PLANT VIROLOGY

PLANT
VIROLOGY
Third Edition

R. E. F. Matthews

Department of Cellular and Molecular Biology
The University of Auckland
Auckland, New Zealand

ACADEMIC PRESS, INC.
Harcourt Brace Jovanovich, Publishers
San Diego   New York   Boston   London
Sydney   Tokyo   Toronto

This book is printed on acid-free paper. ∞

Academic Press, Inc.
San Diego, California 92101

United Kingdom Edition published by
Academic Press Limited
24–28 Oval Road, London NW1 7DX

Library of Congress Cataloging-in-Publication Data

Matthews, R. E. F. (Richard Ellis Ford), Date.
      Plant virology / R.E.F. Matthews. -- 3rd ed.
         p.   cm.
      Includes index.
      ISBN   0-12-480553-1   (alk. paper)
      1. Plant viruses. 2. Virus diseases of plants. 3. Plant viruses-
   -Control.   I. Title.
   3B736.M37   1991
   632'.8--dc20   90-40409
                                                        CIP

Printed in the United States of America
91 92 93 94   9 8 7 6 5 4 3 2 1

Contents

3 Isolation

4 Components

5 Architecture

6 *Replication I: Introduction to the Study of Virus Replication*

7 Replication II: Viruses with Single-Stranded Positive Sense RNA Genomes

8 Replication III: Other Virus Groups and Families

15 *Ecology*

16 *Economic Importance and Control*

17 *Nomenclature, Classification, Origins, and Evolution*

Preface

In the 10 years since the second edition was written there have been major developments in the field of plant virology. Most of these developments stem from the application of gene manipulation technology. Two groups of plant viruses have DNA genomes, but the great majority contain RNA. Thus the major influence in the 1980s has been the development of procedures whereby DNA copies of RNA genomes can be made and DNA sequencing techniques applied to them. Complete nucleotide sequences are now known for representatives of over half the groups of plant viruses with RNA genomes. The ability to prepare *in vitro* infectious RNA transcripts of RNA viral genomes from cloned viral cDNA has been of particular importance. For example it has allowed techniques such as site-directed mutagenesis to be applied to the study of genome function. Nucleotide sequence information has had, and continues to have a profound effect on our understanding of many aspects of plant virology. We now know the number, location, size and functions of the genes in many representative viruses.

The ability to introduce DNA copies of single viral genes into the genome of host plants has opened up new possibilities for the control of some important viral diseases. Methods for the assay and detection of viruses and the diagnosis of diseases have improved greatly in sensitivity, specificity and convenience. These advances have been due mainly to advances in serological procedures and in the application of nucleic acid hybridization technology.

The deluge of nucleotide sequence information becoming available opens up the possibility that we can develop a classification system for viruses based on their evolutionary relationships. This sequence information emphasizes more than ever the essential unity that exists between viruses infecting all groups of organisms.

These various developments have led to a very substantial rewriting of most chapters in the text. Some are entirely new. The extent of the changes can be gauged from the following data. Approximately 52% of the written text is new. Much of the remainder has been revised; of the 251 illustrations, 126 are new; of the 21 tables, 20 are new; and approximately 2,000 of the 3,000 references in the bibliography were not in the second edition.

While much of the older material has been deleted for reasons of space, I have retained many older references to interesting but unexplained biological phenomena, because the technologies for exploring and understanding many of these at the molecular level are now becoming available.

As did the first edition, this volume covers all aspects of the subject from molecular to ecological. The bases for experimental procedures are discussed, but detailed protocols are not provided. The volume is intended primarily for graduate students in plant virology, plant pathology, general virology and microbiology, and for teachers and research workers in these fields. It should also prove useful to some people in related fields such as molecular biologists, biochemists, plant physiologists and entomologists.

I am much indebted to the following colleagues who critically read and commented on sections of the manuscript: P. J. G. Butler, T. W. Dreher, the late R. I. B. Francki, R. L. S. Forster, R. C. Gardner, J. Marbrook, E. Mayr, B. A. Morris, L. R. Nault, D. L. Nuss, Y. Okada, R. A. Owens, D. Penny, A. W. Robards, D. D. Shukla, R. H. Symons, J. M. Thresh, and M. Zaitlin. I thank the following for most useful discussions: J. Berriman, R. I. B. Francki, R. C. Gardner, J. M. Kaper, D. Lane, D. L. Nuss, and J. W. Randles. I also thank the many colleagues in various countries who provided information by correspondence, who sent manuscripts prior to publication, and who provided photographs for illustrations. Figures are acknowledged individually in the text. I also thank editors and publishers for permission to reproduce figures and photographs. I thank M. Gibbs for preparing computer-generated figures; P. R. Fry for help with the bibliography, and last, but not least, Jean Parrott for typing the manuscript.

Preface to the Second Edition

There have been substantial developments in many areas of plant virology since the first edition was published.

Advances have been made in all branches of the subject, but these have been most far reaching with respect to the structure of viruses and of their components, and in our understanding of how viral genomes are organized and how viruses replicate in cells. Significant developments have also occurred in our understanding of how viruses are transmitted by invertebrates and in the application of control measures for specific diseases. The taxonomy of viruses has advanced significantly, and there are now 25 internationally approved families and groups of plant viruses. All these developments have required that most sections be entirely rewritten. The extent of the changes can be gauged from the fact that 1881 of the 2667 references in the bibliography did not appear in the first edition.

As did the first edition, this volume is written to cover all aspects of the field, and is intended primarily for graduate students in plant pathology, plant virology, general virology, and microbiology, and for teachers and research workers in these fields. It should also prove useful to some people in related disciplines—molecular biologists, biochemists, plant physiologists, and entomologists.

Preface to the First Edition

As in many other areas of biology, there has been rapid growth over the past few years in our knowledge of plant viruses and the diseases they cause. Thus there was a substantial need for a new text covering all aspects of the subject.

This book was written primarily for graduate students in plant pathology, plant virology, general virology, and microbiology and for teachers and research workers in these fields. I hope that it will also prove useful as a reference work for those in disciplines related to plant virology—molecular biologists, biochemists, plant physiologists, and entomologists.

I have attempted to cover, to some degree at least, all aspects of the subject, a difficult task in view of the wide range of disciplines involved. There is a brief historical account of the development of plant virology in the first chapter, but the general approach is not a historical one. Those interested will find this aspect well covered in earlier texts.

Topics dealt with include the structure of viruses and viral components; the replication of viruses; their macroscopic, cytological, and biochemical effects on the host plant; the nature of virus mutation; relationships with invertebrate vectors; and a discussion of ecology and control. Throughout I have attempted to indicate how progress in any particular area has been dependent on the development and application of appropriate experimental methods. Specific details of methodology have not been given since these are available elsewhere.

The subject has grown to the extent that it would be impossible to quote all papers on any given topic in a book of this size. In general I have referred to important early papers and to the most important or most suitably illustrative recent papers. From these the reader should be able to gain rapid access to the literature on any relevant topic.

In a text on a subject that draws a wide range of scientific disciplines, I believe that illustrative material is most important, particularly for students or newcomers to the field. For this reason I have gone to some pains, and have had the support of many colleagues, in selecting graphs and photographs to highlight and supplement the text.

In certain areas, particularly the molecular biology of viral replication, our knowledge of plant viruses lags behind that of animal and bacterial viruses. I have therefore drawn on information about these viruses where it seemed appropriate to set the stage for considering more fragmentary facts about plant viruses.

One recent development that created problems was the discovery that many diseases previously thought to be caused by unstable viruses are very probably caused by mycoplasma-like organisms. Although, in general, I have not included diseases in which the probability of a mycoplasma-like organism being involved is high, one chapter on agents causing virus-like diseases is devoted mainly to a consideration of such organisms in plant disease. Other recent work of considerable general interest has resulted in the discovery that several plant viruses have their genetic material divided up between two or more particles. Thus I have devoted a chapter to the consideration of defective virus particles, dependent viruses, and multiparticle viruses.

I have followed the Commonwealth Mycological Institute list of "Plant Virus Names" (Martyn, 1968). I have not attempted to deal with individual viruses or virus diseases in any systematic or comprehensive way, so that the list of "Plant Virus Names" should be regarded as a valuable companion book for the present text, especially for those interested in the tremendous amount of literature on the plant pathological aspects of virus diseases.

In the last chapter I have outlined the various viewpoints regarding nomenclature and classification. Since, from the long-term point of view, at least, classification of viruses must take origins into consideration, some space is given to speculation on the origins of viruses.

Introduction

I. HISTORICAL

The scientific investigation of plant diseases now known to be caused by viruses did not begin until the late nineteenth century. However, there are much earlier written and pictorial records of such diseases. The earliest known written record describing what was almost certainly a virus disease is a poem in Japanese written by the Empress Koken in 752 A.D., and translated by T. Inouye as follows:

> In this village
> It looks as if frosting continuously
> For, the plant I saw
> In the field of summer
> The colour of the leaves were yellowing

The plant, identified as *Eupatorium lindleyanum,* has been found to be susceptible to the tobacco leaf curl *Geminivirus,* which causes a yellowing disease (Osaki *et al.,* (1985).

In Western Europe in the period from about 1600 to 1660, many paintings and drawings were made of tulips that demonstrate flower symptoms of virus disease. These are recorded principally in the Herbals of the time (e.g., Parkinson, 1656). During this period, blooms featuring such striped patterns were prized as special varieties. One of the earliest written accounts of an unwitting experimental transmission of a virus is that of Lawrence (1714). He described in detail the transmission of a virus disease of jasmine by grafting. This description was incidental to the main purpose of his experiment, which was to prove that sap must flow within plants. The following quotation from Blair (1719) describes the procedure and demonstrates, rather sadly, that even at this protoscientific stage experimenters were already indulging in arguments about priorities of discovery.

The inoculating of a *strip'd Bud* into a plain stock and the consequence that the Stripe or Variegation shall be seen in a few years after, all over the shrub above and below the graft, is a full demonstration of this Circulation of the Sap. This was first observed by Mr. *Wats* at *Kensington,* about 18 years ago: Mr. Fairchild performed it

9 years ago; Mr. Bradly says he observ'd it several years since; though Mr. Lawrence would insinuate as if he had first discovered it. (Lawrence, 1714)

In the latter part of the nineteenth century, the idea that infectious disease was caused by microbes was well established, and filters were available that would not allow the known bacterial pathogens to pass. Mayer (1886) described a disease of tobacco that he called *Mosaikkrankheit*. He showed that the disease could be transmitted to healthy plants by inoculation with extracts from diseased plants. Iwanowski (1892) showed that sap from tobacco plants displaying the disease described by Meyer was still infective after it had been passed through a bacteria-proof filter candle. This work did not attract much attention until it was repeated by Beijerinck (1898). Baur (1904) showed that the infectious variegation of *Abutilon* could be transmitted by grafting, but not by mechanical inoculation. Beijerinck and Baur used the term virus in describing the causative agents of these diseases to contrast them with bacteria. The term virus had been used as more or less synonymous with bacteria by earlier workers. As more diseases of this sort were discovered, the unknown causative agents came to be called "filterable viruses."

Between 1900 and 1935, many plant diseases thought to be due to filterable viruses were described, but considerable confusion arose because adequate methods for distinguishing one virus form another had not yet been developed. The original criterion of a virus was an infectious entity that could pass through a filter with a pore size small enough to hold back all known cellular agents of disease. However, diseases were soon found that had viruslike symptoms not associated with any pathogen visible in the light microscope, but that could not be transmitted by mechanical inoculation. With such diseases, the criterion of filterability could not be applied. Their infectious nature was established by graft transmission and sometimes by insect vectors. Thus it came about that certain diseases of the yellows and witches'-broom type, such as aster yellows, came to be attributed to viruses on quite inadequate grounds. Many such diseases are now known to be caused by mycoplasmas and spiroplasmas, and a few, such as ratoon stunting of sugarcane, by bacteria.

An important practical step forward was the recognition that some viruses could be transmitted from plant to plant by insects. Fukushi (1969) records the fact that in 1883 a Japanese rice grower transmitted what is now known to be the rice dwarf *Phytoreovirus* by the leafhopper *Recelia dorsalis*. However, this work was not published in any available form and so had little influence. Kunkel (1922) first reported the transmission of a virus by a planthopper; within a decade many insects were reported to be virus vectors.

During most of the period between 1900 and 1935, attention was focused on the description of diseases, both macroscopic symptoms and cytological abnormalities as revealed by light microscopy, and on the host ranges and methods of transmission of the disease agents. Rather ineffective attempts were made to refine filtration methods in order to define the size of viruses more closely. These were almost the only aspects of virus disease that could be studied with the techniques that were available. The influence of various physical and chemical agents on virus infectivity was investigated, but methods for the assay of infective material were primitive. Holmes (1929) showed that the local lesions produced in some hosts following mechanical inoculation could be used for the rapid quantitative assay of infective virus. This technique enabled properties of viruses to be studied much more readily and paved the way for the isolation and purification of viruses a few years later.

Until about 1930, there was serious confusion by most workers regarding the diseases produced by viruses and the viruses themselves. This was not surprising, since virtually nothing was known about the viruses except that they were very small. Smith (1931) made an important contribution that helped to clarify this situation. Working with virus diseases in potato he realized the necessity of using plant indicators—plant species other than potato, which would react differently to different viruses present in potatoes. Using several different and novel biological methods to separate the viruses, he was able to show that many potato virus diseases were caused by a combination of two viruses with different properties, which he named X and Y. Virus X was not transmitted by the aphid *Myzus persicae* (Sulz.), whereas virus Y was. In this way, he obtained virus Y free of virus X. Both viruses could be transmitted by needle inoculation, but Smith found that certain solanaceous plants were resistant to virus Y. For example, by needle inoculation of the mixture to *Datura stramonium,* he was able to obtain virus X free of virus Y. Furthermore, Smith observed that virus X from different sources fluctuated markedly in the severity of symptoms it produced in various hosts. To quote from Smith (1931): "There are two factors, therefore, which have given rise to the confusion which exists at the present time with regards to potato mosaic diseases. The first is the dual nature, hitherto unsuspected, of so many of the potato virus diseases of the mosaic group, and the second is the fluctuation in virulence exhibited by one constituent, i.e., X, of these diseases."

Another discovery that was to become important was Beale's (1928) recognition that plants infected with tobacco mosaic contained a specific antigen. Gratia (1933) showed that plants infected with different viruses contained different specific antigens. Chester (1935, 1936) showed that different strains of tobacco mosaic *Tobamovirus* (TMV) and potato X *Potexvirus* (PVX) could be distinguished serologically. He also showed that serological methods could be used to obtain a rough estimate of virus concentration.

Since Fukushi (1940) first showed that rice dwarf virus could be passed through the egg of a leafhopper vector for many generations, there has been great interest in the possibility that some viruses may be able to replicate in both plants and insects. It is now well established that plant viruses in the families Rhabdoviridae and Reoviridae and the *Tenuivirus* and *Marafivirus* groups multiply in their insect vectors as well as in their plant hosts.

The high concentration at which certain viruses occur in infected plants and their relative stability turned out to be of crucial importance in the first isolation and chemical characterization of viruses, because methods for extracting and purifying proteins were not highly developed. In 1926, the first enzyme, urease, was isolated, crystallized, and identified as a protein (Sumner, 1926). The isolation of others soon followed. In the early 1930s, workers in various countries began attempting to isolate and purify plant viruses using methods similar to those that had been used for enzymes. Following detailed chemical studies suggesting that the infectious agent of TMV might be a protein, Stanley (1935) announced the isolation of this virus in an apparently crystalline state. At first Stanley (1935, 1936) considered that the virus was a globulin containing no phosphorus. Bawden *et al.* (1936) described the isolation from TMV-infected plants of a liquid crystalline nucleoprotein containing nucleic acid of the pentose type. They showed that the particles were rod-shaped, thus confirming the earlier suggestion of Takahashi and Rawlins (1932) based on the observation that solutions containing TMV showed anisotropy of flow. Best (1936a) noted that a globulinlike protein having virus activity was precipitated from infected leaf extracts when they were acidified, and in 1937 he independently confirmed the nucleoprotein nature of TMV (Best, 1937b).

Electron microscopy and X-ray crystallography were the major techniques used in early work to explore virus structure, and the preeminence of these methods has continued to the present day. Bernal and Fankuchen (1937) applied X-ray analysis to purified preparations of TMV. They obtained accurate estimates of the width of the rods and showed that the needle-shaped bodies produced by precipitating the virus with salt were regularly arrayed in only two dimensions and, therefore, were better described as paracrystals than as true crystals. The isolation of other rod-shaped viruses, and spherical viruses that formed true crystals, soon followed. All were shown to consist of protein and pentose nucleic acid.

Early electron micrographs (Kausche *et al.*, 1939) confirmed that TMV was rod-shaped and provided approximate dimensions, but they were not particularly revealing because of the lack of contrast between the virus particles and the supporting membrane. The development of shadow-casting with heavy metals (Müller, 1942; Williams and Wycoff, 1944) greatly increased the usefulness of the method for determining the overall size and shape of virus particles. However, the coating of metal more or less obscured structural detail. With the development of high-resolution microscopes and of negative staining in the 1950s, electron microscopy became an important tool for studying virus substructure.

From a comparative study of the physicochemical properties of the virus nucleoprotein and the empty viral protein shell found in turnip yellow mosaic *Tymovirus* (TYMV) preparations, Markham (1951) concluded that the RNA of the virus must be held inside a shell of protein, a view that has since been amply confirmed for this and other viruses by X-ray crystallography. Crick and Watson (1956) suggested that the protein coats of small viruses are made up of numerous identical subunits arrayed either as helical rods or as a spherical shell with cubic symmetry. Subsequent X-ray crystallographic and chemical work has confirmed this view. Caspar and Klug (1962) formulated a general theory that delimited the possible numbers and arrangements of the protein subunits forming the shells of the smaller isodiametric viruses. Our recent knowledge of the larger viruses with more complex symmetries and structures has come from electron microscopy using negative-staining and ultrathin-sectioning methods.

Until about 1948, most attention was focused on the protein part of the viruses. Quantitatively, the protein made up the larger part of virus preparations. Enzymes that carried out important functions in cells were known to be proteins, and knowledge of pentose nucleic acids was rudimentary. No function was known for them in cells, and they generally were thought to be small molecules. This was because it was not recognized that RNA is very susceptible to hydrolysis by acid, by alkali, and by enzymes that commonly contaminate virus preparations.

Markham and Smith (1949) isolated TYMV and showed that purified preparations contained two classes of particle, one an infectious nucleoprotein with about 35% of RNA, and the other an apparently identical protein particle that contained no RNA and that was not infectious. This result clearly indicated that the RNA of the virus was important for biological activity. Analytical studies (e.g., Markham and Smith, 1951) showed that the RNAs of different viruses have characteristically different base compositions while those of related viruses are similar. About this time it came to be realized that viral RNAs might be considerably larger than had been thought.

The experiments of Hershey and Chase (1952), which showed that when *Escherichia coli* was infected by a bacterial virus, the viral DNA entered the host cell while most of the protein remained outside, emphasized the importance of the nucleic acids in viral replication. Harris and Knight (1952) showed that 7% of the threonine could be removed enzymatically from TMV without altering the biological activity of the virus, and that inoculation with such dethreonized virus gave rise to normal virus with a full complement of threonine. A synthetic analog of the normal base guanine, 8-azaguanine, when supplied to infected plants was incorporated into the RNA of TMV and TYMV, replacing some of the guanine. The fact that virus preparations containing the analog were less infectious than normal virus (Matthews, 1953c) gave further experimental support to the idea that viral RNAs were important for infectivity. However, it was the classic experiments of Gierer and Schramm (1956), Fraenkel-Conrat and Williams (1955), and Fraenkel-Conrat (1956) that demonstrated the infectivity of naked TMV RNA and the protective role of the protein coat. These discoveries ushered in the era of modern plant virology. The remainder of this section summarizes the major developments of the past 35 years.

The first amino acid sequence of a protein (insulin) was established in 1953. Not long after this event the full sequence of 158 amino acids in the coat protein of TMV became known (Anderer *et al.*, 1960; Tsugita *et al.*, 1960; Wittmann and Wittmann-Liebold, 1966). The sequence of many naturally occurring strains and artificially induced mutants was also determined at about the same time. This work made an important contribution to establishing the universal nature of the genetic code and to our understanding of the chemical basis of mutation.

Brakke (1951, 1953) developed density gradient centrifugation as a method for purifying viruses. This has been one of the most influential technical developments in virology and molecular biology. Together with a better understanding of the chemical factors affecting the stability of viruses in extracts, this procedure has allowed the isolation and characterization of many viruses. The use of sucrose density gradient fractionation enabled Lister (1966, 1968) to discover the bipartite nature of the tobacco rattle *Tobravirus* (TRV) genome. Since that time, density gradient and polyacrylamaide gel fractionation techniques have allowed many viruses with multipartite genomes to be characterized. Their discovery, in turn, opened up the possibility of carrying out genetic reassortment experiments with plant viruses (Lister, 1968; van Vloten-Doting *et al.*, 1968).

Density gradient fractionation of purified preparations of some other viruses has revealed noninfectious nucleoprotein particles containing subgenomic RNAs. Other viruses have been found to have associated with them satellite viruses or satellite RNAs that depend on the "helper" virus for some function required during replication. With all of these various possibilities, it is in fact rather uncommon to find a purified virus preparation that contains only one class of particle.

The 1960s can be regarded as the decade in which electron microscopy was a dominant technique in advancing our knowledge about virus structure and replication. Improvements in methods for preparing thin sections for electron microscopy allowed completed virus particles to be visualized directly within cells. The development and location of virus-induced structures within infected cells could also be studied. It became apparent that many of the different groups and families of viruses induce characteristic structures, or viroplasms, in which the replication of virus components and the assembly of virus particles take place. Improved techniques for extracting structural information from electron microscope images of negatively stained virus particles revealed some unexpected and interesting variations on the original icosahedral theme for the structure of "spherical" viruses.

There were further developments in the 1970s. Improved techniques related to

X-ray crystallographic analysis and a growing knowledge of the amino acid sequences of the coat proteins allowed the three-dimensional structure of the protein
shells of several plant viruses to be determined in molecular detail.

For some decades, the study of plant virus replication had lagged far behind that
of bacterial and vertebrate viruses. This was mainly because there was no plant
system in which all the cells could be infected simultaneously to provide the basis
for synchronous "one step growth" experiments. However, following the initial
experiments of Cocking (1966), Takebe and colleagues developed protoplast systems for the study of plant virus replication (reviewed by Takebe, 1977). Although
these systems had significant limitations, they greatly increased our understanding
of the processes involved in plant virus replication. Another important technical
development has been the use of *in vitro* protein-synthesizing systems such as that
from wheat germ, in which many plant viral RNAs act as efficient messengers.
Their use allowed the mapping of plant viral genomes by biochemical means to
begin.

During the 1980s, major advances have centered on improved methods of
diagnosis for virus diseases, a great increase in our understanding of the organization and strategy of viral genomes, and the development of techniques that promise
novel methods for the control of some viral diseases.

Improvements in diagnosis have centered on serological procedures and on
methods based on nucleic acid hybridization. Since the work of Clark and Adams
(1977), the ELISA technique has been developed with many variants for the sensitive assay and detection of plant viruses. Monoclonal antibodies against TMV
were described by Dietzen and Sander (1982) and Briand *et al.* (1982). Since that
time there has been a very rapid growth in the use of monoclonal antibodies for
many kinds of plant virus research and for diagnostic purposes.

The use of nucleic acid hybridization procedures for sensitive assays of large
numbers of samples has been greatly facilitated by two techniques: (i) the ability to
prepare double-stranded cDNA from a viral genomic RNA and to replicate this in a
plasmid grown in its bacterial host, with the batches of cDNA labeled radioactively
or with biotin to provide a sensitive probe; and (ii) the dot blot procedure, in which a
small sample of a crude plant extract containing virus is hybridized with labeled
probe as a spot on a sheet of nitrocellulose or other material.

The major influence in the 1980s has been the development of procedures
whereby the complete nucleotide sequence of viruses with RNA genomes can be
determined. Of special importance has been the ability to prepare *in vitro* infectious
transcripts of RNA viruses derived from cloned viral cDNA (Ahlquist *et al.,*
1984b). This has allowed techniques such as site-directed mutagenesis to be applied
to the study of genome function. Nucleotide sequence information has had, and
continues to have, a profound effect on our understanding of many aspects of plant
virology, including the following: (i) the location, number, and size of the genes in a
viral genome; (ii) the amino acid sequence of the known or putative gene products;
(iii) the molecular mechanisms whereby the gene products are transcribed; (iv) the
putative functions of a gene product, which can frequently be inferred from amino
acid sequences similarities to products of known function coded for by other viruses; (v) the control and recognition sequences in the genome that modulate expression of viral genes and genome replication; (vi) the understanding of the structure and replication of viroids and of the satellite RNAs found associated with some
viruses, (vii) the molecular basis for variability and evolution in viruses, including
the recognition that recombination is a widespread phenomenon among RNA viruses and that viruses can acquire host nucleotide sequences as well as genes from

other viruses; and (viii) the beginning of a taxonomy for viruses that is based on evolutionary relationships.

In the early 1980s, it seemed possible that some plant viruses, when suitably modified by the techniques of gene manipulation, might make useful vectors for the introduction of foreign genes into plants. Although this has been achieved for several genes in model experiments, the concept has not yet demonstrated any practical significance. However, some plant viruses have been found to contain regulatory sequences that can be very useful in other gene vector systems.

In the early decades of this century, attempts to control virus diseases in the field were often ineffective. They were mainly limited to attempts at general crop hygiene, roguing of obviously infected plants, and searches for genetically resistant lines. Developments since this period have improved the possibilities for control of some virus diseases. The discovery of two kinds of soilborne virus vectors (nematodes, Hewitt *et al.*, 1958; fungi, Grogan *et al.*, 1958) opened the way to possible control of a series of important diseases. Increasing success has been achieved with a range of crop plants in finding effective resistance or tolerance to viruses.

Heat treatments and meristem tip culture methods have been applied to an increasing range of vegetatively propagated plants to provide a nucleus of virus-free material that then can be multiplied under conditions that minimize reinfection. Such developments frequently have involved the introduction of certification schemes. Systemic insecticides, sometimes applied in pelleted form at time of planting, provide significant protection against some viruses transmitted in a persistent manner by aphid vectors. Diseases transmitted in a nonpersistent manner in the foregut or on the stylets of aphids have proved more difficult to control. It has become increasingly apparent that effective control of virus disease in a particular crop in a given area usually requires an integrated and continuing program involving more than one kind of control measure. However, such integrated programs are not yet in widespread use.

Cross-protection is a phenomenon in which infection of a plant with a mild strain of a virus prevents or delays infection with a severe strain. The phenomenon has been used with varying success for the control of certain virus diseases, but the method has various difficulties and dangers. Powell-Abel *et al.* (1986) considered that some of these problems might be overcome if plants could be given protection by expression of a single viral gene. Using recombinant DNA technology, they showed that transgenic tobacco plants expressing the TMV coat protein gene either escaped infection following inoculation or showed a substantial delay in the development of systemic disease. These transgenic plants expressed TMV coat protein mRNA as a nuclear event. Seedlings from self-fertilized transformed plants that expressed the coat protein showed delayed symptom development when inoculated with TMV. Thus a new approach to the control of virus diseases emerged. Since these experiments, the phenomenon has been shown to occur for several viruses and hosts. It may lead to economically useful protection against specific viruses at least for some annual crops.

In recent years considerable progress has been made in the development of a stable and internationally accepted system for the classification and nomenclature of viruses. Three hundred and twelve plant viruses have been placed in two families and thirty two groups. A further 279 probable or possible members have been recognized (Table 17.1). The three virus families and most (but not all) of the groups are very distinctive entities. They possess clusters of physical and biological properties that often make it quite easy to allocate a newly isolated virus to a particular family or group. The rapidly expanding information on nucleotide sequences of viruses infecting plants, invertebrates, vertebrates, and microorganisms is emphasizing, even more strongly than in the past, the essential unity of virology. The time is therefore ripe for plant virologists to adopt the family and genus taxa used by virologists working with all the other host groups.

In spite of our greatly increased understanding of the structure, function, and replication of viral genomes, there is still a major deficiency. We have virtually no molecular understanding of how an infecting virus induces disease in the host plant. The processes almost certainly involve highly specific interactions between viral macromolecules—proteins and nucleic acids—and host macromolecular structures. At present we appear to lack the appropriate techniques to make further progress. Perhaps an understanding of disease processes will be the exciting area of virology in the 1990s. Certain historical aspects of plant virology have been discussed by Bos (1981), Matthews (1987), and Brakke (1988). Hull *et al.* (1989) provide a useful directory and dictionary of viruses and terms relating to virology.

II. DEFINITION OF A VIRUS

In the size of their nucleic acids, viruses range from a monocistronic mRNA in the satellite virus of tobacco necrosis virus (STNV) to a genome larger than that of the smallest cells (Fig. 1.1). Before attempting to define what viruses are we must consider briefly how they differ from cellular parasites on the one hand and transposable genetic elements on the other. The three simplest kinds of parasitic cells are the Mycoplasmas, the Rickettsiae, and the Chlamydiae.

Mycoplasmas and related organisms are not visible by light microscopy. Cells are 150–300 nm in diameter with a bilayer membrane, but no cell wall. They contain ribosomes and DNA. They replicate by binary fission, and some that infect vertebrates can be grown *in vitro*. Their growth is inhibited by certain antibiotics.

The Rickettsiae, for example, the agent of typhus fever, are small, nonmotile bacteria, usually about 300 nm in diameter. They have a cell wall, plasma mem-

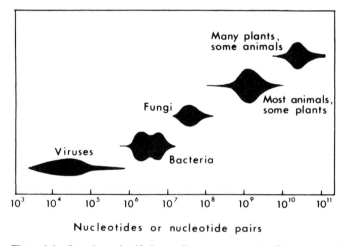

Figure 1.1 Organisms classified according to genome size. The vertical axis gives an approximate indication of relative numbers of species (or viruses) within the size range of each group. (Modified from Hinegardner, 1976.)

brane, and cytoplasm with ribosomes and DNA strands. They are obligate parasites and were once thought to be related to viruses, but they are definitely cells because (i) they multiply by binary fission, and (ii) they contain enzymes for ATP production.

The Chlamydiae, for example, the agent causing psittacosis, include the simplest known type of cell. They are obligate parasites and lack an energy-generating system. They have two phases in their life cycle. Outside the host cell they exist as infectious *elementary bodies* about 300 nm in diameter. These bodies have dense contents, no cell wall, and are specialized for extracellular survival. The elementary body enters the host cell by phagocytosis. Within 8 hours it is converted into a much larger noninfectious *reticulate body*. This is bounded by a bilayer membrane derived from the host. The reticulate body divides by binary fission within this membrane, giving thousands of progeny within 40–60 hours. The reticulate bodies are converted to elementary bodies, which are released when the host cell lyses.

There are several criteria that do not distinguish all viruses from all cells. These include:

1. Size: some poxviruses are bigger than the elementary bodies of Chlamydiae.
2. Nature and size of the genome: many viruses have dsDNA like that of cells, and in some the DNA is bigger than in the Chlamydiae.
3. Presence of DNA and RNA.
4. A rigid cell envelope is absent in viruses and mycoplasmas.
5. Growth outside a living host cell does not occur with viruses or with many groups of obligate cellular parasites, for example, Chlamydiae.
6. An energy-yielding system is absent in viruses and Chlamydiae.
7. Complete dependence on the host cell for amino acids, etc., is found with viruses and some bacteria.

There are three related criteria that do appear to distinguish all viruses from all cells:

1. Lack of a continuous membrane separating viral parasite and host during intracellular replication. Cellular parasites that replicate inside a host cell appear always to be separated from host cell cytoplasm by a continuous bilayer membrane (see Fig. 10.2).
2. Absence of a protein-synthesizing system in viruses.
3. Replication of viruses is by synthesis of a pool of components, followed by assembly of many virus particles from the pool. Even the simplest cells replicate by binary fission.

Plasmids are autonomous extrachromosomal genetic elements found in many kinds of bacteria. They consist of closed circular DNA. Some can become integrated into the host chromosome and replicate with it. Some viruses infecting prokaryotes have properties like those of plasmids and, in particular, an ability to integrate into the host cell chromosome. However, viruses differ from plasmids in the following ways:

1. Normal viruses have a particle with a structure designed to protect the genetic material in the extracellular environment and to facilitate entry into a new host cell.
2. Virus genomes are highly organized for specific virus functions of no known value to the host cell, whereas plasmids consist of genetic material often useful for survival of the cell.

3. Viruses can cause death of cells or disease in the host organism but plasmids do not.

We can now define a virus as follows: *A virus is a set of one or more nucleic acid template molecules, normally encased in a protective coat or coats of protein or lipoprotein, that is able to organize its own replication only within suitable host cells. Within such cells, virus replication is (i) dependent on the host's protein-synthesizing machinery, (ii) organized from pools of the required materials rather than by binary fission, (iii) located at sites that are not separated from the host cell contents by a lipoprotein bilayer membrane, and (iv) continually giving rise to variants through various kinds of change in the viral nucleic acid.*

To be identified positively as a virus, an agent must normally be shown to be transmissible and to cause disease in at least one host. However, the *Cryptovirus* group of plant viruses is an exception. Viruses in this group rarely cause detectable disease and are not transmissible by any mechanism except through the seed or pollen.

The structure and replication of viruses have the following features.

1. The nucleic acid may be DNA or RNA and single- or double-stranded. If the nucleic acid is single-stranded it may be of positive or negative sense. (Positive sense has the sequence that would be used in an mRNA for translation to give a viral-coded protein.)
2. The mature virus particle may contain polynucleotides other than the genomic nucleic acid.
3. Where the genetic material consists of more than one nucleic acid molecule, each may be housed in a separate particle or all may be located in one particle.
4. The genomes of viruses vary widely in size, encoding between 1 and about 250 proteins. Plant viral genomes are at the small end of this range, encoding between 1 and 12 proteins. The viral-coded proteins may have functions in virus replication, in virus movement from cell to cell, in virus structure, and in transmission by invertebrates or fungi.
5. Viruses undergo genetic change. Point mutations occur with high frequency as a result of nucleotide changes brought about by errors in the copying process during genome replication. Other kinds of genetic change may be due to recombination, reassortment of genome pieces, loss of genetic material, or acquisition of nucleotide sequences from unrelated viruses or the host genome.
6. Enzymes specified by the viral genome may be present in the virus particle. Most of these enzymes are concerned with nucleic acid synthesis.
7. Replication of many viruses takes place in distinctive virus-induced regions of the cell, known as viroplasms.
8. Some viruses share with certain nonviral nucleic acid molecules the property of integration into host-cell genomes and translocation from one integration site to another.
9. A few viruses require the presence of another virus for their replication.

2

Methods for Assay, Detection, and Diagnosis

The ability to assay viruses is basic to almost all kinds of virological investigation. Simultaneous application of two or more assay methods that depend on different properties of the virus to separate samples of the same material is useful and is often essential for many kinds of experiment. The problems of detecting viruses and of diagnosing virus disease involve the use of assay techniques, so these topics are also dealt with in this chapter. Some of the methods are used for assay, some for diagnosis, and some for both. Diagnosis of a virus disease on a routine basis is a requirement for the development of satisfactory measures to control that disease (Chapter 16).

We can distinguish three situations where the procedures outlined in this chapter are used: first, when we wish to assay a virus, for example, during the step in a procedure designed to purify the virus, or in experiments to investigate the stability of the virus *in vitro;* second, when we can anticipate from previous experience that one or more known viruses may be present in a crop, and we wish to detect and identify these and assess their prevalence through a growing season; and third, when we isolate a new virus in a particular crop or region, and we wish to determine whether it is a new virus or a variant or strain of a known virus, or identical to a known virus. In the past, descriptions of viruses have often been published without sufficient data to distinguish between these possibilities. Thus the literature has become cluttered with partial descriptions of viruses that are essentially useless.

The problem of delineating strains of viruses is discussed in Chapter 13, while the criteria that can be used to describe and classify an unknown virus are summarized in Chapter 17. Other chapters include material relevant to the problem of disease diagnosis: Chapter 3, viral components; Chapter 4, viral architecture; Chapters 6–8, replication; and Chapter 11, agents causing viruslike symptoms.

The choice of plant material to be sampled is of great importance for successful assay, detection, and diagnosis. The distribution of virus in an infected plant may be very uneven (Chapter 10). Many factors influence the concentration of virus in a plant and therefore its ease of detection (Chapter 12). These problems may be particularly important for crops, where tissue samples are taken from bulbs, corms, or tubers, and for woody perennial crops.

When appraising the relative merits of different methods, the following important factors need to be considered: (i) sensitivity—how small an amount of virus can be measured or detected; (ii) accuracy and reproducibility; (iii) numbers of samples that can be processed in a given time by one operator; (iv) cost and sophistication of the apparatus and materials needed; (v) the degree of operator training required; and (vi) adaptability to field conditions. The last three considerations may be particularly important in the developing countries of the tropics. For this reason, adequate biological tests may be especially important in these countries (Lana, 1981).

Finally, it must always be remembered that diseased plants in the field may be infected by more than one virus. Thus an early step in diagnosis for an unknown disease must be to determine whether more than one virus is involved. This may be done by serological tests, by examination in the electron microscope, or by inoculation to a range of potential host plants with back inoculation to a standard host.

The methods involved in assay, detection, and diagnosis can be placed in four groups according to the properties of the virus upon which they depend: biological activities; physical properties of the virus particle; properties of viral proteins; and properties of the viral nucleic acid.

I. METHODS INVOLVING BIOLOGICAL ACTIVITIES OF THE VIRUS

Biological methods for the assay, detection, and diagnosis of viruses are much more time-consuming than most other methods now available. Nevertheless, they remain very important. Only measurements of infectivity give us relative estimates of the concentration of viable virus particles. As far as diagnosis is concerned, in most circumstances only inoculation to an appropriate host species can determine whether a particular virus isolate causes severe or mild disease. Some practical aspects of biological testing are described by Hill (1984a).

A. Infectivity Assays

1. Quantitative Assay Based on Local Lesions

Holmes (1929) showed that the necrotic local lesions produced in leaves of *Nicotiana glutinosa* following mechanical inoculation with TMV could be used for assay of relative infectivity. The method was more precise and used fewer plants than the older procedure of estimating the number of systemically infected plants in an inoculated group. Since that time, much effort has been devoted to seeking local lesion hosts for particular viruses. Various aspects of mechanical transmission are discussed in Chapter 10.

Whenever possible, hosts that give a clear-cut necrotic or ring spot type of local lesion are used for local lesion assays (Fig. 11.1). Some viruses give reproducible chlorotic lesions, but for others, chlorotic lesions may grade from clear-cut spots to faint yellow areas that require arbitrary and subjective assessment. With such plants it is sometimes possible to take advantage of the fact that the starch content of the cells in the lesion may differ from that in the uninfected cells. At the end of a photosynthetic period, virus-infected cells may contain less starch. At the end of a dark period, they may contain more starch. Leaves are decolored in ethanol and stained with iodine. For satisfactory and reproducible results with starch–iodine lesions, environmental conditions and sampling times need to be carefully controlled. Necrotic local lesions induced by heat treatment of the leaves have been used for certain host–virus combinations (Foster and Ross, 1975).

The nutritional state of the plant may affect the distinctness of the local lesions formed. For example, in nitrogen-deficient Chinese cabbage plants, TYMV may produce well-defined purple local lesions (Diener and Jenifer, 1964).

With due care and an appropriate experimental design, local lesion assays can distinguish a difference of as little as 10–20% between two preparations. However, there are many examples in the literature where unwarranted conclusions are drawn from local lesion assays. The three major aspects to be borne in mind are, first, the wide variation in number of local lesions produced in different leaves by a standard inoculum; second, the general nature of the curve relating dilution of inoculum and lesion number; and third, the statistical requirements for making valid comparisons.

Variation between Leaves

The environmental and physiological factors that influence susceptibility to infection are discussed in Chapter 12. They may include the age of the plant, genetic variation in the host, position of the leaf on the plant, nutrition of the plant, water supply, temperature, light intensity, season of the year, and time of day. Samuel and Bald (1933) recognized that there was much less variation between opposite halves of the same leaf than between different leaves. Since that time most experimental designs for local lesion assays have used half-leaf comparisons. For plants like *Phaseolus vulgaris,* which have two usable primary leaves in the same position on the plant, four fairly equivalent half-leaves are available. Close attention to the inoculation procedure, and to the conditions under which inoculated leaves are held, may improve both the sensitivity and reproducibility of local lesion assays (e.g., McKlusky and Stobbs, 1985).

The number of half-leaf comparisons that are necessary depends on the accuracy required, the uniformity of the test plants, and the number of samples to be compared. With an assay plant grown under fairly standard conditions it is usually possible after some experience to pick individual plants that will have a susceptibility differing markedly from the group, and to discard them. For a single comparison between two samples a minimum number of six to eight leaves should be used. When more than two samples are compared, a variety of experimental layouts are possible. One-half of every leaf can be inoculated with the same standard preparation, and the various test solutions to the other half-leaves. This is a simple design but rather wasteful of plants.

Where appropriate, a latin square design is effective. For example, with a plant like *N. glutinosa* where about four to eight leaves may be available on each plant, it is possible, for a limited number of samples, to arrange that each sample is compared on a leaf at each position on the plant.

Where the number of treatments exceeds the number of leaves, an appropriate design is one in which each test inoculum appears on each leaf position the same number of times. Kleczkowski (1950), Fry and Taylor (1954), and Preece (1967) give some examples of more complex experimental designs. However, there may be a useful limit to the size and complexity of local lesion assays. In large experiments the risk of error in the inoculations or labeling is increased. Because of the long time required to carry out inoculations, changes in the susceptibility of plants with time of day might influence results unless the experimental design is further complicated to take account of this effect, which may be quite large.

The Relation between Dilution of Inoculum and Lesion Number

Best (1937a) examined the nature of the curve relating dilution of TMV inoculum and numbers of local lesions produced in *N. glutinosa*. He found that the curve could be divided into three parts: (i) a section at high concentrations where a change in concentration is accompanied by very little change in local lesion number; (ii) a section in the middle of the curve where a change in concentration is accompanied by a more or less equivalent change in lesion number; and (iii) a section at low virus concentration where change in concentration has little effect on lesion number. This general situation holds for many viruses, but not for all. Two dilutions curves are shown in Fig. 2.1.

Virus dilution curves are interesting from both the theoretical and the practical points of view. Mathematical analyses are considered in relation to the processes of infection in Chapter 10. From the practical point of view, the form of the dilution curve has several important implications. First, comparisons of two samples where local lesion numbers are very high or very low are quite useless. They may be in error by several orders of magnitude. Second, valid comparisons can be made only

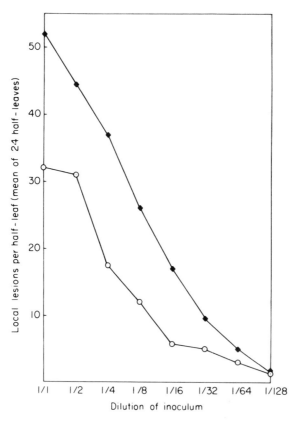

Figure 2.1 Infectivity dilution curves. Effect of dilution of inoculum on number of local lesions produced by two plant viruses. (◆———◆) Tomato bushy stunt *Tombusvirus* TBSV in *N. glutinosa*, (○———○) TMV in *N. glutinosa*. (Data from Kleczkowski, 1950.)

in the region where lesion number is responding more or less proportionally to dilution. Third, the exact slope of the dilution curve is variable and unpredictable from one experiment to another. This means that two samples must be compared at several dilutions (usually two-, five-, or ten-fold). Generally speaking, for leaves about the size of *N. glutinosa* leaves, mean figures in the range of 10–100 local lesions per half-leaf give useful estimates. For much larger or much smaller leaves the range would be different.

Factors that may affect the slope of the dilution curve include: (i) the presence of inhibitors in the inoculum—dilution of the inhibitor may give rise to a curve that is flatter than expected; (ii) virus that is in an aggregated state but becomes disaggregated on dilution—this also would give a flattened curve; (iii) the need for more than one virus particle to give a local lesion—this would give a curve steeper than expected (R. W. Fulton, 1962) and has been found for several viruses that require two or more particles for infectivity (Fig. 10.6); and (iv) changes in susceptibility of test plants during the time taken to carry out inoculations—this could affect the slope of the curve either way depending on order of inoculation and time of day.

Relationship between Lesion Number and Concentration of Infective Virus

Even when account has been taken of leaf-to-leaf variation (by replication and experimental design) and when the actual lesion numbers fall in the middle range of the dilution curve, lesion number cannot be translated directly into relative infective virus content. The simplest practical way to overcome this problem is to arrange the dilutions (on the basis of a preliminary test) so that the samples compared give nearly equal numbers of local lesions (within the useful range of about 10–100 per leaf) in one of the comparisons.

Proper use of local lesion data requires some statistical analysis. Kleczkowski (1949, 1953) drew attention to the fact that neither lesion numbers nor logarithms of these numbers (a transformation used by earlier workers) are satisfactory for statistical analysis. Numbers of local lesions are not normally distributed, and the variance of the mean increases with increasing mean. Kleczkowski derived a transformation that is satisfactory when the mean number of local lesions is greater than about 10. This transformation is

$$y = \log_{10}(X + C)$$

where X is the number of local lesions and C is a constant. C can be assessed for each experiment, but any value of C between about 5 and 20 could be used satisfactorily.

Some General Considerations

The many factors that can influence infectivity of viruses and susceptibility of plants should be borne in mind when devising and using an assay system. When studying the effect of some treatment on the infectivity of a preparation it is important to consider whether the treatment may be altering the state of the medium in some way (e.g., pH) rather than having an effect directly on the virus.

The size and complexity of an assay should be appropriate to the needs of the experiment. It is a waste of labor to set up an elaborate randomized design when a very approximate estimate of infectivity will give the required answer. The more common failing is to draw conclusions from inadequately designed and analyzed experiments. In this connection one factor that has been widely neglected is the influence of time of day on susceptibility of test plants (Chapter 12). The magnitude of this effect varies a great deal and, with assays that take only an hour or so to carry out, probably can be neglected. However, with complex experimental designs and the larger numbers of plants required for maximum accuracy, or with a large number of samples to be assayed, this factor can influence results in a systematic way.

Assay in Insect Vector Cell Monolayers

L. M. Black and colleagues have developed an assay technique for potato yellow dwarf *rhabdovirus* (PYDV) and wound tumor *Phytoreovirus* (WTV) in which the virus is applied to insect vector cells growing as a monolayer on coverslips. The method provides an assay for these two viruses that is in principle the same as the plaque methods available for bacterial and vertebrate viruses. In a comparative study with PYDV, Hsu and Black (1973) found that on the basis of the number of cells per unit area of monolayer, or of leaf epidermis, an assay using insect cells was $10^{3.7}$ times more sensitive than that using local lesions on *Nicotiana rustica*. Assay on vector cells was also much less variable than when leaves were used. More recently, Kimura (1986) developed conditions for infectivity assays of rice dwarf *Phytoreovirus* using a focus counting technique on vector cell monolayers. This method was about 100 times more sensitive than vector injection. Rice gall dwarf *Phytoreovirus* has also been assayed in cell monolayers (Omura *et al.*, 1988b).

2. Quantal Assay Based on Number of Individuals Infected

Before Holmes (1929) introduced the local lesion assay, the only method available for measuring infectivity was to inoculate the sample at various dilutions to groups of plants and record the number of plants that became systematically infected. This type of test takes longer and requires many more plants to obtain a reliable answer. Nevertheless, there are still occasions when this type of test has to be used, for example, with viruses having no suitable local lesion host or with viruses that have to be assayed by the use of insect vectors. Statistical aspects of quantal assays are discussed by Brakke (1970).

Mechanical Inoculation of Whole Plants

Groups of plants are inoculated with a series of dilutions of the inocula to be compared. The dilutions must span the range for which the test plants are neither all infected nor all healthy. Presence or absence of systemic symptoms are subsequently recorded. This procedure is known as a quantal assay, distinct from a quantitative assay based on local lesions. Various means have been used to improve the precision of quantal assays. When large numbers of plants can be grown and inoculated easily, these can give useful data (Fig. 2.2). Probit analysis can be used to estimate the LD_{50} and to give a statistical estimate of precision with such data.

When plants are inoculated with small amounts of virus, systemic symptom development takes longer than with heavy inocula. A record of the time taken for systemic symptoms to appear, combined by some arithmetic manipulation with the proportion of plant infected, may give increased precision to a quantal assay (Diener and Hadidi, 1977).

Incubated Tissue Samples

It is sometimes possible to estimate the rate at which a virus moves from one part of the plant to another by sampling many small pieces of tissue that may contain

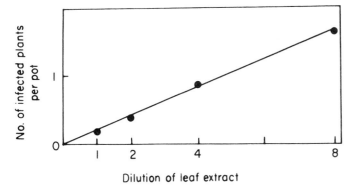

Figure 2.2 Quantal assays. Use of large numbers of plants to obtain precise assays of infectious virus. Four dilutions of crude sorghum leaf extract from plants infected with sugarcane mosaic *Potyvirus* were prepared and each dilution inoculated to 525 sorghum seedlings (25 to a pot). This experiment was repeated seven times. Thus each point is based on 147 pots. 1, 1/160,000; 2, 1/80,000; 4, 1/40,000; 8, 1/20,000. Since the straight line passes through the origin, sugarcane mosaic virus probably has a monopartite genome. (From Dean, 1979.)

no virus or very low amounts of virus. The tissue samples are incubated in isolation to allow any virus present to multiply to give a detectable amount, and then are assayed for the presence or absence of virus either by infectivity or some other method. Such a procedure was used by Fry and Matthews (1963) to determine the time after inoculation at which cells beneath the epidermis of tobacco leaves became infected with TMV.

Insect Vectors

For viruses that are not mechanically transmissible but that have an insect vector, it may be possible to use the percentage of successful insect transmissions to estimate relative amounts of virus. Insects might feed on the test plants or through membranes on solutions containing the virus, or they might be injected with such solutions.

Results obtained by feeding insects on infected tissue may reflect differences in the availability of virus to the insect rather than concentration in the tissue or organ. These methods are all laborious and involve biological variability in the insect as well as in the host plant. For example, the length of time insects are allowed to feed on the test plants may affect the results. Insects may die on the test plants at various times. If more than one insect is used on each test plant, more elaborate statistics are needed. Because of these difficulties the methods have not been widely used, but they have provided valuable information for certain interesting viruses such as WTV (Whitcomb and Black, 1961).

B. Indicator Hosts for Diagnosis

Disease symptoms on plants in the field are almost always inadequate on their own to give a postive identification. This is particularly so when several viruses cause similar symptoms, as do yellowing diseases in beet (Duffus, 1973) when a single virus, such as cucumber mosaic virus (CMV), is very variable in the symptoms it causes (Francki and Hatta, 1980), or when both of these factors are relevant in a single host (Francki *et al.*, 1980a). Thus since the early days of plant virology, searches have been made for suitable species or varieties of host plant that will give clear, characteristic, and consistent symptoms for the virus or viruses being studied, usually under glasshouse conditions. Such indicator hosts provide one of the most basic tools for routine diagnosis. Many good indicator species have been found in the genera *Nicotiana, Solanum, Chenopodium, Cucumus, Phaseolus, Vicia,* and *Brassica.* Sometimes the usefulness of a particular plant can be enhanced by changing the growing conditions (e.g., Lee and Singh, 1972).

Our knowledge of virus structure and replication has increased greatly in recent years, and the taxonomy of plant viruses has developed to the stage where most of the known viruses have been allocated to families or groups. Nevertheless, the search for improved indicator hosts continues. For example, Schmidt and Zobywalski (1984) surveyed a set of 33 French bean cultivars and lines for their ability to discriminate between a range of isolates of bean yellow mosaic *Potyvirus.* Van Dijk *et al.* (1987) screened some 200 species and accessions within the genus *Nicotiana* to find several new and useful indicator hosts.

When comparing the results of viral diagnosis from different laboratories using indicator species, it must always be borne in mind that different lines of the same named variety may vary quite markedly in their symptom response to a given virus (e.g., Van der Want *et al.,* 1975).

Testing large numbers of samples using indicator hosts requires glasshouse facilities that may be occupied for weeks or longer. However, the actual manipulations involved in mechanical inoculation may take less time per sample than many other methods of testing, and many attempts have been made to streamline the procedures for sap extraction and inoculation (e.g., Laidlaw, 1986).

For some groups of diseases, international cooperation has led to the definition of standard sets of indicator plants for particular diseases. Thus the working group on fruit tree virus diseases of the International Society for Horticultural Science has developed a list of indicator hosts for virus and viruslike diseases of eight major woody fruit species. Conditions for field or greenhouse indexing are defined. Of the 137 diseases listed, diagnosis depended solely on the indicator hosts for 87. Antisera were available for testing the remaining 50 by ELISA to complement the indicator results (Dunez, 1983).

C. Host Range in Diagnosis

In earlier work on plant viruses, host range was used as an important criterion in diagnosis. Such information may still be important, or even crucial in certain circumstances. However, apart from the fact that only a relatively small proportion of possible host–virus combinations has been tested, our knowledge of host ranges is limited, or must be qualified, in several ways:

1. In many of the reported host range studies only positive results have been recorded.

2. Absence of symptoms following inoculation of a test plant has not always been followed by back inoculation to an indicator species to test for masked infection. Such infections may not be uncommon. In testing the susceptibility of 456 plant species to 24 viruses, Horváth (1983a) discovered 1312 new susceptible host–virus combinations. Of these, 13% were latent infections.

Where inoculation tests are carried out, an arbitrary decision may have to be made about what constitutes multiplication of virus over and above inoculum remaining on the leaf. Thus, Holmes (1946) took 10 lesions per leaf on back inoculation to *N. glutinosa* to mean that multiplication of TMV or tobacco etch *Potyvirus* (TEV) had occurred in the test species. Multiplication of the virus in one or a few cells of a host near the sites of infection may well go undetected by current procedures. Inoculation with infectious RNA would reduce or eliminate background infectivity due to residual inoculum.

3. The manner of inoculation may well affect the results. Mechanical inoculation has almost always been used in extensive host range studies because of its convenience, but many plants contain inhibitors of infection that prevent mechanical inoculation to the species, or from it, or both.

4. In studying large numbers of species it is usually practicable to make tests only under one set of conditions, but it is known that a given species may vary widely in susceptibility to a virus depending on the conditions of growth (Chapter 12, Section V,C).

5. Quite closely related strains of a virus may differ in the range of plants they will infect. Host range data may apply only to the virus strain studied.

6. Mesophyll protoplasts may be readily infected with a virus that gives little or no infection when applied to intact leaves (e.g., Huber *et al.*, 1977). (See also Table 12.1.)

In spite of these qualifications it may still be worthwhile to seek sets of host and nonhost plants that will distinguish between certain viruses in particular crop species. For example, an international study of persistent aphid-transmitted viruses causing diseases in legumes identified a set of 12 leguminous and nonleguminous test plants that could discriminate between the eight viruses in the study (Johnstone *et al.*, 1984).

Sometimes host range is more discriminating than other tests. For example, raspberry plantings in Scotland have been protected against raspberry bushy dwarf virus by the use of cultivars resistant to the common strain of the virus. A recently discovered resistance-breaking strain could not be distinguished from the common strain either serologically or by indicator hosts (Murant *et al.*, 1986). Different viruses vary widely in the taxonomic breadth of their host range. Host range studies for diagnosis will usually be most useful for those infecting a relatively narrow range of plants.

D. Methods of Transmission in Diagnosis

The different methods of virus transmission, discussed in Chapters 10 and 14, may be useful diagnostic criteria. Their usefulness may depend on the particular circumstances. For example, a virus with an icosahedral particle that is transmitted through the seed and by nematodes is very probably a nepovirus. On the other hand, such a virus transmitted mechanically and by the aphid *Myzus persicae* might belong to any one of several groups.

E. Cytological Effects for Diagnosis

The cytological effects of infection described in Chapter 11 can often be used to assist in diagnosis. Cytological effects detectable by light microscopy can sometimes be used effectively to supplement macroscopic symptoms in diagnosis.

Christie and Edwardson (1986) provide an illustrated catalog of virus-induced inclusions and discuss the problems involved. The light microscope has a number of advantages when the inclusions are large enough to be easily observed: (i) it is a readily available instrument; (ii) specimen preparation techniques can be simple, for example, inclusions seen in epidermal strips can assist in the rapid diagnosis of some virus diseases in red clover (Khan and Maxwell, 1977); (iii) there is a wide field of view, thus allowing many cells to be readily examined; and (iv) a variety of cytochemical procedures are available.

Nevertheless, electron microscopy of thin sections is necessary for some types of inclusion to provide information for use in diagnosis. Hamilton *et al.* (1981) list nine virus groups that induce inclusions that are diagnostic for the group. Presence of characteristic inclusions may be diagnostic for a particular virus when a specific host plant is involved, for example, the citrus tristeza *Closterovirus* in citrus trees (Brlansky, 1987). Individual strains of a virus may produce distinctive cytological effects, as occurs with cauliflower mosaic *Caulimovirus* (CaMV) (Shalla *et al.,* 1980).

F. Mixed Infections

Simultaneous infection with two or even more viruses is not uncommon. Such mixed infections may complicate a diagnosis based on biological properties alone, especially if the host response is variable, as with the internal rib necrosis disease of lettuce in California (Zink and Duffus, 1972). However, several possible differences in biological properties may be used to separate the viruses: (i) if one virus is confined to the inoculated leaf in a particular host, while the other moves systemically; (ii) if a host can be found that only one of the viruses infects; (iii) if the two viruses cause distinctive local lesions in a single host; and (iv) if the two viruses have different methods of transmission, for example, by different species of invertebrate vectors.

On the other hand, certain diseases in the field may be the result of a more or less stable association between two or more viruses. At least three viruses may be implicated in the carrot "motley dwarf" disease complex (Krass and Schlegel, 1974) and in the lettuce speckles disease (Falk *et al.,* 1979b).

G. Preservation of Virus Inoculum

To aid in diagnosis of new diseases and for other virus studies, it is often useful to store virus inocula rather than maintain stock cultures in plants in the glasshouse. Storage saves space and minimizes the risk of cross-contamination or change in the virus isolate. Preservation of purified viruses is discussed in Chapter 3, Section VI.

Many strains of TMV can be stored for long periods in air-dried leaf or in nonsterile aqueous media. Inoculum for most other viruses loses infectivity more or less rapidly unless special conditions are met. Most procedures involve removal of water from the tissue or liquid, the addition of protectant materials, storage at low temperature, or a combination of these procedures. Hollings and Stone (1970) found that a high proportion of the 74 viruses they tested survived at least a year, after lyophilization of infective leaf sap. D-Glucose and peptone were added before lyophilization, and the ampules were stored at room temperature. Some remained infective for over 10 years.

Skim milk has been used as a protectant for lettuce necrotic yellows rhabdovirus (LNYV) and some other unstable viruses stored in a dehydrated state at 4°C (Grivell *et al.*, 1971).

Deep-frozen liquid inocula are satisfactory for some viruses, for example, watermelon mosaic *Potyvirus* (De Wijs and Suda-Bachmann, 1979), but may be unsatisfactory for others, for example, red clover mottle *Comovirus* (Marcinka and Musil, 1977). Many of the more unstable viruses have been stored for periods of years in pieces of chemically dehydrated tissue held at about 10°C (McKinney *et al.*, 1965). Leaf tissue taken from young, actively growing infected plants and held in sealed vials gave the longest storage of infective virus.

For air transport between countries, fresh leaf samples can be placed between pieces of damp blotting paper, sealed in polyethylene bags, and transported at $\simeq$ 4°C in a "cool bag." For particularly perishable material or where 4°C storage is not available, leaves may be soaked in 50% glycerol for a few hours before being sealed in bags (Alhubaishi *et al.*, 1987).

II. METHODS DEPENDING ON PHYSICAL PROPERTIES OF THE VIRUS PARTICLE

A. Stability and Physicochemical Properties

1. Measurement of Stability

Historically, stability of the virus as measured by infectivity (often in crude extracts) was an important criterion in attempting to establish groups of viruses. Such properties as thermal inactivation, aging at room temperature, and the effect of dilution were used. The main justification for their use was the absence of alternative physical data. It is unfortunate that they are still sometimes invoked, since they have been shown to be far too variable to provide a sound basis for the identification of viruses or the placing of viruses into groups (Francki, 1980). Publication of a virus description based solely on biological properties together with stability in a crude extract will usually do no more than clutter the literature. However, useful comparative information can sometimes be obtained from the kinds of reagent that degrade a virus. For example, those whose subunits appear to be held together mainly by ionic bonds are degraded by strong salt, which tends to stabilize those held together mainly by hydrophobic forces. The latter type may be readily degraded by reagents such as urea.

2. Physicochemical Properties

A virus has certain independently measurable properties of the particle that depend on its detailed composition and architecture. These properties can be useful for identification and as criteria for establishing relationships. The most commonly measured properties are

1. *Density,* measured in water, sucrose, or CsCl solutions. For nonenveloped viruses, this property reflects mainly the proportion of nucleic acid in the virus. The buffer used may affect the results obtained in CsCl (Scotti, 1985).
2. *Sedimentation coefficient* and *diffusion coefficient*. These properties reflect the mass, density, and shape of the virus particle.

3. *Ultraviolet absorption spectrum.* This property depends mainly on the ratio of nucleic acid to protein in the virus. Thus many unrelated viruses with similar compositions have similar absorption spectra. Ultraviolet absorption provides a useful assay for purified virus preparations, provided other criteria have eliminated the possibility of contamination with non-viral nucleic acids or proteins, especially ribosomes. Measurements of A_{260} may be unreliable for rod-shaped viruses that vary substantially in their degree of aggregation from one preparation to another, thus leading to changes in the amount of light scattering for a given virus concentration.

4. *Electrophoresis.* Electrophoretic mobility is now usually determined by migration in polyacrylamide gels. Thus this property will depend both on the size of the particle and on the net charge at its surface. While strains of the same virus will usually have very similar mobilities, different viruses within a group may be distinguished as was shown for the *Cucumovirus* group by Hanada (1984).

B. Ultracentrifugation

1. Analytical Ultracentrifugation

Plant viruses fall in a size range that makes them very suitable for assay using the analytical ultracentrifuge. The technique is particularly useful for monitoring the progress of a virus purification procedure, for studying the effects of various treatments on the physical state of a virus, and for assay of amounts of virus in crude tissue samples. With Schlieren optics, virus at 100 μg/ml can usually be measured, and it can be detected down to about 50 μg/ml. The best measurements are made at concentrations of single components in the range 0.5–2.0 mg/ml. One of the big advantages of the analytical centrifuge for virus assay is that, as well as giving a measure of the amount of material present, it can provide a physical criterion of identity, the sedimentation coefficient.

Analytical centrifugation is a valuable method for studying the purity of virus preparations with respect to (i) presence of viruslike particles with different sedimentation properties, (ii) presence of a contaminating virus, (iii) presence of contaminating host macromolecules, and (iv) state of aggregation of the virus.

2. Density Gradient Centrifugation

Density gradient centrifugation, developed by Brakke (1951, 1960), is a method that can be used for both isolation and assay of plant viruses. It has proved to be a highly versatile technique and is widely used in the fields of virology and molecular biology. It has largely replaced the analytical ultracentrifuge in studies on viruses. A centrifuge tube is partially filled with a solution having a decreasing density from the bottom to the top of the tube. For plant viruses, sucrose is commonly used to form the gradient, and the virus solution is layered on top of the gradient. With gradients formed with cesium salts, the virus particles may be distributed throughout the solution at the start of the sedimentation or they may be layered on top of the density gradient.

Brakke (1960) defined three ways in which density gradients may be used. *Isopycnic gradient centrifugation* occurs when centrifugation continues until all the particles in the gradient have reached a position where their density is equal to that of the medium. This type of centrifugation separates different particles on the basis of their different densities. Sucrose alone may not provide sufficient density for isopycnic banding of many viruses. In *rate zonal sedimentation* the virus is layered over a preformed gradient before centrifugation. Each kind of particle sediments as a zone or band through the gradient, at a rate dependent on its size, shape, and density. The centrifugation is stopped while the particles are still sedimenting. *Equilibrium zonal sedimentation* is like rate zonal sedimentation except that sedimentation is continued until most of the particles have reached an approximate isopycnic position. The role of the density gradient in these techniques is to prevent convectional stirring and to keep different molecular species in localized zones. The theory of density gradient centrifugation is complex. In practice, it is a simple and elegant method that has found widespread use in plant virology.

A high-speed preparative ultracentrifuge and appropriate swingout or angle rotors are required. Following centrifugation, virus bands may be visualized due to their light scattering. The contents of the tube are removed in some suitable way prior to assay. The bottom of the tube can be punctured and the contents allowed to drip into a series of test tubes. Fractionating devices based on upward displacement of the contents of the tube with a dense sucrose solution are available commercially. The UV absorption of the liquid column is measured and recorded, and fractions of various sizes can be collected as required. Figure 2.3 illustrates the sensitivity of this procedure.

Since successive fractions from a gradient can be collected, a variety of procedures can be used to identify the virus, noninfectious viruslike components, and host materials. These include infectivity, UV absorption spectra, and examination in the electron microscope. Using appropriate procedures, very small differences in sedimentation rate can be detected (Matthews and Witz, 1985).

With rate zonal sedimentation, if the sedimentation coefficients of some components in a mixture are known, approximate values for other components can be estimated. If antisera are available, serological tests can be applied to the fractions, or antiserum can be mixed with the sample before application to the gradient. Components reacting with the antiserum will disappear from the sedimentation pattern.

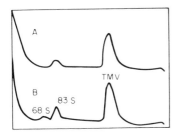

Figure 2.3 Density gradient centrifugation for the assay of viruses in crude extracts. An extract of 30 mg of epidermis (A) or the underlying tissue (B) from a tobacco leaf infected with TMV was sedimented in a 10–40% sucrose gradient at 35,000 rpm for 2 hours. Absorbancy at 254 nm through the gradient was measured with an automatic scanning device. Note absence of detectable 68 S ribosomes in the epidermal extracts. (P. H. Atkinson and R. E. F. Matthews, unpublished.)

C. Electron Microscopy

1. Number of Virus Particles per Unit Volume

A very crude but rapid indication of relative numbers of virus particles can be obtained for reasonably concentrated preparations by mixing a drop of the solutions with an appropriate amount of phosphotungstic acid stain, placing a small amount of the mixture on an electron microscope grid, and examining directly in the microscope for characteristic virus particles. To obtain an accurate estimate of the number of virus particles by electron microscopy it is necessary to know the volume of solution being examined and to be able to count all the particles that were in that volume. Backus and Williams (1950) described a method in which the virus samples were diluted in a solution containing volatile salts (ammonium acetate or ammonium carbonate). The sample was then mixed with a solution containing a known weight of polystyrene latex particles of known and uniform size. The mixture was sprayed onto electron microscope grids using an atomizer.

The number of polystyrene latex particles present can be counted in photographs of drops, and the number present gives an estimate of drop size. The number of characteristic virus particles in the drop also is counted. The ratio of number of virus particles to number of latex particles will vary in different drops. A number of drops must be counted to give a reliable estimate of particle number, and appropriate statistical procedures must be applied (e.g., Williamson and Taylor, 1958). The method is laborious, but gives valuable information in some kinds of experiments.

2. Virus Identification

Knowledge of the size, shape, and any surface features of the virus particle is a basic requirement for virus identification. Electron microscopy can provide this information quickly and, in general, reliably. For the examination of virus particles in crude extracts or purified preparations, a negative-staining procedure is now used almost universally. Commonly used negative stains are sodium phosphotungstate or uranyl acetate, depending on the stability of the virus to these stains.

Approximate particle dimensions can be determined. This is particularly useful for rod-shaped viruses for which particle length distributions can be obtained. Measured particle size may depend very much on how the specimen is prepared and stained (Fig. 2.4). Measurements of particle diameters made on crystalline arrays seen in thin sections of infected cells may also be in error (Hatta, 1976). Diameters of icosahedral viruses are sometimes best determined by measuring linear arrays of particles on the grid. Surface features may be seen best in isolated particles on the grid. Depending on size and morphology, a virus may be tentatively assigned to a particular taxonomic group. However, some small icosahedral viruses cannot be distinguished from members of unrelated groups on morphology alone (Hatta and Francki, 1984). Good results depend very much on optimizing the extraction method and the staining procedure for the particular virus and host plant. Details of current methods are given by Milne (1984a), Roberts (1986), and Christie *et al.* (1987).

For many viruses, examination of thin sections by electron microscopy is a valuable procedure for detecting virus within cells and tissues, but this has its limitations. The large enveloped viruses, the plant reoviruses, and the rod-shaped

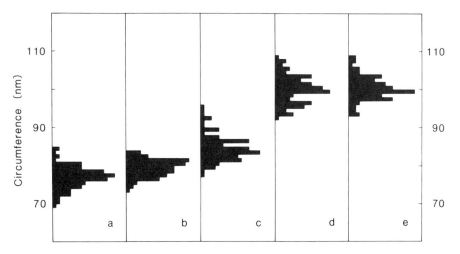

Figure 2.4 Effects of various treatments on the apparent diameter of an icosahedral virus. Histograms of the circumference measurements (nm) of tobacco ringspot *Nepovirus* (TRSV) particles treated in different ways: particles freeze-dried and shadowed after washing grids with ammonium formate (a) and ammonium acetate (b) buffers; grids negatively stained with 2% ammonium molybdate, pH 7.0 (c), 2% uranyl acetate, pH 3.5 (d), and 2% methylamine tungstate, pH 7.0 (e). Only particles negatively stained with ammonium molybdate (c) have dimensions similar to those from freeze-dried preparations. (From Roberts, 1986.)

viruses can usually be readily distinguished because their appearance in thin sections generally differs from that of any normal structures. However, the concentration of virus in the cell and the distribution of particles must be such that there is a reasonable probability of observing virus particles if they are present, in a random section through the cell. It should be remembered that it would require 1000 serial sections 50 nm thick to examine completely the contents of a single cell 50 μm in diameter.

Most of the small isometric viruses have staining properties and apparent diameters that make it very difficult to distinguish scattered individual virus particles from cytoplasmic ribosomes. Some of these viruses form crystalline arrays within the cell and can thus be recognized as virus particles rather than ribosomes.

Some isometric viruses can be induced to form readily recognizable intracellular crystalline arrays if the water content of the tissue is reduced either by wilting (Milne, 1967; Ushiyama and Matthews, 1970) or by plasmolysis with sucrose solutions (Hatta and Matthews, 1974). The pattern of particles seen in the arrays will depend on the relationship between the plane of the section and the crystal lattice (Hatta, 1976).

Ribosomes are susceptible to digestion with RNase while a small isometric virus may be resistant. This characteristic has been developed as a method for detecting such virus particles in cells (Hatta and Francki, 1981a). If isometric virus particles occur in cells or organelles where 80 S ribosomes are absent, scattered virus particles may be recognized, for example, in nuclei (Esau and Hoefert, 1973), plasmodesmata (De Zoeten and Gaard, 1969b), sieve elements (Esau and Hoefert, 1972), and vacuoles. However, nuclei of healthy cells sometimes contain crystalline structures that might be mistaken for viral inclusions (e.g., Lawson *et al.*, 1971).

The identification technique that combines serological diagnosis with electron microscopy is discussed later in this chapter. Electron microscopy in relation to the architecture of virus particles is further discussed in Chapter 5.

D. Chemical Assays for Purified Viruses

Procedures used in chemical analysis can be used for the assay of viruses that can be obtained in a sufficiently purified condition. The simplest procedure, and one that is often overlooked, is to measure the dry weight content of a given volume of the solution. However, dry weights do not distinguish between virus and noninfectious particles containing less than the full amount of nucleic acid. Dry weight measurements form the basis for determining the nitrogen and phosphorus content of the virus and of other components such as ribose or particular amino acids. Measurement of one of these components then could provide a method of virus assay in purified solutions. A solution containing a known weight of purified virus can be used to determine the absorbancy at 260 nm for a given weight per milliliter of the virus, and also factors for making refractive index measurements with a differential refractometer, for converting the areas under Schlieren peaks, or absorbancy peaks from sucrose gradients to weight of virus. These measures can then be used for the assay of other preparations of the virus.

E. Assay Using Radioisotopes

1. *In Vivo* Experiments

When the radioactive isotope ^{32}P as orthophosphate is introduced into tissue where a virus is multiplying, either through the roots of intact plants or in tissue floated on a solution of the isotope, the viral RNA becomes labeled. Similarly, ^{35}S-labeled sulfate can be used to label virus protein. These two isotopes are fairly cheap and are readily available with high specific activities. In certain circumstances they can be used to detect very small amounts of virus. ^{35}S-Labeled methionine is a very convenient material for labeling virus proteins in experiments using protoplasts to study virus replication.

2. *In Vitro* Experiments

For viruses such as TMV and TYMV that can be fairly readily freed of most contaminating material, use of radioactive virus provides a sensitive and accurate assay for certain kinds of *in vitro* experiments. To obtain the best yield of labeled virus, plants are fed the isotope for a period of days during the time of maximum virus increase. For intact plants the highest practicable amount of ^{32}P-labeled orthophosphate or ^{35}S-labeled sulfate is about 10 mC per plant. Using these amounts, a purified virus is obtained containing roughly 1000 cpm/μg counting with an efficiency of about 5% for both labels.

Most methods of virus isolation from infected tissue lead to losses of some virus. The extent of such losses can be estimated by adding a very small known amount of radioactive virus to the starting material and estimating the loss of radioactivity as the isolation proceeds (e.g., Fraser and Gerwitz, 1985).

III. METHODS DEPENDING ON PROPERTIES OF VIRAL PROTEINS

Some of the most important and widely used methods for assay, detection, and diagnosis depend on the surface properties of viral proteins. For most plant viruses this means the protein or proteins that make up the viral coat. Different procedures may use the protein in the intact virus or the protein subunits from disrupted virus. Quite recently nonstructural proteins coded for by a virus have been used in diagnosis.

A. Serological Procedures

1. The Basis for Serological Tests

Antigens

Antigens are usually fairly large molecules or particles consisting of, or containing, protein or polysaccharides that are foreign to the vertebrate species into which they are introduced. Most have a molecular weight greater than 10,000, although smaller peptides can elicit antibody production. There are two aspects to the activity of an antigen. First, the antigen can stimulate the animal to produce antibody proteins that will react specifically with the antigen. This aspect is known as the immunogenicity of the antigen. Second, the antigen must be able to combine with the specific antibody produced. This is generally referred to as the antigenicity of the molecule. Some small molecules with a specific structure such as amino acids may not be immunogenic by themselves but may be able to combine with antibodies produced in response to a larger antigen containing the small molecule as part of its structure. Such small molecules are known as *haptens*.

Large molecules are usually more effective immunogens than small ones. Thus, plant viruses containing protein macromolecules are often very effective in stimulating specific antibody production. The subunits of a viral protein coat are much less efficient. About 15 amino acids at the surface of a protein may be involved in an antigenic site. However, there are difficulties in defining such sites precisely. These are discussed by Van Regenmortel (1989b).

Antibodies

One response of a vertebrate animal to the presence of a foreign antigen is the production of antibody proteins known as immunoglobulins (IgG), which circulate in the body and are capable of binding specifically to the antigen. The structure of an IgG molecule is shown schematically in Fig. 2.5. It consists of two light chains and two heavy chains joined together by disulfide bridges. The two identical antigen binding sites are made from the variable regions of the light and heavy chains.

These variable regions consist of a framework of relatively conserved amino acid sequence interrupted by several regions of highly variable sequence. These hypervariable regions interact to form the two highly specific antigen binding sites. These variable amino acid sequences arise through multiple rearrangements of the DNA of the genes coding for the light and heavy chains within the cells of the B lymphocyte lineage. These are the cells that secrete antibodies in response to the presence of an antigen. There are two general stages in the differentiation of B cell lineages. The first stage, which is independent of the presence of antigen, begins in the fetus and is maintained continuously in the bone marrow of the adult. Each new B cell as it is generated expresses on its surface the IgG molecule of its particular structure and therefore antigen-binding potential. All its progeny express the same IgG. The B cells then migrate to other organs of the immune system such as spleen and lymph nodes, where they remain as resting cells in the absence of an appropriate antigen. The second stage of B cell differentiation occurs when circulating antigen combines with the surface receptor IgG molecule. This stimulates the B cell to

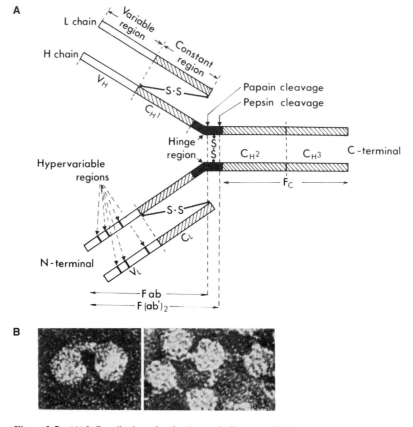

Figure 2.5   (A) IgG antibody molecule. Arrows indicate specific sites where the molecule is cleaved by the enzymes papain and pepsin to give the Fab, Fc, and (Fab')₂ fragments. L, light chains; H, heavy chains. The four polypeptide molecules are joined by disulfide bridges. The two antigen-combining sites are made up from the variable regions of the L and H chains. (From Van Regenmortel, 1982.) (B) Antibody–virus complexes. Electron micrographs showing individual particles of STNV linked together by virus-specific IgG antibody molecules. (Courtesy of S. Höglund.)

proliferate and differentiate into plasma cells that secrete into the bloodstream large amounts of the IgG molecule of exactly the same specificity as that present on the surface of the progenitor B cell. During this proliferation, large numbers of point mutations occur in the variable region DNA. These mutants are under selection pressure from antigen and thus the fit between antibody and antigenic site becomes refined.

An antigenic protein molecule will have several, and perhaps many different structural sites, or *antigenic determinants,* on its surface that can be recognized by the surface receptor IgG molecule of some B cell lineage. Thus for any given protein antigen there may be many different B lymphocytes with slightly differing binding sites that will be stimulated by the antigen. Thus the resulting antiserum will be *polyclonal,* that is, it will contain many different antibodies combining with the antigen, each arising from a different clone of B lymphocytes.

B lymphocytes cannot be cultured *in vitro.* To overcome this problem, Köhler and Milstein (1975) took B lymphocytes from an immunized mouse and fused these *in vitro* with an "immortal" mouse myeloma cell line. Selection of appropriate single fused cells gave "hybridomas" producing only a single kind of antibody—a *monoclonal* antibody. The uses of monoclonal antisera in plant virology are discussed in Section III, A,5.

Another important feature of IgG molecules of many species is the ability of the Fc region (see Fig. 2.5) to bind protein A with very high affinity. Protein A is a molecule isolated from the cell wall of *Staphylococcus aureus*. This binding of protein A is used in several serological procedures.

Production of Antisera

Antisera have been produced against plant viruses in a variety of animals. Rabbits have been used most often since they respond well to plant virus antigens, are easy to handle, and produce useful volumes of serum. Individual animals may vary quite widely in their response to a particular antigen. The amount and specificity of antibody produced in response to a given plant virus antigen may be determined genetically. Unstable plant viruses may have their immunogenicity increased by chemical stabilization (e.g., Francki and Habili, 1972). Protocols for the production of antisera and for the removal of any antibodies reacting with host antigens are given by Van Regenmortel (1982).

Although mice provide relatively small volumes of antiserum, their use may be an advantage when very small amounts of viral antigen are available. Highly inbred strains of mice also minimize variation in response between animals. Mice are of course used in the production of monoclonal antisera, as are rats. Chickens are convenient animals to use for the production of polyclonal antisera. When laying hens are immunized large quantities of purified IgG can be obtained from the egg yolks in a relatively short time (Van Regenmortel, 1982).

Purified IgG from rabbit antisera can be obtained by isolating a specific precipitate of virus and IgG, dissociating virus and IgG at low pH, and removing the virus by high-speed centrifugation. Alternatively, to avoid long acid treatment, antibodies can be bound to virus made insoluble by polymerization with glutaraldehyde. This procedure requires only a short low-speed centrifugation after acid treatment to remove the virus (Maeda and Inouye, 1985).

It has been assumed for a long time that plants could not produce antibodies. However, recent work has shown that transgenic tobacco plants containing the genes for the light and heavy chains of a mouse antibody can produce functional antibody protein in good yield (see Chapter 16, Section II,C,1).

Methods for Detecting Antibody–Virus Combination

A wide variety of methods have been developed for demonstrating and estimating combination between antibodies and antigens. Some of these, such as complement fixation and anaphylaxis, have rarely been used with plant viruses and will not be further discussed. Most traditional methods for using antisera with plant viruses involved direct observation of specific precipitates of virus and antibody, either in liquid media or in agar gels. Over about the past 10 years these methods have been progressively superseded by the use of enzyme-linked immunoabsorbent assay (ELISA), immunoabsorbent electron microscopy, and "dot blots" employing either polyclonal or monoclonal antibodies.

Advantages and Limitations

Serological tests provide rapid and convenient methods for the identification and estimation of plant viruses, the main advantages being the following: (i) the specificity of the reaction allows virus to be measured in the presence of host material or other impurities; (ii) results are obtained in a few hours or overnight compared with days for infectivity assays; (iii) the methods give an answer that is directly proportional to virus concentration over a wide range of concentrations; (iv) some serological detection and assay procedures are more sensitive than infectivity measurements; (v) serological tests are particularly useful with viruses that have no good local lesion host or that are not sap transmissible; and (vi) antisera can be stored and comparable tests made over periods of years and in different laboratories.

It must be borne in mind that serological tests detect and measure the virus protein antigen, not the amount of infective virus. This fact may, of course, be used to advantage in some situations.

The possibility of chance cross-reactions must be considered. Just a few amino acids in a protein sequence are able to elicit an antibody combining with that sequence, provided it is suitably displayed at the surface of the protein. Thus it is theoretically possible that two quite unrelated protein antigens might elicit cross-reacting antibodies purely because of a chance short amino acid sequence similarity. Such similarity has been established between the coat protein of the U1 strain of TMV and the large subunit of the host protein ribulose bisphosphate carboxylase (Dietzgen and Zaitlin, 1986). There was no cross-reaction with distantly related TMV strains or other viruses.

Besides forming the basis for a range of assay and diagnostic methods, serological reactions can be used in a variety of other ways, such as investigating virus structure (Chapter 5), virus relationships (Chapters 13 and 17), and virus activities in cells (Chapter 6). Technical details of methods are given by Van Regenmortel (1982). As of 1984, more than 300 plant viruses had been identified by antisera (Van Regenmortel, 1984a).

2. Precipitation Methods

Precipitation in Liquid Media

The formation of a visible specific precipitate between the antigen and antibody is one of the most direct ways of observing the combination between antibody and virus, but relatively high concentrations of reagents are needed.

Plant viruses are polyvalent, that is, each particle can combine with many antibody protein molecules. The actual number that can combine depends on the size of the virus antigen. Polyvalent antigen combines with divalent antibody to form a lattice-structured precipitate. The precipitation reaction of a virus with a particular antiserum is clearly delineated if a series of tests are made with twofold dilutions of both reagents in all combinations, under standard conditions of temperature and mixing (Matthews, 1957).

In a modification of the precipitation reaction in tubes, known as a ring test, a small volume of undiluted or slightly diluted serum is placed in a small glass tube and overlaid carefully with the virus antigen solution. With time, antibody diffuses into the virus solution and virus diffuses into the antiserum. Somewhere near the boundary, a zone of specific precipitate will form, provided that both reagents are sufficiently concentrated. It is a useful quick test, but is rather insensitive.

When a drop of freshly expressed leaf sap from plants infected with viruses occurring in high concentration is mixed on a microscope slide with a drop of an antiserum, clumping of small particles of host material occurs. This may be seen with the naked eye but is viewed more readily with a hand lens or low-powered microscope. Chloroplasts and chloroplast fragments are prominent in the clumped aggregates. However, some viruses such as wheat streak mosaic *Potyvirus* may undergo rapid antigenic modification in leaf extracts unless glutaraldehyde fixation is used (Langenberg, 1989).

Precipitation Methods Using Virus Bound to Larger Objects

The sensitivity of the precipitation reaction depends on the smallest amount of antigen that will form a visible precipitate. The smaller the antigen, the greater the weight of antigen required. Various methods have been developed in which the virus is adsorbed to particles substantially larger than viruses, such as latex or red blood cells, before reaction with antiserum. Compared to simple precipitation tests these procedures allow the detection of 100- to 1000-fold smaller quantities of virus (Van Regenmortel, 1982). Latex agglutination allowed the detection of between 0.1 and 0.5 μg/ml of six rod-shaped viruses infecting legumes (Demski *et al.*, 1986). A simple field kit using latex agglutination for three viruses infecting potatoes allowed results to be obtained within 3–10 minutes (Talley *et al.*, 1980). Coating of latex particles with protein A has been used to increase the sensitivity of the procedure (Torrance, 1980).

Reverse passive hemagglutination (RPH) is a procedure in which antibody is linked nonspecifically to red blood cells by some appropriate treatment. A solution containing virus is added to a suspension of the antibody-linked red blood cells in round-bottom microtiter plates. In a negative test the cells settle as a compact button. Agglutination occurs in a positive test, giving a shield of cells over the bottom of the tube. Sander *et al.* (1989), using TMV as a model system, have shown that RPH is equal in sensitivity and specificity to double antibody sandwich ELISA. The method involves only one step; readings can be made by eye within 90 minutes without the need of equipment; and appropriately stabilized red blood cells can be stored for long periods.

Immunodiffusion Reactions in Gels

The great advantages of immunodiffusion reactions carried out in gels are: (i) mixtures of antigenic molecules and their corresponding antibodies may be physically separated, either because of differing rates of diffusion in the gel, or because of differing rates of migration in an electric field (in immuno-electrophoresis), or by a combination of these factors; and (ii) direct comparisons can be made of two antigens by placing them in neighboring wells on the same plate. The method may not be as sensitive as the tube precipitation method in terms of a detectable concentration of virus, but of course much smaller volumes of fluid are required. In early work, immunodiffusion tests were carried out in small tubes (i.e., in one dimension). However, this type of test has been superseded by double diffusion tests in two dimensions on glass slides. Ouchterlony (1962) gives a general account of these methods.

Wells are punched in the agar in a defined geometrical arrangement. It is usual in double diffusion tests to have the antiserum in a central well and the antigen solutions being tested in a series of wells surrounding the central well. Antigen and antibody diffuse toward each other in the agar, and after a time a zone will form where the two reagents are in suitable proportions to form a precipitating complex. Both reactants leave the solution at this point and more diffuses in to build up a visible line of precipitation that traps related antigen and antibody. Unrelated antigens or antibodies can pass through the band of precipitation. Bands can be recorded either by direct visual observation with appropriate lighting, by the use of protein stains, or by photography. When radioactive virus is used, radioautography can be used to detect the bands.

When comparing bands formed by two antigens in neighboring wells, several types of pattern have been distinguished. Movement of antigen in the agar gel is strongly dependent on size and shape of the virus. For small isometric plant viruses the methods are satisfactory. Rods may diffuse slowly or not at all. For routine tests a suitable detergent in the immunodiffusion system may allow rapid migration in the gel of virus degradation products. Elongated virus particles can also be made to diffuse by sonication. This has the advantage that the original antigenic surface of the virus survives in the sonicated fragments (e.g., Moghal and Francki, 1976).

A recent modification of gel diffusion uses an agarlike polysaccharide from *Pseudomonas elodea* (Gelrite) instead of agar. Antigens were detected more rapidly in Gelrite, and with about a 100-fold increase in sensitivity. This would make the test almost comparable with ELISA in this respect (Ohki and Inouye, 1987).

Immunoelectrophoresis in Gels

In immunoelectrophoresis, a mixture of antigens is first separated by migration in an electric field in an agar gel containing an appropriate buffer. Antiserum is then placed in a trough parallel to the path of electrophoretic migration and an immunodiffusion test is carried out. This is a powerful method for resolving mixtures of antigens as two independent criteria are involved—electrophoretic mobility and antigenic specificity.

In immuno-osmophoresis two sets of wells are cut in agar buffered near pH 7. Antiserum is placed in one well of each pair and a virus dilution in the other, and a voltage gradient is applied. Antibody protein moves as a result of endosmotic flow. The virus moves because of a net negative charge. Thus the reagents are brought together in the gel much more rapidly than by simple gel diffusion (e.g., John, 1965).

In rocket immunoelectrophoresis, antigen migrates in an electric field in a layer of agarose containing the appropriate antibody. The migration of the antigen toward the anode give rise to rocket-shaped patterns of precipitation. The area under the rocket is proportional to antigen concentration. This procedure has been adapted to the assay of 25-ng to 10-μg amounts of CaMV in leaf extracts from about 25 mg of tissue (Hagen *et al.*, 1982).

3. Radioimmune Assay

Radioimmune assay, using radioactively labeled viral antigens, provides a sensitive and specific method for measuring plant viral antigens (Ball, 1973). In this method, viral antibody is irreversibly attached to the walls of disposable tubes. The amounts of labeled virus that bind to the surface of such tubes can then be measured. In view of the sensitivity now available with ELISA procedures this method of assay is unlikely to be widely used.

4. ELISA Procedures

Clark and Adams (1977) showed that the microplate method of ELISA could be very effectively applied to the detection and assay of plant viruses. Since that time the method has come to be more and more widely used. Many variations of the basic procedure have been described, with the objective of optimizing the tests for particular purposes.

The method is very economical in the use of reactants and readily adapted to quantitative measurement. It can be applied to viruses of various morphological types in both purified preparations and crude extracts. It is particularly convenient when large numbers of tests are needed. It is very sensitive, detecting concentrations as low as 1–10 ng/ml. Detailed protocols are given by Van Regenmortel (1982), Koenig and Paul (1982), Hill (1984b), and Clark and Bar Joseph (1984). Two general procedures have been developed.

Direct Double Antibody Sandwich Method

The principle of the direct double antibody sandwich procedure is summarized in Fig. 2.6. The kinds of data obtained in ELISA tests is illustrated in Fig. 2.7. This is the method described by Clark and Adams (1977). It has been widely used, but suffers two limitations: (i) It may be very strain specific. For discrimination between virus strains this can be a useful feature, however for routine diagnostic tests it means that different viral serotypes may escape detection. This high specificity is almost certainly due to the fact that the coupling of the enzyme to the antibody interferes with weaker combining reactions with strains that are not closely related. (ii) It requires a different antivirus enzyme–antibody complex to be prepared for each virus to be tested.

Indirect Double Antibody Sandwich Methods

In the indirect procedure, the enzyme used in the final detection and assay step is conjugated to an antiglobulin antibody. For example, if the virus antibodies were raised in a rabbit, a chicken antirabbit globulin might be used. Thus one conjugated globulin preparation can be used to assay bound rabbit antibody for a range of viruses. Furthermore, indirect methods detect a broader range of related viruses with a single antiserum (Koenig, 1981).

Many variations of these procedures are possible. Koenig and Paul (1982) studied nine of them with the aim of optimizing the tests for different objectives (Fig. 2.8). Their results emphasized the versatility of the assay for different applications. They concluded that the direct double antibody sandwich method is the most convenient for the routine detection of plant viruses in situations where strain specificity and very low virus concentrations cause no problems. The broadest range of serologically related viruses is detected by indirect ELISA using unprecoated

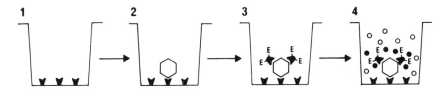

Figure 2.6 Principle of the ELISA technique for plant viruses (direct double antibody sandwich method). (1) The gamma globulin fraction from an antiserum is allowed to coat the surface of wells in a polystyrene microtiter plate. The plates are then washed. (2) The test sample containing virus is added and combination with the fixed antibody is allowed to occur. (3) After washing again, enzyme-labeled specific antibody is allowed to combine with any virus attached to the fixed antibody. (Alkaline phosphatase is linked to the antibody with glutaraldehyde.) (4) The plate is again washed and enzyme substrate is added. The colorless substrate *p*-nitrophenyl phosphate (○) gives rise to a yellow product (●), which can be observed visually in field applications or measured at 405 nm using an automated spectrophotometer. An example of the kind of data obtained is given in Fig. 2.7. (Modified from Clark and Adams, 1977, with permission from Cambridge University Press.)

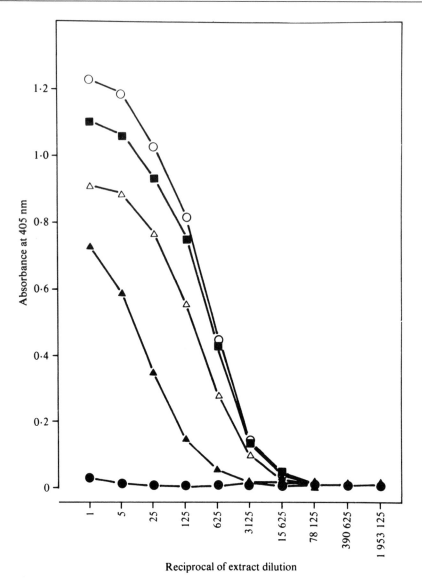

Figure 2.7 Example of data obtained in an ELISA test. Detection of lettuce necrotic yellows rhabdovirus (LNYV) in *N. glutinosa* plants systemically infected for various periods. Uninfected plants, ●——●; and plants infected for 7 days (no systemic symptoms), ▲——▲; 15 days (prominent systemic symptoms), ■——■; 19 days (prominent systemic symptoms), ○——○; and 30 days (severe chlorosis and stunting), △——△. (From Chu and Francki, 1982.)

plates, but this procedure is open to interference by crude plant sap. Precoating of plates with antibodies or their $F(ab')_2$ fragments (Barbara and Clark, 1982) eliminated the interference problem but narrowed the specificity. Other forms of interference may occur. For example, roots of herbaceous plants contain a factor that makes the use of protein A–horseradish peroxidase unsatisfactory as an enzyme conjugate (Jones and Mitchell, 1987). Other viral-coated proteins such as the cylindrical inclusion body protein produced by potyviruses can be used for diagnosis with ELISA methods (Yeh and Gonsalves, 1984).

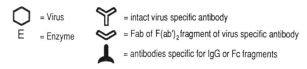

○ = Virus
E = Enzyme

Y = intact virus specific antibody
ᕔ = Fab of F(ab')₂ fragment of virus specific antibody
= antibodies specific for IgG or Fc fragments

(i) Black and white symbols indicate antibodies derived from two different animal species.
(ii) Shaded areas indicate a reaction between the Fc portion of a virus-specific antibody and an Fc-specific antibody.
(iii) Reagents added in order from bottom of diagram except where vertical bars indicate two reagents are preincubated together before addition to the plate.

Direct procedures   Indirect procedures

Short procedures with preincubation of two ingredients.

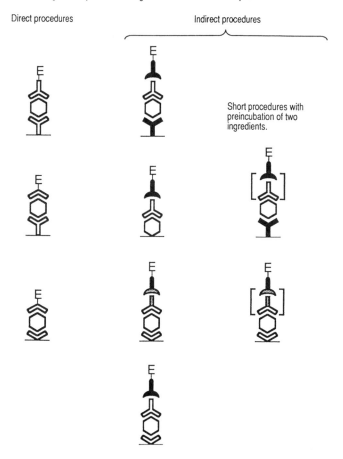

Figure 2.8 Diagrammatic representation of some variations of ELISA. (Modified from Koenig and Paul, 1982.)

Other Modifications

The following examples illustrate the versatility of the basic ELISA procedure.

1. *Biotin avidin*   This detection system is based on the very high affinity of the protein avidin for biotin. The biotin is chemically coupled to the globulin fraction of the antiserum, while the enzyme to be used for detection is coupled to the avidin. Biotin coupling does not interfere with the binding capacity of the antiviral antibody so that a broad range of viral serotypes can be detected using the direct double antibody sandwich procedure, and the assays have increased sensitivity (Zrein *et al.*, 1986). Further advantages are speed of development and versatility (Hewish *et al.*, 1986).

2. *Use of a fluorogenic substrate*   Torrance and Jones (1982) showed that a fluorogenic substrate of alkaline phosphatase (4-methyl-umbelliferyl phosphate) gave a more sensitive ELISA test than the standard chromogenic substrate, but the technique requires an expensive plate-reading machine. Reichenbächer *et al.* (1984) developed an ultramicro ELISA test using the same fluorogenic substrate and plates with a 10-μl test volume. Time-resolved fluoroimmunoassay has been developed to increase sensitivity by reducing background (Siitari and Kurppa, 1987). This is done by using a europium chelate that produces an intense fluorescence with a very long decay time, during which background fluorescence has decayed. Using this procedure with MAbs, Sinijärv *et al.* (1988) were able to detect PVX at 5 pg/ml in potato tuber extracts diluted 7×10^4-fold, and in leaf extracts diluted 2×10^7-fold.

3. *Enzyme amplification*   Torrance (1987) described a procedure for increasing the sensitivity of ELISA assays, in which the enzyme bound to the antibody catalyzes the conversion of NADP to NAD, which then takes part in a second enzyme-mediated cyclic reaction to produce a red-colored end product. The method could be used for rapid diagnosis of a luteovirus occurring in very low concentrations in plants and also for detecting the virus in individual aphid vectors.

4. *Use of F(ab')$_2$ and a protein A–enzyme conjugate*   Barbara and Clark (1982) described a simple indirect ELISA test in which the virus was trapped by F(ab')$_2$ fragments of homologous antibody bound to the plate. Trapped virus was then allowed to react with intact antivirus immunoglobulin. Thus only one viral antiserum is needed. Bound immunoglobulin was then assayed by a conjugate of protein A and enzyme. Protein A will bind only to the Fc portion of the intact immunoglobulin. The procedure has been used effectively with several viruses in small fruits (Converse and Martin, 1982).

5. *Radioimmune ELISA*   Ghabrial and Shepherd (1980) developed a simple and highly sensitive radioimmunosorbent assay using the principle of the direct double antibody sandwich technique in microtiter plates, except that ^{125}I-labeled IgG is substituted for the enzyme-linked IgG in an ELISA assay. The bound and labeled IgG is dissociated from the sandwich before being assayed.

6. *Measurement of specific activity of viruses in crude extracts*   Konate and Fritig (1983) combined two procedures to enable the specific activity of virus radioactively labeled *in vivo* to be determined. Indirect double antibody sandwich ELISA was used to estimate the amount of virus in a well. Then, as a second step, the radioactivity of the virus trapped in the well was assayed.

7. *ELISA on polystyrene beads*   Polystyrene beads 6.5 mm in diameter have been used as the solid phase (one bead per test) instead of microtiter plates (Chen *et al.*, 1982). This system was more discriminating for detecting differences among isolates of soybean mosaic *Potyvirus* than the standard plate procedure.

8. *Dot ELISA*   Several laboratories have used nitrocellulose membranes as the solid substrate for ELISA tests. In one procedure the virus in a plant extract is electroblotted onto the membrane as the first step (Rybicki and von Wechmar, 1982). As an alternative first step, the membrane is coated with antiviral IgG by soaking in an appropriate solution (Banttari and Goodwin, 1985). For the final color development, a substrate is added that the enzyme linked to the IgG converts to an insoluble colored material. The intensity of the colored spot can be assessed by eye or by using a reflectance densitometer. This kind of assay was twice as sensitive as the test carried out in microtiter plates for potato leaf roll *Luteovirus* (PLRV) (Smith and Banttari, 1987). An example of a dot blot assay is given in Fig. 2.9.

The dot immunobinding procedure for TMV used by Hibi and Saito (1985) was only about one-tenth as sensitive as a similar ELISA test in microtiter plates.

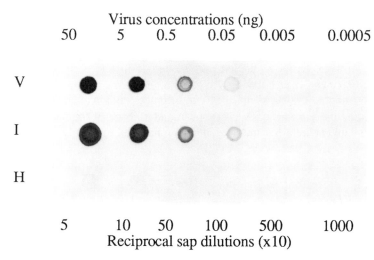

Figure 2.9 The sensitivity of a dot immunobinding assay. V = purified strawberry pseudo mild yellow edge *Carlavirus;* I = crude sap from infected leaves; and H = crude sap from healthy leaves. The conjugated enzyme was alkaline phosphatase and the substrate for color development was Fast Red TR salt. (From Yoshikawa *et al.,* 1986.)

Nevertheless, 1.0 ng of virus could be detected, and its simplicity and speed (a few hours) make the test useful as a practical diagnostic technique.

Graddon and Randles (1986) described a single antibody dot immunoassay for the rapid detection of subterranean clover mottle *Sobemovirus.* The method was found to be 12 times as sensitive as ELISA in terms of total antigen detectable. The main advantages of the method were speed (3 hours to complete a test), low cost, and the small amount of reagents required. Dot blot immunoassays may be particularly useful for routine detection of virus in seeds or seed samples, especially for laboratories where an inexpensive and simple test is needed (Lange and Heide, 1986). However, endogenous insect enzymes may interfere with tests using insect extracts (Berger *et al.,* 1985). Dot blot immunobinding using plain paper has been developed as a diagnostic method for five potato viruses using direct extraction from green leaves (Heide and Lange, 1988).

The technique of dotting the antigen onto nitrocellulose or other papers can be used with methods other than ELISA for detection of the antigen. For example, Hsu (1984) used gold-labeled goat antirabbit IgG to detect rabbit antibodies bound to TMV on nitrocellulose paper. The gold-labeled antibody is directly visible because of its pink color. The method could detect 1–5 pg of TMV protein and is also simple and economical. The use of electrophoretic separation of proteins followed by immunoblotting is discussed in Section III,B,2. Koenig and Burgermeister (1986) discuss the application of various immunoblotting techniques and consider the problem of unexplained or false positive reactions that sometimes occur. These reactions may be due to nonspecific binding of immunoglobulins to coat proteins of certain plant viruses (Dietzgen and Francki, 1987).

Collection, Preparation, and Storage of Samples

As with other diagnostic tests, it is necessary to define and optimize time of sampling and tissue to be sampled to achieve a reliable routine detection procedure (e.g., Torrance and Dolby, 1984; Rowhani *et al.,* 1985; Pérez de san Román *et al.,* 1988). The main factor limiting the number of tests that can be done using ELISA procedures is the preparation of tissue extracts. Various procedures have been tested to minimize this problem (e.g., Mathon *et al.,* 1987). On the other hand, with some viruses at least, it may be possible to avoid the extraction step by placing several small disks of leaf tissue, cut from the leaves to be tested, directly in the well of a microtiter plate (Romaine *et al.,* 1981).

When many hundred of field samples have to be processed it is often necessary to store them for a time before ELISA tests are carried out. Storage conditions may be critical for reliable results. For this reason conditions need to be optimized for each virus and host (e.g., Ward *et al.,* 1987).

5. Monoclonal Antibodies

Since monoclonal antibodies against TMV were described (Dietzen and Sander, 1982; Briand *et al.,* 1982) there has been an explosive growth of interest in this type of antibody for many aspects of plant virus research, and particularly for detection and diagnosis. The term monoclonal antibody has been variously abbreviated as MAb, McAb, MA, and McA. I will use the first of these. By 1986, MAbs had been prepared against more than 30 different plant viruses (Van Regenmortel, 1986), and many additions to the list have been reported since.

The Nature of Antibody Specificity in Relation to MAbs

The binding site of a MAb, or any individual antibody in a polyclonal serum, is able to bind to many different antigenic determinants with varying degrees of affinity, that is, an individual binding site on an antibody is polyfunctional. Furthermore, antibodies against a single antigenic determinant (or epitope) may vary in affinities from one that is scarcely measurable with standard techniques to one that is 100,000-fold higher. Thus it is quite possible for an antibody to bind more strongly to an antigenic determinant differing from the one that stimulated its production. Such a phenomenon may well be obscured by the many different antibodies in a polyclonal antiserum, but with MAbs it will show up clearly. For example, when MAbs were produced following immunization of a mouse with a particular strain of TMV it was found that some of the MAbs reacted more strongly with other strains of the virus (van Regenmortel, 1982; Al-Moudallal *et al.,* 1982). This has important implications for the use of MAbs in the delineation of virus strains (Chapter 13).

Another aspect of the specificity of MAbs must be borne in mind. A particular protein antigen may have only one site for binding a particular MAb. If this single site is shared with another protein, a significant cross-reaction could occur even in the absence of general structural similarity between the two proteins. This may be an uncommon phenomenon. Nevertheless, cross-reactions detected with MAbs cannot be taken to indicate significant structural or functional similarity between two proteins without other supporting evidence (Carter and ter Meulen, 1984).

The Production of MAbs

Technical details for the production of MAbs directed against plant viruses are given by Sander and Dietzen (1984). The use of rats is described by Torrance *et al.* (1986a,b). An outline of the procedure using mice is shown in Fig. 2.10.

The kind of screening test used to test for MAb production is of critical importance (van Regenmortel, 1986). Most workers use some form of ELISA. Multilayered sandwich procedures, especially using chicken antibody in one of the layers, appeared to be the most sensitive (Al-Moudallal *et al.,* 1984). The results

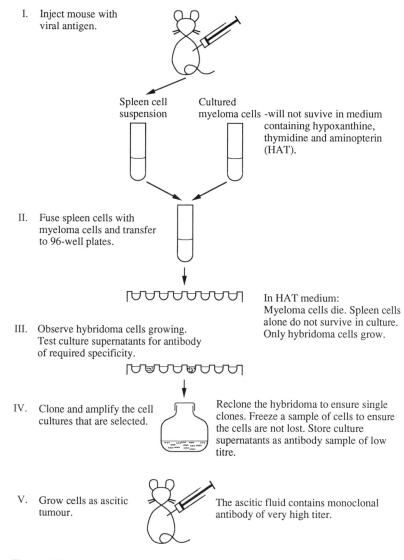

I. Inject mouse with viral antigen.

Spleen cell suspension

Cultured myeloma cells -will not suvive in medium containing hypoxanthine, thymidine and aminopterin (HAT).

II. Fuse spleen cells with myeloma cells and transfer to 96-well plates.

In HAT medium: Myeloma cells die. Spleen cells alone do not survive in culture. Only hybridoma cells grow.

III. Observe hybridoma cells growing. Test culture supernatants for antibody of required specificity.

IV. Clone and amplify the cell cultures that are selected.

Reclone the hybridoma to ensure single clones. Freeze a sample of cells to ensure the cells are not lost. Store culture supernatants as antibody sample of low titre.

V. Grow cells as ascitic tumour.

The ascitic fluid contains monoclonal antibody of very high titer.

Figure 2.10 Steps in the production of monoclonal antibodies.

obtained depend very much on the quality of the reagents used and the exact conditions of pH and other factors in the medium.

Tests Using MAbs

Many MAbs do not provide a precipitation reaction, especially with protein antigen monomers, for which there may be only one antibody combining site per molecule. However, the use of a MAb and an anti-MAb is a convenient way of extracting or precipitating a peptide or protein antigen with a single determinant. For detection and diagnosis, MAbs have been most commonly used in conjunction with ELISA tests or sometimes with radioimmune assays. Again, exact conditions with respect to pH and other factors may be vitally important. For many applications it may be best to use the same ELISA protocol that was used to screen for MAbs during the isolation procedure because quite different MAbs may be selected depending on the ELISA procedure used (van Regenmortel, 1986). If highly concentrated preparations of MAbs are used in ELISA tests, spurious cross-reactions between viruses in different groups may be detected. Such effects can be avoided by the use of milk proteins instead of bovine albumin as a blocking agent and as a diluent (Zimmermann and van Regenmortel, 1989). The reactivity of a MAb with a given antigen may differ greatly depending on the kind of ELISA procedure used Dekker *et al.*, 1989).

Diaco *et al.* (1985) used MAbs in a biotin–avidin ELISA for the detection of soybean mosaic *Potyvirus* in soybean seed, while Omura *et al.* (1986) used MAbs with latex flocculation to detect rice stripe *Tenuivirus* in plants and insects. Sherwood *et al.* (1989) obtained a MAb that reacted in ELISA and dot blot tests with the inner core of tomato spotted wilt virus, a virus with a lipoprotein envelope. For diagnostic purposes it may be useful to select for MAbs that react with a common antigenic determinant on several strains of the virus [e.g., potato virus Y *Potyvirus* (PVY), Gugerli and Fries, 1983; PVX, Torrance *et al.*, 1986a; citrus tristeza *Closterovirus*, Vela *et al.*, 1986]. Alternatively, several strain-specific or virus-specific MAbs may be pooled to give the required reagent (e.g., Dore *et al.*, 1987a; Pérez de san Román *et al.*, 1988).

Advantages of MAbs

1. *Requirements for immunization* Mice and rats can be immunized with small amounts of antigen ($\simeq$100 μg or less). If the virus preparation used is contaminated with host material or other viruses it is still possible to select for MAbs that react only with the virus of interest.
2. *Standardization* MAbs provide a uniform reagent that can be distributed to different laboratories, eliminating much of the confusion that has arisen in the past from the use of variable polyclonal antisera. Furthermore, MAbs can be obtained in almost unlimited quantities in suitable circumstances.
3. *Specificity* MAbs combine with only one antigenic site on the antigen. Thus they may have very high specificity and can therefore provide a refined tool for distinguishing between virus strains (Chapter 13). They can also be used to investigate aspects of virus architecture (Chapter 5) and transmission by vectors (Chapter 14).
4. *High affinity* The screening procedure for detecting MAbs allows for the possibility of obtaining antibodies with very high affinity for the antigen. High-affinity antibodies may be used at very high dilutions, minimizing problems of background in the assays. They can also be used for virus purification by affinity chromatography.
5. *Storage of cells* Hybridomas can be stored in liquid nitrogen to provide a source of MAb-producing cells over a long period.

Disadvantages of MAbs

1. *Preparation* Polyclonal antisera are relatively easy to prepare. The isolation of new MAbs is labor-intensive, time-consuming, and relatively expensive. For any particular project these realities must be weighed against the substantial advantages discussed above.
2. *Specificity* MAbs may be too specific for some applications, especially in diagnosis.
3. *Sensitivity to conformational changes* Because of their high specificity, MAbs may be very sensitive to conformational changes in the antigen,

brought about by binding to the solid phase or by other conditions in the assay.

Summary

Although MAbs provide an excellent tool for many aspects of plant virus research, it is unlikely that they will replace the use of polyclonal antisera in all applications. As far as detection and diagnosis are concerned, it is hoped that a battery of appropriate MAbs will soon become available to diagnostic laboratories in all countries.

6. Serologically Specific Electron Microscopy

A combination of electron microscopy and serology was first used by Larson *et al.* (1950) and also illustrated by Matthews (1957), but the technique of negative staining was needed before the procedure could be developed as an effective and widely applicable diagnostic tool. Derrick (1973) described a procedure in which the support film on an electron microscope grid was first coated with specific antibody for the virus being studied. Grids were then floated on appropriate dilutions of the virus solution for one hour. They were then washed, dried, shadowed, and examined in the electron microscope. Under appropriate conditions, many more virus particles are trapped on a grid coated with antiserum than on one coated with normal serum. Thus the method offers a diagnostic procedure based on two properties of the virus—serological reactivity with the antiserum used and particle morphology. The method was not popular for some years but is now widely used as a diagnostic tool in appropriate circumstances. Various terms have been used to describe the process: serologically specific electron microscopy (SSEM), immunosorbent electron microscopy, solid-phase immune electron microscopy, and electron microscope serology. I will use the first of these terms. The subject has been reviewed by Milne (1990).

The Decoration Technique

Milne and Luisoni (1977) introduced to plant virology a modification of the general procedure in which, after being adsorbed onto the EM grid, virus particles are coated with virus-specific antibody. This produces a halo of IgG molecules around the virus particles that can be readily visualized in negatively stained preparations. The phenomenon was termed "decoration." This procedure probably offers the most convincing demonstration by electron microscopy of specific combination between virus and IgG. For a critical test, a second serologically unrelated virus of similar morphology should be added to the preparation (Fig. 2.11).

Technical details of protocols are given, and difficulties with the procedure discussed by Milne (1984a), Milne and Lesemann (1984), and Katz and Kohn (1984). Particles of some viruses may disintegrate following decoration (Langenberg, 1986b). Nevertheless, the unstable rod-shaped viruslike particles associated with lettuce big vein disease have been recognized using the technique and shown to be related to tobacco stunt virus (Vetten *et al.*, 1987).

The main advantages of the method are: (i) the result is usually clear, in the form of virus particles of a particular morphology, and thus false positive results are rare; (ii) sensitivity may be of the same order as with ELISA procedures and may be one thousand times more sensitive for the detection of some viruses than conventional EM (Roberts and Harrison, 1979); (iii) when the support film is coated with antibody, much less host background material is bound to the grid; (iv) antisera can

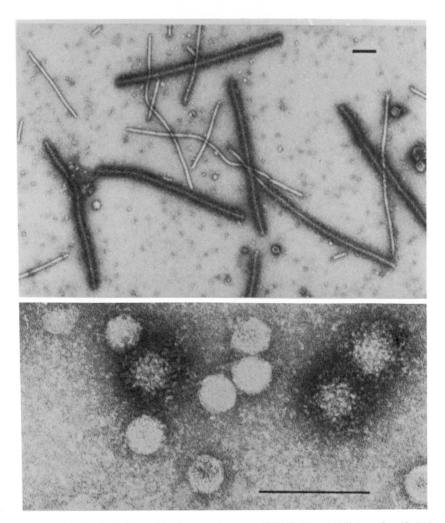

Figure 2.11   Serologically specific electron microscopy (SSEM). (*Bottom*) Mixture of purified tobacco necrosis *necrovirus* (TNV) and tomato bushy stunt *Tombusvirus* (TBSV) adsorbed to the grid and then treated with saturating levels of antibodies to TBSV. TBSV particles are "decorated." TNV particles remain clean, with sharp outlines. Antibody molecules appear in the background. Bar = 100 nm. (*Top*) A natural mixture of two potyviruses from a perennial cucurbit, *Bryonia cretica*. The antiserum used has decorated particles of only one of the viruses in the mixture. One particle near the center is longer than normal and decorated for only part of its length. This particle probably arose by end-to-end aggregation between a particle of each of the two viruses. Bar = 100 nm. (From Milne, 1990.)

be used without fractionation or conjugation, low-titer sera can be satisfactory, and only small volumes are required; (v) very small volumes (≈ 1 μl) of virus extract may be sufficient; (vi) antibodies against host components are not a problem inasmuch as they do not bind to virus; (vii) one antiserum may detect a range of serological variants (on the other hand, the use of monoclonal antibodies may greatly increase the specificity of the test) (e.g., Diaco *et al.*, 1986b); (viii) results may be obtained within 30 minutes; (ix) when decoration is used, unrelated undecorated virus particles on the grid are readily detected; and (x) prepared grids may be sent to a distant laboratory for application of virus extracts and returned to a base laboratory for further steps and EM examination.

Some disadvantages of the procedure are: (i) it will not detect virus structures too small to be resolved in the EM (e.g., coat protein monomers); (ii) sometimes the method works inconsistently or not at all for reasons that are not well understood (Milne and Lesemann, 1984); (iii) it involves the use of expensive EM equipment, which requires skilled technical work and is labor-intensive. For these reasons it cannot compete with, say, ELISA for large-scale routine testing; and (iv) when quantitative results are required, particle counting is laborious, variability of particle numbers per grid square may be high, and control grids are required.

In summary, SSEM cannot replace ELISA tests when large numbers of samples have to be tested. Its main uses are (i) in the identification of an unknown virus, (ii) in situations where only a few diagnostic tests are needed; and (iii) when ELISA tests are equivocal and need direct confirmation.

Immunogold Labeling

Gold particles are highly electron dense and thus show the position on an EM grid of any particle to which they are attached. Protein A forms a reasonably stable complex with gold particles. This complex can then be used to locate any IgG molecules bound to virus coat protein on the EM grid.

Van Lent and Verduin (1985, 1986) prepared gold particles of two average diameters (7 and 16 nm). The size of these particles in relation to virus particles and the specificity of the reaction are illustrated in Fig. 2.12.

Immunogold labeling is particularly valuable in locating viral antigens in thin sections of infected cells (Fig. 7.11). Van Lent and Verduin found that the larger gold particles were more readily seen in stained thin sections. The technique has been adopted for localization of viral antigens by light microscopy (Van Lent and Verduin, 1987) and has been reviewed by Patterson and Verduin (1987).

7. Fluorescent Antibody

The fluorescent antibody method has been applied to the study of the intracellular location and distribution of plant viruses within tissues of the host plant and

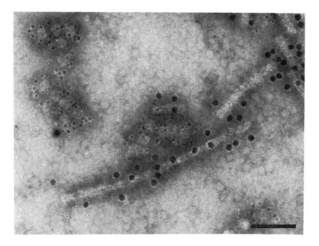

Figure 2.12 Illustration of protein A labeled with gold particles combining with virus-specific IgG near the virus surface. Virus particles were labeled in suspension with protein A–gold complexes. A mixture of purified cowpea chlorotic mottle *Bromovirus* labeled with 7-nm protein A–gold and TMV labeled with 16-nm protein A–gold. Bar represents 100 nm. (From van Lent and Verduin, 1986.)

in insect vectors. Nagaraj (1965) gives the detailed procedures that must be carried out in the preparation and storage of TMV antibody conjugated with fluorescein to avoid nonspecific staining. Appropriate controls are essential. Fluorescent antibody staining is particularly suited to detecting viral antigens in isolated infected protoplasts (Otsuki and Takebe, 1969). Chiu and Black (1969) and others have used standard fluorescent antibody methods very successfully for detecting virus antigens in insect vectors (Chapter 14).

8. Neutralization of Infectivity

When viruses are mixed with a specific antiserum, infectivity is reduced to a greater extent than when mixed with a nonimmune serum. At high dilutions of serum, inactivation may occur only with the specific antiserum (Rappaport and Siegel, 1955). This phenomenon has been little used in plant virology mainly because of the lack of precision in assays for infectivity. Neutralization tests have been used to demonstrate serological relationships between viruses that are not transmissible mechanically (e.g., Rochow and Duffus, 1978). With a sufficiently stable virus such as TMV, infectivity of the virus can be restored by removal of the antibody at low pH. The mechanism of neutralization has not been established. The binding of IgG to the virus protein may prevent effective release of the nucleic acid from the protein coat. Alternatively, it may block virus from attaching to some site within the cell at an early stage of infection. Dietzen (1986) selected a monoclonal antibody specific for a C-terminal antigenic determinant of the TMV coat protein using a chemically synthesized tetrapeptide. This MAb neutralized the infectivity of TMV. A systematic study using these procedures may illuminate the mechanism of neutralization.

B. Electrophoretic Procedures

Electrophoresis in a suitable substrate separates proteins according to size and net electric charge at the pH used. Polyacrylamide gel electrophoresis in a medium containing sodium dodecyl sulfate (SDS-PAGE) is commonly used. The position and amount of the proteins can then be visualized by a nonspecific procedure such as staining or by a specific procedure such as immunoassay.

1. Gel Electrophoresis Followed by Staining

Provided a virus occurs in sufficient concentration, and provided it can be freed sufficiently from any interfering host proteins by some simple preliminary procedure, gel electrophoresis is a rapid method for detecting viral coat proteins (Paul, 1975). Provided they occur in high enough concentration, other viral-coded proteins may be detected. Alper *et al.* (1984) used gel electrophoresis followed by Coomassie brilliant blue staining to detect three potyviruses in leaf extracts of bulbous irises and to differentiate between them. In addition, bands were found in the position expected for the viral-coded inclusion body protein.

2. Electrophoresis Followed by Electroblot Immunoassay

O'Donnell *et al.* (1982) and Rybicki and von Wechmar (1982) were the first to use the protein fractionating power of electrophoresis together with the sensitivity and specificity of solid-phase immunoassay to identify and assay viral proteins. The main stages in the technique are (i) fractionation of the proteins in infected plant sap by SDS–PAGE; (ii) electrophoretic transfer of the protein bands from the gel to activated paper (usually nitrocellulose); (iii) blocking of remaining free binding sites on the paper with protein, usually serum albumin; (iv) probing for the viral coat protein band with specific antiserum; and (v) detection of the antibody–antigen complex with ^{125}I-labeled protein A or by some ELISA procedure. A great advantage of this technique is that it identifies the virus by two independent properties of its coat protein—molecular weight and serological specificity. Shukla *et al.* (1983) found that aminophenylthioether (APT) paper was more efficient than nitrocellulose in binding coat proteins. One-half nanogram of viral protein could be detected compared with 500 ng for Coomassie blue staining. Furthermore, APT paper binds the proteins very stably so that a sheet can be cleaned of antibody and reprobed sequentially with a number of antisera. Electroblot immunoassay can be used following electrophoresis of intact virus particles in agarose gels and transfers of the virus by diffusion blotting to the paper (Koenig and Burgermeister, 1986).

IV. METHODS INVOLVING PROPERTIES OF THE VIRAL NUCLEIC ACID

General properties of a viral nucleic acid, such as whether it is DNA or RNA, double- (ds) or single-stranded (ss), or consists of one or more pieces, are fundamental for allocating an unknown virus to a particular family or group. However, with the exception of dsRNA, discussed in Section IV,D, these properties are usually of little use for routine diagnosis, detection, or assay. As noted in Chapter 6, the ability to make DNA copies of parts or all of a plant viral RNA genome has opened up many new possibilities. The nucleotide sequence of the DNA copy can be determined but this is far too time-consuming to be considered as a diagnostic procedure, except in special circumstances.

Cleaving DNA copies of RNA genomes at specific sites with restriction enzymes and determining the sizes of the fragments by PAGE is a possible procedure for distinguishing viruses in a particular group, but it has not yet achieved any wide acceptance in plant virology. The methods that are being developed depend on the fact that ss nucleic acid molecules of opposite polarity and with sufficient similarity in their nucleotide sequence will hybridize to form a ds molecule. The motivation to develop hybridization methods came first with the viroids, which have no associated proteins that can be used for diagnosis (Palukatis *et al.*, 1981; Allen and Dale, 1981). For the *Geminivirus* group, Haber *et al.* (1987) used labeled DNA from one virus of the group to identify a range of group members by probing restriction enzyme fragments that gave a pattern of bands after PAGE characteristic for each member. Haber *et al.* discuss the possibility of extending this technique to RNA viruses, where restriction enzymes are not directly applicable.

A. The Basis for Hybridization Procedures

The theory concerning nucleic acid hybridization is complex. It is discussed by Britten *et al.* (1974). When a solution containing ds nucleic acid is heated, the secondary bonds holding the strands together are broken and the strands separate.

The ds nucleic acid is said to melt or denature. If the mixture of single strands is then incubated at a lower temperature, the ds structure is re-formed, that is, the nucleic acid renatures. The temperature at which 50% of the sequences are denatured is called the melting temperature or T_m. The denaturation of nucleic acid secondary structure leads to a rise in the absorbance of the solution at 260 nm. This can be used to follow the denaturation and establish the T_m. Figure 9.2 illustrates postulated stages in the denaturation of a small circular, mainly double-stranded RNA. Figure 2.13 shows the curves for two ds nucleic acids with ssRNA for comparison.

The T_m is affected primarily by the following factors: (i) the base composition of the nucleic acid, with a high G + C content giving a higher T_m; (ii) the salt concentration, with higher salt concentration raising the T_m; and (iii) the presence of a hydrogen bond-breaking agent such as formamide, which will decrease the T_m. For most plant viral nucleic acids the T_m in 0.3 M NaCl lies between 88 and 93°C. For hybridization tests, temperature conditions are usually chosen that maximize the rate of association between the complementary strands. This is 22–27°C below the T_m, that is, about 65°C for most plant viral nucleic acids.

Hybridization will take place only between ss nucleic acids that have sequence complementarity. The degree of complementarity required before two sequences can hybridize depends on the temperature of the reaction mixture. As a rule of thumb, reassociation is possible between sequences with 1% mismatch for each 1°C below the T_m. Thus at 65° sequences with 73–78% similarity will anneal. The technique can be used to obtain an estimate of the extent of sequence similarity between two ss nucleic acids. Over a range of conditions, association between strands reaches a dynamic steady state with strands associating and dissociating. Conditions in solution can be chosen (more stringent or less stringent) that allow less or more hybridization to occur. This topic is discussed more fully by Gould and Symons (1983). Hybridization can take place when both nucleic acids are in solution, or when one of the strands is immobilized on a solid matrix. It is the latter kind of procedure that is most readily adapted to the screening or assay of many samples.

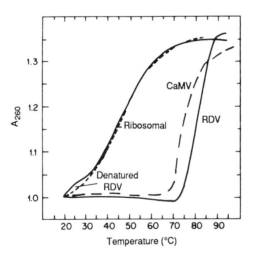

Figure 2.13 Melting profile of rice dwarf *Phytoreovirus* (RDV) dsRNA, ribosomal RNA, and denatured rice dwarf virus RNA all in 0.01 × SSC, and CaMV dsDNA in 0.1 × SSC (SSC = 0.15 M NaCl–0.015 M Na citrate). (Adapted from Miura *et al.*, 1966, and Shepherd *et al.*, 1970.)

B. Dot Blot Hybridization

Dot blot hybridization is now the most commonly used procedure for testing of large numbers of samples. The main steps are: (i) a small amount of sap is extracted from the plant under test; (ii) the viral nucleic acid is denatured by heating if it is ds; (iii) a spot of the extract is applied to a nitrocellulose sheet, or sometimes to a nylon-based membrane; (iv) the nitrocellulose is baked to bind the nucleic acid firmly to it; (v) nonspecific binding sites on the membrane are blocked by incubation in a prehybridization solution containing a protein, usually bovine serum albumin, and small ss fragments of an unrelated DNA, together with salt, etc.; (vi) hybridization of a labeled probe nucleic acid to the test nucleic acid bound to the substrate; and (vii) washing off excess (unhybridized) probe and estimation of the amount of probe bound by a method appropriate to the kind of label used for the probe. The pre-hybridization step (about 2 hours) and the hybridization step (overnight) are carried out in heat-sealable plastic bags in a water bath at about 65°C.

Boulton and Markham (1986) used an adaptation of the procedure they termed "squash-blot" to assay maize streak *Geminivirus* in a single leafhopper vectors squashed directly onto the nitrocellulose filter. Their probe was ^{32}P-labeled and the extent of hybridization was determined by autoradiography. The method is illustrated in Fig. 2.14 (see also Fig. 13.6).

Technical details for blotting methods using labeled nucleic acid probes are given by Owens and Diener (1984) and Hull (1985, 1986). A dot blot technique has been used successfully for screening large numbers of potato plants in a program of breeding potatoes for resistance to several viruses (Boulton *et al.*, 1986). The

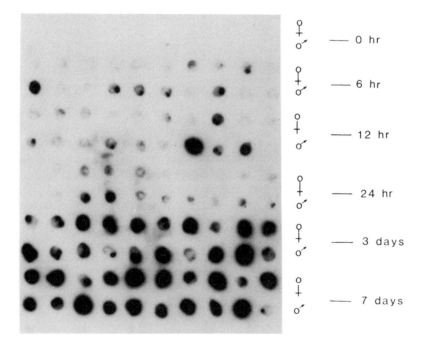

Figure 2.14 Use of the squash-blot nucleic acid hybridization technique to estimate the time course of uptake of maize streak *Geminivirus* (MSV) by a leafhopper vector. Autoradiograph of squash-blots of male (♂) and female (♀) *Cicadulina mbila* showing the time course of MSV acquisition. Each blot is from one insect. (From Boulton and Markham, 1986.)

method has been used to assess relationships between tombusviruses but some unexpected cross-hybridizations were observed (Koenig *et al.*, 1988a).

C. Preparation and Labeling of Probes

Labeled viral RNA has occasionally been used as a probe and has been shown to have several advantages (e.g., Lakshman *et al.*, 1986), but in most experiments, either with DNA or RNA viruses, DNA probes have been used. The probe may be labeled with a radioactive marker (usually ^{32}P for nucleic acids bound to paper or ^{3}H for cytological experiments). Alternatively, a nonradioactive marker such as biotin may be used. Sensitivity of detection may be related to probe size, with larger probes giving more sensitive assays (e.g., Barbara *et al.*, 1987). The use of nucleic acid probes in diagnosis has been reviewed by McInnes and Symons (1989).

Where appropriate information is available, it may be possible to make diagnostic cDNA probes to regions of the viral genome where the extent of sequence similarity has been shown to distinguish between different viruses in a group and to identify strains within a group (see Chapter 13, Section V,C, 4 and 5). Various procedures can be used to radioactively label nucleic acid probes, or to make nonradioactive probes.

1. Reverse Transcription

The viral RNA is first "random" primed with short DNA oligomers. Then DNA copies of the RNA are made using a retrovirus reverse transcriptase. The strand formed is complementary to the viral RNA (copy or cDNA). It can be used directly as a probe if radioactive nucleotides are incorporated during reverse transcription.

2. Cloned Probes

In a more commonly used procedure, a double-stranded cDNA is made, inserted into an appropriate bacterial plasmid, and grown in the bacterial host. (See Sambrook *et al.*, 1989, for practical aspects and Old and Primrose, 1985, for theory.) This procedure can provide a long-term source of a standard probe. Batches of the probe can then be labeled by nick translation, an enzymatic process in which some nucleotides are excised from the dsDNA and the gaps filled in with nucleotides, some of which are labeled. Alternatively, the ds cDNA is made ss by heating and then random primed with short synthetic oligomers. DNA polymerase may then be used to complete new ds molecules, again using some labeled nucleotides. A cloned DNA for plum pox *Potyvirus* was used to transcribe ^{32}P-labeled complementary RNA copies of the viral RNA. In dot blot hybridization assays this probe had a sensitivity of 4 pg of purified virus per spot, some 25–250 times more sensitive than a similar nick-translated DNA probe or ELISA (Varveri *et al.*, 1988).

3. Strand-Specific Probes

For some experiments it is necessary to have DNA probes that are either homologous to or complementary to the viral RNA. By cloning with the bacteriophage M13, which packages ssDNA into virus particles, it is possible to make probes of either positive or negative polarity.

4. Synthetic Probes

When the nucleotide sequence of part or all of the viral genome is known, it is possible to chemically synthesize oligonucleotides of the desired sequence with a chain length of 15–20 bases. In principle at least, this procedure has a number of advantages: (i) large amounts of ss probe can be made and readily end-labeled with a ^{32}P-labeled nucleotide by means of a polynucleotide kinase; (ii) strand-specific probes can be made; (iii) several oligomers 15–20 bases long can be joined in tandem and cloned as outlined above; and (iv) it is possible to construct a library of probes specifically designed to detect particular parts of the viral genome.

Bar-Joseph et al. (1986) discuss details of the application of synthetic probes. One limitation is sensitivity. Using such probes they could detect 4 ng of TMV RNA, whereas Sela et al. (1984), using randomly ^{32}P-labeled TMV cDNA, could detect as little as 25 pg.

5. Nonradioactive Probes

There are two reasons why various workers have sought alternatives to radioactive labels for nucleic acid probes. ^{32}P, the isotope most commonly used, has a short half-life, and many diagnostic laboratories are not well equipped to handle this isotope. The most promising alternative label is biotin. In the first procedure developed, biotin chemically linked to UTP was introduced into the probe by nick translation (Leary et al., 1983). The biotin combines very strongly with the bacterial protein streptavidin. The streptavidin is then conjugated with an appropriate enzyme as in ELISA tests.

Forster et al. (1985) developed a simple chemical procedure for labeling intact nucleic acids with a photoactivatable analog of biotin called photobiotin. When a mixture of photobiotin and nucleic acid in solution is exposed to visible light the photoactivatable group combines covalently with the nucleic acid. Single- or double-stranded, RNA or DNA can be labeled. It is very important that the probe nucleic acid be well purified because photobiotin reacts with any organic material. With one photobiotin molecule bound per 100–400 nucleotides, and alkaline phosphatase as the enzyme conjugated to avidin or streptavidin, 0.5 pg of target DNA could be detected in dot blots (Forster et al., 1985). Habili et al. (1987) used a cloned photobiotin-labeled cDNA probe for the routine diagnosis of barky yellow dwarf Luteovirus (BYDV) in nucleic acid extracts from field samples of cereals using a dot blot procedure. The sensitivity was comparable to that obtained with a nick-translated ^{32}P-labeled DNA probe. The minimum detection level with a biotin-labeled probe for papaya mosaic Potyvirus was 50 pg, about the same as that with a ^{32}P-labeled probe (Roy et al., 1988).

D. Double-Stranded RNAs in Diagnosis

Double-stranded RNAs are associated with plant RNA viruses in two ways: (i) the plant reoviruses (Chapter 8, Section III) and cryptoviruses have genomes consisting of dsRNA pieces and (ii) in tissues infected with ssRNA viruses, a ds form of the genome RNA accumulates that is twice the size of the genomic RNA. This is known as the replicative form (Fig. 6.1). These dsRNAs have been used for diagnosis either following characterization of the ds form by PAGE or by the use of antibodies reacting with dsRNA.

1. Electrophoresis of Isolated dsRNAs

Dodds *et al.* (1984) and Dodds (1986) summarize procedures for the isolation of dsRNAs from infected tissue, and for their separation and characterization by PAGE or an agarose gel, using appropriate molecular weight markers. Bands of dsRNA are normally revealed by staining with ethidium bromide. The dsRNA bands can also be eluted from the gel and used to make radioactive probes for use in dot blot analyses.

In principle, each RNA virus should give rise to a distinctive band of dsRNA. This has been found for some virus groups, for example, a series of rod-shaped viruses with monopartite genomes (Valverde *et al.,* 1986). However, a series of dsRNA molecules smaller than the full-length dsRNA are almost always present in the gels. The pattern of these smaller virus-specific RNAs is characteristic for the virus. Sometimes strains of the same virus can be distinguished by the pattern of smaller bands.

The use of dsRNAs in diagnosis is complicated by the fact that some uninoculated and apparently healthy plants contain a series of dsRNA species. For example, Nameth and Dodds (1985) found that 40 out of 50 glasshouse-grown uninoculated cucurbit cultivars contained readily detectable dsRNA species in the molecular weight range $0.5–11.0 \times 10^6$. Each cultivar had a characteristic pattern, indicating seed transmission. The nature of these dsRNAs has not been established, but in view of the high molecular weight of the largest species, they may be associated with undescribed cryptic viruses. On the other hand, Wakarchuk and Hamilton (1985) described high-molecular-weight dsRNAs form one variety of *Phaseolus vulgaris*. These dsRNAs had sequence similarity to the genome DNA of the bean variety in which it was found, as well as that of other bean varieties that contained no detectable dsRNA.

In spite of these difficulties, the dsRNA method has a role to play in virus diagnosis for some host species. Thus tomato ring spot *Nepovirus,* raspberry bushy dwarf virus, and raspberry leaf spot virus could be readily detected in infected *Rubus* spp. Field samples of diseased plants often contained two or more viruses that could be readily identified from the dsRNA patterns (Martin, 1986). The procedure provided an alternative to diagnosis by grafting or aphid transmission to indicator hosts. Similarly, the method was valuable for diagnosis of citrus tristeza *Closterovirus,* and strains of the virus, in various citrus species, provided the tissue sampled was in optimal condition (Dodds *et al.,* 1987). A low-molecular-weight dsRNA associated with groundnut rosette disease has been used as a diagnostic tool (Breyel *et al.,* 1988).

2. Antibodies against dsRNA

Antibodies reacting nonspecifically against dsRNAs were found in antisera prepared against plant reoviruses (Ikegami and Francki, 1973; Van der Lubbe *et al.,* 1979). Such nonspecific antisera can be prepared using synthetic poly(I):poly(C) antigen (Stollar, 1975), but the procedure has not received wide application for plant viruses.

V. DISCUSSION AND SUMMARY

Various form of ELISA, nucleic acid hybridization tests, and SSEM appear to be emerging as the most favored procedures for detection and diagnosis of plant viruses, but it is not possible to generalize about which method is best. Among many variables, the particular virus and the source material are often the most important. For example, in tests for blueberry shoe string virus in its aphid vector, SSEM was very much less effective than radioimmunosorbent assay or ELISA (Gillet *et al.*, 1982). On the other hand, SSEM was 100 times more sensitive than ELISA for detecting several viruses in roses (Hollings, 1978). For diagnosis, two procedures stand out because they involve two properties of the virus: SSEM (particle morphology and serological specificity) and electrophoresis followed by electroblot immunoassay (coat protein molecular weight and serological specificity).

Where large numbers of samples have to be handled the following factors will be important in choosing a test procedure: (i) specificity; (ii) sensitivity; (iii) ease and speed of operation; and (iv) cost of equipment and consumable supplies. Serological tests now offer a range of specificities from exquisitely specific monoclonal antibodies to broadly cross-reacting polyclonal antisera. However, nucleic acid hybridization tests may sometimes provide a better test across a range of related viruses (Harrison *et al.*, 1983). The sensitivity of many of the newer tests is adequate for most purposes. For ease and speed of operation, and low cost, recent developments in dot blots based on either an immunological test or nucleic acid hybridization have a lot to offer. However, in comparative tests on these methods using MAbs for the routine diagnosis of tomato spotted wilt *Tospovirus* (TSWV) in the field, ELISA was found to be the most suitable (Huguenot *et al.*, 1989).

For many kinds of experiments, for example, studies on virus inactivation, interest centers on the results of measurements made by different methods that depend on different properties of the virus. When two or more methods are used on the same samples, the difficulties of interpretation that may arise because of the different sensitivities of the methods used must always be borne in mind.

Apart from any technical problems associated with any particular testing methods, certain general difficulties may be encountered in routine diagnosis: (i) Various factors that influence virus concentration in the plant can affect reliability. This may relate to variation throughout the plant, or to very localized variations, as with potato mop top *Furovirus* in potato tubers (Mills, 1987). (ii) The possibility of mixed infections with more than one virus. Some recognized disease entities are caused by multiple infections (e.g., the lettuce speckles disease; Falk *et al.*, 1979b). (iii) Sometimes other kinds of agent combine with a virus infection to produce a disease state. Thus the internal browning disorder in tomato appears to depend on infection by TMV at a particular stage of fruit maturity (Taylor *et al.*, 1969).

Perhaps the only generalizations that can be made are that the biological methods of detection will always be important for validating any application of the newer technologies, especially when weak reactions are observed, and that symptoms and host range, tested on an appropriate set of species, still give essential information for a reliable diagnosis.

The newer serological and biochemical screening procedures allow more effective large-scale testing in many programs aimed at the control of virus disease. These include: (i) assays for infection in seed; (ii) testing of stock plants in certification programs; (iii) indexing of commercial crops derived from certification programs; (iv) screening for sources of virus resistance; (v) surveys of virus incidence in crops, weeds, and so on; and (vi) forecasting of epidemics by direct testing of insect vectors.

Finally, with respect to diagnosis, there has been some collaboration between laboratories on three fronts. First, a number of laboratories in the United States maintain many hundreds of virus isolates in viable stored form, which are available for diagnostic comparisons. Similarly, some antisera are available (Hampton,

1983). In this connection it is very unfortunate that the American Type Culture Collection has not been able to maintain an adequate service to plant virologists. Second, ELISA plates have been exchanged between laboratories working on viruses of forage legumes (McLaughlin *et al.*, 1984), and third, a computer-based virus identification data exchange is being developed (Boswell and Gibbs, 1986; Boswell *et al.*, 1986). With cooperation from plant virologists worldwide, this taxonomic database had descriptions for about 750 plant viruses by late 1990 (A.J. Gibbs, personal communication).

3

Isolation

Since the classic studies of Stanley (1935), Bawden and Pirie (1936, 1937), and others in the 1930s, a great deal of effort has been put into devising methods for the isolation and purification of plant viruses. To study the basic properties of a virus it is essential to be able to obtain purified preparations that still retain infectivity. It is not surprising that the first viruses to be isolated and studied effectively (TMV, PVX, and TBSV) were among those that are fairly stable and occur in relatively high concentration in the host plant. Today, interest has extended to a range of viruses that vary widely in concentration in the host and in their stability toward various physical and chemical procedures. There are no generally applicable rules. Procedures that are effective for one virus may not work with another apparently similar virus. Even different strains of the same virus may require different procedures for effective isolation.

A great deal has been written about purity and homogeneity as they apply to plant viruses. In a chemical sense, there is no such thing as a pure plant virus preparation. Even if a preparation contained absolutely no low- or high-molecular-weight host constituents (which is most unlikely), there are other factors to be considered:

1. Most preparations almost certainly consist of a mixture of infective and noninfective virus particles. The latter will probably have one or more breaks in their nucleic acid chains. Most of these breaks will have occurred at different places in different particles.
2. Most virus preparations almost certainly consist of a mixture of mutants even though the parent strain greatly predominates. Such mutants will differ in the base sequence of their RNA or DNA in at least one place. If the mutation is in the cistron specifying the coat protein then this may also differ from the parent strain.
3. Purified preparations of many viruses can be shown to contain one or more classes of incomplete, noninfective particles.
4. The charged groups on viral proteins and nucleic acids will have ions associated with them. The inorganic and small organic cations found in the purified virus preparation will depend very much on the nature of the buffers and so on used during isolation.

5. Some of the larger viruses appear to cover a range of particle sizes having infectivity.

6. A variable proportion of the virus particles may be altered in some way during isolation. Enzymes may attack the coat protein. For example, extracts of bean (*Phaseolus vulgaris*) contain a carboxypeptidaselike enzyme that removes the terminal threonine from TMV (Rees and Short, 1965). Proteolysis at defined sites may give rise, during isolation of the virus, to a series of coat protein molecules of less than full size (e.g., with *Solanum nodiflorum* mottle *Sobemovirus,* Chu and Francki, 1983; and barley yellow mosaic *Potyvirus,* Ehlers and Paul, 1986). Coat proteins may undergo chemical modification when leaf phenols are oxidized (Pierpoint *et al.,* 1977). More complex viruses such as the reoviruses may lose part of their structure during isolation (e.g., Hatta and Francki, 1977).

Thus, for plant viruses, purity and homogeneity are operational terms defined by the virus and the methods used. A virus preparation is pure for a particular purpose if the impurities, or variations in the particles present, do not affect the particular properties being studied or can be taken account of in the experiment. Effective isolation procedures have now been developed for many plant viruses. Rather than describe these in detail, I shall consider, in general terms, the problems involved in virus isolation. Hull (1985) gives detailed protocols for several isolation procedures. Hammond *et al.* (1983) describe the development of a method for isolation of a virus occurring in very low concentration.

I. CHOICE OF PLANT MATERIAL

A. Assay Host

During the development of an isolation procedure, it is usually essential, but not always possible, to be able to assay fractions for infectivity. Of course, this is best done with a local lesion host. Great accuracy usually is not necessary in the preliminary assays, but reliability and rapid development of lesions are a great advantage. If no local lesion host is available, then assays must be done using a systemic host. Assays by the injection of insect vectors sometimes have been used where mechanical transmission is impossible. In some circumstances, dot blot ELISA, dot blot nucleic acid hybridization, or electron microscopy can be used to follow the progress of purification.

B. Starting Material

The choice of host plant for propagating a virus may be of critical importance for its successful isolation. Perhaps the single most important property of the host plant is that it should not contain certain substances in sufficient concentration to inhibit or irreversibly precipitate the virus. These substances include phenolic materials, organic acids, mucilages and gums, certain proteins, and enzymes, particularly ribonucleases. For example, many stone fruit viruses were very difficult to isolate from their natural hosts, since most members of the Rosaceae contain high concentrations of tannins in their leaves. Discovery of alternative nonrosaceaous hosts, for example, cucumber (*Cucumis sativus* L.), has allowed the isolation of several such viruses. A relatively small group of suitable host plants has now been used for the isolation of a range of viruses. These hosts, besides cucumber and related plants, include cowpea [*Vigna sinensis* (Endl.)], *Chenopodium amaranticolor* (Coste and Reyn), *C. quinoa* (Willd.), and *Petunia hybrida* (Vilm).

The plant used, the conditions under which it is grown, and the time at which it is harvested should be chosen to maximize the starting concentration of infectious virus. For many viruses, concentration rises to a peak after a few days or weeks and then falls quite rapidly (Fig. 10.8). Sometimes the distribution of virus within the plant is so uneven that it is worthwhile to dissect out and use only those tissues showing prominent symptoms. Viruses frequently occur in much lower concentration in the midrib than in the lamina of the leaf. If the midrib and petiole is large, it may pay to discard it. In special situations, dissection of tissue is almost essential, for example, with WTV and other viruses causing tumors where the virus is found associated with the tumor tissue. Another reason for harvesting only certain parts of the infected plant may be to avoid high concentrations of inhibitory substances or materials that adsorb to the virus and are later difficult to remove. Such materials frequently occur in lower concentration in new young growth. In certain hosts, virus can only be isolated from such tissue. Similarly, root tissue may sometimes provide more favorable starting material than leaves (e.g., Ford, 1973), although virus concentration is almost always lower in roots.

The possibility that the host used to culture a virus may already harbor another virus or become infected with one must always be borne in mind. Contamination of greenhouse-grown plants with unwanted viruses is not at all uncommon. Strains of TMV, PVX, and TNV may be particularly prevalent, especially in greenhouses that have been used for virus work for some time. It is not necessarily sufficient to use a host that is only a local lesion host for such contaminating viruses (e.g., *N. glutinosa* for TMV). Very small amounts of such a resistant virus may become differentially concentrated during isolation of a second virus.

Freezing of the plant tissue before extraction facilitates subsequent removal of host materials and is a useful step for stable viruses such as TMV. For many other viruses, however, such freezing has a deleterious effect.

II. EXTRACTION MEDIUM

A medium such as serum is a natural one for most animal viruses. There is no equivalent for plant viruses. Once infected plant cells are broken and the contents released and mixed, the virus particles find themselves in an environment that is abnormal. Thus it is often necessary to use an artificial extraction medium designed to preserve the virus particles in an infectious, intact, and unaggregated state during the various stages of isolation. The conditions that favor stability of purified virus preparations may be different from those needed in crude extracts or partially purified preparations (e.g., Brakke, 1963). Moreover, different factors may interact strongly in the extent to which they affect virus stability. The main factors to be considered in developing a suitable medium are as follows.

A. pH and Buffer System

Many viruses are stable over a rather narrow pH range, and the extract must be maintained within this range. Choice of buffer may be important. Phosphate buffers have often been employed, but these have deleterious effects on some viruses.

B. Metal Ions and Ionic Strength

Some viruses require the presence of divalent metal ions (Ca^{2+} or Mg^{2+}) for the preservation of infectivity and even for the maintenance of structural integrity. Ionic strength may be important. Some viruses fall apart in media of ionic strength below about 0.2 M, while others are unstable in media above this molarity. AMV particles may be precipitated by Mg^{2+} concentrations above 0.001 M and degraded by concentrations above 0.1 M (Hull and Johnson, 1968). For some viruses EDTA may be included to minimize aggregation by divalent metals. On the other hand, EDTA disrupts certain viruses.

C. Reducing Agents and Substances Protecting against Phenolic Compounds

Reducing agents such as sodium sulfite, sodium thioglycollate, 2-mercaptoethanol, or cysteine hydrochloride are frequently added to extraction media. These materials assist in preservation of viruses that readily lose infectivity through oxidation. They also may reduce adsorption of host constituents to the virus. Phenolic materials may cause serious difficulties in the isolation and preservation of viruses. Several methods have been used more or less successfully to minimize the effects of phenols on plant viruses during isolation.

Cysteine or sodium sulfite added to the extraction medium both probably act by inhibiting the phenol oxidase and by combining with the quinone (Pierpoint, 1966).

Polyphenoloxidase is a copper-containing enzyme. Two chelating agents with specificity for copper, diethyldithiocarbamate and potassium ethyl xanthate, have been used to obtain infectious preparations of several viruses (e.g., prunus necrotic ring spot *Ilavirus;* Barnett and Fulton, 1971).

Materials that compete with the virus for phenols have sometimes been used. For example, Brunt and Kenten (1963) used various soluble proteins and hide powder to obtain infective preparations of swollen shoot virus from cocoa leaves. Synthetic polymers containing the amide link required for complex formation with tannins have been used effectively to bind these materials. The most important of these is polyvinyl pyrrolidone (PVP).

D. Additives That Remove Plant Proteins and Ribosomes

Many viruses lose infectivity fairly rapidly *in vitro*. One reason for this loss may be the presence of leaf ribonucleases in extracts or partly purified preparations. Dunn and Hitchborn (1965) made a careful study of the use of magnesium bentonite as an additive in the isolation of various viruses. They found that under appropriate conditions, contamination of the final virus product with nucleases was reduced or eliminated. In addition, ribosomes, 19 S protein, and green particulate material from fragmented chloroplasts were readily adsorbed by bentonite, provided Mg^{2+} concentration was $10^{-3} M$ or greater. However, variation occurs in the activity of different batches of bentonite and the material must be used with caution as some viruses are degraded in its presence. Charcoal may be used to adsorb and remove host materials, particularly pigments. Subsequent filtration to remove charcoal may lead to substantial losses of virus adsorbed in the filter cake.

EDTA as the sodium salt at 0.01 M in pH 7.4 buffer will cause the disruption of most ribosomes, preventing their cosedimentation with the virus. This substance can be used only for viruses that do not require divalent metal ions for stability.

E. Enzymes

Enzymes have been added to the initial extract for various purposes. Thus Adomako *et al.* (1983) used pectinase to degrade mucilage in extracted sap of cocoa leaves prior to precipitation of cocoa swollen shoot virus. Improved yield of a virus limited to phloem tissue was obtained when fibrous residues were incubated with Driselase and other enzymes (Singh *et al.*, 1984). This material contains pectinase and cellulase and presumably aids in the release of virus that would otherwise remain in the fiber fraction. The enzymes also digest materials that would otherwise coprecipitate with the virus. Treatment with trypsin at an optimum concentration can markedly improve the purification of turnip mosaic *Potyvirus* (Thompson *et al.*, 1988).

F. Detergents

Triton X-100 or Tween 80 have sometimes been used in the initial extraction medium to assist in release of virus particles from insoluble cell components. Detergents may also assist in the initial clarification of the plant extract. Nonionic detergents dissociate cellular membranes, which may contaminate or occlude virus particles.

III. EXTRACTION PROCEDURE

A variety of procedures are used to crush or homogenize the virus-infected tissue. These include (i) a pestle and mortar, which are useful for small-scale preparations, (ii) various batch-type food blenders and juice extractors, which are useful on an intermediate scale, and (iii) roller mills, colloid mills, and commercial meat mincers, which can cope with kilograms of tissue. For long, fragile, rod-shaped viruses, grinding in a pestle and mortar may be the safest procedure to minimize damage. The addition of acid-washed sand greatly improves the efficiency of extraction. If an extraction medium is used, it is often necessary to ensure immediate contact of broken cells with the medium. The crushed tissue is usually expressed through muslin or cheesecloth.

IV. PRELIMINARY ISOLATION OF THE VIRUS

A. Clarification of the Extract

In the crude extract, the virus is mixed with a variety of cell constituents that lie in the same broad size range as the virus and that may have properties that are similar in some respects. These particles include ribosomes, 19 S (fraction I) protein from chloroplasts, which has a tendency to aggregate, phytoferritin, membrane fragments, and fragments of broken chloroplasts. Also present are unbroken cells, all the smaller soluble proteins of the cell, and low-molecular-weight solutes.

The first step in virus isolation is usually designed to remove as much of the macromolecular host material as possible, leaving the virus in solution. The extraction medium may be designed to precipitate ribosomes, and so on, or to disintegrate them. The extract may be subject to some treatment such as heating to 50–60°C for a few minutes, or the addition of K_2HPO_4, to coagulate much host material. For some viruses, organic solvents such as chloroform give very effective precipitation of host components. For others, the extract can be shaken with *n*-butanol–chloroform, which denatures much host material. The treated extract is then subjected to centrifugation at fairly low speed (e.g., 10–20 minutes at 5000–10,000 g). This treatment sediments cell debris and coagulated host material. With the butanol–chloroform system, centrifugation separates the two phases, leaving virus in the aqueous phase and much denatured protein at the interface. It should be noted that although some viruses can withstand the butanol–chloroform treatment, quite severe losses may occur with others. Chloroform alone gives a milder treatment than does a chloroform–butanol mixture.

For many viruses, it probably would pay to carry out the isolation procedure as fast as possible once the leaves have been extracted. On the other hand, some viruses occur in membrane-bound packets or other structures within the cell. It may take time for the virus to be released from these after the leaf extract is made. A low-speed centrifugation soon after extraction may then result in much virus being lost in the first pellet.

B. Concentration of the Virus and Removal of Low-Molecular-Weight Materials

1. High-Speed Sedimentation

Centrifugation at high speed for a sufficient time will sediment the virus. Provided the particular virus is not denatured by the sedimentation, it can be brought back into solution in active form. This is a very useful step, as it serves the double purpose of concentrating the virus and leaving behind low-molecular-weight materials. However, high-speed sedimentation is a physically severe process that may damage some particles (e.g., some reoviruses, Long *et al.,* 1976). Following high-speed sedimentation, some viruses remain as characteristic aggregates when the pellets are redissolved. The virus particles in these aggregates may be quite firmly bound together (Tremaine *et al.,* 1976). If host membranes are involved in the binding, a nonionic detergent may help to release virus particles. Sedimentation of viruses occurring in very low concentration may result in very poor recoveries. The major process causing losses appears to be the dissolving and redistribution of the small pellet of virus as the rotor comes to rest (McNaughton and Matthews, 1971). Redissolving particles from the surface of the pellet can lead to preferential losses of more slowly sedimenting components (Matthews, 1981) (Fig. 3.1).

2. Precipitation with Polyethylene Glycol

Hebert (1963) showed that certain plant viruses could be preferentially precipitated in a single-phase polyethelene glycol (PEG) system, although some host DNA

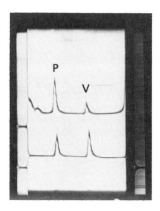

Figure 3.1 Preferential loss of empty protein shells (P) during the purification of okra mosaic *Tymovirus* by high-speed sedimentation. Upper Schlieren trace: extract of infected cucumber leaves diluted one in three in 0.05 *M* NaCl. Lower trace: virus (at approximately 3 mg/ml in 0.05 *M* NaCl) isolated from the same batch of leaves by three cycles of high-speed sedimentation. P, protein shells; V, virus nucleoprotein. The upper trace shows contaminating host macromolecules.

may also be precipitated. Since that time, precipitation with PEG has become one of the commonest procedures used in virus isolation. The exact conditions for precipitation depend on pH, ionic strength, and concentration of macromolecules. Its application to the isolation of any particular virus is empirical, although attempts have been made to develop a theory for the procedure (Juckes, 1971).

PEG precipitation is applicable to many viruses, even fragile ones. For example, the procedure gave a good yield of intact particles of citrus tristeza *Closterovirus,* a fragile rod-shaped virus (R. F. Lee *et al.,* 1987). It has the advantage that expensive ultracentrifuges are not required, although differential centrifugation is often used as a second step in purification procedures.

3. Density Gradient Centrifugation

Many viruses, particularly rods, may form pellets that are very difficult to resuspend. Density gradient centrifugation offers the possibility of concentrating such viruses without pelleting. A density gradient is illustrated in Fig. 2.3. The following modification of the density gradient procedure may sometimes be used even with angle rotors for initial concentration of a virus without pelleting at the bottom of the tube. A cushion of a few milliliters of dense sucrose is placed in the bottom of the tube, this is overlayed with a column of low-density sucrose about 2 cm deep, and the rest of the tube is filled with clarified virus extract. Under appropriate conditions, virus can be collected from the region of the interface between the two sucrose layers. Density gradient centrifugation is used in the isolation procedure for many viruses.

4. Salt Precipitation or Crystallization

Salt precipitation was commonly employed before high-speed centrifuges became generally available. It is still a valuable method for viruses that are not inactivated by strong salt solutions. Ammonium sulfate at concentrations up to about one-third saturation is most commonly used (Fig. 3.2), although many other salts will precipitate viruses or give crystalline preparations. After standing for

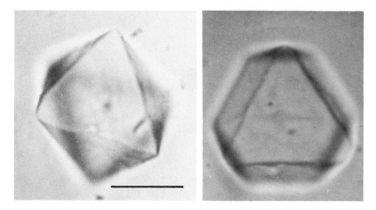

Figure 3.2 Crystals of TYMV formed in ammonium sulfate solutions. Bar = 25 μm. (Courtesy of T. Hatta.)

some hours or days the virus is centrifuged down at low speed and redissolved in a small volume of a suitable medium.

5. Precipitation at the Isoelectric Point

Many proteins have low solubility at or near their isoelectric points. Isoelectric precipitation can be used for viruses that are stable under the conditions involved. The precipitate is collected by centrifugation or filtration and is redissolved in a suitable medium.

6. Dialysis

Dialysis through cellulose membranes can be used to remove low-molecular-weight materials from an initial extract and to change the medium. It is more usually employed to remove salt following salt precipitation or crystallization, or following density gradient fractionation in salt or sucrose solutions, as discussed in the next section.

V. FURTHER PURIFICATION OF THE VIRUS PREPARATION

Virus preparations taken through one step of purification and concentration will still contain some low- and high-molecular-weight host materials. More of these can be removed by further purification steps. The procedure to be used will depend very much on the stability of the virus, the scale of the preparation, and the purpose for which the preparation is required.

Sometimes highly purified preparations can be obtained by repeated application of the same procedure. For example, a preparation may be subjected to repeated crystallization or precipitation from ammonium sulfate, or may be given several cycles of high- and low-speed sedimentation. The latter procedure leads to the preferential removal of host macromolecules because they remain insoluble when the pellets from a high-speed sedimentation are resuspended. Losses of virus often occur with this procedure, either because some or all of the virus itself also becomes insoluble or because insufficient time is allowed for the virus to redissolve. This is a particular difficulty with some rod-shaped viruses. Losses may also occur from virus redissolving before the supernatant fluid is removed. Such losses may be severe with very small pellets, where the surface to volume ratio is high.

Generally speaking, during an isolation it is useful to apply at least two procedures that depend on different properties of the virus. This is likely to be more effective in removing host constituents than repeated application of the same procedure.

A. Density Gradient Centrifugation

One of the most useful procedures for further purification, particularly of less stable viruses, is density gradient centrifugation, which was already discussed as an assay method in Chapter 2, Section II,B,2 where the differences between rate and isopycnic gradients were presented. Sucrose is the most commonly used material for making the gradient. Sucrose density gradient centrifugation is frequently the method of choice for further purification. It is a relatively mild procedure, although there may be a loss of infectivity with some viruses. It can give some indication of the purity of the preparation. It allows a correlation between particles and infectivity to be made, and it frequently reveals the presence of noninfective viruslike particles or multiparticle viruses. Bands of a single component may spread more widely in the gradient than is apparent from a trace of optical density. Several cycles of density gradient centrifugation may be necessary to obtain components reasonably free of mutual contamination. Work with multicomponent viruses has shown that it may be extremely difficult to obtain one component completely free of another, even by repeated density gradient fractionation. Density gradients prepared from the nonionic medium Nycodenz were used effectively for the further purification of several viruses (Gugerli, 1984).

Brakke and Dayly (1965) showed that zone spreading of a major component by nonideal sedimentation can cause zone spreading of a minor component that the major component overlaps. The way in which zones are removed from the gradient may have a marked effect on the extent of cross-contamination.

Strong solutions of salts such as cesium chloride are also effective gradient materials for viruses that are sufficiently stable. Successive fractionation in two different gradients may sometimes give useful results. The effective buoyant density and the stability of a virus in strong CsCl solutions may depend markedly on pH and on other ions present (Matthews, 1974). Viruses that are unstable in CsCl may be stable in Cs_2SO_4 (Hull, 1976).

When a virus preparation is subjected to density gradient centrifugation in CsCl or Cs_2SO_4, multiple bands may be formed. These may be due to the presence of components containing differing amounts of RNA, as is found with TYMV (Keeling et al., 1979). Virus particles containing uniform amounts of nucleic acid may sometimes form multiple bands in CsCl or other salt gradients because of such factors as differential binding of ions (e.g., Hull, 1977; Noort et al., 1982).

B. Gel Filtration

Filtration through agar gel or Sephadex may offer a useful step for the further purification of viruses that are unstable to the pelleting involved in high-speed centrifugation. However, such a step will dilute the virus.

C. Immunoaffinity Columns

Monoclonal antiviral antibodies can be bound to a support matrix such as agarose to form a column that will specifically bind the virus from a solution passed through the column. Virus can then be eluted by lowering the pH. Clarified plant sap may destroy the reactivity of such columns (Ronald *et al.*, 1986), while low pH may damage the virus. To avoid such treatment, De Bartoli and Roggero (1985) developed an electrophoretic elution technique. The use of such columns is probably justified only in special circumstances.

D. Chromatography

Chromatographic procedures have occasionally been used to give an effective purification step for partially purified preparations. McLean and Francki (1967) used a column of calcium phosphate gel in phosphate buffer to purify LNYV. Oat blue dwarf *Marafivirus* was purified using cellulose column chromatography (Banttari and Zeyen, 1969), while Smith (1987) used fast protein liquid chromatography to separate the two electrophoretic forms of CPMV.

E. Concentration of the Virus and Removal of Low-Molecular-Weight Materials

At various stages in the isolation of a virus, it is necessary to concentrate virus and remove salts or sucrose. For viruses that are stable to pelleting, high-speed centrifugation is commonly employed for the concentration of virus and the reduction of the amount of low-molecular-weight material. Ordinary dialysis is used for removal or exchange of salts. For unstable viruses, some other procedures are available. A dry gel powder (e.g., Sephadex) can be added to the preparation, or the preparation in a dialysis bag can be packed in the dry gel powder and placed in the refrigerator for a period of hours, while water and ions are absorbed through the tubing by the Sephadex. Ultrafiltration under pressure through a membrane can be used to concentrate larger volumes and to remove salts. Pervaporation, in which virus solution in a dialysis bag is hung in a draft of air, may lead to loss of virus due to local drying on the walls of the tube, and to the concentration of any salts that are present.

VI. STORAGE OF PURIFIED VIRUSES

Storage of purified preparations of many plant viruses for more than a few days may present a problem. It is often best to avoid long-term storage by using the preparations as soon as possible after they are made. Under the best of conditions, most viruses except TMV lose infectivity on storage at 4°C in solution or as crystalline preparations under ammonium sulfate. Such storage allows fungi and bacteria to grow and contaminate preparations with extraneous antigens and enzymes. Addition of low concentrations of sodium azide or thymol will prevent growth of microorganisms. Preparations may be stored in liquid form at low temperature by the addition of an equal volume of glycerol. By careful attention to the additives used in the medium (a suitable buffer plus some protective protein, sugar or polysaccharide), it may be possible to retain infectivity in deep-frozen solutions or as frozen dried powders for fairly long periods (e.g., Fukumoto and Tochihara, 1984). Preparations to be used for analytical studies on protein or nucleic acid are best stored as frozen solutions, after the components have been separated.

In solutions containing more than about 10 mg/ml of viruses such as TYMV, the virus particles interact quite strongly and spontaneously degrade, especially if the ionic strength is low.

VII. IDENTIFICATION OF THE INFECTIVE PARTICLES AND CRITERIA OF PURITY

The best methods for determining the purity of a virus preparation depend on the purpose of the experiments and the conclusions to be drawn. The three most common purposes for which virus preparations are made are

1. to establish the general properties of a newly isolated virus. This aspect is discussed in the following sections.
2. to aid in the diagnosis of a particular disease. Diagnosis is discussed in Chapter 2.
3. to prepare specific antisera or diagnostic nucleic acid probes.

A. Identification of the Characteristic Virus Particle or Particles

With newly isolated viruses, or during attempts to isolate such viruses, the main objective is to purify the virus sufficiently to identify the infectious particles and characterize them, at least in a preliminary way. The most suitable method is sucrose density gradient centrifugation of the purified preparation. Fractions taken from the gradient can be assayed for infectivity singly, and in various combinations if a multiparticle system is involved. Ultraviolet absorption spectra can be obtained. Samples can be examined by electron microscopy for characteristic particles. Further advantages of the sucrose gradient system are that it requires relatively small quantities of virus, and that an approximate estimate of sedimentation coefficients can be obtained.

B. Criteria of Purity

Many of the criteria for purity that are applied to virus preparations involve the application of the assay methods discussed in Chapter 2. Crystallization used to be taken as an important criterion of purity, but it does not distinguish components that cocrystallize with the infectious virus. A nucleoprotein-type spectrum is often put forward as evidence of purity. This is a very poor measure since contamination with large amounts of host nucleoprotein, or polysaccharides, or significant amounts of host proteins can easily go undetected. Sucrose density gradient analysis or sedimentation in the analytical ultracentrifuge will often reveal the presence of nonviral material, provided the contaminants do not sediment together with the virus and provided they are present in high enough concentrations.

Sedimentation analysis of purified preparations of some spherical viruses may reveal the presence of dimers or trimers of the virus (Markham, 1962). These might be confused with host contaminants unless other tests are applied. Likewise, preparations of rod-shaped viruses often contain a range of particles of various lengths shorter than the virus. These shorter particles and the virus may aggregate to give a range of sizes that sediment over a broad band, or as two or more discrete peaks. Sedimentation analysis is less useful for detecting the presence of nonviral materials with multiparticle viruses since more of the sedimentation profile is occupied by virus particles.

Equilibrium density gradient sedimentation in cesium chloride is a much more sensitive criterion of homogeneity than sedimentation velocity analysis for viruses or viral components that are stable in the strong salt solution. For example, a purified preparation of the B_0 noninfectious nucleoprotein from a TYMV preparation appeared as a single component on sedimentation velocity analysis but after equilibrium sedimentation in cesium chloride it was clearly shown to be a mixture of more than one species (Matthews, 1981) (Fig. 3.3).

Electron microscopy is often useful in a qualitative way for revealing the presence of extraneous material, provided it is of sufficient size and differs in appearance from the virus itself. Fraction I protein, phytoferritin, or pieces of host membrane structures often can be detected in purified preparations. Small proteins and low-molecular-weight materials would not be detected unless they crystallized on the grids or were present in relatively large amounts.

Serological methods, such as immunodiffusion and immunoelectrophoresis, provide very sensitive methods for detecting diffusable host impurities that are antigenic.

Where the gross composition of a particular virus such as TMV is well estab-

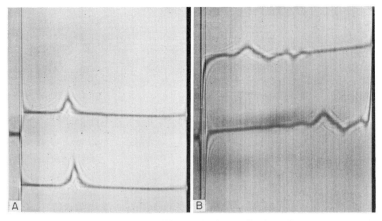

Figure 3.3 Velocity sedimentation and equilibrium density gradient centrifugation as criteria of homogeneity. (A) Sedimentation velocity. Lower pattern: TYMV B_0 component at approximately 1.1 mg/ml in 0.1 M NaCl at pH 7.4 sedimented at 35,600 rpm for 12 minutes. Only one component is apparent. Upper pattern: TYMV B_{00} under the same conditions. (B) Sedimentation equilibrium. TYMV B_0 component (approximately 200 μg) in cesium chloride of $D = 1.35$ g/cm^3 (lower pattern) and 1.39 g/cm^3 (upper pattern) after centrifugation for 42 hours at 35,600 rpm. Schlieren patterns show the presence of more than one component.

lished, chemical analysis of major constituents, for example, phosphorus nitrogen ration, would show up a nitrogen- or phosphorus-containing impurity if it were present in sufficient amount. More sophisticated chemical methods might also be used in certain circumstances, for example, end group amino acid analysis on the protein, which should show only one amino acid. Fingerprinting of the peptides from tryptic digests of well-characterized virus proteins might also be used, but again the sensitivity as a test for purity would not be very high.

Radiochemical methods can be used in various ways to test for purity. For example, a healthy plant can be labeled with ^{32}P orthophosphate or ^{35}S sulfate for several days. This material is then mixed with equivalent unlabeled virus-infected tissue before the leaf is extracted. If millicurie amounts of radioactivity are used, the sensitivity of this method can be rather depressing. Even so, it is not a complete check on purity with respect to compounds containing these two elements. For example, it is possible that virus infection leads to a kind of contaminant not found in healthy tissues. At present, the only general rule for deciding what criteria of purity to apply is that they should ensure that the virus preparation is adequate for the desired purpose.

VIII. VIRUS CONCENTRATION IN PLANTS AND YIELDS OF PURIFIED VIRUS

A. Measurement of Yield

There are many difficulties and ambiguities in obtaining estimates of virus concentration in a plant.

1. Nonextracted Virus

The method used to extract virus from the tissue may frequently leave a variable but quite large proportion of the virus bound to cell debris or retained in vesicles or organelles. Some viruses, particularly those rod-shaped viruses that occur as large fibrous inclusions in the cell, may remain aggregated during the initial stages of isolation. The extent to which this factor leads to virus loss often has not been adequately assessed.

2. Method of Measurement

Infectivity measurements cannot give an absolute estimate of amount of virus and are influenced by many variables. For physical or chemical methods that require purified virus, quite large and variable losses may occur during isolation, before the measurements can be made. For a reasonably stable virus an estimate of such losses can be obtained by adding a small, known amount of radioactively labeled virus to the starting material. The proportion of label recovered in the final preparation gives an estimate of virus loss. This method would not account for virus bound within cell structures, however.

For viruses occurring in high enough concentration, sucrose density gradient analysis on clarified extracts is probably one of the most direct methods for estimating virus concentration in plant extracts.

3. Tissue Sampled

As discussed in Chapter 10, Section IV,H, the concentration of virus in different parts of the plant may vary widely—even in different parts of a leaf showing
mosaic symptoms.

4. Basis for Expressing Results

Virus concentration or virus yield is usually expressed as weight per unit of
fresh weight of tissue or weight per milliliter of extract. However, different tissues
and different leaves, even in the same plant, vary widely in their water content under
the same conditions, reflecting, for example, the size of vacuoles or the amount of
fibrous material.

B. Reported Yields of Virus

There is no entirely satisfactory answer to many of the preceding problems.
Most workers, in reporting the isolation of a virus, if they give estimates of yield at
all, express their results as weight of purified virus obtained from a given weight of
starting material. This is usually whole leaves or leaf laminae without midribs.
Reported yields for the same virus from different laboratories may vary quite widely
because of such factors as host species, growing conditions, and isolation procedure. Yields vary from several mg of virus/g fresh weight of tissue for viruses like
TMV and TYMV down to fractions of μg/g for luteoviruses (e.g., Matsubara *et al.*,
1985).

IX. DISCUSSION AND SUMMARY

Different viruses vary over a 10,000-fold range in the amount of virus that can be
extracted from infected tissue (from about 0.4 to 4000 μg/gm fresh weight). They
also vary widely in their stability to various physical, chemical, and enzymatic
agents that may be encountered during isolation and storage. For these reasons,
isolation procedures have to be optimized for each virus, or even each strain of a
virus.

Important factors for the successful isolation of a virus are (i) choice of host
plant species, and conditions for propagation that will maximize virus replication
and minimize the formation of interfering substances; and (ii) an extraction medium
that will protect the virus from inactivation or irreversible aggregation.

Most viruses can be isolated by a combination of two or more of the following
procedures: high-speed sedimentation, density gradient fractionation, precipitation
using polyethylene glycol, salt precipitation or crystallization, gel filtration, and
dialysis. Positive identification of the infectious virus particle, or particles, can
usually be best achieved by density gradient fractionation of the purified preparation, with combined physical and biological examination of particles in fractions
from the gradient.

When a new virus or strain is being isolated it is essential to back-inoculate the
isolated virus to the original host to demonstrate that it, and it alone, is in fact the
cause of the original disease. Attention must also be paid to the precise conditions

under which purified viruses are stored, as many viruses may lose infectivity and undergo other changes quite rapidly following their isolation.

In a chemical sense there is no such thing as a pure virus preparation. Purity and homogeneity are operational terms. A virus preparation is pure for a particular purpose if the impurities or inhomogeneities in it do not interfere with the objectives of the experiment.

4

Components

Most plant viruses have ssRNA genomes enclosed in either a tube-shaped or an isometric shell made up of many small protein molecules. In these small geometric viruses, there is usually only one kind of protein molecule but some have two. Some viruses have dsRNA genomes while others have ssDNA or dsDNA. A few plant viruses have an outer envelope of lipoprotein. Some of the larger viruses contain several viral-coded proteins, including enzymes involved in nucleic acid synthesis. Thus, the major components of all viruses are protein and nucleic acid. A few contain lipids, while some contain small molecules such as polyamines and metal ions. This chapter deals with the isolation and some properties of these various components.

Over the past few years important developments have taken place with respect to our ability to determine the nucleotide sequences of viral nucleic acids, and from those sequences to deduce the amino acid sequences of viral-coded proteins. There is great interest in comparing the nucleotide and amino acid sequences of different viruses, from the point of view of virus evolution, virus classification, and the functional roles of viral genes and controlling elements. Reeck *et al.* (1987) have drawn attention to the terminological muddle that has arisen through the imprecise use in the literature of the term "homology" in comparing two sequences when in fact "similarity" is what is meant. Similarity is a fact, a measurable property of two sequences, to which measures of statistical significance can be applied. Homology describes a relationship between two things. The word has a precise meaning in biology of "having a common evolutionary origin." Thus a sequence is homologous or it is not. Sequences that are homologous in an evolutionary sense may range from being highly similar to having no significant sequence similarity. In the interest of clarity I will attempt to maintain this distinction.

I. NUCLEIC ACIDS

For many viruses containing ssRNA, the RNA can act directly as an mRNA upon infection. Such molecules are called positive sense or plus strand RNAs. The

Table 4.1

Types of Nucleic Acids Comprising Plant Viral Genomes[a]

	Number of viruses	Percentage
Plus-stranded RNA	484	76.6
Minus-stranded RNA	82	13.0
Double-stranded RNA	27	4.3
Single-stranded DNA	26	4.1
Double-stranded DNA	13	2.0

[a]From Zaitlin and Hull (1987).

complementary sequences are called negative sense or minus strand RNAs. The types of nucleic acid found in plant viruses are listed in Table 4.1.

In the following sections, most attention is given to ssRNAs since most plant viruses contain this type of nucleic acid. The RNAs of plant viruses contain adenylic, guanylic, cytidylic, and uridylic acids—as are found in cellular RNAs. Except for the methylated guanine in 5' cap structures (Section C,3) no minor bases have been demonstrated to be present in plant viral RNAs.

The plant reoviruses, containing dsRNA, have regularities in base composition like those found in dsDNA, namely, guanine = cytosine and adenine = uracil. For the viruses containing ssRNA there is no expectation that guanine should equal cytosine or that adenine should equal uracil, and many viruses show a wide deviation from any base-pairing rule. In plant viruses containing DNA, deoxyribose replaces ribose, and thymine is present instead of uracil as with other DNAs. If the DNA is ds, guanine = cytosine and adenine = thymine. A good account of the properties of nucleic acids in general is given by Adams *et al.* (1981).

A. Isolation

The usual aim in nucleic acid isolation is to obtain an undegraded and undenatured product in a state as close as possible to that existing in the virus particle. A variety of physical and chemical agents can be used to remove the protein from viruses and give infectious nucleic acid provided that (i) the nucleic acid was infectious within the intact particle, (ii) extremes of pH are avoided, and (iii) the nucleic acid is protected against attack by nucleases.

Much of the pioneering work on viral RNA was carried out with TMV. However, the RNA of this virus while it is within the protein coat has exceptional stability. For most other viruses the RNA within the intact virus particle is probably subject, in varying degrees, to some degradation, with loss of infectivity. This occurs both in the intact plant and during isolation and storage of the virus. Best conditions for maintaining intact RNA within the virus particle differ from one virus to another.

Isolation of viral RNA will usually involve the use of phenol. Gierer and Schramm (1956) used a phenol procedure to isolate infectious TMV RNA. Phenol is an effective protein denaturant and nuclease inhibitor. There are many variations of the basic procedure. Particular variations that work well for one virus may not be effective for another. However, phenol is not satisfactory for some viruses (e.g., wheat streak mosaic *Potyvirus*, Brakke and van Pelt, 1970; BYDV, Brakke and Rochow, 1974).

Destruction of nucleases and effective release of the viral nucleic acid may be achieved by a preliminary incubation with pronase in the presence of 1% SDS. Phenol extraction is then used to remove pronase and protein digestion products. In most currently used RNA isolation methods, SDS is included during the phenol extraction. If this is done, good nucleic acid preparations can be obtained with most viruses.

Various procedures can be used to fractionate or to purify further the nucleic acids isolated from viruses or from infected cells. These include sucrose density gradients; equilibrium density gradient centrifugation in solutions of CsCl or Cs_2SO_4; fractionation on columns of hydroxylapatite (Bernardi, 1971); cellulose chromatography (e.g., Jackson et al., 1971); and electrophoresis in polyacrylamide gels (e.g., Symons, 1978). This last is the method of choice for many purposes since the procedure can be fast and simple, requires very small amounts of nucleic acids, and can detect heterogeneity that would not be seen by other methods (e.g., Fowlks and Young, 1970).

In summary, for any particular virus, a protocol for nucleic acid isolation has to be devised that maximizes the required qualities in the final product. These qualities may include yield, purity, structural integrity, infectivity, or the ability to act with fidelity in in vitro translation systems (e.g., Brisco et al., 1985).

B. Methods for Determining Size

The size of a virus nucleic acid is perhaps the most important property of the virus that can be expressed as a single number. There is some uncertainty for any method of measurement of nucleic acid size except with a full sequence analysis.

1. Sequence Analysis

The nucleotide sequences for many viral RNAs and DNAs and viroid RNAs have been determined. These allow chemically precise determinations of MW.

2. Polyacrylamide Gel Electrophoresis

Polyacrylamide gel electrophoresis (PAGE) is now a widely used procedure for estimating the MW of a viral RNA, by reference to the mobilities of standard RNAs of known MW. Standard "RNA ladders" are now available commercially. Any secondary structure in the RNAs will affect the mobility and, therefore, the MW estimate. Even formaldehyde treatment may not eliminate all secondary structure. If the analysis is performed in a Tris–EDTA buffer at pH 7.5 containing 8 M urea at 60°C, more reliable estimates of MW may be obtained (Reijinders et al., 1974). Glyoxal has also been used as an effective denaturant (Murant et al., 1981). Estimates of MW using gels will be approximate whatever method and markers are used.

3. Renaturation Kinetics

Where it appears necessary to check whether a virus is monopartite or whether it might have, say, two RNAs of different base sequence but of the same size, then renaturation of the RNA with complementary DNA can be particularly useful. The sequence complexity of the nucleic acid can give a clear indication as to whether one, two, or more different molecules exist (e.g., Gould, *et al.*, 1978).

4. Electron Microscopy

In recent years the size of ds nucleic acids has often been estimated from length measurements made on electron micrographs of individual molecules. This method may give erroneous results with ss nucleic acids because of doubt as to the internucleotide distances (Reijinders *et al.*, 1973). Errors are considerable and reliable marker molecules must be used.

5. Physicochemical Methods

Gierer (1957, 1958) estimated the size of isolated TMV RNA by measuring the sedimentation coefficient and the intrinsic viscosity of the RNA. He found the general relationship between molecular weight (m) and sedimentation coefficient (s) for this RNA to be $m = 1100s^{2.2}$. With the s equal to 31 he determined that the MW of the isolated intact TMV RNA was 2.1×10^6. Empirical relationships between MW (m) and $s_{20.w}$ of the general form $m = ks^d$ have been published for RNAs by several workers (Spirin, 1961; Hull *et al.*, 1969a). Such a relationship is very useful but it should be remembered that methods depending on the determination of s are much more laborious than those using PAGE and are no more accurate.

C. ssRNA Genomes

1. Heterogeneity

When RNA is prepared from a purified virus preparation and subjected to some fractionation procedure that separates RNA species on the basis of size, RNAs of more than one size are usually found. Apart from degradation during RNA isolation and storage, there are several possible reasons for heterogeneity.

Multipartite Genomes

Of the 28 established groups of nonenveloped plant viruses with ssRNA genomes, 12 have their RNA in two or more separate pieces of different size, as discussed in detail in later chapters.

Other Sources of Heterogeneity

1. Some degradation of the viral RNA may have occurred inside the virus before RNA isolation.
2. Less than full length copies of the viral RNA (subgenomic pieces) synthesized as such in infected cells may be encapsidated (Palukaitis, 1984).
3. Some host RNA species may become accidentally encapsidated. For example, the empty protein shells found in preparation of egg plant mosaic virus and several other tymoviruses contain an average of two to three small RNA molecules with tRNA activity (Bouley *et al.*, 1976). These are very probably encapsidated host tRNAs. Similarly host RNAs may become encapsidated in TMV coat protein (Siegel, 1971).
4. A number of viruses contain small satellite RNAs unrelated to the viral genome in nucleotide sequence (Chapter 9, Section II,B).

5. Small defective interfering RNAs made up of viral sequences are found in some virus infections (Chapter 6, Section VIII,B).

6. As noted in the next Section (2,b), ssRNAs in solution may have a substantial degree of secondary and tertiary structure. A single RNA species may assume two or more distinct configurations under appropriate ionic conditions. These "conformers" may be separable by electrophoresis or centrifugation. Thus they give rise to additional heterogeneity that is not based on differences in length (Dickerson and Trim, 1978).

7. The RNA may migrate as a transient dimer under certain conditions (e.g., Asselin and Zaitlin, 1978).

2. Some Physical Properties of ssRNAs

Ultraviolet Absorption

Like other nucleic acids, plant viral RNAs have an absorption spectrum in the ultraviolet region between 230 and 290 nm that is largely due to absorption by the purine and pyrimidine bases. The absorption spectra of the individual bases average out to give a strong peak of absorption near 260 nm with a trough near 235 nm. Ultraviolet absorption spectra are usually of little use for distinguishing between one viral RNA and another.

The absorbance of an RNA solution at 260 nm measured in a cell of 1-cm path length is a convenient measure of concentration. The absorbance per unit weight varies somewhat with base composition. At 1 mg/ml for TMV RNA, $A_{260} \simeq 29$; for TYMV RNA, $A_{260} \simeq 23$ (in 0.01 M NaCl at pH 7).

Secondary Structures

In the intact virus particle, the three-dimensional arrangement of the RNA is partly or entirely determined by its association with the virus protein or proteins (Chapter 5). Here I shall consider briefly what is known about the conformation of viral RNAs in solution. dsDNA has a well-defined secondary structure imposed by base-pairing and base stacking in the double helix. Single-stranded viral RNAs have no such regular structure. However, it has been shown by a variety of physical methods that an RNA such as that of TMV in solution near pH 7 at room temperature in, say, 0.1 M NaCl does not exist as an extended thread. Under appropriate conditions ssRNAs contain numerous short helical regions of intrastrand hydrogen-bonded base-pairing interspersed with ss regions. They behave under these conditions as more or less compact molecules. The degree of secondary structure in the molecule under standard conditions will depend to some extent on the base sequence and base composition of the RNA.

The helical regions in the RNA can be abolished, making the molecule into a random, extended, disorganized coil by a variety of changes in the environment (Spirin, 1961). These include heating, raising or lowering the pH, or lowering the concentrations or changing the nature of the counterions present (e.g., Na^+, Mg^{2+}) (Boedtker, 1960; Eecen *et al.*, 1985). The change from helical to random coil alters a number of measurable physical properties of the viral RNA. The absorbance at 260 nm and viscosity are increased, while the sedimentation rate is decreased.

A polyribonucleotide lacking secondary structure has about 90% the absorbancy of the constituent nucleotides found on hydrolysis. A fully base-paired structure (e.g., a synthetic helical polyribonucleotide composed of poly(A) plus poly(U) has about 60% of the UV absorbancy of the constituent nucleotides. The changing absorption characteristics of RNA with changing pH are due in part to changes in the extent of the base-paired helical regions in the molecule, and in part to shifts in the absorption spectrum of individual bases due to changes in their ionization state. Methods for probing the structure of RNAs in solution are discussed by Ehresmann *et al.* (1987). Viral RNAs with amino acid accepting activity almost certainly have a three-dimensional tRNA-like configuration near the 3′ terminus when they are in solution under appropriate conditions (see Fig. 4.2).

Effective Buoyant Density in Solutions of Cesium Salts

When a dense solution of CsCl or Cs_2SO_4 is subjected to centrifugation under appropriate conditions, the dense Cs ions redistribute to form a density gradient in the tube. If a nucleic acid is present it will band at a particular position in the density gradient. This density, known as the effective buoyant density, provides a useful criterion for characterizing virus nucleic acids and for distinguishing between RNAs and DNAs. dsDNA forms a band at about 1.69–1.71 g/cm^3 in a solution of CsCl with starting density of about 1.70 g/cm^3. The exact banding position depends on the G + C content of the DNA. Thus the method can be used to discriminate between DNAs with differing base composition. Under the same conditions RNA is pelleted from the gradient. Cs_2SO_4 solutions form gradients in which all the various kinds of nucleic acid can form bands. In a gradient formed from a solution with a starting density of about 1.56 g/cm^3, dsDNA bands at 1.42–1.44 g/cm^3; ssDNA at about 1.49 g/cm^3; DNA–RNA hybrids at about 1.56 g/cm^3; dsRNAs at about 1.60 g/cm^3; and ssRNAs at about 1.65 g/cm^3.

3. End-Group Structures

Many plant viral ssRNA genomes contain specialized structures at their 5′ and 3′ termini. This section summarizes the nature of these structures; while their biological functions are discussed in Chapter 6, Section VI,A.

The 5′ Cap

Many mammalian cellular messenger RNAs and animal virus messenger RNAs have a methylated blocked 5′-terminal group of the form

$$m7G5'ppp5'X^{(m)}pY^{(m)}p. \ldots$$

where $X^{(m)}$ and $Y^{(m)}$ are two methylated bases.

Some plant viral RNAs have this type of 5′ end, known as a "cap," but in the known plant viral RNAs the bases X and Y are not methylated. Virus groups with 5′ capped RNAs are the tobamoviruses, tobraviruses, tymoviruses, bromoviruses (also cucumoviruses and AMV), carmoviruses, furoviruses, potexviruses, and hordeiviruses.

5′ Linked Protein

Members of several plant virus groups have a protein of relatively small size ($\approx$3500–24,000 MW) covalently linked to the 5′ end of the genome RNA. These are known as VPg's (short for *v*irus *p*rotein, *g*enome linked). For some viruses the viral gene coding for the VPg has been identified (e.g., for cowpea mosaic *Comovirus* (CPMV), see Chapter 7, Section V,A). All VPg's are coded for by the virus concerned. If a multipartite RNA genome possesses a VPg, all the RNAs will have the same protein attached. For CPMV and VPg is attached to the genomic RNAs by

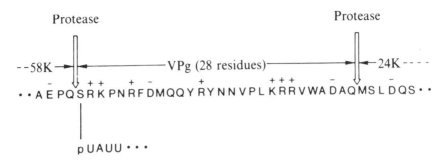

Figure 4.1 Primary structure and properties of CPMV VPg. VPg is released from longer precursors by the 24K protease at Gln/Ser and Gln/Met cleavage sites, respectively. Basic and acidic amino acid residues are indicated by + and − signs, respectively. (From Jaegle *et al.*, 1987.)

a phosphodiester bond between the β-OH group of a serine residue located at the − NH$_2$ terminus of the VPg and the 5′-terminal uridine residue of the two genomic RNAs (Jaegle *et al.*, 1987). Figure 4.1 shows how the CPMV VPg is excised from a longer precursor by a protease. The basic nature of the protein is also indicated.

Daubert and Bruening (1984) discuss methods for the detection and characterization of VPg's, which have been found in viruses belonging to the following groups: potyviruses, comoviruses, nepoviruses, sobemoviruses, and luteoviruses.

3′ Poly(A) Tracts

Polyadenylate sequences have been identified at the 3′ terminus of the messenger RNAs of a variety of eukaryotes. Such sequences have been found at the 3′ terminus of several plant viral RNAs that can act as messengers. Members of the following groups have 3′-terminal poly(A) sequences: potyviruses, potexviruses, capilloviruses, comoviruses, nepoviruses, and furoviruses. The length of the poly(A) tract may vary for different RNA molecules in the same preparation, and such variation appears to be a general phenomenon. For example, in clover yellow mosaic *Potexvirus* (ClYMV) RNA the range was bout 75–100 A residues (AbouHaidar, 1983). For CPMV, numbers of residues ranged from about 25 to 170 for B RNA and 25–370 for M RNA (Ahlquist and Kaesberg, 1979).

3′ tRNA-like Structures

Pinck *et al.* (1970) reported that TYMV RNA, when incubated with various ^{14}C-labeled amino acids in the presence of appropriate cell-free extracts from yeast or *E. coli*, bound valine, which became attached to the 3′-terminal adenosine by an ester linkage. Later work showed that several other tymoviruses accepted valine (Pinck *et al.*, 1975). The accepting activity was also present in the ds replicative form of the viral RNA, and in this state was resistant to RNase attack. These experiments demonstrated that the amino acid accepting activity was an integral part of the viral positive sense ssRNA. tRNA-like structures are also known for the 3′ termini of tobamoviruses, hordeiviruses, and bromoviruses (also cucumoviruses).

Subsequent work has shown that the 3′ terminus of the RNAs of the type strain of TMV and some other tobamoviruses accept histidine (Litvak *et al.*, 1973; García-Arenal, 1988), those of BMV RNAs and BSMV RNAs accept tyrosine (Loesch-Fries and Hall, 1982), and CMV RNAs accept tyrosine (Kohl and Hall, 1974). BSMV RNAs are unusual in having an internal poly(A) sequence between the end of the coding region and the 3′ tRNA-like sequence that accepts tyrosine (Agranovski *et al.*, 1981, 1982, 1983). Methylated purines and pseudouracil occur in cellular tRNAs and some of these bases have cytokinin activity. Claims for cytokinin activity in TMV RNA have not been confirmed (Whenham and Fraser, 1982).

Various tRNA-like secondary and tertiary structures have been proposed for these 3′-terminal sequences. For example, Joshi *et al.* (1985) discuss the conformational requirements for the aminoacylation and adenylation of the tRNA-like structure in TMV. To varying degrees all of these structures are speculative. Recently Dumas *et al.* (1987) have inferred a tRNA-like structure for the 3′ end of TYMV RNA using three-dimensional graphics modeling, together with previously known biochemical results and sequences comparisons (Fig. 4.2).

A key feature of the model is a "pseudoknot" based on a set of three GC base pairs involving the triple G's in loop I (Fig. 4.2A) and the triple C's in the single-stranded sequence between loops I and II. This base-pairing together with the coaxial stacking of stems I and II gives rise to an amino acid acceptor arm of 12 stacked base pairs as found in classical tRNAs (Figs. 4.2C and F). Additional evidence for the presence of the pseudoknot came from a study of the thermal unfolding of the structure (van Belkum *et al.*, 1988). Pseudoknots are further discussed in Chapter 6, Section IV,A,1.

4. Infectivity

Historical

Gierer and Schramm (1956) used several criteria to establish that phenol-extracted TMV RNA isolated from the virus was infectious. The infectivity in their RNA preparations was only about 0.1% that of the same amount of RNA in intact virus. Fraenkel-Conrat and Williams (1955) had shown that while the infectivity of isolated TMV RNA was very much lower than in intact virus, the infectivity of the RNA could be greatly enhanced by allowing viral protein subunits to reaggregate around the RNA to reform virus rods. When TMV RNA from one virus strain was reconstituted *in vitro* with protein from a distinct strain to form "hybrid" rods, and these particles were inoculated onto plants, the progeny virus had protein of the type naturally found associated with the RNA used (Fraenkel-Conrat, 1956). This was strong evidence that the RNA alone specified the virus coat protein structure. Since these classic experiments infectious RNA has been prepared from many plant viruses. Most viral RNA preparations are much less infectious than the intact virus on an equal RNA basis and the infectivity of the RNA relative to intact virus varies considerably with method of preparation, with different viruses, and with inoculation medium (e.g. cowpea chlorotic mottle *Bromovirus,* Wyatt and Kuhn, 1977).

Structural Requirements

From experiments with various inactivating agents it is has been established that the complete intact TMV RNA molecule is necessary for infectivity. A single break in the polynucleotide chain destroys infectivity. Chemical modification of only the terminal base at either the 5′ or the 3′ terminus may inactivate a viral RNA (Kohl and Hall, 1977). On the other hand, the polyadenylic acid covalently linked to the 3′ terminus of some viral RNAs may not be necessary for infectivity. For BSMV (Agranovsky *et al.,* 1978) and tobacco etch *Potyvirus* (TEV) (Hari *et al.,* 1979), RNA molecules lacking poly(A) were as infectious as the polyadenylated molecules.

The VPg's found at the 5′ ends of certain viruses are necessary for the infec-

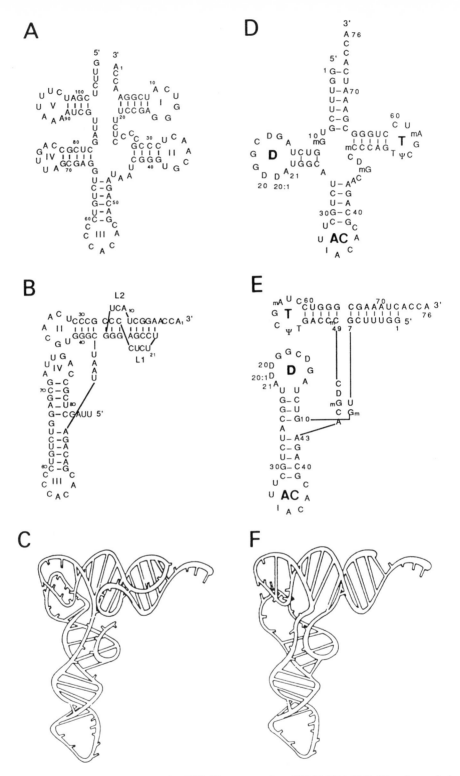

Figure 4.2 Comparison between the tRNA-like structure from TYMV RNA (A,B,C,) and canonical tRNA (D,E,F). (A) Secondary structure of the 106 3′ terminal nucleotides of TYMV RNA. (B) Scheme of folding of the tRNA-like structure (86 nucleotides). (C) Three-dimensional wire model of the tRNA-like structure (86 nucleotides). (D) Secondary structure of yeast tRNA[Val]. (E) Scheme of the folding of yeast tRNA[Val] (to be compared with (B)). (F) Three-dimensional model of canonical tRNA. (From Dumas *et al.*, 1987.)

tivity of some RNAs (e.g., those of nepoviruses; Mayo *et al.*, 1982a; Hellen and Cooper, 1987) but not for others (e.g., the comoviruses and tomato bushy stunt *Tombusvirus* (TBSV); Mayo *et al.*, 1982a).

Full-length transcripts of TMV RNA from cDNA lacking the 5′ cap were much less infectious than capped molecules (Meshi *et al.*, 1986). Similarly BMV RNA transcripts without a cap were much less infectious than capped molecules (Janda *et al.*, 1987).

The ds replicative RNAs of TMV (Jackson *et al.*, 1971) and CPMV (Shanks *et al.*, 1985) are not infectious. Heat denaturation renders such dsRNA preparations infectious because the infectious positive sense strands are released from the ds structure.

The ability to produce infectious viral RNA copies from cloned viral DNA (Ahlquist *et al.*, 1984b) has made it possible to examine the effects of adding extra nonviral nucleotides to either end of the genomic RNA. Addition of a single extra 5′ G residue reduced the infectivity of BMV transcripts more than threefold, while 7 or 16 base extensions caused a marked fall in infectivity (Janda *et al.*, 1987). On the other hand, addition of nonviral bases at the 3′ ends of BMV or TMV had much less effect, and they were eliminated in progeny virus (Ahlquist *et al.*, 1984b; Meshi *et al.*, 1986).

The Rhabdoviridae and Tomato Spotted Wilt Virus

Members of the Rhabdoviridae possess an ss negative sense RNA genome, and virus particles contain a viral RNA polymerase (transcriptase) essential for initiating infection. Thus the isolated RNAs of these viruses are not infectious. Similarly the isolated RNAs of tomato spotted wilt virus (TSWV), being mainly negative sense, are not infectious.

Viruses with Multipartite ssRNA Genomes

All of these viruses require a complete set of intact genomic RNAs to establish a full infection giving rise to progeny virus. However, in several groups with bipartite genomes the RNA-dependent RNA polymerase is coded for on the larger genome RNA. Thus this RNA alone can infect cells and reproduce itself, while not being able to elicit virus particles because the coat protein gene is on the other genomic RNA. This applies to tobraviruses (Harrison and Nixon, 1959) and nepoviruses (Robinson *et al.*, 1980). While the B component of CPMV can replicate alone when inoculated to protoplasts (Goldbach *et al.*, 1980), it cannot produce symptoms or move from cell to cell when inoculated onto leaves (Rezelman *et al.*, 1982). Similarly RNAs 1 and 2 of BMV and CMV can replicate without the presence of RNA3, but cannot form virus particles or move from cell to cell (Chapter 7, Sections VI,A and B).

AMV has a genome made up of three RNAs (1, 2, and 3). A fourth smaller RNA containing the coat protein gene is usually encapsidated. The coat gene is also found in RNA3. The three isolated genomic RNAs alone are not infectious, because they require the presence of some coat protein molecules or RNA4 in the inoculum. Each of the three RNAs needs to bind a coat protein molecule at a specific site to initiate infection. This phenomenon is discussed further in Chapter 7, Section VI,C,4.

D. dsRNA Genomes

The genomes of plant reoviruses consist of 10 or 12 pieces of dsRNA (e.g., Hibi *et al.*, 1984). The cryptoviruses that have been studied contain two segments of

dsRNA (Accotto and Boccardo, 1986). The fully base-paired double-helical structure is most stable in solutions of high ionic strength ($>0.1\ M$), at pH values near
neutrality, and at low temperatures. At low ionic strengths, extremes of pH, or high
temperatures, the ds structure is lost (i.e., the structure melts) as described in
Chapter 2, Section IV,A. These dsRNAs can be distinguished from ssRNAs by a
number of physical and chemical properties. Many of these differences result from
the fact that dsRNAs have a much more highly ordered structure than ssRNAs.

The following physical and chemical properties are commonly used.

1. *Base composition*   Base-pairing of the Watson–Crick type is present, thus
   base analyses show that adenine = uracil and guanine = cytosine in these
   RNAs.
2. *X-ray diffraction*   Because of their highly ordered structure it was possible
   to prove the ds nature of these nucleic acids by X-ray diffraction analysis
   (Tomita and Rich, 1964).
3. *Melting profile*   Native rice dwarf *Phytoeovirus* (RDV) dsRNA has a sharp
   DNA-like melting curve, with $T_m = 80°C$. RDV RNA that had been denatured by heating following by rapid cooling behaved like ribosomal and
   transfer RNA (Miura *et al.*, 1966) (Fig. 2.13).
4. *Electron microscopy*   dsRNA has a general appearance like that of dsDNA
   when shadowed or stained specimens are examined in the electron
   microscope.
5. *Resistance to RNase*   The ds structure makes the phosphodiester backbone
   of the RNA strands resistant to attack by pancreatic RNase provided the
   reaction is carried out at an appropriate pH and ionic strength. If dsRNA is
   denatured by heating followed by rapid cooling, the individual RNA strands
   become susceptible to RNase. The increase in absorbance at 260 nm gives a
   convenient measure of the extent to which a given sample of nucleic acid is
   attacked by the enzyme.
6. *Buoyant density in Cs_2SO_4*   dsRNAs have a characteristic buoyant density
   in Cs_2SO_4 gradients (about $1.60\ g/cm^3$).
7. *Reactivity with formaldehyde*   Formaldehyde reacts with the free amino
   groups of the nucleic acid bases. The amino groups in dsRNA are involved
   in the hydrogen-bonded base pairs and do not react with formaldehyde under
   conditions where ssRNA does. This difference can be used as an additional
   criterion to identify dsRNA (e.g., Ikegami and Francki, 1975).
8. *Infectivity*   The isolated dsRNAs of reoviruses are not infectious. This is
   because the virus particles contain a viral RNA polymerase that functions at
   an early stage in infection to produce positive sense strands from negative
   sense strands in the infecting virus.

E. dsDNA Genomes

Until recently, members of the *Caulimovirus* group were the only plant viruses
known to have a genome consisting of dsDNA. These genomes are about 8000 base
pairs in length. Working with CaMV, Shepherd *et al.* (1968) provided the first good
evidence that a plant virus contained DNA. The nucleic acid was identified as DNA
by the following properties: (i) a positive diphenylamine reaction; (ii) the presence
of thymine; (iii) infectivity of nucleic acid isolated from the virus was abolished by

pancreatic DNase, but not by RNase; and (iv) when purified virus was subject to equilibrium density gradient centrifugation in cesium chloride, the DNA as measured by the diphenylamine test (or material absorbing at 260 nm) and the infectivity all banded at the same position in the gradient. Further work showed that the DNA was ds, because it had a melting profile typical of a dsDNA (Fig. 2.13) and because it had a base composition in which G = C and A = T (Shepherd et al., 1970).

DNA isolated from CaMV, subjected to gel electrophoresis, and examined in the electron microscope reveals a number of conformations (Menissier et al., 1983). Open circles are the most abundant, but singly and multiply knotted forms also occur (Fig. 4.3).

Both strands of the DNA contain ss discontinuities. In one strand there is a unique discontinuity found in all caulimoviruses. In the second strand different viruses or strains have one, two, or three ss discontinuities. The strand with a single discontinuity is the α or minus strand. At each of the three discontinuities the 5′ and 3′ extremities of the interrupted strand overlap one another to a variable extent over a range of about 8–20 nucleotides (Richards et al., 1981).

A small ssDNA fragment is also found in virus particles with an RNA species of 75 nucleotides covalently linked to the 5′ end (Turner and Covey, 1984). The RNA had the properties of host plant tRNA^{met}. Many of the ss strands making up the dsDNA genome also possess one or more ribonucleotides covalently bound to their 5′ termini (Guilley et al., 1983). These RNAs are primers or remnants of primers involved in replication (Chapter 8, Section I). Isolated CaMV DNA infects turnip plants (e.g., Howell et al., 1980) and protoplasts (Yamaoka et al., 1982a).

Commelina yellow mottle (CoYMV) is the type member of a new group of plant viruses approved by the I.C.T.V. but not yet given an official name. The genome consists of one molecule of open circular dsDNA of ≃7500 base pairs with a single-strand discontinuity at one site in each strand (Lockhart, 1990).

F. ssDNA Genomes

One group of plant viruses, the geminiviruses, has ssDNA genomes, as was first shown by Goodman (1977a,b) and Harrison et al. (1977). In some geminiviruses the genome is a single covalently closed circle of about 2.5–3.0 kb. The genome of others consists of two such circles of about the same size, generally known as DNAs 1 and 2 (e.g., Bisaro et al., 1982).

The single circle of genomic DNA encapsidated in maize streak virus is partially ds, having a short (70–80 bases) strand base paired to it. The short strand is capped with one or a few ribonucleotides (Howell, 1984). African cassava mosaic virus (ACMV) has a genome consisting of two ssDNA circles. In addition to these two molecules, DNA preparations made from the virus contain minor DNA populations of twice and approximately half the length of genomic DNA (Stanley and Townsend, 1985). The dimeric DNA is involved in the replication cycle (Chapter 8, Section II,E). The half-size DNA, which is derived from only one of the genome strands, may represent DI molecules.

Goodman (1977a) showed that isolated *Geminivirus* DNA was infectious. Stanley (1983), using cloned DNA, demonstrated that both DNAs of ACMV are necessary for infectivity and movement in plants. However, one of the DNAs (DNA1) can replicate alone in protoplasts but cannot move systemically in plants (Davies et al., 1987a). It has been demonstrated, using cloned DNA, that the single

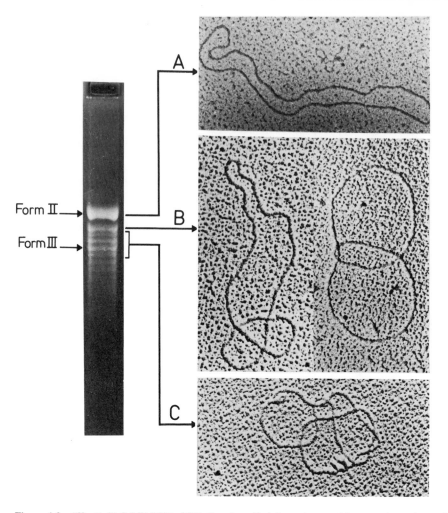

Figure 4.3   "Knotted" CaMV DNA. DNA (1 μg) purified from virus particles was electrophoresed on a 1% agarose gel at 30 V for 16 hours. The gel was then stained with ethidium bromide (0.5 μg/ml) and observed under uv illumination. DNA in different sections of the gel was electroeluted and prepared for electron microscopy. Band A corresponds to open circular DNA molecules. Band B corresponds to singly knotted DNA molecules. Band C mainly corresponds to high-knotted circular DNA. Linear DNA molecules, sometimes knotted, were also observed (not shown). (From Menissier *et al.*, 1983.)

ssDNA of maize streak virus is infectious and constitutes the complete genome (Lazarowitz, 1988).

G. Nucleotide Sequences

The nucleotide sequences at the 5' and 3' termini were determined for several plant viruses in the 1970s by sequencing RNA fragments directly. The first plant viral genome to be fully sequenced was the DNA of CaMV (Franck *et al.*, 1980). The ability to generate cDNA libraries covering the full extent of viral RNA genomes allowed DNA sequencing methods to be applied to determining the complete nucleotide sequence of such RNAs. The first such sequence to be determined was that of TMV (Goelet *et al.*, 1982). Since that time many other plant viral RNA genomes have been fully sequenced (Chapter 7). Of particular importance has been the development of a system in which cDNA clones of an RNA virus can be transcribed *in vitro* to produce RNA copies that are fully infectious *in vivo* (Ahlquist *et al.*, 1984b).

Many plant viral nucleotide sequences are available in sequence data banks (e.g., Bilofsky *et al.*, 1986). Details of sequencing procedures are given by Ausubel *et al.* (1987, 1988). Functional aspects of nucleotide sequences are discussed in Chapters 6–8.

II. PROTEINS

Most of the small geometric plant viruses with ssRNA as their genetic material have only one kind of protein subunit in the virus particle. The ssRNA *Comovirus* group contains two. The plant reoviruses, the *Caulimovirus* group, and the much larger enveloped plant viruses contain several different proteins, some of which are enzymes.

The same 20 amino acids, or a selection of them, are found in plant virus coat proteins as are found in other living material, and the average proportion of amino acids shows the same general trend. For example, cysteine, methionine, tryptophan, histidine, and tyrosine usually occur in low amounts.

A. Isolation from Virus Preparations

Many of the conditions and reagents used to isolate viral RNA can also be used to give a preparation of the virus protein that is more or less free of RNA. However, the best method for obtaining intact RNA from a particular virus is usually different from that which gives satisfactory protein preparations.

1. Detergents

With the advent of electrophoresis in polyacrylamide gels, disruption of virus samples in SDS has become a popular procedure to prepare the monomers from noncovalently linked polypeptides in the virus. Under these conditions mobility in the gel may depend almost entirely on size.

2. Acids

Fraenkel-Conrat (1957) described an elegant and simple method for the isolation of TMV protein subunits by treating the virus with cold 67% acetic acid. The ease with which the virus is split varies with different strains of the virus. TMV protein isolated by this procedure is in a native state and can be used to reconstitute virus rods (see Chapter 7, Section III,A,9). Unfortunately the acetic acid method is not applicable to most other viruses.

3. Strong Salt Solutions

One molar $CaCl_2$ dissociates and solubilizes the protein coats of broad bean mottle *Bromovirus* (BBMV) (Yamazaki and Kaesberg, 1963a) and BMV (Yamazaki and Kaesberg, 1963b). For BMV the product obtained appeared from sedimentation analysis to be a relatively stable dimer of the chemical subunit. A simple procedure using 1 m NaCl produced protein subunits that could be reconstituted with RNA to give infective virus particles (Hiebert *et al.*, 1968). LiCl is useful as an alternative to CaCl$_2$ for some viruses and may give a superior product (e.g., see Moghal and Francki, 1976).

B. Nature of the Protein Product

The chemical and physical state of the isolated protein varies widely depending on the virus and the treatment used. The types of product are discussed in the following sections.

1. Viral Protein Shells

Virus particles from which the RNA has escaped (e.g., TYMV).

2. Subunits

Subunits in native form, and in various stages of aggregation may be present. For example, minor components (a few percent of the total protein) were found in CCMV and STNV (Rice, 1974). These components had the size and composition expected for a dimer of the coat protein. The chemical evidence suggested that the dimers were joined by covalent bonds other than disulfide bonds. The coat protein of TRSV is a monomer of MW $\simeq$ 13,000; but following dissociation of the virus in SDS, urea, and mercaptoethanol, a polymeric series of five polypeptides was obtained with MWs up to 110,000 (Chu and Francki, 1979).

3. Reversibly Denatured Subunits

For example, phenol treatment of TMV gives denatured protein, but Anderer (1959) found that if the denatured protein is precipitated from the phenol by methanol it could be renatured to give fully functional subunits by heating at 60°C and pH 7.0–7.5.

4. Irreversibly Denatured Aggregates

Irreversibly denatured aggregates are insoluble in aqueous media. A major factor leading to the production of insoluble aggregates is the formation of cross-linking disulfide bridges between cysteine residues due to oxidation of the sulfhydryl groups. Such cross-linking may be blocked by various procedures.

5. Partly Degraded Protein

Some of the protein subunits in a virus particle may be cleaved at specific sites during isolation from the plant or on subsequent storage. For example, Koenig *et al.* (1978) described how PVX protein can be partially degraded from the N terminus in the intact virus by reducing agent-dependent proteases in crude plant extracts, and at the C terminus by reducing agent-independent proteases that occurred in some virus preparations.

Ezymatic degradation could lead to ambiguity as to the "true" size of the protein subunit in the virus particle. For example, most sobemoviruses have a single species of coat protein. However, proteins of more than one size are found in some members of the group. Thus dissociated velvet tobacco mottle *Sobemovirus* yields proteins of $M_r \simeq 37,000, 33,000$, and 31,500. Chu and Francki (1983) considered that the 33,000 protein was an unstable intermediate resulting from the loss of about 40 amino acids from the 37,000 protein. Loss of about 15 more gave rise to the 31,500 protein. The related *Solanum nodiflorum* mottle virus has a single coat protein of $M_r \simeq 31,000$. It is not clear whether the reduction in size from 37,000 to 31,500 is a biologically significant event or merely an artifact of the isolation process (Kibertis and Zimmern, 1984).

6. Coat Protein Covalently Linked to a Host Protein

A more unusual situation has been described for the U_1 strain of TMV. A protein of about 26,500, occurring at about one molecule per virus particle, has most of the amino acid sequences found in TMV coat protein together with unrelated sequences (Collmer and Zaitlin, 1983). The isolated protein, called protein H, reassociated into U_1 TMV rods when coat protein, H protein, and RNA were mixed *in vitro* (Collmer *et al.*, 1983). The host-derived portion of H protein has been shown by amino acid sequencing and immunological cross-reactivity to be ubiquitin, probably linked to the TMV coat protein at lysine 53 (Dunigan *et al.*, 1988). Ubiquitin is a small protein (76 amino acids) found in all eukaryotes with a high degree of amino acid conservation between groups of organisms. The significance of its binding to TMV is not known, but Dunigan *et al.* (1988) suggest that it may be a stress response by the plant.

C. Size Determination

Most plant viral coat polypeptides have MWs in the range 17,000–40,000. This size distribution is at the lower end of the range of MWs of most polypeptides found in higher plants. Sedimentation–diffusion measurements are difficult to apply to molecules as small as the coat proteins. SDS–PAGE and calculations from amino acid sequence are now the methods most commonly used to determine MW.

1. Polyacrylamide Gel Electrophoresis

The most popular method in recent years has been electrophoresis of viral proteins in polyacrylamide gels containing SDS and urea. The MWs are estimated by reference to the mobility of marker proteins of known MW. Outstanding advantages of the method are its simplicity, the small quantities of proteins required, and the fact that mixtures of proteins are readily resolved.

However, there are several limitations that must be taken into account. First, there is an intrinsic variability of about 5–10% in estimate of MW. Second, the method is based on several assumptions: (i) that the unknown proteins have the same conformation in the SDS gels as the marker proteins (perhaps a hairpinlike structure); (ii) that all the proteins bind the same amount of SDS (about 1.4 g/g protein); and (iii) that as a consequence of uniform binding of SDS all the proteins have a similar charge/mass ratio. Different proteins do not bind uniform amounts of

SDS. Anomalous behavior is revealed when estimates of MW made at different gel concentrations give different answers (e.g., Ghabrial and Lister, 1973; Shepherd, 1976).

Coat proteins that are certainly single species sometimes form double bands (e.g., TYMV protein with apparent MWs of 20,100 and 21,700; Matthews, 1974). Substitution of certain groups on a protein may lead to unexpected changes in apparent MW. For example, after carboxymethylation, TYMV protein had an apparent MW of 23,200 (Matthews, 1974). Such anomalies are presumably due to the existence of different conformational states of the polypeptide that bind SDS differently, but the details are not understood.

2. Chemical Methods

Where the full sequence of amino acids is established the MW is known with a high degree of precision. A combination of amino acid analysis and tryptic peptide mapping, mapping of cyanogen bromide fragments, or end-group analysis can give estimate of size in the absence of a full sequence analysis. However, these procedures alone may give erroneous results (e.g., Gibbs and McIntyre, 1970).

D. Amino Acid Sequences

The primary structure of TMV coat protein was determined by Fraenkel-Conrat and colleagues (Tsugita *et al.*, 1960) and by Anderer *et al.* (1960). There were many difficulties associated with techniques and interpretation of results over a period of years, and the elucidation of the full sequence of 158 amino acids represented a substantial achievement. The full sequence of the coat protein amino acids is now known for several strains of TMV and for other tobamoviruses. The sequences for two different tobamoviruses are shown in Fig. 4.4.

Knowledge of the sequence of amino acids in TMV coat protein was of great value in studies that correlated base changes in the RNA with changes in amino acid

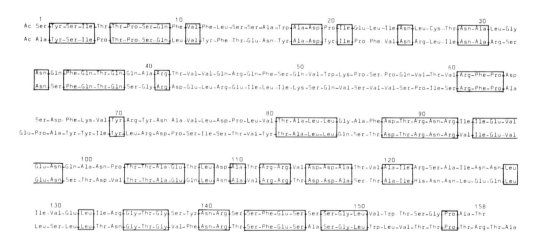

Figure 4.4 The amino acid sequences of the type strain of TMV protein (*top line*) and that of sunn-hemp mosaic *Tobamovirus* (*bottom line*). Numbers written above amino acid residues indicate the sequential numbers from the NH$_2$ terminus of the protein. Residue 65 in the sunnhemp virus protein has been moved to the left one place to obtain the best fit. (From Rees and Short, 1975.)

sequence in the coat proteins of virus mutants. These studies confirmed the nature and universality of the genetic code.

During the 1970s the partial or full amino acid sequences of several coat proteins were determined by direct chemical methods, and such procedures are sometimes still used (e.g., Rees and Short, 1982; Short *et al.*, 1986). However, with the development of procedures for copying RNA genomes into DNA, and the rapid sequencing of such DNA, it has proved simpler and quicker to determine the amino acid sequence of proteins from the nucleotide sequence of the genes coding for them. Use of this strategy for plant virus proteins began in the early 1980s (e.g., for STNV coat protein, Ysebaert *et al.*, 1980; and BMV, Ahlquist *et al.*, 1981a). The amino acid sequences of many coat proteins and other viral-coded proteins have now been determined by these procedures. Identification of the coat protein gene in the RNA sequence can sometimes be made from its size and its position in the genome compare with that of known related viruses. However, proof of identity can be obtained by determining chemically the sequence of 20–30 amino acids somewhere in the coat protein and matching this sequence with the appropriate nucleotide sequences (e.g., Carrington *et al.*, 1987, for turnip crinkle *Carmovirus* coat protein).

E. Secondary and Tertiary Structure

A variety of physical and chemical techniques that give some information about secondary and tertiary structure have been applied to plant virus coat proteins. The essential requirements for the determination of a full three-dimensional structure are a knowledge of the primary sequence of amino acids and a crystalline form of the protein that allows the techniques of high-resolution X-ray crystallography to be applied. A knowledge of the chemistry of the amino acid sequence is essential for the model building that assists in the three-dimensional interpretation of the data obtained by X-ray crystallographic analysis. In addition, where sequences for several related proteins are known, residues of particular importance can be identified (see Fig. 17.4). Other physical and chemical techniques can give additional guiding or confirmatory information.

1. X-Ray Crystallography

X-ray analysis has been the key technique in elucidating the three-dimensional structure of the coat proteins of several viruses. These structures are discussed in Chapter 5 in relation to the virus as a whole. Where the primary sequence of the coat proteins of two or more related viruses is known, computer-based methods are available for making predictions about tertiary structures (Sawyer *et al.*, 1987). Such predictions may give useful information but they are no substitute for detailed X-ray analysis.

2. Other Physical Techniques

A variety of other physical techniques have been applied to the study of viral coat protein structure. These techniques may give information on which low-resolution models can be based. For example, Kan *et al.* (1982) used data from proton magnetic resonance to propose a low-resolution model for AMV coat protein consisting of a rigid core and flexible N-terminal arm of approximately 36 amino acids.

3. Residues Exposed at the Surface of the Protein or Intact Virus

It is possible to obtain evidence as to whether particular amino acids are at the surface of the protein subunit or the intact virus by using specific enzymes, chemical reagents, or antibodies. For example, carboxypeptidase A removes the C-terminal threonine residue from TMV which must, therefore, be exposed in the intact virus (Harris and Knight, 1955). Of 15 side chain carboxylic acid groups in TMV protein, only three (on residues 64 and 66 and the C terminus) are readily available in the virus for reaction with a carbodiimide, and therefore are on or near the surface (King and Leberman, 1973).

Biochemical evidence together with immunological studies showed that the hydrophilic N-terminal region of the potyvirus coat protein is at the surface of the virus (Allison *et al.*, 1985) (see Chapter 13, Section III,B,4).

4. Groups Added to the Protein Structure

Many coat proteins appear to consist entirely of unmodified polypeptides. In others, various groups are added subsequent to protein synthesis. For example, the coat protein of CaMV may be both phosphorylated (Hahn and Shepherd, 1980) and glycosylated (Du Plessis and Smith, 1981). Lee *et al.* (1972) presented indirect evidence that neuraminic acid or some related compound is present at the surface of sowthistle yellow vein rhabdovirus.

Partridge *et al.* (1976) showed that carbohydrate was associated with the coat protein of BSMV. Its composition included glucose, mannose, xylose, galactosamine, and glucosamine. Pronase digestion of the coat protein of BSMV yielded a tripeptide, Gly-Asp-Ala, attached to carbohydrate through an amide bond with asparagine (Gumpf *et al.*, 1977).

5. The Problem of Polypeptide Chain Folding

Following biosynthesis, the primary polypeptide chain must fold correctly to give the functionally correct three-dimensional structure. Anfinsen (1973) has pointed out that a chain of 149 amino acids would theoretically be able to assume about 4^{149}–9^{149} different conformations in solution. Thus folding to give the correct structure must proceed from a limited number of initiating events. It is not yet possible to predict a three-dimensional structure just from the amino acid sequence, except where the structure of a protein with a very similar amino acid sequence is known. Nevertheless, progress is being made with this general problem of protein structure (Creighton, 1988a,b).

F. Enzymes and Other Noncoat Proteins in Virus Particles

1. Caulimoviruses

CaMV particles contain a DNA polymerase activity. The enzyme–DNA template complex was protected within the virus particle. The enzyme had a MW $\simeq$ 76,000 (Ménissier *et al.*, 1984). CaMV also contains an endogenous protein kinase activity. The main proteins phosphorylated were the coat protein and two larger proteins. Serine and threonine were the residues phosphorylated (Ménissier de Murcia *et al.*, 1986; Martinez-Izquierdo and Hohn, 1987).

2. Reoviridae

By analogy with reoviruses infecting animals, members of the family infecting plant would be expected to contain an RNA-dependent RNA polymerase within the virus particle. Such activity has been detected in preparations of several plant reoviruses, including rice ragged stunt and rice black-streaked dwarf viruses (Uyeda *et al.*, 1987b). In addition, these viruses, for example, rice ragged stunt virus (Hagiwara *et al.*, 1986), contain five or possibly more structural polypeptides. These are discussed in relation to the structure of the viruses in Chapter 5, Section IV.

3. Rhabdoviridae

RNA-dependent RNA polymerase that transcribe viral minus strand RNA into mRNA have been found in animal rhabdoviruses. Similar enzymes should also be present in plant members of the family. Such activity has been shown in preparations of LNYV (Francki and Randles, 1972) and broccoli necrotic yellows virus (Toriyama and Peters, 1981). Like members of the Rhabdoviridae infecting animals, plant rhabdoviruses contain several structural proteins (see Table 5.2).

4. Tomato Spotted Wilt *Tospovirus*

The tomato spotted wilt virus appears to contain four major structural proteins (Mohamed, 1981).

5. Tenuiviruses

The filamentous nucleoprotein particles of rice stripe *Tenuivirus* have an RNA-dependent RNA polymerase associated with them (Toriyama, 1986a). A minor polypeptide of MW $\simeq 23,000$, which may be a polymerase, was found in preparations of another tenuivirus (Toriyama, 1987). It is not yet established whether these enzyme activities are due to viral-coded proteins.

6. Cryptoviruses

Carnation cryptic *Cryptovirus* contains an RNA-dependent RNA polymerase activity, which appears to be a replicase that catalyzes the synthesis of copies of the ds genomic RNAs (Marzachi *et al.*, 1988).

III. OTHER COMPONENTS IN VIRUSES

A. Polyamines

Johnson and Markham (1962) reported the presence of a polyamine in preparations of TYMV. On dissociating TYMV into RNA and protein, about two-thirds of the amine was found with the RNA. Johnson and Markham considered that in the intact virus it was primarily associated with the RNA. Beer and Kosuge (1970) identified the polyamine as a mixture of spermidine (I), making up about 1.0% of the TYMV by weight, and spermine (II), making up only 0.04%. This would be sufficient to neutralize about 20% of the RNA phosphate groups in the virus.

$$\text{I: } H_2N\text{-}CH_2\text{-}CH_2\text{-}CH_2\text{-}NH\text{-}CH_2\text{-}CH_2\text{-}CH_2\text{-}CH_2\text{-}NH_2$$
$$\text{II: } H_2N\text{-}CH_2\text{-}CH_2\text{-}CH_2\text{-}NH\text{-}CH_2\text{-}CH_2\text{-}CH_2\text{-}CH_2\text{-}NH\text{-}CH_2\text{-}CH_2\text{-}CH_2\text{-}NH_2$$

These amines have been identified as occurring widely in plant and animal tissues. Spermidine occurs in uninfected Chinese cabbage and the amount present is increased by TYMV infection (Beer and Kosuge, 1970). Although the protein shell of TYMV can adsorb small amounts of these polyamines reversibly, the shell is impermeable to them. The virus does not exchange or leak the polyamines from within the virus particle (Cohen and Greenberg, 1981). The polyamines in the virus become associated with the RNA before or at the time it is packaged into virus particles.

Not all small plant viruses contain these polyamines. CPMV contained about 1% by weight of spermidine, whereas in BSMV and three bromoviruses no polyamines were detected (Nickerson and Lane, 1977).

B. Lipids

It can be safely assumed that all viruses with an outer bilayer envelope contain lipid. A few enveloped plant viruses have been analyzed for this component. Preparations of TSWV from *N. glutinosa* contained about 19% lipid (Best and Katekar, 1964). Potato yellow dwarf rhabdovirus isolated from *N. rustica* contained more than 20% of lipid (Ahmed *et al.*, 1964). Lipid has also been reported in LNYV (Francki, 1973).

Selstam and Jackson (1983) found that sonchus yellow net rhabdovirus contained about 18% of lipid composed of 62% phospholipids, 31% sterols, and 7% triglycerides. Detailed analysis of these fractions showed that the lipid composition of this virus differed in several respects from that of rhabdoviruses infecting vertebrates. The detailed composition fits with the view that the virus acquires most of its lipid from a host membrane. Ahmed *et al.* (1964) considered that no more than one-fifth of the lipid in their PYDV preparations was due to contaminating host materials, but more detailed chemical characterization is required.

C. Metals

Since early work on the composition of plant viruses (Stanley, 1936) it has been recognized that the ash fraction of purified virus preparations contains various metals. We can distinguish two kinds of metal binding by viruses.

First, there are the metals bound as a consequence of the medium and method used for virus isolation. For example, Johnson (1964) found that the metal content of TYMV varied widely according to the method of preparation. Such variable binding may be trivial as far as the biological activity of the virus is concerned, but it may affect physical properties. Thus the effective buoyant density of TYMV in CsCl gradients depends on the amount of divalent metal ions associated with the virus. In this connection, Durham *et al.* (1977) have pointed out that most virologists are not aware how many trace ions are present even in the purest distilled water.

Second, there are bound metals that may have important biological functions. Durham and colleagues have shown that several viruses (e.g., papaya mosaic *Potexvirus*, Durham and Bancroft, 1979) bind H^+ near pH 7.0 in some way not predictable from the amino acid composition of the protein. Ca^{2+} displaces H^+, suggesting that coordinating oxygens are involved in the binding site. Carboxylate groups constrained in an electronegative environment are probably involved in this Ca^{2+} binding. There are probably three such binding sites per subunit for papaya mosaic virus. Durham *et al.*, have proposed that Ca^{2+} binding in particular may be important in the disassembly of some viruses (Durham *et al.*, 1977).

TMV in unbuffered solutions binds two Ca^{2+} ions per subunit. The two sites are nonidentical and titrate independently (Gallagher and Lauffer, 1983). The stability of some viruses depends on the presence of certain ions. For example, the stability of SBMV appears to depend critically on the presence of Ca^{2+} and Mg^{2+} (Hsu *et al.*, 1976). With the structure of SBMV now established at high resolution, the binding sites for Ca^{2+} have been established (Abdel-Meguid *et al.*, 1981). The Ca^{2+} binding sites in TBSV are shown in Fig. 5.22.

Some plant viruses such as tobacco streak *Ilarvirus* (TSV) have a "zinc finger" binding domain that specifically binds an atom of zinc in a protein involved in nucleic acid binding (see Chapter 7, Section VI,D).

D. Water

The importance of water in living cells is well recognized but the water content of viruses and its significance have been somewhat neglected. The exact meaning to be attached to the term hydration varies with the method of measurement, a problem discussed by Jaenicke and Lauffer (1969). Measurement of various hydrodynamic properties of a virus in solution allow the degree of hydration to be calculated. The use of intensity fluctuation spectroscopy of laser light has allowed the molecular weights, particle dimensions, and solvation of viruses to be measured with increased ease and accuracy (Harvey, 1973; Camerini-Otero *et al.*, 1974). However, errors concerning solvation may be large even with modern techniques.

Typical prokaryotic cells contain about 3–4 ml of water/g dry matter. The few enveloped viruses for which data are available have water contents in the same range (Matthews, 1975). By contrast, the small geometric viruses contain much less (e.g., 0.75 ml/g for TBSV, Camerino-Otero *et al.*, 1974; 0.78 ml/g for TYMV and about 1.1 ml/g for TSV, J. D. Harvey, personal communication). These amounts are similar to those reported for proteins such as bovine serum albumin (0.6 ml/g) (Kuntz and Kauzmann, 1974). 0.6 ml/g corresponds approximately to an average of three water molecules per amino acid residue.

CaMV, which has two protein shells, had 1.7 ml/g, somewhat more than the smaller geometric viruses. Reovirus has 1.5 ml water/g (Harvey *et al.*, 1974). It is probable that plant reoviruses will be found to be hydrated to a similar extent. These viruses, with two protein shells, may have intermediate amounts of bound water, but more examples of this type need to be studied.

It seems likely that water is associated with virus particles in several different ways. The geometric viruses, and probably enveloped viruses as well, have water bound to their protein components ($\simeq 0.5$–1.0 ml/g) in the same way as water binds to isolated proteins. Those viruses with osmotically functional bilayer membranes have, in addition, 2–3 ml/g water that is probably retained in the same way as cells retain water. Structures of viral coat proteins analyzed to atomic resolution show that some water molecules are bound at specific sites within the protein.

IV. DISCUSSION AND SUMMARY

A knowledge of the components found in a virus and details of their properties and structure are essential to our understanding of virus architecture. For example, the detailed symmetry and the size of the protein shell of isometric viruses depend on the properties of the coat protein. The length of rod-shaped viruses depends on the length of the viral RNA, as well as the protein subunit, which determines the pitch and diameter of the helix.

The genomes of most plant virus groups consist of positive sense ssRNA, which can act directly as mRNA upon infecting a cell. For such viruses the isolated RNA is infectious. However, for a few viruses of this sort, a small protein covalently bound to the 5' terminus (VPg) is necessary for infectivity. If the complete genome is split between two or three pieces of RNA then all must be inoculated to produce a complete infection. The isolated dsDNA genomes of caulimoviruses and the ssDNA genomes of geminiviruses are infectious. However, the isolated negative sense ssRNA genomes of rhabdoviruses and the dsRNA genomes of reoviruses are not. This is because viral mRNA synthesis requires a viral-coded polymerase that is present in the virus particle. The full nucleotide sequence has been determined for the genome of at least one member of many plant virus families and groups. This information provides the basis for our understanding of viral genome strategy and replication.

Most of the small geometric viruses contain a single kind of coat protein. These proteins range in size from about $M_r = 17,000$ to $40,000$. In earlier work the amino acid sequences of several coat proteins were determined by direct chemical methods. However, today the simplest procedure is to derive the amino acid sequence from the nucleotide sequence of the coat protein gene.

The plant virus families and groups with large particles (rhabdoviruses, reoviruses, CaMV, and TSWV) contain more than one kind of protein, some of which are enzymes. Others have a structural role. The two kinds of enveloped plant viruses (rhabdoviruses and TSWV) contain lipid as part of an outerbounding lipoprotein envelope. This envelope is probably derived from a host bilayer membrane. All viruses contain water, with enveloped viruses containing more than geometric ones. Many viruses contain divalent metal ions, especially Ca^{2+}, and a few contain polyamines. These components may be important for stability of the intact virus particle.

5

Architecture

A knowledge of the detailed structure of virus particles is an essential prerequisite to our understanding of many aspects of virology, for example, how viruses survive outside the cell, how they infect and replicate within the cell, and how they are related to one another. Knowledge of virus architecture has increased greatly in recent years, due both to more detailed chemical information and to the application of more refined electron microscopic, optical diffraction, and X-ray crystallographic procedures.

The term *capsid* has been proposed for the closed shell or tube of a virus and the term *capsomere* for clusters of subunits seen in electron micrographs. The mature virus has been termed the *virion* (Caspar *et al.*, 1962). In membrane-bound viruses the inner nucleoprotein core has been called the *nucleocapsid*. These names appear to me not to be necessary, and at times confusing. For example, what is the capsid or the virion in a multicomponent virus like AMV? Which is the capsomere in a virus like TBSV, where different parts of the same protein subunit are arranged with twofold and threefold symmetry? However, the term *encapsidation* is now widely used to refer to the process by which a viral genome becomes encased in a shell of viral protein. This term now has an established meaning, and I will use it. I shall also use *protein subunit* or structural subunit to refer to the covalently linked peptide chain. The term *morphological subunit* will refer to the groups of protein subunits revealed by electron microscopy and X-ray crystallography.

The implications of virus structure for virus self-assembly are discussed in Chapters 7 and 8 and for the classification of viruses in Chapters 13 and 17. Figure 5.1 indicates the range of sizes and shapes found among plant viruses.

I. METHODS

A. Chemical and Biochemical Studies

As already pointed out in Chapter 4, a knowledge of the size and nature of the viral nucleic acid and of the proteins and other components occurring in a virus are

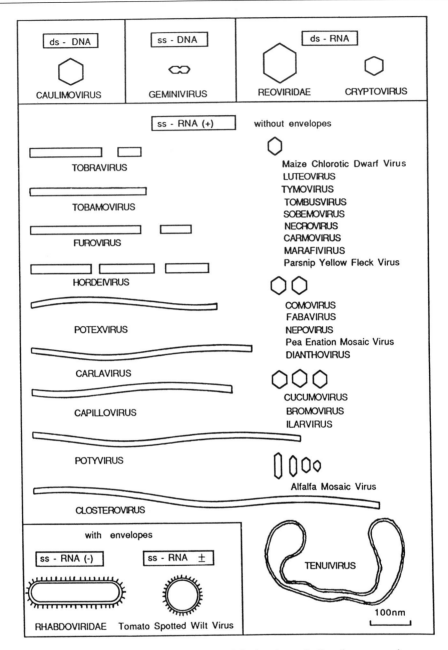

Figure 5.1 The families and groups of viruses infecting plants. Outline diagrams are drawn approximately to scale. (Courtesy R. I. B. Francki.) One additional group, the commelina yellow mottle virus group with dsDNA and bacilliform particles, was approved by the I.C.T.V. in August 1990.

essential to an understanding of its architecture. Chemical and enzymatic studies may give various kinds of information about virus structure. For example, the fact that carboxypeptidase A removed the terminal threonine from intact TMV indicated that the C terminus of the polypeptide was exposed at the surface of the virus (Harris and Knight, 1955). Methyl picolinimidate reacts with the exposed amino group of lysine residues. Perham (1973) used this reagent with several TMV mutants to determine which lysine residues were at the surface of the virus and which were buried. For viruses with more complex structures, partial degradation by chemical or physical means—for example, removal of the outer envelope—can be used to establish where particular proteins are located within the particle (e.g., Jackson, 1978) (see Section VI,A).

Studies on the stability of a virus under various pH and ionic or other conditions may give clues as to structure and the nature of the bonds holding the structure together. Many studies of this sort have been carried out but results are often difficult to interpret in a definitive way (see Kaper, 1975). Some examples of this approach are given later in this chapter.

B. Size of Viruses

1. Hydrodynamic Measurements

The classic procedure for hydrodynamic measurements is to use the Svedberg equation (see Schachman, 1959). This requires a knowledge of s, the sedimentation coefficient, D, the diffusion coefficient, and $\bar{v}$, the partial specific volume for the virus being studied. D and $\bar{v}$ are rather troublesome to determine by classic procedures. s can be determined readily in the analytical ultracentrifuge. For this reason many workers go no further than determining s for a new virus. The term molecular weight is widely used with reference to viruses. I will follow this usage but particle weight would be strictly a more appropriate term.

s may be conveniently determined in a preparative ultracentrifuge by sedimenting the virus in a linear sucrose density gradient and comparing the distance it sediments to internal markers of known s. The markers and unknown should have similar sedimentation rates. If not, other methods of calculation can be used (Clark, 1976).

Laser light scattering has been used to determine the radii of several approximately spherical viruses with a high degree of precision (Harvey, 1973; Camerino-Otero et al., 1974). This method is of particular interest since it gives an estimate of the hydrated diameter whereas the Svedberg equation and electron microscopy give a diameter for the dehydrated virus.

2. Electron Microscopy

Measurements made on electron micrographs of isolated virus particles, or thin sections of infected cells, offer very convenient estimates of the size of viruses. For some of the large viruses and for the rod-shaped viruses such measurements may be the best available, but they are subject to significant errors. There are various magnification errors inherent in measurements on electron micrographs. These may be overcome in part, but not entirely, by using a well-characterized, stable, and distinctive virus, such as TMV as an internal standard (Bos, 1975). However, Markham et al. (1964) pointed out that TMV rods may not be absolutely uniform in length. The width of the rods may offer a better standard. Flattening of particles on the supporting film may cause significant errors. Flattening can be detected with a tilting stage (Serwer, 1977). If films not coated with carbon are used there may be significant distortion of individual virus particles probably due to stretching of the film (Ronald et al., 1977).

When length distributions of rods are being prepared it is necessary to combine individual measurements into size classes. If the size classes chosen are too large then the presence of more than one length of particle may be obscured. There are additional difficulties in measuring the contour lengths of flexuous rods (De Leeuw, 1975).

Hatta (1976) pointed out that measurements of interparticle distances in arrays of particles seen in thin sections of crystals of purified small isometric viruses will always be less than the true diameter of the particles because particles overlap in the arrays.

3. X-Ray Crystallography

For viruses that can be obtained as stable crystalline preparations, X-ray crystallography can give accurate and unambiguous estimates of the radius of the particles in the crystalline state. The technique is limited to viruses that are stable or can be made stable in the salt solutions necessary to produce crystals.

4. Neutron Small-Angle Scattering

Neutron small-scale scattering (see Section I,E) can give accurate estimates of the radius of iscosahedral viruses in solutions containing relatively low salt concentrations. The method has confirmed that particles of some small icosahedral viruses such as BMV, in solution under conditions where they are stable, may vary in size (Chauvin *et al.*, 1978).

C. Electron Microscopy to Determine Fine Structure

Horne (1985) gives an account of the application of electron microscopy to the study of plant virus structure.

1. Metal Shadowed Preparations

In early electron microscope studies of virus particles, shadowing with heavy metals was used to enhance contrast. Such shadowing tended to obscure surface details but gave information on overall size and shape of the dry particle. Much more information can be obtained if specimens are freeze-dried and then shadowed in a very high vacuum (e.g., Hatta and Francki, 1977). For example, Roberts (1988), using this kind of procedure, was able to show that tobacco ringspot *Nepovirus* had a structure resembling that of models made up of 60 subunits in clusters of five arranged in a $T = 1$ structure (see Section II,B).

2. Freeze Etching

As at present developed, this technique could give useful information about the surfaces and substructure of larger viruses—particularly those with lipoprotein bilayer membranes. Although Steere (1969) illustrated arrays of freeze-fractured potato yellow dwarf rhabdovirus particles within an infected cell, no novel structural information has yet been obtained for plant viruses using this procedure.

3. Negative Staining

The use of electron-dense stains has proved of much greater value than metal shadowing in revealing the detailed morphology of virus particles. Such stains may be positive or negative. Positive stains react chemically with and are bound to the virus [e.g., various osmium, lead, and uranyl compounds and phosphotungstic acid (PTA) under appropriate conditions]. Chemical combination may lead to alteration in or disintegration of the virus structure. In negative staining, on the other hand, the electron-dense material does not react with the virus but penetrates available spaces on the surface or within the virus particle. This is now the preferred technique for examining virus structure by electron microscopy. Common negative stains are potassium phosphotungstate (KPT) at pH 7.0, and uranyl acetate or formate used near pH 5.0. The virus structure stands out against the surrounding electron-dense material. The carbon substrates used to mount stained specimens have a granular background. Structural detail can often be seen most clearly in images of particles in a film of stain suspended over holes in the carbon grid. The stain may or may not penetrate any hollow space within the virus particle.

With helical rod-shaped particles, the central hollow is frequently revealed by the stain. With the small spherical viruses, which often have associated with them empty protein shells, it was assumed by some workers that stained particles showing a dense inner region represented empty shells in the preparation. However, staining conditions may lead to loss of RNA from a proportion of the intact virus particles, allowing stain to penetrate, while stain may not enter some empty shells. Stains differ in the extent to which they destroy or alter a virus structure, and the extent of such changes depends closely on the conditions used.

Even in the best electron micrographs, the finer details of virus structure tend to be obscured, first, by noise due to minor irregularities in the actual virus structure and other factors such as irregularities in the stain, and second, by the fact that contrast due to the stain is often developed on both sides of the particle to varying degrees. To overcome these difficulties and to extract more reliable and detailed information from images of negatively stained individual virus particles, several methods have been used.

The main procedures have been photographic superposition (Markham *et al.*, 1963, 1964; Finch and Holmes, 1967), shadowgraphs (Finch and Klug, 1966, 1967), and optical transforms (Klug and Berger, 1964; Crowther and Klug, 1975; Steven *et al.*, 1981). Vogel and Provencher (1988) developed a computational procedure for three-dimensional reconstruction from projections of a disordered collection of single particles and have applied it to tomato bushy stunt *Tombusvirus* (TBSV). With recent developments facilitating the application of X-ray crystallographic analysis to virus structure these procedures will probably not be widely used, except for viruses to which X-ray analysis cannot yet be applied.

4. Thin Sections

Some aspects of the structure of the enveloped viruses, particularly bilayer membranes, can be studied using thin sections of infected cells or of a pellet containing the virus (see Figs. 5.32 and 5.33).

5. Cryo-electron Microscopy

Jeng *et al.* (1989) used TMV as a test object to develop a procedure based on cryo-electron microscopy (that is, extremely rapid freezing in an aqueous medium) for determining structural detail in objects with helical symmetry. They obtained data on TMV with spacing better than 10 Å from electron images of hydrated TMV frozen in vitreous ice. Butler *et al.* (1990) have used this procedure to demonstrate the presence of double disks of TMV under rod assembly conditions.

D. X-Ray Crystallographic Analysis

Where it can be applied, X-ray crystallography provides the most powerful means of obtaining information about structures that are regularly arrayed in three dimensions. Finch and Holmes (1967) give an introductory account of the methods involved.

Over about the past 10 years there have been significant advances that have allowed the application of X-ray crystallographic analysis to more viruses, and at higher resolutions. With the definition of structures at the atomic level it has become possible to interpret structure in more detail in relation to biological function. The major advances have been: (i) An increase in the capacity and speed of computers, together with a reduction in cost. (ii) Noncrystallographic symmetry averaging. Icosahedral viruses such as TBSV and STNV have 60-fold symmetry in the virus particle. The degree to which this coincides with the symmetry of the crystal used for analysis depends on the crystal form that happens to be obtained. For example, in TBSV the smallest crystal repeating unit (the assymetric unit) consists of five trimers of the coat protein. These are exactly related by crystallographic symmetry. Thus symmetry averaging for this virus takes place around a chosen fivefold axis. Successive recalculations from an initial electron density map remove noise and enhance detail in the map. An example of the procedure is given by Olson *et al.* (1983) for TBSV. (iii) The development of computer graphics. This technology can now largely replace the laborious manual model building of the past, which was necessary for the refinement of a structure (Olson *et al.*, 1985). Computer graphics are also extremely useful for exploring particular aspects of a three-dimensional structure and for communicating structural ideas in a more readily comprehensible form, by such means as color coding and selective omission of detail (Namba *et al.*, 1984, 1988). (iv) The fourth advance, not yet widely applied to plant viruses, is the introduction of site-directed mutagenesis to protein crystallography. The possibility of exchanging one amino acid for another in any chosen site in a protein changes crystallography from a passive technique to one in which the relation between structure and function can be studied in a systematic and rational fashion.

X-ray crystallography analysis has two main limitations as far as virus structures are concerned: (i) For nearly all icosahedral viruses most of the nucleic acid is not arranged in a regular manner in relation to the symmetries in the protein shell. It is these symmetries that determine the orientation of virus particles in the crystal. Therefore little detailed structural information about the nucleic acid can be obtained. Some density attributed to the RNA within STNV has been recorded, but no meaningful data were obtained by X-ray analysis (see Bentley *et al.*, 1987, for discussion). (ii) Larger viruses such as members of the Rhabdoviridae cannot be crystallized. In addition, it is probable that many features of their structure are not strictly regular. For such viruses electron microscopy provides the best information.

E. Neutron Small-Angle Scattering

Neutron scattering by virus solutions is a method by which low-resolution information can be obtained about the structure of small isometric viruses and in particular about the radial dimensions of the RNA or DNA and the protein shell. The effects of different conditions in solution on these virus dimensions can be determined readily. The method takes advantage of the fact that H_2O–D_2O mixtures can be used that match either the RNA or the protein in scattering power. Analysis of the neutron diffraction at small angles gives a set of data from which models can be built (e.g., Fig. 5.17).

F. Serological Methods

The reaction of specific antibodies with intact viruses or dissociated viral coat proteins has been used to obtain information that is relevant to virus structure. Monoclonal antibodies are proving to be particularly useful for this kind of investigation, although there are significant limitations.

Some workers distinguish two types of antigenic determinant in proteins. A *continuous determinant* is a continuous sequence of amino acids exposed at the surface of a protein that possesses a distinctive conformation. A *discontinuous determinant* is made up of amino residues that are close together on the protein surface but that are located at distant positions in the primary structure. It has been suggested that if the surface area of a protein that is recognized by an antibody molecule is about 2.0×2.5 nm then all protein antigenic determinants are likely to be discontinuous (Barlow *et al.*, 1986). Antigenic determinants specific for quaternary structure in the virus particle have been demonstrated for all the groups of viruses (Van Regenmortel, 1982). The existence of such determinants in the TMV rod was confirmed by the finding that, out of 18 MAb's raised against the virus, 8 did not react with viral subunits in ELISA tests (Altschuh *et al.*, 1985).

The molecular dissection of antigens by monoclonal antibodies has been discussed by Van Regenmortel (1984b). Three methods have been used to localize antigenic determinants: (i) peptides obtained from the protein by chemical or enzymatic cleavage are screened for their reactivity with antibodies; (ii) short synthetic peptides representing known amino acid sequences in the protein can be similarly screened; and (iii) immunological cross-reactivity is assayed between closely related proteins having one or a few amino acid substitutions at known sites.

There are substantial limitations for all these procedures. Peptides derived by cleavage or synthesis may not maintain in solution the conformation that the sequence possessed in the intact molecule. Furthermore, such peptides will only rarely represent all the amino acids in the original antigenic determinant. For these reasons reactivity of peptides with antibodies to the intact protein is usually very low.

The limitations of the peptide approach can be further illustrated from work with TMV. Tests with a set of synthetic and cleavage peptides indicated that the dissociated TMV protein possesses seven antigenic determinants (Altschuh *et al.*, 1983). More recent work with a set of synthetic peptides representing almost the full length of the protein showed that almost the entire sequence possesses antigenic activity (Al-Moudallal *et al.*, 1985).

Structural interpretation of the results of cross-reactions between closely related proteins can be confused by the fact that the conformation of any antigenic determinant may be changed by an amino acid substitution occurring elsewhere in the protein. Furthermore, the ability of different MAbs to detect residue exchanges may be extremely variable as was found for a set of TMV mutants (Al-Moudallal *et al.*, 1982). Nevertheless, serological techniques have produced some structural information, especially with viruses for which the detailed protein structure has not been determined by X-ray crystallography. For example, at least three different antigenic determinants were distinguished on the coat protein of PVX with a set of MAbs

(Koenig and Torrance, 1986). The results are summarized in Fig. 5.2. The surface location of the N terminus was confirmed by Söber *et al.* (1988).

The N-terminal and C-terminal regions of the coat protein of TYMV were interpreted to be at the surface of the virus (Quesniaux *et al.*, 1983a,b). Similarly immunological evidence confirmed that these N-terminal regions are also at the virus surface of potyviruses (Shukla *et al.*, 1988b). A surface location for both N and C termini appears to be a common feature in other rod-shaped viruses.

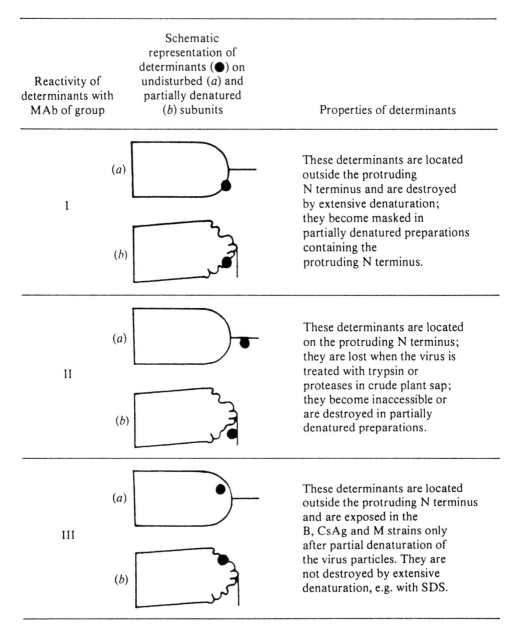

Reactivity of determinants with MAb of group	Schematic representation of determinants (●) on undisturbed (*a*) and partially denatured (*b*) subunits	Properties of determinants
I	(*a*) (*b*)	These determinants are located outside the protruding N terminus and are destroyed by extensive denaturation; they become masked in partially denatured preparations containing the protruding N terminus.
II	(*a*) (*b*)	These determinants are located on the protruding N terminus; they are lost when the virus is treated with trypsin or proteases in crude plant sap; they become inaccessible or are destroyed in partially denatured preparations.
III	(*a*) (*b*)	These determinants are located outside the protruding N terminus and are exposed in the B, CsAg and M strains only after partial denaturation of the virus particles. They are not destroyed by extensive denaturation, e.g. with SDS.

Figure 5.2   Schematic representation of the three kinds of antigenic determinants distinguished in the coat protein of PVX by their differential reactivity to a set of MAbs (From Koenig and Torrance, 1986.)

Caution must be used in interpreting the results of ELISA tests in structural terms when the whole virus is the antigen, as the following results demonstrate. Antibodies against a synthetic peptide of TBSV making up part of the flexible amino-terminal arm (amino acids 28–40) reacted with whole virus as antigen in ELISA tests (Jaegle *et al.*, 1988). Since the amino-terminal arm is in the interior of a very compact shell (Section III,B,2) this result must mean that virus structure was opened up sufficiently on the ELISA plate to allow antibodies to react with the normally buried arm. Similar results have been obtained for SBMV (MacKenzie and Tremaine, 1986).

As was indicated by Dore *et al.* (1987b), a definitive delineation of antigenic sites requires a knowledge of the three-dimensional structure of the polypeptides. However, if the three-dimensional structure is known, what is the point, in relation to their structure, of determining the antigenic sites for plant viruses? The situation is quite different for viruses infecting vertebrates, where a knowledge of such sites may be very important for vaccine development.

Dore *et al.* (1988) developed an elegant procedure involving ELISA reactions on electron microscope grids and gold-labeled antibody. Using this procedure they showed that anti-TMV MAb's that reacted with both virus particles and the coat protein subunits bound to the virus rods only at one end (Fig. 5.3). Further studies have shown that the MAb's bind to the surface of the protein subunit that contains the right radial and left radial α-helices (see Fig. 5.10) (Dore *et al.*, 1989). This result confirms in a graphic way a fact already known from X-ray crystallographic analysis, namely, that the coat protein in the TMV rod presents a chemically different aspect on its upper and lower surfaces.

Using a similar technique, Lesemann *et al.* (1990) distinguished three groups of MAbs reacting with beet necrotic yellow vein *Furovirus*. One reacted along the entire length of the particles. The other two groups reacted with antigenic sites on opposite ends of the particles.

G. Methods for Studying Stabilizing Bonds

The primary structures of viral coat proteins and nucleic acids depend on covalent bonds. In the final structure of the simple geometric viruses these two major components are held together in a precise manner by a variety of noncovalent

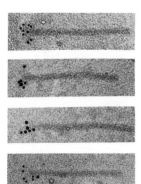

Figure 5.3 Electron micrographs showing the binding of a gold-labeled MAb to one extremity of TMV rods. Bar = 100 nm. (From Dore *et al.*, 1988.)

bonds. Three kinds of interaction are involved: protein–protein, protein–RNA, and RNA–RNA. In addition, small molecules such as divalent metal ions (Ca^{2+} in particular) may have a marked effect on the stability of some viruses. A knowledge of these interactions is important for understanding the stability of the virus in various environments, how it might be assembled during virus synthesis, and how the nucleic acid might be released following infection of a cell. The stabilizing interactions are hydrophobic bonds, hydrogen bonds, salt linkages, and various other long- and short-range interactions. A variety of physical and chemical methods have been used in attempts to refine our understanding of the role of these bonds in virus structure.

1. X-Ray Crystallographic Analysis

Models built to about 2.8 Å, based on X-ray crystallographic analysis and a knowledge of the primary structure of the coat protein, provide a detailed understanding of the bonds in the secondary and tertiary structure of the protein subunit. In addition, the bonds between subunits that make up the quaternary structure of the virus, and the role of accessory molecules and ions, mainly water and Ca^{2+}, can be defined. A virus structure examined to atomic resolution provides a vast amount of detail that cannot be encompassed in a book of this size. The highlights of several such structures are discussed in later sections.

2. Stability to Chemical and Physical Agents

The effects of pH, ionic strength, kind of ion, temperature, compounds such as phenol and detergents, and hydrogen bond-breaking agents such as urea on the stability of viruses have been studied in many laboratories (Kaper, 1975). Such experiments can give us information only of a general kind about the bonds involved in virus stability.

Among the small isometric viruses there is a wide range of stabilities. Viruses like TYMV with strong protein–protein interactions are the most stable. The other end of the spectrum is illustrated by CMV, where protein–RNA interactions predominate (Kaper, 1975).

Protein–protein interactions are obviously very important for the stability of TYMV since empty protein shells are quite stable. Hydrophobic bonding between subunits is an important factor in this stability because (i) TYMV is very stable at high ionic strengths, (ii) it is readily degraded by phenol and ethanol, and (iii) degradation of the virus and empty protein shell by urea, organic mercurials, and other chemicals can be interpreted on the basis that the predominant protein–protein interaction is via hydrophobic bonding (Kaper, 1975). Extreme conditions such as high pH (Keeling and Matthews, 1982) and freezing and thawing (Katouzian-Safadi and Haenni, 1986) have been used to study release of RNA from TYMV particles.

Alkaline conditions have been used to demonstrate the removal of protein subunits from the TMV rod, beginning at the 5' end, and to reveal intermediates in the stripping process, these being attributed to regions of unusually strong interaction between the protein and RNA (Perham and Wilson, 1978).

3. Chemical Modification of the Coat Protein

Particular amino acid residues in the coat protein can be modified by the chemical addition of a side chain, and the effects of such substitution on stability of the virus can be examined (e.g., Wilson and Perham, 1985, for TMV).

4. Removal of Ions

For small isometric viruses whose structures are partially stabilized by Ca^{2+} ions, removal of these ions, usually by EDTA, leads to swelling of the virus particle. A study of the swelling phenomenon can give information about the kinds of bonds important for stability. A variety of techniques have been used for monitoring swelling. For example, Krüse et al. (1982) used small-angle X-ray and neutron scattering, analytical centrifugation, and fluorescence techniques to study swelling of TBSV.

Swelling of STNV in the presence of EDTA was demonstrated by ultracentrifugation and X-ray crystallography (Unge et al., 1986). In swollen virus an amino-terminal peptide that was normally well buried in the structure became susceptible to trypsin. The kinetics of swelling of SBMV was studied using photon correlation spectroscopy (Brisco et al., 1986b).

Durham et al. (1984) used hydrogen-ion titration curves to follow the reversible swelling and contraction of particles of TBSV, CCMV, and TCV. They drew conclusions regarding the semipermeability of various protein shells.

5. Circular Dichroism

Circular dichroism spectra can be used to obtain estimates of the extent of α-helix and β structure in a viral protein subunit (e.g., Odumosu et al., 1981, for SBMV).

6. Methods Applicable to Nucleic Acid within the Virus Particle

Several methods are available to give approximate estimates of the degree of helical base-pairing or other ordered arrangement of the nucleic acid within the virus. These include relative absorbancy at 260 nm (e.g., Haselkorn, 1962, for TYMV), laser Raman spectroscopy (e.g., Hartman et al., 1978, for TYMV), circular dichroism spectra (e.g., Odumosu et al., 1981, for SBMV), and magnetic birefringence (e.g., Torbet et al., 1986, for CaMV).

II. PHYSICAL PRINCIPLES IN THE ARCHITECTURE OF VIRUSES

Crick and Watson (1956) put forward a hypothesis concerning the structure of small viruses, which has since been generally confirmed. Using the knowledge then available for TYMV and TMV, namely, that the viral RNA was enclosed in a coat of protein and (for TMV only, at that stage) that the naked RNA was infectious, they assumed that the basic structural requirement for a small virus was the provision of a shell of protein to protect its ribonucleic acid. They considered that the relatively large protein coat might be made most efficiently by the virus controlling production in the cell of a large number of identical small protein molecules, rather than in one or a few very large ones.

They pointed out that if the same bonding arrangement is to be used repeatedly in the particle, the small protein molecules would then aggregate around the RNA in a regular manner. There are only a limited number of ways in which the subunits can be arranged. All the geometric viruses are either rods or spheres.

There is no theoretical restriction on the number of protein subunits that can pack into a helical array in rod-shaped viruses. From crystallographic considerations, Crick and Watson concluded that the protein shell of a small "spherical" virus could be constructed from identical protein subunits arranged with cubic symmetry, for which case the number of subunits would be a multiple of 12.

Crick and Watson (1956) pointed out that cubic symmetry was most likely to lead to an isometric virus particle. There are three types of cubic symmetry: tetrahedral $(2:3)$, octahedral $(4:3:2)$, and icosahedral $(5:3:2)$. Thus, an icosahedron has fivefold, threefold, and twofold rotational symmetry. For a virus particle the three types of cubic symmetry would imply 12, 24, or 60 identical subunits arranged identically on the surface of a sphere. These subunits could be of any shape.

Klug and Caspar (1960) pointed out that many viruses probably had shells made up of subunits arranged with icosahedral symmetry. Horne and Wildy (1961) discussed possible models for the arrangement of protein subunits in icosahedral shells with $5:3:2$ symmetry. They considered possible packing arrangements for clusters of five and six protein subunits. Pawley (1962) enumerated the plane groups that can be fitted on polyhedra, and he suggested that these may have application in the study of virus structure. Caspar and Klug (1962) further developed the principles of virus construction, particularly with respect to the isometric viruses based on icosahedral structures. A shell made up of many small identical protein molecules makes most efficient use of a virus' genetic material.

Figure 5.4 shows a regular icosahedron (20 faces). With three units in identical positions on each face, this icosahedron gives 60 identical subunits. This is the largest number of subunits that can be located in identical positions in an isometric shell. Some viruses have this structure, but many have much larger numbers of subunits, so that all subunits cannot be in identical environments.

A. Quasiequivalence

Caspar and Klug (1962) with their theory of quasiequivalence laid the basis for our further understanding of the way the shells of many smaller isometric viruses are constructed. In general terms, they assumed that all the chemical subunits in the shell need not be arrayed in a strictly mathematically equivalent way, but only quasiequivalently. They assumed that the shell is held together by the same type of bonds throughout, but that the bonds may be deformed in slightly different ways in different nonsymmetry-related environments. They calculated that the degree of deformation necessary would be physically acceptable. Quasiequivalence would occur in all icosahedra except the basic structure (Fig. 5.4).

However, more recent detailed information on the arrangement of subunits in the shells of small viruses has altered this view. Considering the relatively undistortable nature of protein domains, the possible flexion permitted by side chain motion is, for many viruses, too small to accommodate the required angular shifts (Harrison, 1983; Rossmann, 1984). Viruses such as TBSV have evolved multidomain subunits that, to a large extent, adjust to the different symmetry-related positions in the shell by means other than distortion of intersubunit bonds (Section III,B,2).

Other viruses have evolved different variations on the icosahedral theme. For example, comoviruses (Section III,B,4) and members of the Reoviridae (Section IV) have different polypeptides in different symmetry environments within the shell.

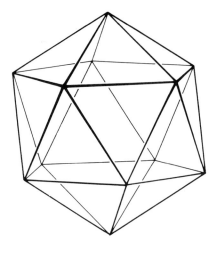

Figure 5.4 The regular icosahedron. This solid has 12 vertices with fivefold rotational symmetry; the center of each triangular face is on a threefold symmetry axis, and the midpoint of each edge is on a twofold symmetry axis. There are 20 identical triangular faces. Three structural units of any shape can be placed in identical positions on each face, giving 60 structural units. Some of the smallest viruses have 60 subunits arranged in this way.

B. Possible Icosahedra

Caspar and Klug (1962) enumerated all the possible icosahedral surface lattices and the number of structural subunits involved. The basic icosahedron (Fig. 5.4), with $20 \times 3 = 60$ structural subunits, can be subtriangulated according to the formula

$$T = P \cdot f^2$$

where T is called the *triangulation number*.

1. The Meaning of Parameter f

The basic triangular face can be subdivided by lines joining equally spaced divisions on each side.

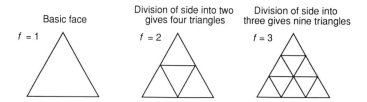

Thus f is the number of subdivisions of each side and f^2 is the number of smaller triangles formed.

2. The Meaning of Parameter P

There is another way in which subtriangulation can be made and this is represented by P. It is easier to consider a plane network of equilateral triangles.

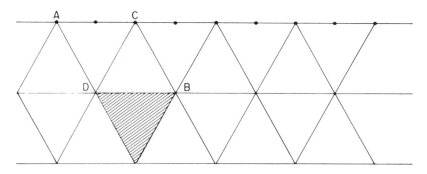

Such a sheet can be folded down to give the basic icosahedron by cutting out one triangle from a hexagon (e.g., cross-hatching) and then joining the cut edges to give a vertex with fivefold symmetry.

However, if each vertex is joined to another by a line not passing through the nearest vertex, other triangulations of the surface are obtained. In the simplest case the "next but one" vertices are joined.

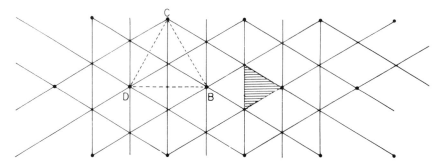

This gives a new array of equilateral triangles. This plane net can be folded to give the solid in Fig. 5.5C by removing one triangle from each of the original vertices (e.g., the shaded triangle) and then folding in to give a vertex with fivefold symmetry. It can be shown by simple trigonometry that each of the small triangles has one-third the area of the original faces. This can be seen by inspection by noting that there are six new half triangles within one original face (dashed lines C-B-D.). In this example $P = 3$. In general

$$P = h^2 + hk + k^2$$

where h and k are any integers having no common factor.

$$\text{For } h = 1, k = 0, P = 1$$
$$\text{For } h = 1, k = 1, P = 3$$
$$\text{For } h = 2, k = 1, P = 7$$

Where $P \geq 7$ the icosahedra are skew, and right-handed and left-handed versions are possible. The physical meaning of h and k in a virus structure is illustrated in the following section.

Since each of the triangles formed with the P parameter can be further subdivided into f^2 smaller triangles, T gives the total number of subdivisions of the original faces and $20T$ the total number of triangles. Figure 5.5 gives some examples.

Thus the number of structural subunits in an icosahedral shell $= 20 \times 3 \times T = 60T$.

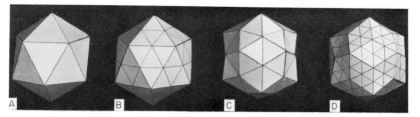

Figure 5.5 Ways of subtriangulation of the basic icosahedron shown in Fig. 5.4 to give a series of deltahedra with icosahedral symmetry (icosadeltahedra). (A) The basic icosahedron, with $T = 1$ ($P = 1, f = 1$). (B) With $T = 4$ ($P = 1, f = 2$). (C) With $T = 3$ ($P = 3, f = 1$). (D) With $T = 12$ ($P = 3, f = 2$). (From Caspar and Klug, 1962.)

C. Clustering of Subunits

The actual detailed structure of the virus surface will depend on how the physical subunits are packed together. For example, three clustering possibilities for the basic icosahedron are

Pentamers   Trimers   Dimers

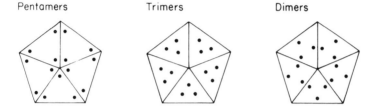

In fact many smaller plant viruses are based on the $P = 3$, $f = 1$, $T = 3$ icosahedron. In this structure, the structural subunits are commonly clustered about the vertices to give pentamers and hexamers of the subunits. These are the morphological subunits seen in electron micrographs of negatively stained particles (e.g., Fig. 5.16).

Since there are always 12 vertices with fivefold symmetry in icosahedra we can calculate the number of morphological subunits (M) (assuming clustering into pentamers and hexamers) as follows:

$$M = \frac{(60T - 60)}{6} \text{ hexamers } \frac{+60}{5} \text{ pentamers}$$

$$= 10 (T - 1) \text{ hexamers } + 12 \text{ pentamers}$$

In photographs of virus particles where the pentamers and hexamers can be unambiguously recognized (e.g., one-sided images of negatively stained particles or freeze-fracture replicas of the outer faces of larger viruses) the parameters h and k might be used to establish the icosahedral class of the particle. This procedure would be particularly useful for shells containing large numbers of hexamers. h and k represent the numbers of hexamers that must be traversed to move by the shortest route from one pentamer to the next. Thus we must identify two adjacent pentamers. For example, the following was observed on the surface of freeze-etch replicas of phage λ (Bayer and Bocharov, 1973).

$$h = 4, \ k = 1, \ T = 21$$

This indicates a skew icosahedron with "right-handed" skewness.

D. "True" and "Quasi" Symmetries

In the basic icosahedron (Fig. 5.4) a feature located halfway along any edge of a triangular face is positioned on an axis of rotation for the whole solid. Thus it is on a "true" or icosahedral symmetry axis. This is a true dyad. In any more complex icosahedron ($T > 1$) there is more than one kind of twofold symmetry. For example, in a $P = 3$, $T = 3$ shell (Fig. 5.5C), the center of one edge on each of the 60 triangular faces is on a true dyad axis relating to the solid as a whole. The center positions of the other two edges of a face have only local twofold symmetry. These are called "quasi" dyads.

III. GEOMETRIC VIRUSES WITH ssRNA

A. Helical Rods

1. *Tobamovirus* Group

General Features

The particle of TMV is a rigid helical rod. It is an extremely stable structure, having been reported to retain infectivity in nonsterile extracts at room temperature for at least 50 years (Silber and Burk, 1965). The stability of naked TMV RNA is no greater than that of any other ssRNA. Thus, stability of the virus with respect to infectivity is a consequence of the interactions between neighboring protein subunits and between the protein and the RNA.

X-ray diffraction analyses have given us a detailed picture of the arrangement of the protein subunits and the RNA in the virus rod. The subunits are closely packed in a helical array. The pitch of the helix is 2.3 nm, and the RNA chain is compactly coiled in a helix following that of the protein subunits (Fig. 5.6). There are three nucleotides of RNA associated with each protein subunit, and there are 49 nucleotides and $16\frac{1}{3}$ protein subunits per turn. The phosphates of the RNA are at about 4 nm from the rod axis. By tilting negatively stained TMV particles in the electron beam and noting changes in the edge appearance of the rods, Finch (1972) established that the basic helix of TMV is right-handed. In a proportion of negatively stained particles one end of the rod can be seen as concave, and the other end is convex. The 3' end of the RNA is at the convex end and the 5' at the concave end (Wilson *et al.*, 1976; Butler *et al.*, 1977). A central canal with a radius of about 2 nm becomes filled with stain in negatively stained preparations of the virus (Fig. 5.7)

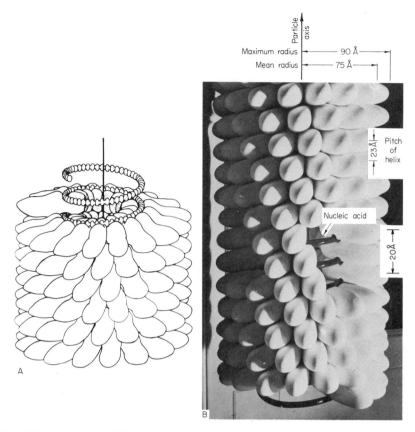

Figure 5.6 Structure of TMV. (A) Drawing to show the relationship of the RNA and the protein subunits. The RNA shown free of protein could not maintain this configuration in the absence of the protein subunits. There are 3 nucleotides per protein subunit, or 49 per turn of the major helix spaced about 5 Å apart. (B) Photograph of a model of TMV with major dimensions indicated. The structure repeats after 69 Å in the axial direction, the repeat containing 49 subunits distributed over three turns of the helix. (From Klug and Caspar, 1960, and based largely on the work of R. E. Franklin.)

Short Rods

Most purified preparations of TMV contain a proportion of rods that are of variable length and less than 300 nm. Study of these short rods may be complicated by the problem of aggregation. Both intact TMV rods and the shorter rods have a tendency to aggregate end to end under appropriate conditions, giving a wide range of lengths. The length distribution observed is very dependent on the precise history of the virus preparation. Some factors affecting the length distribution are pH, ionic strength, nature of ions present, temperature, and length of storage, but not all the factors involved are as yet fully understood.

Many of the shorter rods are probably of no special significance. They may be virus particles that were only partially assembled at the time of isolation, or parts of rods fractured during isolation. However, in beans infected with sunnhemp mosaic *Tobamovirus* (formerly known as the cowpea strain of TMV) the RNA containing the coat protein gene (MW $\simeq 3 \times 10^5$), which is synthesized as a discrete species during virus replication, is assembled into a rod about 40 nm long (Higgins *et al.*, 1976) (see Fig. 17.2).

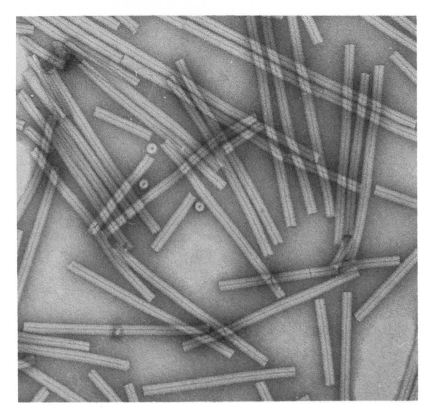

Figure 5.7 Electron micrograph of negatively stained TMV particles. The axial hole filled with stain can be seen clearly. A few protein disks showing the central hole are also present. ×230,000. (Courtesy of R. W. Horne.)

Properties of the Coat Protein

The protein monomer can aggregate in solution in various ways depending on pH, ionic strength, and temperature. The major forms are summarized in Fig. 5.8. A kind of aggregate known as the stacked disk (not illustrated in Fig. 5.8) consists of three or more pairs of rings of subunits (Raghavendra *et al.*, 1986). Experiments with MAbs show that both ends expose the same protein subunit surface in stacked disks. Thus each two-layer unit in the stack must be bipolar (i.e., facing in opposite directions) (Dore *et al.*, 1989). The existence of these various aggregates has been important both for our understanding of how the virus is assembled (Chapter 7, Section III,A,9) and also for the X-ray analysis that has led to a detailed understanding of the virus structure. The helical protein rods that are produced at low pH are of two kinds, one with $16\frac{1}{3}$ subunits per turn of the helix, as in the virus, and one with $17\frac{1}{3}$. In both of these forms the protein subunit structure is very similar to that in the virus. The RNA is replaced by at least one anion binding near a phosphate binding site (Mandelkow *et al.*, 1981). Constraints on the permissible amino acid exchanges found in the coat proteins of different tobamoviruses are discussed in Chapter 17, Section II,B.

Structure of the Double Disk

The double disk containing two rings of 17 protein subunits is of particular interest. Under appropriate conditions the disks form true three-dimensional

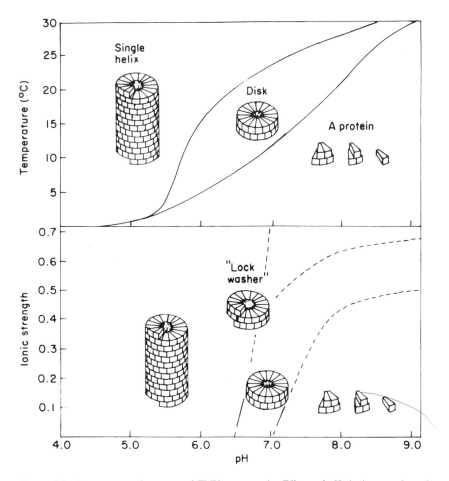

Figure 5.8 Some aggregation states of TMV coat protein. Effects of pH, ionic strength, and temperature. Upper figure modified from Richards and Williams (1976); lower figure modified from Durham *et al.* (1971, reprinted by permission from *Nature (London), New Biol.* **229**, 37–42. Copyright © Macmillan Magazines Ltd.). The helical rod of protein, the A protein, and the 20 S disk have been well characterized (Champness *et al.*, 1976). The lockwasher is the proposed intermediate in the initiation of the assembly of TMV (Durham *et al.*, 1971) (Fig. 7.9). Although the lockwasher form has never been isolated, the "nicked" protein helices formed when the pH is lowered rapidly give direct support to the existence of such a form. (Durham and Finch, 1972.) (See also Chapter 7, Section III,A,9.)

crystals. Although the repeating unit is very large, X-ray crystallographic procedures can be applied. This was the approach taken by Klug and colleagues, which after 12 years work led to an elucidation of the structure of the protein subunit and the double disk to 2.8-Å resolution (Bloomer *et al.*, 1978).

In the disk the polypeptide chain is in a flexible or disordered state below a radius of about 4 nm so that no structure was revealed in this region. There are four main helices in the rest of the molecule as illustrated in Fig. 5.9. These helices take up about 60 out of the 158 amino acids in the molecule. The distal ends of the helices are braced together by a region of β-sheet. Both N and C termini are at the circumference of the disk. There are strong polar and hydrophobic interactions between adjacent subunits in the disk (Fig. 5.9).

Viewed in section the subunits of the upper ring in the disk are flat. Those of the lower ring are tilted downward toward the center of the disk (Fig. 5.10). There are

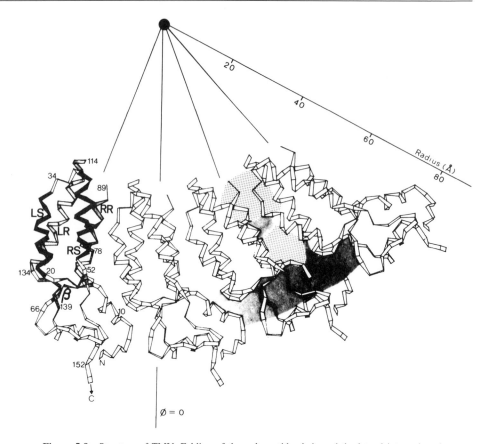

Figure 5.9 Structure of TMV. Folding of the polypeptide chain and the lateral interactions between subunits. Four adjacent subunits within a ring are shown viewed from above. The N and C termini of the chain are marked on the extreme left subunit together with the four main helices, left and right slewed (LS and RS) in the upper half of the molecule with right and left radial (RR and LR) below, and the β sheet that connects all four helices at their higher radius ends. The other subunits show the overall shape of the subunit as a scimitar with a pronounced slew relative to the radius of the ring. The interface contains alternating patches of polar residues, indicated by stippling, and of hydrophobic residues, indicated by solid shading, which are all contributed to by residues from both the subunits forming the interface. The hydrophobic patch at higher radius is continuous with the hydrophobic girdle, which extends circumferentially across the whole width of a subunit and across the interface, forming a continuous belt around the ring. The hydrophobic girdle includes residues from the high radius side of the β sheet, the upper loop that joins the RS and RR helices, and the two lower loops leading to the termini of the chain. [From Bloomer *et al.*, 1978, reprinted by permission from *Nature (London)* **276**, 362–368. Copyright © 1978 Macmillan Magazines Ltd.]

three regions of contact vertically between subunits as detailed in Fig. 5.10. The two disks, in contact at the outer part, open toward the center like a pair of jaws. The flexible inner parts of the folded chains are indicated by dotted lines in Fig. 5.10. Other physical studies show that there is thermal motional disorder in this region rather than static disorder (Jardetsky *et al.*, 1978).

Virus Structure

Intact TMV does not form three-dimensional regular crystalline arrays in solution. For this reason structural analysis has not yet proceeded to the detail available for the double disk. However, using fiber diffraction methods, Stubbs *et al.* (1977)

solved the structure to a resolution of 4 Å. Using this, together with the other data already available, they produced a model for the virus (Figs. 5.11 and 5.12).

Namba and Stubbs (1986) established a structure at 3.6 Å that has refined the structure of Stubbs *et al.* (1977) in certain details that are particularly relevant to our understanding of virus assembly. Namba *et al.* (1989) have produced an even more refined structure at 2.9 Å that has generated a great deal of detailed information. The following discussion is based mainly on the structure of Namba and Stubbs (1986).

The following are important features.

1. *The inner surface*   The presence of the RNA stabilizes the inner part of the protein subunit in the virus so that its position can be established. The highest peak in the radial density distribution is at about 2.3 nm. This is the region occupied by the vertical chains containing the V helices (Namba and Stubbs, 1986). These chains fill a space 0.9 nm wide by 2.3 nm high, and they are packed closely together to form a dense wall around the axial hole, protecting the RNA from the medium.
2. *The outer surface*   The $-NH_2$ terminus ends at the virus surface. The $-COOH$ terminus is also at the surface but residues 155–158 were not located on the density map and are therefore assumed to be somewhat disordered (Namba and Stubbs, 1986).
3. *The RNA binding site*   The binding site is in two parts, being formed by the top of one subunit and the bottom of the next. The three bases associated with each protein subunit form a claw that grips the left radial helix of the top subunit as viewed in Fig. 5.12. The left radial helix has a large number of aliphatic residues between positions 117 and 128, which form three faces for the bases. The bases lie flat against the hydrophobic side chains and each face can accommodate any base. The other part of the RNA binding site is mostly on the right slewed helix.

Two phosphate groups form ion pairs with Arg[90] and Arg[92]. The third phosphate appears to form a hydrogen bond with Thr[37] (Namba and Stubbs, 1986). Arg[41] extends toward the same phosphate as Arg[90] but does not approach as closely. Namba *et al.* (1989) suggest that the electrostatic

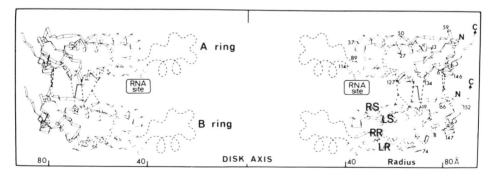

Figure 5.10   Side view through the disk showing the relative disposition of subunits in the two rings and the regions of contact between them. There are three regions of contact indicated by a solid line for the hydrophobic contact of Pro 54 with Ala 74 and Val 75, by dashed lines for the hydrogen bonds between Thr 59 and Ser 147 and 148, and further dashed lines for the extended salt-bridge system. The low radius region of the chain, which has no ordered structure in the absence of RNA, is shown schematically with an indication of an additional one to two turns extending the LR helix before it turns upward into the vertical column. (Courtesy of A. Klug and A. C. Bloomer.)

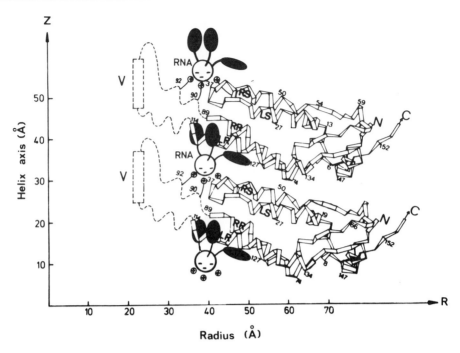

Figure 5.11 Structure of TMV. Schematic outline of the arrangement of the RNA and protein in the virus as seen in a side view of two subunits. The radial helices are lettered as in Figs. 5.9 and 5.10. There is, in addition, the vertical column (v) represented as a box. This is not present in the disk (Fig. 5.10). The triplet of RNA bases interacting with each subunit is shown in black. The bases form a "claw" grasping the LR helix. The ribose 2'-OH of the two bases at 40 Å radius probably hydrogen bond with the two aspartate residues present at positions 115 and 116 in all strains of the virus. The middle base of each triplet lies at a higher radius and may hydrogen bond to serine 123. The three negatively charged phosphate groups have been thought to form salt bridges with three arginine residues as indicated, but see text. (Courtesy of K. C. Holmes.)

interactions between protein and RNA are best considered as complementarity between the electrostatic surfaces of protein and RNA, rather than as simple ion pairs between arginines and phosphates.

4. *Electrostatic interactions involved in assembly and disassembly* Caspar (1963) and Butler *et al.* (1972) obtained results indicating that pairs of carboxyl groups with anomolous pK values (near pH 7.0) were present in TMV. Namba and Stubbs (1986) identified two intersubunit pairs of carboxyl groups in their model that may be those proposed by Caspar: Glu^{95}–Glu^{106} at 2.5 nm radius and in a side-to-side interface, and Glu^{50}–Asp^{77} at 5.8 nm radius in the top-to-bottom interface. These groups may play a critical role in assembly and disassembly of the virus (Chapter 7, Section III,A).

Namba *et al.* (1989) identified three sites where negative charges from different molecules are juxtaposed in subunit interfaces. These create an electrostatic potential that could be used to drive disassembly.

—a low-radius carboxyl–carboxylate pair appears to bind calcium;

—a phosphate–carboxylate pair that also appears to bind calcium;

—a high-radius carboxyl–carboxylate pair in the axial interface. This could not bind calcium but could bind a proton and thus titrate with an anomolous pK.

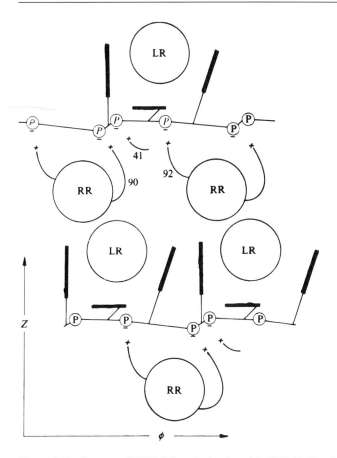

Figure 5.12 Structure of TMV. Schematic drawing of the RNA binding site viewed along a radius and in the plane of the RNA. LR, left radial helix; RR, right radial helix; P, phosphate. Bases are indicated by thin rectangles. Numbers and signs indicate arginines, which may form salt bridges with the phosphates. [From Stubbs *et al.*, 1977, reprinted by permission from *Nature (London)* **267**, 219. Copyright © 1977 Macmillan Magazines Ltd.]

5. *Water structure*   Namba *et al.* (1989) showed that water molecules are distributed throughout the surface of the protein subunit, both on the inner and outer surfaces of the virus and in the subunit interfaces.

6. *Specificity of TMV protein for RNA*   Gallie *et al.* (1987d) showed that TMV protein does not assemble with DNA even if the origin of assembly sequence is included. Namba *et al.* (1989) concluded that this specificity must involve interactions made by the ribose 2′ hyroxyl groups, because all three base-binding sites could easily accommodate thymine.

2. *Tobravirus* Group (Tobacco Rattle Virus)

TRV is a rigid cylindrical rod built on the same general plan as TMV. TRV requires two particles, a long rod and a shorter one, to cooperate in the production of intact virus particles. This aspect is discussed in Chapter 7. Here I shall consider the structure of the long rods. The short rods are constructed on the same plan using the same protein subunit.

Most information about the structure of this virus has been obtained by electron microscopy of virus particles stained in uranyl formate (Offord, 1966) (Fig. 5.13). Some information was obtained by optical diffraction. For the strains studied by Offord, length of the infectious particle was approximately 191 nm and its diameter was 25.6 nm. The pitch of the helix was 2.55 nm with 76 subunits in three turns. The radius of the central hole was about 2.7 nm. An annular feature was observed at a radius of about 8.2 nm, which might represent the RNA chain. A densely stained ring has been observed at about this radius in cross sections of rods (Tollin and Wilson, 1971). The strain studied by Offord had about 72 turns of the helix giving a MW for the protein coat $\simeq 44 \times 10^6$. There are probably 4 nucleotides per protein subunit, giving 7100 nucleotides with an average MW $\simeq 322$. Thus for the RNA the MW $\simeq 2 \times 10^6$ and for the long rod the MW $\simeq 46 \times 10^6$.

Harrison and Woods (1966) noted that in many of their electron micrographs of different TRV isolates, the two ends of the particle had different shapes. One end was slightly convex, while at the other end of the axial canal was slightly flared. They suggest that this appearance of the ends of the rods might be due to the protein subunits being banana-shaped with the long axes not inclined exactly 90° to the long axis of the particle.

Like TMV coat protein, TRV protein can exist in various discrete states of aggregation in solution. In particular a double disk structure, with $s = 35$, has been observed that may well play a part in virus assembly (see Gugerli, 1976).

3. Other Helical Viruses

The structures of other viruses with rigid rod-shaped particles—the hordeiviruses and BNYVV—have not been studied in the same detail as TMV. Likewise the particle structure for most of the five virus groups with flexuous rods is known only to the extent of establishing overall parameters for the helices formed by the protein subunits.

Some flexuous rods are illustrated in Fig. 5.14. Papaya mosaic *Potexvirus* (PapMV) is the best understood structure among the flexuous rods (AbouHaidar and

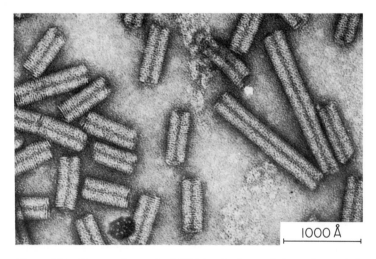

Figure 5.13 Electron micrograph of TRV showing long and short rods negatively stained with uranyl formate. (From Offord, 1966.)

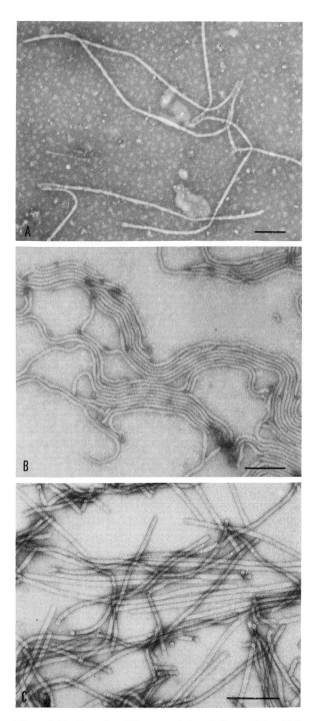

Figure 5.14 Examples of flexuous rod-shaped virus particles. (A) Beet yellow's *Closterovirus* showing pairing of particles at one end. (Courtesy of G. J. Hills.) (B) Lilac chlorotic leaf spot *Closterovirus* (Courtesy of R. W. Horne.) (C) PVX. (Courtesy of R. W. Horne.) Bars = 0.1 μm.

Erickson, 1985). The molecular basis for flexibility is of particular interest. Flexibility probably depends on the existence of a gradient of axial interactions between protein subunits in the rod, with the highest density of interactions occurring at low radius. At high radius there is probably a "spongy" quaternary structure with sparse axial intersubunit contacts and water-filled cavities. There is, as yet, no structural proof for this idea, but various lines of evidence support it (AbouHaidar and Erickson, 1985). For example, PapMV particles in gels can be seen to interpenetrate whereas TMV rods do not.

Mention should also be made of viruses in the rice stripe virus group that show folded, branched, or coiled threadlike particles in electron micrographs (Toriyama, 1986b). Preparations of these particles are infectious.

B. Small Icosahedral Viruses

At present we can distinguish four kinds of structure among the protein shells of small icosahedral plant viruses whose architecture has been studied in sufficient detail. However, high-resolution structures are available only for the second cluster of viruses. When more detailed information is available for the other groups, the arrangement suggested here may change.

1. Tymoviruses, bromoviruses, and cucumoviruses, which have 180 somewhat banana-shaped protein subunits clustered into pentamers and hexamers to form a $T = 3$ protein shell.
2. Tombusviruses, sobemoviruses, and STNV. The first two of these groups have 180 subunits in a $T = 3$ lattice, while STNV has 60 subunits in a basic $T = 1$ icosahedron. However, the S (shell or surface) domain of all three subunits has a similar polypeptide topology. They also have an amino-terminal domain that interacts with the RNA. In addition, the tombusviruses have a P (protruding) domain that forms surface projections on the virus particle.
3. AMV. This virus has several particles, the smallest being a $T = 1$ icosahedron, with a series of larger particles being prolate extensions of the $T = 1$ shell.
4. Comoviruses, which have two different proteins in the shell occupying different symmetry-related positions. Icosahedral RNA virus structure has been reviewed by Rossmann and Johnson (1989).

1. Tymoviruses, Bromoviruses, and Cucumoviruses

Tymovirus *Group*

The Protein Shell Using negative-staining procedures, Huxley and Zubay (1960) and Nixon and Gibbs (1960) showed that the protein shell of TYMV is made up of 32 protuberances, occupying two structurally distinct sites in the shell. Klug and colleagues used TYMV extensively in developing X-ray diffraction and electron microscopy as tools for the study of smaller isometric viruses. Klug *et al.* (1966) and Finch and Klug (1966) concluded that the protein shell has 180 scattering centers lying at a radius of about 14.5 nm. These points were identified with protuberances of the protein structure units at the surface of the particle.

Each individual protein subunit is somewhat banana shaped. Within the intact

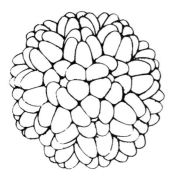

Figure 5.15 Structure of TYMV. Drawing of the outside of particle showing clustering of protein subunits into groups of five and six. (From Finch and King, 1966.)

virus each protein subunit is made up of about 9% α-helix and 43% β-sheet. About 48% of the polypeptide is in an irregular conformation (Hartman *et al.*, 1978). This conclusion was broadly confirmed by Tamburro *et al.* (1978). The orientation of the helical regions within the protein is not established.

The X-ray data on the virus gave good agreement for a model of the protein shell with 32 scattering centers lying at a radius of about 12.5 nm. The numerical relationships of the protein subunits in the TYMV shell are as follows:

Class: $P = 3$; $f = 1$; giving $T = 3$; number of structural subunits $= 60 \times 3 = 180$; M (number of morphological subunits) $= (10 \times 2)$ hexamers $+ 12$ pentamers $= 32$.

Figure 5.15 summarizes the knowledge of the external arrangement of the protein subunits while Fig. 5.16 shows views of these particles in three different orientations obtained by the three-dimensional image reconstruction technique.

Location of the RNA Finch and Klug (1966) considered that folds of the RNA were intimately associated with each of the 32 morphological protein units. They thought that it was the presence of this RNA in and around these positions that enhanced the appearance of the 32 morphological subunits seen in electron micrographs of TYMV, as compared to TYMV empty protein shells. However, neutron

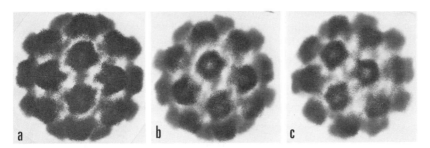

a   b   c

Figure 5.16 Structure of TYMV. Views of the reconstructed density-distribution plotted section by section on to photographic plates, which have been stacked to form a glass block. Only the top three-quarters of the model is shown so that the top morphological units are not superimposed on other units at the bottom of the model. The views are approximately down a twofold axis in (a), a threefold axis in (b), and a fivefold axis in (c). The central morphological units in (b) and (c) are clearly a hexamer and a pentamer, respectively. The morphological units viewed sideways on are seen to be slightly conical, splaying out at their bases to form connections with neighboring units. (From Mellema and Amos, 1972.)

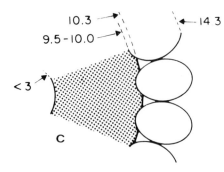

10.3

9.5 - 10.0

← 14 3

< 3

C

Figure 5.17 Schematic drawing of the diameters of the RNA and protein within TYMV based on data from neutron small-angle scattering experiments. Numbers indicate radii (nm). The shaded area corresponds to the RNA, which on this model does not penetrate deeply into the protein shell. [From Jacrot *et al.*, 1977, reprinted by permission from *Nature (London)* **266**, 420. Copyright © 1977 Macmillan Journals Ltd.]

small-angle scattering shows that there is very little penetration of the RNA into the protein shell (Fig. 5.17) and that the protein subunits are densely packed.

Classes of Particle Purified TYMV preparations can be fractionated on CsCl density gradients into a large number of components. There are three classes of particle:

1. *The empty protein shell* About one-third to one-fifth of the particles found in a TYMV preparation isolated from infected leaves are empty protein shells (the top or T component). These contain no RNA but otherwise are identical in structure to the protein shell of the infectious virus (B_1 component).

2. *Infectious virus nucleoprotein (B or bottom components and particles derived from them)* The infectious virus fractionates to form two density classes in CsCl gradients (B_{1a} and B_{1b}) that are equally infectious (Matthews, 1974). A third B_1 fraction, B_{1c}, more dense than B_{1b}, has been characterized. B_{1b} and B_{1c} both contain copies of the coat protein mRNA as well as a molecule of genome RNA. These B_1 components can be converted in strong solutions of CsCl to a B_2 series with higher densities, especially if the pH > 6.5. These are designated B_{2a}, B_{2b}, and B_{2c}. Their formation is prevented by the presence of 0.1 M $MgCl_2$ in the CsCl.

3. *Nucleoprotein particles containing subgenomic RNAs and having densities in CsCl intermediate between that of the T and B components* Mellema *et al.* (1979) and Keeling *et al.* (1979) isolated a series of eight minor components. The coat mRNA and a series of eight other subgenomic RNAs of discrete size have been isolated from these particles. These eight subgenomic RNAs have not yet been firmly allocated to particular nucleoprotein species. If the coat mRNA and the eight others are encapsidated in various combinations, and numbers of copies per particle, there may in fact be a very large number of particles of slightly differing density.

Some properties of the minor noninfective nucleoprotein fractions have been determined by Mellema *et al.* (1979) and Keeling *et al.* (1979). Their RNA content ranges from about 5% for fraction 1 to 28% for fraction 8. The proportion of total minor nucleoproteins relative to the infectious nucleoproteins is about 5% on a particle number basis. The coat protein cistron is found in most of the minor nucleoproteins along with other subgenomic RNAs (Pleij *et al.*, 1977; Higgins *et al.*, 1978).

When TYMV B_1 is taken to pH 11.6 in 1 M KCl the particles swell from 14.6

to 15.2 nm within 30 seconds (Keeling *et al.*, 1979). RNA escapes from these particles in 3–10 minutes in a partially degraded state. In addition, an amount of protein is lost within 1–3 minutes that is equivalent to the loss of one pentamer or hexamer of subunits from each particle. No such loss of protein occurs with the minor nucleoproteins (Keeling and Matthews, 1982). On return to pH 7.0 the radius of the result empty shells returns to normal. The nucleoproteins containing less than the full genome RNA do not swell at pH 11.6 under the same conditions and their RNA does not escape, although it is also degraded within the particle.

Bromovirus *Group*

Particle Structure   The structure of all three members of this group has been studied: BMV (Bancroft *et al.* (1967), broad bean mottle virus (Finch and Klug, 1967), and cowpea chlorotic mottle virus (Bancroft *et al.*, 1967).

The protein shell of these viruses is 25–28 nm in diameter, made up of 180 protein subunits of MW $\simeq$ 19,400. The 180 subunits show pentamer–hexamer clustering to give 32 morphological subunits arranged at the vertices of an icosahedral lattice with $T = 3$. Thus, the surface structure is similar to that of TYMV.

Finch *et al.* (1967b) found that the morphological units of broad bean mottle virus protrude at least 1.5 nm from the body of the particle. The negative stain appeared to penetrate into the center of the virus particles, suggesting the presence of an appreciable central hole, which is not found in TYMV. The presence of this hole, about 5.5 nm in radius, was confirmed by X-ray diffraction studies (Finch *et al.*, 1967b), and for BMV by small-angle neutron scattering (Fig. 5.18). Finch and Klug (1967) suggested that the absence of a central hole in TYMV may be due to the need to pack a higher proportion of RNA into the particle.

Stability of the Virus   Although bromoviruses are very similar to tymoviruses in the arrangement of their viral RNA and protein shells, they are much less stable particles. As the pH is raised between 6.0 and 7.0, particles of BMV swell from a radius of 13.5 nm to over 15 nm (Zulauf, 1977). The consequences of this swelling depend on many factors and particularly on the ionic conditions. The particles are readily disrupted in 1 *M* NaCl at pH 7.0. When swollen virus is dissociated the protein subunits can reassemble under a variety of conditions to form a range of products, described by Bancroft and Horne (1977). Of particular interest is the fact that, in the presence of trypsin, the coat protein of BMV loses 63 amino acids from the amino terminus, and can then self-assemble into a $T = 1$ empty shell (Cuillel *et al.*, 1981).   BMV can be fully dissociated into subunits by high pressures.

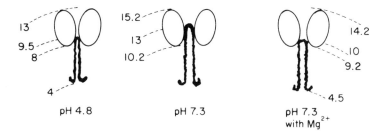

Figure 5.18   Structure of BMV. Relationship between BMV RNA and the protein shell under three ionic conditions, as deduced by neutron small-angle scattering. The degree of penetration of the RNA was experimentally determined. The folding as shown is hypothetical. Dimensions given are the radii in nanometers. Experiments were made on a mixture of the three BMV particles and, therefore, represent average dimensions. (From Jacrot *et al.*, 1976.)

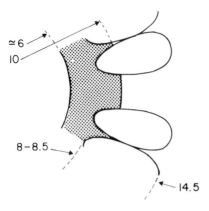

Figure 5.19 Structure of CMV. Sketch of the relationship between RNA and protein subunits in the virus as deduced from neutron small-angle scattering experiments. The size of the holes in the protein shell was calculated assuming 32 large holes, one in the center of each pentamer and hexamer in the shell. Dimensions are in nanometers. [From Jacrot *et al.*, 1977, reprinted by permission from *Nature (London)* **266,** 420. Copyright © Macmillan Magazines Ltd.]

The formation of $T = 1$ particles is a step in the disassembly process (Silva and Weber, 1988).

The structural polymorphism of BMV has been investigated using neutron small angle scattering (Jacrot *et al.*, 1976; Chauvin *et al.*, 1978b). At low pH (around 5.0) the virus is in a compact state under a range of ionic conditions. Near pH 7.0 at moderate ionic strength the virus particle swells and the RNA penetrates more deeply between the protein subunits. Mg^{2+} suppresses this effect (Fig. 5.18).

Titration studies have shown that BMV contains several (probably two) cation binding sites per subunit. Both sites titrate together at about pH 6.7. It is probable that both carboxylate groups of the protein and RNA phosphates contribute to this binding (Pfeiffer and Durham, 1977). When both sites bind H^+, Ca^{2+}, or Mg^{2+} the virus is compact. As these ions are released the virus swells.

The N-terminal 25 amino acids of the coat protein are rich in basic amino acids, and structure prediction methods indicate that this sequence may interact in a helical form with the RNA (Argos, 1981).

Cucumovirus *Group*

Finch *et al.* (1967a), using electron microscope methods similar to those employed in their study of broad bean mottle virus, showed that CMV is very similar, both in surface structure and in the fact that there is a central hole in the particle, to the bromoviruses. CMV is an unstable particle when compared to TYMV. However, CMV does not swell at pH 7.0 under conditions where BMV does. CMV has 180 subunits grouped in 32 morphological subunits (12 pentamers and 20 hexamers). The diameter of the negatively stained particle is 30 nm.

Jacrot *et al.* (1977) showed that there is substantial penetration of the RNA into the shell of protein (Fig. 5.19). The packing of the protein subunits is such that about 15% of the surface (at a radius of 11.7 nm) could be made up of holes, which would expose the RNA to inactivating agents and could explain the sensitivity of this virus to RNase.

Purified virus preparations contain four RNA species housed in three particles in the same arrangement as the bromoviruses. Some isolates of cucumoviruses have associated with them a small satellite RNA (Chapter 9, Section II,B,1).

2. Tombusviruses, Sobemoviruses, and STNV

The S domains (Fig. 5.21) of the protein subunit in the viral shells for TBSV, SBMV, and STNV are compared in Fig. 5.20. These protein subunits have in

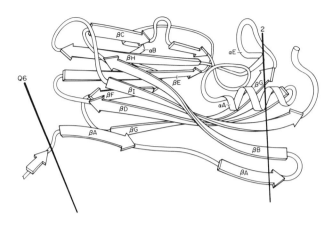

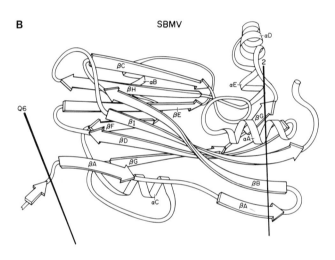

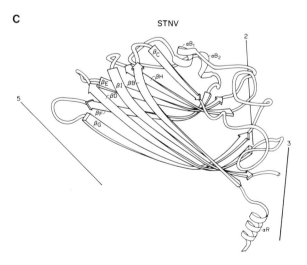

Figure 5.20 Diagrammatic representation of the backbone folding of the coat protein of (A) TBSV, (B) SBMV, and (C) STNV shown in roughly comparable orientations. (From Rossmann *et al.*, 1983.)

common a β-barrel structure. These are more or less wedge-shaped structures whose sides are made up of β-strands, which are maximally stretched polypeptide chains. One can envisage side-by-side positioning of the β-strands to generate a β-sheet, stabilized by noncovalent forces. Bending the shell into the wedgelike structure produces a type of β-barrel.

The S domain of TBSV can be described as an eight-stranded antiparallel β-barrel consisting of two back-to-back four-stranded shells (Fig. 5.20a). The domain is narrower at one end, making it wedge-shaped and thus allowing close approach of the subunits in the shell.

The S domain of SBMV is very similar to that of TBSV (Fig. 5.20b), while that of STNV (Fig. 5.20c) has somewhat less similarity to the other two structures. In the virus shell the wedge-shaped end of the S domain in all three viruses is close to the five- or sixfold vertices.

TBSV

The structure of TBSV has been determined crystallographically to 2.9-Å resolution (Harrison *et al.*, 1978; Olson *et al.*, 1983). It contains 180 protein subunits of MW = 43,000 arranged to form a $T = 3$ icosahedral surface lattice, with prominent dimer clustering at the outside of the particle, the clusters extending to a radius of about 17 nm.

The essential features of the structure are summarized in Fig. 5.21. The protein subunit is made up of two distinct globular parts (domains P and S) connected by a flexible hinge involving five amino acids (Fig. 5.21b). Each P domain forms one-half of the dimer-clustered protrusions on the surface of the particle. The S domain forms part of the icosahedral shell. Each P domain occupies approximately one-third of one icosahedral triangular face. The shell is about 3 nm thick and from it protrude the 90 dimer clusters formed by the P domain pairs. The β sheet of the P domain has what is known as a jelly-roll conformation (Gibson and Argos, 1990). In addition to domains P and S, each protein subunit has a flexibly linked N-terminal arm containing 102 amino acid residues. This has two parts, the R domain and the connecting arm called a (Fig. 5.21b).

The coat protein assumes two different large-scale conformational states in the shell that differ in the angle between domains P and S by about 20° (Fig. 5.21d). The conformation taken up depends on whether the subunit is near a quasi dyad or a true dyad in the $T = 3$ surface lattice. The state of the flexible N-terminal arm also depends on the symmetry position.

The N-terminal arms originating near a strict dyad follow along in the cleft between two adjacent S domains. On reaching a threefold axis, such an arm winds around the axis in an anticlockwise fashion (viewed from outside the particle). Two other N-terminal arms originating at neighboring strict dyads will be at each threefold axis. The three polypeptides overlap with each other to form a circular structure called the β-annulus around the threefold axis (Fig. 5.22). Thus 60 of the 180 N-terminal arms form an interlocking network that, in principle, could form an open $T = 1$ structure without the other 120 subunits. The annulus is made up of 19 amino acids from each arm. The R and a domains arising from the 120 subunits at quasi dyad positions hang down into the interior of the particle in an irregular way so that their detailed position cannot be derived by X-ray analysis. The RNA of TBSV is tightly packed within the particle and these N-terminal polypeptide arms very probably interact with the RNA (see Section VII,B,1).

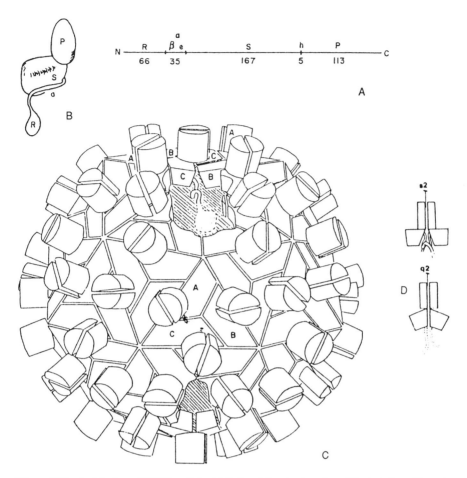

Figure 5.21 Architecture of TBSV particle. (a) Order of domains in polypeptide chain from N terminus to C terminus. The number of residues in each segment is indicated below the line. The letters indicate R domain (possible RNA binding region), arm a (the connector that forms the β-annulus and extended arm structure on C subunits and that remains disordered on A and B), S domain, hinge, and P domain. (b) Schematic view of folded polypeptide chain, showing P, S, and R domains. (c) Arrangement of subunits in particle. A, B, and C denote distinct packing environments for the subunit. S domains of A subunits pack around fivefold axes; S domains of B and C alternate around threefold axes. The differences in local curvature can be seen at the two places where the shell has been cut away to reveal S domain packing near strict (top) and quasi (bottom) dyads. (d) The two states of the TBSV subunit found in this structure, viewed as dimers about the strict (s2) and local (q2) twofold axes. Subunits in C positions have the interdomain hinge "up" and a cleft between twofold-related S domains into which fold parts of the N-terminal arms. Subunits in the quasi-twofold-related A and B positions have hinge "down," S domains abutting, and a disordered arm. [a, b, and c from Olson *et al.*, 1983; d from Harrison *et al.*, 1978; d reprinted by permission from *Nature (London)* **276**, 370. Copyright © Macmillan Journals Ltd.]

Two divalent cation binding sites are a feature of each of the three trimer contacts between the S domains (Fig. 5.22). If the divalent cations are removed above pH 7.0 the particle swells.

The structure of another *Tombusvirus*, TCV, has been determined at 3.2-Å resolution. The overall structure is very similar to that of TBSV (Hogle *et al.*, 1986).

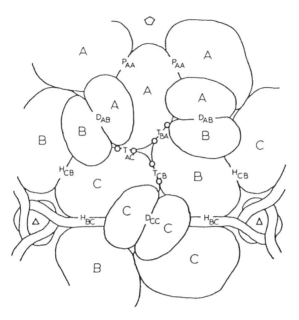

Figure 5.22 Packing of subunits in the icosahedral asymmetric unit of TBSV and notation for inter-faces. The different kinds of subunit contact are labeled D (dimer), T (trimer), P (pentamer), and H (hexamer), with a subscript showing the types of subunit interacting across the contact in question. The positions of Ca^{2+} at the T interfaces are shown by small circles. A fivefold axis is indicated at the top of the diagram. The β-annulus with threefold symmetry is shown twice in the lower part of the diagram. (From Olson *et al.*, 1983.)

SBMV

SBMV has an icosahedral particle about 30 nm in diameter, with 180 protein subunits in a $T = 3$ lattice. There is a single coat protein of MW = 28,200. There is a single piece of RNA with MW of 1.4×10^6. There appear to be no empty shells, minor nucleoproteins, or satellites associated with viruses of this group. The virus particle shows little surface detail in negatively stained preparations.

The structure of SBMV has been determined at 2.8-Å resolution (Abad-Zapatero *et al.*, 1980; Silva and Rossmann, 1987). The folding of the S domain of the protein subunit is very similar to that of TBSV (Fig. 5.20b). There is also an amino-terminal R domain. SBMV lacks the P domain found in TBSV, which accounts for the smaller subunit (260 amino acids compared with 380), the smaller outside radius, and the smoother appearance in electron micrographs.

The arrangement of the three quasiequivalent subunits (A, B, and C) is very similar to that in TBSV. As with TBSV, part of the R domain of the C subunits is in an ordered state forming a β-annulus around each icosahedral threefold axis. The R domain is shorter than in TBSV. The first 64 residues of the A and B subunits (the R domain) and the first 38 residues of the C subunit are disordered and associated with the RNA. Cation binding sites (Ca^{2+}) are near the external surface of the shell. The virus is strongly dependent on Mg^{2+} and Ca^{2+} for its structural integrity. About 200 of these ions are bound firmly to each virus particle. Their removal with EDTA causes a reversible destabilization of the structure. Treatment with trypsin removes the N-terminal 61 amino acids from the isolated protein subunit. The resulting MW = 22,000 fragment can assemble into a 17.5-nm-diameter $T = 1$ particle (Erickson and Rossmann, 1982). Rossmann (1985) gives a detailed account of SBMV structure.

STNV

STNV is the smallest known virus. It has a shell made of 60 protein subunits of MW = 21,300 arranged in a $T = 1$ icosahedral surface lattice (Fig. 5.23). The structure of the protein subunit has been determined crystallographically with refinement to 2.5-Å resolution (Jones and Liljàs, 1984). The general topology of the polypeptide chain is like that of the S domains of TBSV and SBMV (Fig. 5.20c) but the packing of the subunits in the $T = 1$ icosahedral structure is clearly different (Rossmann et al., 1983) and there is no P domain like that in TBSV. In the amino terminus of STNV, there are only 11 disordered residues followed by an ordered helical section (residues 12–22) buried in the RNA. Three different sets of metal ion binding sites (probably Ca^{2+}) have been located. These link the protein subunits together. A more detailed structure for the protein shell has been proposed by Montelius et al. (1988).

3. AMV and *Ilarvirus* Groups

The structure of the coat protein of AMV is not yet known at atomic resolution. Proton magnetic resonance studies provide a low-resolution model in which the protein (MW = 24,250) consists of a rigid core and a flexible amino-terminal part of about 36 amino acid residues (Kan et al., 1982). The coat protein behaves as a water-soluble dimer stabilized by hydrophobic interactions between the two molecules. This dimer is the morphological unit out of which the viral shells are constructed.

Under appropriate conditions of ionic strength, ionic species, pH, temperature, and protein concentration, the protein dimer forms a $T = 1$ icosahedral structure built from 30 dimers (Fig. 5.24) (Driedonks et al., 1977). This structure has been confirmed by X-ray crystallographic analysis (Fukuyama et al., 1983).

Purified preparations of AMV contain four nucleoproteins present in major amounts (bottom, B; middle, M; top b, Tb; and top a, Ta). They each contain an RNA species of definite length. The genome is split between the Tb, M, and B RNAs. Three of the four major components are bacilliform particles, 19 nm in diameter. The fourth (Ta) is normally spheroidal with a diameter slightly larger

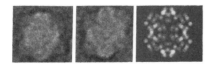

Figure 5.23 STNV. Electron micrographs of two virus particles oriented close to the twofold axis, and a model of 12 morphological subunits shaped to approximate to the same pattern. (Courtesy of J. T. Finch.)

Figure 5.24 $T = 1$ model of the 30 S AMV particle composed of 30 dimers of coat protein. The twofold symmetry axes of the dimers coincide with the dyad positions of an icosahedron leaving large holes at five- and threefold positions. For clarity the rear-facing subunits are omitted from the model. [With permission from Driedonks *et al.*, 1977. Copyright © Academic Press, Inc. (London) Ltd.]

than 19 nm (Fig. 5.25). However, two forms of Ta component have been recognized (Heijtink and Jaspars, 1976). Ta^t is spheroidal and soluble in 0.3 M MgSO$_4$. Ta^b is a rodlet and insoluble in 0.3 M MgSO$_4$. These two particles appear identical in other properties. The Ta components have 120 protein subunits, and Cusack *et al.* (1983) have raised the possibility that these may have a nonicosahedral structure, an idea not yet proven.

From a careful study of the MWs of the RNAs, the protein subunit, and the virus particles, Heijtink *et al.* (1977) concluded that the number of coat protein monomers in the four major particles $= 60 + (N \times 18)$, n being 10, 7, 5, or 4.

The details of the arrangement of the protein monomers in the particles have not been established. However, from analysis of electron micrographs of negatively stained AMV particles, Mellema and van den Berg (1974) concluded that the virus coat forms two kinds of lattice—a stacked and a helical type. In both, the coat protein forms dimers and these are associated with the twofold symmetry axes. In the stacked lattice the morphological units are arranged in staggered rings with two rings in one repeat distance. In the helical type of lattice, the rings of unit cells are transformed into a double helix. The molecular basis for this transition is not understood.

A small-angle neutron scattering study has given a model for the B compound showing the distribution of RNA and protein at low resolution (Cusack *et al.*, 1981) (Fig. 5.25).

The hemispherical ends of the virus rods are believed to be "caps" formed by a switch from sixfold to fivefold icosahedral arrangement of the protein subunits (R. E. Hull *et al.*, 1969b; Mellema and van den Berg, 1974). The virus is unstable with regard to high ionic strength and SDS, and is sensitive to RNase. This sensitivity may be explained by the holes in the protein coat illustrated in Fig. 5.25. Conformational changes occur at mildly alkaline pH (Verhagen *et al.*, 1976).

AMV nucleoproteins are readily dissociated into protein and RNA at high salt concentrations. Bacilliform particles can be reformed under appropriate conditions. Thus the virus is mainly stabilized by protein–RNA interactions (see van Vloten-Doting and Jaspars, 1977). Under a wide range of solvent conditions, AMV particles do not show the phenomenon of swelling (Oostergetel *et al.*, 1981). Compared with the small isometric viruses, AMV may be regarded as being in a permanently swollen state.

Purified preparations of some AMV strains can be fractionated by polyacrylamide gel electrophoresis to reveal the presence of at least 17 nucleoprotein components (Bol and Lak-Kaashoek, 1974). These are all made up from the single viral coat protein (van Beynum *et al.*, 1977). Besides the four major RNA species there are at least 10 minor RNA species of different lengths. The nucleoproteins occurring in minor amounts contain both major and minor species of RNA. Particles of somewhat different size may contain the same RNA, while particles of the same size may contain different RNAs. Thus a small variation is possible in the amount of RNA encapsidated by a given amount of viral protein (Bol and Lak-Kaashoek, 1974).

Most of the minor component particles are smaller than the B component, but one was found to be 87 nm long and contained two molecules of middle component RNA (Heijtink and Jaspars, 1974). Jaspars (1985) gives a full account of AMV structure.

The ilarviruses such as TSV have isometric particles of four different size classes (van Vloten-Doting, 1975). Their particle structure has not been established in detail but they share many properties in common with the AMV group (van

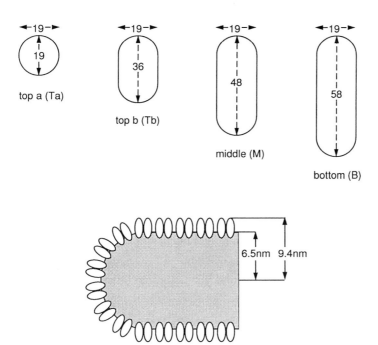

Figure 5.25 AMV particles. (*Top*) Sizes (in nm) of the four main classes of particle. (*Bottom*) A schematic representation of the distribution of protein and RNA in AMV bottom component. RNA is indicated by the hatched area and the protein molecules are represented by ellipsoids. The model is derived from the analysis of both the 30 S and the bottom component by small-angle neutron scattering. (Bottom reprinted from Cusack *et al.*, 1981.)

Vloten-Doting, 1976). In fact, many workers consider that AMV should be a member of the *Ilarvirus* group.

4. *Comovirus* Group

CPMV has a diameter of about 28 nm and an icosahedral structure with an unusual arrangement of subunits. The virus shell contains two proteins, a large and small one (MWs = 42 and 22 × 10^3) with different amino acid compositions (Wu and Bruening, 1971). From the sizes of the protein and the virus shell, Geelen *et al.* (1972) calculated that there should be 60 of each of these proteins in the shell. Using this information and data they obtained from a three-dimensional image reconstruction, Crowther *et al.* (1974) proposed a model in which the 60 larger proteins are clustered at the 12 positions with fivefold symmetry; the 60 smaller proteins form 20 clusters about the positions with threefold symmetry (Fig. 5.26).

Although the particle of CPMV is clearly a $T = 1$ icosahedral structure, the fact that it has two proteins in different symmetry environments gives it some overall resemblance to a $T = 3$ structure (Schmidt and Johnson, 1983). The structure of CPMV at 3.5 Å shows that the two coat proteins produce three distinct β-barrel domains in the icosahedral asymmetric unit. Although two of these β-barrels are covalently linked, their relationship is essentially like the β-barrels found in the three major separate coat proteins of the picornaviruses (Stauffacher *et al.*, 1985). A similar structure has been deduced for bean pod mottle *Comovirus* (Chen *et al.*, 1989).

Purified preparations of CPMV contain three classes of particle, with identical protein shells, that can be separated by centrifugation. Early studies on this virus were complicated by the essentially trivial fact that the smaller protein is susceptible *in situ* to attack by proteolytic enzymes both in the host plant and after isolation of

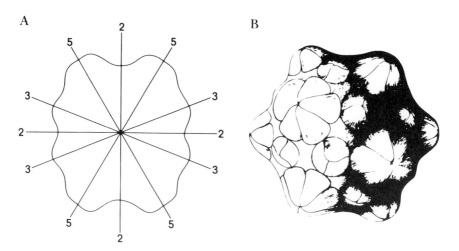

Figure 5.26 Structure of CPMV based on image reconstruction and on known chemical composition. (A) Relation between symmetry axes and contour at the virus surface. Trace of the equator of the image reconstruction in a plane normal to a twofold axis. The positions of the twofold, threefold, and fivefold axes lying in the plane are indicated. The radii of the contour at the fivefold, threefold, and twofold positions are approximately 12.0, 10.6, and 9.3 nm, respectively. The indentations at the twofold positions are approximately twice as long (between two fivefold axes) as they are wide (between two threefold axes). (B) Schematic diagram of the model. The large protein subunits are clustered about the fivefold positions while the small subunits are clustered about the threefold positions. (From Crowther *et al.*, 1974.)

the virus. CPMV contains about 200 spermidine molecules and a trace of spermine per particle.

5. Ourmia Melon Virus

A recently described virus infecting melons in Iran has such novel structural properties that it almost certainly needs to be placed in a new virus group, or family (R. G. Milne, personal communication). The single structural protein has an $M_r =$ 25,000. Three ss positive sense RNAs of MW 0.91, 0.35, and 0.32×10^6 were detected. The particles resemble both geminiviruses and AMV in some respects. They are cylindrical, with an 18.5-nm diameter, and have pointed ends.

The protein subunits cluster into dimers or trimers. The pointed ends may be formed from icosahedra cut through threefold axes for a dimer or twofold axes for a trimer. The tubular body does not form a continuous geometrical net as with AMV, but contains discontinuities marked by fissures between double disks of the protein (Fig. 5.27). Particles consisting of 2, 3, 4, or 6 double disks have been observed,

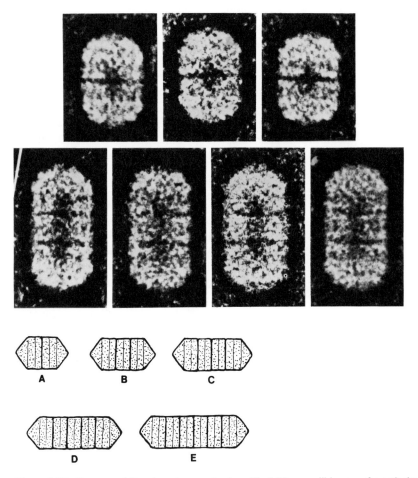

Figure 5.27 Structure of Ourmia melon mosaic virus. (*Top*) "Averaged" images of negatively stained particles. Each image was built up photographically by equal superimposed exposures of 10 original particle images. The top row has two double disks and the bottom row has three. (*Bottom*) Sketches showing the suggested arrangement of double disks in particles of different length. Particles of type D have not yet been observed. (Courtesy of R. G. Milne.)

with 4- and 6-disk particles being rare. Figure 5.27B illustrates these structures diagrammatically.

IV. GEOMETRIC VIRUSES WITH dsRNA

A. Reoviridae

The basic structure of the plant reoviruses is similar to that of *Reovirus*, a genus of the family Reoviridae that infects vertebrates. Particles are isometric and about 70 nm in diameter as seen in negatively stained preparations. By analogy with *Reovirus* they probably have a substantially greater diameter in solution (Harvey *et al.*, 1974).

Two genera have been established for the plant reoviruses: *Phytoreovirus* for those with 12 pieces of dsRNA, no spikes on the outer coat, type member WTV; and *Fijivirus* for those with 10 pieces of RNA, spikes present on the outer coat, type member FDV.

Almost all the information on the structure of plant reovirus particles has been derived from electron microscope observations on individual negatively stained particles and on subviral structures produced by partial breakdown of the intact virus. Reoviruses contain an RNA-dependent RNA polymerase as part of the particle. This enzyme activity has not yet been identified with any of the polypeptides revealed by gel electrophoresis for the plant reoviruses. However, it is located in the subviral particle, not in the outer envelope (e.g., Black and Knight, 1970; Ikegami and Francki, 1976).

1. *Phytoreovirus*

There are seven different proteins in the particle of WTV (Reddy and MacLeod, 1976) (see Table 8.1). The particles consist of an outer shell of protein and an inner core containing protein and the 12 pieces of dsRNA. However, there is no protein in close association with the RNA. The particles are readily disrupted during isolation, by various agents. Under suitable conditions subviral particles can be produced, which lack the outer envelope and which reveal the presence of 12 projections at the fivefold vertices of an icosahedron.

By controlled degradation and study of the products it has been possible to locate some of the proteins within the particle (Table 8.1). If the outer shell is removed by a gentle treatment (e.g., enzyme digestion) the resultant cores have unimpaired infectivity for insect vector monolayer cultures (Reddy and MacLeod, 1976).

The location of the RNA within rice dwarf virus, another member of the *Phytoreovirus* genus, has been studied by small-angle neutron scattering (Inoue and Timmins, 1985). The RNA is located within a radius of 23 nm, and the protein shell lies between 23 and 36 nm from the center of the particle. Thus there appears to be no major interpenetration of RNA and protein. The location of the proteins within the virus have been established for two other members: rice gall dwarf virus (Omura *et al.*, 1985) and rice ragged stunt virus (Hagiwara *et al.*, 1986). The particle

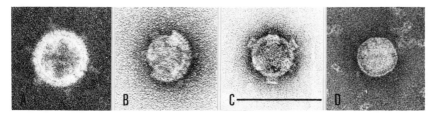

Figure 5.28 Structure of maize rough dwarf *Fijivirus* as revealed by images of negatively stained particles. (A) Complete particle fixed in glutaraldehyde and negatively stained in uranylacetate (UA) showing the external A spike. (B) Partially stripped particles negatively stained with UA. (C) Inner core plus B spikes, stained in UA. (D) Inner core alone, stained in potassium phosphotungstate. (Courtesy of R. G. Milne.)

structure of this latter virus appears to differ from typical members of either the *Phytoreovirus* or *Fijivirus* genera (Kawano *et al.*, 1984). It might appropriately belong in a third genus.

2. *Fijivirus*

Maize rough dwarf virus (MRDV) is one of the best characterized members of the *Fijivirus* genus (Milne and Lovisolo, 1977). Milne *et al.* (1973) detected 12 spikes projecting from the surface of intact MRDV particles, one at each fivefold symmetry axis. These A spikes were about 11 nm long. Beneath each A spike was a B spike about 8 nm long, revealed when the outer coat was removed. The B spikes were associated with some differentiated structure in the inner core, which they termed a baseplate. Detached B spikes could be seen to be made up of five morphological units surrounding a central hole. Figure 5.28 illustrates the kind of images from which the structure has been deduced.

Cores without spikes (smooth cores) contain 136,000 and 126,000 MW polypeptides. The spiked cores contained, in addition, a 123,000 MW polypeptide that is, therefore, probably located in the B spikes (Boccardo and Milne, 1975). None of the other polypeptides has been unequivocally associated with particular structures seen in negatively stained images of particles (Boccardo and Milne, 1975; Luisoni *et al.*, 1975). FDV has a very similar structure to that of maize rough dwarf virus, as revealed by electron microscopy (Fig. 5.29).

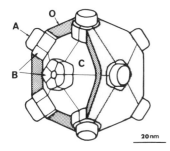

Figure 5.29 A scale model of the Fiji disease virus. A = A spike; B = B spike. Part of the outer shell (O) of the particle and one of the A spikes have been removed to expose the core (C) and the structure and arrangement of the B and A spikes. The arrangement of distinct morphological subunits within the A spikes, outer shell, and core has not been established. (From Hatta and Francki, 1977.)

V. GEOMETRIC VIRUSES WITH DNA

A. *Caulimovirus* Group

CaMV has a very stable isometric particle about 50 nm in diameter containing dsDNA. In spite of the great amount of information available on the genome organization and replication of this important virus, its detailed structure remains unknown. Electron microscopy reveals a relatively smooth protein shell with no structural features. Calculations from MW of the coat protein and the amount of protein in the virus suggest that it may have a $T = 7$ icosahedral structure (Krüse *et al.*, 1987). The distribution of protein and DNA in the particle has been determined by neutron small-angle scattering. Data from the work of Krüse *et al.* (1987) are summarized in Table 5.1.

The particle consists of an outer protein shell. Within this is a zone where both DNA and protein are present. Most of the DNA is in a zone with very little protein, while the center region of the particle (one-eighth of the volume) is occupied only by solvent. The DNA does not appear to be associated with any significant amount of histonelike protein (Al Ani *et al.*, 1979b). Exposure of the virus to pH 11.25 leads to the release of DNA tails without total disruption of the protein shell (Al Ani *et al.*, 1979a).

B. *Geminivirus* Group

Geminiviruses contain ssDNA and one type of coat polypeptide. Particles in purified preparations consist of twinned or geminate icosahedra (Fig. 5.30).

From a study of negatively stained particles together with models of possible structures it is probable that chloris striate mosaic virus consists of two $T = 1$ icosahedra joined together at a site where one morphological subunit is missing from each, giving a total of 22 morphological units in the geminate particle. This structure receives support from a study of the size of the DNA and protein subunit of this virus, indicating that each geminate particle shell, consisting of 110 polypeptides with MW of 2.8×10^4 arranged in 22 morphological units, contains one molecule of ssDNA with MW of 7.1×10^5 (Francki *et al.*, 1980b).

Table 5.1

The Spherically Averaged Radial Organization of CaMV and the Volume Fraction Occupied by Protein, DNA, and Water in Each Shell[a]

	Shell (Å)	Protein	DNA	Water
I	120–150	0.27 (11.0)[b]	0.46 (6.4)[b]	0.74
II	150–185	0.04 (2.9)	0.18 (51.6)	0.78
III	185–215	0.29 (25.8)	0.12 (42.0)	0.59
IV	215–250	0.43 (60.3)	0.00 (0)	0.57

[a]Volume fractions are accurate to 0.05 for large and 0.02 for small values.

[b]Number in parentheses indicates the percentage of total protein or DNA present in the corresponding shell. (From Krüse *et al.*, 1987).

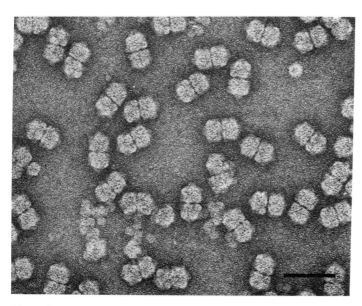

Figure 5.30 Purified geminivirus virus from *Digitaria* negatively stained in 2% aqueous uranyl acetate. Bar = 50 nm. (From Dollet *et al.*, 1986.)

VI. ENVELOPED VIRUSES

A. Rhabdoviridae

Rhabdoviridae is a family of viruses whose members infect vertebrates, invertebrates, and plants. They have a complex structure. Rhabdoviruses from widely differing organisms are constructed on the basic plan shown in Fig. 5.31.

Some animal rhabdoviruses may be bullet shaped, but most and perhaps all plant members are rounded at both ends to give a bacilliform shape. Electron microscopy on thin sections and negatively stained particles, has been used to determine size and details of morphology (Fig. 5.32). The size measurements can be only approximate and are probably too small because of shrinkage taking place during dehydration for electron microscopy. Particles of these viruses readily deform and fragment *in vitro* unless pH and other conditions are closely controlled (e.g., Francki and Randles, 1978) (Fig. 5.32E).

The proteins present in plant rhabdoviruses are detailed in Table 5.2. The location of the various proteins within the particles has been established by fractionation of subviral structures, by successive removal of superficially located proteins by detergent or enzyme treatment, and by the labeling of the exposed proteins with ^{125}I (e.g., Ziemiecki and Peters, 1976). The N protein appears to be firmly bound to the RNA to form the ribonucleoprotein core (RNP). This is usually seen in thin sections as having cross-striations 4.5–5.5 nm apart. The RNP is a single continuous strand arranged in helical fashion (e.g., Hull, 1976). The surrounding membrane contains two proteins: the matrix or M protein (or proteins), which is in an hexagonal array, and the G protein (glycoprotein), which forms the surface projections (e.g., Ziemiecki and Peters, 1976).

The detailed fine structure remains to be determined. In particular it is not clear

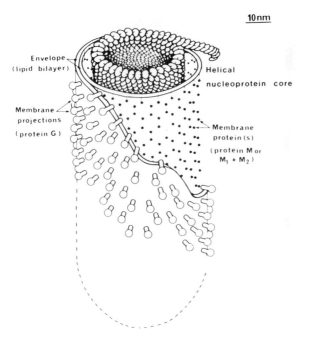

Figure 5.31 Generalized diagram for the structure of rhabdoviruses. (From Francki and Randles, 1980, reprinted with permission from "The Rhabdoviruses," Vol. 3. Copyright © CRC Press, Inc., Boca Raton, Florida.

how the hexagonally arrayed M protein is related spatially to the helical RNP. Likewise the way the rounded ends are formed and the arrangement of RNP within them is not established. The RNA-dependent RNA polymerase activity found in the rhabdoviruses has not yet been unequivocally assigned to any particular polypeptide or set of polypeptides but the L and NS proteins are involved. The enzyme is found in the RNP of LNYV (Francki and Randles, 1973).

Table 5.2
Proteins in Four Plant Rhabdoviruses

Name of protein[a]	MW ($\times 10^{-3}$)			
	LNYV[b]	PYDV[c]	SYVV[d]	SYNV[e]
L	160–180	+[f]	150 (CHO)	140 and 120
G	78 (CHO)	78 (CHO)	83 (CHO)	76.8
N	57	56	60	63.8
NS	−	+	−	−
M_1	+	33	44	45.5
M_2	19	22	36	39.5

[a]The nomenclature of Wagner *et al.* (1972) for the structural proteins of vesicular stomatitis virus has been used. For this virus, L (large) protein is associated with the nucleoprotein core; G is glycosylated and forms the membrane spikes; N is the major structural protein of the nucleoprotein core; NS is nonstructural protein; M is the matrix protein (M_1 and M_2) a major component of the viral membrane but buried deeply compared with the G protein.

[b]From Francki and Randles (1978) and Dietzgen and Francki (1988).

[c]Potato yellow dwarf virus, from Knudson and MacLeod (1972).

[d]Sow thistle yellow vein virus, from Ziemiecki and Peters (1976).

[e]Sonchus yellow net virus, from Jackson and Christie (1977) and Jackson (1978).

[f]+, Small amounts of protein detected; −, no protein detected; CHO, carbohydrate. Proteins are allocated to each group according to their size and, for some, according to their determined location in the particle.

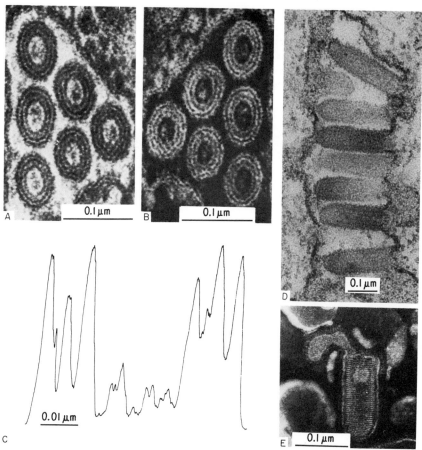

Figure 5.32 Rhabdoviruses. (A, B) Transverse sections through particles of potato yellow dwarf virus showing the lamellar nature of the viral envelope. Particles show some lateral compression. (B) is a photographic reversal of (A). (C) Microdensiometer trace through the top particle in (A). (From MacLeod *et al.*, 1966.) (D) Section through a cell of *N. rustica* showing a group of potato yellow dwarf particles enclosed in a membrane-bound vesicle. (Courtesy of R. MacLeod.) (E) A particle of sowthistle yellow vein virus in a negatively stained leaf dip preparation. (From Richardson and Sylvester, 1968.)

Hull (1970b) suggested that there were two subgroups of plant rhabdoviruses, those with narrower particles (60–70 nm diameter), for example, LYNV, and those with wider particles (80–100 nm diameter), for example, potato yellow dwarf virus.

B. Tomato Spotted Wilt *Tospovirus*

The tomato spotted wilt virus is very unstable and difficult to purify (e.g., Joubert *et al.*, 1974). Its structure has been studied in thin sections of infected cells, and in partially purified preparations. Isolated preparations contain many deformed and damaged particles.

As seen in thin sections particles are about 100 nm in diameter (Milne, 1970) (Fig. 5.33). By analogy with an enveloped animal virus of similar size (Harvey, 1973) they are probably much larger in the living cell in the hydrated state. The central core of the particle contains the RNA. Outside the core is a layer of dense material, surrounded by a membrane about 7.5 nm thick (Milne, 1970). The internal

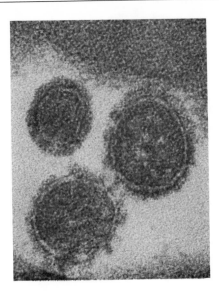

Figure 5.33 Tomato spotted wilt virus in a thin section of tomato. (Courtesy of R. G. Milne.)

structure has not been defined. The virus contains lipid (Best, 1968). The outer membrane is a typical lipoprotein bilayer. The RNAs and structural proteins of TSWV are discussed in Chapter 8, Section V.

VII. THE ARRANGEMENT OF NUCLEIC ACID WITHIN ICOSAHEDRAL VIRUSES

In the viruses with helical symmetry the RNA is more or less tightly embedded in the protein helix. For rods such as TMV, which are accessible to study by X-ray crystallographic methods, the position and bonding of the RNA may be determined in atomic detail. Other methods have given us some indications about the arrangement of the nucleic acid and its relationship to the protein shell in icosahedral viruses, and high-resolution X-ray crystallographic methods are providing molecular detail for a few of these.

A. Order in the Nucleic Acid

1. STNV

The crystal structure of STNV has been studied at 16-Å resolution using neutron diffraction in H_2O/D_2O (Bentley *et al.*, 1987). At 40% D_2O, scattering arises largely from the RNA component. The two main RNA motifs are shown in Fig. 5.34. These are in fact connected by regions of weaker RNA density. Each spherical motif (II) is connected to five symmetry-related extended motifs (I). They form a continuous network of RNA density on the inside surface of the protein coat. The I motifs form the edges of an icosahedron, leaving triangular holes centered on the threefold axes. It is into these holes that triple-helical arms of amino-terminal

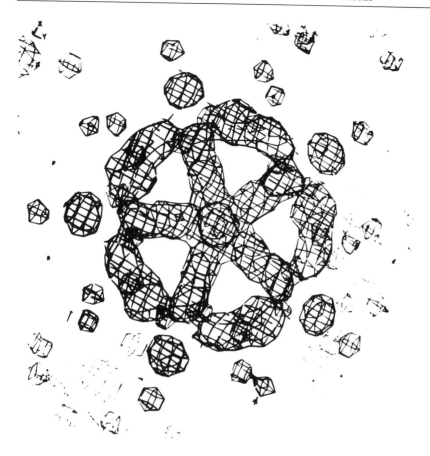

Figure 5.34 Low-resolution structure of the RNA within STNV determined by neutron diffraction. Positive density at 40% D_{20} looking down the fivefold axis from the center of the virus. I is the RNA density motif, which lies along the edges of each triangle and forms the edges of an icosahedron. Its length is about 45 Å and its diameter is 22–25 Å. The RNA density of motif II lies along each fivefold axis at a distance of 67 Å from the virus center. Minor regions of density (III) at higher radii correspond to positive fluctuations of protein density. (From Bentley *et al.*, 1987.)

regions of the protein subunits penetrate, and make close contact with the RNA (Fig. 5.35). Basic amino acids are well placed to make contact with the RNA.

Except for this protein–RNA interface the inner face of the protein shell is separated from the RNA by a thin layer of solvent. The cross section of motif I fits well with the idea that it represents a double helix in the RNA. If this is so, then 72% of the total RNA would be in double-helical form. The fact that STNV RNA has thermal denaturation kinetics like that of a transfer RNA indicating a high degree of secondary structure (Mossop and Francki, 1979a) confirms this.

2. BPMV

Nearly 20% of the RNA in BPMV particles binds to the interior of the protein shell in a manner displaying icosahedral symmetry. The RNA that binds is single-stranded, and interactions with the protein are dominated by nonbonding forces with few specific contacts (Chen *et al.*, 1989).

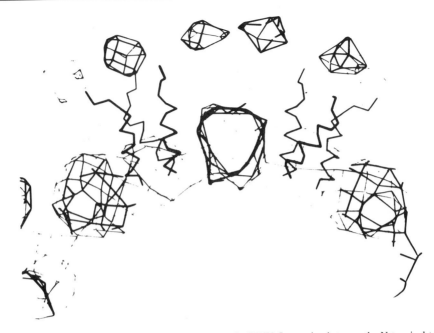

Figure 5.35   Relation between protein and RNA in STNV. Interaction between the N-terminal triple-helical arms and the RNA density viewed along the length of RNA motif I of Fig. 5.34, that is, perpendicular to the icosahedral axis. Positive density at 40% D_{20}. αcarbon atoms from amino acid residues 12 to 27 from each subunit form a pair of triple-helical arms. (From Bentley *et al.*, 1987.)

3. Other Small Icosahedral Viruses

Percentage of Double-Helical Structure

It is probable that the RNA inside many small icosahedral viruses has some double-helical structure. For example, Haselkorn (1962) concluded from hypochromicity studies that at least two-thirds of the bases in TYMV were in an ordered structure. Laser Raman spectroscopy indicated that about 77% of the RNA was in α-helical form (Hartman *et al.*, 1978). Circular dichroism studies confirmed that the RNA has a considerable amount of base-pairing and/or base stacking, and that the configuration of isolated RNA is very similar to that inside the virus (Tamburro *et al.*, 1978).

From hypochromicity experiments on intact CMV and isolated RNA, Kaper *et al.* (1965) concluded that the RNA inside the virus has a high degree of secondary structure, equivalent to a helical content of about 70%.

Circular dichroism studies on turnip rosette *Sobemovirus* suggest that the RNA within the particle has considerable base-pairing (Denloye *et al.*, 1978). In CCMV the RNA has a high degree ($\approx$95%) of ordered secondary structure as determined by laser Raman spectroscopy (Verduin *et al.*, 1984).

Magnetically induced birefringence in solution for several isometric plant viruses suggests that at least part of the RNA core has a symmetry differing from that of the protein shell (Torbet, 1983). [31]P nuclear magnetic resonance studies indicated that at 10°C the RNA within particles of AMV lacked movement, but that at 25°C there was some mobility in this component (Kan *et al.*, 1987).

Contributions from Base Sequence

The question as to whether the base sequence in an RNA plays a role in the folding of RNA within an icosahedral virus has been approached by Yamamoto and Yoshikura (1986), using a computer program for calculating RNA secondary structure. Calculations for 20 viral RNAs indicated that genomes of icosahedral viruses had higher folding probabilities than those of helical viruses. When the folding probability of a viral sequence was compared with that of a random sequence of the same base composition, the viral sequences were more folded. These results suggest that base sequence plays some part in the way in which ssRNA genomes fold within the virus.

4. Rice Dwarf Virus

Mizuno *et al.* (1986) have presented electron microscopic evidence suggesting that the ds genomic RNA segments of rice dwarf *Phytoreovirus* are packed within the viral core as a supercoiled structure complexed within protein.

5. CaMV

The dsDNA of CaMV is not thought to be complexed with any packaging protein within the virus. Magnetic birefringence measurements have been interpreted to suggest that the DNA is wound in a spool-like structure (Torbet *et al.*, 1986). Taking the neutron scattering data into account (Section V,A), the DNA may be packed into a spool composed of two annuli following the contours of the inside of the protein shell.

B. Interactions between RNA and Protein in Small Isometric Viruses

As has already been indicated in earlier sections, current knowledge suggests that there may be two types of RNA–protein interaction in the small isometric viruses, depending on whether basic amino acid side chains or polyamines neutralize charged phosphates of the RNA.

1. Viruses in Which Basic Amino Acids Bind RNA Phosphates

The best studied virus in which basic amino acids bind RNA phosphates is TBSV (Olson *et al.*, 1983). A total of 21 lysine and arginine residues are available to bind phosphates (11 on the R domain, 4 on the arm, and 6 on the inner surface of the S domain). Since there are approximately 26 nucleotides in the RNA for each protein subunit, most of the RNA charges can be neutralized by these amino acids. The flexible connection of the R arm allows it to conform to irregularities in RNA packing.

A similar situation exists for the SBMV (Silva and Rossmann, 1987). STNV has a shorter amino-terminal arm associated with the RNA (Jones and Liljàs, 1984). The positive charges on the inner surface of the protein shell and the amino-terminal arm are sufficient to neutralize 70% of the RNA phosphates. The remaining 30% are probably neutralized by bound Mg^{2+} ions (Liljàs *et al.*, 1982).

BMV coat protein has a basic amino-terminal region that is predicted to interact in a helical form with the RNA (Argos, 1981). A coat protein mutant in which the first 25 N-terminal amino acids in the basic arm of BMV coat protein had been deleted failed to direct the packaging of RNA (Sacher and Ahlquist, 1989). A mutant giving rise to a protein lacking only the first seven amino acids packaged viral RNA *in vivo*.

Argos also predicted a similar helical domain for the amino-terminal region of the AMV coat protein. About 10–15 amino acids penetrate the RNA (Oostergetel *et al.*, 1983). These helices are considered to be stabilized by neutralization of the positive charges by RNA phosphates. The composition of these N-terminal arms bears a similarity to histones (Argos, 1981). None of these viruses contains polyamines as part of their structure.

2. TYMV

There is no evidence for significant penetration of protein into the RNA of TYMV, nor is there any accumulation of basic amino acids at the protein–RNA interface. Ehresmann *et al.* (1980) used two *in situ* cross-linking procedures to identify three regions of the coat protein that lie close to the RNA. Cytosine was the base most prominently involved in cross-linking but there was no enrichment of basic amino acids in the cross-linked peptides. An examination of the coat protein sequences of TYMV and another *Tymovirus*, eggplant mosaic virus, showed that neither protein possessed an accumulation of basic residues able to form a strong ionic interaction with the RNA (Dupin *et al.*, 1985).

The RNA within TYMV may be stabilized by two kinds of interaction: first by glutamyl or aspartyl side chains hydrogen bonding to cytosine phosphate residues as suggested by Kaper (1972), and second by the spermine and spermidine found in this virus (Chapter 4, Section III,A), which could neutralize a significant proportion of the charged RNA, together with divalent cations. It is probable that all tymoviruses are stabilized in part by polyamines, but some members of the group may lose these components quite readily during isolation. Thus belladonna mottle virus isolated in the absence of CsCl contained about 100–200 polyamine molecules and 500–900 Ca^{2+} ions per virus particle. The polyamines could readily be exchanged with other cations such as Cs, leading to a loss of particle stability (Savithri *et al.*, 1987).

3. Comoviruses

CPMV may be similar to TYMV with respect to neutralization of charged phosphates, since particles containing RNA also contain polyamines, and there is no evidence for a mobile protein arm (Virudachalum *et al.*, 1985).

Chen *et al.* (1989) resolved protein–RNA interactions in bean pod mottle *Comovirus* at 3.0-Å resolution. They suggested that seven ribonucleotides that can be seen as ordered in the structure lie in a shallow pocket on the inner surface of the protein shell formed by the two covalently linked domains of the large coat protein.

VIII. DISCUSSION AND SUMMARY

Advances in the capacity, speed, and availability of computers, the use of non-crystallographic symmetry averaging, and developments in computer graphics have allowed resolution down to atomic detail for the proteins and protein coats of several geometric plant viruses, both rod shaped and isometric. X-ray crystallographic analysis can, of course, be applied only to viruses or virus coat proteins that can be obtained in crystalline form. For larger viruses that have not been crystallized, other techniques, especially electron microscopy remain important for studying virus architecture.

Serological methods, even those employing monoclonal antibodies, have proved to be of limited use in delineating the structures of the small plant viruses, because of ambiguities in interpretation of the results, unless they can be related to a protein structure that is already established by other methods.

The idea of quasiequivalence in the bonding between subunits in icosahedral protein shells with $T > 1$ as put forward by Caspar and Klug in 1962 is still useful but has required modification in the light of later developments. Viruses have evolved at least two methods by which a substantial proportion of the potential nonequivalence in bonding between subunits in different symmetry related environment can be avoided: (i) by having quite different proteins in different symmetry-related positions as in the reoviruses and the comoviruses, and (ii) by developing a protein subunit with two or more domains that can adjust flexibly in different symmetry positions, as in TBSV and SBMV.

The many negatively charged phosphate groups on the nucleic acid within a virus are mutually repelling. To produce a sufficiently stable virus particle these charges, or most of them, need to be neutralized. Structural studies to atomic resolution have revealed three solutions to this problem. (i) In TMV the RNA is closely confined within a helical array of protein subunits. Two of the phosphates associated with each protein subunit are close to arginine residues. However, the electrostatic interactions between protein and RNA are best considered as complementarity between two electrostatic surfaces. Similar arrangements may hold for other rod-shaped viruses, but no detailed data are available for these. (ii) In TBSV and a number of other icosahedral viruses, a flexible basic amino-terminal arm with a histonelike composition projects into the interior of the virus interacting with RNA phosphates. Additional phosphates are neutralized by divalent metals, especially Ca^{2+}. (iii) In TYMV, where there is little interpenetration of RNA and protein, the RNA phosphates are neutralized by polyamines and Ca^{2+} ions.

X-ray analysis has located Ca^{2+} ions in specific locations between protein subunits in several virus shells. All this detailed structural information has gone some distance in explaining the relative stability of different viruses to various agents such as chelating compounds and changes in pH or ionic strength.

The organization of the nucleic acid within icosahedral virus particles has been difficult to study because the nucleic acid makes no contribution to the orientation of virus particles within a crystal. However, indirect methods show that the nucleic acids are quite highly ordered within the virus, most viral RNAs having substantial double-helical structure. In the simplest virus, STNV, neutron diffraction data show that much of the RNA is arranged in an icosahedral network just inside the protein shell. A high-resolution density map of BPMV shows that about 20% of the RNA is arranged inside the protein shell in an icosahedral manner.

For the near future, the most fruitful development in studies on virus architecture will probably be the application of site-directed mutagenesis to change specific amino acids at known sites in a virus such as TMV or TBSV and to study the consequences of such changes by various methods, including X-ray crystallographic analysis.

6

Replication I: Introduction to the Study of Virus Replication

Our understanding of the ways in which plant viruses replicate has increased remarkably over the past few years. This is mainly because new techniques have allowed the complete nucleotide sequences of many plant viral genomes to be established. This in turn has allowed the number, size, and amino acid sequence of putative gene products to be determined. We now have this information for representatives of many plant virus groups and families. Developments in gene manipulation technology have permitted an artificial DNA step to be introduced into the life cycle of RNA viruses, thus infectious genomic RNA transcripts with a uniform nucleotide sequence can be produced *in vitro*. In turn, this has allowed the application of site-directed mutagenesis in experiments to determine functional regions in the noncoding regions of genomic nucleic acid, as well as the functions of gene products and putative gene products. Determination of the functions and properties of gene products has also been greatly assisted by the use of well-established *in vitro* translation systems.

Studies on virus replication *in vivo* using isolated protoplast preparations have provided new information on many aspects of virus replication. The ability to produce transgenic plants in which every cell contains a functional DNA copy of a single viral gene or genome segment is opening up new possibilities for the study of virus replication *in vivo*. In this chapter I will discuss, in general outline, the methods used to establish viral genome structure, expression strategy, and the details of *in vivo* replication. Most attention will be given to viruses with ss positive sense RNA genomes, because 28 of the 35 recognized groups and families of plant viruses have genomes of this type. In Chapters 7 and 8, details concerning the replication of representatives of the best studied virus groups and families are discussed.

I. GENERAL PROPERTIES OF PLANT VIRAL GENOMES

A. Information Content

In theory, the same nucleotide sequence in a viral genome could code for up to 12 polypeptides. There could be an open reading frame (ORF) in each of the three reading frames of both the positive and negative sense strands, giving six polypeptides. If each of these had a leaky termination signal they could give rise to a second read-through polypeptide. However, in nature, there must be severe evolutionary constraints on such multiple use of a nucleotide sequence, because even a single base change could have consequences for several gene products. However, two overlapping genes in different reading frames do occasionally occur, as do genes on both positive and negative sense strands. Read-through proteins are quite common.

The number of genes found in plant viruses ranges from 1 for the satellite virus, STNV, to 12 for some reoviruses. Most of the ss positive sense RNA genomes code for about four to seven proteins. In addition to coding regions for proteins, genomic nucleic acids contain nucleotide sequences with recognition and control functions that are important for virus replication. These control and recognition functions are mainly found in the 5' and 3' noncoding sequences of the ssRNA viruses, but they may also occur internally, even in coding sequences.

B. Economy in the Use of Genomic Nucleic Acids

Viruses make very efficient use of the limited amount of genomic nucleic acids they possess. Eukaryote genomes may have a content of introns that is 10–30 times larger than that of the coding sequences. Like prokaryote cells, most plant viruses lack introns. Plant viruses share with viruses of other host groups several other features that indicate very efficient use of the genomic nucleic acids: (i) Coding sequences are usually very closely packed, with a rather small number of noncoding nucleotides between genes. (ii) Coding regions for two different genes may overlap in different reading frames (e.g., in BNYVV, see Fig. 7.33) or one gene may be contained entirely within another in a different reading frame (e.g., the *Carmovirus* genome, see Fig. 7.15). (iii) Read-through of a "leaky" termination codon may give rise to a second, longer read-through polypeptide that is coterminal at the amino end with the shorter protein. This is quite common among the virus groups with ss positive sense RNA genomes. In a few viruses there may be a second read-through protein giving rise to a set of three proteins coterminal at the amino end. "Transframe" proteins in which the ribosome avoids a stop codon by switching to another reading frame have a result that is similar to a "leaky" termination signal. (iv) A single gene product may have more than one function. For example, the coat protein of AMV has a protective function in the virus particle and a function in the initiation of infection by the viral RNAs. (v) A functional intron has been found in wheat dwarf *Geminivirus* (Schalk *et al.*, 1989) and a similar situation exists in *Digitaria* streak *Geminivirus* (Accotto *et al.*, 1989). Thus mRNA splicing may be a feature common to all geminiviruses infecting members of the Poaceae. This process can increase the diversity of mRNA transcripts available, and therefore the number of gene products. (vi) A functional viral enzyme may use a host-coded protein in combination with a viral-coded polypeptide (e.g., the replicase of TYMV). (vii) Regulatory functions in the nucleotide sequence may overlap with coding sequences (e.g., the signals for subgenomic RNA synthesis in TMV). (viii) In the 5′ and 3′ noncoding sequences of the ssRNA viruses a given sequence of nucleotides may be involved in more than one function. For example, in genomic RNA, the 5′-terminal noncoding sequences may provide a ribosome recognition site, and at the same time contain the complementary sequence for a replicase recognition site in the 3′ region of the minus strand. In members of the *Potexvirus* group the origin of assembly is also at the 5′ end of the genome (Chapter 7, Section II,A,4).

C. The Functions of Viral Gene Products

The known functions of plant viral gene products may be classified as follows:

1. Structural Proteins

Coat proteins of the small viruses; the matrix or core proteins of the reoviruses and those viruses with a lipoprotein membrane; and proteins found within such membranes.

2. Enzymes

Proteases

Proteases coded for by those virus groups in which the whole genome or a segment of the genome is first transcribed into a single polyprotein. The properties of these enzymes have been reviewed by Wellink and van Kammen (1988).

Enzymes Involved in Nucleic Acid Synthesis

It is now highly probable that all plant viruses, except satellite viruses, code for one or more proteins that have an enzymatic function in nucleic acid synthesis, either genomic nucleic acid or mRNAs or both. The general term for these enzymes is *polymerase*. There is some inconsistency in the literature in relation to the terms used for different polymerases. I will use the various terms with the following meanings. Polymerases that transcribe RNA from an RNA template have the general name *RNA-dependent RNA polymerase*. If such an enzyme makes copies of an entire RNA genome it is called a *replicase*. Replicase enzymes may also be involved in the synthesis of subgenomic mRNAs. If an RNA-dependent RNA polymerase is found as a functional part of the virus particle as in the Rhabdoviridae and Reoviridae it is called a *transcriptase*. The enzyme coded by caulimoviruses, which copies a full-length viral RNA into genomic DNA, is called an *RNA-dependent DNA polymerase* or *reverse transcriptase*.

Two features have made the RNA-dependent RNA polymerase enzymes difficult to study. First, they are usually associated with membrane structures in the cell, and on isolation the enzymes often become unstable. They are, therefore, difficult to purify sufficiently for positive identification of any viral-coded polypeptides. Second, tissues of healthy plants may contain, in the soluble fraction of the cell, low amounts of an enzyme with similar activities. The amounts of such enzyme activity may be stimulated by virus infection.

Conserved amino acid sequence motifs have now been identified in the RNA polymerase genes of many ssRNA plant viruses. Indeed the presence of such a motif is used to identify putative polymerase genes in newly sequenced viral genomes. Recently, the discovery of another series of amino acid motifs from a wide range of organisms has led to the definition of a large group of proteins, the nucleic acid helicases (Gorbalenya *et al.*, 1989b). Their functions are believed to include nucleic acid unwinding and acting in the replication, recombination, repair, and expression of DNA and RNA genomes. These helicase sequence motifs have been identified in a wide range of ssRNA plant viral genomes (Habili and Symons, 1989) and can be assumed to function in plant viral RNA replication. In some viruses the polymerase and helicase motifs appear to be in the same protein. In others they are separate. They are discussed in relation to virus classification and evolution in Chapter 17, Section V.

Two kinds of RNA structures have been isolated from viral RNA synthesizing systems. One, known as *replicative form* (RF), is a fully base-paired ds structure, whose role is not certain. For example, it may represent RNA molelcules that have ceased replicating. The other, called *replicative intermediate* (RI), is only partly ds and contains several ss tails (nascent product strands) (Fig. 6.1). This structure is closely related to the one actually replicating the viral RNA. It is thought that the RI as isolated may be derived from a structure like that in Fig. 6.1c by annealing of parts of the progeny strands to the template. Cytological evidence suggests that the RI of TYMV is essentially single-stranded (Chapter 7, Section IV,A,5).

3. Virus Movement and Transmission

For several groups of viruses, a specific viral-coded protein has been identified as an essential requirement for cell-to-cell movement and for systemic movement within the host plant. Other gene products have been identified as essential for successful transmission by invertebrate vectors. Viral gene products may also be involved in transmission by fungi.

4. Nonenzymatic Role in RNA Synthesis

The 5′ VPg protein found in some virus groups is thought to act as a primer in RNA synthesis.

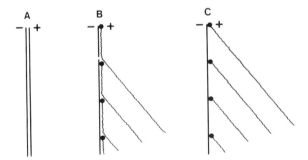

Figure 6.1 Forms of association between positive and negative sense strand viral RNA. (A) Replicative form (RF). A base-paired structure with full-length positive and negative sense strands. (B) Replicative intermediate (RI). A partially base-paired structure with polymerase molecules (●) and ss tails of nascent progeny positive sense strands. (C) Probable true state of the RI *in vivo*. The progeny positive sense strands and the template negative sense strand are almost entirely ss.

5. Coat Protein of AMV

The AMV coat protein and the corresponding protein in ilarviruses have an essential role in the initiation of infection by the viral RNA, possibly by priming negative strand synthesis.

6. Protein Recognizing Host Cells

Bacterial cells that can be infected by viruses usually exist in a liquid medium. Cells in an animal body usually have no cytoplasmic connections between them. Thus viruses infecting bacterial or animal cells normally release virus from infected cells to infect further cells of an appropriate type. To identify markers on the surface of appropriate host cells, most viruses infecting bacteria and animals have recognition proteins on their surface. Such recognition proteins are lacking for most plant viruses, which is probably related to two properties of the host organism. A recognition protein would be of no use to a plant virus for recognizing the surface of a suitable host plant because the virus cannot penetrate the surface layers unaided. Furthermore, a plant virus, once it infects a host cell, say in a leaf, can move from cell to cell via plasmodesmata and vascular tissue throughout almost the entire plant. It therefore has no need for a special recognition protein inside the host organism. Possible exceptions among the plant viruses are those that replicate in their invertebrate vectors.

7. Summary

On present knowledge, only two functions appear to be common to all plant viruses except satellite viruses. These are the coat proteins and the enzymes involved in genome replication. Increasing numbers of viruses are being shown to require a protein with a cell-to-cell movement function. This may turn out to be another universal requirement for viruses that can move freely through various plant tissues, replacing, in a sense, the surface recognition proteins of bacterial and animal viruses noted earlier.

There is no doubt that other functions, important for replication, will be discovered for the viral genes that have already been identified structurally. This will be an important area for study in the future. For example, some viruses alter the balance of host-cell metabolism in a manner advantageous to virus replication. Some cause morphological alterations within the cell, or cell organelles, that are closely connected with virus replication. These aspects are discussed in more detail in Chapters 7 and 8.

II. HOST FUNCTIONS USED BY PLANT VIRUSES

Like all other viruses, plant viruses are intimately dependent on the activities of the host cell for many aspects of replication.

A. Components for Virus Synthesis

Viruses use amino acids and nucleotides synthesized by host-cell metabolism to build viral proteins and nucleic acids. Certain other more specialized components found in some viruses, for example, polyamines, are also synthesized by the host.

B. Energy

The energy required for the polymerization involved in viral protein and RNA synthesis is provided by the host cell, mainly in the form of nucleoside triphosphates.

C. Protein Synthesis

Viruses use the ribosomes, tRNAs, and associated enzymes and factors of the host cell's protein-synthesizing system for the synthesis of viral proteins using viral mRNAs. All plant viruses appear to use the 80 S cytoplasmic ribosome system. There is no authenticated example of the chloroplast or mitochondrial ribosomes being used. Most viruses also depend on host enzymes for any posttranslational modification of their proteins, for example, glycosylation.

D. Nucleic Acid Synthesis

Almost all viruses code for an enzyme or enzymes involved in the synthesis of their nucleic acids, but they may not contribute all the polypeptides involved. For example, in the first phase of the replication of caulimoviruses, the viral DNA enters the host-cell nucleus and is transcribed into RNA form by the host's DNA-dependent RNA polymerase II. In TYMV, and probably some other viruses, the RNA-dependent RNA polymerase is composed of two subunits, one coded for by the virus and one contributed by the host.

E. Structural Components of the Cell

Structural components of the cell, particularly membranes, are involved in virus replication. For example, viral nucleic acid synthesis usually involves a membrane-bound complex. Viral protein synthesis involves endoplasmic reticulum. In *Tymovirus*-infected cells, viral RNA synthesis takes place in vesicles that are invaginations of the two chloroplast membranes.

F. Movement within the Plant

There is no doubt that many viruses code for a specific protein that is necessary for cell-to-cell movement. Nevertheless, since they lack any form of independent motility, they must depend on host mechanisms for movement within infected cells and for long-distance transport in the phloem.

III. GENERALIZED OUTLINE FOR THE REPLICATION OF A SMALL ssRNA VIRUS

As an introduction to the subject I shall consider a brief outline of the probable main stages in the replication of a small positive sense strand RNA virus. There are many variations in detail in these stages.

1. The virus particle enters the cell. At the time of entry or shortly afterward the RNA is released from the protein coat.
2. The infecting RNA becomes associated with host ribosomes and is translated to give a viral RNA polymerase, and perhaps other virus-specific proteins.
3. The RNA polymerase uses the parental positive sense strand to transcribe negative sense strand copies.
4. These are then used by the polymerase to produce progeny strands via a replicative intermediate form of the RNA or a structure related to it (see Fig. 6.1). These steps may take place at some specific site in the cell that is characteristic of the virus or group of viruses.
5. Partial copies of the RNA may be made as well as full-length molecules. A monocistronic coat protein messenger is often made and is used to produce large quantities of this protein.
6. Coat protein subunits and RNA strands are assembled to give new virus particles, which accumulate within the cell, usually in the cytoplasm.

In the second edition of this book I enumerated nine basic questions about this replication process that at that stage were essentially unanswered. In the intervening years we have obtained significant answers to four of the questions, which related essentially to the gene content of the viruses and their genome strategies. The remaining questions still require effective answers of general application:

1. How, when, and where does the infecting particle become uncoated?
2. How does the parental RNA establish and organize a specific replication site within the cell?
3. Is the synthesis of individual virus-specific proteins a controlled process with respect to time of synthesis and amount synthesized? If so, what are the control mechanisms? We are beginning to obtain an understanding of some of these mechanisms.
4. How and where are progeny virus particles assembled from the RNA and coat protein? We now have some knowledge of this process for a few viruses.
5. How is the overall quantity of virus produced in a cell controlled?

IV. THE STRATEGIES OF PLANT VIRAL GENOMES

Genome strategy is a useful but rather vague term (see Wolstenholme and O'Connor, 1971), which could be extended to include almost every aspect of virus structure, replication, and ecology. I take the term to include (i) the structure of the genome (DNA or RNA; ds or ss; if ss whether it is positive or negative sense); (ii) the question as to whether the nucleic acid alone is infectious; (iii) general aspects of the enzymology by which the genome is replicated (e.g., the presence of nucleic acid polymerases in the virus particle, and any other enzymes concerned in nucleic acid metabolism that are coded for by the virus); and (iv) the overall pattern whereby the information in the genome is transcribed and translated into viral proteins (not the detailed molecular biology of this process).

There are five families and groups that have genome structures and strategies that are unique among the plant viruses and that are now well understood. These are the plant members of the Reoviridae and Rhabdoviridae families, the *Caulimovirus* and *Geminivirus* groups, and TSWV. These strategies are discussed in Chapter 8. In this section I will outline the kinds of strategy that have evolved among the groups of plant viruses with positive sense RNA genomes. Several selection pressures have probably been involved in this evolution.

A. The Eukaryotic Protein-Synthesizing System

In eukaryote protein-synthesizing systems, in most circumstances, mRNAs contain a single ORF. Thus the 80 S ribosomes are adapted to translate only the ORF immediately downstream from the 5′ region of an mRNA. ORFs beyond this point in a viral mRNA normally remain untranslated. Much of the variation in the way gene products are translated from viral RNA genomes appears to have evolved to meet this constraint.

On current knowledge there are five main strategies by which RNA viral genomes ensure that all their genes are accessible to the eukaryotic protein-synthesizing system.

(i) *Subgenomic RNAs*   The synthesis of one or more subgenomic RNAs enables the 5′ ORF on each such RNA to be translated.

(ii) *Polyproteins*   Here the coding capacity of the RNA for more than one protein, and sometimes for the whole genome, is translated from a single ORF. The polyprotein is then cleaved at specific sites by a viral-coded protease, or proteases, to give the final gene products.

(iii) *Multipartite genomes*   The 5′ gene on each RNA segment can be translated.

(iv) *Read-through proteins*   The termination codon of the 5′ gene may be "leaky" and allow a proportion of ribosomes to carry on translation to another stop codon downstream from the first, giving rise to a second longer functional polypeptide.

(v) *Transframe proteins*   Another mechanism by which two proteins may commence at the same 5′ AUG is by a switch of reading frame near the termination codon of the 5′ ORF to give a second longer "transframe" protein.

Another possible mechanism of this sort is premature termination of translation in the absence of a termination codon. In two plant virus groups it has been suggested that an internal ORF may be translated by internal initiation even though the system is a eukaryotic one. These are the *Luteovirus* and *Comovirus* groups, but the idea has not been substantiated. In TYMV, two polypeptides initiate at two closely spaced 5′ AUGs in different reading frames. The five main strategies are illustrated in Fig. 6.2.

A few plant virus groups and families have made use of only one of these devices to develop a successful genome strategy, but most use two or three different strategies in combination. The strategies used by the virus groups with well-studied representatives are summarized in Table 7.1.

All of these various combinations of translation strategies must have survival value.

On the other hand, it is difficult to visualize how the polyprotein strategy of the potyviruses can be efficient. The coat protein gene is at the 3′ end of the genome (see Fig. 7.1). Thus for every molecule of the 20,000 MW coat protein produced by TEV, a molecule of all the other gene products has to be made, totaling about 320,000 MW. Since about 2000 molecules of coat protein are needed for each virus particle, this appears to be a very inefficient procedure. Indeed large quantities of

1. Subgenomic RNAs

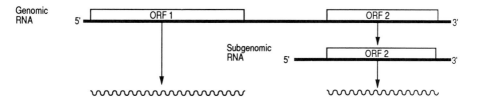

2. Multipartite genome

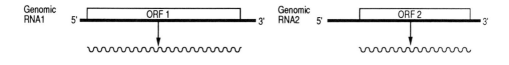

3. Polyprotein

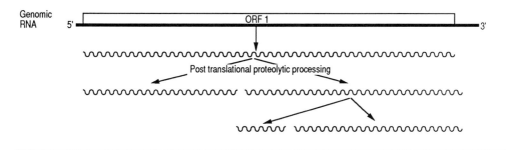

Post translational proteolytic processing

4. Read through protein

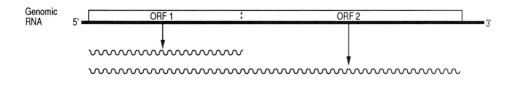

5. Translational frameshift

The ribosome bypasses a stop codon in Frame 0 by switching back one nucleotide to Frame -1 at a UUUAG sequence before continuing to read triplets in Frame -1 to give a fusion or transframe protein.

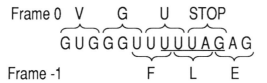

Figure 6.2 Five strategies used by plant viruses to allow protein synthesis in a eukaryotic system from a positive sense RNA genome containing more than one gene.

several gene products, apparently in a nonfunctional state, accumulate in infected cells (Chapter 7, Section I,A). Nevertheless, the potyviruses are a very successful group. There are many member viruses, and they infect a wide range of host plants. Other viruses using polyprotein processing have additional devices that can avoid this problem. Comoviruses have their two coat proteins on a separate genome segment. There does not appear to be any massive accumulation of noncoat gene products in cells infected with these viruses.

With respect to the problem of eukaryote mRNA processing there is one potential adaptation that no plant viruses appear to have exploited. Chloroplasts contain a prokaryotic protein-synthesizing system but there is no authenicated example of viral protein synthesis taking place in chloroplasts. This suggests that there may be some fundamental barrier to viral mRNAs entering chloroplasts and becoming functional there. However, TMV RNA is know to enter chloroplasts *in vivo* (Schoelz and Zaitlin, 1989).

B. Other Selection Pressures

Other selection pressures have almost certainly influenced the development of the strategies illustrated in Fig. 6.2. For example, all these small RNA viruses require much more of the coat protein than any other gene product. Most of them use a coat protein subgenomic mRNA originating from the 3′ region of a genomic RNA. These small monocistronic RNAs are highly efficient messengers for coat protein. Furthermore, transcription of the mRNA from a minus strand template is an additional mechanism for increasing the production of coat protein relative to other gene products.

Subgenomic RNAs might facilitate production of gene products at different sites or at different stages during virus replication. Overlapping genes in different reading frames, read-through, and transframe proteins give more gene products from a small genome.

In theory, the viruses with multipartite genomes might be at a disadvantage in that the probability of establishing successful infection may be reduced, because two or more particles must cooperate—presumably by infecting the same cell or at least by infecting cells that are close neighbors having protoplasmic connections. In fact, this does not appear to be a difficulty. All the groups of viruses with bi- and tripartite genomes can be transmitted, most of them very readily, by mechanical means. Most have insect vectors as well. Two groups with tripartite genomes (*Ilarvirus* and *Hordeivirus*) have no known insect vectors. On the other hand, members of the *Luteovirus* group with a monopartite genome are confined to the phloem and are transmitted only by aphids. Thus there is no indication that bi- and tripartite genomes are at a disadvantage with respect to transmission in nature.

Apart from allowing the 5′ gene on each genome segment to be translated effectively in a eukaryote system, multipartite genomes may have some other selective advantages, such as: (i) Genetic flexibility. In viruses with several pieces of RNA in one particle, genetic reassortment can take place during virus replication. In addition to reassortment at this stage, the existence of the RNA in several separate particles allows for the possibility of selection and reassortment at other stages as well, for example, during transmission, entry into the cell, and movement through the plant. However, genome segments from different strains of the same virus are not always compatible (see Fulton, 1980). (ii) To separate early and late gene functions. (iii) To separate functions that are required to be carried out in different parts of the host cell. (iv) To separate insect and plant functions. For viruses replicating in both plant and insect hosts, a function or functions required only in one kind of host may be segregated on a particular genome segment (see Fig. 14.15). (v) To facilitate encapsidation of the genome by keeping particles, especially icosahedral ones, relatively small. (vi) Multiplicity reactivation. Reanney (1987) suggested that genome subdivision is an adaptive response to the high error rates in RNA virus replication, resulting mainly from the lack of error-correcting mechanisms.

V. METHODS FOR DETERMINING GENOME STRATEGY

The actual sequence of events that has led to the understanding of viral genome strategies has varied widely for different viruses. This is because of the rather haphazard manner in which a particular branch of science tends to develop. To take two examples: all four of the definite TMV gene products, three of which are nonstructural, were identified before the nucleotide sequence of the genome of that virus had been established. At the other extreme, the full nucleotide sequence of BYDV is known, while only two of seven potential gene products have been characterized, and these are both proteins found in the virus.

I shall attempt to present a summary of the various methods involved in a more logical sequence than that in which they have actually been applied to many viruses. First, we must understand the structure of the genome, in particular, the number of genome pieces, the arrangement of the ORFs in the genome, the deduced amino sequences for those ORFs, and the positions of any likely regulatory and recognition nucleotide sequences. As a second stage we must define the ORFs that are actually functional by both *in vitro* and *in vivo* studies. We need to recognize the gene functions of the virus, either by direct studies on any viral proteins that can be isolated from infected cells, or by classic genetic studies that may reveal various biological activities. Finally, we need to match these viral gene activities with the functional ORFs. It is here that the modern techniques of reverse genetics can play a very important role. In reverse genetics an alteration (base change, insertion or deletion of bases) is made at any preselected position in the genome. The consequences of the change are then studied with respect to its effect on the gene product and the product's biological function, a function that may not have been previously recognized by traditional methods.

A. Structure of the Genome

There are several steps in determining the structure of a viral genome. The starting material is almost always nucleic acid isolated from purified virus preparations.

1. Kind of Nucleic Acid

Whether the nucleic acid is ss or ds, DNA or RNA, or linear or circular can be established by the various chemical, physical, and enzymatic procedures outlined in Chapter 3.

2. Number of Genome Pieces

When virus particles housing separate pieces of a multipartite genome differ sufficiently in size or density, they may be fractionated on density gradients and the nucleic acids isolated from the fractions. Alternatively, nucleic acids of differing size may be separated on density gradients or by gel electrophoresis. When the two pieces of a bipartite genome are of very similar size, as with some geminiviruses, the existence of two distinct parts of the genome may be inferred from hybridization experiments estimating sequence complexity. However, formal proof that the genome is in two pieces of nucleic acid can best be obtained by cloning the full length of each piece separately and demonstrating that both are required for infectivity (e.g., the geminiviruses, Hamilton *et al.,* 1983).

3. Terminal Structures

Chemical and enzymatic procedures can be used to establish the nature of any structures at the 5′ and 3′ termini of a linear nucleic acid.

4. Nucleotide Sequence

A knowledge of the full nucleotide sequence of a viral genome is essential for understanding genome structure and strategy. The methods used are detailed in the relevant publications referred to in Chapters 7 and 8.

5. Open Reading Frames

With the help of an appropriate computer program, the nucleotide sequence is searched for ORFs in each of the three reading frames of both positive and negative sense strands. All ORFs are tabulated, as is illustrated in Fig. 6.3 for a *Tymovirus.*

As shown in Fig. 6.3, a large number of ORFs may be revealed. Those ORFs that could code for polypeptides of MW less than 7000–10,000, or that would give rise to proteins of highly improbable amino acid composition, are usually not given further consideration. However, sequence similarity between small ORFs in several

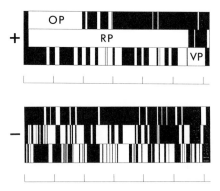

Figure 6.3 Diagram of the three triplet codon phases of the plus and minus strand RNAs of ononis yellow mosaic *Tymovirus* genomic RNA. White boxes indicate all ORFs that begin with an AUG and terminate with UGA, UAG, or UAA. There are three significant ORFs labeled OP (overlapping protein), RP (replicase protein), and VP (coat protein). (From Ding *et al.,* 1988.)

viruses may indicate that they are functional (e.g., the 7000 MW ORFs of potex-viruses and carlaviruses).

ORFs of significant size representing possible proteins of 100 amino acids or more occur in the negative sense strands of CPMV RNA2 (Lomonossoff and Shanks, 1983), TMV (Goelet *et al.*, 1982), AMV RNAs 1 and 2 (Cornelissen *et al.*, 1983a,b), papaya mosaic *Potexvirus* (AbouHaidar, 1988), and one strain of WCMV(M), but not PVX or the 0 strain of WCMV (Harbison *et al.*, 1988) or clover yellow mosaic *Potexvirus* (AbouHaidar and Lai, 1989). There is no evidence that any of these have functional significance. However, there is no reason, in principle, why functional ORFs should not occur in the negative sense strand. Such ORFs are found in the geminiviruses and TSWV (Chapter 8, Sections II and V).

6. Amino Acid Sequence

The amino acid sequence and MW of the potential polypeptide for each ORF of interest can then be determined from the nucleotide sequence and the genetic code.

7. Regulatory Signals

Various parts of the genome and particularly the 5′ and 3′ noncoding sequences are searched for relevant regulatory and recognition signals as will be discussed in Section VI.

8. mRNAs

The genomes of DNA viruses must be transcribed into one or more mRNAs. These must be identified in nucleic acids isolated from infected tissue and matched for sequence with the genomic DNA (e.g., CaMV; see Chapter 8, Section I). Many plant viruses with ss positive sense RNA genomes have some ORFs that are trans-lated only from a subgenomic RNA. These too must be identified to establish the strategy of the genome.

In RNA preparations isolated from virus-infected tissue or from purified virus preparations, subgenomic RNAs may be present that can be translated *in vitro* to give polypeptides with a range of sizes that do not correspond to ORFs in the genomic RNA. For example, Higgins *et al.* (1978) detected RNAs of eight discrete lengths in RNA isolated from preparations of TYMV. Mellema *et al.* (1979) were able to associate five of these with particular polypeptides synthesized in the re-ticulocyte system. The full-length translation products of these RNAs and the gen-omic RNA overlapped with one another and shared a common amino terminus. Mellema *et al.* concluded that these RNAs share a common translation initiation site near their 5′ termini.

In vitro translation of TMV RNAs isolated from TMV-infected tissue gave rise to a series of products with MWs of 45,000, 55,000, 80,000, and 95,000 (Goelet and Karn, 1982). These formed a nested set of proteins sharing C-terminal se-quences and having staggered N-terminal amino acids. Goelet and Karn suggested that viral RNA may be transcribed into a set of incomplete negative sense strands that are in turn transcribed into a set of incomplete mRNAs. Since no function in viral replication has yet been ascribed to such N-terminal or C-terminal families of proteins, they will not be further discussed.

Thus it may be a difficult task to establish whether a viral RNA of subgenomic size is a functional mRNA or merely a partly degraded or partly synthesized piece of genomic RNA. One criterion is to first isolate an active polyribosome fraction and then isolate the presumed mRNAs. The RNAs may then be fractionated by gel electrophoresis, and those with viral-specific sequences identified by the use of appropriate hybridization probes.

Not infrequently genuine viral subgenomic mRNAs are encapsidated along with the genomic RNAs. These can then be isolated from purified virus preparations and characterized. When the sequence of the genomic nucleic acid is known there are two techniques available to locate precisely the 5' terminus of a presumed subgenomic RNA. In the S1 nuclease protection procedure, the mRNA is hybridized with a complementary DNA sequence that covers the 5' region of the subgenomic RNA. The ss regions of the hybridized molecule are removed with S1 nuclease. The DNA that has been protected by the mRNA is then sequenced. In the second method, primer extension, a suitable ss primer molecule is annealed to the mRNA. Reverse transcriptase is then used to extend the primer as far as the 5' terminus of the mRNA. The DNA produced is then sequenced. Carrington and Morris (1986) used both these procedures to locate the 5' termini of the two subgenomic RNAs of CarMV (see Fig. 7.15). A sequence determination that reveals a single termination nucleotide rather than several is a good indication that the subgenomic RNA under study is a single distinct species and not a set of heterogeneous molecules (e.g., Sulzinski *et al.*, 1985).

B. Defining Functional ORFs

Some of the ORFs revealed by the nucleotide sequence will code for proteins *in vivo*, whereas others may not. The functional ORFs can be unequivocally identified only by *in vitro* translation studies using viral mRNAs and by finding the relevant protein in infected cells.

1. *In Vitro* Translation of mRNAs

The monocistronic RNA of STNV was translated with fidelity in the prokaryotic *in vitro* system derived from *Escherichia coli* (Lundquist *et al.*, 1972). However, results with other plant viral RNAs were difficult to interpret. Three systems derived from eukaryotic sources have proved useful with plant viral RNAs. The general outline of the procedure for these three systems follows.

1. The RNA or RNAs of interest are purified to high degree, using density gradient centrifugation and/or polyacrylamide gel electrophoresis. For viruses whose particles become swollen under the conditions of the *in vitro* protein-synthesizing system, the RNA associated with the virus may act effectively as mRNA (e.g., Brisco *et al.*, 1986a).
2. The RNAs are then added to the protein-synthesizing system in the presence of amino acids, one or more of which is radioactively labeled.
3. After the reaction is terminated, the polypeptide products are fractionated by electrophoresis on SDS–PAGE, together with markers of known size.
4. The products are located on the gels by means of the incorporated radioactivity.

The three systems are:

1. *Toad oocytes*   These, strictly do not constitute an *in vitro* system. Intact oocytes of *Xenopus* or *Bufo* are injected with the viral mRNA and incubated in a labeled medium.

2. *The rabbit reticulocyte system*   The cells from anemic rabbit blood are lysed in water and centrifuged at 12,000*g* for 10 minutes. The supernatant fluid is then used. This is a useful system because of the virtual absence of RNase activity. Figure 6.4 illustrates the use of this system.

3. *The wheat embryo system*   In this system, the viral RNA is added in the presence of an appropriate label to a supernatant fraction from extracted wheat embryos from which the mitochondria have been removed.

Jagus (1987a,b) gives technical details of these methods. Many papers in which the methods have been used are referred to in Chapters 7 and 8. Not infrequently, when purified viral genomic RNAs are translated in cell-free systems, several viral-specific polypeptides may be produced in minor amounts in addition to those expected from the ORFs in the genomic RNA. It is unlikely that such polypeptides have any functional role *in vivo*. They are probably formed *in vitro* by one or more of the following mechanisms: (i) endonuclease cleavage of genomic RNA at specific sites; (ii) proteolytic cleavage of longer products during the incubation; (iii) misreading of sense codons as termination signals; and (iv) secondary structure of the genomic RNA, formed under the *in vitro* conditions, which prevents translocation of a proportion of ribosomes along the RNA. For example, multiple polypeptides of MW below 11,000 are synthesized in the rabbit reticulocyte system with TMV RNA (Wilson and Glover, 1983).

When the I_2 subgenomic RNA of TMV is translated *in vitro*, a family of polypeptides besides the 30,000 protein is produced, but only the 30,000 protein

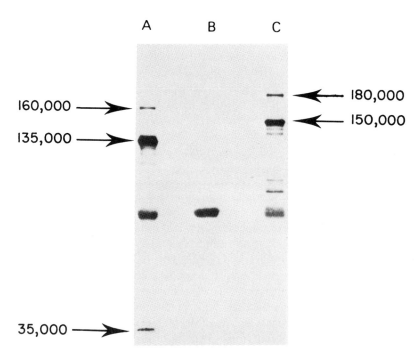

Figure 6.4   Translation of plant viral RNAs in the rabbit reticulocyte system. The polypeptide products were fractionated by electrophoresis in a polyacrylamide gel and located by radioautography of the gel. (A) Products using TMV RNA as message. (B) Control with no added RNA. (C) Products using TYMV RNA. The unmarked smaller polypeptides may be incomplete transcripts of the viral message, or due to endogenous mRNA. Note that no protein the size of viral coat protein (MWs ≈17,500 or 20,000) is produced by either viral RNA. (From Briand, 1978.)

could be detected *in vivo* (Ooshika *et al.*, 1984). Such polypeptides will not be discussed further in relation to virus replication.

What criteria can be used to "optimize" conditions for *in vitro* translation? Measurement of total radioactivity incorporated is not particularly informative. Measurements of radioactivity in individual polypeptides separated on poly-acrylamide gels are much more useful. One might aim for conditions producing (i) the greatest number of polypeptides, (ii) the fewest, (iii) the longest, or (iv) the most of a particular known gene product. It thus becomes apparent that to obtain defini-tive mapping of the genome from studies on the polypeptides produced *in vitro*, we must also know what polypeptides are actually synthesized *in vivo* by the virus.

2. Methods for Identifying ORFs That Are Functional *in Vivo*

Viral-coded proteins other than those found in virus particles may be difficult to detect *in vivo* especially if they occur in very low amounts, and only transiently during a particular phase of the virus replication cycle. However, a battery of methods is now available for detecting viral-coded proteins *in vivo* and matching these with the ORFs in a sequenced viral genome. In particular, the nucleotide sequence information gives a precise estimate of the size and amino acid composi-tion of the expected protein, and knowledge of the expected amino acid sequence can be used to identify the *in vivo* product either from a partial amino acid sequence of that product or by reaction with antibodies raised against a synthetic polypeptide that matches part of the expected amino acid sequence.

Proteins Found in the Virus

Coat proteins are readily allocated to a particular ORF based on several criteria: (i) amino acid composition compared with that calculated for the ORF; (ii) amino acid sequence of part or all of the coat protein; (iii) serological reaction of an *in vitro* translation product with an antiserum raised against the virus; and (iv) for a few viruses such as TMV, assembly of an *in vitro* translation product into virus particles when mixed with authentic coat protein. The rhabdoviruses may be exceptional in that five of the six gene products, corresponding to five of the ORFs in the genome, are found in purified virus preparations (Chapter 8, Section IV).

Direct Isolation from Infected Tissue or Protoplasts

Healthy and virus-infected leaves or protoplasts are labeled with one or more radioactive amino acids. Cell extracts are fractionated by appropriate procedures and proteins separated by gel electrophoresis. Protein bands appearing in the sam-ples from infected cells and not from healthy cells may be identified with the expected product of a particular ORF by comparing its mobility and its pattern of tryptic peptides with that of an *in vitro* translation product of the ORF (e.g., Bujarski *et al.*, 1982). With appropriate *in vivo* labeling, partial amino acid se-quencing of the isolated protein may allow the precise location of its coding se-quence in the genome to be established (e.g., Wellink *et al.*, 1986). In potyvirus infections, large amounts of several virus-coded nonstructural proteins accumulate in infected cells, facilitating the allocation of each protein to its position in the genome.

Serological Reactions

Antisera provide a powerful set of methods for recognizing viral-coded proteins produced *in vivo* and identifying them with the appropriate ORF in the genome.

1. *Antisera against synthetic peptides* A synthetic peptide can be prepared corresponding to part of the amino acid sequence predicted from an ORF. An antiserum is raised against the synthetic peptide and used to search for the expected protein in extracts of healthy or infected tissue or protoplasts. For example, Kibertis *et al.* (1983) synthesized a peptide corresponding to the C-terminal 11 amino acids of a 30,000 MW ORF in the TMV genome. They were able to show that a polypeptide corresponding to this ORF was synthesized in infected protoplasts.

2. *Antisera against* in vitro *translation products* If an mRNA is available that is translated *in vitro* to give a polypeptide product clearly identified with a particular ORF or genome segment, antisera raised against the *in vitro* product can be used to search for the same protein in extracts of infected cells or tissue. For example, such antisera have been used to identify nonstructural proteins coded for by AMV RNAs in tobacco leaves. The antisera were used in conjunction with a very sensitive immunoblotting procedure.

3. *Antisera against recombinant proteins* Antibodies can be raised against recombinant protein derived *in vitro* from RNA transcripts of a cloned gene and used to search for the corresponding protein in extracts of infected plant. For example, this procedure was used to establish the position of the nucleocapsid protein gene in the rhabdovirus SYNV (Zuidema *et al.*, 1987).

4. *Immunogold labeling* Antibodies produced against a synthetic peptide corresponding to part of a particular ORF in the genomic nucleic acid and labeled with gold can be used to probe infected cells for the presence of the putative gene product. This was done for the 30,000 MW gene product of TMV (Tomenius *et al.*, 1987). As an alternative procedure, antibodies can be produced against the protein product of a plant viral gene expressed in *E. coli*. This was combined with immunogold labeling to demonstrate that the product of CaMV gene 1 is expressed in infected leaves (Linstead *et al.*, 1988).

Comparison with Genes Known to Be Functional in Other Viruses

Size, location in the genome, and nucleotide sequence similarities with known functional genes may give a strong indication that a particular ORF codes for a functional protein *in vivo*. For example, no nonstructural viral proteins have yet been isolated from BYDV-infected tissue. However, the 60,000 MW ORF revealed by the nucleotide sequence of the BYDV genome would give rise to a protein containing an amino acid sequence that is highly conserved among known and putative viral RNA-dependent RNA polymerases (Miller *et al.*, 1988b) (Section V,D,2). The 60,000 MW ORF in BYDV is therefore very likely to be a functional gene.

Presence of a Well-Characterized Subgenomic RNA

Occasionally a viral subgenomic RNA has been well characterized but no *in vivo* protein product has been detected. Thus, the I_1 subgenomic RNA of TMV is very probably a functional mRNA because (i) it is located in the polyribosome fraction from infected cells and (ii) it has a precisely defined 5' terminus (Sulzinski *et al.*, 1985). Thus it is reasonable to suppose that the 5' ORF of this subgenomic RNA is functional *in vivo* (see Fig. 7.5).

Presence of Appropriate Regulatory Signals in the RNA

AUG triplets that are used to initiate protein synthesis may have a characteristic sequence of nucleotides nearby (Section VI,A,2). Upstream of the AUG triplet there may be identifiable ribosome recognition signals. Presence of these sequences would indicate that the ORF is functional.

Codon Usage

Frequency of codon usage has sometimes been used to indicate whether an ORF revealed in a genomic nucleotide sequence is likely to produce a functional protein (e.g., Morch *et al.*, 1988).

Reoviruses

The reoviruses are a special case with respect to establishing functional ORFs. Each ds genome segment is transcribed *in vitro* to give an mRNA that gives a single protein product (Nuss and Peterson, 1981). On this basis it was reasonable to assume that each genome piece has a single functional ORF.

C. Recognizing Activities of Viral Genes

Before information on the sequence of nucleotides in viral genomes became available and before the advent of *in vitro* translation systems, there were two ways of recognizing the activities of viral genes—the identification of proteins in the virus particle and classic virus genetics.

1. Gene Products in the Virus

Fraenkel-Conrat and Singer (1957) reconstituted the RNA of one strain of TMV in the protein of another strain that was recognizably different. The progeny virus produced *in vivo* by this *in vitro* "hybrid" had the coat protein corresponding to the strain that provided the RNA. Since this classic experiment, it has been universally assumed that viral genomes code for coat proteins. Likewise it has usually been assumed that other proteins found as part of the virus particle are also virus coded, for example, those found in reoviruses and rhabdoviruses.

2. Classic Viral Genetics

Two kinds of classic genetical study have identified many biological activities of viral genomes, and both of these procedures are still useful in appropriate circumstances.

Allocation of Functions in Multiparticle Viruses

The discovery of viruses with the genome divided between two or three particles opened up the possibility of locating specific functions on particular RNA species. The requirements for and stages in this kind of analysis are:

1. Purification of the virus.
2. Fractionation of the genome components, either as nucleoprotein particles (on density gradients of sucrose or cesium salts) or as isolated biologically active RNA species (usually by electrophoresis on polyacrylamide gels).
3. Definition of the set of RNA molecules that constitute the minimum viral genome.
4. Identification and isolation of natural strains or artificial mutants differing in some defined biological or physical properties, which will provide suitable

experimental markers. For example, Dawson (1978a) isolated a set of *ts* mutants of CMV. One group of mutations mapped on RNA3 and the rest on RNA1.

5. *In vitro* substitution of components from different strains or mutants in various combinations. These are inoculated to appropriate host plant species. The relevant biological or physical properties of the various combinations are determined. A particular property may then be allocated to a particular genome segment or segments.

6. Back-mixing experiments. In such experiments the parental genome pieces are isolated from the artificial hybrids, mixed in the original combinations, and tested for appropriate physical or biological properties. Such tests are necessary to show that the RNAs of the hybrids retain their identity during replication.

7. Supplementation tests. These tests provide an alternative procedure to *in vitro* reassortment. Individual wild-type genome segments are added to a defective (mutant) inoculum. Restoration of the wild-type character in a particular mixture will indicate which segment controls the character (e.g., Dawson, 1978a). Transgenic plants expressing a single genome segment can also be used in supplementation tests.

8. Mixing of mutants. Unfractionated preparations of two different mutants may be mixed and tested. If the wild-type property is restored it can be assumed that the two mutations are on different pieces of RNA.

Supplementation tests and mixing of mutants do not require purification and fractionation of the mutant viruses. They can provide independent confirmation of results obtained by *in vitro* substitution experiments (de Jager, 1976).

Various factors may complicate the analysis of reassortment experiments:

1. A particular property may be determined by more than one gene, located on the same or on separate pieces of RNA.

2. Some genes may be pleiotropic (i.e., have more than one effect).

3. If certain parts of the RNA are used to produce two proteins with different functions (e.g., by the read-through mechanism) then a single base change might induce changes in the two different functions.

4. Some amino acid replacements might be "silent" with respect to one property of the protein but not another.

However, by using these procedures, many activities of viral genes have been attributed to one or more of the genome segments in a multipartite virus. Local or systemic symptoms in particular hosts, host range, and proteins found in the virus particle are activities that have commonly been studied. It must be recognized that these reassortment experiments have two limitations: (i) where more than one gene is present on the RNA or DNA segment they cannot allocate an activity to a particular gene, and (ii) except for structural proteins, they do not prove that the gene product is responsible for the activity. In principle, the activity could be due to some direct effect of the nucleic acid itself.

Virus Mutants

The study of naturally occurring or artificially induced mutants of a virus has allowed various virus activities to be identified. Again, many of the activities involve biological properties of the virus.

Mutants that grow at a normal (permissive) temperature but that replicate abnormally or not at all at the nonpermissive (usually higher) temperature are particularly useful. Such temperature-sensitive (*ts*) mutants are easy to score and manipulate, and most genes seem to be potentially susceptible to such mutations. They arise when a base change (or changes) in the viral nucleic acid gives rise to an amino acid substitution (or substitutions) in a protein, which results in defective function at the nonpermissive temperature. Alternatively, the base change might affect the function of a nontranslated part of the genome—a control element, for example. The experimental objective is to collect and study a series of *ts* mutants of a particular virus. To be useful for studies on replication, *ts* mutants must possess certain characteristics: (i) they must not be significantly "leaky" at the nonpermissive temperature and (ii) the rate of reversion to wild type must be low enough to allow extended culture of the mutant at both the permissive and nonpermissive temperatures.

If the mutation studied occurred in the gene for the coat protein and the amino acid sequence of the coat protein was known, it was possible to locate the mutation within that protein. The location of mutations in other viral genes had to await a knowledge of nucleotide sequence and genome structure. The *ts* strain of TMV known as LS1 is a good illustration. This strain replicates normally at 22°C, but at the nonpermissive temperature (32°C) there is very little replication compared with the parent virus (TMV-L). Nishiguchi *et al.* (1980) studied the replication and movement of the two strains at the two temperatures, using fluorescent antibody staining to identify cells where virus had replicated. The results, illustrated in Fig. 6.5, showed clearly that LS1 was defective in a cell-to-cell movement function. However, these results did not locate the cell-to-cell function involved. This had to await a knowledge of the genome structure of TMV and the site of the mutation within that structure (Chapter 7, Section IIIA). Viral mutants and other variants are discussed further in Chapter 13.

D. Matching Gene Activities with Functional ORFs

A variety of methods are now available for attempting to match the *in vivo* function of particular viral gene products with a particular ORF. A few of these give unequivocal proof of function, whereas others are more or less strongly indicative of a particular function. There are two kinds of method. In the first, which may not be generally applicable, the natural gene product produced in infected tissue is isolated, and its activity is established by direct methods. The second group of methods involves, directly or indirectly, the use of recombinant DNA technology.

1. Direct Testing of Protein Function

Some viral-coded proteins besides coat proteins have functions that can be identified in *in vitro* tests.

On In Vitro *Translation Products*

Carrington and Dougherty (1987a) prepared an *in vitro* plasmid expression vector that allowed cell-free synthesis of particular segments of the TEV genome. The RNAs obtained were translated in rabbit reticulocyte lysates to give polypeptides that could be assessed for protease activity and the ability to act as protease substrates. In this way they showed that the 40,000 MW protein was a viral protease.

Figure 6.5   Example of a ts mutant of TMV defective in a cell-to-cell transport function. Fluorescent antibody staining of epidermal cells indicates the distribution of coat protein antigen 24 hours after inoculation. Tomato leaflets were infected at the permissive or nonpermissive temperature with a *ts* strain (LSI) or a wild-type strain (L) of TMV. Inoculated leaflets were cultured for 24 hours (*a*) LSI at 32°C; (*b*) LSI at 22°C; (*c*) L at 32°C; (*d*) L at 22°C. (From Nishiguchi *et al.*, 1980.)

On a Protein Isolated from Infected Cells

Thornbury *et al.* (1985) purified the protein helper component for aphid transmission from leaves infected with PVY. They showed that it has an MW of 58,000. They isolated the corresponding protein from another potyvirus and produced antisera against the two proteins. Using an *in vitro* aphid feeding test, they showed that the antisera specifically inhibited aphid transmission of the virus that induced formation of the corresponding protein in infected plants. These tests demonstrated that the polypeptides were essential for helper component activity. Various workers have obtained evidence for the role of viral-coded polypeptides in RNA-dependent RNA polymerase activity using extracts from infected tissues (e.g., for bromoviruses, Miller and Hall, 1984).

2. Approaches Depending on Recombinant DNA Technology

Location of Spontaneous Point Mutations

Knowledge of nucleotide sequences in natural virus variants allows a point mutation to be located in a particular gene, even if the protein product has not been isolated. In this way the changed or defective function can be allocated to a particular gene. The *ts* mutant of TMV known as LS1 can serve again as an example. At the nonpermissive temperature it replicates and forms virus particles normally in protoplasts and infected leaf cells, but is unable to move from cell to cell in leaves. A nucleotide comparison of the LS1 mutant and the parent virus showed that the LS1 mutant had a single base change in the 30,000 MW protein gene that substituted a serine for a proline (Ohno *et al.*, 1983b). This was a good indication, but not proof, that the 30,000 MW protein is involved in cell-to-cell movement.

Introduction of Point Mutations, Deletions, or Insertions

Ahlquist *et al.* (1984b) designed a template viral cDNA clone in the transcription vector pPM1 containing a modified λ P_R promoter that allowed *in vitro* transcription to initiate precisely at the 5′ terminus of the viral sequence. Using this technique, they were able to show that a complete infectious BMV genome consisting of the three viral RNAs could be generated *in vitro*. The technique is generally applicable to RNA viruses. The introduction of a DNA stage in an RNA virus life cycle allows the application of recombinant DNA technology to RNA viruses. Furthermore, it allows the production of genetically well-defined virus isolates. One important application of this technique is the introduction of defined changes in particular RNA viral genes to study their biological effects, and thus define gene functions, an approach commonly known as reverse genetics.

For example, Meshi *et al.* (1987) introduced, into the parent TMV strain, the same mutation in the 30,000 MW protein as found in the *ts* LS1 strain of TMV discussed in the last section. The generated mutant showed the same phenotype as the natural LS1. Thus the mutation is sufficient to account for the LS1 phenotype. This result was confirmed by various frameshift mutations placed in the 30,000 MW gene. Such mutants could replicate in protoplasts but could not move from cell to cell in tobacco plants.

Recombinant DNA techniques may also be used to introduce deletions into specific genes. For example, Woolston *et al.* (1983) introduced a 126-base-pair deletion into ORFII of a strain of CaMV that was transmitted by aphids. They showed that virus containing the deletion was not aphid-transmitted and did not produce the 18,000 MW product coded by gene II.

Expression of the Gene in a Transgenic Plant

Using a modified Ti plasmid in *Agrobacterium tumefaciens* it is possible to introduce a single plant viral gene into the host plant genome. In this state, it is possible to study the gene's function in the absence of the expression of other viral genes and of other elements of the viral genomic RNA. Deom *et al.* (1987) introduced into tobacco cells a chimeric gene containing a cloned complementary DNA for the 30,000 MW protein of TMV. Transgenic plants regenerated from transformed tobacco cells expressed the 30,000 MW protein mRNA and accumulated 30,000 MW protein. Inoculation of transgenic plants expressing the normal 30,000 MW protein with strain LS1 of TMV (discussed in the preceding section) allowed the LS1 strain to cause systemic disease at the nonpermissive temperature. This experiment proves that the TMV 30,000 MW protein potentiates movement from cell to cell. Graybosch *et al.* (1989) produced transgenic tobacco plants expressing the TVMV cylindrical inclusion body protein. Such plants were quite symptomless, showing that expression of this polypeptide is, by itself, unable to induce disease.

In another example involving transgenic plants, Rogers *et al.* (1986) used the Ti plasmid vector to introduce the two DNA components of tomato golden mosaic virus separately into the genomic DNA of petunia plants. Plants containing either DNA alone were symptomless but one-quarter of the progeny from crosses between plants containing each of the DNAs produced symptoms. These experiments showed that natural or modified genome components can be introduced into every cell of a plant, whether or not any active virus can be produced. Such transformed plants provide a new method for physical dissection of the viral genome and assignment of functions to the ORFs identified by computer.

Recombinant Viruses

Recombinant DNA technology can be used to construct viable viruses from segments of related virus strains that have differing properties, and thus to associate that property with a particular viral gene. For example, Saito *et al.* (1987a) constructed various viable recombinants containing parts of the genome of two strains of TMV, only one of which caused necrotic local lesions on plants such as *Nicotiana sylvestris,* which contain the N′ gene. Their results indicated that the viral factor responsible for the necrotic response in N′ plants is coded for in the coat protein gene. This response is discussed further in Chapter 12.

Another example of the use of recombinants constructed *in vitro* is given by the work of Woolston *et al.* (1983) with CaMV. They infected plants with hybrids constructed from the genomes of an aphid-transmissible and an aphid-non-transmissible strain of the virus. The results showed that aphid transmission and the synthesis of a protein with MW = 18,000 were located in either ORFI or ORFII. Tests with a deletion indicated that ORFII was the gene involved.

A third example involves BMV and CCMV, both members of the *Bromovirus* group. RNA replication requires products of both RNAs 1 and 2 of these viruses. Heterologous combinations of RNAs 1 and 2 do not support replication. Construction of hybrids by precise exchange of segments between BMV and CCMV RNA2 has made possible the mapping of virus-specific replication functions in this RNA (Traynor and Ahlquist, 1990). The sequence of the 5′ and 3′ segments of RNA2 appeared to be involved in the interaction of the RNA2-coded protein with RNA1 or its protein. It is these terminal segments that are less strongly conserved among different viruses than the central region.

It is also possible to construct viable recombinant hybrids between different viruses. Sacher *et al.* (1988) used biologically active cDNA clones to replace the natural coat protein gene of BMV RNA3 with the coat protein gene of sunnhemp mosaic *Tobamovirus*. In this virus the origin of assembly lies within the coat protein gene (see Fig. 17.2). In barley protoplasts coinoculated with BMV RNAs 1 and 2, the hybrid RNA3 was replicated by *trans*-acting BMV factors, but was coated in TMV coat protein to give rod-shaped particles instead of the normal BMV icosahedra.

Baculoviruses as Vectors for Plant Viral Genes

Foreign genes can be inserted into the genome of baculoviruses—large-DNA viruses that infect insects. Maule *et al.* (1988) obtained expression of the nonstructural genes I and II of CaMV in insect cells by incorporating these genes into a baculovirus vector. This system should allow the CaMV proteins to be isolated in sufficient quantity and purity for studies to be made on their functions.

Hybrid Arrest and Hybrid Select Procedures

Hybrid arrest and hybrid selection procedures can be used to demonstrate that a particular cDNA clone contains the gene for a particular protein. In hybrid arrest the cloned cDNA is hybridized to mRNAs, and the mRNAs are translated in an *in vitro* system. The hybrid will not be translated. Identification of the missing polypeptide defines the gene on the cDNA.

In the hybrid select procedure, the cDNA–mRNA hybrid is isolated and dissociated. The mRNA is translated *in vitro* to define the protein coded for. In appropriate circumstances these procedures can be used to identify gene function. For example, Hellman *et al.* (1988) used a modified hybrid arrest procedure to obtain evidence identifying the protease gene in TVMV.

Sequence Comparison with Genes of Known Function

As noted in Section V,B,2 sequence comparisons can be used to obtain evidence that a particular ORF may be functional. The same information may also give strong indications as to actual function. For example, the gene for an RNA-dependent RNA polymerase has been positively identified in poliovirus. The study by Kamer and Argos (1984) revealed amino acid sequence similarities between this poliovirus protein and proteins coded for by several plant viruses. This similarity implies quite strongly that these plant viral-coded proteins also have a polymerase function. The conserved amino acid sequence is $GXXXTXXXN(X)_{20-40}GDD$, where X represents any amino acid. The sequence is found in the proposed RNA-dependent RNA polymerases of all plant and most animal viruses (Hamilton *et al.*, 1987).

Functional Regions within a Gene

Spontaneous mutations and deletions can be used to identify important functional regions within a gene. However, mutants obtained by site-directed mutagenesis, and deletions constructed *in vitro* can give similar information in a more systematic and controlled manner. For example, the construction and transcription of cDNA representing various portions of the TEV genome, and translation *in vitro* and testing of the polypeptide products, showed that the proteolytic activity of the

49,000 MW viral protease lies in the 3′-terminal region. The amino acid sequence in this region suggested that it is a thiol protease related in mechanism to papain (Carrington and Dougherty, 1987a). García *et al.* (1989) found viral proteolytic activity expressed in *E. coli* harboring plasmids with a cDNA insert of the plum pox *Potyvirus* genome. They used site-directed mutagenesis to study the amino acid sequences involved in proteolytic activity.

VI. THE REGULATION OF VIRUS PRODUCTION

There is ample evidence for many plant viruses that overall virus production usually proceeds under some form of control. For example, TRV long rods accumulate earlier than short rods in infected protoplasts, but both classes reach stable plateau values (Harrison *et al.*, 1976). In chronically infected young Chinese cabbage leaves, the production of TYMV is as closely controlled as that of cytoplasmic ribosomes (Faed and Matthews, 1972). On the other hand, production of viral protein may not always be under effective control. The very large production of empty viral protein shells by okra mosaic *Tymovirus* in cucumber may be an example of an unbalanced synthesis of RNA and coat protein (Marshall and Matthews, 1981). In an interesting series of experiments, Föglein *et al.* (1975) showed that the upper limit of virus component production in cells may not be fixed. When protoplasts were isolated from TMV-infected tobacco leaves in which virus synthesis had slowed down or stopped, there was substantial renewed viral RNA synthesis.

As far as the virus is concerned, there will usually be regulatory controls involving genomic nucleic acid synthesis, mRNA transcription from the genome, translation of the mRNAs, and assembly of virus particles. Regulatory and recognition signals in both the viral nucleic acids and proteins will be involved. Our present understanding of these processes is discussed in this section.

Host-cell factors must also interact with viral regulatory processes to modulate virus replication, perhaps through the supply of necessary metabolites and in other ways of which we are, at present, quite ignorant.

A. Regulatory and Recognition Signals in RNA Viral Genomes

Regulatory control of eukaryote translation has been reviewed by Kaempfer (1984).

1. The Role of Terminal Structures

The kinds of structure found at the 5′ and 3′ termini of RNA viruses were outlined in Chapter 4, Section I,C,3. Their effects on the infectivity of RNAs were discussed in Chapter 4, Section I,C,4. Some of these terminal structures have regulatory or recognition functions in virus replication. Noncoding or coding nucleotide sequences in other parts of the RNA genome may also have such functions. These are discussed in the following section. Regulatory and recognition functions in DNA plant viruses are discussed in Chapter 8 in the sections on the *Caulimovirus* and *Geminivirus* groups.

The 5' Cap

Most eukaryotic mRNAs have a 5' cap in which the first two nucleotides following the 5'-terminal m^7Gppp are methylated on the ribose. A number of plant virus RNAs have this structure without the methylation of the first two nucleotides. The role of the cap structure for RNA viruses can be considered in relation to three properties—infectivity of the virus, stability of the RNA, and efficiency of translation. Different capped mRNAs vary in their dependence on the cap structure for effective functioning. There is some evidence to suggest that mRNAs with little ability to form stable base pairs at the 5' leader sequence have a decreased dependence on the cap structure for efficient translation (Gehrke *et al.*, 1983).

1. *Infectivity* cDNA transcripts of the RNA4 of AMV have been synthesized. The transcripts can then be capped *in vitro*. Assays in either protoplasts (Loesch-Fries *et al.*, 1985) or bean leaves (Langereis *et al.*, 1986a) showed that the cap structure was essential for infectivity. With two other viruses, TMV (Meshi *et al.*, 1986) and BMV (Janda *et al.*, 1987), transcripts without a cap had some infectivity, but very much less than capped molecules.

2. *Stability* In *in vitro* systems, the cap protects mRNAs generally from degradation by exonucleases. It has been suggested by several authors that increased stability *in vivo* may be the reason why the cap is needed for infectivity.

3. *Efficiency of translation* Capped and uncapped transcripts of RNA4 of AMV were translated in the reticulocyte lysate system with about equal efficiency (Langereis *et al.*, 1986a). Smirnyagina *et al.* (1989) used RNase H to cleave BMV RNA3 between the two genes for which it codes (see Fig. 7.26). The RNA fragment containing the coat protein gene had no cap structure and contained 20–22 more nucleotides at its 5' terminus than the natural coat protein subgenomic RNA. Nevertheless, coat protein was efficiently translated from it in an *in vitro* system.

Other experiments suggest that the cap structure may be involved in the mechanism by which different viral mRNAs are translated at different rates. A binding factor, or factors, is involved in the initiation of protein synthesis. In a vertebrate system one of these factors is a three-subunit protein that is involved in the recognition of cap structures and in the binding of mRNAs to ribosomal subunits that already carry the initiating $tRNA^{met}$ (Clemens, 1987). Discrimination among naturally capped mRNAs in *in vitro* systems has been shown to involve competition by mRNAs for the initiation factor implicated in the binding of ribosomes. Godefroy-Colburn *et al.* (1985a) measured the translational discrimination between the four RNAs of AMV. Taking the affinity of RNA3 = 1.0, affinities of the other RNAs were: RNA1 = 10; RNA2 = 60; and RNA4 = 150. These ratios were similar in the reticulocyte lysate and the wheat germ systems. Godefroy-Colburn *et al.* (1985b) proposed that this translational discrimination between naturally capped mRNAs may be related to the "breathing" of hydrogen-bonded structures near the 5' end of the RNA. Whatever the exact mechanism, there appears to be a powerful regulatory role for the 5' nucleotide sequence of capped mRNAs. On the other hand, STNV RNA, which is naturally uncapped, has been shown to be an effective mRNA *in vitro* (e.g., Leung *et al.*, 1976). In fact STNV RNA is one of the most efficient mRNAs known.

The VPg

In picornaviruses the 5' VPg plays a role as a primer for the replication of both minus strand and plus strand RNA. Viral RNA isolated from polyribosomes in poliovirus-infected cells does not contain the VPg. Removal of the VPg is not essential for mRNA activity *in vitro* (reveiwed by Rueckert, 1985). Among the plant viruses with a 5′ VPg (comoviruses, nepoviruses, and sobemoviruses), the protein is essential for the infectivity of some RNAs but not others (Chapter 4, Section I,C,4). Likewise the effect of removing the VPg enzymatically on messenger function *in vitro* varies with the virus. For most viruses, it had no effect, but for SBMV, RNA messenger activity was greatly reduced (reviewed by van Vloten-Doting and Neeleman, 1982). Unlike the VPg of poliovirus, that of CPMV is not enzymatically removed from the RNA during incubation in the rabbit reticulocyte lysate system. The linked RNA protected the VPg from proteolysis, with as few as 17 nucleotides being sufficient for such protection (de Varennes *et al.*, 1986).

The VPg of CPMV has not been found in free form in infected cells. It occurs either in its 60,000 MW precursor form or covalently linked to the 5′ terminus of the viral RNA. This suggests that initiation of viral RNA synthesis may be directly coupled to release of VPg molecules from its precursor (van Kammen, 1985). Although proof is lacking, it seems most likely that the VPg's of plant viral RNAs are involved in the initiation of RNA synthesis. However, if this is their only function, why does the primer remain bound to the genomic RNA after their function is completed?

The 3′ Poly(A) Sequence

The structure, synthesis, and possible roles of the 3′ poly(A) sequence in cellular mRNAs have been reviewed by Sachs and Davis (1989). Poly(A) is added to the 3′ end of mRNA transcripts, synthesized by RNA polymerase II in the cell nucleus, by a multiprotein complex. This complex cleaves transcripts to produce the mature mRNA, and then adds polyadenylate. The role or roles of poly(A) in cellular mRNAs have not yet been clearly defined, but a poly(A) binding protein is involved.

For one plant virus at least the 3′ poly(A) sequence appears to be coded for by a poly(U) sequence in the minus strand. Poly(U) sequences are present at the 5′ termini of both the M and B negative sense RNAs of CPMV (Lomonossoff *et al.*, 1985). Neither of these RNAs has an AAUAAA polyadenylation signal preceding the 3′ poly(A) sequence. However, most plant viruses with a 3′ poly(A) sequence have such a polyadenylation signal. Thus they are assumed to have the poly(A) sequence added enzymatically after synthesis of the positive strand, perhaps by a host-cell poly(A) polymerase. This idea has been confirmed experimentally for BNYVV RNA3, which has a 3′ poly(A) sequence in the genomic RNA. Jupin *et al.* (1990) constructed a transcription vector for RNAs in which the poly(A) sequence was eliminated. When transcripts were inoculated to *C. quinoa* along with normal transcripts of genomic RNAs 1 and 2, biological activity was lower than with normal RNA3. Nevertheless, among the local lesions produced, some contained RNA3 and these had acquired long 3′ poly(A) sequences. These experiments also show that the 3′ poly(A) sequence is necessary for RNA3 replication. Although most eukaryotic mRNAs have a 3′ poly(A) sequence, not all plant viruses have this structure. Some viruses that do not contain a 3′ poly(A) sequence in the genomic RNA do not appear to acquire such a sequence after infection. For example, it has been shown that the AMV RNAs present in polyribosomes of infected tobacco leaves do not bind to oligo (dT) columns (Bol *et al.*, 1976). Thus if these mRNAs had any poly(A) sequence they must have been very short.

Various studies have shown that the 3′ poly(A) sequence increases the stability of eukaryotic mRNAs. Evidence suggests that this is also true for some plant viral RNAs. Huez *et al.* (1983) compared the *in vitro* translation of mRNA of CPMV (3′

poly(A)), RNA4 of BMV (tRNA-like), and RNA4 of AMV (no special structure). They compared the translational stability of the natural RNAs in *Xenopus* oocytes with the CPMV RNA that had been adenylated, and with the BMV and AMV RNAs to which about 40 A residues had been added enzymatically to the 3′ terminus. The results showed clearly that the poly(A) sequence substantially increased the translational stability of all three RNAs. It was also shown that the stability of nonadenylated RNAs depends on the nature of the cytoplasm used in the tests. Thus natural AMV RNA was quite stable in oocytes, but poorly stable in HeLa cells. This difference raises the question of how relevant experiments carried out in animal cells are in relation to the natural plant hosts of these viruses.

Experiments with globin mRNA in rabbit reticulocyte lysates or the wheat germ system showed that globin synthesis was directly proportional to the rate of loss of A units from the poly(A) sequence (Rubin and Halim, 1987). Under nontranslating conditions no A units were lost. The cleavage of A units was not due to ribonuclease in the extracts. The relevance of this finding for plant virus mRNA remains to be determined.

3′ tRNA-like Structures

A variety of functions have been proposed for the tRNA-like structures at the 3′ termini of some viral RNAs (reviewed by Florentz *et al.*, 1984). The following four possible functions and origins are not necessarily mutually exclusive: (i) accepting and donating an amino acid in some aspect of protein synthesis; (ii) facilitating translation; (iii) acting as the replicase recognition site; and (iv) representing molecular fossils from a very early stage of evolution.

1. *An amino acid accepting function*   The addition of an amino acid to the 3′ end of the RNA implies that several enzymes must be able to recognize specifically some structure or structures in the RNA. Direct evidence has been obtained to show that the tRNA-like structure at the 3′ end of the TYMV RNA is recognized by several enzymes and factors: (i) tRNA nucleotidyltransferase from *E. coli* that can add an AMP residue to the $-CpC_{OH}$ end of the RNA (Litvak *et al.*, 1970); (ii) *E. coli* valyl-tRNA synthetase, which catalyzes the binding of valine in an ester linkage to the terminal adenosine (Yot *et al.*, 1970); (iii) *N*-acylaminoacyl-tRNA hydrolase, which cleaves the ester linkage between *N*-acyl amino acids and tRNA (Yot *et al.*, 1970); TYMV-val RNA forms a complex with the peptide elongation factor Tu in the presence of GTP and is protected by such binding from nonenzymatic deacylation and nuclease digestion (R. L. Joshi *et al.*, 1984); and (iv) TYMV RNA charged with valine is cleaved by an *E. coli* tRNA maturation endonuclease releasing a 3′-terminal fragment of about 3.5 S bearing valine (Prochiantz and Haenni, 1973). The 4.5 S terminal fragment isolated before acylation could be acylated with valine, showing that the rest of the TYMV molecule was not necessary. TYMV RNA has a strong affinity for yeast valyl-tRNA synthetase, the rate constant for the addition of valine being only slightly lower than for tRNA^{val} (Giegé *et al.*, 1978).

TYMV RNA is valylated and consequently adenylated *in vivo* in a natural host, Chinese cabbage (Joshi *et al.*, 1982). Full-length unencapsidated RNAs of both BMV and BSMV are aminoacylated in barley protoplasts, but the RNAs found in isolated virus are not (Loesch-Fries and Hall, 1982). However, *in vitro* studies indicate that charged RNAs do not act as amino acid donors (e.g., Chen and Hall, 1973).

Bujarski *et al.* (1985) and Dreher and Hall (1988b) used oligonucleotide-directed site-specific mutagenesis to introduce numerous mutations into a cDNA clone of the 3' 200 nucleotides of BMV RNAs. They studied the effect of each mutation on adenylation and tyrosylation *in vitro*. For both tRNA-like activities, a pseudoknot permitting the formation of an aminoacyl acceptor stem was very important. The ability of several of these mutants that were defective *in vitro* to replicate *in vivo* indicates that tyrosylation is not essential for RNA replication (Dreher *et al.*, 1989).

TRV RNAs contain a 3' tRNA-like structure about 40 nucleotides long that can be adenylated but that cannot be aminoacylated, in contrast to the tRNA-like structures found in the viruses listed earlier (van Belkum *et al.*, 1987). This example reduces the probability that aminoacylation of other viral RNAs with a tRNA-like structure is necessary for the replication of such viruses.

Rao *et al.* (1989) studied the effects of various single or double base changes in the $-CCA_{OH}$ terminus of the tRNA 3' structure of BMV RNA3. Although these constructs had previously been shown to be defective in aminoacylation and replication *in vitro*, *in vivo* they were fully viable from a very early time after inoculation. In progeny RNAs the terminal triplet was rapidly restored to wild type, indicating a rapid turnover and correction of the 3' termini *in vivo*, probably by the plant tRNA nucleotidyltransferase. For this reason Rao *et al.* (1989) suggest tha the 3' $-CCA_{OH}$ termini may function like the telomeres of chromosomal DNA (i.e., to protect the coding sequences at the end of the genomic nucleic acid).

2. Facilitating translation Florentz *et al.* (1984) noted that the 5' and 3' terminal sequences of TYMV RNA are partially complementary, and that a hydrogen-bonded circular molecule could form in which the initiating AUG for translation is buried. They proposed that interaction of the tRNA-like structure with valyl-tRNA synthetase opens up the structure, allowing translation to begin.

3. A replicase recognition site The 3' region of all genomic viral RNAs must contain the replicase recognition site for initiating minus strand synthesis. Detailed evidence for the role of tRNA-like structures in such recognition is given in Section VI,A,2,b,i. The problem remains, however, why some viral replicases require a complex three-dimensional recognition site and others do not.

4. Molecular fossils Weiner and Maizels (1987) proposed that the tRNA-like structures present in certain modern RNA viruses are survivors of similar structures from the original RNA world, where they tagged genomic RNA for replication and also ensured that nucleotides were not lost from the 3' terminus during replication. In other words, they are molecular fossils. This idea is discussed further in Chapter 17.

Other Pseudoknots in the 3' Noncoding Region

A pseudoknot structure is formed in RNA when a single-stranded loop region forms standard Watson–Crick base pairs with a complementary sequence outside this loop. Thus a pseudoknot will have at least two stems. The various possible kinds of structure are discussed by Pleij (1990).

Pleij *et al.* (1987) have shown that the 3' noncoding region of TMV RNA can be folded to form a series of pseudoknots between the end of the coat protein gene and the tRNA-like structure (Fig. 6.6). The sequence that can form these pseudoknots is conserved among various TMV strains (e.g., García-Arenal, 1988). They further found that one or more such structures could exist in almost all capped nonpolyadenylated plant viral RNAs. The function of these regions is unknown. Their position in TMV makes it unlikely that they are involved in virus assembly. Pleij *et al.* (1987) raise the possibility that they might be involved in recombination

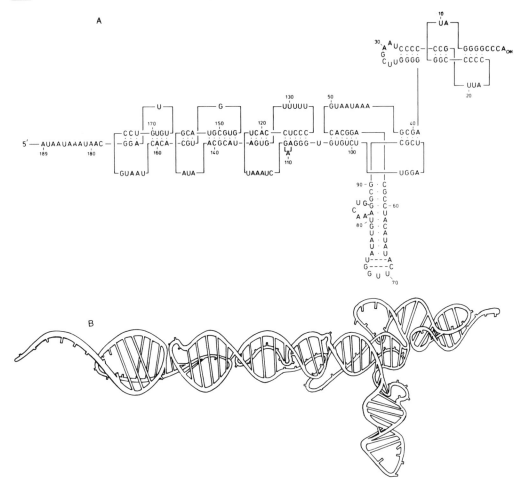

Figure 6.6 Pseudoknots. Proposed structure for the 3′ noncoding region of TMV RNA (Vulgare strain). The tRNA-like structure is at the right. Between this and the end of the coat protein gene is a stalklike structure containing four more pseudoknots. (A) Schematic representation. (B) An artist's view. (From Pleij *et al.*, 1987.)

events. Takamatsu *et al.* (1990) introduced various deletions into the pseudoknot region of TMV RNA (see Fig. 6.6). All of the helical regions except that closest to the tRNA-like structure could be deleted without affecting virus multiplication. Point mutations which destabilized the helical region closest to the tRNA-like structure greatly reduced or prevented virus multiplication.

Discussion

Many questions remain unanswered concerning the role of terminal RNA structures in plant viral genomes. The central question is why different viruses with ss positive sense RNA genomes have such different structures at their 5′ and 3′ termini. For example, if the 3′-terminal tRNA-like structures function as a replicase recognition site, how is the recognition site formed for viruses without such structures? Why do these viruses have such different structures as a cap or a VPg at the 5′ terminus? Until we have answers to such questions we will not fully understand the significance of these terminal structures in the virus life cycle.

2. The Role of Nucleotide Sequences

Complete genome nucleotide sequences are now known for representatives of many plant virus groups. However, regulatory and recognition functions in these sequences have been studied for only a few plant viruses. Nevertheless, the examples discussed in this section are probably relevant for many other viruses. Specific nucleotide sequences play regulatory and recognition roles in both protein and nucleic acid synthesis.

In Protein Synthesis

Both coding and noncoding nucleotide sequences may have recognition and regulatory roles in the translation of viral RNAs.

1. *Ribosome recognition sequences*   As discussed in Section IV, plant virus RNAs, with rare exceptions, behave like eukaryotic mRNAs in that they are translated as monocistronic mRNAs. Thus they are translated only from an AUG near the 5′ terminus and not from internal initiation sites, as happens in prokaryotes. Therefore we expect to find one or more ribosome recognition sites between the 5′ terminus and the first functional AUG. The 5′ untranslated region of viral RNAs usually has a low G + C content, suggesting a low degree of secondary structure. A number of plant viruses have been shown to bind two or three ribosomes along this sequence, forming disome or trisome initiation complexes. The best studied is TMV.

The 5′ region of TMV RNA is not polymorphic (Dawson *et al.*, 1986), indicating an important role for this sequence of nucleotides. Under conditions where protein elongation is inhibited, the 5′ leader sequence can bind two wheat germ ribosomes *in vitro*, one occupying the AUG initiation site and the other a site 5′ to this, in the leader sequence (Filipowicz and Haenni, 1979). The site for initiating binding of the second ribosome is probably an AUU triplet situated at the 3′ end of the following sequence in the TMV 5′ leader: –UAUUUUUACAACAAUU– (Tyc *et al.*, 1984). There is a 51-nucleotide sequence separating the AUU from the AUG, which provides sufficient space for the two ribosomes to bind without steric hindrance. While the isolated linear form of the 5′ leader sequence binds two ribosomes, a circular form ligated *in vitro* does not, suggesting that the free 5′ end is required for the initial interaction with 80 S ribosomes (Konarska *et al.*, 1981).

There is some evidence that ribosome binding may be initiated by binding between a nucleotide sequence in the viral leader and a complementary sequence at the 3′ terminus of the 18 S ribosomal RNA of the small ribosomal subunit. An example of such a complementary sequence is shown in Fig. 6.7. Twelve base pairs could form out of 18 residues.

2. *Selection of AUG initiation codons*   Since there are usually AUG triplets in an mRNA other than the one used for initiation of translation, there must be a mechanism whereby the correct AUG is recognized by the ribosome and/or associated factors. The

```
                        30              40
                        .               .
SBMV     5' . . .UG A U U U U C C U A C C U U U G U G U U U C AUG
                   . . . .       . . . .   . . . . .
18S rRNA 3'HO- A U U A C U A G G A A G G C G U C C . . . . .
```

Figure 6.7 Complementarity between part of a viral 5′ leader nucleotide sequence and the consensus sequence of 18 S ribosomal RNA 3′ termini derived from several organisms (Hagenbückle *et al.*, 1978). The viral sequence is from SBMV. The first AUG is indicated. (From Wu *et al.*, 1987.)

recognition lies in the nucleotides immediately surrounding the correct AUG triplet with an A in position -3 and a G in position $+4$ perhaps being particularly important as shown by Kozak (1981). Lütcke *et al.* (1987), from a study of a large number of mRNAs, concluded that the consensus sequence around the AUG differed significantly for plants and animals. The plant consensus sequence was AACAAUGGC. For 41 ORFs in plant viral RNAs I found the most frequent nucleotides to be CACAAUGGC. However the significance of such agreement is open to question. A more detailed analysis is needed on a larger number of ORFs grouped (i) according to the role of their gene products or (ii) on the basis of the "super-families" discussed in Chapter 17.

In general, plant viral RNAs have not been translated effectively in *in vitro* systems derived from prokaryotes, usually *E. coli.* However, Glover and Wilson (1982) showed that the genomic RNA of TMV could be translated efficiently in such a system to produce authentic TMV coat protein, except that it had an N-terminal methionine. They pointed out that, in the 36 nucleotides 5′ to the coat initiation AUG, there were two regions of homology with the idealized Shine–Dalgarno 16 S RNA binding site. These sites were 5′ to the beginning of the coat protein mRNA.

3. *Sequences enhancing translation rate* Sequences from the 5′ leaders of several plant viruses will act in *cis* to enhance the translation of foreign RNAs both *in vitro* and *in vivo,* and in prokaryote or eukaryote systems (Gallie *et al.*, 1987b).

In an attempt to define more closely the sequences involved in enhancement, a mutational analysis of the 5′ leader sequence of TMV was carried out that involved deletion mutants, a base substitution, and many base replacements. The results were difficult to interpret, because particular changes affected the enhancer activity differently depending on the test system (mesophyll protoplasts, *Xenopus* oocytes, or *E. coli*) (Gallie *et al.*, 1988a). On the basis of *in vitro* experiments with the 5′ leader sequence linked to a reporter gene, Sleat *et al.* (1988a) suggested that the enhancement of translation by the sequence may be due to reduced secondary structure either within the leader sequence or between this sequence and downstream coding regions. Such secondary structure could inhibit or delay translation initiation by interfering with the binding of ribosome subunits or other protein factors.

Using various constructs and a reporter gene, Carrington and Freed (1990) showed that a 144 base AU-rich 5′ nontranslated nucleotide sequence in TEV RNA had substantial translation enhancement activity in a cell-free system.

4. *Suppressor tRNAs*   As discussed in Chapter 7, in several ss positive sense RNA plant viruses, the first ORF from the 5′ terminus of the genome, or genome segment, may terminate in a stop codon that is sometimes suppressed. This suppression allows the ribosome to read to the next stop codon, giving rise to a read-through protein of greater length. The read-through protein includes all the amino acids of the shorter protein. The synthesis of a read-through protein depends primarily on the presence of appropriate suppressors tRNAs. Thus two normal tRNAs[tyr] from tobacco plants were shown to promote UAG read-through during the translation of TMV RNA *in vitro* (Beier *et al.*, 1984a). The tRNA[tyr] must have the appropriate anticodon to allow effective read-through. A tRNA[tyr] from wheat germ, with an anticodon different from that in tobacco, was ineffective (Beier *et al.*, 1984b).

The proportion of read-through protein produced may be modulated by nucleotide sequences around the termination codon (Bouzoubaa *et al.*, 1987; Miller *et al.*, 1988b) (Fig. 6.8). This opens up the possibility that the sequence context may be important for the interaction between the particular suppressor tRNA and the termination codon.

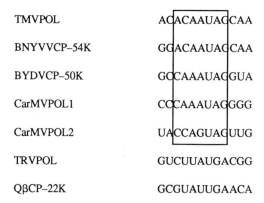

TMVPOL	AC|ACAAUAC|CAA
BNYVVCP–54K	GG|ACAAUAC|CAA
BYDVCP–50K	GC|CAAAUAG|GUA
CarMVPOL1	CC|CAAAUAG|GGG
CarMVPOL2	UA|CCAGUAC|UUG
TRVPOL	GUCUUAUGACGG
QβCP–22K	GCGUAUUGAACA

Figure 6.8   Alignment of nucleotide sequences flanking stop codons that are known to be read-through during translation, compared with the sequence flanking the BYDV coat protein gene stop codon. Boxed region indicates similar sequences shared by some plant viruses, including CarMV and BNYVV, which have amino acid sequence similarities with BYDV read-through sequences. The abbreviation POL indicates a stop codon in a putative polymerase gene, and CP indicates one at the end of a coat protein gene. Two stop codons occur in the CarMV polymerase gene. (From Miller *et al.*, 1988b.)

5. *Sequences controlling transframe proteins*   In some animal viruses, fusion proteins between two continuous genes located in different reading frames can be synthesized. This comes about by a translational frameshift event that allows a tRNA to bypass the stop codon at the 3′ end in one reading frame and switch to the second reading frame so that translation can continue to the next stop codon in that reading frame (Hizi *et al.*, 1987). In all known examples of such transframe proteins a −1 frameshift is involved. In addition, the sequence in the frameshift region is either AAAAAAC or UUUA. This sequence is followed by a variable stem loop structure. A frameshift is illustrated in Fig. 6.2.

Miller *et al.* (1988b) have proposed that the 60,000 MW ORF in BYDV (see Fig. 7.13) is expressed by such a translational frameshift at the end of the 30,000 MW ORF to produce a 39,000–60,000 MW fusion, or transframe protein. This proposed frameshift would be −1, and there is a UUUA sequence at the end of the 30,000 MW ORF of BYDV, followed by a stem-loop structure. There is also a stem-loop structure 5′ to the proposed frameshift (Fig. 6.9). *In vitro* translation experiments or identification of the transframe protein *in vivo* are necessary to verify this proposal. If the idea is correct it will provide the first example of a plant viral transframe protein.

In RNA Synthesis

1. *Promoter sequences*   The term promoter was first used to designate nucleotide sequences in cellular DNA, just upstream of a gene, that were recognized by a DNA-dependent RNA polymerase as the initial event in transcription of mRNA for the gene. Proposed promoter sequences for a DNA virus are illustrated in Fig. 8.8.

With RNA viruses, the term promoter is used to indicate nucleotide sequences recognized by the viral RNA-dependent RNA polymerase for the synthesis of both genomic RNAs and subgenomic mRNAs. There are three kinds of site to consider: (i) promoter sites most likely to be near the 3′ terminus of the plus strand genomic RNA used to initiate synthesis of a minus strand template; (ii) promoter sites most likely to be near the 3′ terminus of the minus strand used to initiate synthesis of plus

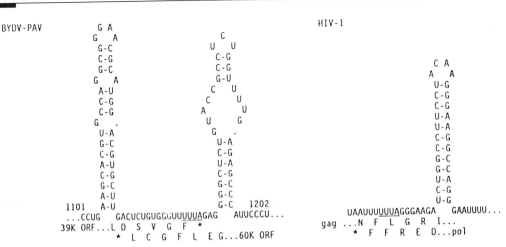

Figure 6.9   Right: The known frameshift region of human immunodeficiency virus (HIV-1) (Jacks *et al.*, 1988). Left: The proposed frameshift region between the 39K and 60K ORFs of BYDV strain PAV. Numbering of nucleotides indicates position in the genome. The conserved frameshift sequence is underlined. The amino acid sequences near the site of the frameshift are shown for the two reading frames. (Asterisks indicate in-frame stop codons. (From Miller *et al.*, 1988b.)

strand genomic RNA; and (iii) internal promoters on the minus strand used to initiate synthesis of subgenomic mRNAs.

Three technical developments with the BMV system have allowed rapid progress in our understanding of some of the elements in the genome that control viral RNA replication. These are (i) the availability of a template-dependent and template-specific replicase system. This system remains unique among plus strand RNA viruses (Marsh *et al.*, 1990). (ii) the ability to transcribe cDNA clones *in vitro* (Ahlquist and Janda, 1984; Dreher *et al.*, 1984), and (iii) appropriately constructed cDNA clones of all three RNAs that can be transcribed *in vitro* to produce a set of infectious RNAs (Ahlquist *et al.*, 1984b). This allows the three RNA components to be produced free of any cross-contamination. It also allows for defined mutations to be introduced via the cDNA. Because of these developments we know more about the control of RNA synthesis for BMV than for any other plant virus. There are three kinds of promoter to consider.

Minus strand promoters   In viruses with segmented genomes the 3′-terminal region usually shows a high degree of sequence homology. Figure 6.10 illustrates such homology for BNYVV. The 3′ sequences of related viruses usually show more sequence homology than the genome as a whole. For example, the 155-nucleotide sequence from the 3′ end of two rather distantly related tobamoviruses is highly conserved (Fig. 6.11). Upstream from residue 156 and into the coat protein gene the extent of homology decreased markedly. This marked conservation of sequence undoubtedly reflects an important function or functions. One function must be the viral replicase recognition sequence.

Possible roles for the tRNA-like structure at the 3′ terminus of some viruses have been discussed in Section VI,A,1. Ishikawa *et al.* (1988) constructed chimeric viruses containing the protein coding region from TMV L strain and 3′ noncoding regions from different TMV strains. Some had

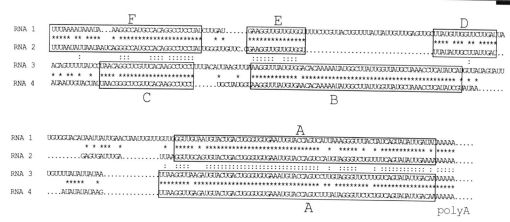

Figure 6.10 Sequence homologies near the 3′ termini of the four BNYVV RNAs. Boxes mark domains of homology. A region of extensive homology between all four RNAs is marked A; B and C indicate extensive homology between RNAs 3 and 4; D, E, and F indicate extensive homology between RNAs 1 and 2. Asterisks indicate positions in which RNAs 1 and 2 or 3 and 4 are identical. Colons indicate positions in which all four RNAs are identical. (Modified from Bouzoubaa *et al.*, 1986.)

tRNA-like structures accepting histidine, while another accepted valine. All chimeric viruses could replicate but in some combinations the extent of replication was reduced, indicating significant specificity between the L-strain replicase and its putative promoter.

More detail is known about the promoter for minus strand synthesis in the BMV system from *in vitro* experiments. Using a hybrid-arrest replication reaction, Ahlquist *et al.* (1984c) showed that the promoter for synthesis of negative strand RNA molecules was located in the 3′ tRNA-like structure

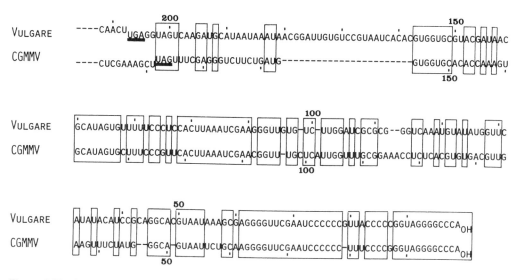

Figure 6.11 Conservation of nucleotide sequences in the 3′ terminal region of two distantly related tobamoviruses. Alignment of the nucleotide sequences of the 3′ noncoding regions of CGMMV and the common strain of TMV (Vulgare). Common nucleotides between them are boxed and termination codons of the coat protein cistrons are underlined. (From Meshi *et al.*, 1983.)

common to all the BMV RNAs. This was confirmed by the observations that short 3' fragments are active templates for minus strand synthesis (Miller *et al.*, 1986) and that single base changes can decrease promoter activity (Dreher *et al.*, 1984). Miller *et al.* (1986) established that initiation of minus strand synthesis occurs at the second nucleotide from the 3' end, and that the 3'-terminal A is not necessary. They also showed that the 134-nucleotide 3'-terminal sequence common to all four BMV RNAs contains all the signals required to initiate minus strand synthesis. Sequences 5' to this region appear to play no role in template selection.

Dreher and Hall (1988a) have defined much more closely the structural requirements for promoter activity in the 3' 134-nucleotide tRNA-like structure. They did this by introducing, via cDNA, numerous mutations (either substitutions or small deletions) throughout the sequence, but designed to have localized effects. Promoter activity was assayed *in vitro*. The main results were: (i) destabilization of the pseudoknot structure in the aminoacyl acceptor stem gave lower promoter activity (a pseudoknot is shown in Fig. 6.6), which demonstrated the importance of a three-dimensional tRNA-like conformation; (ii) substitution of the C residue used for initiation lowered promoter function; and (iii) deletions in one particular arm of the tRNA-like structure, particularly in the loop of unpaired bases, also gave low activity. Thus much of the tRNA-like structure appears to be involved in promoter activity.

3' BMV RNA sequences are capable of assuming secondary structures more extensive than the tRNA-like feature that include a pseudoknot. The pseudoknot is rather easily denatured. A suggested structure is shown in Fig. 6.12. The actual replicase recognition event presumably involves such secondary structure, with subsequent linearization of the molecule to allow transcription of the sequence to begin.

Promoters near 5' termini   If the same viral replicase initiates minus strand synthesis and plus strand synthesis using the same recognition sequence, then we would expect the replicase recognition sequences at the 5' and 3' ends of the genomic RNA to be complementary. For most RNAs there does not appear to be any such significant complementary relationship. However, tandem regions showing sequence similarity to the BMV plus RNA1 5' terminus were found on the minus strand 5' termini of other viruses possessing aminoacylatable tRNA-like structures (i.e., bromo-, cucumo-, tobamo-, and tymoviruses) (Marsh *et al.*, 1989).

Since far more plus strand than minus strand RNA is synthesized during virus replication, the 3' promoter of the minus strand must be used much more frequently than that of the plus strand. It is not known how this imbalance is maintained, but there are several possibilities: (i) Two different viral-coded replicases might contain different recognition sites. In this connection, many plant viruses have two proteins that might be involved in RNA recognition. On the other hand, only one region with amino acid sequence similarity to RNA-dependent RNA polymerases is present, suggesting a single core polymerase for BMV (Section V,D,2). (ii) One viral RNA polymerase may recognize both 3' minus strand and 3' plus strand promoter sequences either with the same or with different active sites. (iii) Some host or viral-coded protein factor may affect the specificity of a single recognition site in the replicase. Support for the involvement of a host factor comes from the fact that the 5' terminal sequences of the three BMV RNAs

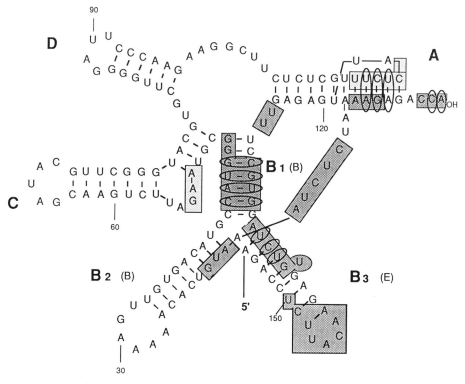

Figure 6.12 Proposed tRNA-like structure of the 3' end of BMV RNA, derived from studies in which portions of the structure were protected by the synthetase from enzymatic attack. Strongly protected regions are indicated by heavily shaded boxes; weakly protected regions by lightly shaded boxes. The CCA terminus is also boxed. No apparent anticodon triplet exists in the structure. Loci found by mutational analysis to be important for tyrosylation are circled. (From Perret *et al.*, 1989.)

closely resemble the internal control regions of tRNA genes (Marsh and Hall, 1987). (iv) The promoter specificity of the polymerase may be altered by different subunit associations. Evidence from the BMV system shows that, for this virus, a viral-coded product influences the balance of plus : minus strand synthesis. When BMV RNAs 1 and 2 were inoculated into barley protoplasts the ratio was 1.6 : 1. Inclusion of RNA 3 in the inoculum shifted the ratio to >100 : 1 (Marsh *et al.*, 1990).

For viruses with multipartite genomes, roughly equal amounts of genomic RNA segments are usually synthesized. ssRNA viruses with segmented genomes show some sequence homology in the segments near the 5' terminus but it is usually less extensive than at the 3' end. For example, the homologous sequence at the 5' terminus between RNAs 1 and 2 of the PSG strain of TRV is 5'-AUAAAACAUU–, presumably corresponding to part of the replicase recognition signal at the 3' end of the minus strand (Cornelissen *et al.*, 1986a). In another example, BMV genomic RNAs 1 and 2 show substantial sequence homology in the 5' region but RNA3 shows much less (Fig. 6.13).

The detailed structural requirements for the promoter near the 3' terminus of the minus strand genomic RNAs of BMV have not yet been delineated. The synthesis of plus strand genomic RNAs *in vitro* using the BMV

BMV1   m^{7}GpppGUAGACCACGGAACGAGGUUCAAUCCCUUGUCGACC ACGGUUCUGC
BMV2   m^{7}GpppGUAAACCACGGAACGAGGUUCAAUCCCUUGUCGACCCACGGUU UGCGCAACACACAUCUG
BMV3   m^{7}GpppGUAAAAUAC CAACUAAUUCUCGUUCGAUUCCGGCGAACA UUCUAUUUUUACCAACAUCGG
BMV4   m^{7}GpppGUAUUA

Figure 6.13 Sequence homologies in the 5′ nucleotide sequences of the RNAs of a virus with a tripartite genome. The three genomic RNAs of BMV are shown together with subgenomic RNA4. Agreement between two or more sequences is shown by heavy type. (From Ahlquist *et al.*, 1984a.)

replicase from infected tissue and minus strand templates has not yet been achieved (Ahlquist, 1987).

Internal promoters   The subgenomic mRNAs used by many RNA plant viruses for coat protein synthesis are not necessary for infectivity. Therefore the subgenomic RNA must be transcribed from a genomic RNA. The mechanism for this transcription was a matter for speculation until Miller *et al.* (1985) showed that the subgenomic RNA4 of BMV arises by internal initiation on the negative sense strand of RNA3. Marsh *et al.* (1988) used modified cDNA clones and *in vitro* transcription to define the sequence requirements for *in vitro* promoter activity in the intercistronic region of BMV RNA3. French and Ahlquist (1988) used biologically active cDNA clones to investigate sequences controlling RNA4 synthesis *in vivo*. Essentially similar results were obtained in the two studies. There is a core promoter sequence of about 20 nucleotides immediately upstream of the initiation nucleotide (G) (Fig. 6.14). A smaller domain overlaps the 5′ untranslated end of RNA4. There is an oligo(A) tract 16–22 nucleotides long immediately upstream of the core sequence. Elimination of this tract reduced RNA4 synthesis 10-fold in protoplasts. 5′ to the oligo(A) tract is a nine-nucleotide sequence required for high levels of promoter activity *in vitro*. This promoter bears no structural resemblance to the promoter at the 3′ terminus of the positive sense genomic RNAs. This raises the question of whether proteins other than the replicase used for genomic negative strand synthesis are involved in subgenomic RNA plus strand replication. There may be a core polymerase with other transcriptional factors that give full promoter specificity.

2. *Other sequence requirements for efficient RNA synthesis*   French and Ahlquist (1987) used expressible cDNA clones to construct partial deletions throughout BMV RNA3 and to define the *cis*-acting sequences involved in RNA3 replication. The effects on RNA3 synthesis were tested in protoplasts that were coinfected with unaltered RNAs 1 and 2. Both 5′ and 3′ noncoding sequences are required, but either of the two coding regions can be deleted, or both changed by frameshift mutations, without markedly affecting RNA synthesis. However, if both are deleted in the same construction, RNA synthesis is markedly suppressed, suggesting the need for a spatial separation of the 5′ and 3′ termini. In addition, a 150-base sequence in the intercistronic region was necessary for effective RNA3 synthesis. This element contained a sequence of about 12 nucleotides that was substantially conserved near the 5′ end of RNAs 1, 2, and 3, and in the analogous positions in CMV RNAs. The sequence may have a role in plus strand synthesis. However, it is not part of the internal promoter illustrated in Fig. 6.14. With respect to the 3′ noncoding region, a sequence of 163–200 nucleotides is required for normal RNA3 accumulation *in vivo*. This is significantly larger than the 134 nucleotides necessary for efficient minus strand initiation *in vitro*. It is consistent with the very close conservation of the 3′-terminal 193 nucleotides in all three BMV RNAs.

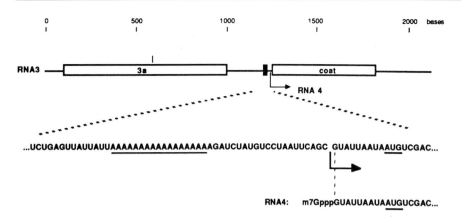

Figure 6.14   The internal promoter for BMV subgenomic RNA4 synthesis situated in RNA3. In the schematic map of RNA3, coding regions are shown as boxes and noncoding regions as lines. The nucleotide sequence of the intercistronic promoter region is shown, and below that the 5′ terminal sequence of RNA4, which overlaps with the 3′ region of the promoter. The coat protein AUG initiation site is underlined. The positive sense sequence packaged in virus particles is shown. The actual promoter sequence for RNA4 plus strand synthesis is complementary to this sequence. (From French and Ahlquist, 1988.)

The *cis*-acting sequences required for replication of the related *Bromovirus* CCMV RNA3 are organized differently. Normal levels of CCMV RNA3 synthesis required only between 125 and 220 bases from the 3′ noncoding region and no more than the first 89 bases of the 238-base 5′ noncoding region (Pacha *et al.*, 1990). Thus the entire intercistronic region of RNA3 is not needed for RNA replication. The smallest replicon tested that possessed all the *cis*-acting elements for RNA3 replication consisted of only 454 bases.

3. *A signal for polyadenylation*   In eukaryotic mRNAs with a 3′ poly(A) sequence there is frequently a polyadenylation signal (AUAAA) preceding the 3′ terminus by about 10–15 nucleotides. This signal is present in some, but not all, viral RNAs with a poly(A) sequence.

In Virus Assembly

There are two general mechanisms whereby viral coat proteins could package only viral nucleic acids into particles: (i) if synthesis and assembly of virus particles takes place in a cell compartment that does not contain free host nucleic acids or (ii) if virus assembly is initiated by a specific recognition process between coat protein and viral RNA. The second process is known to operate for some viruses and is probably widespread. The initiation of TMV rod assembly involves a specific internal nucleotide sequence in the RNA (Chapter 7, Section III,A,9). *Potexvirus* rod assembly is initiated by a sequence near the 5′ terminus of the RNA (Chapter 7, Section II,A,4). It has been proposed that specific nucleotide sequences are involved in the correct selection and packaging of *Phytoreovirus* RNA components (see Fig. 8.11).

B. Regulatory and Recognition Roles for Viral Proteins

It seems certain that viral-coded proteins must have roles concerning regulation and recognition in relation to the virus life cycle and to the effects of infection on the host cell. These functions will involve interactions between viral-coded proteins, between such proteins and viral nucleic acids, and between viral proteins or nucleic acids and host components. At present this is an area where little meaningful information is available. However, some examples, and possible examples, are emerging.

1. AMV Coat Protein

A molecule of the coat protein of AMV combines at a specific site on the viral RNAs to initiate viral RNA replication. This phenomenon is discussed in Chapter 7, Section VI,C,4.

2. The 32,000 MW Protein of CPMV

The experiments of Vos *et al.* (1988b) show that two viral-coded proteins are involved in the proteolytic processing of the polyproteins of CPMV. One is a 24,000 MW protease encoded in the B-RNA (see Fig. 7.23). The second is a 32,000 MW protein also encoded in B-RNA. This protein is not itself a protease, but it has a regulatory role. In the presence of the 32,000 MW protein, the 24,000 MW protease cleaves the M polyprotein at a glutamine–methionine site, thus releasing coat protein precursor. Without the 32,000 MW protein the protease cleaves only at the glutamine–glycine site. Thus the 32,000 MW protein appears to ensure that the coat protein precursor polypeptide is formed first from the B polyprotein, before cleavage to give the two coat proteins.

3. The Protease of CPMV

The VPg of CPMV is considered to act as a primer for the initiation of viral RNA synthesis (Section VI,A,1). The VPg is part of the 202,000 MW polyprotein translated from B-RNA. This protein has to undergo several successive proteolytic processing steps before the VPg is released (see Fig. 7.23). Thus the protease that processes the 202,000 MW protein appears to play an important regulatory role in the replication of the viral RNA (van Kammen, 1985).

4. The TYMV Replicase

The TYMV replicase consists of a viral-coded 115,000 MW polypeptide and a 45,000 MW host protein. In spite of nucleotide sequence conservation in the replicase gene of a range of tymoviruses (Chapter 7, Section IV,A), Candresse *et al.* (1986) found that there was substantial antigenic variability between the 115,000 MW proteins of different tymoviruses. They suggested that this variability may result from the need for the 115,000 MW viral subunit of different tymoviruses to recognize the smaller protein from different hosts to form a functional replicase. Indirect support for this idea came from the fact that the only *Tymovirus* 115,000 MW protein to cross-react significantly with the TYMV 115,000 MW protein was that of *Erysimum* latent virus, which, like TYMV, has a host range in the Brassicaceae.

VII. EXPERIMENTAL SYSTEMS FOR STUDYING REPLICATION *IN VIVO*

The tissue that has been most commonly used in the study of virus replication is the green leaf blade. This tissue constitutes approximately 50–70% of the fresh weight of most experimental plants, and final virus concentration in the leaf blade is often 10–20 times higher than in other parts of the plant. We can distinguish four types of systems *in vivo:* the intact plant, surviving tissue samples, cells or organs in tissue culture, and protoplasts. The advantages and difficulties of these systems are discussed next. Some plant viruses also replicate in their insect vectors. This topic is discussed in Chapter 14.

A. The Intact Plant

In Chapter 11, Section IV, some of the variables involved in sampling intact plants are discussed. It should be borne in mind that in spite of these difficulties there are certain aspects of virus replication that can be resolved only by study of the intact developing plant, for example, the relationship between mosaic symptoms and virus replication.

1. Inoculated Leaves

Inoculated leaves have several advantages. Events can be timed precisely from the time of inoculation. A fairly uniform set of leaves from different plants can be selected, and half-leaves may be used as control material. There are two major disadvantages.

1. A typical leaf such as a tobacco leaf with a surface area of 200 cm^2, for example, contains about 3×10^7 cells. The upper limit for the proportion of epidermal cells that can be infected by mechanical inoculation under the best conditions is not known precisely, but is probably not more than about 10^4 cells per leaf. Thus, at the beginning of an experiment, only about 1 in 10^3 of the cells in the system has been infected. Even for those that are directly infected, the synchrony of infection may not be very sharp, especially if whole virus is used as inoculum. Thus, early changes in the small proportion of infected cells will probably be diluted out beyond detection by the relatively enormous number of as yet uninfected cells. Then, as infection progresses, a mixed population of cells at different stages of infection will be produced.

2. The second major disadvantage of inoculated leaf tissue, at least for studying events over the first few hours, is that mechanical inoculation itself is a severe shock to the leaf, causing changes in respiration, water content, and probably many other things as well, including nucleic acid synthesis. Thus, the use of appropriately treated control leaf is essential. A third difficulty applies to experiments in which radioactively labeled virus is used as inoculum. Most of the virus applied to the leaf does not infect cells, and a substantial but variable proportion cannot be washed off after inoculation. The fate of the infecting particles may well be masked by the mass of potentially infective inoculum remaining on or in the leaf.

For particular kinds of experiments, two modifications in the use of the inoculated leaf have proved useful. With some leaves grown under appropriate conditions it is relatively easy to strip areas of epidermis from the leaf surface. Very limited amounts of tissue can be harvested in this way, but the method increases by a factor of about 8 the proportion of cells infected at times soon after inoculation (Fry and Matthews, 1963).

Several workers have used micromanipulation methods to infect single cells on a leaf—usually leaf hair cells—and then to follow events in the living cells as they can be observed by phase or ultraviolet microscopy or in preparations stained with fluorescent antibody. This procedure, while it has given useful information, is limited to microscopical examination and cannot at present be used for biochemical investigations.

2. Systemically Infected Leaves

Moving from the inoculated leaf, a virus may invade the youngest leaves first, then successively infect the older and older leaves (Fig. 10.7). Thus, systemically infected leaves may be in very different states with respect to virus infection. Furthermore, the time at which infectious material moves from inoculated leaves to young growth may vary significantly between individual plants in a batch. Nevertheless, it is probable that in young systemically infected leaves (perhaps about 4 cm long at the time virus enter for plants like tobacco and chinese cabbage) most of the cells in a leaf become infected over a period of 1–2 days. Such a leaf has been used to study the replication of TMV (Nilsson-Tillgren *et al.*, 1969) and TYMV (Hatta and Matthews, 1974; Bedbrook *et al.*, 1974).

The synchrony of infection in the young systemically infected leaf can be greatly improved by manipulating the temperature. The lower inoculated leaves of an intact plant are maintained at normal temperatures ($\approx25-30°$) while the upper leaves are kept at 5–12°. Under these conditions, systemic infection of the young leaves occurs, but replication does not. When the upper leaves are shifted to a higher temperature, replication begins in a fairly synchronous fashion (W. O. Dawson *et al.*, 1975; Dawson and Schlegel, 1976a,b,c). This procedure has not yet been widely applied, but it should provide a very useful system that complements, in several reports, the study of virus replication in protoplasts, discussed subsequently. The technique uses intact plants, is simple, and can provide substantial amounts of material. The main requirement is for a systemic host with a habit of growth that makes it possible for upper and lower leaves to be kept at different temperatures. Its use has been extended to some other viruses, for example CMV in tobacco (Roberts and Wood, 1981).

B. Surviving Tissue Samples

1. Excised Leaves

These are useful when fairly large quantities of leaf tissue are required. Petioles may be placed in water or a nutrient solution. Under these conditions, leaves vary widely in the amount of fluid they take up, and may wilt unpredictably. Tissue near the cut end of the petiole acts as a "sink" for radioactively labeled metabolites (Pratt and Matthews, 1971). On the other hand, the method minimizes the problem of the growth of microorganisms in the tissue during incubation. More commonly, leaves are placed in dishes covered with glass under moist conditions. Growth of bacteria, fungi, and protozoa is then likely to be a problem.

2. Disks of Leaf

Disks of tissue 5–20 mm in diameter cut from leaves with a cork borer and floated on distilled water or some nutrient salt solution have the advantage that pieces from many leaves can be combined in one sample to smooth out leaf-to-leaf variations. The physiological state of the leaves from which disks are taken affects uptake and metabolism of radioactively labeled materials (e.g., Kummert and Semal, 1969). There may be two serious disadvantages: (i) microorganisms grow on the surface of the disks and in the intercellular spaces, so addition of antibiotics may not block all microorganisms and may well alter the biochemical situation in the cells of interest, and (ii) excised disks are not uniform in several ways (Pratt and Matthews, 1971). First, there is a "geographical" gradient from the cut edge to the center of the piece of tissue. Differences involve the uptake of labeled precursors and their utilization for nucleic acid synthesis. Second, excised tissues change with time in a complex fashion in their ability to accumulate substances from the medium. There may be a differential accumulation of labeled precursors in the cut ends of veins. Third, further variables are introduced when the excised tissue is treated with a drug such as actinomycin D, which may be distributed very unevenly in the tissue.

3. Epidermal Strips

Dijkstra (1966) explored the possibility of studying TMV replication in strips of epidermis removed from leaves immediately after inoculation with TMV and floated on nutrient solutions or distilled water, but no significant progress has been made with this system.

C. Tissue Culture

Plant cells can be grown in tissue culture in several ways, either as whole organs (e.g., roots or stem tips) or as solid masses of callus tissue growing in solid or liquid culture, or as cell suspensions. Amounts of virus produced in cultured tissue or cells are usually very much less than in intact green leaves, although tobacco callus cells disrupted in the presence of TMV inoculum produced high yields of virus (Murakishi *et al.*, 1971; Pelcher *et al.*, 1972). Various methods have been tested in the study of virus replication, but except for some microscopical studies, results have been disappointing. White *et al.* (1977) and Wu and Murakishi (1979) have adapted the low-temperature preincubation procedure of W. O. Dawson *et al.* (1975) to callus cultures infected with plant viruses. The virus growth curves obtained for TMV in tobacco callus cells were comparable to that obtained with protoplasts.

D. Cell Suspensions and Tissue Minces

In principle, suspensions of surviving but nondividing cells offer considerable advantages in the study of virus replication. Dissociated cells from callus tissue grown in culture and leaf cells separated enzymatically have been used. For example, Jackson *et al.* (1972) successfully used separated leaf cells to study the replication of TMV RNA.

E. Protoplasts

Protoplasts are isolated plant cells that lack the rigid cellulose walls found in intact tissue. Cocking (1966) showed that protoplasts could be made from tomato fruit by using enzymes to degrade the cell wall. Takebe *et al.* (1968), Takebe and Otsuki (1969), and Aoki and Takebe (1969) showed that metabolically active protoplasts could be isolated from tobacco leaf cells; that such protoplats could be synchronously infected with TMV or TMV RNA; and that virus replication could be studied in such protoplasts. Since then, protoplasts have been prepared from many species and infected with a range of viruses. Progress has been reviewed by Murakishi *et al.* (1984) and Sander and Mertes (1984).

In outline, protoplasts are prepared as follows: the lower epidermis is stripped from the leaf tissue, which is then vacuum infiltrated with a solution of a commercial pectinase (polygalacturonidase) preparation called Macerozyme from *Rhizopus* sp. The medium contains $0.4-0.7$ M mannitol plus 0.5% potassium dextran sulfate. The leaf pieces are then shaken on a waterbath. Early fractions of cells released from the tissue may be discarded. The veins and so on are removed by filtration and the cells collected by centrifugation. They are then treated with a cellulase preparation (from *Trichoderma viride*). On complete removal of the cellulose wall, the cells, now bounded only by the plasma membrane, assume a spherical shape (Fig. 6.15). About 10^7 palisade cells can be obtained from 1 g of tobacco leaf in 2 hours. Many minor variations on the procedure have been developed (e.g., Kassanis and White, 1974; Beier and Bruening, 1975, 1976; Motoyoshi and Oshima, 1976; Shepherd and Uyemoto, 1976; Kikkawa *et al.,* 1982). The ability to infect protoplasts synchronously enables plant virologists to carry out one-step virus growth experiments (Fig. 7.7), an important kind of experiment that has long been available to those studying viruses of bacteria and mammals. Besides improved synchrony of infection, protoplasts have several other advantages: (i) close control of experimental conditions; (ii) uniform sampling can be carried out by pipetting; (iii) the high proportion of infected cells (often 60–90%); (iv) the relatively high efficiency of infection; and (v) organelles such as chloroplasts and nuclei can be isolated in much better condition from protoplasts than from intact leaves.

However, a number of actual or potential limitations and difficulties must be borne in mind.

1. Protoplasts are very fragile, both mechanically and biochemically, and their fragility may vary markedly, depending on the growing conditions of the plants, season of the year, time of day, and the particular age of leaf chosen. Defined plant growth conditions may improve the quality and reproducibility of the isolated preparations (e.g., Kubo *et al.,* 1975b).

2. Under culture conditions that favor virus replication, protoplasts survive only for 2–3 days and then decline and die.

3. To prevent growth of microorganisms during incubation, antibiotics may be added to the medium. These may have unexpected effects on virus replication (e.g., gentamycin, Kassanis *et al.,* 1975).

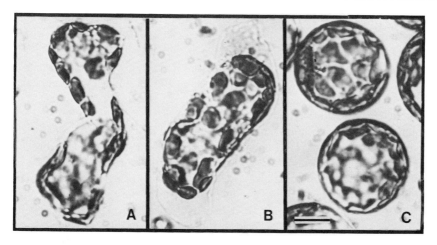

Figure 6.15 Isolation of protoplasts from chinese cabbage leaves. (A) A separated spongy mesophyll cell and (B) a separated palisade cell, following pectinase treatment. The cells still retain the cellulose wall. Following treatments of isolated cells with cellulase, spherical protoplasts are produced (C). Bar = 10 μm. (Courtesy of Y. Sugimura.)

4. Compared to intact tissue, relatively small quantities of cells are made available.

5. Cytological effects observed in thin sections of infected leaf tissue may not be reproduced in protoplasts—probably because of the effects of changed osmotic conditions on cell membranes, for example, with TMV in tobacco (Otsuki *et al.*, 1972a), TYMV in Chinese cabbage (Sugimura and Matthews, 1981), Festuca leaf streak rhabdovirus in cowpea (van Beek *et al.*, 1985), and CaMV in turnip (Yamaoka *et al.*, 1982b).

6. The isolation procedure and the medium in which they are maintained must drastically affect the physiological state of the cells. Physical and chemical disturbances include (i) partial dehydration; (ii) severing of plasmodesmata; (iii) loss of the cell wall compartment, which is not metabolically inert; (iv) reversal of the cell's electrical potential; (v) inhibition of leucine uptake; (vi) a large increase in RNase activity; and (vii) cellulose synthesis and wall regeneration, which begin very soon after the protoplasts are isolated. In addition, tobacco mesophyll protoplasts have been shown to synthesize six basic proteins that are undetectable in tobacco leaf. Three of these are like a 1,3-β-glucanase and two chitinases found in TMV-infected tobacco leaves (Grosset *et al.*, 1990) (Chapter 12, Section III).

As a consequence of these changes protoplasts vary with time in many properties during the period that they survive after isolation. Although little systematic study has yet been made of the changes, it is known that some features of virus replication differ in intact leaves and in protoplasts. Thus Föglein *et al.* (1975) showed that when protoplasts are prepared from leaves fully infected with TMV, vigorous viral RNA synthesis is reinitiated. Tobacco protoplasts containing the *N* gene escape necrotic cell death when infected with TMV (Otsuki *et al.*, 1972b).

In many studies using protoplasts it has been reported that yields of virus (virus particles per cell) are very similar to that found in intact plants. For example, Renaudin *et al.* (1975) found that Chinese cabbage protoplasts infected *in vitro* produced about 10^6 TYMV particles per cell. This figure is similar to the published yields of TYMV obtained with extracts of intact leaf. These estimates were based on the assumption that all the cells in the leaf were infected, and that they were of the same size. If, however, the estimates are made on the same class of cell as used for the *in vitro* studies (i.e., palisade mesophyll), and if only infected cells are considered, then yields per cell in the intact leaf are about 10 times higher (Sugimura and Matthews, 1981). Despite these limitations, protoplast systems have contributed to our knowledge and will contribute more in the future.

Many efforts have been made to improve the process of infection of protoplasts with viruses or viral RNAs. For example, Watanabe *et al.* (1982) inoculated tobacco protoplasts successfully with TMV RNA encapsulated in large unilamellar vesicle liposomes. Another technique known as electroporation is being increasingly used to introduce viral nucleic acids into protoplasts. It involves the application of a brief high-voltage pulse to a mixture of cells and nucleic acid. The pulse renders the cells transiently permeable to the nucleic acid. It is a widely applicable procedure but the mechanism is not well understood, and optimum conditions need to be determined for each system. For example, Nishiguchi *et al.* (1986) infected tobacco leaf protoplasts with TMV and CMV RNAs using this procedure. The method has been adopted to facilitate infection of protoplasts by infectious TMV RNA produced in low concentrations by transcription from TMV cDNA (Watanabe *et al.*, 1987a). A positively charged virus (BMV) could be readily induced to infect by electroporation but the negatively charged CCMV gave only a poor rate of infection (Watts *et al.*, 1987).

A polycation such as poly(L) ornithine is necessary in the medium for infection using viral RNAs or viruses with an acidic isoelectric point. Takanami *et al.* (1989) reported a greatly increased efficiency of infection using polyethyleneimine in the medium.

F. Radioisotopes

The use of radioactively labeled virus precursors is essential for many studies on virus replication. There are substantial difficulties and limitations in the effective use of tracer compounds for studying the replication of plant viruses. Various ways have been used to introduce the labeled material into the tissue being studied. Whole plants can be removed from their pots, the roots carefully washed free of soil, and the isotope applied to the roots. This procedure is useful for ^{32}P-labeled orthophosphate and ^{35}S-labeled sulfate. Provided there is no delay in applying the isotope after washing the roots, uptake is rapid and efficient. With plants such as actively growing Chinese cabbage, ^{32}P may be detected in leaves within minutes of application, and uptake into the plant may be more or less complete within a few hours. With these two isotopes, uptake into leaves through the roots is much more effective than floating intact disks of leaf tissue on solutions of the isotope, even if the disks are sliced to expose more vein ends. Placing leaves with their cut petioles in the solution can lead to a highly variable and irregular uptake of isotope. However, by careful timing and attention to growth conditions, quite high specific activities can be obtained (e.g., 1 mCi ^{32}P/mg viral RNA; Bastin and Kaesberg, 1975). Kopp *et al.* (1981) describes a procedure in which pieces of leaf from which the lower epidermis has been stripped are floated on a solution containing the radioactive precursor. No systematic study of the best ways to introduce such precursors as amino acids and nucleotides appears to have been made. Devices are available for injecting solutions into leaves (e.g., Hagborg, 1970; Konate and Fritig, 1983).

Most plant leaves have rather large reserves of low-molecular-weight phosphorus compounds. By various manipulations, it is possible to lower or raise the overall concentration of phosphorus compounds not more than two- to threefold. Thus, in leaf tissue it has not been possible to carry out effective pulse–chase type experiments with phosphorus. With most organic compounds that can be used as labeled virus precursors, active leaves are continually providing an endogenous source of supply. Furthermore, plant tissues have the capacity to metabolize carbon compounds in many different ways, so that the labeled atom may soon appear in a wide range of low-molecular-weight compounds. For certain kinds of experiments it is useful to be able to label purified virus chemically *in vitro* to high specific activity. A variety of procedures are available (e.g., Frost 1977; Montelaro and Rueckert, 1975).

G. Metabolic Inhibitors

Inhibitors of certain specific processes in normal cellular metabolism have been widely applied to the study of virus replication. Three have been of particular importance: (i) actinomycin D, which inhibits DNA-dependent RNA synthesis but not RNA-dependent RNA synthesis; (ii) cycloheximide, which is used as a specific inhibitor of protein synthesis on 80 S cytoplasmic ribosomes; and (iii) chloramphenicol, which inhibits protein synthesis on 70 S ribosomes (e.g., in chloroplasts, mitochondria, and bacteria).

Results with these inhibitors must always be treated with caution, as they may have other diverse subsidiary effects in eukaryotic cells, which may make it difficult to interpret results. For example, actinomycin D may affect the size of nucleotide pools (Semal and Kummert, 1969), can cause substantially increased uptake of metabolites by excised leaves (Pratt and Matthews, 1971), may reduce uptake by infiltrated disks (Babos and Shearer, 1969), and may not suppress synthesis of certain species of host RNA (e.g., Antignus *et al.*, 1971).

Synthesis of the large polypeptide of ribulose bisphosphate carboxylase takes place in the chloroplasts on 70 S ribosomes, while the small polypeptide is synthesized on 80 S ribosomes in the cytoplasm. Owens and Bruening (1975) used these two polypeptides as an elegant internal control in their examination of the effects of chloramphenicol and cycloheximide on the synthesis of CPMV proteins.

H. Metabolic Compartmentation

If we count a membrane as a compartment, eukaryotic cells have at least 20 compartments. In their replication, plant viruses have adapted in a variety of ways to the opportunities provided by this intracellular metabolic diversity. In thinking about experiments on virus replication (particularly those involving the use of radioisotopes and/or metabolic inhibitors) we must take account of the fact that processes take place in cells that have a high degree of metabolic compartmentation. This exists in several forms: (i) in different cell types, which are metabolically adapted for diverse functions; (ii) in membrane-bound compartments within individual cells, for example, nuclei, mitochondria, chloroplasts, lysosomes, peroxisomes, and vacuoles; (iii) in isolatable stable complexes of enzymes; and (iv) in microenvironments created without membranes, by means of weakly interacting proteins, or unstirred water layers near a surface.

I. Sites of Synthesis and Assembly

Two general kinds of procedure have been used in attempts to define the intracellular sites of virus synthesis and assembly: (i) fractionation of cell components from tissue extracts followed by assay for virus or virus components in the various fractions and (ii) light and electron microscopy.

There are many difficulties involved in using cell fractionation procedures to locate sites of virus assembly.

1. Chloroplasts are fragile organelles, and a proportion of these are always broken. Chloroplast fragments cover a wide range of sizes and will contaminate other fractions.
2. Viruses such as TMV, occurring in high concentration, will almost certainly be distributed among all fractions, at least in small amounts.
3. Virus-specific structures may be very fragile and unable to withstand the usual cell breakage and fractionation methods.
4. If virus-specific structures are stable, they may fractionate with one or more of the normal cell organelles.
5. Virus infection may alter the way in which certain cell organelles behave on fractionation.

Considerable progress has been made with some viruses using cell fractionation procedures. However, in recent years we have learned more from ultrastructural studies, and most where both kinds of technique have been applied.

Viruses belonging to many different groups induce the development in infected cells of regions of cytoplasm that differ from the surrounding normal cytoplasm in staining and ultrastructural properties. These are not bounded by a clearly defined membrane but usually include some endoplasmic reticulum and ribosomes. They vary widely in size and may be visible by light microscopy. In varying degrees for different viruses there is evidence that these bodies are sites of synthesis of viral components and the assembly of virus particles. I use the term *viroplasm* to describe such inclusions. Some of the amorphous and "X-body" inclusions described in the older literature are of this type. The detailed structure of the viroplasms may be highly characteristic for different virus groups (Chapters 7, 8, and 11), and sometimes even for strains within a group. Proteins coded for by the viral genome probably bring about the formation of these characteristic structures. How this is accomplished is quite unknown. Immunocytochemical methods are being increasingly used to locate viral coded proteins within cells and tissues (e.g., Stussi-Garaud *et al.*, 1987) (see Fig. 7.12).

VIII. ERRORS IN VIRUS SYNTHESIS

The mechanisms by which genetic variability is generated in viruses are discussed in Chapter 13. Here we will discuss two different kinds of process that give rise to incorrect or defective virus particles.

A. Mixed Virus Assembly

Mixed virus assembly can be shown to take place *in vitro* between the RNA of one strain of a virus and the coat protein of another (e.g., Okada *et al.*, 1970;

Okada, 1986b), between RNA and protein from unrelated viruses (e.g., Matthews, 1966), and between one kind of RNA and two different coat proteins (e.g., Wagner and Bancroft, 1968; Talianski *et al.*, 1977). Of more interest is the formation of mixed virus particles *in vivo*.

When two viruses multiply together in the same tissue, some progeny particles may be formed that consist of the genome of one virus housed in a particle made partially or completely from the structural components of the other virus. Among enveloped viruses infecting animals, mixed infections may lead to the production of nucleoprotein cores of one virus enclosed in an envelope of the other. Such mixed particles, called *pseudotypes,* have not been observed with enveloped plant viruses. They will probably be found among the plant Rhabdoviridae.

Other kinds of mixed particle may be formed. Where the genome of one nonenveloped virus is encased in a protein shell made entirely of subunits of another virus (or strain) the phenomenon has been called *genomic masking.* When the protein coat consists of a mixture of proteins from the two viruses, it has been called *phenotypic mixing.*

I shall use the term phenotypic mixing for the process that gives rise *in vivo* to any virus particle consisting of components from two distinct viral parents. Dodds and Hamilton (1976) give an account of the methods used to study the phenomenon. Various studies on phenotypic mixing have been carried out with defective mutants of TMV whose protein will not form rods with the RNA when plants are grown at high temperature. When such strains are grown in mixed infections with type TMV (or some other strain able to form virus rods at the higher temperature) then a proportion of the progeny contain the mutant strain RNA in a rod made with the protein of the competent strain (Schaskolskaya *et al.*, 1968; Sarkar, 1969; Atabekov *et al.*, 1970b).

Such mixing may take place in leaves only under conditions where two viral RNAs are present and one functional coat protein is made (Atabekova *et al.*, 1975). On the other hand, Otsuki and Takebe (1978) showed that when protoplasts were inoculated with TMV together with tomato mosaic *Tobamovirus,* some of the individual progeny rods were coated with a mixture of the two coat proteins.

Strains of BYDV show aphid vector specificity (Chapter 14, Section III,G). When a strain of the virus normally transmitted in a particular vector was grown in oats in a double infection with a serologically unrelated strain not normally transmitted by the aphid, this latter strain was transmitted. Rochow (1970) showed that this transmission was due to the fact that some of the RNA of the second strain had been assembled into protein shells of the normally transmitted strain. A novel immunohybridization procedure has been developed to demonstrate directly that, in mixed infections in the field, a non-aphid-transmitted strain of BYDV became encapsidated in the protein of an aphid-transmitted strain (Creamer and Falk, 1990). Phenotypic mixing can occur between two unrelated helical viruses with different dimensions (TMV and barley stripe mosaic virus in barley) as shown by Dodds and Hamilton (1974). It has even been found between a helical virus (barley stripe mosaic virus) and an icosahedral one (BMV) (Peterson and Brakke, 1973).

Lettuce speckles mottle virus is associated with beet western yellows virus in an aphid-transmitted disease complex of sugar beet and other hosts. It appears to consist of naked ssRNA, which when in a mixed infection with beet western yellows virus becomes coated with the protein of this virus (Falk *et al.*, 1979a,b).

The formation of distinctive inclusion bodies has been used to confirm that two unrelated viruses can replicate in the same cell, for example, TMV and TEV in tobacco (Fujisawa *et al.*, 1967), TuMV and CaMV in *Brassica perviridis* (Kamei *et al.*, 1969), and soybean mosaic *Potyvirus* and bean pod mottle *Comovirus* in soybean (Lee and Ross, 1972).

The existence of phenotypic mixing also suggests that two unrelated viruses or two related strains can replicate together in the same cell at least under some conditions. Other kinds of evidence support this view (Chapter 13).

Nevertheless, in tobacco leaves doubly infected with TMV plus PVX or PVY plus PVX, no assembly of one viral RNA in the coat protein of another could be detected (Goodman and Ross, 1974b). A likely reason for this is that whereas closely related strains of a virus might replicate in the same region of the cell, different viruses may be assembled from components accumulated in separate sites or viroplasms in the same cell. Such separation may not always be complete.

Efficient and specific virus assembly would be favored by the localization of the RNA and protein subunits in a compartment within the cell. There are several reasons for this. First, if *in vivo* assembly is due to random meeting between protein subunits, then maintenance of a high local concentration of these would favor efficient assembly. Second, since subunits can pack around nonviral RNA of appropriate size, and since insignificant amounts of nonviral RNA are usually present in virus particles, free host RNA must be largely excluded from the assembly sites. Third, *in vitro* studies show that aggregation of subunits is markedly dependent on ionic environment and pH. These specific conditions differ *in vitro* for different viruses; and fourth, uncoated RNA must be protected from attack by nucleases.

Nevertheless, with some viruses significant amounts of host RNA may be incorporated into viruslike particles or pseudovirions. Such particles have been reported as making up to 2.5% of preparations of various strains of TMV (Siegel, 1971; Rochon *et al.*, 1986). Most of the encapsidated host RNA is the 5' region of 18 S ribosomal RNA (Rochon *et al.*, 1986). The site for initiation of this packaging has been located within a 43-nucleotide region beginning at position 157 from the 5' terminus of the rRNA. This sequence has limited similarity to the TMV assembly initiation sequence (see Fig. 7.8) but it can fold to give a stem-loop structure (Gaddipati *et al.*, 1988).

The mRNA for the large subunit of ribulose 1,5-bisphosphate carboxylase, which is coded by chloroplast DNA, is encapsidated in TMV coat protein in infected cells. This mRNA was found to contain at least three sites that are capable of reacting with coat protein aggregates *in vitro* to initiate rod formation (Atreya and Siegel, 1989). The most reactive site had significant sequence similarity to the initiation site in TMV.

B. Defective Interfering Particles

Defective interfering RNAs (DI RNAs) and DI particles are widespread in animal virus infections. These DI particles have the following properties: (i) they usually are derived from the viral genome, that is, consist mainly or entirely of genomic nucleotide sequences; (ii) they reduce the yield of helper virus; and (iii) they cause milder disease symptoms. Adam *et al.* (1983) described a population of DI-like particles associated with a plant rhabdovirus that arose after 30 passages. Some cultures of soilborne wheat mosaic *Furovirus* contained an additional short rod. The RNA in this rod was derived from genomic RNA2 (the shorter genomic RNA) (Shirako and Brakke, 1984). The sizes and proportions of the RNAs shorter than RNA2 varied from plant to plant, suggesting that they arise by continued spontaneous deletion mutation. Shorter than normal dsRNA segments are associated with transmission-defective isolates of WTV. Nuss and Summers (1984) showed that these RNAs are formed by the deletion of up to 85% of a genomic RNA segment, giving rise to terminally conserved RNAs that are functional with respect to transcription, replication, and packaging. These isolates interfere with standard virus production in leafhopper cell monolayers (Reddy and Black, 1977)

MacDowell *et al.* (1986) characterized a less than full-sized class of DNA from preparations of tomato golden mosaic *Geminivirus*. These DNAs, of slightly varying length, were derived only from genomic DNA B, by a deletion of about half the DNA. These DNAs delayed symptom expression and to this extent were like the DI nucleic acids of animal viruses.

Hillman *et al.* (1987) and Morris and Hillman (1989) described an abnormal RNA from a culture of TBSV that met all the criteria noted earlier for a DI RNA. The RNA was about 396 nucleotides long and was derived from the genomic RNA by six internal deletions, the 5' and 3' sequences being conserved. Two of the deletions were large (1180 and 3000 nucleotides) while the others were much smaller (Fig. 6.16). Coinoculation of the small RNA depressed virus synthesis in whole plants and attenuated disease symptoms. Although the DI RNA could represent 60% of viral specific RNA in leaf extracts, only about 3–4% of the encapsidated RNA was DI RNA. The DI RNA probably replicates by the same mechanism as the viral RNA. Experiments in protoplasts showed that the DI RNA suppresses replication of genomic TBSV RNA (Jones *et al.*, 1990). A similar DI RNA has been described from a culture of *Cymbidium* ring spot *Tombusvirus* (Burgyan *et al.*, 1989).

Ismail and Milner (1988) isolated DI particles from *Nicotiana edwardsonii* plants chronically infected with sonchus yellow net rhabdovirus. Most of the DI particles were 73–86% as long as the standard virus. Alone they were noninfectious, but when coinoculated with complete virus, they were replicated to a greater extent than the infectious particles. A DI RNA associated with TCV was shown to be a mosaic molecule containing sequences derived from TCV together with a block of nucleotides corresponding to a 5' sequence found in satellite RNAs associated

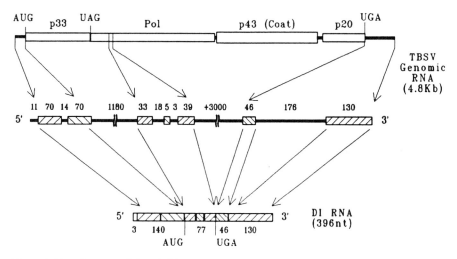

Figure 6.16 Alignment of the TBSV genomic (top) and DI RNA (bottom) sequences. The middle diagram shows the size (nucleotides) and position of the regions of genomic sequence deleted to produce the DI RNA species. (From Morris and Hillman, 1989.)

with TCV (Li *et al.*, 1989). This DI RNA was also unusual in that its presence made the disease caused by TCV more severe.

It is generally believed that such DI RNAs interfere with the helper virus by competing for the viral RNA-dependent RNA polymerase, and for coat protein during virus assembly. The generation of DI RNAs is probably by a "copy choice" mechanism in which the polymerase occasionally makes an error during replication by jumping to another RNA template molecule or to a different part of the same molecule. These various DI RNAs should be useful in exploring the molecular basis of disease modulation.

IX. FUTURE STUDIES ON VIRUS REPLICATION

I conclude this chapter with brief mention of some technical developments that are playing or may play an important role in the future in developing our further understanding of virus replication. They all depend on recombinant DNA technology.

A. *In Vitro* Mutagenesis and Recombinant Viruses

The ability to introduce, *in vitro,* base changes, deletions, or insertions at defined sites in a viral genome and the construction of hybrid viruses *in vitro* open up many possibilities for experiments that will illuminate the functions of viral genomes and the proteins for which they code. The methods are already being used widely, and some examples are given in Chapters 7 and 8. The possibilities have been reviewed by Botstein and Shortle (1985).

B. Transgenic Organisms

Expression of single viral genes in transgenic plants, as discussed in Section V,D,2,opens up many possibilities for studying various aspects of virus replication *in vivo*. In addition, a vector system has been developed that would also allow the expression of plant viral genes, singly or in combination, in a variety of mammalian systems. Dougherty *et al.* (1986) constructed a chimeric transcription unit consisting of a vaccinia virus promoter linked to a 2400-base-pair dsDNA copy of the 3′ end of TEV RNA. This was introduced into the vaccinia virus genome. The TEV genetic information was transcribed into RNA molecules that were correctly processed. The protein products were recognized by antisera to the appropriate TEV proteins. However, there must always be a question mark concerning the relevance of results obtained in such heterologous systems.

C. *In Situ* Transcription

Tecott *et al.* (1988) describe a procedure in which mRNA can serve as a template for reverse transcriptase *in situ,* within fixed tissue sections. An oligonucleotide complementary to a sequence in the mRNA of interest is prepared and hybridized *in situ*. This provides the primer for reverse transcriptase action.

During this step, radioactive nucleotides are included to allow for subsequent radioautographic localization of the mRNA of interest. The transcripts may then be eluted from the section and used in other procedures, for example, in cloning.

D. The Polymerase Chain Reaction

The polymerase chain reaction (PCR) is a major recent development in DNA technology. A limitation of cloning techniques using restriction endonucleases has been the difficulty in detecting DNA sequences occurring as a very low proportion of the DNA sample, for example, single-copy genes or nucleic acid associated with a low level of virus infection or integration. The PCR overcomes this difficulty by allowing a DNA sequence of interest to be amplified several million times, so that the target DNA can be visualized on a gel. An introductory account of the procedure is given by Marx (1988).

To apply the technique, the nucleotide sequence of the DNA of interest, or sequences near it, must be known. The PCR is a mechanism for amplifying the DNA between two oligonucleotide primers. Two primers are synthesized to be complementary to known sequences in the target DNA on opposite strands usually about 150–500 base pairs apart. Design of the primers is a very important aspect of the process. The dsDNA is first heated to separate the strands in the presence of the primers. When the temperature is reduced the primers anneal. At a selected temperature a thermophilic DNA polymerase extends the primers to copy the DNA. Successive cycles of denaturation and synthesis achieve logarithmic amplification of the DNA. In principle each new cycle doubles the amounts of product DNA. However, the precise conditions in the incubation medium can have a marked effect on yield of product. Technical details can be found in Saiki *et al.* (1988). A major technical problem is the possibility that nontarget DNA may be amplified.

7

Replication II: Viruses with Single-Stranded Positive Sense RNA Genomes

Twenty-eight groups of plant viruses with ss positive sense RNA genomes have been established, and we have a significant body of information about the replication of 18 of these. Thus it might be thought sufficient to describe the replication of a few representative groups. There are two main reasons why this would not be satisfactory. First, study of the different groups has proceeded very unevenly, both for technical reasons and because of particular interests in various laboratories. Thus the kind and depth of information we have for different groups varies quite widely. For example, as far as virus assembly is concerned, we know most about TMV. When it comes to regulatory mechanisms in the viral genome, we know most for BMV; while for the tymoviruses, interest has centered, until recently, on the relationship between cytological effects and virus replication.

The second reason why it is difficult to choose a few representative groups is that these viruses have evolved a variety of strategies for gene expression. Figure 6.2 summarized 5 mechanisms by which viral RNA genomes achieve efficient translation in a eukaryotic system. Here I have included translational frameshifting as having the same effect as read-through proteins. In principle, these 4 adaptations, in various combinations, could give rise to 15 different genome strategies. On present knowledge, the groups of ss positive sense RNA plant viruses have adopted 7 of these. They are summarized in Table 7.1. The table includes 16 groups for which genome maps are available plus 2 other groups that can be included: *Cucumovirus* by analogy with *Bromovirus,* and *Ilarvirus* by analogy with alfalfa mosaic virus. If we include the plant reoviruses (one strategy—multipartite genome) and the plant rhabdoviruses (probably one strategy—subgenomic RNAs) and tomato spotted wilt *Tospovirus* (three strategies—multipartite genome, subgenomic RNAs, and read-through protein), then 21 RNA plant virus groups and families have evolved 8 different strategies for survival in eukaryotic cells.

The organization and expression of the genomes of representative viruses belonging to the groups listed in Table 7.1 are summarized in this chapter, in the order

Table 7.1

Summary of Genome Strategies Adopted by 18 Single-Stranded Positive Sense RNA Plant Virus Groups

Number	Strategy (see Fig. 6.2)		Virus group	Number of ORFs	Number of proteins coded
I	One strategy	Polyprotein	*Potyvirus*	1	8
II	One strategy	Subgenomic RNA	*Potexvirus*	5	4–5
			Tombusvirus	5	5
III	Two strategies	Subgenomic RNA plus read-through or frameshift protein	*Tobamovirus*	5	4–5
			Luteovirus	6	6–7
			Carmovirus	5	5–7
IV	Two strategies	Subgenomic RNAs and polyprotein	*Tymovirus*	3	3–5
			Sobemovirus	4	4–5
V	Two strategies	Multipartite genome and polyprotein	*Comovirus*	2	≈9
			Nepovirus	2	≈6
VI	Two strategies	Subgenomic RNAs and multipartite genome	*Bromovirus*	4	4
			Cucumovirus	4	4
			Alfalfa mosaic virus	4	4
			Ilarvirus	4	4
			Hordeivirus	7	7
VII	Three strategies	Subgenomic RNAs, multipartite genome, and read-through protein (or frameshift)	*Tobravirus*	5	5
			Furovirus	9	6–9
			Dianthovirus	4	4

shown in the table, along with other aspects of the replication process. One additional virus, pea enation mosaic virus, is discussed because of its apparent association with the nucleus. Some other virus groups are described in Chapter 9 in relation to satellite viruses and satellite RNAs.

Organization and expression of the genome is summarized in a map for each group. The information contained in these maps comes from the relevant references in the text. The conventions used in the maps are summarized in Table 7.2.

There is no doubt that further experimental work will lead to changes in particular aspects for most of these genome maps. In general, the nucleotide sequence data are soundly based. Doubts usually revolve about which ORFs actually give rise to functional proteins *in vivo,* and about details of the translation mechanisms. These question marks are usually noted in the text rather than the diagrams.

The arrangement of groups by genome strategy in Table 7.1 and this chapter brings certain groups together that, on other grounds, would be in separate categories. The arrangement I have used here is for convenience and should not be regarded as a proposed classification. Problems concerning virus classification are discussed in Chapter 17.

I. ONE STRATEGY: POLYPROTEIN

A. *Potyvirus* Group

The potyviruses are a large and economically very important group of viruses. A great deal has been learned over the past few years about the organization of the

Table 7.2
Conventions Used in the Genome Maps for RNA Viruses

Scale for genome length in 1000s of nucleotides

0   1   2   3 kb

Genomic RNA
2200 nt

Genome or subgenomic RNAs with length indicated in nucleotides

End groups, and structures near termini

5' cap = me^{7}Gppp

VPg = covalently bonded protein

5' = no special structure

? = not known

poly(A) 3' = a run of As usually of variable length (not included in quoted genome length)

tRNA^{tyr}3' = 3' terminal nucleotides can fold to give a tRNA like structure accepting the amino acid indicated

3' OH = no special structure at 3' termini

AUG   UAA
3495   4917

= open reading frame 1, with a coding potential of Mr ≈ 54000
position of initiation and termination codons indicated by number of nucleotides from 5' terminus of
the genomic RNA

ORFs are numbered in one series for the whole genome

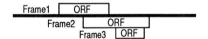

indicates a leaky termination codon giving rise to a read-through protein of 80K

Frame1   ORF
  Frame2   ORF
    Frame3   ORF

Where more than one reading frame is involved they are arranged as indicated

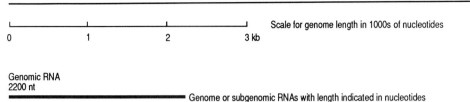 = Protein product

genome and the nature and functions of the gene products through the application of molecular biological techniques. In contrast to most other plant viruses, some stable, nonstructural gene products may accumulate in substantial amounts in infected cells, greatly facilitating their study. Nucleotide sequences have become available for TEV (Allison *et al.*, 1985, 1986), TVMV (Domier *et al.*, 1986;

Hellman *et al.*, 1988; Shahabuddin *et al.*, 1988), plum pox virus (Maiss *et al.*, 1989; Lain *et al.*, 1989), and PVY (Robaglia *et al.*, 1989). The best known members of the group are tobacco etch virus (TEV) and tobacco vein mottling virus (TVMV). The expression and function of their gene products have been reviewed in detail by Dougherty and Carrington (1988). At present most is known about the gene products of TEV and their expression. Thus the following sections deal mainly with this virus. Future studies will be greatly facilitated by the ability to produce infectious genomic RNA transcripts *in vitro* from cDNA copies (Domier *et al.*, 1989).

1. Genome Structure

At the 5' terminus there is a covalently linked VPg. The main features of the TEV genome are: (i) a 5' noncoding region of 144 nucleotides rich in A and U; (ii) a single large ORF of 9161 nucleotides initiating at residue 145–147, which could code for a polyprotein with about 3000 amino acids (about 340,000 MW); and (iii) a 3' untranslated region of 190 bases terminating in a poly(A) tract (Fig. 7.1). The coding sequence shows amino acid sequence similarities with several other virus groups (Domier *et al.*, 1987) (Chapter 17, Section V,A). The genomes of the other potyviruses that have been sequenced have the same plan with some variations in length.

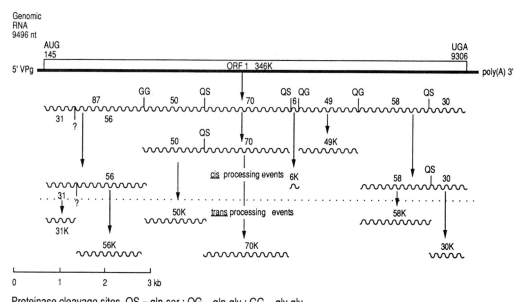

Figure 7.1   Organization and expression of a *Potyvirus* genome (TEV).

2. Gene Products

Proteins in the Virus

About 2000 molecules of a coat protein of $M_r \simeq 20,000$ (20K) make up the rod-shaped particle of TEV. The coat protein is probably also involved in successful aphid transmission of potyviruses (Shaw *et al.*, 1990). In addition, there is a single molecule of a VPg linked covalently to the 5′ terminus. Early work suggested that this is about 6K in size for TEV (Hari, 1981) and very much larger (24K) for TVMV (Siaw *et al.*, 1985). However, more recent work suggests that the VPg of TEV is much larger (Shaw *et al.*, 1990). Thus a role for the 6K polypeptide has not been established.

Cytoplasmic Pinwheel Inclusion Protein

A protein monomer that aggregates to form characteristic pinwheel inclusion bodies is found in plants infected by all potyviruses (Edwardson, 1974) (see Fig. 11.23), including those transmitted by mites (Brakke *et al.*, 1987a). *In vitro* translation studies have shown that this protein is coded by the virus (Dougherty and Hiebert, 1980). For TEV it has a MW $\simeq$ 70K. It has been suggested that it is involved in cell-to-cell movement of virus (Langenberg, 1986a), but this seems unlikely. From nucleotide sequence similarities with similar proteins in picornaviruses and comoviruses it is likely to be involved in RNA replication (Dougherty and Carrington, 1988). In particular, it contains the highly conserved NTP binding motif.

Cytoplasmic Amorphous Inclusion Protein
and the Helper Component Protein

The cytoplasmic amorphous inclusion body protein has been detected in some but not all potyviruses. The helper component protein is a viral-coded protein necessary for aphid transmission of the virus (Chapter 14, Section III,D,6). Several lines of evidence indicate they are the same proteins, in whole or in part, and that the gene is located in the 5′ region of the genome: (i) they comigrate in gel electrophoresis; (ii) amorphous inclusion body protein antibody reacts with the main protein in purified helper protein extracts; (iii) antiserum to a partially purified helper component blocked the biological activity of homologous helper component and also reacted with a cell-free 75K translation product from homologous RNA; and (iv) amorphous inclusion protein and 5′ cell-free translation products have similar partial proteolysis patterns (de Mejia *et al.*, 1985; Thornbury *et al.*, 1985). The helper component protein of all the potyviruses that have been sequenced contains a potential "zinc finger" metal binding site of unknown function (Robaglia *et al.*, 1989). Helper component activity is located in the 56K protein coded near the 5′ terminus (Fig. 7.1). This protein is bifunctional and probably has two domains. The other activity is a proteinase which cleaves the Gly-Gly site. On the basis of site-directed mutagenesis studies, this proteinase is of the cysteine class of proteolylic enzyme. (Oh and Carrington, 1989).

Nuclear Inclusion Proteins

The nuclear inclusion bodies found in TEV and some other *Potyvirus* infections are composed of two proteins of $M_r \simeq 58K$ and 49K. These are coded for by all potyviruses but only some produce the inclusions. The 49K protein of TEV is a viral-coded proteinase related to the trypsin superfamily of serine proteinases, that excised from the polyprotein by autoproteolysis (Carrington and Dougherty, 1987a,b). It is also responsible for four other cleavages in the 3' two-thirds genome (Fig. 7.1). The 58K nuclear inclusion protein has some sequence similarity with polymerases of some other viruses (Allison *et al.*, 1986; Domier *et al.*, 1987), and therefore may have this function.

Other Proteins

Computer analysis indicates that a 31K protein arising at the 5' end of the polyprotein has sequence similarity with the TMV 30K cell-to-cell movement protein (Domier *et al.*, 1987). Another expected protein, of 50K, has not yet been found in infected plants but based on sequence similarities with a poliovirus proteinase, Dougherty and Carrington (1988) predict that it has a proteinase function.

3. Organization of the Genes

Understanding of gene organization in potyviruses has come from a knowledge of the nucleotide sequence of the genome, from cell-free translation studies, and from analysis of potyviral proteins from infected plants, as outlined by Dougherty and Carrington (1988) (Fig. 7.1).

4. Expression and Processing of the Gene Products

Five sites that are cleaved by the 49K proteinase have been identified in the 346K polyprotein (Fig. 7.1). These cleavage sites (gene product boundaries) are defined by the following sequence of seven amino acids in the polyprotein:

P6 P5 P4 P3 P2 P1   P'1
-Glu-Xaa-Xaa-Tyr-Xaa-Gln-Ser (or Gly)

This sequence was conserved in 10 geographical isolates of TEV (Dougherty *et al.*, 1988). The cleaved bond is between Gln–Ser (or Gln–Gly). Site-directed mutagenesis showed that Gln is essential for cleavage by the 49K proteinase, and that the naturally conserved amino acids were strongly preferred (Dougherty *et al.*, 1988; Carrington *et al.*, 1988). To determine whether the seven-amino-acid sequence was sufficient to provide a cleavage site, Carrington and Dougherty (1988) inserted the sequence into a protein and showed that the protein could then be cleaved at the Gln–Ser site. However, additional amino acid sequences outside the seven-amino-acid sequence affected the efficiency of cleavage. Within the seven-amino-acid sequence the P6, P3, and P1 determinants are essential in defining the cleavage site (Dougherty *et al.*, 1989a). The P2, P4, and perhaps P5 positions appear to regulate the rate of cleavage. The polyprotein method of gene expression prevents potyviruses from using transcriptional control to vary the production of different gene products. Thus the TEV 49K proteinase may represent a means of posttranslational gene regulation (Dougherty *et al.*, 1990).

Data from experiments using site-directed mutagenesis suggest that the catalytic triad of amino acids in the 49K proteinase of TEV is probably His[234], Asp[269], and Cys[339] (Dougherty *et al.*, 1989b).

Cell-free protein synthesis studies have shown that the 49K proteinase is autocatalytically cleaved in *cis* (intramolecularly) from the polyprotein. Subsequently it carries out other cleavages in *trans* (intermolecularly) (Dougherty and Carrington,

1988). It is not certain whether any complete polyprotein molecules are synthesized or whether all are cleaved at least once before synthesis is complete. Figure 7.1 summarizes a proposed scheme for processing of the polyprotein. Studies in transgenic plants have revealed that a third proteinase activity is necessary for cleavage between the 31K and 56K proteins indicated in Fig. 7.1 (Carrington *et al.*, 1990). This activity may be partly or wholly host encoded. An analysis of the *in vitro* translation products of bean yellow mosaic virus RNA (Chang *et al.*, 1988b) and peanut mottle virus (Xiong *et al.*, 1988) indicate a genome organization similar to that in Fig. 7.1.

5. Other Studies on Protein Synthesis *in Vitro*

Translation in a reticulocyte system of full-length genomic RNA of TEV isolated from infected tobacco leaves gave several large polypeptides reacting with antisera to known TEV-coded proteins, but the sizes were mostly larger than the proteins found *in vivo* (Dougherty, 1983).

It has recently been shown that the initiation of protein synthesis on poliovirus RNA involves binding of a ribosome to an internal sequence within the 5' noncoding region (Pelletier and Sonenberg, 1988). In view of other similarities (Chapter 17, Section V,A) it may be that a similar mechanism is involved with the potyviruses.

Shields and Wilson (1987) examined the polypeptides formed *in vitro* by two systems programmed with turnip mosaic *Potyvirus* (TuMV) RNA. Wheat germ extracts yielded predominantly a range of labeled products of 55K or less, probably due to an endogenous proteinase. In the rabbit reticulocyte system there was clear evidence for early synthesis of a large protein followed by subsequent cleavage, but precursors and products could not be related to a scheme such as that in Fig. 7.1.

6. Subgenomic RNAs

Shields and Wilson (1987) found no evidence for the presence of subgenomic RNAs in their preparations of TuMV. Likewise, Dougherty (1983) could find no evidence for authentic subgenomic RNAs in total RNA preparations made from TEV-infected leaves.

7. *In Vivo* RNA and Protein Synthesis

Some of the proteins found *in vivo* have been discussed in the preceding Section 2. Donofrio *et al.* (1986) have isolated an RNA-dependent RNA polymerase activity from corn infected with maize dwarf mosaic *Potyvirus*. The activity was solubilized and attributed to a MW $\simeq$ 160K protein. The subunit structure of the polymerase has not been characterized. Vance and Beachy (1984) found a genomic-length RNA associated specifically with active polyribosomes in extracts of soybean leaves infected with soybean mosaic *Potyvirus*. They concluded that this RNA is the only viral RNA translated *in vivo*. Full-length viral RNA, the complementary strand, and a ds viral RNA have been found associated with the chloroplast fraction in tissue extracts (Gadh and Hari, 1986). However, there was no evidence that viral RNA synthesis took place in the chloroplasts.

II. ONE STRATEGY: SUBGENOMIC RNA

A. *Potexvirus* Group

The genomes of four potexviruses have been sequenced. Genomic structure is highly conserved within the group. The assembly *in vitro* of papaya mosaic *Potexvirus* (PapMV) from its RNA and protein subunits is the best understood among the plant viruses with flexuous rods.

1. Genome Structure and Strategy

Potexviruses have an ss positive sense RNA genome with a MW in the range 2.0–2.5×10^6. The 5′ end has an m^7 Gppp cap and the 3′ end is polyadenylated. The genome of white clover mosaic virus (WCMV) has been sequenced (Harbison *et al.*, 1988; R. L. S. Forster *et al.*, 1988). It has five ORFs, with the arrangement and coding capacity shown in Fig. 7.2. PVX strains Russian (Skyrabin *et al.*, 1988a) and X3 (Huisman *et al.*, 1988), narcissus mosaic virus (Zuidema *et al.*, 1989), and PapMV (Sit *et al.*, 1989) have also been sequenced and all show the same genome organization.

Forty of the most 5′-terminal nucleotides show strong homology among the potexviruses. The 147K ORF begins 108 nucleotides from the 5′ terminus. There are three spaced termination codons indicating two possible read-through proteins of MW $\simeq$ 150K and 151K. No evidence has been found for the presence of these proteins *in vivo*. The protein encoded by the 147K ORF has domains of sequence similarity with the presumed polymerase genes of other RNA viruses. It corresponds to a 160K protein translated *in vitro* from the genomic RNA (Forster *et al.*, 1987). A full-length cDNA clone of PVX has been constructed and infectious RNA

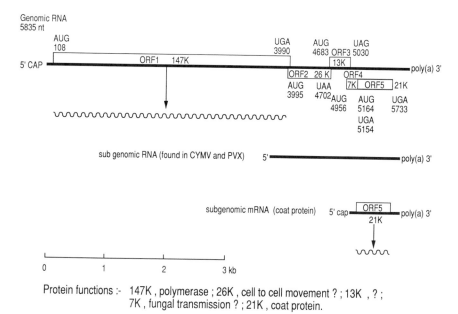

Protein functions :- 147K , polymerase ; 26K , cell to cell movement ? ; 13K , ? ; 7K , fungal transmission ? ; 21K , coat protein.

Figure 7.2 Organization and expression of a *Potexvirus* genome (white clover mosaic virus).

transcripts obtained (Hemenway *et al.*, 1990). This will facilitate detailed analysis of genome functions.

The nucleotide sequence in the 3'-terminal region of PapMV (AbouHaidar, 1988) and clover yellow mosaic virus (AbouHaidar and Lai, 1989) contains the coat protein genes, which are similar in structure to those of other potexviruses. There is a polyadenylation signal (AAUAAA) at various distances from the 3' terminus in different potexviruses.

The genome of potato M *Carlavirus* has been partially sequenced from the 3' terminus (Rupasov *et al.*, 1989). The 25K, 12K, and 7K ORFs showed significant amino acid sequence similarities with several nonstructural proteins of potexviruses. Sequence similarities with other virus groups are discussed in Chapter 17, Section V,B.

2. RNA and Protein Synthesis

Subgenomic RNAs for the coat protein gene have been found for narcissus mosaic virus (Short and Davies, 1983), clover yellow mosaic virus (Bendena *et al.*, 1987), and white clover mosaic virus (Forster *et al.*, 1987). For these last two viruses the coat protein mRNA was found in polyribosomes, indicating that they in fact function *in vivo*. Forster *et al.* (1987) also found the genomic RNA in association with polyribosomes. In the rabbit reticulocyte system the genomic RNA directed mainly the synthesis of a protein with MW $\simeq$ 160K. No dsRNA corresponding to the coat protein subgenomic RNA could be detected in tissues infected with PapMV or foxtail mosaic *Potexvirus*, but genome-sized dsRNAs were detected (Mackie *et al.*, 1988).

Bendena and Mackie (1986) showed that the genomic RNAs of three potexviruses template for a 160–182K nonstructural protein in the reticulocyte cell-free system. In the same system, clover yellow mosaic virus coded for one major polypeptide with MW = 182K, while in pea protoplasts the 182K protein and a 22K coat protein were produced (Brown and Wood, 1987). On the other hand, Guilford and Forster (1986) detected a series of five subgenomic RNAs in leaves infected with daphne X *Potexvirus*. The significance of these remains to be determined.

3. Site of Replication *in Vivo*

When protoplasts were infected by PVX, particles absorbed end-on to the plasma membrane and were taken into the cell in pinocytotic vesicles (Honda *et al.*, 1975). The virus had a time course of multiplication in tobacco protoplasts similar to that found for TMV (Otsuki *et al.*, 1974). Cycloheximide completely inhibited virus production, indicating that viral protein is synthesized on 80 S ribosomes. Large aggregates of virus particles developed in the cytoplasm as infection proceeded, but the nucleus, chloroplasts, and other organelles remained structurally normal.

In infected leaves, a depolymerized form of PVX protein has been detected serologically (Shalla and Shepard, 1970). Large masses of viral antigen have been observed at an early stage in the cytoplasm and vacuole of *Vicia* cells infected with clover yellow mosaic virus (Schlegel and Delisle, 1971). Shalla and Petersen (1973) detected PVX antigen in infected protoplasts many hours before virus particles were seen, but the significance of these observations has been questioned by Honda *et al.* (1975).

Strains of PVX produce amorphous inclusions, which are variable in size and number in different host species (Bawden and Sheffield, 1939). Viewed by light microscopy they appear rather similar to inclusions induced by viruses belonging to some other groups, but their ultrastructure is substantially different (e.g., Shalla and Shepard, 1972). At present there is no convincing evidence to show that they are the sites of virus synthesis and assembly.

The inclusion bodies contain a range of normal components (mitochondria, dictyosomes, endoplasmic reticulum, and vacuoles) as well as many virus particles stacked in arrays or randomly oriented. Laminate sheets, often in a scroll-like form, are a prominent feature. Some sheets are heavily beaded on both sides with small densely staining bodies, with a diameter smaller than that of 80 S ribosomes. Other sheets are smooth (Shalla and Shepard, 1972) (Fig. 7.3). The beads and the sheets are unrelated antigenically either to the virus or to the depolymerized viral protein.

4. Virus Assembly

Among the plant viruses with flexuous rod-shaped particles, most is known about the assembly *in vitro* of the potexviruses and in particular PapMV (reviewed by AbouHaidar and Erickson, 1985). The process shows both similarities to and differences from the assembly of TMV. Substantially less detail has been revealed, because the three-dimensional structure of the coat protein has not been established at high resolution. As with TMV, the coat protein of PapMV forms various kinds of aggregates in the absence of RNA, depending on such factors as pH, ionic strength,

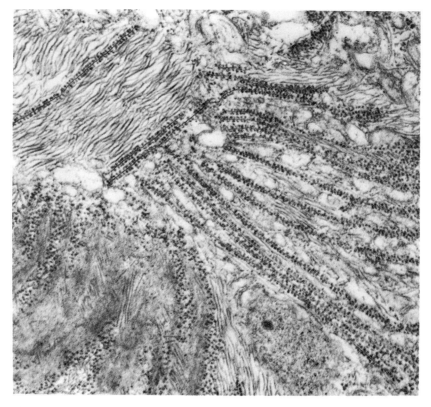

Figure 7.3 Cytoplasmic inclusion induced by PVX in a tobacco cell. Section showing beaded sheets and regularly arranged flexuous virus rods. (Courtesy of T. Shalla.)

and temperature. At pH values below neutrality, helical rods are formed. At pH 7, 14 S double disks are produced. These contain two layers of nine subunits.

The subassembly of subunits that initiates rod formation has not been unequivocally established, but it is probably the 14 S double disk. Five nucleotides are associated with each subunit in the virus rod. The site for nucleation on the RNA has been established as being at the 5' end (AbouHaidar and Bancroft, 1978).

The sequence of the first 158 nucleotides at the 5' end of PapMV is known (Lok and AbouHaidar, 1986). These workers implicated a set of eight pentamer repeats (N(C or A)AAA) at the 5' terminus as having a role in the initiation site. However, no such repeating sequence was found in some other potexviruses (white clover mosaic virus, narcissus mosaic virus, and PVX, R. L. S. Forster *et al.*, 1988). Nevertheless, all the available 5' sequences contain 5' cap GAAAA as the first nucleotides (Zuidema *et al.*, 1989). This might be the initiation site for the coating of genomic RNA. *Potexvirus* subgenomic RNAs do not contain this 5' sequence and most are not encapsidated. However, the narcissus mosaic virus subgenomic RNA is encapsidated (Zuidema *et al.*, 1989). The specificity of assembly of PapMV is markedly pH dependent. At pH 8.0 only the homologous RNA and that of a closely related virus are encapsidated. At lower pH, selectivity is much less stringent.

In vitro studies on the association of coat protein with viral RNA to form rods have been made with clover yellow mosaic *Potexvirus* (AbouHaidar, 1981). The site for initiating assembly is within 150–200 nucleotides of the 5' end of the RNA. Most efficient assembly took place at neutral pH in a low ionic strength medium.

B. *Tombusvirus* Group

The particle structure of the type member of this group, tomato bushy stunt virus (TBSV), is now well understood (Chapter 5, Section III,B,2). The genome consists of one molecule of linear positive sense ssRNA about 4.7 kb in length that has neither poly(A) nor a tRNA-like structure at the 3' terminus. The complete nucleotide sequence has been established for cucumber necrosis virus (Rochon and Tremaine, 1989), cymbidium ring spot virus RNA (Grieco *et al.*, 1989), and TBSV (Hayes *et al.*, 1988e; Hillman *et al.*, 1989; Hearne *et al.*, 1990). The organization is similar for all these viruses. That for TBSV is shown in Fig. 7.4.

Three RNA species were found both in purified virus particles of cymbidium ring spot virus and in virus-infected tissue (Russo *et al.*, 1988). The largest was genomic RNA. The others were two subgenomic RNAs 3' coterminal with the genomic RNA. Infectious TBSV RNA has been synthesized *in vitro* from cDNA clones (Hearne *et al.*, 1990). Other small viral specific dsRNAs that appear to be unrelated to the replication cycle have been found for TBSV (Hayes *et al.*, 1984). *In vitro* translation studies showed that the only product of the smaller (1 kb) subgenomic RNA is a 22K protein that is not the coat protein. The 43K coat protein was translated from the 5' end of the larger subgenomic RNA and is therefore internally located in the genome. Satellite RNAs may also be present (Chapter 9, Section II,B,3). DI RNAs may also be present (see Fig. 6.16).

Both ss and dsRNAs corresponding to the genomic and two subgenomic RNAs have been found in TBSV-infected tissue (Hayes *et al.*, 1988f). ssTBSV RNAs isolated from polysomes was translated *in vitro* to give a MW = 37K protein from genomic RNA, 40K from the 2.0-kb RNA, and 22K from the 0.9-kb RNA. A

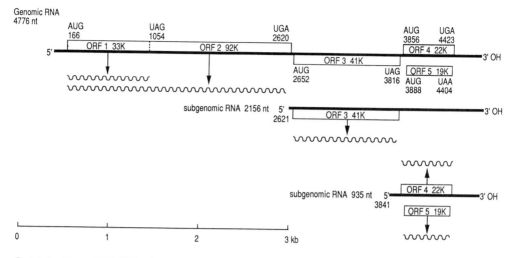

Figure 7.4 Organization and expression of a *Tombusvirus* genome. (TBSV, cherry strain).

protein of MW = 90K was also detected. This may be the read-through product of the 37K protein. It has not yet been established whether proteins of ≈22K and 19 are both translated from the smaller subgenomic RNA.

In cells infected with tombusviruses, large numbers of virus particles are scattered randomly through the cytoplasm and sometimes in crystalline arrays. Characteristic multivesiculate membranous bodies that contain dsRNA are found in the cytoplasm. They are characteristic for the group (Martelli and Russo, 1981). In infected cells the chloroplasts develop large flask-shaped invaginations that always face a multivesiculate body. These bodies together with the invaginations are probably the sites of viral RNA synthesis. Another type of inclusion known as dense granules consists mainly of coat protein (Bassi *et al.*, 1986; Appiano *et al.*, 1986).

III. TWO STRATEGIES: SUBGENOMIC RNAs PLUS READ-THROUGH PROTEIN

A. *Tobamovirus* Group

The replication of TMV has been studied more than any other plant virus. For this reason it is discussed in more detail than for the other groups. In particular, a great deal of experimental work has been carried out on the assembly of the virus *in vitro* from its components.

1. Genome Structure

The ss positive sense RNA of TMV has been sequenced (Goelet *et al.*, 1982). The sequence revealed several closely packed ORFs. An m⁷Gppp cap is attached to the first nucleotide (guanylic acid). This is followed by an untranslated leader sequence of 69 nucleotides. This initiates an ORF that codes for a MW = 126K protein. Experiments described next show that the termination codon for this protein (UAG) is "leaky" and that a second larger read-through protein is possible. This read-through protein has a MW = 183K. The terminal five codons of this read-through protein overlap a third ORF coding for a protein of MW = 30K. This ORF terminates two nucleotides before the initiation codon of the fourth ORF, which is closest to the 3' terminus. It codes for a MW = 17.6K coat protein. As discussed in the following, the two smaller ORFs at the 3' end of the genome are translated from subgenomic RNAs. The 3' untranslated sequences can fold in the terminal region to give a tRNA-like structure that accepts histidine. These relationships are summarized in Fig. 7.5. A comparison of other sequenced or partially sequenced *Tobamovirus* genomes suggests that the genome structure outlined is common to all members of the group (Solis and García-Arenal, 1990).

A third subgenomic RNA called I_1 RNA, representing approximately the 3' half of the genome, has been isolated from TMV-infected tissue. S1 mapping showed that this RNA species had a distinct 5' terminus, at residue 3405 in the genome (Sulzinski *et al.*, 1985). These workers proposed a model for the translation of I_1 RNA. There is an untranslated region of 90 bases followed by an AUG codon initiating a protein of MW = 54K terminating at residue 4915. Thus the amino acid sequence of the 54K protein is the same as the residues at the carboxy terminus of the 183K protein.

2. Proteins Synthesized *in Vitro*

TMV genomic RNA has been translated in several cell-free systems. Two large polypeptides were produced in reticulocyte lysates (Knowland *et al.*, 1975) and in the wheat germ system (e.g., Bruening *et al.*, 1976), but no coat protein was found by these or later workers. The TMV genome is not large enough to code independently for the two large proteins that were produced. Using a reticulocyte lysate system, Pelham (1978) showed that the synthesis of these two proteins is initiated at the same site. The larger protein is generated by partial read-through of a termination codon. The two proteins are read in the same phase, so the amino acid sequence of the smaller protein is also contained within the larger one, a conclusion fully confirmed by the nucleotide sequence discussed in the previous section. Increased production of the larger protein occurred *in vitro* at lower temperatures (Kurkinen, 1981). The extent of production of the larger protein may also depend on the kind of $tRNA^{tyr}$ present in the extract (Beier *et al.*, 1984a,b) (Chapter 6, Section VI, A,2,a,iv).

The I_1 subgenomic RNA isolated from infected tissue is translated *in vitro* in the rabbit reticulocyte system to produce a polypeptide of MW = 54K (Sulzinski *et al.*, 1985).

The I_2 genes coding for the 30K and 17.6K proteins are not translated from genomic RNA but from two subgenomic RNAs. The I_2 RNA isolated from infected tissue has been studied in *in vitro* systems by many workers. It is translated to produce a protein of MW = 30K. It is uncapped (Joshi *et al.*, 1983) and appears to terminate in 5' di- and triphosphates (Hunter *et al.*, 1983). The initiation site for transcription has been mapped at residue 1558 from the 3' terminus (Watanabe *et al.*, 1984a). The I_2 RNA also contains the smaller 3' gene (Fig. 7.5) but this is not translated in *in vitro* systems.

The smallest TMV gene (the 3' coat protein gene) is translated *in vitro* only from the monocistronic subgenomic RNA (Fig. 7.5) (Knowland *et al.*, 1975;

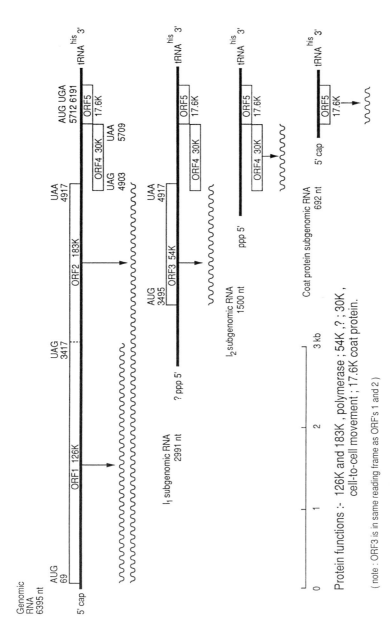

Genomic
RNA
6395 nt

AUG
69

ORF1 126K

UAG
3417

ORF2 183K

UAA
4917

AUG UGA
5712 6191

ORF4 30K
ORF5
17.6K

UAA
5709

5' cap

tRNA his 3'

I₁ subgenomic RNA
2991 nt

? ppp 5'

AUG
3495

ORF3 54K

UAA
4917

UAG
4903

ORF4 30K
ORF5
17.6K

tRNA his 3'

I₂ subgenomic RNA
1500 nt

ppp 5'

ORF4 30K
ORF5
17.6K

tRNA his 3'

Coat protein subgenomic RNA
692 nt

5' cap

ORF5
17.6K

tRNA his 3'

0   1   2   3 kb

Protein functions :- 126K and 183K, polymerase ; 54K, ? ; 30K,
cell-to-cell movement ; 17.6K coat protein.

(note : ORF3 is in same reading frame as ORF's 1 and 2)

Figure 7.5 Organization and expression of a *Tobamovirus* genome (TMV, Vulgare strain).

Beachy and Zaitlin, 1977). This gene can also be translated efficiently *in vitro* by prokaryotic protein-synthesizing machinery from *Escherichia coli* (Glover and Wilson, 1982).

3. Proteins Synthesized *in Vivo*

Proteins corresponding approximately in size to the 183K and 126K proteins have been found in infected tobacco leaves (e.g., Scalla *et al.*, 1976) and in infected protoplasts (e.g., Sakai and Takebe, 1974). Cyanogen bromide peptide analysis on the 110K protein from infected leaves showed it to be the same as the *in vitro* translation product of similar size (Scalla *et al.*, 1978). The 30K protein has been detected in both infected tobacco protoplasts (Beier *et al.*, 1980) and leaves (Joshi *et al.*, 1983).

Kibertis *et al.* (1983) and Ooshika *et al.* (1984) raised antibodies against a synthetic peptide with the predicted sequence for the 11 or 16 C-terminal amino acids of the 30K protein. The 30K protein from TMV-infected protoplasts was precipitated by these antibodies, positively identifying it as the I_2 gene product.

Many workers have demonstrated the synthesis of TMV coat protein *in vivo*. Determination of the nucleotide sequence in the 3' region of the TMV genome readily located the gene for this protein since the full amino acid sequence was already known (Tsugita *et al.*, 1960; Anderer *et al.*, 1960). *In vitro* protein synthesis primed with alkali-treated virus is discussed in Section III,A,7.

No protein has yet been detected *in vivo* that corresponds to the 54K *in vitro* translation product of the I_1 subgenomic RNA. Thus the TMV genome codes for four gene products, with a probable fifth yet to be established.

4. Functions of the Viral-Coded Proteins

Coat Protein

The major function of the coat protein is obvious. However, recent work shows that it is multifunctional. *Nicotiana* species containing the N' gene react to the common strain of TMV with systemic mosaic disease whereas tomato mosaic virus induces necrotic local lesions without mosaic disease. Saito *et al.* (1987a) showed, by constructing recombinants between the two virus strains, that the viral factor causing the necrotic response lies in the coat protein gene. Using recombinants between wild type TMV and mutants induced by nitrous acid, Knorr and Dawson (1988) identified a single point mutation at nucleotide 6157 in the coat protein gene that leads to a substitution of phenylalanine for serine at position 148 and that is responsible for the necrotic local lesion response. Other studies with mutants in which the coat protein gene was partially or completely deleted showed that the coat protein can also influence symptoms in other ways (Dawson *et al.*, 1988). The coat protein must also play some role in virus movement since the mutants established systemic infection less effectively than did wild-type virus.

The 126K and 183K Proteins

One or both of these proteins are involved in viral RNA replication. They have significant amino acid sequence homology with known viral RNA-dependent RNA polymerases (replicases) (Kamer and Argos, 1984). Mutation in the 126K gene caused a reduction in the synthesis of the 30K protein and its mRNA, suggesting that the 126K (and/or 183K) proteins are involved in the synthesis of I_2 RNA

(Watanabe *et al.*, 1987b). The most convincing evidence is the fact that a mutant constructed by Meshi *et al.* (1987) in which both the 30K and the coat protein genes were deleted gave rise, in infected protoplasts, to a shortened viral RNA.

In vitro mutagenesis at or near the leaky termination codon of the 126K gene indicates that both proteins are necessary for normal TMV replication in tobacco leaves (Ishikawa *et al.*, 1986).

The actual molecule of the TMV replicase has not yet been characterized. However, Young *et al.* (1987) have been able to solubilize the replication complex containing the 126K and 183K polypeptides. By analogy with some other viruses it may possibly involve host-coded as well as viral-coded polypeptides.

The 30K Protein

Genetic studies with *ts* mutants demonstrated the existence of a virus-coded function required for cell-to-cell movement of TMV (e.g., Taliansky *et al.*, 1982b; Atabekov *et al.*, 1983). Studies on the four proteins coded for by a TMV mutant *ts* for cell-to-cell movement (LsI) showed a difference from the normal virus only in the 30K protein (Leonard and Zaitlin, 1982). The mutation leading to the *ts* state in this mutant was shown to cause a change from a serine to a proline in the 30K protein (Ohno *et al.*, 1983b). Two different kinds of experiment have further demonstrated the role of the 30K protein in cell-to-cell movement. Transgenic tobacco seedlings expressing the normal 30K protein complemented the LsI mutant, allowing it to spread from cell to cell and move systemically (Deom *et al.*, 1987). In a reverse genetics approach the point mutation found in the LsI mutant was introduced into the parent virus. The resultant virus had the LsI phenotype. Various frameshift mutations in the 30K gene also gave a defective phenotype (Meshi *et al.*, 1987). These mutations could replicate in protoplasts but none showed infectivity for tobacco, that is, local lesions or systemic infection. Comparison of the amino acid sequences of several tobamoviruses showed two well-conserved regions in the middle region of the molecule, which are therefore probably important for the movement function (Saito *et al.*, 1988). The 30K protein has been shown to bind ssRNA or DNA strongly in a non-sequence-specific manner. The binding domain lies within amino acid residues 65–86 (Citovsky *et al.*, 1990). These authors suggest that this binding might play a role in the movement function.

5. Controlling Elements in the Viral Genome

Five controlling elements have been recognized or inferred in TMV RNA: (i) The nucleotide sequence involved in initiating assembly of virus rods. This is discussed in Section III,A,9. (ii) The replicase recognition site in the 3' noncoding region (Chapter 6, Section VI,A). (iii) The start codon sequence context may be one form of translational regulation. The context differs for each of the four known gene products (Chapter 6, Section VI,A,2). For example, in strain U_1 the contexts are as follows:

126K:	ACA<u>AUG</u>G	54K:	GAU<u>AUG</u>C
30K:	UAG<u>AUG</u>G	coat protein:	AAU<u>AUG</u>U

The context for the 30K protein is least optimal according to Kozak's model (Kozak, 1981, 1986). This might be the reason why so little 30K protein is produced compared with coat protein. However, changing the start codon context for the 30K AUG to the optimal strong context as defined by Kozak (1986) did not increase expression of the gene in tobacco plants (Lehto and Dawson, 1990). Insertion of sequences containing the coat protein subgenomic RNA promoter and leader upstream from the 30K ORF did not lead to increased production of the 30K product (Lehto *et al.*, 1990a). In fact the production of 30K protein was delayed and virus movement impaired, suggesting that different sequences influence timing of the expression of different genes. (iv) RNA promotors presumably have a role in regulating the amounts of subgenomic RNAs produced but the exact sequences for these subgenomic promotors have not been identified. That for the coat protein lies within 100 nucleotides upstream of the ORF. (v) There is a further possible control mechanism operating on the infecting virus particle. Studies using pH 8.0 treated virus in an *in vitro* protein-synthesizing system suggest that the strong coat protein–RNA interactions occurring at the origin-of-assembly nucleotide sequence may be a site where translocation of 80 S ribosomes is inhibited during uncoating of the virus rod. However, the effects of such a control mechanism would differ with different strains of the virus (Wilson and Watkins, 1985). The regulation of *Tobamovirus* gene expression is discussed by Dawson and Korhonen-Lehto (1990).

6. Viral RNA Synthesis *in Vitro*

In vitro RNA synthesis has been detected only in more or less crude fractions from infected cells. A virus-specific RNA-dependent RNA polymerase is associated with the membrane fraction (Zaitlin *et al.*, 1973; Romaine and Zaitlin, 1978). This can catalyze the synthesis of both positive and negative sense strands.

Young and Zaitlin (1986) analyzed the RNA structures synthesized *in vitro* by a crude 30,000*g* pellet fraction from TMV-infected tobacco leaves, using DNA probes of positive or negative sense. The major products labeled with radioactive nucleotides had the expected properties for TMV RF and RI (see Fig. 6.1). Much more label was incorporated into plus strands than into minus strands in the RF fraction. In the RI fraction only labeled plus strands could be detected. The system has a number of limitations: synthetic activity is short-lived; the system does not respond to added template RNA; and few if any free progeny viral RNA molecules are formed. These results are very similar to those reported by Watanabe and Okada (1986). They used a 2500*g* pellet and showed that coat protein subgenomic RNA could also be synthesized by this fraction. They detected RF of genomic size but not of coat protein subgenomic size. Furthermore they detected no synthesis of minus strand RNA of this size. Thus the subgenomic coat protein RNA may be synthesized by internal initiation on the negative strand genomic RNA rather than by transcription from a subgenomic negative sense strand. The cDNA cloning of the complete genome of TMV and the demonstration that infectious transcripts could be made *in vitro* was an important step as far as future studies on the virus are concerned (Dawson *et al.*, 1986).

7. Early Events Following Infection

Disassembly of the Virus *in* Vitro

To initiate infection TMV RNA must be uncoated, at least to the extent of allowing the first ORF to be translated. Most *in vitro* experiments on the disassembly of TMV have been carried out under nonphysiological conditions. For example, alkali or detergent (1% SDS) cause the protein subunits to be stripped from TMV RNA beginning at the 5' end of the RNA (the concave end of the rod) (e.g., Perham and Wilson, 1976). Controlled disassembly by such reagents yields a series of subviral rods of discrete length (e.g., Hogue and Asselin, 1984). Various cations slow down or prevent the stripping process at pH 9.0 (Powell, 1975). Durham *et al.* (1977) suggested that Ca^{2+} binding sites might act as a switch controlling disassembly of TMV in the cell. Removal of Ca^{2+} would result in a change in the conformation of protein subunits leading to their disaggregation. Durham (1978) proposed that TMV (and other small viruses) may be disassembled at or within a cell membrane. The virus might be in a medium roughly $10^{-3}\,M$ with respect to Ca^{2+} outside the cell, while inside the cell Ca^{2+} is about $10^{-7}\,M$. The ion dilution would provide free energy to help break intersubunit bonds. These ideas remain speculative.

Wilson (1984a) found that treatment of TMV rods briefly at pH 8.0 allowed some polypeptide synthesis to occur when the treated virus was incubated in an mRNA-dependent rabbit reticulocyte lysate. He suggested that the alkali treatment destabilizes the 5' end of the rod sufficiently to allow a ribosome to attach to the 5' leader sequence and then to move down the RNA, displacing coat protein subunits as it moves. He called the ribosome–partially-stripped-rod complexes "striposomes" and suggested that a similar uncoating mechanism may occur *in vivo*. In contrast to the reticulocyte system, in the wheat germ system virus treated at pH 8.0 gave rise to three times as much polypeptide synthesis as isolated TMV RNA, presumably due to protection of the RNA in the rod from nucleases before it was uncoated (Wilson, 1984b). In an *in vitro* protein-synthesizing system from *E. coli*, pH 8.0 treated virus gave rise to significant amounts of the 126K protein, whereas TMV RNA gave polypeptides of 50K or less, with a substantial amount of coat protein size (Wilson 1986).

Treatment of TMV *in vitro* with sodium dodecyl sulfate (SDS) for $\simeq 15$ seconds exposed a sequence of nucleotides from the 5' terminus to beyond the first AUG codon. No more 5' nucleotides were exposed during a further 15 minutes in SDS. Incubation of SDS-treated rods with wheat germ extract, or rabbit reticulocyte lysate, led to the binding of one or two ribosomes in $\simeq 20\%$ of the particles (Mundry *et al.*, 1990). Structure predictions suggest that the exposed sequence up to the first AUG exists in an extended single-stranded configuration, which would assist in the recruitment of ribosomes.

Experiments with Protoplasts

To obtain infection of a reasonable proportion of protoplasts it is necessary to treat the virus and/or the protoplasts in one of several ways (Chapter 6, Section VII, E). Electron microscopy has been used to study the entry process, and it has been suggested that polyornithine stimulates entry of TMV either by damaging the plasmalemma (Burgess *et al.*, 1973) or by stimulating endocytotic activity (Takebe *et al.*, 1975). Estimates of the extent of uncoating of the adsorbed TMV inoculum vary from 5% (Wyatt and Shaw, 1975) to about 30% one hour after inoculation (Zhuravlev *et al.*, 1975), but the proportion of fully stripped RNA has not been determined. In view of the abnormal state of the cells and particularly the nature of the suspension medium, the relevance of studies in protoplasts to the infection process in leaves following mechanical inoculation is open to question.

Uncoating in Oocytes

Xenopus oocytes microinjected with TMV produced at least as much immunoreactive 126K protein as did oocytes injected with TMV RNA (Ph.C. Turner *et al.*, 1987). This experiment appears to rule out a specific role for the cellulose wall in the uncoating of TMV in leaves. Whether it also rules out a role for the plasma membrane depends on whether intact virus particles contaminating the outside of the needle used for injection could have entered the oocytes via the cell membrane, being uncoated on the way.

Early Events in Intact Leaves

The nature of the leaf surface, the requirement for wounding, the efficiency of the process, and other aspects of infection in intact leaves are discussed in Chapter 10. The uncoating process has been examined directly by applying TMV radioactively labeled in the protein or the RNA or in both components (e.g., Shaw, 1973; Hayashi, 1974). The following conclusions can be drawn from such experiments: (i) within a few minutes of inoculation, about 10% of the RNA may be released from the virus retained on the leaf; (ii) much of the RNA is in a degraded state but some full-length RNAs have been detected; (iii) *in vivo* stripping of the protein from the rod begins at a minimum of two and probably many more sites along the rod (Shaw, 1973); (iv) the early stages of the process do not appear to depend on preexisting or induced enzymes (Shaw, 1969); and (v) the process is not host specific, at least in the early stages. There is a fundamental difficulty with all such experiments. Concentrated inocula must be used to provide sufficient virus for analysis, but this means that large numbers of virus particles enter cells rapidly (Fig. 7.6). It is impossible to know which among these particles actually establish an infection.

There is some evidence suggesting that TMV particles are uncoated *in vivo* in epidermal cells as an 80 S ribosome moves along the RNA from the 5' end (Shaw *et al.*, 1986) in the manner suggested from *in vitro* studies (the preceding Section III, A,7). Translation complexes with the expected properties of striposomes have been isolated from the epidermis of tobacco leaves shortly after inoculation (Plaskitt *et al.*, 1987). However, more definitive experiments are needed to establish that this uncoating mechanism operates *in vivo*. In particular, the intracellular mechanism equivalent to SDS or alkali treatment is unknown. The topic has been reviewed by Wilson *et al.* (1989).

Transgenic tobacco plants in which the TMV coat protein gene is expressed are resistant to TMV infection (Chapter 16, Section II,C,1). One mechanism proposed for this resistance involves the endogenously produced coat protein shifting the kinetics of the initial uncoating step in favor of fully coated rods (Register *et al.*, 1988).

8. RNA and Protein Synthesis *in Vivo*

The data of Siegel *et al.* (1978) indicate that viral protein synthesis does not suppress total host-cell protein synthesis but occurs in addition to normal synthesis. Two days after infection, viral coat protein synthesis accounted for about 7% of total protein synthesis. Synthesis of the 126K protein was about 1.4% and that of the 183K protein about 0.3% as much as coat protein.

Much of the TMV-induced RNA polymerase in tobacco leaves was in bound form, being present in the pellet following centrifugation at 31,000g for 20 minutes (Zaitlin *et al.*, 1973). The TMV–RNA replication complex is not associated with chloroplasts, nuclei, or mitochondria, but rather is found in a membranous complex bound to cytoplasmic ribosomes (e.g., Ralph *et al.*, 1971).

Several workers have reported the association of both full-length TMV RNA and the coat protein mRNA with cytoplasmic polyribosomes in infected tobacco

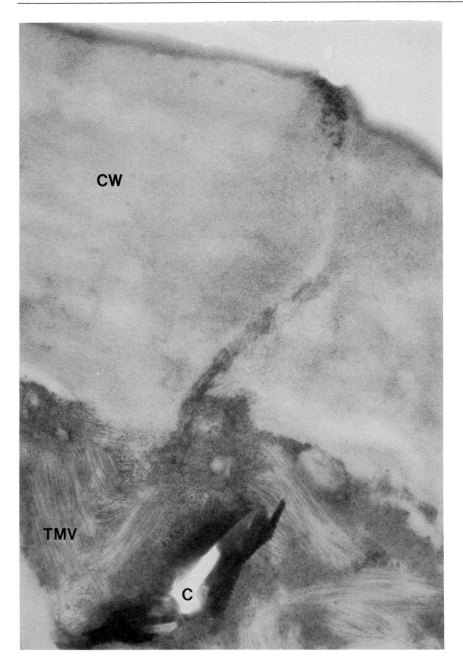

Figure 7.6 TMV particles that have entered a tobacco leaf lower epidermal cell through a wound caused by abrasive (celite). Tissue was excised and fixed immediately after inoculation. Large numbers of virus rods (TMV) are visible in the cytoplasm. CW, cell wall; C, celite. (From Plaskitt *et al.*, 1987.)

leaves (e.g., Beachy and Zaitlin, 1975). Beachy and Zaitlin also found TMV dsRNAs to be associated with the membrane-bound polyribosome fraction. Confirmation that TMV proteins are synthesized on 80 S ribosomes comes from the fact that cycloheximide completely inhibits TMV replication in protoplasts, whereas chloramphenicol does not (Sakai and Takebe, 1970).

The ss and ds RNAs produced by TMV infection of tobacco leaves have been investigated in Zaitlin's laboratory (Zelcer *et al.*, 1981; Palukaitis *et al.*, 1983; Sulzinski *et al.*, 1985). Four TMV-specific dsRNAs were found in the fraction of nucleic acids soluble in 2 *M* LiCl. All four contained the 3' plus strand sequences. In decreasing order of size they were identified as genomic, I_1, I_2, and coat protein gene dsRNAs. Four mRNAs corresponding in size to these ds forms were found associated with the polyribosome fraction from infected cells.

Dorokhov *et al.* (1983, 1984a) isolated a ribonucleoprotein fraction from infected tobacco tissue in CsCl density gradients, which had a higher buoyant density than TMV. This material can be released from polyribosomes by EDTA treatment. Genomic, I_1, I_2, and coat protein RNAs and polypeptides of various sizes were identified as components of the ribonucleoprotein complex.

From studies on viral protein synthesis in tobacco leaves using wild-type virus and a *ts* mutant held at different temperatures and a protein synthesis inhibitor, Dawson (1983) concluded that the synthesis of 183K, 126K, and 17.5K proteins was correlated with dsRNA synthesis rather than that of ssRNAs. It is possible that nascent ssRNAs from replication complexes function as mRNAs. Dawson's results suggest that TMV mRNA is relatively transitory *in vivo*. However, Dawson and Boyd (1987b) showed that synthesis of TMV proteins in tobacco leaves was not translationally regulated under conditions of heat shock, as were most host proteins. Thus under appropriate conditions most of the protein being synthesized was viral coded.

Watanabe and Okada (1986) detected low levels of viral-specific RNA synthetic activity in protoplast extracts, 3 hours after inoculation. Activity had increased markedly by 6 hours. Figure 7.7 illustrates the time course of appearance of various TMV-related macromolecules.

As might be expected, RI and RF appear early in infections as does the 126K

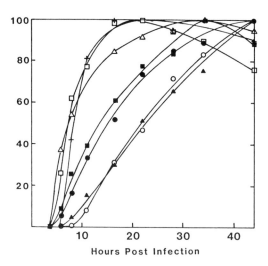

Figure 7.7   Time course of production of TMV-related RNAs, proteins, and progeny virus particles in synchronously infected protoplasts. One-half of a batch of protoplasts was incubated with [¹⁴C]uridine in the presence of actinomycin D from the time of inoculation. The other half was incubated with [¹⁴C]leucine under the same conditions. Samples were taken for analysis at the times indicated. Data are expressed as the percentages of the maximum values attained for each component during the time course studies. (□) RI; (△) RF; (●) TMV RNA; (■) coat protein mRNA; (+) 140K protein; (▲) coat protein; (○) progeny virus particles. (From Ogawa and Sakai, 1984.)

protein (estimated as 140K in Fig. 7.5). Coat protein mRNA and genomic RNA are early products. At a later stage virus production follows closely that of coat protein synthesis. Thus it appears that the amount of coat protein available may limit the rate at which progeny virus is produced. Kibertis *et al.* (1983) found that the 30K protein was synthesized only between 8 and 16 hours after inoculation of protoplasts. However, in intact leaves, production of the 30K protein continued for some days, but was maximal around 24 hours (Lehto *et al.*, 1990b). Watanabe *et al.* (1984b) first detected the 183K, 126K, 30K, and coat proteins 2–4 hours after infection, and before infectious virus could be found. The 183K, 126K, and coat protein were synthesized over a period of many hours but synthesis of the 30K protein and its mRNA was detected between 2 and 9 hours after inoculation. It is possible that the 54K protein expected to be synthesized from I_1 mRNA has not yet been detected because its synthesis is even more transient than that of the 30K protein.

Using whole tobacco leaves inoculated with TMV RNA and held at 27°C, Fry and Matthews (1963) determined the following sequence of events: Movement of infectious material from the inoculated epidermis to the underlying mesophyll was first detected by a very sensitive method at 4 hours. At 7 hours, infectious progeny virus was detected. At about 8 hours there was a rise in the total RNA content in inoculated epidermis. The logarithmic increase in infectivity in young tobacco leaves in which infection of the cells had been substantially synchronized began 8 hours after plants were moved to 15°C (Dawson *et al.*, 1975). In the leaves pretreated at low temperature, synchrony of infection approached that of protoplast systems. Given differences in experimental conditions, the time lag between infection and the detection of progeny virus is very similar in protoplasts and whole-leaf tissue. This supports the view that the more detailed studies on replicative events carried out in protoplasts may be relevant for events in the intact leaf.

Yamaya *et al.* (1988a) incorporated a full-length cDNA copy of TMV genomic RNA into the genome of tobacco plants using a disarmed Ti plasmid vector. Virus production and symptom development took place in transformed plants, opening the way for further studies on TMV replication.

9. Assembly of the TMV Rod

Assembly in Vitro

In their classic experiments, Fraenkel-Conrat and Williams (1955) showed that it was possible to prepare TMV coat protein and TMV RNA and to reassemble these into intact virus particles. TMV RNA alone had an infectivity about 0.1% that of intact virus. Reconstitution of virus rods gave greatly increased specific infectivity (about 10–80% that of the native virus) and the infectivity was resistant to RNase attack. Since these early experiments, many workers have studied the mechanism of assembly of the virus rod, for there is considerable general interest in the problem. The three-dimensional structure of the coat protein is known in atomic detail (see Fig. 5.10) and the complete nucleotide sequence of several strains of the virus and related viruses is known. The system therefore provides a useful model for studying interactions during the formation of a macromolecular assembly from protein and RNA.

There are four aspects of rod assembly to be considered: the site on the RNA where rod formation begins; the initial nucleating event that begins rod formation; rod extension in the 5' direction; and rod extension in the 3' direction. There is a general consensus concerning most of the details of the initiation site and the initial event, but the nature of the elongation processes remains controversial.

A central problem has been the fact that the coat protein monomer can exist in a variety of aggregation states, the existence of which is closely dependent on conditions in the medium (see Fig. 5.8). Equilibria exist between different aggregates so that while one species may dominate under a given set of experimental conditions, others may be present in smaller amounts. The subject has been reviewed recently by Butler (1984) and Okada (1986b).

1. *The assembly origin in the RNA*   Coat protein does not begin association with the RNA in a random manner. Zimmern and Wilson (1976) located the origin of assembly between 900 and 1300 nucleotides from the 3′ end. Butler *et al.* (1977) and Lebeurier *et al.* (1977) produced evidence showing that the longer RNA tail loops back down the axial hole of the rod. Lebeurier *et al.* used electron microscopy to show that in partially assembled rods, a long and a short tail of RNA both protrude from one end of the rod. As the rod lengthens the longer tail disappears. The short tail is then incorporated, completing rod formation. These and other experiments (e.g., Otsuki *et al.*, 1977) demonstrated an internal origin of assembly. The nucleotide sequence of the origin of assembly has been established by various workers for the common strains of TMV (e.g., Zimmern, 1977; Jonard *et al.*, 1977). Nucleotide sequences near the initiating site can form quite extensive regions of internal base-pairing as is illustrated in Fig. 7.8. The origin of assembly is located elsewhere in some tobamoviruses (see Fig. 17.2).

Loop 1 in Fig. 7.8 with the sequence AAG, AAG, UCG combines first with a 20 S aggregate of coat protein. Steckert and Schuster (1982) assayed the binding of 25 different trinucleoside diphosphates to polymerized TMV coat protein at low temperatures and pH. 5′-AAG-3′ bound most strongly. Under conditions used for virus reconstitution, longer oligomers ((AAG)$_2$ or AAGAAGUUG) were required for strong binding (Turner *et al.*, 1986). The importance of various aspects of loop 1 have been established in more detail by site-directed mutagenesis (Turner and Butler, 1986; Turner *et al.*, 1988). These studies have shown that specific assembly initiation occurs in the absence of loops 2 and 3 of Fig. 7.8, but loop 1 is essential. Deletion or alteration of the unpaired sequence in loop 1 abolishes rapid packaging. The binding of loop 1 is mainly due to the regularly spaced G residues. The sequences (UUG)$_3$ and (GUG)$_3$ are as effective as the natural sequence. However, sequences such as (CCG)$_3$ and (CUG)$_3$ reduced the assembly initiation rate. Thus there is some, but not complete, latitude in the bases in the first two positions.

Base sequence in the stem of loop 1 is not critical, except that shortening the stem reduces the rate of protein binding, as do changes that alter the RNA folding close to the loop (Turner *et al.*, 1988). Overall stability of loop 1 is important because base changes that made the stem either more or less strongly base-paired were detrimental to protein binding. The small loop at the base of the stem is also important, but its phasing to the top loop is not critical.

The preceding discussion refers to the vulgare strain of TMV. Other tobamoviruses have somewhat different sequences in the stem-loop structure. From comparing these sequences, Okada (1986b) suggested a different target sequence in the loop, namely, GAAGUUG.

2. *The initial nucleating event*   Since Butler and Klug (1971) showed that a 20 S polymer of coat protein was responsible for initiating TMV rod assembly, the finding has been confirmed in various ways by many workers. The structure of the 20 S double disk is known to atomic resolution (see Fig. 5.10). For many years it

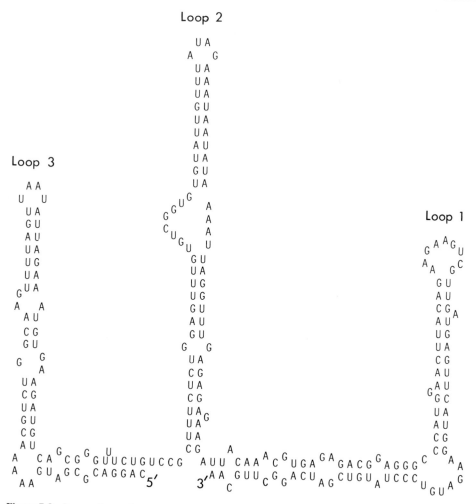

Figure 7.8 Proposed secondary structure for the TMV origin of assembly extending from bases 5290 to 5527 of the viral RNA sequence. (From Zimmern, 1983b.)

has been assumed that this was the configuration that initiated assembly, and that on interaction with the origin of assembly sequence in the RNA, the disk converts to a protohelical form. The structure in Fig. 5.10 was derived from 20 S aggregates crystallized in solutions of high ionic strength, and it has been assumed that the double disk was also the favored structure under reconstitution conditions. Studies using sedimentation equilibrium (Correia *et al.*, 1985), near-UV circular dichroism (Raghavendra *et al.*, 1985), and electron microscopy (Raghavendra *et al.*, 1986) under salt and temperature conditions used in rod reconstitution experiments suggested that the 20 S species observed is a helical aggregate of 39 ± 2 subunits. However, using the rapid-freeze technique, it has been shown directly that under conditions in solution favoring most rapid TMV assembly, about 80% of the coat protein is in a structure that is compatible only with an aggregate of two rings (P. J. G. Butler, personal communication). Thus the predominant structure in solution must be very similar to that found in the crystal (Fig. 5.10).

Certainly, the model proposed by Champness *et al.* (1976) appeals on functional grounds. The disks can form an elongated helical rod of indefinite length at lower pH values. The transition between helix and double disk is mainly controlled by a switching mechanism involving abnormally titrating carboxyl groups. At low pH the protein can form a helix on its own because the carboxyl groups become protonated. When the protein is in the helical state either at low pH or in combination with RNA in the virus, the interlocking vertical helices are present (Fig. 5.11). In this condition the abnormally titrating carboxyl groups are assumed to be forced together. Champness *et al.* proposed that the inner part of the two-layered disk acts as a pair of jaws (Fig. 5.10) to bind specifically to the origin-of-assembly loop in the RNA (Fig. 7.8), in the process converting to a protohelix. A model for the early steps in the assembly of TMV is shown in Fig. 7.9.

3. *Rod extension in the 5' direction*   Following initiation of rod assembly there is rapid growth of the rod in the 5' direction (e.g., Zimmern, 1977) until about 300 nucleotides are coated. Zimmern (1983b) has proposed a model for this initial rapid assembly involving two additional hairpin loops located 5' to the origin of assembly loop (Fig. 7.8). The spacing of these loops is consistent with the idea that they interact successively with three double disks. Various workers have isolated partially assembled rods of definite length classes. At least some of these classes are probably due to regions of RNA secondary structure that delay rod elongation at particular rod lengths (Godchaux and Schuster, 1987).

All workers agree that rod extension is faster in the 5' direction than in the 3' but disagreement remains as to other aspects. Butler's group believed that the 20 S aggregate is used in 5' extension and the A protein in the 3' direction with complete rods being formed in 5–7 minutes. (Aggregation states of the coat protein monomer are illustrated in Fig. 5.8.) Okada's group (e.g., Fukuda and Okada, 1985, 1987) claimed that 5' extension uses 4 S protein while 3' extension uses 20 S aggregates, with complete rods taking 40–60 minutes to form. Some of the earlier relevant experiments are discussed in detail by Butler (1984), Lomonossoff and Wilson (1985), and Okada (1986b). The experiments of Fukuda *et al.* (1978) show that for the strain of virus and assembly conditions they used, full-length rods were formed only after 40–60 minutes. Fukuda and Okada (1985) found that rod elongation in the 5' direction is complete in 5–7 minutes giving rise to a 260-nm rod, which subsequently elongates in the 3' direction to give the 300-nm rod. They used a Japanese strain of TMV and a higher ionic strength than Butler's group, which may account for some of the differences.

Recent work from Butler's laboratory lends strong support to the views that under optimum assembly conditions, growth of the TMV rod in the 5' direction is mainly by the addition of double disks, or sometimes single disks of coat protein (Turner *et al.*, 1989). They prepared *in vitro* RNA transcripts containing various heterologous nonviral RNAs 5' to the TMV origin of assembly sequence, instead of the natural TMV sequence. There was no sequence 3' to the origin of assembly. They then determined the lengths of RNA fragments protected from nuclease attack after allowing a short period for rod assembly. They found a series of lengths in steps of slightly over 100 nucleotides, or occasionally of just over 50 nucleotides. These are approximately the lengths expected to be protected by two (or one) turns of protein helix. They found such steps in length whatever heterologous RNA was used. This experiment rules out the possibility that the steps are due to some regularity in the RNA sequence that slows assembly at certain distances along the molecule.

4. *Rod extension in the 3' direction*   Fairall *et al.* (1986) studied reassembly with RNAs that had been blocked at various sites by short lengths of hybridized and cross-linked cDNA probes. They found that even when 5' extension was incomplete

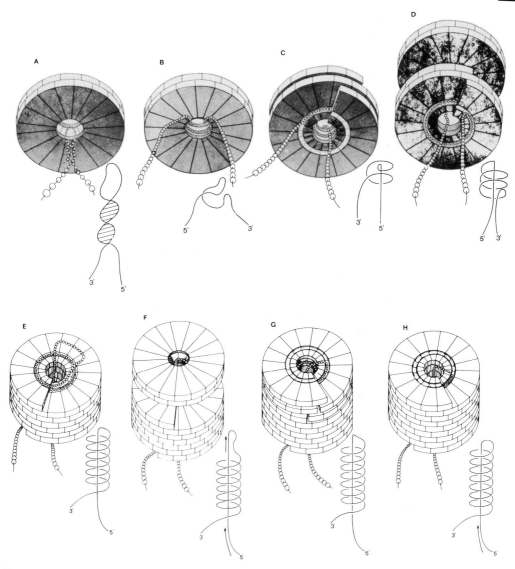

Figure 7.9 Model for the assembly of TMV; (A–C) initiation; (D–H) elongation. (A) The hairpin loop inserts into the central hole of the 20 S disk. This insertion is from the lower side of the disk as viewed in Fig. 5.10. It is not yet apparent how the correct side for entry is chosen. (B) The loop opens up as it intercalates between the two layers of subunits. (C) This protein RNA interaction causes the disk to switch to the helical lockwasher form (a protohelix). Both RNA tails protrude from the same end. The lock-washer–RNA complex is the beginning of the helical rod. (D) A second double disk can add to the first on the side away from the RNA tails. As it does so it switches to helical form and two more turns of the RNA become entrapped. (E–H) Growth of the helical rod continues in the 5′ direction as the loop of RNA receives successive disks, and the 5′ tail of the RNA is drawn through the axial hole. In each drawing the three-dimensional state of the RNA strand is indicated. (Courtesy of P. J. G. Butler. Copyright © Medical Research Council.)

due to the blocked sequence, 3′ extension was completed. However, lengths of rods were determined after 20 minutes of incubation so the data are not relevant to the question of whether 3′ elongation is completed in 5–7 minutes.

Fukuda and Okada (1987) prepared an ss cDNA probe that extended from the origin of assembly to the 3′ terminus and that was complementary to TMV RNA.

They used this to determine the length of rod extension in the 3' direction. The results showed that significant rod extension in the 3' direction did not occur at least in the first 4 minutes. It was first observed at 8 minutes and was still increasing between 15 and 40 minutes. At 4 minutes there was substantial encapsidation of RNA in the 5' direction. Fukuda and Okada (1987) found a series of discrete sizes of RNA in the RNA protected in the 3' direction and suggest that these differed by about 100 nucleotides in length. However, the results illustrated in their paper indicate a very uneven increment in length from about 55 to 135. This scatter in lengths may be due to the effects of nucleotide sequence on assembly using A protein and on nuclease specificity rather than to the addition of double disks. Turner et al. (1989) carried out assembly experiments on RNA with heterologous sequences inserted 3' to the TMV origin of assembly sequence. They found no evidence for banding in the protected RNA, giving strong support to earlier work indicating that extension of the rod in the 3' direction is by the addition of small A protein aggregates.

The assembly of other strains of TMV and other tobamoviruses has not been studied as intensively as that of the type strain, but the available evidence indicates that the same basic assembly mechanism operates for all these viruses. However, the structures involved may differ in detail.

Studies on reassembly in vitro in which the TMV origin of assembly was embedded in various positions support the ideas that rod extension is much faster in the 5' than the 3' direction, that 5' extension is probably completed before 3' extension begins, and that the two extension reactions are different from each other (Gaddipatti and Siegel, 1990).

Artificial RNAs have been constructed that combine the TMV origin of assembly and the mRNA for a foreign gene. This RNA has been assembled into a rod with TMV coat protein and used to introduce the mRNA into plant cells (Chapter 8, Section VI,D,3).

Assembly in Vivo

Very short rods have been seen in electron micrographs of infected leaf extracts but the 20 S aggregate has not been definitively established as occurring in vivo. Nevertheless, the following evidence shows that the process involved in the initiation of assembly outlined earlier is almost certainly used in vivo. (i) The conditions under which assembly occurs most efficiently in vitro (pH 7.0, 0.1 M ionic strength, and 20°C) can be regarded as reasonably physiological. (ii) The correlation between the location of the origin of assembly in different tobamoviruses (see Fig. 17.2) and the encapsidation, or not, of short rods containing the coat protein mRNA (Chapter 5, Section III,A,1) strongly suggests that the origin of assembly found in vitro is used in vivo. (iii) The mutant Ni2519, which is ts for viral assembly in vivo, has a single base change that is at position 5332 (Zimmern, 1983b). This change weakens the secondary structure near the origin-of-assembly loop. (iv) Tobacco plants transgenic for the CAT gene with the TMV origin of assembly inserted next to the 3' terminus give rise to RNA transcripts which can be assembled into viruslike rods with the TMV coat protein when the plants are systemically infected with TMV (Sleat et al., 1988b).

It is known from in vitro experiments that TMV coat protein can form rods with other RNAs. In vivo there appears to be substantial specificity in that most rods formed contain the homologous RNA. In vivo this specificity may be due, first, to specific recognition of the correct RNA by the 20 S disk, and second, because rods are assembled at an intracellular site where the homologous viral RNA predominates. Nevertheless, fidelity in *in vivo* assembly is not total (Chapter 6, Section VIII,A).

There is no evidence that establishes the method by which the TMV rod elongates *in vivo*, but there is no reason to suppose that it differs from the mechanism that has been proposed for *in vitro* assembly.

10. Intracellular Location of Virus Synthesis and Accumulation

The existence of amorphous inclusions or X bodies in cells infected with TMV has been known for many years (Fig. 7.10). In tobacco cells, electron microscopy has shown that these structures consist of an assemblage of endoplasmic reticulum, ribosomes, virus rods, and wide filaments that may be bundles of tubules (Fig. 7.11). They occur in all types of cells infected with TMV. The weight of evidence suggests that TMV components are synthesized and assembled into virus in or near these cytoplasmic viroplasms.

On the basis of cell fractionation experiments, various claims have been made for the chloroplasts as a site of TMV assembly. These claims were probably based

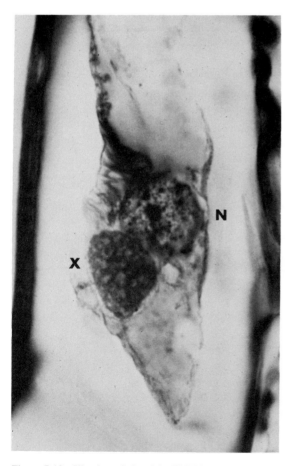

Figure 7.10   Viroplasm induced by TMV in a tobacco cortical parenchyma cell. Light microscope, ×1500. N, nucleus; X, X-body. (From Esau and Cronshaw, 1967. Reproduced from the *Journal of Cell Biology,* **33,** 673, by copyright permission of the Rockefeller University Press.)

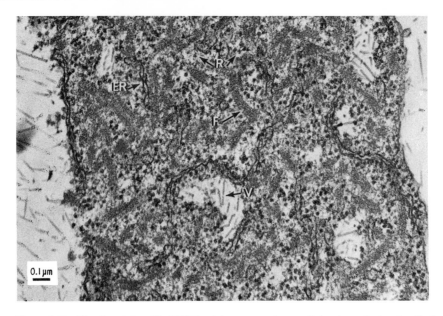

Figure 7.11   Viroplasm induced by TMV in a tobacco parenchyma cell showing endoplasmic reticulum (ER), ribosomes (R), virus rods (V), and wide filaments (F); ×53,000. The inclusion has no delimiting membrane. (From Esau and Cronshaw, 1967. Reproduced from the *Journal of Cell Biology, **33***, 673, by copyright permission of the Rockefeller University Press.)

on contamination of the isolated organelles with either free virus or virus contained within cosedimenting structures. Viruslike rods have been observed in thin sections of chloroplasts and of nuclei under certain conditions. Again, these observations are unlikely to mean that these organelles are sites for TMV synthesis. Reinero and Beachy (1986) showed that TMV coat protein could be isolated from chloroplasts of systemically infected leaves. Full-length TMV genomic RNA has been shown to be present in chloroplasts in infected tissue (Schoelz and Zaitlin, 1989). This is unlikely to be relevant for the main virus replication process, since no subgenomic RNAs could be detected. It is possible that the genomic RNA could be translated on chloroplast ribosomes to give rise to some coat protein in these organelles.

The 126K polypeptide coded by TMV was found to be associated with host chromatin in extracts of infected leaves (van Telgen *et al.*, 1985), but this has been shown to be due to the association of cytoplasmic viroplasms containing the protein with the nuclei (Wijdeveld *et al.*, 1989). These results illustrate the caution needed in interpreting the results of experiments involving cell fractionation procedures. Using hybridization with probes specific for TMV RNA, Okamoto *et al.* (1988) showed that TMV minus strand RNA was localized in the cytoplasm of infected protoplasts.

Following inoculation of tobacco leaf protoplasts, Otsuki *et al.* (1972a) first detected TMV rods in the cytoplasm in thin sections after 6 hours. Their appearance was soon followed by aggregates of rods, which grew rapidly in size. These aggregates were not associated with nuclei, mitochondria, or chloroplasts, and no virus particles were seen within these organelles. Electron-opaque filaments and tubules as observed in leaf tissue (Fig. 7.11) were not seen in the protoplasts.

Saito *et al.* (1987b) used specific antisera and immunogold labeling to localize the components of the TMV RNA-dependent RNA polymerase (130K and 180K proteins) in granular inclusions (Fig. 7.12). These granular inclusions are almost certainly early stages in the development of the viroplasms illustrated in Fig. 7.11. Label present in other regions including nuclei, chloroplasts, mitochondria, and the general cytoplasm appeared to be nonspecific.

Similarly Hills *et al.* (1987), using an antiserum specific for the 126K protein, found the immunogold label almost exclusively in viroplasms and in pockets of TMV particles at the periphery of the viroplasms. Within the viroplasms they found much of the gold label associated with the viroplasmic tubules (or wide filaments) of the kind seen in Fig. 7.11. This evidence strongly suggests that TMV RNA is synthesized in these viroplasms and that viral protein synthesis also occurs at these sites, with virus rods being assembled in the interior or near the periphery of the viroplasms. Using light microscopy, Sheffield (1939) described how these bodies go through a developmental process as they enlarge during the course of infection. It is probable that young viroplasms are the most active synthetically and that the fully developed structures (Fig. 7.11) represent mainly an accumulation of virus-related proteins and cell ribosomes at the end of the synthetic process.

As infection in a cell progresses, virus particles accumulate in the cytoplasm and vacuole and may form large hexagonal plates consisting of virus rods in regular arrays (see Fig. 11.21). About 10^6-10^7 virus particles per cell are formed.

Tomenius *et al.* (1987) used antibodies raised against a C-terminal peptide of the 30K protein together with immunogold labeling to show that this protein accumulates in plasmodesmata, with maximum labeling about 24 hours after inoculation. This observation is interesting in view of the proposed role of the 30K protein in cell-to-cell movement (Chapter 6, Section V,D,2).

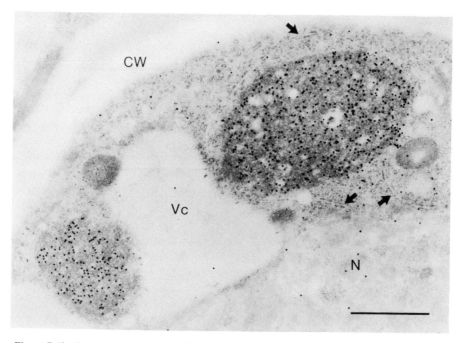

Figure 7.12 Immunocytochemical localization of the 130K–180K TMV proteins in a cell from a young infected tobacco leaf. Leaf sections were treated with an anti-130K–anti-180K antiserum, and then with a protein A–gold complex. Gold label is strongly localized in the viroplasm regions of the cytoplasm. CW = Cell wall; Vc = vacuole; N = nucleus; arrows indicate ER. Bar = 1 μm. (From Saito *et al.*, 1987b.)

B. *Luteovirus* Group

The *Luteovirus* group includes some economically very important viruses such as the type member, barley yellow dwarf virus (BYDV), which infects cereals, and potato leaf roll virus (PLRV). The group has been difficult to study for two reasons: (i) in general, these viruses are confined to the phloem and therefore occur at a very low concentration in the plant as a whole and (ii) they are not transmitted by mechanical inoculation. Thus very little is known about the proteins coded for by members of this group apart from the coat protein. There is also very little experimental information about the way in which the genes are expressed. Thus when Miller *et al.* (1988b) determined the nucleotide sequence of BYDV, they made interesting predictions concerning the genes and genome strategy of BYDV based on sequence similarities with the genes of viruses whose strategies are more firmly established.

1. Genome Structure

The genome of PLRV consists of a single molecule of positive sense ssRNA with a 5' linked VPg and no poly(A) sequence (Mayo *et al.*, 1982b). The nucleotide sequence of the BYDV genome has been determined except for the 5'-terminal base (Miller *et al.*, 1988a,b). A diagram of the deduced genome organization is shown in Fig. 7.13. Six ORFs are suggested for the plus strand, together with a possible transframe protein. The minus strand has five ORFs, which could encode proteins of M_r 10–15K. However, the unusual amino acid compositions suggest that none of them are functional genes. If the 6.7K ORF shown in Fig. 7.13 is the one nearest to

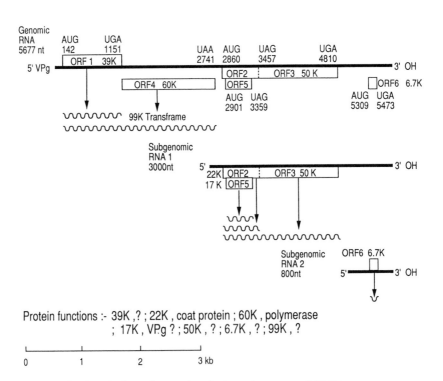

Figure 7.13 Organization and expression of a *Luteovirus* genome (BYDV).

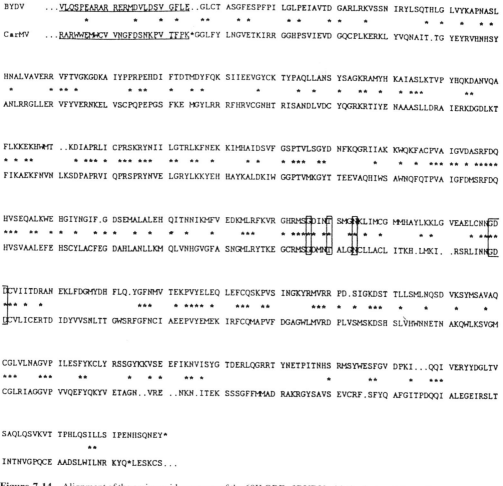

```
BYDV     ...VLQSPEARAR RERMDVLDSV GFLE..GLCT ASGFESPFPI LGLPEIAVTD GARLRKVSSN IRYLSQTHLG LVYKAPNASL
              *          *   *  *    **      * *       * *    * *    *          *   *   *   * *
CarMV    ...RARHWEMICV VNGFDSNKPV TFPK*GGLFY LNGVETKIRR GGHPSVIEVD GQCPLKERKL YVQNAIT.TG YEYRVHNHSY


HNALVAVERR VFTVGKGDKA IYPPRPEHDI FTDTMDYFQK SIIEEVGYCK TYPAQLLANS YSAGKRAMYH KAIASLKTVP YHQKDANVQA
   *       * ** *       * **      *   * *              * *      *   ** *   * ***       **
ANLRRGLLER VFYVERNKEL VSCPQPEPGS FKE MGYLRR RFHRVCGNHT RISANDLVDC YQGRKRTIYE NAAASLLDRA IERKDGDLKT


FLKKEKHWMT ..KDIAPRLI CPRSKRYNII LGTRLKFNEK KIMHAIDSVF GSPTVLSGYD NFKQGRIIAK KWQKFACPVA IGVDASRFDQ
* *  **       * *** *  *** ***   **  **  *      * *    * ***  **        *    *   *  *** ** * *****
FIKAEKFNVN LKSDPAPRVI QPRSPRYNVE LGRYLKKYEH HAYKALDKIW GGPTVMKGYT TEEVAQHIWS AWNQFQTPVA IGFDMSRFDQ


HVSEQALKWE HGIYNGIF.G DSEMALALEH QITNNIKMFV EDKMLRFKVR GHRMSGDINT SMGNKLIMCG MMHAYLKKLG VEAELCNNGD
*** **  *  *   * ** *  *   *    *   *   *        ***   * ***** *  *** *  *      * *     * ****
HVSVAALEFE HSCYLACFEG DAHLANLLKM QLVNHGVGFA SNGMLRYTKE GCRMSGDMNT ALGNCLLACL ITKH.LMKI. .RSRLINNGD


DCVIITDRAN EKLFDGMYDH FLQ.YGFNMV TEKPVYELEQ LEFCQSKPVS INGKYRMVRR PD.SIGKDST TLLSMLNQSD VKSYMSAVAQ
*** *  * *               ***    * **** *    *** **     ***  *   *** *    *    *       *    *
DCVLICERTD IDYVVSNLTT GWSRFGFNCI AEEPVYEMEK IRFCQMAPVF DGAGWLMVRD PLVSMSKDSH SLVHWNNETN AKQWLKSVGM


CGLVLNAGVP ILESFYKCLY RSSGYKKVSE EFIKNVISYG TDERLQGRRT YNETPITNHS RMSYWESFGV DPKI...QQI VERYYDGLTV
***      *** ** *** ** *   * *  ** *               *   ** *  *   ***
CGLRIAGGVP VVQEFYQKYV ETAGN..VRE ..NKN.ITEK SSSGFFMMAD RAKRGYSAVS EVCRF.SFYQ AFGITPDQQI ALEGEIRSLT


SAQLQSVKVT TPHLQSILLS IPENHSQNEY*
             **
INTNVGPQCE AADSLWILNR KYQ*LESKCS...
```

Figure 7.14 Alignment of the amino acid sequence of the 60K ORF of BYDV with the first read-through region of the putative polymerase gene of CarMV. Asterisks indicate matches, dots indicate spacing required to optimize the fit, underlined sequences are from the upstream reading frames, and boxed amino acids are highly conserved among RNA-dependent RNA polymerases. (From Miller *et al.*, 1988b.)

the 3′ terminus to be expressed then the 3′ untranslated sequence is unusually long, with 568 nucleotides. There is no indication that a tRNA-like structure can be formed.

The 5′ ORF in the plus strand encodes for a potential protein of 39K. This ORF overlaps by 13 bases with the next ORF of $M_r = 60K$. The 60K ORF appears to encode the viral RNA-dependent RNA polymerase because it contains an amino acid sequence shared by the putative RNA-dependent RNA polymerases of all plant and most animal viruses (Kamer and Argos, 1984; Hamilton *et al.*, 1987). The sequence is $GXXXTXXXN(X)_{20-40}$ GDD, where X represents any amino acid. The amino acid sequence of the 60K ORF shows striking sequence similarity to the RNA polymerase of carnation mottle virus (Fig. 7.14). No such sequence similarities were found between the 60K ORF and the RNA-dependent polymerase genes of other viruses.

Following the 60K ORF is a sequence of 116 noncoding nucleotides. The next ORF of 22K encodes the coat protein gene. This was shown by partial amino acid sequence of the coat protein and by the fact that a fragment of the 22K gene expressed in *E. coli* gave a product that was recognized by viral antibodies (Miller *et al.*, 1988a). There is a 17K ORF contained entirely within the 22K gene in a different reading frame. This might be the ORF coding for the viral VPg.

Immediately after the coat protein gene, in the same reading frame, is an ORF that could encode a 50K protein, and following this in the same frame as the 17K ORF is a 6.7K ORF (Fig. 7.13), which may be expressed.

2. Gene Expression

The only positively identified gene product is the coat protein. However, on the basis of the sequence data, Miller *et al.* (1988b) have proposed several expression strategies.

Translational Frameshift for the 60K ORF

Miller *et al.* (1988b) defined the reading frame for the 60K ORF as −1 with respect to the 39K ORF by the amino acid sequence similarities seen in Fig. 7.14. The 5′ limit of the ORF is then defined by a UGA termination codon. The first AUG downstream from this position is some 285 nucleotides away. However, there is evidence from amino acid sequence conservation in mutants, and sequence similarities with carnation mottle virus, that this 285-nucleotide sequence is translated. In the light of these features, Miller *et al.* (1988b) proposed a translational frameshift that allows ribosomes reading the 39K ORF to bypass the stop codon at the 3′ end of this ORF and continue to translate the 60K ORF to give a 99K fusion or transframe protein (see Fig. 6.9).

The Coat Protein and 17K and 6.7K ORFs

The coat protein and 6.7K ORFs may be expressed from subgenomic RNAs. RNAs of suitable size (3.0 and 0.8 kb) and appropriate map positions were detected in RNAs from infected tissue (Gerlach *et al.*, 1987a) (Fig. 7.13). It is possible that the 17K ORF is expressed by internal initiation on the proposed coat protein mRNA, because the sequence context of the AUG for the 17K ORF is in good agreement with the consensus for plant initiator AUGs. In fact it is closer than that of the coat protein initiator—plant initiation consensus sequence: AACAAUGGC (Lütcke *et al.*, 1987); 17K ORF: CAAAAUGGC; coat protein: GUGAAUGAA.

The 50K ORF

The 50K ORF is in the same reading frame as the coat protein and separated from it only by a UAG stop codon. Miller *et al.* (1988b) suggested that the 50K ORF may be translated as a read-through protein from the coat protein for the following reasons: (i) A short nucleotide sequence 5′ to the stop codon is similar in several viruses in which that codon is read through. The BYVD sequence at the end of the coat protein gene is similar to these (see Fig. 6.8). (ii) The amino acid sequence of the read-through protein of BNYVV shows some similarity to the 50K ORF of BYVD. Further sequence similarities between BYDV and other viruses are mentioned in Chapter 17, Section V,B.

3. Other Luteoviruses

Beet western yellows virus (BWYV) RNA (Veidt *et al.*, 1988) and PLRV RNA (van der Wilk *et al.*, 1989; Mayo *et al.*, 1989) have also been fully sequenced. All three viruses have a similar arrangement of ORFs. PLRV is about 200 nucleotides longer than the other two viruses. However, on nucleotide sequence homology, PLRV and BWYV are more closely related to each other than they are to BYDV. Two subgenomic RNAs of 2.9 and 0.72 kb have been detected in tissue infected with BWYV (Falk *et al.*, 1989). This virus also contained a 3.1-kb RNA, unrelated in sequence to genomic RNA, which may be a large satellite RNA.

C. *Carmovirus* Group

The type member of this group, carnation mottle virus (CarMV), and others that have been studied resemble members of the *Tombusvirus* group in several properties, including the size of the single genomic RNA, size and three-dimensional structure of the coat protein, and organization and size of the protein shell. The group is distinguished from the tombusviruses on the basis of genome structure and strategy. There is no marked nucleotide sequence similarity between the members of the two groups. The properties of the group have been reviewed by Morris and Carrington (1988). The two most studied viruses are CarMV and turnip crinkle virus (TCV). Although the genome of CarMV is only about half the size of the TMV genome, it shows many similarities in gene arrangement and expression.

1. Genome Structure

The genome consists of a single piece of ss positive sense RNA. The complete sequence of the nucleotides in CarMV has been established (Guilley *et al.*, 1985). There is a 69-nucleotide 5′ leader sequence before the first AUG. There are two UAG termination codons followed by a UAA arranged as shown in Fig. 7.15. This

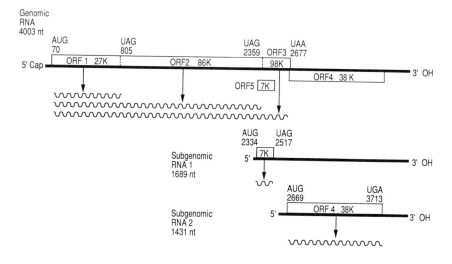

Protein functions :- 27K , ? ; 86K , polymerase ; 98K ,? ; 38K , coat protein ; 7K , cell-to-cell transport

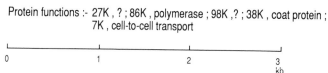

Figure 7.15   Organization and expression of a *Carmovirus* genome (CarMV).

could give rise to a protein of 27K terminating at the first UAG. Read-through of this UAG to the second one or to the UAA codon would give proteins of 86K and 98K. ORF4 in the second reading frame shown in Fig. 7.15 would give a protein of MW = 38K. A small ORF in reading frame 3 would give a polypeptide of 7K. This is followed by a 3′ noncoding region of 290 nucleotides.

Besides genomic RNA, two subgenomic RNA species of 1.6 and 1.75 kb are produced *in vivo* and encapsidated in virus particles (Carrington and Morris, 1984). They appear to be 3′ coterminal with the genomic RNA. Their 5′ termini have been precisely mapped (Carrington and Morris, 1986). The TCV genome has been sequenced, and its organization shows it to be closely related to CarMV (Carrington *et al.*, 1989b). Infectious TCV RNA has been transcribed *in vitro* from full-length cDNA clones (Heaton *et al.*, 1989b). Maize chlorotic mottle virus has a similar genome organization to, and amino acid sequence similarities with, CarMV and TCV. It may also belong in this group (Nutter *et al.*, 1989).

2. The Proteins Produced and Their Functions

In the *in vitro* reticulocyte lysate system the 1.6-kb subgenomic RNA 2 efficiently produces a 38K protein that is the coat protein, based on serological reactivity with an anti-viral serum (Harbison *et al.*, 1984; Carrington and Morris, 1985). This is confirmed by the predicted amino acid composition. The genomic RNA gave rise to two major polypeptides corresponding to the 27K ORF and the 86K read-through ORF revealed by the nucleotide sequence (Fig. 7.15) (Carrington and Morris, 1985). Harbison *et al.* (1985) identified the 98K double read-through protein in lysates optimized for stop codon read-through. The coat protein gene is silent in the genomic RNA. The sequences that are highly conserved in the genes of several plant viruses thought to code for viral RNA-dependent RNA polymerases (Kamer and Argos, 1984) are present in the read-through portion of the 86K protein, suggesting that this protein encodes a polymerase activity. The role of the 1.75-kb subgenomic RNA1 has not been established, but a 7K polypeptide is coded for in *in vitro* translation (Carrington and Morris, 1986).

It has been suggested that, in view of the other genomic similarities between TMV and CarMV, the product of the 7K ORF may have a cell-to-cell transport function like that of the 30K TMV protein. *In vivo*, only the 38K coat protein and the 98K double read-through protein have been detected (Harbison *et al.*, 1985).

3. *In Vitro* Assembly

Dissociation of TCV at high pH and ionic strength gives rise to coat protein dimers and a ribonucleoprotein complex made up of the genomic RNA, six coat protein subunits, and an 80K protein that is a covalent coat protein dimer. TCV particles can be reassembled *in vitro* using coat protein and either the free RNA or the ribonucleoprotein complex. For both forms of the RNA the process is selective for viral RNA, and proceeds by continuous growth of a shell from an initiating structure (Sorger *et al.*, 1986).

4. *In Vivo* Studies

Very little is known about the replication of carmoviruses *in vivo*. Kluge *et al.* (1983) infected *Dianthus* protoplasts with CarMV and showed that most virus production occurred between 12 and 24 hours after inoculation.

IV. TWO STRATEGIES: SUBGENOMIC RNA AND POLYPROTEIN

A. *Tymovirus* Group

The first *Tymovirus* genomes were fully sequenced only recently, and the complete genome strategy is not yet established. A notable feature of the tymoviruses is the intimate association of viral RNA synthesis with small vesicles induced by the virus at the periphery of the chloroplasts. This phenomenon has enabled a quite detailed sequence of cytological effects to be related to stages in virus replication.

1. Genome Structure

The full nucleotide sequence has been determined for five tymoviruses: European TYMV (Morch *et al.*, 1988), Australian TYMV (Keese *et al.*, 1989), eggplant mosaic virus (Osario-Keese *et al.*, 1989), Ononis yellow mosaic virus (Ding *et al.*, 1989), and Kennedya yellow mosaic virus (Ding *et al.*, 1990a). The five genomes vary slightly in length but all have a very similar structure with three ORFs of significant length. However, the coat protein gene in different tymoviruses may be in different reading frames with respect to the long ORF (Ding *et al.*, 1990b).

The genomic organization is very compact. For example, with the European TYMV only 192 (3%) of the 6318 nucleotides are noncoding. The largest ORF initiates at position 95 and ends at position 5627 with a UAG (to give a protein of 206K and 1844 amino acids). A second ORF runs from the first AUG on the RNA (beginning at nucleotide 88) and terminates with a UGA at 1972 (69K; 628 amino acids). It overlaps the 206K gene is in the same reading frame as the coat protein gene. The coat protein gene is in reading frame −1 compared with the 206K gene for the virus illustrated in Fig. 7.16. There is a 105-nucleotide noncoding 3′ region, which contains a tRNA-like structure (Fig. 7.16).

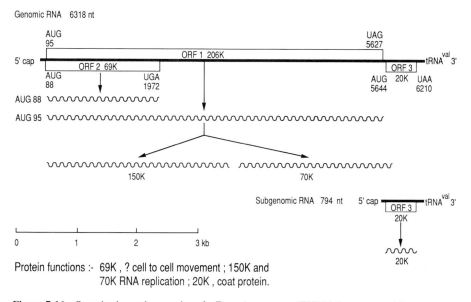

Protein functions :- 69K , ? cell to cell movement ; 150K and
                      70K RNA replication ; 20K , coat protein.

Figure 7.16   Organization and expression of a *Tymovirus* genome (TYMV, European strain).

2. Subgenomic RNA

A small subgenomic RNA is packaged with the genomic RNA and in a series of partially filled particles (e.g., Pleij *et al.*, 1977). No dsRNA corresponding in length to this subgenomic RNA could be detected *in vivo*. Gargouri *et al.* (1989) detected nascent subgenomic coat protein plus strand RNAs on dsRNA of genomic length. Thus the coat protein mRNA is synthesized *in vivo* by internal initiation on minus strands of genomic length. Ding *et al.* (1990b) compared the available *Tymovirus* nucleotide sequences around the initiation site for the subgenomic mRNA. They found two conserved regions, one at the initiation site and a 16-nucleotide sequence on the 5′ side of it. This longer sequence, which they called the tymobox, may be an important component of the promoter for subgenomic RNA synthesis.

3. *In Vitro* Translation Studies

The small subgenomic RNA is very efficiently translated in *in vitro* systems to give coat protein (e.g., Pleij *et al.*, 1977; Higgins *et al.*, 1978). The coat protein gene in the genomic RNA is not translated *in vitro*.

Weiland and Dreher (1989) obtained infectious TYMV RNA transcripts from cloned full-length cDNA. By making mutants in the initiation codons they showed that the 69K protein is expressed from the first AUG beginning at nucleotide 88, while the much larger gene product is expressed from an AUG beginning at nucleotide 95. Antibodies raised from a synthetic peptide corresponding to the C terminus of the ORF are specific for the 69K protein, demonstrating *in vitro* expression (C. Bozarth, J. Weiland, and T. Dreher, personal communication).

The large ORF with the AUG beginning at nucleotide 95 appears to be expressed in a variety of *in vitro* systems and is then cleaved *in vitro* to give a larger 5′ and a smaller 3′ product (Morch *et al.*, 1982, 1988, 1989; Zagorski *et al.*, 1983). The protease concerned and the cleavage site or sites have not been characterized. The posttranslational processing shown in Fig. 7.16 is based on recent work by K. Bransom and T. Dreher (personal communication). It is not yet known whether the 206K protein is fully processed *in vitro*. Neither has the full complement of mature polypeptides formed *in vitro* been firmly defined.

4. Functions of the Proteins

The 206K polypeptide contains sequences conserved in a number of RNA viral gene products that are involved in nucleotide binding (Fig. 7.22) and in replicase function. The latter sequence is located toward the 3′ terminus of the 206K structure in a 70K cleavage product. However, the TYMV replicase isolated from infected plants contains a 115K (=120K) viral-coded polypeptide (Mouches *et al.*, 1984). This situation is not yet resolved.

5. RNA Synthesis

In extracts from Chinese cabbage plants infected with TYMV, the RNA polymerase activity is associated with the chloroplast fraction (Bové *et al.*, 1967; Ralph and Wojcik, 1966). The enzyme has been detected in the bounding membrane fraction of the chloroplasts that contains the small virus-induced vesicles (Laflèche *et al.*, 1972; Bové *et al.*, 1972). Mouches *et al.* (1984) purified the TYMV replicase from infected Chinese cabbage leaves. The enzyme was template-dependent and showed specificity for TYMV RNA. It consisted of two subunits. One of 115K was virus coded as shown by serological cross-reaction between this protein and the 115K *in vitro* translation product of genomic RNA (Candresse *et al.*, 1986). The other, of 45K, is host encoded, because antibodies raised against it do not react with any TYMV *in vitro* translation products (Bové and Bové, 1985). *In vivo* the viral replicase is tightly bound in a replication complex involving the chloroplast envelope membrane. By contrast the host-encoded RNA-dependent RNA polymerase found in healthy and infected leaves is a soluble enzyme of different size and with different template specificity (Mouches *et al.*, 1984).

Garnier *et al.* (1980), using a combination of electron microscopy and other techniques, were able to demonstrate that, *in vivo,* the RNA replicative intermediate is mainly in ss form and that the dsRNA isolated from infected leaves (Bockstahler, 1967) is mainly an artefact generated during isolation (see Fig. 6.1). However, late in the infection cycle some dsRF may accumulate *in vivo.*

The idea that most of the viral RNA synthesis occurs in the small peripheral vesicles found in diseased chloroplasts (Fig. 7.17) is supported by the following additional evidence: (i) when observed in thin sections many vesicles can be seen to contain stranded material with the staining properties expected for a ds nucleic acid (Fig. 7.17) (Ushiyama and Matthews, 1970); (ii) the ds form of TYMV RNA is associated with the chloroplast fraction in extracts of infected tissue rather than with the nuclei or soluble fraction (Bové *et al.*, 1967; Ralph and Wojcik, 1966); (iii) using autoradiography with ³H-labeled uridine, Laflèche and Bové (1968) showed that radioactivity accumulated in the spaces between clumped chloroplasts in diseased cells; and (iv) immunogold labeling located the viral replicase near the periphery of clumped chloroplasts in infected cells (Garnier *et al.*, 1986). In spite of the close association of TYMV with the chloroplasts, the presence of chlorophyll is not necessary for TYMV replication. The virus multiplied in chlorophyll-less protoplasts from etiolated Chinese cabbage hypocotyls (Fernandez-Gonzalez *et al.*, 1980).

TYMV infection also stimulates nucleic acid synthesis in the nucleus. Laflèche and Bové (1968) noted that ³H-labeled uridine accumulated in or near the nucleoli in TYMV-infected cells treated with actinomycin D, as well as near the margins of the clumped chloroplasts. Bedbrook *et al.* (1974) showed that radioactively labeled

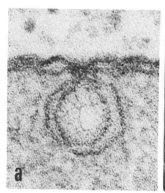

Figure 7.17 Fine structure of TYMV-induced peripheral vesicles in the chloroplasts of infected Chinese cabbage cells. (A) Thin section showing continuity of inner chloroplast and outer vesicle membranes and stranded material inside the vesicle with the staining properties of ds nucleic acid (X235,000). (Courtesy of S. Bullivant.) (B) Fine structure of vesicle membranes revealed by freeze-fracturing of isolated chloroplasts (X92,000). (From Hatta *et al.*, 1973.)

nuclear nucleic acids prepared from TYMV-infected tissue contained two components not present in nucleic acids from equivalent healthy leaves. The significance of these events in the nuclei for TYMV replication remains to be established.

The unusually high cytidylic acid content of TYMV RNA made it possible to distinguish fairly readily between ss and ds TYMV RNA by nucleotide ratios of RNA labeled *in vivo* with ^{32}P. The results fitted with the view that the viral RNA is produced by an asymmetric semiconservative process in which the positive sense strands are produced more frequently than the negative sense strands (Ralph *et al.*, 1965). More detailed studies indicated that at very early and very late stages of infection, negative sense strand synthesis predominates (Bedbrook and Matthews, 1976).

The subgenomic RNA containing the coat protein gene becomes labeled *in vivo* with ^{32}P more rapidly than genomic RNA, presumably reflecting more rapid synthesis of the coat mRNA (Matthews *et al.*, 1963). However, the site of production of this mRNA is not established.

6. Viral Proteins Synthesized *in Vivo*

Coat Protein

Biochemical studies indicate that coat protein is synthesized on 80 S cytoplasmic ribosomes. Renaudin *et al.* (1975) showed that viral protein synthesis in Chinese cabbage protoplasts is inhibited by cycloheximide but not by chloramphenicol. The cytoplasmic site for viral protein synthesis has been confirmed by cytological evidence (see the following Section 8), but viral protein also accumulates in the nucleus.

The uracil analog, 2-thiouracil, blocks TYMV RNA synthesis but not protein synthesis (Francki and Matthews, 1962), implying that the mRNA for coat protein is a relatively stable molecule. The kinetics of labeling with [^{35}S]methionine of empty protein shells and viral nucleoprotein in infected protoplasts shows that these two protein shells are assembled from different pools of protein subunits (Sugimura and Matthews, 1981).

Other Proteins

The 69K protein has been detected *in vivo* in both Chinese cabbage and *Arabidopsis thaliana*. It is expressed at a 500× lower level than coat protein and appears to be an early nonstructural protein. The 70K C-terminal fragment shown in Fig. 7.16 is also found *in vivo* (C. Bozarth, J. Weiland, and T. Dreher, personal communication).

Mouches *et al.* (1984) showed that the TYMV replicase consisted of viral-coded polypeptide of MW = 115K and a host-coded protein of 45K. Using an immunoblotting technique, Candresse *et al.* (1987b) showed that the viral-coded replicase subunit appears very soon after inoculation of plants or protoplasts. The 115K protein has not yet been reconciled with the data summarized in Fig. 7.16. Other proteins of 80K and 150K were detected using the same antiserum (Bové and Bové, 1985), and these may represent other functional products expected from the genome structure, but more definitive evidence is needed.

7. Early Events following Inoculation

In vitro studies show that under various nonphysiological conditions the RNA can escape from TYMV without disintegration of the protein shell. Thus at pH 11.5

in 1 *M* KCl the RNA is rapidly released together with a cluster of 5–8 protein subunits from the shell (Keeling and Matthews, 1982). A hole corresponding to 5–7 subunits is left in the protein shell following release of RNA by freezing and thawing (Katouzian-Safadi and Berthet-Colominas, 1983). Treatment of TYMV with 3–7% butanol at pH 7.4 leads to the rapid release of RNA and 5 or 6 protein subunits in monomer form (S. Deroles and R. E. F. Matthews, unpublished).

In Chinese cabbage leaves, Kurtz-Fritsch and Hirth (1972) found that about 2% of the retained TYMV inoculum was uncoated after 20 minutes. They showed that empty shells and low-MW protein were formed following RNA release.

Matthews and Witz (1985) confirmed these findings and showed that a significant proportion of the retained inoculum was uncoated within 45 seconds, and that the process was complete within 2 minutes. At least 80–90% of this uncoating takes place in the epidermis. Approximately 10^6 particles per epidermal cell can be uncoated (see also Chapter 10, Section III,B and C). The process gives rise to empty shells that have lost about 5–6 protein subunits and to low-MW protein. At the high inoculum concentrations used, most of the released RNA must be inactivated, presumably in the epidermal cells. Celite was used as an abrasive, so that the mechanism of entry proposed for TMV (Fig. 7.6) could account for the large numbers of particles entering each cell. The uncoating process just described was not confined to known hosts of TYMV.

8. TYMV Replication in Relation to Cytological Effects

The *Tymovirus* group produces very characteristic cytological alterations in infected cells. Some of these changes are intimately associated with virus replication and will be discussed here. Other, secondary cytological effects are described in Chapter 11.

The small peripheral vesicles that contain the RI and viral RNA-dependent RNA polymerase are a consistent feature of *Tymovirus* infection (Fig. 7.17). They are present in all infected host species, and they occur in all parts of the plant that have been examined (evidence summarized by Matthews, 1973). How the vesicles are formed is not established. In a formal sense they can be considered to be derived from invaginations of the chloroplast membranes but it is not known whether preexisting membrane is modified or whether additional chloroplast membrane is synthesized.

Using the presence of vesicles as a marker of infection it was possible to follow cytologically the invasion of a single systemically infected leaf as the infection spread out from the phloem elements of the small veins (Hatta and Matthews, 1974). The more important stages are illustrated in Figs. 7.18 and 7.19 and described in the legends. The earliest stage at which we could detect TYMV particles was at stage D, when microcrystalline arrays were seen in the cytoplasm next to the electron-lucent zones. (Fig. 7.19A). We concluded that the clumping of the chloroplasts is a secondary effect of infection, not essential for virus synthesis. In successive stages later than C, more and larger arrays of virus particles appeared in the cytoplasm while the electron-lucent material disappeared. We concluded that the electron-lucent material is an accumulation of viral coat protein, mainly or entirely in the form of pentamers and hexamers, and that these are used for the assembly of virus particles.

The role of the ER that appears transiently at stage B (Fig. 7.18) is not

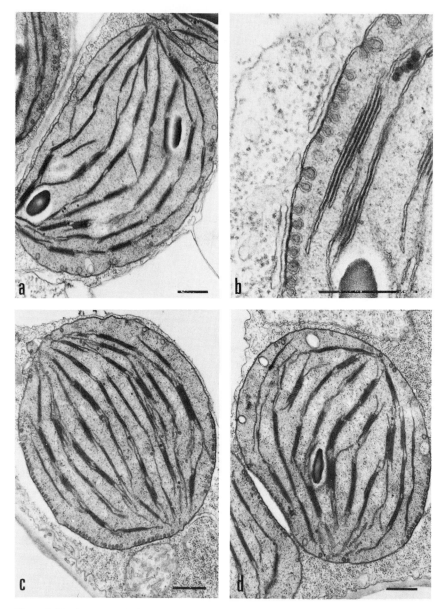

Figure 7.18   Some steps in the sequence of cytological changes induced by TYMV infection in Chinese cabbage leaf cells. (a) Scattered small peripheral vesicles; chloroplasts are otherwise normal. (b) Chloroplasts are now swollen. Scattered vesicles are still present, but many clusters of vesicles are also present. In freeze-fracture preparations these can be seen as hexagonal arrays. ER is present in the cytoplasm over the clustered vesicles. The zones of ER over each cluster of vesicles are connected by strands of ER, making a network over the chloroplast surface (Hatta and Matthews, 1976). (c) Electron-lucent areas have appeared in the cytoplasm over the clustered vesicles. This material reacts with antibodies against TYMV. Freeze-fracturing indicates that it may be in the form of pentamers and hexamers. (d) The chloroplasts have become clumped with electron-lucent areas in contact. At a later stage the electron-lucent material is replaced by virus particles. (From Hatta and Matthews, 1974.)

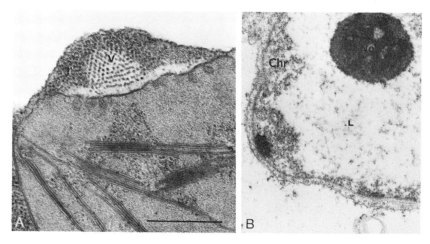

Figure 7.19 The intracellular sites of TYMV and TYMV protein accumulation. (A) Small crystalline array of TYMV particles associated with electron-lucent area in the cytoplasm of a stage C cell subjected to plasmolysis. Bar = 500 nm. r, ribosomes; v, TYMV particles. (From Hatta and Matthews, 1974.) (B) Accumulation of TYMV coat protein (electron-lucent material, L) in the nucleus of a stage D cell not subjected to plasmolysis. Bar = 500 nm. (Modified from Hatta and Matthews, 1976.)

established. From the timing of its appearance, and its very specific position in the cell, it could be involved in any or all of the following processes: (i) the synthesis of a protein or proteins needed to form the peripheral vesicles; (ii) synthesis of viral RNA polymerase; and (iii) synthesis of viral coat protein.

Viral coat protein in the form of empty protein shells accumulates in large quantities in the nuclei of cells infected with TYMV and other tymoviruses (Hatta and Matthews, 1976). No virus particles have been identified in nuclei. The protein can be detected in nuclei of TYMV-infected Chinese cabbage cells before the electron-lucent zones appear in the cytoplasm. The amount increases with time until the nuclei are almost filled with empty protein shells (Fig. 7.19B). We must assume that the viral protein is synthesized in the cytoplasm and moves into the nucleus either via the nuclear pores or directly through the nuclear membrane.

Okra mosaic virus produces large quantities of empty viral protein shells in inoculated cucumber cotyledons (Marshall and Matthews, 1981). At 1 day after inoculation, about one-half the total viral protein shells were found in the nucleus. This accumulation occurred in the presence of virus particles that were found exclusively in the cytoplasm. The most likely explanation for this active and preferential accumulation is that viral coat protein enters the nucleus in the form of monomers or pentamers and hexamers and is assembled into shells once inside. The accumulation of normal plant-coded proteins in the nucleus is an extremely selective process, depending on a signal sequence of amino acids in the proteins (Dingwall, 1985). Thus the mechanism by which *Tymovirus* coat proteins accumulate in the nucleus in very large amounts is worth further study. It may be relevant to this problem that the TYMV coat protein hydropathic profile (Kyte and Doolittle, 1982) looks very much like that of a membrane protein (R. E. F. Matthews, unpublished).

9. Virus Assembly

TYMV particles have not yet been reassembled *in vitro* from RNA and protein subunits. Figure 7.20 illustrates a possible model for the assembly of TYMV parti-

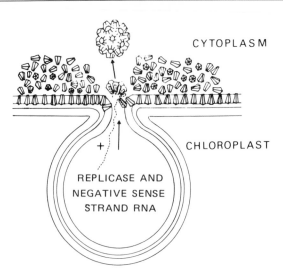

Figure 7.20   A model for the assembly of TYMV. (i) Pentamer and hexamer clusters of coat protein subunits are synthesized by the ER and accumulate in the cytoplasm overlying clustered vesicles in the chloroplast. (ii) These become inserted into the outer chloroplast membrane in an oriented fashion, that is, with the hydrophobic sides that are normally buried in the complete protein shell lying within the lipid bilayer, with the end of the cluster that is normally inside the virus particle at the membrane surface. (iii) An RNA strand synthesized or being synthesized within a vesicle begins to emerge through the vesicle neck. (iv) At this site a specific nucleotide sequence in the RNA recognizes and binds a surface feature of a pentamer cluster lying in the outer chloroplast membrane near the vesicle neck, thus initiating virus assembly. (v) Assembly proceeds by the addition of pentamers and hexamers from the uniformly oriented supply in the membrane. (vi) The completed virus particle is released into the cytoplasm.

cles *in vivo,* based on the information in the preceding section and Figs. 7.18 and 7.19. This is discussed in more detail by Matthews (1981). The model predicts that there is an accumulation of coat protein just before virus assembly begins, and that unlike TMV replication, there would be no accumulation of complete uncoated viral genomes. As the pentamer and hexamer clusters are depleted in the membrane they would be replaced by others from the electron-lucent layer until the supply was exhausted. Empty protein shells presumably represent errors in virus assembly, which take place in the absence of RNA. They can form because of the strong protein–protein interactions in the shell of this virus.

The model readily explains the effect of 2-thiouracil on TYMV replication noted in the preceding Section 6. When genome RNA synthesis in the vesicles is blocked by the analog, empty protein shells are made in increased amounts from the accumulated coat protein and from further protein being synthesized on preexisting coat mRNA.

10. Virus Increase in Protoplasts and Leaves

In infected *Brassica* protoplasts virus production probably begins somewhat earlier than 12 hours after inoculation. Renaudin *et al.* (1975) found that $1-2 \times 10^6$ virus particles per protoplast were produced 48 hours after infection, which is similar to estimates made for cells in intact leaves (Matthews, 1970). However, these latter estimates were made on extracts of whole infected leaves, without any account being taken of variation in size and virus productivity of different cell types or of the percentage of cells actually infected in the leaf. On the basis of virus particles per infected mesophyll cell, protoplasts support only about one-tenth of the virus produced by similar cells in the intact leaf. In protoplasts, empty protein shell production is not reduced as much as that of virus production. The two types of particle are produced in almost equal amounts compared with about 20% of empty shells in leaves. The approximate maximum rates of particle production per infected protoplast are about the same for cells infected *in vitro* or *in vivo*. In protoplasts, virus increase ceased after about 48 hours, but continued for about 10 days in intact leaves. Disease development in protoplasts is not synchronous as judged by rounding and clumping of chloroplasts (Sugimura and Matthews, 1981).

11. Polyamine Synthesis

In protoplasts derived from infected leaves, or in healthy protoplasts infected *in vitro,* newly formed virus particles contained predominantly newly synthesized spermidine and spermine (Balint and Cohen, 1985a,b). When a specific inhibitor of spermidine was present there was increased synthesis of spermine and an increase in the spermine content of virus particles. Thus there is some flexibility in the way in which the positive charge contributed by the polyamines is conserved. The biosynthesis of polyamines and their possible roles in plants are discussed by Smith (1985).

B. *Sobemovirus* Group

The genome of one member of this group has been fully sequenced, but the gene products coded for have not yet been fully delineated. The properties of the group have been reviewed by Hull (1988).

1. Genome Structure

The genome of southern bean mosaic virus (SBMV) consists of a single strand of positive sense ssRNA. Strain C has four ORFs (Wu *et al.*, 1987) (Fig. 7.21). The

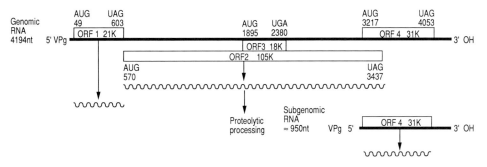

Protein functions :- 21K,? ; 105K, polymerase, ATP binding,VPg ; 18K, not known to be translated ; 31K, coat protein .

Figure 7.21 Organization and expression of a *Sobemovirus* genome (SBMV strain C).

5' terminus is covalently linked to a small VPg (Mang *et al.*, 1982). The 48-nucleotide leader contains a sequence that is partially complementary to the 3' end of 18 S ribosomal RNA, suggesting a possible role in ribosome binding. This site is seven bases 5' to the first AUG codon (see Fig. 6.7). There is a 3'-terminal untranslated region 153 nucleotides long that does not have the potential to form a tRNA-like structure.

2. Gene Products

The four ORFs are in three reading frames. As discussed by Wu *et al.* (1987), *in vitro* translation studies have given a complex series of translation products. ORF4 encodes the coat protein that is translated from a subgenomic RNA (RNA2) that is 3' coterminal with genomic RNA and that has the same VPg as genomic RNA at the 5' terminus (Ghosh *et al.*, 1981).

Genomic RNA isolated from virus is translated to give a 105K protein, corresponding to ORF2. This product may be a polyprotein cleaved in *cis* by a viral protease. Amino acid sequence similarities with other viruses suggest that the 105K ORF may code for a viral polymerase, the VPg, and an ATP binding function (Wu *et al.*, 1987). The proposed ATP binding domain is shown in Fig. 7.22. A group of smaller proteins is also translated from the RNAs of various sizes isolated from virus preparations. Some of these correspond to the protein expected for ORF1. No evidence has been found for the translation of ORF3 *in vitro*.

RNAs of several other members of the group have been translated in *in vitro* systems or protoplasts to give a high-MW product of about 105K plus smaller polypeptides, for example, lucerne transient streak virus (Morris-Krsinich and Forster, 1983), *Solanum nodiflorum* virus (Kiberstis and Zimmern, 1984), and turnip rosette virus (Morris-Krsinich and Hull, 1983). In summary, SBMV appears to produce several proteins from an RNA genome that is smaller than average.

3. Viral RNA Synthesis

A viral-specific RNA-dependent RNA polymerase activity has been isolated from tissue infected with velvet tobacco mottle virus (Rohozinski *et al.*, 1986).

```
                            •           • •
                      •     •         • • •
CONSENSUS        I - L R G K A G - G K S L - S N - I A R
                        • • •       • •       • • •
SBMV(1626-1682)  I E N R G K V K L G K R E F A W . V P K


CONSENSUS        Q N P D G A M S
                 • •   • • • • •
SBMV(1773-1796)  E S F D G A L P


                      •         •
CONSENSUS        M A S - L E E K
                 • • •   • •   •
SBMV(1923-1940)  L A S . L E . K
```

Figure 7.22   The putative ATP binding domain corresponding to a region of ORF2 of SBMV RNA between nucleotides 1626 and 1940. A consensus sequence was obtained from an alignment of 18 picornaviral 2C regions and an alignment of AMV, BMV, and TMV. The letters in the consensus lines represent the most prevalent amino acid with a similarity of at least 60% and the hyphens show where the similarity is less than 60%. The dots between the lines mark identical or equivalent residues between the two sequences. The periods in the SBMV lines indicate where gaps have been introduced. A double asterisk indicates that all aligned residues in the consensus sequence at that position are identical. A single asterisk indicates that all the aligned residues are equivalent. (From Wu *et al.*, 1987.)

Only one species of dsRNA, corresponding to the genomic RNA, has been found in tissue infected with several viruses of the group. This suggests that the subgenomic mRNA is derived by internal initiation on the minus strand genomic RNA, as it is for several other virus groups (Hull, 1988).

4. Assembly of Virus Particles

SBMV particles can be assembled *in vitro* at low ionic strength from isolated RNA and coat protein. The components assembled into $T = 1$ or $T = 3$ particles depending on the size of the viral RNA used and the pH (Savithri and Erickson, 1983).

5. *In Vitro* Translation

When swollen SBMV particles were incubated with a wheat germ extract containing [^{35}S]methionine, sucrose density gradient analysis showed that 80 S ribosomes were associated with intact or almost intact virus particles (Shields *et al.*, 1989). The data suggested that translation of the viral RNA begins before it is fully released from the virus particle.

V. TWO STRATEGIES: MULTIPARTITE GENOME AND POLYPROTEIN

A. *Comovirus* Group

CPMV is the type member of the *Comovirus* group. The following discussion centers on this virus since it has been used for most experimental work. The structure, replication, and expression of the genome have been reviewed by Goldbach and van Kammen (1985) and van Kammen and Eggen (1986). The structure and organization of the bipartite CPMV genome is remarkably similar to that of the animal picornaviruses, which have a monopartite genome. This aspect is discussed in Chapter 17, Section V,A.

1. Genome Structure

The genome of CPMV consists of two strands of positive sense ssRNA called M- and B-RNAs, with MW = 1.22×10^6 and 2.04×10^6. Both RNAs have a small VPg covalently linked at the 5' end, and a poly(A) sequence at the 3' end. The full nucleotide sequence of the M-RNA was determined by van Wezenbeek *et al.* (1983) and that of B-RNA by Lomonossoff and Shanks (1983). There is a single long open reading frame in both RNAs. In M-RNA there are three AUGs beginning at nucleotides 161, 512, and 524. Using site-directed mutagenesis, Holness *et al.* (1989) confirmed that the AUG at position 161 is used *in vitro* to direct the synthesis of a 105K product. Both the 150K protein initiating at nucleotide 161 and a 95K product initiating at nucleotide 512 have been detected in infected protoplasts, as have their cleavage products as illustrated in Fig. 7.23 (Rezelman *et al.*, 1989). As discussed in Section A,5, these polyproteins are cleaved to give a series of functional products (Fig. 7.23).

The first 44 nucleotides in the 5' leader sequences of M- and B-RNA show 86%

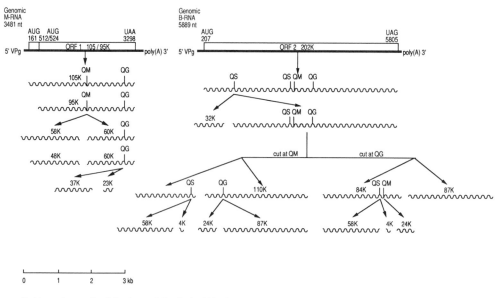

Proteinase cleavage sites Q-S = gln-ser ; Q-G = gln-gly ; Q-M = gln-met

Protein functions :- M encoded , 58K and 48K , cell-to-cell movement ; 37K and 23K , coat proteins
                    B encoded , 32K , Modulation of protease ; 4K , VPg ; 24K , protease ; 58K , RNA replication complex ; 110K replicase

Figure 7.23   Organization and expression of a *Comovirus* genome (CPMV).

homology. The last 65 nucleotides preceding the poly(A) sequence show 83% homology. The seven nucleotides before the poly(A) sequence are the same in the two RNAs. This conservation suggests that these sequences are involved in recognition signals for viral or host proteins. Eggen *et al.* (1990) constructed a series of mutants in B-RNA involving the seven 3'-terminal nucleotides and the first four A's of the poly(A) sequence and tested their effects on B-RNA replication in protoplasts and plants. Only mutants with minor modifications were able to replicate. Mutant transcripts reverted stepwise to the wild-type sequence during replication in plants.

Neither RNA has an AAUAAA polyadenylation signal preceding the 3' poly(A) sequence. The presence of poly(U) stretches at the 5' end of negative sense strands in RF molecules indicate that the poly(A) sequences are transcribed in the RNA replication process (Lomonossoff *et al.*, 1985).

2. The Partial Independence of B-RNAs

Both the B and M virus particles, or the RNA they contain, are necessary for an infection producing progeny virus particles (van Kammen, 1968; De Jager, 1976). However, inoculation of protoplasts with B component leads to production of B-encoded polypeptides and to replication of B component RNA. M-RNA is not replicated in the absence of B-RNA. Thus B-RNA must code for proteins involved in replication, while M-RNA codes for the structural proteins of the virus. B-RNA inoculated to intact leaves cannot move to surrounding cells in the absence of M-RNA (Rezelman *et al.*, 1982).

3. Proteins Coded by B-RNA

B-RNA is translated both *in vitro* and *in vivo* into a 200K polyprotein that is subsequently proteolytically cleaved at specific sites, in a series of steps, to give the functional proteins. The relations between the various polypeptides synthesized in protoplasts have been established by comparison of the proteolytic digestion patterns of the isolated proteins and by immunological methods (e.g., Rezelman *et al.*, 1980; Goldbach *et al.*, 1981, 1982; Zabel *et al.*, 1982; Goldbach and Rezelman, 1983; Franssen *et al.*, 1984a). More recently, radioactively labeled CPMV proteins synthesized in infected protoplasts were isolated and partial amino acid sequences determined (Wellink *et al.*, 1986). This allowed precise location of the coding sequence for each protein on the B-RNA, precise determination of molecular weights, and determination of the cleavage sites at which they are released from the polyprotein precursor. There are three types of cleavage site: glutamine–serine (two sites); glutamine–methionine (one site); and glutamine–glycine (one site). The scheme for proteolytic processing derived from this and other work is shown in Fig. 7.23.

4. Proteins Coded by M-RNA

A scheme for the translation of M-RNA shown in Fig. 7.23. It is based on the *in vitro* translation and processing studies of Franssen *et al.* (1982) and later work. The two mature coat proteins are known and the 60K precursor of these proteins has been detected in infected protoplasts (Wellink *et al.*, 1987b). These workers, using antisera raised against synthetic peptides, detected a 48K M-RNA-encoded protein in the membrane fraction of infected cells. A 58K polypeptide initiated at the AUG at nucleotide 161 and sharing a C-terminal sequence with the 48K protein was detected in protoplasts inoculated with CPMV (Wellink and van Kammen, 1989). Some other proteins indicated in Fig. 7.23 have not yet been detected *in vivo*, including the 95K primary transcript, presumably because it rapidly cleaved. A 95K protein was detected *in vitro* initiating from either of the AUGs at positions 512 or 524 (Holness *et al.*, 1989). The 95K protein did not appear to be necessary for M-RNA replication in protoplasts.

5. Functions of the Viral-Coded Proteins

Coat Proteins

The *in vitro* studies of Franssen *et al.* (1982) demonstrated directly that the two coat proteins are coded on M-RNA. Other evidence was given in the preceding sections. This virus can spread through the plant only if the RNAs are in the form of virus particles (Wellink and van Kammen, 1989).

Viral Protease

It has been known for some time that proteolytic activity was associated with a product or products of B-RNA that acted in *cis* on the B-RNA polyprotein and in *trans* on the M-RNA polyprotein. Recently this activity has been shown by two different procedures to be carried by the 24K product. Wellink *et al.* (1987a) synthesized a peptide corresponding to an amino acid sequence in the 200K polyprotein that showed similarity to picornaviral 3C proteases. Antibodies to the peptide reacted with a 24K protein found in CPMV-infected protoplasts and leaves.

Verver *et al.* (1987) constructed a full-length cDNA copy of B-RNA. RNA transcribed *in vitro* could be efficiently translated and proteolytically cleaved. Introduction of an 87-bp deletion into the coding region of the 24K protein abolished cleavage activity, demonstrating this protein to be the viral protease. Vos *et al.* (1988a,b) extended this work to include a cDNA copy of M-RNA. They constructed a series of deletion mutants in the 32K and 24K reading frames of B-RNA, and the results showed that the 24K protein is the protease responsible for all cleavages in both B and M polyproteins. For efficient cleavage of the glutamine–methionine site in the M-RNA polyprotein, a second B-encoded protein (32K) is essential, although this protein does not itself have proteolytic activity. Thus it is very probable that the 32K protein has a regulatory function in the processing of M-RNA-encoded proteins. Studies on amino acid sequence variation around the cleavage sites in different comoviruses suggest that protease specificity depends on some three-dimensional structural feature of the substrates (Shanks and Lomonossoff, 1990).

Replicase

Two lines of evidence indicate that the 110K protein encoded by B-RNA is involved in RNA replication. First, it is part of the replicase complex isolated from infected leaves and cosediments precisely with RNA polymerase activity (Dorssers *et al.*, 1984). Second, the polypeptide shows significant sequence similarity with the replicase of picornaviruses (see Chapter 17, Section V,A).

Membrane Attachment

The B-RNA-encoded 60K and 58K polypeptides may be involved in membrane attachment in the membrane-bound replication complex (Goldbach and van Kammen, 1985).

The VPg

Partial amino acid sequencing of the VPg allowed it to be precisely mapped on the B-RNA (Zabel *et al.*, 1984). The functions of VPg's in general are discussed in Chapter 6, Section VI,A,1.

Transport Function

As noted in Section V,A,2, B-RNA is dependent on a function of M-RNA for cell-to-cell movement. Besides a requirement for the coat proteins, experiments using insertion and deletion mutants have shown that both the 58K and 48K proteins encoded by M-RNA are needed for cell-to-cell movement (Wellink and van Kammen, 1989). A 43K protein coded by red clover mosaic *Comovirus* (corresponding to the 48K CPMV protein) was found by immunogold labeling to be associated with the plasmodesmata (Shanks *et al.*, 1988, 1989).

Cells infected with CPMV develop tubular structures that protrude from or penetrate cell walls and that contain viruslike particles. Using immunogold labeling, van Lent *et al.* (1990) showed that the 58K and/or 48K protein were located in or on these tubular structures, supporting a role in cell-to-cell movement for these proteins.

6. Viral RNA Synthesis

A host-encoded RNA-dependent RNA polymerase activity increases some 20-fold following infection of leaves with CPMV. No such increase occurs in infected protoplasts. The activity is associated with a 130K polypeptide. Although this enzyme is active in the crude membrane fraction isolated from infected leaves, it plays no role in CPMV replication (Dorssers *et al.*, 1983, 1984; van der Meer *et al.*, 1984).

A CPMV RNA replication complex, containing a 110K viral-coded polypeptide, has been isolated from infected cowpea leaves (Dorssers *et al.*, 1984). Other, host-specific polypeptides were also present in the complex but their role, if any, in RNA replication is not established. However, actinomycin D inhibits RNA synthesis if applied immediately after inoculation, but not if supplied later. (de Varennes *et al.*, 1985). This suggests that some host-coded protein may be needed for RNA replication. This replication complex is capable of elongating nascent viral RNA strands *in vitro* to full-length RNA. Possible models for RNA replication are discussed by van Kammen and Eggen (1986). The initiation of viral RNA replication may be linked to processing of the polyprotein, and the VPg may play some role, but further work is needed. The position of the 110K polypeptide in the polyprotein (Fig. 7.23) requires that for every molecule of this protein produced, the other B-RNA-encoded polypeptides must also be formed, if not fully cleaved. Other B-RNA-encoded proteins are in fact present in these cytopathic structures (Wellink *et al.*, 1988) but their identity has not been established.

Only a small fraction of the nonstructural proteins detected in infected cowpea leaves was active in RNA replication (Eggen *et al.*, 1988). Thus virus replication proteins may be used only once, perhaps because of a strong coupling between polyprotein processing and replication.

7. General Features of CPMV Replication

In inoculated cowpea protoplasts there was a rapid increase in infectious virus between 9 and 24 hours after inoculation and over 10^6 progeny virus were produced per protoplast (Hibi *et al.*, 1975). Two dsRNA species have been characterized from infected *Vigna* leaves corresponding to the two ss genome RNAs (van Griensven *et al.*, 1973).

From experiments using cycloheximide and chloramphenicol, with the polypeptides of ribulose bisphosphate carboxylase as an internal control, Owens and Bruening (1975) concluded that both the proteins found in CPMV are synthesized in the cytoplasm. Progeny virus particles in crystalline arrays have been seen in cytoplasm and vacuoles, but not within nuclei, chloroplasts, or mitochondria (Langenberg and Schroeder, 1975). CPMV-infected cells develop quite large cytopathological structures (Fig. 7.24). These contain groups of membranous vesicles forming a reticulum. The vesicles contain stranded material with the staining appearance of a ds nucleic acid. Virus particles are also present. From a detailed study combining electron microscopy and cell fractionation procedures, De Zoeten *et al.* (1974) concluded that the viral dsRNAs are located in these structures, which can, therefore, be regarded as viroplasms. These viroplasms have also been observed in infected cowpea and tobacco protoplasts (Hibi *et al.*, 1975).

B. *Nepovirus* Group

The two most studied members of this group are tobacco ring spot virus (TRSV) and tomato black ring virus (TBRV). Although the nepoviruses show many similarities to the *Comovirus* group, their exact genome strategy has not yet been

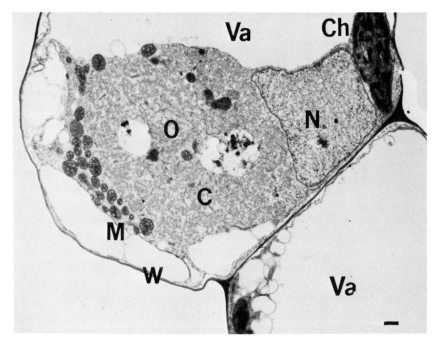

Figure 7.24   Intracellular site of CPMV synthesis. An electron micrograph of a CPMV-infected meso-phyll cell. Note the cytopathological structure (C) associated with mitochondria (M) and the nucleus (N); the large vacuoles (Va); wall (W); and chloroplasts (Ch). Bar = 0.5 μm. (From De Zoeten *et al.*, 1974.)

established. Pseudorecombinants produced by exchanges of the two nucleoprotein components are discussed in Chapter 13, Section II,F.

1. Genome Structure

The genome of nepoviruses is bipartite, consisting of two strands of ss positive sense RNA. The larger RNA (RNA1) has about 8100–8400 nucleotides depending on the virus, while RNA2 is much more variable with about 3400–7200 nucleotides (Murant *et al.*, 1981). Both RNAs have a 3′ poly(A) tail, and both have a VPg linked to the 5′ terminus, which is essential for infectivity but not for *in vitro* translation (Chu *et al.*, 1981; Mayo *et al.*, 1982a).

The nucleotide sequence of TBRV RNA2 has been established (M. Meyer *et al.*, 1986). Most of the sequence is taken up by a single ORF with the capacity to code for a polypeptide of MW = 220K (Fig. 7.25). The coat protein, from its known amino acid sequence, was tentatively assigned to the 3′ region of RNA2. The sequence of the 5′-terminal part of the polypeptide resembled parts of the 30K polypeptides of tobamoviruses, suggesting a transport function for this region. In TRSV, which has a much longer RNA2 than TBRV, there is a 1.9-kb sequence homology at the 3′ termini of RNAs 1 and 2 (Rott *et al.*, 1988).

The nucleotide sequence of TBRV RNA1 has been established (Grief *et al.*, 1988). It contains a single long ORF (Fig. 7.25). The 3′ noncoding sequences for RNAs 1 and 2 are identical. The amino acid sequence in the RNA1 polyprotein shows three regions each having about 60% sequence similarity with the CPMV proteins thought to be involved in RNA replication. The GDD sequence found in all RNA polymerases is present, as is the consensus sequence around a phosphate-

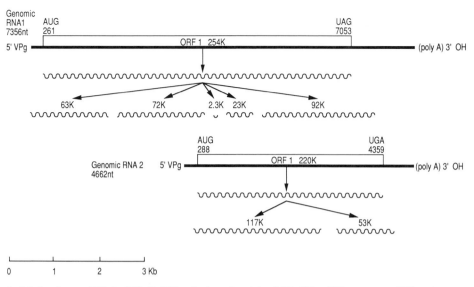

Protein functions :- RNA 1: 63K , ? ; 72K , attachment protein ; 2.3K , VPg ; 23K , protease ; 92K , polymerase
RNA 2: cell-to-cell transport , coat protein.   The indicated cleavage site is hypothetical

Figure 7.25   Organization and expression of the *Nepovirus* genome (TBRV).

binding pocket found in many nucleoside-triphosphate-binding proteins (see also
Fig. 17.8).

2. *In Vitro* Translation Products

RNAs 1 and 2 of TBRV are translated in the wheat germ and rabbit reticulocyte
systems to give polypeptides of MW = 220K and 160K representing about 80 and
90% of the coding capacity of the RNAs (Fritsch *et al.*, 1980). Polypeptides of
similar sizes are produced from RNAs 1 and 2 of grapevine fanleaf virus (Morris-
Krsinich *et al.*, 1983). *In vitro* translation studies with TRSV suggest that the two
polyproteins are posttranslationally cleaved by a protease coded for by RNA1 (Forster and Morris-Krsinich, 1985; Jobling and Wood, 1985). A 53K protein precipitated with an antiviral serum was assumed to be the coat protein, but no clear pattern
of proteolytic cleavage has yet been established.

3. *In Vivo* Studies

RNA1 can replicate in the absence of RNA2 and therefore must carry the RNA
polymerase and VPg genes. The replication of nepoviruses has not been studied in
detail. In cucumber cotyledons, TRSV-induced RNA polymerase activity rises
rapidly to a maximum at about 3 days after inoculation and then falls (Peden *et al.*,
1972). TRSV presumably replicates by means of an RI form of the RNA, but only
heterogeneous low-MW dsRNA has been isolated from infected plants (Rezaian and
Francki, 1973).

The viral coat protein appears to be synthesized on cytoplasmic ribosomes, and
both electron microscope and cell fractionation experiments indicate that TRSV
replicates in the cytoplasm (Rezaian *et al.*, 1976), probably in association with
characteristic membranous vesicles. Virus particles can often be observed in rows in

long tubules in *Nepovirus* infections (e.g., Walkey and Webb, 1970; Saric and Wrischer, 1975).

VI. TWO STRATEGIES: SUBGENOMIC RNA AND MULTIPARTITE GENOME

A. *Bromovirus* Group

The *Bromovirus* group contains four definitive members. Most experimental work has used BMV, the type member, and the following account mainly concerns this virus. The organization of this genome has been discussed by Dougherty and Hiebert (1985), Kaesberg (1987), and Ahlquist *et al.* (1987). Exploitation of several technical developments, discussed in the following Section 3, has led to very rapid progress in our understanding of the sequence elements in the BMV genome that regulate RNA replication. Many of the features revealed for BMV will probably be found to operate for other positive sense ssRNA plant viruses. The possible use of BMV RNAs as gene vectors is discussed in Chapter 8, Section VI,C,1.

1. Genome Structure

BMV has a tripartite genome totaling 8243 nucleotides. In addition, a subgenomic RNA containing the coat protein gene is found in infected plants and in virus particles. This coat protein gene is encoded in the sequence toward the 3' end of RNA3. Each of the four RNAs has a 5' cap and a highly conserved 3'-terminal sequence of about 200 nucleotides. The terminal 135 nucleotides of this sequence can be folded into a three-dimensional tRNA-like structure (Perret *et al.*, 1989), which accepts tyrosine, in a reaction similar to the aminoacylation of tRNAs, but this reaction probably needs the 3'-terminal 155 nucleotides.

The genome of BMV has been completely sequenced (Ahlquist *et al.*, 1984a). RNAs 1 and 2 encode single proteins as indicated in Fig. 7.26. RNA3 encodes a protein of predicted MW = 32,480. Between this cistron and the coat protein cistron there is an intercistronic noncoding region approximately 250 nucleotides long. An internal poly(A) sequence of heterogeneous length (16–22 nucleotides) occurs in this intercistronic region ending 20 bases 5' to the start of the coat protein gene. The first 9 bases of the RNA4 consist of the last 9 bases of the intercistronic region. The internal poly(A) sequence is not essential for RNA3 replication. However, during replication, *in vivo*, of constructs lacking this sequence, restoration of the poly(A) took place (Karpova *et al.*, 1989).

The genome of cowpea chlorotic mottle *Bromovirus* (CCMV) has been sequenced (Allison *et al.*, 1989). There is extensive sequence similarity with the proteins of BMV and also CMV. However, the CCMV RNA3 5' noncoding region contains a 111-base insertion not found in the other viruses.

2. Proteins Encoded and Their Functions

BMV RNAs are efficient mRNAs in *in vitro* systems, and in particular in the wheat germ system that is derived from a host plant of the virus. In this system, RNA1 directs the synthesis of a single polypeptide of M_r = 110K (109K in Fig.

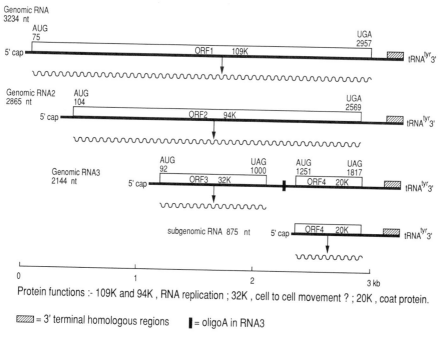

Protein functions :- 109K and 94K , RNA replication ; 32K , cell to cell movement ? ; 20K , coat protein.

▨ = 3' terminal homologous regions   ▮ = oligoA in RNA3

Figure 7.26 Organization and expression of a *Bromovirus* genome (BMV).

7.26) and RNA2 directs a single polypeptide of $M_r = 105K$ (94K in Fig. 7.26) (Shih and Kaesberg, 1976; Davies, 1979). RNA3 directs the synthesis of a 35K protein (32K in Fig. 7.26) (Shih and Kaesberg, 1973). The coat protein cistron in RNA3 is not translated. RNA4 directs synthesis of the 20K coat protein very efficiently. It is preferentially translated in the presence of the other viral RNAs in part, at least, because of more efficient binding of ribosomes (Pyne and Hall, 1979). *In vitro* the coat protein inhibits RNA synthesis by the BMV replicase, in a specific manner, possibly by partial assembly of nucleoprotein (Horikoshi *et al.*, 1987).

Four new proteins were observed in tobacco protoplasts after infection with BMV. These had MWs of 20K (coat protein), 35K, 100K, and 107K (Sakai *et al.*, 1979). Four BMV-induced proteins with the same MWs (within the error of estimation in gels) were also found by Okuno and Furusawa (1979). They were found in infected protoplasts prepared from three plant species—a systemic host, a local lesion host, and a nonhost. These four proteins account for over 90% of the viral genome. They correspond well in size with the *in vitro* products noted earlier and the precise MWs in Fig. 7.26.

RNAs 1 and 2 inoculated alone to protoplasts replicate, showing that they must contain a replicase function (Kiberstis *et al.*, 1981). French *et al.* (1986) demonstrated with cloned material, where no contaminating RNA3 could be present, that RNAs 1 and 2 by themselves could replicate. Further evidence that the proteins from RNAs 1 and 2 are involved in RNA replication comes from the amino acid sequence similarities with proteins of other viruses known to function as viral RNA polymerases, or genetically implicated in replication functions (see Fig. 17.19). An active replicase can be isolated from BMV-infected tissue. Bujarski *et al.* (1982) demonstrated the presence of a 110K protein in such a replicase fraction that had the same electrophoretic mobility and tryptic polypeptide pattern as the *in vitro* translation product of RNA1. Quadt *et al.* (1988) synthesized C-terminal peptides derived from the polypeptides coded by RNAs 1 and 2. Antibodies raised against these peptides could recognize the corresponding native proteins in replicase preparations. Antibodies against the polypeptide of RNA1 but not that of RNA2 completely blocked replicase activity. Similar results were obtained by Horikoshi *et al.* (1988).

Kroner *et al.* (1989) introduced defined mutations resulting in substitutions between amino acids 451 and 484 of the BMV RNA2 protein (94K), a region conserved in many positive strand RNA viruses. Four mutants had unconditional blocks in RNA synthesis while five others showed *ts* defects in RNA replication. Two of the *ts* mutants also showed a preferential reduction in the synthesis of genomic RNA relative to subgenomic RNA, at both permissive and nonpermissive temperatures. Thus there is no doubt that the gene product of RNA 2 is involved in RNA replication, probably with several functions. Other studies using directed mutagenesis have shown that each of the three conserved domains in BMV gene products of RNAs 1 and 2 (109K and 94K) are involved in RNA replication (Ahlquist *et al.*, 1990).

On the basis of amino acid sequence similarities it has been suggested that the protein translated from RNA3 may have a role in cell-to-cell movement analogous to that of the 30K protein of TMV (Section III, A,4) (Haseloff *et al.*, 1984). This possible function is supported by the fact that RNAs 1 and 2 will replicate in protoplasts in the absence of RNA3 (Kiberstis *et al.*, 1981) but this RNA is needed for systemic plant infection. Thus a function has been assigned or tentatively indicated for all four proteins known to be translated from the BMV genome.

3. Elements Controlling RNA Replication in the BMV Genome

From a practical point of view, Janda *et al.* (1987) greatly improved the efficiency with which infectious transcripts of BMV RNAs could be synthesized *in vitro* with T7 RNA polymerase by fusion of BMV cDNA directly to the initiation site of the canonical $\Phi10$ promoter of phage T7. These and other developments have greatly increased our understanding of replication of the BMV genome (see Chapter 6, Section VI,A,2).

4. BMV Replication *in Vivo*

The Process of Infection

In the experiments of Kurtz-Fritsch and Hirth (1972), 17% of the BMV inoculum was uncoated after 20 minutes. Uncoating led to the appearance of low-MW viral protein rather than empty protein shells. However, *in vitro* experiments with CCMV indicate that cotranslational disassembly may occur with these viruses (Roenhorst *et al.*, 1989). In translational mixes with CCMV, up to four ribosomes were associated with each virus particle. However, cotranslational disassembly could not be demonstrated to occur *in vivo* (Roenhorst, 1989).

Time Course of Events

In barley protoplasts infected *in vitro,* a membrane-bound RNA polymerase activity resistant to actinomycin D was increased up to 30-fold over mock inoculated protoplasts (Okuno and Furusawa, 1979). In the same system, RNAs 1 and 2 were detected 6 hours after infection. All four RNAs were present at 10 hours. Production of ds forms of all four ssRNAs followed a time course similar to that of the ssRNAs (Loesch-Fries and Hall, 1980). Maximum RNA synthesis was from 16 to 25 hours. Virus particle formation was greatest between 10 and 25 hours after inoculation.

In young cowpea leaves in which infection with CCMV was synchronized by differential temperature treatment, the three largest RNAs were synthesized at relatively constant ratios throughout the infection (Dawson, 1978b), but very little RNA4 was produced early in infection. As the infection progressed the proportion of RNA4 continued to increase. The RFs of components 1, 2, and 3 were produced with kinetics similar to that for the corresponding ssRNAs.

Assembly of the Virus

The pioneering work of Bancroft and colleagues showed that the protein subunits of several bromoviruses could be reassembled *in vitro* to give a variety of structures (reviewed by Bancroft and Horne, 1977). In the presence of viral RNA, the protein subunits could reassemble to form particles indistinguishable from native virus. However, the conditions used were unphysiological in several respects. The assembly mechanism of CCMV is thought to involve carboxyl–carboxylate pairs (Jacrot, 1975).

Of considerable interest in relation to possible *in vivo* mechanisms is the assembly of infectious CCMV under mild conditions (Adolph and Butler, 1977). A 3 S protein aggregate dimer at pH 6.0, ionic strength 0.1–0.2, and 25°C, in the absence of added Mg^{2+} combines with the RNA to form infectious virus. The reassembled particles cannot be distinguished from native virus by physicochemical or structural means.

The competition experiments of Cuillel *et al.* (1979) with various foreign RNAs showed that under appropriate conditions BMV protein can recognize its own RNA molecules to some extent. Using neutron and X-ray scattering, Cuillel *et al.* (1983) found that the assembly of BMV empty protein shells *in vitro* was a very rapid process, with forward scattering reaching half-maximum in under 1 second. The oligonucleotides $(Ap)_8A$ and $(A–T)_5$ interact with empty shells of CCMV. They bind specifically to the arginine and lysine residues of the N-terminal arm of the polypeptide within the shell (Chapter 5, Section III,B,1). The binding studies support a model for virus assembly in which a random coil to α-helix transition occurs, induced by neutralization of the basic amino acid side chains (Vriend *et al.*, 1986).

Various *in vitro* studies with BMV and CCMV indicate that the highly basic N-terminal region of the coat protein is involved in reactions with the viral RNA. Experiments with BMV variants containing known deletions in the coat protein gene confirm this view. The first seven amino acids are completely dispensible for packaging of the RNA *in vivo* but if 25 N-terminal amino acids are missing no virus particles are produced (Sacher and Ahlquist, 1989).

Sites of Accumulation within the Cell

Using immunoradioautography, Lastra and Schlegel (1975) detected BBMV antigen in the nucleus and cytoplasm of broad bean leaves 1 day after inoculation. The amount of antigen associated with the nucleus remained more or less constant while massive amounts accumulated in the cytoplasm. No antigen was associated with chloroplasts or mitochondria. Amorphous membranous inclusion bodies, which are probably viroplasms, and filamentous inclusions have been described for cells infected with CCMV (Kim, 1977).

Ultrastructural changes induced by BMV (and CCMV) involve a proliferation and modification of parts of the endoplasmic reticulum. In addition, the nuclear membrane appears to give rise to small cytoplasmic vacuoles that appear to contain nucleic acid (Burgess *et al.*, 1974a).

B. *Cucumovirus* Group

The cucumoviruses have many properties in common with the bromoviruses although they have not been studied in such detail. The genome organization is very similar to that of the bromoviruses except that each of the three genomic RNAs and the subgenomic coat protein RNA are slightly longer. Expression of the genome gives four protein products. Although much work has been carried out, a viral-coded replicase has not yet been adequately characterized (Khan, *et al.*, 1986). However, *in vivo* experiments suggest that replicase activity is coded for by RNAs 1 and 2 (Nitta *et al.*, 1988). The ability of strain Fny of CMV to induce rapid and severe disease in zucchini squash is a function of RNA1 (Roossinck and Palukaitis, 1990). Infectious RNA transcripts have been obtained from full-length cDNAs of all three CMV RNAs (Rizzo and Palukaitis, 1990). The nucleotide sequences of the CMV RNAs have similarities to those of BMV, AMV, and some other viruses (Chapter 17, Section V,B). A major difference is the occurrence of satellite RNAs in many CMV isolates (Chapter 9, Section II,B,1). Genome structure and expression of cucumoviruses have been reviewed by Dougherty and Heibert (1985). Membrane-bound vesicles associated with the tonoplast may be the site of CMV RNA synthesis (Hatta and Francki, 1981b).

C. Alfalfa Mosaic Virus

AMV is one of the most studied plant viruses, especially with respect to the role of the coat protein in the virus life cycle. Our knowledge of the virus has been reviewed by Jaspars (1985).

1. Genome Structure

The genome of AMV is tripartite with a fourth, subgenomic coat protein mRNA. The arrangement is very similar to that of BMV. The genome of strain 425 was the first tripartite RNA genome to be fully sequenced (Cornelissen *et al.*, 1983a,b; Barker *et al.*, 1983; Brederode *et al.*, 1980). All RNA species have an m^7 G^5ppp cap structure and a 3′ terminus ending with –AUGC–OH. They cannot be aminoacylated. RNAs 1 and 2 each have a long ORF as indicated in Fig. 7.27. RNA3 encodes a protein of 300 amino acids and the coat protein gene. The coat protein gene is repeated in RNA4. Each RNA has a 5′ leader sequence as indicated in Fig. 7.27. The leader of RNA3 is longer than the others and contains a sequence of 28–30 nucleotides repeated three times. This repeat is found in all strains of the virus (Langeris *et al.*, 1986b). With the exception of a few nucleotides, the last 145 at the 3′ terminus are identical in all four RNA species.

2. Proteins Encoded by the Genome

The four RNAs have all been translated in various cell-free systems. RNA1 directs the synthesis of a 115K protein at low mRNA concentrations in the rabbit

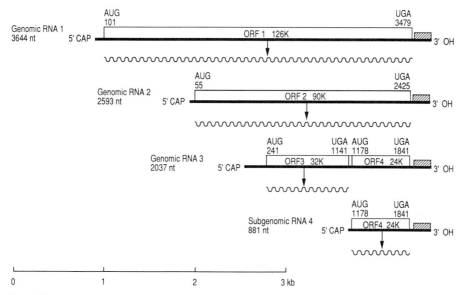

Protein functions :- 126K and 90K, polymerase ; 32K, cell-to-cell transport ; 24K, coat protein

▨▨ = 3' terminal 145 nt homologous region

Figure 7.27 Organization and expression of the alfalfa mosaic virus genome.

reticulocyte system, or a mixture of 58K and 62K proteins at high mRNA concentrations (van Tol and van Vloten-Doting, 1979; van Tol *et al.*, 1980). It remains to be determined whether these latter products are functional proteins. However, they have the same N terminus as the 115K protein (126K in Fig. 7.27). In various cell-free systems RNA2 was translated to give a protein of MW = 84K (Mohier *et al.*, 1976) corresponding to the 90K protein predicted in Fig. 7.27. In a variety of *in vitro* systems the primary translation product from RNA3 has a MW = 35K (Dougherty and Hiebert, 1985) corresponding to the expected 32K protein. RNA4 is efficiently translated *in vitro* to produce the viral coat protein (e.g., Mohier *et al.*, 1976).

All three nonstructural proteins indicated in Fig. 7.27 have been detected in a crude membrane fraction from AMV-infected tobacco leaves using antisera raised against synthetic peptides corresponding to the C-terminal sequences (Berna *et al.*, 1986).

3. Functions of the Nonstructural Proteins

The 126K and 90K Proteins

The following evidence indicates that the 126K and 90K proteins are involved in RNA replication, presumably as part of a membrane-bound replicase complex: (i) RNAs 1 and 2 can replicate independently of RNA3 in protoplasts (Nassuth *et al.*, 1981). (ii) The time course of accumulation of these two proteins following inoculation corresponded to the rise and fall in replicase activity in infected tobacco leaves (Berna *et al.*, 1986). In infected protoplasts the 126K and 90K proteins were detected early and disappeared when virus production reached a plateau (van Pelt-Heerschap *et al.*, 1987b). (iii) The effects of mutations in RNAs 1 and 2 suggest that they function cooperatively (Sarachu *et al.*, 1983). (iv) RNA1 and RNA2 *ts* mutants were defective in the synthesis of viral minus strand RNAs (Sarachu *et al.*, 1985). A period of 6 hours immediately after inoculation at the permissive temperature was sufficient to allow normal minus strand synthesis (Huisman *et al.*, 1985). (v) There are strong amino acid sequence similarities between three regions of the 90K protein and regions in the RNA polymerase of some other viruses (Kamer and Argos, 1984) (see Fig. 17.9).

Complementation experiments between *ts* mutants show the presence of two complementation groups on RNA2, suggesting that the product directed by this RNA has two functional domains (Roosien and van Vloten-Doting, 1982). No relationship between such domains and the structure of the 90K protein has yet been delineated.

The 32K Protein

Three lines of evidence suggest that the 32K protein has a cell-to-cell transport function: (i) Studies with *ts* mutants of RNA2 showed that some can multiply in protoplasts but cannot move from cell to cell in tobacco leaves (Huisman *et al.*, 1986). (ii) The 32K protein is largely confined to the cell wall fraction of infected tobacco leaves (Godefroy-Colburn *et al.*, 1986). Using an immunocytochemical technique, Stussi-Garaud *et al.* (1987) showed that the 32K protein was localized almost entirely in the middle lamella of recently infected cells. The 32K protein was not detected in fully infected cells. However, no relationship of the 32K protein to plasmodesmata was observed in contrast to the situation with TMV. (iii) Virus particles containing only RNAs 1 and 2 can replicate the RNAs in protoplasts but cannot move from cell to cell in leaves.

4. Requirement of Coat Protein for RNA Replication

Besides being the structural protein of virus particles, the coat protein of AMV (and also of ilarviruses) plays an essential role in viral genome replication. The nature of this role is not yet understood, but a great deal has been learned about the interactions between the coat protein and the four viral RNAs. The main features are as follows: (i) Some coat protein or coat mRNA, as well as the three genomic RNAs, is necessary to initiate an infection (van Vloten-Doting and Jaspars, 1977). (ii) AMV infection starts with some coat protein bound to each of the three genomic RNAs (Smit *et al.*, 1981). (iii) Coat protein has a high affinity for viral RNA. Free viral RNA added to a virus preparation will remove coat protein subunits from the nucleoprotein (van Vloten-Doting and Jaspars, 1972). (iv) The binding and genome activation are specific to AMV coat protein and the coat proteins of the related ilarviruses (e.g., van Vloten-Doting, 1975). AMV protein does not bind specifically to RNAs of unrelated viruses (Zuidema and Jaspars, 1985). (v) Transgenic tobacco plants expressing AMV coat protein will support virus replication following inocu-lation with AMV RNAs 1, 2, and 3 or TSV RNAs 1, 2, and 3. Transgenic plants expressing TSV coat protein also support replication of the three AMV RNAs (van Dun *et al.*, 1988b). (vi) The coat protein binds to a specific site near the 3' terminus of RNA4 (Fig. 7.28). There is a homologous sequence at the 3' termini of the three genomic RNAs. The RNA binds either one or three coat protein dimers. The proposed binding sites are indicated in Fig. 7.28. Internal coat protein binding sites are also found on all the genomic RNAs, but these do not appear to have such strong affinity for the protein (Jaspars, 1985). (vii) Removal of the N-terminal 25 amino acids of the coat protein destroys effective binding to RNA and the ability to make the genome RNAs infectious (Zuidema *et al.*, 1983).

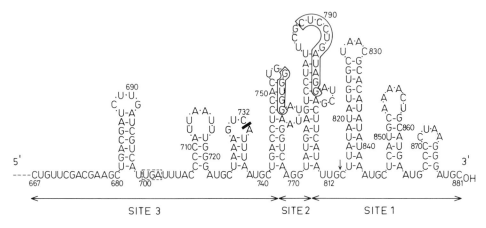

Figure 7.28   Nucleotide sequence of the 3'-terminal coat protein binding region of RNA4 of AMV 425. The region can be subdivided into three sites, numbered 1 to 3 starting from the 3' end, and separated by two short sequences (encircled with uninterrupted line) that were not detected in the assay used to define the binding site. The thick bar indicates the end of the 3'-terminal region homologous in all RNA species of the virus. The encircled UGA at positions 700 to 702 is the termination codon of the coat protein cistron. (Reprinted with permission from Houwing and Jaspars, *Biochemistry*, **21**, 3408. Copyright © 1982, American Chemical Society.)

Several lines of evidence suggest that specific coat protein binding to the 3'-terminal site in genomic RNAs is necessary for recognition of the plus strand RNAs by the replicase (e.g., Smit *et al.*, 1981; Nassuth and Bol, 1983; Houwing and Jaspars, 1986, 1987).

However, van der Kuyl (1990) carried out a deletion analysis which indicated that a sequence located between nucleotides 133–163 from the 3' end of AMV RNA3 was sufficient to direct the synthesis of minus strand products by a replicase isolated from AMV-infected beans. Thus the high affinity coat protein binding sites at the 3' termini of AMV RNAs are not involved in RNA recognition by the replicase, at least *in vitro*. Further experiments are needed to obtain definitive evidence on the recognition and regulatory roles of the coat protein early and late in the infection cycle.

5. Other Aspects of RNA Synthesis

Three ds RF RNAs corresponding to RNAs 1, 2, and 3 have been isolated from plants in which AMV was multiplying (Pinck and Hirth, 1972; Mohier *et al.*, 1974). Only mixtures of components that were infectious gave rise to any detectable RF RNA. Mohier *et al.* (1974) could detect no RF corresponding to RNA4.

Dore and Pinck (1988) showed that plasmid DNA containing a copy of RNA3 can substitute for RNA3 when mixed with RNAs 1, 2, and 4 in an inoculum. Progeny RNA3 derived from the circular plasmid DNA was of normal length. Presumably the incoming plasmid is transcribed in the nucleus.

RNA4 is not able to influence the genetic properties of the progeny of heterologous RNAs (Bol and van Vloten-Doting, 1973). No RNA4 was found among the RNAs produced in protoplasts inoculated with virus particles containing only RNAs 1, 2, and 4 (Nassuth *et al.*, 1981). van der Kuyl *et al.* (1990) prepared RNA transcripts from cDNA of AMV RNA3 using a virus-specific RNA-dependent RNA polymerase from infected plants. They used deletion analysis to show that a sequence located between −8 and −55 upstream of the initiation site for RNA4 synthesis was sufficient for coat protein subgenomic RNA synthesis *in vitro*. Thus, like BMV, AMV RNA4 synthesis appears to be initiated at an internal replicase recognition site on the minus strand of RNA3 as for BMV (Nassuth *et al.*, 1983a).

Host factors may be involved in viral RNA synthesis because actinomycin D added early after inoculation of protoplasts substantially inhibited both minus and plus strand synthesis (Nassuth *et al.*, 1983b). The time course of RNA synthesis in protoplasts is shown in Fig. 7.29.

6. Viral Protein Synthesis *in Vitro* and *in Vivo*

The four AMV RNAs vary widely in their relative affinities for the initiation complex in the *in vitro* wheat germ protein-synthesizing system (Godefroy-Colburn *et al.*, 1985a). Taking the affinity of RNA3 as 1.0, the other RNAs had affinities as follows: RNA1 = 10; RNA2 = 60; and RNA4 = 150. Similar relationships were found in the reticulocyte lysate system. Strains with different secondary structures in the 5′ leader sequence had similar affinities, indicating that the secondary structure was not an important influence. Other studies showed that accessibility of the cap structure may be a factor determining relative affinities for the initiation of protein synthesis (Godefroy-Colburn *et al.*, 1985b).

The time course for the production of the four AMV gene products in protoplasts irradiated to suppress host protein synthesis is shown in Fig. 7.30. Coat protein production goes on over a longer period than that of the nonstructural proteins, but the relative amounts of the proteins produced do not correlate at all closely with the *in vitro* data.

7. Assembly of Virus Particles

The dimer of the coat protein is a very stable configuration in solution (Driedonks *et al.*, 1977) but no clear picture has yet emerged concerning the factors

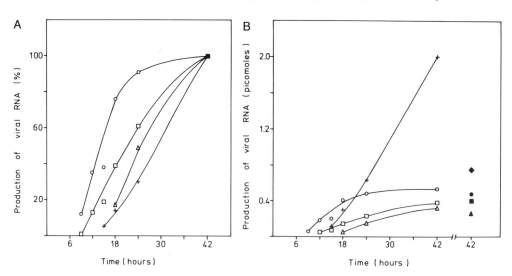

Figure 7.29   Time course of plus-strand AMV RNA synthesis in infected cowpea protoplasts. (A) The relative increase of viral RNA with the amount of each RNA present at 42 hours postinoculation taken as 100%; (B) the increase of viral RNA in picomoles. RNA1 (□), RNA2 (△), RNA3 (○), and RNA4 (+). Solid symbols at the right of (B) indicate the amounts of RNAs 1 (■), 2 (▲), 3 (●), and 4 (◆) present in virions 42 hours after inoculation of the protoplasts. (From Nassuth *et al.*, 1983a.)

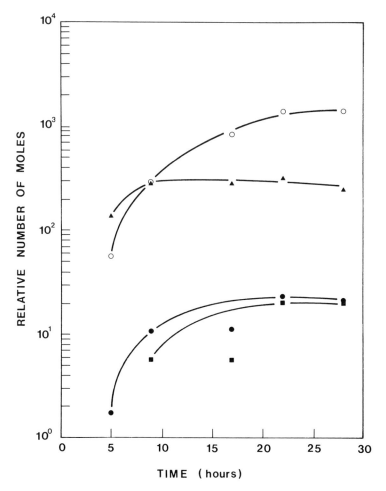

Figure 7.30 Time course of the synthesis of AMV gene products P1 (●), P2 (■), P3 (▲) and coat protein (○) in lucerne mesophyll protoplasts irradiated with UV light prior to inoculation. (From Samac *et al.*, 1983.)

controlling the morphogenesis of AMV. Driedonks *et al.* (1980) followed polymerization in the analytical ultracentrifuge. They postulated four stages in assembly of the virus: (i) an initiation stage; (ii) initial cap formation; (iii) cylindrical elongation; and (iv) a cap formation or closure stage. However, no well-defined class of particle resembling a natural virus particle in size was observed.

8. Site of Virus Synthesis and Assembly

The replication complex is found in the cytoplasmic fraction. Coat protein was present in the cytoplasm 6 hours after inoculation and appeared in the nucleus after 48 hours (van Pelt-Heerschap *et al.*, 1987a). Masses of AMV particles were seen in the cytoplasm of AMV-infected cells (De Zoeten and Gaard, 1969a; Hull *et al.*, 1970) and sometimes in the nucleus (Hull *et al.*, 1970). While the evidence is not conclusive, it is most likely that all AMV components are synthesized and assembled in the cytoplasm.

9. Time Course of Virus Production

Little is known about the site or mechanism of uncoating of AMV nucleoproteins following infection. In infected protoplasts, progeny virus appeared 12 hours after inoculation and the amount produced rose rapidly until 24 hours, when the rate of production slowed (Alblas and Bol, 1977).

D. *Ilarvirus* Group

Tobacco streak virus (TSV) and other members of the *Ilarvirus* group are very similar to AMV. The genome consists of three positive sense ssRNA segments. The 3' region of the smallest genomic RNA (3) is repeated in a smaller subgenomic RNA (RNA4) that encodes the coat protein. As with AMV, the three genomic RNAs are not infectious unless either RNA4 or coat protein is present in the inoculum. The coat proteins of AMV and TSV are interchangeable in this respect. There is little structural similarity between the two proteins (van Vloten-Doting, 1975). There is virtually no sequence similarity between the 3'-terminal regions of TSV and AMV RNAs but they have in common a sequence of stable hairpin structures flanked by the tetranucleotide sequence AUGC. These structures probably provide the common recognition site for the coat proteins (Koper-Zwarthoff and Bol, 1980). TSV virus particles contain about one zinc atom per four protein subunits. One zinc per two subunits were found for AMV, while other viruses tested contained none (Sehnke *et al.*, 1989). There is a sequence (Cys X2 Cys X10 Cys X2 His) between residues 28 and 45 of the TSV coat protein, and this sequence is found in several zinc-binding proteins involved in nucleic acid binding and gene regulation. Thus the "zinc finger" motif in TSV coat protein and a similar motif in AMV may play a role in binding to the RNA (Sehnke *et al.*, 1989).

The complete nucleotide sequence of TSV RNA3 has been determined (Cornelissen *et al.*, 1984). Its overall structure is similar to that of AMV. There are no significant sequence similarities between the coat protein of TSV and that of AMV, CMV, or BMV. However, some similarity exists between TSV and AMV in the 5' gene coded by RNA3, although there was no serological relationship (van Tol and van Vloten-Doting, 1981). The RNAs of TSV have been translated in both the wheat germ (Davies, 1979) and the rabbit reticulocyte systems (van Tol and van Vloten-Doting, 1981). The pattern of translation from a mixture of all four RNAs is similar to that of AMV. Four proteins with MW ≃ 120K, 100K, 34K, and 25K (coat protein) are produced. There have been few *in vivo* studies of *Ilarvirus* replication.

E. *Hordeivirus* Group

For several years it appeared that some strains of barley stripe mosaic virus (BSMV), the type member of the group, contained two genome pieces while others had three. This was because RNAs 2 and 3 of some strains were difficult to separate by gel electrophoresis. Full-length genomic cDNA clones of two strains of BSMV were transcribed *in vitro* in the presence of 5' cap analogs. A combination of three transcribed RNAs was infectious for barley, demonstrating the tripartite nature of the genome (Petty *et al.*, 1989). The biology of hordeiviruses has been reviewed by Carroll (1986). Genome organization is discussed by Jackson *et al.* (1989).

1. Genome Structure and the Proteins Encoded

The three genomic RNAs of BSMV are sometimes designated α, β, and γ. I will use the terms RNA 1, 2, and 3 (in decreasing order of size) for consistency with the other viruses having tripartite genomes. These genomic RNAs have an m^7 Gppp cap at the 5' end (Agranovsky *et al.*, 1979) and a t-RNA-like structure at the 3' end, which can be aminoacylated with tyrosine (Agranovsky *et al.*, 1981; Loesch-Fries and Hall, 1982). There is an internal poly(A) sequence of 8–30 residues, approximately 210 nucleotides from the 3' terminus (Agranovsky *et al.*, 1982). Genome structure of hordeiviruses has been reviewed by Jackson *et al.* (1989).

RNA1

A single protein of MW = 120K is translated *in vitro* from RNA1 (Dolja *et al.*, 1983). Complete sequencing data show that this is the only polypeptide likely to be coded for by this RNA with a true MW $\simeq$ 130K (Gustafson *et al.*, 1989). A MW $\simeq$ 120K virus-specific protein, probably the 130K product, has been found *in vivo*. The 130K gene product shows amino acid sequence similarities with the large proteins of BMV RNA1, AMV RNA1, and the 126K protein of TMV. The main regions of similarity are located near the amino and carboxy 3' termini (see Fig. 17.9). The carboxy-terminal region contains six motifs commonly found in enzymes involved in DNA and RNA replication and recombination. Similar conserved sequences are found in the 58K ORF of RNA2 and also in RNA3 (Gustafson *et al.*, 1987).

RNA2

RNA2 has been fully sequenced (Gustafson and Armour, 1986). There are four major ORFs as illustrated in Fig. 7.31. Three lines of evidence show that RNA2, and ORF2 in particular, codes for the BSMV coat protein: (i) RNA2 is translated efficiently *in vitro* to give coat protein (Dolja *et al.*, 1979); (ii) the amino acid composition deduced from the nucleotide sequence is very close to that determined experimentally for the coat protein; and (iii) the predicted sequence of the 30 N-terminal amino acids in the ORF2 translation product is identical to 29 of the corresponding amino acids in BSMV coat protein determined by direct sequencing.

ORF3 could code for a protein of MW = 58K, ORF4 for one of 17K, and ORF5 for one of 14K. ORF5 overlaps the last 29 nucleotides of ORF3 and the first 188 of ORF4. None of these potential proteins has yet been positively identified as *in vitro* or *in vivo* translation products. However, Brakke *et al.* (1988) have described a protein of MW $\simeq$ 60K occurring in large amounts in infected barley leaves that is probably viral-coded and that could be an ORF3 product. Its association with large amounts of viral RNA make it a candidate for a replicase component. ORFs 4 and 5 have significant sequence similarity with proteins encoded by RNA2 of BNYVV (K. Richards, personal communication). Analysis of nucleotide sequence in potential translation initiation sites and codon usage suggest that ORFs 3 and 4 are probably translated while ORF5 is less likely to be used. Under certain ionic conditions *in vitro*, RNA2 directs the synthesis of a MW = 25K polypeptide in addition to coat protein (Dolja *et al.*, 1983), but such a protein has not been detected *in vivo*, and its relation to the ORFs in RNA2 is not established. ORFs 3 and 4 are presumably translated from subgenomic RNAs, but these have not yet been identified.

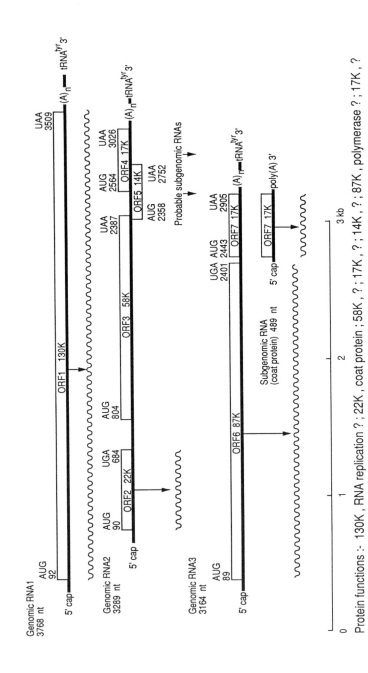

Figure 7.31 Organization and expression of a *Hordeivirus* (BSMV, type strain).

Protein functions :- 130K , RNA replication ? ; 22K , coat protein ; 58K , ? ; 17K , ? ; 14K , ? ; 87K , polymerase ? ; 17K , ?

RNA3

Unlike RNAs 1 and 2, which are relatively constant in length between different strains of BSMV, RNA3 varies significantly between some strains. Gustafson *et al.* (1987) sequenced two RNA3's, of different length, from the type and ND18 strains. There are two ORFs separated by an intercistronic region. The larger size of the type strain is due to an in-frame 366-nucleotide direct tandem repeat located within ORF6. The structure of the RNA3 type strain is shown in Fig. 7.31.

The predicted size of the translation product for type strain ORF6 is 87K and for the ND18 strain it is 74K. This strain-specific difference is readily seen in the polypeptides translated *in vitro* from these two strains (Gustafson *et al.*, 1981). The amino acid sequence predicted in ORF6 contains similarities with the proteins known to be involved in RNA replication for BMV and AMV. The homologies are located in the carboxy-terminal regions of the peptides. This BSMV protein may therefore also have a polymerase function.

The polypeptide predicted from ORF7 of RNA3 is translated from a subgenomic RNA (e.g., Jackson *et al.,* 1983). The initiation point for this RNA is 27 nucleotides upstream of the ORF7 initiation codon in the intercistronic region. Unlike subgenomic RNAs of some other viruses, the ORF7 RNA is not coterminal with the 3′ end of the genomic RNA. Instead it contains a poly(A) terminus of variable length up to 150 nucleotides (Stanley *et al.,* 1984). Synthesis of the subgenomic RNA is presumably initiated at an internal site on minus strand RNA3. The protein coded by ORF7 is rich in sulfur-containing amino acids, with no sequence similarity to any proteins coded for by other viruses with tripartite genomes. This protein has not yet been detected *in vivo,* and no function has been proposed.

In summary, the genome of BSMV differs from that of the other viruses with plus sense tripartite RNA genomes in several ways: (i) it codes for six (or perhaps seven) proteins instead of four; (ii) the coat protein gene is at the 5′ end of RNA2 and is translated from the genomic RNA; and (iii) the RNA3 subgenomic RNA is not 3′ coterminal with the genomic RNA.

2. Virus Replication *in Vivo*

In barley protoplasts inoculated with BSMV, progeny virus was first detected after about 12 hours, increasing to a maximum of about 2×10^6 particles per protoplast at 48 hours (Ben-Sin and Po, 1982). In chronically infected younger leaves of barley in which virus is replicating, free viral RNA is difficult to detect. By contrast, during acute systemic infections, large amounts of uncoated RNA accumulate and this RNA is associated with a 60K protein (Brakke *et al.,* 1988).

Using an immunogold cytochemical technique in which tissue was stained after sectioning, Lin and Langenberg (1985) localized dsRNA in vesicles at the periphery of proplastids in cells of infected barley roots. By following sequential cell-to-cell infection in the meristematic region of infected roots and shoots, Lin and Langenberg (1984a) established a series of cytological events. The earliest event (Stage 1) was the appearance of peripheral vesicles in the proplastids of the root meristem cells. Viral protein but not virus particles was detected late in Stage 1. Stage 2 was marked by the appearance of rod-shaped virus particles oriented vertically and attached to the proplastid membrane. In Stage 3 rods were also seen attached to the endoplasmic reticulum. In Stage 4 viral protein (but very few rods) was found in the nuclei. No virus protein was seen in heterochromatin. The sequence of events bears some similarities to that found for TYMV (Section IV,A,8). It is to be hoped that the immunogold procedure will be sensitive enough to locate some of the nonstructural BSMV proteins within this sequence of infected cells.

VII. THREE STRATEGIES: SUBGENOMIC RNAs, MULTIPARTITE GENOME, AND READ-THROUGH PROTEIN

A. *Tobravirus* Group

Tobraviruses have a bipartite ss positive sense RNA genome. The group comprises three distinct viruses based on nucleic acid hybridization experiments (Robinson and Harrison, 1985). The three viruses are TRV, pea early browning virus, and pepper ring spot virus (PepRSV) (formerly known as the CAM strain or the Brazilian isolate of TRV). There is extensive sequence homology between the RNAs of strains within each virus category, and some homology between the RNAs of the three viruses.

TRV was the first virus shown to have a genome split between two particles. Using sucrose density gradients, Lister (1966, 1968) fractionated preparations of TRV rods into two size classes. The RNA in long rods could replicate alone but could not form virus rods. Short rods could not replicate at all. Lister concluded that long rods code for the enzyme necessary for RNA replication, while short rods code for coat protein. These results were later confirmed using protoplasts (Kubo *et al.*, 1975a).

1. Genome Structure

RNA1

The RNA1 of TRV (strain SYM) has been fully sequenced (Boccara *et al.*, 1986; Hamilton *et al.*, 1987). The RNA is 6791 nucleotides long and all tobraviruses have an RNA1 about this length. There are four ORFs arranged as shown in Fig. 7.32. There is a one-base intergenic region between the 194K and 29K ORFs. There is a 3' noncoding region of 255 nucleotides. Although TRV RNA1 cannot be aminoacylated, there is a tRNA-like feature in the 3'-terminal region (van Belkum *et al.*, 1987).

A cDNA clone of TRV RNA1 was infectious for tobacco leaves and produced the expected subgenomic RNA species (Hamilton and Baulcombe, 1989).

Pea early browning virus (PEBV) RNA1 has been fully sequenced (MacFarlane *et al.*, 1989) and shows a similar gene arrangement to that of TRV (Fig. 7.32). However, the protein coded by ORF4 is somewhat smaller (12K). The difference in size is due to a continuous sequence of 35 amino acids absent from PEBV but present in TRV. In both proteins there is a duplicated motif of cysteine and histidine residues. Thus this *Tobravirus* gene product may be a "zinc finger" protein, and thus may be capable of zinc-dependent nucleic acid binding.

RNA2

The RNA2 of tobraviruses has widely different lengths in the range of 1800–4000 nucleotides. Different strains of TRV show substantial diversity in the sequences of their RNA2. The complete sequence of the nucleotides in RNA2 of the PSG strain of TRV is known (Cornelissen *et al.*, 1986), as is that of PepRSV (Bergh *et al.*, 1985).

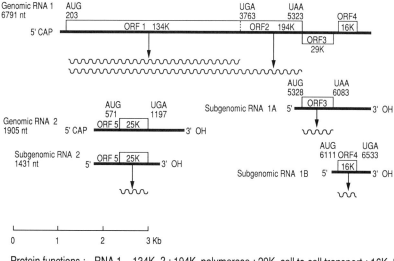

Protein functions :- RNA 1   134K, ? ; 194K, polymerase ; 29K, cell to cell transport ; 16K, ?
RNA 2   25K, coat protein.

Figure 7.32   Organization and expression of a *Tobravirus* genome (TRV) (RNA1 = strain SYM; RNA2 = strain PSG).

The 3′ 497 nucleotides of PSG RNA2 are identical in sequence to the 3′-terminal sequence of strain PSG RNA1. Therefore part of the ORF for the 16K protein of RNA1 is also present in RNA2 on the 3′ side of the coat protein gene. The sequence that is unique to RNA2 has only one significant ORF, for a protein of MW ≃ 23K, with a 5′ noncoding sequence of 570 nucleotides and one at the 3′ end of 708 nucleotides. The 5′ leader sequence contains eight AUG codons. A subgenomic RNA2 found in strain PSG preparations was found to be derived from genomic RNA2 (Fig. 7.32). Three subgenomic RNAs like those shown in Fig. 7.32 have been found in tissues infected with PepRSV (Bergh and Siegel, 1989).

The RNA2 of TRV strain TCM contains an ORF for a 29K protein unrelated in sequence to any other gene on RNAs 1 or 2. This ORF is in addition to a complete 16K ORF, the coat protein gene, and the 3′-terminal sequence in common with RNA1 (Angenent *et al.*, 1986). Thus the wide variability in the length of the RNA2's of TRV strains is due to (i) a variation in the length of the 3′-terminal sequences in common with RNA1 and (ii) the presence or absence of an additional RNA2-specific ORF. No protein product has been found for the 29K ORF on RNA2 of strain TCM.

All the natural TRV isolates have a 3′ region of identical nucleotide sequence between RNAs 1 and 2. However, viable pseudorecombinants can be formed between RNAs 1 and 2 with differing 3′ sequences. After 25 passages in tobacco, each RNA retained its distinctive sequence. Thus in this virus perfect 3′ homology is not a prerequisite for a stable genotype (Angenent *et al.*, 1989b).

2. Proteins Encoded

RNA1

The RNA1 of the SYM strain encodes a protein of MW ≃ 134K, together with a read-through protein of ≃194K. The 194K protein shows sequence similarities with the putative replicase genes of TMV, AMV, BMV, and CMV (Hamilton *et al.*, 1987), confirming Lister's conclusions for RNA1. The similarities are particularly strong with TMV. The 29K protein shows some sequence similarity to the TMV 30K protein and thus may be involved in virus movement from cell to cell. No similarities were found between the 16K ORF and other viral proteins.

RNA1 is translated in the rabbit reticulocyte system to give two polypeptides of MW 170,000 and 120,000 (Pelham, 1979). These two polypeptides are also produced together with many smaller products in the wheat germ system containing added spermidine (Fritsch *et al.*, 1977). These two proteins correspond to the 194K and 134K ORFs shown in Fig. 7.32. Similar results have been reported for PEBV (Hughes *et al.*, 1986). A protein product for the 29K ORF is translated from a subgenomic RNA (1A) (Robinson *et al.*, 1983). The 16K protein is also translated from a subgenomic RNA (1B), which is not required for replication or cell-to-cell transport in leaves (Guilford, 1989). A 16K protein product was found in infected protoplasts (Angenent *et al.*, 1989a), which was incorporated into a high-MW cellular component.

RNA2

RNA2 of the PRN strain is translated *in vitro* to give the coat protein identified by serology, peptide mapping, and specific aggregation with authentic coat protein to form disk aggregates (Fritsch *et al.*, 1977). A second unrelated protein of 31K is also translated. Different strains and viruses appear to differ in the products translated *in vitro* from RNA2 preparations, perhaps in part due to variable contamination with subgenomic RNAs. No messenger activity has been detected for TRV strain SYM RNA2 in *in vitro* tests (Robinson *et al.*, 1983). However, a subgenomic mRNA derived from RNA2 was shown to be the mRNA for coat protein.

In summary, current evidence suggests that the RNA1 of tobraviruses gives rise to two large gene products involved in RNA replication, one being a read-through protein. A smaller ($\approx$29K) protein that may be involved in cell-to-cell movement is translated from subgenomic RNA1A. A 16K protein of unknown function has been found in protoplasts. RNA2 has one known gene product—the coat protein. However, if coat protein is the only gene product it is not clear why a subgenomic RNA should be required. Certain types of disease symptoms are specified by the short rods even when these give rise to identical coat proteins (e.g., Robinson, 1977). Thus there may be a second protein coded for by RNA2.

3. Assembly of Virus Rods

Like TMV, TRV coat protein can form a series of stable aggregates (Gugerli, 1976). Detailed electron microscopic studies showed that the 40 S disk aggregate consists of three layers of protein subunits arranged in a helix (Roberts and Mayo, 1980).

PepRSV can be reconstituted from its isolated protein and RNA. Reconstitution was most effective at pH 4.7 and at low temperature (AbouHaidar *et al.*, 1973). The initiation step (binding of a disk to the RNA) took place over a wide range of conditions, but efficient elongation required closely defined conditions of pH, ionic strength, and temperature. The short-rod RNA reconstituted more efficiently than long-rod RNA, and the binding site for the disk was at or near the 5' terminus of the short RNA (AbouHaidar and Hirth, 1977).

4. Events *in Vivo*

Cycloheximide inhibits TRV replication, indicating that 80 S ribosomes are used for virus protein synthesis (Harrison and Crockatt, 1971). An outline of the events leading to TRV replication in tobacco mesophyll protoplasts has been obtained using a variety of techniques (Harrison *et al.*, 1976). At 22–25°C, infective long-rod RNA was detected 7 hours after inoculation. No other changes were seen at this stage. Nucleoprotein rods were detected by electron microscopy, fluorescent antibody, and infectivity at 9 hours. There appeared to be no accumulation of coat protein but some accumulation of viral RNA, which was incorporated into rods about 4–5 hours after synthesis. Infectious RNA synthesis was largely complete by 12 hours, and infectious nucleoprotein by 24 hours.

Although some short rods appeared at early times, their synthesis lagged behind that of long rods, but by 40 hours both species had reached plateau values. A second difference between the two kinds of rod is their distribution in the cell. Short rods occur mainly scattered in the cytoplasm, whereas long rods are associated with mitochondria in a characteristic fashion. The ends of the rods are closely appressed to the mitochondrial membrane, but do not penetrate it. No rods were associated with other cell organelles. There was no apparent association of the Californian strain of the virus with mitochondria in *N. tabacum* (De Zoeten, 1966).

B. *Furovirus* Group

The type member of the *Furovirus* group is soilborne wheat mosaic virus with a genome consisting of two ssRNA molecules. However, little is known about the genomic structure of this virus. The genome of beet necrotic yellow vein virus (BNYVV), has been fully sequenced but it appears to differ from other members in having four genomic RNAs, all of which have a 3' poly(A) sequence. BNYVV causes a severe disease of sugar beet called rhizomania, which has spread through Europe and to the United States in recent decades. The protein products of the genomic RNA of sorghum chlorotic spot virus, a proposed *Furovirus,* have been described (Kendall *et al.*, 1988). The coat protein gene is in a similar position to that of BNYVV (Fig. 7.33). The biology and molecular biology of the furoviruses has been reviewed by Brunt and Richards (1989). *Nicotiana velutina* mosaic virus RNA shows some similarities to the *Furovirus* group (Randles and Rohde, 1990), but its status has not yet been established.

1. Genome Structure

The genome of BNYVV consists of four molecules of ss positive sense RNA numbered 1–4 in decreasing order of size. The four RNAs have a 5' cap and a 3' poly(A) sequence. The nucleotide sequences of all four RNAs have been established (Bouzoubaa *et al.*, 1985, 1986, 1987). The 5' sequences all begin with an AAA sequence following the cap. The 3'-terminal 60 nucleotides of the four RNAs show extensive sequence homology. RNA1 has an ORF for one large polypeptide. RNA2 has an ORF for a protein of 21K at the 5' end with a UAG termination followed by a long ORF in the same reading frame. Four additional ORFs are present as indicated in Fig. 7.33. RNAs 3 and 4 each contain a single ORF with a relatively long 5' untranslated sequence. Biologically active transcripts have been prepared from cloned cDNA for RNAs 3 and 4 (Ziegler-Graff *et al.*, 1988) and for RNAs 1 and 2 (Quillet *et al.*, 1989).

Jupin *et al.* (1989) isolated an RNA about 600 nucleotides long from infected leaves that has the properties of a subgenomic RNA from RNA3. It is colinear with the 3' region of RNA3, with its 5' terminus at about nucleotide 1230. It always

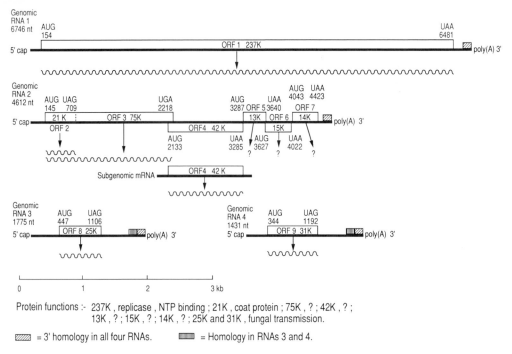

Figure 7.33   Organization and expression of a possible *Furovirus* genome (BNYVV).

appears in infections where RNA3 is present. A small ORF able to encode a 4.6K polypeptide is present. The existence of this 4.6K product has not been established.

2. Functions of the Encoded Proteins

RNA1

The large ORF in RNA1 has regions of sequence homology with the genes known to code for proteins involved in RNA synthesis in some other viruses such as TMV. In particular it contains the purine NTP binding consensus sequence near the middle of the molecule, and embedded in the longer sequence characteristic of helicases. Toward the 3′ end it contains motifs found in the putative replicases of all plus strand RNA viruses (see Fig. 17.9). In the wheat germ system RNA1 is translated to give a polypeptide of MW $\simeq$ 200K and another somewhat larger, which may represent the full ORF of RNA1 (Bouzoubaa *et al.*, 1987).

In vitro translation of RNA1 in the wheat germ system gave two large polypeptides: 240K initiating at AUG (154) and 220K initiating at an internal AUG (496). In rabbit reticulocyte lysate only the 220K protein was made. However, the nucleotide context around AUG (154) is much closer to the plant consensus sequence (Chapter 6, Section VI,A,2) than that near AUG (496) (Jupin *et al.*, 1988).

RNA2

The predicted polypeptide from ORF2 (Fig. 7.33) has the amino acid composition expected for the viral coat protein (Bouzoubaa *et al.*, 1987). *In vitro* translation studies showed that the coat protein gene is on RNA2 (Ziegler *et al.*, 1985). In addition, a 75K polypeptide is produced as a read-through product of the coat protein gene, corresponding to the two in-frame ORFs in the 5′ region (Fig. 7.33).

A similar situation has been found for the RNA2 of soilborne wheat mosaic virus (Hsu and Brakke, 1985). There is some evidence that the 42K polypeptide of the RNA2 of BNYVV is expressed via a subgenomic RNA (Bouzoubaa *et al.*, 1986).

RNAs 3 and 4

The pattern of four BNYVV RNAs isolated from root extracts of plants infected in the field from various countries was quite constant. By contrast, in leaf extracts of various mechanically inoculated species there was great variation in the number and size of the two smaller RNAs (Koenig *et al.*, 1986). Lemaire *et al.* (1988) found an explanation for this phenomenon that at the same time indicates a function for RNAs 3 and 4. BNYVV is transmitted in nature by the fungus *Polymyxa betae* and is usually confined to the roots. The four RNAs can always be isolated from the roots of naturally infected beets. However, when inoculated to *Chenopodium quinoa* leaves, RNAs 3 and 4 undergo deletions or may become undetectable in isolated RNA preparations. Inoculations back to sugar beet roots by means of the fungal vector showed that successful transmission was always associated with the reappearance of full-length RNAs 3 and 4. Thus it seems likely that the gene products of these RNAs are involved in fungal transmission. The reappearance of the two small RNAs was probably due to their being present in quantities below the level of detection in RNA preparations made from leaves (Lemaire *et al.*, 1988).

3. Other *In Vivo* Studies

Jupin *et al.* (1990) reported studies in which polyclonal antisera raised against polypeptides specific for each gene of BNYVV were used in Western blot experiments with extracts of infected *Chenopodium quinoa* leaves. Polypeptides corresponding to all the viral ORFs were detected except the 15K ORF6 of RNA2 (Fig. 7.33). The 25K and 31K ORFs of RNAs 3 and 4 were expressed in spite of the fact that neither RNA is essential for virus replication in leaves. In cell fractionation studies none of the detected polypeptides was found to be tightly associated with the cell wall. Thus there is no evidence so far for a cell-to-cell movement protein for BNYVV. However, encapsidation of the RNA appears to be necessary for long-distance movement, because a mutant defective in packaging did not move systemically (Quillet *et al.*, 1989).

Using RNA transcripts from cDNA clones, Quillet *et al.* (1989) and Jupin *et al.* (1990) showed that while RNA 1 or 2 alone was not infectious, RNAs 1 and 2 together were so. This experiment formally proves that traces of RNAs 3 and 4 are not necessary for infection of leaves. However, the presence of RNA3 has a marked effect on symptom expression, producing bright yellow local lesions.

C. *Dianthovirus* Group

Red clover necrotic mosaic *Dianthovirus* (RCNMV) has a bipartite ss positive sense genome. Both RNA1 (Xiong and Lommel, 1989) and RNA2 (Lommel *et al.*, 1988) have been fully sequenced.

1. Genome Organization

The nucleotide sequences show three large ORFs in RNA1 and one in RNA2 (Fig. 7.34). In RNA1 there is only a single intervening nucleotide between the

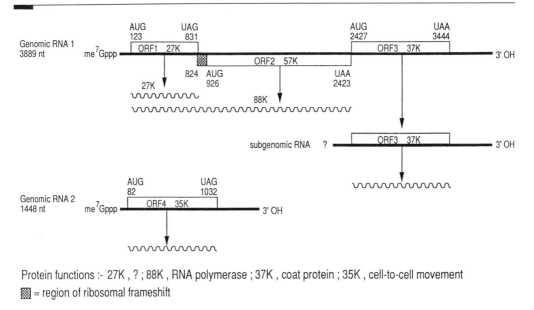

Figure 7.34   Organization and expression of a *Dianthovirus* genome (RCNMV).

termination triplet of the 57K ORF and the AUG of the 37K ORF. There is a 95-nucleotide intervening sequence between the 27K and 57K ORFs. However, the 57K ORF can be extended in the 5′ direction through this 95-nucleotide sequence terminating in a UGA at position 824. This extension is shaded in Fig. 7.34. A ribosomal frameshift is assumed to occur within the 7-nucleotide overlap to give rise to an 88K fusion protein consisting of the 27K ORF, the intergenic region, and the 57K ORF. The frameshift is thought to occur within a 5-nucleotide uracil sequence composed of two phase-shifted phenylalanine codons, as has been proposed for BYDV (Section III,B,2).

There is a region of strong sequence homology in the 3′ untranslated region of RNAs 1 and 2. This homologous region can give rise to a highly stable stem-loop structure, which is probably involved in polymerase recognition.

2. Proteins Synthesized *in Vitro*

RNA2 directs the synthesis of a 35K protein *in vitro* (Lommel *et al.*, 1988). In the rabbit reticulocyte lysate system, RNA1 directed the synthesis of 90K, 50K, 39K, and 27K proteins. The 90K protein was only a minor component among the *in vitro* translation products.

3. Subgenomic RNAs

A single 1.5-kb RNA species homologous to RNA1 was detected in RNAs from infected tissue. This RNA mapped to sequences near the 3′ end of RNA1 and is assumed to be a subgenomic RNA for the coat protein (Xiong and Lommel, 1989).

No subgenomic RNA could be detected *in vivo* for the 50K protein. *In vitro* there must be a start codon accessible to ribosomes for the 50K protein (57K ORF). This may be an artefact of the *in vitro* system. It remains to be determined whether this protein is produced *in vivo*.

4. Functions of the Gene Products

The 39K *in vitro* product was the only one reacting with a coat protein antiserum. Striking amino and sequence similarities were found with the CarMV and TCV coat proteins (Xiong and Lommel, 1989). The 35K protein coded by RNA2 shows some sequence similarity with the 32K protein of BMV, which is considered to be involved in cell-to-cell movement.

Studies on pseudorecombinants between different RNAs 1 and 2 showed that the ability to systemically invade cowpea resides in RNA2 (Osman *et al.*, 1986). *In vivo* studies using protoplasts and whole plants also support a role for the 35K protein in cell-to-cell movements (Paje-Manalo and Lommel, 1989). It may also have a role in determining host range.

Significant amino acid sequence similarity was found between the 88K ORF of RCNMV and the product of similar size and position in the genome, for CarMV, TCV, MCMV, and BYDV (Xiong and Lommel, 1989). The region of highest sequence similarity was downstream from the read-through or frameshift positions and contained the GDD motif characteristic of many RNA-dependent RNA polymerases. RNA1 could replicate alone in protoplasts but failed to cause systemic disease in plants, supporting the view that the polymerase gene is on this RNA (Paje-Manalo and Lommel, 1989). Pseudorecombinant studies with three spontaneous mutants of RCNMV indicate that symptom expression is determined mainly by RNA1 (Osman and Buck, 1989).

VIII. PEA ENATION MOSAIC VIRUS GROUP

There is only one virus currently recognized in this group, but it is of particular interest, being an RNA virus whose replication is intimately associated with the nucleus. No nucleotide sequences have been reported.

A. Genome Structure

The genome consists of two molecules of ss positive sense RNA. The 5' end of both RNAs has a VPg attached (Reisman and De Zoeten, 1982).

B. *In Vitro* Translation

RNA1, the larger RNA, was translated *in vitro* to yield two major products of MW 88K and 36K with no common tryptic peptides; that is, the 88K protein is not a read-through product of the 36K (Gabriel and De Zoeten, 1984). RNA2 coded for one product of MW = 45K. An antiserum against the virus reacted with the 88K protein, indicating that this contained coat protein sequences.

C. *In Vivo* Studies

Tests with fluorescent antibody showed that antigen is located in both nucleus and cytoplasm at early stages after inoculation of tobacco protoplasts but at later stages detectable antigen was mainly confined to nuclei (Motoyoshi and Hull,

1974). In another electron microscope study with protoplasts the first visible signs of PEMV infection at 17 hours were cytoplasmic membrane-bound bodies enclosing a series of vesicles containing fibrils. Some of these appeared to fuse with the nuclear membrane. Virus particles were seen only in the nucleus (Burgess *et al.*, 1974b). Thus, the antigen detected in the cytoplasm by Motoyoshi and Hull (1974) may have been a subviral form of the viral protein.

By a variety of techniques De Zoeten *et al.* (1976) established that PEMV dsRNA is localized in the nuclei of infected cells. The PEMV-induced RNA polymerase activity is also associated with the nuclei as well as with virus-induced vesicles in the cytoplasm (Powell *et al.*, 1977).

Powell and De Zoeten (1977) have shown that nuclei isolated from healthy pea plants can support the initiation of PEMV RNA replication when PEMV RNA was added to them *in vitro*. This was shown by (i) an increase in actinomycin D-resistant polymerase activity with a maximum after about 10 hours incubation and (ii) hybridization experiments, which showed that at least some of the polymerase activity led to PEMV-specific RNA synthesis. Most of the RNA made was negative sense strand in ds form but some ss positive sense strand was also made. This priming reaction took place with viral RNA but not intact PEMV, strongly suggesting that in the intact cell the virus RNA is uncoated before it reaches the nucleus. Several other viral RNAs were tested but only PEMV RNA stimulated viral RNA synthesis in the nuclei.

Replication III: Other Virus
Groups and Families

This chapter summarizes knowledge concerning five kinds of plant viruses that do not have positive sense ssRNA genomes. These are the *Caulimovirus* group with dsDNA; the *Geminivirus* group with ssDNA; the Reoviridae family with dsRNA; the Rhabdoviridae family with negative sense ssRNA; and tomato spotted wilt virus, which has the properties of a bunyavirus. Little is known about the replication of cryptoviruses, a second group with dsRNA genomes. A final section deals with the use of plant viruses as gene vectors.

I. CAULIMOVIRUS GROUP

The caulimoviruses are the only plant viruses with a dsDNA genome (Chapter 4, Section I, E). In 1979 very little was known about the replication of this group, but since then progress has been very rapid. There have been two main motivating factors. First, it was hoped that these viruses, because of their dsDNA genomes, might be effective gene vectors in plants. This aspect is discussed in Section VI,A. Second, the realization that the DNA is replicated by a process of reverse transcription, similar to that of the animal retroviruses, made their study a matter of wide interest. Most experimental work has been carried out on CaMV. During virus replication, viroplasms, which have an appearance characteristic for the group, appear in infected cells. Recent reviews include Hull and Covey (1985), Hohn *et al.* (1985), Hull *et al.* (1987), Pfeiffer *et al.* (1987), and Mason *et al.* (1987).

A. Genome Structure

The DNA nucleotide sequences of three isolates of CaMV (Franck *et al.*, 1980; Gardner *et al.*, 1981; Balázs *et al.*, 1982), figwort mosaic virus (Richins *et al.*, 1987), and soybean chlorotic mottle virus (Hasegawa *et al.*, 1989) have been completely determined. They all consist of a circular dsDNA molecule of about 8 kb. The circular dsDNA of CaMV has a single gap in one strand and two in the complementary strand. Gap 1, the single discontinuity in the α-strand, which is the strand used for transcription, corresponds to the absence of only one or two nucleotides compared with the complementary strand. The two discontinuities in the complementary strand have no missing nucleotides. They are regions of overlap where a short sequence is displaced from the double helix by an identical sequence at the other boundary of the discontinuity. All caulimoviruses have a single gap in the α- (or minus) strand. Viruses in the group other than CaMV may have one, two, or three discontinuities in the plus strand. In CaMV it has been found that the terminal 5' deoxyribonucleotide is at a fixed position and often has one or more ribonucleotides attached. The DNA encodes six and possibly eight genes. These are closely spaced but with very little overlap except for the possible gene VIII. The arrangement of the ORFs in relation to the dsDNA is shown in Fig. 8.1.

B. Proteins Encoded and Their Functions

The eight ORFs illustrated in Fig. 8.1 have been cloned, enabling *in vitro* transcription and translation experiments to be carried out. All eight ORFs could be translated *in vitro* (Gordon *et al.*, 1988). However, not all the protein products have been detected *in vivo* and not all have had functions unequivocally assigned to them. Gordon *et al.* (1988) tested translation products from all eight ORFs with an antiserum prepared against purified dissociated virus. Gene products of ORFs III, IV, and V reacted with this antiserum. The others did not.

1. ORF I

The MW of the gene product expected from the DNA sequence is 37K. The gene product has been produced in expression vectors and antiserum prepared. Such an antiserum detected viral-specific proteins of 46K, 42K, and 38K only in replication complexes isolated from infected tissue (Harker *et al.*, 1987b). In a similar study Martinez-Izquierdo *et al.* (1987) found a viral-specified protein of $M_r \simeq 41K$. A 45K product was found by M. J. Young *et al.* (1987). Given the ambiguities in MW determinations in gels, the major polypeptide found by all three groups is probably the expected 37K protein, especially as the *in vitro* translation product had an $M_r = 41K$ (Gordon *et al.*, 1988).

The ORF I product was detected immunologically in an enriched cell wall fraction from infected leaves (Albrecht *et al.*, 1988). At early times after infection the protein was detected in replication complexes, from which it later disappeared. Immunogold cytochemistry indicated an accumulation of the ORF I protein at or in plasmodesmata between infected cells (Linstead *et al.*, 1988). It appeared to have a close association with the cell wall matrix associated with modified plasmodesmata, suggesting a role in cell-to-cell movement. Such a role is supported by some degree of sequence similarity between the ORF I protein and the TMV P30, which is involved in viral movement (Hull *et al.*, 1986) (Chapter 7, Section III,A,4).

2. ORF II

A protein with a $M_r \simeq 19K$ is translated from ORF II mRNA *in vitro* (Gordon *et al.*, 1988) and an 18K ORF II product can be isolated from infected leaf tissue in

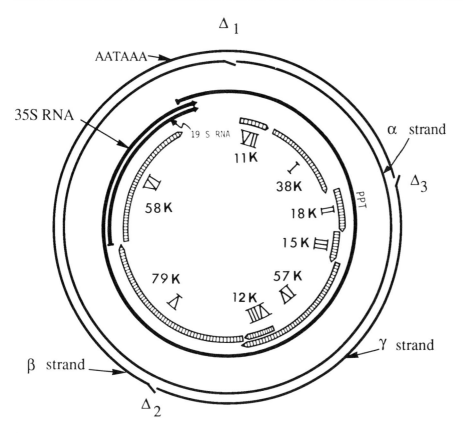

Figure 8.1 Genetic organization of CaMV. The 8-kbp viral DNA features three single-stranded interruptions, one (Δ1) in the α (or coding) strand and two (Δ2 and Δ3) in the noncoding strand, defining the β and γ DNA species. The DNA encodes eight potential ORFs. The MWs of the predicted proteins are indicated. The capped and polyadenylated 19 S and 35 S RNAs have different promoters, but share the same 3' termini. The two mRNAs are translated from a fully ds form of the DNA and not from the gapped form shown here. (From Pfeiffer *et al.*, 1987.)

association with the viroplasms. Some naturally occurring strains of CaMV are transmitted by aphids, whereas others are not. Deletions made within ORF II led to loss of aphid transmissibility (Armour *et al.*, 1983; Woolston *et al.*, 1983). The Campbell isolate is a natural isolate of CaMV that is not aphid transmissible. The 18K product is produced by this strain (Harker *et al.*, 1987a) but appears to be inactive in the aphid transmission function because of an amino acid change, glycine to arginine, in position 94 (Woolston *et al.*, 1987). Experiments in which various recombinant genotypes were produced between various CaMV strains showed that aphid transmissibility mapped to ORF II (Daubert *et al.*, 1984). Collectively these results demonstrate a function in aphid transmission for the product of ORF II. The ORF II protein also appears to increase the extent to which virus particles are held within the viroplasms (Givord *et al.*, 1984).

3. ORF III

ORF III mRNA is translated *in vitro* to give a protein of the expected M_r (14K) (Gordon *et al.*, 1988). A protein of $M_r = 15K$ has been detected *in vivo* using an antiserum raised against an NH_2-terminal 19-amino-acid peptide with a sequence corresponding to that of the ORF III (Xiong *et al.*, 1984). A fusion protein expressed in bacteria and consisting of the N terminus of β-galactosidase and the complete gene III protein (15K) showed DNA-binding activity with a preference for dsDNA (Giband *et al.*, 1986). Mesnard *et al.* (1990) showed that the gene III product is a non-sequence-specific DNA binding protein. They suggested that it is a structural protein within the virus particle.

4. ORF IV

The product of ORF IV is a 57K precursor of the 42K protein subunit of the icosahedral shell of the virus (Franck *et al.*, 1980; Daubert *et al.*, 1982). The 42K protein is assumed to be derived from the 57K molecule by proteolysis after formation of the virus shell (Hahn and Shepherd, 1982). The coat protein is phosphorylated on serine and threonine residues by a protein kinase that is firmly bound to the virus (Ménissier de Murcia *et al.*, 1986; Martinez-Izquierdo and Hohn, 1987). The coat protein is also glycosylated to a limited extent (e.g., Du Plessis and Smith, 1981).

5. ORF V

This is the largest ORF in the genome. mRNA derived from cloned DNA is translated *in vitro* to give a protein of the expected MW of 79K (Gordon *et al.*, 1988). The gene can also be expressed in *E. coli*. Antibodies raised against the protein produced in *E. coli* reacted with an 80K protein in extracts from infected leaves as well as a series of smaller polypeptides (Pietrzak and Hohn, 1987). The following lines of evidence have established that ORF V is the viral reverse transcriptase gene: (i) A protein of the expected size is present both in replication complexes (Pfeiffer *et al.*, 1984) and in virus particles (Ménissier *et al.*, 1984; Gordon *et al.*, 1988). Furthermore the 80K polypeptide associated with the replication complexes is recognized by antibodies against a gene V translation product (Laquel *et al.*, 1986). (ii) There are regions of significant sequence similarity between gene V of CaMV and the reverse transcriptase gene of retroviruses (see Fig. 17.12). (iii) Takatsuji *et al.* (1986) cloned gene V and expressed it in *Saccharomyces cerevisiae*. Yeast expressing the gene accumulated significant levels of reverse transcriptase activity.

It has been suggested that the CaMV reverse transcriptase gene may be translated as a large fusion protein involving the neighboring coat protein gene IV. This idea comes from analogy with the retroviruses, where there is a gene known as *gag* next to the *pol* (polymerase) gene. A *gag* polyprotein gives rise to four small structural proteins of the virus. The *pol* gene may encode a complex of four enzyme activities—reverse transcriptase, RNase-H, endonuclease, and protease. The *pol* gene is translated as a large fusion polyprotein with the *gag* gene (reviewed by Mason *et al.*, 1987). The fusion protein suggestion for CaMV is supported by the fact that there is a domain in CaMV gene V with sequence similarity to the retrovirus protease. Using a dicistronic mRNA for genes IV and V transcribed from a cloned DNA construct, Gordon *et al.* (1988) could detect no such fusion protein, suggesting that the transcriptase gene is translated from its own AUG. However, Toruella *et al.* (1989) showed directly that the N-terminal domain of ORF V produces a functional aspartic protease that can process several CaMV polyproteins.

6. ORF VI

The product of ORF VI (58K) has been identified as the major protein found in CaMV viroplasms (e.g., Odell and Howell, 1980; Xiong *et al.*, 1982). An RNA transcribed from cloned DNA was translated to give a polypeptide of appropriate size (Gordon *et al.*, 1988). The ORF VI product was detected immunologically in a cell fraction enriched for viroplasms (Harker *et al.*, 1987b). Various studies show that the ORF VI product plays a major role in disease induction, in symptom expression, and in controlling host range, but that other genes may also play some part (Schoelz and Shepherd, 1988; Baughman *et al.*, 1988; Shepherd *et al.*, 1988). These topics are discussed further in Chapter 12, Section II,B,1. Gene VI also functions in *trans* to activate translation of other viral genes (Section I,C).

7. ORFs VII and VIII

The two smallest ORFs (VII and VIII) are not present in another caulimovirus (carnation etched ring virus) so their significance is doubtful. However, they can be translated *in vitro* and there is indirect evidence for their existence *in vivo* (reviewed by Givord *et al.*, 1988). Gene VII has no role in aphid transmission. They have a high content of basic amino acids and maybe DNA binding proteins (Givord *et al.*, 1988). Site-directed mutagenesis of ORF VIII, in which start and stop codons were removed or a new stop codon introduced, did not affect infectivity (Schultze *et al.*, 1990).

8. Some Unresolved Questions

Although much has been learned about CaMV gene products, problems remain. These include (i) the possible roles of ORFs VII and VIII; (ii) the possible roles of the less than full size products found for several genes *in vivo;* (iii) the origins of kinase and protease activities; and (iv) the significance of the DNA-binding properties of the gene III product.

C. RNA and DNA Synthesis

There are a large number of publications concerned with CaMV nucleic acid replication and the phenomenon of reverse transcription. Many of these are referred to in the review articles noted at the beginning of this section. Here, only a few key or recent references will be noted.

By 1983 various aspects of CaMV nucleic acid replication led three groups to propose that CaMV DNA is replicated by a process of reverse transcription involving an RNA intermediate (Guilley *et al.*, 1983; Hull and Covey, 1983; Pfeiffer and Hohn, 1983). Some of the observations that led to the model were: (i) the fact that a full-length RNA transcript is produced that has terminal repeats (Covey and Hull, 1981); (ii) the fact that DNA in virus particles has discontinuities (Fig. 8.1) while that found in the nucleus does not, but is supercoiled and is associated with histones as a minichromosome (Ménissier *et al.*, 1982; Olszewski *et al.*, 1982); (iii) the existence of dsDNA in knotted forms (see Fig. 4.3); and (iv) the existence of other forms of CaMV DNA in the cell that are not encapsidated, such as an ss molecule of 625 nucleotides with the same polarity as the α-strand covalently linked to about

100 ribonucleotides (Covey *et al.*, 1983). Since 1983 there has been much further evidence confirming the reverse transcription model, including the partial characterization of the viral-coded reverse transcriptase discussed in the previous section.

It is now clear that there are two phases in the nucleic acid replication cycle. In the first, the dsDNA of the infecting particle moves to the cell nucleus, where the overlapping nucleotides at the gaps are removed, and the gaps are covalently closed to form a fully dsDNA. These minichromosomes form the template used by the host enzyme, DNA-dependent RNA polymerase II, to transcribe two RNAs of 19 S and 35 S, as indicated in Fig. 8.2. At 31 nucleotides upstream from the initiation site of the 35 S RNA there is a TATATAA sequence, and a similar TAT^A_TTAAA is upstream of the 19 S RNA initiation site. The functional regions of the 35 S promoter have been analyzed in more detail (Section VI,D,1). The two promoters are very active, particularly that for the 35 S RNA. It is highly active when used in constructs to express genes in a variety of plant cells (Section VI,D,1). An enhancer sequence of 338 base pairs has been identified in the region upstream from the 35 S TATA sequence (Odell *et al.*, 1988).

The two polyadenylated RNA species migrate to the cytoplasm for the second phase of the replication cycle that takes place in the viroplasms (e.g., Mazzolini *et al.*, 1985).

The 19 S RNA is the mRNA for the viroplasm protein that is produced in large amounts. The 35 S RNA must be used to produce the other gene products since no functional smaller transcripts for most of the ORFs have been found. The 600-nucleotide leader sequence can fold into a large stem and loop structure, centered around a stretch of 36 nucleotides conserved in different viruses (Fütterer *et al.*, 1988). This leader region inhibits translation of downstream genes in an *in vivo* transient expression system (Baughman and Howell, 1988).

There is an AUG at the beginning of every ORF, and the ORFs are close together in the genome. These observations led to a "relay race" model for the translation of the 35 S RNA (Dixon and Hohn, 1984). In this model, a ribosome binds first to the 5' end of the RNA and translates to the first termination codon. At this point it does not completely leave the RNA but reinitiates protein synthesis at the nearest AUG whether just downstream or upstream from the termination. Support for the model came from site-directed mutagenesis studies in ORF VII and the region between ORF VII and ORF I, regions that are not essential for infectivity under laboratory conditions. Insertion of an AUG into either of these regions rendered the viral DNA noninfectious unless the inserted AUG was followed by an in-frame termination codon (Dixon and Hohn, 1984). However, the model is not proven, and a suppression or frameshift model would still be a possibility.

Gene VI is the only *Caulimovirus* gene to be transcribed as a separate transcript from its own promoter, suggesting that it may have an important role at an early stage following infection (Gowda *et al.*, 1989). Reporter genes in various plasmid constructions have been used to show gene VI functions in *trans* in the post-transcriptional expression of closely spaced genes of the full-length RNA transcript (Gowda *et al.*, 1989; Bonneville *et al.*, 1990). Mutagenesis of the coding part of gene VI showed that it was the protein product rather than the mRNA that was responsible for transactivation. The mechanism by which this protein product enhances substantially the translation of the other viral genes on a polycistronic mRNA is not understood.

To commence viral DNA synthesis on the 35 S RNA template, a plant methionyl tRNA molecule forms base-pairs over 14 nucleotides at its 3' end with a site on the 35 S RNA corresponding to a position immediately downstream from the Δ1 discontinuity in the α-strand DNA. The viral reverse transcriptase commences synthesis of a DNA minus strand and continues until it reaches the 5' end of the 35 S RNA. At this point a switch of the enzyme to the 3' end of the 35 S RNA is needed to complete the copying. The switch is made possible by the 180-nucleotide direct repeat sequence at each end of the 35 S RNA. When the template switch is completed, reverse transcription of the 35 S RNA continues up to the site of the tRNA primer, which is displaced and degraded. The Δ1 gap is present in the newly synthesized DNA.

The used 35 S template is then removed by an RNase H activity. It is not certain whether this activity is a function of the reverse transcriptase or a host enzyme. In this process two purine-rich tracts of the RNA are left near the position of gaps Δ2 and Δ3 in the second DNA strand (plus strand). Synthesis of the second (plus) strand of the DNA then occurs, initiating at these two RNA primers. The growing plus strand has to pass the Δ1 gap in the minus strand, which again involves a template switch.

Extracts from infected cells contain CaMV DNA replication complexes that are active in the synthesis of both plus and minus strand DNA (e.g., Thomas *et al.*, 1985). Various forms of viral DNA isolated from these complexes fit with the model for dsDNA synthesis outlined earlier (e.g., hairpin structures, Turner and Covey, 1988). Some of the replicating DNA appears to be in structures with the properties of incomplete virus particles (e.g., Marsh and Guifoyle, 1987; Füetterer and Hohn, 1987). An outline of the preceding model for DNA synthesis is summarized in Fig. 8.2.

In summary, the mechanism of CaMV DNA replication is reasonably well understood. The pathway for the production of the gene VI product is clear, but the way in which the other gene products are produced remains to be firmly established.

D. Recombination in CaMV DNA

The fact that CaMV DNA is converted to a covalently closed ds circle to allow transcription shows that there must be an early involvement of host plant DNA repair enzymes following infection. This idea is reinforced by the fact that cloned DNA, excised from the plasmid in linear form, is infectious and that the progeny DNA is circular. Coinfection of plants with nonoverlapping defective deletion mutants usually led to the production of viable virus particles (Howell *et al.*, 1981). Analysis of the progeny DNA showed that the rescue was by recombination rather than complementation. Lebeurier *et al.* (1982) showed that pairs of noninfectious recombinant full-length CaMV genomes integrated with a plasmid at different sites regained infectivity on inoculation to an appropriate host. In the progeny virus DNA all the plasmid DNA had been eliminated and the viral DNA had a normal structure. Walden and Howell (1982) provided further evidence for intergenomic recombination.

On the basis of experiments with pairs of heterologous genomes, Geldreich *et al.* (1986) proposed a model for recombination in CaMV mediated by the 35 S RNA. In this model, just after inoculation two different DNAs with identical cohesive ends can be ligated together to give a dimer DNA. This dimer is then

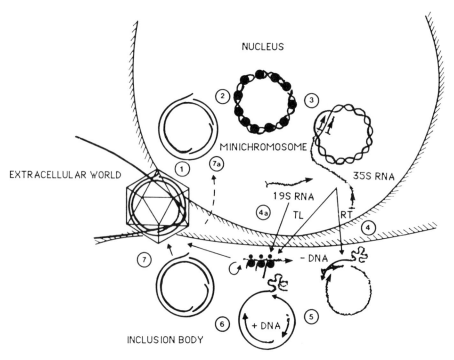

Figure 8.2   A current model of the life cycle of CaMV. Upon infection, the CaMV virus particles release their DNA (1), which gets repaired and sealed in the nucleus (2), where it associates with histones to generate a transcribing minichromosome (3). The 19 S and 35 S RNA are exported to the cytoplasm for translation (4a). Some of the 35 S RNA goes to the viroplasms, where it associates with a met-tRNA molecule (4) to get reverse transcribed into DNA. After digestion of the RNA template and synthesis of the second strand of DNA (5), the viral DNA (6) is packaged into virions (7). Two pathways may coexist: the 35 S RNA may be encapsidated as early as step 4 and DNA synthesis occur in (pre)particles, or it may produce replication complexes that generate free viral DNA that is directed back into the nucleus for amplification (shunted pathway 7a). (From Pfeiffer *et al.*, 1987.)

transcribed to generate a hybrid 35 S RNA that is responsible for the formation of the recombinant genome by reverse transcription.

E. Replication and Movement *in Vivo*

As noted in an earlier section, CaMV (and other caulimoviruses) induce characteristic viroplasms in the cytoplasm of their host cells (Fig. 8.3). These viroplasms are the site for progeny viral DNA synthesis and for the assembly of virus particles. Viral coat protein appears to be confined to them. Most virus particles are retained within the viroplasms.

At an early stage in their development the viroplasms appear as very small patches of electron-dense matrix material in the cytoplasm, surrounded by numerous ribosomes. Larger viroplasms are probably formed by the growth and coalescence of the smaller bodies. The mature viroplasms vary quite widely in size from about 0.2 to 20 μm in diameter. They are usually spherical and are not membrane bound. They often have ribosomes at the periphery and consist of a fine granular

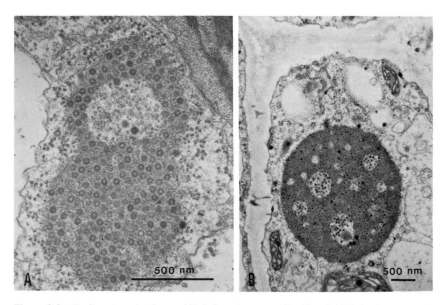

Figure 8.3 *Caulimovirus* viroplasms. (A) A *Brassica rapa* cell with a CaMV viroplasm, containing virus particles. (Courtesy of G. P. Martelli.) (B) Cell of infected *Brassica perviridis* with a viroplasm (V) containing virus particles. (From Martelli and Castellano, 1971, by permission from Cambridge University Press.)

matrix with some electron-lucent areas not bounded by membranes. Virus particles are present in scattered or irregular clusters in the lucent areas and the matrix.

Little is known about the way CaMV particles are assembled. No empty virus shells and very little unencapsidated DNA of the sort found in virus particles are found in infected tissue. These observations suggest that encapsidation may be closely linked to DNA synthesis. The role of glycosylation and phosphorylation of the coat protein remains to be determined. The form in which infectious material moves from cell to cell has not been established, but virus particles have been observed in plasmodesmata (Kitajima and Lauritis, 1969).

Protoplasts from various *Brassica* species can be infected with either virus or isolated DNA. The time course of viral DNA production varied with different species over the 4 days tested. Virus synthesis in turnips reached a maximum 65–96 hours after inoculation (Maule, 1983).

II. GEMINIVIRUS GROUP

The geminiviruses are the only recognized group of plant viruses with a genome consisting of ssDNA. Since Goodman (1977a,b) established the DNA nature of a virus in *Phaseolus,* progress has been quite rapid in our understanding of the organization and replication of the genomes of these viruses. Like the caulimoviruses, interest in members of this group was stimulated by the possibility that, because they contained DNA, they might be developed as gene vectors for plants. Recent reviews include Stanley (1985), Stanley and Davies (1985), Davies *et al.* (1987a,b), Davies and Stanley (1989), and Bisaro *et al.* (1989). Geminiviruses as potential gene vectors in plants are discussed in Section VI,B.

A. Genome Structure

Three subgroups within the *Geminivirus* group have been approved by the I.C.T.V. They are:

Subgroup	Type member	Hosts	Vectors	Genome structure
I	Maize streak virus (MSV)	Monocotyledons	Leafhoppers	One circular ssDNA
II	Beet curly top virus (BCTV)	Dicotyledons	Leafhoppers	One circular ssDNA
III	Bean golden mosaic virus (BGMV)	Dicotyledons	Whiteflies	Two circular ssDNA

Thus, Subgroup II combines some properties of Subgroups I and III.

However, all geminiviruses have a conserved genome sequence in common. There is a large (~200 base) noncoding intergenic region in the two-component geminiviruses. This region has sequences capable of forming a hairpin loop. Within this loop is a conserved sequence found in all geminiviruses (Lazarowitz, 1987). Because it is so highly conserved it has been considered to be involved in RNA transcription (see Section II,D) and in DNA replication.

1. ACMV

The full nucleotide sequence of ACMV DNA showed that the genome consists of two ssDNA circles of similar size (Stanley and Gay, 1983). Stanley (1983) demonstrated that both DNAs are needed for infectivity. Similar evidence has been provided for TGMV and bean golden mosaic virus (BGMV) (Morinaga *et al.*, 1988). Among all well-characterized DNA viruses, these and similar geminiviruses are the only type with a bipartite genome. The nucleotide sequences of the two DNAs are very different except for a 200-nucleotide noncoding region, called the common region, which is almost identical on each DNA. The two DNAs have been arbitrarily labeled 1 and 2. (Some workers use the terminology A and B.) The strands that are found in virus particles are designated virus or plus, and the complementary strands are minus. For these viruses this is an arbitrary distinction because both plus and minus strands contain coding sequences. Comparison of the sequences of ACMV with other bipartite geminiviruses that have been sequenced, such as TGMV (Hamilton *et al.*, 1984) and BGMV (Howarth *et al.*, 1985), revealed that six of these ORFs are in a conserved arrangement (Fig. 8.4). Insertion and deletion mutagenesis has shown that both the ORFs of DNA2 of ACMV are essential for infectivity (Etessami *et al.*, 1988). The coat protein gene in DNA1 is not required (Etessami *et al.*, 1989). It is noteworthy that the two small ORFs in the negative sense DNA1 strand overlap substantially in different reading frames.

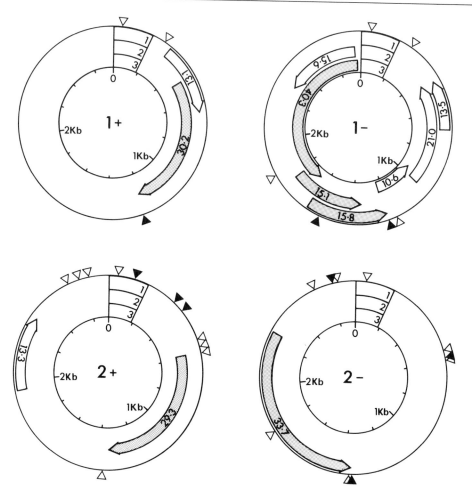

Figure 8.4 Genome organization of ACMV. Arrows indicate ORFs and direction of transcription. Coding capacity of each ORF is indicated. The shaded ORFs are the six found in other bipartite geminiviruses, and those for which RNA transcripts have been identified. The boxed region to the right of the O kb position, showing reading frames 1, 2, and 3, is the 200-base common intergenic region. △ = Possible promoter sequences (TATA boxes); ▲ = possible polyadenylation signals (AATAAA sequences). (From Townsend *et al.*, 1985.)

2. MSV

The nucleotide sequence of the MSV genome (Nigerian strain) revealed a single DNA circle of 2687 nucleotides that had no detectable overall sequence similarity with genomes of the two-circle type (Mullineaux *et al.*, 1984). Analysis of ORFs with potential for polypeptides of MW ≥ 10K revealed seven potential coding regions arranged as shown in Fig. 8.5. There are two intergenic regions. A slightly different map than that found for the Nigerian isolate was determined for a South African isolate of MSV, on a clone that was infectious for maize using an *A. tumefaciens* delivery system (Lazarowitz, 1988). The genome of two other leafhopper-transmitted viruses, chloris striate mosaic virus (Anderson *et al.*, 1988) and WDV (MacDowell *et al.*, 1985), have been sequenced.

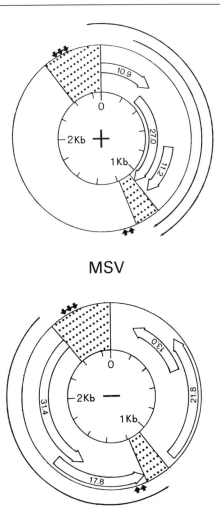

Figure 8.5 Organization of the genome of MSV (Nigerian) showing the coding and potential coding regions of the virion sense (+) DNA and its complement (−). The open reading frames (⇒) assume that the protein synthesis initiates at the first ATG and that potential products are those of at least 10K molecular weight. The narrow-headed outer arrows (→) represent the approximate mapping positions of the RNA transcripts. Those on the (+) DNA are the 1.05- and 0.9-kb RNAs, overlapping at their 5′ termini (nucleotides 163 and 2682) and coterminating at nucleotide 1114. The thick arrows (↓) indicate the positions of five of the nine possible stem-loops, and the stippled regions represent the large and small intergenic regions. (From Mullineaux *et al.*, 1984.)

3. BCTV

The sequence of 2993 nucleotides in an infectious clone of BCTV has been established (Stanley *et al.*, 1986). The organization of the single-component genome resembles the DNA1 of the whitefly-transmitted group (Fig. 8.6). The four conserved coding regions of DNA1 have very similar counterparts in BCTV, except for the coat protein gene, whose sequence is more closely related to those of the leafhopper-transmitted group.

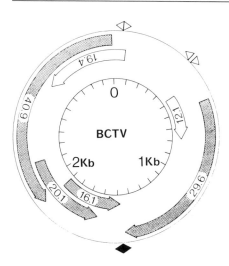

Figure 8.6 Organization of the genome of BCTV showing the potential coding regions. The shaded ORFs represent those that have equivalent homologous ORFs in other geminiviruses. On the basis of comparison with other geminiviruses, the 29.6K ORF is the coat protein gene. Consensus TATA sequences are indicated ($\triangle$) and also potential polyadenylation signals ($\blacktriangle$). (From Davies *et al.*, 1987a.)

B. Functions of the ORF Products

1. Coat Protein

In ACMV, the coat protein gene has been mapped to the 30.2K ORF of DNA1 (see Fig. 8.4). (Townsend *et al.*, 1985). The coat protein gene also occurs in this position in TGMV (Kallender *et al.*, 1988). In MSV this gene has been mapped to an equivalent position, that is, in the 27K ORF of the plus sense strand (Morris-Krsinich *et al.*, 1985). It might be expected that the properties of the coat protein might play a role in insect vector specificity. This idea is supported by the sequence homologies between the coat proteins of viruses transmitted by *Bemisia tabaci*. By contrast, the coat protein ORFs of BCTV, MSV, and WDV, each of which has a different leafhopper vector, show much less homology (MacDowell *et al.*, 1985; Stanley *et al.*, 1986). Briddon *et al.* (1990) constructed chimeric clones in which the coat protein gene in DNA 1 of ACMV (whitefly transmitted) was replaced with the coat protein gene of BCTV (leafhopper transmitted). The BCTV gene was expressed in plants and gave rise to particles containing the ACTV DNAs. These chimeric particles were transmitted by the BCTV leafhopper vector, whereas normal ACMV was not when injected into the insects. These results demonstrated that vector specificity is a function of the coat protein.

Genetic analysis of TGMV (whitefly transmitted) has shown that the coat protein is not essential for infectivity, for systemic spread, or for symptom development (Gardiner *et al.*, 1988). Unlike the situation with the bipartite TGMV, the coat protein gene of monopartite viruses is essential for infectivity, as shown for MSV by Boulton *et al.* (1989a), Lazarowitz *et al.* (1989), and for WDV (Woolston *et al.*, 1989) and BCTV by Briddon *et al.* (1989).

2. ORFs Related to DNA Replication

Using ACMV, or cloned copies of the genome, Townsend *et al.* (1986) showed that DNA1 could replicate independently in protoplasts from *Nicotiana plumbaginifolia,* but DNA2 could not. Similarly, in transgenic petunia plants, DNA1 of TGMV on its own gave rise to virus particles containing DNA1 (Sunter *et al.,* 1987). Thus DNA replication functions must be associated with one or more of the ORFs in DNA1 minus since the DNA1 plus ORF encodes the coat protein. Using mutations created *in vitro,* Elmer *et al.* (1988a) showed that the largest gene in the minus strand of DNA1 of TGMV known as AL1 ($=40.3$K in Fig. 8.4) is required for DNA synthesis. AL1 is the only viral gene product that is absolutely required for single- and double-stranded replication of TGMV DNA, but accumulation of ssDNA requires the AL2 ORF ($=15.1$K ORF in DNA1$-$ of Fig. 8.4). Maximal replication of both ss and ds DNA depends on the presence of a functional AL3 gene product ($=15.8$K ORF in DNA1$-$ of Fig. 8.4) (Rogers *et al.,* 1989; Hayes and Buck, 1989; Hanley-Bowdoin *et al.,* 1990).

3. Virus Movement

Since DNA1 of ACMV can replicate in protoplasts, but does not infect plants (Stanley, 1983), DNA2 must encode a function for cell-to-cell movement of the virus.

4. Insect Transmission

Since DNA2 is found only with whitefly-transmitted geminiviruses it has been suggested that products of this DNA must be involved in insect transmission. Insertion and deletion mutagenesis experiments show that both coding regions of ACMV DNA2 must have other functions in addition to any role in insect transmission (Etessami *et al.,* 1988). The experiments described in the preceding Section 1 show that the coat protein, a product of a DNA1 gene, controls transmission by insect vectors.

5. A Gene Product with a Possible Function in Poaceae

A polypeptide of 11K encoded by ORF 10.9K of MSV (Fig. 8.5) has been detected among the products of *in vitro* translation using RNAs from maize plants infected with MSV (Mullineaux *et al.,* 1988). It was also found in protein extracts from infected leaves, but not in virus particles. Two other geminiviruses infecting Poaceae had sequences in common with the ORF 10.9 of MSV, in equivalent ORFs, while viruses infecting dicotyledons did not. Thus the product of ORF 10.9 may have a function related to infection of Poaceae.

C. mRNAs

Five virus-specific polyadenylated RNA transcripts found in ACMV-infected plants were mapped to either the plus or minus DNA strands, demonstrating that transcription is bidirectional in both DNAs (Townsend *et al.,* 1985). Some possible polyadenylation signals for ACMV are shown in Fig. 8.7.

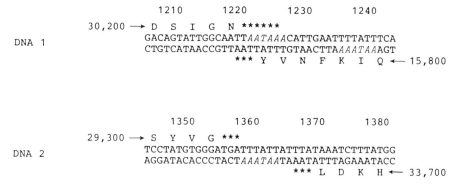

Figure 8.7   Possible polyadenylation signals (AATAAA) in ACMV DNAs. The DNAs are shown in ds form. Termination codons for four ORFs are indicated by *** with calculated MWs of the polypeptides shown to each side (see also Fig. 8.4). Amino acids at the C termini are in the one-letter code. Potential polyadenylation signals are shown in italics. (From Stanley and Davies, "Molecular Plant Virology". Vol. II. Copyright © 1985, CRC Press, Inc., Boca Raton, Florida.)

Three virus-specific polyadenylated RNA transcripts have been found in maize leaves infected with MSV. Two mapped to the plus strand and one to the minus strand, demonstrating bidirectional transcription (Morris-Krsinich *et al.*, 1985). The position of the possible promoter sequences relative to the ORFs fits with this mechanism. A DNA-dependent RNA polymerase type II reaction was active in virus-specific RNA synthesis in nuclei isolated from *Nicotiana* tissue infected with TGMV (Coutts and Buck, 1985). Both DNA components were transcribed in a bidirectional fashion but relationships between the transcripts and the ORFs in the DNAs were not established. In later work, an mRNA isolated from infected tissue was identified as that for the coat protein (Petty *et al.*, 1988). Schalk *et al.* (1989), using infected protoplasts, identified an intron in the genome of WDV involved with two distinct overlapping reading frames. The bipartite geminiviruses infecting dicotyledons did not have an intron in the gene for the corresponding protein. A 92-base intron has been identified in another monopartite *Geminivirus, Digitaria* streak virus (DSV) (Accotto *et al.*, 1989).

The detailed transcription mapping carried out on DSV RNAs formed *in vivo* by Accotto *et al.* revealed that all except 18 nucleotides of the genome are transcribed either from the viral strand (two RNA species) or from the complementary strand (up to five RNA species). Thus this virus can be assumed to produce more than one RNA species from the same transcription unit, and therefore more than one polypeptide.

Sunter *et al.* (1989) identified six polyadenylated virus-specific RNAs in preparations from *Nicotiana benthamiana* plants agroinfected with TGMV. They corresponded in size, polarity, and map location to the six ORFs in the genome (as illustrated for ACMV in Fig. 8.4). However, with more detailed primer extension and S1 nuclease protection experiments, Sunter and Bisaro (1989) showed that leftward transcription of the TGMV genome is more complex than previously thought. The 1.3-kb (DNA2−) and 1.6-kb (DNA1−) RNA species were shown to be composed of a family of similar-sized transcripts that appear to be 3′ coterminal but that have different 5′ termini.

Thus the genome organization of all the groups of geminiviruses may be more complex, and the number of proteins coded for greater than is indicated in Figs. 8.4 and 8.5.

D. Control of Transcription

The genomes of ACMV, MSV, and BCTV have a number of consensus promoter sequences (TATAA/TAA/T) lying outside or just within the common untranslated regions. Two such promoter sequences are illustrated in Fig. 8.8. In the vertebrate dsDNA virus SV40, gene expression is regulated in time by transcriptional polarity. Howarth *et al.* (1985) suggested that a similar timing control may occur with geminiviruses.

Using a maize protoplast transient expression system, Fenoll *et al.* (1988) defined further the structure of the MSV rightward promoter that drives transcription of the RNA for coat protein (the 27K ORF in the plus strand) (Fig. 8.5). They identified a 122-bp sequence upstream from the start site for transcription that enhances promoter activity. This "upstream activating sequence" lies in the large intergenic region and includes a region that is common to all geminiviruses. The 122-bp sequence activates the MSV core promoter in a position-dependent, but orientation-independent fashion. The activating sequence specifically binds proteins in maize nuclear extracts.

E. DNA Replication

The details of *Geminivirus* DNA replication are not well established. *Geminivirus* particles usually accumulate in the nucleus, and with some, such as MSV, large amounts of virus accumulate there. In some infections, fibrillar rings, which must be part of a spherical structure, appear in the nucleus (Francki *et al.*, 1985c) but their composition and significance are not known. Nuclei isolated from *Nicotiana* tissue infected with TGMV synthesized variable amounts of plus and minus strands of both DNAs 1 and 2 (Coutts and Buck, 1985).

The presence of the same 200-nucleotide sequence in the intergenic regions of both DNAs 1 and 2 of ACMV indicates that it plays an important part in viral replication. A common region with the capacity to form a stable loop containing a conserved nanonucleotide (TAATATTAC) sequence has been found in all geminiviruses. This loop structure has been proposed as a binding site for the host enzymes responsible for priming complementary strand synthesis (see Davies *et al.*, 1987b). By inserting short sequences into TGMV B-DNA at sites within and just outside the common region, Revington *et al.* (1989) showed that the conserved hairpin loop structure is necessary for DNA replication. The loop may be part of the viral origin of replication. However, a possible second mechanism is indicated by the fact that MSV preparations contain a population of small DNA molecules, in particular one of ≈80 nucleotides complementary to genomic DNA, with ribonucleotides covalently linked to the 5′ terminus. These molecules can prime the

Figure 8.8 Comparison of the sequences bordering the two putative promoter regions on ACMV DNA1 (+) upstream of the coat protein and 13K ORFs. The sequences are aligned to show the homology. (From Davies *et al.*, 1987b.)

synthesis of a complementary strand *in vitro,* at one site in the genome (Donson *et al.,* 1984). This small DNA does not map to the loop with the conserved nanonucleotide sequence in the large intergenic region, but rather in the small intergenic region (Fig. 8.5). Similar small DNA molecules have been found in a geminivirus from *Digitaria* that is related to MSV (Donson *et al.,* 1987) in WDV (Hayes *et al.,* 1988b; MacDonald *et al.,* 1988) and chloris striate mosaic virus (Anderson *et al.,* 1988).

Extracts of *Nicotiana* leaves infected with TGMV or ACMV contain a variety of viral DNA forms in addition to ssDNA. Three forms of unit-length dsDNA have been found: closed circular supercoiled, relaxed circular, and linear. Other high-MW species appeared to be concatamers consisting of two or more unit-length genomes (Hamilton *et al.,* 1982; Sunter *et al.,* 1984; Slomka *et al.,* 1988). The existence of these ds forms and the concatamers indicates that the virus replicates via a circular ds replicative form, possibly by a rolling circle mechanism (Stanley and Davies, 1985). Support for this mechanism comes from the observation of Rogers *et al.* (1986) that there is strong similarity between the highly conserved sequence TAATATTAC found in the DNA components of all geminiviruses that have been sequenced and the cleavage site for the gene A protein of the ssDNA bacterial virus ΦX174, which acts as an origin for rolling circle replication of the DNA.

Rogers *et al.* (1986) showed that the DNA1 component of TGMV could replicate independently of DNA2, in keeping with the strong similarities between the DNA1 component and the single DNA of geminiviruses such as MSV. It is possible that these DNAs code for a protein with functions like that of ΦX174 protein A. A rolling circle mechanism for DNA replication could also account for the fact that infection of *Nicotiana benthamiana* with uncut cloned tandem dimers of TGMV DNA gave rise to normal monomeric unit-length DNAs soon after inoculation (Hayes *et al.,* 1988a). Similar results have been obtained with bean golden mosaic virus (Morinaga *et al.,* 1988).

Subgenomic ss and ds forms of DNA2 have been characterized for ACMV (Stanley and Townsend, 1985) and for TGMV (MacDowell *et al.,* 1986). Their significance is unknown, but they may be similar to the defective interfering DNAs of some animal viruses.

Recombination is known to occur during *Geminivirus* DNA replication. For example, insertion or deletion mutagenesis of the two large ORFs of ACMV DNA2 destroyed infectivity but was restored by coinoculation of constructs that contained single mutations in different ORFs (Etessami *et al.,* 1988). Frequent intermolecular recombination produced dominant parental-type virus. Infection of *Nicotiana benthamiana* with uncut cloned tandem dimers of TGMV DNA components gave rise to genome-length ssDNA species of both components (Hayes *et al.,* 1988a). It has not been established whether intermolecular recombination or some replicative event leads to the production of these unit-sized DNAs.

F. Agroinfection with Geminiviruses

Grimsley *et al.* (1987) showed that *Agrobacterium* containing tandem repeats of MSV DNA inoculated to whole maize plants led to symptoms of MSV infection. Since MSV DNA is not mechanically transmissible, and intact virus can infect only by means of an insect vector, this experiment provided a very sensitive demonstration that *Agrobacterium* could infect a monocotyledon. Elmer *et al.* (1988b) adapted the agroinfection procedure of Grimsley *et al.* (1987) to provide a simple and efficient assay for TGMV replication. They produced transgenic *Nicotiana benthamiana* plants containing multiple tandem copies of TGMV B-DNA. They found that an inoculum containing as few as 2000 *Agrobacterium* cells containing TGMV A-DNA could produce 100% virus infection. Agroinfection is also a highly efficient way of introducing DNAs A and B together into *N. benthamiana* (Hayes *et al.*, 1988e). The technique has also been extended to *Digitaria* streak virus (Donson *et al.*, 1988), MSV in various species of Poaceae (Boulton *et al.*, 1989a), and ACMV (B. A. M. Morris *et al.*, 1988; Klinkenberg *et al.*, 1989).

Hayes *et al.* (1989) noted that constructs based on TGMV with DNAs longer than normal viral DNA were stable when introduced into the plant cell genome by agroinfection, but rapidly reverted to viral size when introduced by other means. The absence of deleted molecules in agroinfected plants may be because vector molecules replicate in single cells with little or no cell-to-cell movement. This would give deletion mutants much less opportunity to compete.

III. PLANT REOVIRIDAE

Plant members of the Reoviridae family are placed in two genera: *Phytoreovirus* with 12 dsRNA genome segments, the type member being wound tumor virus (WTV); and *Fijivirus*, with 10 dsRNA genome segments, the type member being Fiji disease virus (FDV). The molecular biology of WTV has been reviewed by Nuss (1984) and Nuss and Dall (1989).

A. Genome Structure

The sizes of the 12 genome segments of WTV are listed in Table 8.1. The full nucleotide sequence of the smallest segment was established by Asamizu *et al.*

Table 8.1
Interrelationship of WTV Genome Segments and Gene Products[a]

Segment	Mol. wt. of genome segment ($\times 10^6$)	Mol. wt. of gene products		Location in virus	Nomenclature
		In vivo	*In vitro*		
1	2.90	155,000	155,000	Nucleoprotein core	P1
2	2.40–2.48	130,000	130,000	Outer protein coat	P2
3	2.20–2.25	108,000	108,000	Nucleoprotein core	P3
4	1.78–1.80	72,000	74,000	—	Pns4[b]
5		76,000	76,000	Outer protein coat	P5
6	1.10–1.15	57,000	57,000	Nucleoprotein core	P6
7	1.05–1.15	52,000	52,000	—	Pns7
8	0.83–0.90	42,000	42,000	Capsid	P8
9	0.57–0.63	41,500	41,500	Capsid	P9
10	0.55–0.61	39,000	39,000	—	Pns10
11	0.54–0.60	35,000	35,000	—	Pns11
12	0.32–0.35	19,000	19,000	—	Pns12

[a]From Nuss (1984).

[b]Pns = nonstructural polypeptide.

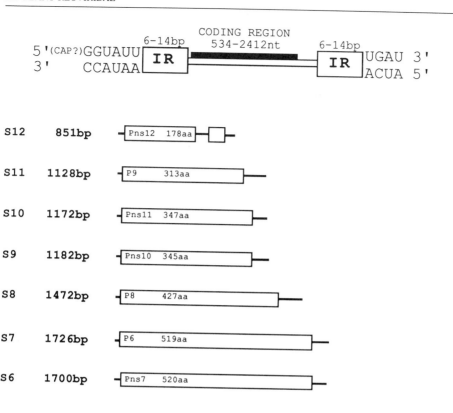

Figure 8.9 Organization of the WTV genome. The general organization of WTV genomic segments is illustrated at the top of the figure. The terminal conserved hexanucleotide and tetranucleotide sequences are indicated in large letters while the positions of the 6- to 14-base pair (bp) segment specific terminal inverted repeats (IR) are indicated by the large open boxes. The solid rectangle indicates the general position of the coding region (534–2412 nucleotides). The remainder of the figure illustrates specific information regarding the size of individual genomic segments (bp) and their encoded polypeptides (aa) as well as the coding assignments. There is no evidence for expression of the second small coding region present in genomic segment S12. (From Nuss and Dall, 1990.)

(1985). It consists of 851 nucleotides with a single ORF sufficient to code for the expected gene product Pns12 (Fig. 8.9; Table 8.1). Segments 5 (Anzola *et al.*, 1987) 4, 6, 9, 10 (Anzola *et al.*, 1989a), 7 (Anzola *et al.*, 1989b) 8 (Xu *et al.*, 1989b), and 11 (Dall *et al.*, 1989) have also been sequenced. Each contains a single functional ORF as is illustrated in Fig. 8.9.

The nucleotide sequence of genome segment 10 of rice dwarf *Phytoreovirus* (RDV) has been determined (Uyeda *et al.*, 1987a; Omura *et al.*, 1988a). This segment has 1319 nucleotides. The first AUG is at residues 27–29 and a single ORF extends for 1059 nucleotides. The predicted protein product has 352 amino acids and a MW $\simeq$ 39K, similar in size to the product of WTV segment 10 (Table 8.1). The amino acid sequence showed no strong similarities to other proteins, including polymerases. The sequences at the 5' and 3' termini are the same as those in WTV (Fig. 8.9). The sequence of RDV genome segment 9 has also been determined (Uyeda *et al.*, 1989). It consists of 1305 nucleotides with an ORF coding for a putative polypeptide of 351 amino acids (MW = 38,600). The six 5'-terminal nucleotides and three at the 3' terminus were the same as in segment 10.

B. RNA Transcription and Translation

The plant reoviruses, like their counterparts infecting vertebrates, contain a transcriptase that uses the RNA in the particle as template to produce ssRNA copies. The WTV enzyme has an optimum at pH 8.2 and a temperature optimum at 25°C (Black and Knight, 1970). Similar enzyme activities have been reported for rice dwarf virus (Kodama and Suzuki, 1973), FDV (Ikegami and Francki, 1976), and rice ragged stunt virus (S. Y. Lee *et al.*, 1987). In *Reovirus*, the enzyme is active only after the outer protein shell of the virus is removed to give "cores." With the plant reoviruses, the enzyme activity can be detected directly on purified virus preparations. This is almost certainly due to the fact that the plant reoviruses readily lose their outer protein shell, and purified virus preparations may consist largely or entirely of cores. WTV has been shown to contain a methylase that catalyzes the incorporation of methyl groups from *S*-adenosyl-*L*-methionine into the RNA strands synthesized *in vitro* (Rhodes *et al.*, 1977), giving the 5'-terminal structure $m^7G^{5'}ppp^{5'}Ap$---.

The transcriptase in purified WTV synthesizes *in vitro* 12 ssRNA products corresponding to the 12 ds segments of the genome (Reddy *et al.*, 1977). Nuss and Peterson (1981) were able to assign each of the 12 mRNA transcripts synthesized *in vitro* to their corresponding genome segment. Similar results have been obtained for RDV (Uyeda and Shikata, 1984) and for rice gall dwarf virus (Yokoyama *et al.*, 1984).

The existence of the active transcriptase within the virus particles has facilitated the identification of the protein products of the 12 genes. Thus Nuss and Peterson (1980) identified 12 polypeptides synthesized in a cell-free system using mRNAs synthesized *in vitro*. They also used a cell-free system in which WTV transcription was coupled with translation. Furthermore they found that the same 12 virus-specific polypeptides were synthesized in cell monolayers of the vector *Agallia constricta*. All this information is summarized for WTV in Table 8.1. Genome segments 2 and 5 play a vital role in insect transmission (see Chapter 14, Section IV, D,2). Many of the assignments of protein products to genome segments listed in Table 8.1 have been confirmed by *in vitro* expression of tailored full-length cDNA clones (Xu *et al.*, 1989a).

C. Intracellular Site of Replication

Plant reoviruses replicate in the cytoplasm as do those infecting mammals (Wood, 1973). Following infection, densely staining viroplasms appear in the cytoplasm (Fig. 8.10). Viroplasms were present in cells of various tissues of leafhopper vectors infected with WTV as well as infected plant cells (Shikata and Maramorosch, 1967). Immunofluorescence demonstrated the presence of viral antigen in the cytoplasm of cultured leafhopper cells (Chiu *et al.*, 1970). It is not yet

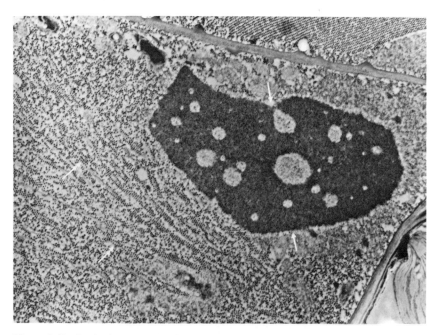

Figure 8.10   Leaf vein tumor cells of maize experimentally infected with maize rough dwarf *Fijivirus*. The three different kinds of inclusions caused by MRDV are easily recognizable: viroplasm (arrows), cytoplasmic tubules along and inside which the virus particles are aligned (double arrows), and part of a virus crystal (top right). (From Bassi and Favali, 1972. Reprinted with the permission of Cambridge University Press.)

possible to relate the *in vitro* studies on the replication of WTV to the structures seen cytologically.

Enzyme digestion experiments and radioautographic assay of the incorporation of ³H-labeled uridine into maize cells infected with maize rough dwarf virus indicate that much of the viroplasm is made up of protein—probably viral proteins. Viral RNA appears to be synthesized in the viroplasm, where the mature particles are assembled. The mature particles then migrate into the cytoplasm where they may (i) remain as scattered particles, (ii) form crystalline arrays, or (iii) become enclosed in or associated with tubelike proteinaceous structures (Fig. 8.10) (Bassi and Favali, 1972; Favali *et al.*, 1974). The autoradiographic studies failed to implicate the nucleus, mitochondria, or chloroplasts in virus replication.

A detailed electron microscopic study supported the view that the viroplasms caused by FDV in sugarcane are the sites of virus component synthesis and assembly (Hatta and Francki, 1981c). The viroplasms are composed mostly of protein and dsRNA. Some areas contained numerous isometric particles 50–60 nm in diameter. Some appeared to be empty shells while others contained densely staining centers of dsRNA. These particle types appeared to be incomplete virus particles or cores. Complete virus particles were seen only in the cytoplasm.

D. RNA Selection during Virus Assembly

Every WTV particle appears to contain one copy of each genome segment because (i) RNA isolated from virus has equimolar amounts of each segment (Reddy and Black, 1973) and (ii) an infection can be initiated by a single particle

(Kimura and Black, 1972). Thus there is a significant problem with this kind of virus. What are the macromolecular recognition signals that allow one, and one only, of each of 10 or 12 genome segments to appear in each particle during virus assembly? For example, the packaging of the 12 segments of WTV presumably involves 12 different and specific protein–RNA and/or RNA–RNA interactions. The first evidence that may be relevant to this problem comes from the work of Anzola *et al.* (1987) with WTV. They established the structure of a defective (DI) genomic segment 5, which was only one-fifth the length of the functional S5 RNA because of a large internal deletion. However, this DI RNA was packaged at one copy per particle, as for the normal sequence. Thus they established that the sequence(s) involved in packaging must reside within 319 base pairs of the 5′ end of the plus strand and 205 from the 3′ end. Reddy and Black (1974, 1977) showed that an increase in the DI RNA content in a virus population led to a corresponding decrease in the molar proportion of the normal fragment, that is, the DI fragment competes only with its parent molecule. Thus there must be two recognition signals, one that specifies a genome segment as viral rather than host, and one that specifies each of the 12 segments.

Anzola *et al.* (1987) sequenced the 5′- and 3′-terminal domains of all 12 genome segments (Fig. 8.11). They suggested that a fully conserved hexanucleotide sequence at the 5′ terminus and a fully conserved tetranucleotide sequence at the 3′ terminus might form the recognition signals for viral as opposed to host RNA. They also found segment-specific inverted repeats of variable length just inside the conserved segments (Fig. 8.11). They suggested that these might represent the specific recognition sequence for each individual genome segment. A similar inverted repeat was found for segment 9 of RDV (Uyeda *et al.*, 1989).

Other members of the Reoviridae family each have conserved 5′ and 3′ sequences (Asamizu *et al.*, 1985). Influenza viruses, which have segmented ssRNA genomes, also have comparable conserved sequences at the 5′ and 3′ termini of each genome segment (Stoeckle *et al.*, 1987). These similarities strengthen the idea that the 5′ and 3′-terminal sequences have a role in the packaging of these segmented RNA genomes. However, the problem is by no means solved. If only RNA–RNA recognition is involved, how is this brought about to give a set of 12 dsRNAs for packaging? If protein–RNA recognition is important, how are 12 specific sites constructed out of the three proteins known to be in the nucleoprotein viral core?

In *Reovirus* the negative sense strands are synthesized by the viral replicase on a positive sense template that is associated with a particulate fraction (Acs *et al.*, 1971). These and related results led to the proposal that dsRNA is formed within the nascent cores of developing virus particles, and that the dsRNA remains within these particles. If true, this mechanism almost certainly applies to the plant reoviruses. It implies that the mechanism that leads to selection of a correct set of 12 genomic RNAs involves the ss plus strand. Thus the base-paired inverted repeats illustrated in Fig. 8.11 could be the recognition signals. It may be that other virus-coded "scaffold" proteins transiently present in the developing core are involved in RNA recognition rather than, or as well as, the three proteins found in mature particles.

Xu *et al.* (1989b) constructed a series of transcription vectors that allowed production of an exact transcript of S8 RNA and of four analogs that differed only in the immediate 3′ terminus. Their experiments provided three lines of evidence supporting the view that the 5′- and 3′-terminal domains interact in a functional way: (i) nuclease T1 sensitivity assays showed that even a slight change in the 3′-terminal sequence can affect the conformation of the 5′ terminus; (ii) translation *in*

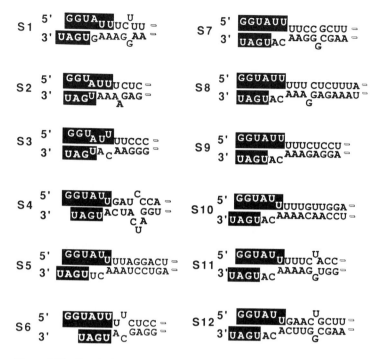

Figure 8.11 Terminal sequence domains of the positive sense strands of the 12 genome segments of WTV. The segment-specific inverted repeats near the 5′ and 3′ termini are oriented to indicate potential base-pairing interactions. The conserved 5′ hexanucleotide and 3′ tetranucleotide shared by all 12 genome segments are shown in white on black. (From Anzola *et al.*, 1987.)

vitro is slightly decreased by alterations in the 3′ terminus, which extends the potential for 3′–5′ terminal base-pairing, and is increased by changes that reduce potential base-pairing; and (iii) computer modeling for minimal energy structures for six WTV transcripts predicted a conformation in which the terminal inverted repeats were base-paired.

Dall *et al.* (1990) developed a gel retardation assay with which they demonstrated selective binding of WTV transcripts by a component of extracts from infected leafhopper cell cultures. Using terminally modified and internally deleted transcripts they established that the segment-specific inverted repeats present in the terminal domains were necessary but not sufficient for optimal binding. Some involvement of internal sequences was also necessary. There was no evidence for discrimination in binding between transcripts from different segments. The binding component or components present in extracts of infected cells, which are not present in those of healthy cells, have not yet been characterized.

IV. PLANT RHABDOVIRIDAE

The rhabdoviruses have the largest particles among the plant viruses. They are one of only two groups whose particles are bounded by a lipoprotein membrane. Knowledge about the rhabdoviruses has been reviewed on a number of occasions, for example, by Francki and Randles (1980), Francki *et al.* (1985d), and Jackson *et al.* (1987). They have been placed in two groups depending on their site of maturation and accumulation within the cell.

A. Genome Structure

Plant rhabdoviruses, like those infecting vertebrates, possess a genome consisting of a single piece of ss negative sense RNA, with a length in the range of 11,000–13,000 nucleotides. A discrete mRNA is transcribed from the negative sense genome for each protein coded for. Gene sequencing techniques have only recently been applied to one plant member of the family—*Sonchus* yellow net virus (SYNV).

Tobacco plants infected with SYNV contain transcripts 139–144 nucleotides long that are complementary to the 3′-terminal region of the genomic RNA. This region and the whole of the first gene from the 3′-terminal region of the genomic RNA have been sequenced (Zuidema *et al.*, 1986, 1987). This ORF encodes a 475-amino-acid polypeptide of MW ≃ 53.6K. The next gene, in a 5′ direction on the genomic RNA, has also been sequenced (Heaton *et al.*, 1987). This second ORF encodes a protein of 345 amino acids and a MW ≃ 38.3K. The arrangement of these two genes is summarized in Fig. 8.12.

Heaton *et al.* (1989a), using hybridization with a library of cDNA clones, have established that the order of the six genes in the SYNV genome is 3′-N-M2-sc4-M1-G-L-5′. The intergenic and flanking sequences are highly conserved as is shown in Fig. 8.13. All the SYNV RNAs begin with the sequence AACA or AACU (positive sense).

B. Proteins Encoded

Five viral-coded proteins are found in SYNV virus particles (see Table 5.2). The proteins coded by the first two mRNAs to be sequenced have been identified by raising antibodies against the product of the cloned gene and showing that they react

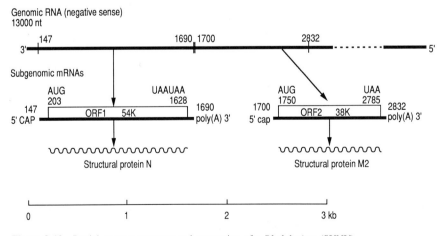

Figure 8.12 Partial genome structure and expression of a *Rhabdovirus* (SYNV).

leader	**UUUCUUUUU**	**GG**	**UUGUA**	N
N	**AUUCUUUUU**	**GG**	**UUGUC**	M2
M2	**AUUCUUUUU**	**GG**	**UUGUC**	sc4
sc4	**AUUCUUUUU**	**GG**	**UUGAA**	Ml
Ml	**AUUCUUUUU**	**GG**	**UUGAA**	G
G	**AUUCUUUUU**	**GG**	**UUGUA**	L
	1	2	3	

Figure 8.13 Nucleotide sequences in the leader and at the gene junctions of genomic SYNV RNA. The 3′ direction of the genomic RNA is to the left. 1 = The putative polyadenylation signal at the 5′ terminus of each gene; 2 = an untranslated intergenic sequence; and 3 = the transcription site at the 3′ terminus of each gene. (From Heaton *et al.*, 1989a.)

with a structural component of the purified virus. On this basis the first gene from the 3′ end of the genomic RNA codes for the nucleocapsid (N) protein (Zuidema *et al.*, 1987), while the second gene encodes another structural protein M2 (Heaton *et al.*, 1987), confirmed by the later work noted above. When proteins from highly purified LYNV were fractionated on acrylamide gels, many minor protein bands were detected. These were mostly degradation or aggregation products of the G and N proteins (Dietzgen and Francki, 1988).

C. The Viral Transcriptase

Variable amounts of transcriptase activity have been found in different plant rhabdoviruses, but the enzyme is not yet well characterized. Lettuce necrotic yellows virus (LNYV) transcriptase is located in the internal nucleoprotein core of the virus and uses the viral RNA as a template (Francki and Randles, 1972; Randles and Francki, 1972). The enzyme has a requirement for Mg^{2+}. The *in vitro* product of this transcriptase was almost entirely ssRNA (Francki and Randles, 1973). The product was heterogeneous in size, being much smaller than complete viral RNA. The product strands are rapidly released from the cores. The transcription complex could be separated into two inactive fractions by centrifugation. Activity was restored by recombining the fractions (Toriyama and Peters, 1980). Transcriptase activity has also been reported in SYNV (Flore, 1986), and for wheat rosette stunt, a probable rhabdovirus (Sun and Gong, 1988).

D. Genome Replication

Replication of the genome of plant rhabdoviruses is assumed to take place on a full-length positive sense RNA template. The detailed mechanism for genome replication has not been established for any rhabdoviruses (Masters and Banerjee, 1988). In particular the question remains as to how the transcriptase can ignore the intergenic junctions that delineate the mRNAs to produce a complete positive sense strand.

E. mRNA Synthesis

Wolanski and Chambers (1971) used radioautography of tissue infected with LNYV and labeled with [³H]uridine in the presence of actinomycin D to follow the appearance and location of virus-specific RNA. At early times this was mostly in the nucleus, but at later stages synthesis in the cytoplasm predominated. Viral protein synthesis probably occurs in the cytoplasm.

From patterns of hybridization with cDNA clones, Heaton *et al.* (1989a) showed that the SYNV genome is transcribed into a short 3′-terminal "leader" RNA and six mRNAs. Stenger *et al.* (1988) detected four size classes of polyadenylated RNAs in extracts of tissue infected with sowthistle yellow vein virus. Using hybridization with randomly labeled LYNV genomic RNA, Dietzgen *et al.* (1989) detected five polyadenylated RNAs in extracts from infected *N. glutinosa*. More discriminating procedures will presumably reveal six mRNAs for all the plant rhabdoviruses.

F. Protein Synthesis

Four major viral proteins could be detected by immunoprecipitation in protoplasts infected with SYNV (G, N, M, and M_2) (van Beek *et al.*, 1986). In addition, several larger proteins were detected that were probably aggregates of unglycosylated G protein. Structural viral proteins first appeared between 8 and 12 hours after inoculation.

G. Cytological Observations on Replication

Because of their large size and distinctive morphology the rhabdoviruses are particularly amenable to study in thin sections of infected cells. They appear to fall into three groups: (i) Those that accumulate in the perinuclear space with some particles scattered in the cytoplasm. With some viruses of this group, structures resembling the inner nucleoprotein cores have been seen within the nucleus. The envelopes of some particles in the perinuclear space can be seen to be continuous with the inner lamella of the nuclear membrane. Figure 8.14 illustrates this group. Immunogold labeling with an antiserum against the five structural proteins of PYDV showed that viral proteins accumulate mainly in the nucleus (Lin *et al.*, 1987).

(ii) In the second group, for example, LNYV, maturation of virus particles occurs in association with the endoplasmic reticulum, and particles accumulate in vesicles in the reticulum. Figure 8.15 gives a graphical representation for this group. Biochemical evidence suggests that the nucleus is involved in the early stages of infection by members of this group. (iii) In this group, exemplified by barley yellow striate mosaic virus, infection leads to the formation of large viroplasms in the cytoplasm of infected cells. These consist of electron-dense granular or fibrous material. Mature virus particles appeared in membrane-bound sacs within the viroplasm particularly near its surface (Conti and Appiano, 1973). There is no nuclear phase associated with the replication of this group (Bassi *et al.*, 1980). Festuca leaf streak virus appears to belong here. The ability to infect protoplasts with this virus (van Beek *et al.*, 1985) may be a useful advance for the study of the morphogenesis of these viruses.

V. A PLANT MEMBER OF THE BUNYAVIRIDAE

Tomato spotted wilt *Tospovirus* can infect 360 angiosperm species belonging to 50 families. The genome consists of three RNA segments: L ($\approx$8200 nucleotides), M

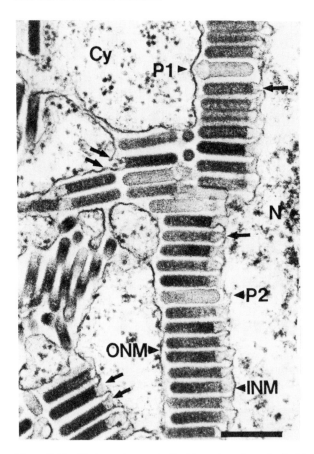

Figure 8.14 Electron micrograph of a thin section of a maize leaf infected with a Hawaiian isolate of maize mosaic rhabdovirus. Virus particles apparently budding through the inner nuclear membrane (INM) and through intracytoplasmic extensions (double arrows) of the outer nuclear membrane (ONM); single arrows indicate constriction of the INM. Alignment of particles at P1 and P2 suggests budding on the ONM. Cy, cytoplasm; N, nucleus. Bar = 0.3 μm. (From McDaniel, Ammar, and Gordon, 1985).

($\approx$5400), and S ($\approx$3000). There are four structural polypeptides. Two are glycosylated (G1 and G2) and are at the surface. The N protein binds to the RNA, and there is a large protein occurring in a minor amount. On the basis of these and other properties, Milne and Francki (1984) suggested that TSWV was possibly a member of Bunyaviridae, a large family of viruses replicating in vertebrates and invertebrates. This suggestion has been amply confirmed by recent work in which the M and S RNAs have been sequenced (de Haan *et al.*, 1990a,b; P. de Haan, personal communication). Figure 8.16 summarizes the main features of these two genome segments. Both RNAs contain inverted complementary repeats at their termini. These are probably involved in RNA replication and in the formation of the circular configuration found in virus particles.

M RNA is negative sense and has nine-base-long complementary sequence at the 5′ and 3′ termini. The complementary strand of mRNA contains a long ORF with a UGA termination codon that is probably leaky. Thus a 35K protein and a read-through protein of 185K may be produced.

The S RNA has an ambisense coding strategy. In other words, it is positive sense for one gene and negative sense for the other (Fig. 8.16). The positive sense

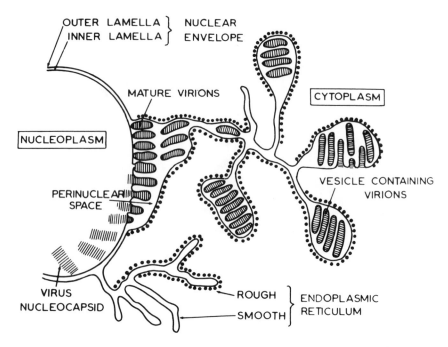

Figure 8.15 Diagrammatic representation of morphogenesis and cellular distribution of a rhabdovirus such as LNYV. (From Francki and Hatta, "Atlas of Plant Viruses," Vol. I. Copyright © 1985, CRC Press, Inc., Boca Raton, Florida.)

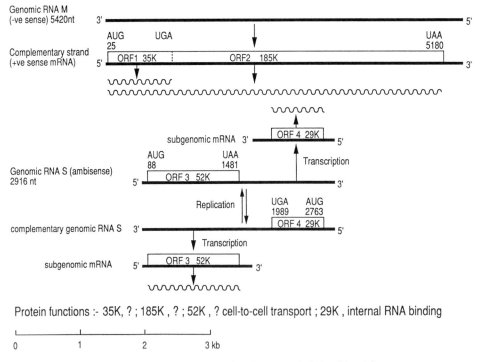

Figure 8.16 Organization and expression of TSWV genomic RNAs M and S.

ORF codes for a protein with a predicted size of 52K. The ORF in the complementary sense encodes a 29K protein. Both proteins are translated from subgenomic RNAs, which have been identified in extracts from TSWV-infected cells (de Haan *et al.*, 1990b). There is a stable hairpin loop structure in the intergenic region of S RNA. The two subgenomic RNAs probably terminate at this loop. These various features clearly identified TSWV as a member of the Bunyaviridae family, with the status of a new genus. Little is known about the replication of TSWV *in vivo* except that it almost certainly takes place in the cytoplasm.

VI. POSSIBLE USES OF VIRUSES
FOR GENE TRANSFER

In the early 1980s there was considerable interest in the possibility of developing plant viruses as vectors for introducing foreign genes into plants. At first, interest centered on the caulimoviruses, the only plant viruses with dsDNA genomes, because cloned DNA of the viruses was shown to be infectious (Howell *et al.*, 1980). Interest later extended to the ssDNA geminiviruses, and then to RNA viruses when it became possible to reverse transcribe these into dsDNA, which could produce infectious RNA transcripts.

The main potential advantages of a plant virus as a gene vector were seen to be: (i) the virus or infectious nucleic acid could be applied directly to leaves, thus avoiding the need to use protoplasts and the consequent difficulties in plant regeneration; (ii) it could replicate to high copy number; and (iii) the virus could move throughout the plant, thus offering the potential to introduce a gene into an existing perennial crop such as orchard trees.

Such a virus vector would have to be able to carry a nonviral gene (or genes) in a way that did not interfere with replication or movement of the genomic viral nucleic acid. Ideally, it would also have the following properties: (i) inability to spread from plant to plant in the field, providing a natural containment system; (ii) induction of very mild or no disease; (iii) a broad host range, which would allow one vector to be used for many species, but would be a potential disadvantage in terms of safety; and (iv) maintenance of continuous infection for the lifetime of the host plant.

The major general limitations in the use of plant viruses as gene vectors are: (i) they are not inherited in the DNA of the host plant, and therefore genes introduced by viruses cannot be used in conventional breeding programs; (ii) plants of annual crops would have to be inoculated every season, unless there was a very high rate of seed transmission; (iii) by recombination or other means the foreign gene introduced with the viral genome may be lost quite rapidly with the virus reverting to wild type; and (iv) it would be necessary to use a virus that caused minimal disease in the crop cultivar. The virus used as vector might mutate to produce significant disease, or be transmitted to other crops that were susceptible. Infection in the field with an unrelated second virus might cause very severe disease.

In recent years technological advances in the use of the modified Ti plasmid of *Agrobacterium tumefaciens* for gene transfer have made this the method of choice for most dicotyledon plant systems (reviewed by Schell, 1987). For monocotyledons, methods involving DNA introduction into protoplasts, or direct introduction of DNA into leaf cells by a particle gun, have been developed. The present status of plant viruses as useful gene vectors can be gauged from the fact that they are not mentioned in a recent review of genetic engineering applied to crop improvement (Gasser and Frayley, 1989). However, plant viruses have been very useful in one respect. They have acted as a source of control elements for use in other vector systems. The 35 S promoter of CaMV has proved particularly versatile. In the following section I will review briefly the experiments aimed at developing various types of plant viruses as gene vectors. Aspects of the topic have been discussed by Wilson (1989).

A. Caulimoviruses

Howell *et al.* (1981) inserted an eight-base-pair *Eco*RI linker molecule into the large intergenic region of cloned CaMV DNA and showed there was no impairment of infectivity. Gronenborn *et al.* (1981) reported the successful propagation of foreign DNA in turnip plants using CaMV as a vector. However, they found that the size of the DNA insert that could be successfully propagated was limited to about 250 base pairs or less.

Brisson *et al.* (1984) reported the first successful expression of a foreign gene in plants using CaMV as a vector. They used the 234-bp dihydrofolate reductase gene from *E. coli,* which confers methotrexate resistance in the bacterium. They inserted the gene into a derived strain of CaMV from which most of gene II had been deleted. The chimeric DNA was stably propagated in turnip plants and the bacterial gene was expressed, as shown by assays for methotrexate resistance. De Zoeten *et al.* (1989) replaced the ORF II of CaMV DNA with the human interferon (IFNαD) gene. They obtained a stable strain of CaMV that replicated in *Brassica rapa* and led to the production of IFNαD. The interferon was located in the CaMV viroplasms and it had antiviral activity in an animal cell assay.

Paszkowski *et al.* (1986) constructed a hybrid CaMV genome containing the selectable marker gene neomycin phosphotransferase type II, which replaced the gene VI coding region. This construct was not viable in plants and could not be complemented in *trans* by wild-type CaMV. However, inoculation of *Brassica campestris* protoplasts under DNA uptake conditions gave rise to stable cell lines genetically transformed for the marker gene. This occurred only when the hybrid was coinoculated with wild-type CaMV. The mechanism for this effect is not understood. CaMV as a probe for studying gene expression in plants is discussed by Pfeiffer and Hohn (1989).

B. Geminiviruses

Much attention has been focused on the geminiviruses as potential gene vectors because of their DNA genomes and because the small size of the genomes makes them convenient for *in vitro* manipulations (reviewed by Coutts *et al.,* 1987; Davies and Stanley, 1989). Nevertheless, this small size may restrict the amount of viral DNA that can be deleted (see Davies *et al.,* 1987b). However, this is counterbalanced by the fact that for some geminiviruses a viable coat protein and encapsidation are not necessary for successful inoculation by mechanical means, or for systemic movement through the plant (Gardiner *et al.,* 1988).

There are other potential difficulties (Davies *et al.,* 1987b). Recombination can occur to give parental-type molecules. Most geminiviruses are restricted mainly to the phloem and associated cells. However, the wide host range of the geminiviruses (compared with the caulimoviruses) makes them of considerable interest. The fact that some members infect cereal crops would be particularly useful, except for the fact that they are not seed transmitted and are mechanically transmitted only with difficulty. In any event, inoculations on the scale needed for cereal crops would be impractical.

Nevertheless, model experiments have shown that a geminivirus can be used to introduce and express foreign genes in plants. Various workers have shown that an intact coat protein gene is not essential for TGMV replication and movement. Hayes *et al.* (1988d) constructed a chimeric TGMV DNA in which most of the coat protein gene had been deleted and replaced by the bacterial neomycin phosphotransferase gene (neo gene). They used *Agrobacterium* inoculation (agroinfection) to introduce this construct into tobacco plants. The TGMV DNA was replicated in the transgenic plants and the neo gene expressed. Ward *et al.* (1988) isolated a coat protein mutant of ACMV DNA1 in which 727 nucleotides in the coat protein gene were deleted. This mutant was noninfectious. However, when the coat protein ORF was replaced by the coding region of the bacterial chloramphenicol acetyltransferase (CAT) gene under control of the coat protein promoter, infectivity was restored, virus spread through the plant, and the CAT gene was expressed. Nevertheless, systemic movement of infectious constructs with larger than normal DNAs in the coat protein position exerted a strong selection pressure favoring derivatives of normal size (Elmer and Rogers, 1990).

Transient expression systems allow for the rapid screening of DNA constructs designed to study the activities of promoter sequences, RNA processing signals, and so on, in cells, and as a preliminary screen in the construction of transgenic plants. In principle, viral DNA might be altered to provide plasmid-type vectors for high copy number and rapid expression of modified or foreign genes. The work of Elmer *et al.* (1988a) with a geminivirus progressed some distance toward the construction of such a plant plasmid. Deletion derivatives of TGMV DNA A defined the minimal DNA fragment capable of self-replication as about 1640 bp or 60% of the A sequence.

Hanley-Bowdoin *et al.* (1988) showed that following agroinoculation of petunia leaf disks with TGMV DNA A, the viral coat protein gene was transiently expressed, beginning about 2 days after agroinfection. Constructs were then made in which the coat protein ORF was replaced by the bacterial CAT or β-glucuronidase (GUS) genes, under the control of the coat protein promoter. Following agroinoculation, these foreign genes were also transiently expressed in petunia leaf disks.

Finally, Gröning *et al.* (1987) found that the DNA of a geminivirus infecting *Abutilon sellovianum* was located in the chloroplasts, raising the possibility of constructing a chloroplast-specific transformation vector.

C. RNA Viruses

The ability to manipulate RNA virus genomes by means of a cloned cDNA intermediate has opened up the possibility of using RNA as well as DNA viruses as gene vectors. In principle the known high error rate in RNA replication (Chapter 17, Section IV,B,4) might place a limitation on the use of RNA viruses (van Vloten-Doting *et al.*, 1985; Siegel, 1985). The experimental evidence to date suggests that mutation may not be a major limiting factor, at least in the short term.

1. *Bromovirus* Group

French *et al.* (1986) constructed variants of BMV RNA3 in which the coat protein gene was replaced with the bacterial CAT gene, or in which the CAT gene was inserted near the 5′ end of the coat gene. When inoculated onto barley protoplasts together with normal BMV RNAs 1 and 2, these RNA3 constructs replicated and produced subgenomic RNAs equivalent to the normal coat protein subgenomic RNA (see Fig. 7.26). When the CAT gene was inserted in-frame with the upstream coat protein initiation codon, CAT expression exceeded that in plant cells transformed by Ti plasmid-based vectors. This work has been reviewed by Ahlquist *et al.* (1987).

Pacha *et al.* (1990) showed that all the *cis*-acting elements required for the replication of the 2.1-kb CCMV RNA3 can be contained in a 454-base replicon made of 5′- and 3′-terminal sequences. Thus it may be possible to express a foreign gene of significant size using this replicon.

2. TMV

Takamatsu *et al.* (1987) prepared a TMV cDNA construct in which most of the coat protein gene was removed and that had the CAT gene in its place. When *in vitro* transcripts from this construct were inoculated onto tobacco leaves the local lesions that formed were smaller than normal, but biologically active CAT was produced, and this activity increased in the inoculated leaves for 2 weeks. RNA was extracted from these leaves and reinoculated onto tobacco after being encapsidated in TMV protein *in vitro*. CAT activity was again detected, indicating that this replicating RNA had some degree of stability *in vivo*. Yamaya *et al.* (1988a), using a disarmed Ti plasmid vector, introduced a cDNA copy of the TMV genome under the control of the 35 S CaMV promoter into the genomic DNA of tobacco plants. This experiment demonstrated that a non-seed-transmitted RNA virus can be made seed transmissible.

Dawson *et al.* (1989) constructed a hybrid TMV in which the CAT gene was inserted between the coat protein and the 30K genes. This construct replicated efficiently, produced an additional subgenomic RNA and CAT activity, and assembled into 350-nm virus rods. However, during systemic infection the insert was precisely deleted, giving rise to wild-type virus. Takamatsu *et al.* (1990b) constructed, via cDNA, a TMV RNA in which an additional sequence encoding Leu-enkephalin, a pentapeptide with opiate-like activity, was incorporated just 5′ to the termination codon for coat protein. The pentapeptide was expressed in protoplasts as a fusion product with coat protein.

3. STNV

Mechanical inoculation of cowpea with cloned cDNA full-length copies of STNV RNA in the presence of helper TNV resulted in the production of viable STNV particles (van Emmelo *et al.*, 1987). Short linker segments could be introduced at various sites in the STNV genome and these were stably transmitted to progeny, thus raising the possibility that STNV could be used as a gene vector.

4. *Tobravirus* Group

Angenent *et al.* (1989c) found that sequences of 340 nucleotides at the 5′ end and 405 at the 3′ end of the RNA2 of TRV (strain PLB) were sufficient for replication. Constructs in which the deleted viral nucleotides were replaced with a 1401-nucleotide sequence from plasmid DNA were replicated in protoplasts coin-oculated with the complete genome of strain PLB, indicating that this virus may have potential as a gene vector.

D. Viruses as Sources of Control Elements for Transgenic Plants

Certain plant viral nucleic acid sequences have been found to have useful activity in chimeric gene constructs as promoters of DNA transcription and as enhancers of mRNA translation.

1. Promoters

Transcription of CaMV DNA gives rise to a 19 S and a 35 S mRNA (Section I, C). Shewmaker *et al.* (1985) introduced a full-length copy of CaMV DNA into the T DNA of the *Agrobacterium* Ti plasmid, and this was integrated into various plant genomes. They showed that the 19 S and 35 S promoters were functional in this form in several plant species. They are both strong constitutive promoters and have found wide application for the expression of a range of heterologous genes (e.g., Balázs *et al.*, 1985; Bevan *et al.*, 1985; Nagy *et al.*, 1986; Odell *et al.*, 1985, Lloyd *et al.*, 1986; reviewed by Weising *et al.*, 1988).

The 35 S promoter has been found to be much more effective than the 19 S in several systems. For example, expression of the α-subunit of β-conglycinin in petunia plants under control of the 35 S promoter was 10–50 times greater than from the 19 S promoter (Lawton *et al.*, 1987). The 35 S promoter was also found to be 10–30 times more effective than the nopaline synthase promoter from *Agrobacterium tumefaciens* (García *et al.*, 1987a; Saunders *et al.*, 1987).

Odell *et al.* (1985) found that correct initiation from the 35 S promoter required proximal elements that included a TATA sequence. Rate of transcription was deter-mined by sequences that were dispersed over 300 bp of upstream DNA. Ow *et al.* (1987) made a more detailed analysis of the functional regions of the 35 S promoter using the firefly luciferase gene as a reporter of promoter activity. Deletion analysis revealed that the promoter is a composite structure within the 150-bp DNA sequence upstream from the start of 35 S RNA transcription. There are at least three func-tional domains: a proximal region with the TATA box; a middle region containing a CCAAT-like box element; and a distal region with three elements similar to that of the SV40 core enhancer

$$
\begin{array}{c}
\text{AAA} \\
\text{GTGG} \qquad \text{G} \\
\text{TTT}
\end{array}
$$

Fang *et al.* (1989) have also defined multiple *cis* regulatory elements for the max-imal expression of the CaMV promoter in transgenic plants.

Kay *et al.* (1987) constructed a variant 35 S promoter that contained a tandem duplication of 250 bp of upstream sequences. This modification gave about a 10-fold increase in transcriptional activity. A "34 S" promoter from figwort mosaic *Caulimovirus* had about the same activity as the 35 S CaMV promoter (Sanger *et al.*, 1990).

The 35 S promoter is nominally a constitutive one. However, Benfey and Chua

(1989) showed that there was marked histological localization of expression of GUS activity in petunia under control of this promoter. By contrast, the 19 S promoter directed expression of the CAT gene in a range of tissues in tobacco plants (C. Morris *et al.*, 1988). Benfey *et al.* (1989) showed that the 35 S CaMV promoter contains at least two domains (A and B) that can confer different gene expression patterns in different tissues and at different stages of plant development. Modularity in promoters generally is discussed by Dynan (1989). In further work, Benfey *et al.* (1990a) showed that domain B has a modular organization, with at least five domains that are all able to confer distinct expression patterns when fused with domain A. Various combinations of subdomains conferred different expression patterns both early in development and in mature transgenic plants (Benfey, 1990a,b).

2. Untranslated Leader Sequences as Enhancers of Translation

Untranslated leader sequences of several viruses have been shown to act as very efficient enhancers of mRNA translational efficiency both *in vitro* and *in vivo* and in prokaryotic and eukaryotic systems. AMV RNA4 is known to be a well-translated message for AMV coat protein. Jobling and Gehrke (1987) replaced the natural leader sequences of a barley and a human gene with AMV RNA4 leader sequence. These constructs showed up to a 35-fold increase in mRNA translational efficiency in the rabbit reticulocyte and wheat germ systems. Sleat *et al.* (1987) made similar constructs involving the uncapped mRNAs for two vertebrate genes and the bacterial GUS gene with or without a 5′-terminal 67-nucleotide sequence derived from the untranslated region of TMV RNA (the Ω sequence). These were tested *in vitro* in the rabbit reticulocyte, wheat germ, and *E. coli* systems. The TMV leader sequence enhanced translation of almost every mRNA in every system. Gallie *et al.* (1987a,b) extended these results to show that the 67-nucleotide sequence was also a potentially useful enhancer *in vivo* in mesophyll protoplasts and *Xenopus* oocytes. A deletion derivative of Ω appeared to be functionally equivalent to a Shine–Dalgarno sequence in several bacterial systems (Gallie and Kado, 1989). The translational enhancement brought about by the TMV Ω sequence is mediated by the ribosomal fraction of the *in vitro* system used (Gallie *et al.*, 1988b). The experiments of Sleat *et al.* (1988c) support the view that the untranslated viral leader sequences reduce RNA secondary structure, making the 5′ terminus more accessible to scanning by ribosomal subunits or by interaction with initiation factors.

3. *In Vitro* Studies on Transcription

In vitro translation systems which faithfully reproduce *in vivo* gene expression have proved very useful in animal and other systems for developing an understanding of nucleotide sequences and protein factors involved in the control of transcription. Cooke and Penon (1990) have taken a step in this direction for plant systems. They obtained a partially purified extract from tobacco cell suspensions that contained all the factors necessary for transcription from the 19 S promoter.

4. Use of the TMV Origin of Assembly (OAS) for the Introduction of Foreign RNAs

Sleat *et al.* (1986) showed that a correctly oriented OAS sequence located 3′ to a foreign RNA sequence could initiate the efficient encapsidation of the foreign

RNA *in vitro*. In an extension of these experiments, Gallie *et al.* (1987c) prepared transcripts that encoded the OAS together with the RNA sequence for the CAT enzyme. When these were encapsidated in TMV coat protein *in vitro* and inoculated to a wide range of cell types the CAT mRNA was transiently expressed. Immunogold labeling located the site of disassembly and transient gene expression in epidermal cells of inoculated tobacco leaves (Plaskitt *et al.*, 1987).

Sleat *et al.* (1988c) constructed a plasmid derivative containing the 5′ leader sequence of TMV followed by the CAT sequence and the OAS. This was introduced into the DNA of tobacco plants using an *Agrobacterium* vector. Transcripts from this nuclear DNA were encapsidated into TMV-like rods when the transgenic plants were infected with TMV. These experiments demonstrated an efficient complementation between functions encoded in the host genome and those of the infecting virus.

5. Other Possible RNA Virus Sequences

The subgenomic promoters for some RNA viruses have now been defined. These may prove useful for gene amplification, as might 5′- or 3′-terminal sequences that have not yet been tested. The development of simple and specific RNA enzymes from satellite RNA sequences is discussed in Chapter 9, Section II,B,6. In principle, such ribozymes, integrated into the plant DNA, might be used to suppress a particular host mRNA in conjunction with the introduction of new genes by means of vectors.

9

Viroids, Satellite Viruses, and Satellite RNAs

I. VIROIDS

A variety of important viruslike diseases in plants have been shown to be caused by pathogenic RNAs known as viroids, a term first introduced by Diener (1971) to describe the infectious agent causing the spindle tuber disease of potato. They are small circular molecules, a few hundred nucleotides long, with a high degree of secondary structure. They do not code for any polypeptides and replicate independently of any associated plant virus. Viroids are of practical importance as the cause of several economically significant diseases and are of general biological interest as the smallest known agents of infectious disease. Work on viroids has been reviewed many times, for example, by Semancik (1987) and Diener (1987b). The most studied viroid is potato spindle tuber viroid (PSTVd). Viroid names are abbreviated to initials with a d added to distinguish them from abbreviations for virus names.

A. Structure of Viroids

1. Circular Nature of Viroid RNA

The circular nature of many viroids was first shown directly by electron microscopy. Under nondenaturing conditions the molecules appear as small rods with an axial ratio of about 20 : 1, and for PSTVd an average length of about 37 nm. When spread under denaturing conditions the molecules can be seen to be covalently closed circles (Sänger *et al.*, 1976; McClements and Kaesberg, 1977; Riesner *et al.*, 1979). All viroid preparations contain a variable proportion of linear molecules. The nucleotide sequence and proposed secondary structure for PSTVd are shown in Fig. 9.1.

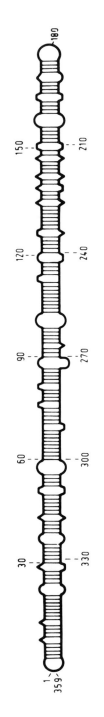
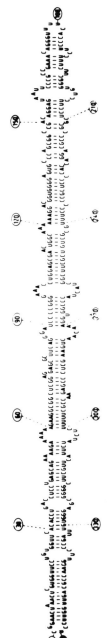

Figure 9.1 Potato spindle tuber viroid. Lower: Nucleotide sequence as determined by Gross *et al.* (1978); center: proposed secondary structure in outline; upper: three-dimensional representation of the viroid molecule. (Courtesy of H. Sanger.)

2. Nucleotide Sequences

The nucleotide sequences of about 50 viroids and viroid sequence variants are now known. These sequences have confirmed the circular nature of the molecules (Fig. 9.1). All viroids have some degree of sequence similarity. On the basis of degree of overall sequence similarity, 16 distinct viroids have been recognized, and these have been placed in five groups (Table 9.1). Other distinct viroids continue to be discovered. An alternative classification is shown in Fig. 9.4.

The use of cDNA clones to sequence field isolates of viroids has revealed that a single isolate may contain a range of closely related sequence variants (e.g., Visvader and Symons, 1985).

3. Secondary Structure

From the primary sequence it is possible to predict a secondary structure that maximizes the number of base pairs. This gives rise to rodlike molecules with base-paired regions interspersed with unpaired loops (Fig. 9.1).

Table 9.1
Viroids Arranged in Five Groups Based on Overall Similarities in Nucleotide Sequence[a,b]

Group	Viroid	Number of nucleotides	Selected references
PSTVd	Potato spindle tuber viroid (PSTVd)	359	Gross *et al.* (1978)
	Chrysanthemum stunt viroid (CSVd)	356, 354	Haseloff and Symons (1981)
	Citrus exocortis viroid (CEVd)	370–375	Visvader and Symons (1985); Duran-Vila *et al.* (1988)
	Citrus cachexia viroid (CCaVd)	?	Semancik *et al.* (1988)
	Hop stunt viroid (HSVd)	297	Shikata (1987); Sano *et al.* (1989)
	Tomato apical stunt viroid (TASVd)	360	Kiefer *et al.* (1983); Candresse *et al.* (1987a)
	Tomato planta macho viroid (TPMVd)	360	Kiefer *et al.* (1983)
	Columnea latent viroid (CLVd)	370	Hammond *et al.* (1990)
CCCVd	Coconut cadang-cadang viroid (CCCVd)	246, 247 plus larger forms	Haseloff *et al.* (1982)
	Coconut tinangaja viroid (CTiVd)	254	Keese *et al.* (1988a)
ASBVd	Avocado sunblotch viroid (ASBVd)	247	Symons (1981)
ASSVd	Apple scar skin viroid (ASSVd)	330	Hashimoto and Koganezawa (1987)
	Grapevine yellow speckle viroid (GYSVd)	367	Koltunow and Rezaian (1988)
	Grapevine viroid 1B (GVd1B)	363	Koltunow and Rezian (1989a)
HLVd	Hop latent viroid (HLVd)	256	Puchta *et al.* (1988)

[a]Modified from Puchta *et al.* (1988).

[b]Sequence variants of many of these viroids, derived from field isolates, have been described (e.g., Sano *et al.*, 1989). Sequence variants of a known viroid (or viroid strain) have been defined as having greater than 90% sequence homology with that of a published viroid sequence (Keese and Symons, 1985).

4. Structural Transitions in PSTVd

Melting curves for ds nucleic acids were discussed in Chapter 4, Section I,C,2. Detailed studies of the melting curves of PSTVd, together with the sequence data, allowed Riesner *et al.* (1979) to develop a model for the stages in the denaturation of PSTVd RNA as the temperature is raised. A more refined model has been proposed by Steger *et al.* (1984) (Fig. 9.2).

5. Structural Domains in Viroids

Viroids in the PSTVd group (Table 9.1) have been the most studied. From an examination of sequence similarity among more than 40 PSTVd-like viroid sequences, a model of secondary structure involving five structural domains has been

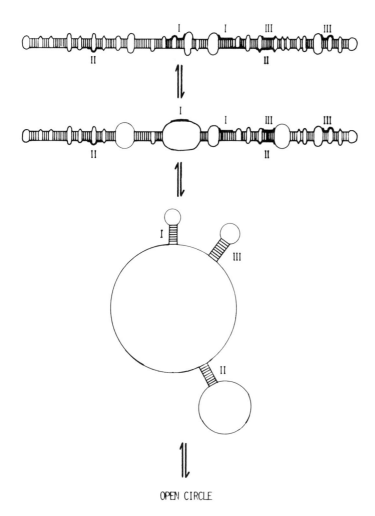

OPEN CIRCLE

Figure 9.2 Mechanism of denaturation and renaturation of PSTVd. Schematic structures are represented at temperatures of ca. 25, 70, 75, and above 95°C. The formation of two or three stable hairpins (I, II, III) is possible with PSTVd during the main transition. The depicted hairpins are formed by base-pairing from the following regions: (I) 79–87/110–102, (II) 227–236/328–319, and (III) 127–135/168–160. Analogous hairpins to hairpins I and II of PSTVd are found in CEVd and CSVd. In CCCVd only hairpin I is found. (From Steger *et al.*, 1984.)

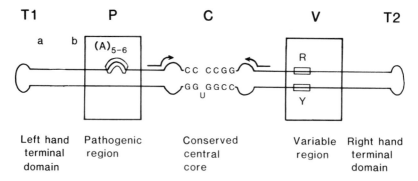

Figure 9.3 Model of viroid domains. Five viroid domains (T1, P, C, V, T2) were determined from sequence homologies between PSTVd-like viroids. The arrows depict an inverted repeat sequence. R, Y represent a short oligopurine, oligopyrimidine helix. (From Keese and Symons, 1985.)

developed for this group (Keese and Symons, 1985) (Fig. 9.3). This model also applies to CCCVd and CTVd.

The Conserved Central Domain

The central domain, of about 95 nucleotides, is the most highly conserved region among most viroids. It contains a 9-nucleotide inverted repeat that can form a stem-loop structure corresponding to loop I in Fig. 9.2. The repeats flank a strictly conserved bulged helix:

$$\begin{array}{cc} \text{UCC—CCGG} \\ |\,|\,|\ \ \ |\,|\,|\,| \\ \text{AGG\ \ GGCC} \\ \text{U} \end{array}$$

The stem loop and bulged helix cannot exist at the same time. It was therefore proposed that the two structures serve different functions (Keese and Symons, 1985).

The Pathogenic Domain

The pathogenic domain contains an adenine-dominated purine-rich sequence in one strand and an oligo(U) sequence in the opposite strand. It has been implicated in pathogenesis (Section I,D).

The Variable Domain

The variable domain is the most variable region in the molecule and may show less than 50% homology between otherwise closely related viroids.

The Terminal Domains

The main sequence homologies between PSTVd-like viroids are in the two terminal domains shown in Fig. 9.3. They have been implicated in viroid replication (Section I,B).

6. A Classification of Viroids Based on the Conserved Central Domains

Puchta *et al.* (1988) proposed a classification of viroids like that shown in Table 9.1 based on percentage of sequence similarity. A similarity of 65–100% was taken

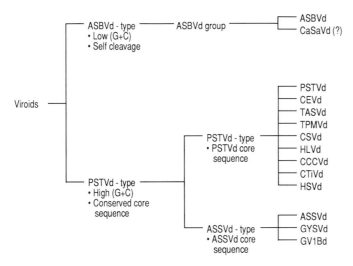

Figure 9.4 Proposed classification of viroids into three groups based on conserved core sequences and the ASBVd self-cleavage site. Name abbreviations are the same as in Table 9.1, with the addition of carnation stunt associated viroid (CaSaVd). (From Koltunow and Rezaian, 1989b.)

to indicate viroids belonging to the same group. This need to choose arbitrary percentage similarity values is an unsatisfactory basis for classification into groups. Koltunow and Rezaian (1989b) have proposed a classification of viroids into three distinct groups based on the distinct conserved core sequences of PSTVd and ASSVd and on the self-cleavage site of ASBVd (Fig. 9.4). This appears to be a sounder basis for placing viroids into groups. Sequence similarity remains important for distinguishing between distinct viroids and strains or sequence variants of a single viroid, as defined by Keese *et al.* (1988b). They proposed that sequence variants with more than 90% sequence similarity be regarded as members of a single viroid (species).

B. Replication of Viroids

Studies on viroid replication have been reviewed in detail by Sänger (1987) and Robertson and Branch (1987).

1. Translation of Viroid RNAs

Even if it was assumed that the three out-of-phase potential reading frames were fully utilized, viroids do not contain enough information to code for an RNA replicase. In principle, however, they could code for a relatively small polypeptide that could combine with host proteins to form a viroid-specific replicase, in the way that the small bacteriophage QB operates in its host. Nevertheless, it is now generally accepted that viroids are not translated to give any polypeptides. The main lines of evidence are (i) A few viroids contain one or more AUG codons, but many, such as PSTVd and its variants, contain none. (ii) Attempts to demonstrate mRNA function for viroid RNAs *in vitro* using various systems under a range of conditions have all failed. (iii) Analysis of the leaf proteins present in healthy and viroid-infected tissues have failed to reveal any new viroid-specified polypeptides.

2. Possible Pathways for Viroid RNA Replication

Given that viroids code for no polypeptides of their own, we must assume that they use preexisting host nucleic acid-synthesizing enzymes. Healthy plants contain two kinds of RNA-synthesizing enzymes: DNA-dependent RNA polymerases I, II, and III, and an RNA-dependent RNA polymerase. Early experiments on viroid replication, particularly the finding that actinomycin D inhibited viroid replication in leaf tissue, suggested that viroids might replicate via a DNA template. However, no viroid-specific DNA has been found in infected tissue (Zaitlin *et al.*, 1980; Branch and Dickson, 1980; Hadidi *et al.*, 1981) and it is now accepted that viroids replicate via a minus strand RNA template.

3. Kinds of Viroid RNA Found in Infected Tissue

RNA strands complementary to viroid RNA (defined for viroids as minus strands) were first described by Grill and Semancik (1978) in tissue infected with CEVd. Owens and Cress (1980) showed that the minus PSTVd strands present in RNase-treated dsPSTVd RNA had the same electrophoretic mobility as linear 359-nucleotide plus strands. Zelcer *et al.* (1982) found that PSTVd-infected tissues contained RNA complementary to the entire viroid RNA. Later studies showed that the minus strand PSTVd RNA existed as a tandem multimer of several unit-length monomers (e.g., Branch *et al.*, 1981). These were present as complexes with extensive ds regions. Further studies showed that monomeric PSTVd plus strands, both circular and linear, are complexed with long multimeric minus strands. Synthesis of the ds complex increases simultaneously with that of ss positive sense RNA (Owens and Diener, 1982). Time course studies with PSTVd-infected protoplasts showed that synthesis of oligomeric linear minus strand RNA preceded that of the plus strand oligomers, suggesting that the minus strand oligomers represent transient intermediates (Faustmann *et al.*, 1986).

4. Models for Viroid RNA Replication

The experiments briefly noted in the previous section and many others (e.g., Steger *et al.*, 1986) demonstrated that viroids replicate via an RNA template. The existence of closed circular monomers complexed to long linear multimers makes it highly probable that a rolling circle type of replication gives rise to progeny viroid RNA. Several models of the rolling circle mechanism have been put forward (e.g., Branch *et al.*, 1981; Owens and Diener, 1982; Ishikawa *et al.*, 1984; Branch and Robertson, 1984; Hutchins *et al.*, 1985). Figure 9.5 illustrates two rolling circle models.

In Fig. 9.5A the infecting circular plus strand acts as the template for synthesis of linear minus strand multimers. These then act as a template for linear plus strand multimers that are subsequently cleaved to give plus strand monomers, which are circularized in the final step. The scheme in Fig. 9.5B is a variation in which the multimeric minus strands are cleaved to monomers, which are circularized, and can then act as a rolling circle template to produce plus strand multimers. A distinguishable feature of the two models in Fig. 9.5 is that minus strand circular monomers are present in B but not in A. Branch *et al.* (1988a) could find no minus strand circular monomers among the RNAs from PSTVd, suggesting that model A is more probable for this viroid. While it seems virtually certain that viroids replicate via a RNA

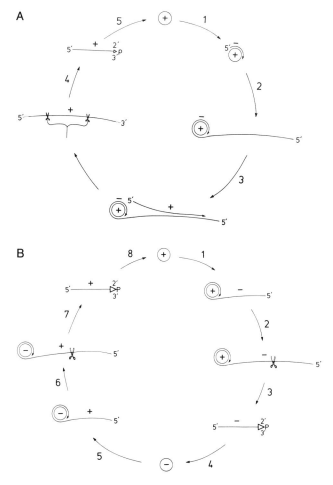

Figure 9.5 Asymmetric and symmetric models of rolling-circle replication. The replication models begin at the top of each diagram, with an infecting circular plus strand (+) which becomes a template for minus-strand (−) synthesis (step 1). In the asymmetric cycle (A), a multimeric linear minus strand is copied into a multimeric plus-strand precursor (step 3), and the plus-strand precursor is cleaved (step 4) and circularized (step 5). In the symmetric cycle (B), the multimeric minus strand is first cleaved to unit length (step 3) and circularized (step 4). The circular minus strand then serves as a rolling-circle template for the synthesis of a plus-strand precursor (step 5), and this precursor is cleaved (step 6) and ligated (step 8). (From Branch *et al.*, 1988b).

rolling circle mechanism, none of the models can be taken as proven in all their details. Indeed, different viroids might use somewhat different pathways.

5. Enzymes Involved in Viroid Replication

Enzyme activity is required for several steps in viroid replication—copying of plus and minus strands, which could involve different enzymes; cleavage of multimers at a precise site to give linear monomers; and ligation to give the closed circle progeny.

RNA Polymerases

The host polymerases involved in viroid RNA replication have not been identified beyond doubt. However, studies with isolated nuclei strongly suggest that the transcription of the infecting monomeric plus strand into multimeric minus strands is carried out by the DNA-dependent RNA polymerase II found in the nucleoplasm.

Mühlbach and Sänger (1979) studied the effect of α-amanitin on replication of cucumber pale fruit viroid in tomato protoplasts. They found that viroid synthesis was inhibited by concentrations around 10^{-8} M, which is characteristic for the inhibition of DNA-dependent RNA polymerase II by this drug. Rackwitz *et al.* (1981) purified the enzyme from healthy plant tissues and showed that it is capable of synthesizing full-length linear minus strand viroid copies from plus strand templates *in vitro*. Since other enzymes can copy viroid RNAs *in vitro* this work does not demonstrate a role for the polymerase II enzymes (Symons *et al.*, 1985). Polymerase II initiates synthesis of different viroid minus strands at specific sites (Sänger, 1987). Various workers have shown that viroids replicate and accumulate in the nucleus, a location that would fit with a role for DNA polymerase II (e.g., Takahashi and Diener, 1975). In addition, Rivera-Bustamente and Semanick (1989) prepared a subnuclear fraction from nuclei of CEVd-infected tissue that was able to synthesize CEV RNA.

Other work suggests a possible role for DNA-dependent RNA polymerase I, which normally carries out ribosomal RNA synthesis in the nucleolus. Riesner *et al.* (1983) showed that PSTVd RNA accumulates in the nucleolus while Palukaitis and Zaitlin (1987) described regions of sequence similarity between regions of PSTVd and an rDNA promoter sequence. However, DNA-dependent polymerase III transcribes viroid RNA *in vitro* into complete copies whereas polymerase I produced low yields of smaller products (Sänger, 1987).

As a third possibility, it was found that an RNA-dependent RNA polymerase from healthy tomato plants can use PSTVd RNA as a template to produce full-length copies in low yield (Boege *et al.*, 1982). Synthesis was not inhibited by relatively high concentrations of α-amanitin. The viroid copy is the first well-defined homogeneous product reported to be synthesized *in vitro* by an RNA-dependent RNA polymerase from healthy plants (Sänger, 1987).

Cleavage of Multimers to Give Monomers

Hutchins *et al.* (1986) prepared tandem dimeric cDNA clones of ASBVd, and RNA dimers were transcribed from these clones. They found that self-cleavage of both plus and minus RNA transcripts occurred at specific sites in each transcript to give plus and minus viroid monomers. They proposed a hammerheadlike secondary structure around the cleavage site (Fig. 9.6a) similar to those proposed for satellite RNAs (Fig. 9.10). However, the ASBV hammerhead structure proposed is unlikely to be sufficiently stable. A. C. Forster *et al.* (1988) therefore proposed a double hammerhead structure involving the association of two plus strand or two minus hammerheads of ASBVd (Fig. 9.6b). This model has two advantages: (i) the proposed cleavage site has sufficient stability and (ii) monomer-length RNAs could not cleave further. Only dimers or multimers could form the double hammerhead structure and thus be self-cleaved. As with the satellite RNAs discussed in Section II,B,7 self-cleavage requires a divalent ion and gives rise to 5′ hydroxyl and 2′,3′ cyclic phosphate 3′ termini. The sequences required to form the plus and minus strand self-cleavage structures in ASBVd are situated side by side in the middle of the molecule. About one-third of the total viroid sequence appears to be devoted to the self-cleavage function (Symons *et al.*, 1987).

Secondary structures such as those proposed for the processing of ASBVd and satellite RNAs have not been found in other viroids such as PSTVd and CEVd. Furthermore, Tsagris *et al.* (1987a) tested oligomeric linear PSTVd RNAs under a

A

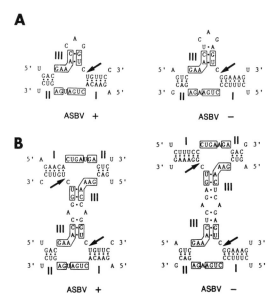

Figure 9.6 Hammerhead structures proposed for self-cleaving ASBVd RNA. (A) Single hammerhead structures proposed by Hutchins *et al.* (1986), which are similar to those proposed for self-cleaving satellite RNAs (Fig. 9.10). (B) Double hammerhead structure proposed for ASBVd RNA by A. C. Forster *et al.* (1988). Sites of cleavage are indicated by arrows. Nucleotides conserved between ASBVd, newt self-cleaving RNA and plant satellite RNAs are boxed. (From A. C. Forster *et al.*, 1988.)

wide range of conditions *in vitro,* but could find no evidence for autolytic processing as has been found for ASBVd. On the other hand, such oligomers were accurately processed to monomers when incubated with a nuclear extract from healthy potato cells (Tsagris *et al.*, 1987b). Various treatments of the extracts showed that excision of the linear monomers required the presence of catalytically active protein in the extract. All viroid multimer RNAs except those of ASBVd are processed by a host nuclear enzyme but this has not yet been characterized. The central conserved domain plays an important role in the generation of monomers from multimers for viroids like PSTVd, but other less efficient sites may be used (Hammond *et al.*, 1989). Hecker *et al.* (1988) discuss the roles of transient and stable secondary structures in viroid processing.

Ligation of Linear Monomers

The last step in viroid replication must be the ligation of linear monomers. Linear PSTVd monomers were efficiently converted to the circular form by an RNA ligase isolated from wheat germ (Branch *et al.*, 1982). The nuclear extract from potato cells shown by Tsagris *et al.* (1987b) to cleave PSTVd multimer RNAs into linear monomers also ligated these monomers to give plus strand circles. The ligation process is presumably assisted by the 5′ hydroxyl and 2′,3′ cyclic phosphate termini formed during cleavage of the multimers. Kikuchi *et al.* (1982) described a novel type of RNA ligase from wheat germ that circularized linear viroid RNA to form 2′ phosphomonester and 3′,5′ phosphodiester bonds.

Infectivity of Cloned Viroid DNA

Various workers have cloned full-length, greater than full-length, or dimeric viroid cDNA and have tested the infectivity of the DNA and of RNA transcripts made *in vitro* (Cress *et al.*, 1983; Ohno *et al.*, 1983a). In general, plasmid DNAs and their RNA transcripts were shown to be infectious. Tsagris *et al.* (1987b) provided direct evidence supporting the idea that infectivity of a viroid construct depends on its ability to be processed *in vivo*. In deletion experiments with TASVd cDNA, Candresse *et al.* (1990) obtained evidence confirming the importance of the conserved central domain (see Fig. 9.3) in viroid replication.

Evidence for Recombination between Viroids

The available nucleotide sequence data make it highly probable that recombination events have taken place in the past between different viroids, presumably during replication in mixed infections (Keese and Symons, 1985). For example, TASVd appears to be a recombinant viroid made up of most of the sequence of a CEVd-like viroid but with the T2 domain replaced by a T2 domain from a PSTVd-like viroid. Other examples are known (Keese and Symons, 1985; Hammond *et al.*, 1990). A newly studied viroid, Australian grapevine viroid, appears to have originated from extensive RNA recombination between four other viroids. Its sequence can be divided into regions each with high sequence similarity with sequences of CEVd, PSTVd, ASSVd, or GYSVd. The central conserved domain is that of ASSVd (Rezaian, 1990).

C. Biological Aspects of Viroids

1. Macroscopic Disease Symptoms

As a group, there is nothing that distinguishes the disease symptoms produced by viroids from those caused by viruses, which are described in Chapter 11. They include stunting, mottling, leaf distortion, and necrosis. From an agricultural crop point of view symptoms cover a wide range from the slowly developing lethal disease in coconut palms caused by the CCCVd viroid (e.g., Haseloff *et al.*, 1982) to the worldwide symptomless infection of HLVd latent viroid (Puchta *et al.*, 1988). It is probable that many more symptomless viroid infections remain to be discovered. A graft-transmissible dwarfing of citrus that is almost certainly due to viroid infection can, in principle, be used to advantage commercially because of lower costs of spraying, harvesting, and other factors (Schwinghamer and Broadbent, 1987).

2. Cytopathic Effects

Various effects of viroid infection on cellular structures have been reported. For example, in some infections, changes have been observed in membranous structures called "plasmalemmasomes." Several workers have described pronounced corrugations and irregular thickness in cell walls of viroid-infected tissue (e.g., Momma and Takahashi, 1983). A variety of degenerative abnormalities have been found in the chloroplasts of viroid-infected cells (e.g., da Graca and Martin, 1981).

3. Biochemical Changes

Viroid infection appears to cause no gross changes in host nucleic acid metabolism. By contrast, marked changes in the amounts of various host proteins have been described in infected tissue, for example, a 14K protein (Diener, 1987a).

Perhaps the most dramatic effect is the increase in a 140K host protein in tomatoes infected with several different viroids (Camacho-Henriquz and Sänger, 1984). Induction of this protein was not specific to viroid infection. Increases in other pathogenesis-related proteins have also been noted following infections with viroids and some of these proteins appear to bind to viroid molecules *in vivo* (Grannel *et al.*, 1987; Hadidi, 1988). Hiddinga *et al.* (1988) identified a host-encoded 68K protein that is differentially phosphorylated in extracts from viroid-infected tissue and also mock-inoculated tissue. The protein appeared to be a dsRNA-dependent protein kinase immunologically related to a similar protein in mammalian cells that has been implicated in the regulation of virus synthesis. Significant changes in the composition of cell walls have been found in viroid-infected tissues (Wang *et al.*, 1986).

4. Movement in the Plant

Viruses with defective coat proteins and naked RNAs move slowly through the plant by cell-to-cell movement. By contrast viroids move rapidly through a host plant in the manner of competent viruses, almost certainly through the phloem (Palukaitis, 1987). The relative resistance of viroid RNA to nuclease attack probably facilitates their long-distance movement. It is also possible that viroid particles undergo translocation while bound to some host protein.

5. Transmission

Viroids are readily transmitted by mechanical means in most of their hosts. Transmission in the field is probably mainly by contaminated tools and similar means. This ease of transmission in the presence of nucleases is probably due to viroid secondary structure and to the complexing of viroids to host components during the transmission process.

PSTVd is transmitted through the pollen and true seed of potato plants (Grasmick and Slack, 1986) and can survive in infected seed for long periods. However, this route of transmission is not likely to be of great commercial importance in potato. Several viroids have been shown to be pollen and seed transmitted in tomato (Kryczynski *et al.*, 1988).

PSTVd was reported to be transmitted at low frequency in a nonpersistent manner by the aphid *Macrosiphum euphorbiae* (De Bokx and Piron, 1981). However, it is doubtful if aphid transmission is of any significance in the field.

6. Interference between Viroids

Fernow (1967) showed that inoculation of tomato plants with a very mild strain of PSTVd gave substantial protection against a second inoculation with a severe strain applied 2 weeks after the first. It was later shown that cross-protection occurs not only between strains of a particular viroid but also between different viroids (Niblett *et al.*, 1978). In experiments in which a mild and a severe strain of PSTVd were inoculated simultaneously, the severe strain dominated and most plants developed severe disease, even when the mild strain was in 100-fold excess in the inoculum (Branch *et al.*, 1988b). In other experiments, PSTVd RNA transcripts from a cloned PSTVd DNA were inoculated together with HSVd RNA. PSTVd reduced the level of HSVd RNA in infected plants. Plants inoculated with dual transcripts—two copies of a severe PSTVd strain linked to two of HSVd—developed PSTVd symptoms and only PSTVd progeny RNA could be detected. The molecular basis for this interference is not understood (see also Chapter 13, Section IV,A).

7. Epidemiology

The main methods by which viroids are spread through crops are by vegetative propagation, mechanical contamination, and through pollen and seed. The relative importance of these methods varies with different viroids and hosts. For example, vegetative propagation is dominant for PSTVd in potatoes and CSVd in chrysanthemums. Mechanical transmission is a significant factor for others such as CEVd in citrus and HSVd in hops. Seed and pollen transmission are factors in the spread of ASBVd in avocados.

For most viroid diseases the reservoir of inoculum appears to be within the crop itself, which raises the question as to where the viroid diseases came from. The evidence suggests that many viroid diseases are of relatively recent origin (Diener, 1987a). As Diener pointed out, none of the recognized viroid diseases was known to exist before 1900. In fact many of them were first described since 1940. The sudden appearance and rapid spread of a new viroid disease can probably be accounted for by the following factors. Viroids are readily transmitted by mechanical means. Many modern crops are grown as large-scale monocultures. Thus from time to time a viroid present in a natural host and probably causing no disease might escape into a nearby susceptible commercial crop and spread rapidly within it. If the viroid and crop plant had not evolved together, disease would be a likely outcome. There is direct evidence for such a sequence of events with the tomato planta macho disease in Mexico (Diener, 1987a). A study of HLVd in the United Kingdom suggests that the current prevalence of this viroid in hops is a consequence of infection becoming established in the hop propagation system during the late 1970s (Barbara *et al.*, 1990).

D. Molecular Basis for Biological Activity

Because of their very small size, their autonomous replication, the known structure of many variants, and the lack of any viroid-specific polypeptides, it has been a hope of many workers that viroids might provide a simple model system that would provide insights as to how variations in the structure of a pathogen modulate disease expression. This hope has not yet been realized. As discussed in the following, very small changes in nucleotide sequence may give rise to dramatic changes in the kind of disease induced by a viroid. Therefore disease induction must involve specific recognition of the viroid sequence by some host macromolecule. Until the nature of this host macromolecule (or molecules) is known the interpretation of correlations between nucleotide sequence and biological properties of viroids will remain speculative. At present the only practicable biological properties that can be observed are infectivity and severity of disease. Four approaches are being used and these will be briefly summarized.

1. Sequence Comparisons of Naturally Occurring Viroid Variants

The sequence comparison approach takes advantage of the fact that many sequence variants of a particular viroid may exist in nature, which provide a range of viable mutants. A major difficulty is that many natural infections contain a mixture of viroid variants. This problem has been overcome by the preparation of cDNA clones of isolates each with a single defined nucleotide sequence. The work of Visvader and Symons (1985) is an example of this approach. By cDNA cloning they isolated 11 new variants of CEVd. These variants, together with 6 already known, fell into two classes, severe and mild, with respect to symptom severity in tomatoes. Sequences differences lay in the pathogenicity and variable domains indicated in Fig. 9.3. Later work using cDNA chimeric clones located pathogenicity in the P_L domain (Visvader and Symons, 1986).

For PSTVd only the P domain has been shown to be involved in pathogenicity (Schnölzer et al., 1985). They placed PSTVd variants in four classes with respect to disease severity in tomato: mild, intermediate, severe, and lethal. Increasing severity of disease was correlated with decreasing stability in vitro of the secondary structure in the virulence-modulating domain. Such a correlation was not observed with the CEVd variants described earlier (Visvader and Symons, 1985).

2. Site-Directed Mutagenesis

In principle the ability to prepare infectious cDNA clones of viroids should allow the application of site-directed mutagenesis to the study of the effects of structural changes on disease induction. In practice it has been found that many base changes in a viroid sequence render it noninfectious, even those that might be expected to have a minimal effect on secondary structure. When a PSTVd cDNA containing a single base change was introduced into the host plant in an *Agrobacterium* vector, wild-type viroid was recovered from the inoculated plant (Owens et al., 1986; Hammond and Owens, 1987). This reversion may well be a consequence of the high error frequency characteristic of RNA copying (Chapter 13, Section III,A,1,a). Because of this reversion, the progeny arising from inoculations with mutant viroid cDNAs must be carefully examined to determine whether or not they have maintained the alteration(s) present in the cDNA inoculum.

3. Chimeric Viroids

Construction of chimeric viroid cDNAs provides another possible approach to the study of disease induction, but this has so far met with little success. For example, Owens et al. (1986) constructed mixed tandem dimers of full-length PSTVd and TASVd cDNAs. These were infectious but analysis of the progeny showed that individual plants contained only normal PSTVd or TASVd. None of the expected chimeras was found. In another approach Owens et al. (1986) constructed chimeric cDNA monomers between parts of the PSTVd and TASVd cDNA molecules, but these were not infectious. Hammond et al. (1989) constructed a chimera between PSTVd and TPMVd, which produced infectious progeny. Owens et al. (1990) constructed infectious chimeras between tomato apical stunt viroid (TASVd) and CEVd. Symptoms in tomato were milder than those produced by TASVd, and resembled those produced by CEVd, which was the source of the pathogenicity domain in the construct.

4. Sequence Comparisons with Host RNAs

Another possible way to attempt to identify viroid nucleotide sequences that function in pathogenesis is to search for possible host RNA sequences to which viroid sequences might bind. For example, during ribosomal RNA processing there is an intronlike sequence between 5.8 S RNA and 25 S RNA that is excised. U3 small nuclear RNA is thought to bind to the intronlike sequence in the reaction leading to excision. Jakab *et al.* (1986) found sequence similarities between two domains of PSTVd and broad bean U3 snRNA. On this basis they suggested that viroids (except ASBVd) exert their pathogenic effects by interfering with normal pre-rRNA processing in nucleoli. Haas *et al.* (1988) showed that the 7 S RNA from tomato leaf tissue resembles a sequence recognition particle RNA. It also has a notable sequence complementarity to the part of PSTVd RNA known to modulate virulence. Thus this 7 S RNA is another possible host target for viroid RNA.

5. Summary

All of these approaches are a long way from explaining how pathogens as small as viroids cause such a variety of diseases ranging from symptomless to lethal. Further significant progress in this important area will have several prerequisites, including the following: (i) specific *in vitro* and *in vivo* assays for different viroid functions; (ii) a knowledge of viroid structure (or structures) *in vivo;* and (iii) identification of the host macromolecules with which viroids interact during the processes of infection, replication, and movement in the plant.

An approach to this latter question has been made by Klaff *et al.* (1989). They recognized a 43K host protein that bound to PSTVd when the viroid was mixed *in vitro* with nuclear extracts. A similar complex was found *in vivo* in the nuclei of infected cells.

E. Diagnostic Procedures for Viroids

Since viroids produce no specific proteins, the immunological methods applied so successfully to viruses cannot be used for the diagnosis of diseases caused by viroids. Similarly, because no characteristic particles can be detected, electron microscopic techniques are inappropriate. For these reasons diagnostic procedures have been confined to biological tests, gel electrophoresis, and more recently nucleic acid hybridization tests.

1. Biological Tests

Biological tests for viroid detection and diagnosis have been important where suitable diagnostic test plants have been identified, and they remain important for some viroids, for example, where strains of different severity exist. Some indicator hosts have been found that give severe symptoms with strains causing both mild and severe disease in the crop of interest (e.g., Singh, 1984). However, no suitable indicator hosts have been found for some viroid diseases such as cadang-cadang disease. Mild isolates of viroids may produce barely detectable symptoms. Environmental conditions may affect markedly the disease produced by other isolates. For such reasons *in vitro* tests based on the properties of the viroid RNA have assumed considerable importance.

2. Gel Electrophoresis

Viroids generally occur in very low concentration in the infected host. Thus some partial purification and concentration procedure must be used before the nucleic acids are run in an appropriate PAGE system. Various modifications have dramatically increased the sensitivity of viroid detection by electrophoresis. For example, Schumacher et al. (1986) devised a "return" gel technique in which the samples were first subjected to electrophoresis under nondenaturing conditions. Denaturing conditions to produce open viroid circles are then applied, and the polarity of the electric field is reversed. Viroid molecules then lag behind, well clear of all host RNA species. Such a procedure has been used to detect PSTVd in single true seeds of potato (R. P. Singh et al., 1988).

3. Nucleic Acid Hybridization

Methods for detection and diagnosis based on nucleic acid hybridization are becoming increasingly important. For application in routine testing it was necessary to develop methods for the large-scale production of cloned, highly labeled viroid cDNA and to develop procedures where clarified tissue extracts could be used in the tests (e.g., Owens and Diener, 1981). Methods for preparing labeled nucleic acid probes were discussed in Chapter 2 (Section IV,C). RNA probes are now used for routine detection of PSTVd in tuber flesh and sprouts (Salazar et al., 1988a). Sano et al. (1988b) used synthetic oligonucleotide probes to diagnose HSVd strains and CEVd. The dot blot hybridization procedure is now being widely used (e.g., Candresse et al., 1988, for CSVd). This procedure has been coupled with the use of photobiotin-labeled DNA probes to provide a nonradioactive and sensitive procedure for the routine diagnosis of viroids in plant extracts (McInnes et al., 1989; Roy et al., 1989).

4. General

With present procedures, it is relatively simple to verify a positive test result for viroid infection. It is much more difficult to ensure that a negative result means a viroid-free plant. This may be an important practical issue where vegetative crops are concerned. Methods for the control of viroid diseases are discussed in Chapter 16, Section II.

II. SATELLITE VIRUSES AND SATELLITE RNAs

Purified virus preparations isolated from infected plants may contain a variety of RNAs other than the genomic RNAs. Some of these, such as subgenomic RNAs and defective RNAs, have already been discussed in Chapters 6–8. In addition, some isolates of certain plant viruses contain satellite agents. Two classes of these agents can be distinguished according to the source of the coat protein used to encapsidate the RNA. In *satellite viruses* the satellite RNA codes for its own coat protein. In *satellite RNAs* the RNA becomes packaged in protein shells made from coat protein of the helper virus. Satellite viruses and satellite RNAs have the following properties in common:

1. Their genetic material is an ssRNA molecule of small size. The RNA is not part of the helper virus genome and usually has no sequence similarity to it.
2. Replication of the RNA is dependent on a specific helper virus.
3. The agent affects disease symptoms, at least in some hosts.

4. Replication of the satellite interferes to some degree with replication of the helper.
5. Satellites are replicated in the cytoplasm on their own RNA template.

In recent years several satellite RNAs associated with a particular group of viruses have been shown to have viroidlike structural properties. These agents have been termed virusoids but this term is no longer favored. Satellite viruses and satellite RNAs have been reviewed by Murant and Mayo (1982) and Francki (1985). This section summarizes the properties of satellite viruses and satellite RNAs. Their methods of replication, and the ways in which they induce or modulate expression of disease, are also discussed.

A. Satellite Plant Viruses

Two definite satellite plant viruses have been described, together with two probable examples. They are the smallest known viruses.

1. Satellite Tobacco Necrosis Virus (STNV)

STNV is the most studied of the satellite viruses. The helper virus, TNV, is a typical small icosahedral virus with a diameter of about 30 nm. It replicates independently of other viruses and normally infects plant roots in the field. It had been known for some time that certain cultures of TNV contained substantial amounts of a smaller viruslike particle (see Fig. 5.23). It was shown by Kassanis and his colleagues (e.g., Kassanis, 1962) that the smaller particle with a diameter of about 18 nm, now know as STNV, depended for its replication on the larger virus. There is significant specificity in the relationship between satellite and helper. Strains of both viruses have been isolated, and only certain strains of the helper virus will activate particular strains of the satellite (Uyemoto *et al.*, 1968). The ability of particular TNV isolates to activate the satellites was correlated with the ability of TNV to infect bean or tobacco plants but not with their serological relatedness (Kassanis and Phillips, 1970).

Both STNV and TNV are transmitted by the zoospores of the fungus *Olpidium brassicae* (Chapter 10, Section II,B). Transmission depends on an appropriate combination of four factors: satellite and helper virus strains, race of fungus, and species of host plant (e.g., Kassanis and Macfarlane, 1968).

The complete nucleotide sequence of STNV RNA was one of the first viral sequences to be determined (Fig. 9.7A). The arrangement of the RNA within the virus is shown in Fig. 5.34. The amino acid sequence of the coat protein was deduced from the nucleotide sequence (Fig. 9.7B) and later confirmed by direct sequencing of the STNV coat protein. These results demonstrated that STNV codes for only one gene product—its coat protein. The topology of the coat protein molecule is illustrated in Fig. 5.20C.

STNV RNA has no significant sequence similarity with the TNV genome. The function of the long 3′ noncoding region in STNV RNA is unknown. However, the RNA has a high degree of secondary structure and it has been suggested that this region may be involved in the stability of the molecule (Ysebaert *et al.*, 1980). STNV RNA is remarkably stable *in vivo*, having been shown to survive in inoculated leaves for at least 10 days in the absence of helper virus (Mossop and Francki, 1979a). This stability may have evolved to allow the satellite to survive a period

A

AGUAAAGACAGGAAACUUUACUGACUAACAUGGCAAAACAACAGAACAACAGGCGAAAAUCCGCAACAAUGCGUGCAGUGAAGCGCAUGAUAAAUACACA 100

CUUGGGAGCAUAAAAGGUUUGCACUGAUCAACUCAGGGAACACCAAUGCAACUGCUGGUACAGUACAAAAUCUGUCCAACGGUAUAAUCCAAGGAGAUGAU 200

AUCAACCAGAGAAGUGGUGAUCAAGUGCGUAUAGUUUCACAUAAACUUCACGUACGAGGCACUGCCAUCACCGUCAGCCAGACCUUUAGAUUUAUCUGGU 300

UUCGUGAUAACAUGAACCGUGGGACCACUCCCACAGUUCUUGGAGGUGUUGAACACUGCGAAUUUCAUGUCGCAGUAUAACCCAAUCACGUUGCAGCAAAA 400

GAGAUUUACAAUACUCAAGGAUGUAACUCUCAAUUGUUCGCUGACAGGGGAGAGCAUUAAAGAUCGGAUAAUUAACCUUCCAGGACAACUGGUGAACUAU 500

AAUGGAGCGACGGCUGUAGCAGCCUCCAAUGGUCCCGGCGCAAUAUUUAUGUUGCAGAAUUGGCGACUCCUUUGGUUGGUCUGUGGGGACUCCUCUUAUGAGG 600

CUGUGUACACAGAUGCAUAAUCCCAGAGGUUCACAAUGUUGAGUAGGUGGCGCUGAAAGAUGCGUAGCUUCUUCUCUUCUGGAGCCCACUUCCUGGUGGUGGUAAGC 700

AGAAAUCCAAGGGUACGGUGGUACGGUGGAAAAGCAGUCCCAGCUCUGCAUUGGGGAACCGGCUUUUACACCCAGCUUAGGGCUAAAGUGUACUACUACUCUCAU 800

UUGUAGUCUAAAAUGAGACGUUGGCCUCGACGUGUCGAGGUGGCCUAAGGGAUUUGGAACCCCUGAUGGUCGUAGUCGAAUUUCCCGUGUUUCAUUCCGAGU 900

CUCUUGGGUCAUAAUGCCAUUAGUAGGUCUAGCACUCAACGUAACUUCAAAGAUAAUCCUCCUUUGCAACAAGAAUAUGUGCGCCGUCGUGUGUUUAAAGCGGU 1000

AUAUAUUAAGUGCGCCGGCAUAUCGUUGUUUGGACCAGGGGCCCCACGCCGGUUUGGGUACCCGGGUGGGCUUCCCCUCGUUCACAGGGCUUUAGGGAGAUGAUAAAG 1100

GUAUAGUUAUUAUUAGACAAAUGCGGACAAACCUGAAAAGCUCGCUAGUGUGGUGGGCCUGGCCAAGCGAAGAACCUCAUCCAGGUAUAGUUCUACAUGGGAAAUU 1200

UGGUACCAUCCCAAACUUCUAUGAAGUCCUCUCGACUACCCCC

B

ALA-LYS-GLN-GLN-ASN-ASN-ARG-ARG-LYS-SER-ALA-THR-MET-ARG-ALA-VAL-LYS-ARG-MET-ILE- 20

ASN-THR-HIS-LEU-GLU-HIS-LYS-ARG-PHE-ALA-LEU- ILE-ASN-SER-GLY-ASN-THR-ASN-ALA-THR- 40

ALA-GLY-THR-VAL-GLN-ASN-LEU-SER-ASN-GLY- ILE- ILE-GLN-GLY-ASP-ASP- ILE-ASN-GLN-ARG- 60

SER-GLY-ASP-GLN-VAL-ARG- ILE-VAL-SER-HIS-LYS-LEU-HIS-VAL-ARG-GLY-THR-ALA- ILE-THR- 80

VAL-SER-GLN-THR-PHE-ARG-PHE- ILE-TRP-PHE-ARG-ASP-ASN-MET-ASN-ARG-GLY-THR-THR-PRO- 100

THR-VAL-LEU-GLU-VAL-LEU-ASN-THR-ALA-ASN-PHE-MET-SER-GLN-TYR-ASN-PRO- ILE-THR-LEU- 120

GLN-GLN-LYS-ARG-PHE-THR- ILE-LEU-LYS-ASP-VAL-THR-LEU-ASN-CYS-SER-LEU-THR-GLY-GLU- 140

SER- ILE-LYS-ASP-ARG- ILE- ILE-ASN-LEU-PRO-GLY-GLN-LEU-VAL-ASN-TYR-ASN-GLY-ALA-THR- 160

ALA-VAL-ALA-ALA-SER-ASN-GLY-PRO-GLY-ALA- ILE-PHE-MET-LEU-GLN- ILE-GLY-ASP-SER-LEU- 180

VAL-GLY-LEU-TRP-ASP-SER-SER-TYR-GLU-ALA-VAL-TYR-THR-ASP-ALA 195

Figure 9.7 (A) Nucleotide sequence of STNV RNA. The sequence, represented here as the viral RNA, combines information obtained from cloned copy DNA and the 5′-terminal sequence reported by Leung *et al.* (1976). The initiation and termination codons of the STNV coat protein gene are boxed. (From Ysebaert *et al.*, 1980.) (B) Amino acid sequence of STNV coat protein as deduced from the nucleotide sequence. (From Ysebaert *et al.*, 1980.)

within a cell after uncoating, but before the cell becomes infected with helper virus. Little is known about the replication of STNV *in vivo,* but it is widely assumed that STNV RNA replication must be carried out by an RNA-dependent RNA polymerase coded for, at least in part, by the helper virus.

Replication of STNV substantially suppresses TNV replication (I. M. Jones and Reichmann, 1973), and it is possible that this may involve competition for the replicase. The presence of STNV in the inoculum reduces the size of the local lesions produced by the helper virus. This could be due to the reduction in TNV replication.

STNV is readily translated *in vitro* in both prokaryotic and eukaryotic systems, the only product being the coat protein. Early experiments with prokaryotic systems were noted in Chapter 6, Section V,B. In the wheat germ system, where both TNV and STNV RNAs were translated in a mixture, STNV RNA was preferentially translated even in the presence of an excess of TNV RNA (Salvato and Fraenkel-Conrat, 1977). Using mRNAs transcribed from recombinant plasmids containing DNA copies of various lengths of the 5′ region of STNV RNA, sequences that affect the requirement for various initiation factors have been delineated in the wheat germ protein synthesis system (Browning *et al.,* 1988).

The 5′ terminus of STNV is unlike that of most ss positive sense RNA plant viruses. There is neither a 5′ cap nor a VPg. The 5′ termination is 5′-ppApGpUp– (Ysebaert *et al.,* 1980). The 3′-terminal region can be folded to give a tRNA-like structure with an AUG anticodon, but there is no evidence that this structure can accept methionine. The fact that STNV RNA, compared with other plant viral

RNAs, is efficiently translated *in vitro* in prokaryotic systems may be explained by the 5′ noncoding region containing a sequence –AGGA–, which is part of the Shine–Dalgarno prototype sequence complementary to a region at the 3′ end of the 16 S ribosomal RNA of *E. coli* (Ysebaert *et al.*, 1980). A cDNA copy of STNV RNA inserted into a plasmid gave rise to efficient translation of STNV coat protein in *E. coli*, in line with its efficient translation in a bacterial system *in vitro* (van Emmelo *et al.*, 1984).

Van Emmelo *et al.* (1987) described experiments in which the direct infection of plant leaves with plasmids containing a full-sized DNA copy of STNV RNA in the presence of helper virus led to the production of infectious STNV particles. Progeny virus particles were also obtained with cDNAs that carried oligonucleotide insertions at various sites in the genome. STNV has been considered as a possible gene vector in plants (Chapter 8, Section VI).

2. Satellite Panicum Mosaic Virus

A relationship very similar to that between TNV and STNV exists between panicum mosaic virus (PMV) and a smaller satellite virus (SPMV) (Buzen *et al.*, 1984). There is no nucleotide sequence or serological relationship between any of these four viruses. Like STNV, SPMV codes for a single protein that is the coat protein for the satellite virus particle (Masuta *et al.*, 1987).

3. Maize White-Line Mosaic Satellitelike Virus

Maize white-line mosaic virus is an ungrouped virus similar in size to TNV but with no evidence of relationship to it. Small satellite virus particles have been found in the roots of infected plants (Gingery and Louie, 1985). However, the satellite relationship has not been fully documented because the viruses do not appear to be transmissible by sap inoculation.

4. Satellite Virus of Tobacco Mild Green Mosaic Virus

Valverde and Dodds (1986) described a small RNA, unrelated in sequence to TMV, that they found in some natural isolates of TMV strain U5 (now known as tobacco mild green mosaic virus, TMGMV). At first they considered this to be a satellite RNA. However, when the TMGMV isolation procedure was altered, a 17-nm-diameter icosahedral virus particle was isolated that contained the satellite RNA (Valverde and Dodds, 1987). The satellite virus had no serological relationship with the helper virus or with other satellite viruses. These experiments are a good example of the need for care when attempting to delineate a newly isolated satellite RNA. The satellite RNA has been sequenced (Mirkov *et al.*, 1989). Comparison of this sequence with that of TMGMV showed that similarity ($\approx 60\%$) was limited to the 3′ noncoding region of TMGMV (Solis and García-Arenal, 1990). The 3′ end of the satellite RNA can fold to give a tRNA-like structure involving two pseudoknots as has been proposed for TMV and TMGMV. These similarities are assumed to reflect dependence of the satellite on a replicase coded for by the helper virus.

B. Satellite RNAs

White and Kaper (1989) describe a simple method for the detection satellite RNAs in small samples of plant tissue. Such RNAs have been found in association with members of four plant virus groups.

1. Satellite RNAs of Cucumoviruses

CMV

In 1972 a devastating outbreak of a lethal necrotic disease of field-grown tomatoes occurred in the Alsace region of France. It was realized that CMV was involved in this outbreak but it was not clear why necrosis occurred instead of the usual fern-leaf symptoms (e.g., Putz *et al.*, 1974). Kaper and his colleagues (e.g., Lot and Kaper, 1976) described the presence of a fifth small RNA component present in some isolates of CMV in addition to the three genomic RNAs and the subgenomic coat protein mRNA and that was not part of the viral genome. Kaper and Waterworth (1977) showed that the additional small RNA present in cultures of CMV strain S caused lethal necrotic disease in tomatoes when added to the CMV genomic RNAs. They called this satellite RNA CARNA5 and proposed that it had been responsible for the lethal necrotic disease in Alsace. Similar recent outbreaks in tomatoes in southern Italy have been shown to be due to a necrogenic isolate of CARNA5 (Kaper *et al.*, 1990).

Several different satellite RNAs have been described but CARNA5 is the best characterized (Fig. 9.8). It is found in all CMV density fractions following equilibrium density gradient centrifugation. Therefore, it is probably packaged in particles containing RNAs 1 and 2, and also, in varying numbers of copies per particle, in particles not containing any CMV RNA (Kaper *et al.*, 1976). This RNA is dependent on, but not significantly related in overall nucleotide sequence to, the CMV genome. Packaging in CMV protein enables the satellite RNA to be transmitted by aphids that transmit CMV (Mossop and Francki, 1977).

Many natural variants related in nucleotide sequence to CARNA5 have been described (e.g., Palukaitis and Zaitlin, 1984a; García-Luque *et al.*, 1984; Hidaka *et al.*, 1988). They vary in length from about 330 to 390 nucleotides. It is probable that most satellite RNA preparations consist of populations of molecules with closely related sequences.

Like the helper genomic CMV RNAs they are all capped with m^7Gppp at their 5' termini and have a hydroxyl group at the 3' end. Ten residues appear to be conserved at the 5' terminus and six at the 3' terminus, with other conserved regions in between. The satellite RNAs have a potential for substantial secondary structure, and models have been proposed (e.g., Gordon and Symons, 1983). A 3'-terminal tRNA-like structure is possible but attempts to aminoacylate the RNA have been unsuccessful. One to three ORFs are present (e.g., Fig. 9.8) and these can be translated *in vitro*. The satellite-encoded polypeptides do not appear to play a role in pathogenesis. The high degree of secondary structure may account for the stability of the RNA both *in vitro* and *in vivo* and for its relatively high specific infectivity (Mossop and Francki, 1978).

With many CMV isolates the amount of satellite RNA is small, and some workers could not eliminate the satellite from CMV cultures (e.g., Kaper *et al.*, 1976). This may result from the high stability and specific infectivity of the small

<pre>
                          met   glu   asn   cys   ala   glu   gly   leu   tyr
m⁷G_{ppp}GUUUUGUUUG AUG GAG AAU UGC GCA GAG GGG UUA UAU

     leu   arg   glu   asp   leu   ser   leu   gly   gly   val   gly   tyr
     CUG CGU GAG GAU CUG UCA CUC GGC GGU GUG GGA UAC

     leu   pro   ala   lys   ala   gly
     CUC CCU GCU AAG GCG GGU UGAGUGAUGUUCCCUCGGACUGG

                                              met   ser   ala   thr   leu   ser   thr
     GGACCGCUGGCUUGCGAGCU AUG UCC GCU ACU CUC AGU ACU

     thr   leu   ser   phe   glu   pro   pro   leu   ser   leu   leu   ala
     ACA CUC UCA UUU GAG CCC CCG CUC AGU UUG CUA GCA

     glu   pro   gly   thr   trp   phe   ala   asp   thr   met   asp   phe
     GAA CCC GGC ACA UGG UUC GCC GAU ACU AUG GAU UUU

     leu   lys   lys   his   ser   val   arg   trp   tyr   glu   ser
     CUA AAG AAA CAC UCU GUU AGG UGG UAU GAG UCA UGA

     CGCACGCAGGGAGAGGCUAAGGCUUAUGCUAUGCUGAUCUCCGUGAA

     UGUCUAUCAUUCCUCUGCAGGACCC_{OH}
</pre>

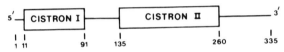

Figure 9.8 The sequence of 335 nucleotides in CARNA 5 (strain 7). It is not yet established beyond doubt that this RNA is an mRNA *in vivo*. The possible coding scheme shown alongside the base sequence is that proposed by Richards *et al.* (1978).

RNA. The effects of the presence of satellite RNA on disease symptoms are discussed in Section II,B,7.

Using RNA transcripts from a cDNA clone of a CMV satellite inducing yellowing symptoms (strain Y), Masuta *et al.* (1988b) showed that extra nonsatellite nucleotides at the 5′ terminus greatly reduced infectivity. This parallels the effect seen with transcripts from cloned viral RNAs.

The amounts of satellite RNA produced are highly variable depending on both the host species used and the strain of helper virus. RNAs 1 and 2 of CMV have highly conserved 5′ leader sequences that form part of a hairpin structure. This conserved 5′ sequence appears in the complementary sequence of CMV satellite RNA. Base-pairing of this sequence to the 5′ leaders of genomic RNAs 1 and 2 may be one mechanism whereby the satellite RNA influences both viral and satellite RNA production (Rezaian *et al.*, 1985). On the other hand, it has been shown that a CMV satellite RNA binds specifically to a 33-nucleotide region in the coat protein gene of CMV, probably forming an unusual knotlike structure (Rezaian and Symons, 1986). This finding provides another possible mechanism whereby the satellite RNA could regulate CMV replication.

Peanut Stunt Virus

Peanut stunt virus isolates may contain a satellite RNA known as PARNA5. It is more than 15% longer than any CMV satellite RNA (393 nucleotides), has the same 5′- and 3′-terminal structures as CMV satellites, and shows short stretches of sequence similarity near these termini (Collmer *et al.*, 1985). However, PARNA5 has several regions of high sequence similarity with some viroids. There are also striking similarities with host intron sequences essential for correct RNA processing.

2. Satellite RNAs of Nepoviruses

Several nepoviruses support the replication of satellite RNAs, which become encapsidated in particles made of the helper virus coat protein. Some of these satellite RNAs appear to be like the CMV satellites and are probably replicated by the helper virus replicase. Others appear to replicate by a quite different rolling circle mechanism.

Satellite RNA of Tobacco Ringspot Virus (STRSV)

STRSV consists of a small RNA species about 359 nucleotides long (Buzayan *et al.*, 1986d) in a protein shell identical to that of the helper virus. Twelve to twenty-five satellite RNA molecules become packaged in a single particle (Schneider *et al.*, 1972a). Different field isolates of the satellite produce different lesion types (Schneider *et al.*, 1972b). The satellite cannot replicate on its own, and it interferes with the replication of TRSV. There is an absence of ORFs, and the RNA is not translated *in vitro* (Owens and Schneider, 1977). Schneider and Thompson (1977) isolated a population of dsRNA molecules from leaves infected with STRSV that were absent from leaves infected with TRSV alone, or from healthy tissue. The MWs found in this multicomponent dsRNA preparation were much greater than that expected from dsSTRSV RNA. Electrophoretic analysis showed that the monomer RNA was the smallest and most abundant of at least 10 size classes with the STRSV sequence. A dsRNA fraction from infected tissue, when denatured, gave a similar series of up to 12 zones containing both positive and negative sense strands. (The polarity of the encapsidated ss satellite RNA is defined as positive sense.) The MW increment of each zone corresponded approximately to the size of the monomer satellite RNA (Kiefer *et al.*, 1982). No polyadenylate was detected at the 3′ terminus and no VPg at the 5′ terminus. (These are present in the helper virus.) The 3′ terminus was a cytosine 2′,3′-cyclic phosphate, and adenosine was at the 5′ terminus. Circular monomeric and dimeric forms of STRSV RNA have also been isolated (Linthorst and Kaper, 1984). STRSV RNA may have some base-paired viroidlike secondary structure (Buzayan *et al.*, 1987).

Satellite RNA of Arabis Mosaic Virus

A small satellitelike RNA has been found in some isolates of arabis mosaic virus from hops. The evidence suggests that, when present together with the helper virus, it may be the cause of the important nettlehead disease of hops (Davies and Clark, 1983).

Satellite RNA of Tomato Black Ring Virus (STBRV)

Some isolates of TBRV have associated with them a small RNA of MW ≃ 0.5 × 10^6 that has the properties of a satellite RNA (Murant *et al.*, 1973). The satellite RNA is packaged in varying numbers in particles made of the helper virus coat protein, giving rise to a series of components of differing buoyant density in solutions of CsCl. Like the helper virus the satellite RNA is transmitted through the seed, and also by the nematode vector of the virus (Murant and Mayo, 1982). Various strains of the satellite have been isolated in the field, and there is some specificity between strains of helper virus and strains of the satellite RNA. STBRV has a VPg at the 5' terminus and is polyadenylated at the 3' end. The satellite does not appear to affect replication of the helper virus or to modify symptoms except that the number of local lesions on *Chenopodium amaranticolor* is reduced (Murant *et al.*, 1973).

The first complete nucleotide sequence of an isolate of STBRV revealed a molecule 1375 nucleotides long. The sequence contained one ORF, with 5' and 3' noncoding sequences. The 3' region contained a sequence resembling a poly-adenylation signal (Meyer *et al.*, 1984). On the basis of sequence homology, five STBRV isolates fell into two groups with about 60% homology between groups. Within groups, homology was much stronger (Hemmer *et al.*, 1987). All isolates contained a single ORF with a coding capacity for a protein of 419–424 amino acids. Several regions of potential amino acid sequence were identical in all isolates.

Satellite RNAs with general properties quite similar to those of STBRV have been described for two other nepoviruses—strawberry latent ring spot virus (Mayo *et al.*, 1982c) and grapevine fanleaf virus (Pinck *et al.*, 1988; Fuchs *et al.*, 1989).

Satellite RNAs Associated with Chicory Yellow Mottle Virus (ChYMV)

An interesting situation with respect to satellite RNAs has recently been de-scribed for ChYMV, a nepovirus from southern Italy. Strain ChYMV-T contains two major RNA components with the properties of satellite RNAs. Their MWs ≃ 500K and 170K and hybridization experiments showed no sequence homology between the two RNAs, or between these RNAs and the helper virus genomic RNAs (Piazzolla *et al.*, 1989).

Both satellite RNAs have been sequenced (Rubino *et al.*, 1990). The 500K satellite has only linear molecules and is 1145 nucleotides long with a 3' poly(A) sequence. There is one ORF capable of coding for a polypeptide of MW ≃ 40K. There is a 5' leader sequence of 16 nucleotides and a 3' noncoding region of 40 nucleotides. The ORF is translated *in vitro* to give the expected product. The 170K satellite is 457 nucleotides long. There is no ORF of significant length and no mRNA activity *in vitro*. Both linear and circular forms are found in infected tissue. There are large regions of sequence similarity with other small *Nepovirus* satellites.

3. Satellite RNAs of Tombusviruses

Particles of most members of the *Tombusvirus* group contain a small satellite RNA. These RNAs have extensive sequence similarity with each other but not with the helper virus genomic RNA (Gallitelli and Hull, 1985). In most other systems, replication of satellite RNAs is strain-specific with respect to the helper virus. By contrast, replication of these *Tombusvirus* satellite RNAs can be supported by heterologous viruses that are not normally associated with the satellite. Presence of the satellite RNA may alter disease expression (e.g., Hillman *et al.*, 1985).

4. Satellite RNAs of Carmoviruses

Turnip crinkle virus (TCV) supports a family of satellite RNAs (Altenbach and Howell, 1981). Two of the satellites (D and F) do not affect symptoms, while C is virulent, intensifying TCV symptoms in turnip. Simon and Howell (1987) synthesized RNA copies of the virulent satellite (RNA C) *in vitro* using full-length cDNA copies in an expression vector. The RNA was infectious only when inoculated together with helper TCV. cDNA or negative sense RNA copies of the satellite were not infectious. Using a cloned cDNA probe of the satellite RNA, Altenbach and Howell (1984) detected a series of at least six multimeric forms of the RNA. Most of this RNA was of the same polarity as the encapsidated monomer, suggesting a role in RNA replication for the multimers. They also found some sequence relatedness between the satellite and host DNA, but the significance of this remains to be determined.

The virulent satellite C of TCV is of particular interest since, unlike other known satellites, it has a substantial region of homology with the helper TCV genome (Simon and Howell, 1986; Simon *et al.*, 1988). The 189 bases at the 5' end of RNA C are homologous to the entire sequence of a smaller nonvirulent TCV satellite RNA. The rest of the RNA C molecule (166 bases in the 3' region) is nearly identical in sequence to two regions at the 3' end of the TCV helper genome. This interesting composite molecule is illustrated in Fig. 9.9 along with the functional domains in the molecule established by Simon *et al.* (1988) by means of insertion and deletion analysis. Thus in a structural sense at least, TCV satellite RNA C is a hybrid between a satellite RNA and a DI virus particle (Chapter 6, Section VIII,B). When plants were inoculated with a mixture of TCV, satellite RNA D, and satellite RNA C transcripts containing nonviable mutations in the 5' domain, recombinant satellite RNAs were recovered (Cascone *et al.*, 1990). Sequence analysis around 20 recombinant junctions supported a copy-choice model for this recombination. In this model, while in the process of replicating minus strands, the replicase can leave the template together with the nacent plus strand and can reinitiate synthesis at one of two recognition sequences on the same or a different template.

5. Viroidlike Satellite RNAs Associated with Four Sobemoviruses

Four viruses described from Australia, which belong in the *Sobemovirus* group, have been shown to contain small RNAs that occur in both circular and linear forms. Their properties are somewhat like those of viroids (reviewed by Francki, 1985; Symons, 1990). However, their thermodynamic properties and their ability to self-cleave are quite different from those ot viroids, except possibly ASBVd. They do not show the cooperativity during melting that is found in viroids (Fig. 9.2), but behave like random base sequences (Steger *et al.*, 1984). Two of the viruses, velvet tobacco mottle virus and *Solanum nodiflorum* mottle virus, infect Solanaceae while the other two, lucerne transient streak virus and subterranean clover mottle virus,

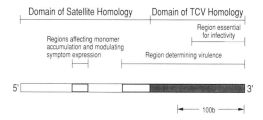

Figure 9.9 Domains in the virulent composite satellite RNA C of TCV. (Modified from Simon *et al.*, 1988.)

infect legumes. The small RNAs have been shown to have the biological properties of satellite RNAs. The helper virus can replicate independently, while the small RNA cannot (Jones *et al.*, 1983; Jones and Mayo, 1984; Francki *et al.*, 1986a). The satellite RNAs, which range in size from about 325 to 390 nucleotides, are encapsidated in both linear and closed circular forms. They show no sequence similarity with the helper RNAs. All four satellite RNAs share a common sequence GAUUUU, which is also in the same position and loop structure for all four RNAs (Keese *et al.*, 1983). Canadian and Australian isolates of the lucerne transient streak satellite have 80% sequence homology (AbouHaidar and Paliwal, 1988).

While in biological properties these RNAs are undoubtedly satellites, in some properties, particularly their small size, circularity, high degree of base-pairing, and lack of mRNA activity, they are like viroids (Francki, 1985). A sequence GAAAC is found in the central region of viroids and also in a similar position in the satellites associated with viruses infecting solanaceous plants (Keese and Symons, 1987).

6. A Satellite RNA of PEMV

In addition to the two genomic RNAs, a third small RNA ($M_r = 0.3 \times 10^6$) is found in some PEMV preparations and has been characterized as a satellite RNA (Demler and De Zoeten, 1989). The presence of the RNA does not affect symptom expression, aphid transmission, or particle morphology.

7. Replication of Satellite RNAs

In the previous sections the various satellite RNAs were classified according to the kind of helper virus involved. This may not be the most functional way to consider these small RNAs. It is becoming increasingly apparent that satellite RNAs differ in the ways in which they are replicated, and that most of them may fall into two groups on this basis. Piazzolla *et al.* (1989) suggested that one group may consist of satellite RNAs that have 5'- and 3'-terminal structures like those of the helper virus, and whose replication depends on the helper virus RNA replicase. The second group uses a rolling circle means of replication and depends on some other function of the helper virus. This grouping cuts across the classification based on the taxonomic position of the helper virus. Indeed, Piazzolla *et al.* (1989) and Rubino *et al.* (1990) have described one satellite of each type, both of which are dependent on ChYMV (Section II,B,2).

Satellite RNAs with Terminal Structures like Those of the Helper Virus

This group is typified by the satellite RNAs of CMV and TBRV. Their properties with respect to replication may be summarized as follows.

1. *Terminal structures* CMV satellite RNAs have a 5' cap structure and a 3' hydroxyl group as does CMV. STBRV has a 5' VPg and 3' polyadenylation as does TBRV. These similarities to the corresponding helper virus suggest a common method for replication of the satellite and helper RNAs.

2. *The presence of ORFs* Satellite RNAs in this group that have been sequenced contain one or two ORFs. Some satellites of CMV contain two ORFs (Fig. 9.8) while others contain one (Avila-Rincon *et al.*, 1986a). Five variants of STBRV have been sequenced (Hemmer *et al.*, 1987). All contained a single large ORF coding for a putative protein containing 419–424 amino acids with several regions

of identical amino acid sequence. These features strongly suggest a functional role for the ORF, but it is doubtful whether they play any role in symptom induction.

3. *Translation of the ORFs*   The isolate of CARNA5 used by Owens and Kaper (1977) was translated in the *in vitro* wheat germ system to give two polypeptides of MW $\simeq$ 5200 and 3800. These polypeptides have not yet been found *in vivo* so their significance is uncertain. This *in vitro* translation is inhibited by 7-methylguanosine-5′-monophosphate, which would be expected since the 5′ end of CARNA5 is capped. The possible coding scheme shown in Fig. 9.8 could fit reasonably with the size of the two polypeptides translated *in vitro* from CARNA5. On the other hand, two polypeptides were produced from the single ORF in strain S of CARNA5 by a mechanism that has not yet been established (Avila-Rincon *et al.*, 1986b). Different strains of CARNA5 are translated *in vitro* with differing efficiencies. Steen *et al.* (1990) showed that this variation can be related to the predicted secondary structures of their first 55 nucleotides.

Satellite RNAs of TBRV were translated in the wheat germ system to give a single polypeptide of MW $\simeq$ 48,000 requiring most of the satellite's coding potential (Fristch *et al.*, 1980). The protein was produced in both the absence and the presence of helper viral RNAs 1 and 2 (Fristch *et al.*, 1978). A protein of the same size was detected in extracts of protoplasts infected with TBRV preparations containing the satellite, but not with isolates that lacked it.

4. *RNA replication via a unit-length negative sense template*   P. Palukaitis (personal communication) suggested a replication mechanism involving autocatalytic cleavage and ligation to account for the presence *in vivo* of multimer forms of ds satellite RNA. However, the weight of present evidence does not favor such a model. The following lines of evidence strongly suggest that the CMV satellite RNAs replicate via a unit-length negative sense template:

   a. Kaper and Diaz-Ruiz (1977) isolated from infected tobacco four dsRNA species corresponding to the four CMV ssRNA species and in addition a ds species of MW-220,000 corresponding to an RF of CARNA5. Similar results were obtained in protoplasts by Takanami *et al.* (1977), where large amounts of unit-length dsRNA accumulated.

   b. The kinetics of labeling of the two strands in the dsRNA with ^{32}P fit with the production of an excess of positive sense strands early in infection, with more ds satellite RNA accumulating later (Piazzolla *et al.*, 1982).

   c. The ds forms of both CARNA5 and genomic CMV RNA3 contain an unpaired guanosine at the 3′ end of the minus strand. This is a feature of the RFs of some other viruses and suggests that the viral and satellite RNAs share a common replicative mechanism (Collmer and Kaper, 1985).

   d. Baulcombe *et al.* (1986) and Masuta *et al.* (1989) constructed plasmids containing a single cDNA copy of the RNA of a CMV satellite and introduced this into tobacco plants using *Agrobacterium tumefaciens*. Transformed plants could produce biologically active satellite RNA when inoculated with CMV. Following CMV inoculation there was no delay in symptom appearance in plants transformed with a monomer compared with those containing a dimer DNA copy. Thus the monomer form appeared to be as effective in replication as the dimer.

5. *The nature of dependence on the helper virus*   Linthorst and Kaper (1985) showed that CARNA5 did not replicate in protoplasts unless the helper virus was present. This result suggests that the satellite depends on the helper for some replication function, rather than just for encapsidation or movement through the

plant. Inoculation of tobacco mesophyll protoplasts with RNAs 1 and 2 of CMV induces the synthesis of viral RNA replicase activity. Satellite RNA was replicated when added to such protoplasts (Nitta *et al.*, 1988).

The preceding evidence strongly suggests that replication of satellite RNAs of this type depends on a replicase function of the helper virus. This raises the question of the function of the polypeptide potentially coded for by these satellites. Fritsch *et al.* (1978) suggested that this polypeptide might modulate the activity of the host replicase so that it preferentially replicated the satellite RNA.

Satellite RNAs with a Viroidlike Replication

This group is typified by STRSV, the satellite RNA of arabis mosaic virus, and by the viroidlike satellite RNAs. STRSV has no detectable mRNA activity (Owens and Schneider, 1977) nor do the nucleotide sequences indicate ORFs of significant length. There is no clear evidence as to why satellite RNAs of this group are dependent for their replication on a helper virus. However, there is quite strong evidence that they are replicated by a viroidlike rolling circle mechanism involving intermediates that are multimeric tandem repeats of the STRSV RNA (Kiefer *et al.*, 1982; Branch and Robertson, 1984).

The circular and multimer forms of STRSV RNA found in infected tissue have been described in Section II,B,2. Large amounts of the circular monomer form may be present in nucleic acids extracted from infected tissue (Linthorst and Kaper, 1984) but only linear molecules are packaged in virus particles.

In vitro, autolytic processing of dimeric and trimeric forms of STRSV takes place to give rise to biologically active monomers of 359 nucleotides (Prody *et al.*, 1986). The reaction is promoted by Mg^{2+} and some other divalent ions. The autolytic processing appears to be a phospho-transfer reaction between an adenylate residue and the 2' hydroxyl of a neighboring cytidylate, to give rise to a 5' hydroxyl terminal adenosine and a 3' cytosine 2'3' cyclic phosphodiester group. As noted earlier, these termini are characteristic of the monomeric RNA. Under appropriate conditions *in vitro*, the process is reversible in a nonenzymatic reaction. Ligation restores the original bonds. RNA self-cleavage by satellite RNAs has been reviewed by Sheldon *et al.* (1990).

Multimeric polyribonucleotides with STRSV minus RNA sequences are also able to process autolytically at an A–G bond 48 residues distant from the autolytic site in the plus strand (Buzayan *et al.*, 1986b). Nonenzymatic ligation *in vitro* of STRSV minus RNA also occurs, and at a far higher rate than ligation of the plus strand. The ligation reaction is the precise reversal of autolytic processing (Buzayan *et al.*, 1986c).

The cleavage reaction of the plus STRSV RNA was shown to be a property of a core sequence no longer than 64 satellite nucleotides that contains the bond that is cleaved (Buzayan *et al.*, 1986a). Comparisons have been made of the nucleotide sequences of related satellite RNAs with viroidlike replication, for example, several isolates of STRSV RNA (Buzayan *et al.*, 1987) and the satellite RNA of arabis mosaic virus with a STRSV sequence (Kaper *et al.*, 1989). These show that the sequences near the 5' and 3' termini of the linear forms that are involved in ligation, and in cleavage of circular forms, are highly conserved. In the more central region of the sequence various differences are apparent, including single-base deletions and insertions, base substitutions, and inverted blocks of sequences indicating past recombinational events.

The kinetics of labeling of the circular and linear forms of the satellite RNA of

VTMoV indicates that the linear form is a precursor of the circular form (Hanada and Francki, 1989). A large pool of unencapsidated satellite RNA appears to be present in infected cells.

8. The Development of Simple and Specific RNA Enzymes

The fact that the self-cleavage activity of circular satellite RNAs was confined to a relatively short and conserved part of the molecule led several groups to define more closely the structural requirements for specific phosphodiester bond cleavage. Thus Forster and Symons (1987a,b) showed that a 55-nucleotide sequence in the viroidlike satellite RNA associated with lucerne transient streak virus self-cleaves at a unique site. They proposed a hammerhead-shaped structural model for the sequence containing the cleavage site. The cleavage site in the plus and minus strands showed strong sequence similarities (Fig. 9.10). Sheldon and Symons (1989) have explored the sequence requirements for self-cleavage *in vitro* by this RNA, using insertion, deletion, and base substitution mutations. There was some flexibility in the sequence requirements, but alterations usually reduced the efficiency of self-cleavage.

Uhlenbeck (1987) and Haseloff and Gerlach (1989) have further defined the minimum sequence and structural requirements for RNA enzyme or "ribozyme" activity in the hammerhead structure. Haseloff and Gerlach took the single self-cleaving domain of the plus strand of STRSV and physically separated the sequences responsible for the enzyme and substrate activities. Three essential aspects were revealed, as illustrated in Fig. 9.11.

Haseloff and Gerlach (1989) proposed that the highly conserved sequence in region B may provide a metal ion binding site that could precisely position a metal ion in close proximity to the 2′ OH group next to the phosphodiester bond to be cleaved. The ability to synthesize highly specific tailor-made ribozymes opens up many possibilities, including a novel approach to the control of virus infection (Chapter 16, Section II,C,1). Dzianott and Bujarski (1989) have used the self-cleavage sequence of STRSV RNA to develop a cDNA system in which viral RNA transcripts can be automatically processed at the 3′ terminus to produce infectious viral RNAs. Symons (1989) has reviewed knowledge about the self-cleavage of small pathogenic RNAs.

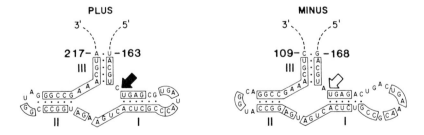

Figure 9.10 Location of the self-cleavage sites, indicated by arrows, in the hammerhead-shaped self-cleavage domains proposed for the viroidlike satellite RNA of lucerne transient streak virus. Boxed nucleotides are conserved between the plus and minus hammerhead structures. (From Forster and Symons, 1987a.)

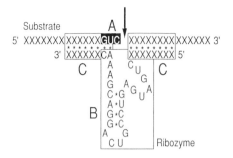

Figure 9.11 Essential sequences in the self-cleaving domain derived from STRSV RNA. A = GUC triplet in the substrate RNA immediately 5′ to the bond to be cleaved; B = catalytic region consisting of a highly conserved sequence with a base-paired stem and a loop; C = flanking regions without necessary sequence conservation, but base-paired to the substrate RNA and that allow precise positioning of the ribozyme relative to the cleavage site. (From Haseloff and Gerlach, 1988.)

9. The Molecular Basis for Symptom Modulation

As already noted in Section II,B,1, satellite RNAs may be responsible for outbreaks of severe disease in the field. In addition, because of their small size and the availability of many sequence variants, either isolated from nature or produced in the laboratory, they are amenable to detailed molecular study. For these reasons it has been hoped that studies correlating molecular structure of the satellite RNAs with their effects on disease in the plant would give us some insight into the molecular basis of disease induction. Results so far have been difficult to interpret. This is not surprising, since, as discussed next, multiple factors are involved. The subject has been reviewed by Kaper and Collmer (1988).

Variation in the Disease Symptoms Induced by Satellite RNAs

The biological effects of CARNA5 depend on the host. In tobacco, large amounts of satellite RNA are produced, and the yield of CMV is markedly depressed, and symptoms become milder (Kaper and Tousignant, 1977). Disease symptoms may also be suppressed in other hosts (Waterworth *et al.*, 1979). By contrast, the addition of CARNA5 to CMV in tomato leads to a lethal necrotic disease as noted in Section II,B,1,a. Figure 9.12 shows the contrasting effects of CARNA5 in tomato and pepper.

Another satellite RNA associated with CMV in the field causes the tomato white-leaf disease (Gonsalves *et al.*, 1982). A different satellite RNA of CMV caused disease attenuation when inoculated together with some strains of CMV (Mossop and Francki, 1979b). With other strains of CMV this satellite had no effect on symptoms. Yet another CMV satellite RNA (Y strain) caused a bright yellowing of the leaves when inoculated with the helper CMV to some, but not all, of the *Nicotiana* species tested (Takanami, 1981). On the other hand, in tomato it induced lethal necrosis in combination with one strain of CMV but not with another (Masuta *et al.*, 1988a).

White and Kaper (1987) made a systematic study of the effects of CMV strain D and CARNA5 on 52 *Lycopersicon* spp. accessions. This combination, which causes lethal necrosis in *L. esculentum* cv. Rutgers, caused only mild mosaic symptoms in many species. There was wide variation in different accessions of the same species. Thus while most accessions of *L. hirsutum* gave no necrotic plants, one

Figure 9.12 Modulation of CMV symptoms by its satellite CARNA 5. Upper row of plants was infected by virus alone; lower row of plants by virus plus CARNA 5. In tabasco pepper plants on the left, CARNA 5 attenuates disease symptoms; in tomato plants on the right, CARNA 5 induces lethal necrosis. (From Kaper and Collmer, in "RNA Genetics" (Holland and Ahlquist, eds.), Vol. III. Copyright © 1988, CRC Press, Inc., Boca Raton, Florida.)

gave rise to 100% necrotic plants. Symptom modulation is not confined to satellite RNAs associated with CMV. For example, a satellite RNA of groundnut rosette virus is largely responsible for the symptoms of groundnut rosette disease (Murant *et al.*, 1988a). This is also true for VTMoV and its satellite. The groundnut rosette disease has two main forms, "chlorotic mottle" and "green rosette." Sometimes mild chlorosis or mottle symptoms are found. These variations depend on the satellite RNAs or mixture of satellite RNAs, that are present in the plant (Murant and Kumar, 1990).

In summary, these and other results demonstrate that, where satellite RNAs occur, the disease outcome depends on an interaction between the helper virus strain, the strain of satellite RNA, and the species and cultivar of host plant. Environmental conditions no doubt provide a fourth factor.

Interference between Satellite RNAs

Jacquemond (1982) showed that infection of young tomato plants with CMV together with a satellite RNA giving mild symptoms protected plants against a second inoculation with a satellite RNA that normally caused lethal necrosis. Similar results have been reported by Yoshida *et al.* (1985). The use of satellite RNAs in disease control is discussed in Chapter 16, Section II.

Nucleotide Sequences in Relation to Disease Modulation

Kaper *et al.* (1988) compared the nucleotide sequences of nine CARNA5 variants that were necrogenic in tomato with five that were not. The sequence of ORF1 (Fig. 9.8) was completely conserved in all nine necrogenic strains but not in any of the nonnecrogenic strains, suggesting a possible role for the putative product of ORF1 in causing necrosis. However, in a direct approach to this question, Collmer and Kaper (1988) used site-directed mutagenesis to eliminate the AUG initiation codon for ORF1. Such mutants retained the ability to cause necrosis, demonstrating that a protein product of ORF1 is not involved. Comparison of the nucleotide sequences of six CMV satellite RNAs suggested that only a few nucleotide changes may be necessary to change the host response and that different kinds of disease response (e.g., yellowing or necrosis) may be associated with different domains of the satellite sequence (Palukaitis, 1988).

Several workers have used cDNA clones to construct recombinant RNA genomes *in vitro* to investigate functional domains in satellite RNAs. Thus Kurath and Palukaitis (1989a) constructed six recombinant satellite RNA genomes from infectious cDNA clones of three CMV satellite RNAs. The results showed clearly that the domain for chlorosis in tomato is in the 5' 185 nucleotides of these RNAs, while the domain for necrosis is in the 3' 150 nucleotides. Devic *et al.* (1989), using a similar approach, showed that the determinant for symptom production by the CMV satellite RNA Y lay between nucleotides 1 and 219, while the domain for necrosis in tomato lay in the 3' region, beyond nucleotide 219. The experiments of Masuta and Takanami (1989) suggest that the formation of a secondary structure between nucleotides 100 and 200 in the Y satellite may be involved in the induction of chlorosis.

The sequence of the WL1-sat, which attenuates CMV symptoms on tomatoes, differs from all necrogenic satellite RNAs at three nucleotides positions in the conserved 3' region of the RNA. Sleat and Palukaitis (1990a) used site-directed mutagenesis to show that when all three nucleotide positions in the attenuating strain were mutated to the nucleotides found in necrogenic strains, lethal necrosis was induced in tomato. If only two of the three nucleotides were mutated to those present in a necrogenic satellite, no necrosis ensued.

A satellite RNA of particular interest in relation to disease modulation is the virulent satellite RNA of TCV illustrated in Fig. 9.9. Other, avirulent strains of the satellite (e.g., strain F) lack the 3' sequence derived from the helper TCV genome. To demonstrate that this domain determines virulence, Simon *et al.* (1988) constructed a chimeric satellite composed mostly of the 5' sequence of strains F and the 3' domain of the virulent satellite C. Other constructs contained small insertions or deletions at various sites. Test with these RNAs led to the conclusion that the 3' sequence derived from TCV contains a region essential for infectivity and a larger overlapping region determining virulence (see Fig. 9.9). The domain of satellite homology contains regions affecting monomer accumulation and modulating symptom expression.

The only general conclusion to be drawn from studies of this kind is that changes in a disease induced by the presence of a satellite RNA depend on changes in nucleotide sequence in the RNA. So far there is no convincing evidence that such changes are mediated by a polypeptide translated from the satellite RNA. Indeed, differences in disease modulation occur between RNAs that have no significant ORFs. Disease modulation is most probably brought about by specific macromolecular interactions between the satellite RNA and (i) helper virus RNAs, (ii) host RNAs, (iii) viral-coded proteins, (iv) host proteins, or (v) any combination of (i)–(iv). Symptom modulation of CMV in tomato by CARNA5 involves the function in the helper virus that supports satellite replication as well as some function of the satellite itself (J. M. Kaper, personal communication). A similar conclusion can be drawn concerning CMV satellite RNAs causing systemic chlorosis in tobacco (Sleat and Palukaitis, 1990b). Nucleic acid hybridization studies have enabled CMV strains to be placed in two groups. Induction of chlorosis by satellite RNAs occurred only with subgroup II strains of the virus, and also appeared to be associated with CMV RNA 2.

Until the kinds of interaction that are important in disease induction have been established on a molecular basis, differences in nucleotide sequence between related satellites will remain largely uninterpretable.

10

Transmission, Movement, and Host Range

One could envisage a virus surviving for hundreds of years in an individual tree of a long-lived species. However, being obligate parasites, viruses will usually depend for survival on being able to spread from one susceptible individual to another fairly frequently. A knowledge of the ways in which viruses are transmitted from plant to plant is important for several reasons:

1. From the experimental point of view we can recognize a particular disease as being caused by a virus only if we can transmit the virus to healthy individuals by some means and reproduce the disease.
2. Viruses are important economically only if they can spread from plant to plant fairly rapidly in relation to the normal commercial lifetime of the crop.
3. A knowledge of the ways in which a virus maintains itself and spreads in the field is usually essential for the development of satisfactory control measures.
4. The relationships between viruses and their invertebrate and fungal vectors are of considerable general biological interest.
5. Certain methods, particularly mechanical transmission, are very important for the effective laboratory study of viruses.

Viruses cannot penetrate the intact plant cuticle. This problem is overcome either by avoiding the need to penetrate the intact outer surface (e.g., in seed transmission or by vegetative propagation) or by some method involving penetration through a wound in the surface layers, as in mechanical inoculation and transmission by insects. Our knowledge about virus transmission is far from complete. For many viruses where only one or two methods of transmission are known at present, others will almost certainly be found.

Section IV of this chapter deals with movement of viruses through the host plant, and their final distribution in various organs and tissues. These topics are related both to the replication of viruses and to the methods by which they are transmitted from plant to plant. In recent years it has become apparent that the ability of a virus to infect a particular plant species successfully may be closely related to its ability to move through the plant following inoculation. For this reason the last section deals with factors affecting the host range of viruses.

I. DIRECT PASSAGE IN LIVING HIGHER PLANT MATERIAL

A. Through the Seed

About one-fifth of the known plant viruses are transmitted through the seed of infected host plants. Seed transmission provides a very effective means of introducing virus into a crop at an early stage, giving randomized foci of infection throughout the planting. Thus, when some other method of transmission can operate to spread the virus within the growing crop, seed transmission may be of very considerable economic importance. Viruses may persist in seed for long periods so that commercial distribution of a seed-borne virus over long distances may occur.

Table 10.1 lists the approximate frequency with which seed transmission has been found among viruses of various groups. Seed transmission occurs for some viruses not yet assigned to groups and also for viroids (e.g., Kryczynski *et al.*, 1988).

Two general types of seed transmission can be distinguished. With TMV in tomato, seed transmission is largely due to contamination of the seedling by mechanical means. This type of transmission may occur with other tobamoviruses. The external virus can be readily inactivated by certain treatments eliminating all, or almost all, seed-borne infection. A small but variable proportion of the seed may be infected in the endosperm where virus may persist for many years. No TMV has been detected in the embryo (Broadbent, 1965).

In the second and more common type of seed transmission the virus is found within the tissues of the embryo. The embryo may become infected through the ovary or via the pollen. Reports that a virus is not seed transmitted in a particular host are quite often based on inadequate testing. Rigorous testing may be a substantial task.

Some infections cause disease symptoms in the seed (Chapter 11) but there is not necessarily a correlation between seeds showing symptoms and those transmitting the virus. Small seeds have been found to transmit pea seed-borne mosaic virus at a much higher rate than larger seeds (Khetarpal *et al.*, 1988).

1. Factors Affecting the Proportion of Infected Seed

Virus and Virus Strain

The proportion of infected seed from infected plants varies quite widely with different viruses. It may be as much as 100%–for example, TRSV in soybean (Athow and Bancroft, 1959) or strawberry latent ring spot *Nepovirus* in celery (Walkey and Whittingham-Jones, 1970). By contrast, 1% transmission was found for Andean potato latent *Tymovirus* in potato (Jones and Fribourg, 1977), and lettuce plants infected with lettuce mosaic *Potyvirus* may produce about 3–15% of seed giving rise to infected plants (Couch, 1955). Seed transmission of AMV varied

Table 10.1

Relative Importance of Seed Transmission for Viruses That Are Members or Possible Members of Various Virus Groups and Families[a]

Virus group	Number of members		Type of potential injury[b]					
	In group	Seed-borne	A	B	C	D	E	F
Alfalfa mosaic	1	1	+	+	+			
Bromovirus	5	1	+	+	+			
Capillovirus	4	1						
Carlavirus	50	2						
Carmovirus	16	0						
Caulimovirus	14	0						
Closterovirus	12	1						
Comovirus	16	6	+	+	+			
Cryptovirus	26	26						
Cucumovirus	4	4	+	+	+			
Dianthovirus	3	0						
Fabavirus	3	0						
Furovirus	7	1						
Geminivirus	25	1						
Hordeivirus	4	2	+	+	+		+	
Ilarvirus	13	8				+		
Luteovirus	27	0						
Maize chlorotic dwarf	2	0						
Marafivirus	5	0						
Necrovirus	2	1						
Nepovirus	34	22	+	+	+			
Pea enation mosaic	1	1						
Plant reovirus	9	0						
Potexvirus	39	4						
Potyvirus	117	16	+	+	+	+	+	+
Rhabdovirus	77	1						
Tenuivirus	7	0						
Sobemovirus	10	2						
Tobamovirus	15	7	+	+	+			
Tobravirus	3	3	+	+	+			
Tomato spotted wilt	1	1						
Tombusvirus	13	1						
Tymovirus	19	3						
Viroids	15	5				+	+	

[a]Modified from Stace-Smith and Hamilton (1988) with data provided by R. I. Hamilton.

[b]Viroids are included for comparative purposes. A, survival of inoculum; B, dispersal of inoculum; C, primary inoculum source; D, contamination of germ plasm lines; E, contamination of virus-free planting material; and F, direct crop losses due to plants arising from infected seed. The information in this table needs qualifying in two ways: (i) the numbers and the types of damage will change as further information becomes available and (ii) many viruses that are seed transmitted are transmitted in this way only in certain hosts.

considerably for different strains (Frosheiser, 1974) as did peanut mottle *Potyvirus* in peanuts (Adams and Kuhn, 1977).

Reassortment or pseudorecombination experiments with the two RNAs of some nepoviruses showed that seed transmissibility in *Stellaria media* was markedly dependent on some virus function carried by RNA1. RNA2 had an additional but smaller influence (Hanada and Harrison, 1977).

Host Plant

Some viruses that are seed transmitted are transmitted in this way by a wide range of host species. For example, tomato black ring *Nepovirus* was seed transmitted by all nine species tested in six families (Lister, 1960). Other viruses may be seed transmitted in one host but not in another. Thus dodder latent mosaic virus was transmitted through 5% of seed from infected *Cuscuta campestris,* but not through seed from cantaloupe, buckwheat, or pokeweed.

Different varieties of the same host species often vary widely in the rate at which seed transmission of a particular virus occurs. For example, lettuce mosaic virus does not appear to be transmitted through the seed in the variety Cheshunt Early Giant (Couch, 1955). Grogan and Bardin (1950) found rates of transmission ranging from 1 to 8% in other varieties. Seed transmission rates reported for BSMV in different barley cultivars have varied from 0 to 75% (Carroll and Chapman, 1970). There is also variation between plants of a given variety.

Time at Which Plant Is Infected

Generally speaking, the earlier a plant is infected, the higher the percentage of seed that will transmit the virus (e.g., Owusu *et al.,* 1968). One exception to this trend appears to be BSMV in barley, where percentage of infected seed rose steadily as time of infection was delayed, reaching a maximum at 10 days before heading. After this time the percentage declined (Eslick and Afanasiev, 1955).

Crowley (1959) examined the effect of time of flowering in relation to time of inoculation on the proportion of bean seeds (*Phaseolus vulgaris* L.) infected with SBMV. In the bean plant, self-fertilization takes place as the flower opens. Timing of this event in relation to inoculation time was recorded. Embryos were removed from the seed before they were fully mature and were tested for infectivity. The virus was able to infect the embryo both by infecting the gametes and by infecting the embryo, but only during the early stages of its development (up to about 4 days after fertilization). Crowley found a similar timing for TRSV in soybean and BSMV in barley.

Location of Seed on Plant

There appears to be no consistent pattern in the way infected and noninfected seeds are distributed on the plant. For example, Athow and Laviolette (1962) found that position of seed in the pod and location of pod on the plant did not affect the proportion of soybean seeds that were infected with TRSV. In these experiments the plants that were used had been infected for a long time. For plants that set seed in succession over a period, infection occurring near or during the flowering period might give a distribution in which older seeds would have less infection than younger seeds.

Age of Seed

Some viruses appear to be lost quite rapidly from seed on storage, while others persist for years. Fulton (1964) described a loss of virus from the seed of *Prunus pennsylvanica* carrying cherry necrotic ring spot virus. For the first 4 years of storage at 2°C the percentage remained fairly constant at 60–70%. By the sixth year, less than 5% of the seed was infected, while a loss of viability in the seed was minor. There was little loss of AMV in infected alfalfa seeds after 5 years at minus 18°C or room temperature (Frosheiser, 1974).

High Temperature

Well-dried seed is much more resistant to high temperatures than most other plant parts. Some seed-borne viruses appear to tolerate about as high a temperature as the seed they infect, for example, a mosaic virus in muskmelons, which was probably squash mosaic *Comovirus* (Rader *et al.,* 1947). TRSV survived for 5 years in soybean seed at 16–32°C as effectively as at 1–2°C, although seed germination was greatly reduced at the higher temperature (Laviolette and Athow, 1971). The reason for the resistance of these viruses in the seed is quite obscure. It may be due to the general stabilizing effect on intact virus particles of low water and high protein content.

2. Transmission through Infected Pollen

Most seed-transmitted viruses are probably also transmitted through pollen from infected plants, though not all have been adequately tested. Conversely there appears to be no example of a pollen-transmitted virus that is not also seed transmitted. AMV is more efficiently transmitted through pollen than the ovaries (Frosheiser, 1974). In contrast, while there was 5% transmission of lettuce mosaic virus through the ovule in lettuce, there was less than 0.5% infected seed produced by pollen transmission (Ryder, 1964). Self-pollination of infected plants presumably can result in a higher percentage of infected seed than when only one of the gametes comes from an infected individual.

The extent to which infected pollen is a significant factor in the spread of viruses in the field has not been thoroughly assessed. It may well be more important economically with cross-pollinated woody perennials than with annual crops. With certain viruses, infected pollen may cause only the resulting seed to become infected when a healthy plant is pollinated. With others, however, the plant itself may become infected. For example, Gilmer (1965) records an experimental tree-to-tree transmission of sour cherry yellows virus by pollen. The pollen may be carried by humans, wind, or honey bees (e.g., Converse and Lister, 1969). Natural transmission in the field may be via infected pollen, and no other means, as was found for raspberry bushy dwarf ?*Ilarvirus* in raspberry by Murant *et al.* (1974). Francki and Miles (1985) showed that if leaves onto which pollen from plants infected with sowbane mosaic *Sobemovirus* had fallen were subjected to mechanical abrasion, then virus transmission occurred. However, this is unlikely to be an important factor in the field.

Transmission through the gametophyte is probably more efficient for most host–virus combinations than transmission through the pollen. For example, cross-pollination experiments showed that the principle route for seed transmission of TRSV in soybean was by infection of the megagametophytes. Infected pollen may be able to compete only poorly with normal pollen during fertilization.

BSMV (Carroll, 1974) and TRSV particles (Yang and Hamilton, 1974) have been found by electron microscopy within infected pollen grains. Particles of BSMV were seen within the sperm cells in both the cytoplasm and nuclei and were found to be widely distributed in cells and tissues associated with embryogenesis (Brlansky *et al.*, 1986). In another study, immunogold labeling showed AMV particles to be widely distributed in ovules, pollen, and anthers of alfalfa (Pesic *et al.*, 1988). For such viruses the way in which the egg cell is infected is presumably via the infected sperm, or perhaps the sperm nucleus alone. On the other hand, the exine of mature pollen from infected plants has been shown to carry several viruses, such as TMV, occurring in high concentration (Hamilton *et al.*, 1977). These observations point to a second mechanism for pollen transmission. The germ tubes growing from the infected pollen grain may pick up virus particles, or actually become infected by mechanical means, and thus carry active virus to the ovule. Members of the *Cryptovirus* group are unusual in that they are transmitted with high efficiency through pollen and seed, but not by mechanical transmission, grafting, or invertebrate vectors (Lisa *et al.*, 1986).

3. Distribution of Virus within the Seed

Some seed-transmitted viruses have been shown to be in the embryo, for example, AMV (Gallo and Ciampor, 1977). Others are assumed to be there. Bean common mosaic virus was detected in blossoms, young pods, and seeds of *Phaseolus*. It was present in cotyledons and embryos but not detected in seed coats (Ekpo and Saettler, 1974).

4. Theories Concerning Seed Transmission

It is easy enough to see why seed transmission should occur. The virus may infect the embryo through either the gametophyte, the pollen, or both. The problem is to understand why certain viruses, particularly stable ones, are not seed transmitted.

It is generally accepted that for true seed transmission to occur the virus must enter and survive in the embryo. If a virus was unable to enter the embryo, or if it entered but was later eliminated, seed transmission would not take place. Carroll (1972) detected a seed-transmissible strain of BSMV in pollen and in embryos and endosperm of barley seed. A nontransmitted strain was not detected at these sites. However, it would be technically very difficult to demonstrate beyond doubt that complete exclusion had occurred. A few viruses that are confined to vascular tissues may be unable to enter the ovule, which has no vascular connection with the parent.

Whether or not a particular virus is seed transmitted in a particular host and the efficiency of the transmission may depend on the rate of loss of virus from embryos that were initially all, or almost all, infected. Inhibitors of virus infection have been found in seed extracts. Gunnery and Datta (1987) isolated a low-MW RNA from barley embryos that specifically inhibited the initiation of protein synthesis. Such molecules might play a role in inhibiting virus synthesis and seed transmission in some species.

B. By Vegetative Propagation

Vegetative propagation is an important horticultural practice, but it is unfortunately a very effective method for perpetuating and spreading viruses. Economically important viruses spread systemically through most vegetative parts of the plant. A plant once systemically infected with a virus usually remains infected for its lifetime. Thus, any vegetative parts taken for propagation, such as tubers, bulbs, corms, runners, and cuttings, will normally be infected. There are many instances where every individual of a particular cultivar tested has been found infected with a particular virus, for example, some potato cultivars infected with PVX. However, when a healthy plant of a vegetatively reproducing species is infected, even at a fairly early stage of growth, the virus may not move throughout the plant in the first growing season.

C. By Grafting

Grafting is essentially a form of vegetative propagation in which part of one plant grows on the roots of another individual. Once organic union has been established the stock and scion become effectively a single plant. Where either the rootstock or the individual from which the scion is taken is infected systemically with a virus, the grafted plant as a whole will become infected if both partners in the graft are susceptible. Early descriptions of graft transmission are noted in Chapter 1. Since the early days of work with plant viruses, the demonstration that a disease was transmissible by grafting, together with the absence of a pathogen visible by light microscopy, has been taken as an indication that the disease is due to a virus. Many viruses once thought to be transmissible only by grafting are now known to be transmitted by other means as well.

A wide variety of grafting techniques are used in horticultural work, and most of these have been applied in experimental virus transmission (e.g., Fig. 10.1).

Other forms of grafting or pseudografting not normally used in horticulture have also been found effective for transmitting some viruses. For example, the insertion of pieces of infected leaf under the bark can lead to virus transmission (Garnsey and Whidden, 1970). This method may be successful between species where an organic union would not be expected to form and persist. The method may be useful for transmitting viruses of woody fruit trees to herbaceous hosts and at seasons when normal budwood is not available (e.g., Fulton, 1964; Stouffer, 1969). Cores of stem tissue removed with a 2-mm-diameter canula and inserted into slightly smaller holes in the stem of the test plant have also been used successfully (Dimock *et al.*, 1971).

Grafting transmission may lead to a different disease from that appearing after, say, mechanical inoculation. For example, *N. glutinosa* normally gives necrotic local lesions with no systemic movement of virus following mechanical inoculation with TMV. However, healthy plants grafted with tobacco plants infected systemically with TMV die of a systemic necrotic disease. This effect of grafting is probably due to the introduction of virus into the vascular elements of the hypersensitive host (Zaitlin, 1962).

Grafting may succeed in transmitting a virus where other methods fail. Nevertheless, it is not always an efficient process. This may sometimes be due to lack of complete systemic invasion in the plant supplying the supposedly diseased scions. Where tissue of the healthy plant material being grafted gives a necrotic local

Figure 10.1 Transmission by cleft grafting. Scion of potato carrying virus Y (arrowed) cleft grafted onto a potato plant, cultivar Dakota Red. Note death of apices in the lateral shoots of the stock plant. (Courtesy of M. J. Dahlberg.)

reaction to the virus, transmission of the disease may not be accomplished. For example, Chamberlain *et al.* (1951) found that when healthy buds from Burbank and Sultan plums were grafted onto plum rootstocks infected with the veinbanding type of plum mosaic, the buds reacted with necrosis and died. This is probably the reason why these varieties were very rarely found infected with the virus in the field. Under standard conditions different viruses are transmitted in different minimal times after grafting. Minimum times of bud contact to give 100% transmission varied from 74 to 152 hours for 12 *Prunus* viruses studied by Fridlund (1967b). Unaided graft formation is a rather uncommon occurrence, and this method of virus transmission is not of wide importance in nature.

D. By Dodder

Dodder (*Cuscuta* spp.) (Convolvulaceae) is a vine that is parasitic on higher plants and that lacks leaves or chlorophyll. There are many different species with different host ranges, some of which are extensive. Bennett (1940b) showed that dodder would transmit viruses from plant to plant. The parasite forms haustoria, which connect with the vascular tissues of the host. Viruses are probably transmitted via the plasmodesmata that transiently connect the parasite's hyphal tips with host-cell cytoplasm.

Transmission by dodder is in some respects similar to grafting. However, graft compatibility is limited to quite closely related plants—usually within a genus. Dodder, on the other hand, can be used to transmit a virus between distantly related plants (e.g., Desjardins *et al.*, 1969). The virus being transmitted experimentally may not multiply in the dodder, which then appears to act as a passive pipeline connecting two plants. Transmission of TMV was substantially increased by conditions (such as pruning the dodder and shading the healthy plant) that might be expected to lead to a flow of food materials through the dodder from the diseased to the healthy plant (Cochran, 1946). Bennett (1940b) was able to separate CMV from TMV because it persisted in the dodder when the parasite was grown on hosts immune to both viruses, whereas the TMV was lost.

Dodder used in transmission studies may sometimes harbor an unsuspected virus. Thus, Bennett (1944) found that symptomless *C. californica* was frequently infected with a virus he called dodder latent mosaic virus, which caused serious disease in several unrelated plant species.

One of the main experimental uses of dodder transmission has been to transfer viruses from hosts where they are difficult to study to useful experimental plants. Dodder is probably an insignificant factor in the transmission of economically important viruses in the field, and it has rarely been used in experimental work in recent times.

II. TRANSMISSION BY ORGANISMS OTHER THAN HIGHER PLANTS

A. Invertebrates

Many plant viruses are transmitted from plant to plant in nature by invertebrate vectors. This major topic is considered separately in Chapter 14.

B. Fungi

Several viruses have been shown to be transmitted by soil-inhabiting fungi. The known vectors are members of the class Plasmodiophoromycetes in the division Myxomycota, or in the class Chytridiomycetes in the division Eumycota. Both classes include endoparasites of higher plants. Species in the chytrid genus *Olpidium* transmit viruses with isometric particles, while species in two plasmodiophorus genera (*Polymyxa* and *Spongospora*) transmit rod-shaped or filamentous viruses (Table 10.2).

The most studied fungal vector is the chytrid *Olpidium brassicae* (Wor) Dang., which is a soilborne obligate parasite infecting the roots of many plants. In root cells the fungus forms resting spores, which are released into the soil when the root disintegrates. The resting spores, under appropriate conditions in the root or soil, release numerous zoospores into the soil water. These can then infect fresh roots. The ultrastructure of the zoospores and of the infection process have been described by Temminck and Campbell (1969a,b). There are two phases in the infection process. During the encystment phase, the zoospore attaches firmly to the root, and the single flagellum is withdrawn and the axonemal fibrils are seen coiled within the

Table 10.2
Viruses Transmitted by Soil-Inhabiting Parasitic Fungi

Fungal species	Virus	Survival of infectivity in air-dried soil	Selected references
Olpidium brassicae	TNV	?	Teakle (1962); Kassanis and Macfarlane (1964a); Fry and Campbell (1966); Campbell and Fry (1966)
Olpidium brassicae	STNV	?	Kassanis and Macfarlane (1968)
Olpidium brassicae	Lettuce big vein[a]	8 years	Campbell and Grogan (1964); Lin *et al.* (1970)
Olpidium brassicae	Tobacco stunt	?	Hiruki (1987)
Olpidium cucurbitacearum	Cucumber necrosis	2 months	Dias (1970)
Polymyxa graminis	Barley yellow mosaic	?	Adams *et al.* (1988)
Polymyxa graminis	Soilborne wheat mosaic	10 months at 17°C 20 months at 5°C	Canova (1966); Estes and Brakke (1966); Rao (1968)
Polymyxa graminis	Wheat spindle streak mosaic	?	Slykhuis and Barr (1978)
Polymyxa graminis	Indian peanut clump	?	Mayo and Reddy (1985)
Polymyxa graminis	Oat golden stripe	?	Plumb and Macfarlane (1977)
Polymyxa graminis	Peanut clump	3 months	Thouvenel and Fauquet (1981)
Polymyxa graminis	Rice stripe necrosis	?	Fauquet and Thouvenel (1983)
Polymyxa betae	Beet necrotic yellow vein	?	Guinchedi and Langenberg (1982); Abe and Tamada (1986)
Polymyxa betae	Beet soil borne	?	Ivanovic *et al.* (1983)
Spongospora subterranea	Potato mop top	9 months	Jones and Harrison (1969)
Spongospora subterranea	Watercress chlorotic leaf spot	≈12 months	Tomlinson and Hunt (1987)

[a]The virus assumed to cause this disease has not been definitively characterized.

zoospore body. The zoospore then secretes a cyst wall. In the second phase, the host cell wall is breached and the cyst cytoplasm enters the host cell, leaving behind it the ectoplast and the tonoplast (Fig. 10.2). The fungal cytoplasm becomes surrounded by a new tonoplast inside the host cell. A virus may be carried to the root on the surface of the zoospore (e.g., TNV, STNV, or cucumber necrosis virus) or within the zoospores (e.g., lettuce big vein agent and tobacco stunt virus).

Campbell and Fry (1966) suggested that the zoospores are not carrying TNV as they are released from the roots, but that they pick up TNV also released into the

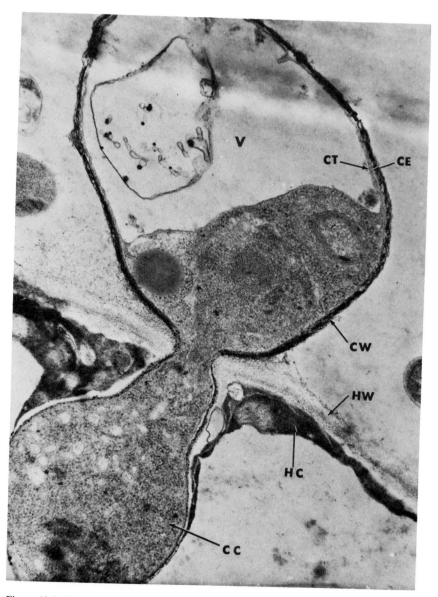

Figure 10.2 Fungal transmission. Infection of a root cell by *Olpidium*. Electron micrograph showing contents of an encysted zoospore entering the host cell. CC, cyst cytoplasm; HC, host cytoplasm; HW, host wall; CW, cyst wall; CT, cyst tonoplast; CE, cyst ectoplast; V, vacuole. (Courtesy of J. H. M. Temminck, reproduced by permission of the National Research Council of Canada from the *Canadian Journal of Botany*, Vol. 47, p. 421, 1969.)

soil water. *Olpidium* appears to be a very efficient vector since transmissions occur with liquids containing only about 50–100 zoospores/ml (Kassanis and Macfarlane, 1964a; Fry and Campbell, 1966). Zoospores of the fungus can transmit TNV to tobacco callus tissue in culture with more dilute inocula than are effective with mechanical inoculation of such tissues (Kassanis and Macfarlane, 1964b).

The fact that specific antiserum blocks infectivity of zoospores suggested that the virus is on or near the surface of the zoospore. On the other hand, washing by repeated centrifugation of zoospores carrying TNV did not prevent TNV transmission, suggesting that the virus is firmly bound to the zoospore (Campbell and Fry, 1966). Different isolates of *Olpidium* vary in their ability to transmit TNV. The ability to transmit is correlated with the ability of the zoospores to adsorb TNV to the surface membranes of both their body and their axonemal sheath as has been shown directly by electron microscopy (Temminck *et al.*, 1970).

When the axonemal sheath is withdrawn inside the zoospore during encystment, the attached virus is assumed to enter the fungal cytoplasm (Temminck and Campbell, 1969b). This virus can then enter the root epidermal cell in the fungal cytoplasm (Fig. 10.2). Presumably it is then released into epidermal cytoplasm, where it replicates. The properties of fungus-transmitted labile rod-shaped plant viruses have been discussed by Brunt and Shikata (1986).

C. *Agrobacterium tumefaciens*

DNA copies of viral genomes can be inserted into the Ti plasmid carried by *Agrobacterium tumefaciens*. Infection of a plant by *Agrobacterium* carrying such an engineered plasmid offers a novel experimental method for transmission of plant viruses. Thus Grimsley *et al.* (1986) showed that systemic infection of turnips occurred following inoculation with *A. tumefaciens* containing more than one copy of CaMV DNA tandemly inserted into the Ti plasmid. Therefore, following infection by the bacterium, a complete CaMV genome must be excised to replicate and move through the plant.

Agrobacterium tumefaciens is thought not to infect monocotyledons. However, Grimsley *et al.* (1987) showed that, when cultures of the bacterium containing a plasmid with tandemly repeated copies of MSV DNA were inoculated to whole maize plants, the plants developed symptoms caused by MSV. Since this *Geminivirus* is not transmitted by mechanical means they could conclude that *Agrobacterium* can transfer DNA to maize. Agroinfection with geminiviruses was discussed in Chapter 8, Section VI,B.

III. MECHANICAL TRANSMISSION

Mechanical inoculation involves the introduction of infective virus or viral RNA into a wound made through the plant surface. When virus establishes itself successfully in the cell, infection occurs. This method of transmission is of great importance for many aspects of experimental plant virology, particularly for the assay of viruses (Chapter 2) and in the study of the early events in the interaction between a virus and susceptible cells (Chapter 6). Viral RNA as inoculum was discussed in Chapter 4, Section I,C,4. When intact virus is used as inoculum the viral nucleic acid must be partly or entirely uncoated at an early stage. This process is discussed in Chapter 7 for TMV in Section III,A,7 and for TYMV in Section IV,A,7.

A. Applying the Inoculum

1. Method of Application

In early work, drops of the inoculum were placed on the leaf and the leaf surface was scratched or pricked with a needle to cause wounding. This was very inefficient and it was later found that gently rubbing the leaf surface with some suitable object wetted with the inoculum gave more efficient transmission. A wide variety of objects have been used, depending on the preference of the operator and the volume of inoculum available. The objective in mechanical inoculation is to make numerous small wounds in the leaf surface without causing death of the cells. This process is greatly enhanced by the use of certain additives (Section III,A,2). The pressure required to do this depends on many factors such as plant species, age and condition of leaf, and additives present in the inoculum. Macroscopic areas of dead tissue appearing on the inoculated leaf within a day or so indicate that the wounding was excessive. With a few viruses and hosts severe abrasion is more effective (Louie and Lorbeer, 1966). Citrus tristeza virus, which is probably confined to the phloem, is transmitted by a knife-cutting procedure, while ordinary mechanical inoculation is ineffective (Garnsey *et al.,* 1977). Citrus exocortis viroid is also transmitted efficiently by cutting stems with a contaminated blade (Garnsey and Whidden, 1973). Multiple slashing of stems was also the most effective way of transmitting tomato ring spot *Nepovirus* to *Prunus* seedlings (Bitterlin *et al.,* 1987).

Polson and von Wechmar (1980) used a process of electro-endosmosis to introduce MSV into leaves through cut petioles in a way that gave rise to infection even though this virus is normally transmitted only by leaf-hoppers. Konate and Fritig (1984) described an efficient microinoculation procedures that allowed early events following mechanical inoculation to be studied at predetermined individual infection sites on the leaves. Injection of infective material into petioles or stems with a hypodermic syringe has been used occasionally, but it is generally a very inefficient method. However, a high-pressure medical serum injector was used successfully to transmit beet curly top *Geminivirus* (Mumford, 1972).

Inoculating large numbers of plants can be a time-consuming process, and various procedures have been adopted to reduce the time involved. For example, dipping and moving the leaves of seedling plants in the inoculum may provide a rapid method for inoculating large numbers of seedlings at the time of transplantation. Where large numbers of leaves are to be inoculated with the same inoculum, airbrushes of the type used by artists may prove useful. Alternatively the inoculum may be applied in a solid stream (Louie *et al.,* 1983). A sensitive and rapid method using a specially designed air gun has been described by Laidlaw (1987). To avoid washing pestles and mortars or other glassware where large numbers of individual tests have to be made, it may be possible to rub a piece of diseased leaf directly on a leaf of the test plant (Murakishi, 1963).

2. Additives That Increase Efficiency

The efficiency of mechanical inoculation is greatly increased when some abrasive material is added to the inoculum or sprinkled over the leaves before inoculation. The most commonly used abrasives are carborundum (400–500 mesh) or diatomaceous earths such as Celite. The increase in number of local lesions obtained by the use of abrasives varies with different hosts and viruses but may be 100-fold or more. The time of addition of these materials may be important. Celite added after grinding and dilution was much more effective than when added before grinding (Yarwood, 1968).

Abrasives are the only generally used additives, but for certain viruses and host plants other substances may give spectacular increases in the number of local lesions, particularly dipotassium phosphate (Yarwood, 1952) and potassium sulfite (Yarwood, 1969). Other additions have included divalent metal ions, bentonite, and sucrose. Buffer mixtures with additives may be especially designed for successful inoculation from particular hosts (e.g., Martin and Converse, 1982).

3. Frozen and Freeze-Dried Inoculum

Lawson and Taconis (1965) described a method for transmitting dahlia mosaic *Caulimovirus* to *Verbesina encelioides* (Cav.) Benth. and Hook. Transmission by usual mechanical procedures failed or was very inefficient. When leaves from mosaic-infected dahlias were frozen in liquid nitrogen and ground in a precooled mortar, and the frozen tissue powder applied directly to *Verbesina* leaves with a brush, efficient transmission was obtained. Presumably the low temperature reduces inactivation of the virus by substances present in the extract. Water may not be necessary in the inoculum to achieve effective mechanical transmission. Freeze-dried powders from infected tissue caused efficient transmission without the need for additional abrasive (Ragetli *et al.*, 1973).

4. Other Components in the Inoculum

There are many substances that, when present in the inoculum, may affect the number of successful infections produced by a virus. These are host constituents in crude extracts, or substances added to extracts or purified virus preparations. They are discussed in Chapter 2. The kinds of material added to leaf tissue to facilitate isolation of a virus may also be used to allow transmission by mechanical inoculation of crude extracts.

5. Washing Leaves

Holmes (1929) considered that washing inoculated leaves with water immediately after inoculation increased the number of local lesions formed, and this has become a fairly widespread practice. However, washing leaves after inoculation, spraying with water, or dipping leaves in water may substantially reduce the number of local lesions produced by several viruses or have variable effects depending on other conditions (Yarwood, 1973). The effect of washing or dipping leaves in water on the number of lesions probably depends on many factors, and particularly on whether inhibitors of infection are present in the inoculum. If such inhibitors are present, washing may minimize their effect.

If the leaves are dried rapidly after inoculation either by blotting or with an air jet there may be a marked increase in the number of local lesions, but again the effect is variable (Yarwood, 1973).

6. The Plant Part Inoculated

Sometimes a virus can be transmitted mechanically by inoculating the cotyledon leaves, but not the first true leaves, for example, with a virus of sweet potato (Alconero, 1973). Mechanical inoculation of roots is inconvenient and is often less successful than with leaves. However, transmission by this means has been achieved for several viruses (e.g., Moline and Ford, 1974).

7. The Plant Part Used as Inoculum

Virus inhibitors are often present in highest concentration in leaves and may be present in low concentrations or absent in other organs, for example, petals. Thus extracts from petals have been used successfully to transmit several viruses mechanically (see Fulton, 1964).

8. Viral RNA as Inoculum

For any virus that has positive sense ssRNA as its genetic material it should be possible, in principle, to prepare an extract of total nucleic acids from infected tissue and use this to inoculate healthy plants. This procedure may allow mechanical transmission when whole-leaf extracts are ineffective. Success may be due to removal of virus inhibitors into the phenol phase or into the material that accumulates at the interface between phases, or it may be due to the existence of unstable or incomplete virus that is inactivated by nucleases unless these are removed by the phenol. With some tissues and viruses, grinding or blending in the presence of phenol may release virus from sites where it remains bound in normal sap extracts. Any infectivity in phenol extracts or RNA preparations will be fully susceptible to nucleases on contaminated glassware, once the phenol is removed, as it must be, before inoculation using the extract.

B. Nature and Number of Infectible Sites

1. Nature of the Leaf Surface

On the upper surface of leaves commonly used for mechanical inoculation there will be about $1–5 \times 10^6$ cells of various types. Most will be epidermal cells, with a smaller number of stomatal guard cells. In different plants the surface of the guard cells may be raised above, at the same level as, or sunken below the level of the epidermis. In many species there are two subsidiary cells surrounding the guard cells. These cells are smaller than the typical epidermal cells. Many leaves possess trichomes (Fig. 10.3). There are numerous kinds of trichomes and more than one kind may occur on the same leaf. For example, in tobacco there are ordinary trichomes or hairs and glandular hairs with a multicellular secretory head. The whole of the leaf surface is covered with a series of protective layers. The amounts and arrangement of layers vary widely in different species and under differing growth conditions. A typical arrangement of the outer layers is shown in Fig. 10.4.

The outermost layer is the epicuticular wax, which in different species may be very different in amount, fine structure, and chemical constitution. The epicuticular wax layer will have a strong influence on the wettability of the leaf surface during mechanical inoculation. The cuticle extends through the stomatal apertures and

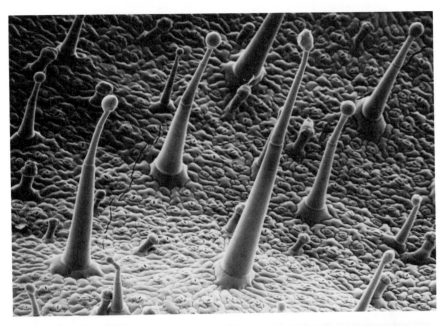

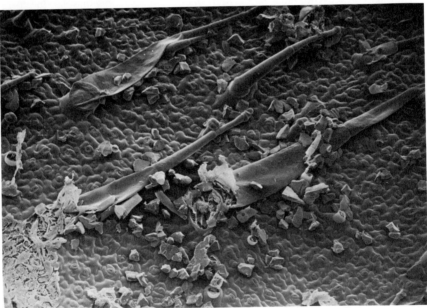

Figure 10.3 Scanning electron micrographs of the surface of *Nicotiana glutinosa* leaves before and after mechanical inoculation. (*Top*) An untreated leaf showing intact leaf hairs and epidermis; (*Bottom*) the broken hairs following mechanical inoculation. The particles of the carborundum abrasive may be clearly seen. Bar = 0.1 nm. (Courtesy of M. J. W. Webb.)

covers the inner walls of epidermal cells where these are exposed to internal air spaces. Plasmodesmata are believed to occur in the outer cell wall of leaf epidermal cells. These are called ectodesmata. They probably extend out to, but not through, the cuticle. The nature of plant surfaces has been reviewed by Juniper and Jeffree, 1983).

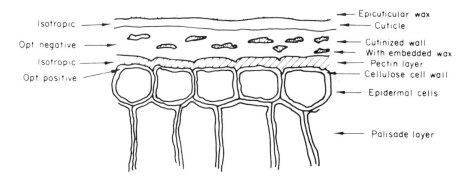

Figure 10.4 The barrier to virus infection. Diagrammatic representation of the epicuticle of the plant seen in cross section. The lines dividing the layers above the epidermal cells indicate regions of major change in the construction of components rather than sharp boundaries. Individual plant species may depart greatly from this general arrangement. (B. E. Juniper; from Eglinton and Hamilton. Copyright © 1967 by the AAAS.)

2. Nature of the Infectible Sites

Virus placed on an intact leaf surface cannot infect. Wounds must be made which break through the inert leaf surface. It is possible that some types of cell on the leaf surface are more susceptible to wounding than others. It has been shown directly, by microsurgical methods, that virus can be introduced into leaf hairs, but the infection rate was low (Zech, 1952). Although trichomes may be heavily damaged during mechanical inoculation (Fig. 10.3), there is no convincing evidence that they play any special role in virus entry during mechanical transmission. Likewise no special role for guard cells or ectodesmata has been established.

In summary, it seems probable that all the cell types making up the epidermis are potentially capable of being infected by mechanical inoculation. The cuticle is probably the major barrier to infection. Wounds that penetrate right through the cuticle and the cell wall are probably effective in allowing virus to enter, but such wounds may not be necessary if ectodesmata in fact play a role. It may be necessary merely to break the cuticle to allow access to the ectodesmata or a cellulose wall with altered properties. However, it has not been demonstrated that a virus particle can enter a cell in the leaf through an intact plasmolemma, in the way viruses infect isolated protoplasts. Nevertheless, it is very probable that such a process can occur. Experiments with several viruses indicate that, following mechanical inoculation, some underlying mesophyll cells may be infected directly with the inoculum (Salinas Calvete and Wieringa-Brants, 1984; Matthews and Witz, 1985).

3. Evidence from the Infectivity Dilution Curve

The relation between numbers of local lesions produced and the dilution of the inoculum was discussed in relation to the assay of viruses in Chapter 2. Various theoretical models have been developed in attempts to explain the nature of the dilution curve. These models involve assumptions about the number of infectible sites. For example, Furumoto and Mickey (1967) put forward a modification of an earlier hypothesis. They point out that for animal and bacterial viruses the idea that one particle can infect one cell appears to be valid. Thus, it is likely to be true for plant viruses as well. In their formulation they assume the following: (i) a susceptible region is in fact an individual epidermal cell; (ii) all or almost all cells are potentially infectible, thus they consider that the number of infectible sites is very much larger than had been thought—up to about 10^6 sites on a leaf; and (iii) the probability of penetration and infection of the cell by a given virus particle is the same for all virus particles, but this probability may vary randomly from cell to cell.

A simple approximation to their formulation is

$$Y = N\alpha \log_e (1 + cV/\beta)$$

where Y is expected average number of lesions, N is number of susceptible cells per half-leaf, V is virus concentration in mg/ml, c is a constant relating the virus concentration to the average number of virus particles outside a cell, and α and β are constants determining the distribution of cell penetration probabilities. Figure 10.5 shows the graphical fit of the theoretical model for the combined data of more than 30 experiments with TMV. For these composite results $N\alpha = 4.5$ and $c/\beta = 380$. (In individual experiments these parameters varied widely, and the fit was not as good as shown in Fig. 10.5, especially at high virus dilutions.)

The formulation cannot be used to predict the dilution curve in any given experiment. The close agreement between the theoretical curves and experimental points proves nothing with respect to the actual biological situation. The fact that some equations fit better than others may result merely from the trivial fact that the number of adjustable parameters is greater. A simple growth curve model has been proposed by Gokhale and Bald (1987). With multiparticle viruses two or more particles cooperate to produce a single infection. This phenomenon should lead to a change in the shape of the infectivity dilution curve. R. W. Fulton (1962) described such curves for two unstable viruses, *Prunus* necrotic ring spot *Ilarvirus* and prune dwarf *Ilarvirus*. The dilution curves were significantly steeper than expected for a theoretical one-hit curve. The extent of the deviation varied with assay host used. Fulton concluded that the abnormally steep curves were probably due to the need for two or more virus particles to initiate infection.

Dilution curves for AMV (requiring three particles) and TNV (assumed to require only one particle) are shown in Fig. 10.6 for data obtained in the same host species.

Although the difference between the two curves in Fig. 10.6 is quite clear, a dilution curve on its own cannot be used to decide whether a viral genome is housed in one, two, or three particles. There are many other factors that cause the slope of the curve to vary from experiment to experiment (Chapter 12, Section V,C,1).

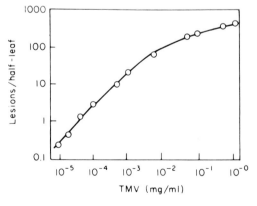

Figure 10.5 Graphical fit of theoretical model for the infectivity dilution curve, and composite experimental results from more than 30 experiments with TMV. (From Furumoto and Mickey, 1967.)

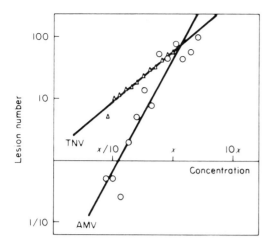

Figure 10.6 Comparison of dilution curves for a single-particle virus (TNV) and a multiparticle virus (AMV). Local lesions in *Phaseolus vulgaris* for both viruses. For TNV, $X = 0.15$ μg/ml; for AMV, $X = 9$ μg/ml. (From van Vloten-Doting, 1968.)

4. Lifetime of Infectible Sites following Wounding

The time for which wounds remain infectible has usually been studied by dipping leaves into inoculum at various times after abrasion and counting the number of lesions that subsequently develop. Generally speaking, the number of infectible sites falls off very rapidly after abrasion. For example, Furumoto and Wildman (1963a) found that about 70% of the sites on *N. glutinosa* leaves that were susceptible to infection with TMV by dipping 2 seconds after wounding had lost their susceptibility by 90 seconds. The remaining sites lost their susceptibility much more slowly over a period of about 1 hour. However, not all viruses and hosts follow this pattern. Jedlinski (1964) found that during the first 10 minutes after abrasion, the number of infectible sites could either increase, decrease, or remain constant, depending on the host–virus system tested. There are many treatments that, applied to the leaf before or after wounding, can alter the course of events (Chapters 12, Section V).

C. Number of Particles Required to Give an Infection

There are two aspects to consider concerning the number of particles required for infection. First, how many particles have to be applied to a leaf to give one successful infection? Second, is one particle from among those applied sufficient to give an infection?

1. Number of Particles Applied

Regarding the first point, in the absence of an efficient method of inoculation it is not possible to obtain unequivocal estimates of the proportion of infectious particles in a virus preparation, but Furumoto and Wildman (1963b) concluded that at least one in ten of the TMV particles in a purified preparation was infectious.

Mechanical inoculation of leaves has been widely regarded as a very inefficient process. Various estimates suggested that between about 10^4 and 10^7 virus particles have to be applied to the leaf for each local lesion that subsequently develops (e.g., Walker and Pirone, 1972a).

With multiparticle viruses the requirement for more than one particle makes the inoculation process even less efficient than for single-particle viruses. For CPMV 10^6–10^8 particles were applied mechanically for each lesion produced (van Kammen, 1968) and for AMV the figure is 10^8–10^{10} (van Vloten-Doting, 1968). The major factor leading to the requirement for a large number of virus particles for each successful infection is almost certainly the inefficiency of the process as it is usually performed. There are probably several reasons for this: (i) it is very likely that only a very small proportion of epidermal cells have potentially infectible wounds made in the leaf surface above them; (ii) the lifetime of the infectible wounds is short, and many of the virus particles in the fluid above a wound never make contact with the site; (iii) the distribution of the virus applied over the leaf surface is probably very uneven in terms of surface areas of the size of single cells; (iv) much of the virus may be adsorbed to inactive sites on the leaf surface and remain trapped there; and (v) some virus particles may enter potentially infectible cells, but not become successfully established, unless "rescued" in some way. The existence of such centers has been shown for some host–virus combinations (Rappaport and Wu, 1963).

In most estimates of the efficiency of mechanical inoculation relatively large volumes of inoculum were applied to the test leaves, and the calculation involved determining the number of particles applied per lesion produced. However, if a very small volume of inoculum is applied using a series of dilutions of the virus, the limiting dilution at which an infection is obtained gives a much lower estimate. Thus Walker and Pirone (1972b) found that about 450 TMV particles in 2.5 µl of inoculum were sufficient to infect a tobacco plant. Using even smaller volumes of inoculum (0.1–1.0 µl) as few as 10–30 particles of TYMV were required to produce a single local lesion in Chinese cabbage (Fraser and Matthews, 1979a). These very substantial increases in efficiency of the mechanical inoculation process may be due to three effects: (i) for a given number of virus particles applied, the smaller the volume, the higher the virus concentration; (ii) a virus particle in a very small volume will have a greater probability of finding an infectible site in a short time than one in a large volume; and (iii) the fact that rapid drying increases the number of local lesions (Section III,A,5) may also be a factor. Two other methods have been used to determine efficiencies of infection—inoculation of protoplast suspensions and microinjection of cells.

We can distinguish three quantities in these inoculation procedures:

1. Total number of virus particles applied per infection obtained. This is the only measurement that can be made for leaf inoculation. (Under limiting conditions in inoculated leaves it is reasonable to assume that each lesion arises from the infection of a single cell.)
2. Number of particles adsorbed to each viable cell, for each successful infection. This can be measured for inoculated protoplasts and is usually about 0.1–1.0% of the applied inoculum. Efficiency of infection of protoplasts can be expressed as the average number of virus particles adsorbed per protoplast to give infection in one-half of the protoplasts (ID_{50}). Values of the ID_{50} calculated from published data for several viruses fell in the range 50–500 (Fraser and Matthews, 1979a). The numbers of particles actually supplied in the inoculum were of the order of 100–1000 times these numbers. Thus, the

efficiency of inoculation of protoplasts may not be intrinsically any higher than mechanical inoculation of leaves.

3. Number of particles actually entering a cell for each successful infection. This can be measured only with microinjection. For example, Halliwell and Gazaway (1975) obtained an ID_{50} of 310 TMV particles injected in 1 picoliter into single tobacco cells.

2. Is One Particle Sufficient?

It is generally agreed that for many viruses only a single particle is needed actually to infect a cell and give rise to the visible lesion (e.g., Boxall and MacNeil, 1974). Theoretical consideration of the dilution curve discussed in the last section is consistent with these data. Experiments with protoplasts indicate that about one TMV particle is sufficient to infect one protoplast (Takebe, 1977). Reddy and Black (1973) found that two to three WTV particles were sufficient to infect a cell in insect vector cell monolayers.

D. Mechanical Transmission in the Field

Compared to transmission by invertebrate vectors or vegetative propagation, field spread by mechanical means is usually of minor importance. However, with some viruses it is of considerable practical significance. TMV can readily contaminate hands, clothing, and implements and be spread by workers in tobacco and tomato crops. This is particularly important during the early growth of the crop, for example, during the setting out of plants. Plants infected early act as sources of infection for further spread either during cultural operations as disbudding or by rubbing together of healthy and infected leaves by wind. TMV may be spread mechanically by tobacco smokers, because the virus is commonly present in processed tobacco leaf. For example, all 37 brands of cigarette sold in West Germany contained TMV (Wetter, 1975).

PVX can also be readily transmitted by contaminated implements or machines and by workers or animals that have been in contact with diseased plants (Todd, 1958). On some materials the virus persists for several weeks.

PVX can spread either by direct leaf contact between neighboring plants (Clinch *et al.*, 1938) or between neighbors when leaves are not in contact. This has been assumed to be due to mechanical inoculation by contact between roots, but a soil-inhabiting vector has not been excluded. Some viruses of fruit trees have been shown to be spread in orchards by cutting tools (e.g., Wutscher and Shull, 1975). Field trials have suggested that red clover mottle *Comovirus* can be readily transmitted by mowing machines (Rydén and Gerhardson, 1978). On the other hand, Heard and Chapman (1986) concluded that in mown ryegrass swards most of the local spread of ryegrass mosaic *Potyvirus* was due to transmission by mites rather than the mowing operation.

E. Abiotic Transmission in Soil

1. Aboveground

Allen (1981) showed that, in glasshouses, TMV could be transmitted from soil containing the virus coming into contact with leaves. Some other stable viruses may well be transmitted in this way.

2. Belowground

Several viruses may infect roots from virus-contaminated soil apparently without fungi or arthropods being involved, for example, TBSV (Kleinhempel and Kegler, 1982), carnation ring spot *Dianthovirus* (Brown and Trudgill, 1984), and SBMV (Teakle, 1986).

IV. MOVEMENT AND FINAL DISTRIBUTION IN THE PLANT

Some viruses are confined to the inoculated leaf while others may move systemically through the plant. If a virus replicates in the initially infected cell but cannot move to neighboring cells, its replication may remain undetected. This has implications in considering the host range of viruses (Section V). In many host–virus combinations, cell-to-cell movement is limited, giving rise to a relatively small zone of infected tissue that may be visible as a local lesion, often consisting of dead cells (see Fig. 11.1). In still others movement within the inoculated leaf is not limited but systemic spread may not occur.

The experiments summarized next have shown that viruses spread through the plant in two ways—a slow cell-to-cell spread through the plasmodesmata from the site of inoculation and a much more rapid movement through the vascular tissues, usually through the phloem. In these processes there are three possible barriers to movement: (i) movement from the first infected cell; (ii) movement out of parenchyma cells into vascular tissues; and (iii) movement out of vascular tissue into the parenchyma cells of an invaded leaf.

The final distribution of virus through tissues and organs may be very uneven. Many factors, including host genes, viral genes, and environmental factors, affect the rate and extent of virus movement through the plant. The effects of viral genes are discussed in Section IV,G; host genes in Chapter 12, Sections III and V,B,2; and environmental factors in Chapter 12, Section V,C. Maule (1991) has reviewed recent advances in our understanding of virus movement in plants.

A. Methods

The classic experiments on movement and distribution of viruses involved dissection of plants into appropriate pieces at various times after inoculation. Extracts of these parts were then inoculated to suitable assay hosts either immediately or after an incubation period to allow very small amounts of virus that might be present to increase and give detectable amounts (see Fig. 10.7) Fluorescent antibody methods have been applied to the detection of viruses in various parts of the plant. This method is of course much less sensitive than infectivity. Various other procedures used for assay and diagnosis could be used to follow movement (Chapter 2).

In recent experiments, discussed next, a variety of techniques have been used to characterize the form in which infectious virus moves and the role of viral gene products in such movement.

The luciferase gene from the firefly has been used as a reporter of gene expression by light production from transgenic plants (Ow *et al.*, 1986). The same gene has been integrated into vaccinia virus and used as a sensitive marker to follow dissemination

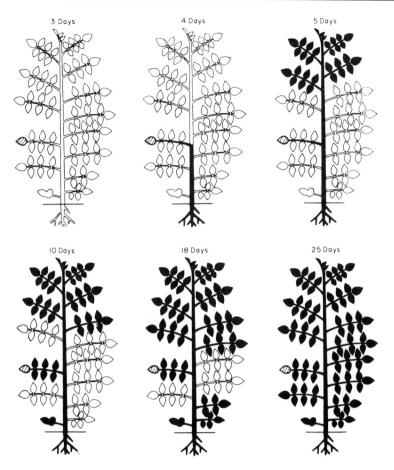

Figure 10.7 Diagram showing the spread of TMV through a medium young tomato plant. The inoculated leaf is shaded, and systematically infected tissues are shown in black. (From Samuel, 1934.)

of the virus in infected animals (Rodriguez *et al.*, 1988). If the gene could be similarly integrated into a plant virus it might be useful for a similar purpose.

B. The Form in Which Virus Moves

Since both intact virus and isolated positive sense ssRNA can infect cells, both of these structures may be involved in cell-to-cell spread. Virus particles have been visualized in plasmodesmata and sieve pores (see Section IV,C). However, such observations by no means demonstrate that intact virus particles are the form, or the only form, in which a virus moves from cell to cell. There is clear evidence that viral RNA can move into and infect neighboring cells. For example, the long rods of TRV inoculated alone are incapable of forming complete rods and, therefore, must migrate from cell to cell as naked RNA. Similarly, defective mutant strains of certain viruses (e.g., TMV) are unable to produce complete virus particles. Infective viral RNA and defective coat protein are produced and yet the infection moves from cell to cell. Following inoculation with type TMV, the underlying mesophyll becomes infected some 3 hours before mature virus particles are produced in the epidermis (Fry and Matthews, 1963). As far as long-distance transport within the plant is concerned,

Dawson *et al.* (1975), using the differential temperature treatment discussed in Chapter 6, Section VII,A,2, showed that TMV could move from an inoculated tobacco leaf into the upper parts of the plant in some form other than mature virus particles. On the other hand, the results of Saito *et al.* (1990) indicate that intact TMV particles may be necessary (Section IV,G,1).

In eukaryotic cells mRNAs are thought to be contained, under some conditions at least, in cytoplasmic ribonucleoprotein particles made up of cell proteins complexed with the mRNAs and called "informosomes" (e.g., Ajtkhozhin *et al.*, 1976). Atabekov and his colleagues investigated the possibility that plant viral RNAs may move through the plant in similar complexes (reviewed by Atabekov and Dorokhov, 1984). They used the mutant Ni118 of TMV, which is temperature sensitive for coat protein production.

At the nonpermissive temperature virtually no virus particles are produced but infection can spread systemically. From plants infected with Ni118 at the nonpermissive temperature they isolated ribonucleoprotein particles composed of genomic and subgenomic TMV RNAs and both TMV-specific and host-cell proteins. They called these complexes virus-specific RNP or vRNP. These vRNP complexes have the following properties (Dorokhov *et al.*, 1983, 1984a,b): (i) the buoyant density of vRNP in CsCl or Cs_2SO_4 is considerably higher than that of TMV particles, and like that of plant or animal informosomes; (ii) the RNA is susceptible to RNase; (iii) vRNPs appear as flexuous rods in the electron microscope; (iv) on the basis of pulse–chase experiments vRNP does not appear to serve as a precursor of mature TMV; (v) polypeptide analysis of vRNP from tobacco showed the presence of several proteins, the most prominent being 39K, 37K, 31k (unrelated to TMV coat protein), and TMV coat protein; (vi) vRNP from PVX-infected tobacco tissue contains a set of proteins of different size, while TMV vRNP from various hosts had the same sizes as found in infected tobacco; and (vii) vRNP is found in the free state in cells and also in association with polyribosomes, indicating a role in the synthesis of viral-coded proteins.

The main evidence implicating vRNP in cell-to-cell and long-distance transport is as follows: (i) The I_2 RNA of TMV, which is the template for the 30K transport protein (see Fig. 7.5), is present in vRNP as is the 30K protein itself. (ii) There is a correlation between the ability of TMV mutants to move systemically through the plant and the presence of vRNP in leaves distant from the site of inoculation. For example, Dorokhov *et al.* (1984b) studied two groups of mutants. Ni118, as an example of one group, contains a *ts* mutation in the coat protein gene, but is *tr* (normal) for transport. It can spread very slowly because of the defective coat protein. Ls1 is an example of the other group of mutants that has normal coat protein but is *ts* for the transport function (see Chapter 6, Section V,C,2). Dorokhov *et al.* (1984b) showed that vRNP can be produced at the restrictive temperature by *tr,* but not *ts,* mutants. *ts* mutants produce vRNP only at the permissive temperature. (iii) The mutants LS1 and Ni118 complement each other with respect to long-distance transport at the nonpermissive temperature, that is, mixed infections permitted long-distance transport and the appearance of vRNP. This result suggests a role for TMV coat protein in transport, a view not supported by other evidence (Section IV,G). (iv) Replication, but not movement, of type TMV is inhibited at 5°C. When inoculated leaves of tobacco plants were held at 25°C and the rest of the plant at 5°C infective material and vRnP but no virus particles accumulated in the upper leaves (Dorokhov *et al.*, 1981, 1984b). This evidence suggests, but by no means proves, that vRNPs are the form in which TMV and other viruses move both from cell to cell and over long distances. The presence of characteristic virus particles in sieve tubes or phloem exudate suggests that, in some circumstances at least, it is the intact virus that is translocated (e.g., for TMV, Esau and Cronshaw 1967; Oxelfelt, 1975). See also Section IV,G.

C. The Role of Plasmodesmata

Plasmodesmata consist of two concentric cylinders of plasma membrane and endoplasmic reticulum that traverse the cellulose walls between adjacent plant cells. The annulus between the two membrane cylinders gives continuity of the cytosol between cells. This continuity is probably regulated and therefore is not permanent. It is functionally equivalent to the gap junction of animal cells (Gunning and Overall, 1983). Plasmodesmata, together with the cytoplasm of individual cells, form a continuous three-dimensional network, the symplast.

There are two reasons for believing that plasmodesmata must be the route by which infectious virus—in whatever form—moves from cell to cell. (i) Kinetic data indicate that even a small molecule like sucrose is transported from mesophyll cells to the small-vein network via the symplastic route (i.e., through the plasmodesmata; Cataldo, 1974). Thus there appears to be no other reasonable route for very much larger virus particles. However, TMV antigen has been found associated with cell walls of infected tissue (De Zoeten and Gaard, 1984). (ii) Direct evidence that it is possible for viruses to move through the plasmodesmata comes from electron microscopic observations showing the presence of various kinds of virus particles within these intercellular channels (e.g., Esau *et al.*, 1967; De Zoeten and Gaard, 1969b; Kitajima and Lauritis, 1969) (see also Fig. 10.10). Plasmodesmata vary considerably in diameter, for example, from about 20 to 200 nm in young tobacco leaves (Shalla, 1959). However, the effective channel for movement may be much smaller. For example, Tucker (1982) used fluorescence-labeled peptides to assess the upper limit of size for movement through *Setcreasea* staminal hair cells. He established a cutoff at a MW of 700–800, a figure similar to that found by other workers in various species and tissues. Given these dimensions, it is not surprising that ribosomes do not move through plasmodesmata (Gunning and Overall, 1983). Thus we must conclude that viruses alter the properties of plasmodesmata in a manner that allows their passage through them. This idea is supported by the existence of specific viral-coded proteins that are necessary for virus movement and by virus-induced alterations in plasmodesmata (Section IV, G). On the other hand, some virus infections may reduce the availability of plasmodesmata. Shalla *et al.* (1982) showed that the restricted movement of a *ts* mutant of TMV in tobacco leaves was associated with a substantial reduction in the numbers of plasmodesmata.

D. Time of Movement from First Infected Cells

Uppal (1934) stripped epidermis from *N. sylvestris* at various times after inoculation with TMV and assayed the underlying mesophyll in another host. He found that virus moved into the mesophyll in 4 hours at 24–30°C. Estimates for the time taken for virus to move into the underlying mesophyll using similar methods have been made for several host–virus combinations: 2 hours at 28°C for CMV in cowpea (Welkie and Pound, 1958); 10 hours at 20°C for TMV in *N. glutinosa* (Dijkstra, 1962); and 4 hours at 27°C for TMV in tobacco (Fry and Matthews, 1963).

E. Rate of Cell-to-Cell Movement

The rate of movement from cell to cell has been estimated in several ways. The radius of necrotic local lesions produced by three tobamoviruses in *N. glutinosa* increases in a linear fashion with time (Rappaport and Wildman, 1957). Each of the viruses examined had a different rate of spread ranging from about 6 to 13 μm/hour. Uppal (1934) estimated that TMV moved from the upper epidermis to the lower at the rate of about 8 μm/hour, a value in agreement with the estimate of lateral spread. Rates of cell-to-cell spread may vary with leaf age, with different cell types, and in different directions within the leaf. TYMV infection spreads outward from the phloem of the small veins of young systemically infected Chinese cabbage leaves at about 4 μm/hour, that is, roughly one cell every 3 hours (Hatta and Matthews, 1974). Rate and extent of cell-to-cell spread in inoculated Chinese cabbage leaves differed widely with different strains of TYMV (Matthews, 1981). Experiments with *ts* mutants of TMV show that rate of cell-to-cell movement of this virus is also influenced by the viral genome (Nishiguchi *et al.*, 1978).

There are generally fewer plasmodesmata per unit area on the vertical walls of mesophyll cells than on the walls that are more or less parallel to the leaf surface. Furthermore, there tend to be trains of mesophyll cells linked efficiently together and ending in contact with a minor vein. Viruses may spread more rapidly along such routes than in other directions within the mesophyll. In local lesions caused by TYMV in Chinese cabbage not all the cells become infected even after many days. For these various reasons, estimates of rates of cell-to-cell movement based on numbers of infected cells and those based on estimates of virus production can give only average values.

F. Long-Distance Movement

Because economically important disease is usually produced only by systemic infection, most attention has been given to viruses that move out from the point of entry and invade many parts of the plant. A few viruses appear to be restricted to parenchyma tissues. These can spread only by slow cell-to-cell movement, presumably through the plasmodesmata. Viruses with a defective coat protein and, therefore, having uncoated RNA appear to move in this way. However, most viruses can move fairly rapidly through conducting tissue.

The time at which infectious material moves out of the inoculated leaf into the rest of the plant varies widely depending on such factors as host species and virus, age of host, method of inoculation, and temperature. After transmission by aphids, BYDV may move out of the inoculated leaf within 12 hours (Jensen, 1973). TMV moved out of tobacco leaves 32–48 hours after mechanical inoculation (Oxelfelt, 1970). Sonchus yellow net virus moved systemically into leaves and roots within 24 hours of inoculation (Ismail *et al.*, 1987).

That viruses may move rapidly over long distances was first shown by the classic experiments of Samuel (1934) with TMV in tomatoes. He inoculated one terminal leaflet and then followed the spread of virus with time by cutting up sets of plants into many pieces at various times. He incubated the pieces to allow any small amount of virus present to increase and then tested for the presence of virus by infectivity. Figure 10.7 shows the course of virus spread in a medium-sized tomato plant. Virus moved first to the roots, then to the young leaves. It was some time before the middle-aged and older leaves became infected. In very young plants, older leaves did not become infected, even after several months.

Evidence that viruses can move over long distances in the phloem comes from several kinds of experiments: (i) virus spread is influenced by the flow of metabolites in the plant (Bennett, 1940a); (ii) when a section of stem is killed or ringed, movement is prevented or substantially delayed (e.g., Helms and Wardlaw, 1976); (iii) presence of virus particles in sieve elements has been demonstrated for such phloem-restricted viruses as BYV and beet western yellow *Luteovirus* in *Beta vulgaris* (Esau *et al.*, 1967; Esau and Hoefert, 1972), beet yellow stunt *Closterovirus* in *Sonchus* (Esau and Hoefert, 1981), and also for TRSV in soybean, which is not phloem restricted (Halk and McGuire, 1973); and (iv) when TYMV is translocated into a young Chinese cabbage leaf the first mesophyll cells to show cytological signs of infection are always next to phloem elements and not xylem (Hatta and Matthews, 1974).

Once virus enters the phloem, movement may be very rapid. Values of about 1.5 cm/hour have been recorded for TMV (Bennett, 1940a). Capoor (1949) obtained a value of about 8 cm/hour for the movement of TMV and PVX in tobacco stems. Helms and Wardlaw (1976) reported rates of up to 3.5 cm/hour for TMV in *N. glutinosa* compared to 60 cm/hour for ^{14}C assimilates. SBMV appears to move in the phloem of pinto beans, and in these cells it moves 10–100 times as fast as the cell-to-cell movement through parenchyma (Worley, 1965). Worley drew attention to the possibility that the rapid rates in phloem cells may be due to the very directional nature of protoplasmic streaming in these cells, compared to rather haphazard movements in parenchyma cells. The distribution of P protein within mature sieve tubes (Dempsey *et al.*, 1975) indicates that there would be little physical impediment to the movement of virus particles through either the pores of the sieve plate or the sieve tube itself.

Long-distance translocation does not appear to depend on a concentration gradient of virus along the route of translocation, but rather on the rapid and random transport of infective material. In the early stages of systemic infection, virus may pass through infectible tissues without causing infection there (e.g., Capoor, 1949).

BYDV appears to be primarily confined to the phloem. Following introduction by aphid vectors there may be some local multiplication perhaps in phloem parenchyma before translocation begins (Gill, 1968). Sugar beet curly top *Geminivirus*, transmitted by leafhoppers and confined to the phloem, can move rapidly over long distances with no initial delay. Bennett (1934) recorded rates of 2.5 cm/minute.

The time at which unrelated viruses move from the inoculated leaf of the same individual host plant may be different. For example, Smith (1931) found that in tobacco plants inoculated with a mixture of PVX and PVY, PVY moved ahead of PVX, and could be isolated alone from the tip of the plant.

The characteristic particles of LNYV were observed in young xylem cells of leaf veins (Chambers and Francki, 1966). Infective virus could also be recovered from xylem sap in the stem. However, this virus is not confined to the xylem, as particles have been observed in mesophyll, epidermal cells, and leaf hairs. There was no evidence that LNYV particles found in xylem played any role in systemic movement. Potato mop top *Furovirus* is defective, in that the protein helices of its particles are mostly uncoiled at one end (Harrison and Jones, 1970). It moves rather slowly through tobacco plants. The pattern of movement is consistent with transport in the xylem (Jones, 1975). The defective TMV mutants PM_1 and PM_2 may move by the same route. Some viruses normally found in other tissues may in certain circumstances be able to move in the xylem. Thus, Schneider and Worley (1959a,b) found that SBMV, introduced by grafting or injection into the stem, was able to move either upward or downward through pieces of pinto bean stem that had been killed by steaming. Direct injection of virus into killed sections of stem led to infection at a distance, showing that local multiplication was not essential for movement of this virus. For several viruses, movement in the xylem was correlated with transmission by beetles (Gergerich and Scott, 1988b).

Systemic invasion of a leaf may result in a much more even movement of virus into mesophyll cells than that following mechanical inoculation. The differential temperature system that leads to infection of the cells in a young tobacco leaf with TMV under conditions where TMV replication does not occur (Chapter 6, Section VII,A,2) indicates that in this leaf, cell-to-cell spread from the vascular elements is not dependent on the complete cycle of virus replication.

G. Role of Viral Gene Products in Virus Spread within the Plant

Jokusch (1966b) first suggested that viral genes may play a role in the cell-to-cell movement of viruses. In recent years this idea has been confirmed experimentally. There are two kinds of gene products to be considered—coat protein genes and specific transport genes. The role of viral genes in transport has been reviewed by Atabekov and Talianisky (1990).

1. The Coat Protein

Various lines of evidence indicate that the coat protein may not be required for cell-to-cell movement of viruses (Section IV,B). However, long-distance systemic movement may require a functional coat protein. Experiments with designed mutants in which deletions were made in the N-terminal arm of the BMV coat protein showed that packaging of the RNA of this virus into particles was necessary for systemic infection. If the first seven amino acids were missing, packaging and systemic infection were normal. If 25 amino acids were deleted, no packaging and no systemic infection occurred, even though the RNA could replicate normally as shown in protoplasts (Sacher and Ahlquist, 1989). In a similar approach using TMV, Saito et al. (1990) found that certain mutations in the coat protein and the assembly origin in the RNA gave rise to greatly reduced ability to move systemically. Pseudorecombinants formed from infectious RNAs transcribed from cDNAs showed that the inability of strain M of CMV to systemically infect squash is a function of the coat protein (Shintaku and Palukaitis, 1990). A single amino acid substitution in the coat protein of BNYVV interfered with packaging of the RNAs and long-distance spread of the virus in spinach (Jupin et al., 1990).

The mutant DT-IG of TMV produces no coat protein and does not move from the inoculated leaf (Sarkar, 1986). However, when the RNA of this mutant is inoculated onto transgenic tobacco plants expressing viable TMV coat protein, the RNA can move systemically. Rod-shaped nucleoprotein particles were present in the systemically invaded leaves. Homogenates of such leaves retained their infectivity for Xanthi nc tobacco after nuclease digestion (J. Osbourne, S. Sarkar, and T. M. A. Wilson, personal communication). These experiments demonstrate a requirement for coat protein in long-distance transport of TMV.

Observations on TGMV replication in transgenic plants showed that DNA A codes for all the functions necessary for replication and production of virus particles, but that some function of DNA B is required for cell-to-cell movement (Sunter *et al.*, 1987). A genetic analysis of TGMV, in which the coat protein gene was modified or almost entirely deleted, showed that this gene is not essential for systemic spread although development of systemic symptoms was delayed (Gardiner *et al.*, 1988). Similar results have been obtained for ACMV (Ward *et al.*, 1988). However, the coat protein gene is essential for systemic movement of MSV, a whitefly-transmitted geminivirus (Lazarowitz *et al.*, 1989).

2. Genes with a Specific Transport Function

Various kinds of genetic experiments have shown that some viral gene (or genes) other than the coat protein gene is involved in virus movement. Experiments involving the use of recombinant DNA technology, discussed in Chapter 6, Section V,D,2, have shown beyond doubt that the 30K protein coded for by TMV is involved in cell-to-cell movement of the virus. Further details concerning this protein are given in Chapter 7, Section III,A,4,c. A study of TMV mutants with altered production of the 30K protein has shown that only those producing less than 10% of wild type had a reduced ability to move. Mutants with increased production of the 30K protein moved no faster than wild type. Early production of the 30K protein was necessary for effective cell-to-cell movement (Lehto *et al.*, 1990a). The timing of 30K protein synthesis in relation to virus accumulation (early rather than late) appears to be more important than the amount of protein produced (Lehto and Dawson, 1990).

The fact that several mutants of AMV were *ts* for replication in the plant at 30°C but could replicate in protoplasts at this temperature suggested that they were defective in some cell-to-cell transport function (Huisman *et al.*, 1986). In cowpea protoplasts, the RNA1 of red clover necrotic mosaic *Dianthovirus* replicates independently of RNA2. However, RNAs 1 and 2 are necessary for infection of intact plants, suggesting that RNA2 codes for a cell-to-cell movement function (Osman and Buck, 1987).

B component RNA of red clover mottle *Comovirus* (RCMV) is not transported between cells in an inoculated leaf in the absence of M RNA. However, it could spread in leaves previously infected with TMV Ni118 and held at the nonpermissive temperature. Ni118 is a mutant that is *ts* for coat protein. Thus TMV coat protein was not functioning to enable cell-to-cell movement of the B component comovirus RNA (Malyshenko *et al.*, 1988). Immunogold cytochemistry located the 43K protein coded by the M component of RCMV in the plasmodesmata of infected pea tissue, supporting a role in cell-to-cell movement for this protein (Shanks *et al.*, 1989).

Genes with a cell-to-cell movement function have been identified positively or tentatively for many groups of viruses, as summarized in Chapters 7 and 8. The cell-to-cell movement proteins of several viruses have amino acid sequence similarities. Thus in BMV, RNA 3 codes for a protein of 32K, required for systemic movement, which has sequence similarities with the TMV 30K protein, as does the 31K protein at the 5' end of the potyvirus polyprotein, and the ORF1 protein of CaMV. Cell-to-cell movement proteins may not be found in all virus groups. Luteoviruses are almost entirely confined to the phloem and the nucleotide sequence of PLRV shows no ORF with a sequence similar to the movement proteins of other groups (Harrison *et al.*, 1989).

More than one gene may be involved in the movement of some viruses. For example, while the ORF1 product of CaMV is involved in cell-to-cell movement, gene VI is also necessary for systemic spread (Schoelz and Shepherd, 1988). In MSV the coat protein gene, as well as a plus strand ORF upstream of the coat gene, coding for a 10.9K protein is essential for systemic movement (Lazarowitz *et al.*, 1989). Cell-to-cell transport of CPMV requires both coat protein and the 58K/48K protein (Wellink and van Kanamen, 1989).

The mechanism by which these gene products enable virus movement to occur has not been established. There are several possibilities that are not mutually exclusive: (i) The role of plasmodesmata was discussed in Section IV,C. The viral-coded movement protein may alter the properties of plasmodesmata in some way that allows virus to move through them. This possibility is supported by the facts that the 30K TMV movement protein accumulates inside plasmodesmata (Tomenius *et al.*, 1987), and that it accumulates, and is stable, in the cell wall fraction of transgenic tobacco plants expressing the 30K gene (Deom *et al.*, 1990). Similarly the ORF1 gene product of CaMV is associated with a cell wall fraction (Albrecht *et al.*, 1988) and is located near modified plasmodesmata between infected mesophyll cells, and in the end walls of phloem parenchyma cells (Linstead *et al.*, 1988). There is some direct evidence that viruses may modify plasmodesmata. Kitajima and Lauritis (1969) showed that in zinnia leaf tissue infected with dahlia mosaic *Caulimovirus* there was no tubular structure within the modified plasmodesmata and no ER associated with them. The modified plasmodesmata had consistently larger openings and were of uniform diameter throughout their length. They were large enough to accommodate readily the caulimovirus particles observed within them. Wolf *et al.* (1989) used fluorescein-labeled dextrans of various sizes to measure movement from an initially injected cell. Cells of a transgenic line of tobacco expressing the coat protein gene allowed movement of particles that were about 10 times as large as those that could move in control cells, demonstrating directly an effect of the coat protein gene on plasmodesmata. (ii) The movement protein may suppress some host response that limits virus movement from the initially infected cells. There is no evidence to support this idea. (iii) The movement protein may form part of a virus transport complex of the sort described for TMV in Section IV,B. (iv) The movement protein may make it possible for virus particles to leave the vascular tissue and enter the mesophyll cells of systemically infected leaves. This could involve mechanism (i). It does not appear possible to infect Chinese cabbage leaves with TYMV by standing cut petioles in solutions of the virus, even though virus moves into the leaf (Matthews, 1970). Similar results were obtained for TMV in tobacco and tomato (Taliansky *et al.*, 1982b). However, plants inoculated on the upper leaves in the normal way with type TMV, and later excised allowed systemic infection through the cut stems by the mutant *LS*, which produces a defective cell-to-cell movement protein. Thus it appears that the functional movement protein of the type TMV allowed the defective mutant to move systemically from the vascular tissue into mesophyll cells.

In summary, viral-coded movement proteins may act in one or more than one way. The proteins coded by different viruses may act by different mechanisms, but the fact that there are amino acid sequence similarities between proteins coded by different viruses would favor a similar mechanism for some, at least. On current evidence this is most likely to involve structural and functional alterations to the plasmodesmata. Godefroy-Colburn *et al.* (1990) discuss this hypothesis.

H. Final Distribution in the Plant

Whether or not a virus will move systemically at all depends on events in or near the local infection. It is often assumed that viruses that do move systemically become fairly evenly distributed throughout the plant, but in fact this hardly ever happens. There are several factors that usually result in a very unequal distribution.

1. Limitation of Local Lesion Size

Necrotic Local Lesions

With many host–virus combinations, virus moves into and multiplies in only a small group of cells around the point of infection. It has been suggested that where infection results in rapid necrosis and death of cells, further spread of the virus is limited by the fact that no movement of virus from the dead cells can occur. However, examples are known where limited local infection occurs without death of cells (e.g., PVY in potato, Gebre Selassie et al., 1985). Limitation of spread may be a complex phenomenon associated with development of resistance to virus invasion in surrounding tissue (Chapter 12).

A virus that normally produces only necrotic local lesions with no systemic spread when inoculated by mechanical means may spread through the plant systemically if infection is achieved by grafting. For example, SBMV, which on mechanical inoculation to *Phaseolus vulgaris* var. Pinto gives only necrotic local lesions, moves both upward and downward in the stem if introduced by grafting, and causes scattered limited areas of infection (Schneider and Worley, 1959a). Environmental factors as they affect extent and rate of virus movement are discussed in Chapter 12, Section V,C.

Chlorotic Local Lesions

For many host–virus combinations giving rise to chlorotic local lesions, increase in lesion diameter does not continue indefinitely. This limitation may be due in part to increasing age of the inoculated leaf, but it bears no necessary relation to the question of systemic movement.

TYMV moves out of inoculated Chinese cabbage leaves about the time local lesions first become visible, usually 4 to 5 days. Local lesions do not increase significantly in size after about 8 days. With TYMV it is possible to assess the proportion of infected cells using light microscopy of protoplasts to determine whether the chloroplasts are rounded and clumped (see Fig. 11.20). Using this procedure we have established that even in the central zone of a well-established yellow lesion there are three kinds of cell: (i) Those showing full infection at the time protoplasts were prepared. (ii) Those showing no sign of infection by light microscopy when first prepared, but that within about 6 hours, developed rounding and clumping of chloroplasts. Electron microscopic examination showed that this group was infected in the intact lesion, as judged by the presence of characteristic peripheral vesicles (see Fig. 7.17), but further development of cytological effects had been suppressed in the intact leaf. (iii) Those that showed no sign of infection by light or electron microscopy at any stage. These constituted 25–50% of cells in different experiments (R. E. F. Matthews and I. Trembath, unpulished). Tissue samples further from the center of lesions showed progressively higher proportions of cells that appeared to be uninfected. Whether the cells showing no evidence of infection in a local lesion are equivalent to dark green tissue in the systemic mosaic remains to be determined.

2. Escaping Infection

The efficiency with which localized infection resulting from mechanical inoculation leads to systemic invasion varies quite widely for different viruses and hosts.

Thus, in a rapidly growing tobacco or tomato plant a single local infection with TMV is sufficient to give systemic invasion in a high proportion of plants. In contrast, Bennett (1960) found that only about 10% of *Chenopodium capitatum* plants with 25 lesions or less became infected systemically with BYV. Two hundred or more local lesions were needed to ensure systemic infection in every plant.

Certain viruses may move with difficulty into particular leaves. SBMV and TRSV both multiply in an inoculated primary leaf of Black Valentine bean, and both readily move from this leaf to younger leaves on the plant. However, while SBMV readily moves to the uninoculated primary leaf in as short a time as 4 days, TRSV rarely moves into this leaf (Schneider, 1964). Schneider suggested that TRSV for some unknown reason becomes dependent on slow cell-to-cell invasion in the petiole of the uninoculated primary leaf.

Asymmetric infection may be even more marked with viruses that move systemically rather inefficiently. Inoculation of one primary leaf of soybean seedlings with TNV results in an asymmetric infection of the entire plant. In most trifoliate leaves, half the center leaflet and the leaflet on the same side as the inoculated primary leaf became infected. This asymmetry extended to the root system (Resconich, 1963). Some viruses that move rapidly into the root system when inoculated into leaves may not move out of the root system when the roots are inoculated, or may move only after a long delay (Roberts, 1950).

With viruses infecting woody perennials, distribution may be very uneven through the plant. For example, Gilmer and Brase (1963) found a very nonuniform distribution of prune dwarf *Ilarvirus* in mature sweet and sour cherry trees that had been infected for at least 5 years.

3. Rise and Fall in Virus Concentration with Age

The amount of some viruses rises rapidly in infected leaves and then falls off. Considering the plant as a whole, this can lead to a very uneven distribution of virus in different leaves at any given time during growth. Figure 10.8 illustrates the distribution of potato A *Potyvirus* in leaves of field-grown tobacco plants over a 21-week period.

4. Uneven Concentration in Different Organs and Tissues

In a plant infected for some time with a systemic virus, the concentration of virus may not be uniform in different organs. With most mosaic-type viruses that have been investigated, virus reaches a much higher concentration in the leaf lamina than in other parts of the plant. For example, TYMV concentration in the stem and root system and the midrib and petiole of expanded leaves in only one-tenth to one-twentieth of that found in the leaf lamina.

5. Viruses Confined to Certain Tissues

Citrus tristeza *Clostervirus* is not mechanically transmitted, but is transmitted by a phloem-feeding aphid. Price (1966) has found phloem cells in infected lime plants to be packed with viruslike rods. These rods were absent from parenchyma tissues, and it seems likely that the virus is confined to the phloem. BYDV, like other luteoviruses, is found only in phloem parenchyma, companion cells, and sieve

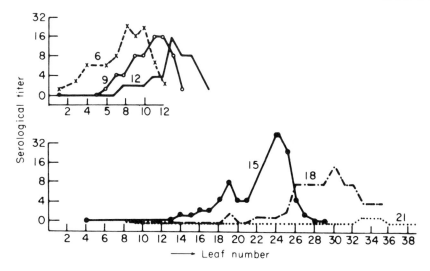

Figure 10.8   Change in concentration of potato A *Potyvirus* with age in the leaves of a single Samsun tobacco plant inoculated on leaf number 1. Virus concentration was measured serologically. From week 15 onward, some older leaves had died. Numbers beside graphs show weeks after infection. (From Bartels, 1954.)

elements (e.g., Gill and Chong, 1975). BYV appears to be confined to phloem at early stages of infection but at later stages it is found throughout the mesophyll and in the epidermis (Esau *et al.*, 1967). As judged by staining with specific fluorescent antibody, the viral antigen (protein) of WTV is confined to the pseudophloem tissue of root and stem tumors in sweet clover and to a few thick-walled cells in the xylem region (Nagaraj and Black, 1961). Similarly, curly top virus antigen was detected only in the phloem of several hosts (Mumford and Thornley, 1977). Some other viruses that are confined to the phloem were noted in Section IV,G.

6. Uneven Distribution in Leaves Showing Mosaic Patterns

Dark green areas in the mosaic patterns of diseased leaves usually contain very little virus compared with yellow or yellow-green areas. This has been found for viruses that differ widely in structure, and infecting both mono- and dicotyledonous hosts. The phenomenon may, therefore, be a fairly general one for diseases of the mosaic type. Figure 10.9 illustrates TMV distribution at the junction between a dark green and a yellow-green area in a leaf showing a mosaic pattern. Possible reasons for the low concentration of virus in dark green areas are discussed in Chapter 12, Section IV,D. This uneven distribution of virus in mosaic diseases may also extend to the petals to broken flowers. Thus TMV produces white sectors in pink tobacco flowers to which virus appears to be confined.

7. Distribution of Virus near Apical Meristems

There is evidence that some viruses may invade the primary meristematic tissues. For example, Walkey and Webb (1968) examined squashes of excised apical meristem tissue by electron microscopy and were able to show the presence of several viruses in the apices of various hosts. Virus particles have been observed in

Figure 10.9 Distribution of TMV in the region of a junction between a dark green area (right) and a yellow-green area (left) of a tobacco leaf showing mosaic, determined by electron microscopic examination of a thin section across the macroscopically visible junction. To the left of the heavy line, all cells showed at least one crystalline inclusion of TMV. To the right, no crystalline inclusions were seen, and cells appeared cytologically normal. Numbers indicate the numbers of viruslike rods seen in the section of each cell. X, crystalline TMV inclusion seen; ●, plasmodesmata observed between two cells. (Atkinson and Matthews, 1970.)

the apical initials of tobacco shoots (Roberts *et al.*, 1970). PepRSV was identified in meristematic cells of both root and shoot apices (Kitajima and Costa, 1969). Rods of PVX have been detected within or close to the potato shoot apex (Appiano and Pennazio, 1972). Rods of *Odontoglossum* ring spot *Tobamorivus* were observed in mitotically active cells in the apical meristem of *Cymbidium* (Toussaint *et al.*, 1984). Clusters of BSMV-infected cells were distributed quite unevenly in both infected root and shoot apices of wheat (Lin and Langenberg, 1984b).

However, with many virus–shoot combinations, there appears to be a zone of variable length (usually about 100 μm but up to 1000 μm) near the shoot or root tip that is free of virus or that contains very little virus (e.g., Mori *et al.*, 1982; Faccioli *et al.*, 1988). This situation has been exploited to obtain virus-free clones by growing excised shoot tips in tissue culture (Chapter 16, Section II,A,3). However, with some virus-infected plants, one can start with meristems containing virus, and after a period in tissue culture these may give rise to virus-free plantlets (Walkey *et al.*, 1969).

Root tips of plants infected with one of several viruses have been found to be free of detectable virus (e.g., Appiano and D'Agostino, 1983). Smith and Schlegel (1964) studied the distribution of clover yellow vein mosaic *Tymovirus* in root tip of *Vicia faba*. They cut serial sections and assayed for infectivity (Fig. 10.10). Within the limits of the assay method, the first 400 μm of the root tip, which included the root cap and the meristem, were virus-free.

The reasons why the meristematic zone of cells frequently fails to support virus growth are quite unknown. Normal protein synthesis is much more active in dividing than nondividing cells so that the concentration of host mRNA may be much

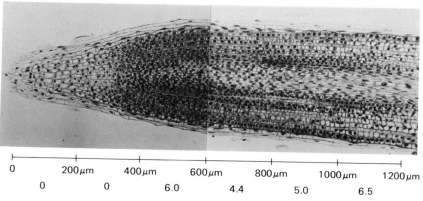

Figure 10.10 Distribution of clover yellow vein *Potyvirus* in the root tips of *Vicia faba*. Longitudinal section, showing the number of local lesions obtained on inoculation with extracts of 200-μm sections. (Smith and Schlegel, 1964.)

higher. The repertoire of mRNAs in use may also be different. Thus in dividing cells normal mRNAs may compete more effectively with viral RNAs for the translation apparatus. It is also possible that in some host–virus combinations the viral cell-to-cell movement protein is unable to function in meristematic cells. On the other hand, some viruses can invade the root meristem. Crowley *et al.* (1969), using electron microscopy, found particles of TRSV in dividing cells of bean root tips and in the cells of the inner root cap. Using the same virus and host, Atchison and Francki (1972) showed that actual virus replication took place in the root tip cells for only about 1 day following invasion. Virus concentration in the root tip must have been maintained after this initial period by translocation to the tip of virus synthesized elsewhere.

V. THE HOST RANGE OF VIRUSES

Since the early years of this century, plant virologists have used host range as a criterion for attempting to identify and classify viruses. In a typical experiment the virus under study would be inoculated by mechanical means to a range of plant species. These would then be observed for the development of viruslike disease symptoms. Back inoculation to a host known to develop disease might be used to check for symptomless infections. In retrospect it can be seen that reliance on such a procedure gives an oversimplified view of the problem of virus host ranges. Over the past few years our ideas of what we might mean by "infection" have been considerably refined, and some possible molecular mechanisms that might make a plant a host or a nonhost for a particular virus have emerged.

A. Limitations in Host Range Studies

1. General

Almost all the plant viruses so far described have been found infecting species among the angiosperms. Only a minute proportion of the possible host–virus combinations has been tested experimentally. The following arithmetic indicates the scale of our ignorance. Horváth (1983a) tested the host range of 24 viruses on 456 angiosperm species. He found 1312 new host–virus combinations, that is, 12% of those he tested. There may be about 250,000 species of angiosperms (Heywood, 1978) and about 500 plant viruses have been recorded. If the 12% rate applied on average to all these plants and viruses, then there may be about 15×10^6 new compatible host–virus combinations awaiting discovery. In relation to this figure the number of combinations already tested must be almost negligible.

Our present knowledge of the occurrence and distribution of viruses among the various groups of plants is both fragmentary and biased. There are three probable reasons for this. First, plant virologists working on diseases as they occur in the field have been primarily concerned with viruses causing economic losses in cultivated plants. They have usually been interested in other plant species only to the extent that they might be acting as reservoirs of a virus or its vector affecting a cultivated species. Thus, until fairly recently all the known plant viruses were confined to the angiosperms. Within this group, most of the known virus hosts are plants used in agriculture or horticulture or are weed species that grow in cultivated areas.

Most of the world's population is fed by 12 species of plant (three cereals, rice, wheat, and corn; two sugar plants, beet and cane; three root crops, potato, sweet potato, and cassava; two legumes, bean and soybean; and two tree crops, coconut and banana). Seventy-one of the 587 viruses recognized by the I.C.T.V. (R. I. B. Francki, personal communication) have names derived from these 12 host plants. The number of angiosperm species is commonly reported to be about 250,000. Thus about 12% of the viruses have been described from about 0.005% of the known species. Tobacco and tomato provide another example. Both these annual crops are of high value commercially on a per hectare basis. Twenty-five viruses are listed with tobacco and tomato as part of the vernacular name. Not many of the species in lower phyla are used commercially. It is not surprising, therefore, that the first virus found in these groups was one associated with cultivated mushrooms.

The second reason for our incomplete knowledge is a more speculative one. It seems likely that widespread and severe disease in plants due to virus infection is largely a consequence of human agricultural manipulations. Under natural conditions viruses are probably closely adapted to their hosts and cause very little in the way of obvious disease. Thus, casual inspection of plants growing in their natural habitat may give little indication of the viruses that might be present. Adequate testing of a significant number of such species by means of inoculation tests, both by mechanical transmission and with possible invertebrate vectors, would be laborious and time-consuming. Very little systematic testing of this sort has been carried out.

Third, the genera and species chosen for a host range study may not form a taxonomically balanced selection. For example, Watson and Gibbs (1974) pointed out that most virologists, working in the north temperate zone, use mainly festucoid grasses in host range studies; whereas in other parts of the world nonfestucoid groups predominate in the flora and in agricultural importance.

2. Technical

There are a number of technical difficulties and pitfalls in attempting to establish the host range of a virus. These were discussed in relation to diagnosis in Chapter 2, Section I,C. Another kind of potential limitation in the use of mechanical inoculation is illustrated by experiments with the viroid PSTVd. On the basis of mechanical inoculation tests, a particular accession of *Solanum acaule* has been considered to be immune to PSTVd. However, *Agrobacterium*-mediated inoculation of this plant led to the systemic replication of the viroid as did grafting with viroid-infected tomato (Salazar *et al.*, 1988b).

B. Patterns of Host Range

In spite of the preceding limitations some general points can be made. Different viruses may vary widely in their host range. At one extreme BSMV is virtually confined to barley as a host in nature (Timian, 1974). At the other, CMV, AMV, TSWV, TMV, and TRSV have very wide host ranges.

The host range of one virus may sometimes fall completely within the host range of another apparently unrelated virus. Thus, Holmes (1946) found that among 310 species tested, 83 were susceptible to both TMV and tobacco etch *Potyvirus* (TEV), 116 were susceptible to TMV but not TEV, none to infection by TEV but not TMV, and 111 were not susceptible to either virus.

Bald and Tinsley (1970) showed that eight cocoa virus isolates could be placed in a host range containment and divergence series. The most virulent isolate infected 21 of the 26 species examined, and this host range contained the host ranges of the other seven isolates.

Bald and Tinsley (1967) reexamined data of earlier workers by statistical procedures. There was a tendency for phylogenetically more advanced groups of plants to contain a higher proportion of species susceptible to the viruses examined than the lower phylogenetic groups. Generally speaking, viruses can infect a higher proportion of species in the family of the common field host and closely related families than in distantly related families.

C. The Molecular Basis for Host Range

Biological and statistical studies of the sort described in the previous section cannot lead us to an understanding of reasons why a virus infects one plant species and not another. For this we need biochemical, molecular biological, and genetical information of the sort that is beginning to become available. There is no doubt that very small changes in the viral genome can affect host range. For example, Evans (1985) described a nitrous acid mutant of CPMV that was unable to grow in cowpea but that could grow in *Phaseolus vulgaris*. The mutation was in the B-RNA.

On the basis of present knowledge there are four possible stages where a virus might be blocked from infecting a plant and causing systemic disease: (i) during initial events—the uncoating stage; (ii) during attempted replication in the initially infected cell; (iii) during movement from the first cell in which the virus replicated; and (iv) by stimulation of the host's cellular defenses in the region of the initial infection. These stages will be considered in turn.

1. Initial Events

Recognition of a Suitable Host Cell or Organelle

Bacterial viruses and most of those infecting vertebrates have specific proteins on their surface that act to recognize a protein receptor on the surface of a susceptible host cell. The surface proteins on plant rhabdoviruses may have such a cell recognition function, but this has not been established. Such a function may be unlikely with LNYV, as virus particles whose outer membranes have been removed with detergent are infectious (Randles and Francki, 1972). However, the G protein spikes (see Fig. 5.31) may be important in recognizing membranes of insect vectors. Surface proteins on plant reoviruses almost certainly have a recognition role in their insect vectors. There is no evidence for cell recognition receptors on the surface of any of the ssRNA plant viruses. The evidence available for these small viruses suggests that host range is usually a property of the RNA rather than the protein coat. When it has been tested, the host range of a plant virus is the same whether intact virus or the RNA is used as inoculum. Attempts to extend host ranges by using infectious RNA rather than virus have generally been unsuccessful. Hiebert *et al.* (1968) showed that artificial hybrid *Bromovirus* particles, consisting of CCMV RNA in a coat of BMV protein, could still infect cowpea, which is immune to BMV. The hybrid could not infect barley, the normal host of BMV. These and similar tests showed that the host range of a viral RNA cannot be extended by coating the RNA in the protein of a virus that can attack the host. Atabekov and colleagues (Atabekov, 1975) have studied the host ranges of many "hybrid" viruses and found examples where a heterologous coat protein limited host range. Thus BMV RNA in a coat of TMV was unable to infect its normal hosts (Atabekov *et al.*, 1970a). On balance it appears that viral coat proteins play little if any positive part in cell recognition. This view is supported by the fact that viral uncoating following inoculation appears not to be host specific (Section C,1,b). Surface recognition proteins may be of little use to a virus in the process of infecting a plant because of the requirement that they enter cells through wounds on the plant surface.

Leaf-hair cells have been infected with TMV by introducing the virus directly into the cell with a microneedle (Zech, 1952), thus presumably bypassing any virus–cell surface interaction. Similarly, intact virus particles or other infectious materials appear to be able to pass from cell to cell through the plasmodesmata and cause infection while remaining within the plasma membrane.

From experiments using model membranes and plant viral coat proteins in various aggregation states, Datema *et al.* (1987) concluded that hydrophobic lipid–coat protein interactions do not occur. Experiments on the binding and uptake of CCMV by cowpea protoplasts gave no evidence to support an endocytic uptake of virus mediated by specific receptors (Roenhorst *et al.*, 1988).

As discussed in Chapters 7 and 8, stages in the replication of many viruses take place in association with particular cell organelles. Recognition of a particular organelle or site within the cell by a virus (or by some subviral component or product) must be a frequent occurrence. Plant viruses may have evolved a recognition system basically different from that of viruses that normally encounter and recognize their host cells in a liquid medium or at a plasma membrane surface.

Lack of Specificity in the Uncoating Process

Various lines of evidence suggest that there is no host specificity in the process. Thus for TMV (Kiho *et al.*, 1972) and TYMV (Matthews and Witz, 1985), virus was uncoated as readily in nonhosts as in host species. Gallie *et al.* (1987c) showed that when mRNA coding for the enzyme chloramphenicol acetyltransferase was packaged into rods with TMV coat protein and inoculated to protoplasts or plant leaves, the RNA became uncoated. Francki *et al.* (1986b) showed that when velvet tobacco mottle virus and the viroid PSTVd were inoculated together onto a susceptible host, the viroid was incorporated into virus particles. When such virus was inoculated to a species that was a host only for the viroid, viroid infection occurred, indicating that the nucleic acids had been uncoated. However, these experiments did not eliminate the possibility that velvet tobacco mottle virus replicated only in the initially infected cells of the presumed nonhost species.

2. Replication

There is no recorded example where a particular step in the viral replication cycle has been proved to be a determinant of host range. Following inoculation of TMV to plant species considered not to be hosts for the virus, viral RNA has been found in polyribosomes (Kiho *et al.*, 1972). Furthermore, TMV particles uncoat and express their RNA in *Xenopus* oocytes (P. C. Turner *et al.*, 1987). However, there is some evidence that specific viral genes involved in replication may also be involved as host range determinants. Mouches *et al.* (1984) and Candresse *et al.* (1986) obtained evidence that the replicase of TYMV consists of a 115K viral-coded polypeptide and possibly a 45K subunit of plant origin. The 115K polypeptide from different tymoviruses showed great serological variability. They suggested that this variability might be a consequence of the need for the viral-coded polypeptide specifically to recognize the host subunit in different host species in order to form a functional replicase. Thus the viral-coded peptide might be directly involved in defining host specificity.

Most strains of CaMV infect only Brassicaceae. The few strains that also infect some solanaceous species can be divided into three types according to which species they infect. To determine which CaMV genes determine this host range, Schoelz *et al.* (1986) made recombinant viruses by exchanging DNA segments between the cloned strains. The resulting hybrids were then tested on the relevant solanaceous species. These experiments indicated that the first half of gene VI, which codes for the inclusion body protein (see Fig. 8.1), determines whether the virus can systemically infect *Datura stramonium* and *Nicotiana bigelovii*.

Squash leaf curl disease in the United States is caused by two distinct but highly homologous bipartite geminiviruses. The host range of one virus is a subset of the other (Lazarowitz, 1990). Analysis of agroinfected leaf disks indicated that virus replication was involved in the host restriction of one of the viruses. Replication of the restricted virus was rescued in *trans* by coinfection with the nonrestricted virus. Sequence analysis revealed that the restricted virus had a 13-nucleotide deletion in the common region. In other respects the sequences of the two common regions were almost identical. Lazarowitz (1990) suggested that this deletion may have been involved in the host range restriction.

As discussed in Chapters 6 and 7, some viruses synthesize "readthrough" proteins. Successful readthrough depends on the presence of an appropriate suppressor tRNA in the host plant (Chapter 6, Section VI,A,2). The presence of such a tRNA may, in principle at least, be a factor determining host range for viruses depending on the readthrough process.

3. Cell-to-Cell Movement

Two lines of evidence strongly support the view that possession of a compatible and functional cell-to-cell movement protein is one of the factors determining whether a particular virus can give rise to readily detectable virus replication in a given host species or cultivar.

First, the experiments summarized in Section IV,G show that many viruses contain a gene coding for a cell-to-cell movement protein. As discussed earlier, viruses cannot usually infect leaves through the cut end of petioles or stems. However, Taliansky et al. (1982a,b,c) showed that if the upper leaves were already infected with a helper virus, a virus in solution could pass into the upper leaves and infect them. This happened with mutants defective in the transport function, and also with viruses that were not normally able to infect the species of plant involved. For example, they showed that BMV can be transported from conducting tissues in tomato plants preinfected with TMV. Tomato is not normally a host for BMV. The same occurred with TMV introduced into wheat plants already infected with BSMV. A similar situation appears to exist in field infections in sweet potato, where prior infection with sweet potato feathery mottle *Potyvirus* allows CMV to infect this host (Cohen et al., 1988). Thus, when the resistance of an apparent nonhost species is due to a block in the transport function, this block can be overcome by a preinfection with a virus that has a transport protein compatible with the particular host. This appears to be a fairly general phenomenon.

The second line of evidence concerning the importance of cell-to-cell movement in host range is that a number of viruses have been shown to be able to infect and replicate in protoplasts derived from species in which they show no macroscopically detectable sign of infection following mechanical inoculation of intact leaves. For example, Beier et al. (1977) attempted to infect 1031 lines of cowpea by mechanical inoculation with CPMV. Sixty-five lines were defined as operationally immune because no disease developed and no virus could be recovered. Protoplasts could be prepared from 55 of the "immune" lines. Fifty-four of these could be infected with CPMV. Similar results have been obtained with other hosts and viruses.

Sulzinski and Zaitlin (1982) mechanically inoculated cowpea and cotton plants with TMV. They isolated protoplasts at intervals and determined the proportion of infected cells at various times using fluorescent antibody. Only about 1 in $5–15 \times 10^4$ protoplasts were infected and this number remained unchanged for at least 11 days. These results show that subliminally infected plants can support virus replication in individual cells, but that the virus cannot move out of the initially infected cells. Viruses that can be induced to invade and replicate systemically by a helper virus, as described in the first part of this section, can presumably infect the host concerned on their own in a similar subliminal fashion. This idea is supported by the observation that the *Geminivirus* BGMV, normally confined to phloem tissue, moved to cells of many types in bean leaves doubly infected with BGMV and TMV (Carr and Kim, 1983).

However, not all sequence differences in the movement proteins can be involved in defining host range. Solis and García-Arenal (1990) showed that there were major differences in the C-terminal region of the movement proteins of TMV, another *Tobamovirus*, TMGMV, and TRV, and yet all of these viruses share the same range of natural hosts in the Solanaceae.

A functional cell-to-cell movement protein may not be the only requirement for systemic movement of a virus. CCMV infects dicotyledons (cowpea) and the related BMV replicates in monocotyledons (barley). Both viruses replicate in protoplasts of both hosts. If the movement protein was the only factor controlling systemic movement, transferring RNA3 between the two viruses should switch the host range. This does not occur. Therefore systemic infection must also involve factors coded for by RNAs 1 or 2 or both (Allison et al., 1988). Since BMV must be encapsidated for systemic movement (Section IV,B) it is probable that all four genes are involved in systemic movement and host specificity.

4. Stimulation of Host-Cell Defenses

In some host–virus combinations the virus stimulates host-cell defense mechanisms that may be a factor limiting virus replication and movement to cells near the first infected cell, giving rise to local lesions, without subsequent systemic spread. This phenomenon, which makes the plant resistant to the virus from a practical point of view, is discussed in Chapter 12, Section III.

5. Host Genes Affecting Host Range

Host genes must be involved in all interactions between a virus and the plant cell following inoculation, whatever the outcome may be. The effects of certain host genes that limit virus infection to local lesions have been studied in some detail. These are discussed in Section V,C,5 and Chapter 12, Section III.

6. Summary

Concerning the molecular basis for viral host ranges, present evidence suggests that: (i) There is usually no virus–host cell receptor recognition system and little host specificity in the initial uncoating process. (ii) In some plant–virus combinations molecular mechanisms may exist for blocking virus at some stage in the replication process in the cells where virus first gained entry. This forms the basis for true immunity to infection. (iii) In many host–virus combinations replication can occur in the first infected cells but movement out of these cells is not possible because of a mismatch between a viral movement protein and some host-cell structure. This situation gives rise to subliminal infection and, for practical purposes, a resistant species or cultivar. (iv) In other virus–host combinations movement may be limited by host responses to cells in the vicinity of the initially infected cell, giving rise to a "local lesion host," which from a practical point of view is "field resistant." Using these ideas, the different kinds of host response to inoculation with a virus are defined in Table 12.1

VI. DISCUSSION AND SUMMARY

The ability to be transferred to healthy host plant individuals is crucial for the survival of all plant viruses. Viruses cannot, on their own, penetrate the undamaged plant surface. For this reason each kind of virus has evolved ways to bypass or overcome this barrier. Viruses that are transmitted from generation to generation with high frequency via pollen and seed or in plant parts in vegetatively reproducing species can avoid the need to penetrate the plant surface.

Many groups of plant viruses are transmitted by either invertebrate or fungal vectors, which penetrate the plant surface during the process of feeding or infection, carrying infectious virus into the plant cells at the same time.

Mechanical transmission is a process whereby small wounds are made on the plant surface in the presence of infectious virus. For many groups of viruses, if conditions are favorable, infection follows. Mechanical inoculation is a very important procedure in experimental virology. In the field, it is not of great importance except for a few groups such as tobamoviruses and potexviruses, which appear to have no other means of transmission. These viruses are, however, well adapted to this form of transmission as they are relatively stable and occur in high concentration in infected leaves. These features make transmission possible when leaves of neighboring plants abrade one another.

It is probable that successful entry of a single virus particle into a cell is sufficient to infect a plant, but in the normal course of events many may enter a single cell. Following initial replication in the first infected cell, virus moves to neighboring cells via the plasmodesmata, this being a relatively slow process. Virus then enters the vascular tissue, usually the phloem, movement in this tissue being much more rapid. Some viruses code for specific gene products that are necessary for the virus to move out of the initially infected cell and through the plant.

The final distribution of virus through the plant may be quite uneven. In some host–virus combinations virus movement is limited to local lesions. In others, some leaves may escape infection, while in mosaic diseases dark green islands of tissue may contain little or no virus. Some viruses are confined to certain tissues. For example, luteoviruses are confined mainly to the phloem. Some viruses penetrate to the dividing cells of apical meristems. Others appear not to do so.

There is clear evidence that both virus particles and infectious viral RNA can move from cell to cell, but this may not be the form in which all viruses move. Evidence suggests that TMV may move in the form of nucleoprotein complexes made up from viral RNA, viral-coded proteins, and some host components. Viral-coded movement proteins may exert their effect by altering the properties of the plasmodesmata to facilitate viral passage, but this has not been proven.

The molecular basis for the host range of viruses is not understood. Most plant viruses do not have a mechanism by which intact particles can recognize a host cell that is suitable for replication. Uncoating of the viral nucleic acid appears to occur in hosts and nonhosts alike. No particular step in the replication of a virus has yet been implicated in limiting host range. However, there is good evidence that in "nonhosts" of some viruses, the virus can replicate in the first infected cell but cannot move to neighboring cells. This is because the viral-coded cell-to-cell movement protein does not function in the particular plant. For practical purposes such a plant is resistant to the virus in question.

11

Disease Symptoms and Effects on Metabolism

Viruses are economically important only when they cause some significant deviation from normal in the growth of a plant. For experimental studies we are usually dependent on the production of disease in some form to demonstrate biological activity. Symptomology was particularly important in the early days of virus research, before any of the viruses themselves had been isolated and characterized. Dependence on disease symptoms for identification and classification led to much confusion, because it was not generally recognized that many factors can have a marked effect on the disease produced by a given virus (Chapter 12, Section V).

Most virus names in common use include terms that describe an important symptom in a major host or the host from which the virus was first described. There is a vast literature describing diseases produced by viruses. This has been summarized by Smith (1937, 1972), Bos (1978) Holmes (1964), and in the C.M.I./A.A.B. Descriptions of Plant Viruses issued from 1970 onward and edited by B. D. Harrison and A. F. Murant. Some viruses under appropriate conditions may infect a plant without producing any obvious signs of disease. Others may lead to rapid death of the whole plant. Between these extremes a wide variety of diseases can be produced.

Virus infection does not necessarily cause disease at all times in all parts of an infected plant. We can distinguish six situations in which obvious disease may be absent: (i) infection with a very mild strain of the virus; (ii) a tolerant host; (iii) nonsterile "recovery" from disease symptoms in newly formed leaves; (iv) leaves that escape infection because of their age and position on the plant; (v) dark green areas in a mosaic pattern (discussed in Chapter 12, Section IV,D; and (iv) plants infected with cryptic viruses (Section I,D).

I. MACROSCOPIC SYMPTOMS

A. Local Symptoms

Localized lesions that develop near the site of entry on leaves are not usually of any economic significance but are important for biological assay (Chapter 2). Infected cells may lose chlorophyll and other pigments, giving rise to chlorotic local lesions (Fig. 11.1A). The lesion may be almost white or merely a slightly paler shade of green than the rest of the leaf. In a few diseases, for example, in older leaves of tomato inoculated with TBSV, the lesions retain more chlorophyll than the surrounding tissue. For many host–virus combinations, the infected cells die, giving rise to necrotic lesions. These vary from small pinpoint areas to large irregular

Figure 11.1 Types of local lesion. (A) Chlorotic lesions caused by beet mosaic *Potyvirus* in *Chenopodium amaranticolor*. (Courtesy of P. R. Fry.) (B) Ring spot lesions due to AMV in tobacco. (Courtesy of S. A. Rumsey.) (C) Necrotic lesions due to tomato spotted wilt virus in tobacco. (Courtesy of E. E. Chamberlain.)

spreading necrotic patches (Figs. 11.1B and C). In a third type, ring spot lesions appears. Typically these consist of a central group of dead cells. Beyond this develop one or more superficial concentric rings of dead cells with normal green tissue between them (Fig. 11.1B). Some ring spot local lesions consist of chlorotic rings rather than necrotic ones. Some viruses in certain hosts show no visible local lesions in the intact leaf, but when the leaf is cleared in ethanol and stained with iodine, "starch lesions" may become apparent.

Viruses that produce local lesions when inoculated mechanically onto leaves may not do so when introduced by other means. For example, BYV produces necrotic local lesions on *Chenopodium capitatum,* but does not do so when the virus is introduced by the aphid *Myzus persicae* feeding on parenchyma cells (Bennett, 1960). However, AMV does produce local lesions following aphid transmission.

B. Systemic Symptoms

The following sections summarize the major kinds of effects produced by systemic virus invasion. It should be borne in mind that these various symptoms often appear in combination in particular diseases, and that the pattern of disease development for a particular host–virus combination often involves a sequential development of different kinds of symptoms.

1. Effects on Plant Size

Reduction in plant size is the most general symptom induced by virus infection (e.g., Figs. 11.2 and 12.7). There is probably some slight general stunting of growth even with "masked" or "latent" infections where the systemically infected plant shows no obvious sign of disease. For example, mild strains of PVX infecting potatoes in the field may cause no obvious symptoms, and carefully designed experiments were necessary to show that such infection reduced tuber yield by about 7–15% (Matthews, 1949d). The degree of stunting is generally correlated with the severity of other symptoms, particularly where loss of chlorophyll from the leaves is concerned. Stunting is usually almost entirely due to reduction in leaf size and internode length. Leaf number may be little affected.

In perennial deciduous plants such as grapes there may be a delayed initiation of growth in the spring (e.g., Gilmer *et al.,* 1970). Root initiation in cuttings from virus-infected plants may be reduced, as in chrysanthemums (Horst *et al.,* 1977).

In vegetatively propagated plants, stunting is often a progressive process. For example, virus-infected strawberry plants and tulip bulbs may become smaller in each successive year.

Stunting may affect all parts of the plant more or less equally, involving a reduction in size of leaves, flowers, fruits, and roots and a shortening of petioles and internodes. Alternatively, some parts may be considerably more stunted than others. For example, in little cherry disease, fruits remain small due to reduced cell division, in spite of apparently ample leaf growth. A reduction in total yield of fruit is a common feature and an important economic aspect of virus disease. The lower yield may sometimes be due to a reduction in both size and number of fruits (e.g., Hampton, 1975). In a few diseases, for example, prune dwarf, fruits may be greatly reduced in number but of larger size than normal. Healthy cherry trees pollinated with pollen from trees infected with this virus (Way and Gilmer, 1963) or necrotic ring spot virus (Vértesy and Nyéki, 1974) had a reduced fruit set. Seed from infected

Figure 11.2 Stunting effect of rice dwarf *Phytoreovirus* infection in rice. Healthy plant on right. Plants infected with a standard strain (O) and a severe mutant (S). (From Kimura *et al.*, 1987.)

plants may be smaller than normal, germination may be impaired, and the proportion of aborted seed may be increased (Walkey *et al.*, 1985).

2. Mosaic Patterns and Related Symptoms

One of the most common obvious effects of virus infection is the development of a pattern of light and dark green areas giving a mosaic effect in infected leaves. The detailed nature of the pattern varies widely for different host–virus combinations. In dicotyledons, the areas making up the mosaic are generally irregular in outline. There may be only two shades of color involved—dark green and a pale or yellow-green, for example. This is often so with TMV in tobacco. Or there may be many different shades of green and yellow, as with TYMV in Chinese cabbage. The junctions between areas of different color may be sharp and such diseases resemble quite closely the mosaics produced by inherited genetic defects in the chloroplasts (Fig. 11.14). *Abutilon* mosaic *Geminivirus* is a good example of this type. TYMV in Chinese cabbage may approach genetic variegation in the sharpness of the mosaic pattern it produces (Fig. 11.3).

The borders between darker and lighter areas may be diffuse. If the lighter areas differ only slightly from the darker green, the mottling may be difficult to observe as with some of the milder strains of PVX in potato. In mosaic diseases infecting herbaceous plants there is usually a fairly well-defined sequence in the development of systemic symptoms. The virus moves up from the inoculated leaf into the growing shoot and into partly expanded leaves. In these leaves the first symptoms are a "clearing" or yellowing of the veins. This may be very faint or may give striking emphasis to the pattern of veins (Fig. 11.4A). Vein-clearing may persist as a major feature of the disease.

In leaves that are past the cell division stage of leaf expansion when they become infected (about 4–6 cm long for leaves such as tobacco and Chinese cabbage) no mosaic pattern develops. The leaves become uniformly paler than normal. In the oldest leaves to show mosaic a large number of small islands of dark green tissue usually appears against a background of paler color. The mosaic areas may be confined to the youngest part of the leaf blade, that is, the basal and central region. Although there may be considerable variation in different plants, successively younger systemically infected leaves show, on the average, mosaics consisting of fewer and larger areas. The mosaic pattern is laid down at a very early stage of leaf development and may remain unchanged, except for general enlargement, for most of the life of the leaf. In some mosaic diseases the dark green areas are associated mainly with the veins to give a vein-banding pattern (Fig. 11.4B).

Figure 11.3 Mosaic disease in Chinese cabbage caused by TYMV. Leaf from a plant infected with the Cambridge stock culture. Inoculation with extracts from different colored blocks of tissue in this leaf gave plants with different symptoms. (Courtesy of J. Endt.)

Figure 11.4   (A) Vein-clearing symptoms in lettuce due to lettuce big vein agent. Healthy leaf on right. (Courtesy of J. Endt.) (B) Vein banding due to CaMV in cauliflower. (Courtesy of S. A. Rumsey.)

In monocotyledons a common result of virus infection is the production of stripes or streaks of tissue lighter in color than the rest of the leaf. The shades of color vary from pale green to yellow or white, and the more or less angular streaks or stripes run parallel to the length of the leaf (Fig. 11.5).

The development of the stripe diseases found in monocotyledons follows a similar general pattern to that found for mosaic diseases in dicotyledons. One or a few leaves above the inoculated leaf that were expanded at time of inoculation show no stripe pattern. In the first leaf to show striping, the pattern is relatively fine and may occur only in the basal (younger) portion of the leaf blade. In younger leaves, stripes tend to be larger and occur throughout the leaf. The patterns of striping are laid down at an early stage and tend to remain unchanged for most of the life of the leaf. Yellowed areas may become necrotic as the leaf ages.

A variegation or "breaking" in the color of petals commonly accompanies mosaic or streak symptoms in leaves. The breaking usually consists of flecks, streaks, or sectors of tissue with a color different from normal (Fig. 11.6). The breaking of petal color is frequently due to loss of anthocyanin pigments, which reveals any underlying coloration due to plastid pigments. In a few instances, for example, in tulip color-adding virus, infection results in increased pigmentation in some areas of the petals. Nectar guides in petals are often invisible to humans (but visible to bees) because they involve pigments that absorb strongly only in the ultraviolet region (Thompson *et al.*, 1972). The effects of virus infection on these nectar guides and on the behavior of honey bees do not appear to have been studied.

Infected flowers are frequently smaller than normal and may drop prematurely. Flower breaking may sometimes be confused with genetic variegation, but it is usually a good diagnostic feature for infection by viruses producing mosaic symptoms. In a few plants virus-induced variegation has been valued commercially. Thus, at one time virus-infected tulips were prized as distinct varieties. As with the development of mosaic patterns in leaves, color-breaking in the petals may develop only in flowers that are smaller than a certain size when infected. Thus, in tulips inoculated with tulip breaking virus less than 11 days before blooming no break symptoms developed even though virus was present in the petals (Yamaguchi and Hirai, 1967). Virus infection may reduce pollen production and decrease seed set, seed size, and germination (e.g., Hemmati and McLean, 1977) (see also Fig. 11.17).

Fruits formed on plants showing mosaic disease in the leaves may show a mottling, for example, cucumbers infected with CMV (Fig. 11.7A). In other diseases, severe stunting and distortion of fruit may occur (Fig. 11.7B). Seed coats of infected seed may be mottled.

Figure 11.5 Narcissus mosaic *Potexvirus* in daffodil. Healthy leaf on right. (Courtesy of S. A. Rumsey.)

Figure 11.6 Flower symptoms. (A) Turnip mosaic *Potyvirus* in stock. (Courtesy of R. I. Hughes.) (B) CMV in gladiolus. (Courtesy of S. A. Rumsey.) (C) CMV in violet. Healthy flower on left. (Courtesy of S. A. Rumsey.)

3. Yellow Diseases

Viruses that cause a general yellowing of the leaves are not as numerous as those causing mosaic diseases, but some, such as the viruses causing yellows in sugar beet, are of considerable economic importance. The first sign of infection is usually a clearing or yellowing of the veins in the younger leaves followed by a

Figure 11.7 Fruit symptoms. (A) CMV in cucumber. Healthy fruit on left. (Courtesy of H. Drake.) (B) Pear stony pit virus in pear. Healthy fruit on left. (Courtesy on L. H. Wright.) (C) Apple, var. Granny Smith with concentric ring and russett patterns due to apple ring spot virus. (Courtesy of S. A. Rumsey.)

general yellowing of the leaves. This yellowing may be slight or severe. No mosaic is produced, but in some leaves there may be sectors of yellowed and normal tissue. In strawberry yellow edge disease, yellowing is largely confined to the margins of the leaf (Fig. 11.8). When severe, a yellows disease may lead to a total loss of the crop (e.g., Weidemann *et al.*, 1975).

4. Ring Spot Diseases

The major symptom in many virus diseases is a pattern of concentric rings and irregular lines on the leaves and sometimes also on the fruit (Fig. 11.7C). The lines

Figure 11.8   Strawberry yellow edge disease in strawberry. (Courtesy of S. A. Rumsey.)

may consist of yellowed tissue or may be due to death of superficial layers of cells, giving an etched appearance. In severe diseases, complete necrosis through the full thickness of the leaf lamina may occur. Figure 11.9 illustrates ring and line patterns. With the ring spot viruses, such as TRSV, there is a marked tendency for plants to recover from the disease after an initial shock period. Leaves that have developed symptoms do not lose these, but younger growth may show no obvious symptoms in spite of the fact that they contain virus. Ring spot patterns may also occur on other organs, for example, bulbs (Asjes *et al.,* 1973).

5. Necrotic Diseases

Death of tissues, organs, or the whole plant is the main feature of some diseases. Necrotic patterns may follow the veins as the virus moves into the leaf (Fig. 11.10). In some diseases the whole leaf is killed. Necrosis may extend fairly rapidly throughout the plant. For example, with PVX and PVY in some varieties of potato, necrotic streaks appear in the stem. Necrosis spreads rapidly to the growing point, which is killed (see Fig. 10.1), and subsequently all leaves may collapse and die. Such systemic necrotic disease is often preceded by wilting of the parts that are about to become necrotic.

6. Developmental Abnormalities

Besides being generally smaller than normal, virus-infected plants may show a wide range of developmental abnormalities. Such changes may be the major feature of the disease or may accompany other symptoms. For example, uneven growth of

Figure 11.9   Systemic ring and line patterns induced by a yellow strain of Oregon TRV in tobacco (left) and tomato (right). (Courtesy of R. M. Lister.)

Figure 11.10   A pattern of systemic necrosis in tobacco following invasion by tomato spotted wilt virus. Left: Leaf photographed in visible light. Right: Photographed in ultraviolet light, showing fluorescence due to accumulation of the fluorescent compound scopoletin in the diseased tissue. (From Best, 1936b.)

the leaf lamina is often found in mosaic diseases. Dark green areas may be raised to give a blistering effect, and the margin of the leaf may be irregular and twisted (Fig. 11.11A). In some diseases, the leaf blade may be more or less completely suppressed, for example, in tomatoes infected with CMV and/or TMV (Fig. 11.11B) (Francki *et al.*, 1980a). In others, the leaf blade may be rolled either upward or

Figure 11.11 Growth abnormalities. (A) A mosaic virus in runner bean. (Courtesy of H. Drake.) (B) A mixed infection with TMV and CMV in tomato. (Courtesy of E. E. Chamberlain.)

Figure 11.12   Galls on stem of white sweet clover, caused by WTV. (Courtesy of L. M. Black.)

downward. Pronounced epinasty of leaf petioles may sometimes be a prominent feature.

Some viruses cause swellings in the stem, which may be substantial in woody plants, for example, in cocoa swollen shoot disease. Another group of growth abnormalities is known as enations. These are outgrowths from the upper or lower surface of the leaf usually associated with veins. They may be small ridges of tissue or larger, irregularly shaped leaflike structures, or long filiform outgrowths. Conversely, normal outgrowths may be suppressed. For example, a potyvirus infection in *Datura metel* causes the production of fruits lacking the normal spines (Rao and Yaraguntiah, 1976).

A variety of tumorlike growths may be caused by viruses. The tumor tissue is less organized than with enations. Some consist of wartlike outgrowths on stems or fruits. The most studied tumors are those produced by WTV. The tumors are characteristic of this disease (Fig. 11.12). In a systemically infected plant, external tumors appear on leaves or stems where wounds are made. In infected roots they appear spontaneously, beginning development close to cells in the pericycle that are wounded when developing side roots break through the cortex. The virus may also cause many small internal tumors in the phloem of the leaf, stem, and root (Lee and Black, 1955).

7. Wilting

Wilting of the aerial parts frequently followed by death of the whole plant may be an important feature (e.g., in virus diseases of chick-pea, Kaiser and Danesh, 1971).

8. Recovery from Disease

Not uncommonly a plant shows disease symptoms for a period and then new growth appears in which symptoms are milder or absent, although virus is still present. This commonly occurs with *Nepovirus* infections. Many factors influence this recovery phenomenon. Environmental factors are discussed in Chapter 12, Section V. The stage of development at which a plant is infected may have a marked effect on the extent to which symptoms are produced. For example, tobacco plants inoculated with sugar beet curly top *Geminivirus* develop disease symptoms. This stage is frequently followed by a recovery period. If very young seedling plants are inoculated, a proportion of these might never show clear symptoms, even though they can be shown to contain virus (Benda and Bennett, 1964).

9. Reduced Nodulation

Various workers have described a reduction in the number, size, and fresh weight of nitrogen-fixing *Rhizobium* nodules induced by virus infection in legumes (e.g., AMV in alfalfa, Ohki *et al.*, 1986; CMV in pea, Rao *et al.*, 1987). In general, overall nitrogen fixation is reduced by virus infection. With AMV infection in *Medicago*, nitrogen fixation per unit of nodule fresh weight is the same as in healthy plants, but infected plants produce less nodule tissue. Hence nitrogen fixation per plant is reduced (Dall *et al.*, 1989b).

10. Genetic Effects

Infection with BSMV induces an increase in mutation rate in *Zea mays* and also a genetic abnormality known as an aberrant ratio (AR) (Sprague *et al.*, 1963; Sprague and McKinney, 1966, 1971). For example, if a virus-infected pollen parent with the genetic constitution A_1A_1, *PrPr*, *SuSu* was crossed with a homozygous recessive line $(a_1a_1$, *prpr*, *susu*,) resistant to the virus, a low frequency of the progeny lines gave significant distortion from the expected ratios for one or more of the genetic markers. $(A_1-a_1$ = presence or absence of aleurone color; *Pr–pr* = purple or red aleurone color; *Su–su* = starchy or sugary seed.) This AR effect was observed only when the original pollen parent was infected and showing virus symptoms on the upper leaves. The AR effect is inherited in a stable manner, with a low frequency of reversion to normal ratios. It is inherited in plants where virus can no longer be detected. Wheat streak mosaic *Potyvirus* also induces the AR effect (Brakke, 1984). The genetics of the AR effect are complex and probably more than one phenomenon is involved. There is no evidence that a cDNA copy of part or all of the viral genome is incorporated into the host genome. The AR effect has been reviewed by Brakke (1984). More is known about the genetics of *Zea mays* than any other angiosperm, which may account for the discovery of the AR effect in this species. Further work may show that similar phenomena occur in other virus-infected plants.

C. Agents Inducing Viruslike Symptoms

Disease symptoms similar to those produced by viruses can be caused by a range of physical, chemical, and biological agents. Such diseases may have interesting factors in common with virus-induced disease. The activities of such agents have sometimes led to the erroneous conclusion that a virus was the cause of a particular disease in the field. Further confusion may arise when a disease is caused by the combined effects of a virus and some other agent.

1. Small Cellular Parasites

A group of diseases characterized by symptoms such as general yellowing of the leaves, stunting, witches-broom growth of axillary shoots, and a change from floral to leaf-type structures in the flowers (phyllody) was for many years thought to be caused by viruses. Diseases of this type are not transmissible by mechanical means. They were considered to be caused by viruses because (i) no bacteria, rickettsiae, fungi, or protozoa could be implicated; (ii) they were graft transmissible; (iii) they were transmitted by leafhoppers; and (iv) some, at least, could be transmitted by dodder.

Mycoplasmas and spiroplasmas belong to the Mycoplasmatales. They are characterized by a bounding unit membrane, pleomorphic form, the absence of a cell wall, and complete resistance to penicillin.

Rickettsialike organisms are distinguished from the Mycoplasmatales by the possession of a cell wall and by their *in vivo* susceptibility to penicillin. They have been found in the phloem of various species.

Since the pioneering experiments of Doi *et al.* (1967) and Ishiie *et al.* (1967) a new branch of plant pathology has been opened up by the demonstration that mycoplasmas, spiroplasmas, and rickettsialike organisms can cause disease in plants (e.g., Windsor and Black, 1973). These agents are generally confined to the phloem or xylem of diseased plants. They are too small to be readily identified by light microscopy. The widespread availability of electron microscopy provided the key technique that has allowed their importance in plant pathology to be recognized. Mycoplasma diseases of plants are described in Maramorosch and Raychaudhuri (1988).

A simultaneous virus and mycoplasma infection may give rise to a more severe disease than either agent alone. For example, oat blue dwarf *Marafivirus* and the mycoplasm agent of aster yellows are both confined to the phloem, and both can be transmitted by the leaf hopper *Macrosteles fascifrons* (Stål.). In a mixed infection they cause more severe stunting than either agent alone (Banttari and Zeyen, 1972).

2. Toxins Produced by Arthropods

Insects and other arthropods feeding on plants may secrete very potent toxins, which move systemically through the plant and produce viruslike symptoms. *Calligypona pellucida* (F.) (Homoptera) produces salivary toxins that cause general retardation of growth and inhibition of tillering. Only females produce the toxins (Nuorteva, 1962).

Eryophyid mites feeding on clover may induce a mosaiclike mottle in the younger leaves. One mite is sufficient to induce such symptoms (Fig. 11.13). These mites are small and may be overlooked since they burrow within the leaf. A viruslike mosaic disease of wax myrtle was found to be caused by a new species of eryophyid mite (Elliot *et al.*, 1987).

Maize plants showing wallaby ear disease are stunted and develop galls on the underside of the leaves, which are sometimes darker green than normal. This

Figure 11.13 Viruslike symptoms induced in lucerne 3 weeks after a single ereophyid mite was placed on the plant. (Courtesy of L. Stubbs.)

viruslike condition has been shown to be due to a toxin produced by the leaf hopper *Cicadulina bimaculata* (Ofori and Francki, 1983).

3. Genetic Abnormalities

Numerous cultivated varieties of ornamental plants have been selected by horticulturists because they possess heritable leaf variegations or mosaics. These are often due to maternally inherited plastid defects. The variegated patterns produced sometimes resemble virus-induced mosaics quite closely. However, in mosaics due to plastid mutation, the demarcation between blocks of tissue of different colors tends to be sharper than with many virus-induced mosaics (Fig. 11.14).

Other virus symptoms may be mimicked by genetic abnormalities. For example, Edwardson and Corbett (1962) described a "wiry" mutant in Marglobe tomatoes that gave an appearance similar to that of plants infected with strains of CMV and TMV. All the leaves of the mutant were like the upper leaves lacking a lamina shown in Fig. 11.11B. The disease could not be transmitted by grafting. Genetic experiments suggested that the mutant phenotype is controlled by a pair of recessive genes.

4. Nutritional Deficiencies

Plants may suffer from a wide range of nutritional deficiencies that cause abnormal coloration, discoloration, or death of leaf tissue. Some of these conditions

Figure 11.14 Mosaic due to a defective chloroplast mutation in tobacco. Right: A leaf from a tobacco plant of the mutant described by Burk *et al*. (1964). Left: Mosaic due to type TMV. (Courtesy of J. Fields.)

can be fairly easily confused with symptoms due to virus infection. For example, magnesium and iron deficiency in soybeans leads to a green banding of the veins with chlorotic interveinal areas. The yellowing, however, is more diffuse than is usual in virus infection. In sugar beet, yellowing and necrosis due to magnesium deficiency may be similar to the disease produced by BYV. Potassium and magnesium deficiency in potatoes produces marginal and interveinal necrosis similar to that found in certain virus diseases.

5. High Temperatures

Growing plants at substantially higher temperatures than normal may induce viruslike symptoms. When *N. glutinosa* plants were held at 37.8°C for 4–8 days and then returned to 22°C, new leaves displayed a pattern of mosaic, vein-clearing, chlorosis, and other abnormalities with a resemblance to virus infection (Fig. 11.15). These symptoms gradually disappeared in newer leaves, but could be induced in the same plants again by a second treatment at high temperature (John and

Figure 11.15 Viruslike symptoms induced in *N. glutinosa* following a period of growth at high temperature (37.8°C). (From John and Weintraub, 1966.)

Weintraub, 1966). Mechanical inoculation and grafting tests to various hosts and electron microscopy failed to reveal the presence of a virus in heated plants.

6. Hormone Damage

Commercially used hormone weed killers may produce viruslike symptoms in some plants. Tomatoes and grapes are particularly susceptible to 2,4-di-chlorophenoxyacetic acid (2,4-D). Growth abnormalities in the leaves caused by 2,4-D bear some resemblance to certain virus infections in these and other hosts. Compounds related to 2,4-D can cause almost complete suppression of mesophyll development, giving a plant with a "shoestring" appearance. Alternatively, vein growth may be retarded more than the mesophyll. Mesophyll may then bulge out between the veins to give an appearance not unlike leaf curl diseases.

7. Insecticides

Certain insecticides have been reported to produce leaf symptoms that mimic virus infection (e.g., Woodford and Gordon, 1978).

8. Air Pollutants

Many air pollutants inhibit plant growth and give rise to symptoms that could be confused with a virus disease. For example, chimney gases from a cement factory caused *Zea mays* seedlings to become stunted and yellowed, with necrotic areas and curled leaf margins (Cireli, 1976).

D. The *Cryptovirus* Group

The viruses now placed in the *Cryptovirus* group escaped detection for many years because most of them cause no visible symptoms, or in a few situations very mild symptoms. They are not transmissible mechanically or by vectors, but are transmitted efficiently in pollen and seed. They occur in very low concentrations in infected plants (reviewed by Boccardo *et al.*, 1987). Nevertheless, they have molecular characteristics that might be expected of disease-producing viruses.

The genome consists of two dsRNA segments. The RNA can be translated *in vitro* after denaturation to give two polypeptides (Accotto *et al.*, 1987). The polyhedral virus particles contain an RNA-dependent RNA polymerase activity (Boccardo and Accotto, 1988; Marzachi *et al.*, 1988). Thus they share some properties with the reoviruses. There is no indication, other than the low concentration at which they occur, as to why they cause symptomless infection.

II. HISTOLOGICAL CHANGES

The macroscopic symptoms induced by viruses frequently reflect histological changes within the plant. These changes are of three main types—necrosis, hypoplasia, and hyperplasia—that may occur singly or together in any particular disease. For example, all three are closely linked in the citrus exocortis disease (Fudl-Allah *et al.*, 1971).

A. Necrosis

Necrosis as the major feature of disease was discussed in Section I,B,5. In other diseases necrosis may be confined to particular organs and tissues and may be very localized. It commonly occurs in combination with other histological changes. It may be the first visible effect or may occur as the last stage in a sequence. For example, necrosis of epidermal cells or of midrib parenchyma may be caused by lettuce mosaic virus in lettuce (Coakley *et al.*, 1973). Necrosis caused by TNV is usually confined to localized areas of the roots (e.g., Lange, 1975). Late infection of virus-free tomato plants with TMV may give rise to internal necrosis in the immature fruits (e.g., Taylor *et al.*, 1969).

In the potato leaf roll disease, the phloem develops normally but is killed by the infection. Necrosis may spread in phloem throughout the plant, but is limited to this tissue (Shepardson *et al.*, 1980).

In *Pelargonium* infected with tomato ring spot *Nepovirus*, histological effects seen by light microscopy were confined to reproductive tissues (Murdock *et al.*, 1976). Pollen grain abortion and abnormal and aborted ovules were common. The symptoms could be confused with genetic male sterility.

B. Hypoplasia

Leaves with mosaic symptoms frequently show hypoplasia in the yellow areas. The lamina is thinner than in the dark green areas, and the mesophyll cells are less differentiated with fewer chloroplasts and fewer or no intercellular spaces (Fig. 11.16).

In stem-pitting disease of apples, pitting is shown on the surface of the wood when the bark is lifted. The pitting is due to the failure of some cambial initials to differentiate cells normally, and a wedge of phloem tissue is formed that becomes embedded in newly formed xylem tissue (Hilborn *et al.*, 1965). The affected phloem becomes necrotic.

The major anatomical effect of apple stem grooving *Closterovirus* in apple stems is the disappearance of the cambium in the region of the groove. Normal phloem and xylem elements are replaced by a largely undifferentiated parenchyma (Pleše *et al.*, 1975).

Reduced size of pollen grain and reduced growth of pollen tubes from virus-infected pollen may be regarded as hypoplastic effects (Fig. 11.17). A variety of other effects on pollen grains have been described (e.g., Haight and Gibbs, 1983).

C. Hyperplasia

1. Cells Are Larger Than Normal

Vein-clearing symptoms are due, with some viruses at least, to enlargement of cells near the veins (Esau, 1956). The intercellular spaces are obliterated, and since there is little chlorophyll present the tissue may become abnormally translucent.

2. Cell Division in Differentiated Cells

Some viruses such as PVX may produce islands of necrotic cells in potato tubers. The tuber may respond with a typical wound reaction in a zone of cells around the necrotic area. Starch grains disappear and an active cambial layer develops (Fig. 11.18).

Similarly, in a white halo zone surrounding necrotic local lesions induced by TMV in *N. glutinosa* leaves, cell division occurred in mature palisade cells (Wu, 1973).

3. Abnormal Division of Cambial Cells

The vascular tissues appear to be particularly prone to virus-induced hyperplasia. In the diseased shoots found in swollen shoot disease of cocoa, abnormal amounts of xylem tissue are produced but the cells appear structurally normal (Posnette, 1947). In plants infected by beet curly top *Geminivirus,* a large number of abnormal sieve elements develop, sometimes associated with companion cells. The arrangement of the cells is disorderly and they subsequently die (Esau, 1956; Esau and Hoefert, 1978). Oat blue dwarf *Marafivirus* causes abnormalities in the development of phloem in oats, involving hyperplasia and limited hypertrophy of the phloem procambium (Zeyen and Banttari, 1972).

In crimson clover infected by WTV, there is abnormal development of phloem cambium cells. Phloem parenchyma forms meristematic tumor cells in the phloem of leaf, stem, and root (Lee and Black, 1955).

Galls on sugarcane leaves arise from *Fijivirus*-induced cell proliferation. This gives rise in the mature leaf to a region in the vein where the vascular bundle is grossly enlarged (Fig. 11.19). Two main types of abnormal cell are present—

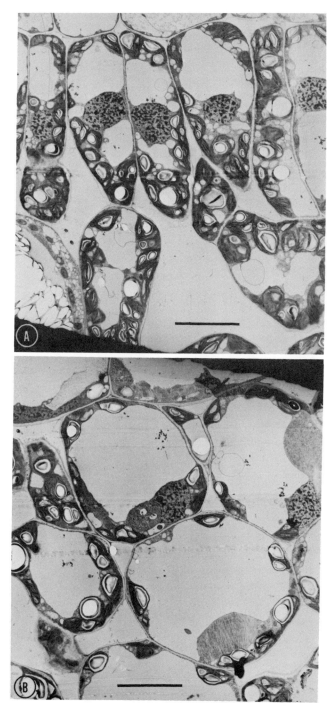

Figure 11.16   Histological and cytological effects of TMV in tobacco. (A) Section through palisade cells of a dark green area in a leaf showing mosaic. Cells are essentially normal. (B) Section through a nearby yellow-green area. Cells are large and undifferentiated in shape. Nuclei are not centrally located as in the dark green cells. Bar = 20 μm. (Courtesy of P. H. Atkinson.)

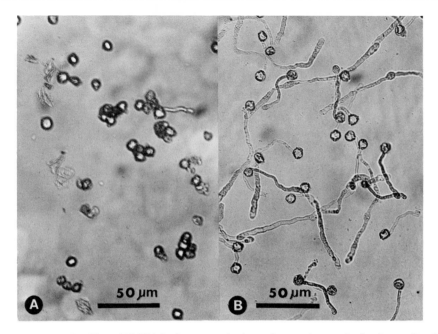

Figure 11.17 Effect of TRSV infection on germination and germ tube growth of soybean pollen. The pollen grains were germinated overnight in 30% sucrose. (A) Infected; (B) healthy. (From Yang and Hamilton, 1974.)

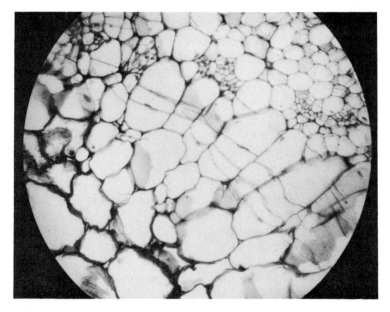

Figure 11.18 Section through a potato tuber infected with PVX, showing a cork cambial layer being developed near a group of necrotic cells (bottom left).

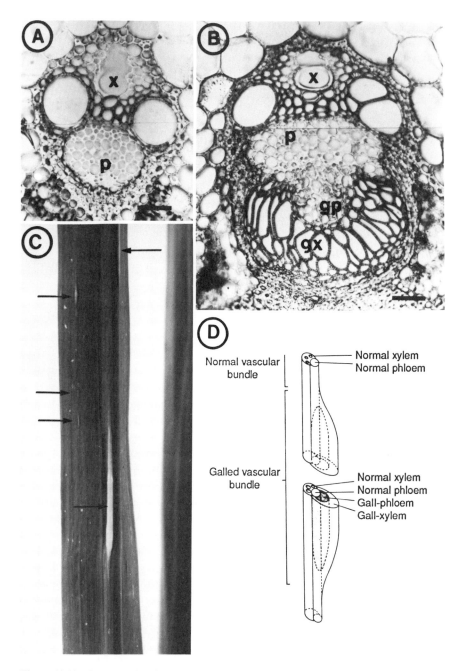

Figure 11.19 Structure of leaf galls on sugarcane infected with Fiji disease virus (FDV). (A) Transverse section of vascular tissue in a leaf vein from a healthy sugarcane plant, showing the xylem (x) and phloem (p) tissues. (B) A transverse section of vascular tissues of a vein on a galled leaf of an FDV-infected sugarcane plant, showing the gall phloem (gp) and gall xylem (gx), in addition to normal phloem (p) and xylem (x) tissues. (Bars = 0.5 μm.) (C) Part of a sugarcane leaf infected with FDV, showing small and large gall (arrows). (D) A diagram of the tissue distribution in the vein of an FDV-infected sugarcane leaf showing normal and gall tissues. (From Egan *et al.*, 1989.)

lignified gall xylem cells and nonlignified gall phloem (Hatta and Francki, 1976). Hyperplastic growth of phloem was marked in plum infected with plum pox *Potyvirus* (Buchter *et al.*, 1987).

III. CYTOLOGICAL EFFECTS

The cytological effects of viruses have been a subject of interest ever since the early searches with light microscopes for causative organisms in diseased tissues. About the beginning of the century these studies led to the discovery of two types of virus inclusion—amorphous bodies or "X bodies" and crystalline inclusions. The X bodies had resemblances to certain microorganisms. Some workers erroneously considered that they were in fact the parasite, or a stage in the life cycle of the parasite causing the disease. These early conclusions were not entirely wrong since many of the X bodies are in fact virus-induced structures in the cell where the components of viruses are synthesized and assembled. These viroplasms are discussed in Chapters 7 and 8.

A. Methods

Light microscopy is still important in the study of cytological abnormalities for several reasons: (i) much greater areas of tissue can be scanned, thus ensuring that samples taken for electron microscopy are representative; (ii) it may allow electron microscopic observations to be correlated with earlier detailed work on the same material using light microscopy; and (iii) observations on living material can be made using both phase and bright field illumination.

Improvements in procedures for the fixing, staining, and sectioning of plant tissues over the past 30 years and the widespread availability of high-resolution electron microscopes has led to a substantial growth in our knowledge about the cytological effects of viruses in cells. As in other fields of biology, electron microscopy is providing a link between macroscopic and light microscope observations on the one hand and molecular biological and biochemical studies on the other.

Examination of stained thin sections remains the standard procedure, but freeze-fracturing can give useful information on virus-induced membrane changes (Hatta *et al.*, 1973). Scanning electron microscopy will probably be of less value in the study of virus diseases but is sometimes useful (Hatta and Francki, 1976). It must be remembered that small differences in conditions under which plants are grown before sampling and in the procedure used to prepare tissue for electron microscopy can have a marked effect on the appearance and stability of organelles and virus-induced structures (e.g., Langenberg, 1982). Chilling of tissue before fixation may improve the preservation of very fragile virus-induced structures (Langenberg, 1979).

To relate any observed cytological effects to virus replication it is very useful to be able to follow the time course of events in infected cells. Some examples of the procedures used were given in Chapter 7. In principle, protoplasts infected *in vitro* should provide excellent material for studying the time course of events. However, various limitations are becoming apparent: (i) some ultrastructural features seen in TMV-infected tobacco leaf cells were not observed during TMV replication in protoplasts (Otsuki *et al.*, 1972a); (ii) crystalline inclusion bodies in TMV-infected leaf cells were degraded in protoplasts made from such leaves (Föglein *et al.*, 1976); and (iii) the development of cytological changes in protoplasts infected *in vitro* may be much less synchronous than indicated by growth curves.

B. Effects on Cell Structures

1. Nuclei

Cytopathological effects of virus infection have been well illustrated by Francki *et al.* (1985a,b). Many viruses have no detectable cytological effects on nuclei. Others give rise to intranuclear inclusions of various sorts and may affect the nucleolus or the size and shape of the nucleus, even though they appear not to replicate in this organelle.

Shikata and Maramorosch (1966) found that in pea leaves and pods infected with PEMV, particles accumulate first in the nucleus. During the course of the disease, the nucleolus disintegrates. Masses of virus particles accumulate in the nucleus and also in the cytoplasm. PEMV also causes vesiculation in the perinuclear space (DeZoeten et al., 1972).Virus particles of several small isometric viruses accumulate in the nucleus (as well as the cytoplasm). They may exist as scattered particles or in crystalline arrays (e.g., SBMV, Weintraub and Ragetli, 1970, TBSV, Russo and Martelli, 1972. Masses of viral protein or empty viral protein shells have been observed in nuclei of cell infected with several tymoviruses (Hatta and Matthews, 1976) (see Fig. 7.19).

Crystalline platelike inclucions were seen by light microscopy in cells infected with severe etch in tobacco (Kassanis, 1939). The plates were birefringent when viewed sideways and were a very regular feature of severe etch infection. Intranuclear inclusions have been described for some other potyviruses.

The viral-coded proteins involved in potyvirus nuclear and cytoplasmic inclusions were discussed in Chapter 7, Section I,A,2. Electron-lucent lacunae appeared in the nucleolus of *Nicotiana* cells infected with potato A *Potyvirus* (Edwardson and Christie, 1983). For some rhabdoviruses, viral cores appear in the nucleus and accumulate in the perinuclear space (see Fig. 8.14).

Geminiviruses cause marked hypertrophy of the nucleolus, which may come to occupy three-quarters of the nuclear volume. Fibrillar rings of deoxyribonucleoprotein appear, and masses of virus particles accumulate in the nucleus (e.g., Rushing *et al.*, 1987). An isolate of CaMV has been described in which nuclei become filled with virus particles and greatly enlarged (Gracia and Shepherd, 1985). The virus particles were not embedded in the matrix protein found in cytoplasmic viroplasms.

2. Mitochondria

The long rods of TRV may be associated with the mitochondria in infected cells (Harrison and Roberts, 1968). The mitochondria in cells of a range of host species, and various tissues, infected with cucumber green mottle mosaic *Tobamovirus* develop small vesicles bounded by a membrane and lying within the perimitochondrial space and in the cristae (Hatta *et al.*, 1971).

Aggregated mitochondria have been observed in *Datura* cells infected by a *Potyvirus* (Kitajima and Lovisolo, 1972) but there was no indication that these aggregates were involved in virus synthesis. The development of abnormal membrane systems within mitochondria has been described for several virus infections (Francki, 1987). They have no established relation to virus replication and are probably degenerative effects. For example, in some *Tombusvirus* infections multivesiculate bodies appear in the cytoplasm. These have been shown to develop from greatly modified mitochondria (Di Franco *et al.*, 1984; Di Franco and Martelli, 1987). In infections with other tombusviruses, multivesiculate bodies originate from modified peroxisomes (Martelli *et al.*, 1984). In *Sonchus* infected with beet yellow stunt *Closterovirus*, virus particles are found in phloem cells. The flexious rod-shaped particles are frequently inserted into the cristae of the mitochondria (Esau, 1979).

3. Chloroplasts

The small peripheral vesicles and other changes in and near the chloroplasts closely related to TYMV replication were discussed in Chapter 7, Section IV,A,8. TYMV infection can cause many other cytological changes in the chloroplasts, most of which appear to constitute a structural and biochemical degeneration of the organelles. The exact course of events in any mesophyll cell depends on (i) the developmental stage at which it was infected, (ii) the strain of virus infecting, (iii) the time after infection, and (iv) the environmental conditions (Matthews, 1973; Hatta and Matthews, 1974).

In inoculated leaves the chloroplasts become rounded and clumped together in the cell. There is little effect on grana or stroma lamellae. The chloroplasts become cup-shaped, with the opening of the cup generally facing the cell wall. Starch grains accumulate. "White" strains of the virus cause degeneration of the grana in inoculated leaves. In expanded leaves above the inoculated leaf that become fully infected without the appearance of mosaic symptoms, the effects of infection on chloroplasts are similar to those seen in inoculated leaves.

In leaves that were small at time of infection and that develop the typical mosaic, a variety of different pathological states in the chloroplasts can readily be distinguished by light microscopy in fresh leaf sections. Islands of tissue in the mosaic showing various shades of green, yellow, and white contain different strains of the virus, which affect the chloroplasts in recognizably distinct ways (Fig. 11.20). In dark green islands of tissue, which contain very little virus, chloroplasts appear normal.

The most important changes in the chloroplasts seen in tissue types other than dark green are (i) color, ranging from almost normal green to colorless; (ii) clumping to a variable extent; (iii) presence of large vesicles; (iv) fragmentation of chloroplasts; (v) reduction in granal stack height; (vi) presence of osmiophilic globules; and (vii) arrays of phytoferritin molecules. Some of these abnormalities are illustrated in Fig. 11.20.

Different strains of TYMV produce particular combinations of abnormalities in the chloroplasts. In blocks of tissue of one type, almost all cells show the same abnormalities and these persist at least for a time as the predominating tissue type when inoculations are made to fresh plants.

In contrast to the small peripheral vesicles, which appear to be induced by all tymoviruses in the chloroplasts of infected cells, none of the changes noted is an essential consequence of tymovirus infection; nor can they be regarded as diagnostic for the group. For example, no clumping of the chloroplasts occurs in cucumbers infected with okra mosaic virus. On the other hand, clumping of chloroplasts is

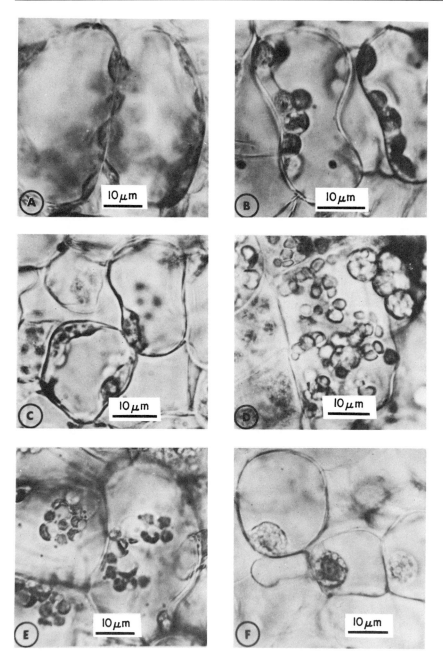

Figure 11.20 Chloroplast abnormalities in different tissue types from Chinese cabbage leaves showing mosaic caused by TYMV. (A) Healthy leaf. (B) Pale green. (C) Yellow-green. (D) Yellow-green tissue with fragmenting chloroplasts. (E) Yellow-green tissue with highly vesiculate choloroplasts. (F) White tissue with a single spherical vesiculate mass of chloroplast material in each cell. (From Chalcroft and Matthews, 1967b.)

induced by turnip mosaic virus in *Chenopodium* (Kitajima and Costa, 1973). Several viruses outside the *Tymorivus* group induce small vesicles near the periphery of the chloroplasts. These vesicles differ from the *Tymorivus* type in that they do not appear to have necks connecting them to the cytoplasm. In *Datura* leaves infected with TBSV the thylakoid membranes undergo varied and marked rearrangements

(Bassi *et al.* 1985). For most of these viruses the vesicles or other changes appear to be degenerative consequences of infection. However, for BSMV, the vesicles appear to be associated with virus replication (Lin and Langenberg, 1984a).

In many infections, the size and number of starch grains seen in leaf cells are abnormal. In mosaic diseases there is, generally speaking, less starch than normal, but in some diseases (e.g., sugar beet curly top and potato leaf roll) excessive amounts of starch may accumulate. Similarly in local lesions induced by TMV in cucumber cotyledons, chloroplasts become greatly enlarged and filled with starch grains (Cohen and Loebenstein, 1975; see also Section IV,D).

4. Cell Walls

The plant cell wall tends to be regarded mainly as a physical supporting and barrier structure. In fact it is a distinct biochemical and physiological compartment containing a substantial proportion of the total activity of certain enzymes in the leaf (Yung and Northcote, 1975). Three kinds of abnormality have been observed in or near the walls of virus-diseased cells:

1. Abnormal thickening, due to the deposition of callose, may occur in cells near the edge of virus-induced lesions (e.g., Hiruki and Tu, 1972). Chemical change in the walls may be complex and difficult to study (Faulkner and Kimmins, 1975).
2. Cell wall protrusions involving the plasmodesmata have been reported for several unrelated viruses. The protrusions from the plasmodesmata into cells may have one or more canals. They may be quite short or of considerably length. They appear to be due to deposition of new wall material induced by the virus, and they may be lined inside and out with plasma membrane (e.g., Bassi *et al.*, 1974). The function of these protuberances has not been established.
3. Depositions of electron-dense material between the cell wall and the plasma membrane may extend over substantial areas of the cell wall (as with oat necrotic mottle *Potyvirus*, Gill, 1974) or may be limited in extent and occur in association with plasmodesmata (as with BSMV, McMullen *et al.*, 1977). They have been called *paramural bodies*.

The major cytopathic effect of citrus exocortis viroid is the induction of numerous small membrane-bound bodies near the cell wall with an electron density similar to that of the plasma membrane (Semancik and Vanderwoude, 1976). They are found in all cell types.

5. Bacteroidal Cells

The first phase in the infection of soybean root cells by *Rhizobium* (i.e., development of an infection thread and release of rhizobia into the cytoplasm) appears not to be affected by infection of the plant with soybean mosaic *Potyvirus*. In the second stage a membrane envelope forms around the bacterial cell to form a bacteroid. Structural differences such as decreased vesiculation of this membrane envelope were observed in virus-infected roots (Tu, 1977).

6. Myelinlike Bodies

Myelinlike bodies consisting of densely staining layers that may be close-packed in a concentric or irregular fashion have been described for several plant virus infections (e.g., Kim *et al.*, 1974). They probably reflect degenerative changes in one or more of the cell's membrane systems. They may be associated with osmiophilic globules, which are thought to consist of the lipid component of cell membranes. Kim *et al.* suggested that myelinlike bodies in bean leaf cells infected with comoviruses may be formed from the osmiophilic globules.

7. Cell Death

Drastic cytological changes occur in cells as they approach death. These changes have been studied by both light and electron microscopy, but they do not tell us how virus infection actually kills the cell.

C. Virus-Induced Structures in the Cytoplasm

The specialized virus-induced regions in the cytoplasm that are, or appear likely to be, the sites of virus synthesis and assembly (viroplasms) were discussed in Chapters 6–8. Here other types of inclusions will be described. These are usually either crystalline inclusions consisting mainly of virus, or the pinwheel inclusions characteristic of the potyviruses.

1. Crystalline Inclusions

Virus particles may accumulate in an infected cell in sufficient numbers and exist under suitable conditions to form three-dimensional crystalline arrays. These may grow into crystals large enough to be seen with the light microscope, or they may remain as small arrays that can be detected only by electron microscopy.

The ability to form crystals within the host cell depends on properties of the virus itself and is not related to the overall concentration reached in the tissue or to the ability of the purified virus to form crystals. For example, TYMV can readily crystallize *in vitro*. It reaches high concentration in infected tissue but does not normally form crystals there. By contrast, STNV occurring in much lower concentrations frequently forms intracellular crystals.

TMV

In tobacco leaves showing typical mosaic symptoms caused by TMV, leaf-hair and epidermal cells over yellow-green areas may almost all contain crystalline inclusions, while those in fully dark green areas contain none. The junction between yellow-green and dark green tissue may be quite sharp, so that there is a zone where neighboring leaf hairs either have no crystals, or almost every cell has crystals. Warmke and Edwardson (1966) followed the development of crystals in leaf-hair cells of tobacco. Virus particles were first seen free in the cytoplasm as small aggregates of parallel rods with ends aligned. These aggregates increase in size. The growing crystals are not bounded by a membrane, and as they become multilayered they may sometimes incorporate endoplasmic reticulum, mitochondria, and even chloroplasts between the layers.

The platelike crystalline inclusions are very unstable and are disrupted by pricking or otherwise damaging living cells. They are birefringent when viewed edge-on, but not when seen on the flat face. They contain about 60% water and otherwise consist mainly of successive layers of closely packed parallel rods oriented not quite perpendicularly to the plane of the layers. Rods in successive layers are tilted with respect to one another. This herringbone effect can be visualized in freeze-fractured preparations for some strains of TMV (Fig. 11.21). Sometimes long, curved, fibrous inclusions, or spikelike or spindle-shaped inclusions made up largely of virus particles, can be seen by light microscopy. Different strains of the virus may form different kinds of paracrystalline arrays. Most crystalline inclusions have been found only in the cytoplasm but some have been detected in nuclei (e.g., Esau and Cronshaw, 1967).

BYV

The inclusions found in plants infected with BYV occur in phloem cells and also appear in other tissues, for example, the mesophyll. By light microscopy, the inclusions are frequently spindle-shaped and may show banding (Fig. 11.22A). Electron microscopy reveals layers of flexuous virus rods (Fig. 11.22B). Most of the viral inclusions occur in the cytoplasm. Smaller aggregates of viruslike particles were seen in nuclei and chloroplasts (Cronshaw *et al.*, 1966).

Other Helical Viruses

Rod-shaped viruses belonging to other groups may aggregate in the cell into more or less ordered arrays. These can frequently be observed only by electron microscopy, and other material besides virus rods may be present in the arrays. As well as such aggregates of virus rods, red clover vein mosaic *Carlavirus* induces the appearance of large crystals in the cytoplasm (Khan *et al.*, 1977). These contain RNA and protein and consist of a crystalline array of polyhedral particles about 10 nm in diameter. No virus rods were present. The significance of these unusual crystals in unknown.

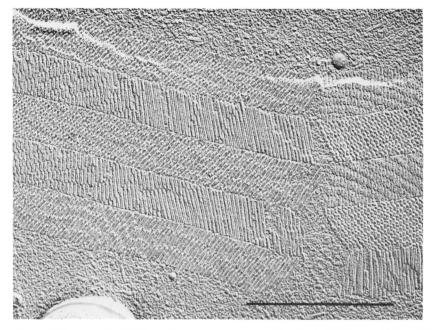

Figure 11.21   Crystal of TMV rods in a freeze-etched preparation. Part of TMV crystal lying within a mesophyll cell that has been penetrated by glycerol. The crystal has retained its herringbone structure and the lattice spacing of about 24 nm, despite the fact that the tonoplast was ruptured. Bar = 1 μm. (From Willison, 1976.)

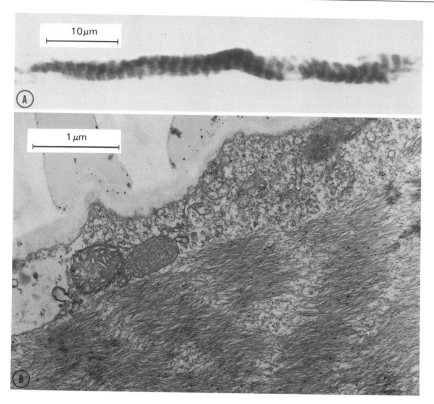

Figure 11.22 Inclusion bodies caused by beet yellow *Closterovirus* in parenchyma cells of small veins of *Beta vulgaris*. The banded form of these inclusions is shown (A) by light microscopy and (B) by electron microscopy. Bands are made up of flexuous virus particles in more or less orderly array. (From Esau *et al.*, 1966.)

Small Icosahedral Viruses

Many small icosahedral viruses form crystalline arrays in infected cells. Sometimes these are large enough to be seen by light microscopy (e.g., TNV, Kassanis *et al.*, 1970). Icosahedral viruses that do not normally form regular arrays may be induced to do so by heating or plasmolyzing the tissue to remove some of the water (Milne, 1967; Hatta, 1976).

Many strains of broad bean wilt *Fabavirus* induce cylindrical tubules of virus particles. Russo *et al.* (1979) described a strain that formed unusual hollow tubules that are square or rectangular in section. The walls are made up of two parallel rows of virus particles.

Reoviruses and Rhabdoviruses

Plant cells infected with viruses belonging to the Reovividae or Rhabdoviridae frequently contain masses of virus particles in regular arrays in the cytoplasm. With some rhabdoviruses the bullet-shaped particles accumulate in more or less regular arrays in the perinuclear space (see Fig. 8.14) (see also Francki *et al.*, 1985d).

2. Pinwheel Inclusions

Potyviruses induce the formation of characteristic cylindrical inclusions in the cytoplasm of infected cells (e.g., Hiebert and McDonald, 1973). The most striking feature of these inclusions viewed in thin cross section is the presence of a central tubule from which radiate curved "arms" to give a pinwheel effect. Reconstruction from serial sections shows that the inclusions consist of a series of plates and curved scrolls with a finely striated substructure with a periodicity of about 5 nm. The bundles, cylinders, tubes, and pinwheels seen in section are aspects of geometrically complex structures. The general structure has been confirmed by examination of freeze-etched preparations (McDonald and Hiebert, 1974) and by the use of a tilting stage together with computer-assisted analytical geometry (Mernaugh *et al.*, 1980) (Fig. 11.23). For some potyviruses the pinwheels are tightly curved. For others they are more open. The inclusions induced by members of the group may differ consistently in various details.

Pinwheel inclusions originate and develop in association with the plasma membrane at sites lying over plasmodesmata (Lawson *et al.*, 1971; Andrews and Shalla, 1974). The central tubule of the pinwheel is located directly over the plasmodesmata and it is possible the membranes may be continuous from one cell to the next. The core and the sheets extend out into the cytoplasm as the inclusion grows. Later in infection they may become dissociated from the plasmodesmata and come to lie free in the cytoplasm. Virus particles may be intimately associated with the pinwheel arms at all times and particularly at early stages of infection (Andrews and Shalla, 1974). The viral-coded protein that is found in these inclusions was discussed in Chapter 7, Section I,A,2.

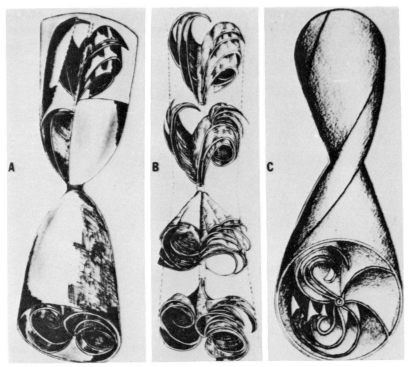

Figure 11.23 Potyvirus pin wheel inclusions. (A) A model for a WSMV inclusion that has a smooth, elliptic, hyperboloid shape. (B) Another model for a WSMV inclusion with a porous surface that may fit the confines of an elliptic hyperboloid. (C) A twisted hyperboloid model depicting the configurations that some TEV pinwheel inclusions may assume. (From Mernaugh *et al.*, 1980.)

D. Cytological Structures Resembling Those Induced by Viruses

Some normal structures in cells could be mistaken for virus-induced effects—for example, crystalline or membrane-bound inclusions in plastids (e.g., Newcomb, 1967). Prolamellar bodies in chloroplasts give the appearance in section of a regular array of tubes. These bodies may be induced by certain chemical treatments (Wrischer, 1973).

Phosphorus-deficient bean leaves (Thompson *et al.,* 1964) showed degenerative changes in the chloroplasts like those seen in some virus infections. Similarly, in sulfur-deficient *Zea mays,* chloroplasts contained many osmiophilic granules and small vesicles (Hall *et al.,* 1972).

Bundle sheath chloroplasts of C4 plants contain numerous small vacuoles near their periphery and vacuoles also have been described for chloroplasts in certain tissues of C3 plants (e.g., Marinos, 1967). In a spontaneous plastid mutant of *Epilobium hirsutum,* the degenerate grana and vacuolation of the mutant plastids seen by electron microscopy (Anton-Lamprecht, 1966) bear some resemblance to the pathological changes induced by TYMV infection in Chinese cabbage.

Nuclei of healthy cells sometimes contain crystalline structures that might be mistaken for viral inclusions (e.g., Lawson *et al.,* 1971).

Such virus-induced effects as disorganization of membrane systems, presence of numerous osmiophilic granules, and disintegration of organelles are similar to normal degenerative processes associated with aging or degeneration induced by other agents.

IV. EFFECTS ON PLANT METABOLISM

All the various macroscopic and microscopic symptoms of disease discussed in Sections I–III must originate in biochemical aberrations induced directly or indirectly by the virus. Many early workers described differences in composition or in rate of some process between healthy and virus-infected tissues. Aspects commonly investigated were total carbohydrates and sugars; total nitrogen and various nitrogen fractions; the carbon–nitrogen ratio; total ash or various ash components; and rates of photosynthesis, respiration, and transpiration. Estimations were often made on fractions containing many different compounds, for example, "soluble nitrogen." Most analytical work has been carried out on fully infected plants, perhaps many weeks after inoculation. In such plants, we may well expect to observe changes in the amounts of many substances and in the rates of major biochemical processes. For obtaining any detailed understanding of the effects of virus infection on host metabolism and the initiation of disease processes, this earlier work is almost useless.

A. Experimental Variables

The experimental systems available for studying effects on host plant metabolism are very much the same as those described in Chapter 6 for virus replication. An effective cell culture system would be most useful for investigating some aspects of the metabolism of infected cells. However, there is increasing evidence to show that following their isolation, leaf protoplasts are in a disturbed and changing metabolic state (Chapter 6, Section VII,E). The results of metabolic studies on the effects of virus infection on such cells need to be interpreted with caution.

There are many variables to be taken into account when using intact plants or organs. The following discussion is concerned mainly with changes taking place in leaves, because these constitute most of the herbaceous host plant. More virus is usually produced in them, and they are most often used for experimental work. There are three kinds of variable factor to consider when designing experiments to study the effects of virus infection on host metabolism: (i) the basis on which results are to be expressed; (ii) factors in the material that vary with time; and (iii) non-uniformity in the material being sampled at any given time.

1. Basis for Expressing Results

Measurements or units that can be used as a basis for expressing results include fresh weight of tissue, dry weight, protein nitrogen, DNA, leaf area, or on a per plant, per leaf, or per cell basis. Fresh weight is the most convenient measurement and has been most widely used. It may be satisfactory where big differences are being sought at fairly early stages after infection. However, virus infection may alter water content of tissues soon after infection. At later times, when stunting of growth may have occurred, serious difficulties arise in the use of fresh weight. Virus-infected leaves, because they are smaller than healthy leaves of the same age, may have a higher concentration of many components per unit fresh weight. On the other hand, when compared with healthy leaves of the same size, they may have a lower concentration of these same components per unit fresh weight. Dry weight measurement may likewise introduce serious ambiguities. The virus itself may come to represent about 10% of the dry weight of a leaf. Stunting may increase the proportion of inert cell wall materials, and starch accumulation due to virus infection may increase dry weight. Similarly, protein nitrogen content may be significantly affected by the presence of the virus and by general stunting. Where DNA content per cell remains constant, it forms a satisfactory basis for making comparisons.

Leaf area is a satisfactory basis for comparison where virus infection has not altered the area of the leaves—for example, at early stages after inoculation. Used on stunted leaves, it could be quite misleading. Measurements on a per plant basis may be quite satisfactory where chronically infected material is examined, such as potatoes grown from infected tubers. Where the information sought is relevant to the economic aspects of disease, the whole plant, or the parts used commercially, will be the most relevant unit. However, when whole inoculated plants are used, the sample will include tissues that have been affected for different periods of time.

For many purposes, the leaf is the most satisfactory basis for calculating results, particularly where measurements can be begun before infection and continue as the disease progresses, and where leaf expansion is taking place. The cell may be the unit of choice in experiments involving tissue minces or cells grown in tissue culture. There may be no ideal method for expressing results, so that, generally speaking, it is highly desirable to use two or more different bases on the same experimental material.

2. Factors That Vary with Time

Leaf Age

For most of the herbaceous species used in experiments with viruses, individual leaves are never in a steady state. They normally pass through four stages, each of which merges into the next. The first stage, up to about 2 cm long for leaves like those of tobacco, is one of intensive cell division. This is followed by a stage in which cell expansion and protein synthesis become the predominant activities. As the leaf approaches full size, photosynthesis and the export of metabolites become the major activity (although it has been going on from an early stage), and finally the processes of senescence take over. Virus infecting leaves at different stages in their development will thus meet different conditions and may have different effects on cell processes.

Diurnal Variations

Many basic processes in leaves of plants grown in daylight follow a diurnal rhythm. For example, the proportion of ribosomes in polyribosome form, which presumably reflects the rate of protein synthesis, follows a diurnal cycle with a minimum at the end of the night period and a maximum near the middle of the day (Clark *et al.*, 1964). Infection by virus at different times of day might produce different immediate effects.

Seasonal Variations

Plants grown at different times of the year under uncontrolled or partly controlled greenhouse conditions vary in many properties and may be affected differently by virus infection. This source of variation can be eliminated by growing plants from seed under fully controlled environmental conditions.

Wound Responses

Metabolic changes, which begin as soon as healthy tissue is excised, alter the rate of uptake of isotopically labeled metabolites (Pratt and Matthews, 1971) and the pathways concerned with respiration (Macnicol, 1976).

3. Nonuniformity at a Given Sampling Time

Between Leaves

Variation is minimal between the two primary leaves, and these form useful experimental material in such plants as *Phaseolus vulgaris*. Apart from the primary leaves, no two leaves on many of the plants used in virus studies will be in exactly the same stage of development. This variation can be overcome to some extent by using groups of plants and selecting one leaf of a particular size for study. Even here, very few leaves in a group will be just at the same stage, since leaf initiation does not proceed in step in different plants. In fast-growing plants such as Chinese cabbage, where young leaves may double in size in 30–40 hours, it may require selection of a few plants from a large group to give a reasonably uniform set of younger leaves.

Within Leaves

Positions in the leaf blade symmetrically placed with respect to the midrib are usually very similar, so that half-leaves cut longitudinally down each side of the midrib give useful samples for comparative studies. Within half-leaves, however, tissue is not uniform. The tip of the leaf is older than the base. Lamina thickness and venation pattern may vary. Rates of uptake of radioactively labeled compounds may not be constant between different parts of the blade.

Cell Type

Mesophyll cells usually make up a high proportion of leaf tissue, but epidermal cells and epidermal appendages may constitute 10–15% of the fresh weight. If the midrib and major veins are large, they are usually discarded in biochemical studies on virus infection, unless they are of special interest.

Mosaic Symptoms

As discussed in Chapter 12, Section IV,D,2, blocks of tissue in leaves showing mosaic symptoms may be very heterogeneous with respect to virus infection and such heterogeneity may be detectable only by microscopic examination. With mosaic patterns such as those induced by TYMV, the dark green areas can be used with advantage as control tissue in studying some of the effects of chronic virus infection on cell components and processes. A dark green area is dissected from one-half of a leaf and a corresponding virus-bearing area is taken from the opposite half-leaf.

Cell Organelles

In higher plants, isolated and apparently intact chloroplasts frequently fractionate into two or more bands in density gradients. In some species of higher plants, two morphological types of chloroplast have been observed that fix carbon by different routes (Slack, 1969). Thus normal variation between chloroplasts could complicate observations on the metabolic effects of virus infection.

Changes in Excised Leaf Pieces

After short periods of incubation the marginal areas of excised leaf disks, even very small ones, vary markedly from the central region in various metabolic activities (Pratt and Matthews, 1971; Macnicol, 1976).

B. Nucleic Acids and Proteins

The synthesis of viral nucleic acids and proteins was discussed in Chapters 6–8. Here I will consider the general effects of virus infection on host-cell nucleic acids and protein synthesis. Induction of specific host proteins following infection is discussed in Chapter 12, Section III.

1. DNA

It is widely assumed that the small RNA viruses have little effect on host-cell DNA synthesis, but there are very few, if any, definitive experiments bearing on the question. Virus infection may well have some effect on host-cell DNA synthesis, but such effects are likely to be fairly small and difficult to establish because (i) DNA content per cell may increase for some time in a normal expanding leaf; (ii) minor DNA fractions, which might be affected by virus infection, may be difficult to isolate and identify; and (iii) any effect might be very transitory and, therefore, difficult to detect in asynchronous infections. Using a radioautographic technique to assay for DNA synthesis in individual cells, Atchison (1973) found that there was a drop in DNA synthesis in the terminal 1 mm of French bean roots about the time they were invaded by tobacco ring spot virus. This was soon followed by a transient drop in the mitotic index.

2. Ribosomes and Ribosomal RNA

Effects of virus infection on ribosomal RNA synthesis and the concentration of ribosomes may differ with the virus, strain of virus, time after infection, and the host and tissue concerned. In addition, 70 S and 80 S ribosomes may be affected differently.

In TMV-infected leaves viral RNA may come to represent about 75% of the total nucleic acids without having any marked effects on the main host RNA fractions except to cause a reduction in 16 S and 23 S chloroplast ribosomal RNAs (Fraser, 1987b). However, under some conditions cytoplasmic ribosomal RNA synthesis is also inhibited. A reduction in chloroplast ribosomes without a marked effect on cytoplasmic ribosomes is a fairly common feature for mosaic diseases (e.g., BSMV in barley, Brakke *et al.*, 1987b; TYMV, see the following discussion).

The variables that affect the outcome of TMV infection on cytoplasmic and chloroplast ribosomes have been discussed by Fraser (1987b). Experiments with TYMV will illustrate the difficulties in making generalizations.

In Chinese cabbage leaves chronically infected with TYMV the concentration of 70 S ribosomes in the yellow-green islands in the mosaic is greatly reduced compared to that in dark green islands in the same leaf (Reid and Matthews, 1966). There is little effect on the concentration of cytoplasmic ribosomes in such yellow-green islands of tissue. The extent of this reduction depends very much on the strain of TYMV, and it also becomes more severe with time after infection. Loss of 70 S ribosomes more or less parallels the loss of chlorophyll, "white" strains causing the most severe loss.

A somewhat different result is obtained if the effect of TYMV infection with time in a young systemically infected leaf is followed. Chloroplast ribosome concentration falls markedly about the time virus concentration reaches a maximum. About the same time there is a significant increase in cytoplasmic ribosome concentration, which is mainly due to the stunting effect of infection. On the other hand, if the effects of virus infection on these components for the plant as a whole are considered, a different picture emerges. Infection reduces both cytoplasmic and chloroplast ribosomes.

These results emphasize the fact that infection of a growing plant with a virus introduces an additional time-dependent variable into a system in which many normal interacting components are changing with time. Analyses made on only one or two components of the system, or at some particular time, are unlikely to give much insight into virus replication and the nature of the disease process. Very little is known about any effects of virus infection on host tRNAs, nuclear RNAs, or mitochondrial ribosomal RNAs. Effects on host mRNAs are discussed in the next section in relation to protein synthesis.

3. Proteins

The coat protein of a virus such as TMV can come to represent about half the total protein in the diseased leaf. This can occur without marked effects on the overall content of host proteins. Many other viruses multiply to a much more limited extent. Effects on host protein synthesis are not necessarily correlated with amounts of virus produced. A reduction in the amount of the most abundant host protein—ribulose bisphosphate carboxylase-oxygenase (rbcs)—is one of the commonest effects of viruses that cause mosaic and yellowing diseases (e.g., TYMV,

Reid and Matthews, 1966; wheat streak mosaic *Potyvirus,* White and Brakke, 1983).

Fraser (1987b) estimated that TMV infection reduced host protein synthesis by up to 75% during the period of virus replication. Infection did not alter the concentration of host polyadenylated RNA, nor its size distribution. This suggested that infection may alter host protein synthesis at the translation stage rather than interfering with transcription. Many viruses infecting vertebrates inhibit host-cell translation by a variety of mechanisms, bringing about conditions that favor translation of viral mRNAs (Schneider and Shenk, 1987). The mechanisms used by plant viruses are beginning to be studied. For example, Stratford and Covey (1988) found that there were changes in the levels of specific translatable mRNAs in response to infection of turnip leaves with CaMV. More such changes were found with a severe strain. In particular the mRNA encoding the precursor to the small subunit of rbcs was markedly decreased following infection with the severe strain.

It is known that the coat protein of TMV, and some other viruses, can encapsidate some host RNAs *in vivo* (Chapter 6, Section VIII,A). Sleat *et al.* (1988b) transformed tobacco seedlings to express chloramphenicol acetyltransferase (CAT) mRNA. Transformed plants that also contained the TMV origin of assembly sequence (see Fig. 7.8) 3' to the CAT gene showed a threefold suppression of CAT activity compared with plants without the origin of assembly sequence. Thus it is possible that the coating of host mRNAs in viral coat protein may be a mechanism for the shutting off of specific host mRNAs during virus replication.

Saunders *et al.* (1989) used another approach to the same problem. They generated a library of cDNA clones corresponding to the host RNAs isolated from turnip leaves infected with CaMV during the early vein-clearing stage. Hybridization was used to select clones that represented RNAs whose levels had been raised or lowered by infection. For example, one RNA that was greatly reduced in amount was identified as the mRNA for the ribulose 1,5-bisphosphate carboxylase small subunit polypeptide. Overall, the findings of Stratford and Covey (1988) and Saunders *et al.* (1989) suggest that there are few major changes in host gene expression during infection with CaMV.

C. Lipids

The sites of virus synthesis within the cell almost always contain membrane structures (Chapters 6–8). TYMV infection alters the ultrastructure of chloroplast membranes, and rhabdovirus particles obtain their outer membrane by budding through some host-cell membrane. There have been a few studies of the effects of virus infection on lipid metabolism (e.g., Trevathan *et al.,* 1982) but none of these has illuminated the mechanisms by which viruses change and use plant membrane systems.

D. Carbohydrates

Some viruses appear to have little effect on carbohydrates in the leaves, while others may alter both their rate of synthesis and rate of translocation. These changes may be illustrated in a simple manner.

Leaves that have been inoculated several days previously with a virus that does not cause necrotic local lesions are harvested in the morning or after some hours in darkness, decolorized, and treated with iodine. The local lesions may show up as dark-staining areas against a pale background, indicating a block in carbohydrate translocation. On the other hand, if the inoculated leaves are harvested in the afternoon after a period of photosynthesis, decolorized, and stained with iodine, the local lesions may show up as pale spots against the dark-staining background of uninfected tissue (Holmes, 1931). Thus, virus infection can decrease the rate of accumulation of starch when leaves are exposed to light.

From the few diseases that have been examined in any detail, it is not possible to make very firmly based generalizations about other carbohydrate changes, but the following may be fairly common effects: (i) a rise in glucose, fructose, and sucrose in virus-infected leaves; (ii) a greater rise in these sugars caused by mild strains of a given virus compared with severe strains; and (iii) effects of infection on mesophyll cells, not yet understood, may reduce translocation of carbohydrates out of the leaves.

E. Cell Wall Compounds

Although cytological studies have demonstrated ultrastructural changes in the cell walls in many virus infections, the biochemical basis of such changes would be difficult to study. Future work may show that virus infection has effects on various activities in the cell wall compartment, which is not metabolically inert. Eighty-five percent of detectable peroxidase activity and 22% of the acid phosphatase are located in the cell wall of healthy tobacco leaves (Yung and Northcote, 1975). Elevated peroxidase activity has been reported as a response of tobacco and many other hosts to virus infection (see Matthews, 1981).

F. Respiration

Many studies have been made of the effects of virus infection on rates and pathways of respiration, but it is not possible to relate the results to the processes involved in virus replication. In summary, for many host–virus combinations where necrosis does not occur, there is a rise in respiration rate, which may begin before symptoms appear and continue for a time as disease develops. In chronically infected plants, respiration is often lower than normal. In the one systemic disease so far examined in detail, there is no detectable change in the pathway of respiration. In host–virus combinations where necrotic local lesions develop, there is an increase in respiration as necrosis develops. This increase is accounted for, at least in part, by activation of the hexose monophosphate shunt pathway (see Matthews, 1981; Fraser, 1987b).

G. Photosynthesis

In a tobacco mutant in which some islands of leaf tissue had no chlorophyll, TMV replication occurred in white leaf areas in the intact plant. However, replication did not occur if the white tissue was detached and floated on water immediately after inoculation (R. E. F. Matthews, unpublished). Detached white tissue supplied with glucose supported TMV replication, indicating that the process of photosynthesis itself is not necessary for replication of this virus. Nevertheless, virus infection usually affects the process of photosynthesis. Reduction in carbon fixation is the most commonly reported effect in leaves showing mosaic or yellows diseases. This reduction usually becomes detectable some days after infection.

Photosynthetic activity can be reduced by changes in chloroplast structure, by reduced content of photosynthetic pigments or ribulose bisphosphate carboxylase, or by reduction in specific proteins associated with the particles of photosystem II (Naidu *et al.*, 1986). However, such changes appear to be secondary, occurring some time after infection when much virus synthesis has already taken place. In tobacco plants infected with various strains of TMV, electron transport rates were reduced when loss of chlorophyll occurred. In inoculated leaves, photosystem II appeared to be irreversibly damaged in inoculated leaves even when no macroscopic symptoms were apparent (van Kooten *et al.*, 1990). A variety of effects of localized and systemic TMV infection in tobacco were observed in experiments with isolated chloroplasts. However, some enzyme activities were little affected (Montalbini and Lupattelli, 1989).

Some effects on photosynthesis are known that appear to be closely linked in time to the early period of maximum virus production. In chloroplasts isolated from Chinese cabbage leaves infected with TYMV, the Hill reaction and cyclic and noncyclic photophosphorylation were all increased compared with healthy leaves during the phase of active virus multiplication (on an equal chlorophyll basis) (Goffeau and Bové, 1965). At a late stage of infection, photosynthetic activity was lower than in controls measured on chloroplasts isolated from whole plants. In young Chinese cabbage leaves infected with TYMV there was a substantial diversion of a the products of photosynthetic carbon fixation away from sugars and into organic acids and amino acids. This change was most marked during the period of rapid virus increase and returned to the normal pattern when virus replication was near completion (Bedbrook and Matthews, 1973). An increase in the activity of the enzymes phosphoenolpyruvate carboxylase and aspartate aminotransferase followed a similar time course.

Magyarosy *et al.* (1973) found a similar shift from the production of sugars to amino acids and organic acids in squash plants systemically infected with squash mosaic *Comovirus*. They isolated chloroplasts from healthy and diseased leaves and showed that both produced a similar pattern of carbon fixation products and that the total carbon fixed was about the same. They concluded that the virus-induced production of amino acids was taking place in the cytoplasm.

In summary, during the period of rapid replication, virus infection may cause a diversion of the early products of carbon fixation away from sugars and into pathways that lead more directly to the production of building blocks for the synthesis of nucleic acids and proteins. The most general result of virus infection is a reduction in photosynthetic activity. This reduction arises from a variety of biochemical and physical changes. The relative importance of different factors varies with the disease.

H. Transpiration

In chronically virus-infected leaves transpiration rate and water content have been found to be generally lower than in corresponding healthy tissues. The reported effects over the first 1–2 weeks after inoculation vary. Results are difficult to compare and interpret because different viruses and host species have been used together with different conditions of growth and different tissue sampling procedures.

Bedbrook (1972) used the cobalt chloride paper method (Stahl, 1894) to estimate relative transpiration rates and to give a measure of stomatal opening. He compared, in intact Chinese cabbage plants, dark green islands in leaves showing mosaic patterns due to TYMV infection and various islands of tissue fully invaded by the virus. In darkness or low light intensity, stomata in darker green and pale green islands were closed, while those in islands of more severely affected lamina were open. In plants that had been held in full daylight the dark and pale green islands were transpiring rapidly. Transpiration from severely affected islands was much less. These and other experiments showed that TYMV infection lowers the responsiveness of the stomata to changes in light intensity, the lowered response being most marked with strains causing the greatest reduction in chlorophyll. Because of diminished transpiration, the temperature of sugar beet leaves in susceptible plants infected with BNYVV was 2–3°C higher than that of a tolerant variety (Keller *et al.*, 1989).

I. Activities of Specific Enzymes

Much of the work dealing with the effect of virus infection on specific enzymes is difficult to interpret for the following reasons: (i) where differences have been found, it has usually been assumed that virus infection alters the amount of enzyme present and little consideration has been given to the possibility that infection may affect enzyme activities through changes in the amount of enzyme inhibitors or activators released when cells are disrupted; (ii) the difficulty of deciding on an appropriate basis for expressing enzyme activity has often been ignored; and (iii) much of the work was done before the widespread existence of isoenzymes was recognized.

There have been many studies involving the use of polyacrylamide gel electrophoresis to fractionate and assay isoenzymes and to study the consequences of virus infection on these patterns. It is relatively easy to generate data by this means. It is much less easy to provide meaningful interpretations of any observed changes. There are several reasons for these difficulties:

1. In the healthy plant there may be a continually changing developmental sequence of isoenzymes (Denna and Alexander, 1975).
2. Electrophoretically distinct isoenzymes may be determined genetically or may be different conformational forms derived by posttranslational modification from the same primary structure.
3. The pattern of isoenzymes in the normal host may differ in closely related genotypes, and the effects of virus infection may differ with these (e.g., Eşanu and Dumitrescu, 1971).
4. Isoenzymes may be distributed in several subcellular sites. For example, a different set of peroxidase isoenzymes was associated with the cell wall and with the soluble fraction in extracts of normal maize root tips (Parish, 1975). Virus infection may affect various sites in different ways.
5. Virus-induced cell death may lead to changes in isoenzyme patterns that do not differ significantly from those induced by entirely unrelated causes of necrosis.

6. The observed effect of virus infection may depend on the substrate used for isoenzyme assay.
7. Aggregation states (e.g., monomer $\rightleftharpoons$ dimer) may affect the kinetic properties of an enzyme, for example, aspartate aminotransferase (Melander, 1975).

Studies have been reported that involve representatives of all the major groups of enzymes (oxidoreductases and so on), but none of these has taken the preceding variables adequately into account. Examples have been discussed by Matthews (1981) and Fraser (1987b). Enzymes involved in the phenomenon of acquired resistance are discussed in Chapter 12, Section III.

J. Hormones

There is little doubt that virus infection influences hormone activities in infected plants and that hormones play some part in the induction of disease. This aspect is discussed in Chapter 12, Section IV,B. Quantitative effects of infection on concentration have been shown for all the major groups of plant hormones (reviewed by Fraser, 1987b). Virus infections tend to decrease auxin and gibberellin concentrations and increase that of abscisic acid. Stimulation of ethylene production is associated with necrotic or chlorotic local responses.

K. Low-Molecular-Weight Compounds

There are numerous reports on the effects of virus infection on concentration of low-molecular-weight compounds in various parts of virus-infected plants. The analyses give rise to large amounts of data, which vary with different hosts and viruses, and which are impossible to interpret in relation to virus replication. In this section some of these effects will be briefly noted.

1. Amino Acids and Related Compounds

The most consistent change observed has been an increase in one or both of the amides, glutamine and asparagine. The imino acid pipecolic acid has been reported to occur in relatively high concentrations in several virus-infected tissue (e.g., Welkie *et al.*, 1967). A general deficiency in soluble nitrogen compounds compared with healthy leaves may occur during periods of rapid virus synthesis.

2. Compounds Containing Phosphorus

Phosphorus is a vital component of all viruses and as such may come to represent about one-fifth of the total phosphorus in the leaf. In spite of this we still have no clear picture for any virus of the source of virus phosphorus, or the effects of infection on host phosphorus metabolism.

In Chinese cabbage leaves infected with TYMV, sampled 12–20 days after inoculation, a rise in virus phosphorus was accompanied by a corresponding fall in nonvirus-insoluble phosphorus, suggesting that this virus uses phosphorus at the expense of (but not necessarily directly from) some insoluble source of phosphate in the leaf (Matthews *et al.*, 1963).

3. Leaf Pigments

Virus infection frequently involves yellow mosaic mottling or a generalized yellowing of the leaves. Such changes are obviously due to a reduction in leaf pigments. Many workers have measured the effects of virus infection on the amounts of pigments in leaves. Frequently it appears to involve a loss of the chlorophylls, giving the yellowish coloration due to carotene and xanthophyll, but the latter pigments are also decreased in some diseases. Changes in chloroplast pigments are probably often secondary changes, since many viruses appear to multiply and accumulate in other parts of the cell, and since closely related strains of the same virus may have markedly different effects on chloroplast pigments even though they multiply to the same extent.

The reduction in amount of leaf pigments can be due either to an inhibition of chloroplast development or to the destruction of pigments in mature chloroplasts. The first effect probably predominates in young leaves that are developing as virus infection proceeds. The rapidly developing chlorosis frequently observed in local lesions when mature green leaves are inoculated with a virus must be due to destruction of pigments already present. In systemically infected leaves, TYMV reduced the concentration of all six photosynthetic pigments to a similar extent. This was due to a cessation of net synthesis, and subsequent dilution by leaf expansion (Crosbie and Matthews, 1974a).

Dark green islands of tissue in Chinese cabbage leaves showing mosaic symptoms had essentially normal concentrations of pigments (Crosbie and Matthews, 1974a). Small leaves near the apex of large Chinese cabbage plants are shielded from light and contain little chlorophyll. When such cream-colored leaves, about 2 cm long, were excised and exposed to light, those from healthy plants became uniformly dark green, while those from TYMV-infected plants developed a prominent mosaic pattern of dark green islands and yellow areas within 24 hours. Thus, in young expanding leaves chlorophyll synthesis is inhibited in those islands of tissue in which TYMV is replicating.

4. Flower Pigments

In view of the work that has been done on the genetics and biochemistry of normal flower coloration, surprisingly little is known about the biochemistry of the flower-breaking process, which is such a conspicuous feature of many virus diseases. In tobacco plants infected with TMV, the normal pink color of the petals may be broken by white stripes or sectors. We have found the virus present only in the white areas. However, presence or absence of virus may not be the only cause for color breaks. In sweet peas (*Lathrus odoratus*) infected with what was presumably a single virus—bean yellow mosaic *Potyvirus*—a pale pink flower sometimes became flecked with both darker pink and white areas.

Virus infection usually appears to affect only the vacuolar anthocyanin pigments. The pigments residing in chromoplasts may not be affected. For example, the brown wallflower (*Cheiranthus cheiri*), which contains an anthocyanin, cyanin, and a yellow plastid pigment (Gairdner, 1936), breaks to a yellow color when infected by turnip mosaic *Potyvirus* (TuMV). A preliminary chromatographic examination of broken and normal parts of petals infected with several viruses showed that the absence of color was due to the absence of particular pigments rather than to other factors, such as change in pH within the vacuole (R. E. F. Matthews, unpublished).

Kruckelmann and Seyffert (1970) examined the effect of TuMV infection on several genotypes of *Matthiola incana* R. Br. Infection brought about both white stripes and pigment intensification. Observations on a set of known host genotypes have shown that virus infection affected only the activities of genes controlling the quantities of pigments produced. It appeared to have no effect on the activities of genes modifying anthocyanin structure.

L. Summary

The physiological and biochemical changes most commonly found in virus-infected plants are (i) a decrease in rate of photosynthesis, often associated with a decrease in photosynthetic pigments, chloroplast ribosomes, and rbcs; (ii) an increase in respiratory rate; (iii) an increase in the activity of certain enzymes, particularly polyphenoloxidases; and (iv) decreased or increased activity of plant growth regulators.

There is no reason to suppose that major disturbances of host plant metabolism are necessarily determined by the major processes directly concerned with virus replication. Some minor initial effect of virus invasion and replication may lead to profound secondary changes in the host cell. Such changes may obscure important primary effects even at an early stage after infection.

Many of the changes in host plant metabolism noted earlier are probably secondary consequences of virus infection and not essential for virus replication. A single gene change in the host, or a single mutation in the virus, may change an almost symptomless infection into a severe disease. Furthermore, metabolic changes induced by virus infection are often nonspecific. Similar changes may occur in disease caused by cellular pathogens or following mechanical or chemical injury. In many virus diseases, the general pattern of metabolic change appears to resemble an accelerated aging process. Because of these similarities, rapid diagnostic procedures based on altered chemical composition of the virus-infected plant must be used with considerable caution. Metabolic changes are discussed further in Chapter 12 in relation to disease processes as a whole.

12

Induction of Disease

In various earlier chapters I have summarized present knowledge about virus rep- lication and the symptoms both macroscopic and microscopic, physiological and biochemical, that are associated with this replication. In this chapter I will examine the extent to which we understand how viruses cause disease. Ultimately we would like to explain, in the terms of molecular biology and biochemistry, how a single virus particle containing genetic material sufficient to specify a few polypeptides can infect and cause disease in its host plant. This is a major task when we remember that the host plant is growing continually and is organized into a variety of tissues and organs with specific structures and functions (see Goldberg, 1988).

We are far from attaining this objective for any host–virus combination. Never- theless, the knowledge about replication provided by molecular biology allows us to think more constructively than in the past about possible ways in which viruses might induce disease. Furthermore, various new methods based on recombinant DNA technology have become available for studying the role of viral gene products in disease induction. Important procedures are site-directed mutagenesis of the viral genome; switching genes between viruses and virus strains; and construction of transgenic plants that express only one or a few viral genes. These procedures are providing new information on the role of viral genes in disease induction, but they have limitations, as discussed in the following sections. Various factors that influ- ence the course of infection and disease are discussed in the last section of this chapter. Induction of disease has been reviewed by van Loon (1987) and Zaitlin and Hull (1987).

I. THE KINDS OF HOST RESPONSE TO INOCULATION WITH A VIRUS

The terms for describing the various kinds of response made by plants to inoculation with a virus have been used in ambiguous and sometimes inconsistent ways. Thus, over many years, there has been confusion about the meaning of certain terms. Cooper and Jones (1983) discussed this problem and suggested a standardized

Table 12.1

Types of Response by Plants to Inoculation with a Virus

IMMUNE (nonhost)   Virus does not replicate in protoplasts, nor in cells of the intact plant, even in the initially inoculated cells. Inoculum virus may be uncoated, but no progeny viral genomes are produced.

INFECTIBLE (host)   Virus can infect and replicate in protoplasts.

   Resistant (*extreme hypersensitivity*)   Virus multiplication is limited to initially infected cells because of an ineffective viral-coded movement protein, giving rise to a *subliminal infection*. Plants are *field resistant*.

   Resistant (*hypersensitivity*)   Infection limited by a host response to a zone of cells around the initially infected cell, usually with the formation of visible necrotic local lesions. Plants are *field resistant*.

   Susceptible (*systemic movement and replication*)

      Sensitive   Plants react with more or less severe disease.

      Tolerant   There is little or no apparent effect on the plant, giving rise to *latent* infections.

---

usage. In particular they introduced the term *infectible* to mean the opposite of *immune*.

To a significant degree, the confusion in terminology has been due to a lack of molecular biological knowledge concerning the different kinds of virus–plant interactions. However, the work summarized in Chapter 10, Section V, allows some of the terms to be defined with greater precision.

In Table 12.1 I have defined the relevant terms in the light of current knowledge, taking account of the suggestion made by Cooper and Jones (1983). The use of some of these terms differs from that in other branches of virology. For example, *latent* used in reference to bacterial or vertebrate viruses usually indicates that the viral genome is integrated into the host genome.

It has been assumed for some years that virus–host cell interactions must involve specific recognition or lack of recognition between host and viral macromolecules. Recently one example has been described where such specific recognition must be involved (Section V,B,2).

Interactions might involve activities of viral nucleic acids, or specific viral-coded proteins, or host proteins that are induced or repressed by viral infection. A novel method for supression of host protein synthesis is suggested by the experiments of Sleat *et al.* (1988b). They produced young tobacco seedlings transformed to express a bacterial gene product. If the TMV origin of assembly nucleotide sequence was coupled to the bacterial gene DNA, and the plants inoculated with TMV, much less bacterial gene activity was detected. This suggests that the formation of natural pseudovirions might be a mechanism for shutting off specific host RNA functions.

II. THE ROLE OF VIRAL GENES
IN DISEASE INDUCTION

A. Some General Considerations

We know that a single base change, say a cytosine converted to a uracil residue by nitrous acid, is sufficient to produce a mutant virus, giving changed symptoms. It is most improbable that such a change in one out of, say, 7000 bases in a viral RNA could directly bring about altered disease symptoms. Any change is much more likely to be mediated by some protein product of the RNA. However, nucleotide changes in viroid RNAs, which do not code for any proteins, can bring about changed symptoms (Chapter 9, Section I,D). Therefore we cannot rule out the possibility that nucleotide sequences in viral genomes may sometimes play a direct role in disease induction.

Various lines of evidence have shown that viral genes must be involved in quite specific ways in the induction of disease. For example, closely related strains of TYMV produce two distinct and mutually exclusive pathways of change in the chloroplasts of TYMV-infected cells. In one, the chloroplasts first become rounded and clumped and then develop a large vacuole. In the other, they become angular before clumping, and then fragment to yield many small pieces. These different pathways must be activated by a viral gene or genes (Fraser and Matthews, 1979b).

The existence of viruses with genomes consisting of two or more pieces of nucleic acid has allowed the production of pseudorecombinants in the laboratory. Reassorting genomic segments has allowed determinants of specific disease symptoms to be located on particular genome segments. Many such experiments have been reported. For example, it was established that at least some determinants of symptom production are located on DNA1 of ACMV (Stanley *et al.*, 1985). Experiments with only a few virus strains tested on only a few hosts may not reveal the full range of genome segments involved. Thus Rao and Francki (1982) constructed 18 pseudorecombinants *in vitro* by exchanging the three genomic RNA segments between pairs of three strains of CMV. These were tested on 10 host plant species. Some host reactions were determined by RNA2 or RNA3 alone, but others resulted from an interaction between both RNAs. Others probably involved all three genome segments. While such reassortment experiments have given useful information, they do not always pinpoint the gene or genes involved since genome segments often code for more than one protein product.

In considering the induction of disease it will be useful to make a distinction between the *functions* of viral genes in the virus life cycle and the *effects* of the genes on the host. In normal circumstances, all the gene products of a viral genome will have one or more functions in the complete virus life cycle. There is no evidence for the existence of viral genes that have a role only in inducing disease.

If we consider a range of host species and environmental conditions (Section V), a given virus can cause far more different kinds of symptoms than it has different gene products or combinations of two or more gene products acting together. We must conclude that a particular viral gene may have a variety of effects on the kind of disease produced depending on the host plant involved, the environmental conditions, and possible interactions with other viral genes. At present we can distinguish three kinds of effect of viral genes: (i) Those based on a specific requirement for an essential virus function. The small vesicles induced by TYMV in chloroplast membranes (see Fig. 7.16) may be an example of this sort. (ii) Those based on a defect in a viral gene, for example, a cell-to-cell transport function, that results in limitation of infection to the point of entry. (iii) Those based on effects of virus infection that are quite inconsequential as far as virus replication or movement are concerned, unless they so damage host cells that further replication is inhibited.

Many virus infections show no observable macroscopic disease symptoms. However, apart from the cryptic viruses, most produce observable and often characteristic cytological effects. Thus most macroscopic disease induced by viruses may be inconsequential in the sense just noted (iii), while some cytological effects may represent essential requirements for virus replication and movement. Perhaps macroscopic disease symptoms should be regarded as effects of the virus on its cellular environment, rather than as part of the viral phenotype.

B. Specific Viral Gene Products

All viral genes may not be involved in disease induction. For example, Graybosch *et al.* (1989) introduced the cytoplasmic inclusion protein gene of a potyvirus into tobacco plants. The gene was expressed in the transformed plants but no disease was induced. However, the possibility remains that this gene product could modulate disease expression during virus infection. Several specific viral gene products have now been clearly implicated in the induction of disease. These are discussed in the sections that follow.

1. CaMV

Experiments discussed in Chapter 8, Section I, showed that gene VI of CaMV contains a host range determinant. The product of this gene (MW = 66K) is also involved in determining the kind of disease that develops. Daubert *et al.* (1984) constructed recombinants from strains of CaMV differing markedly in biological properties. The experiments showed that typical disease expression, consisting of leaf chlorosis and mottling, mapped to a genome segment containing gene VI. Other regions of the genome influenced the disease pattern. This was presumably due to effects of the gene product, because in other experiments in developing leaves of infected turnip plants, severity of vein-clearing symptoms was not correlated with the concentration of CaMV DNA (Maule *et al.*, 1989).

In a more fruitful approach, Baughman *et al.* (1988) transferred a segment of CaMV DNA containing gene VI to tobacco plants using the *Agrobacterium tumefaciens* Ti plasmid. The resulting transgenic plants showed viruslike symptoms. Gene VI from two different virus isolates produced different symptoms—either mosaiclike or a bleaching of the leaves. Symptom production was blocked by deletions or frameshift mutations in gene VI. Production of symptoms was closely correlated with the appearance in the leaves of the 66K gene product, as shown by immunoblots. However, they estimated that the amount of the 66K protein produced in transgenic tobacco plants was only about one-twentieth of that found in infected turnips.

The effects of gene VI may depend on the plant species tested. Thus Goldberg *et al.* (1990) transformed three solanaceous species with gene VI of CaMV and FMV. In these nonhosts there was a positive relationship between the induction of a prominent mottling symptom and the extent of gene VI expression. However, systemically susceptible hosts of CaMV or FMV showed no prominent chlorosis or mottling whatever the level of expression of gene VI. Stratford and Covey (1989) constructed a series of hybrid CaMV genomes between two strains that cause severe or mild disease in turnips. A variety of loci affecting disease development were detected. For example, determinants for the degree of leaf chlorosis were located in a domain consisting of part of gene VI together with the large intergenic region and nucleotides 6103–6190 of gene VII. Plant stunting was affected by at least two separate loci, one containing parts of genes I and II and the second within the reverse transcriptase gene (V). Thus different aspects of the disease process can be assigned to specific parts of the genome, and much of the viral genome appears to be involved.

2. TGMV

Disease induction by TGMV appears to involve genes on both DNA segments of the viral genome, including that of the coat protein (Rogers *et al.*, 1986; Gardiner *et al.*, 1988). Further work is needed to clarify the role of particular genes.

3. TMV

Coat Protein

Various studies, using recombinant viruses, that demonstrated that TMV coat protein plays a role in disease induction were discussed in Chapter 6, Section V, C,2, and Chapter 7, Section III,A,4. A single amino acid change in the coat protein may be the only difference detected between strains causing different diseases (e.g., Sarkar, 1986). Reinero and Beachy (1986) found TMV coat protein in the chloroplasts of infected cells in both the stroma and membrane fractions. Chloroplasts isolated from leaves infected with a symptomless strain contained 10–50 times less coat protein than those from leaves with mosaic symptoms.

Dawson *et al.* (1989) specifically modified the coat protein gene in various ways. Deletions in the gene that retained the C terminus gave rise to a marked yellowing disease resulting from degradation of the chloroplasts.

Tests with a variety of mutant constructs containing insertions, deletions, frameshifts, or single-base changes at defined sites demonstrate that a number of sites in the coat protein gene affect the necrotic response of TMV to *Nicotiana* varieties containing the N' genotype (Saito *et al.*, 1989; Culver and Dawson, 1989a). The point mutations in the coat protein created by Culver and Dawson also influenced systemic symptoms in *N. tabacum* with the n genotype. In principle, it could be the altered coat protein or the changed mRNA that caused these symptom changes. Culver and Dawson (1989b) removed the translational start signal (AUG → AGA) for the coat protein in a mutant eliciting the hypersensitive response. Infectious transcripts failed to induce the response, showing that it is the coat protein itself that is involved. Furthermore, the mutant lacking a coat protein moved slowly out of inoculated leaves, but produced systemic mosaic symptoms.

The 130K and 180K Proteins

Comparison of the nucleotide sequences of two almost symptomless mutants of TMV (LII and LIIA) with the parent-type strain showed that a change from cysteine to tyrosine at amino acid position 348 of the 130K and 180K proteins was involved in loss of symptom production. Two other amino acid changes in these proteins may also have been involved (Nishiguchi *et al.* 1985).

4. RDV

Since members of the plant reovirus group cannot be cloned by single local lesion selection, isolation of mutants has been difficult. However, Kimura *et al.* (1987) injected dilute inoculum of the type strain (0) of RDV into insect vectors. They repeatedly selected for rice plants showing unusually severe symptoms. By these means they obtained a severe strain (S) (see Fig. 11.2). The fourth largest genome segment of strain S had an apparent MW about 20K larger than that of strain 0. The corresponding gene product in strain 0 had an $M_r \simeq 43K$, and in strain S the $M_r \simeq 44K$. This protein is located in the outer envelope of the virus. The idea that this gene product is involved in producing the more severe symptoms receives support from the fact that neurovirulence in a reovirus infecting mice has been shown to be controlled by the outer envelope protein (Weiner *et al.*, 1977).

5. CMV

Strain M of CMV induces severe systemic chlorosis in tobacco. Pseudorecombinants with another strain located the gene responsible on M CMV RNA3. Further experiments with recombinant RNA3's transcribed from engineered cDNAs showed that the symptom in tobacco was controlled by the coat protein gene. There were only eight amino acid differences between the M CMV coat protein and that of the other strain used (Shintaku and Palukaitis, 1990).

6. Viroids and Satellite RNAs

The molecular basis for disease induction by viroids and satellite RNAs was discussed in Section I,D and II,B,7 of Chapter 9.

C. Defective Interfering Particles

The properties of DI particles and RNAs were discussed as a manifestation of faulty virus replication in Chapter 6, Section VIII,B. Their presence in the plant tends to make disease symptoms milder. Thus Hillman *et al.* (1987) described a DI RNA derived from a culture of TBSV (see Fig. 6.16). In *Nicotiana clevelandii,* TBSV alone causes lethal necrosis. Addition of increasing amounts of the DI RNA to the inoculum resulted in increasingly milder symptoms, which was accompanied by a reduced production of TBSV. Ismail and Milner (1988) isolated a DI particle of *Sonchus* yellow net virus whose RNA was 77% as long as standard virus. Particles containing this RNA were only about 80% as long as standard virus. They were noninfectious on their own, but when mixed with standard virus the resulting symptoms in *N. edwardsonii* were chlorotic mottling instead of the normal vein-clearing.

D. Future Studies on the Role of Viral Genes in Disease Induction

The role of viral genes in disease induction is currently one of the most biologically interesting and practically important areas of plant virus research. Site directed mutagenesis and other techniques together with *in vitro* or protoplast systems, and sometimes whole plants, can identify the main function or functions of a gene product, for example that it is a replicase or a protease. However, to study disease induction intact plants or parts of plants must be used, and a full virus replication cycle studied. For these reasons there may be significant difficulties ahead.

1. Establishing the *in Planta* Roles of Particular Viral Gene Products

The *in planta* roles of some of the protein products of viral genes involved in disease induction may be very difficult to study for several reasons:

1. The proteins may be present in very low concentration, as a very few molecules per cell of a virus-specific protein could block or derepress some host-cell functions.
2. It is quite possible that such proteins would be present in the infected cells for a short period relative to that required for the completion of virus synthesis.

3. The virus-specified polypeptide may form only part of the active molecule in the cell.

4. The virus-specified polypeptide may be biologically active only *in situ,* for example, in the membrane of some particular organelle.

5. As noted in Section IIA, many macroscopic disease symptoms may be due to quite unexpected side effects of virus replication. The following hypothetical example illustrates this kind of possibility. Consider the 49K proteinase coded for by TEV (Carrington and Dougherty, 1988). The amino acid sequences that function as substrate recognition signals have been identified. In the usual host, under normal conditions this proteinase can accumulate to high levels within infected cells without causing cell death. This must mean that the proteinase does not significantly deplete the amount of any vital host protein. Suppose that we change to another host species, or to different environmental conditions. In the new situation, some host-coded protein essential for cell function might be sensitive to cleavage, with a new pattern of disease developing as a consequence. In such circumstances it might be difficult to identify the host protein involved, and its functions. In attempting to understand disease induction in molecular terms it may be most profitable to concentrate initially on effects that are known to be a direct consequence of, and essential for virus replication, or for virus movement from cell to cell.

2. Approaches Using Gene Manipulation Technology

Although molecular approaches will no doubt continue to increase our understanding of the role of viral genes in the induction of disease difficulties are emerging. For example, site-directed mutagenesis, which is in principle, a powerful technique, has several limitations. (i) The number of possible permutations and combinations of base sequence alteration is enormous. (ii) It will be difficult to find changes that are not lethal for the virus when a full infectious cycle is required. (iii) Because of the high mutation rate in RNA viruses and the existence of recombination (Chapter 13) it is often necessary to check the complete base sequence of the engineered mutant culture to ensure that spontaneous changes have not caused reversion to the original sequence or altered the sequence in some other way. (iv) A more general difficulty may be that, as increasingly detailed experiments are carried out with a particular virus, most or all viral genes may be found to have interacting roles in various aspects of disease induction.

III. HOST PROTEINS INDUCED BY VIRUS INFECTION

Some viruses that cause necrotic local lesions, induce a non-specific host response that includes the *de novo* synthesis of host-coded proteins. The development of both localized and systemic resistance to superinfection follows the development of the necrotic local lesions. The localization of virus replication in tissue near the site of infection, the hypersensitive response (Table 12.1), is important in agriculture and horticulture as the basis for field resistance to virus infection. These phenomena have been the subject of many studies and are reviewed by Ponz and Bruening (1986), Bol (1988), Bol and van Kan (1988), and van Loon (1989).

A. Biological Aspects of Local and Systemic Acquired Resistance

1. Localized Acquired Resistance

Some varieties of tobacco respond to infection with certain strains of TMV at normal temperatures by producing necrotic local lesions and no systemic spread, instead of the usual chlorotic local lesions followed by mosaic disease. In several *Nicotiana* varieties the reaction is under the control of a single dominant gene, the *N* gene, found naturally in *N. glutinosa* (Holmes, 1938). This has been incorporated into *N. tabacum* cultivar Samsun NN (Holmes, 1938), *N. tabacum* cultivar Burley NN (Valleau, 1952), and *N. tabacum* cultivar Xanthi nc (Takahashi, 1956). In very young seedlings, plants containing the *N* gene (either *NN* or *Nn*) developed systemic necrosis. In older plants only two responses to infection are usually observed— localized necrotic lesions in *NN* and *Nn,* and systemic mosaic disease in *nn* (Holmes, 1938). However, systemic necrosis may occur in some older plants with the *N* gene (Dijkstra *et al.,* 1977).

Another gene in tobacco for local necrosis has been described by Weber (1951). With this gene the heterozygote always developed systemic necrosis (Fig. 12.8). A mutant that shows a clear gene dose effect is *N. tabacum* var. Samsun EN. At temperatures of 23°–26°C much more virus was produced in local lesions of *EN/en* plants than in *EN/EN* (Jockusch, 1966a).

Most experimental work has been carried out with tobacco varieties containing the *N* gene. Ross (1961a) showed that a high degree of resistance to TMV developed in a 1 to 2-mm zone surrounding TMV local lesions in Samsun NN tobacco (Fig. 12.1). The zone increased in size and resistance for about 6 days after inoculation. Greatest resistance developed in plants grown at 20°–24°C. Resistance was not found in plants grown at 30°C.

Genes that induce a hypersensitive necrotic reaction in intact plants or excised leaf pieces fail to do so when isolate protoplasts are infected. This has been found for the *NN* gene in tobacco (Otsuki *et al.,* 1972b) and for *Tm2* and *Tm2²* in tomato (Motoyoshi and Oshima, 1975, 1977). Protoplasts from plants carrying these genes in the homozygous condition allowed replication of TMV without death of the cells. It has been suggested that this effect might be due to the epidermis being involved in the necrotic response (Motoyoshi and Oshima, 1975, 1976), or that cell-to-cell connections are necessary for the hypersensitive phenotype to be expressed (Motoyoshi and Oshima, 1977). Another, and possibly more likely, explanation could be that the presence of the cell wall is necessary for expression of the *N* gene. Peroxidase enzymes appear to be involved in the necrotic response. These enzymes may be located mainly in the cell wall compartment. Thus, removal of the wall may cause a gap in the chain of biochemical events, leading to cell death.

2. Systemic Acquired Resistance

Our knowledge about systemic acquired resistance has come mainly from the work of Ross and colleagues (e.g., Ross, 1961b, 1966) on TMV in the tobacco

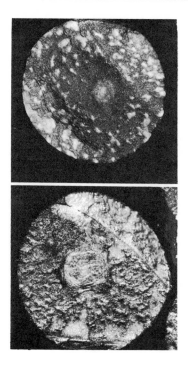

Figure 12.1 Acquired resistance to infection. Upper: A disk cut from a Samsun NN tobacco leaf inoculated first with TMV (large lesion in center) and 7 days later challenge inoculated with a concentrated TMV inoculum. Note absence of lesions from the second inoculation in a zone around the original lesion. Lower: A similar experiment in which PVX was the challenge virus. No zone free of lesions is present. (From Ross, 1961a.)

variety Samsun NN and TNV in pinto bean. In tests with tobacco, lower leaves are inoculated with TMV and then some days later the same leaves or upper leaves may be challenged by a second inoculation with TMV (Fig. 12.2). Acquired resistance is measured by the reduction in diameter of the lesions (and with some viruses, reduction in number). With bean, one primary leaf is inoculated and the opposite primary leaf challenged by inoculation some days later.

Lesions were about one-fifth to one-third the size found in control leaves, but lesion number was not reduced with TMV in Samsun NN tobacco. Resistance was detectable in 2–3 days, rose to a maximum in about 7 days, and persisted for about 20 days. Leaves that developed resistance were free of virus before the challenge inoculation. No conditions have been found that would give complete resistance. In plants held at 30°C no resistance developed. Mechanical or chemical injury that killed cells did not lead to resistance, nor did infection with viruses that do not cause necrotic local lesions. On the other hand, many other nonspecific agents applied to leaves will induce the phenomenon (e.g., Gupta *et al.*, 1974). In such experiments it is not possible to be sure that the same phenomenon is being studied since many treatments affect lesions size.

The resistance induced by TMV was not specific for TMV, but was effective for TNV and several other viruses. A similar lack of specificity in the resistance acquired following the development of necrotic local lesions was found with various other host–virus combinations giving the hypersensitive response. However, virus–specific factors may regulate the extent of the resistance (van Loon and Dijkstra, 1976).

Ross (1966) pointed out that effects on lesion number tend to be more variable and develop later. He considered that a fall in lesion number merely means that in highly resistant leaves, lesions do not become large enough to be countable.

Figure 12.2 Resistance acquired at a distance from the site of inoculation. Right: Samsun NN leaf inoculated first on the apical half with TMV and 7 days later given a challenge inoculation over its whole surface with TMV. Left: Control leaf given only the second inoculation. (From Ross, 1961b.)

B. The Host Proteins Induced in the Hypersensitive Response

Gianinazzi *et al.* (1970) and van Loon and van Kammen (1970) showed that changes in the pattern of soluble leaf proteins occurred in tobacco leaves responding hypersensitively to infection with TMV. Since then these proteins have been studied in many laboratories. They were termed pathogenesis-related proteins or "PR proteins" by Antoniw *et al.* (1980). There are two classes: acidic PR proteins and their basic homologues (Table 12.2). PR proteins have been reviewed by Carr and Klessig (1989) and Bol *et al.* (1990).

In Samsun NN tobacco leaves infected with TMV, at least 14 acidic PR proteins appear. These are soluble at pH 3.0, resistant to proteases, and accumulate in the intercellular spaces (e.g., Ohashi and Matsuoka, 1987; Dumas *et al.,* 1988). These 14 proteins can be placed in five groups based on MW, amino acid composition, and serology (van Loon *et al.,* 1987; Hooft van Huijsduijnen *et al.,* 1987; Kauffmann *et al.,* 1987, 1989). Basic homologues for three of these groups have been characterized (Table 12.2). Eight of the proteins, purified to a substantial degree, contained no carbohydrate (Jamet and Fritig, 1986).

Group 1 contains the acidic proteins known as 1a, 1b, and 1c. These have over 90% sequence similarity and each have a signal peptide of 38 amino acids. When this is removed, 138 amino acids remain. One basic protein, 173 amino acids long,

Table 12.2
Extracellular Proteins Induced in Samsun NN Tobacco by TMV Infection[a]

		Acidic PR proteins		Basic homologues		
Group	Name	MW $\times 10^{-3}$	Estimated number of genes	MW $\times 10^{-3}$	Estimated number of genes	Function
1	1a,1b,1c	15	8	19	8	Unknown
2	2,N,O	40	?	33	?	β-1,3-Glucanases
3	P,Q	29/30	4	32/34	4	Chitinases
4	R,S	23–24	?	—	—	Unknown
5	r_1,r_2,s_1,s_2	13–14.5	?	—	—	Unknown

[a]From Bol and van Kan (1988) and Kauffmann *et al.* (1989).

has been identified, with 67% sequence similarity to the acidic proteins of group 1 (Cornelissen *et al.*, 1987). No function has yet been assigned to the group 1 proteins.

Group 2 contains three acidic proteins known as 2, N, and O. These are acidic β-1,3-glucanases. There is also a serologically related basic enzyme in this group (Kauffmann *et al.*, 1987).

Group 3 proteins, named P and Q, are chitinases (Legrand *et al.*, 1987). There is about 67% amino acid sequence similarity between the acidic and basic chitinases (Hooft van Huijsduijnen *et al.*, 1987; Shinshi *et al.*, 1987).

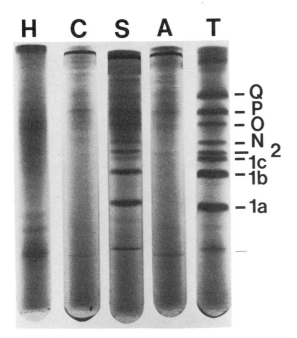

Figure 12.3 Accumulation of PR proteins in the intercellular fluid of tobacco after various treatments. Plants were sprayed with water (H), *p*-coumaric acid (C) or salicylic acid (S) or inoculated with AMV (A) or TMV (T). Samples of the intercellular fluid were electrophoresed in nondenaturing polyacrylamide gels. The position of the major PR proteins is indicated in the margin. (From Bol and van Kan, 1988.)

Groups 4 and 5 each contain acidic proteins with no basic homologues. R and S are serologically related (Kauffman *et al.*, 1989). Protein S shows a 65% sequence similarity to thaumatin, an intensly sweet-tasting protein from the fruits of a West African shrub, *Thaumatococcus danielli* Beuth (Cornelissen *et al.*, 1986b), but the tobacco protein does not taste sweet and is not serologically related to thaumatin (Kauffmann *et al.*, 1989). Two tobacco genes encoding thaumatinlike proteins have been sequenced (van Kan *et al.*, 1989). The functional significance of this sequence similarity is unknown. Protein S also has extensive sequence similarity to a maize bifunctional protein that is an α-amylase-proteinase inhibitor (Richardson *et al.*, 1987) but has no inhibitory activity (Kauffmann *et al.*, 1989). PR proteins of groups 1 and 4 are probably enzymes concerned with the production of various metabolites such as ethylene and phytoalexins.

These various PR proteins are not induced following infection with a virus that does not cause necrotic local lesions in the host used (Fig. 12.3, lane 4).

C. Lack of Specificity in PR Protein Induction

Several lines of evidence demonstrate that the induction of PR proteins is a generalized defense reaction in plants rather than a specific response to virus infections that cause necrosis: (i) studies with various host genotypes and TMV strains showed that the differences in the soluble proteins are associated with a particular symptom type and are not confined to the NN genotype; (ii) proteins homologous to the tobacco PR proteins have been found in at least 20 plant species, for example, tomato, cowpea, maize, and barley (White *et al.*, 1987; van Loon, 1989; Bol, 1988), and amino acid sequences are substantially conserved in different species; (iii) chitin and β-1,3-glucan, the substrates of group 2 and 3 PR proteins, are found in fungal cell walls, not in plants; and (iv) PR proteins are elicited by other viruses causing necrotic lesions in particular hosts, for example, by TNV (Pennazio *et al.*, 1987; Roggero and Pennazio, 1989) by viroids (Granell *et al.*, 1987; Vera and Conejero, 1989), by other invading organisms, for example, *Agrobacterium* (Antoniw *et al.*, 1983), and by the fungi *Thielaviopis basicola* (Gianinazzi *et al.*, 1980), and by a variety of chemicals. Salicylic acid, polyacrylic acid, and ethepon in particular have been studied (e.g., van Loon and Antoniw, 1982; White *et al.*, 1986; Hooft van Huijsduijnen *et al.*, 1986a).

D. Induction of mRNAs for PR Proteins

By the use of cDNA clones as probes, Hooft van Huijsduijnen *et al.* (1986b) and Bol *et al.* (1990) were able to show that, following spraying of Samsun NN tobacco leaves with salicylic acid or inoculation with TMV, a series of RNAs corresponding to PR protein mRNAs were induced (Fig. 12.4). All the mRNAs shown were strongly induced by TMV infection. Immunoprecipitation of hybrid-selected translation products confirmed the identity of some of the mRNAs. The virus-inducible tobacco genes respond differently to various chemical inducers. They show differences in their tissue specific expression, and they are expressed in a light-dependent manner. It can be inferred that the promoter regions of these genes must be composed of a mosaic of *cis*-acting elements interacting with different *trans*-acting factors (Bol *et al.*, 1989). As indicated in Table 12.2, group 1 and

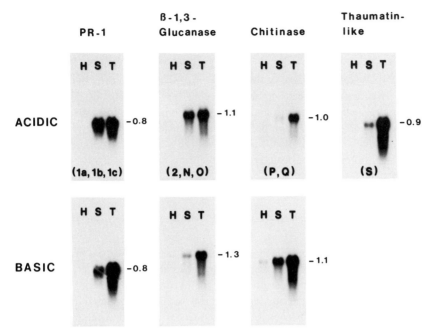

Figure 12.4   Induction of mRNAs for PR proteins in Samsun NN tobacco by TMV infection (T) or salicylate treatment (S) (H, healthy plants). Seven Northern blots were hybridized to ^{32}P-labeled cDNA clones corresponding to the acidic and basic isoforms of the proteins indicated. The estimated size of the mRNAs (in kb) is indicated to the right of the margin for each blot. (Reprinted with permission from Bol *et al.*, 1990.)

group 3 PR protein genes exist as multigene familes (e.g., Matsuoka *et al.*, 1987; Pfitzner and Goodman, 1987; Cornelissen *et al.*, 1987).

E. Synthesis of PR Proteins *in Vitro* and *in Vivo*

Translation *in vitro* of polyribosomal or total poly(A) RNAs from tobacco leaves revealed four translation products specific to hypersensitively responding leaves (Smart *et al.*, 1987). *In vivo* pulse-labeling experiments detected PR protein synthesis in tobacco leaves about 8 hours after treatment with salicylic acid, and more thn 18 hours after infection with TMV. For both treatments, synthesis declined rapidly after 50 hours (Matsuoka and Ohashi, 1986).

The PR proteins that are β-1,3-glucanases and chitinases accumulate mainly in the vacuole (Mauch and Staehelin, 1988). On the other hand, the b_1 protein and those related to it serologically were located both in the cytoplasm and in the intercellular spaces (Dumas *et al.*, 1988).

F. Other Host Proteins Induced during the Hypersensitive Response

Increases in the activity of a variety of enzymes besides those among the PR proteins have been observed during the hypersensitive response to viruses. Peroxidase, polyphenoloxidase, and ribonuclease activities are increased (e.g., Wagih and Coutts, 1982). Peroxidases are not usually considered to be PR proteins but they have similar properties. They are soluble at pH 3.0 and are protease resistant, and acidic and basic isozymes accumulate in the intercellular spaces (Bol and van Kan, 1988). The metabolism of phenylpropanoid compounds is strongly activated by infection with various pathogens, including viruses that induce a hypersensitive response. This activation leads to the accumulation of compounds derived from phenylalanine, such as flavonoids and lignin. Activation of the pathway involves *de novo* enzyme synthesis, for example, *O*-methyltransferase (Collendavelloo *et al.*, 1982). Van Kan *et al.* (1988) identified two genes in Samsun NN tobacco coding for a glycine-rich protein that is strongly induced during the hypersensitive response to TMV infection.

G. Steps in the Induction of Host Proteins

One of the earliest detectable events in the interaction between a plant host and a pathogen that induces necrosis is a rapid increase in the production of ethylene, which is a gaseous plant stress hormone. In the hypersensitive response to viruses there is an increased release of ethylene from leaves (e.g., Gáborjányi *et al.*, 1971). The fact that ethepon (a substance releasing ethylene) introduced into leaves with a needle can mimic the changes associated with the response of Samsun NN to TMV is good evidence that ethylene is involved in the initiation of this hypersensitive response (van Loon, 1977). An early burst of ethylene production is associated with the virus-localizing reaction. The increase in ethylene production is not determined by the onset of necrosis but by a much earlier event (De Laat and van Loon, 1983). Ethylene applied to carrot tissue induced mRNAs for a range of known host defense proteins (Ecker and Davis, 1987). However, experiments with various factors that inhibit or stimulate ethylene production showed that stress ethylene developing during a hypersensitive response to virus infection does not affect the localizing mechanism (Pennazio and Roggero, 1990).

H. Other Biochemical Changes during the Hypersensitive Response

Many other biochemical changes have been observed during the hypersensitive response. For example, Uegaki *et al.* (1988) detected 19 sesquiterpenoids that were considered to be stress-induced compounds in *Nicotiana undulata* inoculated with TMV. An enhanced NADPH-dependent oxygen-generating system was found in a membrane-rich cellular fraction from tobacco leaves reacting hypersensitively to TMV infection. Early electrolyte leakage occurred from cells of cowpea leaves during a hypersensitive response to virus infection and other stress stimuli (Pennazio and Sapetti, 1981). Free abscisic acid concentration was raised up to 18-fold of normal in tissue near or within necrotic local lesions caused by TMV in tobacco (Whenham and Fraser, 1981). The cytokinin content of Xanthi NN tobacco leaves with systemic acquired resistance was increased (Balázs *et al.*, 1977).

I. Roles of Host-Coded and Viral-Coded Proteins

There are four phenomena to be considered in relation to the hypersensitive response: (i) the limitation of virus spread; (ii) necrosis of cells near the initial infection; (iii) resistance to superinfection near the local lesions, and (iv) systemic acquired resistance. In spite of all the experimental work that has been carried out, it is still not entirely clear whether these are four aspects of one process or two or more independent phenomena. Necrosis is not essential for the limitation of virus spread. For example, no necrosis follows inoculation of cucumber cotyledons with TMV, but virus is limited to local infections.

It is not yet established whether any of the PR proteins play a role in the limitation of virus spread and acquired resistance. The PR proteins with known enzymatic activities appear to have a defense role against fungi, bacteria, or insects rather than viruses. If any of the PR proteins are, in fact, active in the antiviral response, they may be found among those proteins for which a function has not yet been established. However, when expressed to high levels in transgenic tobacco plants with the NN constitution, individual PR proteins did not affect the necrotic response to inoculation with TMV (e.g., the 1b gene, Cutt *et al.*, 1989; the 1a and s genes, Linthorst *et al.*, 1989).

PR proteins of group 1 are produced in *Nicotiana rustica* leaves developing necrotic local lesions after inoculation with TMV, but the virus moves systemically (Dumas and Gianinazzi, 1986). In tobacco, Xanthi nc inoculated with TMV synthesis of group 1 PR proteins was not observed when plants were placed in the dark, but necrotic lesions formed and virus spread was restricted (Abad *et al.*, 1986). On the other hand, changes with time in the distribution of virus and PR protein 1a around a single local lesion of TMV-infected tobacco were those that might be expected for a substance closely associated with the restriction of virus spread (Antoniw and White, 1986).

Other experiments suggest that viral-coded proteins may be involved in the local lesion response. Thus a block in synthesis of the 30K transport protein during the hypersensitive reaction may be the reason why TMV becomes localized in tobacco varieties with the *N* gene (Moser *et al.*, 1988).

Whereas the *N* gene results in necrotic local lesions following inoculation with all strains of TMV, the *N'* gene brings about this result with only some strains of the virus, and lesion sizes vary. Differences in abilities of TMV strains to multiply and cause lesions of different sizes on *N'* gene plants were strongly correlated with differences in the thermal stabilities of their isolated coat protein subunits (Fraser, 1983). More direct evidence for involvement of the coat protein comes from experiments with recombinant viruses (Saito *et al.*, 1987a). A common strain of TMV produces systemic infection in *Nicotiana sylvestris*, which contains the *N'* gene. Strain L produces a hypersensitive response. By switching the coat protein genes between the two strains it was shown that the hypersensitive response was a property of the coat protein gene of strain L.

J. Systemic Acquired Resistance

The action at a distance involved in systemic acquired resistance presumably involves the translocation of some substance or substances. Ross (1966) has presented good evidence that transport of a resistance-inducing material is involved. For example, when the midrib of an upper tobacco leaf was cut, resistance did not develop in the portion of the lamina distal to the cut. Similarly, killing sections of petiole of inoculated leaves with boiling water while allowing the leaf to remain turgid prevented development of resistance in other leaves. Other experiments showed that in large tobacco plants the material moved equally well both up and down the stems.

The nature of the material that migrates is unknown, as is the actual mechanism of resistance in the resistant uninfected leaves. This mechanism may or may not be the same as that in the zone of tissue around necrotic lesions. Systemic acquired resistance can be induced by nonnecrotic localized viral infection (Roberts, 1982). Systemic acquired resistance is not effective when the challenge virus is one that moves systemically (Pennazio and Roggero, 1988).

When PR proteins are induce chemically, resistance to virus infection is usually confined to an area near the site of treatment. Ethylene is the exception in that it induces both PR proteins and resistance in leaves other than the treated leaf (van Loon and Antoniw, 1982). This suggests that ethylene could be the trigger that stimulates production of the mobile compound noted earlier.

Fraser (1982) made a quantitative study of the appearance of group 1 PR proteins and of systemic acquired resistance in tobacco responding to TMV with necrotic local lesions. He found no quantitative or temporal relationship between the degree of systemic resistance and the presence of PR proteins. For example, resistance could be detected in leaves before the PR proteins.

With a strain of PYV causing systemic necrosis in tobacco there was an increase in ethylene production and an accumulation of PR proteins but these had no effect on the development of systemic disease (Roggero and Pennazio, 1988). These and other results suggest that PR proteins do not have a significant role in systemic acquired resistance.

K. Wound Healing Responses

Wounds involving necrosis caused by mechanical injury, insects and various pathogens, including viruses, frequently result in a series of wound healing responses by the plant. These responses must involve the nonspecific induction of many host-coded proteins. The most complex response is the development of a wound periderm and cell wall changes, including lignification, suberization, and the deposition of callose. Virus-induced necrosis may lead to such wound responses. Figure 11.18 illustrates a wound–periderm response. A periderm was formed in young bean leaves inoculated with the VM strain of TMV, which gives very small lesions, but not in old leaves or in leaves inoculated with the U_1 strain giving large lesions (Wu, 1973).

Various workers have noted a deposition of callose in cells around necrotic local lesions, leading to thickening of cell walls and probably blocking of plasmodesmata (e.g., Hiruki and Tu, 1972; Wu, 1973). Stobbs et al. (1977) found that callose deposition in live cells extended beyond the margin of detectable virus, while remaining within the zone of fluorescent metabolite accumulation in pinto bean leaves infected with TMV. Cell wall glycoproteins were determined chemically following extraction from leaves by Kimmins and Brown (1973). They found an accumulation of glycoproteins following inoculation of hypersensitive hosts with TMV or TNV. An identical response occurred in leaves that had been sham inoculated.

Other observations suggest that cell wall modifications may not be a factor limiting spread. Appiano et al. (1977) considered that the conspicuous cell wall lignification seen in lesions caused by TBSV in *Gomphrena* leaves was not a barrier to spread of virus because lignification did not follow the whole cell perimeter, and because virus could be detected beyond the cells with modified walls.

L. Antiviral Factors

Sela and his colleagues have partially purified and characterized an "antiviral factor" (AVF) from *Nicotiana* cultures with the *N* gene (reviewed by Fraser, 1987a). Parallels have been drawn between the AVF and human interferons (e.g., Sela *et al.*, 1987). However, the significance of this work in relation to host plant resistance remains to be established.

IV. PROCESSES INVOLVED IN DISEASE INDUCTION

In the previous sections we have seen that viral genes play a part in the induction of disease and that host-coded proteins are induced as part of the disease process. In this section I will discuss some of the processes involved in the induction of disease. At the present time we are unable to implicate specific viral- or host-coded proteins with the initiation of any of these processes.

A. Sequestration of Raw Materials

The diversion of supplies of raw materials into virus production thus making host cells deficient in some respect is an obvious mechanism by which a virus could induce disease symptoms. This mechanism is very probably a factor when the host plant is under nutritional stress. For example, in mildly nitrogen-deficient Chinese cabbage plants the local lesions produced by TYMV have a purple halo, the purple coloration being characteristic of nitrogen starvation (Diener and Jenifer, 1964). Specific nutritional stress can also be induced by environmental conditions. For example, Ketellapper (1963) showed that adverse effects of unfavorable temperature on plant growth could be partly or completely prevented by providing the plants with some essential metabolites, which applied under normal temperature conditions had no effect. The effects were to some extent species specific. Thus, at least part of the effect of adverse temperature on plant growth may be due to deficiencies in specific metabolites. Virus infection may well aggravate such deficiencies.

Sometimes an increase in severity of symptoms is associated with increased virus production. Thus, with bean pod mottle virus in soybean, Gillaspie and Bancroft (1965) observed flushes of severe symptoms, followed by a recovery period at 1.5 and 4.5 weeks after inoculation. Assays on leaf extracts showed that highest infectivity per unit of tissue occurred at the first symptom flush. Infectivity then fell, but rose again during the second symptom flush. However, in well-nourished plants there is no general correlation between amount of virus produced and severity of disease (Palomar and Brakke, 1976). Similarly there is not necessarily a correlation between the severity of macroscopic disease and various physiological and biochemical changes brought about by infection (Ziemiecki and Wood, 1976), although some correlations of this sort have been found.

Using 26 different strains of TMV in tobacco and tomato, Fraser *et al.* (1986) used multiple regression analysis to show that variation in virus multiplication and symptom severity could account for most of the variation of growth in tobacco. The same situation did not apply in tomato. Furthermore it is in the nature of such experiments that they cannot demonstrate a cause and effect relationship.

Except under conditions of specific preexisting nutritional stress, as indicated, it is unlikely that the actual sequestration of amino acids and other materials into virus particles has any direct connection with the induction of symptoms. The following considerations support this view:

1. Viruses are made up of the same building blocks in roughly the same proportions as are found in the cell's proteins and nucleic acids. Even with viruses, such as TYMV, which reach a relatively high concentration in the diseased tissue (>1.0 mg/g fresh weight of lamina), the amount of virus formed may be quite small relative to the reduction in other macromolecules caused by infection. In Chinese cabbage plants infected with TYMV the reduction in normal proteins and ribosomes was more than 20 times as great as the amount of virus produced (Crosbie and Matthews, 1974b).
2. Closely related strains of the same virus may multiply in a particular host to give the same final concentration of virus, and yet have markedly different effects on host cell constituents (e.g., strains of TYMV in the stock culture).
3. The type strain of TMV multiplying in White Burley tobacco produced chlorotic lesions at 35°C but none at 20°C. About one-tenth as much virus is made at 35°C as at 20°C (Kassanis and Bastow, 1971b).
4. A single gene change in the tobacco plant may result in a change from the typical mosaic disease produced by TMV to the hypersensitive reaction. F_1 hybrids between the two genetic types may respond to TMV with a lethal systemic necrotic disease with greatly reduced virus production (Fig. 12.8).

In the preceding discussion I have been considering the amount of virus produced as measured in tissue extracts. However, the *rate* of virus replication in *individual cells* could be an important factor in influencing the course of events. Very high demand for key amino acids or other materials over a very short period perhaps of a few hours, could lead to irreversible changes with major long-term effects on the cell, and subsequently on tissues and organs. There is no unequivocal experimental evidence for such effects, measured on individual cells.

B. Effects on Growth

1. Stunting

There appear to be three biochemical mechanisms by which virus infection could cause stunting of growth—a change in the activity of growth hormones, a reduction in the availability of the products of carbon fixation, and a reduction in uptake of nutrients.

Changes in Growth Hormones

There is little doubt that a virus-induced change in hormone concentration is one of the ways in which virus infection causes stunting.

It should be remembered that plant hormones have been defined chemically and biologically in controlled growth tests on excised tissues or organs. In the intact plant, however, each of the groups of hormones induced many growth and physiological effects. Their functions overlap to some extent and their interactions are complex. For a given process their effects may be similar, synergistic, or antagonistic. In the intact plant, members of all or most of the groups are involved in any particular developmental process. There are many possible ways in which virus infection could influence plant growth by increasing or decreasing the synthesis, translocation, or effectiveness of these various hormones in different organs and at different stages of growth. Thus it is not surprising that our understanding of the interactions between viruses and hormones is extremely sketchy.

In most situations, virus infection decreases the concentration of auxin and gibberellin and increases that of abscisic acid. Ethylene production is stimulated by virus-induced necrosis and by development of chlorotic local lesions, but not where virus moves systemically without necrosis. Reported effects on cytokinins are variable (Fraser and Whenham, 1982; Whenham, 1989). Three examples of studies on virus–hormone interactions will be considered.

Ethylene production was enhanced in cucumber cotyledons infected with CMV, and the increase began just before hypocotyl elongation rate slowed down (Marco *et al.,* (1976). Artificially reducing the ethylene content enhanced elongation of infected seedlings, but not as much as it did healthy seedlings, suggesting that some other factors were involved. This was borne out by further work showing that suppression of hypocotyl elongation was also accompanied by a reduction in gibberellinlike substances and an increase in abscisic acid (Aharoni *et al.,* 1977). This work illustrates well the complex effects of infection on hormone balance and the difficulty in establishing cause and effect relationships.

Whenham *et al.* (1985) found that TMV infection of tobacco increased the concentration of abscisic acid outside the chloroplasts, while having little effect on the amount of the hormone inside the chloroplasts. The increase in abscisic acid concentration was strongly and positively correlated with increasing severity of disease for six strains of TMV. However, infection of tomatoes with TMV did not increase abscisic acid (Fraser and Whenham, 1989).

Fraser and Matthews (1981) found that inoculation of the cotyledon leaves of small Chinese cabbage seedings with TYMV was followed by a rapid transient inhibition of leaf initiation in the shoot apex, long before infectious virus moved out of the inoculated leaves. The idea that abscisic acid might be responsible for this effect was supported by the fact that it could reproduced by applying abscisic acid in a single dose to the cotyledon leaves. The magnitude of the effect was independent of concentration over the range 10^{-5} to 3×10^{-9} M abscisic acid (Fraser and Matthews, 1983). Further support for the involvement of abscisic acid came from the following experiment. When the petioles of freshly inoculated Chinese cabbage leaves are placed in water a substance diffuses from the cut ends that can mimic the rapid effects of virus inoculation on leaf initiation. This effect of the eluate can be abolished by the addition of gibberellic acid (GA_3 at 10^{-4} M) (Fraser *et al.,* 1984). Thus it seems probable that virus inoculation brings about a rapid change in the balance of abscisic acid and GA_3 in the apical dome.

Reduction in the Availability of Fixed Carbon

Apart from any effects on hormone balance, plants will become stunted (on a dry weight basis, at least) if the availability of carbon fixed in photosynthesis is limiting. A reduction in available fixed carbon could be brought about in several ways.

Direct Effects on the Photosynthetic Apparatus This is the most obvious and perhaps the most common way by which infection reduces plant size.

Infection of Chinese cabbage leaves with TYMV reduced the three major chloroplast components, chlorophyll a, ribulose bisphosphate carboxylase, and 68 S ribosomes, to about the same extent on a per plant basis (Crosbie and Matthews, 1974b). These changes took place after most of the virus replication was completed, and reduction in the first two began before the ribosomes were affected.

The initial events that lead to reduced carbon fixation in the chloroplasts are not known for any host–virus combination. Hormones may play a role in the initiation of chlorophyll degradation. In leaves of *Tetragona expansa* inoculated with bean yellow mosaic *Potyvirus,* chlorotic local lesions develop. Shortly after their appearance there was a substantial increase in ethylene release from the leaves (Gáborjányi *et al.*, 1971). In cucumber cotyledons infected with CMV, Marco and Levy (1979) have shown that ethylene is produced before local lesions appearance. They were able to delay the appearance of chlorotic local lesions by removing the ethylene by hypobaric ventilation.

Starch Accumulation in Chloroplasts   The accumulation of starch in chloroplasts commonly seen in virus infections must deprive the growing parts of the plant of some newly fixed carbon. This accumulation may be due to reduced permeability of the chloroplast membrane or to changes in enzyme activities within the chloroplast.

Stomatal Opening   Lowered photosynthesis in yellows-infected sugar beet could be accounted for in part by a virus-induced reduction in stomatal opening (Hall and Loomis, 1972). The reduced responsiveness of stomata to changes in light intensity might be a factor limiting carbon fixation in TYMV-infected leaves during the earlier part of the day.

Translocation of Fixed Carbon   Any effect of virus infection such as necrosis that reduces the efficiency of phloem tissues must limit translocation of fixed carbon from mature leaves to growing tissues. However, reduced permeability of leaf cells to the migration of sugars into the phloem may be more commonly the limiting factor.

Leaf Posture   Leaf posture may affect the overall efficiency of photosynthesis in the field. For example, a wheat variety with an erect habit fixed more CO_2 than one with a lax posture (Austin *et al.*, 1976). Virus infection can affect growth habit, but no studies have been made of this factor in relation to carbon fixation per plant.

Reduced Leaf Initiation

Reduction in leaf number is a very small factor in the stunting of herbaceous plants following virus infection. For example, 100 days after infection, tobacco plants infected with TMV possessed about one leaf less than healthy plants (Takahashi, 1972b). Careful analysis of leaf initiation following inoculation of very young Chinese cabbage plants with TYMV has shown that there is a rapid transient reduction in the rate of leaf initiation (the preceding Section 1,a). The rate of leaf initiation is reduced to about one-third to one-quarter that of control plants for a period of 1–2 days. The rate of initiation in inoculated plants then recovers. Plants then have approximately one leaf less than healthy plants for the rest of their life (Fraser and Matthews, 1981). The same effect is produced by turnip mosaic *Potyvirus* and by CaMV in Chinese cabbage. Thus the effect may be a general one.

Reduction in Uptake of Nutrients

Very little experimental work has been done on the effects of virus infection on the capacity of roots to take up mineral nutrients from the soil and to transport them to other parts of the plant. However, nitrogen fixation may be adversely affected. For example, TRSV may impair growth of soybean, at least in part by suppressing leghemoglobin synthesis and nitrogen fixation during the early stages of growth (Orellana *et al.*, 1978). Nodulation of soybean plants by *Rhizobium* was reduced by infection with soybean mosaic *Potyvirus,* bean pod mottle *Comovirus,* or both viruses. Greatest reduction (about 80%) occurred when plants were inoculated with virus at an early stage of growth (Tu *et al.*, 1970). The first phase in the infection of soybean root cells by *Rhizobium* (i.e., development of an infection thread and release of rhizobia into the cytoplasm) appears not to be affected by infection of the plant with soybean mosaic virus (Tu, 1973). In the second stage of infection a membrane envelope forms around the bacterial cell to form a bacteroid. Structural differences such as decreased vesiculation of this membrane envelope were observed in virus-infected roots (Tu, 1977).

Growth Analysis of the Whole Plant

Analysis of the stunting process in a herbaceous plant during the development of a mosaic disease is an extremely complex problem, which has not been adequately investigated. A beginning has been made with Chinese cabbage plants infected with a "white" strain of TYMV. Such strains cause a very marked reduction in chlorophyll content in diseased tissue (Crosbie and Matthews, 1974b).

Healthy plant growth was approximately logarithmic over the period we studied (as was the amount of chlorophyll per plant). Virus infection caused a virtual cessation of chlorophyll a production for several weeks. Growth of the diseased plant was almost linear over this period. We can assume that this was because chlorophyll content increased very little for several weeks.

The diseased plant appeared to respond to the reduction in chlorophyll in two ways. First, a higher proportion of the material available for growth was used to make leaf lamina rather than midrib, petiole, stem, or roots. Second, the proportion of dark green island tissue increased in later formed leaves, so that chlorophyll in these islands came to form a significant proportion of the total.

The "white" strain of TYMV used in these experiments causes a much more severe disease than mild "pale green" strains of the virus when judged by eye inspection. Closer analysis showed there was little difference in total fresh weight of aerial parts or of roots between plants infected with the two kinds of strain. The gross differences were mainly due to (i) different distribution of total chlorophyll within the plant and (ii) a more extreme alteration in growth form with the severe strain (i.e., reduction in stem, petiole, and midrib). Takahashi (1972a,b) studied the effect of TMV infection on tobacco leaf size and shape and internode length, but he did not relate the effects to chlorophyll loss.

2. Epinasty and Leaf Abscission

The experiments of Ross and Williamson (1951) indicated that the epinastic response and leaf abscission in *Physalis floridiana* infected with PVY were associated with the evolution of a physiologically active gas, which was probably ethylene.

Beginning about 48 hours after inoculation of cucumber cotyledons with CMV, there was a decrease in ethylene emanation compared to controls (Levy and Marco, 1976). Over the same period there was an increase in the internal concentration of ethylene. Infection gives rise to a marked epinastic response in the cotyledons. The following evidence strongly suggests that the epinasty was due to the increased ethylene concentration:

1. Ethylene reached a peak concentration before the epinastic response began.
2. Treatment of healthy cucumbers with ethylene caused epinasty.
3. Epinasty could be prevented in infected cotyledons by hypobaric ventilation, which facilitates removal of the internal ethylene.

This is the best evidence showing that a hormone is directly involved in the production of symptoms induced by virus infection. How virus infection stimulates ethylene production and how the hormone produces its effects are not known.

3. Abnormal Growth

Virus-Induced Tumors

All the plant viruses belonging to the Reoviridae, except rice dwarf virus, induce galls or tumors in their plant hosts, but not in the insect vectors in which they also multiply.

There is a clear organ or tissue specificity for the different viruses. For example, tumors caused by WTV predominate on roots and, to a lesser extent, stems. FDV causes neoplastic growths on veins of stems and leaves. Thus, we can be reasonably certain that some function of the viral genome induces tumor formation, but we are quite ignorant as to how this is brought about. Wounding plays an important role as an inducer or promoter of tumors caused by WTV. Hormones released on wounding may play some part in this process. Microscopic tumor initials, which normally do not develop into macroscopic tumors, occur in the stem apices of sweet clover infected with WTV. Application of IAA stimulates the growth of these tumors to macroscopic size (Black, 1972).

Distortion of Tissues

In leaves showing mosaic disease, the dark green islands of tissue frequently show blistering or distortion. This is due to the reduced cell size in the surrounding tissues, and the reduced size of the leaf as a whole. The cells in the dark green island are much less affected and may not have room to expand in the plane of the lamina. The lamina then becomes convex or concave to accommodate this expansion. Other effects on leaf shape are discussed in Chapter 11.

C. Effects on Chloroplasts

1. Chlorotic Local Lesions

Marco and Levy (1979) have suggested that the chlorotic local lesions formed in cucumber cotyledons following inoculation with CMV may be caused by the virus-induced release of endogenous ethylene. They found a substantial rise in ethylene production in the few hours before local lesions appeared. Furthermore, application of ethylene caused yellowing of the leaves, but suppression of ethylene production delayed lesion appearance.

2. Structural Changes

TYMV infection in Chinese cabbage provides particularly favorable material for the study of virus-induced structural changes of chloroplasts. This is because virus-induced clumping of chloroplasts and some other changes such as the formation of large vesicles (sickling) and fragmentation can be monitored in large numbers of cells by light microscopy. Protoplasts isolated from leaves inoculated with TYMV can be used to study the effects of various factors on these virus-induced structural changes.

For example, in the sickling process a large clear vesicle appears in chloroplasts of an infected protoplast, the chlorophyll-bearing structures being confined to a crescent-shaped fraction of the chloroplast volume. The vesicle is bounded by a membrane that appears to arise from stroma lamellae (Fraser and Matthews, 1979b). Red and blue light are equally effective inducers of sickling, which does not occur in the dark (Matthews and Sarkar, 1976). At 6–8 days after inoculation of leaves very few protoplasts show sickling when freshly isolated, although the light intensity received by the leaves is more than adequate to induce sickling in isolated protoplasts. Thus, some factor partially represses the sickling process in intact leaves.

D. Leaf Ontogeny and Mosaic Disease

Chalcroft and Matthews (1966) and Reid and Matthews (1966) suggested that the mosaic patterns commonly found in virus-infected plants are laid down in the shoot apex. We assumed that in a plant infected with a mixture of virus strains the first virus particle to establish itself in a dividing cell preempts that cell and all or almost all its progeny, giving rise in the mature leaf to a macroscopic or microscopic island of tissue occupied by the initial strain. Since that time evidence has accumulated to support this general view.

Much of our understanding of the ontogeny of the dicotyledon leaf has come from a study of homoplastidic periclinical chimeras. Knowledge of cell lineages in such chimeras in relation to mosaic diseases was summarized by Matthews (1981). Plant chimeras in general are discussed by Tilney-Bassett (1986).

Cell lineages have not been studied in the same detail in monocotyledons. However, it is well established that once the developing leaf primordium reaches a certain size, new cells contributing to elongation arise from a basal intercalary meristem (e.g., Kirchanski, 1975). In chimeras containing a plastid mutation this form of development gives rise to longitudinal stripes of mutant and normal tissue, a situation commonly found in ornamental monocotyledons. As one proceeds acropetally along a monocotyledon leaf the cells become older.

1. Role of Virus Strains and Dark Green Tissue in Mosaic Disease

In some infections such as TMV in tobacco, the disease in individual plants appears to be produced largely by a single strain of the virus. However, it has been known for many years that occasional bright yellow islands of tissue in the mosaic contain different strains of the virus. Such strains probably arise by mutation, and during leaf development come to exclude the original mild strain from a block of tissue. In Chinese cabbage plants infected with TYMV there may be many islands of tissue of slightly different color from which different strains of the virus can be isolated (Chalcroft and Matthews, 1967a,b).

Dark green islands of tissue that superficially appear normal occur in these two diseases and in many other mosaics. They are a prominent component of the mosaics and may be important in the growth of the diseased plant.

2. Evidence That Mosaic Development Depends on Leaf Ontogeny

Leaf Age at Time of Infection

By inoculating plants at various stages of growth it has been demonstrated for TYMV (Chalcroft and Matthews, 1966) and TMV (Atkinson and Matthews, 1970; Takahashi, 1971) that the type of mosaic pattern developing in a leaf at a particular position on the plant depends not on its position but on its stage of development when infected by the virus. There is a critical leaf size at the time of infection above which mosaic disease does not develop. This critical size is about 1.5 cm (length) for tobacco leaves infected with TMV (Atkinson and Matthews, 1970; Gianinazzi *et al.*, 1977). Symptoms of African cassava mosaic were determined early in leaf development and subsequently change little (Fargette *et al.*, 1987).

Gradients in Size of Islands

Although the size of macroscopic islands in a mosaic are very variable, there tends to be a definite gradient up the plant. Leaves that were younger when systemically infected tend to have a mosaic made up of larger islands of tissue. Even within one leaf there may be a relationship between both the number and size of islands and the age of different parts of the lamina as determined by frequency of cell division.

Presence of Virus, and the Mosaic Pattern in Very Young Leaves

Many viruses can invade the apical dome. The mosaic pattern is already laid in very small TYMV-infected Chinese cabbage leaves that have not yet developed significant amounts of chlorophyll (Matthews, 1973).

Patterns in the Macroscopic Mosaic

Patterns in the macroscopic mosaic are often so jumbled that an ontogenetic origin for them cannot be deduced. Occasionally patterns that are clearly derived from the apical initials have been observed. For example, Chinese cabbage leaves (or several successive leaves) have been observed to be divided about the midrib into two islands of tissue—one dark green and the other containing a uniform virus infection (Matthews, 1973). These observations show that individual apical initials must have been infected either with virus or with the agent that induces dark green tissue.

The Microscopic Mosaic

The existence of microscopic mosaics provides the strongest evidence that mosaic patterns depend on leaf ontogeny. In Chinese cabbage leaves infected with TYMV the microscopic mosaic develops only on leaves that were less than about 1–2 mm long at the time of infection. In such leaves, areas in the mosaic pattern that macroscopically appear to be a uniform color may be found on microscopic examination to consist of mixed tissue in which different horizontal layers of the meso-phyll have different chloroplast types. A wide variety of mixed tissues can be found. For example, in some areas, both palisade and the lower layers of the spongy tissue may consist of dark green tissue while the central zone of cells in the lamina is white or yellow green (Fig. 12.5A). This situation may be reversed, with the central layer being dark green and the upper and lower layers consisting of chlorotic cells (Fig. 12.5B).

As seen in fresh leaf sections these areas of horizontal layering may extend for several millimeters, or they may be quite small, grading down to islands of a few cells or even one cell of a different type. The junction between islands of dark green cells and abnormal cells in the microscopic mosaic is often very sharp.

Such arrangements are not confined to dark green versus virus-infected layers. The microscopic mosaic also includes cell layers infected by different strains of the virus with distinctive effects on chloroplasts. These arrangements can all be interpreted in terms of the ontogenetic processes described for healthy leaves from observations on chimeras.

How Many Initially Infected Cells Give Rise to an Island in the Mosaic?

Stewart and Dermen (1975) showed that at any stage during leaf ontogeny one cell may divide only a few more times while a sister cell may give rise to a substantial area of tissue. The simplest assumption is that in cells still undergoing division the first virus particle to establish itself in a cell preempts that cell and its progeny. The progeny give rise in the mature leaf to a macroscopic or microscopic island of tissue occupied by that particular strain of virus. Alternatively, the cell is converted to a dark green state and gives rise to a dark green island in the mature leaf.

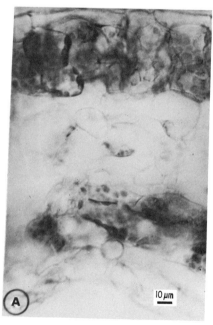 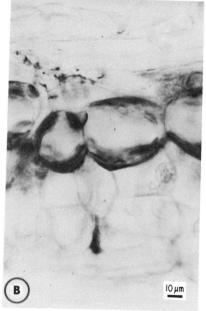

Figure 12.5 Examples of layers in the microscopic mosaic in Chinese cabbage leaves infected with TYMV. (A) Upper and lower cell layers (LII) dark green; central zone (LIII), white. (B) Reverse of (A). (From Chalcroft and Matthews, 1967b.)

In some circumstances cell-to-cell spread and infection by the virus in the shoot apex must be possible, for example, early after infection of the apex. Following such spread some islands of tissue in the mature mosaic will arise from the progeny of a group of neighboring cells in the apex, rather than as a clone from a single cell.

Islands of dark green tissue that extend through all the histological layers of a leaf must have arisen by cell-to-cell spread of an agent during leaf ontogeny (Atkinson and Matthews, 1970).

3. Genetic Control of Mosaics

In spite of the fact that the most striking effects of TYMV infection in *Brassica* spp. are on the chloroplasts, the response of these organelles to infection is under some degree of nuclear control. Certain varieties of *Brassica rapa* respond to all the strains isolated from *Brassica pekinensis* with a mild diffuse mottle. In reciprocal crosses between *B. rapa* and *B. chinensis* all the progeny gave a *B. chinensis* type of response to the strains (Matthews, 1973).

4. Mosaic Disease in Monocotyledons

The macroscopic islands of tissue in mosaic diseases in monocotyledons almost always consist of stripes, streaks, or elongated blocks of tissue lying parallel to the axis of the leaf rather than in an irregular pattern. This arrangement is almost certainly due to the fact that these leaves elongate by means of a basal meristem, producing files of cell.

5. The Nature of Dark Green Tissue

Dark green islands in the mosaic pattern are cytologically and biochemically normal as far as has been tested. They contain low or zero amounts of infectious virus and no detectable viral protein or viral dsRNA. In pumpkin leaves showing mosaic disease caused by watermelon mosaic *Potyvirus*, the cylindrical inclusion and amorphous inclusion proteins as well as viral coat protein were absent from dark green areas of the mosaic, or present in trace amounts (Suzuki *et al.*, 1989).

Various lines of evidence show that dark green islands are resistant to superinfection with the same virus or closely related viruses (e.g., Chalcroft and Matthews, 1967a,b; Atkinson and Matthews, 1970; Loebenstein *et al.*, 1977).

Various factors can influence the proportion of leaf tissue that develops as green islands in a mosaic. These include leaf age, strain of virus, season of the year, and removal of the lower leaves on the plant (Crosbie and Matthews, 1974b). The dark green islands of tissue may not persist in an essentially virus-free state for the life of the leaf. "Breakdown" leading to virus replication usually takes place after a period of weeks, or after a sudden elevation in temperature (Atkinson and Matthews, 1970; Matthews, 1973; Loebenstein *et al.*, 1977).

There is no convincing evidence for any of the various theories that have been put forward to explain the nature of dark green islands. They certainly do not consist merely of tissue that has escaped infection. There is no evidence for the presence of very mild strains of virus in dark green islands. It is possible that the cells in dark green islands are occupied by defective strains, but such strains would have to produce little or no intact virus or viral antigen, and would have to replicate without detectable cytological effects. These cannot be ruled out at present, but on the other hand, there is no positive evidence for their presence.

Gera and Loebenstein (1988) isolated an inhibitor of virus multiplication from dark green island tissue of tobacco leaves infected with CMV. However, they do not appear to have assayed for the presence of inhibitors in the yellow virus-bearing tissues. The role of such inhibitors in dark green tissue remains to be established. When transgenic tobacco plants expressing the 30K movement protein are infected with TMV, the mosaic disease which develops appears normal in all respects (R. N. Beachy, personal communication). Thus we can assume that the 30K protein is not involved in the formation or maintenance of dark green islands.

Dark green islands may be part of a more general phenomenon in plant virus infections. In local lesions caused by TYMV in Chinese cabbage a proportion of cells show no evidence of virus infection and remain in that state for considerable periods even though neighboring cells are fully infected (Chapter 10, Section, IV,H,1). It may be that these cells are in a resistant state like that found in dark green islands of the mosaic. Perhaps a proportion of cells in all infected leaves develop the resistant state, but only cells still retaining the potential to divide can give rise to microscopic or macroscopic islands of dark green tissue.

The phenomenon of lysogeny has been known among bacterial viruses for a long time. Noting that dark green islands of tissue appeared in the TYMV-induced mosaic only in leaves that were still undergoing cell division at the time of infection, Reid and Matthews (1966) speculated that some process like lysogeny might be taking place to give dark green tissue its resistance to infection. With the advent of the polymerase chain reaction it should be possible to test this idea by searching for TYMV nucleotide sequences in DNA form in dark green tissue. This technique should be sensitive enough to detect one copy of the TYMV genome, or part of it, per cell.

6. An Experimental System for Studying the Origin of Dark Green Tissue

There is no doubt that mosaic patterns are laid down very early in leaf ontogeny. However, these early stages have been very difficult to study because of the great variability and the unpredictability of the detailed mosaic pattern that will develop in any particular leaf. We have developed a system that may have sufficient reproducibility to allow investigation of early events in the developing leaf (A. Ferguson and R. E. F. Matthews, unpublished). Chinese cabbage seedlings 6–11 days old bear two cotyledon leaves and 4–8 small true leaves and leaf primordia. When such seedlings are inoculated on the cotyledons with TYMV, a simple mosaic pattern consisting of a single yellow infected area in the center of the leaf, with a continuous dark green zone around the leaf margins, develops reproducibly in about 40% of the fifth oldest true leaves. The dark green tissue has all the properties described for such tissue in the irregular mosaics formed in larger plants. The fifth leaf was about 1 mm long at the time virus moved out of the inoculated leaves. At this stage it already had a rudimentary midrib and lamina. Cells in the central zone of the leaf were partially differentiated and developed cytological signs of virus infection. Cells in the marginal zone of leaf 5 were still fully meristematic at the time virus moved from the inoculated leaf. Thus induction of the dark green condition appears to depend on the presence of dividing cells in the invaded leaf. With this system it should be possible to apply such techniques as immunogold cytochemistry and *in situ* nucleic acid hybridization to explore further the nature of dark green tissue as the leaf develops.

E. Role of Membranes

Typical leaf cells are highly compartmentalized and there is an increasing awareness of the vital roles that membranes play in the functioning of normal cells. Virus replication involves the induction of new or modified membrane systems within infected cells (Chapters 7 and 8). It is known that both animal and bacterial viruses alter the structure and permeability properties of host cell membranes. Infection of insect vector cells by plant viruses may alter the physical properties of the plasma membrane (Hsu, 1978).

Various virus-induced responses, some of which were discussed in earlier sections, indicate ways in which changes in membrane function may have far-reaching consequences for the plant. For example:

1. Virus infection may affect stomatal opening. Net photosynthesis may be limited by stomatal opening, which in turn depends on guard cell membrane function.
2. A different consequence of virus-induced stomatal closure is indicated by the work of Levy and Marco (1976). A virus-induced increase in stomatal resistance led to increased internal ethylene concentration followed by an epinastic response. The change in stomatal opening was probably induced by changes in properties of the guard cell plasma membranes.
3. Sucrose accumulation in infected leaves appears to depend on membrane permeability changes rather than altered phloem function.
4. Virus-induced starch accumulation in chloroplasts may be due in part at least to alterations in permeability of chloroplast membranes.
5. There is little doubt that ethylene is involved in the induction of some symptoms of virus disease. Ethylene has been found to cause a very rapid decrease in the incorporation of $[1-^{14}C]$glycerol into phospholipids of pea stems (Irvine and Osborne, 1973).
6. Living cells maintain an electrochemical potential difference across their plasma membrane, which is internally negative. There is an interdependence between this potential difference and ion transport across the membrane. Stack and Tattar (1978) showed that infection of *Vigna sinensis* cells by TRSV altered their transmembrane electropotentials.

Thus a more detailed knowledge of the ways in which viruses alter the properties of membranes will be a prerequisite to our fuller understanding of disease induction.

V. FACTORS INFLUENCING THE COURSE OF INFECTION AND DISEASE

In Sections II and III of this chapter the roles of virus genes and induced host gene products in the induction of disease were discussed. The kinds of process involved in disease induction were considered in Section IV. For a given species of plant and a given virus there are many factors that can influence the course of infection and the disease that develops. The way in which a virus is inoculated into the plant may be important (Chapter 10). Sometimes many different strains of a virus occur and these may cause quite different kinds of disease in the same host plant under the same conditions (Chapter 13). In this section I shall discuss inherent variables in the host plant itself, environmental factors, and interactions between unrelated viruses and between viruses and some other agents of disease.

A. Virus Concentration in the Inoculum

The amount of infecting virus may influence the extent to which growth is subsequently depressed (compare A and B in Fig. 12.7). This may in fact be an age-of-plant effect (Section V,B) since it is probable that virus moves systemically through the plant sooner following a heavy inoculation either mechanically or by insects. However, other workers have not found a decrease in plant yield with increasing numbers of aphids (e.g., Skaria *et al.*, 1984). The apparent discrepancy may be in the number of aphids used. To obtain effects like that shown in Fig. 12.7, 100 or more aphids were used per plant for the heavy inoculation.

B. Plant Factors

1. Age

Susceptibility to Infection

Most commonly, small young leaves and old leaves are less susceptible than well expanded younger leaves. There may be a marked gradient of susceptibility with age (Fig. 12.6). The curves in Fig. 12.6 can be taken to indicate the changes in susceptibility that individual leaves undergo with time.

The gradient may not always be of the form shown in Fig. 12.6. Thus, on *N. glutinosa* plants with 8–10 well developed leaves, TMV will produce more local lesions on the middle and lower leaves than on the younger ones. In contrast, TBSV

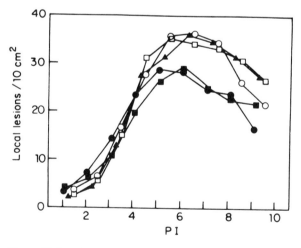

Figure 12.6 Effect of leaf age on number of local lesions produced by TMV in leaves of Samsun NN tobacco. Since younger leaves are expanding the number is expressed per unit leaf area at the time of inoculation. A *plastochron* is the time interval between corresponding developmental stages of successive leaves. The *leaf plastochron index* (PI) is an arbitrary measure that allows plants at slightly different developmental stages to be compared. Each curve represents successive leaves from a single plant. PI ●, 15.10; ■, 15.12; ▲, 15.30; ○, 15.52; □, 15.53. (From Takahashi, 1972b.)

may produce no lesions on the oldest leaves and most on the youngest (Bawden, 1964).

Abscisic acid accelerates senescence processes in leaves. Exogenously applied abscisic acid increased susceptibility of Samsun tobacco leaves to TMV (Balázs *et al.*, 1973), indicating a possible role for this hormone in age-dependent changes in leaf susceptibility.

Systemic Spread

In a fairly mature tobacco or tomato plant inoculated with TMV in the younger leaves, the lower leaves may never become infected with virus unless directly inoculated. With increasing age there is a greater tendency for infection to remain localized. Physiological age rather than actual age is the significant factor. When potato plants are infected with PVX, either naturally in the field or by inoculation, some tubers may be free of the virus at the end of the first season. Beemster (1966) found that the proportion of tubers and eyes infected with PVY is closely related to the time at which the plants are inoculated. For example, the percentage of infected tubers was 100% for plants 8 weeks old on inoculation and only 25% for plants 13 weeks old when inoculated. Time of infection is often an important factor in determining loss of yield in economically important crops. For example, loss of yield is generally more severe when cereals are infected at an early stage of growth with BYDV (Smith, 1967) (Fig. 12.7).

In field experiments with *Capsicum annuum,* plant growth and fruit yield improved almost in direct proportion to the lateness of inoculating the plants with CMV (Agrios *et al.*, 1985). The term "mature plant resistance" has been used for these age-of-plant effects but the mechanism for them is unknown. Many metabolic changes occur as leaves age. For example, ribosome content decreases (Venekamp and Beemster, 1980), but no causal relationship has been established between such changes and increased resistance to the effects of inoculation with a virus. Barker and Woodford (1987) describe a longer-term effect of late infection. Potato plants grown from tubers of plants infected with PLRV late in the previous season developed unusually mild symptoms, although they contained as much virus as plants with severe symptoms. Perhaps a strain selection process was at work.

2. Plant Genotype

The Kinds of Host Response

The genetic makeup of the host plant has a profound influence on the outcome following inoculation with a particular virus. The kinds of host response are defined in Table 12.1. Figure 12.8 illustrates the dramatic effect that a single Mendelian gene can have on disease response. Defining the response of a particular host species or cultivar to a particular virus must always be regarded as provisional. A new mutant of the virus may develop in the stock culture, or a new strain of the virus may be found in nature that causes a different response in the plant. This applies particularly to nonhost immunity. For example, for many decades the potato seedling USDA 41956 was considered to be immune to PVX. However, in due course a strain of this virus was discovered that can infect this genotype (Moreira *et al.*, 1978).

Nevertheless, immunity with long-term durability must occur frequently in nature. It has been suggested with respect to cellular parasites that long-lived plants such as trees, and species naturally dominant over large areas, such as prairie

Figure 12.7 Interaction between time of infection and number of infecting aphids on subsequent depression in size of barley plants by BYDV. (A) About 8 aphids per plant. (B) About 150 aphids per plant. The inoculative feed was 2 days, so that feeding damage was insignificant. Stages of growth when infected: (left to right) control; flowering; boot; tillering; three-leaf; one-leaf. (From Smith, 1967.)

Figure 12.8 Reactions of two varieties of tobacco to a strain of TMV from tomato. Left: Var. White Burley showing typical systemic mosaic. Center: Var. Warne's, necrotic local lesions with no systemic spread of virus. Right: An F_1 hybrid (Warnes ♀ × White Burley ♂)) showing necrotic local lesions and a severe systemic necrosis and stunting.

grasses, must have developed greater resistance to lethal parasites than other species. Similar considerations may apply as far as virus infection is concerned, but little relevant experimental evidence is available.

It is probable that many of the plants described in the past as immune to a particular virus were in fact infectible, but resistant and showing extreme hypersensitivity as defined in Table 12.1 The viruses concerned may have had cell-to-cell movement proteins that were defective in the particular plant, resulting in multiplication only in the initially infected cells. This phenomenon may be detectable only by special procedures (e.g., see Fig. 6.5).

The following points concerning the effects of host genes on the plant's response to infection emerge from many different studies: (i) both dominant and recessive Mendelian genes may have effects. However, while most genes known to affect host responses are inherited in a Mendelian manner, cytoplasmically transferred factors may sometimes be involved (Nagaich *et al.*, 1968); (ii) there may or may not be a gene dose effect; (iii) genes at different loci may have similar effects; (iv) the genetic background of the host may affect the activity of a resistance gene; (v) genes may have their effect with all strains of a virus, or with only some; (vi) some genes influence the response to more than one virus; (vii) plant age and environmental conditions may interact strongly with host genotype to produce the final response; and (viii) route of infection may affect the host response. Systemic necrosis may develop following introduction of a virus by grafting into a high-resistant host that does not allow systemic spread of the same virus following mechanical inoculation (e.g., tomato mosaic *Tobamovirus* in resistant tomato lines; Stobbs and McNeill, 1980).

As pointed out by Fraser (1988), there are three main types of resistance and immunity to a particular virus, considered from the point of view of the complexity of the host population involved: (i) *immunity* involves every individual of the species; little is known about the basis for immunity, but it is related to the question of the host range of viruses discussed in Chapter 10, Section V,C; (ii) *cultivar re-*

sistance describes the situation where one or more cultivars or breeding lines within a species show resistance while others do not; and (iii) *acquired or induced resistance* is present where resistance is conferred on otherwise susceptible individual plants following inoculation with a virus. This last phenomenon was discussed in Section III. Some authors have considered that immunity and cultivar resistance are based on quite different underlying mechanisms. However, recent studies with a bacterial pathogen, in which only one pathogen gene was used, show that for this class of pathogen at least, the two phenomena have the same basis (Whalen *et al.*, 1988).

Genetics of Resistance to Viruses

This section briefly outlines the kinds of genetics involved in cultivar resistance to virus infection and summarizes possible mechanisms for such resistance. Resistance and immunity for the control of virus diseases are discussed in Chapter 16. Plant resistance to viruses has been reviewed by Ponz and Bruening, (1986), Fraser (1987a, 1988), and Evered and Harnett (1987).

The summary in Table 12.3 shows that resistance to viruses in most crop virus combinations is controlled by a single dominant gene. However, this may merely reflect the fact that most resistant cultivars were developed in breeding programs aimed at the introduction of a single resistance gene. There have not been many studies of the inheritance of resistance in wild species. Some specific examples of dominant, incompletely dominant, and apparently recessive genes for resistance are given in Table 12.4.

Sometimes response to virus infection is associated with resistance to some other kind of disease agent. Thus the necrotic response of tobacco to infection by a strain of PVY may be a pleiotropic effect of the gene controlling resistance to a root-knot nematode (Rufty *et al.*, 1983).

The Gene-for-Gene Hypothesis

Gene-for-gene relationships are well known between host plant and fungal or bacterial pathogens. They have been established primarily on the basis of genetic analyses of both plants and pathogens. With these parasites each allele in the host that confers resistance may be reflected in a complementary virulence locus in the parasite that can overcome the resistance. Thirty or more such pairs of genes have been recognized genetically in a particular host–parasite relationship. Recently some avirulence genes have been isolated. For example, cloned avirulence genes

Table 12.3
Genetics of Resistance to Viruses in Crop Species[a]

Genetic basis	Number of host–virus combinations
Single dominant gene	29
Incompletely dominant (gene-dosage dependent)	10
Apparently recessive	11
Subtotal: monogenic	50
Possibly oligogenic	5
Monogenic, with possible modifier genes or effects of host genetic background	8
Subtotal: oligogenic (?)	13
Total number of host–virus combinations in sample	63

[a]Data from Fraser (1987a).

Table 12.4

Examples of Host Genes for Resistance to Plant Viruses

Name of gene	Host species	Virus	Virulent virus isolates known	Selected references
		Controlled by dominant genes		
N	Nicotiana glutinosa	TMV	No	Holmes (1929)
N^1	N. sylvestris	TMV	Yes	Melchers et al. (1966)
Zym	Cucurbita moschata	ZYMV	?	Paris et al. (1988)
Tm-2	Lycopersicon esculentum	TMV	Yes	Pilowski et al. (1981)
$Tm\text{-}2^2$	L. esculentum	TMV	Yes	Cirulli and Alexander (1969)
Nx, Nb	Solanum tuberosum	PVX	Yes	Jones (1982)
By, By-2	Phaseolus vulgaris	BYMV	Yes	Schroeder and Provvidenti (1968)
Rsv_1, Rsv_2	Glycine Max	SbMV	Yes	Buzzell and Tu (1984)
		Controlled by incompletely dominant genes		
Tm-1	L. esculentum	TMV	Yes	Fraser et al. (1980)
L^1, L^{+a}	Capsicum chinense	TMV	Yes	Boukema (1980)
2 genes	Hordeum vulgare	BSMV	Yes	Sisler and Timian (1956)
Multiple genes	Vigna sinensis	SBMV	?	Hobbs et al. (1987)
		Apparently recessive genes		
By-3	P. vulgaris	BYMV	No	Provvidenti and Schroeder (1973)
SW_2, SW_3, SW_4	L. esculentum	TSWV	Yes	Finlay (1953)

from the tomato pathogen *Pseudomonas syringae* pv tomato, when transferred to the soybean pathogen *P. syringae* pv *glycinea,* caused a hypersensitive response in some soybean cultivars (Kobayashi *et al.,* 1989).

Viruses and virus strains may be described in relation to the various host responses defined in Table 12.1. Thus if a particular plant species or cultivar shows immunity or resistance to a virus, that virus is said to be *nonpathogenic* for that species or cultivar. A virus is *pathogenic* if it usually causes systemic disease in a species or cultivar. A gene for immunity or resistance introduced into such a species or cultivar may make the virus *avirulent.* However, a mutant strain may overcome the host gene resistance and it would then be *virulent.*

Gene-for-gene relationships have been proposed by some authors for virus–host interactions (e.g., Fraser, 1987a). A well-studied example of the gene-for-gene hypothesis applied to a plant virus and its host is the resistance of tomato to ToMV (Table 12.5). There are three resistance genes, *Tm-1, Tm-2,* and *Tm-2².* The virus has evolved variants that can overcome all three host resistance genes. The virulent virus strains are numbered according to the host genes they can overcome. For example, strain 0 is avirulent. Strain 2 overcomes gene *Tm-2;* strain *1.2²* overcomes genes *1* and *2².* No virus strains are yet known that overcome host genes *2* plus *2²* or *1, 2,* plus *2.2².* The host genotypes differ in their "durability" in the field. Thus strain 1 isolates appeared within a year in commercial crops containing only the *Tm-1* gene. By contrast most commercial ToMV-resistance tomato cultivars now contain the genes *Tm-1/Tm-1 : Tm-2/Tm-2²* or *Tm-1/ + : Tm-2/Tm-2²,* and these appear to be highly durable in their resistance to TMV (Fraser, 1985).

The relationships summarized in Table 12.5 bear a superficial similarity to the

Table 12.5

Genetic Interactions between ToMV-Resistant Tomato Plants
and Strains of the Virus[a]

Host genotype[b]	Virus genotype[c]					
	0	1	2	2^2	1.2	1.2^2
Wild type	M	M	M	M	M	M
Tm-1	R	M	R	R	M	M
Tm-2[a]	R	R	M	R	M	R
Tm-2[2 a]	R	R	R	M	R	M
Tm-1/Tm-2	R	R	R	R	M	R
Tm-1/Tm-2[2]	R	R	R	R	R	M
Tm-2/Tm-2[2]	R	R	R	R	R	R
Tm-1/Tm-2/Tm-2[2]	R	R	R	R	R	R

[a]Modified from Fraser (1985), with permission from Martinus Nijhoff/Dr. W. Junk, Dordrecht.

[b]Plant with genotypes marked [a] may show local and variable systemic necrosis rather than mosaic when inoculated with virulent strains.

[c]M = systemic mosaic; R = resistance.

gene-for-gene relationships seen between bacterial and fungal pathogens and their hosts. Such pathogens contain large numbers of genes and could easily maintain or develop a suite of genes that either allow or overcome host resistance. By contrast ToMV contains four genes all with functions involved in virus replication or movement. The change to virulence by the virus must involve mutational events in one or more of the four genes or in the controlling elements of the viral genome. An example is provided by the work of Meshi *et al.* (1988). They examined the nucleotide sequence of a resistance breaking strain of ToMV, Ltal, which is able to replicate in tomatoes with the *Tm-1* gene (i.e., a strain of genotype 1 in Table 12.5). They found two base substitutions resulting in amino acid changes Gln 979 → Glu and His 984 → Tyr in the 130K and 180K viral proteins. They demonstrated that these were indeed the changes responsible for resistance breaking by introducing these mutations singly or together into the parent strain. All three constructed mutants replicated in tomato protoplasts with the *Tm-1* gene, but the change His 984 → Tyr did so to a greatly reduced extent.

Another resistance breaking strain of ToMV, Ltbl, is able to multiply in tomatoes with the *Tm-2* gene. Nucleotide sequence analysis revealed two changes in the 30K protein compared to the parent ToMV (Cys 68 → Phe and Glu 133 → Lys) (Meshi *et al.*, 1989). Thus the *Tm-2* resistance in tomato may involve some aspect of cell-to-cell movement.

Mutations in virus genotypes 2 and 2^2 of Table 12.5 very probably involve nucleotide changes in the TMV gene coding for the movement protein, as has been found for the *ts* mutant LsI (Chapter 6, Section V,D,2) (Nishiguchi and Motoyoshi, 1987). Thus for viruses, virulence or avirulence appears to be controlled by single point mutations in genes essential for virus replication or movement, rather than some change in genes dedicated to maintaining or overcoming host resistance.

Mechanisms of Host Immunity and Resistance

Resistance of Cowpea to CPMV   Seedlings of the cowpea cultivar Arlington are resistant to CPMV, and isolated protoplasts are also resistant. This cultivar is therefore immune as defined in Table 12.1. This immunity is governed by a single dominant gene as determined by crosses with the susceptible Black Eye variety. Ponz *et al.* (1988a) found three inhibitory activities in extracts of Arlington cowpea protoplasts that were at higher levels than in Black Eye extracts. These were (i) inhibitor(s) of the translation of CPMV RNAs; (ii) proteinase(s) that degrade CPMV proteins; and (iii) an inhibitor of proteolytic processing of a CPMV polyprotein. The proteinases were not specific for CPMV proteins and were not coinherited with the immunity to CPMV.

The inhibitor of polyprotein processing was specific for CPMV and had the coinheritance expected for an agent mediating the immunity to CPMV. It can be reasonably concluded that the proteinase inhibitor is the host-coded gene product responsible for immunity to CPMV. This is the first such gene product identified for plant viruses and the first example of immunity for which a clear molecular mechanism has been established. The inheritance of the translation inhibitor activities was complex but one or more of these may play an accessory role in immunity. In another study, Ponz *et al.* (1988b) found that the Arlington cowpea line also had a single dominant gene for resistance to TRSV, but that this was quite distinct from the gene for immunity to CPMV.

Impaired Systemic Movement of the Virus   In some cultivars that have a genotype conferring resistance to a particular virus, movement through the vascular tissue is affected in a manner that appears not to involve a viral-coded cell-to-cell movement protein in the way discussed in a following section. For example, the spread of PLRV infection within the phloem of resistant potato cultivars appears to be impaired, leading to much less efficient acquisition of the virus by the aphid vector *Myzus persicae* (Barker and Harrison, 1986). Similarly, in corn (*Zea mays*) resistance to MDMV, the pattern of virus spread in inoculated leaves suggested that the plant inhibited the virus from moving through the vascular system (Lei and Agrios, 1986). The molecular mechanism for such host effects is unknown.

The expression of the TMV 30K movement protein is strongly and selectively enhanced in tobacco protoplasts by actinomycin (Blum *et al.*, 1989). It was suggested that the drug may act by selectively inhibiting a host factor that normally suppresses the expression of the 30K viral protein.

Nonspecific Virus Inhibitors   Extracts of many plants contain substances that inhibit infection by viruses when mixed with the inoculum. Some may inhibit virus replication in experimental systems, and many are known to be proteins or glycoproteins. The relevance of any of these substances to plant resistance to viruses has not been established. One of the most studied is a protein of MW = 29K isolated from *Phytolacca americana*. This protein is known to inhibit a wide range of eukaryotic ribosomes. Ready *et al.* (1986) showed that the *Phytolacca* protein is mainly located in the cell wall. They suggested that when injury to a cell occurs during the process of virus infection, the inhibitor enters the cytoplasm and shuts off virus synthesis.

Ineffective Viral Genes   As discussed in Chapter 10, Section IV,G, many viruses code for a gene product that is necessary for cell-to-cell movement. If a movement protein is ineffective in a particular host plant then virus replication is confined to the initially infected cells. Consequently the plant shows extreme hypersensitivity (Table 12.1) and is effectively resistant to the virus. A virus that does not normally move systemically in a particular plant species may do so in the presence of an unrelated virus that does infect systemically (Section V,D,1). This indicates that a mismatch between viral movement protein and plant species may be a quite common mechanism for plant resistance.

A Mechanism for Resistance Breaking by a Virus

It is not yet clear how the ToMV strain Ltal overcomes the resistance of tomato cultivars containing the *Tm-1* gene. However, Meshi *et al.* (1988) made a suggestion: They found a strong correlation between ability to overcome resistance and a decrease in total net charge on the viral proteins due to the amino acid substitutions noted in a preceding section. Thus an electrostatic interaction between the viral 130K and 180K proteins and a host resistance factor may be involved.

Protoplasts in the Study of Host Resistance

As noted earlier, if protoplasts from a plant species that shows no sign of infection following mechanical inoculation sustain normal virus replication following inoculation *in vitro*, then a defective or mismatched virus movement protein is probably involved in the plant resistance. However, if protoplasts from a resistant cultivar are also resistant to infection *in vitro*, then the species is immune as defined in Table 12.1. For example, this situation was found to apply for protoplasts from the Kyoto cultivar of cucumber, which had been described as resistant to CMV (Maule *et al.*, 1980). Using the definition in Table 12.1, this cultivar would be immune.

Tolerance

The classic example of genetically controlled tolerance is the Ambalema tobacco variety. TMV infects and multiples through the plant, but in the field, infected plants remain almost normal in appearance. This tolerance is due to a pair of independently segregating recessive genes r_{m1} and r_{m2}, and perhaps to others as well with minor effects.

Other examples are known where either a single gene or many genes control tolerance. For example, tolerance of a set of barley genotypes to BYDV was controlled by a single major gene probably with different alleles giving differing degrees of tolerance (Catherall *et al.*, 1970). From a study of the relative abundance of DNA forms and viral RNAs in plants infected with CaMV, Covey *et al.* (1990) concluded that the expression of the CaMV minichromosome is a key phase in the replication cycle that is regulated by the host. Kohlrabi is a host that is tolerant of infection compared with turnip. In this host, high levels of supercoiled DNA accumulated, with very little generation of RNA transcripts, viral products, and virus.

Kinds of Symptoms

The symptoms that develop in plants that are neither resistant nor tolerant will be influenced in many ways by the host genotype. For example, the difference in the mosaic disease induced by strains of TYMV in *Brassica pekinensis* and *B. rapa* is under the control of a nuclear gene (Section IV,D,3). As another example, the host genome influences the color pattern in infected petals.

The time taken for symptoms to develop following inoculation may be influenced by the host variety. Such effects on incubation period may be an important factor influencing the effects of viruses in the field, for example, BCTV in sugar beet (Duffus and Skoyen, 1977). There may be considerable variation within a single cultivar in its response in the field to a particular virus (e.g., Tomlinson and Ward, 1978).

The genotype of a particular cultivar may interact with the infecting virus to cause a characteristic disease condition. For example, a tomato leaf roll disease was caused by interaction of the wilty gene in tomato and TMV infection (Provvidenti and Hoch, 1977).

C. Environmental Factors

The environmental conditions under which plants are grown before inoculation, at the time of inoculation, and during the development of disease can have profound effects on the course of infection. A plant that is highly susceptible to a given virus under one set of conditions may be completely resistant under another. If infection occurs, the plant may support a high or low concentration of virus and develop severe disease or remain almost symptomless, depending on the conditions.

1. Factors Affecting Susceptibility to Infection

Any factor that alters the ease with which the surface of the leaf is wounded will alter the probability of successful entry of virus introduced by mechanical inoculation, while physiological changes in the leaf may make the cell more or less suitable for virus establishment. As a broad generalization, greenhouse-grown plants will have greatest susceptibility when they are grown and used under the following conditions: (i) mineral nutrition and water supply that do not limit growth; (ii) moderate to low light intensities; (iii) a temperature in the range of 18°–30°C, depending on virus and host; and (iv) inoculation carried out in the afternoon.

Light

There are two general situations in which light affects susceptibility—short-term changes in light intensity or deviation over a period of no more than a day or two, and long-term effects that may markedly influence the growth of the plant.

Bawden and Roberts (1947) found that in summer, reducing light intensity or giving plants complete darkness for periods of days increased the susceptibility of several hosts to certain viruses. This effect of shading or darkening plants for about a day before inoculation is often used as a practical measure to increase the susceptibility of test plants with viruses that are difficult to transmit.

Various experiments suggest that light may have two opposing kinds of effects. Excessive illumination may reduce susceptibility. On the other hand, after a period of darkness, even a short burst of light may increase the number of local lesions. For example, when bean plants that had been in the dark for 18 hours were given a 1-minute exposure to 800 foot-candles before inoculation, the number of local lesions produced by a TNV inoculum was double that of plants inoculated under minimal illumination required to see the leaves (Matthews, 1953b). In almost all studies on the effects of light, the experimental plants have been raised under ordinary greenhouse conditions. Thus, any experimentally imposed change will have been superimposed on the natural daily cycle of variation in susceptibility discussed in a following section. The time of day at which an experiment is performed can have a marked effect on results (Matthews, 1953b).

As far as the long-term effects of light on susceptibility are concerned, it is generally found that high light intensities over the period of growth of the test plant give rise to "hard" plants that may have very low susceptibility compared to "soft" plants raised under lower light intensities.

Temperature

In general, preincubation of plants at slightly higher temperatures than normal before inoculation increases susceptibility. The effect of treating plants at a higher temperature after inoculation may vary with the virus or the strain of virus tested (e.g., Kassanis, 1952). A brief shock treatment, for example, at 50°C, after inoculation may affect the kind of lesion that subsequently develops (e.g., Foster and Ross, 1975). Where necrotic ring spot local lesions are produced, light and temperature conditions interact in the production of rings (Harrison and Jones, 1971). Temperature may also affect the speed and efficiency of graft transmission (Fridlund, 1967a).

Water Supply and Humidity

The immediate effect of washing inoculated leaves with water was discussed in Chapter 10, Section III,A,4. Generally speaking, if plants are grown with a minimal supply of water they become more or less stunted, leaf texture is "hard," and the plants will give greatly reduced numbers of local lesions compared to plants raised under conditions where water supply is not limiting growth. For example, Tinsley (1953) found that as many as 10 times more local lesions were produced by several viruses on well-watered *N. glutinosa* and tobacco plants as on poorly watered plants.

Most reports suggest that moderate wilting near the time of inoculation increases susceptibility. For example, when detached bean leaves were wilted before inoculation with TMV, water deficits of 0–15% of the green weight of the leaf increased infection. More severe deficits in the range 15–29% decreased infection. When leaves were wilted after inoculation, numbers of local lesions increased with increasing water deficit over the range 0–35% (Yarwood, 1955a). Maximum increase was four- to eightfold over unwilted leaves.

Nutrition

The nutrition of the host plant may have a marked effect on the numbers of local lesions produced by viruses. As would be expected, interactions between different nutrients are quite complex. For example, Bawden and Kassanis (1950) investigated the effects of various levels of nitrogen, phosphorus, and potassium on susceptibility of tobacco and *N. glutinosa* plants to strains of TMV. The range of fertilizer treatments used produced large effects on plant growth and many significant differences in susceptibility. In general, nutritional conditions that were most favorable for plant growth were also those giving greatest susceptibility. There was no evidence that one particular element increased susceptibility on its own.

Trace elements may also affect transmissibility. For example, increased manganese supply caused a marked increase in the number of local lesions produced by PSTVd in *Scopolia* leaves (Singh *et al.*, 1974).

Time of Day

Since many basic processes in leaves are influenced by the diurnal cycle of night and day, it is not surprising that the susceptibility of leaves to mechanical inoculation varies systematically with time of day. The number of local lesions produced rises to a maximum in the afternoon and falls to a minimum during the night, usually just before dawn (Fig. 12.9).

A diurnal variation has been found for various host–virus combinations. This

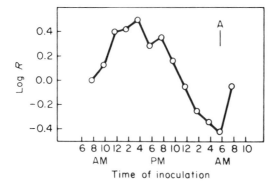

Figure 12.9 Effect of time of day of inoculation on the numbers of local lesions produced by TNV in beans. Log R relates the number produced to the number produced at a base time (8 A.M.). Bar A represents the difference between any two points required for significance at $P = 1\%$. (From Matthews, 1953a.)

diurnal change is not dependent on immediate changes in environmental conditions, but appears to be "built in" as are some other physiological processes in leaves. Thus, bean plants placed in darkness at constant temperature at 4 P.M. and maintained in darkness through the following day showed a very similar cycle of susceptibility to plants that were exposed to the normal daylight cycle (Matthews, 1953b).

Time of Year

Where there are large climatic differences between the seasons, plants may vary widely in their susceptibility to a virus at different times of the year. For example, certain varieties of bean (*Phaseolus vulgaris*) grown at Rothamsted in England during the summer appeared almost immune to CMV. In the winter they produced substantial numbers of necrotic local lesions (Bhargava, 1951). Similar but less extreme differences have been noted with other viruses producing local lesions. The major environmental factor concerned in greenhouse-grown plants is probably light, since Bawden and Roberts (1947) showed that typical winter reactions of plants can be reproduced in summer by appropriate shading.

Chemical and Mechanical Injury

Nematicides and fungicides may increase susceptibility to virus infection. For example, atrazine increased susceptibility of maize hybrids to maize dwarf mosaic *Potyvirus* (MacKenzie *et al.*, 1970).

The mechanical injury involved in transplanting wheat plants led to increased transmission of wheat spindle streak mosaic *Potyvirus* from infected soil (Slykhuis, 1976).

2. Factors Affecting Virus Replication and Disease Expression

Light

Light intensity and duration affect virus production and disease expression in different ways with different viruses, but generally speaking, high light intensities and long days favor replication.

Temperature

Over the range of temperatures at which plants are normally grown, increasing temperature usually increases the rate at which viruses replicate and move through the plant. Like other biological phenomena, however, increase in temperature above a certain point leads to a reduction in the rate of replication. The species of host plant, strain of virus, and age of the leaf in which the virus is replicating may have a major effect on the way the virus behaves with changes in temperature. Figure 12.10 illustrates some effects of temperature on TMV replication.

The temperature at which plants are grown frequently affects the kind of disease that develops. For example, the type of pigmentation developing in subterranean clover infected with red-leaf virus was dependent on the temperature at which the plants were grown (Helms *et al.*, 1985). A severe stem tip necrosis developed in certain soybean cultivars held at 24°C following infection with soybean mosaic virus. At 28°C most plants developed typical mosaic disease (Tu and Buzzell, 1987).

High growth temperatures may annul the effect of some hypersensitivity genes. For example, in *Nicotiana* cultivars with the *N* gene, TMV causes systemic disease at 36°C (Kassanis, 1952). On the other hand, the activity of genes *Nx* and *Nb* in potato, causing hypersensitivity to PVX, were not affected by higher temperatures (Adams, *et al.*, 1986).

Increasing temperature, up to a certain point, increases the rate of systemic movement and decreases the time before the first appearance of systemic symptoms (e.g., Jensen, 1973). The shortest incubation period for systemic invasion does not necessarily occur at the temperature that leads to maximum production of virus.

In plants kept at different temperatures, there may or may not be an approximate correlation between severity of disease symptoms and virus concentration reached. For example, there is such a correlation with TMV and PVX in tobacco in the range 16°–28°C (Bancroft and Pound, 1954).

Changes in temperature at which plants are grown may lead to a selective

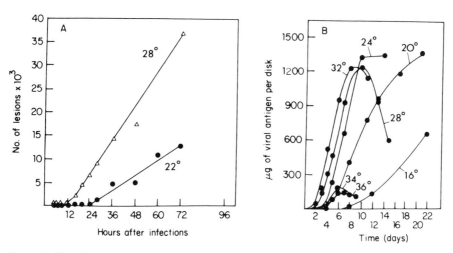

Figure 12.10 Effect of temperature on TMV replication. (A) A linear plot of the time course of TMV replication in tobacco mesophyll protoplasts. (From Dawson *et al.*, 1975.) (B) Effect of temperature on the replication of TMV (common strain) in tobacco leaf disks. Concentration of virus measured serologically. (From Lebeurier and Hirth, 1966.)

multiplication of certain strains adapted to the particular conditions (Chapter 13, Section I,C). Practical applications of heat therapy are discussed in Chapter 16.

Water Supply

The effects of water supply on virus replication have not been systematically studied. A chronic deficiency of water giving stunted "hard" plants will usually give rise to less obvious symptoms. A liberal supply of soil water increases the incidence of internal browning disease in tomatoes due to ToMV infection (Boyle and Bergman, 1967).

Nutrition

A number of investigations have been made into the effects of host plant nutrition on virus replication, particularly of TMV in tobacco plants. Many of the results are conflicting, probably due to variations in methods of growing the plants, the effective concentration of nutrients supplied, strain of virus used, method of estimating virus, and the basis for expressing the results. Supply of the major elements that give best plant growth usually allows for the greatest replication of virus. For example, Bawden and Kassanis (1950) concluded that the effects of nitrogen and phosphorus supply on TMV replication were fairly closely correlated with effects on plant growth. Treatments giving better plant growth led to a greater increase in virus production, both per unit fresh weight and per plant. Pea plants provided with low calcium nutrition developed severe stunting when infected with pea leaf roll *Luteovirus* (Thompson and Ferguson, 1976).

Minor element nutrition (zinc, molybdenum, manganese, iron, boron, and copper) has a variable effect on the capacity of plants to support virus replication. Effects on virus accumulation often parallel effects on plant growth, but exceptions have been observed. For example, manganese deficiency led to an increase in TMV concentration in the leaves. By contrast, toxic concentrations of this element in cowpea led to greatly enhanced replication of CCMV (Dawson, 1972).

Time of Year

The complex factors including day length, light intensity, air and soil temperatures, and water supply that change during the seasonal cycle will affect plant growth, and thus the disease produced by a given virus, and the extent to which a virus replicates. For example, there was a seasonal variation in the concentration of AMV in lucerne, the highest concentrations being recorded in spring and the lowest in autumn (Matisová, 1971). Some viruses, such as potato leaf roll virus in potato and BYV in sugar beet, cause much more distinct disease symptoms in summer than in winter. Others, such as PVX in potato, may cause more severe disease under winter conditions. Chloroplasts in the leaves of *Abutilon* infected with Abutilon mosaic *geminivirus* showed a marked seasonal variation in the severity of ultrastructural changes, being most severe in the summer (Schuchalter-Eicke and Jeske, 1983).

D. Interaction with Other Agents

Most viruses mutate frequently to give new strains. The presence of such strains may have a marked effect on the disease produced (Chapter 13). The disease produced by a particular virus in a given host and the extent to which it replicates are sometimes markedly influenced by the presence of a second independent and unrelated virus or by infection with cellular parasites. These latter effects are discussed next.

1. Interactions between Unrelated Viruses

The interaction between TNV and STNV discussed in Chapter 9, Section II,A,1, involves a complete dependence of one unrelated virus on the presence of another for replication. Another kind of situation exists between two viruses where both are normally associated with a recognized disease in the field. For example, the important tungro disease of rice is normally caused by a mixture of rice tungro bacilliform virus (RTBV), a DNA virus, and rice tungro spherical virus (RTSV) an RNA virus (Jones *et al.,* 1989) and is transmitted by leafhopper vectors (e.g., Cabautan and Hibino, 1984). It is the DNA virus which causes severe disease in rice. The DNA has properties indicating that it replicates in a retrovirus-like manner via an RNA template, as does CaMV (Hull *et al.,* 1990). The speckles disease, which is found in lettuce, sugar beet, and spinach in California, is caused by a complex of two viruses—beet western yellows *Luteovirus* and lettuce speckles mottle virus, the complex being transmitted by an aphid vector. Apart from these kinds of relationship, there are many examples known where unrelated viruses, both capable of infecting and replicating independently, interact together when by chance they infect the same plant.

Effects on Numbers of Local Lesions

Thomson (1961) found that when PVX was inoculated onto tobacco leaves together with TMV or PVY, the PVX lesions were more numerous. The increase was highly variable, ranging from 2- to 80-fold depending on conditions. From studies on the effect of dilution and of changing ratios of the two viruses, Close (1962) concluded that the maximum stimulation by TMV of PVX lesion formation occurs at a definite concentration of TMV in the inoculum and does not depend on the ratio of amounts of the two viruses. Infection by several unrelated viruses normally infecting cowpea was inhibited by CPMV in hosts immune to this virus. However the inhibition occurred only in plants with dominant immunity and not in cultivars with recessive immunity to CPMV (Saayer-Rieps and De Jager, 1988).

Effects on Virus Replication

Leaves of tobacco plants infected during the rapidly expanding stage contained up to 10 times as much PVX if PVY was also present (e.g., Stouffer and Ross, 1961). Ultrastructural studies and fluorescent antibody staining showed that both viruses were replicating in the same cells and that the increased production of PVX was due to an increase in virus production per cell rather than an increase in the number of cells supporting PVX replication (Goodman and Ross, 1974a).

The concentration of maize chlorotic mottle *Sobemovirus* in maize was increased about fivefold in mixed infections with a strain of maize dwarf mosaic *Potyvirus* (Goldberg and Brakke, 1987). A striking effect of this sort was found with black raspberry necrosis virus in herbaceous hosts. In mixed infections with *Solanum nodiflorum* mottle *Sobemovirus,* the normally very low concentration of the raspberry virus was increased 1000-fold (Jones and Mitchell, 1986).

Effects on Virus Movement

As discussed in Section V,B,2 and in Chapter 10, infection and systemic movement by one virus in a particular host may allow the cell-to-cell and systemic movement of an unrelated virus that normally would not move from the initially infected cells in that host.

Double infection with an unrelated virus may allow another virus, normally restricted to a particular tissue, to invade the plant more widely. For example, BGMV, normally confined to the phloem, invades nonphloem tissue in double infections with TMV (Carr and Kim, 1983). Invasion of nonphloem tissue in *Nicotiana clevelandii* by PLRV is enhanced in plants that are also infected by the unrelated PVY (Barker, 1987).

Effects on Macroscopic Disease Symptoms

Not uncommonly, a mixed infection with two viruses produces a more severe disease than either alone. The classic example here is the mixture of PVX and PVY in tobacco which produces a severe veinal necrosis instead of the milder mottling or veinbanding diseases seen with either virus separately (Smith, 1931). In many potato varieties, these two viruses together produce a "rugose mosaic" that is more severe than the disease produced by either virus alone. Many strains of ToMV produce only a mosaic disease in tomatoes, but in combination with PVX a severe "streak" disease ensues and usually kills the plant. Double infections with PVY and the viroid PSTVd caused severe necrotic disease in some potato cultivars in the field (Singh and Somerville, 1987).

Isolated protoplasts have been used to study mixed infections. For example, Otsuki and Takebe (1976a) inoculated tobacco mesophyll protoplasts with TMV and CMV. Seventy to eighty percent of protoplasts supported both viruses without any synergistic or antagonistic effects. On the other hand, in mixed infections of BMV and CCMV in tobacco protoplasts, both types of virus particle are produced, but only those of BMV are infectious. Only RNA3 of CCMV was synthesized in the mixed infections (Sakai *et al.*, 1983).

Cytological Effects in Mixed Infections

Mixed infections leading to necrosis must clearly have marked effects on individual cells. In mixed infections where cell death does not occur, two different viruses may replicate in the same cell producing their characteristic inclusion bodies or arrays of virus particles without any significant indications of mutual interference (e.g., Poolpol and Inouye, 1986; Langenberg, 1987).

2. Interactions between Viruses and Fungi

Effect of Virus Infection on Fungal Diseases

Increased Resistance   Observations in the field and glasshouse tests have shown that infection with *Phytophthora infestans* developed less rapidly in potato plants infected with one of a number of viruses (Müller and Munro, 1951). Numerous other examples are known where virus infection reduces susceptibility to, or development of, fungal and bacterial parasites. For example, infection of a hypersensitive tobacco cultivar with TMV induced systemic and long-lived resistance against *Phytophthora parasitica, Peronospora parasitica,* and *Pseudomonas tabaci* (McIntyre *et al.*, 1981). Similarly, systemic resistance to anthracnose in cucumber was induced by inoculation with TMV (Jenns and Kuć, 1980), as was resistance to *Peronospara tabacina* in tobacco (Ye *et al.*, 1989).

The development of resistance of this sort probably involves the nonspecific host responses discussed in Section III. Indeed fungicidal compounds have been isolated from plants reacting with necrosis to virus infection (e.g., Burden *et al.*, 1985). Generalized defense reactions may not be involved in some other virus–fungus interactions. Thus infection of faba bean with bean yellow mosaic *Potyvirus* decreased pustule density on leaves subsequently inoculated with *Uromyces vicia-fabae*. Changes were most marked on leaves showing yellowing symptoms (Omar *et al.*, 1986).

Increased Susceptibility   Russell (1966) showed that sugar beet plants in the field that were infected with beet mild yellowing *Luteovirus* had greatly increased susceptibility to *Alternaria* infection. BYV had no such effect. Beet mild yellowing virus increased and BYV decreased susceptibility to another fungus, *Erisphye polygoni*. Plants infected with both viruses had about the same susceptibility as healthy plants. The precise extent of the interaction depended on the genetic constitution of the host plant and on the environmental conditions.

Many other instances of increased susceptibility to fungi have been reported following virus infection. For example, prior infection of wheat or barley with BYDV predisposed the ears to infection by *Cladosporium* spp. and *Verticillum* spp. (Ajayi and Dewar, 1983). Sporulation of *Helminthosporium maydis* on corn leaves began sooner and was more abundant in lesions formed on leaves infected with MDMV (Stevens and Gudauskas, 1983). Asparagus infected with asparagus 2 *Ilarvirus* had increased susceptibility to *Fusarium* crown and root rot (Evans and Stephens, 1989).

Effects of Fungal Infection on Susceptibility to Viruses

Infection by a rust fungus may greatly increase the susceptibility of leaves to several viruses (Yarwood, 1951). For example, pinto bean leaves were heavily inoculated with the uredinial stage of *Uromyces phaseoli* on one half-leaf and then later with TMV over the whole leaf. Subsequent estimations of the amounts of TMV showed the presence of up to 1000 times as much virus infectivity in the rusted as in the nonrusted half-leaves. Rust infection partially suppressed the development of visible necrotic lesions. This suppression made it impossible to determine whether the increase in virus content was due to an increase in the number of successful entry points or to increased virus multiplication or to a combination of both of these factors.

Other fungi may induce resistance or apparent resistance to viral infection. Xanthi tobacco plants that had been injected in the stem with a suspension of spores of *Peronospora tabacina* Adam. produced fewer and smaller necrotic local lesions when inoculated with TMV 3 weeks later (Mandryk, 1963). The fungus *Thielaviopsis basicola* causes necrotic local lesions in tobacco. Hecht and Bateman (1964) found that if tobacco plants were infected on the lower leaves with this fungus, upper leaves became resistant to infection with TNV or TMV, as judged by size of the necrotic viral lesions. Inoculation of cucumber with *Colletotrichum lagenarium* or *Pseudomonas lachrymans* induced systemic resistance to CMV in an upper leaf (Bergström *et al.*, 1982). Some of these inhibitory effects at a distance may be due to the nonspecific response involving PR proteins discussed in Section III. On the other hand, a glucan preparation from *Phytophthora megasperma* protected a number of *Nicotiana* species from infection by several viruses without the induction of pathogenesis-related proteins (Kopp *et al.*, 1989).

3. Effects of Virus Infection on Nonvector Insects

Infection by TMV improved the suitability of tomato for survival of Colorado potato beetle larvae (Hare and Dodds, 1987). It was suggested that virus infection may facilitate the adaptation of phytophagous insects to "marginal" host plant species.

4. Interactions between Virus Infection and Air Pollutants

There is increasing concern about the effects of air pollutants such as ozone on plant growth. There have been several investigations of the effect of virus infection on the severity of ozone leaf damage. For most host–virus combinations, virus infection reduced damage. For example, infection with TMV reduced leaf damage in tobacco leaves due to ozone from 11 to 5% (Bisessar and Temple, 1977). The effect was seen in both glasshouse and field trials. Subacute doses of sulfur dioxide caused small but consistent increases in the content of SBMV in beans and MDMV in maize (Laurence *et al.*, 1981).

VI. DISCUSSION AND SUMMARY

Depending on the response to inoculation with a virus, plants can be described as either *immune* or *infectible*. If a plant is immune it is a nonhost for the virus, and the virus does not replicate in any cells of the intact plant or in isolated protoplasts. Immunity usually applies to all the members of a given species, but sometimes it is applicable to just a particular cultivar. The molecular mechanism for one example of such immunity has been established. Cowpeas of the cultivar Arlington are effectively immune to CPMV through the presence of a single dominant Mendelian gene. This gene codes for an inhibitor that blocks the action of the viral-coded proteinase needed for the specific processing of the CPMV polyprotein in the early stages of viral replication.

In infectible species or cultivars, the virus can infect and replicate in isolated protoplasts. The plant may be either resistant or susceptible to infection. We can distinguish two kinds of resistance. In resistance involving *extreme hypersensitivity,* virus multiplication is limited to the initially infected cells because the viral-coded protein necessary for cell-to-cell movement of virus is nonfunctional in the particular host. This gives rise to a subliminal infection. In the past, many examples of this type of resistance were described as immune.

In the second kind of resistance, infection is limited by a host response to a zone of cells around the initially infected cell usually resulting in necrotic local lesions. Uninfected tissue surrounding these lesions becomes resistant to infection. This is called *acquired resistance*. Acquired resistance involves the induction of at least 14 host proteins known as pathogenesis-related (PR) proteins. The appearance of these proteins following virus infection appears to be part of a generalized nonspecific defense reaction against bacterial and fungal parasites, insects, or chemical and mechanical injury, as well as viruses. In many host species the necrotic hypersensitive response (as opposed to a nonnecrotic response leading to systemic infection and disease) is governed by a single dominant Mendelian gene.

There are other kinds of *cultivar resistance* to virus infection. Most of these are determined by one or a few dominant genes. Some are controlled by incompletely dominant genes, and a few by apparently recessive genes.

A virus that does not cause systemic disease in a particular plant is *non-pathogenic* for that plant. If a virus or virus strain causes systemic disease in a particular species or cultivar it is *pathogenic*. A gene for resistance introduced into such a susceptible species or cultivar may make the virus *avirulent*. However, the virus may mutate and overcome the host resistance to become a *virulent* strain. Thus both host and viral genes interact to determine the outcome of inoculation. The change from an avirulent to a virulent virus strain may involve no more than a single-amino acid change in a viral-coded protein.

In species or cultivars that are susceptible, the virus replicates and moves systemically. In a *sensitive* reaction, disease ensues. If the plant is *tolerant* there is no obvious effect on the plant, giving rise to a *latent* infection. The consequences of infection for a susceptible plant are determined by both host and viral genes. For example, a single-base change in the TMV coat protein gene may be sufficient to alter the nature of the resulting disease.

The actual processes involved in the induction of disease are not well understood. Many of the biochemical changes involved may not be directly connected with virus replication. Stunting probably involves changes in the balance of growth hormones. The formation of mosaic patterns in virus-infected leaves involves events that occur in the early stages of leaf ontogeny.

Many environmental factors influence the course of infection and disease. These include light, temperature, water supply, nutrition, and the interactions between these factors during the growing season. Complex interactions may occur when plants are infected with two unrelated viruses or with a virus and a cellular pathogen.

13

Variability

Like other living entities, viruses remain substantially like the parent during their replication, but can change to give rise to new types or "strains." Because RNA viruses lack an error-correcting mechanism during genome replication, they give rise to many mutants involving one or a few nucleotide changes. There are several other mechanisms whereby new variants arise. Our knowledge of the pathways of virus evolution (Chapter 17) is quite fragmentary but there is no doubt that viruses have undergone, and continue to undergo, evolutionary change that is sometimes rapid. New strains provide the raw material for such change.

From the plant pathologist's point of view, the existence of virus strains in the field that cause different kinds of disease is often a matter of considerable practical importance. Reliable criteria are needed for distinguishing and identifying strains. For these reasons the study of virus strains is one of the most interesting and important aspects of plant virology. Aspects of variability have been discussed by van Vloten-Doting and Bol (1988).

I. ISOLATION OF STRAINS

The property of a new strain that first allows it to be distinguished from other known strains of a virus has been almost always biological—usually a difference in disease symptoms in some particular host. There are several ways in which new strains may be obtained. Where the virus is mechanically transmissible it is usual to pass new isolates through several successive single local lesion cultures, if a suitable host is available. This is done to ensure as far as possible that a single strain is being dealt with, and that no unrelated contaminating virus is present in the culture. There is good evidence that a single virus particle can give rise to a local lesion (Chapter 10, Section III,C,2). On the other hand, there is ample evidence that new mutants soon appear.

From a consideration of the mutation rate of viruses, discussed later, it is highly improbable that any virus culture actually consists entirely of a single strain. Many mutants probably arise even during the development of a single local lesion. For example, such mutants have been found in U1 TMV passaged through single lesions (García-Arenal *et al.,* 1984) (see also Section III,A,1). The various chemical and physical methods described in this chapter give useful information only because they are not sufficiently sensitive to detect the small proportion of any particular mutant or variant present in a culture.

A. Strains Occurring Naturally in Particular Hosts

Different strains of a virus frequently occur in nature, either in particular host species or varieties or in particular locations. These can be cultured in appropriate host plants in the greenhouse.

B. Isolation from Systemically Infected Plants

Plants systemically infected with a virus frequently show atypical areas of tissue that contain strains differing from the major strain in the culture. These areas of tissue may be different parts of a mosaic disease pattern (Chapter 12) or they may be merely small necrotic or yellow spots in systemically infected leaves, for example, in tobacco plants infected with mild mottling strains of PVX. When such areas or spots are dissected out and inoculated to fresh plants they may be shown to contain distinctive strains.

The preparation of protoplasts from systemically infected leaves, even when these are showing apparently uniform symptoms, offers the possibility of a very fine "dissection" of the leaves. Natural mutants may be revealed among such protoplasts either by regenerating plants from them (Shepard, 1975) or by inoculating test plants with virus from a single cell (Fraser and Matthews, 1979a).

C. Selection by Particular Hosts or Conditions of Growth

A particular strain may multiply and move ahead of others in a certain plant. Such a host can be used to isolate the strain. Similarly, strains may differ in the rate at which they multiply and move at different temperatures in a given host. Holmes (1934) found that tomato and tobacco stems inoculated with a severe strain of TMV and incubated at 35°C subsequently contained mild strains, which were able to multiply readily at 35°C, although the original severe strain was not. This result has since been amply confirmed by others. For example, Lebeurier and Hirth (1966) used serial passage in tobacco at successively higher temperatures to isolate a strain of TMV that grew effectively at 36°C.

Low temperatures have also been used to isolate strains (McGovern and Kuhn, 1984). It seems reasonable to suppose that selection of strains by particular hosts or conditions of growth may sometimes involve selection of a strain with a cell-to-cell movement protein that is better adapted to the host or conditions than the previously dominant strain. A small subpopulation of virus exists in a U1 TMV culture that can cause the hypersensitive reaction in *Nicotiana sylvestris*. This subpopulation moves upward rapidly and is selected for in the upper parts of the plant during rapid growth (bolting) of the shoot axis (Khan and Jones, 1989).

D. Isolation by Means of Vectors

Vectors may be used in three ways to isolate strains. First, by using short feeding periods on the plants to be infected, only one strain out of a mixture may be transmitted. This occurred with BCTV transmitted by a leafhopper vector following a 15-minute infection feeding (Thomas, 1970). Second, particular vectors may preferentially transmit certain strains of a virus (see Section III,C,3). Third, inoculation of insect vectors with diluted inoculum followed by repeated selection of infected plants for a particular type of symptom can result in the isolation of a variant virus (Kimura *et al.*, 1987).

E. Isolation of Artificially Induced Mutants

Experimentally induced mutants have been used for two important kinds of investigation in plant virology. Nitrous acid-induced mutations in the coat protein gene of TMV were of considerable importance in determining the nature of the genetic code and in confirming the nature of mutation. Nitrous acid has also been used to induce temperature-sensitive (*ts*) mutants, mainly in TMV, to aid in the delineation of the *in vivo* functions carried by the viral genome. Nitrous acid mutants have been used as a source of mild virus strains to give disease control by cross-protection (Chapter 16, Section II,C,2).

There is a spontaneous background mutation rate, but many of the natural mutants will be suppressed in inoculated leaves by a highly infectious inoculum containing mainly wild-type virus. Mutagens such as HNO_2 inactivate much of the treated virus. Thus, to show that a mutagen is increasing the mutation rate, it is necessary for treated and untreated virus to be assayed at fairly high dilutions that give about the same number of total infections for treated and control samples (e.g., Melchers, 1968).

1. Coat Protein and Other Mutants of TMV

To isolate mutants, the reactions of different hosts containing the N or N' genes have been exploited. One convenient method for isolating mutants is to inoculate a necrotic local lesion host under conditions where most lesions will have arisen from infection with single virus particles. Sometimes mutants give a recognizably different necrotic local lesion, very frequently smaller than normal (Siegel, 1960). To detect other symptom differences, single lesions are dissected out and inoculated to hosts giving systemic symptoms. Mutants may then be recognized by the different symptoms they produce.

Another method for isolating mutants and estimating their frequency is to use a parent strain of virus in a host that gives systemic symptoms without necrotic or other conspicuous local lesions. One then looks for mutants producing necrotic, yellow, or other characteristic local lesions in the host. This method selects one class of mutants, but it is sometimes difficult to eliminate parent virus from the culture, even by repeated single local lesion culture (Fig. 13.1).

A third method for isolating mutants depends on diluting the inoculum to a point where only about one-half (or less) of the plants inoculated become infected. A host giving systemic infection is used. Under these conditions it can be assumed that most of the infected plants were infected by a single virus particle.

Figure 13.1 Tobacco leaf inoculated with a mutant strain of TMV that produces large slowly developing yellow local lesions. The isolate is not completely freed of the type strain (causing no obvious local lesions). The sharp boundaries to the yellow (light) areas in the lower left quarter of the leaf delineate areas in which type TMV has multiplied and prevented invasion by the yellow strain. (Courtesy of B. Kassanis.)

2. *ts* Mutants

In isolating *ts* mutants following treatment of virus with nitrous acid, the objective is to obtain a range of independent mutants that are defective in one virus function at the nonpermissive temperature. The procedure used to select for such mutants should not select against mutants in any particular cistron.

Early methods selected for *ts* mutants that had amino acid substitutions giving coat proteins that were defective at the nonpermissive temperature (Jockusch, 1964; Wittmann and Wittmann-Liebold, 1966).

Dawson and Jones (1976) described a selection procedure for *ts* mutants of TMV based on the idea that all *ts* mutants should infect and begin replication more slowly at the nonpermissive temperature (32°C). U1 TMV replicates in Xanthi nc leaves at 32°C but no necrotic lesions appear. On return to 25°C necrotic lesions develop rapidly. If an infection was due to a *ts* mutant, development of necrosis on return to 25°C would be significantly delayed. Using this procedure, Dawson and Jones screened approximately 50,000 lesions formed by nitrous acid treated TMV. They finally obtained 25 *ts* mutants. The reversion to wild type by many of these mutants is a sufficiently rare event that they can be used for biochemical experiments at 25°C (Jones and Dawson, 1978).

In an attempt to obtain *ts* mutants of TRV with the mutation located in the long-rod RNA rather than the coat protein, Robinson (1973) added a large excess of untreated short rods to nitrous acid-treated virus. The mixture was inoculated to a local lesion host at 20°C. Single lesions were isolated and inoculated to a local lesion host at 20° and 30°C as a preliminary screen for mutants.

II. THE MOLECULAR BASIS OF VARIATION

A. Mutation (Nucleotide Changes)

1. Chemical Mutagens

Mutations involving single nucleotides consist of the replacement of one base by another at a particular site, or the deletion or the addition of a nucleotide. Single-base changes occurring in a coding region may lead to replacement of one amino acid by another in the protein product or to the introduction of a new stop codon that results in early termination of translation and a shorter polypeptide. Deletion or addition of a single nucleotide in a coding region will give rise to a frameshift, with consequent amino acid changes downstream of the deletion or addition. Such deletions or additions will usually be lethal unless compensated for by a second change (addition or deletion) that restores the original reading frame. Nucleotide changes in noncoding regions will vary in their effects depending on the regulatory or recognition functions of the sequence involved.

Treatments that cause mutation also inactivate viruses. Quantitative studies on inactivation and on the appearance of necrotic local lesions in a culture normally giving chlorotic lesions fitted quite well with a theoretical curve based on the assumptions that a single chemical event can result in inactivation and that a single event can result in mutation.

Gierer and Mundry (1958) first demonstrated the high efficiency of nitrous acid as an *in vitro* mutagen for TMV. The mutagenic action was considered to be through deamination of cytosine to give uracil and deamination of adenine to give hypoxanthine. The hypoxanthine acts like guanine during replication of the treated RNA and base-pairs with a cytosine. In the next copying event this will pair with a guanine. Thus, an adenine is replaced by a guanine at the original site. The amino acid exchanges found in the coat protein of mutants of TMV induced by nitrous acid confirmed these deaminations as the basis for the induced mutations. Such studies on TMV mutants made an important contribution to our understanding of the genetic code. They confirmed that the code is nonoverlapping and degenerate and gave strong support for the idea that the code is universal (Wittmann and Wittmann-Liebold, 1966; Sarkar, 1986). The single-base changes brought about by nitrous acid have also been amply confirmed by later amino acid and nucleotide sequence data (e.g., Rees and Short, 1982; Knorr and Dawson, 1988).

5-Fluorouracil is an analog of uracil that is incorporated into the RNA of viruses that are replicating in plants supplied with the analog. It replaces uracil residues in the RNA and can lead to the changes uracil → cytosine and adenine → guanine (Wittmann and Wittmann-Leibold, 1966). Some other chemicals have less clearly defined mutagenic effects on RNA viruses. For the experimental induction of mutations, nitrous acid will usually be the most useful mutagen.

2. X Irradiation and Ultraviolet Irradiation

Ionizing radiation and UV irradiation inactivate viruses containing dsDNA, and these agents also cause mutations in such viruses. Temperature-sensitive mutants of AMV have been isolated following irradiation of purified Tb component of AMV with UV light (van Vloten-Doting *et al.*, 1980).

3. Elevated Temperature

Several workers have noted that an increased number of variant strains could be isolated when plants were grown at higher temperatures (e.g., Mundry, 1957). However, there is good evidence, discussed later in this chapter, that the multiplication of and invasion by particular strains may be favored by certain temperatures. It seems probable that a major effect of heat treatment is to favor certain types of spontaneous mutant.

4. Natural Mutations

There is no doubt that a single-base change giving rise to a single-amino acid substitution in the protein concerned is a frequent source of virus variability under natural conditions *in vivo*. Thus, of 16 spontaneous TMV mutants listed by Wittmann and Wittmann-Liebold (1966), six had one amino acid exchange in the coat protein, one had two, and one had three such exchanges. Many of the coat protein exchanges represented base changes different from those induced by nitrous acid. Eight of the mutants had normal coat proteins and, therefore, must have had one or more base changes in the RNA outside the coat protein cistron. Most of the 15 amino acid substitutions found between the coat proteins of two naturally occurring strains of AMV could be explained by single point mutations (Castel *et al.*, 1979). Presumably, most of these mutants would have arisen from copying errors made by the RNA-dependent RNA polymerase during viral RNA replication. The differences between naturally occurring sequence variants of viroids usually consist mainly of a series of base substitutions (e.g., Visvader and Symons, 1985), although additions and deletions of nucleotides also occur.

B. Recombination

For many years it was considered that recombination was a genetic mechanism confined almost entirely to organisms with DNA as their genetic material. It is now known that recombination occurs in plant viruses with genomes consisting of either DNA or RNA.

1. DNA Viruses

Recombination *in vivo* was demonstrated for CaMV by inoculating plants with pairs of defective viral genomes with subsequent recovery of fully competent virus (Walden and Howell, 1982; Lebeurier *et al.*, 1982). Recombinants were characterized by restriction enzyme mapping. Figure 13.2 depicts two mechanisms proposed by Walden and Howell for recombination *in vitro*. In their *in vivo* studies they found that rescue probably occurred by a double recombination event (Fig. 13.2B). However, they considered that complementation between the two defective genomes may occur during the initial stages following coinfection of a cell.

DNA sequencing of CaMV strain CM_4-184 showed that it contained regions related to two other strains—strains S and CM1841 (Dixon *et al.*, 1986). The site of recombination between the two strains giving rise to CM_4-184 was located in the region R, where intergenomic switching between two RNA templates usually occurs during DNA minus strand synthesis by reverse transcription (Fig. 13.3). The repeated sequences (R) facilitate a template switch to another template, which may be of the same or a different virus strain.

Other studies of recombination between mutant CaMV DNAs indicate that recombination may sometimes occur by crossover events between double-stranded DNAs (Choe *et al.*, 1985). These authors favored a mechanism involving heteroduplex repair. Noncomplementing mutant DNAs would form heteroduplex loops at the mutant loci. Repair of these loops using mutant DNA as the template would result in the conversion of wild type to noninfectious mutant. Repair using the wild-type DNA as template gives rise to a viable genome.

Site-directed mutations in various ORFs of the two DNA molecules of TGMV rendered the genome noninfectious. Appropriate mixtures of noninfectious mutants were infectious, primarily as the result of complementation. However, a low proportion of wild-type DNA A molecules were detected. These could have arisen by either recombination or reversion (Brough *et al.*, 1988).

In experiments of a similar sort with the DNA2 of another geminivirus, ACMV, Etessami *et al.* (1988) found that recombination between certain pairs of mutants was a frequent occurrence, with the resulting wild-type virus rapidly becoming dominant. On the other hand, some mutants were retained within the virus population, indicating that rescue can occur by means of *trans*-acting gene products. Mutants of ACMV with deletions in the coat protein were competent for replication in the deleted form in protoplasts. However, systemic movement in a host plant was accompanied by reversion of DNA1 to a size comparable to that of native DNA1. This was accomplished by the generation of an extensive tandem repeat in the

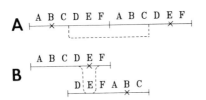

Figure 13.2 Proposed mechanism of recombination taking place between mutant viral genomes *in vivo*. (A) Single recombination event. (B) Double recombination or internal gene conversion event. ×, virus-lethal modification sites. (From Walden and Howell, 1982, by permission of Raven Press, Ltd.)

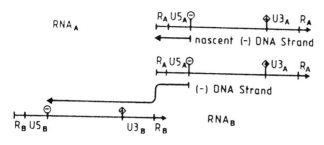

Figure 13.3 Schematic representation of a model for recombination in CaMV involving an intergenomic template switch between two RNA templates (A and B) during reverse transcription of a nascent DNA minus strand. This recombination event would give rise to recombinants with the U5 and part of the R region of one strain (A), the remainder of the genome being derived from the other strain (B). R = Terminal repeats; U5 = sequence between R and the primer binding site for minus strand DNA synthesis (⊖); U3 = the sequence between R and the primer site for plus strand DNA synthesis. (From Dixon *et al.*, 1986.)

mutant DNA (Etessami *et al.*, 1989). The reason for this phenomenon is not clear. When two mutants of MSV were inoculated together, *trans*-acting gene products allowed replication to occur and eventually intermolecular recombination resulted in the appearance of wild-type virus (Boulton *et al.*, 1989b).

2. RNA Viruses

The existence of DI RNA molecules with internal deletions (Chapter 6, Section VIII,B) indicated that RNA plant viruses have a mechanism for recombination. Direct evidence for recombination in a complete viral genome was provided by Bujarski and Kaesberg (1986). The RNAs 1, 2, and 3 of BMV have a very similar sequence of 200 nucleoties at the 3' terminus. A mutant was constructed in which a 20-base hairpin in the tRNA-like structure of RNA3 was deleted. When this mutant was inoculated together with wild-type RNAs 1 and 2, progeny RNAs contained both RNAs 3 and 4 with the mutant deletion. Analysis of virus with time in successive leaves revealed the appearance of increasing quantities of RNA3 that was similar to wild-type RNA3. One revertant RNA3 had the last 267 bases of wild-type RNA3 replaced with 307 bases from the 3' end of RNA2. Other revertants had similar structures differing slightly in length. Thus recombination must have occurred with a relatively high frequency at various positions in the RNA. The recombination events may also occur quite rapidly, because Rao and Hall (1990) found similar recombinants of mutant BMV RNA3 24 hours after inoculation of barley protoplasts. A mixture containing wild type BMV RNAs 1 and 3 and designed mutants of RNA2 with alterations in the pseudoknot region were inoculated to *Chenopodium hybridum*. Analysis of virus from single lesions showed that RNA2 had been restored to normal by homologous recombination within the tRNA-like region (Rao *et al.*, 1990). The experiments of Robinson *et al.* (1987) and Angenent *et al.* (1989b) with natural isolates of tobraviruses suggest that recombination of the sort described by Bujarski and Kaesberg (1986) also occurs in nature.

Deletions in either of the genes coded for by CCMV RNA3 blocked systemic movement of the virus. However, coinoculation with the two kinds of deletion mutant rapidly regenerated wild-type RNA3, giving rise to systemic infection (Allison *et al.*,

1990). Progeny RNA from such infections contained a full-length RNA 3 with both genes intact. Nucleotide substitutions introduced into the coat protein deletion mutant were recovered in the full-length progeny, demonstrating that recombination had occurred.

Simon and Howell (1986) described three satellite RNAs of TCV. One of these had a substantial 3′ domain homologous with the 3′ domain of the helper virus (see Fig. 9.7). The regions of homology did not have identical sequences, suggesting that exchanges of sequences between the satellites or between satellite RNAs and helper viruses, perhaps by recombination, are not very frequent occurrences. Recombination is also known to occur in viroids (Chapter 9, Section I,B,5).

The mechanism of recombination in RNA viruses is not firmly established but three diverse mechanisms have been suggested: (i) enzymatic cutting and religation; (ii) a copy choice model between two templates that by chance lie close together during replication (of these two, the latter mechanism is more widely favored); and (iii) there are several lines of evidence showing that nonviral RNAs can be reverse-transcribed and incorporated into the eukaryote genome (Baltimore, 1985). Thus it is possible that rearrangement of viral RNA sequences might occur when the RNA is reverse-transcribed into DNA form, for example, by a transcriptase belonging to another virus or a cellular retrotransposon (Haseloff *et al.*, 1984). Recombination in RNA viruses has been reviewed by King (1988). The rate of recombination in RNA viruses varies from uncommon to very frequent, a feature that is not yet understood.

C. Deletions

1. DNA Viruses

Howarth *et al.* (1981) described a naturally occurring deletion mutant of CaMV in which 421 base pairs were missing from ORF II. This mutant is infectious but is not transmitted by aphids. These authors found short direct repeats at the borders of the 421-bp deletion. These were probably involved in generation of the deletion as has been found in various nonviral systems.

Naturally occurring subgenomic DNAs arising by substantial deletions have been described for the geminiviruses ACMV (Stanley and Townsend, 1985), TgMV (MacDowell *et al.*, 1986), and WDV (MacDonald *et al.*, 1988). These various deletion mutants were each somewhat variable in length, but close to half the size of the normal DNA component. The mechanism by which they arise has not been established. Intramolecular recombination in a dsDNA form is one possibility. However, generalized recombination would not be possible because of the lack of sequence homology between the deletion borders. MacDowell *et al.* (1986) proposed that these DNAs might be generated by topoisomerase I-mediated non-homologous recombination. This enzyme shows some preference for the sequences CTT and GTT. These occur close to the deletion boundaries in the DNAs.

2. RNA Viruses

A deletion in RNA2 appeared to be the cause of a defective form of TSWV in which no enveloped particles were produced (Verkleij and Peters, 1983). Substantial deletions of genomic nucleic acid may give rise to the defective interfering RNAs discussed in Chapter 6, Section VIII,B. The first authentic DI RNA associated with a plant virus (TBSV) was described by Hillman *et al.* (1987). Sequence analysis showed that this small RNA consisted of a mosaic of seven fragments of different sizes derived from, and arranged in the same order as found in, the TBSV genome (see Fig. 6.16). Two mechanisms have been proposed to explain the origin of animal virus DI RNAs: (i) autocatalytic or enzymatic splicing reactions, which would involve sequence-specific splice sites, or (ii) occasional errors in which the RNA polymerase jumps to a different part of the same viral template RNA or to another template. The second mechanism is favored for the DI RNA of TBSV since no sequence conservation was found on either side of the deletion junctions (Hillman *et al.*, 1987).

When WTV, a phytoreovirus, is cultured for long periods in plants without leafhopper transmission, ability to be transmitted is lost in some isolates. Transmission-defective isolates have been shown to be formed by a substantial deletion (up to 85%) of genomic RNA segment numbers 2 and 5 giving rise to 5' and 3' terminally conserved remnants that are functional with respect to transcription, replication, and packaging (Nuss and Summers, 1984) (see Chapter 8, Section III,D, also Fig. 14.15). A somewhat similar situation has been found in the *Furovirus* BNYVV. The genome structure of this virus is illustrated in Fig. 7.33. Isolates from field-infected sugar beet roots showed that the two smaller RNAs (3 and 4) consistently had the standard length. By contrast, these two RNAs from isolates obtained from leaves of sugar beet or other hosts showed great variation in length and relative amounts, being shorter than the standard RNAs 3 and 4 (Koenig *et al.*, 1986).

Bouzoubaa *et al.* (1985) showed by sequence analysis that some isolates of RNAs 3 and 4 from virus cultured in *Chenopodium quinoa* had internal deletions (Fig. 13.4). With natural isolates of a virus it is difficult to estimate how frequently deletions occur in a virus stock. However, as pointed out by Jupin *et al.* (1990), synthetic RNA transcripts offer two major advantages for studying the deletion process: (i) they provide completely defined starting materials and (ii) they provide a

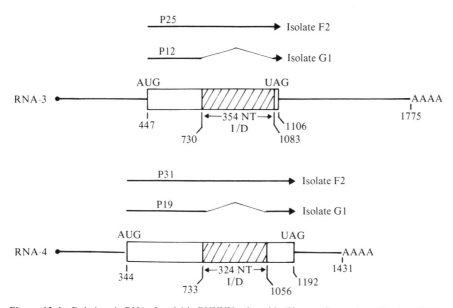

Figure 13.4 Deletions in RNAs 3 and 4 in BNYVV cultured in *Chenopodium quinoa.* The long ORFs in the F2 isolate are indicated by boxes. The hatched areas represent the deletions found in isolate G1. The sizes of the potential translation products are shown on the arrows. Nucleotide numbers from the 5' end are indicated. (From Bouzoubaa *et al.*, 1985.)

way of "setting the clock to zero" with respect to deletion events. Using a set of such transcripts for BNYVV RNAs 1, 2, 3, and 4, they isolated, after only one passage in *C. quinoa*, two specific deleted forms of RNA3 and one of RNA4.

Lemaire *et al.* (1988) showed that isolates of BNYVV containing no detectable RNAs 3 and 4 were very poorly transmitted by the fungal vector *Polymyxa betae*. Thus it is reasonable to assume that these two RNAs code for functions necessary for fungal transmission to roots, but not for viral replication in leaves. Lemaire *et al.* found that when successful fungal transmission of isolates with no detectable RNAs 3 and 4 did take place, full-length RNAs 3 and 4 reappeared. Presumably these RNAs were present in the cultures from leaves, but at concentrations below the level of detectability by physical methods.

Subgroups among CMV strains were characterized by both deletions and additions of nucleotide sequences (Quemada *et al.*, 1989).

D. Additions

In the literature many examples are described of mutant viruses with additional base sequences that have been generated by *in vitro* modification of recombinant DNA plasmids. Naturally occurring examples of such additions are much less common. In a group of naturally occurring variants it may sometimes be difficult to establish whether a difference in length is due to an addition or a deletion of nucleotides. Kimura *et al.* (1987) repeatedly selected rice plants for severe symptoms after inoculation with rice dwarf *Phytoreovirus* by leafhoppers injected with a dilute inoculum of the stock culture (strain 0). This procedure allowed the isolation of a severe strain (S). The fourth largest RNA of strain S had an M_r about 20K greater than that of strain 0. The MW of the protein corresponding to the 43K protein of strain 0 was 44K in strain S. It has not been demonstrated that S was derived from 0 by an addition of nucleotides. Strain 0 could have been derived from S by a deletion event, with the parent strain maintained at a low level in the culture.

Among the bromoviruses, the CCMV RNA3 5' noncoding sequence contains a clearly demarcated 111-base insertion not present in BMV, which must represent a sequence rearrangement in one of the two viruses (Allison *et al.*, 1989).

Repetition of blocks of sequences is known in some RNA viruses. For example, sequences of 56 and 75 nucleotides are duplicated in the leaders of the RNAs 3 of strain S and strain L of AMV (Langereis *et al.*, 1986b). The duplications are next to one another in the leader sequences and do not appear to be essential for replication. Langereis *et al.* suggested that these duplications may have been generated by polymerase molecules releasing prematurely from a minus strand RNA3 template and reinitiating again on the same template with the nascent strand still attached.

E. Nucleotide Sequence Rearrangement

An example of nucleotide rearrangement has been described among satellite RNAs of TRSV (Buzayan *et al.*, 1987). STRSV RNAs from strains 62L and NC-87 of TRSV have the same 360-residue sequence. The budblight satellite RNA with 359 residues differs from these mainly in the nucleotides from 100 to 140. The differences in sequences in this region are consistent with rearrangements of blocks of nucleotide residues as indicated in Fig. 13.5.

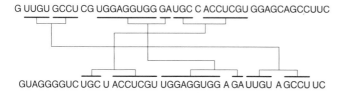

Figure 13.5 Possible nucleotide sequence rearrangements between two satellite RNAs of TobRSV. Upper line: strain 62L, nucleotides 100–145. Lower line: budblight strain, nucleotides 100–143. Heavy underlines indicate blocks of sequences that may have been rearranged. Connecting lines indicate identical sequences in the two strains. (Courtesy of G. Bruening.)

F. Reassortment of Multiparticle Genomes

Since the classic experiments with the long and short rods of strains of TRV (Lister, 1966, 1968; Frost *et al.*, 1967), genetic assortment experiments have been carried out with most of the known multipartite viruses. These experiments have demonstrated beyond doubt that new variants can arise by reassortment of the preexisting segments of the viral genome, both in the laboratory and in nature (Fulton, 1980) (see also Chapter 6, Section III,C,2).

Successful reassortment may not always be mutually effective. For example, Rao and Hiruki (1987) found with two strains of red clover necrotic mosaic *Dianthovirus* that a mixture of RNA1 of strain TpM34 plus RNA2 of strain TpM48 was infectious, while the reciprocal mixture was not.

Although BMV and CCMV have been regarded as distinct bromoviruses, their individual RNA components will complement one another in certain combinations (Allison *et al.*, 1988). Capped *in vitro* transcripts were made from complete cDNA copies of genomic RNAs 1, 2, and 3 of each virus. No viral replication was detected with any heterologous combination of RNAs 1 and 2, which code for proteins involved in RNA replication. By contrast, heterologous RNA3 was viable in both combinations, replicating in protoplasts and giving local lesions in *Chenopodium*. However, neither hybrid virus systemically infected the natural parental host plants.

G. The Origin of Strains in Nature

All the kinds of molecular change noted earlier in Section II will contribute to the evolution of strains in nature. A single-base change resulting in a single-amino acid change in a protein is probably one of the most common events giving rise to natural variation. The primary structure of the coat proteins of naturally occurring strains of TMV supports this view (Wittmann and Wittmann-Liebold, 1966).

Viruses appear to vary quite widely in the rate at which they give rise to new strains. Large numbers of TMV strains are known, while only a few have been isolated for TBSV. Different strains of a virus may also vary quite markedly in the rates at which they give rise to mutants of a particular symptom type. Thus, some strains of PVX producing chlorotic local lesions in tobacco frequently gave rise to ring spot local lesion strains, and other strains gave none at all (Matthews, 1949a). Strains of TNV that produced white lesions on cowpea frequently gave rise to strains giving red lesions (Fulton, 1952). These red lesions always first appeared as spots resembling sectors in association with a white lesion. Various white strains produced red mutants at different rates over a fivefold range. The reverse process (red strains giving rise to white) was not observed.

These various differences in the rate of appearance of strains may not necessarily be due to differences in actual mutation rate. Some viruses or strains may produce a much higher proportion of defective or completely nonviable mutants than others. Some of the apparent differences probably reflect our ability to detect mutants.

We can envisage strains diverging further and further from the parent type as changes brought about by various mechanisms accumulate in the various proteins specified by the virus. The survival and spread of such strains will often depend on their competitive advantage within the host in which they happen to arise, and in others which they subsequently infect. Selection of strains by host species is discussed in Section IV,B.

The survival of new strains may not always depend on immediate selective advantage, although continued survival would of course require an adequate combination of properties. For example, Reddy and Black (1977) point out how deletion variants of WTV could by chance come to dominate in particular shoots of a growing clover plant. If the deletion event occurred in a cell near the apical meristem where virus concentration is low, the mutant may come to replace the parent virus entirely in a particular shoot. They point out that the process can be regarded as an example of evolution by geographic isolation in miniature, with the branches of the plant providing the geographic isolation. The evolution of virus strains is further discussed in Chapter 17, Section IV,B,3.

III. CRITERIA FOR THE RECOGNITION
OF STRAINS

A virus might be defined simply as a collection of strains with similar properties. Sometimes we wish to ask whether two similar virus isolates are identical or not; on other occasions we will have to decide whether two isolates are different viruses or strains of the same virus. Two kinds of properties are available for the recognition and delineation of virus strains—structural criteria based on the properties of the virus particle itself and its components, and biological criteria based on various interactions between the virus and its host plant and its insect or other vectors. These criteria are further discussed in Chapter 17 in relation to the general problem of virus classification. Serological properties are based on the structure of the viral protein or proteins, but because of their practical importance, serological criteria are considered here in a separate section.

A. Structural Criteria

1. Nucleic Acids

Heterogeneity in Viral RNA Genomes

All the RNA genomes that have been examined have been found to exist not as a single nucleotide sequence but as a distribution of sequence variants around a consensus sequence. Thus sequence microheterogeneity is always present in natural populations (Holland *et al.*, 1982; Domingo *et al.*, 1985; Morch *et al.*, 1988). This nonuniformity in genome sequence has led to the concept of a "quasi species" for many viruses. The high mutation rate can be attributed to the high error rate in RNA-dependent RNA polymerases, which is thought to be due to the lack of any error-correcting mechanism for these polymerases. They have error rates in the range of $10^{-3}-10^{-5}$ misincorporations per site, which is three to four orders of magnitude higher than the rate for DNA polymerases (see Domingo and Holland, 1988). This fact, coupled with the very high rate of virus replication that is possible, can potentially lead to rapid change in virus populations. Most of the variants in a culture of a particular virus strain will normally consist of base substitutions at various sites perhaps with some deletions or additions of nucleotides. However, more substantial variation can occur when one or more segments of a multipartite genome are not under selection pressure. Variants may arise quite rapidly, and these could lead to confusion in strain identification. For example, genomic RNAs 3 and 4 of BNYVV code functions required for fungal transmission in the soil (Section II,C,2). When virus was isolated from leaves of infected sugar beets, where there is presumably no selection pressure to maintain the integrity of these genes, a wide range of deletion mutants was recovered (Burgermeister *et al.*, 1986). This potential for rapid change must be borne in mind when considering nucleic acid differences as criteria for recognizing virus strains.

Methods for Assessing Nucleic Acid Relationships

Nucleotide Sequences Nucleotide sequences give valuable information about the extent of relationships between viruses. Determination of even relatively short sequences can give useful information. However, to study effectively the relationships between strains of a virus it is necessary to determine the full nucleotide sequence of at least one, or preferably several isolates. In earlier studies on nucleotide sequences, where sequence was determined directly on the labeled viral RNA, the microheterogeneity noted earlier was not detected. The sequence determined would normally be the consensus sequence for that particular culture (e.g., Swinkels and Bol, 1980). When cDNA clones of RNA viruses are used for sequencing, the resulting sequence may, by chance, contain sequences that are not typical of the consensus sequence. This possibility is checked by the sequencing of several clones covering the same section of the genome. To ensure that a functional genome has been sequenced it is necessary to use a cloned DNA that has been shown to be infectious (e.g., Lazarowitz, 1988) or a cloned DNA that can be transcribed into infectious RNA (e.g., Dawson *et al.*, 1986).

In considering nucleotide sequences as a measure of relatedness between virus strains it must be remembered that the extent and distribution of such differences may vary quite widely depending on the virus and the part of the genome examined. To take some examples: (i) The coat protein gene and 3' untranslated region of three strains of PVY and two strains of WMV2 had 83 and 92%-identity, respectively (Frenkel *et al.*, 1989). There were a few clusters of nonidentical nucleotides but most were distributed more or less randomly throughout the sequence studied. (ii) By contrast, the coat protein and 3' untranslated regions of four strains of plum pox *Potyvirus* and two strains of TEV had 97–99% identity (Frenkel *et al.*, 1989). Similarly, the 39-nucleotide 5' leader sequences of the RNA4 of seven strains of AMV were identical except at position 26, where A, G, or U residues were found (Swinkels and Bol, 1980). Sequences with one or more recognition functions may often be highly conserved. For example, on the basis of sequence homology in the 3' nontranslated region it has been proposed that pepper mottle *Potyvirus* should be regarded as a strain of PVY (van der Vlugt *et al.*, 1989).

Hybridization Nucleic acid hybridization experiments can give valuable information concerning the degree of base sequence homology between the nucleic acids of different virus isolates, but interpretation of the data may not be straightforward.

The basis for nucleic acid hybridization was discussed in Chapter 2, Section IV,A. A significant advantage of hybridization procedures is that a comparison can be made between total genome RNAs, or RNA segments. This contrasts with serological tests, for example, which usually compare antigenic sites on the coat protein, which, for many viruses, represents only about one-tenth of the total genome. However, the degree of nucleic acid homology estimated between different strains will depend on various experimental factors such as the stringency of the hybridization conditions and the conditions for enzymatic removal of unpaired nucleotides (see Fig. 13.6). Figure 13.6 also illustrates the use of the dot blot method in a semiquantitative manner.

Koenig *et al.* (1988a) found that quantitative dot blot hybridization provided a very sensitive method for distinguishing closely related viruses that could barely be distinguished in serological tests. On the other hand, they observed some unexpected cross-hybridization between viruses belonging to different taxonomic groups. These cross-reactions may have resulted from real base sequence similarities as has been found between some viruses in separate groups and may also have resulted from the use of random-primed cDNAs.

Heterogeneity Mapping Heterogeneity mapping is a method based on RNA hybridization that allows the detection of single point mutations provided they occur in a significant proportion of the molecules. The method takes advantage of the

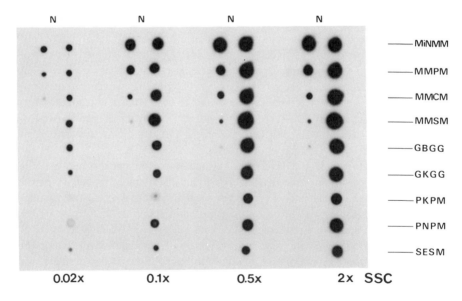

Figure 13.6 Use of the dot blot technique to estimate degree of relationship between strains of a virus. Autoradiograph showing the extent of sequence homology between MSV-MNM and other MSV strains. Four identical filters were each spotted, on the right with 2 ng of DNA from the different MSV strains under test and on the left with doubling dilutions (2 ng → 7.8 pg) of MSV-MNM (N) as controls. Filters were hybridized with a full-length nick-translated clone of MSV-MNM (N) prior to washing under conditions of different stringency (0.02, 0.1, 0.5, and 2 [SSC] at 65°C). (From Boulton and Markham, 1986.)

ability of RNase A to recognize and excise single-base mismatches in RNA hetero-duplexes (Winter *et al.*, 1985). Labeled minus sense RNA probes are transcribed from cDNA clones of the RNAs. Following hybridization with the test RNA, and digestion by RNase A and T, fragments are separated and sized by PAGE. The method has been used to assess the extent and location of heterogeneity among strains of CMV (Owen and Palukaitis, 1988) and satellite RNAs of this virus (Kurath and Palukaitis, 1989b). The method makes possible the assessment of entire populations of RNA molecules for major sites of heterogeneity. By contrast, the sequencing of individual clones gives precise and detailed information on a few molecules that may make up a tiny fraction of the virus population.

Other Properties of the Genomic Nucleic Acid   Size of the virus nucleic acid may differ with different virus strains, for example, with various TRV isolates (e.g., Cooper and Mayo, 1972). Restriction endonuclease mapping has been used to characterize the dsDNA genomes of some caulimoviruses (e.g., Hull and Donson, 1982).

Additional Genes

The RNA2 of the TCM strain of TRV is considerably larger than that of other strains that have been sequenced, for example, PSG. Angenent *et al.* (1986; 1989b) showed that the greater length was due partly to a repetition of 1099 3' nucleotides from RNA1, which includes a 16K ORF, and partly to a 29K ORF that was unique to this RNA2 (Fig. 13.7). The 29K ORF has no significant homology with the 28.8K ORF of PSG RNA1, and its origin and function, if any, are not established.

Subgroups of Strains

When a sufficient number of strains have been examined, nucleotide sequence relationships may be used to delineate subgroups of strains of a virus. For example,

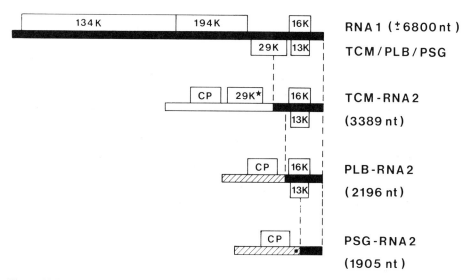

Figure 13.7   Variation in length of RNA2 in strains of TRV. Schematic representation of the genome structure of strains TCM, PLB, and PSG. RNA1 is illustrated at the top. RNA1 sequences in the RNAs are indicated by solid bars. Sequences that are homologous in RNAs 1 and 2 are connected by dashed lines. Sequences that are identical in PLB RNA2 and PSG RNA2 are indicated by hatched boxes. The asterisk in TCM RNA2 indicates the sequence that is unique to TCM RNA2. CP = Coat protein. The MWs of other gene products are indicated. (From Angenent *et al.*, 1989b.)

on the basis of competition hybridization tests, 30 strains of peanut stunt *Cucumovirus* could be divided into two groups with little homology between them, but extensive homology within groups (Diaz-Ruiz and Kaper, 1983).

Some Limitations Concerning Base Sequence Data

A given base substitution, deletion, or addition may have very different effects in the protein coded for depending on a number of circumstances. The following factors may be important:

1. Because the genetic code is degenerate, many base substitutions cause no change in the amino acid being coded for. For example, in the TMV strain L11A, derived from the L strain, there were 10 base substitutions, 7 of which occurred in the third position of in-phase codons and did not influence amino acid sequence (Nishiguchi *et al.*, 1985).

2. A given base substitution may result in change to an amino acid of very similar properties, which causes very little change in the protein. Alternatively, the change may be to a very different kind of amino acid (e.g., from an aliphatic side chain to an aromatic one), giving rise to a viable protein with changed physical properties or to a nonfunctional mutant that does not survive.

3. A single-base deletion or addition will cause a frameshift mutation with greater or lesser effect depending on whether it is near the beginning or the end of the gene, whether a second change (addition or deletion) brings the reading frame back to the original, and how many proteins are coded for by the section of nucleic acid in question.

4. A more general problem in using base sequence data for classification is that some parts of the genome, and some products may have multiple functions. Some parts of the genome may code for a single polypeptide, but others may code for more than one; and some polypeptides may have more than one function. Some parts of the genome may have both coding and control or recognition functions. Furthermore, mutations in one gene may affect the production of another. For example, mutations in the presumed polymerase gene of TMV mutant *LIIA* cause reduced synthesis of the cell-to-cell movement protein (30K), thus reducing efficiency of movement of this strain (Watanabe *et al.*, 1987b).

Thus, even if we knew the full base sequences for the nucleic acids of a set of virus strains, we would be unwise to use these sequences to establshi degrees of relationship without other information. It was once thought that a virus classification scheme based only on nucleotide sequnce would be the ultimate aim (Gibbs, 1969). It is now apparent that the significance to be placed on nucleic acid base sequence data can be judged from a biological point of view only in conjunction with a knowledge of the organization of the genome and the functions of its parts and products.

2. Structural Proteins

The coat protein or proteins and other structural proteins found in viruses are very important, both for the viruses and for virologists wishing to delineate viruses and virus strains. The coat proteins of the small RNA viruses must have evolved to give a satisfactory balance between three important functions.

1. The ability to self-assemble around the RNA; mutants are known in which this function is defective even at normal temperatures. For example, strain PM2 of TMV cannot form virus rods with RNA. The protein aggregates *in vitro* at pH 5.2 to form long, open, flexuous, helical structures rather than compact rods (Zaitlin and Ferris, 1964);

2. Stability of the intact particle inside the cell, and during transmission to a fresh host plant.

3. The ability to disassemble to the extent necessary to free the RNA for transcription and translation. A variant that could not carry out this function would not survive in nature. For example, Bancroft *et al.* (1971) described a mutant of CCMV induced by nitrous acid that was unable to be uncoated in the cell, and was, therefore, noninfectious in spite of the fact that RNA isolated by the phenol procedure from the virus was highly infectious.

For the small RNA viruses the coat protein is of particular importance for the delineation of viruses and virus strains. Besides the intrinsic properties of this protein (size, amino acid sequence, and secondary and tertiary structure), many other measurable structural properties of the virus depend largely or entirely on the coat protein. These include serological specificity, architecture of the virus, electrophoretic mobility, cation binding, and stability to various agents. Thus, ideas on relationships within groups of virus strains, based on properties dependent on the coat protein, may be rather heavily biased. On the other hand, if mutations in the noncoat protein genes have occurred more or less at the same rate as in the coat protein during the evolution of strains in nature, then such views on relationships may be reasonably well based.

The *Potyvirus* group will serve to illustrate the use of the properties of coat proteins in the delineation of virus strains. Potyviruses have been one of the most difficult virus groups to study taxonomically. The group contains about one-quarter of all the known plant viruses. The viruses infect a wide range of host plants and exist in nature as many strains or pathotypes differing in biological properties such as host range and disease severity. It has been considered by some workers that strains of potyviruses may form a continuous spectrum between two or more otherwise distinct viruses, making delineation of viruses and groups of strains difficult or impossible. However, recent comparisons between the amino acid sequences of the coat proteins of several viruses and many strains indicate that this approach may provide a useful basis for taxonomy within the group (Shukla and Ward, 1989a,b). Analysis of the 136 possible pairings between a set of viruses and strains revealed a clear-cut bimodal distribution, with distinct viruses having an average sequence homology of 54%, while strains averaged 95% (Fig. 13.8). These data give no support for the "continuous spectrum" idea among the potyviruses. Distinct viruses showed major differences in length of their coat proteins (Fig. 13.9). Major differences in amino acid sequence were near the N termini, with high homology in the C-terminal half of the proteins. On the other hand, strains have very similar N termini.

Two exceptions to this pattern appear to reflect the misplacing of certain *Potyvirus* isolates on the basis of previous data. Serological tests suggested that PeMV and PVY were only distantly related. However, the sequence data shown in Fig. 13.9 clearly indicate that PeMV should be considered a strain of PVY. SMV-N and SMV-V, formerly considered to be strains of SMV, but now shown to have a sequence homology of 58%, should be considered as two distinct viruses (Shukla and Ward, 1988). Similar analyses for other potyviruses are suggesting revised

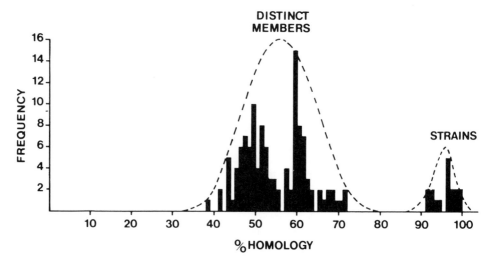

Figure 13.8 Demarcation between the extent of amino acid sequence homologies in coat proteins among distinct individual potyviruses (left-hand distribution) and between strains of the same virus (right-hand peak). The 136 possible pairings between 17 strains of eight distinct viruses were analyzed. The homologies between distinct viruses had a mean value of 54.1% and a standard deviation of 7.29%, while the homologies between strains of individual viruses showed a mean of 95.4% and standard deviation of 2.56%. The dashed curves show that all values for distinct viruses and strains fall within ±3 standard deviations from their respective mean values. (From Shukla and Ward, 1988.)

relationships, for example, that watermelon mosaic virus 2 and SbMV-N are strains of the same potyvirus (Yu *et al.*, 1989). Similarly, JgMV is not a strain of SCMV but rather a distinct virus (Shukla *et al.*, 1987) as confirmed by serology (Shukla *et al.*, 1989a).

High-performance liquid chromatography of tryptic peptides may be useful in differentiating potyviruses and their strains (Shukla *et al.*, 1988a). This technique does not provide the detailed information obtained by amino acid sequencing, but its greatest value may lie in the ease with which the method can be applied.

The projecting (P) domain of *Tombusvirus* coat proteins has a more variable amino acid sequence than the structural (S) domain (Hearne *et al.*, 1990). Thus greater variability at the exposed surface may be a feature of both rod-shaped and icosahedral plant viruses.

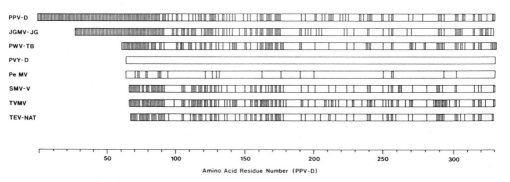

Figure 13.9 Schematic diagram showing the location of amino acid sequence differences between seven distinct members of the *Potyvirus* group and PeMV. The sequences were compared with strain D of PVY, the type member. PeMV is very similar to PVY in its coat protein sequence. (From Shukla and Ward, 1988.)

3. Nonstructural Proteins

Insufficient amino acid sequence data are available at present for the nonstructural viral-coded proteins to be of much use in delineating strains. Mayo *et al.* (1982) could detect no difference in the tryptic peptides obtained from the VPg of different strains of raspberry ring spot *Nepovirus* or TBRV. Some strains of TMV may differ widely in the amino acid sequence of their 30K proteins, as discussed by Atabekov and Dorokhov (1984). Four structural proteins of the two serotypes of PYDV—SYDV and CYDV—fell into two groups based on peptide mapping. Proteins M and N differed little between the strains, whereas M_2 and G were significantly different (Adam and Hsu, 1984).

4. Other Structural Features

Architecture of the Virus Particle

Related viruses will be expected to have very similar size, shape, and geometrical arrangement of subunits. However, significant differences in particle morphology have been found within groups of related strains. Differences in rod length are frequent between strains of helical viruses such as TRV (e.g., Cooper and Mayo, 1972) and BSMV (Chiko, 1975). Sometimes the variation in architecture appears to be "abnormal" even though the strain of virus is a viable one. Thus, the packing of the coat protein of the Dahlemense strain of TMV involves a periodic perturbation of the helix (Caspar and Holmes, 1969). Some AMV strains contain abnormally long particles that have the normal diameter but contain more than one RNA molecule (Heijtink and Jaspars, 1974).

Electrophoretic Mobility

The electrophoretic mobility of a virus depends in the first place on the amino acid composition of the protein and second on the three-dimensional structure, which affects the availability of ionizing groups. Mobility is also dependent on the ions present in the buffer used.

Stability and Density

Among the small RNA viruses, differences in stability and density have been used to differentiate virus strains. The RNA content of the virus may vary with strain and thus affect buoyant density in strong salt solutions (e.g., Lot and Kaper, 1976). However, differences in the coat protein most commonly lead to a difference in stability or density.

B. Serological Criteria

The nature of antigens and antibodies, the basis for serological tests, and their advantages and limitations were discussed in Chapter 2, Section III,A. This section considers the use of serological criteria to delineate viruses and virus strains.

1. Some General Considerations

Presence or Absence of Serological Relationship

Serological tests provide a useful criterion for establishing if two virus isolates are related or not. Any of the tests described in Chapter 2 can be applied, but most commonly some modification of the precipitation reaction or ELISA tests is used. Provided adequate precautions are taken, serological tests can be valuable for placing viruses into groups.

If two virus isolates show some degree of serological relationship it is highly probable that they will have many other properties in common and belong in the same virus group. There are a few unexplained exceptions. Various examples are known of viruses that undoubtedly belong in the same group but that show no serological cross-reactivity—for example, TYMV and EMV in the *Tymovirus* group. In making tests for serological relationships there are several potential sources of error.

1. Presence in viral antisera of antibodies reacting with host constituents such as the abundant protein ribulose 1′,5′-bisphosphate carboxylase.
2. Nonspecific precipitation of host materials in crude extracts.
3. Nonspecific precipitation of viral antigens, especially at high concentrations.
4. Contamination of antigen preparations with other viruses.
5. Virus altered during isolation. It should always be borne in mind that virus may be altered during isolation in a way that can affect its serological specificity.
6. Nonreciprocal positive reactions. To demonstrate that two viruses are serologically unrelated, reactive antisera must be prepared against each of the viruses under test. It must be shown that each reacts with its own antiserum, but gives no reaction with the heterologous antiserum. This reciprocal test is necessary since the viruses might in fact be related, but one may occur in too low a concentration in the extracts to give any positive reaction. Negative one-way tests are of little value. As discussed in the next section, it is preferable to use high-titer antisera to demonstrate a lack of serological relationship.
7. Isolates taken from the field may be mixtures of several different serotypes (e.g., Dekker *et al.*, 1988).

These considerations apply particularly to the use of polyclonal antisera. The use of MAb's avoids several of these problems, but they have limitations of their own as discussed in Section III,B,3.

Degrees of Serological Relationship among a Group of Virus Strains

A considerable amount of experimental work has been directed toward determining degrees of relatedness within groups of strains and in attempts to correlate serological properties with other biological and chemical characteristics. Delineation of virus strains is a particularly important aspect of any program designed to produce resistant varieties of a host species.

If two isolates of a virus are identical they will react identically when cross-reacted with each other's antisera, whatever form of serological test is applied. If, however, they are related but distinct, some degree of cross-reaction will be observed, at least with polyclonal antisera, but the reactions will not be identical. Various types of serological test can be used to identify and distinguish virus strains. Examples are given in table 13.2.

When a group of only two or three virus isolates is to be considered, it is a relatively simple matter, provided technical precautions are observed, to determine whether the isolates are unrelated serologically, whether they are identical, or whether they show differing degrees of relationship. Using the same set of isolates and the same antisera, quite reproducible results can be obtained, to indicate, for example, that strains A and B are closely related and that both are more distantly related to strain C. However, when large numbers of related strains are tested the situation may become quite complex and less and less meaningful as more strains are considered in relation to one another.

Experimental Variables

There are a number of important experimental variables that can affect the estimated degree of serological relationship between viruses and strains. These include the following: (i) A major source of experimental variation is the variability in antisera, both in successive bleedings from the same animal and in sera from different individuals. The proportion of cross-reacting antibody present in a series of bleeding taken over a period of months from a single animal may vary widely (Koenig and Bercks, 1968). (ii) The extent to which antisera to two virus strains cross-react is usually correlated with the antibody content of the serum. Sera of low titer show lower cross-reactivity, and those with high titers show greater cross-reactivity. Thus to detect serological differences between closely related strains using polyclonal antisera it is preferable to use antisera of fairly low titer. To demonstrate distant serological relationships, it may be necessary to use high-titer antisera. (iii) Many virus preparations used for immunization and for antibody assay may contain varying amounts of free coat protein or coat protein in various intermediate states of aggregation or in a denatured state. Coat protein in the intact virus may lose amino acids through proteolysis. Antibodies reactive with coat protein in these various forms may or may not indicate the same sort of relationships as antibodies against intact virus. An example of this sort is discussed in relation to the *Potyvirus* group in Section III,B,4. The method used to detect and assay cross-reacting and strain-specific antibodies may affect the apparent degree of relationship. Examples are given in the references listed in Table 13.2.

The Serological Differentiation Index

In spite of all the variables, useful assessment of degrees of serological relationship can be obtained by testing successive bleedings from many animals and pooling the results. Most quantitative measurements of degrees of serological relationship have been carried out using precipitation titers. In such tests the extent of serological cross-reactivity can be expressed by a serological differentiation index (SDI) (van Regenmortel and von Wechmar, 1970). The SDI is the number of twofold dilution steps separating homologous and heterologous titers. The SDI values are equal to the difference in those titers expressed as negative Log 2. For example, such replicated comparisons have been made for sets of tobamoviruses (Van Regenmortel, 1975) and tymoviruses (Koenig, 1976).

ELISA tests can also be used to calculate SDIs as a measure of the extent of serological relatedness between viruses or virus strains (Jaegle and Van Regenmortel, 1985; Clark and Barbara, 1987). Table 13.1 shows a comparison of the SDIs obtained from ELISA and precipitin tests. There were differences in the reciprocal SDIs found by ELISA for pairs of viruses, and the average of these values did not correspond closely to those found by precipitin SDIs. Nevertheless, both kinds of test show clearly that CGMMV is substantially different from the other tobamoviruses.

Clark and Barbara (1987) describe a more refined statistical procedure for calculating SDIs from ELISA tests that is capable of discriminating reliably among virus strains that differ by as little as 0.2 SDI.

Table 13.1

Serological Differentiation Indices (SDI) for Pairs of Tobamoviruses Calculated from ELISA
and from Precipitin Tests[a]

Tobamoviruses		SDI from ELISA			Average SDI from precipitin tests	% Sequence similarity in the viral coat proteins
x	y	y-anti-x	x-anti-y	Average value		
TMV –ToMV		1.4	0.5	0.9	1.2	82
U2 –ToMV		0.6	1.9	1.3	1.9	70
TMV –RMV		2.0	1.2	1.6	2.1	44
TMV –U2		2.4	1.9	2.1	2.7	74
ToMV–RMV		1.9	0.7	1.3	4	47
U2 –RMV		3.5	2.5	3.0	4.5	46
RMV –CGMMV		7.0	6.4	6.7	5	
U2 –CGMMV		2.6	8	5.3	6	
TMV –CGMMV		5.4	6.9	6.2	6.8	36
ToMV–CGMMV		5.7	6.9	6.3	7	

[a]From Jaeyle and Van Regenmortel (1985).

2. The Role of Virus Components in Serological Reactions

There is no good evidence that ss plant viral RNA can elicit RNA-specific antibodies. Antibodies formed in response to injection with a plant virus react only with the virus protein, either in the intact virus or as various partial degradation products of the intact protein shell or rod. A formal demonstration of the role of the protein was made by Fraenkel-Conrat and Singer (1957). They carried out mixed reconstitution experiments between serologically distinct strains of TMV. The artificial hybrid virus had the serological type of the protein used to coat the RNA, but the progeny following infection had protein of the type from which the RNA was obtained.

Nevertheless, the viral RNA may play some secondary role in stimulating production of antibodies against the viral protein shell. Intact TYMV is substantially more immunogenic than the apparently identical empty protein shell, which contains no RNA (Marbrook and Matthews, 1966). This difference was found in rabbits and mice using several routes of injection. The difference persisted throughout the time course of the primary response and was also found in the response to a second injection. Isolated TYMV RNA injected at the same time did not augment the immunogenicity of the empty protein shells. Artificial empty protein shells produced from the infectious virus *in vitro* were no more immunogenic than the natural empty shells.

Noninfectious TYMV nucleoprotein was just as immunogenic as infectious virus. Thus, we concluded that the enhanced immunogenicity of the nucleoprotein must be due to the physical presence of the RNA inside the particle. Whether this enhanced immunogenicity of the viral nucleoprotein is a general feature of plant viruses remains to be determined. TMV appears to be more immunogenic than either protein rods or subunits (Marbrook and Matthews, 1966; Loor, 1967). The mechanism by which the ssRNA within the virus stimulates immunogenicity is not yet understood.

dsRNAs can be immunogenic. dsRNA antisera react with dsRNA but not dsDNA or ss nucleic acids. The antisera lack specificity for particular ds nucleic acids (see Chapter 2, Section IV,D,2).

There are several reasons why we would expect intact viruses (such as TMV and TYMV) and protein subunits or subviral aggregates prepared from them to differ in the antigenic sites they possess:

1. Some antibody-combining sites on the intact virus may be made up of parts of the exposed surface of two or more subunits. Such a site would not exist in isolated subunits.
2. Subunits probably have characteristic combining sites, which are masked when the subunits are packed into the intact shell.
3. Conformational changes occur when the subunits aggregate, so that the configuration of the exposed surface of the packed subunit may not be the same as when it exists as a monomer.

Examples of all these phenomena are known among plant viruses and their protein subunits.

3. Procedures Used for Delineating Viruses and Strains

Assay Methods

The various serological methods that are used in the detection and assay of viruses were discussed in Chapter 2, Section III,A. Most of these procedures have been used for delineating viruses and virus strains. Some recent examples are listed in Table 13.2. This list reflects the fact that ELISA procedures have become the most popular for the delineation of viruses and strains.

Monoclonal Antibodies

The advantages and disadvantages of using MAbs for assay, detection, and diagnosis of viruses were summarized in Chapter 2, Section III,A,5. The outstanding value of MAbs in the delineation of virus strains is that their molecular homogeneity ensures that only one antigenic determinant is involved in a particular reaction.

Table 13.2

Some Serological Procedures Used for the Delineation of Viruses and Virus Strains

Procedure	Virus or virus group	References
Indirect ELISA	Tymo-, tombus-, como-, tobamo-, potex-, carla-, and potyviruses	Koenig (1981)
F(ab′)$_2$ ELISA	Carlaviruses	Adams and Barbara (1982)
Radial double diffusion in agar, ELISA, and SSEM	Cucumoviruses	Rao *et al.* (1982)
Quantitative rocket immunoelectrophoresis	Panicum mosaic ? *Sobemovirus* strains	Berger and Toler (1983)
Electroblot immunoassay	Tymo-, tombus-, como-, nepo-, tobamo-, potex-, carla-, and potyviruses	Burgermeister and Koenig (1984)
Direct or indirect ELISA	Ilaviruses and AMV	Halk *et al.* (1984)
Various ELISA procedures	Grapevine fanleaf *Nepovirus* strains	Huss *et al.* (1987)
Indirect protein A–sandwich ELISA	Tobamoviruses and virus strains	Hughes and Thomas (1988)
Indirect ELISA	Maize streak *geminivirus* isolates	Dekker *et al.* (1988)

The high specificity of this single interaction is not swamped in a large number of other interactions as with a polyclonal antiserum. Provided a MAb can be found that recognizes a small antigenic change between two virus strains, then very fine distinctions can be made in a reproducible manner. Although the ability of different MAbs to distinguish single-amino acid exchanges may vary widely, some may be able to do so (Al-Moudallal *et al.*, 1982).

However, there are several limitations in the use of MAbs. (i) There is usually no immunoprecipitation between MAbs and viral protein monomers. (ii) MAbs are often sensitive to minor conformation changes in the antigen such as may be caused by detergent or by binding of antigen to an ELISA plate (e.g., Dekker *et al.*, 1987). (iii) Among a set of virus strains the relative reactivity of different MAbs may vary considerably. For example, TMV strain 06 differs from TMV by residue exchanges at positions 9, 65, and 129 of the coat protein. In the tests shown in Fig. 13.10, this strain reacted like TMV with MAb a, more strongly than TMV with MAb c, and not at all with MAb b. (iv) MAbs may be heterospecific, that is, they may frequently react more strongly with other antigens than with the virus used for immunization (see Chapter 2, Section III,A,5). The reaction of strain 06 with MAb c, which is stronger than that with TMV, the strain used as the immunogen, illustrates this phenomenon (Fig. 13.10).

If, during the selection of hybridoma clones for the isolation of MAbs, the clones are tested only with the strain of virus used as immunogen, MAbs with low affinity for this strain may go undetected and be discarded. Among such MAbs may be clones that would be very useful for the detection of other strains. Thus when searching for strain-specific MAbs it is important to screen clones against a panel of structural relatives of the immunogen.

Another potential limitation of MAbs in the delineation of strains can occur if two strains have an identical antigenic determinant in common. If, by chance, the MAb specificity is directed against this determinant, the strains will appear identical even though they may have substantially different determinants elsewhere in the molecule. For example, strain Y-TAMV is a member of the ToMV group of strains that has an 18% difference in coat protein amino acid sequence compared with TMV. The two viruses are readily distinguished by polyclonal antisera but not by some MAb's (e.g., antibody C in Fig. 13.10).

These limitations highlight the importance of generating diverse panels of MAbs for the delineation of viruses and strains. Experiments with ToMV will further illustrate the problem. Strains of this virus are considered to be serologically quite uniform. Ten MAbs raised against the virus reacted in an identical manner with 15 ToMV strains and isolates. However, two of them cross-reacted with TMV and ribgrass mosaic virus (Dekker *et al.*, 1987).

4. Antigenic Sites Involved in the Serological Delineation of Viruses and Strains

Because of the crucial role they play in intersubunit bonding, the sides of protein subunits that make up the shells or rods of a particular virus might be expected to be fairly constrained in the extent to which amino acid replacements would allow the subunit to remain functional. This would be particularly so with rod-shaped viruses in which RNA–protein interactions are also important. One might expect much less constraint on that part of the protein subunit that makes up the surface of the virus, and that would therefore also provide the antigenic sites of the intact virus. This expectation has been confirmed for members of the *Potyvirus* group.

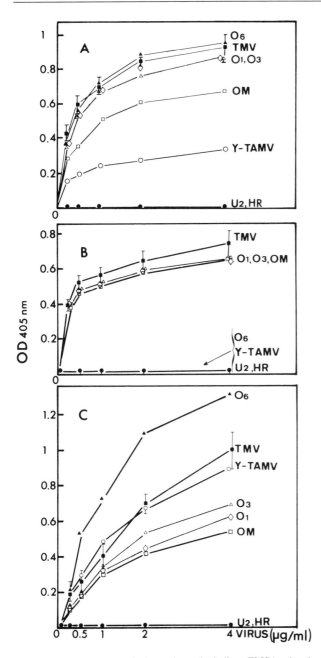

Figure 13.10 Detection of tobamoviruses by indirect ELISA using three MAb's, a, b, and c, obtained using TMV as immunogen. Strains 01, 03, and 06 are orchid strains and 0M is the common Japanese strain of TMV; Y-TAMV is a strain of tomato mosaic virus; U2 and HR are two distinct tobamoviruses. (From Van Regenmortel, 1984c.)

Biochemical and immunological evidence suggested that the N-terminal 29 amino acids of TEV were hydrophilic and located at or near the virus surface (Allison *et al.*, 1985). Mild proteolysis by trypsin of the particles of six distinct potyviruses showed that the N- and C-terminal regions of the coat proteins are exposed at the particle surface. Trypsinization removed 30–67 residues from the N terminus and 18–20 from the C terminus, the length removed depending on the virus (Shukla *et al.*, 1988b). This proteolysis left a fully assembled, infectious virus particle containing protein cores consisting of 216 or 218 amino acids. Electroblot immunoassays with polyclonal antisera showed that *Potyvirus* group-specific antigenic sites are located in the trypsin-resistant core protein region. Thus antibodies to the dissociated core protein should react with most potyviruses. On the other hand, the surface-located N terminus is the only large region in the coat protein that is unique to a particular *Potyvirus*, and most virus-specific antibodies should react with this region. This fits with the amino acid sequence data, which showed that the N-terminal region is the most variable in *Potyvirus* coat proteins (Fig. 13.9). It has been known for some time that potyviruses become partly degraded on storage. The use of partially degraded virus as an immunogen or in antigenic analyses may account for many of the contradictory reports in the literature concerning serological relationships among the potyviruses (Shukla *et al.*, 1988b). (See also Fig. 13.13.)

Shukla *et al.* (1989a) developed the following simple procedure to remove cross-reacting group-specific antibodies: The virus-specific N-terminal region of the coat protein of one *Potyvirus* was removed using lysylendopeptidase. The truncated coat protein was then coupled to cyanogen-bromide-activated Sepharose. By passing antisera to different potyviruses through such a column, the cross-reacting antibodies were bound. Antibodies that did not bind reacted only with the homologous virus and its strains, as judged by electroblot immunoassays.

The practical utility of this procedure has been demonstrated for a group of 17 *Potyvirus* isolates infecting maize, sorghum, and sugarcane in Australia and the United States whose taxonomy was in a confused state (Shukla *et al.*, 1989b). The results demonstrated that the 17 strains belong to four distinct potyviruses, for which the names Johnsongrass mosaic virus, maize dwarf mosaic virus, sugarcane mosaic virus, and sorghum mosaic virus were proposed.

Electroblot immunoassays using native and truncated coat proteins (minus the N terminus) can be used to screen MAb's to determine whether they are group specific or virus specific. This procedure was used to distinguish MAb's that were virus specific from those that reacted with 2 or more, and sometimes all 15, of the potyviruses tested (Hewish and Shukla, reported in Shukla and Ward, 1989b).

5. Production of Antibodies against Defined Antigenic Determinants

Geysen *et al.* (1984) described a procedure for the rapid concurrent synthesis of hundreds of peptides on solid supports. These had sufficient purity to be used in ELISA tests. Using sets of such peptides and antisera against a virus, immunologically important amino acid sequences on the virus could be closely defined. This procedure has been successfully applied to the analysis of both polyclonal sera and MAb's raised against potyviruses (Shukla *et al.*, 1989c). This work opens up the possibility of using synthetic peptides corresponding to defined antigenic sites as immunogens to generate group-specific, virus-specific, and perhaps some strain-specific serological probes. Intrinsic and extrinisc factors affecting antigenic sites in relation to the prediction of important amino acid sequences are discussed in Berzofsky (1985).

6. Antibodies against Nonstructural Proteins

Chang *et al.* (1988a) used the serological reactions of nuclear inclusion proteins to study relationships between a set of potyviruses and *Potyvirus* strains.

7. Other Uses of Strain-Specific Antisera

Besides the use of serological methods for establishing relationships between plant viruses, strain-specific antisera provide very useful reagents for various kinds of experiments.

For example, antisera specific for ToMV strains have been used to monitor the effectiveness of the protection given by infection of tomatoes with mild strains of ToMV to superinfection with wild strains (Cassells and Herrick, 1977a), and to study the mechanism of cross-protection (Barker and Harrison, 1978).

Strain-specific antisera were used to show that when tobacco leaf protoplasts were doubly infected with two TMV strains, some progeny rods contained a mixture of both coat proteins (Otsuki and Takebe, 1976b). Antibodies specific for TMV strains were used to study the conditions under which phenotypically mixed rods of TMV could be formed *in vivo* and *in vitro* (Atabekova *et al.*, 1975; Taliansky *et al.*, 1977). Purcifull *et al.* (1973) used strain-specific antisera for several potyviruses to show that the protein found in the inclusion bodies induced by each strain (Fig. 11.23) was distinct, unrelated to the viral coat protein, and independent of host species in which the virus was grown. The site of initiation and direction of TMV assembly were elegantly confirmed by Otsuki *et al.* (1977) using strain-specific antibody.

C. Biological Criteria

1. Symptoms

Macroscopic Symptoms

As noted earlier, symptom differences are of prime importance in the recognition of mutant strains. However, the extent of differences in disease symptoms may be a quite unreliable measure of the degree of relatedness between different members of a group of strains.

Symptoms produced by different virus strains in the same species and variety of host plant may range from the symptomless "carrier" state, through mosaic diseases of varying degrees of severity, to lethal necrotic disease. Figure 13.11 illustrates the range of systemic symptom types produced by four strains of TSV in tobacco. The strains are sufficiently closely related that experimental reassortment experiments are possible between them.

The diseases produced by a given set of strains in one host plant may not be correlated at all with the kinds of disease produced in another host species. Most viruses, including many of widespread occurrence such as TMV, PVX, PVY, AMV, and CMV, occur as numerous strains in nature. Many "new" viruses have been described primarily on the basis of symptoms and other biological properties, which have turned out later to be a strain of one of these commonly occurring viruses. Some viruses appear to have given rise to relatively few strains as judged by symptoms, for example, PLRV in potato varieties.

A set of defined cultivars that give differential local lesion responses may provide a particularly useful and rapid method for delineating strains among field isolates of a virus. However, the important influence of environmental conditions on local lesion responses must be controlled.

A virus causing severe disease is often said to be more "virulent" than one causing mild disease. From what has been said in other sections, it should be

Figure 13.11 Control of disease expression by the viral genome. Variation in chronic disease symptom type caused by four TSV isolates in tobacco (A) The "Standard" North American strain. Tobaccos became more or less symptomless. (B) A strain causing toothed margins on the leaves. (C) A strain in which tobaccos continue to show mosaic and necrotic symptoms. (D) A strain causing severe chronic stunting. These symptom types can be artificially reassorted by making crosses between top, middle, and bottom components of the various strains (see Fulton, 1975). (Courtesy of R. W. Fulton.)

apparent that the description can only be applied to a given strain of the virus inoculated into a particular variety of host plant in a specific manner and growing under particular environmental conditions.

A named variety of host plant, especially a long-established one, may come to vary considerably in its reaction to a given strain of virus, due, for example, to the fact that seed merchants in different localities may make different selections for propagation. This may add a further complication to the identification of strains by means of symptoms produced on named cultivars. Nevertheless, a systematic study of symptoms produced on several host species or varieties under standard conditions may help considerably to delineate strains among large numbers of field isolates of a virus.

Cytological Effects

The cytological changes induced by different strains of a virus are often readily distinguished. Differences are of three kinds: (i) in the effects on cell organelles; (ii) in the virus-induced structures within the cell; or (iii) in the distribution or aggregation state of virus particles within the cell. Such differences may be of increasing importance in the delineation of viruses and virus strains. However, other factors may cause variation in the extent of differences between strains. For example, various strains in the stock culture of TYMV have markedly different effects on chloroplasts in cells of systemically infected leaves (see Fig. 11.20), but these differences may be much less marked or nonexistent in the infected cells of local lesions.

Different strains of TuMV show differences in the morphology of their cylindrical inclusions (McDonald and Hiebert, 1975).

Ultrastructural changes in both nucleus and cytoplasm of oat cells infected with BYDV strains differed between strains that were specific for a particular aphid vector and those that were not (Gill and Chong, 1979).

Different strains of AMV may differ markedly in the way in which virus particles form aggregates within infected cells (e.g., Wilcoxson *et al.*, 1975). The characteristic viroplasms found in cells infected with caulimoviruses (see Fig. 8.3) may vary with different strains (Givord *et al.*, 1984; Stratford *et al.*, 1988). The variation may be associated with differences in gene II. Mixed infections with two variants of BYDV in oats gave rise to altered patterns of effects in vascular tissue, including a predisposition for the xylem to become infected (Gill and Chong, 1981).

2. Host Range and Host Plant Genotype

Host ranges of viruses generally are discussed in Chapter 10, Section V. Many strains of a virus may have very similar host ranges, but others may differ considerably.

Similar responses of a set of host plant genotypes to two viruses may provide good evidence that they are related strains (e.g., Schroeder and Provvidenti, 1971). On the other hand, a loss in ability to infect a particular host may be brought about by a single mutation. Dahl and Knight (1963) studies 12 mutants isolated from ToMV that had been treated with nitrous acid. One of these strains had lost the capacity to infect tomato.

Strains of a virus that have different host ranges often produce different disease symptoms on some common host. This is not always so. For example, four strains of TMV that were not clearly distinguishable by symptoms on *N. tabacum* or on common varieties of *Lycopersicon esculentum* could be differentiated by their host ranges on a set of *Lycopersicon* hosts, including two varieties of *L. esculentum* and three selections of *L. peruvianum* (McRitchie and Alexander, 1963). Strains of PVX have been grouped according to their reactions to a range of host plant genotypes (Cockerham, 1970).

3. Methods of Transmission

Different arthropod vector species or different races of a single species may differ in their transmission of various strains of the same virus. Differences may be of the following kinds.

1. In the percentage of successful transmissions, for example, maize dwarf mosaic *Potyvirus* strains by aphid species (Louie and Knoke, 1975).
2. In minimum acquisition time by the vector, for example, maize dwarf mosaic virus in aphid vectors (Thongmeearkom *et al.*, 1976).
3. In the length of the latent period, for example, strains of PEMV in its aphid vector (Bath and Tsai, 1969).

4. In the time the vectors remain infective (e.g., Thongmeearkom *et al.,* 1976).

5. Some strains may not be transmitted at all by particular vectors, for example, strains of PYDV and leafhopper species (Black, 1941).

Patterns of transmissibility by three aphid species have allowed large numbers of field variants of BYDV found in North America to be placed into five main groups (Rochow, 1979). The quite stable groupings have facilitated studies on the distribution of virus variants both geographically and in successive seasons.

If one strain of a virus is transmissible by mechanical means all others usually are too. However, there are reports of marked variation in mechanical trans-missibility depending on both host clone and virus strain, for example, AMV in alfalfa (Frosheiser, 1969). Defective strains may occur in which the RNA is not coated or is incompletely coated with protein. Such strains will not be mechanically transmissible except under conditions where they are protected from attack by nucleases.

4. Cross-Protection

The mechanism of cross-protection is discussed in Section IV,A.

It was shown by McKinney (1929) that tobacco plants infected with a green mosaic virus (TMV) developed no further symptoms when inoculated with a yellow mosaic virus.

Salaman (1933) found that tobaccos inoculated with a mild strain of PVX were immune from subsequent inoculation with severe strains of the virus, even if inocu-lated after only 5 days. They were not immune to infection with the unrelated viruses, TMV and PVY. This phenomenon, which has been variously called cross-protection, antagonism, or interference, was soon found to occur very commonly among related virus strains. It is most readily demonstrated when the first strain inoculated causes a fairly mild systemic disease and the second strain causes nec-rotic local lesions. Development of such lesions can be readily observed and a quantitative assessment can be made. Interference between related strains can also be demonstrated by mixing the two viruses in the same inoculum and inoculating to a host that gives distinctive lesions for one or both of the two viruses or strains. Interference by type TMV with the formation of yellow local lesions by another strain is shown in Fig. 13.1.

For a time, cross-protection tests were given considerable weight in establish-ing whether two virus isolates were related strains or not, but subsequent develop-ments have indicated the need for caution. Among a group of strains that on other grounds are undoubtedly related, some may give complete cross-protection, while with other combinations protection may be only partial. This is illustrated in Fig. 13.12 for strains of PVX in *Datura.*

Some virus strains do not appear to cross-protect at all. Thus, none of the strains of BCTV protect against one another in water pimpernel *Samolus parviflorus* (Raf) (Bennett, 1955).

Within a set of isolates that are undoubtedly related strains all possibilities may exist—reciprocal cross-protection of varying degrees of completeness, unilateral cross-protection, and no cross-protection as was found for strains of TSV in tobacco (Fulton, 1978). The other factor that may make cross-protection tests ambiguous is that there can be quite strong interference between some unrelated viruses (Bos, 1970).

Figure 13.12   Cross-protection by strains of PVX in *Datura tatula*. (A) Healthy leaf inoculated with a strain giving necrotic local lesions. (B) Leaf previously systemically infected with a very mild strain of the virus, and showing complete protection against inoculation with the necrotic strain. (C) Leaf previously systemically infected with a mottling strain, and showing only partial protection. (From Matthews, 1949b.)

Most experiments on cross-protection have been carried out using mechanical transmission. Cross-protection may also occur in the plant, with viruses transmitted in a persistent manner by insect vectors. Thus, Harrison (1958) found that infection with a mild strain of PLRV protected plants against infection with a severe strain introduced by the aphid vector *Myzus persicae* (Sulz). Cross-protection also occurs in viroids (Chapter 9, Section I,C,6).

5. Productivity

Different strains of a virus may vary widely in the amount of virus produced in a given host under standard conditions. For example, the common strain of TMV was the most productive, and other naturally occurring strains varied over a range down to about one-tenth that of common TMV when productivity was measured as the number of local lesions produced from inocula made from extracts of single local lesions produced in *N. tabacum* cultivar Xanthi nc (Veldee and Fraenkel-Conrat, 1962).

Chemically induced mutants also varied widely in productivity, and all were less productive than common TMV. Some of these strains caused severe symptoms in certain hosts, but there was no correlation between severity of disease and productivity. Productivity appeared to be a genetically stable character since it remained fairly constant for a given mutant when tested after successive transfers. Chemical mutation quite frequently increased the severity of disease produced, but rarely if ever increased the productivity. From a type culture of TMV, B. Kassanis (personal communication) isolated strains causing slowly spreading bright yellow local lesions, usually without systemic spread, in White Burley tobacco (Fig. 13.1). Virus content of these yellow lesions was extremely low. Such strains are difficult to maintain in the laboratory and would never survive in nature.

6. Specific Infectivity

Bawden and Pirie (1956) showed that infectivity per unit weight of purified type TMV was greater than that of a *Datura* strain when they were tested in *N. glutinosa*. There is some evidence suggesting that the protein coat of a virus may be involved in differences in specific infectivity at least between different viruses. Thus Fraenkel-Conrat and Singer (1957) found that the ribgrass mosaic *Tobamovirus* (RMV) had only about 5% of the specific infectivity of common TMV. However, when RMV RNA was reconstituted with common TMV protein, the specific infectivity was about four times higher than the RMV preparation that provided the RNA. Reconstituted TMV usually has a lower specific infectivity than the intact virus. The reason for the increase when the RMV RNA was coated with type TMV protein is not known, but might be due to the relative ease with which intact RMV and the RMV RNA reconstituted with type protein are uncoated *in vivo*.

7. Proportion of Particle Classes

The proportion of particles with differing sedimentation rates found in purified virus preparations or in crude extracts may vary quite widely with different strains of a virus or members of a virus group. Variation of three kinds can be distinguished:

1. In relative amounts of top component (empty protein shells). For example, among the tymoviruses, the proportion of empty shells to viral nucleoprotein is usually in the range of 1 : 2 to 1 : 5 for TYMV and 10 : 1 to 15 : 1 for okra mosaic virus. Even quite closely related strains may vary in this property, for example, strains of red clover mottle *Comovirus* (Oxelfelt, 1976). For some multipartite viruses the proportion of top component has been shown to depend on a function of one RNA species.
2. The proportion of nucleic acid components encapsidated may vary in different strains of viruses with multipartite genomes, for example, AMV (van Vloten-Doting *et al.,* 1968). Again, nucleoprotein proportions may be under the control of a particular RNA species.
3. Abnormal particle classes may be produced by particular strains. Thus Hull (1970a) described an isolate of AMV producing considerable amounts of particles longer than the B component.

It should be remembered that the proportion of particle classes can be affected by factors other than the strain of virus. These include (i) time after infection; (ii) host species; (iii) environmental conditions; (iv) system of culture, for example, the proportion of TYMV empty protein shells is higher in infected protoplasts than in whole leaf tissue; and (v) isolation procedure.

8. Genome Compatibility

The possibility of carrying out viability tests with mixtures of components from different isolates of viruses with multipartite genomes provides a functional biological test of relationship. Such tests were carried out with TRV strains by Sänger (1969). Only 2 of the 20 combinations he tested gave a functional interaction. Members of the *Nepovirus* group show various degrees of compatibility in genetic reassortment experiments (Randles *et al.,* 1977).

Rao and Francki (1981) found that the RNAs 1, 2, and 3 of three strains of CMV were interchangeable in all combinations. However, with tomato aspermy virus, a distinct virus in the *Cucumovirus* group, only RNA3 could be exchanged with those of the CMV strains. Similarly, only RNA3 could be successfully exchanged between two members of the *Bromovirus* group—BMV and CCMV (Allison *et al.*, 1988).

Genome compatibility can be tested in a more direct fashion when the gene products can be isolated and their function is known. For example, Goldbach and Krijt (1982) showed that the protease coded for by CPMV did not process the primary translation products of other comoviruses. The transcriptase activities found in the particles of two rhabdoviruses (LNYV and BNYV) did not carry out transcription with the heterologous virus (Toriyama and Peters, 1981).

9. Activation of Satellites

Particular isolates of TNV will support the replication of some STNVs but not others (Uyemoto and Gilmer, 1972). Similarly, among the cucumoviruses and the small satellite RNAs found in association with them, some viruses support the replication of particular satellites while others do not (Chapter 9, Section II,B).

In considering use of the various possible criteria for the delineation of virus strains, summarized in Section III, we must bear in mind that, from a strictly genetical point of view, complete nucleotide sequence data would be sufficient to establish relationships between strains. Nevertheless, small changes in nucleotide sequence could have very different phenotypic effects. At one extreme a single-base change in the coat protein gene could give rise to changes in several of the phenotypic properties noted earlier. On the other hand, several base changes might give rise to no phenotypic effects at all. For practical purposes, phenotypic characters such as host range, disease symptoms, and insect vectors must usually be given some weight in delineating and grouping virus strains (see Chapter 17, Section VI).

IV. VIRUS STRAINS IN THE PLANT

In the previous sections we have considered ways of isolating virus strains, the molecular mechanisms by which they originate, and the criteria that can be used for distinguishing them. Here certain activities of strains in the infected plant are discussed.

A. The Mechanism of Cross-Protection

It is generally accepted that protection of a plant by one strain of a virus against infection with a second depends on the presence of the protecting virus in the protected tissue. If for any reason a plant or part of a plant becomes freed of the protecting virus, it is then susceptible to reinfection with the first strain or other strains.

Various experiments with intact plants and protoplasts indicate that some early event is involved in the interference or cross-protection phenomenon. When protoplasts were inoculated with one strain of a virus resistance to superinfection with a related strain increased over a period of hours as the time between first and second inoculation was increased, as occurred with strains of raspberry ringspot *Nepovirus* (Barker and Harrison, 1978). The data of Wu (1964) with strains of TMV in tobacco plants are consistent with the idea that interference occurs at a very early stage of infection and that inhibition of one strain by the other on simultaneous inoculation is not due to inhibition of entry into the cell. Several theories have been put forward to account for the phenomenon of cross-protection.

There is increasing evidence that more than one mechanism must be involved. Since it was shown that transgenic plants expressing a coat protein gene may be resistant to infection with intact virus, there has been greatly renewed interest in the phenomenon, especially regarding the possibility of using appropriate transgenic plants for the control of virus diseases (Chapter 16, Section II,C,d). Theories about the mechanism of cross-protection are of three sorts: general; those depending on activities of the viral coat protein; and those depending on activities of the nucleic acid.

1. General Theories

Competition for Replication Sites

Bawden and Kassanis (1945) suggested that virus-specific multiplication sites might exist in the cell. These sites might be limited in number, and if all were occupied by even a low concentration of virus, an incoming related strain requiring the same site could not begin to multiply. Unrelated viruses would have different sites and would, therefore, be able to multiply. This general idea could account for many features of cross-protection. A specific example of this theory might involve the viral replicases. Certain replicases, if not all, involve a combination of host-coded and viral-coded subunits (Chapter 7). Replication of a second strain might be prevented because all or most of the host component had been preempted by the first strain.

Shortage of Essential Metabolites

In this mechanism, the first strain uses the essential metabolites required by the second strain. There are various reasons for thinking that this is most unlikely. All viruses are made from essentially the same selection of amino acids and nucleotides. Thus sequestration of metabolites should not account for the observed specificity. In well-grown plants the amount of virus produced is probably not limited by the availability of amino acids and nucleotides. For example, mixed infections with unrelated viruses may produce much more virus than single infections (Chapter 12, Section V,D,1). Cross-protection between TSV isolates is not related to the concentration of the protecting virus in the tissue (Fulton, 1978).

The Development of Protective Substances

Another possible mechanism involves the development in virus-infected plants of protective substances akin to the antibodies found in animals. It is true that inhibitory materials (that are not infective virus) have been detected in extracts of infected leaves. However, there is no reason to suppose that these bear any similarity whatever to the immune response in animals, nor, in fact, that they have anything to do with the cross-protection phenomenon in the plant. Quite large molecules can circulate in plants, but there appears to be nothing in plants that is remotely comparable with the complex system of organs and cells involved in antibody production in animals. However, transgenic plants expressing appropriate mammalian genes for antibody subunits can produce functional antibodies (Chapter 16, Section II,C,d).

2. Theories Depending on Activities of the Coat Protein

"Adsorptive" Properties of Infected Cells

Kavanau (1949) suggested that aggregates of virus in cells already infected with a virus have specific "adsorptive" properties. An incoming virus particle of the same or of a sufficiently closely related strain becomes adsorbed in one of the aggregates already present. This theory has some appeal but it requires several assumptions. The incoming virus particle presumably must remain intact and be able to circulate within the cell until it meets an aggregate of virus already present. One could easily imagine this for a cell infected with TMV and containing regular aggregates of rods. The fact that cucumber viruses 3 and 4, which have a significantly different particle morphology, do not cross-protect with TMV could readily be explained on this theory.

Encapsidation of the Superinfecting RNA

The incoming RNA might become coated with the viral subunits already present in the cell before it could begin replication. De Zoeten and Fulton (1975) adduced some arguments in favor of this view.

Viral Coat Protein outside the Cytoplasm

De Zoeten and Gaard (1984) suggested that coat protein in the plasmalemma, the apoplast, and/or the cell wall is involved in cross-protection. They suggested that interaction between the superinfecting virus and the extracellular coat protein of the first strain prevents entry or disrupts the superinfecting virus particle.

Prevention of Initial Partial Uncoating

Wilson and Watkins (1985) suggest that an infecting *Tobamovirus* particle loses some protein subunits from the 5′ end, allowing translation to begin on the exposed RNA (Chapter 7, Section III,A,7). In their *in vitro* tests, adding large amounts of purified capsid protein prevented uncoating of the 5′ end of intact virus. *In vivo* the presence of coat protein of the first infecting strain might prevent partial uncoating of the superinfecting strain. This idea would be unlikely to apply where uncoating of infectious RNA is a rapid all-or-nothing event, as it is for virus like TYMV (Chapter 7, Section IV,A,7).

Inhibition of Viral RNA Synthesis

Horikoshi *et al.* (1987) found that the coat protein of BMV inhibited RNA synthesis *in vitro* by the replicase extracted from BMV-infected barley leaves. The BMV coat protein interacted specifically with BMV RNA, since under the same conditions TMV coat protein and bovine serum albumin had no effect. They suggested that this specific binding might be an important mechanism in cross-protection.

Evidence Favoring the Involvement of Coat Protein

1. The results of experiments using transgenic plants discussed in the following Section 4 strongly favor a role for coat protein.
2. The experiments of Sherwood and Fulton (1982) and Zinnen and Fulton (1986) with strains of TMV and two tobamoviruses supported a role for coat protein in cross-protection that involved prevention of uncoating by the challenge virus.

Evidence for Factors Other Than
Coat Protein Being Involved

1. Cross-protection can occur between viroids (Niblett *et al.*, 1978).
2. A TMV mutant with a protein defective for assembly could cross-protect (Zaitlin, 1976). The interpretation of this result has been questioned by Sherwood (1987).
3. The assortment experiments with multiparticle viruses have indicated that the ability of a strain to cross-protect can map on a genome segment other than that coding for the coat protein (Fulton, 1980).
4. A mutant of TMV called DT-IG produces no coat protein that can be detected by highly sensitive ELISA tests. However, provided highly infectious inoculum of this strain is used it protects against superinfection of inoculated leaves of Samsun tobacco with the U1 strain of TMV (Gerber and Sarkar, 1989). Northern blot tests showed that coat protein mRNA was present in leaves inoculated with DT-IG. The mRNA must therefore be nonfunctional (S. Sarkar, personal communication). Cross-protection of leaves inoculated with DT-IG occurs both with intact virus and with RNA as inoculum. This contrasts with the situation for transgenic plants (see the following Section 4).
5. Plants transgenic for the 54K putative protein of TMV are resistant to infection with TMV (see Chapter 16, Section II,C,1).

3. Theories Depending on Activities of the Viral Nucleic Acid

Gene Recombination

In the gene recombination theory, the incoming strain gets lost, in effect, through genetic recombination with the strain already present. Best (1954) put forward this idea to account for the partial protection and the appearance of new strains found with TSWV. This mechanism does not account for all the known facts of the cross-protection phenomenon, although it might be a significant factor with TSWV and some other viruses.

Negative Strand Capture

Zaitlin (1976) and Palukaitis and Zaitlin (1984b) suggested that annealing of the first produced minus strand copies of the superinfecting virus with plus strands of the virus already present in the cell would block further replication of the incoming virus. It is known that fully base-paired dsRNAs of ssRNA viruses are noninfectious. This idea would explain how RNA viruses with defective coat proteins (or none) and viroids can prevent superinfection. It receives support from the fact that antisense RNAs produced in transgenic plants can block the mRNAs for which they have the complementary sequences (e.g., Rothstein *et al.*, 1987). However, it must be remembered that most of the plus strand RNA of the protecting strain will be in the form of virus particles. One may also ask, if negative strand capture is effective, why are the negative strands of the first infecting virus not rendered ineffective by the virus' own plus strands. Perhaps bound replicase molecules prevent this. However, RNA–RNA interaction cannot explain why transgenic plants expressing a coat protein gene can show resistance to superinfection with the same virus or related strains (see the following Section 4).

Translation Competition

Two strains of TMV that cross-protect in the plant were shown to exhibit translational competition in the reticulocyte lysate and wheat germ systems if the second RNA was added after translation of the first had commenced. The phenomenon was virus specific. AMV RNA did not compete with TMV RNA (Salomon and Bar-Joseph, 1982). In principle, translation competition could occur in the plant, but again this mechanism would not explain the resistance shown by transgenic plants expressing coat protein.

4. Transgenic Plants

Resistance in Plants Expressing
a Viral Coat Protein Gene

Resistance to viruses in transgenic plants has been reviewed by Beachy (1988). Beachy *et al.* (1985) and Bevan *et al.* (1985) first reported the expression of the coat protein of TMV in tobacco plants into which a cDNA containing the coat protein gene had been incorporated. Expression of the coat protein gene of AMV has been reported by several groups (e.g., Loesch-Fries *et al.*, 1987), for TRV by van Dun *et al.* (1987) and for PVX by Hemenway *et al.* (1988). Some laboratories used the 35 S CaMV promoter and others the 19 S to obtain expression. The 35 S was considerably more effective.

Powell Abel *et al.* (1986) showed that transgenic plants expressing TMV coat protein either escaped infection following inoculation or developed systemic disease symptoms significantly later than plants not expressing the gene. Plants that showed no systemic disease did not accumulate TMV in uninoculated leaves (Nelson *et al.*, 1987). Transgenic plants produced only 10–20% as many local lesions as controls when inoculated with a strain of TMV causing local lesions. The idea that transgenic plants resist initial infection rather than subsequent replication was suggested by results obtained using transgenic Xanthi-nc tobacco plants, in which fewer local lesions were produced than on control plants. However, the lesions that did develop were just as big as on control leaves, indicating that once infection was initiated there was no further block in the infection cycle. Plants that were expressing AMV coat protein were also resistant to infection with AMV (e.g., van Dun *et al.*, 1987; Loesch-Fries *et al.*, 1987; Tumer *et al.*, 1987).

Transgenic tobacco plants expressing the PVX coat protein gene were significantly protected against PVX infection as shown by a reduced number of local lesions on inoculated leaves, delayed or no systemic symptom development, and a reduction in virus accumulation in both inoculated and systemically infected leaves. The higher the level of coat protein expression, the higher was the level of protection (Hemenway *et al.*, 1988). Plants expressing an antisense coat protein transcript were resistant to infection with PVX, but only with low concentrations of virus in the inoculum.

Possible Mechanisms for Resistance
in Transgenic Plants

In transgenic plants it would be possible, in principle, for the resistance to infection to be due to the coat protein mRNAs transcribed from the cDNA, to the coat protein itself, or to both of these molecules. To test these possibilities, Register *et al.* (1988) constructed a series of cDNAs generating mRNA sequences that would produce no coat protein, or mRNA that lacked the replicase recognition site but that would produce coat protein. These experiments conclusively implicated the coat protein rather than the mRNA in causing resistance to superinfection. Similar results were obtained with a mutated AMV coat protein (van Dun *et al.*, 1988b). Further experiments with TMV confirmed the earlier results and showed that the 3' tRNA-like sequence was not necessary to generate resistance (Powell *et al.*, 1990).

Register *et al.* (1988) and Register and Beachy (1988) showed that protoplasts made from transgenic plants expressing coat protein were specifically protected against infection with TMV. Register and Beachy (1989) showed that when tobacco protoplasts took up coat protein, they were transiently protected from infection with TMV. Thus coat protein outside the cell is probably not involved in coat protein-mediated protection. Pathogenesis-related proteins (Chapter 12, Section III) also do not appear to be involved in this resistance (Carr *et al.*, 1989).

A careful analysis of virus spread in single lesions in transgenic and control tobacco plants showed that when local infection did take place in tissues expressing the coat protein, there was no inhibition in subsequent cell-to-cell movement (Register *et al.*, 1988). On the other hand, when a leaf-bearing stem segment from a transgenic plant was grafted between lower and upper sections of a nontransgenic plant, systemic movement of TMV into the leaves above the graft was inhibited. Transgenic tobacco plants expressing the 30K movement protein of TMV were not protected against infection or disease development (Deom *et al.*, 1987).

When TMV is treated at pH 8.0, translatability *in vitro* is greatly enhanced (Chapter 7, Section III,A,7). Register *et al.* (1988) and Register and Beachy (1988) found that TMV treated in this manner was able to overcome the resistance of transgenic plants in the same way as RNA. This result certainly supports an early event as being important in the resistance of transgenic plants. Osbourn *et al.* (1989b) showed that TMV-like pseudoparticles containing the GUS reporter gene expressed this gene 100 times less efficiently in protoplasts from coat protein-transgenic tobacco plants than in control protoplasts. The data suggested that about 97% of the GUS pseudoparticles remained uncoated. However, other experiments indicated that inhibition of virus disassembly is insufficient to account entirely for coat protein-mediated resistance, and that some later event or events in virus replication must be involved.

The TMV mutant DT-IG produces no coat protein and does not normally move systemically. When this mutant is inoculated into transgenic tobaccos expressing the coat protein gene, rod-shaped particles were found in the systemically invaded leaves (J. Osbourn, S. Sarkar, and T. M. A. Wilson, personal communication). However, the viral rods isolated from the systemic infection were unable to infect fresh transgenic plants, supporting the view that some early uncoating event is involved in resistance.

Osbourn *et al.* (1989a) tested the possibility that coat protein expressed in transgenic plants inhibits virus replication by recoating uncoated viral RNA from a challenge inoculum. They produced double transformed tobacco plants that were expressing TMV coat protein and a reporter gene (CAT) whose transcripts contained a copy of the TMV origin of assembly sequence. No rods could be detected in cell extracts of these plants by electron microscopy, and there was no significant reduction in CAT activity. However, transformed plants retained their ability to resist infection by TMV. Thus it seems unlikely that reencapsidation of uncoated RNA of challenge virus by endogenous coat protein is involved in the resistance of transgenic plants expressing coat proteins.

*Relationship between Natural Cross-Protection and
Resistance in Transgenic Plants*

The question arises as to whether the resistance generated in transgenic plants is in fact related to the natural phenomenon of cross-protection. There are several similarities that would support the idea: (i) In both situations the degree of resistance depends on the inoculum concentration, with high concentrations reducing the observed resistance. (ii) Both are effective against closely related strains of a virus, less against distantly related strains, and not against unrelated viruses. (iii) In some circumstances cross-protection can be substantially overcome when RNA is used as inoculum rather than whole virus (Sherwood and Fulton, 1982; Dodds *et al.*, 1985). Similarly, the resistance of transgenic plants expressing the coat protein is substantially but not completely overcome when RNA is used as inoculum (TMV, Nelson *et al.*, 1987; AMV, van Dun *et al.*, 1987; Loesch Fries *et al.*, 1987; TRV, Angenent *et al.*, 1990). However, plants expressing high levels of PVX coat protein were resistant to infection with PVX RNA (Hemenway *et al.*, 1988). (iv) In classic cross-protection experiments, no cross-protection was observed between two rather similar viruses, AMV and TSV. In experiments with transgenic plants expressing their viral coat proteins, high resistance to infection was observed against the homologous virus and none against the heterologous virus (van Dun *et al.*, 1988b).

On the other hand, there appear to be some differences between natural cross-protection and coat protein-induced resistance. When cross-protection between related strains of a virus is incomplete, the local lesions produced may be much smaller than in control leaves (illustrated for PVX in Fig. 13.12). This indicates reduced movement and/or replication of the superinfecting strain. Local lesions that formed in transgenic tobacco plants expressing the PVX coat were smaller than those of the controls (Hemenway *et al.*, 1988), in line with the result for PVX shown in Fig. 13.12. However, as noted earlier in this section, the reduced number of local lesions that do form on transgenic plants infected with TMV became as large as controls, indicating no block in replication or local movement once infection was successful.

The idea of negative strand capture (Section IV,A,3) as an effective mechanism for cross-protection does not receive support from the fact the RNA as inoculum can do away with much of the resistance in both infected and transgenic plants.

5. Summary and Discussion

It is very probable that the results of experiments with transgenic plants are relevant to our understanding of the mechanism of natural cross-protection. However, clear interpretations are difficult at the present stage, probably for a number of reasons. In natural cross-protection (i) there may be more than one mechanism involved as there is for transgenic plant resistance (Chapter 16, Section II,C,1); (ii) more than one viral gene may be involved; and (iii) the relative importance of various mechanisms may be different with different viruses and host plants. With respect to transgenic plants, different viruses and hosts have been used and different levels of gene expression have been obtained.

Nevertheless, some features regarding transgenic plants are beginning to emerge (i) Plants expressing a viral coat protein gene show resistance to infection with the virus and related strains. (ii) Degree of resistance is related to the extent of expression of the coat protein gene. (iii) The phenomenon is probably a general one since it has been shown for a range of different viruses, including TMV, PVX, AMV, and CMV. However, resistance may not be generated for viruses, such as members of the *Tymovirus* group, where the expressed coat protein might rapidly form empty protein shells as it does in natural infections. (iv) In transgenic plants it is the coat protein itself rather than the mRNA that is responsible for most, or all of the resistance. (v) An early event within the inoculated cell following inoculation is involved in the resistance. This may involve prevention of partial or complete uncoating of the incoming viral genome, at least for rod-shaped viruses. However, later events in replication may also be involved.

The experiments of Golernboski *et al.* (1990) (Chapter 16, Section II,C,1) open up further possibilities concerning the mechanism of natural cross-protection. More than one mechanism may operate for a particular virus; and the mRNA rather than a protein product may sometimes be involved.

The process of virus replication is not usually an uncontrolled one. This aspect of virus replication has not been given much consideration in relation to understanding cross-protection. The amount of virus per cell reaches an upper limit, but the factors controlling this limit are unknown. It may be that production of virus by a challenge strain is confined to a fraction of the total possible for the virus in question. The fraction would become smaller and smaller as replication of the first strain progressed. Current micro methods for the estimation of viral components may make it possible to test this idea.

B. Selective Survival in Specific Hosts

When a virus culture that has been maintained in an apparently stable state in one host species is transferred to another species and then inoculated back to the original host, it is sometimes found that the dominant strain in the culture has been changed. Carsner (1925) showed that a culture of BCTV could be altered by transmission to *Chenopodium murale*. When the virus was returned to beet from this plant it produced only mild symptoms. Lackey (1932) found that the change was reversible and that the virus culture could be retuned to its original condition, with respect to symptoms in beet, by passage through *Stellaria media* (L.). According to Salaman (1938), a strain of PVX, which caused ring spot symptoms, when inoculated into seedling beet, produced small necrotic rings only. When virus from the local lesions was returned to tobacco, only a faint mottle developed. A reverse situation was described by Matthews (1949c). When PVX cultures giving a mild mottle in tobacco were passed through *Cyphomandra betacea* and reinoculation was made to tobacco, only local and systemic ring spot-type disease was produced. *Cyphomandra* was the only one of 19 solanaceous species tested to cause this change. The mild cultures that were used contained a small proportion of ring spot strains that had presumably arisen by mutation. It was suggested that these ring spot strains multiplied more effectively in *Cyphomandra* than did the mild strains and that they thus came to dominate in this host. Ring spot strains alone reached several times the virus concentration in *Cyphomandra* compared to tobacco. Mild strains were four to eight times as concentrated in tobacco as the ring spot strains in this host.

Johnson (1947) found that passage of the ordinary severe-type TMV through sea holly (*Eryngium aquaticum* L.) resulted in mild symptoms on tobacco. He showed that severe strains moved more slowly in sea holly, thus accounting for the "filtering" action of this host.

Inoculation of a culture of CMV that did not cause systemic infection in Blackeye cowpea to the variety Catjang led to the appearance of some abnormally large local lesions. Inoculation from these large local lesions to the variety Blackeye was followed by systemic necrotic disease (Lakshman *et al.*, 1985).

These various selection phenomena may well involve differences in the cell-to-cell movement protein coded for by strains of the virus. For example, the mild cultures of PVX in tobacco just described may have been giving rise continually to mutants that could not compete in tobacco with the parent strain on the basis of their cell-to-cell movement protein. However, the movement protein of some of these mutants may have been better adapted to systemic movement in *Cyphomandra* than the parent strain. The back mutation rate must be very low since the strain or strains selected in *Cyphomandra* appeared quite stable when cultured in tobacco. Now that the nucleotide sequences of the genomic RNAs are known it should be possible to establish whether strain selection by particular hosts involves mutations in the movement protein. Satellite RNAs such as those associated with CMV may undergo differential replication in particular hosts. This may provide another basis for variation in symptoms following culture of a virus in a given species (Waterworth *et al.*, 1978). For example, CARNA5 may exist as a series of closely related sequences (Richards *et al.*, 1978), which could provide for a rapid response to changed conditions for replication.

C. Double Infections *in Vivo*

The existence of phenotypic mixing (Chapter 6, Section VIII,A) suggests that two virus strains can replicate together in the same cell. The following evidence confirms this view:

1. Various observations have demonstrated the presence of two strains of a virus in the same cell in intact plants. Thus Hull and Plaskitt (1970) could recognize characteristic aggregates of particles of two AMV isolates in the same cell.
2. When the *ts* mutant of TMV, Ni118, was inoculated in a mixture to tobacco with common TMV and grown at 35°C, some mutant RNA was found to be coated with common strain protein (Takebe, 1977).
3. Protoplasts inoculated with TMV show a multitarget response to inactivation by UV light (Takebe, 1977), indicating that more than one particle can initiate infection in the same cell.

Thus while proof is lacking, the weight of evidence suggests that two strains can infect and replicate simultaneously in the same cell in the intact plant.

D. Selective Multiplication under Different Environmental Conditions

Temperature is probably the environmental factor that most commonly influences the survival or predominance of strains occurring in nature. Experimental use of temperature to isolate strains is noted in Sections I,C and E. In parts of the world with hot climates, strains of viruses surviving at these temperatures are selected naturally (Chapter 15, Section I,A,4). Some understanding of the basis for the effects of temperature on naturally occurring strains can be gained from the results of the studies on artificially induced *ts* mutants. (Section I,E,2).

E. Loss of Infectivity for One Host following Passage through Another

Loss of infectivity for one host may develop following repeated passage through another species. For example, several strains of PVX lost infectivity for potato during continued propagation in tobacco (Matthews, 1949a). No change in symptoms produced in tobacco, *N. glutinosa,* or *Datura tatula* could be observed over the period that the AP strain lost its infectivity for potato. The nature of the change is not understood, but it presumably reflects gradual selection of a strain or strains better adapted to growth in tobacco.

An immediate loss of infectivity for one host following passage through another can be due to a virus inhibitor that is effective only in certain hosts, rather than to any change in the virus itself.

V. CORRELATIONS BETWEEN CRITERIA FOR CHARACTERIZING VIRUSES AND VIRUS STRAINS

In the preceding sections we have surveyed the various criteria that can be used to delineate variation among virus isolates. How can we use these criteria to decide whether a particular isolate is identical with another isolate or a related variant or strain, or whether it is a distinct virus? This is a question of considerable practical importance, because the recognition and identification of virus strains may be most important for effective virus control. In addition, the virological literature is cluttered with inadequate descriptions of virus isolates. These are frequently described as new viruses or new strains, particularly if they are found in a new host or a different country, when adequate study might show they were very probably identical to some virus already described. The definition of a virus species is given in Chapter 17, Section I,D.

A. Criteria for Identity

There is only one criterion that will establish that two virus isolates are identical—the identity of the complete base sequence of their genome nucleic acids. In spite of recent rapid advances in nucleic acid sequencing techniques this is unlikely to become a routine procedure. For most practical purposes, the following criteria would be sufficient to establish provisional identity of two virus isolates: (i) identity of size, shape, and of any substructure of the virus particle as revealed by appropriate electron microscopy; (ii) serological identity in adequate tests; (iii) identical disease symptoms and host ranges for a set of indicator hosts and genotypes; and (iv) identical transmission, especially with respect to any arthropod, nematode, or fungal vectors. The presumption of identity would be greatly reinforced by information on some aspect of nucleic acid sequence, for example, identical sequences in a particular region of the genome or identity as judged by heterogeneity mapping.

B. Strains and Viruses

The broad questions of virus classification are dealt with in Chapter 17. Here we will consider the problems involved in using the various properties outlined earlier in this chapter to define virus strains, to group them, and to decide whether an isolate is a strain or a different virus.

One method is to take a quantitatively determined set of characteristics such as the amino acid composition of the coat protein. Statistical procedures and computer analysis are then used to derive a classification with degrees of relationship indicated. Computer analysis is particularly useful for handling large amounts of numerical data as was used, for example, to derive Fig. 13.13 from amino acid sequences. However, a classification based on computer analysis is no more objective than other ways of making a classification. It will depend on the personal judgments and selections made by the taxonomist providing the data.

In the Adansonian approach *all* the known characters are given weight in determining groupings. This approach has become popular with the widespread availability of computers but there are significant limitations. For example, as noted earlier, many of the fairly easily measured characters of a small virus depend on properties of the coat protein. Thus differences in the coat protein may be given undue weight. Similarly, symptom differences between two strains could be emphasized, merely by recording differences on an extended host range.

On the other hand, the hierarchical system involves making arbitrary decisions about which characters are the most important. There are serious objections to applying such a system, without some modification, especially when we are considering classification within a group of related viruses. The most useful characters will be different within different virus groups. Thus, the coat protein of STNV, being the only gene product of this virus, should be given more substantial weight than would the coat protein of a virus with, say, 10 genes. Similarly, particle morphology may be most useful for those groups such as the Rhabdoviridae that possess a complex structure.

The best course is probably the pragmatic one of considering all known properties within a group of variants and weighting them in a commonsense manner in relation to the overall properties of the group in question.

When strains arise in the stock culture of a virus in the laboratory as they do with such viruses as TMV and TYMV, we can be reasonably sure that they will be closely related to the parent strain—usually arising from a single mutation. Phenotypic differences in most properties will usually be small, but may sometimes be large as with the TMV strains, such as PM1, that produce defective coat proteins and no intact virus.

Virus isolates collected in the field, perhaps from different host species in different countries, may appear related on the basis of some properties and unrelated on others. The only generalization that can be made at present is that closely related strains will differ in only a few properties, while distantly related strains will differ in many. The extent to which different properties show correlations varies widely in the different groups of viruses.

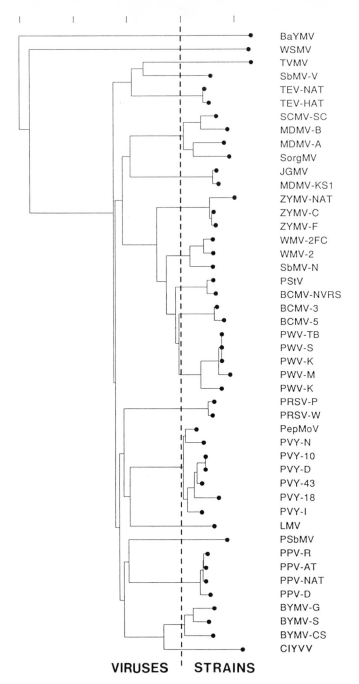

BaYMV
WSMV
TVMV
SbMV-V
TEV-NAT
TEV-HAT
SCMV-SC
MDMV-B
MDMV-A
SorgMV
JGMV
MDMV-KS1
ZYMV-NAT
ZYMV-C
ZYMV-F
WMV-2FC
WMV-2
SbMV-N
PStV
BCMV-NVRS
BCMV-3
BCMV-5
PWV-TB
PWV-S
PWV-K
PWV-M
PWV-K
PRSV-P
PRSV-W
PepMoV
PVY-N
PVY-10
PVY-D
PVY-43
PVY-18
PVY-I
LMV
PSbMV
PPV-R
PPV-AT
PPV-NAT
PPV-D
BYMV-G
BYMV-S
BYMV-CS
CIYVV

VIRUSES   STRAINS

Figure 13.13 Relationships between some potyviruses and their strains. The dendrogram is based on the amino acid sequence of the core of the viral coat proteins, omitting the more variable terminal sequences. The core regions were aligned, and the percentages of different amino acids in each sequence (that is, their distance) were calculated. Gaps were counted as a 21st amino acid. The distances were then converted into the dendrogram using the neighbor joining method of Saitou and Nei (1987). The divisions in the horizontal scale represent 10% differences. The broken vertical line indicates a possible boundary between viral species and strains. Some of the viruses listed do not appear in Appendix 1 (JGMV = Johnson grass mosaic virus; SorgMV = Sorghum mosaic virus; PStV = Peanut stripe virus). The classification suggested here differs from some of the conclusions put forward in the text. The evolutionary distances indicated here fit with known vector relationships. BaYMV is transmitted by fungi, WSMV by mites, while all the others listed have aphid vectors. (Diagram courtesy of A. J. Gibbs.)

C. Correlations for Various Criteria

From a purely genetical point of view the relationships between a set of virus strains can be assessed precisely if we know the differences in nucleotide sequences between their genomes. However, from the virological point of view other factors must be taken into consideration. For example: (i) nucleotide changes that are silent, that is, lead to no change in the structure or function of the virus, are of little interest; (ii) particular gene functions may be of particular ecological and therefore practical significance, for example, mutations in a viral gene that affects insect vector specificity; and (iii) when large numbers of field isolates have to be typed over a short time interval, only rapid diagnostic methods are practicable. The confidence with which particular criteria can be used depends in part on the extent to which they correlate with other criteria. This section gives a brief overview of these problems.

1. Host Responses

Where a group of strains are fairly closely related, host responses may provide the best, or even only practicable, criteria for establishing strain types. For example, Mosch *et al.* (1973) found that 18 isolates of TMV from glasshouse tomato crops could be placed in three groups depending on their pathogenicity for a set of *Lycopersicon esculentum* clones. There were no differences in certain physical properties (buoyant densities and $S_{20,w}$) and only small individual differences in coat protein composition. These did not correlate with the pathogenicity groups.

When a virus of economic importance such as AMV is highly variable, the classification of large numbers of field isolates must usually depend primarily on symptoms and host range on a standard set of indicators (e.g., Crill *et al.*, 1971; Hajimorad and Francki, 1988).

2. Vector Transmission

Among three isolates of BYDV there was a correlation between closeness of serological relationship and transmission by aphid vectors (Aapola and Rochow, 1971). Selected MAb's raised against whitefly-transmitted geminiviruses did not recognize leafhopper-transmitted members of the group (Thomas *et al.*, 1986). Pead and Torrance (1989) found that MAb's could be used to type the three major vector-specific strain groups of BYDV. On the other hand, an isolate of PLRV that was poorly transmitted by aphids was indistinguishable serologically from readily transmitted isolates (Tamada *et al.*, 1984). There was no correlation between serological relatedness and the ability of English populations of *Longidorus attenuatus* to transmit different isolates of TBRV (Brown *et al.*, 1989).

3. Multipartite Genomes

The ability of multiparticle viruses to complement each other successfully provides a powerful functional criterion indicating relationship. However, this property may not correlate closely with the physical properties of the virus particle or other properties of the virus. For example, certain viruses that have been considered as strains of CPMV (Swaans and van Kammen, 1973) did not successfully complement each other in mixed infection experiments (van Kammen, 1968). Successful complementation has been shown to occur not only between already well-recognized strains but also between viruses thought to be distinct members of the same group. Such results further complicate the use of complementation tests as a criterion of relationship. For example, Bancroft (1972) demonstrated successful complementation between BMV and CCMV. These are both in the *Bromovirus* group, but they have almost totally different host ranges and appear unrelated serologically.

At present two groups of viruses are known with a tripartite genome and a separately encapsidated coat protein cistron—the *Ilarvirus* group and the AMV group. With members of both these groups, if the three-genome RNAs are used for infection, the coat protein RNA, or some coat protein itself, is required for infectivity. The coat protein or the coat protein RNA of some ilarviruses, for example, TSV, will activate the RNAs of AMV. The reverse combination is also active. However, mixtures of the three-genome RNAs from the two viruses do not complement each other (Van Vloten-Doting, 1975; Gonsalves and Fulton, 1977), and there is no sequence similarity between the corresponding RNA segments.

Transgenic tobacco plants expressing the TSV coat protein gene were resistant to infection with TSV but susceptible to AMV. They could be infected with AMV RNAs 1, 2, and 3, demonstrating that the endogenously produced TSV coat protein can activate the AMV genome, even though it does not cross-protect against this virus (van Dun *et al.*, 1988b). The coat protein of AMV nucleoproteins is specifically removed by the addition of AMV RNA. Similarly, the nucleoproteins of ilarviruses may lose their protein when free viral RNA is added. There is reciprocity in this reaction between certain ilarviruses and AMV (van Vloten-Doting, 1975; Gonsalves and Fulton, 1977). These results have led some workers to suggest that AMV should be placed in the *Ilarvirus* group.

4. General Nucleotide Sequence Similarities

Using hybridization techniques, there may be complete lack of detectable base sequence homology between viruses that on other grounds, such as morphology of the particle and serology, are certainly related (Zaitlin *et al.*, 1977). At the other extreme, Bol *et al.* (1975) described four strains of AMV with well-characterized differences in biological tests that were virtually indistinguishable in nucleic acid hybridization tests.

Strains of CMV can be divided into two subgroups on the basis of serology, and nucleic acid hybridization, as discussed by Rizzo and Palukaitis (1988). Of 39 strains examined by nucleic acid hybridization, 30 belong in subgroup I and 9 in subgroup II. RNAs belonging to the two subgroups can be reassorted to yield viable recombinants. The RNAs 1 and 2 of representatives of the two groups have been sequenced and compared (Rizzo and Palukaitis, 1988, 1989). Different regions of the RNAs varied in the extent of sequence homology (from 62 to 81%). Strains within the two subgroups cannot be distinguished by the usual nucleic acid hybridization techniques. However, Owen and Palukaitis (1988) used molecular heterogeneity mapping to place 13 of the CMV strains into two groups on the basis of their ability to hybridize to two representative strains. Molecular heterogeneity mapping could distinguish strains within the two groups.

5. 3′ Noncoding Nucleotide Sequences

Another approach for discriminating between distinct potyviruses and strains has been explored by Frenkel *et al.* (1989). They compared the 3′ noncoding nucleotide sequences of 13 potyviruses and found that viruses that were distinct on other grounds had 3' noncoding sequences of different lengths (189–475 nucleotides). The degree of sequence similarity ranged from 39 to 53%. Such values are comparable to that obtained when the 3' untranslated regions from unrelated potyviruses are compared and they are probably in the range expected for chance matching. By contrast, the 3' untranslated regions of sets of viruses recognized on other criteria as related strains were very similar in length and in nucleotide sequence homology (83–99%). WMV-2 and SGMV-N were found to have 78% homology and on this basis were considered to be strains of the same virus.

6. Serological Relationships

Relationships determined by serological methods might, by chance, correlate quite well with any other properties. However, it is reasonable to *expect* that they may show some correlation with those criteria that also depend on some property of the coat protein.

Correlations have been reported between degree of relatedness, measured by cross-protection tests, and serological relatedness (e.g., PVX strains, Matthews, 1949b; BYDV isolates, Aapola and Rochow, 1971). On the other hand, there was no correlation between serological relatedness within a group of TNV isolates and their ability to support the replication of three differing isolates of STNV (Kassanis and Phillips, 1970), nor was there any correlation between serological relatedness and symptoms in tobacco for TRSV (Gooding, 1970).

MAbs raised against strains of PVX reacted in a complex manner with the strains in different groups based on the reaction of host varieties (Torrance *et al.,* 1986a). Nevertheless, a resistance breaking strain could be identified. A series of 10 MAbs raised against PLRV failed to differentiate between strains that caused different symptoms in indicator hosts (Massalski and Harrison, 1987).

Traditionally, viruses and strains within the *Potyvirus* group have been very difficult to delineate. This has been not only because of the large number of viruses involved but because different tests for relationship gave different answers. The work of Shukla and his colleagues discussed in Section III,B,4 has gone some way toward establishing a sound basis for classification of virus isolates belonging to this group. For example, earlier work suggested that cross-protection did not distinguish some isolates that on other criteria such as serology were considered to be separate viruses. Shukla *et al.* (1989b), using antibodies directed against the N-terminal part of the coat protein, showed that potyviruses infecting maize, sorghum, and sugarcane in Australia and the United States comprised four distinct viruses. The earlier cross-protection tests fell neatly into place on this basis (Shukla and Ward, 1989a).

Similarly, the kind of cytoplasmic inclusions found with these isolates supported the idea of four distinct viruses. Thus this set of viruses is beginning to conform with the ICTV definition for a virus species given in Chapter 17, Section I,D, requiring "a set or pattern of correlating stable properties." Difficulties remain, however, because some unexpected serological cross-reactions occur between viruses that on other soundly based criteria are regarded as distant members of the *Potyvirus* group. The antigenic site for these cross-reactions may reside with a few common amino acids close to the N terminus of the coat protein (Shukla and Ward, 1989a,b).

7. Nonstructural Proteins

Yeh and Gonsalves (1984) used antisera raised against the inclusion body proteins of two potyviruses to confirm that they were related strains of one virus rather than two distinct viruses. Thornbury and Pirone (1983) showed that the helper component protein of two different potyviruses were serologically distinct. There was no serological relationship between the 35K protein coded for by AMV and the corresponding proteins of three other viruses with a tripartite genome (van Tol and van Vloten-Doting, 1981).

VI. DISCUSSION AND SUMMARY

The study of variability is one of the most important aspects of plant virology. It is important from the practical point of view because strains vary in the severity of disease they cause in the field, and because strains can mutate to break crop plant resistance to a virus. It is important also for developing our understanding of how viruses have evolved in the past, and how they are evolving at present.

A range of procedures is available for isolating virus variants either from nature or following some form of mutagenesis or other manipulation outside the plant. Mutants of the *ts* type have been particularly useful in studying various aspects of virus structure and replication.

Because of the very high mutation rate it is probable that all cultures of plant viruses consist of a mixture of numerous strains even after single lesion passage. However, a "consensus" genome sequence will usually dominate in the culture and many variants will not be detected.

The molecular mechanisms by which variation within a virus population is produced are like many of those found in cellular organisms, except that for many plant virus groups the material upon which variation operates is RNA rather than DNA. Mechanisms include mutations involving single-nucleotide changes or the addition or deletion of one or a few nucleotides, recombination, deletions or additions of blocks of nucleotide sequences, rearrangement of nucleotide sequences, and reassortment among multipartite genomes.

A range of structural, serological, and biological criteria is available for delineating viruses and virus strains within a group or family of viruses. The kind of criteria to be used will depend on the purpose of the study. If we are studying evolutionary relationships within a virus group or family, or among the variants of a single virus, then the full nucleotide sequences of the viruses concerned will be of prime importance, but a knowledge of the functional products of the viral genome will often be needed as well. If the full nucleotide sequences are known for representative viruses, then other methods, such as various forms of nucleic acid hybridization, can be usefully interpreted for additional viruses and strains. If we are interested in developing methods for reliably and rapidly diagnosing viruses and virus strains from the field, then other methods will be appropriate. Dot blot serological assays using some form of ELISA are emerging as an important type of test. Polyclonal antisera of wide specificity or MAb's of very narrow specificity can be used in such tests as appropriate. Biological criteria such as disease symptoms, host range, methods of transmission, and cross-protection may be important in defining viruses and virus strains.

There is renewed interest in the subject of cross-protection between virus strains, because it has been shown for several viruses that transgenic plants expressing the coat protein gene are resistant to superinfection, and that this phenomenon resembles natural cross-protection in several respects. The mechanism of resistance in transgenic plants is not fully understood but it certainly involves the coat protein in some way. In natural cross-protection it is likely that there is more than one mechanism.

The extent to which virus species have been clearly delineated varies widely among the different groups and families of viruses. There are dangers in formalizing virus species or virus groups before a sufficient number and diversity of strains have been investigated. For example, at a stage when only about seven tymoviruses were known, two subgroups were suggested on the basis of serological relationships and RNA base composition (Gibbs, 1969; Harrison *et al.,* 1971). Since then further tymoviruses have been discovered with intermediate characteristics (Koenig and Givord, 1974). For some groups, such as the potyviruses, "a common set or pattern of correlating stable properties" is emerging that can allow the grouping of virus strains into species with some degree of confidence.

The relative importance, or weight, to be placed on different properties of a virus for purposes of classification remains a difficult problem. An adequate understanding of the significance to be placed on the various properties may come only when we have a detailed knowledge of the structure of the viral genome, the polypeptides it codes for and their functions, and the regulatory or other roles of any translated or untranslated regions in the genome.

Even with such knowledge difficulties will remain. For example:

1. Disease induction, which is a complex process, has been shown for some viruses to depend on the functions of two or more viral genes.
2. Various possible mechanisms are now known whereby a single mutation could have effects on two or more functions.
3. A single gene product may have two or more functions, differing in importance, for the virus life cycle.

Thus from a practical point of view it may be an oversimplification to establish relationships between viruses and strains within a family or group solely on the basis of nucleotide sequences.

14

Relationships between Plant Viruses and Invertebrates

The transmission of viruses from plant to plant by invertebrate animals is of considerable interest from two points of view. First, such vectors provide the main method of spread in the field for many viruses that cause severe economic loss. Second, there is considerable general biological interest in the relationships between vectors and viruses, especially as some viruses have been shown to multiply in the vector. Such viruses can be regarded as both plant and animal viruses. Even with those that do not multiply in the animal vector, the relationship is usually more than just a simple one involving passive transport of virus on some external surface of the animal. Transmission by invertebrate vectors is usually a complex phenomenon involving the virus, the vector, the host plant, and the environmental conditions. An enormous amount of work has been done on these problems. In this chapter I shall consider the groups involved as virus vectors and outline the kinds of relationship that have been found to exist between virus and vector. Vectors are considered in relation to the ecology of viruses in Chapter 15 and in relation to disease control in Chapter 16.

I. VECTOR GROUPS IN THE INVERTEBRATES

Of some 22 phyla in the invertebrates, only 2 have many members that feed on living green land plants. These are the Nematoda and the Arthropoda. Both of these phyla contain vectors of plant viruses. Two additional phyla, the Annelida and Mollusca, have a few plant feeders, and it may be that these contain potential vectors of a strictly mechanical sort.

A. Nematoda

There are 10 orders in the Nematoda (Goodey, 1963). Most of the nematodes parasitic in living green plants belong to the Tylenchida, but none from this group has yet been found to be a virus vector. Vectors known so far are confined to the Dorylaimida group containing only a few plant parasites.

B. Arthropoda

I have followed the taxonomy used by Borror *et al.* (1981). Of the 14 classes in this phylum, 2 contain members feeding on living green land plants—the Insecta (Hexapoda) and the Arachnida. Both these groups contain virus vectors.

1. Insecta

Among 28 orders in the living Insecta there are 10 with members feeding on living green land plants and that might, therefore, be possible vectors. These are listed below with the approximate number of vector species at present known:

1. Collembola—chewing insects; some feed on greed plants (0).
2. Orthoptera—chewing insects; some feed on green plants (27).
3. Dermaptera—chewing insects; a few feed on green plants (1).
4. Coleoptera—chewing insects; many feed on green plants (38).
5. Lepidoptera—chewing insects; many with larvae that feed on green plants (4).
6. Diptera—a few with larvae feeding on green plants (2).
7. Hymenoptera—a few with larvae living on green plants. In a few species adults may eat ripe fruit (0).
8. Thysanoptera (thrips)—some are rasping and sucking plant feeders (4).
9. Hemiptera—feed by sucking on green plants ($\simeq 4$).
10. Homoptera—feed by sucking on green plants; includes aphids, whiteflies, hoppers, and mealybugs ($\simeq 300$).

The first seven orders listed are all chewing insects, and representatives of these orders feed on living green plants as larvae or adults, or both. Vectors of a strictly mechanical sort have been found among the Orthoptera, Dermaptera, and larvae of Lepidoptera and Diptera. Except for a few viruses, vectors in these orders are of minor importance. Important vectors occur in the Coleoptera. The Collembola and Hymenoptera contain relatively few species that are common pests of agricultural plants. There may be potential vectors among them. The Thysanoptera contains vector species for one important plant virus. The Homoptera is numerically the most important order containing plant virus vectors.

2. Arachnida

Only 1 of 12 orders in this class, the Acari (mites and ticks), contains members feeding on living green land plants. The Acari have four families containing mites that are green plant feeders—the Tetranychidae, Tarsonemidae, Eriophyidae, and the Acaridae. Virus vectors are known in the third and possibly the first of these families.

C. Summary

Of about 12 orders of invertebrates that contain at least some members that feed regularly on living green land plants, 9 have been shown to contain vectors of plant viruses. The main exceptions, Tylenchida in the Nematoda and Collembola and Hymenoptera in the Insecta, may well contain some occasional or potential vectors. Of the 9 orders with vectors the Homoptera in the Insecta, which contains a diverse group of plant-feeding insects, is by far the most important.

For almost all viruses that have vectors, such vectors belong to only one of the known vector groups. A few apparent exceptions to this generalization have been reported. For example, there are confirmed reports that tobacco ring spot *Nepovirus* (TRSV) has thrips as well as nematode vectors (J. P. Fulton, 1962; Messieha, 1969). There is no consistent relationship between the morphology of virus particles and the kinds of vectors that transmit them.

II. NEMATODES (NEMATODA)

A. Vector Genera and Viruses Transmitted

Since Hewitt *et al.* (1958) showed that the fanleaf virus of grapes is transmitted by a dagger nematode, several widespread and important viruses have been shown to be transmitted through the soil by nematodes. The three genera found to transmit viruses belong to the order Dorylaimida in which they are the only well-known plant parasites. Two of these genera, *Xiphinema* and *Longidorus*, are closely related and belong in the family Longidoridae. They are large nematodes, the adults being about 3 mm or more long. The third genus containing vector species, *Trichodorus*, belongs to another family, the Trichodoridae, and is smaller, with adults about 1 mm long. The three genera are all ectoparasitic and all have fairly long stylets. They feed on epidermal cells of the root (Fig. 14.1), and feeding punctures occur frequently near the root cap.

Two recognized groups of plant viruses are transmitted by nematodes. Members of the *Nepovirus* group are transmitted by species in the genera *Xiphinema* and *Longidorus*. Members of the *Tobravirus* group are transmitted by species of *Trichodorus*. With the exception of TRSV, none of the viruses in these two groups is known to have vectors other than nematodes.

B. Evidence for Nematode Transmission

Nematodes are difficult vectors to deal with experimentally because of their small size and their rather critical requirements with respect to soil moisture content, type of soil, and, to a lesser extent, temperature. The following kinds of observation and experiment have demonstrated that particular viruses are transmitted by nematodes:

1. The distribution of a particular species of nematode in the field may be quite closely correlated to the occurrence of patches of diseased plants (see Figs. 15.4 and 15.5).
2. Healthy plants grown in the same container as a diseased plant may become infected only when the appropriate vector nematode is added.
3. Hand-picked nematodes from virus-infected plants or from soil that had virus-infected plants in it can be added to sterilized soil in which healthy indicator plants are growing. The indicator plants may then become infected.

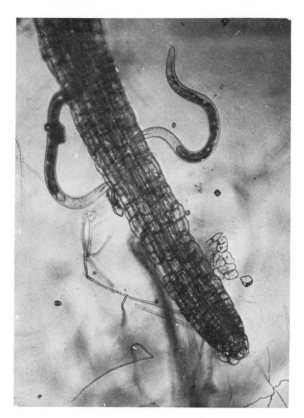

Figure 14.1 Nematodes of the species *Trichodorus christei* feeding on a root of blueberry (*Vaccinium corymbosum*). (Courtesy of B. M. Zuckermann.)

4. Useful, but circumstantial, evidence can be obtained by treating infected soil in different ways and noting the effects on nematodes and the virus-transmitting system in the soil.
5. Electron microscope studies have located specific sites in the nematode gut where a transmitted virus is retained (Figs. 14.2 and 14.3).

A common method for detecting nematode transmission has been to set out suitable "bait" plants (such as cucumber) in a sample of the test soil. These plants are allowed to grow for a time to allow any viruliferous nematodes to feed on the roots and transmit the virus, and for any transmitted virus to replicate. Extracts from the roots and leaves of the bait plants are then inoculated mechanically to a suitable range of indicator species. Various modifications of the procedure can be used. For example, Valdez (1972) described a small-scale procedure for testing individual hand-picked nematodes, while van Hoof (1976) showed that detached tobacco leaves buried in soil could become infected with tobacco rattle *Tobravirus* (TRV) by nematodes within a few hours.

Estimation of the extent of transmission of a particular virus in the field by examination of the nematodes present may be complicated by various factors. For example, the distribution of individuals carrying the virus may be different than that of the population as a whole. Subpopulations within a given area, for example, those in the surface layer and those in subsoil, may be of differing importance (Gugerli, 1977). Populations from different geographical areas may also differ in

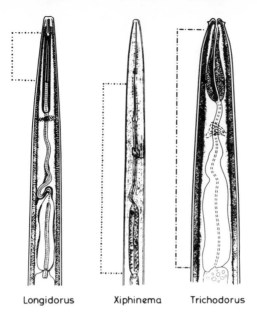

Longidorus   Xiphinema   Trichodorus

Figure 14.2 Localization of viruses within nematode vectors. Diagram of anterior portion of vectors. Broken lines indicate portions of the alimentary tracts where virus particles are retained. (From Harrison *et al.*, 1974b.)

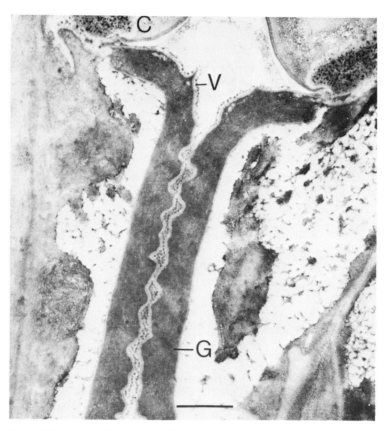

Figure 14.3 Longitudinal section of the buccal region of *Longidorus elongatus* carrying raspberry ring spot *Nepovirus*. Note numerous viruslike particles (V) lining the guide sheath (G) and that none is associated with the stoma cuticle (C). Bar = 1 μm. (From Harrison *et al.*, 1974b.)

efficiency of transmission (Brown, 1986). Immunoabsorbent electron microscopy was about 1000 times more sensitive than conventional electron microscopy in detecting virus in nematode extracts (Roberts and Brown, 1980).

C. Mechanism of Transmission

Nematodes do not infect plants merely by allowing virus present in the soil to gain entry. For example, when soil containing vector nematodes was watered with suspensions containing tomato black ring *Nepovirus* (TBRV), the nematodes did not become infective, whereas they did when the soil contained infected plants (Cadman and Harrison, 1960).

The actual minimum feeding times required for nematodes to acquire a virus and to transmit it to a healthy plant are difficult to establish because it has not been possible to determine when individuals are actually feeding. Times during which the nematodes must have access to the plants are more readily determined. Reported "access times" vary from many hours down to 15 minutes for acquisition of grapevine fanleaf virus by *Xiphinema index* (Das and Raski, 1968).

Once acquired, viruses may persist in transmissible form in starved *Longidorus* for up to 12 weeks, in *Xiphinema* for about a year, and much more than a year in *Trichodorus* (van Hoof, 1970). Transmission does not appear to involve replication of the virus in the vector. Plant virus particles have never been observed within nematode cells. Consistent with this is the fact that no evidence has been obtained for virus transmission through eggs of nematode vectors.

Nevertheless, there is considerable virus–vector specificity. The specificity of nematode genera for virus groups has already been noted. In addition, different species within a genus may transmit different but related viruses. For example, among the nepoviruses, TBRV and raspberry ring spot virus are transmitted by *Longidorus* spp., while arabis mosaic virus and TRSV have *Xiphinema* spp. as vectors. Finer degrees of specificity may occur. For example, the Scottish and English forms of raspberry ring spot virus, which are serologically related but not identical, have different species of *Longidorus* as vectors. On the other hand, the same species of nematode may transmit two viruses showing no serological relationship.

Specificity of transmission does not appear to involve the ability to ingest active virus since both transmitted and nontransmitted viruses have been detected within individuals of the same nematode species (Harrison *et al.*, 1974b). Electron microscope studies have indicated that specificity of transmission involves the adsorption of virus particles to the cuticular lining of the buccal capsule or the esophagus (e.g., McGuire *et al.*, 1970) (Figs. 14.2 and 14.3). Carbohydrate, probably of nematode origin, may be involved in the retention of virus particles on the gut lining (Robertson and Henry, 1986).

Retention of virus particles on the gut lining has been found only for virus–nematode vector combinations (Harrison *et al.*, 1974b). Presumably when a nematode carrying virus begins feeding on a fresh plant and the stylets penetrate a cell, some virus particles are washed from the lining into the plant cell by the passage of salivary secretions.

Genetic reassortment experiments with nepoviruses provide strong evidence that the coat protein of the virus is involved in the specificity of transmission and, therefore, in the binding to the surface of the alimentary tract of vectors. Thus Harrison *et al.* (1974a) showed that in raspberry ring spot *Nepovirus*, both serological specificity and transmissibility by *Longidorus elongatus* were determined by RNA2. Similar results have been obtained with TBRV. Indirect support for the importance of the coat protein comes from the observation that the satellite virus of TRSV (which uses the coat protein of TRSV) is transmitted by *X. americanum* in the same way as TRSV (McGuire and Schneider, 1973).

While the nematode–virus interaction determines specificity of transmission, the efficiency of transmission may be influenced by the plant host (as well as by nematode species and virus strain). For example, TRSV could be acquired by *X. americanum* from soybean as efficiently as from cucumber, but transmission to healthy soybean was much lower than transmission to cucumber (McGuire and Douthit, 1978).

III. APHIDS (APHIDIDAE)

Among insects, the aphids have evolved to be the most successful exploiters of higher plants as a food source, particularly flora of temperate regions. It is, therefore, not surprising that they have also developed into the most important group of virus vectors.

A. Life Cycle and Feeding Habits

1. Life Cycle

In temperate climates, aphids frequently alternate between a primary and a secondary host. They are remarkable for the number of forms produced. A complete cycle follows:

Season	Form	Host
Winter	*Eggs*	Primary
Spring	*Fundatrices* (wingless forms)	host
	Parthenogenetic and viviparous reproduction	
	Fundatrigeniae (wingless forms)	
	Parthenogenetic and viviparous reproduction	
	Migrantes (winged females)	
	Move to secondary host	
	Parthenogenetic and viviparous reproduction	
Summer	*Alienicolae* [alate (winged) and apterous	Secondary
	(wingless) forms]	host
	Many generations of these through the summer reproducing asexually and viviparously	
	Sexuparae (alate viviparous females)	
	Move to primary host	
	Parthenogenetic and viviparous reproduction	
Autumn	*Sexuales* (sexual forms)	Primary
	Mating	host
Winter	*Eggs*	

There are many variations in this cycle, depending on aphid species and on climate. For example, some may overwinter as parthenogenetic viviparous forms. Some may pass through their life cycle on one host species or several species within one genus. *Myzus persicae,* an important vector aphid, has *Prunus* species as its primary host alternating with secondary hosts in over 50 plant families. Figure 14.4 shows a group of feeding aphids.

There are three kinds of variability in aphids, which, as discussed later, may affect their ability to transmit a virus:

1. An aphid species may contain different clones or races, with or without obvious morphological differences.
2. Aphids can exist in different forms, as noted earlier.
3. A particular form has developmental stages known as nymphs. Successive molts by the developing insect define the number of stages or instars.

2. Mouthparts

The mouthparts of aphids consist of two pairs of flexible stylets, held within a groove of the labium. They are extended from the labium during feeding. The maxillary stylets have a series of toothlike projections near their tips (Fig. 14.5). Details of the feeding mechanism such as the sucking pump and the esophageal valve are described by McLean and Kinsey (1984) (Fig. 14.6).

3. Feeding Habits

At the beginning of feeding a drop of gelling saliva is secreted. The stylets then rapidly penetrate the epidermis and, in exploratory probes, the aphid may feed there temporarily. Penetration usually continues into the deeper layers with a sheath of gelled saliva forming during penetration. The stylets usually move between cells until they reach a phloem sieve tube (a process that may take minutes or hours).

Only the maxillary stylets enter the sieve tube. Compression by the cell wall causes the tips to open, exposing the end of the food and salivary canals. Evidence from electronic monitoring of aphids while they feed in the phloem suggests that

Figure 14.4 A group of apterous female adults of the cereal aphid *Rhopalosiphum padi* L. together with various nymphal stages feeding on a leaf of wheat. This species transmits BYDV. (From Lowe, 1964.)

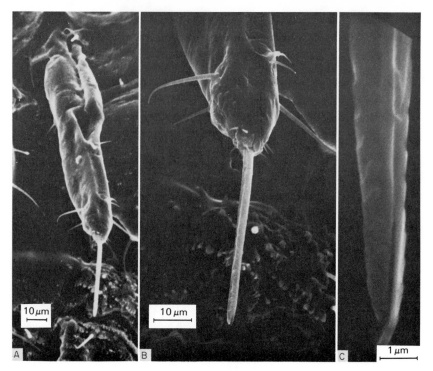

Figure 14.5 Mouthparts of *Myzus persicae* revealed by scanning electron microscopy. (A) Labium with joint area and bristles. Mandibular stylets protrude from the labium. (The aphid was frozen in liquid nitrogen immediately after it had withdrawn its stylets from a leaf.) (B) Tip of labium and mandibular stylets at higher magnification. (C) Tip of madibular stylets showing ridges on both stylets, and the overlap of the tip of one stylet. (From De Zoeten, 1968.)

they salivate, but it is the watery, enzyme-bearing, nongelling saliva. Indeed, if aphids did not salivate in the phloem, the circulative and propagative viruses that are restricted to the phloem would have no access to this tissue from an aphid carrying the virus. No gelling sheath saliva is secreted during feeding in a sieve tube, but on withdrawal such saliva is used to seal the lumen in the salivary sheath that had been occupied by the stylets. This feeding process causes minimal damage to the sieve tube and to surrounding cells in the stylet path. Penetration by aphids has been reviewed by Pollard (1977).

4. Role of the Host Plant

Both physical and chemical features of the plant may markedly affect aphid feeding behavior. For example, resistance of certain brassicas to *Brevicoryne brassicae* (L.) (but not *M. persicae*) has been shown to depend on the physical state of the wax on the leaf surface (Jadot and Roland, 1971). The density of trichomes on soybean leaves influenced probing behavior by several aphid species. Spread of soybean mosaic *Potyvirus* in the field was negatively correlated with density of pubescence (Gunasinghe *et al.*, 1988). Specific chemicals may either attract or inhibit feeding by particular aphid species. For example, sinigrin, a mustard oil glucoside found in the Brassicaceae stimulates feeding by aphids that normally feed on brassicas, but inhibits uptake by species that do not feed on members of this family (Nault and Styer, 1972).

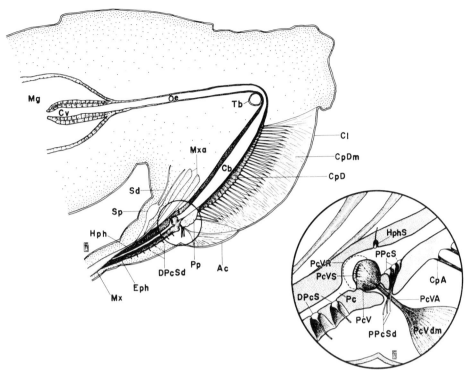

Figure 14.6 Stylized saggital view of an adult *Acyrthosiphon pisum* head, showing the primary structures of the anterior alimentary canal. The epipharyngeal pharynx protuberance (Pp) (actually offset 3 μm from the midline), included here, precisely demonstrates its relative location with respect to the precibarial valve. An enlarged lateral saggital view, at the juncture of the precibarial canal and cibarium (Cb), is shown within the circle. The dashed line that scribes an arc between the precibarial valve piston (PcV) and the precibarial valve receptacle (PcVR) connotes a closed valve. Ac, antecylpeus; Cl, cylpeus; CpA, cibarial pump apodemes; CpD, cibarial pump diaphragm; CpDm, cibarial pump dilator muscle; Cv, cardiac valve; DPcS, distal precibarial sensilla; DPcSd, distal precibarial dendrites; Eph, epipharynx; Hph, hypopharynx; HphS, hypopharyngeal sensillum; Mg, midgut; Mx, maxillary stylets; Mxa, maxillary stylets apodemes; Oe, oesophagus; Pc, precibarium; PcVA, precibarial valve apodemes; PcVdm, precibarial valve dilator muscle; PcVS, precibarial valve suture; PPcS, proximal precibarial sensilla; PPcSd, proximal precibarial dendrites; Sd, salivary duct; SP, salivary pump; Tb, tentorial bar. (From McLean and Kinsey, 1984.)

Several examples are known where infection of a plant with a virus makes the plant more suitable for the insect vector to grow and reproduce. *Aphis fabae* produced more young per mother on beet plants infected with sugar beet mosaic *Potyvirus* than on healthy plants (Kennedy, 1951). Because overcrowding began sooner on virus-infected plants, emigration of aphids began sooner. The cumulative increase of the aphid numbers in a set of plants would be substantial. In other experiments with individual leaves this difference appeared on leaves of all ages on the plant. Baker (1960) found rather similar effects with four species of aphids on beet. *Myzus persicae* preferentially selected plants infected with BYV for feeding, and subsequently bred more rapidly and lived longer than on normal green plants.

Individuals of two aphid species excreted fewer droplets of honeydew when feeding on plants infected with BYDV compared with healthy plants (Ajayi and Dewar, 1982). Maturation of aphids as alatae on oats was favored by BYDV infection (85%) compared with on healthy plants (35%) (Gildow, 1983). On the other hand, aphids reared on BYDV-infected wheat had a shorter life span (Araya and Foster, 1987).

In transmission experiments plants are involved in three ways: (i) for breeding the aphids, (ii) for providing virus-infected material, and (iii) for providing healthy plants to test the ability of aphids to infect. It is common practice to rear virus-free aphids on a plant species "immune" to the virus under study. Thus, when aphids are placed on the virus-infected plant, the change of species may influence their feeding behavior. The species and even the variety of plant used as a source of virus or as a test plant may affect the efficiency of transmission.

Source plants may change in their efficiency with time after inoculation if virus concentration changes. As discussed in Chapter 10 concentration of virus in a systemically infected plant may vary widely even in adjacent areas of tissue. This can affect the efficiency with which an aphid acquires virus. Species and varieties may differ in their susceptibility to a given virus more when one species of aphid is used as a vector than with another.

5. Environmental Conditions

Environmental conditions, particularly temperature, humidity, and wind, may have marked effects on aphid movement and feeding. For example, both transmission and acquisition efficiency for soybean dwarf *Luteovirus* by an aphid vector were greater at 20–22°C than at 10–11°C or 29°C (Damsteegt and Hewings, 1987). High humidity favored transmission of PVY and PLRV by various aphid species (M. N. Singh *et al.*, 1988). Light quality influenced the translocation of two luteoviruses in *Lycopersicon* and affected the recovery of virus by the aphid *Myzus persicae* from the plants (Thomas *et al.*, 1988). These various effects are discussed in relation to virus ecology and control of virus diseases in Chapters 15 and 16. Environmental factors may also affect transmission through effects on plant susceptibility and on the concentration of virus in source plants.

B. The Vector Groups of Aphids

There is no agreed classification for the 3700 aphid species that have been described. Eastop (1977) uses the superfamily Aphidoidea containing three families. The Adelgidae consists of about 45 species living only on Coniferae. They are of no known significance for viruses. The Phyloxeridae with about 60 species have been little studied with respect to viruses except for *Phylloxera* on vines. The Aphididae contain 10 subfamilies of which the Aphidinae contain more than one-half of the species of aphids and most of the important virus vectors. About 66% of the approximately 370 viruses with invertebrate vectors are transmitted by aphids.

Data concerning vector groups can be regarded only as indicative, because the numbers are biased in various ways. For example, in virus–vector studies there has been a marked preference for trials with certain aphid species such as *Myzus persicae* and *Aphis gossipii* that are widespread and easily reared on a range of plant species. Negative results in transmission trials are often not reported. Only a very small proportion of the possible virus–vector combinations have actually been tested. Twenty-two viruses affecting the Solanaceae are transmitted by aphid species most of which had not encountered potatoes until about 400 years ago (Eastop, 1977). Thus, it seems certain that many more actual and potential vectors exist. For example, in a 3-year study in Holland eight new vector species for PVY-N were discovered (Piron, 1986).

C. Types of Aphid–Virus Relationship

Various terms have been used to describe the different ways in which aphids transmit viruses, and the viruses that are aphid-transmitted. I will mainly follow the usage of Harris (1983). Table 14.1 summarizes the main properties of the different kinds of relationships. Some definitions are needed: (i) *Inoculativity* is the ability of an aphid or other insect to deliver infectious virus into a healthy plant. (ii) The *acquisition feed* is the feeding process by which the insect acquires virus from an infected plant. (iii) The *inoculative feed* is the feed during which virus is delivered into a healthy plant. (iv) The *latent period* is the time after an acquisition feed for which the aphid is unable to transmit a virus.

Table 14.1
Categories Describing Aphid Transmission or the Transmitted Viruses

Circulative (persistent)

(i)	Virus acquired through the food canal and translocated.
(ii)	A latent period follows acquisition feeding.
(iii)	Infection feeding involves ejection of virus in saliva from maxillary saliva canal.

Subcategories for circulative

Propagative	**Nonpropagative**
Virus multiplies in insect	Virus not known to replicate in insect

Foregut-borne (noncirculative)

(i)	No detectable latent period.
(ii)	Loss of vector inoculativity following a molt.
(iii)	No evidence for virus in hemocoel or salivary system.

Subcategories for foregut-borne

	Property	**Nonpersistent**	**Semipersistent**
(i)	Mean time for retention of inoculativity following inoculative feeding, without further access to virus	A few minutes	Hours
(ii)	Effect of starving before virus acquisition feeding	Severalfold increase in inoculativity	No effect
(iii)	Time for acquisition and inoculation thresholds	Seconds	Several minutes
(iv)	Effect of continued acquisition feeding	Acquistion probes >1 minute lead to a marked drop in transmission	Inoculative capacity increases for several hours
(v)	Effect following a long inoculative feed	Rarely inoculative after a long feed	No marked effect
(vi)	Effects of starving compared with feeding	Starved aphids remain inoculative longer than those allowed to feed	Inoculativity not affected by starving or feeding

Some workers have used the term *stylet-borne* for nonpersistent viruses but it is by no means certain that inoculum virus is in fact carried on the stylets. As discussed in the next section, the weight of evidence favors the food canal in the maxillae and the foregut as sites of virus retention. I will use the term *foregut-borne* for viruses that do not circulate within the body of the vector. These have sometimes been called noncirculative. It is not known how foregut-borne viruses are released from their site of retention to reenter a plant cell, but vector saliva may play a role. Most aphid-transmitted viruses fall clearly into one of the categories defined in Table 14.1. However, CaMV can be transmitted both nonpersistently and semipersistently. This is known as *bimodal* transmission (Section III,F).

D. Nonpersistent Transmission

Because of its role in the field transmission of many economically important viruses, nonpersistent transmission by aphids has been studied in many laboratories over six decades.

1. The Nonpersistent Viruses

Of the approximately 250 known aphid-borne viruses, most are nonpersistent (see Table 14.3). The virus groups with definite members transmitted in a nonpersistent manner are *Potyvirus, Carlavirus, Caulimovirus* (by *Myzus persicae*), *Cucumovirus*, Alfalfa mosaic virus, and *Fabavirus*. These groups include helical and isometric viruses, DNA and RNA viruses, and viruses with mono-, bi-, and tripartite genomes.

2. Site of Virus Retention

When they alight on a leaf, aphids may make brief probes into the leaf (usually less than 30 seconds). These probes are to test the suitability of the leaf as a food source. The weight of evidence indicates that, at least for aphids like *M. persicae*, the stylets enter the epidermal cells during probing. Thus, the initial behavior of such aphids on reaching a leaf is ideally suited to rapid acquisition of a nonpersistent virus.

Sap sampling on a virus-infected plant will contaminate the stylet tips, the food canal, and the foregut. These sites have been favored for the retention site of virus that will be injected subsequently into a healthy plant following another exploratory probing feed. The weight of evidence now favors the food canal in the maxillae and foregut as the sites where infective virus is retained during nonpersistent transmission.

Pirone and Harris (1977) have reviewed the evidence giving strong support to the idea that aphids can ingest significant quantities of sap during brief feeding probes. Feeding through stretched parafilm membranes on sucrose solutions has been used for many years in the study of aphid feeding behavior and virus transmission. Using this technique, Harris and Bath (1973) showed that during feeding by *M. persicae*, periods of fluid ingestion were followed by regurgitation of alimentary canal contents. Thus the aphid's sucking pump can work in either direction. Regurgitation provides a mechanism whereby aphids could inoculate ingested virus into a healthy plant cell. Many of the results obtained in experimental transmissions are readily explained by a regurgitation mechanism.

As pointed out by Watson and Plumb (1972), the high degree of specificity of transmission found among many of the nonpersistent viruses and their aphid vectors is best accounted for by adsorption phenomena. Specific adsorption sites in the aphid are much more likely to occur on the surface of living cells, for example, in the epipharyngeal region than on exoskeletal structures such as the stylets. (See the following Section 6,a).

3. Retention Time

With a nonpersistent virus, aphids begin to lose the ability to infect immediately after the acquisition feed. The rate at which infectivity is lost depends on many factors, including temperature and whether they are held on plants or under some artificial condition. With many nonpersistent viruses, aphids become noninfective very rapidly when they are allowed to feed on test plants—often in a matter of minutes. The rate at which they become noninfective is about the same on plants that are, and plants that are not, susceptible to the virus (Bradley, 1959).

The main reason for attempting to determine how long aphids retain the ability to infect is in relation to the spread of virus in the field. In most experiments, conditions have not been particularly close to those that might exist under field conditions. However, Cockbain *et al.* (1963) simulated field conditions by allowing tethered aphids to fly for various times in an air current. The infectivity of alatae of *M. persicae* and *Aphis fabae* for pea mosaic *Potyvirus* and sugar beet mosaic *Potyvirus* fell off at about the same rate whether they were flying or kept fasting in a glass container. Figure 14.7 illustrates the rate at which *M. persicae* lost the ability to transmit PVY when held on healthy plants or in a glass container.

4. Virus and Virus Strain Specificity

The V strain of tomato aspermy *Cucumovirus* is transmitted by *Myzus persicae* while the M strain of CMV is not. The RNAs of these two viruses and also that of TMV were reassembled *in vitro* using either V strain or M strain coat protein. All of

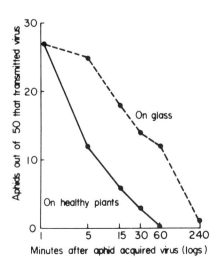

Figure 14.7 Loss of ability of infective *Myzus persicae* to transmit PVY after various periods on healthy tobacco leaves or in a glass dish without food. (From Bradley, 1959.)

the three viral RNAs coated in V strain protein were transmitted by *M. persicae,* following acquisition through membranes. Reassembly with M strain protein did not give aphid transmission (Chen and Francki, 1990). These results show that aphid transmission of cucumoviruses depends solely on the coat protein.

Different strains of the same virus may differ in the efficiency with which they are transmitted by a particular aphid species. Some strains may not be transmitted by aphids at all. Examples are also known where a given virus strain either gained or lost the ability to be transmitted by a particular aphid. Again, properties of the viral coat protein appear to be important in determining specificity of transmission (e.g., Gera *et al.,* 1979).

Different strains of the same nonpersistent virus do not usually interfere with each other's transmission as is sometimes found with propagative viruses. Thus, Castillo and Orlob (1966), in experiments with two strains of CMV and two of AMV, found that each strain was acquired independently of the other. The proportion of aphids transmitting one or both strains was in accord with the probability expected for independent transmission. By contrast, such interference has been reported between strains of PVY (Katis *et al.,* 1986).

5. Aphid Specificity

Aphid Species

The conditions of the aphid culture may affect the efficiency with which individuals transmit and thus the results of experiments on aphid vector specificity. This factor may well account for some of the discrepancies in the literature. Young colonies will provide active individuals, whereas older colonies will give less efficient and more variable transmission.

Aphid species vary widely in the number of different viruses they can transmit. At one extreme, *Myzus persicae* is known to be able to transmit a large number of nonpersistent viruses. Other aphids have been found to transmit only one virus. These differences in part reflect the extent to which different aphid species have been tested, but there is no doubt that real differences in versatility occur.

Among species that transmit a given virus, one species may be very much more efficient than another. In some experiments on this problem adequate care was not taken to ensure that factors such as age and vigor of colony, as well as host plant, were properly controlled. However, even when considerable care is taken in the design of the experiment, big differences in vector efficiency may be observed. Thus, Bradley and Rideout (1953) found marked differences in the efficiency with which PVY was transmitted by different species even when acquisition feed and test feed times were standardized (Fig. 14.8).

Nontransmitters in a Single Population

When individual aphids of a vector species are tested for their ability to transmit a given virus, not all individuals will transmit in a first trial. However, using two vector species for PVY, Gibson *et al.* (1988) showed that individual aphids that did not transmit in a first trial were as likely to transmit virus in a second test as those that did on the first occasion. Thus there was no evidence for nonuniformity in a given vector population. For any given virus, aphid vector, and host plant combination there appears to be a statistical probability that both the acquisition and infection feeds will be successful.

By feeding aphids on solutions containing a virus and a radioisotope marker, Pirone and Thornbury (1988) calculated that individual *Myzus persicae* aphids given

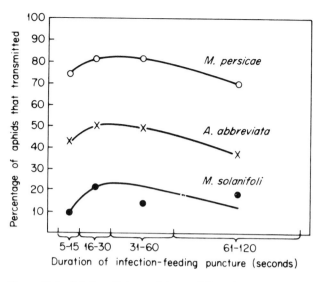

Figure 14.8 Relative efficiency of three aphid species in transmitting PVY after defined acquisition feeding times. (From Bradley and Rideout, 1953, reproduced by permission of the National Research Council of Canada.)

a 10-minute access feeding acquired between 10 and 4000 particles of a *Potyvirus*. Although these estimates were based on an artificial feeding experiment, they indicate the kind of variability involved in virus uptake by individuals in the same populations.

Aphid Form

Alate or apterous viviparous females have been used in most experimental work. The different forms of an aphid species adapted to different seasons and different host plants may vary in the efficiency with which they transmit a virus, although not many species have been studied carefully from this point of view (e.g., Gill, 1970).

Aphid Clones

Aphid species may exist as clones, perhaps showing minor anatomical differences. Such clones may differ in the efficiency with which they transmit. For example, Jurík *et al.* (1980) found that 36 clones of the pea aphid *Acyrthosiphon pisum* varied widely in their ability to transmit BYMV. Successful transmissions that ranged from 3 to 37% were not correlated with the color of the clone or its geographical origin within Czechoslovakia.

Geographically Diverse Populations

Populations of *Acyrthosiphon solani* originating in the United States and New Zealand transmitted strains of soybean dwarf *Luteovirus* less efficiently than did a population from Japan (Damsteeght and Hewings, 1986).

6. Helper Components

It has been known for many years that certain viruses could be transmitted by aphids in a nonpersistent fashion only when another virus was also present in the source plant. For example, transmission of potato aucuba mosaic ?*Potexvirus*

(PAMV) required the presence of PVY. Kassanis and Govier (1971) made the important observation that PAMV need not be present in a mixed infection with PVY in order to be aphid transmitted. Aphids first fed on plants infected with PVY could then be transferred to plants infected with PAMV and then subsequently transmit this virus.

Later work has shown that the helper factor is a specific gene product of the helper virus. Such factors have been found in two kinds of virus, the *Potyvirus* and *Caulimovirus* groups.

Potyvirus *Helper Factors*

The work of Pirone (1981) and others has shown that the existence of helper factors is general among the potyviruses. The helper factor of one potyvirus may or may not permit the aphid transmission of another potyvirus when tested in an *in vitro* acquisition system. Thus there is some specificity in the phenomenon (e.g., Sako and Ogata, 1980; Lecoq and Pitrat, 1985). Thornbury *et al.* (1985) purified the helper components from two potyviruses. These proteins had MWs of 53K (TVMV) and 58K (PVY). Antisera against the proteins specifically blocked aphid transmission in *in vitro* uptake tests. The protein has been shown to be coded for by the virus (Chapter 7, Section I,A,2 and Fig. 7.1). Berger *et al.* (1989) produced transgenic tobacco plants in which the helper component gene of TVMV was expressed. However, expression was at a low level compared with that found in TVMV-infected plants.

Purified TEV in the presence of helper component could be acquired following brief probes through an artificial membrane over a wide range of pH and buffer conditions (Raccah and Pirone, 1984). The way in which the helper protein makes aphid transmission possible has not been established. The most likely kind of effect is that the protein makes it possible for the virus to attach to sites within the aphid in a way that allows it to be transmitted. This idea is supported by experiments in which aphids were allowed to acquire labeled TEV or PVY in the presence of helper component or inactivated helper component. In the presence of active component, virus was selectively associated with the maxillary stylets and with portions of the alimentary canal anterior to the gut (Berger and Pirone, 1986). Coat protein is also involved in successful *Potyvirus* transmission by aphids.

Caulimovirus *Helper Component*

CaMV, and presumably other caulimoviruses, requires a helper component (or aphid transmission factor) when being transmitted in a nonpersistent manner by *Myzus persicae*. The helper component is the product of ORF II in the CaMV genome (Chapter 8, Section I,B,2 and Fig. 8.2). A viable ORF II product is not necessary for virus replication in the plant. Natural isolates are known that are not aphid transmissible. These have a defective ORF II protein product, which may be expressed in the plant (Harker *et al.*, 1987a; Woolston *et al.*, 1987). Non-aphid-transmissible isolates of CaMV are transmitted if the aphids are first fed on plants infected with some other caulimovirus containing a viable aphid transmission protein (Markham and Hull, 1985).

The helper component is associated, at least to a large extent, with the viroplasms, and it was suggested by Rodriguez *et al.* (1987) that viroplasms may be the form in which aphids transmit caulimoviruses. However, Espinosa *et al.* (1988) developed an assay system using *Myzus persicae* feeding through membranes and showed that CaMV is transmitted as free virus particles linked to the helper protein.

E. Semipersistent Transmission

1. Mechanism of Transmission

About 15 aphid-transmitted viruses show transmission properties that are intermediate between nonpersistent and circulative viruses, although as a group the viruses do not show very uniform transmission properties.

The best studied of the semipersistent viruses are BYV and citrus tristeza virus (Raccah *et al.*, 1976). Both of these closteroviruses are found particularly in the phloem, which may account for some of their transmission characteristics. The various features of semipersistent transmission are best accounted for by assuming that the virus particles taken up by the vector aphids are selectively adsorbed to the surfaces of the foregut, resulting in an ingestion—egestion mechanism for transmission (Harris, 1983). Semipersistent viruses such as BYV are confined to the phloem elements and related cells. It takes several minutes for an aphid's stylets to penetrate to the phloem, accounting in part for the longer acquisition feeding time compared with nonpersistent viruses, where the aphid can pick up virus in seconds from an epidermal cell. Aphids must also feed on the phloem for a period of several minutes to become inoculative.

2. Helper Viruses

The aphid *Cavariella aegopodii* transmits both anthricis yellows virus (AYV) and parsnip yellow fleck virus (PYFV) in a semipersistent manner. PYFV is a small isometric RNA virus with a diameter of ≈30 nm, as is AYV (Hemida *et al.*, 1989). The aphid transmits PYFV only when it is carrying AYV, a virus that is not sap transmitted. Unlike the "helper" for some nonpersistent viruses, the AYV itself is the helper agent for PYFV (Elnagar and Murant, 1976). However, PYFV can be acquired from plants infected singly with this virus only if the aphids are already carrying the helper AYV. This suggests that a helper protein may be involved as with some nonpersistent viruses. AYV appears to have a specific retention site in the vector foregut (Harrison and Murant, 1984).

In another helper relationship between two semipersistent viruses, a novel mechanism operates: one closterovirus (*Heracleum* latent virus, with 730-nm rods) depends on another (*Heracleum* virus 6, with 1600-nm rods) for transmission by aphids. In this system the two virus rods become attached to each other at one end (Murant, 1984; Murant *et al.*, 1988b).

F. Bimodal Transmission

Chalfant and Chapman (1962) first showed that a virus (CaMV) could be transmitted both nonpersistently and semipersistently by the same aphid species. Since then, pea seed-borne mosaic *Potyvirus* has been shown to behave in a similar fashion and there are indications that some other viruses may also be bimodally transmitted. The work has been reviewed by Lim and Hagedorn (1977). Aphid species or biotypes that transmit bimodally can acquire virus during short probes (seconds to minutes) and also during long feeding (hours to days). When percentage transmission is plotted against acquisition time, a bimodal curve is obtained (Fig. 14.9).

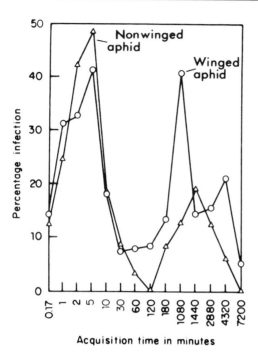

Figure 14.9 Bimodal transmission. Acquisition of pea seed borne mosaic *Potyvirus* by the winged (alatae) and nonwinged (apterae) forms of the New London biotype of the potato aphid, *Macrosiphum euphorbiae* Thomas. (From Lim and Hagedorn, 1977.)

The first peak represents nonpersistent acquisition and inoculation by the aphid. These aphids retained the ability to transmit for only short times. The second peak is semipersistent transmission following longer feeds. Aphids feeding for these times retained their ability to transmit for much longer periods than aphids contributing to the first peak, and their ability to transmit was not increased by preacquisition fasting.

For a given host and virus, various aphid species or biotypes may transmit differently. Thus Chalfant and Chapman (1962) showed that *B. brassicae* transmitted CaMV biomodally while *M. persicae* transmitted only in a nonpersistent manner. Finer distinctions may exist. Thus Lim and Hagedorn (1977) reported that one biotype of the potato aphid transmitted pea seed-borne mosaic bimodally while another transmitted only nonpersistently.

Most viruses that have been reported as bimodally transmitted have many aphid vector species, only one of which transmits bimodally. Among the bimodally transmitted viruses there are no obvious common structural features in the virus particles themselves. The reasons for bimodal transmission are not established. No structural differences could be seen in the stylets of potato aphid biotypes that did or did not transmit bimodally (Lim and Hagedorn, 1977).

Distribution of virus in the tissues of the leaf may be important. Nonpersistent viruses may occur in adequate concentrations only in epidermis and other non-phloem tissues. Semipersistent viruses may occur in phloem, and bimodal viruses may occur in both. However, other factors must operate to account for vector specificity with a particular virus.

In a detailed comparison of CaMV and turnip mosaic *Potyvirus* (TuMV) transmitted by various aphid species in turnips, Markham *et al.* (1987) confirmed a biphasic transmission for CaMV and a nonpersistent one for TuMV. They could find no obvious differences in the distribution of the two viruses within the plants or in the feeding behavior of *Myzus persicae* on plants infected with either virus. They considered that the differences in transmission are probably due to differences in the interaction between virus and vector.

G. Circulative Transmission

The main features of circulative transmission are summarized in Table 14.1. Viruses transmitted in a circulative manner are usually transmitted by one or a few species of aphid. Yellowing and leaf-rolling symptoms are commonly produced by infection with circulative viruses. Viruses that circulate in their aphid vectors may replicate in the vector (propagative) or may not (nonpropagative).

1. Propagative Viruses

Several members of the Rhabdoviridae replicate in their aphid vector. These include SYVV and LNYV, both in the vector *Hyperomyzus lactucae* (L.).

The latent period of SYVV in the vector is long and depends strongly on temperature. Characteristic bacilliform particles have been observed in the nucleus and cytoplasm of cells in the brain, subesophageal ganglion, salivary glands, ovaries, fat body, mycetome, and muscle (Sylvester and Richardson, 1970). Virus particles appear to be assembled in the nucleus. The virus can be serially transmitted from aphid to aphid by injection of hemolymph, and infection is associated with increased mortality of the aphids. Decreased life span varied with different virus isolates. However, since infected aphids lived through the period of maximum larviposition, the intrinsic rate of population growth was little affected (Sylvester, 1973). The virus is transmitted through the egg of *H. lactucae,* about 1% of larvae produced being able to infect plants (Sylvester, 1969). The virus had been shown to multiply in primary cultures of aphid cells (Peters and Black, 1970).

Continuous passage of SYVV in the aphid by mechanical inoculation gives rise to isolates that have lost the ability to infect the plant host (Sylvester and Richardson, 1971).

Similar kinds of evidence have shown that LNYV replicates in its aphid vector (O'Loughlin and Chambers, 1967), as does broccoli yellow net virus (Garrett and O'Loughlin, 1977). It is assumed that virus particles produced in aphid cells are released into the hemolymph, find their way to the salivary glands, and are injected into the plant along with saliva. There is no evidence for interference or cross-protection when aphids are allowed to acquire two different rhabdoviruses (Sylvester and Richardson, 1981).

Strawberry crinkle rhabdovirus has been maintained through six consecutive serial passages by needle inoculation of the aphid vector (*Chaetosiphon jacobi* (Sylvester *et al.,* 1974). Since each transmission involved a 10^{-3} dilution factor of the original inoculum, this is strong evidence for replication. In spite of good survival of and a long life span for injected aphids, their ability to transmit was greatest at 10 days and then declined rapidly. (Fig. 14.10). This rapid falloff might be due to either a drop in concentration of virus arriving at the salivary glands or reduced feeding by aging aphids.

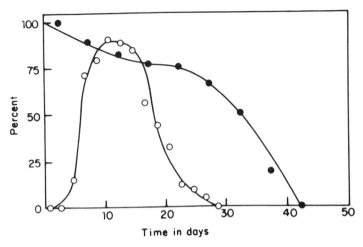

Figure 14.10   Survival and transmission curves of *Chaetosiphon jacobi* aphids injected with strawberry crinkle rhabdovirus during serial passage experiments. After injection, the aphids were caged on healthy strawberry seedlings in a growth chamber at 25°C and constant light and thereafter moved at 24- to 48-hour intervals to fresh seedlings until death. The survival curve represents the percentage of 212 insects that were alive at the beginning of each successive 4-day period. The transmission curve represents the proportion of the test seedlings fed upon by the insects during each successive 48-hour interval that developed syptoms. ○ —— ○, transmission; ● —— ●, survival. (From Sylvester *et al.*, 1974.)

2. Nonpropagative Viruses

Circulative (persistent) viruses for which there is no demonstration of replication in the vector are pea enation mosaic virus (PEMV), representing a monotypic virus group, and the luteoviruses, among which BYDV and PLRV have been the most studied. Some other less well characterized viruses are also transmitted in this manner.

PEMV

Unlike the luteoviruses, PEMV is not confined to the phloem and is sap transmissible. PEMV-like particles have been observed in thin sections of fat body cells, connective tissue cells, and the salivary system, as well as in large quantities in the gut lumen (e.g., Harris and Bath, 1973). However, failure to obtain serial transfer of infectivity in aphids is strong evidence against replication of PEMV in the aphid. Sequential assays of PEMV in aphid vectors using ELISA gave no indication of virus increase in the vector following an acquisition feed (Fargette *et al*, 1982).

Some strains of PEMV are not aphid transmitted. Physical differences between nontransmitted and transmitted strains are due to the presence of a slightly larger RNA and an additional protein associated with the bottom component particle of aphid-transmitted strains (Adam *et al.*, 1979). The bottom component of an aphid-transmitted strain gave multiple banding on electrophoresis, probably due to incremental additions of molecules of a 56×10^3 MW protein into the viral shell.

In aphid-transmitted strains the MW of RNA1 was about 1.2×10^5 larger than in nontransmitted strains. Adam *et al.* (1979) repeatedly transferred an aphid-transmissible isolate of PEMV by mechanical inoculation in *Pisum sativum* L. After 14 such transfers the virus was no longer aphid transmissible. It had also lost the 56×10^3 MW protein and RNA1 was smaller by 1.2×10^5 than in the original isolate.

It seems likely, therefore, that a part of the viral genome allowing aphid transmission resides in RNA1. The amount of RNA deleted would be insufficient to code for the 56×10^3 MW protein, but nevertheless may code for part of its amino acid sequence. This is another example of a viral-coded protein paying a vital role in insect transmission.

BYDV

General Features The median latent period for BYDV in a vector varied between 35 and 65 hours depending on temperature, aphid vector species, and virus isolate (van der Broek and Gill, 1980). BYDV moves from the gut lumen into the hemocoel, probably via a site in the hindgut (Gildow, 1985). Penetration of a sieve element is a necessary prerequisite for transmission of BYDV by inoculative *Sitobion avenae,* but such penetration did not ensure transmission. Penetration of one sieve element gave a 65% chance of transmission, independent of the duration of the phloem contact (Scheller and Shukle, 1986). Wide differences were found in the ability of the five stages of *Schizaphis graminum* to transmit BYDV. Percentage transmission progressively declined from 36% for the first instar to 2% for adults (Zhou and Rochow, 1984).

Dependent Transmission As with certain nonpersistent viruses, some persistent viruses require a helper virus to be present in the plant before aphid transmission can occur. For persistent viruses dependent on another virus, it is the presence of the virus itself in a mixed infection that provides the assistance. BYDV is the best studied of the persistent viruses with dependent transmission.

There is a high degree of vector specificity among the luteoviruses, including BYDV. For example, one form of BYDV, RPV, is transmitted efficiently by *Rhopalosiphum padi* but not by *Sitobion avenae.* For another form, MAV, the reverse is true. Virus can enter the hemolymph of inefficient vectors. Thus the block in transmission appears to lie in the inability of a virus to move from the hemolymph, via the saliva, to the plant. The isolates RPV and MAV differ serologically to a substantial degree.

Antisera specific for the two strains of BYDV can be prepared. *Rhopalosiphum padi* very rarely transmits the MAV isolate from oats when plants are infected only with the MAV strain. However, it regularly transmits both MAV and RPV isolates from doubly infected plants. When aphids are fed through membranes on concentrated mixtures of the two strains purified from separate infections, they transmit only the strain that they transmit specifically from singly infected leaves. Transmission from purified virus is blocked only by homologous antiserum.

When aphids are fed through membranes on virus purified from mixed infections, transmission of MAV by *M. avenae* was blocked by MAV antiserum. From the same mixture, however, *R. padi* infected plants that often developed MAV as well a RPV (Rochow, 1970). Rochow and Muller (1975) injected aphids with strain-specific antisera before they fed on leaves infected with both viruses. Injection of MAV antiserum prevented transmission of MAV by control aphids. However, *R. padi* injected with this antiserum transmitted MAV along with RPV from mixed infections.

These and other experiments are best interpreted by assuming that this type of dependent transmission is due to phenotypic mixing during replication of the two viruses together in the oat plant (Chapter 6, Section VIII,A). Some copies of the MAV genome are presumed to become encapsidated in a coat of RPV protein and are thus protected from inactivation by MAV serum.

Attempts to obtain dependent transmission within *R. padi,* for example, by sequential acquisition of the two strains, have been unsuccessful (Rochow, 1977). This would fit with the view, supported by other evidence, that BYDV does not replicate in its vectors.

Mixed infections with BYDV strains are common in the field. The fact that dependent transmission by *R. padi* occurred with combinations of several MAV-like and RPV-like isolates collected in various parts of the United States and Canada indicate that the phenomenon can influence virus spread in the field.

Phenotypic mixing may occur in other important members of the *Luteovirus* group such as beet western yellows virus.

The MAV and PAV isolates of BYDV are serologically related. When *Sitobion avenae* acquired MAV, fewer aphids transmitted PAV. Various experiments reviewed by Rochow *et al.* (1983) and Gildow (1987) have led to the hypothesis that nonpropagative circulative transmission of luteoviruses by aphids involves specific interactions between the viral-coat protein and receptor sites in the accessory salivary glands of the aphid.

Other Nonpropagative Viruses

Other luteoviruses such as beet western yellows virus and PLRV are transmitted in a circulative but nonpropagative manner. It has been suggested that they are transported through the salivary gland in coated vesicles (Gildow, 1982).

Certain luteoviruses are required as helper viruses for the transmission of other unrelated viruses. About six of these complexes are known. In the following examples the helper virus is named first: carrot red leaf virus and carrot mottle virus (Murant *et al.,* 1985); groundnut rosette virus (see Rajeshwari and Murant, 1988); and beet western yellows virus and lettuce speckles mottle virus (Falk *et al.,* 1979b). These systems have the following characteristics: (i) both viruses are transmitted in a circulative nonpropagative manner; (ii) the dependent virus is sap transmissible, but the helper is not; (iii) the dependent virus is only transmitted by aphids from source plants that contain both viruses. In other words, aphids already carrying helper virus cannot transmit the dependent virus from plants infected only with this virus; and (iv) evidence from a variety of experiments indicates that the dependent virus is transmitted by the aphid vector only when its RNA is packaged in a protein shell made of the helper virus protein (Harrison and Murant, 1984). This phenotypic mixing can take place in doubly infected plants. Groundnut rosette virus depends on its satellite RNA as well as on groundnut rosette assister *Luteovirus* for transmission by *Aphis craccivora* (A. F. Murant, personal communication).

H. Aphid Transmission by Cell Injury

Unstable viruses occurring in low concentration may be readily transmitted by aphid feeding, but some stable viruses such as TMV, TYMV, and SBMV are not. This curious fact is not yet fully explained. Most experimental work on this problem has been done with TMV, with the following results: (i) aphids cannot transmit TMV via their stylets (Harris and Bradley, 1973); (ii) under laboratory conditions they can transmit by making small wounds when they claw the surface of the leaf (Harris and Bradley, 1973); (iii) they can ingest TMV from infected plants and through membranes, and release virus again in an infectious state (Pirone, 1967); (iv) aphid saliva does not inhibit TMV infection; and (v) when purified TMV is mixed with poly(L)-ornithine and KCl, aphids can acquire TMV through a membrane and transmit it to plants via their stylets (e.g., Pirone, 1977). Perhaps the poly(L)-ornithine in some way makes the cell penetrated by the stylets susceptible to infection.

IV. LEAFHOPPERS AND PLANTHOPPERS (AUCHENORRHYNCHA)

A. Structure and Life Cycle

Unlike aphids, leafhoppers have a simple life cycle in which the egg hatches to a nymph, which feeds by sucking and passes through a number of molts before becoming an adult (Fig. 14.11). There may be one or several generations per year. Different species overwinter as the egg, as the adult, or as immature forms.

Only a few comments can be made here on structures important for vector leafhopper relationships with viruses. Figure 14.12 shows in diagrammatic sagittal section the arrangement of major organs in the head and thorax. The salivary glands, which are important in virus transmission, consist of a principal four-lobed gland and an accessory gland. It contains five different types of acini.

The mycetome, a body of unknown function, occurs as an isolated mass on each side of the abdomen. The fat body, which probably has a storage function, surrounds all the organs of the body and is present in head, thorax, and abdomen.

Figure 14.11 *Left to right:* Eggs, nymph, and long- and short-winged adults of the planthopper *Laodelphax striatellus,* vector of several reolike plant viruses. (From Conti, 1984.)

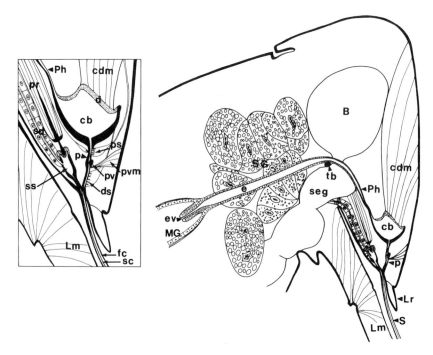

Figure 14.12   Sagittal view of the head of the leafhopper *Graminella nigrifrons* showing the anterior alimentary canal, salivary system, and surrounding structures. (*Inset*) Details of the precibarium and salivary syringe areas. *B*, brain; *cb*, cibarium; *cdm*, cibarial dilator muscle; *d*, cibarial diaphragm; *ds*, distal sensilla; *e*, esophagus, *ev*, esophageal valve; *fc*, food canal; *Lm*, labium; *Lr*, labrum; *MG*, midgut; *p*, precibarium; *Ph*, pharynx; *pr*, piston retractor muscle; *ps*, proximal sensilla; *pv*, precibarial valve; *pvm*, precibarial valve muscle; *S*, stylets; *sc*, salivary canal; *sd*, salivary duct; *seg*, subesophageal ganglion; *SG*, salivary gland; *ss*, salivary syringe; *tb*, tentorial bar. (From Nault and Ammar 1989.)

The alimentary tract is in three main regions: foregut, midgut, and hindgut (Fig. 14.13). Methods for studying the feeding behaviour of hoppers are described by Markham *et al.* (1988).

B. Kinds of Virus–Vector Relationship

No viruses have been found to be transmitted in a nonpersistent manner by hoppers, or by purely mechanical means. Some have the characteristics of semipersistent transmission. Because viruslike particles have been found attached to the cuticular linings of the anterior alimentary canal, Nault and Ammar (1989) introduced the term *foregut borne* to describe this kind of transmission. Hopper-borne viruses are often transmitted by only one or a few closely related species of hopper.

There are about 60 subfamilies in the leafhopper family (Cicadellidae) and two of these, the Agalliinae and the Deltocephalinae, contain species that are virus vectors. The Agalliinae have herbaceous dicotyledonous hosts, while most Deltocephalinae feed on monocotyledons. There are about 15,000 described species of leafhopper in about 2000 genera. Of these only 49 species from 21 genera have been reported as being virus vectors (Nault and Ammar, 1989).

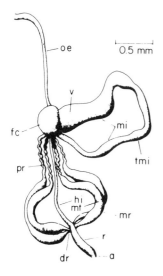

Figure 14.13 Digestive system of *Agallia constricta*. a, anus; r, rectum; dr, distal region; mr, middle region; mt, Malpighian tubules; hi, hind intestine; pr, proximal region; tmi, tubular mid intestine; mi, mid intestine; v, ventriculus; fc, filter chamber; oe, esophagus. (From Gil-Fernandez and Black, 1965.) Modified by Black to conform with the findings of Bharadwaj *et al.* (1966). (Diagram courtesy of L. M. Black.)

There are about 20 families of planthoppers (Fulgoroidea) but only the Delphacidae have definite virus vector species. Members of this family feed on monocotyledons, primarily members of the Poaceae. Thus, all the viruses known to be transmitted by members of this family have hosts in the Poaceae. These cause important diseases of cereal crops, including rice, wheat, and maize. Table 14.2 lists examples of representative hopper-borne viruses and their vectors. Hopper transmission of plant viruses has been reviewed by Nault and Ammar (1989). Evolution of viruses and vectors is discussed in Chapter 17, Section IV,B,5.

C. Semipersistent Transmission

MCDV is transmitted by *Graminella nigrifrons* in a semipersistent manner. The minimum time for acquisition of virus, or inoculation, is 15 minutes (Choudhury and Rosenkranz, 1983). Longer acquisition times increased the frequency of transmission. The minimum time is thought to be the time the hopper takes to penetrate to phloem cells where most virus is found. Infectious virus is lost from the insect in less than 24 hours at 25°C, but is retained for several days at lower temperatures (Nault and Ammar, 1989). Nymphs lose transmissible virus following a molt. Male and female adults are vectors, with females being more efficient (Choudhury and Rosenkranz, 1983). The virus particles attach to the lining of the vector's foregut (Nault and Ammar, 1989), and particles have been observed by electron microscopy adhering to the cuticula lining in the precibarium, cibarium, pharynx, and fore esophagus. None was observed in the alimentary canal beyond the cardiac valve, or in other tissues or the stylets and stylet tips (Childress and Harris, 1989). Hoppers egest material from the foregut from time to time during feeding. Thus transmission of semipersistent viruses may involve an ingestion–egestion mechanism (Harris *et al.*, 1981).

A virus-coded helper protein like that for potyviruses may be needed for the hopper transmission of MCDV, but the evidence is indirect. Hoppers fed MCDV through a membrane do not transmit. However, if hoppers are fed on plants with a

Table 14.2

Some Hopper-Borne Plant Viruses and Vector Genera Grouped according to Mode of Transmission, Virus Taxonomy, and Vector Transmission[a]

Transmission mode and virus group	Vector family	Vector genus	Virus	Selected references
Foregut-borne or semipersistent				
Maize chlorotic dwarf virus	Cicadellidae	*Graminella* (plus 6 others)	Maize chlorotic dwarf virus	Nault and Madden (1988)
Unclassified	Cicadellidae	*Nephotettix*	Rice tungro bacilliform	
Circulative: nonpropagative				
Geminivirus	Cicadellidae	*Cicadulina*	Maize streak virus	Harrison (1985)
	Membracidae	*Micrutalis*	Tomato pseudo-curly top virus	
Circulative: propagative				
Marafivirus	Cicadellidae	*Dalbulus* (plus 3 others)	Maize rayado fino virus	Gamez and Léon (1988)
	Cicadellidae	*Macrosteles*	Oat blue dwarf virus	
Rhabdoviridae	Cicadellidae	*Agallia*[b] *Agalliopis*[b] *Aceratagallia*[b]	PYDV[c]	Jackson *et al.* (1987)
	Cicadellidae	*Graminella*	Oat striate mosaic virus	
	Delphacidae	*Laodelphax*	Barley yellow striate mosaic virus[c]	
	Delphacidae	*Peregrinus*	Maize mosaic virus	
Tenuivirus	Delphacidae	*Laodelphax* (plus 2 others)	Rice stripe virus[c]	Gingery (1988)
	Delphacidae	*Peregrinus*	Maize stripe virus[c]	
Reoviridae				
Phytoreovirus	Cicadellidae	*Agallia*[b] *Agalliopsis*[b] *Aceratagallia*[b]	WTV[c]	Conti (1984)
		Nephotettix *Recelia*	Rice dwarf virus[c]	
Fijivirus	Delphacidae	*Perkinsiella*	Fiji disease virus[c]	Conti (1984)
		Nilaparvata	Rice ragged stunt virus	

[a]Modified from Nault and Amman (1989).
[b]Genera from the subfamily Agalliinae. All other cicadellids are from the subfamily Deltocephalinae.
[c]Transovarial transmission reported in some vectors.

mild strain of the virus and then through a membrane on purified virus of a severe strain, both isolates are transmitted (Hunt *et al.*, 1988). The relationship between vector specificity for MCDV and vector phylogeny is discussed in Chapter 17, Section IV,B,5.

The virus complex causing the important rice tungro disease consists of two viruses, rice tungro spherical virus and rice tungro bacilliform virus. Both are hopper transmitted in a semipersistent manner. The bacilliform virus depends on the spherical virus for its transmission. Several experiments using virus-specific neutralizing antisera suggest that a factor from plants infected with the spherical virus, but not the virus itself, is required by the bacilliform virus for transmission (Hibino and Cabauatan, 1987). The mechanism for vector specificity is not known, but may involve the properties of the cuticular lining of the foregut and/or the nature of salivary secretions (Nault and Ammar, 1989).

D. Circulative Transmission

Circulative transmission involves movement of ingested virus to the salivary glands. As with the aphid vectors, some circulative viruses replicate in the hopper vector (propagative) and some do not (nonpropagative).

1. Nonpropagative Transmission

Some geminiviruses are transmitted by leafhoppers and all of these are transmitted in a nonpropagative manner. The most studied from this point of view are MSV and CSMV, which are monopartite geminiviruses. Acquisition times range from a few seconds to an hour. Longer feeding times give higher transmission rates and longer persistence in the vector (Goodman, 1981; Harrison, 1985). The latent period for 10 nonpropagative viruses was 23 ± 4.1 hours (Nault, 1991). This is presumably the time taken for virus to reach the salivary glands, but the detailed path of transport is not established. The properties of the gut wall appear to regulate vector specificity (Harrison, 1985). The *Geminivirus* coat protein determines whether the virus is transmitted by leafhoppers or whiteflies (Chapter 8, Section II,B,1).

2. Propagative Transmission

Four families and groups of plant viruses have members that replicate in their hopper vectors (Table 14.2). The latent period for 13 propagative viruses from four families and groups of viruses was 368 ± 41 hours (Nault, 1991). The reasons for this very much longer latent period compared with viruses that are merely circulative is not understood. Both enter the plant via saliva.

Reoviridae

The classic experiments demonstrating that some plant viruses could also replicate in their animal vectors were carried out with plant reoviruses and their leafhopper vectors. Early work in this area generated considerable controversy, a subject that has been reviewed by Black (1984). The following were the main kinds of evidence that replication in the hopper vector took place.

Transovarial Transmission and Dilution Experiments   Fukushi (1933) showed that rice-dwarf virus was transmitted through the egg of the leafhopper vector *Nephotettix apicalis* var, *cincticeps* (Uhl). Such transmission occurred when the female parent carried virus, but not when only the male was infective. Fukushi (1940) showed that the virus could be passed through the egg for seven generations derived from one virus-bearing female without access to further virus during the period. He also showed that there was a long incubation period—averaging 30–45 days in the vector.

Serial Injection Experiment   Black and Brakke (1952) injected extracts of the leafhopper *Agallia constricta* infected with WTV into healthy leafhoppers maintained on alfalfa plants that were "immune" to the virus. The insects were allowed to feed for 2 weeks, and then extracts were made from them and injected into healthy hoppers. After seven such serial transfers, the injected hoppers were still infective and contained roughly the same concentration of virus as the first set of insects, as judged by infectivity tests on crimson clover. Since the starting virus would have been diluted about 1 in 10^{18}, multiplication in the insect must have occurred. This work appears to have been the first unequivocal demonstration that a plant virus can replicate in its insect vector.

Growth Curves of Virus and Viral Antigen   In later experiments by various workers, increase in virus or virus antigen was measured following acquisition feeding by the vectors. For example, Reddy and Black (1972) measured the concentration of WTV in hopper vectors using a focus assay in monolayer vector cell cultures. Figure 14.14 illustrates the growth curves they obtained under optimum assay conditions. Kimura (1986) has made similar studies with RDV. The ability to grow insect cells in culture and to infect such cells with propagative viruses has been an important factor in the study of these viruses (reviewed by Black, 1979). A focus assay developed on vector cell monolayers for rice gall dwarf virus was 1000 times as sensitive as ELISA for detecting the virus (Omura *et al.*, 1988b).

Localization of Virus within the Insect   For a virus to multiply, there must be an intimate association between the virus and the cells and organs of the insect host. This association has been revealed for several viruses by electron microscopy, or by the use of specific fluorescent antibody (e.g., Shikata and Maramorosch, 1967).

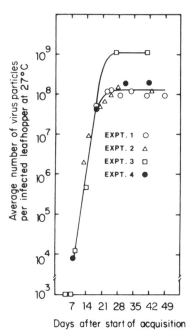

Figure 14.14 Growth curve of WTV in the leafhopper vector *Agallia constricta* after a 1-day acquisition period. 45–50 insects were used for each focus assay. Numbers of virus particles were calculated by reference to an assay carried out at the same time with standard virus. (Courtesy L. M. Black, redrawn from Reddy and Black, 1972.)

Examples of vectors for the two genera of plant reoviruses are given in Table 14.2. The subject has been reviewed by Conti (1984).

The latent, or incubation, period is usually 1–3 weeks. During this period the virus replicates and invades most tissues. Ability to transmit appears to coincide with arrival of the virus in the salivary glands. Hoppers then usually retain infectivity for the rest of their life span. Minimum efficient times for the three stages of transmission are given by Conti (1984). Rates of transmission may be quite low. For example, only 15% of *Perkinsiella saccharicida* reared on infected sugarcane contained FDV antigen and only 6% transmitted the virus (Francki *et al.*, 1986c).

Some, but not all, plant reoviruses have been shown to be transmitted through the eggs of vectors, often with a low frequency Nevertheless, transovarial transmission is important in the ecology of these viruses (Chapter 15).

As discussed in Chapter 13, Section II,C,1, and reviewed by Nuss (1984), when WTV was maintained over a period of years in cuttings made from clover shoots, and without access to the hopper vector, a variety of deletion mutants were obtained. In certain selected isolates, RNA2 or RNA5 was missing (Fig. 14.15). In spite of this, the virus replicated normally in sweet clover and had a full capacity to induce tumors. On the other hand, these isolates were entirely unable to replicate in the leafhopper vector or in vector cell monolayers. Thus segments 2 and 5 do not contain the gene(s) reponsible for tumor induction. They do contain genes required for replication in the insect, but not in the plant. The host cell recognition that initiates *Reovirus* infection in vertebrates is mediated by the Sigma 1 protein located at the icosahedral vertices. It is a tetramer of four 50K subunits. It is a fibrous protein, and in its extended form has a thin fiber conformation with a globular head

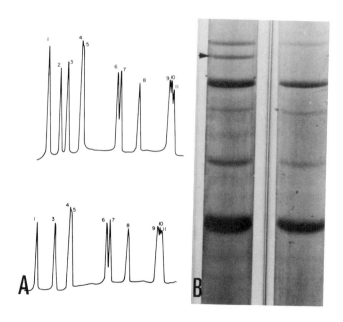

Figure 14.15 The genome of wild WTV and a mutant that had lost the ability to be transmitted by leafhoppers. (A) Scan of a gel electrophoresis separation of the dsRNA genome segments (RNA No. 12 is not shown). Upper scan is of wild-type virus. In the mutant (lower scan), RNA2 is missing. (B) The polypeptides of a wild-type virus are shown on the left. On the right the mutant has one polypeptide missing. This polypeptide is therefore coded for by RNA2. It is involved in leafhopper transmission (see text). (From Reddy and Black, 1977.)

that is assumed to be the recognition structure (Fraser *et al.*, 1990). Proteins coded for by segments 2 and 5 of WTV have MWs of 130k and 76k and are located in the outer protein coat (Table 8.1). It will be of interest to determine whether one of these proteins has a recognition function in insect cells in the manner proposed for Sigma 1 in *Reovirus*.

In cultured vector cells, WTV replication goes through an acute noncytopathic phase for the first 5 days. Following the first cell passage, a persistent phase develops in which viral RNAs and polypeptides decrease to about 5–20% of the acute phase. Viral-specific mRNAs remain at a high concentration but are inefficiently translated (Peterson and Nuss, 1986).

Rhabdoviridae

Work on the plant members of the Rhabdoviridae family has been reviewed by Jackson *et al.* (1987). Eighteen members have hopper vectors. Each virus has cicadellid of delphacid vectors but not both.

The same kinds of experiment described for the Reoviridae have been used to establish that plant members of the Rhabdoviridae replicate in their hopper vectors. Thus Chiu *et al.* (1970) used fluorescent antibody staining to show that soon after infection, PYDV antigen accumulates in the nucleus of *Agallia constricta* cells in culture. Association of maturing virus particles with the nuclear envelope was established by electron microscopy.

The growth curve of PYDV in cultured vector cell monolayers showed an eclipse period of 9 hours at 28°C. Between 9 and 29 hours, virus concentration doubled every 80 minutes. A plateau of virus concentration was then maintained (Hsu and Black, 1974).

Minimum acquisition times range from less than 1 minute to about 15 minutes for different viruses. The longer times are associated with rhabdoviruses confined to the phloem and nearby cells. The latent period may be days or months.

There may be a high degree of vector specificity, even between strains of the same virus. It is probable that the G (glycosylated) protein, which is exposed at the virus surface, is involved in this specificity (Jackson *et al.*, 1987). In rhabdoviruses infecting vertebrates this protein functions in recognition and attachment to a host-cell recognition site. Such a role is supported for SYDV by experiments with its G protein. G-specific antibodies blocked infection of insect vector monolayers (Gaedigk *et al.*, 1986).

Tenuiviruses

ELISA was used to demonstrate that maize stripe virus multiplies in its delphacid planthopper vector, *Peregrinus maidis*. Virus was detected in a wide range of organs, including the salivary glands of all hoppers that were inoculative in a bioassay and in a proportion that were not (Nault and Gordon, 1988).

Using antisera to the 32K coat protein of maize stripe virus and to a major 16.5K viral-coded nonstructural protein, Falk *et al.* (1987) showed that the coat protein could be readily detected in infected maize and *Peregrinus* vectors. However, they could detect the 16.5K protein only in the plant host. Perhaps this protein is required for replication only in the plant.

Reported acquisition thresholds range from 10 minutes to several hours. Inoculation thresholds are much shorter (30 seconds to ≃1 hour) (Gingery, 1988). Latent periods range from 3 to 36 days. Transovarial transmission to a high proportion of progeny has been reported (Table 14.2).

Marafiviruses

The main vector of maize rayado fino virus is the deltocephaline leafhopper *Dalbulus maidis,* in which the virus has been shown to replicate. Although about 80% of insects may be infected as revealed by ELISA tests, only about 10–34% of an ordinary population can transmit the virus. The ability to transmit appears to be under genetic control, because the proportion of transmitters can be rapidly increased by selective breeding (Gamez and Léon, 1988). After a few generations of outcrossing by random mating, transmission reverts to normal rates. Thus active transmitters may be recessive homozygotes for rare alleles that are rapidly diluted in outcrosses.

The virus has been shown to be widespread in the organs of infected insects. There is no evidence for cytopathological effects caused by the virus (Gamez and Léon, 1988).

E. Factors Affecting Multiplication and Transmission

Virus can multiply in hoppers feeding on an immune host. Eggs may overwinter and provide a source of virus for spring crops in the absence of diseased plants. Thus, persistence of virus in the insect and transovarial transmission, and the factors that affect its efficiency, may be of considerable economic importance.

1. Age of Vector When Infected

For several leafhopper vectors under experimental conditions, nymphs are more efficient vectors than adults, and adults decrease in efficiency as they age. For example, Sinha (1967) followed the appearance of WTV antigen in nymphs and adults of *A. constricta* after a 1-day acquisition feed. By 32 days after infection, 50% of the individuals infected as nymphs had antigen in their hemolymph and salivary glands. For individuals infected as adults, only 5% showed antigen at these sites. With the remaining insects, antigen was confined to the primary site of infection in the filter chamber of the intestine. Antigen had spread over a more limited region of intestine than with the nymphs. These experiments suggest that as the insect ages the intestine becomes more refractory to infection, and that virus present in the filter chamber of many adults may not be able to pass into the body cavity.

2. Time after Infection

Slykhuis (1963) showed that the leafhopper *Endria inimica* lost its ability to transmit wheat striate mosaic rhabdovirus after a variable number of days. Some transmitted intermittently and none was infective after 72 days.

3. Temperature

Sinha (1967) examined the effect of elevated temperature on the spread of WTV in *A. constricta*. Groups of nymphs were given an acquisition feed of 1 day at 27°C, and then held at 27°C for 3 further days to allow virus infection to become established in the filter chamber. Virus antigen was found in the filter chamber at this stage. One group continued at 27°C and the other was held at 36°C. High temperature prevented the spread of virus from the intestine to the hemolymph, salivary glands, and other parts of the intestine. By day 6, antigen had disappeared from the filter chamber of 60% of the insects.

Temperature had a marked effect on the rate of transovarial passage of rice stripe *Tenuivirus* in its planthopper vector *Laodelphax striatellus*. At 17.5°C, 83% of viruliferous females passed virus to their progeny at a rate greater than 90%. At 32.5°C, only 12.5% of females reached this frequency of transovarial transmission (Raga *et al.*, 1988).

4. Genetic Variation in the Leafhopper

Different lines or races within a vector species may vary widely in their efficiency as vectors. Thus, to demonstrate with a reasonable degree of confidence that a species is not a vector it may be necessary to test populations from various regions where the virus and insect occur.

5. Change in Properties of the Virus

Long-term culture of a virus in plants without recourse to leafhopper transmission may lead to loss of ability to be transmitted, as noted earlier for WTV.

V. INSECTS WITH BITING MOUTHPARTS

A. Vector Groups and Feeding Habits

A few vectors of a mechanical sort have been reported from the orders Orthoptera and Dermaptera. Important vectors are found in the Coleoptera (beetles). Almost all the vectors belong in a few families. Interest centers on the family Chrysomelidae, which consists of 55,000 species of plant-eating beetles. About 30 of these are known to transmit plant viruses, and each species feeds on a limited range of host plants. Twenty vector species are found in the subfamilies Galerucinae and Halticinae (flea-beetles). Two are in the Crysomelinae and two in Criocerinae. As pointed out by Selman (1973), this distribution almost certainly reflects the interests of investigators rather than the actual situation in nature. Many beetle vectors probably remain to be discovered.

In the Curculionidae and Apionidae a few species of weevils have been shown to be virus vectors.

Leaf-feeding beetles do not have salivary glands. The chrysomelid beetles tend to eat the parenchyma tissues between vascular bundles, thus leaving holes in the leaf, but with heavy infestation damage may be more severe. They regurgitate during feeding, which bathes the mouthparts with plant sap, as well as with viruses, if the plant fed upon is infected. It was once thought that transmission by beetles involved simply a mechanical process of wounding in the presence of virus. This is not so because: (i) some very stable sap-transmissible viruses such as TMV are not transmitted by beetles; (ii) some transmitted viruses may be retained by beetle vectors for long periods; and (iii) there is a substantial degree of specificity between viruses and vector beetles. Beetle transmission has been reviewed by Fulton *et al.* (1987).

B. Viruses Transmitted by Beetles

The viruses transmitted by beetles belong to the *Tymovirus, Comovirus, Bro-movirus,* and *Sobemovirus* groups. Viruses in these groups are not transmitted by members of other arthropod groups and are usually quite stable, reaching high concentrations in infected tissues. They have small isometric particles (25–30 nm diameter) and are readily transmitted by mechanical inoculation. The viruses tend to have relatively narrow host ranges, as do their beetle vectors. TMV has been reported to be transmitted in a purely mechanical way under experimental conditions (e.g., Orlob, 1963), but it is most unlikely that this transmission is of any significance in the field.

C. Beetle–Virus Relationships

Beetles can acquire virus very quickly—even after a single bite—but efficiency of transmission increases with longer feeding, as does retention time (Fulton *et al.,* 1987). Viruses appear quickly in the hemolymph after beetles feed on an infected plant. Insects become viruliferous after injection of virus into the hemocoel. Retention time varies between about 1 and 10 days with different beetles. However, under dormant, overwintering conditions, beetles may stay viruliferous for periods of months. Beetles can transmit the virus with their first bite on a susceptible plant. There is no good evidence for the existence of a latent period following virus acquisition, and no evidence for virus replication in beetle vectors.

It has been established that the regurgitant fluid is a key factor in determining whether a virus will or will not have beetle vectors (Gergerich *et al.,* 1983). This discovery was made possible by a gross wounding technique, which involved cutting disks from a leaf with a glass cylinder contaminated with virus-regurgitant mixture, thus mimicking the kind of wounds made by feeding beetles. When virus was mixed with regurgitant, only viruses normally transmitted by beetles were transmitted by the gross wounding technique. In ordinary mechanical inoculation using abrasives all mixtures were noninfectious (Gergerich *et al.,* 1983; Monis *et al.,* 1986). Regurgitant does not irreversibly inactivate non-beetle-transmitted viruses, because infectious virus could be recovered by dilution of the regurgitant–virus mixture or by isolation of the virus.

Regurgitant from several leaf-feeding beetle species was found to contain an RNase activity equivalent to 0.1–1.0 mg/ml of pancreatic RNase. This enzyme, used in this concentration range, inactivated non-beetle-transmitted viruses such as TMV when inoculated by the gross wounding technique (Gergerich *et al.,* 1986), but the transmission of viruses that are normally transmitted by beetles was not affected. Thus pancreatic RNase could mimic the effect of beetle regurgitant. In further work, three kinds of RNase differing in the way they cleave RNA were found to act with the same discrimination as pancreatic RNase. Other basic proteins did not inhibit transmission of viruses not transmissible by beetles. Thus it appears to be the enzymatic activity of the proteins that affects transmissibility (Gergerich and Scott, 1988a). Why this should be so is not clear. Perhaps the RNase activity affects establishment of the non-beetle-transmitted viruses in the initially inoculated cells as suggested by Gergerich and Scott. Alternatively, RNases may bind more firmly to viruses such as TMV, preventing uncoating in the cell.

Gergerich and Scott (1988b) showed that several beetle-transmitted viruses could move through cut stems. Furthermore, such viruses, inoculated below a steam-killed section of stem in an intact plant, could move and infect the upper parts of the plant, whereas non-beetle-transmitted viruses could not. These results suggest that the ability to be translocated in the xylem and to infect nonwounded tissue is a feature of beetle-transmitted viruses. However, TYMV, which is beetle transmitted, can move into a leaf from a cut petiole but cannot infect the leaf (Matthews, 1970).

When sodium azide was included in the inoculum mixture, cells in a zone around the gross wounding site were rapidly killed but infection by SBMV still occurred. Transmission of a non-beetle-transmitted virus was severely affected (Fulton *et al.*, 1987). It was suggested that the ability of beetle-transmitted viruses to move in the xylem may take them to cells unaffected by the sodium azide treatment.

The apparently simple transmission of viruses by beetles is now seen to be quite a complex process. For example, the experiments summarized in the foregoing shed little light on the problem of specificity among beetle species—why some species are highly efficient vectors of a particular virus and others are not.

VI. OTHER VECTOR GROUPS

A. Mealybugs (Coccoidea and Pseudococcoidea)

Mealybugs are much less mobile on the plant than other groups of vectors such as aphids and leafhoppers, a feature that makes them relatively inefficient as virus vectors. They spread from one plant to another in contact with it, and the crawling nymphs move more readily than adults. Ants that tend the mealybugs may move them from one plant to another. Occasional long-distance dispersal by wind may occur. Mealybugs feed on the phloem. They have been established as the vectors of several viruses affecting tropical plants. The most important economically is the cocoa swollen shoot group of viruses. Six vector species of differing abundance were recorded by Bigger (1981). The relationship between the cocoa swollen shoot virus and mealybugs has some similarities to the nonpersistent aphid-transmitted viruses. Presumably, the virus is carried on or near the stylets of the mealybug. In some species there are transmitting and nontransmitting strains of a mealybug species (Posnette, 1950).

B. Whiteflies (Aleyrodidae)

Whiteflies are known to transmit about 70 disease agents, mainly of tropical and subtropical plants. Many of the agents transmitted by whiteflies cause mosaic disease of a bright yellow or golden nature. Less commonly, the diseases involve marked curling of the leaves or generalized yellowing. The diseases are of substantial importance in tropical regions, but are not confined to these areas. The agents causing many of the diseases have not been characterized. They are transmitted by nonpersistent, semipersistent, and persistent mechanisms. At least three species of whitefly are involved, and it is probable that members of at least seven groups of viruses are transmitted by whiteflies. These are geminiviruses, certain viruses with particles like closteroviruses, carlaviruses, potyviruses, nepoviruses, luteoviruses, and a DNA-containing rod-shaped virus (Duffus, 1985). With respect to whitefly vectors, the best studied group is the geminiviruses.

The most studied vector is *Bemisia tabaci* (Gennadius), which is known to transmit a number of diseases. Only the first instar of the larva is mobile, and it does not move far. Adults are winged, and many generations may be produced in a year. The nymphs of *B. tabaci* are phloem feeders. BGMV is transmitted by *B. tabaci* (Goodman and Bird, 1978). Acquisition and inoculation feeding by *B. tabaci* could occur within 6 minutes, but longer feeding times were more efficient. Individual adults transmitted intermittently for up to 16 days after infection feeding. The same vector species transmits cowpea golden mosaic virus, a probable *Geminivirus*, in a similar manner (Anno-Nyako *et al.*, 1983) and also squash leaf curl *Geminivirus* (Cohen *et al.*, 1983). The result of experiments with the latter virus are consistent with the idea that virus in the hemocoel acts as a reservoir for the salivary gland system where virus-specific sites exist (Cohen *et al.*, 1989).

There is no evidence that the viruses are transmitted through the egg of the whitefly or that they multiply in the vector. The information available suggests that the virus–vector relationship is closest to the circulative nonpropagative situation found with some aphid-transmitted viruses.

Some geminiviruses are transmitted by whiteflies and some by leafhoppers. The leafhopper-borne viruses are serologically unrelated or distantly related to each other and they have different hopper vectors. By contrast, a strong serological relationship was detected between five whitefly-transmitted viruses, all of which are transmitted by *B. tabaci*, suggesting that the viral coat protein may be important in vector transmission (Roberts *et al.*, 1984). The whitefly-transmitted geminiviruses have genomes made up of two circular DNAs (see Fig. 8.4). Insertion and deletion mutagenesis indicates that neither of the DNA2 coding regions of ACMV contributes uniquely to whitefly transmission, other than aiding virus replication and spread through the plant (Etessami *et al.*, 1988).

C. Bugs (Miridae and Piesmatidae)

The mirid bugs feed by means of stylets but their biology and taxonomy are not well understood. *Cyrtopeltis nicotianae* has been shown to be a vector of velvet tobacco mottle virus, SBMV, and several other, but not all, sobemoviruses (Gibb and Randles, 1988). Minimum acquisition time was 1 minute, a characteristic of nonpersistent viruses. However, rate of transmission increased with increasing acquisition feeding time, a property characteristic of semipersistent or circulative transmission. Other characteristics were like those of semipersistent or circulative transmission. There was no evidence for virus replication in the vector. Thus the transmission was like that of beetles in several respects. Mirids are important crop pests, so it is possible that other virus vectors will be found. Beet leaf curl rhabdovirus is transmitted in a persistent, propagative manner by the piesmatid bug *Piesma quadratum*. There is no evidence for transmission through the egg (Proesler, 1980).

D. Thrips (Thysanoptera)

The known vector species of thrips are all in the family Thripidae. The very small size of thrips compared to leafhoppers, or even aphids, makes them difficult to handle experimentally. *Thrips tabaci* (Lindeman) is cosmopolitan, feeding on at least 140 species from over 40 families of plants. It reproduces mainly parthenogenetically. The larvae are rather inactive but the adults are winged and very active. *Thrips tabaci* feeds by sucking the contents of the subepidermal cells of the host plant. Adults live up to about 20 days. Several generations can develop in a year.

At least nine species of thrips can transmit TSWV. *Thrips tabaci* and three species of *Frankliniella* are important vectors of TSWV. Only the larvae can acquire TSWV, with a minimal acquisition time of 15–30 minutes (Sakimura, 1962). The incubation period ranges from 4 to 18 days in *Thrips tabaci* and about 4 to 12 days in *Frankliniella fusca* (Sakimura, 1962). When the incubation period is completed before pupation, larvae become infective. Individuals may retain infectivity for life, but their ability to transmit may be erratic. The virus is not passed through the egg. TSWV has been detected in individual thrips by ELISA (Cho *et al.*, 1988), but no definitive information has been obtained on the question of whether the virus multiplies in the vector.

If TSWV is cultured by successive transfers only in plants, the isolate loses the ability to be transmitted by thrips. Several species of thrips were reported to transmit TRSV (Bergeson *et al.*, 1964). Messieha (1969) confirmed transmission by *T. tabaci*. The virus was acquired within 8 hours and vectors remained infective for 14 days. TRSV is also transmitted by nematodes. TSV has also been reported to be transmitted by species of thrips (Kaiser *et al.*, 1982).

E. Mites (Arachnida)

Members of the mite families Eriophyidae and Tetranychidae feed by piercing plant cells and sucking the contents, but they differ in many other respects. Several members of the first of these families are vectors for three potyviruses. There are a few unsubstantiated claims for vectors in the second family.

1. Eriophyidae

The eriophyid mites are not closely related to other groups of mites. Members are known to transmit at least six plant viruses, and they feed by puncturing plant cells with stylets and sucking in the cell contents. The stylets are held inside the groove of the rostrum, which has two pads that act as ducts for the saliva (Fig. 14.16).

Eriophyid mites are very small arthropods (about 0.2 mm in length) and have limited powers of independent movement. They frequently infest buds and young leaves, where they often cause little damage and may quite easily be overlooked. They are readily killed by desiccation. In spite of this their main method of spread from plant to plant is by wind (Slykhuis, 1955).

Most species are quite specific for the host plant on which they feed, usually being confined to one plant genus, or at most the members of a single family. These mites cannot survive for long periods away from a host plant, and, thus, most of the plant species on which they feed are perennials. They have a relatively simple developmental history that may be completed in 6–14 days. There are two nymphal instars followed by a resting "pseudopupa." Males are not often seen. Some species have two kinds of female, one being specialized for hibernation.

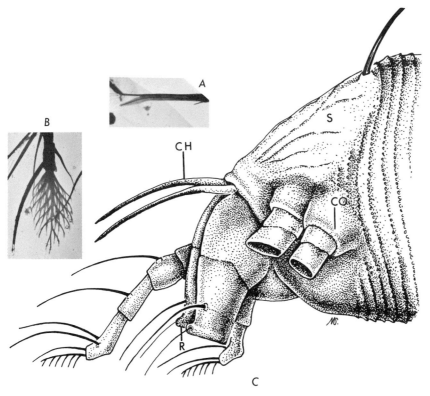

Figure 14.16   Head of an adult mite *Aceria tulipae*. S, shield; CO, coxa; R, rostrum; CH, stylets. (A) Electron micrograph of stylets. A hair lies along the upper side of stylets. Stylets are about 5 μm long and only about one-third of this length penetrates during feeding. (B) Electron micrograph of feather claw. (From Orlob, 1966.)

One of the best-studied mite vectors is *Aceria tulipae* (Keifer), which can transmit two viruses simultaneously—wheat streak mosaic *Potyvirus* (WSMV) and wheat spot mosaic virus (Slykhuis, 1962). Figure 14.17 illustrates the main internal structures of an adult of this species.

The relationship of this vector with WSMV has been studied in considerable detail. Like nonpersistent aphid-transmitted viruses, it is a rod-shaped particle and is readily sap transmitted. *Aceria tulipae* can acquire the virus in a 15-minute period on an infected leaf. A similar minimum period is required to transmit the virus. The mite retains infectivity through molts and may remain infective, at least in the greenhouse, for 6–9 days after removal from an infected plant. On a "virus-immune" host held at 3°C, they remained infective for over 2 months. The mites become infective as nymphs but not as adults when fed on an infected plant. About 30% of mites from a diseased wheat plant tested singly will transmit the virus (Slykhuis, 1965; Orlob, 1966). The uncharacterized wheat spot mosaic virus has a similar relationship with its *A. tulipae* vector (Nault and Styer, 1970).

Virus–vector relationships are difficult to study because of the small size of the mite. There is no good evidence for replication of viruses in mite vectors. Paliwal (1980) found particles of WSMV in the midgut, body cavity, and salivary glands of the mite *Eriophyes tulipae*, suggesting that the virus is circulative in this vector.

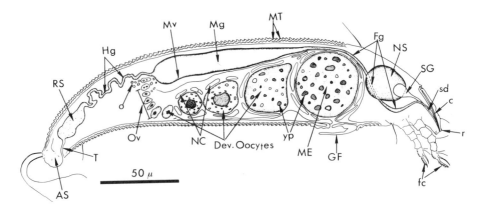

Figure 14.17   Diagram of female adult *Aceria tulipae*. fc, feather claws; r, rostrum; sd, salivary duct; c, chelicera; SG, salivary gland; NS, neurosynganglion; Fg, foregut; GF, genital flap; ME, mature egg; yp, yolk platelets; MT, microtubercles; Mg, midgut; Mv, microvilli; Dev. Oöcytes, developing oöcytes; NC, nurse cells; o, ovariole; Ov, oviduct region; Hg, hindgut; RS, rectal sac; T, rectal tube; AS, anal sucker. (From Whitmoyer *et al.*, 1972.)

2. Tetranychidae

The Tetranychidae consists of medium-sized mites (≈0.8 mm) that are all plant feeders usually with a wide host range. The spider mite *Tetranychus urticae* (Koch) was claimed to be a vector of PVY, but this was not confirmed by Orlob (1968). Orlob showed that *T. urticae* was unable to transmit nine viruses, but could acquire several viruses as revealed by presence of virus within the insect. Mites acquired TMV in 10 seconds but could not transmit it. Mite excretions were infectious, but no infection resulted from this source. Several nontransmitted viruses reached high concentrations in the mite (Orlob and Takahashi, 1971). Robertson and Carroll (1988) described a disease of barley that appears to be caused by a viruslike agent transmitted by the spider mite *Petrobia latens* (Muller).

VII. POLLINATING INSECTS

Raspberry bushy dwarf ?*Ilarvirus* (Murant *et al.*, 1974) and other viruses transmitted through infected pollen, and having insect-pollinated host plants, probably have the infecting pollen distributed by pollinating insects. Field experiments showed that blueberry leaf mottle *Nepovirus* is transmitted via pollen carried by foraging honeybees. About half the honeybees trapped in a field containing infected blueberry bushes had the virus in pollen from their pollen baskets, as determined by ELISA tests (Childress and Ramsdell, 1987). Caging experiments demonstrated that bees and infected pollen were both essential for new infections to occur.

A novel transmission via virus-infected pollen by *Thrips tabaci* was reported by Sdoodee and Teakle (1987). TSV is not transmitted to healthy seedlings by thrips after feeding on infected plants. However, when adults or nymphs were mixed with virus-carrying pollen from TSV-infected tomato, and then placed on *Chenopodium*, the virus was regularly transmitted. Presumably the pollen-borne virus infects via wounds made by the thrips. This sort of transmission might account for the apparently anomalous transmission of TRSV by thrips noted in Section VI,D.

VIII. DISCUSSION AND SUMMARY

Study of the invertebrate vectors that transmit plant viruses is important for two reasons. First, these vectors play a major role in disseminating virus diseases of economic importance in all countries. Second, virus–vector relationships are of considerable biological interest, especially those where the virus replicates in the animal vector as well as its plant host. Only two invertebrate phyla have members that feed on living green land plants and both of these—the Nematoda and Arthropoda—contain vectors of plant viruses.

Three genera of nematodes are vectors. Two, *Xiphinema* and *Longidorus,* contain vectors for the polyhedral viruses of the *Nepovirus* group. Rod-shaped tobraviruses are transmitted by species of *Trichodorus.* There is a substantial degree of specificity in the virus–nematode relationship, almost certainly involving attachment of virus particles to specific zones in the lining of the nematode gut.

Among the classes of the Arthropoda phylum, two have members that feed on living green land plants—the Arachnida and Insecta—and both of these contain virus vectors. The most important group of vectors numerically is to be found in the insect order Homoptera, which includes the aphids, leafhoppers, planthoppers, whiteflies, and mealybugs. Since these groups have sucking mouthparts that penetrate leaf cells and tissues, they are ideally suited to transmit viruses from diseased to healthy plants. Table 14.3 gives an overview of the numbers of viruses transmitted in various ways by the superfamilies (or families) of homopteran vectors. The numbers of viruses known to be transmitted by other invertebrate groups are: beetles (22); nematodes (15); mites (3); thrips (2); and bugs (2). The numbers in Table 14.3 change continually as new viruses and vectors are identified.

About 66% of the arthropod-borne viruses are transmitted by aphids. There are various types of virus–aphid relationship, as summarized in Table 14.1. None of these involves a "flying pin" kind of transmission. Several members of the Rhabdoviridae replicate in their aphid vectors, and vectors may remain infective for their lifetime. Even when the aphid remains infective for a relatively brief period there is specificity in the virus–vector relationship. This specificity may involve viral-coded proteins: (i) Helper factors: potyviruses and caulimoviruses each produce a viral-coded protein that is essential for aphid transmission. They probably act by facilitating the binding of virus to some internal surface of the aphid. (ii) Helper viruses: some semipersistent viruses cannot be transmitted by an aphid vector when the RNA is coated with its own viral-coded coat protein. They can, however, be transmitted when the RNA is encapsidated in the coat protein of some unrelated virus in a plant infected with both viruses. (iii) Viral coat proteins are also involved in the specific retention of particular viruses by particular aphids. (iv) Specificity of aphid vector transmission may involve different strains of the same virus, as occurs with BYDV. Virus strains will be transmitted by a particular vector only if the RNA is in the protein coat appropriate for that vector.

The leafhoppers and planthoppers constitute a second important group of vectors. No viruses are transmitted in a nonpersistent manner by hoppers. One important virus, MCDV, is transmitted in a semipersistent manner but most are circulative and either nonpropagative or propagative. Viruses transmitted by hoppers are not as numerous as those transmitted by aphid vectors, but they include a number of economically very important viruses, especially those infecting food crops belonging to the Poaceae. There is a substantial degree of vector specificity between virus and hopper, and frequently both have rather narrow host ranges.

Table 14.3
Numbers of Plant Viruses Transmitted by Six Homopteran Families among Various Virus Groups and Families[a]

| | Transmission | | Auchenorrhyncha | | | | Sternorrhyncha | |
Virus family or group	Mode	Persistence[b]	Delphacidae (planthoppers)	Cicadellidae (leafhoppers)	Membracidae	Aphididae (aphids)	Aleyrodidae (whiteflies)	Coccoidae (mealybugs)
Tenuivirus	Propagative	P	5					
Rhabdoviridae	Propagative	P	11	7		9		
Reoviridae	Propagative	P	10	3				
Marafivirus	Propagative	P		3				
Geminivirus	Nonpropagative	P		8		(1)	13	
Luteovirus	Nonpropagative	P			(1)[c]	6 (5)	(3)	
Pea enation mosaic	Nonpropagative	P				1		
Maize chlorotic dwarf	Foregut-borne	S		1 (1)				
Closterovirus	Foregut-borne	S				8 (4)	(5)	(1)
Parsnip yellow fleck	Foregut-borne	S				2 (1)		
Caulimovirus	Foregut-borne	N				6 (3)		
Carlavirus	Foregut-borne	N				29 (6)	(3)	
Potyvirus	Foregut-borne	N				68 (87)	(1)	
Potexvirus	Foregut-borne	N				(2)		
Alfalfa mosaic	Foregut-borne	N				1		
Cucumovirus	Foregut-borne	N				3 (2)		
Fabavirus	Foregut-borne	N				4		
Total			26	22 (1)	1	137 (111)	13 (12)	(1)

[a]Modified from Nault (1991).

[b]P = persistent; S = semipersistent; N = nonpersistent.

[c]Numbers in parentheses are viruses considered provisional members of the groups indicated.

Four families and groups—Reoviridae, Rhabdoviridae, and the *Tenuivirus* and *Marafivirus* groups—contain members that replicate in their leafhopper vectors. Such replication usually has little effect on the hoppers. However, from the virus point of view, replication in the vector has two important consequences: (i) once they acquire virus the vectors normally remain infective for the rest of their lives and (ii) replication in the vector is often associated with transovarial passage of the virus, thus giving it a means of survival over winter that is quite independent of the host plant. In the plant reoviruses, particular genome segments code for gene products required for replication in the insect, but not in the plant. With viruses that replicate in their vectors there may be a high degree of specificity between vector and virus or even strains of a virus.

Leaf-feeding beetles have chewing mouthparts and do not possess salivary glands. They regurgitate during feeding, which bathes the mouthparts in sap. This regurgitant will contain virus if the beetle has fed on an infected plant. Beetles can acquire virus after a single bite and can infect a healthy plant with one bite. However, beetle transmission is not a purely mechanical process. There is a high degree of specificity between beetle vector and virus, and some very stable viruses such as TMV are not transmitted by beetles. The viruses that are transmitted belong to the *Tymovirus, Comovirus, Bromovirus,* and *Sobemovirus* groups.

Sometimes one beetle species will transmit a particular virus with high efficiency while a related species does so inefficiently. The reasons for this sort of specificity are not understood. However, we are beginning to understand why some stable viruses are not beetle transmitted. The regurgitant fluid of the beetles contains an inhibitor that prevents the transmission of non-beetle-transmitted viruses but does not affect those that are transmitted. There is good evidence that the inhibitor is an RNase.

Other insect vectors are found among the mealybugs, whiteflies, mirid bugs, and thrips. The viruses transmitted by these groups are not numerous, but the first two are vectors for some viruses causing important diseases in tropical crops. In the Arachnida, eriophyid mites are vectors for several viruses. Lastly, for viruses transmitted through the pollen, pollinating insects can transfer infected pollen to healthy plants, thus transmitting the virus in an indirect manner.

15

Ecology

In order to survive, a plant virus must have (i) one or more host plant species in which it can multiply, (ii) an effective means of spreading to and infecting fresh individual host plants, and (iii) a supply of suitable healthy host plants to which it can spread. The actual situation that exists for any given virus in a particular locality, or on the global scale, will be the result of complex interactions between many physical and biological factors. In this chapter I shall consider briefly the more important of these, with illustrations of the ways they can interact to affect the survival and distribution of plant viruses. An understanding of the ecology of a virus in a particular crop and locality is essential for the development of appropriate methods for the control of the disease it causes. As with most other obligate parasites, the dominant ecological factors to be considered are usually the way viruses spread from plant to plant and the ways that other factors influence such spread. Aspects of ecology and epidemiology are discussed in Maramarosch and Harris (1981).

I. BIOLOGICAL FACTORS

A. Properties of Viruses and Their Host Plants

1. Physical Stability of Viruses and Concentrations Reached

For viruses that depend on mechanical transmission, a virus that is stable, both inside and outside the plant, and that reaches a high concentration in the tissues is more likely to survive and spread than one that is highly unstable. The survival and spread of certain viruses appear to depend largely on a high degree of stability and the large amounts produced in infected tissues. For example, TMV may survive for long periods in dead plant material in the soil, which is then a source of infection for subsequent crops (Johnson and Ogden, 1929). TMV was recovered from 42 brands of cigarettes in Germany by Wetter and Bernard (1977). The virus yield was 0.1–

0.3 mg/g tobacco. TMV from cigarette tobacco had about one-half the specific infectivity of that isolated from fresh leaf.

High concentrations of virus at a particular season may be important. *Stellaria media* (L.) is an overwintering weed host for CMV. Walkey and Cooper (1976) found that CMV reached its highest concentration in this host at low temperatures. Thus, the plant would provide a very good source for acquisition of virus by aphids in the spring.

2. Rate of Movement and Distribution within Host Plants

Viruses or virus strains that move slowly through the host plant from the point of infection are less likely to survive and spread efficiently than ones that move rapidly. Speed of movement will be important as measured relative to the lifetime of individual host plants. Viruses that infect long-lived shrubs or trees can afford to move much more slowly through their hosts than those affecting annual plants. Viruses that can move into the seed and survive there have an important advantage in spread and survival. TNV is confined to the roots in most hosts, and in nature it depends on transmission through the soil and infection of new hosts with the aid of fungal vectors.

3. Severity of the Disease

A virus that kills its host plant with a rapidly developing systemic disease is much less likely to survive than one that causes only a mild or moderate disease that allows the host plant to survive and reproduce effectively. There is probably a natural selection in the field against strains that cause rapid death of the host plant. Leafhoppers living on desert plants of the western United States are infective primarily with strains of beet curly top *Geminivirus* that cause mild symptoms in beet. Virulent strains of curly top virus kill certain desert plant species before they can allow a generation of leafhoppers to mature. Thus, virulent strains in the desert tend to be self-eliminating (Bennett, 1963).

However, disease severity has a very different effect in beet plantings. Sugar beet is a good host for the leafhopper if the plants are small and exposed to full sunlight. They are poor hosts when large, providing a lot of shade. For this reason severe strains of curly top virus facilitate their own spread in beet by producing small, stunted plants, which favor vector multiplication (Bennett, 1963). The spread of severe strains is further assisted by the fact that mild strains do not protect against infection. Leafhoppers overwintering near beet fields carry strains of higher virulence than hoppers found in the desert.

4. Mutability and Strain Selection

The extent to which a virus is able to mutate to produce strains that can cope effectively with changes in the environment may well affect survival and dispersal of the virus. It is difficult to make valid comparisons of rates of mutation between different viruses, but they probably vary quite considerably. For example, PLRV seems relatively stable (although this may merely reflect lack of experimentation or appropriate techniques for this virus), while many others such as TMV and TSWV exist in nature as numerous strains. There is fairly good evidence that different strains of the same virus may vary in the rate at which they produce a certain type of mutant. Mild strains of PVX isolated from potato frequently give rise to ringspot strains in tobacco, but the TBR strain originally isolated from tomato was never observed to do so (Matthews, 1949a). The common TMV can give rise to a wide range of mutant types (Bawden, 1964).

As discussed in Chapter 13, Sections I,C and IV,B various examples are known where particular host plants allow the selective multiplication of certain strains of a virus when presented with a mixture. Likewise, invertebrate vectors have sometimes been shown to transmit some strains of a virus more efficiently than others (Chapter 14). These experiments illustrate ways in which strains might become selected under natural conditions.

Different tobacco species and varieties in the field may be infected naturally with different strains of TMV. It is probable that a major factor leading to the dominance of a particular strain in a particular host is the rate at which it can invade the plant systemically and thus exclude other strains (Chapter 13).

In districts where an annual crop such as tobacco has been farmed for many years, characteristic strains may come to dominate. Thus, Johnson and Valleau (1946) described how different dominant strains of TMV tended to be characteristic of different tobacco farms in old tobacco-growing areas. On a wider scale, geographical isolation may tend to lead to divergence of strains, particularly where climatic conditions are different. Such geographical variation in the dominant strain type is not uncommon (however, see Chapter 17, Section IV,B,4).

One strain of a virus may remain dominant for many seasons in a particular crop and region if there exists a stable natural reservoir host for the strain in that region. In regions with high year-round temperatures, strains of viruses may be adapted to grow at such temperatures. However, where summer temperatures are very high viruses may actually be inactivated *in vivo*. For example, strawberry crinkle rhabdovirus was eliminated from strawberry plants grown in the Imperial Valley, California (Frazier *et al.*, 1965). Season appeared to be a factor in the dominance of two strains of CMV in tomato and pepper crops in southeastern France. A thermosensitive strain dominated in spring, while a thermoresistant strain was prevalent in summer (Quiot *et al.*, 1979). Strains of BYDV can be distinguished by severity of disease and by the species of aphid that transmits them (Chapter 14, Section III,G,2). The dominant strains in New York State have changed gradually over 20 years from isolates of the vector-specific MAV type to those similar to PAV strains (Rochow, 1979). By contrast, in a given crop and season there may be no detectable variation in the pathogenicity of the virus over wide areas, for example, bean pod mottle *Comovirus* in soybean (Ross and Butler, 1985).

Agricultural practices may influence in many different ways the strains of virus that become dominant in a crop. For example, on the estate in Scotland where the Arran varieties of potato were bred, cultural and selection practices unintentionally led to virtually all promising seedlings being infected with PVX at an early stage through contact with old commercial varieties heavily infected with the virus. As seedlings were multiplied by vegetative propagation, individuals showing mosaic symptoms were discarded, ensuring that the variety came to contain a predominantly mild strain of PVX (Matthews, 1949b). An outbreak of rapidly lethal virus disease may well indicate a new and unstable relationship between host and virus. When potato plants free of PVX were inoculated with virus obtained from another variety showing no obvious disease they showed no symptoms in the year of infection. In the following year, the disease produced varied widely, even in different shoots from the same tuber. Some shoots became invaded by strains causing severe necrotic disease and death, while others showed mild mottling or no symptoms (Matthews, 1949a). A period of natural and artificial selection would no doubt have given a line of potatoes again showing only a mild type of infection.

The use of an avirulent strain of TMV (derived from strain 1) for providing cross-protection against ToMV in commercial tomato plantings (Chapter 16, Section II,C,2) led to a marked increase in the appearance of strain 1 infections in susceptible cultivars (Fletcher and Butler, 1975).

Where there is a source of virus for an annual agricultural crop in perennial weeds and wild hosts, successive crops may become infected with strains that never have the opportunity to become well adapted to the crop plant.

When agricultural operations or other factors do not bring about sudden changes, we can envisage, for any particular host species and environment, that selection of strains of a virus will occur until those best adapted for survival will dominate. Factors important for survival of a strain will include (i) efficient transmission by insects or other means, (ii) more rapid multiplication and movement in the plant than any competing strains, and (iii) the production of mild or only moderately severe disease. These factors have been studied experimentally for BSMV in barley by Timian (1974).

5. Plant Host Range of Viruses

The general distribution of viruses in the plant kingdom is considered in Chapter 17. Here I shall consider the ways in which the host range of a virus may affect its survival and distribution in the field.

Viruses vary greatly in the range of species they are able to infect. Some viruses affecting strawberries appear to be confined to the genus *Fragaria*. Other viruses may be able to infect a wide range of plants. For example, the natural hosts of CMV include 476 host species in 67 families. In addition, 536 species in 53 families have been infected experimentally (Horváth, 1983a). Viruses with very narrow host ranges presumably survive either because their host is perennial, or because it is vegetatively propagated, or because the virus is transmitted efficiently through the seed.

A diversity of hosts gives a virus much greater opportunities to maintain itself and spread widely. Viruses that have perennial ornamental plants as hosts as well as other agricultural and horticultural species have become widespread around the world. Important examples among ornamental flowering species are (i) bean yellow mosaic *Potyvirus* and CMV carried in gladioli; (ii) TSWV carried in dahlia, which is frequently infected and is a major reservoir of this virus in many countries (dahlia may also carry CMV, often without symptoms); and (iii) CMV carried in lilies, often without symptoms.

Weeds, wild plants, hedgerows, and ornamental trees and shrubs may also act as virus reservoirs. The actual importance of these various sorts of host for neighboring crops will depend on circumstances, particularly on the presence of active invertebrate vectors. For example, the presence of okra mosaic *Tymovirus* in three malvaceous weeds in Nigeria appeared to be an important source of the virus for crop plants, because infective beetle vectors were shown to be active (Atiri, 1984). On the other hand, although apple chlorotic leaf spot *?Closterovirus* was found in a significant proportion of hedgerow hawthorn plants in England, they appeared to be of little significance for spread of the virus to fruit trees (Sweet, 1980). Among the weed hosts, *Plantago* species may be one of the most important potential virus reservoirs. They are efficient and adaptable perennial weeds with a worldwide distribution. They have been found infected naturally with at least 26 viruses from 19 groups and families (Hammond, 1982). Different strains of a virus may preferentially infect certain weed species. For example, in the southeast of France, thermosensitive strains of CMV were found preferentially in *Rubia peregrina* while thermoresistant strains predominated in *Portulaca oleracea* (Quiot *et al.*, 1979b). The important and complex role of weeds in the incidence of virus diseases is discussed by Duffus (1971).

Many nematodes and the viruses transmitted by them have wide host ranges, including woody perennial plants. Such viruses and their vectors can often survive in woody plants in hedges and forests in the absence of suitable agricultural crop plants. Grape fanleaf virus and its nematode vector *Xiphinema index* are unusual in that they are both largely confined to the grape plant in the field. Since the grape vine is long-lived, alternative hosts may not be necessary for survival. Furthermore, vector and virus can survive for several years in viable roots that may remain in the soil after the grapevine tops have been removed.

B. Dispersal

Dispersal of viruses by airborne or soil-inhabiting vectors, by seed and pollen, and over long distances by human activities plays a key role in the ecology of viruses.

1. Airborne Vectors

Taking plant viruses as a whole, the flying, sap-sucking groups of insect vectors, particularly the aphids, are by far the most important agents of spread and survival. The pattern of spread in the crop and the rate and extent of spread will depend on many factors, including (i) the source of the inoculum—whether it comes from outside the crop, from diseased individuals within the crop arising from seed transmission or through vegetative propagation, from weeds or other plants within the crop, or from crop debris; (ii) the amount of potential inoculum available; (iii) the nature and habits of the vector; for example, with aphids, whether they are winged transients or colonizers; (iv) whether virus is nonpersistent, semipersistent, or persistent in the vector; (v) the time at which vectors become active in relation to the lifetime of the crop; and (vi) weather conditions.

In early work attempting to relate aphid numbers to virus spread, counts were made of aphids actually on plants at different times during the season. Relationships between such numbers and the spread of virus were often not apparent. Doncaster and Gregory (1948) made a significant contribution when they pointed out the importance of migrant winged aphids, particularly those that move through the crop early in the season. The size of the largely static populations that build up on individual plants depends to a great extent on the local weather and other conditions within the crop. Populations can increase at a very rapid rate (about 10-fold in 7 days) and population density may vary widely even in a small region within a crop. Subsequent work has confirmed the general importance of aphid migrations early in the season. For example, Heathcote and Broadbent (1961) placed potatoes growing in pots and infected with PLRV or PVY in plots or in the field for successive periods through the season and noted aphid numbers and subsequent virus incidence. The viruses spread from the infected plants early in the season when there were few aphids colonizing plants, but not later in midseason when wingless aphids were numerous. See Fig. 15.2 for another example of the importance of early infestations.

With different viruses, even those transmitted by members of the same group of vectors, rates and patterns of spread in a crop can vary widely. Two different patterns of spread are illustrated in Fig. 15.1. A virus brought into a crop from outside by vectors does not necessarily spread from the first infected plant in a crop to other plants. Little spread within the crop may occur if the invasion occurs late in the growing season, or if the presence of the vectors in the crop is very transient. This sort of situation is illustrated in Fig. 15.1A. This represents part of a potato field that was initially free of PVY, and was then infected from a source several hundred meters away. There was no evidence of secondary spread. Infected plants were scattered through the field.

Initially the same field had one leaf-roll-infected plant placed at the center. The cluster of leaf-roll-infected plants that developed around the source of infection (Fig. 15.1B) shows that plant-to-plant spread within the crop had taken place with this virus.

Van der Plank (1946) developed a method of testing whether virus was spreading from diseased plants within a crop. It is based on the assumption that virus coming from outside the crop will infect plants at random. On this basis there will be a certain expectation of the frequency with which infected plants will occur side by side as pairs.

$$p = X \frac{(X - 1)}{n}$$

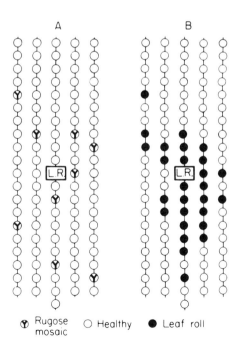

Figure 15.1 Difference in pattern of spread of two aphid-borne viruses in the same plot of potatoes. Diagram A of the plot shows the distribution of plants infected with PVY; diagram B shows PLRV-infected plants. PVY was entering from outside the crop, whereas PLRV was spreading from the infected plant (L.R.). (From Doncaster and Gregory, 1948.)

where p is the expected number of infected pairs, n is the number of consecutive plants examined, and X is the number of infected plants observed. For high values of n, the standard error of p is $\sqrt{p}$.

If the observed number of pairs (counting three plants together as two pairs) is significantly greater than expectation, then spread from infected plants within the field can be assumed. Spread within a crop need not necessarily favor neighboring plants, so that an apparently random distribution cannot exclude the possibility of such spread. The pattern of infection brought about in a field by one vector species may be confused if two forms of the vector are active at the same time, one causing spread to neighbors and the other "jump" transmission to plants at a distance. More sophisticated procedures are now available for analyzing the extent of randomness or clumping (patchiness) in the distribution of infected plants during an epidemic (e.g., Madden and Campbell, 1986; Madden et al., 1987a,b).

Various kinds of traps have been developed for assessing the number of flying aphids, including yellow pan traps, vertical sticky traps, conical nets, or suction traps. The horizontal mosaic green pan trap designed by Irwin is proving useful (Irwin and Ruesink, 1986). Different types of traps may give somewhat different answers with different aphid species, and the height at which the traps are set will affect the data obtained. In general, the frequency of flying aphids decreases with increasing height, and relative numbers of vector species caught in traps may not correspond to relative numbers found on the crop (e.g., Tatchell et al., 1988). Furthermore, trapping alone merely gives an estimate of numbers of aphids of various species and the morphs involved and not the numbers actually infective for particular viruses. Transient winged forms of aphids that do not colonize the crop at all, but move from plant to plant, may be especially important with stylet-borne viruses. Though they may form a small proportion of the total population, they may introduce virus from outside or acquire it from infected plants within the crop and spread it rapidly as they move about seeking suitable food plants. Colonizers will also be important if they move about within the crop.

Ideally, from the point of view of virus transmission, we want to know the numbers of infective aphids that are flying and that will land on the crop of interest. To determine this may be a very tedious and labor-intensive task, especially when nonpersistent viruses are involved. Trapped aphids then have to be placed on healthy test plants as soon as possible after capture. Nevertheless, winged aphids can be collected as they land in a crop, identified, and tested on appropriate plants to determine whether they are carrying particular viruses. For example, Ashby et al. (1979) used wind sock traps to assess numbers of viruliferous vectors of subterranean clover red-leaf virus. On average over the season 1972–1977, 40–50% of alate Aulacorthum solani (Kltb.) were found to be carrying this virus.

Using similar traps in a potato field, Piron (1986) collected 101 aphid species. Twenty-three species transmitted PVY^N, and 22 of these were recorded as new vectors of the virus. This experiment illustrates the fact that a lot more information than presently available may be needed about vectors active in the field in relation to the ecology of particular viruses in particular locations.

The "infection pressure" to which a crop is subjected can also be assessed by placing sets of bait plants in pots within the field for successive periods of a few days. The sets of plants are then maintained in the glasshouse and observed for infection. Figure 15.2 illustrates such a trial in relation to the prevalence of potential vector species. The frequency with which aphids were trapped clearly implicates the early flights of Cavariella aegopodii as being of prime importance in the spread of

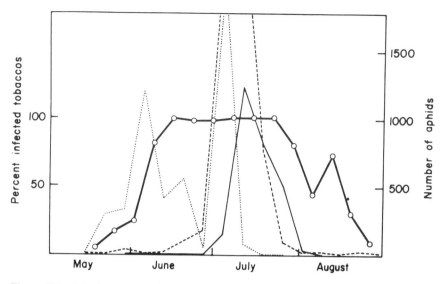

Figure 15.2 Infection pressure of PVY in a field of potatoes. Batches of 100 tobacco plants in pots were set out in the field for 7 days. The left-hand axis and the solid line show the percentage of these plants that subsequently developed infection with PVY in the glasshouse. The number of aphids of three species trapped each week is given by the axis on the right. $\bigcirc$ —— $\bigcirc$, PVY^N in tobacco in %; ——, *Myzus persicae*; ---, *Rhopalosiphum padi*; ······, *Cavariella aegopodii*. (From van Hoof, 1977.)

PVY in this particular field and season. Measurement of vector activity has been discussed by Irwin and Ruesink (1986). If particular virus strains are of interest it is important to know whether the bait plant being used is preferentially infected by one out of two or more stains that are present (Marrou *et al.*, 1979).

Most of the preceding discussion has been concerned with variation in the numbers of plants infected with a virus. In addition to this aspect, the severity of the disease in individual plants may be substantially increased by an increased dose of virus provided by numerous aphids (Smith, 1967) (see Fig. 12.2). Disease is also generally more severe when younger plants are infected.

The number of aphid species known to transmit different viruses varies widely. For example, strawberry crinkle virus has two vectors, while more than 60 vector species have been recorded for CMV.

The fact that the majority of epidemiological studies have been carried out on aphid-borne viruses reflects the importance of this group of vectors. Other groups of airborne vectors such as hoppers and beetles are, of course, also of crucial importance in the ecology of the viruses they transmit. Beetles may differ from aphids and hoppers in the pattern of virus incidence they bring about. For example, flea beetles differ significantly from the aphids in their patterns of movement within a crop. They are generally much more active than aphids. They tend to move from plant to plant over short distances much more frequently than wingless aphids, but they do not usually travel as far as winged forms. Figure 15.3 shows the pattern of spread of TYMV by flea beetles in a field of turnips growing adjacent to a block of plants infected mechanically with the virus. The initially healthy field was divided into strips parallel to the block of infected plants. A strong gradient of infection developed in the field (Markham and Smith, 1949). Similar gradients of infection have been observed for spread by aphids under some conditions.

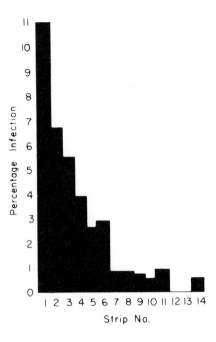

Figure 15.3 Gradient of infection due to spread by flea-beetles. Percentage infection in each strip of plants is plotted against distance from the source of infection, which was next to strip No. 1. Natural spread of TYMV in turnips. Each strip was 2 m wide. (From Markham and Smith, 1949.)

2. Soilborne Viruses

Viruses with No Known Vector

TMV is one of the few viruses transmitted throughout the soil to any significant extent without the aid of any known vector. The stability of this virus allows it to survive from season to season in plant remains, provided conditions are suitable. A susceptible crop planted in contaminated soil will become infected, presumably through small wounds made in the roots during transplanting cultivation, or root growth. Viruses with no known vectors, such as members of the *Potexvirus* and *Tobamovirus* groups, are widespread in soils of forest ecosystems in West Germany (Büttner and Nienhaus, 1989).

Viruses with Fungal Vectors

The ecological implications of transmission by fungal vectors depend on the way in which the fungus carries the virus. Viruses such as soilborne wheat mosaic *Furovirus* and potato mop top *Furovirus* are carried in the resting spores of their plasmodiophoromycete vectors. In these spores the viruses may survive in air-dried soil or in soil that has been stored for many years. Buildup of infection may take several seasons but once established in a field, viruses with this type of vector may persist for many years even in the absence of suitable plant hosts (e.g., Jones and Harrison, 1972). Localized spread of these viruses is by zoospores and resting spores, soil water being a major factor influencing local movement. Viruses with this type of transmission tend to have rather narrow host ranges.

Viruses transmitted by *Olpidium brassicae,* such as TNV, STNV, and cucumber necrosis *Tombusvirus,* are carried on the surface of the spores. Zoospores carrying virus probably survive only a few hours. However, virus that is free in the soil may be picked up and transmitted by newly released zoospores. If the soil is moist, *Olpidium* zoospores can recover TNV from soil for at least 11 weeks.

Viruses transmitted by *Olpidium* do not survive in air-dried soil. In general, viruses of this type have wide host ranges and probably survive in the soil by frequent transmission to successive hosts. Drainage water and movement of soil and root fragments are probably important in the spread of these viruses from one site to another. In model experiments in which viruses were added to a soilless recirculating water supply, several viruses, including some such as TMV with no known vectors, were transmitted (Paludan, 1985).

Polymyxa betae, the fungal vector of BNYVV, forms very persistent cytosores. These were transmitted to infected soil, and both cytosores and BNYVV could pass through the digestive system of the sheep (Heijbroek, 1988).

Viruses with Nematode Vectors

The ecology of viruses with nematode vectors (tobraviruses and nepoviruses) differs considerably from that of viruses with airborne vectors. Nematodes are long-lived, may have wide host ranges, and are capable of surviving in adverse conditions and in the absence of host plants for considerable periods.

Vector nematodes do not have a resistant resting stage, but can survive adverse soil conditions by movement through the soil profile. As soils become dry in summer or cold in winter they move to the subsoil and return when conditions are favorable. Some of the viruses (e.g., arabis mosaic *Nepovirus*) may persist through the winter in the nematode vector.

Nematode-transmitted viruses usually have two other characteristics in common—a wide host range, especially among annual weed species, and seed transmission in many of their hosts. Thus nematode vectors that lose infectivity during the winter may regain it in the spring from germinating infected weeds (Murant and Taylor, 1965).

The spread of nematodes in undisturbed soil may be rather slow. Harrison and Winslow (1961) calculated that a population of *X. diversicaudatum* invaded uncultivated woodland at the rate of about 30 cm per year. Agricultural practices will increase distribution of the vectors during cultivation and probably in drainage or flood water. Farm workers' footwear and machinery contaminated with infested soil may transmit nematodes and their associated viruses over both short and long distances (Boag, 1985). However, the pattern of infection observed in a crop may often depend largely on the vector and virus situation in the soil before the crop was planted. In a field already infected with nematodes and planted with a biennial or perennial crop, it may take 1 or 2 years before the initial pattern of infection becomes apparent, since leaf symptoms may not show until about a year after infection. Subsequent spread may give rise to slowly expanding patches of infected plants within the field (Fig. 15.4).

Individual woody trees or hedges of long-lived woody species have been implicated as a reservoir of vector nematodes (e.g., Fig. 15.5). Sometimes such species may harbor a virus as well.

3. Seed and Pollen Transmission

The occurrence of and mechanisms for seed and pollen transmission were discussed in Chapter 10. These two methods of transmission may be of crucial importance in the ecology of certain viruses. Survival in the seed can be particularly

Figure 15.4 Picture of an outbreak of raspberry ring spot *Nepovirus* in Talisman strawberry. The patchy distribution of stunted plants reflects the patchy distribution of the vector *Longidorus elongatus*. (Courtesy of Scottish Horticultural Research Institute.)

important for viruses that have only annual plants as hosts, and for those that have invertebrate vectors, such as nematodes that normally move only slowly.

CMV persists in seeds of *Stellaria media* buried in soil for at least 12 months (Tomlinson and Walker, 1973). If 10^7 seeds were present per hectare, with 1% infected with CMV, then a 10% emergence of seedlings would give rise to about one infected seedling per square meter. Under such conditions a rapid buildup of CMV infection in a crop might result. Lettuce mosaic *Potyvirus* in lettuce is an outstanding example of the importance of seed transmission in a commercial crop.

Human activities in the dispersal of virus-infected seed are noted in Section I,B,4. Natural dispersal of seeds by wind or water may also be a factor. Seed output per plant may be very large for windborne seeds, but wastage is also high.

Transmission by pollen may be the major, and perhaps only, method of natural spread in the field for certain viruses, for example, *Prunus* necrotic ringspot *Ilarvirus*. On the other hand, pollen dispersal is probably of little or no ecological significance when the host plant is mainly or entirely self-pollinated, for example, barley. However, Shepherd (1972) pointed out that some barley varieties produce large amounts of pollen, which if infected with BSMV could lead to mechanical transmission by abrasion between leaves.

4. Dispersal over Long Distances

Human Dispersal

Even in countries that have had a highly developed agricultural technology for some time, it may be very difficult or impossible to document the arrival and spread of particular viruses. However, there is little doubt that over the past few centuries humans have been mainly responsible for the wide distribution around the world of many viruses that were previously localized in one or a few geographical areas.

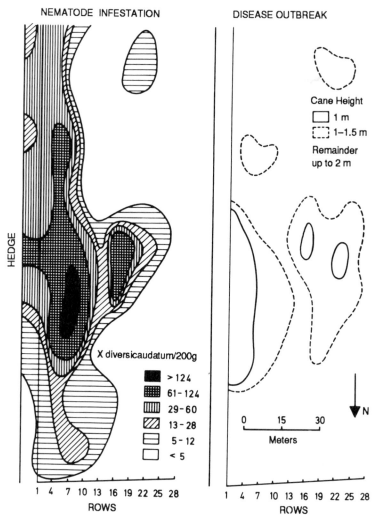

Figure 15.5 Relationship between density of nematode infestation and an outbreak of virus disease. Left: Population contours of *Xiphinema diversicaudatum*. Right: Outbreak of strawberry latent ring spot *?Nepovirus* in a 10-year-old raspberry plantation. The virus causes a reduction in cane height. Plants were spaced 1.8 m between rows and 9.6 m within rows. (From Taylor and Thomas, 1968.)

Viruses have been transported in plants or plant parts and perhaps occasionally in invertebrate vectors. There is little doubt that many of the virus diseases of potato and some of their vectors were brought to Europe with the potato from America and have since been spread to many other countries in tubers (e.g., Jones and Harrison, 1972). Lettuce mosaic virus, because it is seed transmitted, has probably been distributed wherever this crop is grown. The fact that TMV can survive in infectious form in prepared smoking tobacco is probably sufficient to account for its presence wherever tobacco is grown commercially.

The spread of some viruses may be a more complex problem, involving the necessity for spread of a suitable invertebrate or fungal vector as well as the virus and appropriate host plants. For example, Raski and Hewitt (1963) consider that humans have been largely responsible for the dispersal of grapevine fanleaf *Nepovirus* (and its nematode vector) around the world as grape vines were taken from place to place.

In relation to worldwide dispersal of viruses, New Zealand is an interesting example, since it is geographically perhaps the most isolated of the countries that have a diverse and modern agriculture and horticulture. Polynesian immigrants introduced a few food plants, for example, taro (*Colocasia esculenta* Schott) and kumara (*Ipomea batatas* Lam.). Over the past 170 years or so, European immigrants have introduced a wide range of agricultural and horticultural crop plants, together with a large selection of weed species. So far, 139 viruses have been recorded in the country, all present in introduced species. These have almost all been identified as viruses occurring elsewhere, mainly Europe or North America with a few from Australia. It is easy to see how most of these could have arrived in tubers, corms, runners, rootstocks, seed, and so on. For example, 38 of the 139 viruses infect vegetatively propagated fruit trees and vines.

The first virus diseases were discovered in New Zealand (in potatoes) in 1929. Figure 15.6 shows the number known to occur at five-yearly intervals since then. This curve bears little relationship to the time that the viruses arrived in New Zealand. With a few possible exceptions, it mainly reflects the amount of effort that has gone into finding and identifying viruses that have been present for some time. For example, after 1945, systematic work was begun on viruses in fruit trees, and 16 such viruses were found between 1945 and 1965, but most of these had undoubtedly been present for many years.

Most of the vectors for these viruses are introduced species of aphids. New Zealand is very deficient in endemic aphids, their place in the fauna being taken by Psyllidae. It is a curious fact that no viruses with leafhopper vectors are yet known in New Zealand, although representatives of some potential vector genera occur.

There are other examples where the effects of human activities can be more precisely dated. Plum pox *Potyvirus* infects *Prunus* spp. and is transmitted in a nonpersistent manner by aphids. The disease was reported first in Bulgaria in 1915.

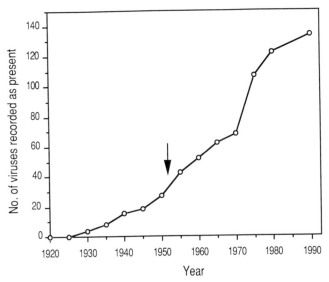

Figure 15.6 Cumulative total number of viruses recorded in New Zealand, plotted at 5-year intervals. Arrow indicates year in which virus quarantine regulations were introduced. (From data supplied by P. R. Fry.)

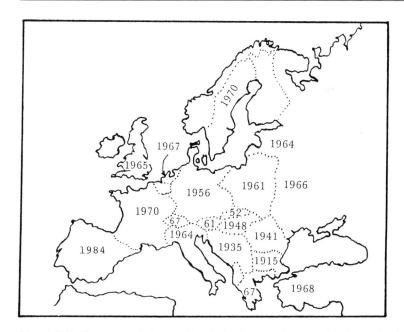

Figure 15.7   First reports of plum pox *Potyvirus* in Europe and western Asia. Dates signify the year the virus was first recorded. (Courtesy of J. M. Thresh.)

It has since spread through Europe as indicated in Fig. 15.7. Aphids spread the virus to nearby orchards. Long-distance spread is most probably by means of vegetative plant material distributed by humans (Walkey, 1985).

The more recent movement of BNYVV, which causes the important rhizomania disease in sugar beet, has been summarized by Hill and Torrance (1989). The disease was first recorded in Italy in 1955. For some years it was confined to southern Europe, but then it moved northward, being discovered in eastern France and Germany in 1974, Austria in 1982, the Netherlands in 1984, and in southeast England in 1987. The disease was first recorded outside Europe in Washington State in 1981 (Al Musa and Mink, 1981), in California in 1983, and then in Texas in 1985 (Duffus and Liu, 1987). A virus arriving in a new environment may sometimes find conditions suitable for very rapid spread. Stubbs (1956) pointed out that carrot motley dwarf virus spread very rapidly in Australia, where the aphid vector *Cavariella aegopodii* was already plentiful as an introduced species. It also lacked natural enemies. By contrast, spread was slow in California, where the aphid vector was heavily parasitized.

Human activities may spread both virus and vector within a country. For example, soils of nurseries in southern England frequently contain nepoviruses and their vectors, which can be transported with the host plants, especially if the infection is latent (Sweet, 1975).

Airborne Vectors

Aphid vectors are important in the long-distance as well as local spread of viruses. This may be true for nonpersistent as well as persistent viruses. For example, detailed studies of aphid movements in the United Kingdom show that many important aphid species recolonize the whole island each year, as long as appropriate host plants are available (Taylor, 1986). This movement involves distances of up to 1000 km, although not usually in one flight. Several successive colonizations may be involved.

On the other hand, under appropriate climatic conditions, a continuous long-distance journeys may not be uncommon. Geostrophic airstreams are air movements at altitudes of about 1000 m or more moving along the isobars of a relatively large-scale weather system. Such airstreams have almost certainly led to the mass transport of winged aphids from Australia to New Zealand, a distance over sea of about 2000 km (Close and Tomlinson, 1975). LNYV and several vector species of aphids have probably been introduced in this way. An example of the potential effects of weather on aphid movement is illustrated in Fig. 15.12. Not only the viruses but also potential aphid vectors are still moving into new continents. Thus *Metopalophium dirhodum,* a vector for BYDV, was first recorded in Australia in 1985 and was almost certainly a fairly recent arrival (Waterhouse and Helms, 1985).

Leafhopper vectors may also travel long distances. For example, large numbers of *Macrosteles fascifrons* may be blown each spring from an overwintering region about 300 km north of the Gulf of Mexico through the midwestern United States and into the prairie provinces of Canada (Chiykowski and Chapman, 1965). BYDV and cereal aphids follow similar routes.

Pollen and Seed Dispersal

Pollen and windborne seed transmission may sometimes be important over shorter distances. Seed-transmitted viruses might be transported over considerable distances by birds, but there is no documented example. Proctor (1968) has shown that viable seeds may persist in the alimentary tract of migratory seabirds for as long as 340 hours, long enough to be transported several thousand kilometers.

Water Dispersal

Over many years, methods have been developed for screening water for the presence of human and animal viruses. Recently these have been applied to a search for waterborne plant viruses, especially in Europe, with some surprising results. The work has been reviewed by Koenig (1986). Most sampling has been done in rivers and lakes. Infectious viruses isolated from such waters include TMV, cucumber green mottle *Tobamovirus,* and other tobamoviruses; a potexvirus; TNV and STNV; TBSV, carnation Italian ringspot, and other tombusviruses; carnation mottle *Carmovirus;* CMV; and carnation ringspot *Dianthovirus.* In addition, carnation mottle virus was isolated from Baltic Sea water (Kontzog *et al.,* 1988). Many of these viruses share several properties. Most of them are very stable and lack airborne vectors that would allow spread over long distances. They occur in high concentrations in infected plants and are released from infected roots and can infect roots without vectors. Many have a wide host range. Most of the infectious virus probably moves in water while adsorbed onto organic and inorganic colloidal particles, especially clays (Koenig, 1986). In this state they would be substantially more resistant to inactivation than as free virus.

The viruses found in water originate by release of virus from living roots, or from decaying plant material, or from sewage. Tomlinson and Faithfull (1984) isolated TBSV from waters of the River Thames in England. This virus and others were isolated from a large number of tomato plants growing in semisolid or dried sludge in the primary settlement beds of various sewage plants. They concluded that these tomato plants and at least some of the viruses were derived from infected tomatoes eaten raw by humans. It is known that TBSV can pass in an infectious state through the human alimentary tract (Tomlinson *et al.*, 1982). Carnation ring spot virus was isolated from water of a canal near a sewage plant in northern Germany, again implicating sewage in possible long-distance dissemination (Koenig *et al.*, 1988b). These water borne viruses may be a factor in the forest decline occurring in Europe, but further research is needed on this question (Koenig, 1986; Büttner *et al.*, 1987).

C. Cultural Practices

The way a particular crop is grown and cultivated in a particular locality and the ways land is used through the year in the area may have a marked effect on the incidence of a virus disease in the crop. This may, of course, offer the opportunity to prevent infection by appropriate cultural practices. Many diverse situations arise. Some examples will illustrate the kinds of factors involved.

1. Planting Date

A clear-cut example of the way time of sowing seed affects virus incidence occurs with the winter wheat crop in southern Alberta. In a normal season, the percentage infection with streak mosaic is markedly dependent on the time at which the seed is sown (Slykhuis *et al.*, 1957). For sowing dates earlier than September, the spring-sown crop, carrying disease, overlaps with the autumn-sown crop. High air temperatures may cause a dramatic reduction in some aphid vector populations. If the planting date for the autumn crop is delayed until such conditions prevail, much less virus spread may take place in a crop. Another example of the importance of planting date is given in Fig. 15.8. The importance of the planting date in relation to vector migrations and control measures is illustrated in Fig. 16.9.

Changes in planting practice that provide an overwintering crop may cause an increased virus incidence. Thus the increased incidence of BYMV in the United Kingdom has been attributed, in part, to the growing of winter malting barley as a profitable spring crop, leading to increased perpetuation of the virus through its fungal vector (Coutts, 1986). The increase in BYMV has also been associated with a switch from spring-sown to autumn-sown barley in various parts of the United Kingdom. In Washington State, the highest incidence of BYDV in winter wheat crops occurred in irrigated areas that supported aphid vectors during the summer, and with early planting dates (Wyatt *et al.*, 1988).

2. Crop Rotation

The kind of crop rotation practiced may have a marked effect on the incidence of viruses that can survive the winter in weeds or volunteer plants. With certain crops, volunteer plants that can carry viruses may survive in high numbers for considerable periods. Doncaster and Gregory (1948) showed that it may take 5 or 6 years to eliminate volunteer potatoes from a field in which potatoes had been grown.

For a crop such as carrots, fields for seed production and for root production together with volunteer plants may form a continuous yearly cycle enabling viruses to be perpetuated (Howell and Mink, 1977). The importance of volunteer plants in the spread of beet viruses is illustrated in Fig. 15.9. With perennial crops, virus incidence will tend to increase as the crop ages.

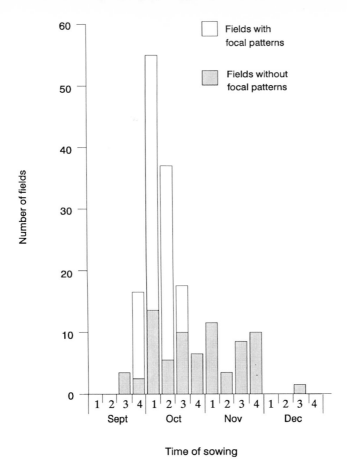

Figure 15.8 Effect of time of sowing on BYDV incidence in wheat fields in the southern United Kingdom in 1976–1977, as estimated from aerial observation of foci of infection. Unshaded columns represent fields with obvious foci of infection; shaded columns are fields apparently free from infection. (From Hill, 1987.)

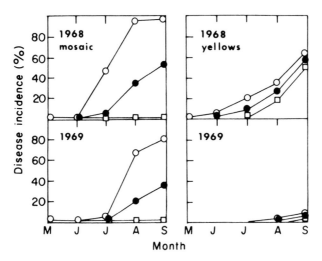

Figure 15.9 The relationship between volunteer sugar beets and the occurrence of beet mosaic *Potyvirus* and beet western yellows *Luteovirus* in beet fields in Washington State during 1968 and 1969. ○, fields containing volunteer plants; ●, fields adjacent to volunteer plants; □, fields at least 1 mile from known volunteer plants. (From Howell and Mink, 1971.)

3. Soil Cultivation

Soil cultivation practices may affect the spread and survival of viruses in the soil or in plant remains. Nematode and fungal vectors may be spread by movement of soil during cultivation. The extent to which soil is aerated and kept moist may affect the survival of TMV in plant remains. Cultivation during frosty weather of fields previously in potatoes may greatly reduce the survival of volunteer potato plants (Doncaster and Gregory, 1948).

4. Field Size

The influence of the size of field on the spread of a virus will depend to a great extent on the source of initial infection. If virus is coming from outside the crop, then, as pointed out by Van der Plank (1948), concentration of the crop into large fields of compact shape will reduce infection from outside the crop to a minimum. For example, this situation has been shown to occur with AMV in lucerne (Gibbs, 1962).

5. Population Density and Plant Size

Airborne vectors bringing a virus into a crop from outside will infect a greater proportion of the plants in a given area when they are widely spaced than when they are close together. For example, the incidence of beet yellows was reduced where the distances between plants or between rows was reduced (Blencowe and Tinsley, 1951). Crop spacing may affect the landing response of flying aphids. Some species were trapped more frequently over widely spaced crops of cocksfoot and kale (A'Brook, 1973).

Large plants in a crop might be expected to become infected more readily with insect-borne viruses than small ones, since they are more likely to be visited by a vector. Broadbent (1957) found that this held in cauliflower seedbeds. In one year, 30% of large seedlings 15% of medium-sized seedlings, and 5% of small seedlings were infected with CaMV.

6. Effects of Glasshouses

It might be expected that the use of glasshouses and polythene tunnels would favor the survival of a stable virus such as TMV, since the structures remain at one site and are in use for intensive cultivation. On the other hand, they might provide some protection against aphid-borne viruses. Thus Conti and Masenga (1977) reported that in the Piedmont, in pepper crops under plastic tunnels, 84% of virus-infected plants contained TMV. The remaining 16% were infected with aphid-borne viruses. By contrast, 88% of virus-infected pepper plants grown outdoors were infected by viruses transmitted by aphids in a nonpersistent manner. Glasshouses are normally used in regions with cool winters and will therefore favor the introduction of viruses more adapted to tropical and subtropical climates, such as with TSWV.

7. Pollination Practices

The horticultural practice of planting mixtures of varieties or pollinators in orchards may favor the spread of pollen-borne viruses. Howell and Mink (1988) present circumstantial evidence suggesting that the annual appearance of new sites of infection with pollen-borne *Prunus* necrotic ringspot virus in a sweet cherry orchard was caused by the practice of moving commercial beehives directly from earlier blooming orchards.

8. Nurseries and Production Fields as Sources of Infection

Nurseries, especially where they have been used for some years, may act for themselves as important sources of virus infection. For example, Hampton (1988) found in several U.S. states no evidence of infection in wild *Humulus lupulus* plants tested for the following viruses: two ilarviruses and three carlaviruses common in cultivated hops, and in particular American hop latent *Carlavirus*. Several hundred hop plants, primarily from breeding nurseries in Oregon, had substantial infection rates with these viruses. Fifty-three non-*Humulus* species growing around infected hop yards and nurseries were found to be free of American hop latent virus. Thus the commercial yards and nurseries appeared to be the only source of infection for this virus.

9. Movement of Crop Plants into New Areas

As well as distributing viruses around the world (Section I,B,4), humans have moved crop species to new countries, often with disastrous consequences as far as virus infection is concerned. Plant species that were relatively virus-free in their native land may become infected with viruses that have long been present in the countries to which they were moved for commercial purposes. It is often difficult to prove a sequence of events of this sort, especially if the movement took place before virologists could investigate and record events. Thresh (1980) details several instances. Cacao swollen shoot virus is a significant example because of the importance of the cacao crop in several West African economies. Cacao was transferred from the Amazonian jungle to West Africa late last century, and since then major commercial production has developed there. The swollen shoot disease, first reported in cacao in 1936, was very probably transmitted from natural West African tree hosts of the virus by the mealybug vectors that are indigenous to the region.

Movement of plant species between countries and continents has been carried out with increasing frequency during the last two centuries. Agriculture in India, North America, and Australia is almost totally dependent on introduced crop plants (Thresh, 1980). Thus there has been ample opportunity for events such as that outlined for cacao to occur.

10. Monocropping

Cultivation of a single crop, or at least a very dominant crop, over a wide area continuously for many years may lead to major epidemics of virus disease, especially if an airborne vector is involved. An example of the development of such an epidemic is shown in Fig. 15.10. Soilborne vectors may also be important from this point of view, for example, with grapevine fanleaf virus in vineyards, where the vines are cropped for many years. Monocropping may also lead to a buildup of crop debris and the proliferation of weeds that become associated with the particular crop.

As pointed out by Diener (1987a), most viroid diseases in crops have appeared

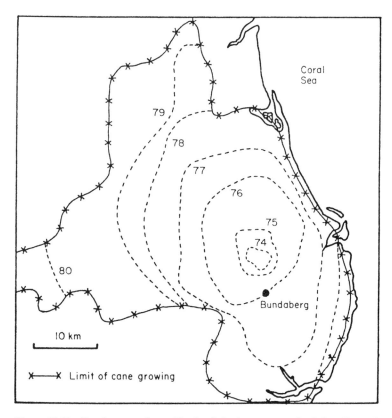

Figure 15.10 Development of an epidemic of planthopper-transmitted virus disease in sugarcane over 7 years. The incidence of Fiji disease *Fijivirus* in the Bundaberg district of Queensland, Australia, 1974–1980. Disease contours indicate the boundary of the area with an average of at least 1% affected cane. The solid line is the outer limit of the main cane-growing area. (Modified from Egan and Hall, 1983.)

only in recent decades. He suggested that viroids themselves have existed for a very long period in wild host species, causing little disease there and escaping from time to time into agricultural crops that were grown on a small scale with varying genetic composition of the crop plants. Under these conditions they would cause no widespread disease. However, with the advent of modern large-scale monocultures of genetically uniform plants, the opportunities for serious outbreaks of viroid diseases have greatly increased. This is particularly so with crops such as chrysanthemums, where propagation of vegetative stocks may be highly centralized.

An outstanding example of the effects of monocropping is the rise in importance of certain virus diseases of rice and of their hopper vectors as a consequence of the "green revolution" that began about 25 years ago. New rice varieties introduced as part of the green revolution in tropical and subtropical areas are a major factor in the greatly improved yields that have been achieved. However, the gains have been seriously impaired by the increased prevalence of several serious virus diseases of rice, and also of their hopper vectors (Thresh, 1988a). An important example is rice grassy stunt *Tenuivirus* (RGSV) and ragged stunt reovirus and their planthopper vector (*Nilaparvata lugens*), known as the rice brown planthopper (Thresh, 1989). As well as transmitting these viruses, the hopper can itself cause serious crop damage. The first new rice cultivars released during 1966–1971 were all susceptible to both vector and RGSV. Effective sources of resistance to both were found and incorporated into new varieties that were released in 1974–1975. These were at first grown widely and successfully in the Philippines, Indonesia, and elsewhere. However, severe infestations with the planthopper were reported within 2–3 years, and in 1982–1983 a resistance breaking strain of RGSV was reported. No new source of virus resistance has been found. A new source of hopper resistance was identified and incorporated into new varieties, and it was successful for a few years until it, too, broke down with the emergence of a new hopper biotype (Thresh, 1989).

Even within one country, the viruses infecting a particular host plant may well be influenced by a new development in agricultural practice. For example, *Solanum laciniatum* Ait., an indigenous plant in New Zealand, had no virus infections recorded from natural habitats. Shortly after commercial plantings were begun, field infections with PVX, PVY, TMV, and CMV were recorded (Thomson, 1976).

II. PHYSICAL FACTORS

A. Rainfall

Rainfall may influence both airborne and soilborne virus vectors. The timing and extent of rainfall may alter the influence it has on vector populations. For example, some rainfall or high humidity is necessary for the buildup of whitefly populations, while continuous heavy rainfall may be a factor in reducing the size of such populations (Vetten and Allen, 1983). Similarly, heavy rainfall just after airborne aphids have arrived in a crop may kill many potential vectors, thus reducing subsequent virus incidence (Wallin and Loonan, 1971).

Potato mop top *Furovirus* has its highest incidence in Scotland in areas with the highest rainfall (Cooper and Harrison, 1973), which was correlated with increased incidence of the fungal vector (*Spongospora subterranea*) in wetter soils. Similarly, in Peru this virus is widely distributed in *Solanum* spp. in the highlands of the Andes, where rainfall and temperature favor the fungal vector (Salazar and Jones, 1975). Symptoms of the virus were not seen in plantings in the dry coastal region.

B. Wind

Wind may be an important factor not only in assisting or inhibiting spread of viruses by airborne vectors, but also in determining the predominant direction of spread. Windbreaks may affect the local incidence of vectors and viruses in complex ways (Lewis and Dibley, 1970). Winged aphids tend not to fly when wind speed is too great, although their direction of flight can be influenced by the prevailing wind. At low wind speeds some species may fly with the wind and others against it (Fig. 15.11).

Freak wind conditions may cause annual epidemics. In 1977 there was a massive and unexpected epidemic of MDMV in the corn crop of the northern U.S. state of Minnesota. This virus had been usually confined to southern states. From studies of the continental weather patterns in 1977, and from the fact that aphid vectors could retain the virus for more than 19 hours, Zeyen *et al.* (1987) proposed that low-

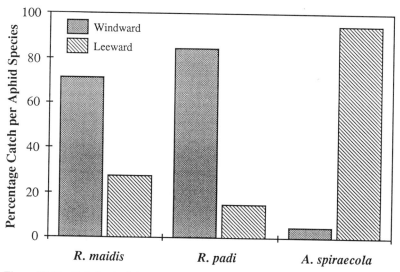

Figure 15.11 Variation in flying direction by different aphid species under low wind velocities. The percentage catch of selected aphid species on the windward and leeward sides of a sticky cylinder set vertically just above a soybean canopy, Urbana, Illinois, U.S.A. 1978. (Modified from Irwin and Ruesink, 1986.)

level jet winds rapidly transported infective aphid vectors from drought-stricken southern areas north to Minnesota (Fig. 15.12).

The direction of movement of leafhoppers may also be markedly influenced by wind speed and direction. The beet hopper *C. tenellus* cannot make progress against a headwind stronger than about 3 km per hour. Movement of whiteflies is also substantially influenced by wind (Fig. 15.13). Mealybugs, which transmit swollen shoot disease of cacao, are relatively inactive and move only short distances on the plant, although they are carried about by ants. They may, however, be carried some distance by wind.

The spread of black currant reversion virus by its gall mite vector (*Phytoptus ribis* Nal.) is influenced to a substantial extent by the prevailing wind over distances of about 15 m from infected plants. However, other factors, such as air turbulence caused by nearby buildings and trees, may influence the passive movement of this vector and spread of the virus (Thresh, 1966). With certain climates and soil types, nematodes of many genera may also be passively carried in windborne dust (Orr and Newton, 1971). However, virus vector species have not yet been identified in such dust.

C. Air Temperature

Air temperature may have marked effects on the rate of multiplication and movement of airborne virus vectors. For example, winged aphids tend to fly only when conditions are reasonably warm. However, very high temperatures may be particularly effective in reducing certain aphid populations.

Exposure of the planthopper vector of MRDV to 36°C prevented virus replication in the vector and suppressed transmission (Klein and Harpaz, 1970). Such air temperatures occur during the hot season in Israel.

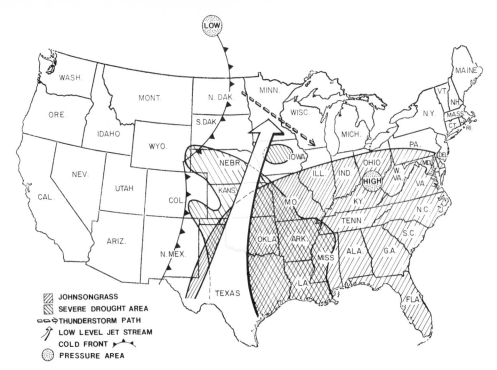

Figure 15.12 Hypothesis to account for long-distance aphid vector movement and a massive outbreak of MDMV in corn in Minnesota in 1977. The continental United States with the midcontinental and southern distributions of johnsongrass, the major wild host for MDMV (western distributions not shown), relative to the severe drought during the months of May and June, 1977. The large arrow indicates the path of low-level jet winds (up to 80 km/hour) on July 2, 1977. The smaller blocked arrow indicates the path of a cold front that caused the low-level jet winds to become diffuse and triggered thunderstorm activity, which moved in a south-southeasterly direction through Minnesota and into Wisconsin in the late evening of July 2 and early morning of July 3, 1977. (From Zeyen *et al.*, 1987.)

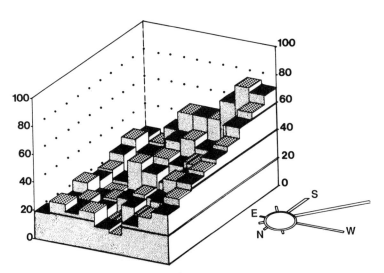

Figure 15.13 The effect of wind direction on the incidence of ACMV infection in a 0.8-ha field of cassava surrounded by a sugarcane windbreak at a coastal site in the Côte d'Ivoire. 0–100 = percentage infection in each block of plants. Wind frequency and direction are indicated by the length of the lines to the right of the block graph. The pattern of catch of whitefly vectors (*Bemisia*) was very similar to that shown for virus incidence. (From Fargette *et al.*, 1985.)

D. Soil

Conditions in the soil can influence the incidence of virus disease in various ways. Highly fertile soils tend to increase the incidence of virus disease. For example, animal manure and several inorganic fertilizers increased the incidence of leaf roll and rugose mosaic diseases in potato crops (Broadbent *et al.*, 1952). Some species of aphid vectors also multiplied faster on such plants. Plant nutrition may also influence the apparent amount of virus infection by accentuating or suppressing symptom expression.

Soil conditions can have a marked effect on the survival of TMV in plant debris. Moist well-aerated soils favor inactivation of the virus compared with dry, compacted, or waterlogged soils (Johnson and Ogden, 1929).

Soil temperature may have a marked effect on the transmission of viruses by nematodes. The optimum temperature and temperature range may vary with different viruses, hosts, and nematode species (e.g., Cooper and Harrison, 1973; Douthit and McGuire, 1975). However, in spite of seasonal fluctuations in temperature, populations of nematodes in the field may remain quite stable for years. Wheat spindle streak mosaic *Potyvirus* in Canada is transmitted by its fungal vector with an optimal soil temperature of 10°C (Slykhuis, 1974). This may be a geographical adaptation since soil borne wheat mosaic virus from Illinois develops optimally at about 16°C.

The physical soil type may also affect vector distribution, and thus the incidence of viral disease. Nepoviruses predominated in cruciferous species on light soils in the German Democratic Republic (Shukla and Schmelzer, 1975). In Scotland, TRV in potatoes was found in freely draining podsols but not on heavy, badly drained soils (Cooper, 1971). Soil type may influence virus incidence in yet another way—by the extent to which moisture is retained. For example, in Tasmania, BYDV infection was consistently higher in pastures on heavier soils, probably because plant growth was better on such soils under drought conditions (Guy, 1988).

E. Seasonal Variation in Weather and the Development of Epidemics

The wide annual variation that can occur in the incidence of virus disease in an annual crop is shown in Fig. 15.14 for beet yellowing viruses in sugar beet. Watson and Heathcote (1965) emphasized the importance of early migration of *M. persicae* for subsequent disease spread. However, other factors must be involved. For example, in 1945, there was a low incidence of *M. persicae* in June followed by a severe outbreak of virus infection. The year 1946 had the highest count of *M. persicae* in June for the 8 years studied, but this was followed by a low incidence of disease. During the 8-year period the extent of infection in nearby seed crops was almost certainly a major factor affecting incidence of virus. Given similar sources of virus inoculum, seasonal variations in virus incidence in a crop like sugar beet are probably due to continuing effects of weather conditions on the aphid vector population through most of the season—affecting timing and size of early migrations into the crop, buildup of population within the crop, and the mobility of this population (Fig. 15.14).

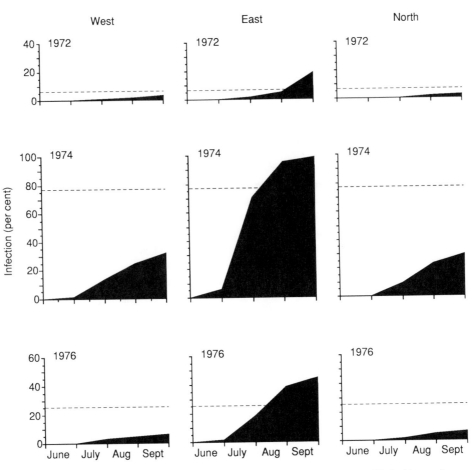

Figure 15.14 Variation in virus incidence with year and location. Mean monthly incidence of sugar beet yellows disease in different regions of England in years with greatly differing amounts of spread. Data supplied by G. D. Heathcote (Brooms Barn Experimental Station) for representative crops sown annually in March or April in eastern England (Suffolk, Essex, and Cambridgeshire), in wetter areas to the west (Shropshire and Worcestershire), and in cooler areas to the north (Yorkshire). The dashed lines indicate the mean incidence of yellows at the end of September as calculated for the entire English crop. (From Thresh, 1983a.)

A sequence of unusual weather conditions can lead to a severe outbreak of a disease that is normally present at fairly low levels in a crop in a particular region each year. The history of a severe epidemic of wheat streak mosaic virus that developed in southern Alberta in the autumn of 1963 will serve to illustrate the kinds of interactions involved. Winter wheat is normally sown in the first 2 weeks of September. This sowing date normally allows the crop to escape infection from maturing spring-sown wheat (see Fig. 16.8). In 1963, an unusual sequence of weather conditions prevailed (Atkinson and Slykhuis, 1963). Spring rainfall was much below normal. Crops of spring-sown grain were sparse and not much seed germinated. This situation was suddenly changed in late June by very heavy rains. In the area that had previously suffered from drought, seed of spring wheat and barley that had lain dormant germinated, now about a month late, and previously stunted plants tillered profusely. Above normal or near normal rainfall in June and July assured further vigorous development of the crop and, by reducing the effectiveness of summer fallow operations, allowed extensive and profuse development of volunteer wheat. In this zone where spring rainfall had been low, extensive acreages of wheat and barley did not mature until late September. The late-developing wheat and barley crops became massive reservoirs of inoculum for the virus and for its mite vector.

Winter wheat sown at the normally recommended time in the first 2 weeks of September then became heavily infected. The autumn weather was also abnormal and facilitated massive spread of virus by the mite. The mean September temperature was 62°F (10° above the 30-year average). These unusually high temperatures continued well into October. The 1964 season was not conducive to spread of the virus, but because of the very extensive spread that had occurred in the winter wheat before the freeze in 1963, very large acreages of grain were lost. The distribution of this epidemic outbreak in 1964 showed a clear relationship with the lack of spring rainfall in 1963 (Atkinson and Grant, 1967).

Three major epidemics of PLRV in potatoes lasting 5–10 years have occurred in the northern seed-growing areas of eastern North America and western Europe during this century. The climax years for these epidemics were 1912, 1944, and 1976 and were near the minima of every third sunspot cycle. The dry warm summers led to a northward movement of the aphid vector *Myzus persicae* (Bagnall, 1988).

An epidemic in which unusual wind patterns were thought to be a factor is illustrated in Fig. 15.12. The problem of forecasting disease outbreaks is discussed in Chapter 16, Section II,D.

III. SURVIVAL THROUGH THE SEASONAL CYCLE

Viruses can survive a cold winter period or a dry summer season in various ways. some have more than one means of survival.

1. Many viruses readily survive from season to season in the same host plant or propagating material from it. These include viruses in perennial hosts, in tubers, runners, and so on, and viruses that are transmitted through the seed. Similarly, crops stored over winter in the field, such as mangolds, may act as a source of both viruses and vectors.
2. Viruses with a wide host range are well adapted to survive, provided their hosts include perennial species, a group of annual species with overlapping growing seasons, or species in which seed transmission occurs.
3. Biennial or perennial wild plants, ornamental trees and shrubs, or weed hosts such as *Plantago* spp. may be important sources for overwintering or oversummering of many viruses (Hammond, 1982).
4. Leafhopper-transmitted viruses that pass through the egg of the vector may overwinter in the egg, or in young nymphs (Fig. 15.15).
5. TMV and a few other viruses can survive through the winter under appropriate conditions in plant refuse, and perhaps free in the soil. TMV can also survive in plant litter left in curing sheds, and so on. In the past, attention has been focused on virus surviving as free virus particles. However, it is known that many viruses infecting vertebrates may adsorb to inanimate surfaces and survive there (Gerpa, 1984). More attention should be paid to the possible

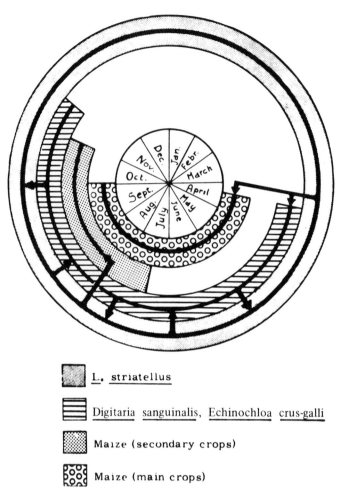

L. striatellus

Digitaria sanguinalis, Echinochloa crus-galli

Maize (secondary crops)

Maize (main crops)

Figure 15.15 Survival of maize rough dwarf *Fijivirus* through the seasonal cycle in northern Italy. *Laodelphax stratellus* overwinters as a young nymph in diapause. This vector carries the virus through the winter when no known plant hosts are available. The black line and arrows indicate movement of the virus to and from its various hosts. (From Conti, 1972.)

survival of plant viruses adsorbed to minerals. Piazzolla *et al.* (1989b) have shown that CMV adsorbs readily to montmonillonite (an expandible layer silicate). In this state it was resistant to inactivation by a strong salt solution.

6. Viruses carried within the resting spores of their fungal vectors may survive for long periods in soil.

7. Agricultural practices may allow the virus to survive in successive crops of the same plant grown in the same locality throughout the year. This may happen where the climate is suitable for production cropping throughout the year, where seed crops of the species are grown through the winter and overlap with production crops, or with wheat, where spring and winter crops are grown in the same area. In recent years substantial increases in the planting of winter oilseed rape in England pose a threat to vegetable crops because of a range of viruses have been identified in these winter crops (Walsh and Tomlinson, 1985).

IV. CONCLUSIONS

We can see from this brief account that all viruses do not possess all characteristics that are generally favorable to survival and spread, which is perhaps just as well. Each virus has some combination of properties that allows it to persist more or less effectively. TMV has no significant invertebrate vector, but survives because of its high resistance to inactivation, the high concentration it generally reaches in its host, the ease with which it can be mechanically transmitted, and its wide host range. By contrast, viruses that multiply in the insect vector and are transmitted through the egg have a complete alternative to the plant host for survival.

TSWV is an example of a very unstable virus that is extremely successful in the field. It can retain infectivity only for periods of hours under normal conditions outside the plant. It has a specialized type of vector (species of thrips that can acquire virus only as nymphs), and yet it is widespread around the world and common within individual countries. Its success probably depends on several factors (Best, 1968): (i) its very wide host range, including many perennial symptomless carriers that provide year-round reservoirs of the virus; (ii) the widespread distribution of species of thrips capable of transmitting the virus; and (iii) the existence of a wide range of strains in strain mixtures, and the probable occurrence of recombination or gene reassortment allowing rapid adaptation of the virus to new hosts to give relatively mild disease.

Many nematode-transmitted viruses are also transmitted through the seed in a range of weed and crop plant hosts. Infected seedlings often appear to be healthy. The combination of nematode and seed transmission may be important for the survival and dispersal of many of these viruses (Murant and Lister, 1967). Germination of infected weed seeds may allow reinfection of the nematode population in a field, and infected seeds may be transferred over longer distances than nematodes to set up new centers of infection. Murant and Lister noted that those viruses such as TBRV, which are commonly found in soils in infected weed seeds, do not persist more than about 9 weeks in their vector *Longidorus elongatus*. In contrast, two nepoviruses (arabis mosaic and grapevine fanleaf), which can persist for 8 months or more in the vectors *Xiphinema* spp., were rarely or never found in seed in infected soil.

Viruses existing in completely natural or almost natural habitats and their ecology have been very little studied. Most of our knowledge is a by-product of the investigation of commercially important diseases. It seems probable that under conditions where natural selection had operated on the virus and the host plant genotype over long periods undisturbed by the activities of humans, viruses would be closely adapted to their plant hosts and invertebrate or other vectors. The effects of virus infection on plant growth would probably be minimal. Severe disease in individual plants might occur occasionally, but epidemics would be rare.

Several investigations suggest that in wild plants, symptomless infection is the most common situation. Thus weed hosts of CMV in both England (Tomlinson *et al.*, 1970) and the German Democratic Republic (Shukla and Schmelzer, 1973) were infected without symptoms.

There may be regional adaptation between virus and host genotype to give mild disease. Tomlinson and Walker (1973) found that *Stellaria media* plants grown from seed from Britain, North America, and Australia were least severely affected by the CMV strain obtained from their country of origin. Thus natural selection of tolerant races of the species and mild strains of the virus had probably taken place.

A passionfruit virus causes disease in the introduced *Passiflora edulis* in the Ivory Coast. However, infections of indigenous *Adenia* spp. in remote areas were often symptomless (De Wijs, 1975). On the other hand, intensive agriculture may bring about complex ecological changes in nearby wild plant areas. Such changes may lead to more severe strains of a virus appearing in the wild flora, for example, with BCTV in California (Magyarosy and Duffus, 1977).

Agricultural practices, even in fairly early primitive stages, must have changed the environment in which viruses existed in several important ways.

1. Plant selection and later plant breeding have given rise to new host genotypes with differing reactions to the existing viruses.
2. Cultivation brought new communities of plants together, both the useful species and the weeds associated with cultivation.
3. These agricultural communities of plants have been transported about the world and have become neighbors of the indigenous floras. Viruses present in these have moved into the agricultural crops.
4. Along with the plants, insect vectors and potential vectors have been transported into the new areas, and insects already present in the new areas may have been able to spread into the agricultural crops.
5. Culture of a single species as the major plant over wide areas may allow the development of enormous populations of insect vectors and the consequent large-scale outbreaks of disease. This possibility may be made more likely by the maintenance of a particular species throughout the year in the same locality (e.g., the growing of seed crops of biennial plants such as sugar beet) or by the introduction of weeds that harbor the virus.
6. Agricultural practices such as cultivation and drainage may change soil conditions so as to markedly affect the population of soil-inhabiting virus vectors.
7. Grafting procedures, apart from their role in spreading viruses, may have allowed the expression of new diseases by selection of virus strains and by the introduction of viruses into previously uninfected species and varieties of plants.

A few agricultural practices may have played a part in unconscious selection of mild strains of a virus but, generally speaking, the foregoing factors combine in various ways continually to produce new and unstable ecological situations in which outbreaks of disease can occur in agricultural and horticultural crops. On the other hand, in regions where a crop has been grown continually for long periods under conditions favoring infection with a virus, resistant cultivars may have developed without a conscious selection procedure. For example, many cultivars of barley from the Ethiopian plateau have been found to have a high degree of resistance to BYDV and BSMV (Harlan, 1976). It is probable that barley has been grown in this region for millenia.

16

Economic Importance and Control

I. ECONOMIC IMPORTANCE

Accurate global figures for crop losses due to viruses are not available. However, some idea of the scale can be obtained by considering that plant disease losses worldwide have been assessed as high as U.S. $60 billion per year (Klausner, 1987). It is generally accepted that of the various plant pathogens, viruses come second only to fungi with respect to the disease losses they cause. Losses differ greatly between and within different countries and regions. Not infrequently, it is developing countries that are dependent on one or a few major crops, and where control measures are inadequate, that are affected most seriously. Crop losses caused by viruses have been discussed by Bos (1982).

Viroids as a group of agents are limited in the range of crop species they affect compared with viruses. Nevertheless, they can cause severe problems in specific crops, for example, cadang-cadang disease of coconuts, potato spindle tuber disease, and chrysanthemum stunt disease. The cadang-cadang viroid has killed over 40 million coconut trees over about 40 years, and losses of about half a million trees each year continue (Garnsey and Randles, 1987).

A. Measurement of Losses

It is often difficult to obtain reliable data either on direct crop losses due to virus disease or on the cost of control. There is wide variation in the extent of direct losses with different crops in different seasons and in various regions, so that average figures are not particularly useful (see Fig. 15.14). We can distinguish three general kinds of situation in which losses occur.

1. Perennial crops, often trees, where a lethal or crippling disease can have very serious consequences because of the time and land invested in such crops. Examples are the tristeza-disease of citrus on most continents, and the swollen shoot disease of cocoa in West Africa.

2. Annual crops sown from seed, for example, many vegetables and grains,

where severity of a particular disease may fluctuate greatly from season to season. For example, virus diseases caused an average annual loss of wheat of about 2% over a decade in the United States, but in Kansas in 1959 losses were estimated at 20% (Losses in Agriculture, 1965).

3. Vegetatively reproduced plants, for example, potatoes and many fruit and ornamental species. In such crops, mild virus infection may be very widespread, often occurring in every individual plant, and may reduce performance every year by a relatively small amount, perhaps of the order of 10%.

The ease with which it is possible to obtain meaningful information on the effects of virus infection on productivity depends on the kind of plant and crop. With annual crops grown from seed it is usually possible to obtain reasonable estimates of yield reduction, provided virus effects are not obscured by such factors as inadequate nutrition or the effects of other diseases or parasites. Accurate yield records are usually available for crops such as wheat or sugar beet that are delivered to silos or factories (Fig. 16.17). When disease incidence is also known in some detail, fairly accurate assessments of losses can be made (e.g., Atkinson and Grant, 1967). Estimates of yield reduction in vegetatively propagated plants have sometimes been impossible because no healthy material was available for comparison. For many years it was thought that potato paracrinkle Carlavirus in King Edward potatoes had no effect on growth. However, when Kassanis (1965) obtained a clone of this variety free of the virus, he was able to show that yield was in fact about 10% better than high-grade commercial stocks of the variety, and the tubers were more uniform in size.

Disease ratings based on symptom severity can be used to assess reduction in yield, provided that the relationship between the ratings and yield reduction has been established. For example, potatoes with slight, moderate, and severe symptoms due to PLRV yielded 65, 80, and 92% less than symptomless plants (Harper *et al.*, 1965). This information was used to estimate yield losses from field observations on numbers of plants in each symptom category.

Aerial photography can be used to assess the extent of infection by a virus that produces symptoms giving a recognizable "signature" in the photograph. Development of an effective procedure requires a knowledge of the most effective wavelength region to use for distinguishing healthy and infected plants (Hilty, 1981). Verification of the technique by ground inspection is necessary.

Field trials to assess crop losses need careful design if the results are to reflect actual losses in commercial plantings. For example, Fargette *et al.* (1988) found that yield reductions in individual cassava plants due to African cassava mosaic *Geminivirus* were much greater in plots where diseased plants were intermixed with healthy ones and subjected to interplant competition than when they were assessed in separate healthy and infected plots.

Losses may be particularly severe in certain regions of the world. For example, it has been estimated that although 11% of the area in the world devoted to maize is in tropical Africa, that region produces only 3% of the total crop. Maize streak virus is often a major contributor to this lower productivity (Connolly, 1987).

B. Biological and Physical Factors Affecting Losses

The extent to which yield is reduced will depend in any particular year and locality on many factors, including variety of host plant and strains of the virus present, the incidence and activity of any vectors, the time at which infection occurs, the nutritional state of the crop, the weather, and the presence of other parasites.

Double infection with two viruses may cause greater than additive effects. Yield of soybean was reduced 8–25% by different strains of SBMV. Beanpod mottle *Comovirus* caused losses up to 40%. Double infection with both viruses caused reductions up to 80% (Ross, 1968). With few exceptions the earlier that viruses infect plants, the more severe the reduction in yield. This is well illustrated by the effects of carrot motley dwarf disease transmitted by the aphid *Caraviella aegepodii* (Scop) (Fig. 16.1).

About 2.5 Tonnes/hectare of carrots were lost for each 10-fold increase in the number of migrants caught on sticky traps in the field. Interpretation of such data may be complicated by the fact that dry periods during spring, which encourage aphid flight and multiplication, may also affect crop yield in other ways, for example, by reducing seed germination.

In addition to the decreased effects of late infection on yield, virus frequently spreads more slowly and less efficiently within old plants. For example, Broadbent *et al.* (1957) found that tubers from potato plants infected late in the season often remained free of PLRV.

The effect of BYDV infection at different growth stages is illustrated in Fig. 12.7 for barley and in Fig. 16.2 for oats.

The effect of virus infection on plant growth and yield may vary with time after infection. The kind of change may be very different for various hosts and viruses. For example, early pickings of cantaloupe were markedly reduced by squash mosaic *Comovirus* infection; but over the whole season the effect of infection was proportionately much less (Fig. 16.3). By contrast, cocoa swollen shoot virus progressively stunted and killed west African cocoa trees (Fig. 16.4).

Apart from overall yield reduction, composition of the commercial end product may be significantly altered by virus infection. For example, total seed oil was decreased and protein increased in soybeans infected with TRSV. The fatty acid composition of the oil was markedly changed (Demski *et al.*, 1971).

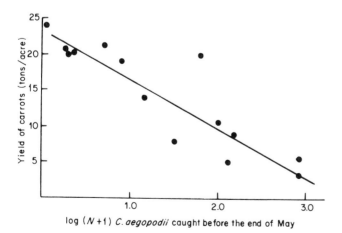

Figure 16.1 Relationship between time of migration of *Cavariella aegopodii*, the vector of carrot motley dwarf virus, and yield of carrots. Data for the period 1959–1965. *N* = Numbers of the vector caught before the end of May, in England. (From Watson and Heathcote, 1965.)

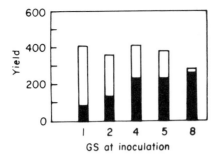

Figure 16.2   Effect of growth stage (GS) at time of inoculation with BYDV on the yield of spring oats (variety 'Condor'). Yield is given as mean weight of grain (g) per plot. ↖, infected; ■ + □, healthy. (Redrawn from Doodson and Saunders, 1970.)

Virus infection may act indirectly to affect various properties of the crop plant that, in turn, cause a reduction in yield. For example, in some varieties of wheat and oats in Canada, BYDV infection severely reduced coldhardiness and the ability to survive after encasement in ice (Paliwal and Andrews, 1979).

When new varieties are introduced into a region they may suffer severe losses compared with varieties grown previously. This is because viruses may be present in the region for which the new variety has no resistance or tolerance. For example, new high-yielding varieties of wheat introduced into central Italy were found to be extremely susceptible to soilborne wheat mosaic *Furovirus,* with yield losses estimated to be as high as 70% (Vallega and Rubies-Antonell, 1985).

Another aspect of variation between cultivars is illustrated for garlic, a crop in which chronic virus infection has been universal. When virus-free bulbs of several cultivars were obtained by a combination of meristem tip culture and heat therapy, some cultivars gave improved yields while one did not (Walkey and Antill, 1988).

Additional complexities in assessing effects on yield are involved when the species concerned is grown as a member of an ecological community containing other species not affected by the virus in question. Some of these complications are

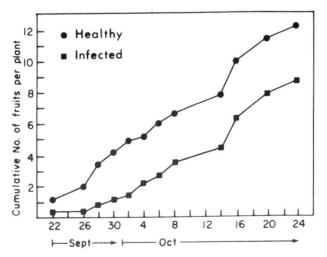

Figure 16.3   Effect of squash mosaic *Comovirus* on the cumulative number of fruits of cantaloupe produced over 12 harvest times. (From Alvarez and Campbell, 1976.)

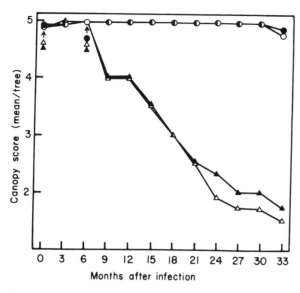

Figure 16.4 The progressive effect of cocoa swollen shoot virus on the canopy condition of West African cocoa trees grown without shade. Canopy scores: 5, dense canopy; 0, defoliated dead tree; 1–4, intermediate states; ○, uninfected; ●, uninfected and receiving fertilizer; △, virus infected; ▲, virus infected and receiving fertilizer. (From Brunt, 1975.)

well illustrated by the work of Catherall and Griffiths (1966a,b) on the effects of a cockfoot virus on cockfoot (*Dactylis glomerata* L). The effects of virus infection on yield of the grass varied depending on whether trials were carried out with plants in pots held outside, trials with plants spaced in the open ground, or trials in a mixed sward with other species.

In another example with grasses, BYDV reduced the aggressiveness of perennial more than Italian ryegrass. The reverse occurred with ryegrass mosaic *Potyvirus* (Catherall, 1987). In mixed swards, perennial ryegrass was suppressed by Italian ryegrass only when BYDV was present, and the status quo was maintained with both viruses present even though overall yield was reduced by 16%.

C. Economic Factors

The loss in actual yield per plant or per acre may not be the only aspect of importance to a grower. In a crop picked over a season, and where the product is graded, the time at which losses occur and the effect of infection on quality may be important. For example, Broadbent (1964) noted that the quality of tomato fruits was severely affected by late infection with TMV. From the prevailing market prices, he calculated financial losses as shown in the following tabulation.

Date plants infected with TMV	Percentage financial loss	
	On a weight basis only	Taking quality into account
March 3	19	19
April 14	18	27
May 31	12	33

In an open market, the relationship between weight loss and financial loss for individual growers is further complicated by the question as to whether most growers supplying a market had similar losses. If losses due to a disease are fairly evenly distributed, price rises will tend to compensate for the losses sustained. Where only a few growers are affected, the financial loss to certain growers may be much more severe for the same loss of crop.

The cost of any control program must be measured in relation to the yield gain. Thus Heathcote (1978) estimated that for the period 1970–1975 in England, the amount of insecticide applied to sugar beet crops cost less than one-third of the resultant increase in yield.

Losses due to virus disease must also be viewed in relation to the economy of the country concerned and the importance of the product in international trade. For example, the production of cocoa in Ghana is a major factor in the economy of the country. From 1947 to 1969, 140 million trees infected with swollen shoot virus were felled during the control campaign (Braudeau, 1969), and production fell dramatically. In 1935–1936, production of dry cocoa was 116,000 tons. This fell to 38,000 tons in 1955–1956, much of the loss being attributable to cocoa swollen shoot virus (Dale, 1962).

II. CONTROL MEASURES

The use of fungicidal chemicals that when applied to crop plants protect them from infection or minimize invasion is an important method for the control of many fungal diseases. No such direct method for the control of virus diseases is yet available. Most of the procedures that can be used effectively involve measures designed to reduce sources of infection inside and outside the crop, to limit spread by vectors, and to minimize the effect of infection on yield. Generally speaking, such measures offer no permanent solution to a virus disease problem in a particular area. Control of virus disease is usually a running battle in which organization of control procedures, care by individual growers, and cooperation among them is necessary year by year. The few exceptions are where a source of resistance to a particular virus has been found in, or successfully incorporated into, an agriculturally useful cultivar. Even here, protection may not be permanent when new strains of the virus arise that can cause disease in a previously resistant cultivar.

I shall consider here the kinds of measures that have been tried for the control of virus diseases. It should be borne in mind that virus infection can sometimes increase the incidence of some other kind of disease. In such situations different sorts of control may be needed. For example, yellowing viruses in sugar beet increase susceptibility to *Alternaria* infection. Spraying with appropriate fungicides reduced this secondary effect of virus infection in some seasons (Russell, 1966).

Correct identification of the virus or viruses infecting a particular crop is essential for effective control measures to be applied. Disease symptoms alone may be very misleading. For example, virus disease in lettuce can be caused by some 14 viruses with an aphid, leafhopper, thrip, nematode, or a fungus vector (Cock, 1968). Many of these viruses produce brown necrotic spots or bronzing on leaves, and later chlorotic stunting. Another example is that of the yellowing diseases of beets in the United Kingdom (Russell, 1958) and the western United States.

It has become apparent that most of the serious virus disease problems around the world are the direct or indirect result of human activities (Thresh, 1982).

Important activities leading to epidemics include (i) introduction of viruses into new areas through transport of infected seed or vegetative material; (ii) introduction of virus vectors into new areas; (iii) introduction of a new variety of a crop into an area when that variety is especially susceptible to a virus already present there; (iv) use of monocultures, that is, the planting of genetically uniform crops over large areas replacing traditional polycultures; (v) use of irrigation to prolong the cropping season; (vi) repeated use of the same fields for the same crop; and (vii) increased use of fertilizer and herbicides or other forms of weed control.

In recent years the most active areas of research into the control of virus diseases have been (i) the breeding of resistant or immune cultivars by classic genetic procedures; (ii) the control of vectors by various strategies; (iii) the production of virus-free stocks of seed and vegetative propagules; and (iv) the production of transgenic plants containing viral genes that confer resistance to the virus. Approaches to control are discussed in Harris and Maramorosch (1982).

Data for many of the control measures discussed in this chapter are derived from laboratory and field trials. Because of the many variables, and the large number of countries involved, it is often difficult to assess the extent to which any particular procedure or set of procedures has actually been adopted on a regular basis into commercial practice.

More and more attention is being given to the possibilities of integrated control involving several strategies. Several of these are noted in this chapter. Again, it is often difficult to assess whether these are actually being effectively used in the field, or whether they remain optimistic dreams. When adequate facilities and expertise are available, a multidisciplinary approach may be useful. The attempts to control diseases caused by TSWV in Hawaii are an example (Cho *et al.*, 1989). Cassava is the third largest plant source of calories in the world, being used for human and animal food and for the production of industrial ethanol. Recently an international project has been set up to attempt to control the virus diseases in this crop on several continents (Fauquet and Beachy, 1989).

A. Removal or Avoidance of Sources of Infection

1. Removal of Sources of Infection in or near the Crop

It is obvious that there will be no virus problem if the crop is free of virus when planted and when there is no source of infection in the field, or none near enough to allow it to spread into the crop.

The extent to which it will be worthwhile to attempt to eliminate sources of infection in the field can only be decided on the basis of a detailed knowledge of such sources and of the ways in which the virus is spreading from them into a crop. Eradication as a control measure has been reviewed by Thresh (1988b).

Living Hosts for the Virus

Living hosts as sources of infection may include (i) perennial weed hosts, annual weed hosts in which the virus is seed transmitted, or annual weed hosts that have several overlapping generations throughout the year; (ii) perennial ornamental plants that often harbor infection in a mild form; (iii) unrelated crops; (iv) plants of the same species remaining from a previous crop; these may be groundkeepers, as with potatoes, or seedling volunteers; and (v) seed crops of biennial plants that may be approaching maturity about the time the annual crop is emerging.

In theory, it should be possible to eliminate most of such sources of infection. In practice, it is usually difficult and often impossible, particularly in cropping areas that also contain private gardens. Private household gardens in temperate and sub-tropical regions often contain a diverse collection of plants, many of which can carry economically important viruses. It is usually difficult or impossible to control such gardens effectively.

The extent to which attempts to remove other hosts of a virus from an area may succeed will depend largely on the host range of the virus. It may be practicable to control alternative hosts where the virus has a narrow host range, but with others, such as CMV and TSWV, the task is usually impossible.

Plant Remains

Plant remains in the soil, or attached to structures such as greenhouses, may harbor a mechanically transmitted virus and act as a source of infection for the next crop. With a very stable virus like TMV, general hygiene is very important for control, particularly where susceptible crops are grown in the same area every year. ToMV may be very difficult to eliminate completely from greenhouse soil using commercially practicable methods of partial soil sterilization (Broadbent et al., 1965). A major development has been the replacement of soil by sand/peat substrates that are renewed every year.

Roguing

Sometimes it may be worthwhile to remove infected plants from a crop. If the spread is occurring rapidly from sources outside the crop, roguing the crop will have no beneficial effect. If virus spread is relatively slow and mainly from within the crop, then roguing may be worthwhile, especially early in the season. Even with a perennial crop, if a disease spreads slowly, roguing and replanting with healthy plants may maintain a relatively productive stand. A study of the distribution of infected plants within the field using the formula developed by Van der Plank (Chapter 15, Section I,B,1) may give an indication as to whether spread within the crop is taking place.

In certain situations roguing may increase disease incidence by disturbing vectors on infected plants (Rose, 1974). In many crops, newly infected plants may be acting as sources of virus for further vector infection before they show visible signs of disease (e.g., Beemster, 1979).

Regular roguing of infected plants has been effective in the control of ACMV in cassava in trials carried out in tropical Africa (Robertson, 1987), but the method is not widely used. One of the most successful examples of disease control by roguing of infected crop plants has been the reduction in incidence of the bunchy top virus in bananas in eastern Australia. Legislation to enforce destruction of diseased plants and abandoned plantations was enacted in the late 1920s. Within about 10 years the campaign was effective to the point where bunchy top disease was no longer a limiting factor in production. Dale (1987) attributes the success of the scheme to the following main factors: (i) absence of virus reservoirs other than bananas, together with a small number of wild bananas; (ii) knowledge that the primary source of virus was planting material, and that spread was by aphids; (iii) cultivation of the crop in small, discrete plantations, rather than as a scattered subsistence crop; and (iv) cooperation of most farmers.

Hygiene

For some mechanically transmitted viruses, and particularly for TMV or ToMV, human activities during cultivation and tending of a crop are a major means by which the virus is spread. Once TMV or ToMV enters a crop like tobacco or tomato, it is very difficult to prevent its spread during cultivation and particularly during such processes as lateraling and tying of plants. Control measures consist of treatment of implements and washing of the hands. For this, Broadbent (1963) recommends a 3% solution of trisodium orthophosphate. Workers' clothing may become heavily contaminated with TMV and thus spread the virus by contact. TMV persisted for over 3 years on clothing stored in a dark enclosed space (Broadbent and Fletcher, 1963), but was inactivated in a few weeks in daylight. Clothing was largely decontaminated by dry cleaning or washing in detergents with hot water.

While TMV is the most stable of the mechanically transmitted viruses, others can be transmitted more or less readily on cutting knives and other tools. These include tulip breaking *Potyvirus,* cymbidium mosaic *Potexvirus,* PVX, and PSTVd. These mechanically transmitted agents may be a particular problem in glasshouse crops, where lush growth, close contact between plants, high temperatures, and frequent handling are important factors in facilitating virus transmission.

Since mechanical transmission is an important means whereby viroids are spread in the field, decontamination of tools and hands is an important control measure. However, this presents difficulties because of the stability of viroids to heat and many decontaminating agents as normally used. Brief exposure to 0.25% sodium or calcium hypochlorite is probably the best procedure (Garnsey and Randles, 1987). One of the earliest and most effective methods for the control of viroids has been the avoidance of sources of infection. This method has been particularly successful for vegetatively propagated crops susceptible to viroids, such as potatoes and chrysanthemums. This success is probably due to the absence of vectors in the field apart from humans.

2. Virus-Free Seed

Where a virus is transmitted through the seed, such transmission may be an important source of infection, since it introduces the virus into the crop at a very early stage, allowing infection to be spread to other plants while they are still young. In addition, seed transmission introduces scattered foci of infection throughout the crop. Where seed infection is the main or only source of virus, and where the crop can be grown in reasonable isolation from outside sources of infection, virus-free seed may provide a very effective means for control of a disease.

Lettuce mosaic *Potyvirus* is perhaps the best example of this. Grogan *et al.* (1952) found that crops grown from virus-free seed in California had a much lower percentage of mosaic at harvest than adjacent plots grown from standard commercial seed. For a time, control was unsatisfactory, until it was realized that even a small percentage of seed infection could give much infection within the crop if the aphid vector was active (Zink *et al.,* 1956).

Tomlinson (1962) obtained similar results under English conditions. To obtain effective control by the use of virus-free or low-virus seed, a certification scheme may be necessary, with seed plants being grown in appropriate isolation. More recent work indicated that even 0.1% seed transmission did not give effective control of lettuce mosaic (Kimble *et al.,* 1975). Only seedstocks that test <0.003% infection are now used in the Salinas Valley area of California. This has given consistent control in practice. Thirty thousand seedlings from each seed lot are grown in a screened greenhouse and observed for mosaic-infected plants. In recent years ELISA tests have replaced other seed-testing procedures.

In the 5 years prior to the introduction of the virus-free seed program in the Salinas Valley area of California, lettuce yields were about 140 cartons/hectare. In the 5 years immediately following introduction of the scheme, average yield was 190 cartons/hectare. Most of this increase can be attributed to the reduction in losses due to lettuce mosaic virus (Kimble *et al.*, 1975).

ELISA tests are now available to test for the presence of a virus in batches of seed. If there is a significant amount of virus in the seed coats that is not involved in seed transmission it may be necessary to remove the seed coats before testing the embryos (Maury *et al.*, 1987). Seed certification schemes against barley stripe mosaic *Hordeivirus* have been credited with avoiding millions of dollars in losses due to this disease in barley crops in some U.S. states (Carroll, 1983).

Tomato seed from ToMV-infected tomatoes carries the virus on the surface of the seed coat. As the seed germinates, virus contaminates the cotyledons and is inoculated into the plant by handling during pricking out. Seed can be cleaned by extraction in HCl, heating dried seed in 0.1 $N/\pm$HCl, or treatment with trisodium orthophosphate or sodium hypochlorite (e.g., Gooding, 1975).

It may be particularly important to attempt to eliminate seed-transmitted viruses from national or international germ plasm collections. Among 207 *Phaseolus vulgaris* accessions in the USDA germ plasm collection, about 60% contained some seed infected with bean common mosaic *Potyvirus* (Klein *et al.*, 1988). In at least one instance, that of pea seedborne mosaic *Potyvirus* in peas, elimination of infected individuals led to a loss of genetic diversity, as judged by seed coat color and isoenzyme genotypes (Alconero *et al.*, 1985). In this circumstance methods that do not involve selective loss of particular plant types might be used to eliminate the virus (see the following section).

3. Virus-Free Vegetative Stocks

For many vegetatively propagated plants, the main source of virus is chronic infection in the plant itself. With such crops, one of the most successful forms of control has involved the development of virus-free clones, that is, clones free of the particular virus under consideration. Two problems are involved. First, a virus-free line of the desired variety with good horticultural characteristics must be found. When the variety is 100% infected, attempts must be made to free a plant or part of a plant from the virus. Second, having obtained a virus-free clone, a foundation stock or "mother" line must be maintained virus free, while other material is grown up on a sufficiently large scale under conditions where reinfection with the virus is minimal or does not take place. These stocks are then used for commercial planting.

Methods for Identification of Virus-Free Material

Visual inspection for symptoms of virus disease is usually quite inadequate when selecting virus-free plants. Appropriate indexing methods are essential. A variety of methods are available, and the most suitable will depend on the host plant and virus. For many viruses, especially those of woody plants, the rather laborious process of graft-indexing to one or more indicator hosts is essential. Distribution of a virus within the tree may be uneven, especially in the early stages after infection, so that tests repeated in successive seasons may be necessary to ensure freedom from virus. For example, Hampton (1966) found, using four buds for indexing per tree, that first year infection with prune dwarf *Ilarvirus* was not detected in a high proportion of cherry trees (29–63%, depending on the variety). The probability of detection improved substantially in trees that had been infected for 3 years. Distribution may also be uneven in herbaceous plants (Chapter 10, Section IV,H). Thus, Beemster (1967) found that in potatoes inoculated with PVY, not all tubers were infected and not all parts of a single tuber might be infected. The heel end of tubers was less frequently infected than the rose end. The rose end was, therefore, a more reliable source material for testing.

Mechanical inoculation to indicator hosts can be used with some viruses, but other methods of diagnosis now rival infectivity tests in their sensitivity for the detection of viruses (Chapter 2).

Methods for Obtaining Virus-Free Plants

Naturally Occurring Virus-Free Material   Occasionally, individual plants of a variety, or plants in a particular location may be found to be free of the virus. If all plants are infected, advantage may sometimes be taken of uneven distribution of the virus in the plant. This is not uncommon with some viruses in fruit trees. Budwood can be taken from uninfected parts of the tree. The shoot tips of rapidly growing stems may sometimes be free of a virus that is systemic through the rest of the plant. Thus, Holmes (1955) was able to obtain dahlia cuttings free of TSWV. This sort of procedure has been used successfully for several viruses in certain hosts. However, many vegetatively reproduced varieties appear to be virtually 100% infected with a virus, and with these, one or more of the special treatments and methods described next have to be used to obtain a nucleus of virus-free material. Some examples have been reported where natural elimination of a virus occurred. When TRV-infected tulip bulbs were grown for several seasons in soil free of the nematode vector, a proportion of the bulbs were found to be free of the virus (van Hoof and Silver, 1976).

Heat Therapy   Heat treatment has been a most useful method for freeing plant material from viruses. Many viruses have been eliminated from at least one host plant by heat treatment (Walkey, 1985).

Two kinds of plant material have been used. Dormant plant parts such as tubers or budwood can generally stand higher temperatures than growing tissues, and the method probably depends on direct heat inactivation of the virus. Temperatures and times of treatment vary widely (35–54°C for minutes or hours). Hot water treatments are often used, as hot air tends to give uneven heating during short treatments. Unless tissues are thoroughly hydrated, dry heat is much less effective than wet heat.

Growing plants are much more generally treated, and hot air rather than hot water is applied. Temperatures in the range of 35 to 40°C for periods of weeks are commonly employed. This form of treatment gives a better survival rate for growing plant material. Details of the treatment vary widely and have to be worked out empirically for each host–virus combination. Very frequently, small cuttings are taken from the shoot tips immediately after the heat treatment, as these may be free of virus when the rest of the plant is not. For example, culture of shoot tips from heat-treated sprouting potatoes gave a useful proportion of plantlets free of PVX (Faccioli and Rubies-Autonell, 1982). The percentage of garlic plants free of three viruses that were regenerated from meristem tip culture increased from 25–50% to 85% when infected plants were treated at 38°C prior to tip culture (Walkey *et al.*, 1987).

Alternatively, apical explants established in culture may be given the heat treatment. Thus strawberry mild yellow edge disease was eliminated from 4-mm-long strawberry stolon explants held at 38°C (Converse and Tanne, 1984). These authors suggest that heat therapy and stolon apex culture contribute independently to the process.

A regime of alternating high ($\approx 40°C$) and normal ($\approx 20°C$) temperatures may help to reduce the damaging effects of high temperatures on the plant tissues while still allowing some shoot tips to be freed of virus. Detailed regimes have to be developed for each host and virus. This procedure has been used for CCMV in cowpea (Lozoya-Saldana and Dawson, 1982) and for grapevine fanleaf and arabis mosaic nepoviruses in grapes (Monette, 1986). Reduction and elimination of virus may depend on the total hours held at the high temperature, as illustrated for CMV in Fig. 16.5. The actual temperature within plant tissues may be several degrees below the measured ambient temperature.

At present there is no basis for predicting that tissue from a certain plant species can, or cannot, be freed of a particular virus. The mechanisms underlying the preferential elimination of virus are not yet understood, but they presumably involve inactivation both of intact virus already synthesized and of the means for making more.

Meristem Tip Culture   The distribution of virus in apical meristems was considered in Chapter 10, Section IV,H,7. Culture of meristem tips has proved an effective way of obtaining vegetatively propagated plants free from certain viruses. Hollings (1965) defined meristem tip culture as aseptic culture of the apical meristem dome plus the first pair of leaf primordia. This piece of tissue is about 0.1–0.5 mm long in different plants. The minimum size of tip that will survive varies with different species. For example, it was necessary to use tips at least 0.3 mm long to obtain survival of rhubarb (Walkey, 1968). The kind of meristem tip usually taken is illustrated in Fig. 16.6 (The smaller the excised tips are at the time of removal the better the chance that they will give rise to virus-free plants, although the more difficult they are to regenerate.) For many plants, at least with the culture methods

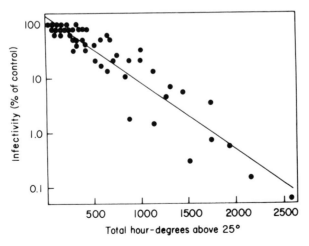

Figure 16.5   Relationship between inactivation of CMV in infected tissue cultures of *Nicotiana rustica* and the total hour-degrees above 25°C resulting from various alternating diurnal temperature regimes. (From Walkey and Freeman, 1977.)

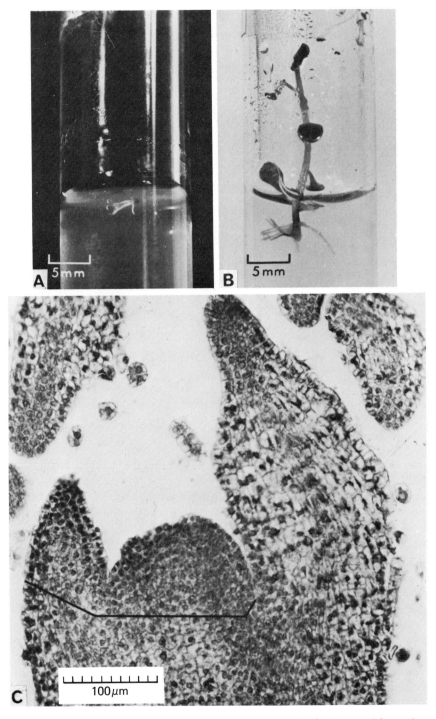

Figure 16.6 Apical meristem tip culture. (A) Stem of potato a week after it emerged from a dormant meristem. (B) Plantlet of potato at a stage ready to be transferred to soil. (C) A histological section along the axis of the apical meristem of a potato sprout showing a two-leaf primordium. The piece including one leaf primordium that is excised for tip culture is shown above the black line. (From Kassanis and Varma, 1967.)

currently used, one leaf primordium needs to be included to get regeneration of a complete plant.

A wide range of nutrient media has been used by different workers. The basic ingredients are an appropriate selection of mineral salts (macro and micro nutrients), sucrose, and one or more growth-stimulating factors such as indole acetic acid or gibberellic acid, sometimes in agar.

Only a proportion of meristem tip cultures yield virus-free plants. It is not always clear to what extent the success of the method depends on (i) the regular absence of virus from meristem tissue, some tips being accidentally contaminated; (ii) some meristem regions in the plant containing virus and others containing none; or (iii) virus present in the meristem being inactivated during culture on the synthetic medium.

Some viruses, for example, those present in members of the Araceae (Hartman, 1974), appear to be readily eliminated by culture in a suitable medium at 20°–25°C. For others, such as TRSV and PVX, most or all such cultures remain infected. In this situation it is now common practice to combine meristem tip culture with heat therapy as discussed in the preceding section. Figure 16.7 illustrates a protocol for the combined use of thermotherapy and meristem tip culture.

As an alternative to direct culture on a synthetic medium, shoot tips up to $\simeq 1.0$ mm long may be grafted aseptically onto virus-free seedling plants growing *in vitro*. Several viruses have been eliminated from peaches by this procedure (Navarro *et al.*, 1983).

Where the virus occurs in the apical meristem it may be difficult to obtain virus-free plants by tip culture (Toussaint *et al.*, 1984). Virus-free plants of several important tropical food crops have been obtained using meristem tip culture. These include cassava (Adejare and Coutts, 1981), sweet potato (Frison and Ng, 1981), and various aroids (Zettler and Hartman, 1987).

Tissue Culture   Culture of single cells or small clumps of cells from virus-infected plants may sometimes give rise to virus-free plants. For example, two viruses were eliminated from *Euphorbia pulcherrima* by cell suspension culture followed by regeneration of plants *in vitro* (Preil *et al.*, 1982). Plants regenerated from shake subculture of small pieces of tobacco tissue originally from TMV-infected plants were free of TMV (Toyoda *et al.*, 1985). A significant proportion of calli obtained from yellow-green areas of TMV-infected *Nicotiana tomentosa* gave rise to regenerated plants that were free of virus (White, 1982). This may mean that some cells in TMV-infected zones of the leaf are in fact free of virus, as appears to be the situation for some cells in TYMV local lesions (Chapter 10, Section IV,H).

Low Temperatures   The effect of holding plants at lower than normal temperatures on virus survival has not been widely investigated. Low temperatures might be expected to have little effect on viruses that are stable *in vitro*. In a few instances, growth at low temperatures has given virus-free plant material. Selsky and Black (1962) grew cuttings from sweet clover plants infected with WTV at 14°C, and no tumors developed even after several vegetative generations. After three generations, cuttings were taken from the plants and grown under normal greenhouse conditions. Ninety-five percent of these gave rise to a second generation of greenhouse-grown cuttings that were 90% free of virus as indicated by absence of tumors. The way in which cold treatment acts is unknown.

Potato and chrysanthemum plants freed from four different viruses were re-

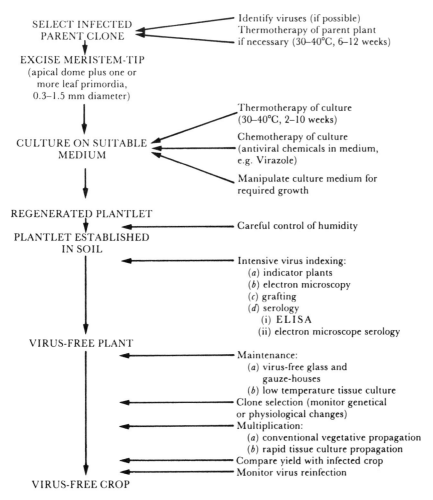

SELECT INFECTED
PARENT CLONE ◄─────── Identify viruses (if possible)
                      Thermotherapy of parent plant
                      if necessary (30–40°C, 6–12 weeks)

EXCISE MERISTEM-TIP
(apical dome plus one or
more leaf primordia,
0.3–1.5 mm diameter)

CULTURE ON SUITABLE ◄─── Thermotherapy of culture
MEDIUM   (30–40°C, 2–10 weeks)

                         Chemotherapy of culture
                         (antiviral chemicals in medium,
                         e.g. Virazole)

                         Manipulate culture medium for
                         required growth

REGENERATED PLANTLET

PLANTLET ESTABLISHED ◄─── Careful control of humidity
IN SOIL

                         Intensive virus indexing:
                         (a) indicator plants
                         (b) electron microscopy
                         (c) grafting
                         (d) serology
                             (i) ELISA
                             (ii) electron microscope serology

VIRUS-FREE PLANT

                         Maintenance:
                         (a) virus-free glass and
                             gauze-houses
                         (b) low temperature tissue culture
                         Clone selection (monitor genetical
                         or physiological changes)
                         Multiplication:
                         (a) conventional vegetative propagation
                         (b) rapid tissue culture propagation
                         Compare yield with infected crop
                         Monitor virus reinfection

VIRUS-FREE CROP

Figure 16.7 A scheme for virus-free plant production by meristem tip culture. (From Walkey, 1985.)

generated from meristem tip cultures that had been held for 6 months in the dark at 6–7°C (Paduch-Cichal and Krycyznski, 1987).

Chemotherapy   Attempts to free infected plant material of a virus solely by the application of antiviral chemicals have been disappointing. There have been several reports of such cures, but they have often been based on very small numbers of plants, or the results have been open to other interpretations. Chemical treatment by itself has not yet found practical use. However, chemical treatment in combination with heat treatment or meristem tip culture may have been an advantage in a few instances.

Several compounds have been used. For example, Hansen and Lane (1985) used ribavirin (also called virazole), an analog of guanosine, in combination with *in vitro* culture to eliminate apple chlorotic leaf spot *?Closterovirus*. 2,4-Dioxohex-ahydro-1,3,5-triazine was used to eliminate potato viruses in potato meristem cultures (Borissenko *et al.*, 1985) and potato stem cuttings (Bittner *et al.*, 1987; for a review, see Tomlinson, 1981).

The Importance of Using Selected Clones

The selection of horticulturally desirable clonal material free of virus infection is an important aspect of any program. Selection may need to be carried out both before and after the material has been rendered virus free by any of the preceding procedures. For example, in New Zealand the planting of apple trees freed from virus infection by heat therapy has resulted in higher yields and better trees, but problems have occurred with poor skin color in colored varieties. This may have been due to poor clone selection prior to treatment (Wood, 1983). The problems caused by variation are even greater if clones are regenerated from single cells.

Similarly there may be horticultural variability in the material after it has been freed of virus. For example, van Oosten *et al.* (1983) noted variation in fruit skin properties in Golden Delicious apples that had been rendered virus free.

The Importance of Adequate Virus Testing

Plants found to be apparently free of the virus at an early stage of growth may develop infection after quite a long incubation, so that in practice it is very important that apparently virus-free plants obtained by meristem tip culture be tested over a period before release. For example, Mullin *et al.* (1974) monitored the progeny of meristem-cultured strawberry plants and found them to be still free of graft-transmissible disease after 7 years.

4. Propagation and Maintenance of Virus-Free Stocks

Once suitable virus-free material of a variety is obtained, it has to be multiplied under conditions that preclude reinfection of the nucleus stock and that allow the horticultural value of the material to be checked with respect to trueness to type. Nuclear stock is then further multiplied for commercial use. This multiplication and distribution phase requires a continuing organization for checking on all aspects of the growth and sale of the material. A classic example is the potato certification scheme in Great Britain, which, over 30 years or so, led to a two- to threefold increase in yield, much of which was due to decreased incidence of virus infection. Tested Foundation Stocks, which are virtually free of virus, are grown in isolation in parts of Scotland that are unfavorable to early aphid migration and colonization (Todd, 1961). High-grade stocks are grown from this seed elsewhere in Britain, in areas selected because of the low incidence of aphid vectors. Health of the stocks is regularly checked. The use of systemic insecticides has extended the areas in which seed potato crops can be grown (For a review, see Ebbels, 1979).

Many such schemes are now in operation around the world for a variety of agricultural and horticultural plants, including stone and pome fruits, grapevines, and berry fruits, as well as potatoes. For example, the Australian Fruit Variety Foundation supplies virus-free trees to Australian growers (Smith, 1983). On the other hand for some groups of plants, particularly those grown for cut flowers and bulbs, lack of cooperation by individual growers may limit the effective application of virus eradication programs.

It may be possible to obtain rapid initial multiplication of virus-free material. For example, Logan and Zettler (1985) describe a procedure, involving repeated rapid shoot proliferation on an appropriate medium, that has the potential to produce more than 50,000 virus-free gladiolus within 30 weeks from a single shoot tip. As virus-free material is introduced into commercial planting, grower cooperation is essential for the implementation of measures to minimize reinfection. Avoidance of soil containing nematode vectors may be important in the propagation of virus-free crops such as hops (Cotten, 1979).

One problem that often arises as a certification scheme develops is that many plants need to be checked for infection before release. Markham *et al.* (1948) suggested a group testing procedure to save labor. Using appropriate sampling conditions, the number of plants infected in a field can be estimated from the proportion of groups found to be infected. The reliability of the test of course increases as the number of plants tested increases.

Any sampling and testing scheme should be considered from two points of view: first, the probability required that a crop of a certain "high" level of infection will be rejected, and second, the probability required that a crop of a certain "low" level of infection will be accepted. It is possible to construct schemes having various probability levels of acceptance of rejection.

This sort of procedure has been developed by Marrou *et al.* (1967) for testing lettuce seeds for freedom from lettuce mosaic *Potyvirus*. In their test, several hundred seeds are extracted together and inoculated to a sensitive indicator host; results are interpreted in relation to graphs or tables based on the binomial distribution.

5. Modified Planting and Harvesting Procedures

Breaking an Infection Cycle

Where one major susceptible annual crop or group of related crops is grown in an area and where these are the main hosts for a virus in that area, it may be possible to reduce infection very greatly by ensuring that there is a period when none of the crop is grown. A good example of this is the control of planting date of the winter wheat crops in Alberta to avoid overlap with the previous spring- or winter-sown crop (Fig. 16.8) This procedure together with elimination of volunteer wheat and barley plants and grass hosts of wheat streak mosaic *Potyvirus* before the new winter crop emerges can give good control in most seasons.

A break during the year where no susceptible plants are grown has proved effective in the control of certain other viruses with limited host ranges; even though they have efficient airborne vectors. Control measures of this sort may be difficult to implement in developing countries where a major food plant is traditionally grown in an overlapping succession. Rice is the major example. The increased use of irrigation in the tropics and protected cropping in cold temperate regions limits the options available to growers for breaking the infection cycle.

Changed Planting Dates

The effect of infection on yield is usually much greater when young plants are infected (see Fig 12.7) Furthermore, older plants may be more resistant to infection and virus moves more slowly through them. Thus, with viruses that have an airborne vector, the choice of sowing or planting date may influence the time and amount of infection. The best time to sow will depend on the time of migration of the vector. If it migrates early, late sowing may be advisable. If it is a late migration, early sowing may allow the plants to become quite large before they are infected. An example of the need for an early sowing date is given in Fig. 16.9.

For any particular crop, the effectiveness of changed planting or harvesting dates in minimizing virus infection has to be considered in relation to other economic

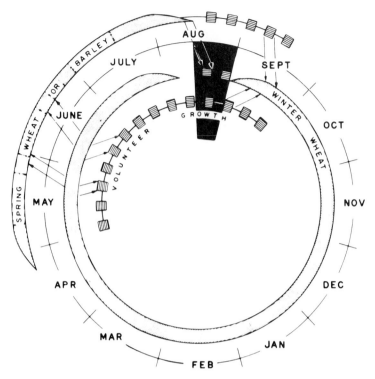

Figure 16.8   Wheat streak mosaic disease cycle. Preventing the infection of winter wheat in the autumn is the key to controlling this disease in southern Alberta. Dark area, period during which effective control can normally be achieved; broken hatched bands, problems presented by volunteer seedlings, early-seeded winter wheat, and/or late-maturing spring wheat or barley; arrows, transfer of virus by wind-blown mites. (Diagram courtesy of T. G. Atkinson.)

factors. Thus, Broadbent *et al.* (1957) found that potatoes planted early and lifted early had reduced virus infection, but a quite uneconomic reduction in yield also resulted.

Plant Spacing

The fact that a higher percentage of more closely spaced plants tend to escape infection than widely spaced ones was discussed in Chapter 15, Section I,C,5. The practical effects of planting density on incidence of a virus and its aphid vectors and on plant yield are well illustrated from the work of A'Brook (1964, 1968) on goundnuts and rosette virus. He tested a wide range of planting densities over several seasons. Aphid densities were higher over well-spaced plants. Figure 16.10 shows the marked reduction in rosette infection with higher plant densities (number of infections were transformed to allow for multiple infections).

Although larger populations decreased rosette incidence, plant competition tended to decrease yield with the very high densities, and seed costs were greatly increased. The objective is to use a planting rate that will achieve complete ground cover as soon as possible without reducing yield due to competition.

Grain yields were significantly higher for some rice varieties planted in a close

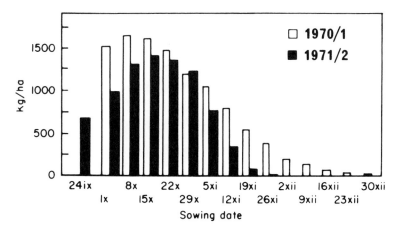

Figure 16.9 Effect of sowing date on yield of field beans (*Vicia faba*) over two seasons in the Sudan. Losses were due to Sudanese broad bean mosaic virus transmitted by aphids. Plots sown after the end of October suffered increasingly severe losses. (From Abu Salih *et al.*, 1973.)

spacing, which reduced the incidence of rice tungro disease (Shukla and Anjaneyulu, 1981). Incidence of yellows in sugar beet was also reduced by increasing plant density (Johnstone *et al.*, 1982). Thus the phenomenon may be a fairly general one. For possible explanations see Chapter 15, Section I,C,5.

Destruction of Aerial Parts of the Plant

To limit virus spread late in the season, some certification schemes for virus-free vegetative stocks require the crop to be harvested before a certain date. This applies to seed potatoes in Holland, where lifting of the crop or killing of the haulms

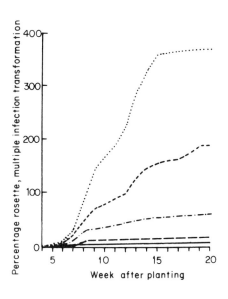

Figure 16.10 Effect of planting density on the percentage of groundnut plants infected with rosette virus. Plants per acre: ⋯⋯, 9600; ---, 19,050; —·—, 38,550; ―――, 78,750; ———, 160,200. (From A'Brook, 1964.) For derivation of multiple-infection transformation, see Gregory, 1948.

is required before a date determined each season from aphid trapping data (De Bokx, 1972).

Isolation of Plantings

Where land availability and other factors permit, isolation of plantings from a large source of aphid-borne infection might give a useful reduction in disease incidence. Thus isolation of beet fields from a large source of beets infected with two viruses markedly reduced infection (Fig. 16.11). Production of virus-free seed potatoes is frequently carried out in areas that are well separated from crops being grown for food. Distances may be controlled by legislation, and the planting of home garden potatoes forbidden within a prescribed area.

6. Prevention of Long-Distance Spread

Most agriculturally advanced countries have regulations controlling the entry of plant material to prevent the entry of diseases and pests not already present. Many countries now have regulations aimed at excluding specific viruses and their vectors, sometimes from specific countries or areas. The setting up of quarantine regulations and providing effective means for administering them is a complex problem. Economic and political factors frequently have to be considered. Quarantine measures may be well worthwhile with certain viruses, such as those transmitted through seed, or in dormant vegetative parts such as fruit trees and bud wood.

There is good evidence indicating that infected rootstocks were important in the worldwide distribution of grapevine viruses (Luhn and Goheen, 1970). Kahn (1976) suggested the use of aseptic plantlet culture for the transfer of vegetatively propagated material between countries. Cooperation between importing and exporting countries or areas may greatly improve the effectiveness of quarantine regulations. Quarantine in relation to plant health is reviewed in Hewitt and Chaiarappa (1977).

The value of quarantine regulations will depend to a significant degree on the previous history of plant movements in a region. For example, active exchanges of ornamental plants between the countries of Europe has been going on for a long period, leading to an already fairly uniform geographical distribution of viruses

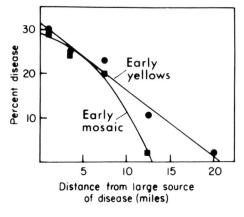

Figure 16.11 Effect of distance from a large source of infection with two aphid-borne viruses on percentage of sugar beet plants infected early in the season. The large source was about 30,000 acres of infected beets in the Sacramento Valley, California. (From Shepherd and Hills, 1970.)

infecting this type of plant (Lovisolo, 1985). On the other hand, the European Plant Protection Organization has found it worthwhile to set up quarantine regulations against fruit tree viruses not already recorded in Europe (Rønde Kristensen, 1983).

It is difficult to obtain any objective assessment as to how effective quarantine regulations have been in limiting long-distance spread of viruses. Because of its long-standing geographical isolation and the fact that almost all crop species have been imported, it is almost certain that all the plant viruses recorded in New Zealand have arrived from other countries within the last 200 years. Viruses were first listed in New Zealand quarantine regulations in 1952. Inspection of Fig. 15.6 suggests, at first sight, that they may have done little to prevent the entry of new viruses. There is no doubt, however, that many of the viruses first described since 1952 were in fact present in the country well before that time.

There is also good evidence in particular instances that quarantine measures can limit further long-distance dispersal of viruses. For example, 37 out of 61 importations of cuttings of *Datura* species from South American grown under quarantine in the United States were found to be infected with a range of viruses (Kahn and Monroe, 1970). South American potato viruses have been found in potatoes in quarantine in Europe. An effective quarantine system requires an effective technological infrastructure that is capable of detecting viruses under a variety of circumstances. Furthermore, quarantine must be concerned not only with important crop species but also with species, unimportant in themselves, that may harbor viruses infecting major crops. However, natural spread of some viruses over very long distances by invertebrate vectors may negate the effects of quarantine measures.

Attempts to prevent viruses spreading into a country may involve more than regulations controlling the importation of particular plant species. For example, since the rhizomania disease of sugar beet reached the Netherlands in 1983, close surveys of British sugar beet crops have been made, especially those in areas closest to mainland Europe. BNYVV was first found in 1987 in a beet crop in Suffolk (Hill and Torrance, 1989). Steps were immediately taken in an attempt to contain the disease. The crop was destroyed, the land was sown in pasture, and strict hygiene measures were imposed. However, in 1989 when some 3000 sugar beet crops were surveyed, two further outbreaks of rhizomania, one near Norwich and one near Kings Lynn, were found. These outbreaks appear to be separate from that of 1987 in Suffolk (Christine Henry, personal communication). In addition, rhizomania was found for the first time in the Western Hemisphere in the early 1980s (Chapter 15, Section I,B,4). Thus, the prospects for preventing worldwide spread of this disease do not appear to be good.

B. Control or Avoidance of Vectors

Before control of virus spread by vectors can be attempted, it is necessary to identify the vector. This information has sometimes been difficult to obtain. Not uncommonly, it is an occasional visitor rather than a regular colonizer that is the main or even the only vector of a virus.

Control or avoidance of invertebrate or fungal vectors is of prime importance for the limitation of crop damage by viruses that have such vectors. For this reason more than one method for control may be used simultaneously for a particular crop, location, virus, and vector.

1. Airborne Vectors

Insecticides

A wide range of insecticides are available for the control of insect pests on plants. To prevent an insect from causing direct damage to a crop, it is necessary only to reduce the population below a damaging level. Control of insect vectors to prevent infection by viruses is a much more difficult problem, as relatively few winged individuals may cause substantial spread of virus. Contact insecticides would be expected to be of little use unless they were applied very frequently. Persistent insecticides, especially those that move systemically through the plant, offer more hope for virus control. Viruses are often brought into crops by winged aphids, and these may infect a plant during their first feeding, before any insecticide can kill them. When the virus is nonpersistent, the incoming aphid when feeding rapidly loses infectivity anyway, so that killing it with insecticide will not make much difference to infection of the crop from outside. On the other hand, an aphid bringing in a persistent virus is normally able to infect many plants, so that killing it on the first plant will reduce spread.

As far as subsequent spread within the crop is concerned, similar factors should operate. Spread of a virus that is nonpersistent should not be reduced as much by insecticide treatment as a persistent virus where the insect requires a fairly long feed on an infected plant. Thus spread of the persistent PLRV in potato crops was substantially reduced by appropriate application of insecticides but spread of the nonpersistent PVY was not (Burt et al., 1964; Webley and Stone, 1972).

Burt et al. (1964) used systemic insecticides, applied as sprays at different times during the season, to estimate the stage at which most spread of PLRV occurred in potatoes in England. Their results emphasized the importance of spread early in the season by winged aphids and the need for plants to be made lethal for aphids as early as possible. Like PVY, other nonpersistent viruses, such as lettuce mosaic Potyvirus, have not been controlled in any useful way by insecticide treatments.

A screen or "trap crop" of plants sprayed with a systemic insecticide may reduce virus infection in plants growing within the screened field (Gay et al. 1973). Use of inappropriate insecticides may cause an increase in virus infection, either by disturbing the aphids present in the crop or by destruction of predators (e.g., Broadbent et al., 1963). This problem also occurs with rice.

Spraying as a farming operation has certain disadvantages. It is an extra operation to be performed, tractor damage to the crop occurs, and spraying may not be practicable at the time required. Drifting spray can also lead to damage in other crops. Persistent systemic insecticides applied in granular form at time of planting overcome many of these difficulties. With a crop such as potatoes, granules can be metered into the furrow through applicators attached to the planting machines. For example, various systemic granular pesticides such as Aldicarb and Thiofanox have given control of aphids and PLRV in potato crops, applied in furrow at planting, or as side bands some weeks later (Woodford et al., 1988). The side band treatments were generally less effective than in furrow.

The photostable synthetic pyrethroid PP321 was effective in field trials in the control of BYDV in winter barley, of rice tungro disease in rice, and of the nonpersistent PVY in potatoes (Perrin and Gibson, 1985), in contrast to the preceding experiments. This same material was effective in controlling beet western yellows Luteovirus in winter oilseed rape (Walsh, 1986).

The effectiveness of systemic insecticide treatment in controlling the nonpersistent PVY in tobacco varied with season, probably depending on climatic factors that affect the numbers and movement of the aphid vector (Pirone *et al.*, 1988).

With leafhopper vectors, the speed at which they are killed after feeding on a treated plant may be an important factor for virus control. For example, several synthetic pyrethroids that killed the vector *Nephotettix virescens* within 7 minutes prevented spread of the rice tungro disease (Anjaneyulu and Bhaktavatsalam, 1987). Slower-acting chemicals were ineffective in reducing virus incidence.

From the point of view of the economics of insecticide use it may be important to forecast whether applications are necessary and, if so, to define optimal times for treatment. For example, in China two crops of rice followed by winter barley and wheat give the planthopper *Laodelphax striatella* a year-round succession of hosts. The best timing for the application of insecticides has been found to be at the winter seedling stage of barley and wheat and at the early paddy stage of the second crop of rice (Yi-Li *et al.*, 1981). In India, where an overlapping cropping pattern is also common, application of systemic insecticides under rice nursery conditions appears to be the most effective for the control of tungro disease (Shukla and Anjaneyulu, 1982).

As discussed in Section II,D, disease forecasting data can be an important factor in the economic use of insecticides. Sometimes a long-term program of insecticide use aimed primarily at one group of viruses will help in the control of another virus. Thus the well-timed use of insecticides in beet crops in England, aimed mainly at reducing or delaying the incidence of yellows diseases, has also been a major factor in the decline in the importance of beet mosaic *Potyvirus* in this crop (Heathcote, 1973).

Oil Sprays

Bradley (1956) and Bradley *et al.* (1962) showed that when aphids carrying PVY probed a membrane containing oil, their subsequent ability to transmit the virus was greatly reduced. These observations led to considerable interest in the possibility of using oil sprays in the field to control viruses spread by aphids. Compared to synthetic insecticides, oil sprays have considerable appeal because of their lack of toxicity for man and animals. However, various limitations have emerged and their commercial use is not widespread.

In some reports, oil sprays have given useful results with field trials against a range of nonpersistent viruses (e.g., Hein, 1971) and also BYV, which is semipersistent. However, oil does not prevent the spread of persistent viruses. Other limitations involve possible plant toxicity, volatility or viscosity of the oil (De Wijs, 1980), adequate coverage of the leaves, and removal of the oil cover by rain or irrigation water.

Mineral oil sprays have proved useful in controlling viruses in bulb crops such as lilies and hyacinths in the Netherlands (Asjes, 1978, 1980). Similar sprays, used routinely on potato seed crops in Brittany, have greatly reduced the spread of PVY (Kerlan *et al.*, 1987). Promising results have been obtained using oil sprays on tomatoes in Sudan to limit the spread of a whitefly-transmitted geminivirus (Yassin, 1983). On the other hand, no control of several nonpersistent viruses was obtained with oil sprays in England (Walkey and Dance, 1979), or with MDMV in sweet corn in Ohio (Szatmari-Goodmann and Nault, 1983). The combined use of mineral oils and pyrethroids gave better control in trials with PVY and beet mosaic virus than either component alone (Gibson and Rice, 1986).

Oil sprays do not appear to affect significantly the susceptibility of the plant, aphid behavior, or virus infectivity. When aphids probe leaves sprayed with oil, the oil spreads readily over the whole length of the stylets. This observation and various experiments summarized by Vanderveken (1977) lead to the idea that oil alters the surface structure or charge on the stylets, thus limiting the ability to adsorb (or release) virus particles.

The normal feeding behavior of the leafhopper *Nephotettix virescens* was disrupted on rice leaves sprayed with oil. There was reduced phloem and increased xylem feeding, with restless behavior, repeated probing, and profuse salivation. These effects may explain the reduced survival time of the insect and the decrease in virus transmission (Saxena and Khan, 1985).

Pheromone Derivatives

Derivatives prepared from the pheromone (E)-β-farnesene, and related compounds interfered with the transmission of PVY by *Myzus persicae* in glasshouse experiments (Gibson *et al.*, 1984). It is possible that these substances act in a manner similar to that of mineral oils since they possess aliphatic carbon chains. While new compounds continue to be developed (Dawson *et al.*, 1988), this type of vector control is still very much at the experimental stage.

Nonchemical Barriers against Infestation

Several kinds of possible barriers to, or repellants against, vector movement into a crop have been investigated.

A tall cover crop will sometimes protect an undersown crop from insect-borne viruses. For example, cucurbits are sometimes grown intermixed with maize. Broadbent (1957) found that surrounding cauliflower seedbeds with quite narrow strips of barley (about three rows 0.3 m apart) could reduce virus incidence in seedlings to about one-fifth. Barley is not attacked by crucifer viruses. Many incoming aphids were assumed to land on the barrier crops, feed briefly, and either stay there or fly off. If they then land on the *Brassica* crop they may have lost any nonpersistent virus they were carrying during probes on the barrier crop.

The reported action against aphids of aluminum strips laid on the ground has been tested for several crops. As the aphids come in to land, the reflected UV light is thought to act as a repellant. Reflective aluminum polythene mulches reduced the incidence of watermelon mosaic *Potyvirus* and increased yields of cucurbits in Western Australia under conditions where both insecticide and oil sprays proved ineffective (McLean *et al.*, 1982). However, these reflective mulches are expensive and where they have come into regular use, difficulties in disposal at the end of the season may occur, at least when disposal by burning is forbidden (Nameth *et al.*, 1986).

A strip of sticky yellow polythene, 0.5 m wide and 0.7 m above the soil, surrounding the trial plots reduced the incidence of aphid-transmitted viruses in peppers (Cohen and Marco, 1973). A yellow polythene mulch significantly delayed the appearance of a yellow vein mosaic virus in okra in India (Khan and Mukhopadhyay, 1985). Nets spread above the crop may also reduce the winged aphid population and virus infection while allowing normal plant growth. Coarse white nets may be the most effective under some conditions (Cohen, 1981). However, most of the findings noted in this section have not moved beyond the experimental stage into commercial practice.

Plant Resistance to Vectors

In recent years there has been substantially increased interest in breeding crops for resistance to insect pests as an alternative to the use of pesticide chemicals. This has been due to various factors, including emergence of resistance to insecticides in insects, the costs of developing new pesticides, and increasing concern regarding environmental hazards and the effects on natural enemies. Along with these developments, there has been increased activity in breeding for resistance to invertebrates that are virus vectors. Some virus vectors are not pests in their own right but others, especially leafhoppers and planthoppers, may be severe pests. In this situation there is a double benefit in achieving a resistant cultivar, sometimes with striking improvements in performance. The subject has been reviewed by Jones (1987).

Sources of resistance have been found among most of the airborne vector groups. Some examples are given in Table 16.1. The basis for resistance to the vectors is not always clearly understood, but some factors have been defined. In general terms, there are two kinds of resistance relevant to the control of vectors. First, *nonpreference* involves an adverse effect on vector behavior, resulting in decreased colonization, and second, *antibiosis* involves an adverse effect on vector growth, reproduction and survival after colonization has occurred. These two kinds of factor may not always be readily distinguished. Some specific mechanisms for resistance are: (i) sticky material exuded by glandular trichomes such as those in tomato (Berlinger and Dahan, 1987); (ii) heavy leaf pubescence in soybean (Gunasinghe *et al.*, 1988); (iii) inability of the vector to find the phloem in *Agropyron* species (Shukle *et al.*, 1987). However, this effect was not operative with an aphid vector of BYDV in barley (Ullman *et al.*, 1988); and (iv) interference with the ability of the vector to locate the host plant. For example, in cucurbits with silvery leaves there was a delay of several weeks in the development of 100% infection in the field with CMV and clover yellow vein *Tymovirus* (Davis and Shifriss, 1983). This effect may be due to aphids visiting plants with silvery leaves less frequently because of their different light-reflecting properties.

Table 16.1

Examples of Plants with Resistance to Airborne Vectors That Have Been Associated with a Decreased Incidence of Virus Infection

Vector	Crop	Virus	References
Aphids			
Aphis gossypii	Musk melon	CMV	Lecoq *et al.* (1981)
Myzus persicae	Potato	PLRV	Rizvi and Raman (1983)
Myzus persicae			
Rhopalosiphum maidis	Soybean	Soybean mosaic virus	Gunasinghe *et al.* (1988)
Aphis citricola			
Leafhopper			
Nephotettix virescens	Rice	Rice tungro viruses	Hibino *et al.* (1987)
Planthopper			
Nilaparvata lugens	Rice	Ragged stunt virus	Parejarearn *et al.* (1984)
Whitefly			
Bemisia tabaci	Tomato	Yellow leaf curl	Berlinger and Dahan (1987)
Thrips			
Frankliniella schultzei	Groundnut	TSWV	Amin (1985)
Mites			
Aceria tulipae	Wheat	WSMV	Martin *et al.* (1984)

Combination of resistance to a vector with some other control measure may sometimes be useful. For example, in field trials, rice tungro disease was effectively controlled by a combination of insecticide application and moderate resistance of the rice cultivar to the leafhopper vectors (Heinrichs *et al.*, 1986). Sprays would be unnecessary with fully resistant cultivars.

There may be various limitations on the use of vector-resistant cultivars: (i) Sometimes such resistance provides no protection against viruses. For example, resistance to aphid infestation in cowpea did not provide any protection against cowpea aphid-borne mosaic *Potyvirus* (Atiri *et al.*, 1984). (ii) If a particular virus has several vector species, or if the crop is subject to infection with several viruses, breeding effective resistance against all the possibilities may not be practicable, unless a nonspecific mechanism is used (e.g., tomentose leaves). (iii) Perhaps the most serious problem is the potential for new vector biotypes to emerge following widespread cultivation of a resistant cultivar, as may happen following the use of insecticides.

This difficulty is well illustrated by the recent history of the rice brown planthopper. With the advent of high-yielding rice varieties in Southeast Asia in the 1960s and 1970s, the rice brown planthopper (*Nilaparvata lugens*) and rice grassy stunt *Tenuivirus*, which it transmits, became serious problems. Cultivars containing a dominant gene (*Bph1*) for resistance to the hopper were released about 1974. Within about 3 years resistance breaking populations of the hopper emerged. A new, recessive resistance gene (*bph2*) was exploited in cultivars released between 1975 and 1983. They were grown successfully for a few years until a new hopper biotype emerged that overcame the resistance (Thresh, 1988a, 1989a,b). Rice cultivars are now being released that contain multiple resistance genes in attempts to develop more durable forms of resistance (Heinrichs, 1988).

In spite of these difficulties, and the problems associated with the identification of plants with resistance to vectors, it seems certain that substantial efforts will continue to be made to improve and extend the range of crop cultivars with resistance to virus vectors. A combination of resistance to the vector and to the virus will frequently be the goal.

Control by Predators or Parasites

Parasites and predators undoubtedly play a major role in limiting the population growth of aphids and other insects. Figure 16.12 shows the results of an experiment by Evans (1954) in Tanzania. He established small colonies of *Aphis craccivora* (one adult and five nymphs) on single plants in various parts of a field of groundnuts. He then followed the growth of the aphid colonies and the appearance of predators.

Under some circumstances, predators might play a part in limiting spread of a virus, but generally they will have little effect if they arrive after the early migratory aphids, which are so important for virus spread. In West Africa, introduction of fungal and hymenopterous parasites of the mealybug vectors of cacao viruses to control swollen shoot virus were unsuccessful, even though the vector is relatively immobile (Thresh, 1958).

2. Soilborne Vectors

Most work on the control of viruses transmitted by nematodes and fungi has centered on the use of soil sterilization with chemicals. However, several factors make general and long-term success unlikely: (i) huge volumes of soil may have to

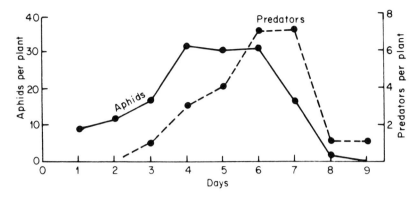

Figure 16.12 Growth of initially small aphid colonies on groundnuts and their subsequent destruction by predators. (From Evans, 1954.)

be treated; (ii) a mortality of 99.99% still leaves many viable vectors; and (iii) use of some of the chemicals involved has been banned in certain countries, and such bans are likely to be extended. In any event, chemical control will be justified economically only for high-return crops, or crops that can remain in the ground for many years.

Nematodes

In principle, the control of viruses transmitted by nematodes should be possible by treatment of infested soil with nematicides. Movement and dispersal of nematodes are generally slow, so that one treatment might be expected to remain effective for longer periods than with airborne vectors. On the other hand, as pointed out by Sol (1963), infective nematodes may occur at considerable depths. Nematodes in soil samples taken up to 80–100 cm deep were able to infect plants with TRV. Thus, it is possible that fumigated soil could be reinfested by nematodes moving up from deep sites where they had escaped the effects of fumigation.

Harrison *et al.* (1963), in field trials at several sites in southern Britain, showed that dichloropropane–dichloropropene (D–D) or methyl bromide at *1* kg/10m², applied in summer, killed over 99% of *Xiphinema diversicaudatum* in the soil. The treatments very largely prevented infection of strawberry crops with arabis mosaic *Nepovirus*. The crops were planted after the fumigation and examined after 1–2 years. Both chemicals killed the nematode down to 70 cm, the greatest depth sampled. Virus incidence following fumigation was closely related to the numbers of surviving nematodes.

Murant and Taylor (1965) and Taylor and Murant (1968) showed that a single soil fumigation with D–D or pentachloronitrobenzene (quintozine) prevented infection of strawberries by tomato black ring and raspberry ring spot nepoviruses transmitted by *Longidorus elongatus*. Quintozine was preferred since it had no marked effect on plant growth. Because *L. elongatus* has a wide host range among cultivated plants and weeds and because it can survive for long periods in the soil without food, its elimination by crop rotation is most unlikely. Chemical control of the vector in the soil probably offers the best solution for viruses transmitted by this nematode.

At present, soil fumigation appears to be the only effective measure for the control of *Xiphinema index,* the vector of grapevine fanleaf virus. If proper care is taken with the fumigation process, vineyards may be grown successfully for 15–20 years (Raski *et al.*, 1983). With vines it is very important to kill all nematodes before replanting. The process may be assisted by treating old vines with a herbicide so that deep roots, and the nematodes they harbor, are killed before fumigation treatment of the upper soil.

Little work has been done on seeking resistance to nematode vectors. However, Bouquet (1981) reported resistance to *X. index* and protection from grapevine fan-leaf infection in *Vitis rotundifolia*.

Fungi

In field experiments in Scotland, decreasing the pH of infective soil to 5.0 by applications of sulfur markedly reduced the infection of potato with potato mop top *Furovirus* and *Spongospora subterranea*, the fungal vector (Cooper *et al.*, 1976). However, neither virus nor vector was eliminated from the soil. Of various fungicides tested only calomel controlled the spread of virus and vector. Cost and toxicity rule out large-scale use of this material.

In general, attempts to control infection with viruses having fungal vectors by application of chemicals to the soil have not been successful. For example, various fungicides failed to control transmission of barley yellow mosaic *Potyvirus* to winter barley by *Polymyxa graminis* (Proeseler and Kastirr, 1988). Several fungicides prevented infection of sugar beet by BNYVV in glasshouse trials, but were ineffective in the field, even at high rates of application (Hein, 1987). The zoospores of *Olpidium brassicae* were susceptible to several fungicides and surfactants in laboratory tests (Tomlinson and Faithfull, 1979).

Some cultivars of wheat show field resistance to soilborne wheat mosaic *Furovirus*. The mechanism for such resistance is not established, but it may involve, in part, resistance to the fungal vector *P. graminis* (Lapierre *et al.*, 1985). Culturing this vector is difficult, which probably accounts for the fact that little work has been done on the development of vector-resistant cereals, in spite of the important viruses this fungus transmits.

C. Protecting the Plant from Systemic Disease

Even if sources of infection are available, and the vectors are active, there is a third kind of control measure available—protecting inoculated plants from developing systemic disease. Genetic protection is the most important means of protecting crops in this way.

1. Genetic Protection

The terms immunity, resistance, and tolerance were defined in Table 12.1. The genetics of plant resistance to viruses and the mechanisms of host resistance were discussed in Chapter 12, Section V,B,2. This section deals with the use of genetic protection in the control of viral diseases.

Immunity

The apparent immunity of seven raspberry cultivars to raspberry vein chlorosis rhabdovirus was confirmed by the fact that they could not be infected by graft inoculation (Jennings and Jones, 1986). Certain cultivars of swede (*Brassica napus*) appear to be immune to turnip mosaic *Potyvirus* (Tomlinson and Ward, 1982), and

certain cultivars of barley appear to be immune to BaYMV. Plants remained uninfected with the virus following root inoculation with virus-bearing cultures of the
fungal vector *Polymyxa graminis*. Transmission by zoospores from the fungus
growing on roots of an immune cultivar was rare or absent. If this is also true in the
field the phenomenon would have significant implications for control of this important virus (Adams *et al.*, 1987).

However, although many searches have been made, true immunity against
viruses and viroids, which can be incorporated into useful crop cultivars, is a rather
uncommon phenomenon. For example, Singh and Slack (1984) found no immunity
to PSTVd among 555 introductions belonging to 81 tuber-bearing *Solanum* species.

Resistance

Where genes for resistance can be introduced into agriculturally satisfactory
cultivars, breeding for resistance to a virus provides one of the best solutions to the
problem of virus disease. Attempts to achieve this objective have been made with
many of the virus diseases of major importance. Genes for resistance or hypersensitivity have often been found, but it has frequently proved very difficult to incorporate such factors into useful cultivars.

A hypersensitive reaction to a virus involving necrotic local lesions with no
systemic spread as discussed in Chapter 12, Section III, can give effective resistance
in the field. For example, the necrotic reaction of *N. glutinosa* to TMV was bred into
some commercial lines of *N. tabacum* (Valleau, 1946). It has since been used very
widely with Virginia-type tobacco cultivars but not with flue-cured types.

Occasionally, very high resistance to a virus has been discovered even in a plant
where most known cultivars were highly susceptible. However, the designation of a
variety or genotype as very highly resistant to a particular virus must always be
provisional, because it is always possible that a mutant of the virus will arise that
can overcome the plant resistance.

The major problem with resistance of any sort as a control measure is its
durability. How long can it be deployed successfully before a resistance breaking
(virulent) strain of the virus emerges? Fraser and Gerwitz (1987) listed 54 host–
virus combinations for which resistance genes have been found. Virulent virus
isolates able to overcome these resistances are known for almost 60% of these.
Fewer than 10% of the resistance genes listed have remained effective when tested
against a wide range of virus isolates over a long period. However, some of the
virulent isolates were found only in laboratory tests rather than field outbreaks.

The costs of a breeding program must be weighed against the possible gains in
crop yield (see Buddenhagen, 1981). Many factors are involved, such as: (i) the
seriousness of the viral disease in relation to other yield-limiting factors; (ii) the
"quality" of the available resistance genes, for example, resistance genes against
CMV are usually "weak" and short-lived, which may be due, at least in part, to the
many strains of CMV that exist in the field; (iii) the importance of the crop (compare, for instance, a minor ornamental species with a staple food crop such as rice);
and (iv) crop quality. Good virus resistance that gives increased yields may be
accompanied by poorer quality in the product, as happened with some TMV-
resistant tobacco cultivars (Johnson and Main, 1983).

Sources of Resistant Genotypes   Quite frequently no resistance could be found
for particular crops and viruses. For example, no resistance to beet western yellows
Luteovirus was found in 70 cultivars and 500 breeding lines of lettuce (Watts,
1975). Russell (1960) found no resistance to BYV in 100,000 beet seedlings. In

such circumstances it may be necessary to search for resistance among wild species. In general, there is certainly a need to broaden the search for resistance genes. For example, in the United Kingdom, 22 horticultural crops have no known source of resistance to a total of 25 viruses that are known to affect them (Fraser, 1988). Nevertheless, many effective sources of resistance have been found and are currently in use.

Sometimes certain existing lines or varieties have been found to have a useful degree of resistance, as with the resistance of Corbett refugee bean to bean common mosaic *Potyvirus* (Zaumeyer and Meiners, 1975). Resistance may not always be uniform, however. For example, even within inbred lines of corn there was variation in resistance to MDMV (Louie *et al.*, 1976). In cabbage, variation in degree of resistance to turnip mosaic *Potyvirus* was found between different lines of the same variety (Polak, 1983).

Sometimes useful sources of resistance can be identified by making initial selections from plants showing good growth in otherwise severely infected fields (e.g., Cope *et al.*, 1978). With important crops, searches for sources of resistance have been made on a worldwide basis. Timian (1975) tested 4889 entries in the Barley World collection and found 44 that showed no symptoms of infection with BSMV in the field. As another example, sources of resistance to cocoa swollen shoot virus in Ghana have been found in the upper Amazon region (Legg and Lockwood, 1977), and attempts have been made to combine resistance from various sources (Kenten and Lockwood, 1977).

Occasionally useful resistance has been found among a collection of mutants induced by physical or chemical means. For example, Ukai and Yamashita (1984) identified such a barley mutant that was highly resistant to BYMV.

In principle, culture of plant cells as protoplasts offers several possibilities for obtaining new sources of resistance to viruses. However, no commercially successful virus-resistant cultivars have yet been derived by this means. The topic has been reviewed by Shepard (1981). When plant cells are grown in culture for a time and then plants regenerated from single cells or small clusters of cells, considerable genetic variation or *somaclonal* variation may be observed in various properties, including resistance to disease. Attempts have been made to increase the frequency and range of variations by treatment of the tissue in culture with mutagens.

Protoplast fusion can take place *in vitro* even between protoplasts belonging to different genera. In principle this offers the possibility of introducing virus resistance genes into a crop cultivar from a quite distantly related species. The donor cells are often irradiated before fusion to fragment the chromosomes. Following fusion, chromosomes and parts of chromosomes are eliminated in a fairly random manner, until a more or less stable set of chromosomes is achieved. The hope is that a cell line will arise that contains a functional set of chromosomes from the crop cultivar, together with a minimal amount of alien genetic material containing the desired resistance gene or genes.

In model experiments, transfer of methotrexate and 5-methyltryptophan resistance from carrot to tobacco was achieved by fusion between leaf mesophyll protoplasts of tobacco and irradiated cell culture protoplasts of carrot (Dudits *et al.*, 1987). Some of the regenerated plants had tobacco morphology and independently segregating genes for the two resistance markers from carrot.

Intergeneric hybridization in a search for resistance to BYDV in wheat has been reviewed by Comeau and Plourde (1987). Substantial resistance has been found in about 20 species in the tribe Triticeae, which could be, or have been, hybridized to cultivated wheat or close relatives.

Low Seed Transmission   The discussion in the previous section was concerned with resistance of the growing plant to virus infection. For those viruses affecting annual crops that are transmitted through the seed, resistance to seed transmission may be an important method for limiting infection in the field. For example, a single recessive gene conditions resistance to seed transmission of BSMV in barley (Carroll *et al.*, 1979). The resistance gene, found in an Ethiopian barley called Modjo, was introduced into a new variety, Mobet, with good agronomic qualities and high resistance to seed transmission of Montana isolates of BSMV (Carroll *et al.*, 1983). Resistance to seed transmission has also been found for lettuce mosaic *Potyvirus* in lettuce and for some legume viruses.

Adequate Testing of Resistant Material   Varieties or lines showing resistance in preliminary trials must then be tested under a range of conditions. Important factors to be considered are strains of the virus, climatic conditions (e.g., Thompson and Herbert, 1970), and inoculum pressure (e.g., Kenten and Lockwood, 1977).

The Need for Resistance to Multiple Pathogens   The difficulties in finding suitable breeding material are compounded when there are strains of not one, but several viruses to consider. Cowpeas in tropical Africa are infected to a significant extent by at least seven different viruses. In such circumstances, a breeding program may utilize any form of genetic protection that can be found. Sources of resistance, hypersensitivity, or tolerance have been found for five of the viruses (D. J. Allen, personal communication). There is of course the further problem of combining these factors with multiple resistance to fungal and bacterial diseases. For example, genetic resistance to TMV, cyst nematodes, root-knot nematodes, and wildfire from *Nicotiana repanda* have been incorporated into *N. tabacum* (Gwynn *et al.*, 1986).

Durability of Resistance and the Emergence of Resistance Breaking Strains   For some crop plants and viruses, resistance has proved to be remarkably durable. Thus the resistance to bean common mosaic *Potyvirus* found in Corbett Refugee bean has been bred into most varieties of dry and snap beans in the United States, and the resistance had not broken down after 45 years (Zaumeyer and Meiners, 1975). One of the most noteworthy examples of durability has been the resistance in sugar beet to the beet curly top *Geminivirus* (Duffus, 1987), the original selections for which were made in the 1920s. The resistance is multigenic and appears to involve a lower concentration of virus in resistant plants that do become infected and a much longer incubation period in resistant varieties. However, over a period of years a series of more aggressive strains of the virus have emerged. Resistance to turnip mosaic *Potyvirus* in lettuce described 20 years ago has been tested with strains of the virus from many countries, but the resistance has not been broken (Duffus, 1987).

A hypersensitive type of resistance to CCMV in cowpeas has recently been shown to be overcome by certain strains of the virus (Paguio *et al.*, 1988). Resistance breaking strains of PVX have been described in the United Kingdom for resistant potato varieties but these have not yet become a serious practical problem (Jones, 1985). Raspberry bushy dwarf ?*Ilarvirus* has been controlled in Scotland by the use of cultivars that are immune to the prevalent strain of the virus. This situation has been threatened by the discovery of a resistance breaking strain (Murant *et al.*, 1986).

Once a substantial population of resistant plants are exposed in the field there is a good probability that a new strain of the virus will evolve, or be introduced, that can overcome the resistance.

The problem of virus strains in the development of a breeding program is well illustrated by the tests carried out by Rast (1967a,b) with 64 different isolates of ToMV on 30 clones of *Lycopersicum peruvianum*. Different isolates (even from the same strain of ToMV as judged by symptoms on tomato or tobacco) differed in the range of *L. peruvianum* clones they would infect. Every clone could be infected by at least one isolate.

Even when an apparently successful resistant variety has been developed it may be important to maintain other measures to minimize contact of the resistant variety with the virus concerned. For example, in tomatoes resistant to ToMV the virus did not move systemically for several weeks and reached only a low concentration (Dawson, 1967). Extracts from infected resistant plants were more infective for healthy resistant plants than virus from susceptible lines. Resistant plants infected with such virus showed more obvious symptoms.

The $Tm2^2$ gene in tomato has been very useful for protection against ToMV for more than 10 years (Chapter 12, Section V,B,2). However, ToMV strains have now been found that can, in the laboratory, overcome the resistance due to the $TM2^2$ gene. It is probably only a matter of time before these become prevalent in commercial glasshouses.

We must conclude that on present knowledge for most crops and most viruses, the search for new sources of resistance and their incorporation into useful cultivars will be a continuing and very long term process, as it is with many fungal and other parasitic agents. The interrelations between host genetic factors and other epidemiological aspects of virus diseases are discussed by Buddenhagen (1983).

Tolerance

Where no source of genetic resistance can be found in the host plant, a search for tolerant varieties or races is sometimes made. However, tolerance is not nearly as satisfactory a solution as genetic resistance for several reasons.

1. The infected tolerant plants may act as a reservoir of infection for other hosts. Thus, it is bad practice to grow tolerant and sensitive varieties together under conditions where spread of virus may be rapid.
2. Large numbers of virus-infected plants may come into cultivation. The genetic constitution of host or virus may change to give a breakdown in the tolerant reaction.
3. Virus infection may increase susceptibility to a fungal disease (Chapter 12, Section V,D,2). However, tolerant varieties may yield very much better than standard varieties where virus infection causes severe crop losses and where large reservoirs of virus exist under conditions where they cannot be eradicated. Thus,tolerance has, in fact, been widely used (see Posnette, 1969). Cultivars of wheat and oats commonly grown in the Midwest of the United States have probably been selected for tolerance to BYDV in an incidental manner, because of the prevalence of the virus (Clement *et al.*, 1986). Tolerance to maize streak *Geminivirus* has been found in maize and rapidly incorporated into high-yielding maize populations for use in tropical Africa (Soto *et al.*, 1982). Walkey and Antill (1989) obtained an unusual result with a variety of garlic called Fructidor. The yield of selected stocks was significantly less than that of unselected infected stocks.

Transgenic Plants

It is now possible to introduce almost any foreign gene into a plant and obtain expression of that gene. In principle, this should make it possible to transfer genes for resistance or immunity to a particular virus, across species, genus, and family boundaries. At present the major problem with this approach is the difficulty in identifying the genes responsible for immunity or resistance. With the exception of those discussed in Chapter 12, Section V,B,2, no such genes active against plant viruses have yet been identified. Nevertheless, several other approaches to producing transgenic plants resistant to virus infection are being actively explored. In particular, expression of coat protein genes or other viral genes may provide transgenic plants with a useful degree of protection against the virus concerned. This and other approaches to control by gene manipulation are discussed in the following sections.

Transgenic Plants Expressing a Viral Coat Protein The development of transgenic plants expressing the coat protein genes for TMV, AMV, TRV, and PVX was discussed in Chapter 13, Section IV,A,4. The mechanism for protection against superinfection with the virus concerned was compared with natural cross-protection. The extent of protection provided by such transgenic plants is correlated with the degree of expression of the coat protein gene. Figure 16.13 shows the striking

Figure 16.13 Protection of transgenic tobacco plants expressing the CMV coat protein gene against mechanical inoculation with CMV at 25 μg/ml. CP$^+$ plants (back row) and CP$^-$ plants (front row) were photographed one month after inoculation. (From Cuozzo *et al.*, 1988.)

protection obtained with the transgenic expression of CMV coat protein. The extent of protection is reduced as inoculum concentration is increased. The question arises as to whether these developments will lead to a practical method of control for some viruses.

Besides the viruses noted, transgenic plants expressing coat protein and showing some resistance to superinfection have been produced for CMV (Cuozzo *et al.*, 1988), TRV (van Dun and Bol, 1988), the *Potexvirus* and *Tombusvirus* groups, and AMV. Thus the phenomenon appears to be fairly generally applicable, at least to these mechanically transmissible viruses. Commercial varieties of potato have been successfully transformed with the PVX coat protein gene without affecting their essential characteristics (Hoekema *et al.*, 1989).

Some important questions remain to be answered, for example: (i) Will resistance be maintained to a useful degree when inoculation in the field is by means of invertebrate or fungal vectors? (ii) Will seed transmission be reduced in transgenic plants that do become infected? (iii) Will resistance be achieved to multiple viruses important in particular crops such as tomatoes and peppers and (iv) will a degree of resistance that protects a useful proportion of plants in an annual crop, be effective in protecting a perennial crop.

Field experiments with tomatoes suggest that transgenic expression of TMV coat protein may be commercially useful in this host (Fig. 16.14). The transgenic plants were partially resistant to TMV and to strains L, 2, and 2^2 of ToMV. In the

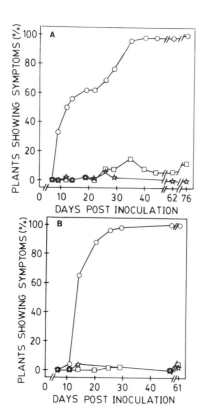

Figure 16.14 Development of systemic symptoms of TMV infection under field conditions in tomatoes transgenic for coat protein. ○ —— ○, nontransformed plants; □——□ and ★——★, plants of two lines of transgenic tomatoes expressing TMV coat protein. Plants were inoculated on terminal leaflets of three successive leaves with TMV strain U1 at 10 μg/ml, 8 days after planting out in the field. Observations were made on 48 plants of each line. (From Nelson *et al.*, 1988.)

field no more than 5% of transgenic plants showed systemic disease symptoms compared with 99% for the control plants. Lack of visual symptoms was correlated with an absence of ToMV. In inoculated control plants, fruit yields were depressed by 26–35%. There was no evidence that expression of the coat protein gene reduced plant growth or fruit yield compared with uninoculated nontransgenic plants (Nelson *et al.*, 1988).

Protection may not always be effective with different strains of a virus. For example, tobacco plants transformed to express the coat protein of strain PLB or TCM of TRV were resistant only to the strain whose coat protein was being expressed (Angenent *et al.*, 1990). On the other hand Nejidat and Beachy (1990) found that transgenic tobacco plants expressing the coat protein gene of TMV were resistant to ToMV and tobacco mild green mosaic virus but not to ribgrass mosaic virus. The protective effect against TMV in transgenic plants expressing coat protein was greatly reduced when tobacco plants were held at 35°C instead of 25°C. However, plants held in a 25°C/35°C night/day cycle retained resistance (Nejida & Beachy, 1989)

The stage has probably been reached where further progress in the commercialization of some transgenic plants resistant to particular viruses will depend mainly on nontechnical aspects, such as regulatory requirements, the cost of the large-scale multisite field tests that are required, and proprietary protection for genetically engineered cultivars (Gasser and Fraley, 1989). Nevertheless, Gasser and Fraley predict that commercial introductions of genetically engineered soybean, cotton, rice, and alfalfa are likely by the mid-1990s and that corn hybrids with resistance to viruses, insects, and herbicides should be available by the year 2000.

Transgenic Plants Expressing Noncoat Viral Genes   When the 30K movement protein gene was expressed in transgenic tobacco plants, movement of the transport-defective LSI mutant of TMV was possible (Deom *et al.*, 1987). Transformed tobacco plants containing cDNA copies of AMV RNAs 1 or 2 showed no resistance to AMV (van Dun *et al.*, 1988a). It is not certain whether this result was due to low levels of expression or whether both genes in the same plant might confer some resistance. No resistance was detectable in tobacco plants transformed with the nonstructural 13K and 16K genes of strain PLB of TRV or the 29K gene unique to strain TCM (Angenent *et al.*, 1990). Much more encouraging results have recently been obtained with TMV. Golemboski *et al.* (1990) transformed tobacco plants with the nucleotide sequence coding for the 54K putative protein of TMV (see Fig. 7.5) The sequence lacked only the three 3′ terminal nucleotides. This sequence encodes a component of the putative replicase complex. The transgenic plants were resistant to infection with TMVU1 or its RNA at high concentrations. It was also resistant to a U1 mutant, but not to two other tobamoviruses or CMV. The transformed plants accumulated the expected sequence—specific transcripts but no protein product could be detected. This situation parallels that found in natural TMV infections (Chapter 7, Sections III,A,3 and 8). Experiments in protoplasts derived from plants transgenic for the 54K protein have shown that a very early event is involved. No 126K or 183K proteins were produced (M. Zaitlin, personal communication).

Transgenic Plants Expressing Satellite RNAs   The general nature of satellite RNAs is described in Chapter 9, Section II,B, and the ability of some satellite RNAs to attenuate the symptoms of the helper virus was discussed there. It has been shown for two satellite RNAs that transgenic plants expressing the satellite RNA are less severely diseased when inoculated with the helper virus. Harrison *et al.* (1987) showed that when transgenic tobacco plants containing DNA copies of a CMV satellite RNA were inoculated with a satellite-free CMV isolate, satellite replication occurred. At the same time, CMV replication was reduced and disease symptoms were greatly attenuated. In untransformed plants the CMV isolate caused mosaic disease and stunting. In the transformed plants no mosaic appeared and plants grew almost as well as healthy ones. These differences persisted for 14 weeks, the longest period tested. Furthermore, the same result was obtained in plants raised from seed of the transformed plants. When transformed plants were inoculated with tomato aspermy *Cucumovirus,* there was a similar attenuation of disease symptoms but without a marked decrease in tomato aspermy viral genome synthesis. Jacquemond *et al.* (1988) showed that tobacco plants transgenic for a CMV satellite RNA were tolerant to infection by aphids, the main method of field transmission for CMV.

Results similar to those of Harrison *et al.* (1987) were reported at the same time for another satellite–virus combination, STRSV and TRSV (Gerlach *et al.,* 1987b). Tobacco plants that expressed full-length STRSV or its complementary sequence as RNA transcripts increased their synthesis of STRSV RNA following inoculation with TRSV, but virus replication was reduced and disease symptoms were greatly ameliorated. This protection was maintained for the life of the plants.

The use of satellite RNAs in transgenic plants to protect against the effect of virus infection has both advantages and disadvantages. The protection afforded is not affected by the inoculum concentration, as it is with viral coat protein transformants. The losses that do occur in transgenic plants because of slight stunting will affect only the plants that become naturally infected in the field, whereas if all plants are deliberately infected with a mild CMV–satellite combination they will all suffer some loss (Section II,C,2). Furthermore, the resistance may be stronger in transgenic plants than in plants inoculated with the satellite. Inoculation is not needed each season, and the mutation frequency is lower.

Nevertheless, these are distinct risks and limitations with the satellite control strategy. The satellite RNA could cause virulent disease in another crop species or could mutate to a form that enhances disease rather than causing attenuation (Chapter 9, Section II,B,7). Another risk is the reservoir of virus available to vectors in the protected plants. Lastly, the satellite approach will be limited to those viruses for which satellite RNAs are known. However, CMV is economically a very important virus.

Transgenic Plants Expressing Antisense RNAs   One method of gene regulation in organisms is by complementary RNA molecules that are able to bind to the RNA transcripts of specific genes and thus prevent their translation. Such RNA has been called antisense or mic RNA (*m*essenger-RNA-*i*nterfering *c*omplementary RNA). Coleman *et al.* (1985) integrated appropriate antisense sequences into the genome of *Escherichia coli* and showed that when these were induced, they could block the replication of the bacteriophage for which they were designed. Appropriate antisense sequences incorporated into a plant genome have been shown to block the activity of specific genes (e.g., van der Krol *et al.,* 1988; Delauney *et al.,* 1988). The possibility of using this strategy for the control of plant viruses is being explored.

Various laboratories have carried out *in vitro* studies with oligonucleotides complementary to some plant virus RNA sequence. Oligodeoxynucleotides complementary to genomic PVX RNA caused translation arrest in a Krebs-2 cell-free system. This was thought to be due to endogenous RNase H activity in the cell-free system (Miroshnichenko *et al.,* 1988). Antisense sequences complementary to sequences near the 5' end of TMV RNA inhibited *in vitro* translation of this RNA in a rabbit reticulocyte lysate. The inhibition was probably due to direct interference with ribosome attachment (Crum *et al.,* 1988). Morch *et al.* (1987) found that the "sense" nucleotide sequences corresponding to the replicase recognition site near the 3' end of genomic TYMV RNA specifically inhibited *in vitro* the activity of the TYMV replicase isolated from virus-infected plants. The relevance of these various experiments to possible virus inhibition *in vivo* remains to be determined.

Among transgenic tobacco plants containing genes for the production of antisense RNAs for three regions of the CMV genome, only one showed some resistance to the virus (Rezaian *et al.,* 1988). Cuozzo *et al.* (1988) compared the extent of protection provided in transgenic tobacco plants by the coat protein gene of CMV or its antisense transcript. Symptom development and virus accumulation were reduced or absent in plants transgenic for the sense gene, and this was unaffected by inoculum concentration over the range used. By contrast, antisense plants were protected only at low inoculum concentrations. Transgenic tobacco plants expressing RNA sequences complementary to the coat protein gene of TMV were not protected as strongly from TMV infection as were plants expressing the coat protein gene itself (Powell *et al.,* 1989). Thus more *in vivo* experimentation will be necessary before the utility of the antisense approach for the control of plant viruses can be properly assessed, but present evidence suggests that it may not be particularly useful.

Transgenic Plants Expressing Ribozymes   The development of simple and specific RNA enzymes was described in Chapter 9, Section II,B,6. Provided the nucleotide sequence of a viral gene is known it should be possible to design a ribozyme specific for that gene and integrate the ribozyme sequence into the plant genome. In principle, such a procedure might specifically inactivate an invading virus.

Transgenic Plants Expressing PR Proteins   The PR host proteins induced following infection with viruses causing necrotic local lesions were discussed in Chapter 12, Section III. These proteins, which are part of a nonspecific host defense reaction, are involved in the phenomenon of local acquired resistance. Treatment of leaves with salicylic acid induces certain PR proteins and inhibits AMV replication in such leaves. Hooft van Huijsduijnen *et al.* (1986b) have isolated and cloned the mRNAs for some of these proteins. In principle it might be possible to provide protection against certain viruses by using "transgenic" plants in which PR protein genes are expressed constitutively under the control of a suitable promoter.

Transgenic Plants Expressing an Insect Toxin   The bacterium *Bacillus thuringiensis* produces a polypeptide that is toxic to insects. Different strains of this bacterium produce toxic polypeptides with specificity for different insect groups. Vaeck *et al.* (1987) produced transgenic plants expressing the toxin gene that were protected against insect attack. Among the virus vector groups of insects, only the Coleoptera appear to be affected by the toxin from some strains of the bacterium. In principle, it might be possible to produce transgenic plants protected against beetle vectors of some plant viruses.

Transgenic Plants Expressing Virus-Specific Antibodies   Plants do not have an immune system like that of animals in which specific antibody proteins are formed in response to an infection, and it has long been assumed that plants could not produce such proteins. However, the work of Hiatt *et al.* (1989) demonstrates that this is possible. They obtained cDNAs derived from mouse hybridoma mRNA, transformed tobacco leaf segments, and regenerated plants. Plants expressing single gamma (heavy) or kappa (light) chains were crossed to produce plants in which both chains were expressed simultaneously. A functional antibody made up over 1% of leaf proteins. It should be possible to adapt this procedure to produce transgenic plants expressing an antibody that might protect them from infection by a specific virus or viruses.

Summary   Transgenic plants offer considerable hope for the production of crop cultivars that are protected genetically against particular viruses. From current knowledge, plants expressing viral coat protein genes, parts of other genes, or satellite RNAs appear to offer the best prospects. Satellite RNAs offer stronger protection but are limited to those viruses known to support a satellite. Expression of coat protein genes offers protection for a range of different kinds of viruses, but the protection is not absolute. Nevertheless, transgenic cultivars of some annual crops with a useful degree of protection derived from a coat protein gene and possibly other viral genes may be available commercially in the near future. The other possibilities noted for transgenic plants are more speculative at this stage.

2. Cross-Protection

Infection of a plant with a strain of virus causing only mild disease symptoms may protect it from infection with severe strains (Chapter 13, Sections III,C,4 and IV,A). Thus plants might be purposely infected with a mild strain as a protective measure against severe disease. While such a procedure could be worthwhile as an expedient in very difficult situations, it is not to be recommended as a general practice, for the following reasons:

1. So-called mild strains often reduce yield by about 5–10%.
2. The infected crop may act as a reservoir of virus from which other more sensitive species or varieties can become infected.
3. The dominant strain of virus may change to a more severe type in some plants.
4. Serious disease may result from mixed infection when an unrelated virus is introduced into the crop.
5. For annual crops, introduction of a mild strain is a labor-intensive procedure.

This subject has been reviewed by Fulton (1986). In spite of these difficulties, the procedure has been used successfully, at least for a time, with some crops. Broadbent (1964) suggested that, because late infection of greenhouse tomatoes with ToMV often causes a severe reduction in quality of the fruit compared with early infection, growers who regularly suffer those losses should inoculate their plants at an early stage with a mild strain of the virus. Rast obtained a nitrous acid mutant of ToMV that was symptomless in tomato (Rast, 1975). Seedling inoculation with this strain was widely practiced for a time in some western European countries. Cross-protection is still used for cherry tomatoes in the United Kingdom. However, cultivars with resistance genes or with multiple genes for tolerance to ToMV have largely replaced seedling inoculation with the attenuated strain in commercial practice.

Citrus tristeza virus provides the most successful example for the use of cross-protection. Worldwide, this is the most important virus in citrus orchards. In the 1920s, after its introduction to South America from South Africa, the virus virtually destroyed the citrus industry in many parts of Argentina, Brazil, and Uruguay. The successful application of cross-protection by inoculation with mild tristeza virus isolates in Brazil is detailed by Costa and Muller (1980). The method has been particularly successful with Pera oranges, with more than eight million trees being planted in Brazil by 1980. Protection continues in most individual plants through successive clonal generations. The search for improved attenuated strains of the virus continues (e.g., Muller and Costa, 1987; Roistacher *et al.*, 1987), and the technique is being adopted in other countries.

Other viruses and crops for which attempts are being made to develop effective cross-protection include papaya ring spot *Potyvirus* in papaya (Yeh *et al.*, 1988) and TMV in pepper (Tanzi *et al.*, 1986). In principle, the cacao swollen shoot disease of cacao in Ghana should be controllable to some degree by cross-protection with mild strains of the virus (Posnette and Todd, 1955), but various difficulties have prevented its effective application (Fulton, 1986). In particular, the use of the technique was incompatible with the objective of treating all known outbreaks by removal of infected trees. This is no longer feasible, so that there is now scope for using mild strain protection in the worst affected areas of Ghana, where other control measures have been abandoned (J. M. Thresh, personal communication).

Tien *et al.* (1987) obtained mild strains of CMV by adding selected satellite RNAs to a CMV isolate. This procedure has been tested, with increased yields, in many areas of China for the control of CMV in peppers. The extent of the protective effect depended on such factors as inoculation time, percentage of plants inoculated, the nature of virulent CMV strains already in the field, and variety of pepper. Yamaya *et al.* (1988b) introduced a cDNA copy of the entire genome of a mild strain of TMV into tobacco plants. The mild strain replicated in the transgenic plants and under experimental conditions protected against superinfection with a severe strain.

Wu *et al.* (1989) compared the degree of protection obtained by preinoculating tobacco and pepper plants with a mild strain of CMV with or without satellite RNA. The presence of satellite RNA increased the protection obtained. Similar results were obtained in greenhouse experiments with CMV in tomato. In field trials, protection was maintained when the virus was introduced by aphids (Montasser *et al.*, 1990). This work has been taken a stage further by Gallitelli *et al.* (1990), who inoculated several hundred young tomato seedlings using several varieties, with CMV strain S carrying a nonnecrogenic satellite called S-CARNA5. These were planted out at the seedling stage in spring on a farm in southern Italy, where a severe epidemic of the tomato necrosis disease was expected. The epidemic occurred, with 100% of plants being destroyed in some fields. In the field containing the inoculated plants, protection against necrosis was almost 100%, while 40% of uninoculated plants developed lethal necrosis. Fruit yields were about doubled in the protected plants. This preliminary trial augurs well for the future of this technique, at least with this host and virus, under conditions where epidemics are frequent.

3. Antiviral Chemicals

Considerable effort has gone into a search for inhibitors of virus infection and multiplication that could be used to give direct protection to a crop against virus infection in the way that fungicides protect against fungi.

There has been no successful control on a commercial scale by the application of antiviral chemicals. The major difficulties are:

1. An effective compound must inhibit virus infection and multiplication without damaging the plant. This is the first, and major, problem. Virus replication is so intimately bound up with cell processes that any compound blocking virus replication is likely to have damaging effects on the host.
2. An effective antiviral compound would need to move systemically through the plant if it were to prevent virus infection by invertebrate vectors.
3. A compound acting systemically would need to retain its activity for a reasonable period. Frequent protective treatments would be impracticable. Many compounds that have some antiviral activity are inactivated in the plant after a time.
4. For most crops and viruses the compound would need to be able to be produced on a large scale at an economic price. This might not apply for certain relatively small-scale, high-value crops, such as greenhouse orchids.
5. For use with many crops, the compound would have to pass food and drug regulations. Many of the compounds that have been used experimentally would not be approved under such regulations.

Many substances isolated from plants and other organisms, as well as synthetic organic chemicals, have been tested for activity against plant viruses. Almost all the substances showing some inhibition of virus infectivity do so only if applied to the leaves before inoculation or very shortly afterward. A recent example is a glucan preparation obtained from *Phytophthora megasperma f.s.p. glycinea,* which appears to inhibit infection by several viruses by a novel, but unknown mechanism (Kopp *et al.,* 1989). Earlier work in this field has been reviewed by Matthews (1981), Tomlinson (1981), and White and Antoniw (1983).

Synthetic analogs of the purine and pyrimidine bases found in nucleic acids have been widely studied, and the search for inhibitory compounds of this type continues (Dawson and Boyd, 1987a). The substituted triazole, 4(5)-amino-1H-1,2,3-triazole-5(4)-carboxyamide, can be regarded as an aza analog of the substituted imidazole compound that is known to be a precursor of the purine ring in some systems. This compound showed some plant virus inhibitory activity but was less effective than 8-azaguanine. The riboside of this compound was suggested as a possible antiviral agent by Matthews (1953c). It was later found to have a broad-spectrum activity in experimental animal virus systems and given the name Virazole (Sidwell *et al.,* 1972). Virazole, also known as ribavirin, has been studied in a variety of plant–virus systems. For example, pretreatment of tobacco plants with the compound delayed or prevented systemic infection with TSWV (De Fazio *et al.,* 1980). Virazole reduced the concentration of CMV and AMV in cultured plant tissues. However, virus-free plants were obtained from meristem tip cultures whether or not Virazole had been in the medium (Simpkins *et al.,* 1981). Other virus-inhibitory compounds have been included in culture media in attempts to improve the efficiency of meristem tip culture for obtaining virus-free plants. None has found widespread use.

4. Suppression of Disease Symptoms by Chemicals

For a time there was considerable interest in the use of certain systemic fungicides to suppress the symptoms of virus infection without necessarily having any effect on the amount of virus produced in the leaves. The fungicides concerned (e.g., Benlate and Bavistin) break down in aqueous solution to give methylbenzimidazole-2-yl-carbamate (MBC or carbendazim). It has been reported that these systemic fungicides possess cytokinin like activity, although at a low level compared to kinetin. Application of MBC as a soil drench caused a substantial reduction in leaf symptoms caused by TMV in tobacco (Tomlinson *et al.*, 1976). However, this compound or others with similar effects have not found commercial application.

D. Disease Forecasting

The problems of monitoring, modeling, and predicting epidemics of plant viral diseases are discussed by McLean *et al.* (1986). For annual crops, forecasting of disease epidemics can be of very great value. For example, appropriate use of organophosphorus insecticides can delay the spread of yellows viruses by aphids within the sugar beet crop (Hull, 1965). Whether such spraying will give an economic return depends on the timing and extent of virus spread in a given season. In Britain, spraying in June gave an economic yield increase when more than 20% of plants on unsprayed plots contract yellows. A spray warning system for growers was developed based on daily counts of aphids in the crops (Hull, 1965). A grower is sent a warning card when infestation in his region reaches one green aphid per four plants. The time of the spray treatment is very important. Between 1962 and 1966, a spray at the time suggested by the warning scheme decreased yellows incidence by 37%. A spray 2 weeks earlier or later decreased the incidence by only 24–25% (Hull and Heathcote, 1967).

Figure 16.15 illustrates the extent to which infection of sugar beet by yellows disease can be predicted from weather factors known to be important for aphid vector and host plant survival during the winter and early spring aphid migration. During the period studied, a succession of three exceptionally cold winters (1962–1964) not only decreased aphid numbers but probably killed weeds and beet plants remaining after harvest. It was several years after 1964 before overwintering hosts built up again.

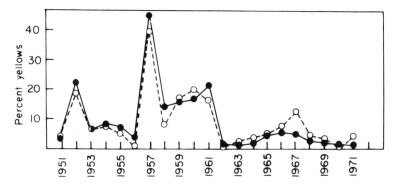

Figure 16.15 Disease forecasting using climatic factors important for aphid vector overwintering and early spring multiplication. ●——●, observed value for percentage yellows disease in sugar beet crops in the United Kingdom; ○---○, values predicted by a regression using the number of frost days (ground temperature < −0.3°C) in January, February, and March and mean April temperatures. (From Watson *et al.*, 1975.)

The mean incidence of yellows in the different sugar beet factory areas declined from south to north. The number of frost days becomes less significant and the mean April temperature more significant on moving north. This was interpreted to mean that few aphids overwinter in the colder areas and that reinfestation takes place each spring from southern England and the European mainland.

Predictions made from winter and spring weather as indicated in Fig. 16.15 have identified years of above-average risk reasonably well, although there was a substantial underestimate of yellows outbreaks in the mid-1970s (Jepson and Green, 1983). Forecasting methods are continually being refined and these, in turn, are dependent on improved techniques for trapping and testing winged vectors (Plumb, 1987; Harrington et al., 1987, 1989; Harrington and Gibson, 1989).

III. DISCUSSION AND CONCLUSIONS

It is impossible to give precise measures of crop losses due to viruses. For a given virus, losses may vary widely with season, crop, country, and locality. Nevertheless, there are sufficient data to show that continuing effort is needed to prevent losses from becoming more and more extensive. Three kinds of situation are of particular importance: (i) annual crops of staple foods such as grains and sugar beet that are grown on a large scale and that under certain seasonal conditions, may be subject to epidemics of viral disease; (ii) perennial crops, mainly fruit trees with a big investment in time and land, where spread of a virus disease, such as citrus tristeza or plum pox, may be particularly damaging; and (iii) high-value cash crops such as tobacco, tomato, cucurbits, peppers, and a number of ornamental plants that are subject to widespread virus infections.

Possible control measures can be classified under three headings: (i) removal or avoidance of sources of infection; (ii) control or avoidance of vectors; and (iii) protecting the plant from systemic disease. In principle, by far the best method for control would be the development of cultivars that resist a particular virus on a permanent basis. Experience has shown that viruses continually mutate in the field with respect to both virulence and the range of crops or cultivars they can infect. Thus breeding for resistance or the development of transgenic plants is unlikely to give a permanent solution for any particular virus and crop. With almost all crops affected by viruses, an integrated and continuing program of control measures is necessary to reduce crop losses to acceptable levels. Such programs will usually need to include elements of all three kinds of control measure noted at the beginning of this paragraph.

The continuing nature of the struggle to keep losses due to viruses to acceptable levels is well illustrated by the history of viruses affecting cucurbits in California (Nameth et al., 1986). Over four decades, the following seven viruses infecting cucurbits have been described from this state: CMV, 1942; squash mosaic *Comovirus*, 1949; papaya ring spot *Potyvirus* (= watermelon mosaic virus 1) and watermelon mosaic 2 *Potyvirus*, 1959; squash leaf curl *Geminivirus*, 1981; lettuce infectious yellows ?*Closterovirus*, 1982; and zucchini yellow mosaic *Potyvirus*, 1983. This list includes both aphid- and whitefly-transmitted viruses. The relative importance of the different viruses fluctuates continually with season and locality.

The following four examples will illustrate some of the kinds of control program that have been developed in various periods and localities to minimize damage by particular viruses.

1. BYDV has been a major disease problem in wheat in New Zealand as it is in most wheat-growing areas of the world. There are two periods of spread in the crop, the first in the autumn (May in the Southern Hemisphere) by winged aphids introducing infection from outside. A small aphid population overwinters in the crop, and the second period of spread (this time within the crop) begins when these aphids multiply in the spring (September). A spray warning system was used for a time, based on the number of aphids in the autumn flight (Close, 1964). With a clear gap between two periods of spread there is considerably more latitude in the time of spraying (July–September) than with sugar beet grown in Europe. Figure 16.16 summarizes the control program that has since been developed for the area. The control of BYDV in cereals is complicated by the very large number of grass species that are alternate hosts and by vector specificity. There are probably at least two distinct viruses involved.

2. Virus-free stocks coupled with an adequate certification scheme offer the best basis for control with many vegetatively propagated crops. Such a scheme must be integrated with other measures aimed at minimizing or delaying reinfection. Thus Pitcher and McNamara (1973) recommend a 2-year interval between the removal of diseased hop vines and replanting with stocks free of arabis mosaic *Nepovirus*.

3. A crop rotation that included several years with barley reduced infection of

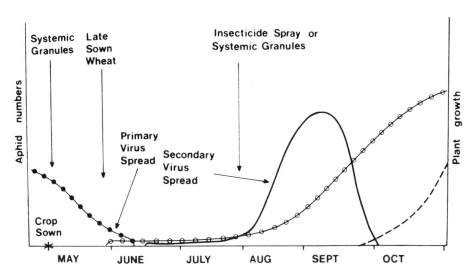

Figure 16.16 Program for the control of BYDV in wheat crops in Canterbury, New Zealand. The recommendation from about 1965 has been to delay sowing until late May/early June. Thus the crop escapes the autumn flight of aphids and consequent virus infection from outside the crop. The spring buildup of aphids within the crop is also minimized. This strategy can be reinforced with two applications of insecticide: the first at sowing to protect the crop until aphids flights have ceased, and the second in August to prevent increase in the numbers of wingless aphids, thus stopping secondary virus spread within the crop. O——O, wheat growth; ---, spring aphid flight; ●——●, autumn aphid flight; ——, wingless aphids within wheat crops. (Reproduced from Burnett, 1976, by permission of Lincoln University.)

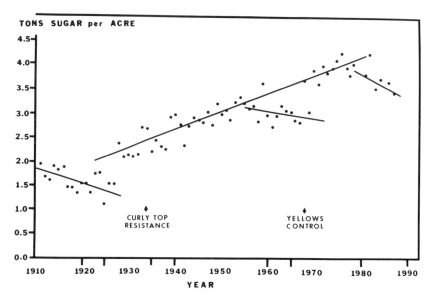

Figure 16.17   Control of virus diseases in sugar beet in California as measured by yield of sugar per acre, for the period 1911 to 1985. Five periods can be delineated. (a) From 1911 to 1925, yields declined because of curly top virus infection. (b) From about 1930, yields improved as a result of removal of badly infected areas from production, and the introduction of curly top-resistant cultivars. (c) From about 1959 to 1969, yields declined again because of more severe strains of curly top virus and the spread of yellows. (d) In 1967, a beet-free period was widely maintained together with a cooperative effort to clean up weed beets. In 1968, varieties with moderate resistance to yellows were introduced, giving a further period of increasing yield. (e) From about 1979, yields have again decreased because of a number of factors: (i) Lettuce infectious yellows virus (Duffus *et al.*, 1986) first occurred in the desert regions of California in 1981, and has since been a limiting factor in sugar production in that region. In susceptible varieties this virus causes a 30% reduction in yield. (ii) In 1983, *Rhizomania* was first found in California and by 1989 it affected over 80,000 hectares. (iii) In recent years, growers in the Delta region have been careless in their management of beet-free periods, resulting in renewed damage by beet yellows virus. (J. E. Duffus, personal communication; figure courtesy of J. E. Duffus.)

potatoes with TRV (French and Wilson, 1976). crops must be monitored in order to detect changing patterns of virus infection, vector behavior, and host response.

4. Viruses in sugar beet. Figure 16.17 illustrates the various measures taken over a 76-year period to control viruses of beet in California.

17

Nomenclature, Classification, Origins, and Evolution

I. NOMENCLATURE AND CLASSIFICATION

A. Historical Aspects

In all studies of natural objects humans have an innate desire to name and to classify. Virologists are no exception. A classification arranges objects showing similar properties into groups, and this becomes an information retrieval system. If made on appropriate criteria, a classification can also give an indication of evolutionary relationships. In theory, it is possible to consider the problems of naming and classifying viruses as separate issues. In practice, however, naming soon comes to involve classification.

Early workers generally gave a virus a name derived from the host plant in which it was found and the most conspicuous disease symptom, for example, tobacco mosaic virus. Viruses were at first thought of as stable entities, and each disease condition in a particular host species was considered to be due to a different virus. However, by the early 1930s three important facts began to be recognized: (i) viruses can exist as different strains, which may cause very different symptoms in the same host plant; (ii) different viruses may cause very similar symptoms on the same host plant; and (iii) some diseases may be caused by a mixture of two unrelated viruses. J. Johnson in 1927 and in subsequent work stressed the need for using some criteria other than disease symptoms and host plants for identifying viruses. He suggested that a virus should be named by adding the word virus and a number to the common name for the host in which it was first found, for example, tobacco virus 1 for TMV. Johnson and Hoggan (1935) compiled a descriptive key based on five characters: modes of transmission, natural or differential hosts, longevity *in vitro*, thermal death point, and distinctive or specific symptoms. About 50 viruses were identified and placed in groups.

K. M. Smith (1937) outlined a scheme in which the known viruses or virus diseases were divided into 51 groups. Viruses were named and grouped according to the generic name of the host in which they were first found. Successive members in a group were given a number. For example, TMV was *Nicotiana* virus 1, and there were 15 viruses in the *Nicotiana* virus group. Viruses that were quite unrelated in their basic properties were put in the same group. Although Smith's list served for a time as a useful catalog of the known viruses, it could not be regarded as a classification.

Holmes (1939) published a classification based primarily in host reactions and methods of transmission. He used a Latin binomial-trinomial system of naming. For example, TMV became *Marmor tabaci,* Holmes. His classification was based on diseases rather than the viruses, and thus 53 of the 89 plant viruses considered by Holmes fell in the genus *Marmor,* which contained viruses known even at that time to differ widely in their properties.

Between 1940 and 1966 various schemes were proposed either for plant viruses only or for all viruses. None of these schemes was adopted by any significant number of virologists. It became increasingly apparent that a generally acceptable system of nomenclature and classification could be developed only through international cooperation and agreement, with the opinions of a majority of working virologists being taken into account.

At the International Congress for Microbiology held in Moscow in 1966, the first meeting of the International Committee for the Nomenclature of Viruses was held, consisting of 43 people representing microbiological societies of many countries. An organization was set up for developing an internationally agreed taxonomy and nomenclature for all viruses. Rules for the nomenclature of viruses were laid down. The subsequent development of the organization now known as the International Committee for Taxonomy of Viruses (I.C.T.V.) has been summarized (Matthews, 1983a, 1985a,b). The main features of the agreed nomenclature and taxonomy as they apply to plant viruses are considered in the following sections.

B. Systems for Classification

There are two general ways in which organisms may be classified. One is the classic monothetic hierarchical system applied by Linnaeus to plants and animals. This is a logical system in which decisions are made as to the relative importance of different properties, which are then used to place a taxon in a particular phylum, order, family, genus, and so on. Maurin *et al.* (1984) have proposed a classification system of this sort that embraces viruses infecting all kinds of hosts. While such systems are convenient to set up and use, there is as yet no sound basis for them as far as viruses are concerned. The major problem with any such system is that we have no scientific basis for rating the relative importance of all the different characters involved. For example, is the kind of nucleic acid (DNA or RNA) more important than the presence or absence of a bounding lipoprotein membrane? Is the particle symmetry of a small RNA virus (helical rod or icosahedral shell) more important than some aspect of genome strategy during virus replication?

An alternative to the hierarchical system was proposed by Adanson (1763). He considered that taxa were best derived by considering all available characters. He made a series of separate classifications, each based on a single character, and then examined how many of these characters divided the species in the same way. This gave divisions based on the largest number of correlated characters. The method is laborious and has not been much used until recently because at least about 60 equally weighted independent qualitative characters are needed to give satisfactory division (Sneath, 1962). However, the availability of computers has renewed interest in this kind of classification. The computer can be used in various ways to give experimental classifications of groups of viruses (e.g., Fig. 13.13).

Comparisons can include numerical information such as nucleic acid base ratios and amino acid composition of the protein. However, computer classification must be regarded as experimental, for it is as limited by lack of information about the viruses as any other method. In practice some weighting of characters is inevitably involved even if it is limited to decisions as to which characters to leave out of consideration and which to include. Its main advantages at present may be to confirm groupings arrived at in other ways and to suggest possibly unsuspected relationships that can then be checked by further experimental work.

The recent rapid accumulation of complete nucleotide sequence information for many viruses in a range of different families and groups is having a profound influence on virus taxonomy in at least three ways: (i) Most of the virus families delineated by the I.C.T.V., mainly on morphological grounds, can now be seen to represent clusters of viruses with a relatively close evolutionary origin. (ii) With the discovery of previously unsuspected genetic similarities between viruses infecting different host groups, the unity of virology is now quite apparent. The stalling approach of some plant virologists to the application of families, genera, and species to plant viruses is no longer tenable. (iii) Genotypic information is now more important for many aspects of virus taxonomy than phenotypic characters. However, there are several limitations to the use of genotype (sequence) data alone. One is the practical problem that even with current technology it is a relatively time-consuming and expensive task to determine the complete nucleotide sequence of a viral genome. Another problem is that with the present state of knowledge, it is very difficult or impossible to predict phenotypic properties of a virus on the basis of sequence data alone. For example, if we have two viral nucleotide sequences differing in a nucleotide at a single site, we could not, at present, deduce from this information alone that one led to mosaic disease and the other to lethal necrosis in the same host. As another example, we could not, at present, decide from the nucleotide sequences alone which of two rhabdoviruses replicated in plants and insects and which in vertebrates and insects.

Any classification of viruses should be based not only on evolutionary history, as far as this can be determined from the genotype, but should also be useful in a practical sense. Most of the phenotypic characters used today in virus classification will remain important even when the nucleotide sequences of most viral genomes have been determined.

C. Families, Groups, and Genera

In 1970 at a meeting in Mexico City, the I.C.T.V. (then called the I.C.N.V.) approved the first taxa for viruses (Matthews, 1983a). There were 2 families and type genera for these, plus 22 genera not placed in families for viruses infecting vertebrates, invertebrates, or bacteria. Sixteen taxa designated as groups were approved for viruses infecting plants. While other virologists subsequently moved quite rapidly to develop a system of families and genera for the viruses, plant virologists have clung to the notion of groups. There are at present two exceptions to this statement. The very obvious and detailed similarities between reoviruses infecting vertebrates, invertebrates, and plants made it inescapable that the plant reoviruses should be placed as members of the already existing Reoviridae family.

This took place in 1978. Similarly, plant rhabdoviruses became members of the Rhabdoviridae family. Two genera of plant reoviruses were delineated—*Phytoreovirus* and *Fijivirus*. This was an important event, because for the first time it brought a few plant viruses into the family and genus structure formed for viruses infecting all other groups of organisms. In August 1990 the I.C.T.V. approved a new genus *Tospovirus* in the family Bunyaviridae to contain TSWV (Chapter 8, Section V). Similarities in nucleic acid sequence and genome organization between viruses infecting various host groups are discussed in Section V.

As noted earlier, these developments strongly reinforce the concept of the unity of virology (see also Kingsbury, 1988). I therefore believe that the time is overdue for plant virologists to rationalize the current "group" taxa into families and genera and thus eliminate some of the serious anomalies and deficiencies in the present state of plant virus taxonomy. The proposal that viral taxonomy should employ only two taxa—groups and viruses—is unlikely to be adopted (Milne, 1984c; Matthews, 1985a,b).

Looked at from the point of view of virology as a whole, most of the plant virus groups clearly have the status of families (e.g., Caulimoviruses, Geminiviruses, Tymoviruses etc). Some other existing groups certainly do not have the properties expected for a family, for example the *Bromovirus* and *Cucumovirus* groups, which are obviously very closely related. They, together with AMV and ilarviruses, should be genera within a single family. There are now clear examples where the genus would be a very useful taxon within a plant virus family, for example a Geminiviridae family, with at least two genera, one for viruses with bipartite genomes and one for those whose genomes are monopartite.

If they were given family status, many plant virus groups would, on present knowledge, contain only one genus. This is no reason against giving such a group family status, if that would be appropriate having regard to the families of viruses affecting other kinds of host.

The delineation of genera within some plant virus families may present difficulties. In such circumstances, decisions should await further information. For example Jellison (1987) applied three different exploratory methods of numerical taxonomy to members of the *Tymovirus* group. The results from one method are shown in Fig. 17.1. The three techniques produced similar but not identical relationships. It might be difficult to place tymoviruses into separate genera on the basis of existing information. This view is supported by other studies (Blok *et al.*, 1987a).

The existing groups and families, listed in Table 17.1, are able to accommodate most newly described viruses. Most of the taxa listed in Table 17.1 have a quite distinctive genome strategy and organization, often accompanied by distinctive morphological characters in their virus particles. Thus most of the clusters of plant viruses delineated by the I.C.T.V. seem to be set for a reasonably stable future. However, no taxonomy is inscribed in stone, and changes will no doubt be needed from time to time as new information becomes available about known viruses and as new viruses are discovered.

For example, Rohde *et al.* (1990) have described a circular single-stranded covalently closed DNA associated with coconut foliar decay disease. This DNA can form a stable stem-loop structure of 10 GC base pairs subtending a loop that contains the sequence TAATATTAC conserved in a similar structure in the *Geminivirus* group (Chapter 8, Section II). However, the virus probably represents a new taxonomic group since it has no other sequence homology with any DNA plant viruses and its size, at 1291 nucleotides, is less than half that of a monopartite *Geminivirus*.

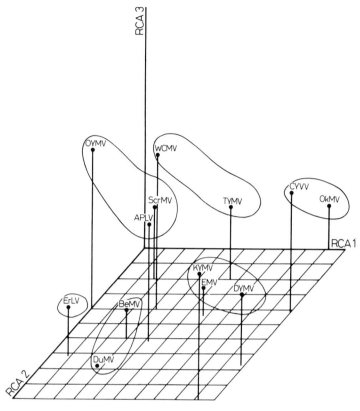

Figure 17.1 The reciprocal averaging (RA) technique of numerical taxonomy applied to an analysis of relationship between members of the *Tymovirus* group. RA was carried out to enable representation of the relationship among the tymoviruses in three dimensions based on both continuous and discontinuous data. Data used to obtain the analysis included: sedimentation coefficients of the top and bottom viral components, percentage nucleic acid in the particles, percentage base composition, particle size, dilution end point, temperature of inactivation, molar percent amino acid composition, and total number of amino acid residues per subunit. Twenty-nine additional binary traits, including host range, geographical distribution, vector species, and host cytological alterations, were also used in the analysis. Individual viruses are circled to indicate pairs or groups of viruses found to be closely related based on the results of another clustering program. (From Jellison, 1987.)

D. Virus Species

1. The Species "Problem" in Relation to Plant Viruses

The species taxon is generally regarded as the most important in any taxonomy, but it has proved the most difficult to apply to viruses. The question of species nomenclature has proved even more difficult.

The central property of viruses is that they consist of a set of genes that code for functional proteins. Through alterations in these genes, viruses undergo evolutionary change. These fundamental properties are like those of cells. Thus the concept of species should be applicable.

The current I.C.T.V. definition of a virus species is: "*A virus species is a concept that will normally be represented by a cluster of strains from a variety of*

Table 17.1

The 35 Families and Groups of Plant Viruses Approved by the ICTV Listed According to Particle Morphology and Type of Nucleic Acid[a]

Characterization	Family or group	Number of members	Number of probable or possible members	Total
dsDNA nonenveloped	*Caulimovirus*	8	6	14
	Commelina yellow mottle virus group[b]	1	0	1
ssDNA nonenveloped	*Geminivirus*	11	14	25
dsRNA nonenveloped	*Cryptovirus*	13	13	26
	Reoviridae	9	0	9
ssRNA enveloped	Rhabdoviridae	11	66	77
	Bunyaviridae	1	0	1
ssRNA nonenveloped				
Monopartite genomes				
Isometric particles	*Carmovirus*	7	8	15
	Luteovirus	11	16	27
	Marafivirus	5	0	5
	MCDV group	1	1	2
	Necrovirus	1	1	2
	Potexvirus	18	21	39
	PYFV group	2	1	3
	Sobemovirus	2	8	10
	Tobamovirus	12	3	15
	Tombusvirus	10	3	13
	Tymovirus	18	1	19
Rod-shaped particles	*Capillovirus*	2	2	4
	Closterovirus	11	1	12
	Carlavirus	27	23	50
	Potyvirus	50	70	120
Bipartite genomes				
Isometric particles	Comovirus	16	0	16
	Dianthovirus	3	0	3
	Fabavirus	3	0	3
	Nepovirus	26	8	34
	PEMV	1	0	1
Rod-shaped particles	*Furovirus*	3	4	7
	Tobravirus	3	0	3
Tripartite genomes				
Isometric particles	*Bromovirus*	4	1	5
	Cucumovirus	3	1	4
	Ilarvirus	11	2	13
Isometric and bacilliform particles	AMV	1	0	1
Rod-shaped particles	*Hordeivirus*	3	1	4
Quadripartite genomes				
Rod-shaped particles	*Tenuivirus*	3	4	7
Total		312	279	590

[a]From data provided by R. I. Hamilton.
[b]New group approved August 1990.

sources, or a population of strains from a particular source, which have in common a set or pattern of correlating stable properties that separates the cluster from other clusters of strains." (Matthews, 1982)

This is a very general definition. It must be recognized that it is impossible for a definition to be sufficiently precise to be a detailed guide in delineating species in any particular virus family and at the same time be relevant for all families.

Van Regenmortel (1989a) proposed a definition of a virus species as *"a polythetic class of viruses constituting a replicating lineage and occupying a particular ecological niche."* This definition incorporates the ideas of biological variability of genomes ("polythetic class") and changes in genomes ("replicating lineage") and of niche occupation.

Personally, I support the pragmatic view of Davis and Heywood (1963) in discussing angiosperm taxonomy: "There is no universally correct definition (of a species) and progress in understanding the species problem will only be reached if we concentrate on the problem of what we shall treat as a species for any particular purpose." The species concept applied to plants is also discussed by Wagner (1984). In day-to-day practice, virologists use the concept of a "virus" as being a group of fairly closely related strains, variants, or pathovars. A virus defined in this way is essentially a species in the sense suggested by Davis and Heywood, and defined by the I.C.T.V.

Some plant virologists believe that the species concept should not be applied to viruses. Their main reason is that because viruses reproduce asexually the criterion of reproductive isolation cannot be used as a basis for defining virus species (e.g., Milne, 1988). In fact the species concept has been usefully applied to many groups of organisms where the criterion of reproductive isolation cannot be applied. As an extreme, the genus *Hieracium* (Compositae) contains some 10,000–20,000 "species" most of which produce seed apomictically. Thus there is no theoretical or practical reason why the concept of species should not be applied to plant viruses. This topic is discussed in more detail by Matthews (1983b, 1985a,b) and Van Regenmortel (1989a).

2. Naming of Viruses (Species)

Questions of virus nomenclature have generated more heat over the years than the much more practically important problems of how to delineate distinct virus species. At an early stage in the development of modern viral taxonomy a proposal was made to use cryptograms to add precision to vernacular names of viruses (Gibbs *et al.,* 1966). The proposal was not widely adopted except, for a time, among some plant virologists. A revision of the idea (Fenner and Gibbs, 1983) is unlikely to stimulate a revival in its use.

When a family, genus, or plant virus group is approved by I.C.T.V., a type species or type virus is designated. However, none of these species (viruses) has yet received an official name. Some virologists favor using the English vernacular name as the official species name. Using part of a widely known vernacular name as the official species name may frequently be a very suitable solution, but it could not always apply (e.g., with newly discovered viruses). Other virologists favor serial numbering for viruses (species). The experience of other groups of microbiologists is that, while numbering or lettering systems are easy to set up in the first instance, they lead to chaos as time passes and changes have to be made in taxonomic groupings. The idea of Latinized binomial names for viruses was supported by the

I.C.T.V. for many years but never implemented for any viruses. The proposal for Latinization has now been withdrawn.

A species name indicates uniqueness, while a genus name can remind us of relatives and their properties. As a mechanism for information retrieval I believe that a simple binomial system, not necessarily Latinized, will turn out to be the most useful and versatile system in the long run. A generic name with a virus (species) epithet, followed by a strain or pathovar designation, should give a complete, unambiguous, and internationally understood designation of a particular virus. In the meantime there is little doubt that the English vernacular names will remain the unofficial specific name for most plant viruses for many years to come.

In successive editions of the Reports of the I.C.T.V. (e.g., Matthews, 1982), virus names in the index have been listed by the vernacular name (usually English) followed by the family, genus, or group name, for example, tobacco mosaic *Tobamovirus*, Fiji disease *Fijivirus,* and lettuce necrotic yellows rhabdovirus. This method for naming a plant virus is becoming increasingly used in the literature. It is, of course, a binomial system but with the virus (species) name coming before the genus name, or what would be the genus name following the rationalization suggested earlier in this section. The fact that the two terms in the binomial are in the reverse order compared with the binomial names of all cellular organisms does not detract from its utility as an information retrieval system. Nevertheless, it becomes somewhat awkward if we wish to designate a strain or pathovar, for example, tobacco mosaic *Tobamovirus* vulgare.

3. Delineation of Viruses (Species)

The delineation of the kinds of virus that exist in nature, that is to say virus species, is a practical necessity, especially for diagnostic purposes relating to the control of virus diseases. For some 20 years, B. D. Harrison and A. F. Murant have acted as editors for the Association of Applied Biologists (A.A.B.) to produce a series of some 339 descriptions of plant viruses and plant virus groups and families. Each description is written by a recognized expert for the virus or group in question. The two editors use commonsense guidelines devised by themselves to decide whether a virus described in the literature is a new virus, or merely a strain of a virus that has already been described. When they publish a new virus description they are in effect delineating a new species of virus. The A.A.B. descriptions of plant viruses are widely used by plant virologists and accepted as a practical and effective taxonomic contribution. The descriptions now cover a substantial proportion of the known plant viruses. I believe that most of them should be adopted officially by the I.C.T.V. as constituting plant virus species.

However, examination of the data summarized in Table 17.1 suggests that at present the delineation of viruses in some groups may be significantly more difficult than in others. Thus two groups, the rhabdoviruses and potyviruses, stand out as having many more members (total) than the other families and groups. Many descriptions of rhabdoviruses rely heavily on electron microscopic data. It is probable that many of the isolates now listed as probable or possible viruses are really strains of a smaller number of distinct viruses. The taxonomy of the potyviruses has also been difficult (Chapter 13, Section V,C,6). The application of new serological procedures (Chapter 13, Sections III,B,4 and 5) and nucleic acid hybridization and amino acid sequence studies may well lead to a reduction in the number of distinct potyviruses that are recognized (see Fig. 13.13).

II. CRITERIA AVAILABLE FOR CLASSIFYING VIRUSES

Properties that are useful for characterizing strains of a virus were discussed in Chapter 13, Section III.

There are two interconnected problems involved in attempting to classify viruses. First, related viruses must be placed in groups (families). For this purpose the more stable properties of the virus, such as amount and kind of nucleic acid, particle morphology, and genome strategy, are most useful. The second problem is to be able to distinguish between related viruses and give some assessment of degrees of relationship within a group. Properties of the virus for which there are many variants are more useful for this purpose. These include symptoms, host range, and amino acid composition of the coat protein. Certain properties, such as serological specificity and amino acid sequences may be useful both for defining groups and for distinguishing viruses within groups.

There are two general problems that must be borne in mind when considering the criteria to be used for virus classification. One is the problem of weighting—the relative importance of one character as compared to another. Some weighting of characters is inevitable, whatever system is used to place viruses in taxa. The second problem concerns the extent to which a difference in one character depends on a difference in another. For example, a single-base change in the nucleic acid may result in an amino acid replacement in the coat protein, which in turn alters serological specificity and electrophoretic mobility of the virus. The same base change in another position in a codon may not affect the gene product at all.

The preceding discussion relates to phenotypic characters. As the complete nucleotide sequences of more and more viruses become known, virus genotype will become increasingly important. What weight is to be given to sequence data in virus classification? A geneticist may say that the evolutionary history of a virus and therefore its relationships with other viruses are completely defined, as far as we will be able to know it, by its nucleotide sequence. This is certainly true in a theoretical sense, but there are also practical aspects to be considered as discussed in Section I,B.

A. Properties of the Viral Nucleic Acid

1. For Delineating Families, Genera, and Groups

The organization and strategy of the viral genome, as discussed in Chapter 6 and detailed in Chapters 7 and 8, is now of prime importance for the placing of viruses into families, genera, and groups, or for the establishment of a new family, genus, or group. Carnation mottle virus provides an instructive example. It shares similar morphological and physicochemical properties with about 27 other viruses, including the definitive members of the *Tombusvirus* group, all having a single isometric particle about 30 nm in diameter. However, following complete sequencing of carnation mottle virus RNA (Guilley *et al.*, 1985) and partial sequencing of the closely related TCV, it became apparent that carnation mottle virus had a genome organization that is quite different from that of the *Tombusvirus* group (Morris and Carrington, 1988). The differences are illustrated in Figs. 7.4 and 7.15.

It was primarily on the basis of these differences that the *Carmovirus* group was established. Once a group has been established on the basis of full nucleotide sequence data and genome strategy for one or a few members, additional members can be allocated to the group on the basis of less complete information about the genome, such as sizes of genomic and any subgenomic RNAs, nucleic acid hybridization data, and information on RNA replication. On the basis of this kind of information, the *Carmovirus* group now has seven definite members.

Nucleic acid hybridization studies may be useful in deciding which of a set of apparently similar viruses belongs in a given group, as with the tombusviruses (Gallitelli *et al.*, 1985). However, it must be borne in mind that "false" hybridization may sometimes occur through guanine-rich tracts in one polymer binding to cytosine-rich tracts in another (Jones and Hyman, 1983).

Physical structure of the genomic nucleic acid may provide a basis for delineating genera within a family as with the plant reoviruses (Francki and Boccardo, 1983) or, in the future, among the geminiviruses (Chapter 8, Section II) (Roberts *et al.*, 1984). Genome organization and nucleotide sequence studies strongly suggest that the four virus groups with ssRNA tripartite genomes and icosahedral particle morphology could usefully be placed in a single family (e.g., Ahlquist *et al.*, 1981b; Rezaian *et al.*, 1984; Davies and Symons, 1988).

2. For Delineating Species

Criteria for delineating species among groups of virus strains have been discussed in Chapter 13, Sections III and V. The use of genome organization to assist in delineating virus species within a group is illustrated in Fig. 17.2 for five tobamoviruses. The figure illustrates the need for considering all available data in delineating species.

Restriction endonuclease maps of a set of geminiviruses indicated the same relationships as those based on full sequence analysis of the same viruses (Kirby *et al.*, 1989). Nucleic acid hybridization studies were employed to delineate clearly three separate viruses within the *Tobravirus* group (Robinson and Harrison, 1985) and the *Hordeivirus* group (Hunter *et al.*, 1989). However, this property may not be equally useful within all families and groups. For example, despite morphological and other similarities between different viruses in the *Potexvirus* group, no nucleic acid similarities could be detected by cross-hybridization using cDNAs (Bendena and Mackie, 1987). However, it is known that there is some base sequence homology between some potexviruses (Harbison *et al.*, 1988; R. L. S. Forster *et al.*, 1988).

B. Viral Proteins

The properties of viral proteins, and in particular the amino acid sequences, are of prime importance in virus classification at all levels: for delineating strains, as discussed in Chapter 13, Sections III,A,2 and 3; for viruses, as discussed in this section; and for indicating evolutionary relationships between families and groups of viruses (Section V).

Coat protein amino acid sequence homologies have been used to distinguish between distinct potyviruses and strains of these viruses (see Fig. 13.8) and to estimate degrees of relationship within the group (see Fig. 13.13). Dendrograms indicating relationships within a virus group have also been based on amino acid composition of the coat proteins, for example, with the tobamoviruses (Gibbs,

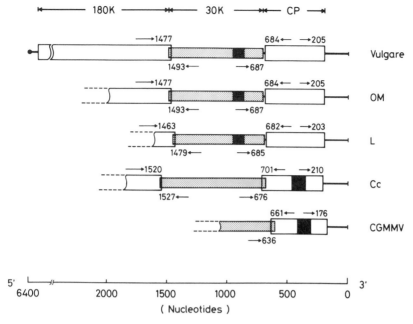

Figure 17.2 Genome organization in the delineation of viruses (species) within a group. Diagrammatic summary of genome structure near the 3' end of several tobamoviruses. Coding regions are indicated by boxes. The number shows the first letter of the initiation codon or the third letter of the termination codon from the 3' end of the genomic RNA. Black areas show the assembly origin. CC = Sunnhemp mosaic virus. (From Takamatsu *et al.*, 1983.) The data suggest that Vulgare OM and L are strains of one species. However, while CC and CGMMV appear to be similar with respect to the position of the assembly origin, from serological and nucleic acid hybridization data they are separate species.

1986). A dendrogram based on peptide patterns obtained from *in vitro* translated 126K proteins of eight tobamoviruses was very similar to that obtained with coat proteins (Fraile and García-Arenal, 1990). Once a set of viruses such as the tobamoviruses has been delineated on the basis of coat protein amino acid composition, new isolates may sometimes be readily placed as a strain of an existing virus (e.g., Creaser *et al.*, 1987).

Using a statistical procedure known as principal component analysis, Fauquet *et al.* (1986) showed that groupings of 134 plant viruses and strains obtained using the amino acid composition of their coat proteins (Fig. 17.3) correlated well with the groups established by the I.C.T.V. (Matthews, 1982).

In recent years it has proved easier to obtain the sequence of nucleotides in genomic nucleic acids than that of the amino acids in the gene products. Thus it is possible to identify amino acid sequence homologies between potential gene products, as has been done for some geminiviruses (Mullineaux *et al.*, 1985) and potyviruses (Domier *et al.*, 1987). Such studies have shown that there may be amino acid sequence similarities between different taxonomic groups of viruses in their nonstructural proteins. Coat proteins, on the other hand, have an amino acid composition that is characteristic of a virus group (Fig. 17.3), and there is usually no significant amino acid sequence similarities between groups. Thus, in spite of the fact that coat proteins represent only a small fraction of the information in most viral genomes, they appear to be the most useful gene products for delineating many distinct viruses and virus groups.

However, functional equivalence of viral coat proteins may not always be

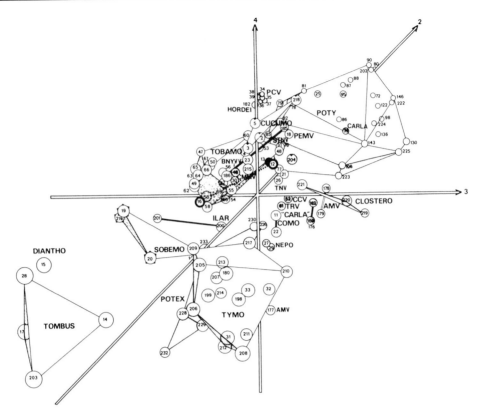

Figure 17.3 Three-dimensional diagram illustrating factors 2, 3, and 4 of a principal components analysis of 122 data sets of plant virus coat proteins compared by their amino acid composition. The three axes contain 62% of the information. The positions of the viruses on axis 2 are indicated by the sizes of the circles. The numbers within the circles are codes for individual viruses. (From Fauquet *et al.*, 1986.)

associated with amino acid sequence similarities. Thus the coat proteins of AMV and TSV are required to activate the genomic RNAs to initiate infection (Chapter 7, Sections VI,C and D). The coat proteins of these two viruses are able to activate each other's genome, recognizing the same sequence of nucleotides in the RNAs, but there is no obvious amino acid sequence similarity between them (Cornelissen and Bol, 1984).

The ultimate goal for using properties of viral proteins in classification will be to know the three-dimensional structure of the protein of at least one member of a group. This would then allow amino acid substitutions in different viruses in a group to be correlated with biological function. This has been achieved for TMV and six viruses related to it (Altschuh *et al.*, 1987). The amino acid sequence homologies of these seven tobamoviruses ranged from about 28 to 82%. Twenty-five residues are conserved in all seven sequences (Fig. 17.4). Twenty of these conserved residues are concentrated in two locations in the molecule: at low radius in the TMV rod near the RNA binding site (36–41, 88–94, and 113–120) and at high radius forming a hydrophobic core (61–63 and 144–145). Where viruses within this set of seven differ in sequence, the differences are often complementary. For example, among buried residues, a change to a large side chain in one position may be compensated by a second change to a smaller one in a neighboring amino acid. This study with

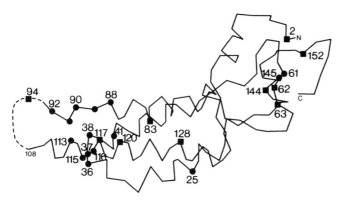

Figure 17.4 Conserved amino acid residues in the coat proteins of seven tobamoviruses for which sequence data are available. The α-carbon chain tracing of one subunit is shown viewed down the disk axis (see Fig. 5.9). The positions are marked for hydrophobic (■) and hydrophilic (●) residues, which are invariant in all seven viruses. (From Altschuh *et al.*, 1987.)

tobamoviruses, while clearly delineating functionally critical regions of the molecule, did not lead to any clear evidence of particular evolutionary relationships within the group.

C. Structure of the Virus Particle

The importance of virus structure in the classification of viruses is summarized in Fig. 5.1. With isometric viruses, particle morphology, as revealed by electron microscopy, has not proved as generally useful as with the rods. This is mainly because many isometric particles lie in the same size range (about 25–30 nm in diameter) and are of similar appearance unless preparations and photographs of high quality are obtained (Hatta and Francki, 1984). Where detailed knowledge of symmetry and arrangement of subunits has been obtained by X-ray analysis or high-resolution electron microscopy, these properties give an important basis for grouping isometric viruses (e.g., Figs. 5.15, 5.16, 5.20, and 5.26). For large viruses with a complex morphology, structure of the particle as revealed by electron microscopy gives valuable information indicating possible relationships (e.g., Figs. 5.28, 5.29, and 5.32).

A mutant of AMV that has mainly spheroidal particles with a strong resemblance to TSV provided supporting evidence for the idea that AMV should be classed as an *Ilarvirus* (Roosien and van Vloten-Doting, 1983).

D. Physicochemical Properties of the Virus Particle

Physicochemical properties and stability of the virus particle have sometimes played a part in classification. These properties are discussed in Chapter 2, Section II. Noninfectious empty viral protein shells and minor noninfectious nucleoproteins are characteristic of certain groups (e.g., tymoviruses). The existence of these components reflects the stability of the protein shell in the absence of the full-length viral RNA.

E. Serological Relationships

Serological methods for determining relationships and their limitations were discussed in Chapter 13, Section III,B. In the past, these methods have been the most important single criterion for placing viruses in related groups. Members of some groups may be all serologically related (e.g., tymoviruses), whereas in others only a few may show any serological relationship (e.g., nepoviruses). In the groups of rod-shaped viruses defined primarily on particle dimensions, many viruses within groups are serologically related, and no serological relationship has been established between groups. It seems probable that serological tests will remain an important criterion upon which virus groups are based. Serological tests may also be used to estimate degrees of relationship between viruses in a group (e.g., Fig. 17.5).

For both tombusviruses and tymoviruses there is no correlation between SDI values as illustrated in Fig. 17.5 and estimates of genome homology. For tobamoviruses, however, there are clear correlations between SDI values, amino acid sequences, and estimates of genome homology (Koenig and Gibbs, 1986). Serological methods can be used to designate a set of virus strains as constituting a new virus within an established group, for example, pepper mild mottle virus in the *Tobamovirus* group (Pares, 1988). When it comes to defining degrees of relationship within a group, borderline situations will sometimes be found. In that circumstance additional criteria will be needed.

Using a broadly cross-reactive antiserum against the viral coat protein core, Shukla *et al.* (1989d) showed that "possible" potyviruses transmitted by mites or whiteflies were serologically related to a definitive potyvirus with aphid vectors. as discussed in Chapter 13, Section III,B, distinct potyviruses have often been difficult to define because serological relationships have been found to be complex and often inconsistent. Shukla *et al.* (1989c) have applied a systematic immunochemical analysis to some members of this group. This method involves the use of overlapping peptide fragments to define the parts of the protein combining with antibodies in particular antisera. This should make it possible to develop both virus-specific and group-specific antisera.

F. Activities in the Plant

The use of biological properties such as host range and symptoms in particular host plants as major criteria led to considerable confusion in the identification and classification of viruses. As knowledge of physical and chemical properties of viruses increased, the biological properties were rated as much less important. However, biological properties must still be given some importance in classifying viruses.

Macroscopic symptom differences will often reveal the existence of a different strain of a virus where no other criterion, except a full nucleotide sequence comparison, will do so. The detailed study of the cytology of infected cells made possible by electron microscopy has shown that many groups of viruses cause characteristic viroplasms or other cytological abnormalities in the cells they infect (Chapter 11, Section III,C,). Ultrastructural changes in cells usually appear to be much more stable characteristics of virus groups than macroscopic symptoms. Certain tombusviruses can be distinguished from other members of the group by their cytopathological effects, especially in *Chenopodium quinoa* (Russo *et al.*, 1987).

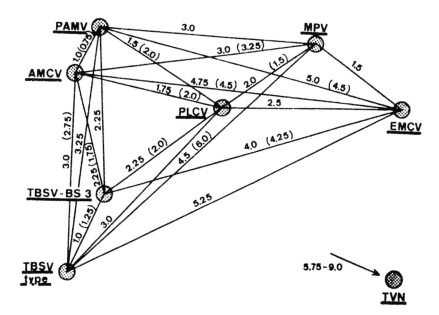

Figure 17.5 The use of serology to estimate degrees of relationship between viruses in a group. The diagram illustrates a classification of 10 tombusviruses with distances representing the mean serological differentiation indices of reciprocal tests (RT-SDI). RT-SDI values have been rounded to the nearest 0.25, and when, in order to represent the relationships in two dimensions, the "observed" and "diagrammatic" (in parentheses) RT-SDI values differ, the two values are shown. CIRV, TVN, and CybRSV are only distantly related to one another and to the other tombusviruses, which form a central cluster. In a multidimensional system these three viruses would have to be arranged in planes above and below that of the central cluster. The arrows indicate the average distance of these viruses from the central cluster as a whole, but not from individual viruses. (From Koenig and Gibbs, 1986.) TVN = *Tombusvirus* Neckar; CIRV = CIRSV in Appendix 2; CybRSV = CyRSV in Appendix 2; PAMV = PeAMV in Appendix 2.

Cross-protection in the plant remains a useful indicator of relationship between viruses (Chapter 13, Section III,C,4).

G. Methods of Transmission

As discussed in Chapter 14, most viruses have only one type of vector, and usually all the viruses within a group have the same type of vector. However, there are exceptions, for example, in the *Potyvirus* and *Geminivirus* groups. In general, the type of vector appears to be a stable character that is useful in delineating major groups of viruses. However, within a group of viruses or virus strains some may be transmitted efficiently by a vector species, some inefficiently, and some not at all. Details of the way in which a virus is transmitted by a vector (e.g., nonpersistent or persistent aphid-transmitted viruses, or on the surface or within the spores of a fungus) may provide further criteria for the grouping of viruses. However, under certain conditions a virus culture may lose the ability to be transmitted by a vector, for example, WTV (Chapter 13, Section II,C,2).

III. SPECULATIONS ON ORIGINS AND EVOLUTION

The most fundamental single property of an organism is the size of its genome. In this respect viruses infecting all kinds of organisms span a range of almost three orders magnitude (see Fig. 1.1). The genome of the smallest virus, STNV, consists of a monocistronic mRNA coding for the coat protein of the virus shell. The largest viruses have genomes about as large as those of the smallest cells. Where and when did this great diversity of agents arise? When and how did they evolve?

Although much relevant information has become available from studies on the structure and replication of viruses and on the molecular biology of normal cells, the origin and evolution of viruses is only now beginning to emerge from the realm of speculation. Nevertheless, the topic is one of general interest, and one that will be relevant to the problem of classification. For a meaningful discussion of these topics we must consider examples from other groups of viruses besides those infecting plants.

There is no compelling reason to suppose that all viruses arose in the same way. Furthermore, it is possible that viruses that originated in one major group of organisms may now exist primarily or solely as agents infecting another group.

A. Origins

Three kinds of origin will be considered in the following sections.

1. Descendants of Primitive Precellular Life-forms

The viruses as we know them today are highly developed obligate parasites. They use the same genetic code as cellular organisms and are dependent for their protein synthesis on ribosomes, tRNAs, and associated enzymes provided by the host cell. Amino acids occur in viral proteins with the same sort of frequency as they do in the globular proteins of other groups of organisms. Thus, although virus particles are relatively *simple* in structure, there is nothing known about their chemistry or mode of replication to show that they are more *primitive* in an evolutionary sense than cellular organisms.

Nevertheless, current views on the origin of life open up the possibility that some RNA viruses at least, or parts of them, may be descended from prebiotic polymers. Until fairly recently, precellular life was considered to involve both proteins and RNAs. However, following the discovery of introns in eukaryotic genes, it was found that the splicing out of introns from RNA transcripts and ligation of the exons could take place in the absence of protein. Furthermore, the ligation reaction can join exons from different RNA transcripts (see Westheimer, 1986) (Chapter 9, Section II,B,8). Thus self-inserting introns can create transposons to shuffle exons. This provides RNA with a vital evolutionary capacity it would otherwise lack—the ability to produce new combinations of genes (see Gilbert, 1986). These observations, and reports of other enzymatic activities for intron RNAs, led Darnell and Doolittle (1986) to propose an evolutionary scheme in which RNAs were the only polymers in the prebiotic stage of evolution. Weiner and Maizels (1987) proposed that the tRNA-like structures found at the 3' terminus of some RNA plant viruses (see Fig. 4.2) are molecular fossils from the original RNA world, retained because of an essential function. They suggested that these structures had several functions including the tagging of genomic RNA molecules for replication. A *Tetrahymena* ribozyme can splice together multiple oligonucleotides aligned on a template strand to give a fully complementary product strand, thus demonstrating the feasibility of RNA–catalyzed RNA replication (Doudna and Szostak, 1989).

On this general theory, RNA viruses might represent greatly modified descendants of prebiotic RNAs that later parasitized the earliest cells. However, there are considerable obstacles from an organic chemical point of view in the initiation of a prebiotic RNA world (Joyce, 1989), and other views are still viable (Waldrop, 1989). Nevertheless, models based on RNA continue to be proposed (e.g., Benner *et al.*, 1989).

2. Development from Normal Constituents of Cells

It has often been considered that viruses may arise from some cell constituent that has escaped from the normal control mechanisms and become a self-replicating entity. Sometimes such a constituent is supposed to escape control in the kind of cell in which it is normally found. More frequently it has been suggested that a normal cell component in one organism may develop into a virus when introduced into the cells of another. For example, some plant viruses might have arisen from normal components of the insects that feed upon plants.

Because viruses carry genetic information it has been proposed that they arose as host genes that escaped from the control mechanisms of the cell. They are considered then to have developed means of being transferred efficiently to other host cells and of replicating independently of cell division. This is a view that is now held by many virologists. Among cellular components, a number of candidates exist as possible progenitors of viruses.

Comparison of Bacterial Plasmids and Viruses

The strongest support for this kind of origin comes from the close parallels between certain extrachromosomal genetic elements in bacteria and some bacterial viruses.

Plasmids are autonomous extrachromosomal genetic elements that are found naturally in many types of bacteria. They consist of closed circular dsDNA, and there may be one or several copies per bacterial cell. They vary in size from about 1500 to 300,000 base pairs. Some can integrate into the bacterial chromosome and are replicated passively along with the chromosomal DNA. An integrated plasmid may become excised from the host DNA and replicate independently again. During this process, some host DNA, including complete genes, may be excised along with the plasmid DNA and subsequently replicate with it.

Several viruses infecting bacteria, and in particular the phages λ, P1, and Mu, have properties like those of plasmids. Phage Mu is perhaps the most interesting in this regard. It has a typical tailed phage morphology. The genome is a linear piece of dsDNA with about 37,000 base pairs. Upon infection, specific sequences in the viral DNA (near its ends) interact with nonspecific sequences of the host DNA leading to integration of the viral genome at many different sites. Integration appears to be necessary for viral replication. Upon excision some host DNA is also excised and packaged into virus particles. Mu can transpose any segment of the bacterial chromosome onto a plasmid. Thus Mu can be regarded as a DNA insertion element.

There are two major differences between plasmids and phages such as λ, P1, and Mu. The phage genomes are highly developed and specialized for virus functions and normally do not contain genetic information that is useful to the host cell. Plasmids frequently contain genetic information that allows the host cell to survive under particular conditions. In some senses, a plasmid can be considered as an additional smaller host chromosome. The other major difference is that viruses in the vegetative state produce virus particles. These structures are elegantly adapted to preserve the viral genome outside the host cell and to both recognize a new appropriate host cell and to inject the DNA into such a cell.

However, it is possible experimentally to convert some viruses (e.g., the phages λ and M13, and vertebrate viruses SV40 and polyoma) into plasmidlike entities by deleting genes except those essential for replication. Conversely, by the addition of appropriate viral genes, such as those required for packaging, it may be possible to convert a plasmid into a viruslike entity. Examples of such constructions are given by Short *et al.* (1988) and Mead *et al.* (1986). Given the existence of modular evolution (Section IVA) it is quite possible to envisage such changes taking place in nature.

Integration of Viral Genes into Eukaryotic Genomes

The fact that the DNA of several groups of viruses can integrate into the host-cell genome makes it easy to believe that some viruses could have originated there.

Among the viruses infecting eukaryotes the RNA tumor virus family (Retroviridae) is of particular interest in relation to viral origins. The genome consists of ssRNA, but upon infection this is copied by a viral-coded, reverse transcriptase into DNA, which is then integrated into the host-cell genome. Various "provirus" and "protovirus" theories have been proposed in which viral genes are considered to be present as normal but repressed components of the cellular genome. Derepression by various forms of stress (e.g., radiation or carcinogens) leads to production of infection virus.

In perhaps its most extreme form, this kind of theory proposes that new viruses are continually entering the biosphere as a consequence of degenerative processes in cells (Smith and Kenyon, 1972). Temin (1976) has proposed that other enveloped RNA animal viruses, and even some small viruses containing DNA, may have evolved from retroviruses. However, it is also quite possible that the viral-related sequences found in some normal individuals represent the surviving remnants of a viral integration into the host genome that took place in the distant past.

Integration of the complete viral genome into the host genome without damage

to the host allows survival and dissemination of the virus through many host generations. This appears to be an almost ideal adaptation for virus survival. If viruses
originated in this way, why is it that so few established virus families and groups
appear to have retained this survival mechanism?

Transposable Elements in Plants

Transposable genetic elements or transposons are now known to occur in many
kind of eukaryotic organisms, including many plant species (reviewed by Saedler *et
al.*, 1987; Nelson, 1988). In the context of viral evolution, the class of transposable
elements known as retrotransposons is of particular interest. Among these, the
copia elements in *Drosophila* and *Ty1* elements in yeast have been the best studied.
These DNA elements are organized like retrovirus DNAs. Baltimore (1985) has
summarized the evidence that they carry out transposition in the cellular genome via
an RNA intermediate and reverse transcription.

Recently, retrotransposonlike mobile elements have been isolated from plants,
including one from tobacco and other solanaceous plants (Grandbastien *et al.*,
1989). This transposon, called *Tnt1*, is a mobile retrotransposonlike element that is
5334 nucleotides long. It contains two 610-base-pair-long terminal repeats and a
single ORF of 3984 nucleotides. Comparison of this ORF with the *Drosophila
copia* retrotransposon, the yeast *Ty* retrotransposon, and vertebrate proretroviruses
shows that *Tnt1* is closely related to *copia*. The ORF encodes a polyprotein containing the functions necessary for autonomous transposition by reverse transcription of
an RNA intermediate, namely, the gag core proteins, a protease, a reverse transcriptase, and a nuclease.

By means of a nucleic acid hybridization procedure, *Tnt1* was found in various
other species in the Solanaceae, but not in species belonging to several other
families. Thus, *Tnt1*-related sequences may have been present in the Solanaceae
during the early stages of speciation in this family. The high degree of amino acid
sequence similarity over the entire length of *Tnt1* and *copia* indicates that the two
elements may share a common origin. This raises the possibility that such elements
can be transmitted horizontally between plants and insects. It is not difficult to
imagine a retrotransposon such as *Tnt1* evolving into a retrovirus with an extracellular phase in the life cycle. However, on present evidence it is just as likely that
these retrotransposons were derived from retroviruses that lost their extracellular
phase and degenerated to become parasites of the cell genome. This view receives
quite strong support from the fact that some of the retrotransposons such as *copia*
have been found packaged into retroviruslike particles inside cells (e.g., Emori *et
al.*, 1985). However, there is no evidence that such particles are infectious.

A *copia*-like transposon called *BS1* with a low copy number was found in maize
following infection with BSMV *Hordeivirus*. The transposon had no sequence
similarity with the viral RNA. The role of this transposon is not known (Johns *et al.*,
1985) but it appears to be the mechanism by which BSMV acts as a mutagen (see
Chapter 11, Section I,B,10).

The promoter and enhancer region of CaMV DNA can be dissected into at least
five domains which can influence plant gene expression in appropriate constructs
(Chapter 8, Section VI,D,1). This, together with the fact that CaMV uses a host
DNA-dependent RNA polymerase for transcription of its DNA suggests the possibility that the promoter and enhancer domains may have been derived from host
plants in a modular fashion in the distant past.

*Amino Acid and Nucleotide Sequence Similarities
between Viruses and Cells*

There are several reports of amino acid or nucleotide sequence similarities between portions of plant viral genomes and some component of cells. Zimmern (1983a) reported that there was 15–23% sequence similarity in a single alignment between the 30K movement protein of two strains of TMV and certain intron-encoded proteins of yeast. On a statistical basis, the significance of the similarities was marginal.

However, similarity in sequences is now more amenable to analysis by "motifs," which are short consensus sequences that appear important for particular functions. As an example, we will consider the internal promoter in BMV genomic RNA3 for subgenomic RNA4 synthesis is illustrated in Fig. 6.14. Comparison of this and other subgenomic promoter sequences (and especially the promoter core sequence) has shown remarkable sequence similarities with the internal control regions 1 and 2 known to be involved in the promotion and control of transcription in eukaryotic tRNAs (Marsh and Hall, 1987). Such sequence similarities were also found in cDNA copies of 5′-terminal sequences of BMV viral RNAs 1, 2, and 3. Thus positive strand synthesis for these viral RNAs uses promoters resembling those used for cellular tRNAs (DNA-dependent RNA polymerase III). However, no such similarities were detected in the tRNA-like sequences at the 3′ end of the viral RNAs.

Gibbs (1987) has summarized several reports of sequence similarities between proteins coded by viruses infecting vertebrates and host proteins. Most of these suggest that the similarities reflect a common evolutionary origin for the genes concerned. However, the possibility of evolutionary convergence is still an open one, especially for shorter lengths of sequence similarity.

The Mobility of Host Genetic Elements

It is now accepted that chloroplasts and mitochondria evolved from a symbiotic relationship between prokaryotic cells and an early eukaryote. However, the three DNA genomes have not remained discrete. There are many "promiscuous" DNA sequences, that is, sequences that are found in more than one of the three genetic compartments (e.g., Timmis and Steele Scott, 1984). Most eukaryotic mRNAs are derived by cleavage from larger primary transcripts. The final mRNAs are composed of base sequences that may not be contiguous in the DNA from which the RNA was transcribed. This RNA "splicing" may have greatly facilitated early eukaryotic evolution. This substantial potential for mobility in host genetic elements would have given an increased probability for the coming together of a chance combination of functions that could give rise to an RNA virus.

3. Origin by Degeneration from Cells

The idea that viruses are extremely degenerate parasitic forms that have evolved from cellular organisms has been put forward on a number of occasions but has lost favor in recent years, mainly because of the similarities between the behavior of certain viruses and plasmids, as outlined earlier.

The properties that distinguish all viruses from all cells reduce to three: (i) lack of a complete membrane separating the virus replication site from the host cytoplasm (or nucleoplasm for some eukaryote viruses); (ii) use of host protein-synthesizing machinery by viruses; and (iii) binary fission in cell reproduction but not with viruses. If any viruses did arise by degeneration from cells, these three properties may have been lost at about the same stage. Absence of a bounding membrane during replication would automatically prohibit binary fission, and at the same time would allow the parasite to become dependent on the host protein-synthesizing machinery.

Fenner (1979) believes that the large size of the poxviruses, their complex structure, including many enzymes within the virus particle, and their ability to replicate in the cytoplasm independently of host nuclear functions make it plausible that these viruses arose from cells. They might form the most degenerate member of the series bacterium → rickettsia → chlamydia → poxvirus. If one family of viruses originated in this way, others could have done so too.

Highly complex symbiotic relationships exist in some living unicellular organisms. Certain viruses might have arisen from DNA in such symbioses by a change to a parasitic mode of intracellular existence and the acquisition of a means of cell-to-cell dissemination. At this point the two kinds of possible origin for viruses—from cells or from cellular DNA—become one and the same. The possible origin of viruses from cells has been discussed by Matthews (1983c).

4. Summary

On present evidence it seems probable that different groups of present-day viruses have originated in different ways. Some of the very large DNA viruses infecting animals are probably descended by a degenerative process from very simple cellular parasites. Other, smaller viruses, especially those with RNA genomes, most probably evolved from host genes via transposons or by other means. Some present-day RNA viruses or parts of them might be direct descendants from a prebiotic RNA world. The true origin of viruses may never be known for certain unless (i) they arose in a modular way from host genes; (ii) amino acid sequences have remained sufficiently conserved in both viruses and hosts; and (iii) in due course sufficient cellular gene modules are sequenced, especially from lower organisms, to enable convincing identification of the modules from which particular viruses arose.

B. Origin of Viroids

1. Possible Origins from Other Nucleic Acids

From Small Nuclear RNAs

Small nuclear RNAs (snRNAs) are believed to play a role in the processing of the primary transcription products of split genes, thus allowing for precise alignment and correct excision of introns. Some, such as U1 snRNA, have been shown to have base complementarity with the ends of introns. Kiss *et al.* (1983) found significant sequence similarity between PSTVd RNA and the U3B snRNA of Novikoff hepatoma cells, suggesting that the viroid might be related to the snRNA.

From Introns

Group 1 introns are widely distributed among the genes for mitochondrial mRNA and tRNA genes, chloroplast tRNA genes, and nuclear tRNA genes. They are characterized by a highly conserved 16-nucleotide sequence and three sets of complementary sequences. The phenomenon of self-splicing has been described for several group 1 introns.

Diener (1981) found a 23-nucleotide region of high base complementarity between the minus strand RNA of PSTVd and the 5'-terminal nucleotides of mammalian U1 snRNA. On this basis he suggested that viroids might have arisen as escaped introns. Dinter-Gottlieb (1986) pointed out that viroids contain a consensus sequence and three sets of complementary bases. In addition, there were stretches of sequence similarity between viroids and a self-splicing intron from *Tetrahymena*. She showed that pairing of these complementary sequences within a viroid could generate a structure resembling the self-splicing group 1 intron from *Tetrahymena*. This would be a more open structure than a maximally base-paired viroid, perhaps stabilized by protein binding. Hadidi (1986) compared sequence similarities between viroids, other RNA pathogens, introns, and exons. As the statistical stringency of the comparisons was increased, only similarities between introns and viroids remained significant.

The escaped intron concept includes the idea of self-processing of the RNAs. However, current evidence for viroids does not fit this model. ASBV is able to self-cleave (Chapter 9, Section I,B) but this viroid does not have the intronlike sequence motifs noted earlier for other viroids. With PSTVd and related viroids, Tsagris *et al.* (1987b) have shown that autocatalytic processing does not operate *in vitro* as with introns. In all the discussions on relationships between snRNA or introns and viroids, it has been assumed that plant snRNAs will be found to have sequences similar to those known from animal systems.

From Hypothetical Signal RNA

Zimmern (1982) proposed that some control elements between cells might consist of small mobile RNAs and that viroids might have arisen from these.

From Transposable Genetic Elements

Sequence analysis of a group of viroids showed some striking similarities with transposable genetic elements, including the proviruses of retroviruses (Kiefer *et al.*, 1983). The similarities included the presence of inverted repeats often ending in the dinucleotides U–G and C–A and flanking imperfect direct repeats. Kiefer *et al.* suggested that viroids might have arisen by deletion of interior nucleotide sequences from transposable elements or retroviruses. ASBVd sequences do not fit this model.

From Satellite RNAs

Although viroids and the viroidlike satellite RNAs show little sequence similarity, they have several properties in common: small size, circular ssRNA, and lack of mRNA activity. In addition, some satellite RNAs probably have a rolling circle model for replication involving greater than unit-length RNAs. On the other hand, viroids replicate in the nucleus, whereas viroidlike satellites do so in the cytoplasm. Nevertheless, Francki *et al.* (1986b) showed that PSTVd could be encapsidated *in vivo* in particles of velvet tobacco mottle virus. Tomato plants, which are immune to this virus, became infected with PSTVd when inoculated with virus containing the viroid. In principle, a viroid might have arisen from a viroidlike satellite by becoming independent of the helper virus for replication.

As noted in Chapter 9, Section II,B,1, the satellite of peanut stunt virus (PAR-

NA5) has several regions of high sequence similarity with various viroids, including sequences of the conserved central region of most viroids. However, the replication of PARNA5 appears to be like that of the helper virus RNA rather than viroidlike. There are also conserved intron sequences in PARNA5 (Collmer *et al.*, 1985).

2. Viroids as Living Fossils of Prebiotic Evolution

It has been proposed by a number of authors (Section III,A,1) that RNA preceded DNA in prebiotic evolution. The recent discovery of RNA molecules with splicing and polymerase activity has led to models for the RNA-catalyzed replication of RNA (Cech, 1986). Thus it is conceivable that modern viroids, because of their small size, circular conformation, inherent stability, RNA → RNA replication cycle, and lack of protein coding capacity, may be descended from a type of RNA molecule that existed before the evolution of DNA. This view is discussed by Diener (1989).

C. Origin of Satellite Viruses and RNAs

1. Satellite Viruses

Apart from their dependence on a helper virus and their small size, satellite viruses appear to belong in the same category of agents as other viruses. They most probably arose from an independent virus by degenerative loss of functions in a mixed infection that provided the helper virus.

2. Satellite RNAs

As discussed in Chapter 9, Section II,B,7, most satellite RNAs fall into one of two groups. The first, typified by satellite RNAs of CMV and TBRV, has terminal structures like that of the helper virus RNAs, replicate via a unit-length negative sense template, depend on the helper virus replicase, and appear to code for one to three proteins. The second group, typified by STRSV, appears to have no mRNA function and is replicated by a viroidlike rolling circle mechanism. The predicted secondary structure of the viroidlike satellite RNAs mimics the rod-shaped configuration of the viroids, but there is little base sequence homology between them except for a conserved GAAAC occurring in some members of each group of agents (Keese and Symons, 1987). Thus it is possible that satellite RNAs have arisen from two different lineages. This idea is shown in a hypothetical scheme in Fig. 17.6. Branch *et al.* (1990) go further, and suggest that viroids, circular plant viral satellite RNAs, and the delta agent associated with human hepatitis B virus should placed in a group of related infectious agents called "circular subviral RNA pathogens."

The satellite RNA C of TCV (see Fig. 9.9) is of particular interest from the evolutionary point of view. Its structure appears to demonstrate that recombination can take place between a satellite RNA and the helper virus genome to generate a new satellite with different biological properties.

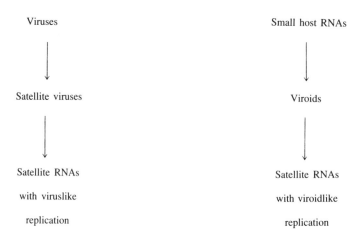

Figure 17.6 Hypothetical scheme for the derivation of satellite RNAs from two separate lineages.

IV. EVOLUTION

A. Mechanisms for Virus Evolution

The molecular mechanisms underlying genetic variation in viruses were summarized in Chapter 13, Section II. They include (i) mutations due to a base change,
(ii) additions or deletion of bases, (iii) deletions or duplications of nucleic acid
sequences, (iv) genetic recombination, (v) reassortment among genome pieces in
viruses where the genetic material is in several pieces of nucleic acid, and (vi) the
acquisition of exogenous host or viral genes. These mechanisms have been discussed by Zimmern (1988).

We can envisage viral evolution proceeding in both a micro and a macro
manner. In microevolution, existing viral genes accumulate small changes by such
mechanisms as nucleotide substitutions, additions, or deletions. In macroevolution
of a virus, a sudden major change may take place by recombination with a related
virus, by duplication of an existing gene, or by acquisition of a gene from the host or
an unrelated virus, processes that have been termed modular evolution. Studies on
the RNAs of two anomalous naturally occurring *Tobravirus* isolates leave little
doubt that recombination takes place in nature (Robinson *et al.*, 1987). The RNA2's
of these two isolates were shown to be recombinant molecules containing sequences
of both TRV and PEBV.

From structural considerations, McLachlan *et al.* (1980) suggested that the
ancestral coat protein of TMV was a dimer of two smaller units. A tandem gene
duplication event was proposed, followed by drift in amino acid sequence, arising
from point mutations, to give rise to the present coat protein subunit. Gorbalenya *et
al.* (1986) have invoked extensive gene duplications to explain internal repeating
homologies in the poliovirus polyprotein. They suggested that this genome was built
up by multistep amplification of a gene coding for an 11-membered peptide followed by divergence in individual copies. The present 22-amino-acid VPg would be
derived from a dimer of the original gene.

Botstein (1980) first applied the concept of modular evolution to bacteriophages. He defined it as "the joint evolution of sets of functionally and genetically
interchangeable elements . . . each of which carries out a particular biological

function." Available evidence suggests that such a process may have occurred in viruses infecting plants. For example, in the genome of BYDV the coat protein shows some distant similarities with those of some other icosahedral viruses such as TBSV and CarMV. The 50K ORF is similar to the read-through protein of the BNYVV coat protein. These various viruses have quite different general properties. Thus BYDV could well be a mosaic of modules and gene expression strategies arranged quite differently from other known viruses (Miller *et al.*, 1988b).

On the other hand, there may be natural mechanisms limiting the extent to which modular evolution can take place. For example, Dawson *et al.* (1989) constructed a hybrid TMV in which the CAT gene was inserted between a 30K and coat protein genes. This construct replicated efficiently in inoculated leaves expressing CAT activity and producing 350-nm rods instead of the natural 300 nm. However, by the time systemic infection had developed, the CAT sequence had been deleted precisely, giving wild-type TMV.

In summary, multiple mechanisms exist for virus evolution, both on a micro and a macro scale with respect to the size of the individual changes involved.

As far as the evolution of viroids is concerned, their viability is readily destroyed by relatively minor sequence alterations (Chapter 9, Section I,D,2). Nevertheless, sequence comparisons of sets of closely related viroids show clearly that single-base changes, insertions, and deletions have played a part in viroid evolution. Of more general interest is the evidence that intermolecular rearrangements may have played a role in viroid evolution as discussed in Chapter 9, Section I,B,5.

B. Evidence for Virus Evolution

No fossil viruses have yet been discovered. Some may await discovery, for example, in insects preserved in amber. In the meantime, evidence for virus evolution must come from the study of present-day viruses in present day hosts.

1. General Properties of Viruses within Families and between Closely Related Groups

On present evidence there is a marked lack of intermediate types between most of the virus families and groups delineated by I.C.T.V. The close similarities in particle morphology, genome strategy, and the three-dimensional structure of proteins leave no room for doubt that the individual viruses within the families and groups delineated by the I.C.T.V., had a common ancestor at some time in the past. A few examples are described briefly in the following sections. In some instances, the evidence suggests that, from an evolutionary point of view, two or more of the existing I.C.T.V. groups of plant viruses would be better placed in a single family.

Tymoviruses

On the basis of particle morphology the tymoviruses form a quite distinct and coherent group of viruses. The genomes of several members have been sequenced (e.g., Osario-Keese *et al.*, 1989). The sequence information clearly supports a common evolutionary origin for members of the group.

Tobamoviruses

Among the definite members of the *Tobamovirus* group, all measures of their relatedness correlate with that indicated by the amino acid composition and amino acid sequence of their coat proteins. Thus they appear to form a distinct group of viruses with a common ancestor (Gibbs, 1986). There is an interesting exception, however. The 3' noncoding region of sunnhemp mosaic virus (SHMV) (formerly known as the cowpea strain of TMV) shows significant sequence similarity with the 3' noncoding region of TYMV RNA. Forty-six out of the 3'-terminal 80 nucleotides are identical (Meshi *et al.*, 1981). Furthermore, SHMV, like TYMV, has a 3'-terminal tRNA-like structure that can be aminoacylated with valine. Other to-bamoviruses can be charged with histidine. Gibbs (1986) suggested that the 3' noncoding region of SHMV was acquired by a recombination event with a *Tymovirus*. On the other hand, gross recombinational events do not appear to have played a major role in the evolution of the *Tobamovirus* group because amino acid sequences in the coat protein and the "125K" protein suggest very similar relationships (Fraile and García-Arenal, 1990).

Geminiviruses

As discussed in Chapter 8, Section II, there are three subgroups of geminivirus. Subgroup I has one circular DNA, is transmitted by leafhoppers, and infects monocotyledons. Subgroup III has two circular DNAs, is transmitted by whiteflies, and infects dicotyledons. BCTV, the only member of Subgroup II, has a single circular DNA and is transmitted by leafhoppers, but infects dicotyledons.

Howarth and Vandemark (1989) constructed phylogenetic trees based on the coat protein amino sequences of 15 geminiviruses, and the replication associated proteins of 16 members of the group. Trees based on coat protein had two main branches matching vector specificity (leafhoppers or whiteflies). Trees based on replication proteins also had two main branches that were very similar to those based on coat protein. BCTV was associated with the group of geminiviruses infecting dicotyledons when the tree was based on replication-associated proteins, and with the whitefly-transmitted group when the coat protein sequences were used. This result goes some way toward defining evolutionary relationships within the *Geminivirus* group. It gives support to the idea that the coat proteins have evolved in response to the needs of vector transmission while replication-associated proteins have evolved with the kind of host (monocotyledons or dicotyledons).

Four Virus Groups with Tripartite Genomes

As discussed in earlier chapters, there are close similarities in genome organization and particle structure between bromoviruses, cucumoviruses, ilarviruses, and AMV. Comparison of the amino acid sequences of the proteins coded for by the 5' ORF of RNA 3 ($\approx$32K) of BMV, CMV, and AMV confirmed the close evolutionary relationships between these plant viruses (Davies and Symons, 1988) and supports the view that they should all be members of a single virus family.

Nepoviruses, Comoviruses, and Fabaviruses

Nepoviruses and comoviruses share a number of properties—a bipartite genome, each segment of which is translated into a single large polypeptide, and similar 5' and 3' structures. They differ in that nepoviruses have one structural protein while comoviruses have two. Strawberry latent ring spot virus is considered to be a possible nepovirus, but with two coat proteins. It may therefore represent a transitional virus in the evolution of these two groups (Dougherty and Hiebert, 1985). Fabaviruses, as represented by broad bean wilt virus, are very similar to comoviruses, but the vector is an aphid. It is possible that the *Nepovirus, Comovirus,* and *Fabavirus* groups should be three genera within a single family.

2. Nucleotide and Amino Acid Sequences

Knowledge of nucleotide sequences in viral genomes, and the corresponding sequences of amino acids in the encoded proteins, has provided powerful confirmation of the evolutionary basis for most of the families and groups of viruses delineated by the I.C.T.V. For example, SYNV is similar but not identical to animal rhabdoviruses in the order of structural genes and in the nucleotide sequences at the gene junctions. The intergenic and flanking-gene sequences are conserved and consist of a central core of 14 nucleotides (3'-UUCUUUUU-GGUUGU/A-5') (see Fig. 8.13). This sequence is similar to the sequence at the gene junctions in vesicular stomatitis and rabies viruses (Heaton *et al.*, 1989a). These conserved features argue strongly for a common origin for all members of the Rhabdoviridae family.

Sequence data can be used to estimate degrees of evolutionary relationship, to develop "trees" indicating possible lines of evolution for viruses within a family (e.g., Fig. 13.13), and to discover unexpected relationships (Section V). While this is a very valuable approach it should be remembered that there are significant difficulties in deriving and interpreting trees constructed from sequence data.

Inherent Difficulties

There are inherent difficulties in constructing meaningful trees, especially when more than four taxa are involved, because of the large number of possible trees. Other difficulties are discussed by Penny *et al.* (1987, 1991).

Nonconstant Rates of Evolution

The molecular clock hypothesis, which assumes a constant rate of change in sequence over evolutionary time, has often been used in the interpretation of "trees." However, comparisons of organelle DNA in both vertebrates (Vawter and Brown, 1986) and plants (Wolfe *et al.*, 1987) reveal extreme rate variation in the molecular clock between nuclear, mitochondrial, and chloroplast DNAs. It is now known that rates of change in different genes and lineages may vary widely (Li *et al.*, 1987). Ubiquitin is an example of a very highly conserved protein. It is composed of 76 amino acids with only three substitutions between animals, higher plants, and yeast (Dunigan *et al.*, 1988). By contrast, there has been very rapid evolution of the reactive center regions of serine protease inhibitors of rodents (Hill and Hastie, 1987). The molecular phylogeny of globin genes is particularly instructive in this respect because amino acid sequences are known for several hundred globin chains (Goodman *et al.*, 1987). There is little doubt that different proteins, or parts of a protein coded by a viral genome, may evolve at different rates. In addition, some noncoding sequences in the genome may be highly conserved, particularly those recognition sequences essential for genome replication.

Sequence Convergence

Sequence similarity between two genes does not necessarily indicate evolutionary relationship (homology). Without other evidence it may be impossible to establish whether sequence similarity between two genes is due to a common evolutionary origin or to convergence. The lysozyme enzymes of foregut fermenters constitute the best studied example of sequence convergence (Stewart and Wilson, 1987). Foregut fermentation has evolved independently twice in placental mammals, first in ruminants and later in colobine monkeys. In both instances lysozyme C was recruited to function in the true stomach. About half the amino acid replacements along the langur lysozyme lineage were in parallel or were convergent with those that had evolved earlier along the cow stomach lysozyme lineage. This convergence was driven by selection for adaptation to the acidic pepsin-containing environment of the stomach. Similar convergence has probably occurred from time to time during the evolution of viral genes. However, with such genes it is difficult to obtain independent evidence demonstrating convergence, as has been possible with the stomach lysozymes. Amino acid sequence similarities that are sufficient to lead to a serological cross-reaction may sometimes arise by chance, as appears to have happened between the TMV coat protein and the large subunit of ribulose bisphosphate carboxylase (Dietzgen and Zaitlin, 1986).

Proteins Approaching the Limits of Change

As two proteins with a common evolutionary origin diverge, they may approach a limit of change that retains common amino acid positions in excess of that expected for random sequences. This limit on change would be due to functional requirements for the protein. When this limit is reached, convergence or back mutations and parallel mutations become as common as divergent mutations. As two diverging proteins approach and remain in this steady-state condition, sequence differences no longer reflect evolutionary distance, and therefore such sequences should not be placed in evolutionary trees. Some bacterial amino acid sequence data are considered to be of this sort (T. E. Meyer *et al.*, 1986). Some coat proteins in certain groups of plant viruses may also have approached this state (e.g., the tobamoviruses). The limit for change may vary even within one protein. Thus the surface domains of some plant viral coat proteins are more variable (= less constrained) than domains with structural roles. The three-dimensional structure of proteins tends to be conserved for longer periods of evolutionary time than primary sequences. Therefore they might be useful for evolutionary comparisons long after amino acid sequences have markedly diverged. Rossmann and Rueckert (1987) pointed out that all the proteins of icosahedral viruses examined up to that time had an 8-stranded β-barrel as their core. By contrast, TMV and probably other rod-shaped viruses have as their core a bundle of four α-helices. This led to the idea that perhaps all icosahedral virus coat proteins had a common origin, and the coat proteins of all rod-shaped viruses had another common origin. An alternative explanation, where no significant amino acid sequence similarity exists, could be that convergent evolution has occurred. Perhaps a functional coat protein for a virus with icosahedral symmetry requires a β-barrel core, and similarly for rod-shaped viruses and α-helices. The small subunit of the icosahedral bean pod mottle *Comovirus* has a 10-strand β-barrel while the large subunit has two 8-stranded β-barrels (Chen *et al.*, 1989). Thus the structure of icosahedral virus coat proteins may not be quite as regular as previously thought.

3. Examples of Evolutionary Change

The Selective Survival of Shorter RNAs

In various experiments both *in vitro* and *in vivo* with derived or constructed viral RNAs, there has been a trend toward survival of shorter RNAs. For example, following infection with BMV, which included RNA3 derivatives carrying multiple subgenomic promoters, smaller subgenomic RNAs always accumulated preferentially (French and Ahlquist, 1988).

Sequence Heterogeneity in a Satellite RNA

Natural populations of the D satellite RNA of CMV have a major site of sequence heterogeneity in which a high proportion of molecules show sequence differences. Kurath and Palukaitis (1990) used infectious transcripts of uniform sequence from a cloned satellite molecule to inoculate various host species. Within two to three passages, and in most of the host species, sequence diversity had been generated at the same site as found in natural populations.

Changes in an Isolate of AMV

The strain ATCC/ 425 of AMV was originally isolated in the United States. For some 20 years separate cultures of strain 425 were maintained in Wisconsin and the Netherlands. During this period several nucleotide changes occurred in RNA3 in both the coat protein cistron and extracistronically. The changes led to five amino acid substitutions in the coat protein (Jaspars, 1985).

The Origin of Assembly in Tobamoviruses

In the Vulgare strain of TMV the OAS sequence lies outside the coat protein gene in a 5′ direction (within residues 876–967 from the 3′ end). In sunnhemp mosaic *Tobamovirus* (SHMV) the OAS lies within the coat protein gene within residues 369–461 (Meshi *et al.,* 1981) (Fig. 17.2). As a consequence, the coat protein subgenomic RNA of SHMV is encapsidated *in vivo* to form short rods, while that of TMV is not (e.g., Fukuda *et al.,* 1980). Guilley *et al.* (1975) isolated an RNA fragment from the coat protein gene of the Vulgare strain of TMV called SERF (*s*pecifically *e*ncapsidated *R*NA *f*ragment). While this RNA fragment can be encapsidated, it does not act as an OAS in the intact RNA. However, it has extensive sequence homology with the Vulgare OAS. The SERF sequence lies at residues 307–388, which is close to that of the SHMV OAS (369–461). Furthermore, this latter virus has a sequence at 831–923 showing extensive homology with the TMV OAS, but that does not function as such (Meshi *et al.,* 1981). Thus the OAS-like sequences are duplicated in both viruses in similar noncontiguous positions, but only one is functional in each virus. Any duplication event presumably took place before divergence of the two viruses occurred during *Tobamovirus* evolution. A *ts* mutant of Vulgare TMV, Ni2519, assembles incompletely *in vitro* at the nonpermissive temperature (32°C). This is because at this temperature both of the OAS-like sequences are functional, giving rise to defective rods (Kaplan *et al.,* 1982).

The Role of Selection Pressure

Functional viral genes will be retained only if they are needed for survival of the virus. For example, maintenance of WTV exclusively in vegetative plant hosts leads to loss of the ability of the virus to replicate in, and be transmitted by, leafhoppers. This functional loss is due to deletions in certain genome segments (see Fig. 14.15). Similarly, RNAs 3 and 4 of BNYVV appear to suffer deletions when not under the requirement to be transmitted by the fungal vector (Chapter 13, Section II,C,2). The reversion of viroid point mutants to wild type provides another example of the effect of selection pressure (Owens *et al.,* 1986).

4. Rates of Evolution

The rate of point mutation for RNA viruses has been estimated to be approximately $10^{-3}–10^{-4}$ per nucleotide per round of replication with some variation between different viruses (reviewed by Holland *et al.,* 1982). This contrasts with estimates of $10^{-7}–10^{-8}$ for DNA polymerases. However, Smith and Inglis (1987) have questioned the idea that the error rate in eukaryotic viral DNA synthesis is significantly lower than that for RNA viruses. In theory, the measured mutation rates would allow for very rapid change in viral genomes either RNA or DNA.

Virus cultures contain large numbers of sequence variants (Domingo and Holland, 1988). Thus there is no doubt concerning the reality of the "quasi species" concept (see Chapter 13, Section III,A,1).

It is very difficult to relate mutation rates to the actual rates of change in viruses that might be occurring in the field at present, or over past evolutionary time. The reasons for this include the following:

Selection Pressure by Host Plants

Conditions within a given host species or variety will exert pressure on an infecting virus against rapid and drastic change. Viral genomes and gene products must interact in highly specific ways with host macromolecules during virus replication and movement. These host molecules, changing at a rate that is slow compared with the potential for change in a virus, will act as a brake on virus evolution. This general idea has led Koonin and Gorbalenya (1989) to propose that RNA viral proteins evolve with an amino acid substitution rate that is similar to that of their host.

Selection pressure in an experimental system is illustrated by the work of Aldaoud *et al.* (1989). They studied the evolution of variants in populations of TMV originating from an *in vitro* transcript of a cDNA clone. One of the markers measured was the ability to cause necrotic lesions in *Nicotiana sylvestris* (nl). The proportion of nl variants in tobacco, tomato, *Solanum nigrum,* and *Petunia hybrida* were similar. However, in *Physalis floridiana* there was strong selection pressure against nl variants, which were reduced to almost undetectable levels. Nevertheless, in all hosts tested there were large and apparently random changes in the proportion of nl variants in individual plants, showing that viral populations can evolve rapidly on a time scale of days. Perhaps these variations in strain population occur during the early stages of infection in new host species. Over a period of many transfers in the same species the populations may stabilize. This idea is supported by the fact that TMV obtained from W. M. Stanley and independently subcultured many times over decades in tobacco varieties in the United States and Germany was shown in both countries to have the same coat protein amino acid sequence (Zaitlin and Israel, 1975).

Further support for stability of the dominant strain in a single host comes from the experiments of Rodriguez-Cerezo and García-Arenal (1989) with tobacco mild green mosaic *Tobamovirus* (TMGMV) in Samsun tobacco, a systemic host. TMGMV was formerly known as U5TMV. They passaged the virus culture twice through single lesions in Xanthi-nc tobacco and then followed heterogeneity in isolates after four series of 20 passages in Samsun tobacco. Ti fingerprint characterization of the population of genomic RNAs showed no detectable variation.

The few field studies that are available support the idea that a given host species tends to stabilize a virus population. Rodriguez-Cerezo *et al.* (1989) studied genome variations in isolates of pepper mild mottle *Tobamovirus* isolated from peppers in the field over the period 1983–1987. They concluded that there was a highly stable virus population from which variants continually arose without replacing the parent population.

The only quantitative study of variation in a virus in a natural host in the wild has been carried out on variants of TMGMV in *Nicotiana glauca* in southern Spain and Australia (Rodriguez-Cerezo *et al.,* 1990). Their two main conclusions were (i) that RNA sequence divergence between isolates was low, no matter what the distance was between the sites where they were isolated, and (ii) there was no correlation between closeness of sequence relationship and geographical proximity of the plants from which they had been isolated. This is difficult to explain since the only known natural method of transmission for TMGMV is by contact between plants.

Variation in Rates of Change in Different Parts of a Viral Genome

As noted in Chapters 6–8, noncoding regions of viral genomes, particularly at the 5' and 3' termini, which function as recognition sites in viral RNA translation and replication, may be highly conserved in the members of a virus family or group. On the other hand, in viruses with multipartite genomes, one genome segment may be conserved and the other highly variable, as with the RNAs 1 and 2 of tobraviruses (see Fig. 13.7). Coat protein genes may be much more strongly conserved in some regions than in others by functional requirements in the proteins, for example, with TMV coat protein (Fig. 17.4) and potyvirus coat proteins (see Fig. 13.9).

Uneven Rates of Change over a Time Period

The environment that dictates the selection pressure on the replication and movement of a virus within a plant consists almost entirely of the internal milieu of the host. Other selection pressures on survival involve transmission from plant to plant, for example, by invertebrate or other vectors. It is possible that a switch to a new host plant species may induce rapid evolution of a virus over quite a short period. Recombination between viral genomes with mutations in different parts of the genome may speed this process. This stage may be followed by a further period of stability for the new virus in the new host (see Chapter 13, Section IV,B). The situation may be somewhat analogous, on a way much shorter time scale, to the episodic evolution or "punctuated equilibria" deduced from the fossil record for the evolution of some groups of invertebrates (Eldredge, 1985).

One way in which a virus may gain a foothold in a new host species is through coinfection with another virus that normally infects that species. For example, TMV does not normally infect wheat, but can do so if the plants are already infected with BSMV (Chapter 10, Section V,C,3).

Lack of Precise Historical Information

Until recently the type member of the *Tymovirus* group, TYMV, had been found occurring naturally only in Western Europe. However, a very closely related virus has been found in an isolated area of Australia infecting the endemic brassicaceous species *Cardamine lilacina* (Keese *et al.*, 1989). From the genomic sequences of the European and Australian TYMV populations, Blok *et al.* (1987b) calculated that the two strains have diverged by about 1%, at most, over the past 10,000 years. This is the only estimate we have of the possible rate of evolution of a plant virus over a geological time span. However, the estimate is based mainly on biogeographical data, concerning the *Cardamine* host, that are themselves subject to considerable error (Guy and Gibbs, 1985). Furthermore, the 1% difference could be due to a very rapid process of adaptation to different host genera over a relatively short period at some time in the past rather than to slow change over thousands of years.

5. Coevolution of Viruses, Host Plants, and Invertebrate Vectors

Fahrenholtz' rule postulates that parasites and their hosts speciate in synchrony (Eichler, 1948). Thus there is a prediction that phylogenetic trees of parasites and their hosts should be topologically identical. Using protein electrophoretic data, Hafner and Nadler (1988) obtained phylogenetic trees for rodents and their ecto-parasites that confirm a history of cospeciation. In view of the known wide host ranges of many present-day plant viruses, it is not to be expected that Fahrenholtz' rule will be followed closely for viruses and their hosts. Nevertheless, it is now widely accepted that viruses have had a long evolutionary history and have co-evolved with their host organisms. In this section I will consider aspects of this idea as they relate to plant viruses. In discussing this idea I do not wish to imply that viruses have evolved to a state of higher complexity following their plant hosts in this respect. On the contrary, the evidence available at present shows that the largest and most complex virus infecting photosynthetic organisms is found in the simplest host—a *Chlorella* like green alga.

Viruses of Prokaryotes and Eukaryotes

Is there a fundamental distinction between viruses infecting prokaryotes and those infecting eukaryotes? There is no authenticated example of a virus of one sort completing its life cycle successfully in cells of the other type. There are a few reports of a short sequence similarity between prokaryote viruses and various eukaryote viruses. Bacteriophage MS_2 had the Asp–Asp sequence in its RNA-dependent RNA polymerase (Kamer and Argos, 1984) (Section V,C,2). Rogers *et al.* (1986) drew attention to similarities between a nucleotide sequence in the stem and loop structure in the common region of various *Geminivirus* DNAs and that in the recognition and cleavage site for the ϕX174 coliphage gene A protein. Seven of eight nucleotides at the two sites are identical. Geminiviruses are the only plant viruses with a ssDNA genome, a feature they also share with bacterial viruses like ϕX174. Rogers *et al.* (1986) have suggested similar mechanisms for the replication of their DNA (see Chapter 8, Section II,C). There is also another observation that may be relevant. It is now well established that chloroplasts evolved from prokaryotic cells. ssDNA from the Abutilon mosaic *Geminivirus* is present in the chloroplasts of infected *Abutilon* (Gröning *et al.*, 1987). However, on present evidence it seems likely that viruses infecting prokaryotes, and most infecting eukaryotes, had separate evolutionary origins, or that if there was a common ancestral virus then it existed in the very distant past.

Viruses Infecting Photosynthetic Eukaryotes below the Angiosperms

Viruslike particle (VLPs) have been observed in thin sections of many eukaryotic algal species belonging to the Chlorophyceae, Rhodophyceae, and the Phaeophyceae. The particles are polygonal in outline and vary in diameter from about 22 to 390 nm. Some have tails reminiscent of bacteriophage. The most studied viruses are those infecting *Chlorella*-like green algae. Members of this group are very diverse biochemically (van Etten *et al.*, 1988). An important technical advance was made when van Etten *et al.* (1983) developed a plaque assay for a virus called PBCV-1 infecting a culturable *Chlorella*-like alga. As a consequence, most is now known about the properties of this virus (reviewed by van Etten *et al.*, 1987). In addition, 19 other plaque-forming viruses have been found infecting this alga in nature (Schuster *et al.*, 1986). They all have large dsDNA genomes and the DNA contains methylated bases. Infection with virus IL 3A induces a type II DNA restriction endonuclease (Xia *et al.*, 1987). The PBCV-1 virus has several properties in common with the Iridoviridae family, typified by Frog virus 3 (Meints *et al.*, 1986): (i) icosahedral morphology; (ii) a large dsDNA genome; (iii) a lipid content of 5–10% by weight; (iv) a cytoplasmic site of assembly; and (v) the presence of methylated bases in the DNA—a very unusual feature in viruses infecting eukaryotes. However, in the *Chlorella* virus, the genome structure is like that of the poxviruses, namely, a linear nonpermuted dsDNA molecule with covalently closed hairpin ends (Rohozinski *et al.*, 1989). The only other virus with these two properties appears to be African swine fever virus. In August 1990, the I.C.T.V. approved family status for the *Chlorella* viruses with the name Phycodnaviridae. The type genus is *Phycodnavirus* with PBCV-1 as the type species.

The *Chlorella*-like algae cultured by van Etten *et al.* (1983) were originally present as hereditary endosymbionts in the protozoan *Paramecium bursaria*. Such *Chlorella*-like algae are also found in *Hydra* and a variety of other organisms. These endosymbiotic relationships may have made possible the appearance of iridoviruses in other more highly evolved animals as well as the green algae. There can be little doubt that the earliest viruses evolved in organisms in an aquatic environment. Bacterial viruses are abundant in unpolluted natural waters (Bergh *et al.*, 1989).

Skotnicki *et al.* (1976b) described a virus infecting the eukaryotic alga *Chara australis*. The virus (CAV) has rod-shaped particles and some other properties like those of tobamoviruses. However, its genome is much larger (11 kb, rather than 6.4 kb for TMV), and about 7 kb of the genome has been sequenced, revealing other relationships (A. J. Gibbs, personal communication): (i) the coat protein of CAV has a composition closer to BNYVV, and closer to TRV than to TMV; (ii) the GDD-polymerase motif of CAV is closest to that of BNYVV; and (iii) the two GKT nucleotide-binding motifs found in CAV are arranged in a manner similar to that found in potexviruses. Thus CAV appears to share features of genome organization and sequences found in several groups of rod-shaped viruses infecting angiosperms. It appears to have strongest affinity with the *Furovirus* group and has no known angiosperm host. For these reasons it is most unlikely that CAV originated in a recent transfer of some rod-shaped virus from an angiosperm host to *Chara*. In its morphology *Chara* is one of the most complex types of Charophyceae, which is a well-defined group with a very long geographical history (Round, 1984). On the basis of cytological and chemical similarities, land plants (embryophytes) are considered to have evolved from a charophycean green alga. *Coleochaete*, another of the more complex types among the Charophyceae, has been shown to contain lignin, a substance thought to be absent from green algae (Delwiche *et al.*, 1989). Recent molecular genetic evidence confirms a charophycean origin for land plants. Group II introns have been found in the tRNA^{ala} and tRNA^{ile} genes of all land plant chloroplast DNAs examined. All the algae and eubacteria examined have uninterrupted genes. Manhart and Palmer (1990) have shown that introns are present in three members of the charophyceae in the same arrangement as in *Marchantia* giving strong support to the view that they are related to the lineage that gave rise to land plants. Tree construction suggests that the Charophyceae may have acquired the introns 400 to 500 million years ago. Thus the virus described in *Chara* is probably the oldest recorded virus infecting a plant on or near the lineage that ultimately gave rise to the angiosperms.

No viruses have been reported from bryophytes. A virus with particles like those of a *Tobravirus* were found in hart's tongue fern (*Phyllitis scolopendrium* (L.) Newn) by R. E. Hull (1968).

A disease of *Cycas revoluta* has been shown to be due to a virus with a bipartite genome and other properties that place it in the *Nepovirus* group (Hanada *et al.*, 1986). The virus was readily transmitted by mechanical inoculation to various *Chenopodium* species. It was also transmitted through the seed of these species.

There have been a few reports of pines being infected experimentally with viruses from angiosperms (see Fulton, 1969). There are a few reports of naturally occurring viruslike diseases in other gymnosperms but the viral nature of the diseases has not been demonstrated (e.g., Schmelzer *et al.*, 1966). However, because of the presence of substances such as tannins, there are technical difficulties in attempting to isolate viruses from gymnosperms.

In summary, the existence of a single-stranded positive sense RNA virus infecting the genus *Chara* suggests an ancient origin for this type of virus. Other than this example, the meager information about viruses infecting photosynthetic eukaryotes below the angiosperms can tell us very little about the age and course of evolution among the plant viruses. The cycads are regarded as living fossils, being in the record since early Mesozoic times. However, the *Nepovirus* found in *Cycas revoluta* is quite likely to have originated in a modern angiosperm, since it readily infects *Chenopodium* spp. The *Phycodnavirus* PBCV-1 infecting a *Chlorella*-like alga is much more likely to be of ancient origin. However, on the basis of structure they do not appear to be primitive viruses. They are much larger and more complex than any known viruses infecting angiosperms, with a genome of about 300,000 base pairs and at least 50 structural proteins (Meints *et al.*, 1986).

The Evolution of Angiosperms and Insects

The most recent major evolutionary explosion, as evidenced in the fossil record, took place about the beginning of the Cretaceous ($\approx$135 million years ago). Monocotyledons and dicotyledons emerged and various orders of mammals and birds appeared. Some orders of insects had evolved by the Devonian (400 million years ago) but several important orders produced large numbers of new types as the higher plants emerged. Many of these coevolved with their angiosperm food plants. Thus some viruses that were already present in insects may have adapted to replicate in the evolving mammals and angiosperms during Cretaceous times.

The angiosperms appear in the fossil record no earlier than the early Cretaceous, that is, about 130 million years ago, with the major early diversification of the group occurring during the mid-Cretaceous (Crane and Lidgard, 1989). However, chloroplast DNA sequence data suggest that the monocotyledons and dicotyledons diverged from a common stock about 200 ± 40 million years ago (Wolfe *et al.*, 1989). Thus over this period of divergence the present-day viruses infecting angiosperms presumably also evolved, at least with respect to their main host specificities.

Horizontal Transmission through Plants of Viruses Infecting Only Insects

Leafhopper A virus (LAV) has particles similar to those of fijiviruses infecting plants. LAV infects and multiplies in the leafhopper *Cicadulina bimaculata* and is transmitted through its eggs. It does not multiply in the maize host plants of the insect. However, when infected hoppers feed on a plant, virus is injected into and circulates transiently in the plant. This virus can infect healthy hoppers that feed simultaneously on the same plant. Thus the maize plant can be regarded as a circulative but nonpersistent vector of an insect virus (Ofori and Francki, 1985).

A virus that infects the aphid *Rhapalosiphum padi* (RhPV) has a 27-nm icosahedral particle containing a single molecule of ssRNA. It is transmitted through the aphid eggs and does not replicate in the host plants of the aphid. However, when virus-infected and healthy aphids were fed simultaneously on the same barley leaf, healthy aphids became infected (Gildow and D'Arcy, 1988). It is not unreasonable to suppose that over a long period of time viruses such as LAV and RhPV might occasionally acquire a gene or genes that would allow them to establish and replicate in the plant as well as the insect.

Affinities of Viruses That Replicate in Both Insects and Plants

The Reoviridae family of viruses has members that infect both vertebrates and invertebrates, and others that infect both invertebrates and plants. This taxonomic distribution of hosts suggests that the more ancient invertebrates may have been the original source of this virus family. A number of features support the view that plant reoviruses originated in the leafhopper vectors (Conti, 1984; Nault, 1987; Nault and Ammar, 1989).

1. Fijiviruses are morphologically similar to viruses such as LAV that replicate only in the insect and to the *Peregrinus maidis* virus (Falk *et al.,* 1988).
2. All known plant reoviruses replicate in their hopper vectors.
3. Plant reoviruses are not seed-borne, nor are they transmitted by mechanical means, except in special circumstances. Thus they are entirely dependent on the hopper vectors for survival.
4. The plant species infected by reoviruses are usually the prime food and breeding hosts of their hopper vectors.
5. The plant reoviruses appear more closely adapted to their hopper hosts because: (i) they replicate to higher titer in the insects; (ii) several plant reoviruses are transmitted through insect eggs but none is transmitted through plant seed; (iii) the percentage of virus-carrying insects in a given vector population is higher than the percentage of plants that can be infected by feeding single hoppers on them; and (iv) some cause cytopathic effects in the hopper vectors, but in general these viruses have less severe pathogenic effects in the insects than in their plant hosts. In fact, most can be considered as causing latent infections.

Like the Reoviridae, members of the Rhabdoviridae family all have a very similar particle morphology and genome strategy and infect either vertebrates and invertebrates or invertebrates and plants. Thus a common origin for this family among the insect vectors is indicated. The viruses are not seed transmitted but are transmitted through the eggs of hopper vectors. They do not appear to cause disease in their insect vectors. The situation is somewhat more complex than with the plant reoviruses since different plant rhabdoviruses have hopper, aphid, piesmid, or mite vectors. Perhaps, as a rare event in the past, one rhabdovirus could transfer from a vector in one family to a vector in another where the vectors had a common host plant (Nault, 1991). It may be relevant to this idea that vesicular stomatitis virus, a rhabdovirus infecting vertebrates, replicates when injected into the leafhopper vector of a plant virus (Lastra and Esparza, 1976).

On the basis of evidence discussed in Chapter 8, Section V, TSWV now belongs in the Bunyaviridae. This family contains over 200 viruses that infect warm- and cold-blooded vertebrates and arthropods. Different viruses are transmitted by mosquitoes, sandflies, or ticks. TSWV, the only known plant member, is transmitted by thrips, but it has not been established that the virus multiplies in these vectors. It seems very probable that like the Reoviridae and Rhabdoviridae families, the bunyaviruses originated in their insect hosts. The tenuiviruses almost certainly originated in insects. The same may be true for the marafiviruses but the evidence is less compelling.

Flock house virus (FHV) is an insect virus belonging to a family of small RNA viruses, the Nodaviridae. It has no known relationship with plants in nature. However, Selling et al. (1990) have shown that the virus replicated to low levels in the leaves of several plant species following mechanical inoculation with FHV RNA. No replication could be detected following inoculation with whole virus. However, inoculation of barley protoplasts with intact FHV resulted in the synthesis of small amounts of progeny virus particles, indicating that the virus could be uncoated in plant cells. FHV particles moved systemically in Nicotiana benthamiana but no replication in systemic leaves could be detected. The virus showed no symptoms in plants it infected. These results suggest that the internal milieu of diverse organisms may be sufficiently similar to be able to support the replication of simple RNA viruses once inside a cell. This may have been a factor in past evolutionary processes.

Adaptation of Plant Viruses to Their Present Invertebrate Vectors

Fossil records indicate that the main taxonomic groups of the Homoptera had diverged by the Upper Triassic about 180 million years ago (Hennig, 1981). As discussed in Chapter 14, there are some aphid species among the vectors of plant viruses that transmit only one virus and others such as Myzus persicae that transmit large numbers. Some viruses are transmitted by only one aphid species, whereas others are transmitted by many. There is some evidence that among a phylogenetically diverse array of aphid vectors some groups are better vectors of a particular virus than others. For example, Zettler (1967) showed that aphids from the subfamily Aphidinae were better vectors of bean common mosaic Potyvirus than were aphids from the subfamilies Callaphidinae or Chaitophorinae. The differences were attributed to differences in probing behavior. Nault and Madden (1988) compared the efficiency of transmission of MCDV by 25 leafhopper species from 13 genera. They concluded that leafhopper species from the tribes Deltocephalini or recent (advanced) Euscelini that use maize as a feeding and breeding host have a higher probability of being MCDV vectors, whereas species from other taxa even if they feed well on maize, have a lower probability of being vectors.

Among the geminiviruses, the whitefly-transmitted viruses all have the same vector species, Bemisia tabaci. In immunoabsorbent electron microscopy tests, strong relationships were detected between five viruses transmitted by Bemisia (Roberts et al., 1984). Similar relationships were found using nucleic acid hybridization, using DNA1, which contains the coat protein gene. No relationship was detected between any whitefly-transmitted and any leafhopper-transmitted virus. The whitefly-transmitted viruses had diverse host ranges, came from different countries, and either were or were not sap transmissible. Roberts et al. (1984) suggested that the relationships found may be due to a key role played by the coat protein in vector transmission by these viruses (see also Section IV,B,1).

Assuming this association by descent, and assuming an insect origin for plant reoviruses and rhabdoviruses, Nault (1987) determined the latest stages during the evolution of vector groups at which the associations could have taken place (Fig.

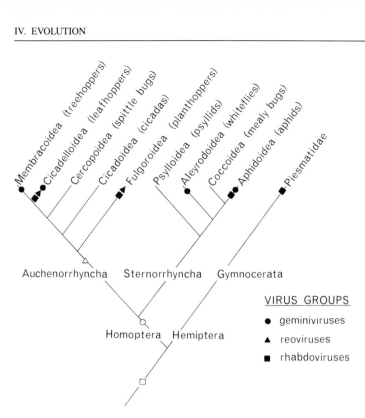

Figure 17.7 Possible coevolution of plant virus families and vector groups. Phylogeny of superfamilies of Homoptera with common names of groups in parentheses. Shown are homopteran groups (and one hemipteran) that are vectors for the geminiviruses, reoviruses, and rhabdoviruses. Assuming these plant viruses are associated by descent with the Homoptera, open symbols show the *latest* time at which associations could have taken place. (From Nault, 1987.) Note: (i) The genus *Phytoreovirus* is transmitted by leafhoppers, and the genus *Fijivirus* by plant hoppers. (ii) Tomato pseudocurly top virus, a possible *Geminivirus,* is transmitted by a treehopper, while another possible *Geminivirus* oat chlorotic stripe virus is transmitted by an aphid. (Thanks to A. C. Wayadande for art work.)

17.7). However, an insect origin for geminiviruses may be doubtful since they appear not to replicate in their vector.

Plant viruses and their beetle vectors may have evolved together even though virus replication in the vector does not occur. The *Tymovirus* group provides an interesting example. Most members of this group have a relatively restricted host range, usually within a single plant family when only natural hosts are considered. The host families do not occur at random among the six subclasses of the dicotyledons. None is found in the Magnoliidae or the Hammaelidae, subclasses that contain the older families among the dicotyledons.

The chrysomelid and curculionid beetle vectors of tymoviruses show specificity with respect to the viruses they transmit. They feed on a fairly narrow range of plant species, usually within the family constituting the host range of the virus. The earliest direct fossil evidence for the Curculionaidae goes back to the Upper Jurassic and for the Crysomelidae to the Middle Eocene (Crowson *et al.,* 1967). Thus although the fossil record is very scanty, it is not inconsistent with the idea that host plants, beetle vectors, and the tymoviruses underwent substantial evolutionary diversification together during the Eocene period.

Adaptation of plant viruses to their invertebrate vectors has probably involved several different processes on different time scales, and these may be very difficult to unravel. These processes include: (i) evolutionary origins by descent on a geological time scale as suggested in Fig. 17.7; (ii) adaptations to particular vector species that may be of quite recent origin, for example, through mutational changes in the viral coat protein or in a helper protein, in a group such as the potyviruses; and (iii) evolutionary origins that involve both coevolution and descent, and direct colonization of new vector groups. For example, such colonizations may explain the fact that among the plant rhabdoviruses there are both insect and mite vectors (Nault, 1987). However, the mite-transmitted coffee ring spot virus has not yet been well characterized as a rhabdovirus.

Geographical Distribution

Another feature favoring an ancient origin for present-day viruses is the geographical distribution of some groups. Members of the *Tymovirus* group of plant viruses have no known means of intercontinental spread except by humans. Different tymoviruses occur in Europe, North and South America, Africa, Asia, and Australia. These continents were still in one landmass at the end of the Jurassic period. The present-day tymoviruses may have diverged from a common ancestral stock since the continents drifted apart between 138 and 80 million years ago. The strain of TYMV found in *Cardamine* in Australia (Section IV,B,4) may be anomalous with respect to this general idea. However, the fact that both the European and Australian TYMVs are found in brassicaceous hosts may have restrained their evolutionary divergence.

Phylogenetic trees based on the amino acid sequences of *Geminivirus* coat proteins or replication-associated proteins suggest that the geographical isolation between Old World and New World geminiviruses may have played a role in the evolution of this group (Howarth and Vandemark, 1989).

There is another aspect of present-day virus and host distribution that is relevant to the problem of virus evolution. An area containing many species and varieties of a plant genus is taken to represent a site where evolution within that genus has occurred. For example, the *Solanum* species related to the potato probably evolved in the Andean region, with a center in the Lake Titicaca area of southern Peru and northern Bolivia, and were first domesticated there (Hawkes, 1967). Most of the viruses infecting potatoes are restricted in nature to the genus *Solanum*, for example, potato s *Carlavirus*, potato x *Potexvirus*, Andean potato mottle *Comovirus*, Andean potato latent *Tymovirus*, and potato moptop *Furovirus*. These viruses have almost certainly been spread from the Andes to other parts of the world in potato tubers. To the extent that they have been examined in detail these viruses show marked variation within the Andean region. For example, 26 isolates of Andean potato latent virus fell into three serological groups. One group was found only in the south of the region, one only in the north, and the third was distributed through the whole region (Koenig *et al.*, 1979). Andean potato latent virus and other viruses such as Andean potato mottle virus that are widely distributed in the Andean region multiply well only under the cool conditions prevailing in the potato areas at 2000–4000/m above sea level (Fribourg *et al.*, 1977.) This diversity of virus strain, climatic adaptation, and geographical restriction of some strains to parts of the region support the idea that the section of the genus *Solanum* comprising the potatoes and the major viruses of potatoes have coevoleved and are still coevolving in the Andean region.

V. GENOME AND AMINO ACID SEQUENCE SIMILARITIES BETWEEN VIRUSES INFECTING PLANTS AND ANIMALS

Given the very similar particle structures of rhabdoviruses and their replication in both plants and animals, it is not surprising that some genome sequence similarities have been revealed. In fact, with this family, there is good reason to believe that the sequence similarities are homologies, that is, have a common evolutionary origin.

It is much more surprising that amino acid sequence similarities in nonstructural proteins have been revealed between various RNA plant virus groups having diverse particle morphology and between these viruses and certain viruses infecting verte-brates (Haseloff *et al.*, 1984; Franssen *et al.*, 1984b; Argos *et al.*,1984; Ahlquist *et al.*, 1985). This has led to the idea that many plant plus strand RNA virus groups may be classified into two major superfamilies and that viruses within these super-families may have a common evolutionary origin (see Goldbach, 1987; Zimmern, 1988). There are also some sequence similarities between members of the two superfamilies. A third supergroup or superfamily centered on the *Luteovirus* group has been proposed by Habili and Symons (1989). Sequence similarities have also been found between caulimoviruses and animal retroviruses.

A. The Picorna-like Plant Viruses

Figure 17.8 summarizes the similarities in genome organization and sequence that have been revealed between the following viruses: *Poliovirus* (a member of the Picornaviridae), CPMV (*Comovirus*), TBRV (*Nepovirus*), and TVMV (*Potyvirus*). These viruses have the following features in common: (i) positive sense ssRNA genomes; (ii) a VPg at the 5' terminus and a poly(A) tract at the 3' terminus; (iii) single long ORFs coding for polyproteins that are processed by viral-coded pro-teases to give the functional gene products; (iv) these viruses encode several non-structural proteins that have similar functions and significant amino acid sequence similarity (>20%); and (v) the genes for these conserved proteins have a similar arrangement for all the genomes.

The B-RNA of CPMV bears a marked resemblance to the P2 and P3 regions of the poliovirus genome. As pointed out by Goldbach (1987), all the polypeptides with greater than 20% amino acid sequence similarity (Fig. 17.8) appear to be involved in membrane-bound replication complexes for both viruses.

Although there is no amino acid sequence similarity between the coat proteins of these viruses, there are some marked similarities in three-dimensional structure. VP1, VP2, and VP3 of poliovirus each has a β-barrel domain. The two coat proteins of CPMV form three β-barrels corresponding to the three for poliovirus. The difference is that, whereas all three proteins are cleaved from the initial poly-protein in poliovirus, only one is cleaved in CPMV, giving rise to the large and the small coat proteins. The single coat protein of nepoviruses with a size of ~60K corresponds in size to the combined MWs of the two comovirus proteins. It will probably be shown to contain three β-barrels.

The potyvirus coat protein is quite different from those of the other Picorna-like viruses, and the gene for it is located elsewhere in the genome, being at the 3' terminus (Fig. 17.8). There are two additional genes that are unrelated to those

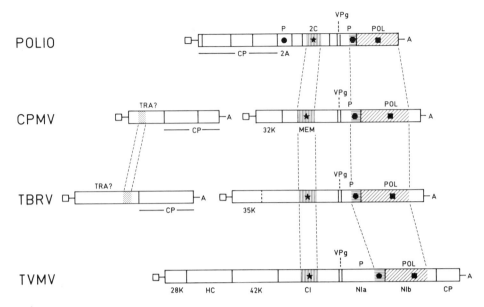

Figure 17.8 Comparison of the genomic RNAs of picornaviruses (POLIO) and the picornalike plant viruses. The "superfamily" of picornalike plant viruses includes comoviruses (CPMV), nepoviruses (TBRV), and potyviruses (TVMV). Open reading frames are represented as open bars, VPg as open squares, and poly(A) tails as A. Regions of significant (>20%) amino acid sequence similarity in the gene products are indicated by similar shading. TRA = Transport function; CP = capsid protein(s); P = protease; MEM = membrane binding; POL = core RNA-dep. RNA polymerase; HC = helper component; ★ = nucleotide binding domain; ● = cysteine protease domain; ■ = polymerase domain. (From Goldbach and Wellink, 1988.)

found in the other Picorna-like viruses. An N-terminal 28K protein of TVMV shows sequence similarity to the 30K protein of TMV and the 29K protein of TRV. These proteins are involved in cell-to-cell movement (Chapter 10, Section IV,G). The aphid transmission factor of TVMV (but not TEV) showed sequence similarity with the aphid transmission factors of two strains of the DNA virus, CaMV (Domier *et al.*, 1987). These similarities indicate that TVMV may have obtained some additional genes by the process of modular evolution.

B. The Sindbis-like Plant Viruses

Several plant RNA viruses have been found to show similarities in amino acid sequence and gene arrangement with Sindbis virus, an *Alphavirus* with a lipoprotein envelope that infects vertebrates. This collection of plant virus groups is quite variable in genome structure and strategy. They are the tobamoviruses (TMV), tombusviruses (CuNV), carmoviruses (CarMV), tobraviruses (TRV), hordeiviruses (BSMV), and the three closely related groups—bromoviruses (BMV), cucumoviruses (CMV), and AMV. Furoviruses (BNYVV) also belong with this superfamily (Bouzoubaa *et al.*, 1987), and potexviruses have sequence similarities that place them in this superfamily (R. L. S. Forster *et al.*, 1988; Skryabin *et al.*, 1988b). All of these viruses have a 5' cap but the 3' termini vary. Nevertheless, most of them specify three proteins with significant sequence similarity to the three nonstructural proteins nsP1, nsP2, and nsP4 of Sindbis virus. These viruses are compared with Sindbis virus in Fig. 17.9. Two of the three proteins (those with

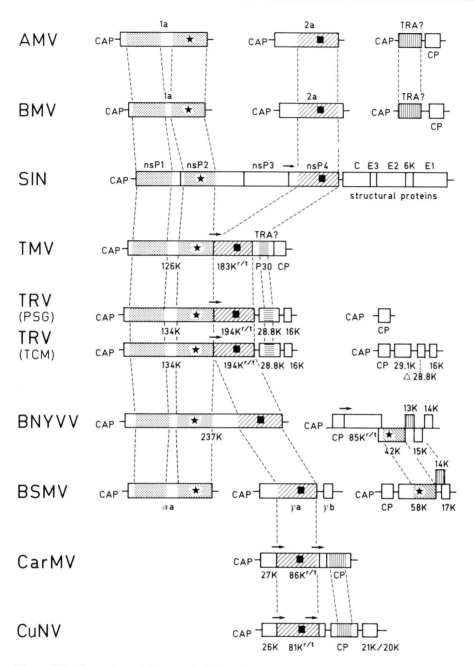

Figure 17.9 Comparison of the genomic RNAs of Sindbis virus (SIN) and the Sindbis-like plant viruses. The "superfamily" of Sindbis-like plant viruses includes the alfalfa mosaic virus group (AIMV), bromoviruses (BMV), tobamoviruses (TMV), tobraviruses (TRV, strains PSG and TCM), furoviruses (BNYVV), hordeiviruses (BSMV), carmoviruses (CarMV), and tombusviruses (CuNV). Open reading frames are represented as open bars, and regions of amino acid sequence similarity in the gene products are indicated by similar shading. For reasons of simplicity, closely adjoining or slightly overlapping genes in the TMV (p30 and CP) and CarMV (CP) genome are drawn contiguously. → = Leaky termination codon; r/t = read-through. For other symbols see the legend to Fig. 17.8. (From Goldbach and Wellink, 1988.)

similarity to nsP1 and nsP2) contain a nucleotide-binding domain sequence and a polymerase domain. Sindbis virus uses a polyprotein strategy so that the conserved domains nsP1 and nsP2 are on separate final polypeptides. In TMV and the tripartite viruses the two corresponding domains remain on a single polypeptide.

The strongest sequence similarity is found between BMV, CMV, and AMV, making it virtually certain that these viruses share a common ancestry. The carmoviruses and tombusviruses have very small genomes encoding only one of the three conserved domains found in the other Sindbis-like viruses. Goldbach and Wellink (1988) suggested that these two groups together with BYDV may turn out to form a third superfamily, a view supported by later work (Table 17.2).

C. Sequence Similarities between Picorna-like and Sindbis-like Viruses and Other Proteins

There is a low level of sequence similarity between certain sites in proteins of viruses belonging to the two RNA virus superfamilies.

1. An NTP Binding Site

Gorbalenya et al. (1985) first pointed out that the 58K protein of CPMV and the protein 2C of poliovirus contain a stretch of amino acids similar to the nucleotide triphosphate binding site of several GTP- or ATP-using enzymes. Such a domain is also found in the proteins of other ssRNA viruses such as SBMV (Wu et al., 1987) (see Fig. 7.22), potyviruses, nepoviruses, and the Sindbis-like viruses (Figs. 17.8 and 17.9). These similarities have been found to extend to a large class of DNA- and RNA-dependent NTPases, at least some of which possess helicase activity. They have been found in E. coli yeast, insects, mammals, poxviruses, and herpes viruses, ssDNA viruses and three groups of positive strand RNA viruses (Hodgman, 1988; Gorbalenya et al., 1989b, 1990). They are assumed to have a role in DNA and RNA replication.

2. RNA-Dependent RNA Polymerase Domains

Kamer and Argos (1984) identified a conserved 14-residue segment consisting of a Gly-Asp-Asp sequence flanked by hydrophobic residues in eight different RNA viruses. These included members of both the Picorna and Sindbis superfamilies (Figs. 17.8 and 17.9). It was also found in BSMV (Gustafson et al., 1987). In Poliovirus, this sequence was identified as being in an RNA-dependent RNA polymerase, implying this identification for the protein in the other viruses. A similar conserved 14-residue sequence consisting of an Asp-Asp sequence flanked by hydrophobic residues was found in retroviral reverse transcriptases, a bacteriophage, influenza virus, CaMV, and hepatitis B virus (see Fig. 17.10), suggesting that the sequence was an active site or a recognition site for polymerases in general. This subject is further discussed by Argos (1988).

A motif 14–15 amino acids long for polymerases was identified by Hodgman (1986). It was found in plant viral RNAs (TMV, AMV, and BMV) and in various DNA vertebrate viruses. It centered about the sequence, -Tyr/Phe-Gly-Asp-Thr-Asp/Glu-.

3. Serine Proteases

A set of three conserved amino acid residues, His, Asp/Glu, and Cys/Ser, has been identified in the chymotrypsinlike serine proteases and in the cysteine proteases of some positive strand RNA viruses (the 3C proteases of picornaviruses and related enzymes in como-, nepo-, and potyviruses) (Gorbalenya *et al.*, 1989a).

D. A Third Possible Superfamily of RNA Viruses

Rochon and Tremaine (1989) pointed out the extensive amino acid sequence homology in the putative replicases of representatives of the *Tombusvirus, Carmovirus,* and *Luteovirus* groups. Habili and Symons (1989) proposed a third supergroup. I will use the term superfamily as I think it is a more appropriate term from the point of view of viral taxonomy as a whole. In their comparative analysis, Habili and Symons used the amino acid sequence motifs in the putative nucleic acid polymerases and helicases of those RNA plant virus groups for which genome sequence data are available. Their proposed arrangement is summarized in Table 17.2.

Their analysis retains the *Sindbisvirus*-like and *Picornavirus*-like superfamilies. Sequence motifs of nucleic acid helicases and RNA polymerases previously considered to be specific for one of these two superfamilies were found to occur together within the new *Luteovirus*-like superfamily. They suggest that this new superfamily provides an evolutionary link between the other two. Arranging the virus groups into three superfamilies based on the RNA polymerase sequence motif alone gave rise to exactly the same arrangement as that based on the helicase motif alone. This lends support to the idea that there may be a real basis for the three superfamilies in Table 17.2.

However, the arrangements within each superfamily may be less suitable. Other criteria cut across these groupings. For example, the 3' end groups in the Sindbis-like superfamily vary, but it contains the plant virus groups with tRNA amino acid acceptor structures. These viruses are found in both the A_1 and A_2 clusters. Marsh *et al.* (1990) have shown that tandem regions showing sequence similarity to the BMV plus strand RNA1 5' terminus occur on the minus strand 5' termini of viruses possessing aminoacylatable tRNA-like 3' termini. The greatest similarity was found between BMV and TYMV, rather than between BMV and CMV. Many more nucleotide sequences will need to be determined before the significance of the proposed superfamilies can be established.

E. CaMV and Retroviruses

The gag–pol polyprotein of retroviruses has five domains: NH_2—structural proteins—protease—ribonuclease H—reverse transcriptase— endonuclease (Fig. 17.10A). Toh *et al.* (1983) described amino acid sequence similarity between retroviral reverse transcriptases and putative polymerases of hepatitis B virus and CaMV. Since 1983, various workers have described sequence similarities between CaMV, hepatitis B virus, and various retroviruses. These are summarized in Fig. 17.10B. Mason *et al.* (1987) have discussed these in relation to CaMV replication.

Table 17.2

Three Superfamilies of Plant Viruses Based on the Similarities in the Amino Acid Sequence Motifs in Their Putative RNA Polymerase and Nucleic Acid Helicase Proteins[a]

Superfamily A (*Sindbisvirus*-like)		Superfamily B (*Luteovirus*-like)		Superfamily C (*Picornavirus*-like)	
A1	*Potexvirus* (PVX) *Tobravirus* (WClMV) *Tymovirus (TYMV)*	B1	*Carmovirus* (CarMV, TCV, MCMV) *Tombusvirus* (CuNV) *Dianthovirus* (RCNMV)[b]	C1	*Nepovirus* (TBRV) *Comovirus* (CPMV)
A2	AMV[c] *Cucumovirus* (CMV) *Tobamovirus* (TMV) *Furovirus* (BNYVV) *Hordeivirus* (BSMV)	B2	*Luteovirus* "Group a" (BYDV, SbDV) "Group b" (PLRV, BWYV)	C2	*Potyvirus* (PPV, TEV, TVMV)
		B3	*Sobemovirus* (SBMV)		

[a]Each superfamily is subdivided on the basis of the closeness of sequence similarities. The viruses whose nucleotide sequence were used are indicated. Modified from Habili and Symons (1989).

[b]I have added *Dianthovirus* because it provides a link between *Luteovirus* and *Carmovirus* (S. A. Lommel, personal communication).

[c]*Bromovirus* and *Ilarvirus* groups would belong in the A2 cateogry.

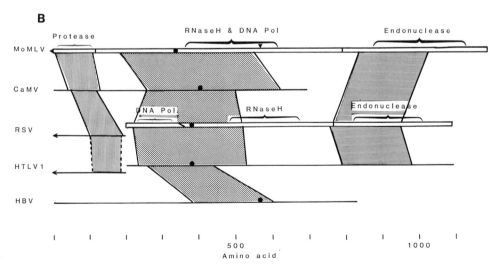

Figure 17.10 Organization of reverse transcriptase domains on retroid virus genomes. (A) Overall distribution of domains in the gag-pol polyprotein of a typical retrovirus. (B) Comparison of the distribution of domains in the pol ORF protein of MoMLV, the gene V product of CaMV, the proteins specified by the pol and carboxy-terminal portion of the gag ORFs of RSV and HTLV I, and the pol ORF of HBV. The positions of the interfaces between domains, where they are known, are shown for MoMLV and RSV. The regions of amino acid similarities are indicated by blocks of shading. ●, the site of Asp-Asp common sequence. (Modified from Mason *et al.,* 1987.) MoMLV = Moloney murine leukemia virus; RSV = Rous sarcoma virus; HTLV1 = Human T-cell leukemia virus 1; HBV = Human hepatitis B virus.

F. Significance for Virus Classification and Evolution

The genome and amino acid sequence similarities discussed suggest relationships that cut across many of the accepted criteria for classifying viruses: (i) host groups—plant and vertebrate viruses; (ii) morphology—viruses with rod-shaped particles, icosahedral particles, and particles with a lipoprotein envelope; (iii) kind of nucleic acid—DNA and RNA; and (iv) numbers of genome segments—viruses with monopartite, bipartite, and tripartite RNA genomes.

The situation we observe among present-day viruses has probably arisen as a consequence of three major processes—divergent evolution of viruses from a common ancestor, convergent evolution of proteins with particular functions, and modular evolution involving the acquisition of genes from other viruses, or the host genome. In the present state of knowledge we can only speculate as to the relative importance of these three processes. Nevertheless, some distinctions may be emerging. The amino acid sequence similarities between the enzymes discussed in Section V,C could well have arisen by convergent evolution, driven by a requirement for particular structures in their active sites. This idea is supported by the following points: (i) the similarities relate to a relatively short sequence of amino acids that are thought to make up the active sites of the enzymes involved and (ii) nonviral enzymes in a range of organisms have similar conserved sequences. By contrast, it would be very difficult to imagine the similarities in amino acid sequence and gene order seen between CPMV and poliovirus (Fig. 17.8) as arising by convergent evolution. Here divergence from a common ancestor seems most probable.

We do not yet know whether all the ss positive sense RNA plant viruses will fit into one of the two or three "superfamilies" already delineated. Whatever the answer to that question, it is unlikely that all this new sequence information will disturb the existing plant virus taxonomy in any major way. In fact, the new information should help to improve existing taxonomy, for example, by providing stronger evidence that the bromoviruses, cucumoviruses, AMV, and ilarviruses should be in a single family, or by supporting the need for genera within the geminiviruses.

There is no doubt that the reoviruses, rhabdoviruses, and bunyaviruses infecting vertebrates, invertebrates, or plants should be in the same families, on the basis of detailed similarities in particle structure and genome organization. By contrast, the relationship between, say, Sindbis virus and TMV is much more remote.

VI. DISCUSSION AND SUMMARY

As far as taxonomy is concerned, most of the groups and families of plant viruses constituted by the I.C.T.V., mainly on the basis of properties of the virus particles, have stood the test of time. Detailed information now available for many groups concerning genome organization and the amino acid sequences of viral-coded proteins has confirmed these groupings and indicates that viruses within a group have a common evolutionary origin. Four of the groups constituted earlier—bromoviruses, cucumoviruses, AMV, and ilarviruses—should now be placed in a single virus family, as suggested some years ago (van Vloten-Doting et al., 1981).

Recent evidence, especially that obtained from nucleotide sequences of viruses from families infecting more than one major host group, confirms that viruses constitute a unified array of biological agents. This makes it imperative that viral taxonomy should reflect that unity. Plant virologists should begin to turn many of the plant virus groups into families, and to establish one or more genera within each family. Most plant viruses will continue to be called by their vernacular names (mainly English) for many years to come. A binomial system for naming plant viruses is emerging in which the vernacular virus name is followed by the group or genus name, for example, wound tumor *Phytoreovirus*.

Structure of the virus particle, strategy of viral replication, nucleotide sequences, and amino acid sequences of viral-coded proteins will remain important for placing viruses into groups and families. Amino acid sequences of viral coat proteins often appear to be more characteristic of individual viruses than nonstructural viral-coded proteins. Thus serological relationships based on coat proteins will remain important for identifying individual viruses and many virus strains.

Three kinds of origin for viruses have been proposed: (i) descendants from a prebiotic RNA world; (ii) from normal host genes; and (iii) by degeneration from simple obligate cellular parasites. The true origin of viruses may never be established, but on present evidence more than one kind of origin is likely. Most speculations on the origin of viroids propose, mainly on the basis of rather limited sequence similarities, that they arose from some small cellular RNA. Satellite viruses probably arose from an independent virus by loss of function. Different satellite RNAs may have had an origin either from viroids or more commonly from satellite viruses.

There is ample experimental evidence from the study of existing viruses to demonstrate that viruses continue to evolve, but rates of evolution for most viruses and virus genes are very difficult to establish with any degree of certainty. As far as mechanisms for virus evolution are concerned we can envisage two kinds of processes: *microevolution* by means of mutations such as single-base changes, deletions, or additions and gene recombination or reassortment of genome pieces between related viruses; and *macroevolution,* involving a major change such as gene exchange with, or acquisition from, an unrelated virus, duplication of an existing gene, or acquisition of a gene from the host. These latter kinds of change give rise to what has been called modular evolution. As has been pointed out by Ding *et al.* (1990b), if modular evolution has been widespread in the evolutionary history of viruses, then it will not be possible to develop a taxonomy based on evolutionary history above the level of genus that is represented by a single nested hierarchy. From this idea, superfamily, and perhaps family, relationships may be properly represented only by two or more hierarchies each for a particular part of the viral genome. In spite of this possibility, as the sequences of more viral genomes and host genes become known it may be possible to develop a true systematics for viruses.

Many features of viruses in the Reoviridae family indicate that they evolved originally in invertebrates, subsequently adapting to both plant and vertebrate hosts. The Rhabdoviridae and Bunyaviridae families also have members infecting invertebrates, vertebrates, or plants. Thus there is much circumstantial evidence to show that the existing virus families and groups are of ancient origin and that they have coevolved with their host organisms and arthropod vectors. By contrast, particular individual viruses may be quite new in evolutionary time.

Full nucleotide sequences for representatives of many ssRNA plant virus groups and vertebrate virus families are now known, which has enabled amino acid sequences of viral-coded proteins to be compared. Many ss plant RNA virus groups can be placed in one of two "superfamilies" on the basis of similarities in amino acid sequence and gene arrangement with certain ssRNA vertebrate viruses. These are Picorna-like viruses, such as CPMV, TBRV, and TVMV, which have similarities in amino acid sequences and gene arrangement with poliovirus, and the Sindbis-like plant viruses, which include TMV, BMV, CMV, and AMV. A third superfamily based on the *Luteovirus* group has been proposed. It remains to be seen whether all the ss positive sense RNA plant viruses fall into one or the other of these three superfamilies. Where gene arrangement parallels sequence similarities, a common evolutionary origin for viruses within these superfamilies at some time in the distant past seems likely. Where amino acid sequence similarities are confined entirely or mainly to the active sites of enzymes, it is possible that evolutionary convergence may have been involved. When further nucleotide sequences are available it may be possible to develop a true systematics for viruses, that is, a classification based on evolutionary descent.

I conclude with the following speculations concerning the origins of the various plant virus groups and families. If the possible rate of mutation in viruses, especially RNA viruses, and the processes of modular evolution had had a free rein in past times, most clues as to relationships between groups and families would have been obliterated by now. However, clear evidence of such relationships exists. Therefore we can assume that the environment presented to viruses by the milieu of their host cells acts as a powerful restraining influence on virus evolution. Thus the theoretical potential for rapid change in a virus lineage is no impediment to postulating an ancient origin. On current evidence we can perceive four clusters of plant viruses with respect to their possible origins.

1. *The Single-Stranded Positive Sense RNA Viruses*   Sindbis virus has an icosahedral core containing the RNA genome surrounded by a tightly applied lipoprotein envelope with two glycosylated proteins. TMV is a helical rod with a single coat protein. These two viruses could hardly be more different in morphology, and yet they show clear similarities in their genomes. Therefore, the ss positive sense RNA viruses may represent the most ancient viruses, not necessarily originating in a single evolutionary event, but evolving slowly to give rise to the existing superfamilies. Relationships between members within these superfamilies infecting plants or vertebrates may be real, but distant. There is no compelling evidence for the origin and evolution of the proposed ssRNA superfamilies in their insect vectors. The ancestral viruses may have predated the evolutionary separation of plants and animals some 10^9 years ago.

2. *Plant Reoviruses, Rhabdoviruses, and Bunyaviruses*   It is a curious fact that the three virus families known to have members infecting vertebrates, invertebrates, and plants—Reoviridae, Rhabdoviridae, and Bunyaviridae—are the three families containing the structurally most complex viruses that infect plants. They are also the only plant viruses with RNA genomes that do not consist of ss positive sense RNA. It is noteworthy that members of these families infecting plants or vertebrates are much more closely related to each other than are the ss positive sense RNA viruses within each superfamily discussed earlier. Therefore they are likely to have had a much more recent common origin. It is reasonable to suppose that the immediate ancestors of these three virus families were present in the early insects and that as insects, angiosperms, and mammals coevolved during Cretaceous times, viruses within each family diverged sufficiently to replicate in one or two of these three major host groups.

3. *Caulimoviruses*   Caulimoviruses might represent ancient molecular fossils of a world in transition from RNA to DNA genomes, as suggested by Weiner (1987) for retroviruses in general.

4. *Geminiviruses*   The similarities between geminiviruses and the bacterial virus φX174 were noted earlier in Section IV,B,5. Geminiviruses may be descended without great change over a long time period from a simple type of DNA virus that first evolved in prokaryotes. Apart from the viral-coded A protein, DNA synthesis of φX174 is dependent on host enzymes (Baas and Jansz, 1978). Thus the fact that present-day geminiviruses replicate in the nucleus may reflect a dependence on host DNA enzymes that goes back to an origin in prokaryotic cells.

Appendix 1

Literature on Plant Virology

The following list includes the major publications that regularly contain original papers or reviews relevant to plant virology. The most important publications are marked with an asterisk. Many other publications are covered by the abstracting service noted at the end of this list.

Core journals

    Intervirology
  *Journal of General Virology
    Journal of Virological Methods
  *Virology

Journals publishing papers mainly on applied aspects

  *Annals of Applied Biology
    Annals of the Phytopathological Society of Japan
    Netherlands Journal of Plant Pathology
  *Phytopathology
    Plant Disease
    Plant Pathology

Journals publishing papers from time to time mainly in basic plant virology

    Archives of Virology
    Biochemical and Biophysical Research Communications
  *Cell
  *EMBO Journal
    European Journal of Biochemistry
    FEBS Letters
  *Journal of Molecular Biology
    Journal of Virology
    Molecular Plant Microbe Interactions
    Nature
  *Physiological and Molecular Plant Pathology

Plant Molecular Biology
Proceedings of the National Academy of Science, U.S.A.
Science
Virus Research

Review Journals

Advances in Disease Vector Research
Advances in Virus Research
Annual Review of Biochemistry
Annual Review of Microbiology
Annual Review of Phytopathology

Abstracting Service

Papers of interest to plant virologists are published in a much wider range of journals than those listed above, particularly national and regional journals covering plant pathology. An excellent abstracting service is provided by the Netherlands Circle for Plant Virology (C. J. Asjes, c/o Bulb Research Centre, Vennestraat 22, Postbox 85, 2160 AB Lisse, The Netherlands).

Appendix 2

Standard Acronyms for Selected Plant Viruses and Viroids

Virus names, subgroups, and acronyms were taken fom Hull *et al.* (1991). The group or genus names approved by the I.C.T.V. have been placed immediately following the virus names. This provides a binomial identification of the virus concerned. Tentative allocations of a virus to a group or genus are indicated by ?. Certain virus names are referred to in the text which do not appear in this list. These are usually from the older literature, and positive identification in relation to the present list may not be possible.

The italics for group or genus names indicate names approved by the I.C.T.V. Rhabdovirus indicates that a virus has been allocated by the I.C.T.V. to the family Rhabdoviridae, but this is not an approved generic name for plant viruses and is therefore not italicized. A dash after the virus name indicates that the virus has not yet been allocated to a group or genus by the I.C.T.V. In the last column a dash indicates that no A.A.B. description has been published. A few important synonyms are included. For others, refer to the A.A.B. descriptions.

Subgroups for some viruses are indicated by a letter, number, or vector type after the family or group name. *Geminivirus:* I, II, or III, as defined in Chapter 8, Section II. *Cryptovirus:* (A) = two dsRNA molecules of MW 1.20 and 0.97×10^6; (B) = two dsRNA molecules of MW 1.49 and 1.38×10^6. Rhabdovirus: (1) mature and accumulate in the cytoplasm; (2) bud at the inner membrane of the nuclear envelope and accumulate in the perinuclear space. *Potyvirus:* (aphid) (mite) (fungus) indicate the kind of vector for viruses currently allocated to this group.

Table 1

Viruses

Virus name	Acronym	A.A.B. Description No.
Abutilon mosaic *Geminivirus* III	AbMV	—
African cassava mosaic *Geminivirus* III	ACMV	297
Agropyron mosaic *Potyvirus* (mite)	AgMV	118
Alfalfa cryptic 1 *Cryptovirus* (A)	ACV1	—
Alfalfa latent *Carlavirus*	ALV	211
Alfalfa mosaic AMV group	AMV	229
American hop latent *Carlavirus*	AHLV	262
American plum line pattern *Ilarvirus*	APLPV	280
American wheat striate mosaic rhabdovirus (1)	AWSMV	99
Andean potato latent *Tymovirus*	APLV	—
Andean potato mottle *Comovirus*	APMV	203
Anthoxanthum latent blanching *Hordeivirus*	ALBV	—
Apple chlorotic leaf spot ? *Closterovirus*	ACLV	30
Apple mosaic *Ilarvirus*	ApMV	83
Apple stem grooving *Capillovirus*	ASGV	31
Arabis mosaic *Nepovirus*	ArMV	16
Arracacha A *Nepovirus*	AVA	216
Arracacha B ? *Nepovirus*	AVB	270
Artichoke Italian latent *Nepovirus*	AILV	176
Artichoke mottle crinkle *Tombusvirus*	AMCV	—
Artichoke vein banding *Nepovirus*	AVBV	285
Artichoke yellow ringspot *Nepovirus*	AYRSV	271
Asparagus virus 1 *Potyvirus* (aphid)	AV1	—
Asparagus virus 2 *Ilarvirus*	AV2	288
Asparagus virus 3 *Potexvirus*	AV3	—
Bamboo mosaic *Potexvirus*	BaMV	—
Barley stripe mosaic *Hordeivirus*	BSMV	68
Barley yellow dwarf *Luteovirus*	BYDV	32
Barley yellow mosaic *Potyvirus* (fungus)	BaYMV	143
Barley yellow striate mosaic rhabdovirus (1)	BYSMV	312
Bean common mosaic *Potyvirus* (aphid)	BCMV	337
Bean distortion dwarf ? *Geminivirus* III	BDDV	—
Bean golden mosaic *Geminivirus* III	BGMV	192
Bean mild mosaic ? *Carmovirus*	BMMV	231
Bean pod mottle *Comovirus*	BPMV	108
Bean rugose mosaic *Comovirus*	BRMV	246
Bean yellow mosaic *Potyvirus* (aphid)	BYMV	40
Bean yellow vein banding—	BYVBV	—
Bearded iris mosaic *Potyvirus* (aphid)	BIMV	338
Beet cryptic virus 1 *Cryptovirus* (A)	BCV1	—
Beet cryptic virus 2 *Cryptovirus* (A)	BCV2	—
Beet curly top *Geminivirus* II	BCTV	210
Beet leaf curl rhabdovirus (2)	BLCV	268
Beet mild yellowing *Luteovirus*	BMYV	—
Beet mosaic *Potyvirus* (aphid)	BtMV	53
Beet necrotic yellow vein *Furovirus*	BNYVV	144
Beet pseudo-yellows ? *Closterovirus*	BPYV	—
Beet soilborne ? *Furovirus*	BSBV	—
Beet western yellows *Luteovirus*	BWYV	89
Beet yellow net ? *Luteovirus*	BYNV	—
Beet yellow stunt *Closterovirus*	BYSV	207
Beet yellows *Closterovirus*	BYV	13
Belladonna mottle *Tymovirus*	BeMV	52
Bermuda grass etched-line ? *Marafivirus*	BELV	—

Virus name	Acronym	A.A.B. Description No.
Bidens mottle *Potyvirus* (aphid)	BiMoV	161
Blackeye cowpea mosaic *Potyvirus* (aphid)	BlCMV	305
Blackgram mottle ? *Carmovirus*	BMoV	237
Black raspberry latent *Ilarvirus*	BRLV	106
Black raspberry necrosis—	BRNV	333
Blueberry leaf mottle *Nepovirus*	BLMV	267
Blueberry red ringspot *Caulimovirus*	BRRV	327
Blueberry shoestring *Sobemovirus*	BSSV	204
Broad bean mottle *Bromovirus*	BBMV	101
Broad bean necrosis *Furovirus*	BBNV	223
Broad bean stain *Comovirus*	BBSV	29
Broad bean true mosaic *Comovirus*	BBTMV	20
Broad bean wilt *Fabavirus*	BBWV	81
Broccoli necrotic yellows rhabdovirus (1)	BNYV	85
Brome mosaic *Bromovirus*	BMV	180
Burdock yellows *Closterovirus*	BuYV	—
Cacao necrosis *Nepovirus*	CNV	173
Cacao swollen shoot—	CSSV	10
Cacao yellow mosaic *Tymovirus*	CYMV	11
Cactus X *Potexvirus*	CVX	58
Canna yellow mottle—	CaYMV	—
Cardamine latent *Carlavirus*	CaLV	—
Carnation cryptic *Cryptovirus* (A)	CCV	315
Carnation etched ring *Caulimovirus*	CERV	182
Carnation Italian ringspot *Tombusvirus*	CIRSV	—
Carnation latent *Carlavirus*	CLV	61
Carnation mottle *Carmovirus*	CarMV	7
Carnation necrotic fleck *Closterovirus*	CMFV	136
Carnation ringspot *Dianthovirus*	CRSV	308
Carnation vein mottle *Potyvirus* (aphid)	CVMV	78
Carnation yellow stripe ? *Necrovirus*	CYSV	—
Carrot mottle—	CMoV	137
Carrot red leaf *Luteovirus*	CaRLV	249
Carrot thin leaf *Potyvirus* (aphid)	CTLV	218
Carrot yellow leaf *Closterovirus*	CYLV	—
Cassava brown streak ? *Carlavirus*	CBSV	—
Cassava common mosaic *Potexvirus*	CsCMV	90
Cassava green mottle *Nepovirus*	CGMV	—
Cassava latent virus = African cassava mosaic virus		
Cassava vein banding ? *Caulimovirus*	CVBV	—
Cassava virus C—	CVC	—
Cassava virus X *Potexvirus*	CsVX	—
Cassia yellow blotch *Bromovirus*	CYBV	334
Cauliflower mosaic *Caulimovirus*	CaMV	243
Celery mosaic *Potyvirus* (aphid)	CeMV	50
Cereal chlorotic mottle rhabdovirus (2)	CeCMV	251
Chara australis—	CAV	—
Cherry leaf roll *Nepovirus*	CLRV	306
Cherry rasp leaf *Nepovirus*	CRLV	159
Chicory virus X *Potexvirus*	ChVX	—
Chicory yellow mottle *Nepovirus*	ChYMV	132
Chloris striate mosaic *Geminivirus* I	CSMV	221
Chrysanthemum B *Carlavirus*	CVB	110

(continued)

Table 1 (*Continued*)

Virus name	Acronym	A.A.B. Description No.
Citrus leaf rugose *Ilarvirus*	CiLRV	164
Citrus tatter leaf ? *Closterovirus*	CiTLV	—
Citrus tristeza *Closterovirus*	CTV	33
Citrus variegation *Ilarvirus*	CVV	—
Clitoria yellow vein *Tymovirus*	CYVV	171
Clover yellow mosaic *Potexvirus*	ClYMV	111
Clover yellow vein *Potyvirus* (aphid)	ClYVV	131
Cocksfoot mild mosaic ? *Sobemovirus*	CMMV	107
Cocksfoot mottle *Sobemovirus*	CoMV	23
Cocksfoot streak *Potyvirus* (aphid)	CSV	59
Coffee ringspot rhabdovirus (2)	CoRSV	—
Cowpea aphid-borne mosaic *Potyvirus* (aphid)	CABMV	134
Cowpea chlorotic mottle *Bromovirus*	CCMV	49
Cowpea green vein banding ? *Potyvirus*	CGVBV	—
Cowpea mild mottle *Carlavirus*	CPMMV	140
Cowpea mosaic *Comovirus*	CPMV	197
Cowpea mottle ? *Carmovirus*	CPMoV	212
Cowpea ringspot ? *Cucumovirus*	CPRSV	—
Cowpea severe mosaic *Comovirus*	CPSMV	209
Crimson clover latent *Nepovirus*	CCLV	—
Cucumber green mottle mosaic *Tobamovirus*	CGMMV	154
Cucumber leaf spot ? *Tombusvirus*	CLSV	319
Cucumber mosaic *Cucumovirus*	CMV	213
Cucumber necrosis *Tombusvirus*	CuNV	82
Cucumber soilborne *Carmovirus*	CSBV	—
Cucumber yellows ? *Closterovirus*	CYV	—
Cycas necrotic stunt *Nepovirus*	CNSV	—
Cymbidium mosaic *Potexvirus*	CyMV	27
Cymbidium ringspot *Tombusvirus*	CyRSV	178
Cynosurus mottle ? *Sobemovirus*	CnMoV	—
Dahlia mosaic *Caulimovirus*	DMV	51
Dandelion yellow mosaic—	DaYMV	—
Daphne X *Potexvirus*	DVX	195
Dasheen mosaic *Potyvirus* (aphid)	DsMV	191
Dendrobium vein necrosis ? *Closterovirus*	DVNV	—
Desmodium yellow mottle *Tymovirus*	DYMV	168
Digitaria streak *Geminivirus* I	DSV	—
Digitaria striate rhabdovirus (1)	DStV	—
Dioscorea latent *Potexvirus*	DLV	335
Dulcamara mottle *Tymovirus*	DuMV	—
Eggplant mosaic *Tymovirus*	EMV	124
Eggplant mottled crinkle *Tombusvirus*	EMCV	—
Eggplant mottled dwarf rhabdovirus (2)	EMDV	115
Elderberry carla *Carlavirus*	ECV	263
Elderberry latent ? *Carlavirus*	ELV	127
Elm mottle *Ilarvirus*	EMoV	139
Epirus cherry—	EpCV	—
Erysimum latent *Tymovirus*	ErLV	222
Euphorbia mosaic *Geminivirus* III	EuMV	—
European wheat striate mosaic *Tenuivirus*	EWSMV	—
Fern mottle ? *Furovirus*	FMoV	—
Figwort mosaic *Caulimovirus*	FMV	—
Fiji disease *Fijivirus* (1)	FDV	—
Foxtail mosaic *Potexvirus*	FoMV	264

Virus name	Acronym	A.A.B. Description No.
Frangipani mosaic *Tobamovirus*	FrMV	196
Galinsoga mosaic *Carmovirus*	GaMV	252
Ginger chlorotic fleck *Sobemovirus*	GCFV	328
Glycine mosaic *Comovirus*	GMV	—
Glycine mottle ? *Carmovirus*	GMoV	—
Grapevine Algerian latent *Tombusvirus*	GALV	—
Grapevine Bulgarian latent *Nepovirus*	GBLV	186
Grapevine chrome mosaic *Nepovirus*	GCMV	103
Grapevine fanleaf *Nepovirus*	GFLV	28
Groundnut rosette—	GRV	—
Groundnut rosette assistor *Luteovirus*	GRAV	—
Guinea grass mosaic *Potyvirus* (aphid)	GGMV	190
Helenium S *Carlavirus*	HVS	265
Henbane mosaic *Potyvirus* (aphid)	HMV	95
Heracleum latent ? *Closterovirus*	HLV	225
Heracleum virus 6 ? *Closterovirus*	HV6	—
Hibiscus chlorotic ringspot *Carmovirus*	HCRSV	227
Hibiscus latent ringspot *Nepovirus*	HLRSV	233
Hippeastrum mosaic *Potyvirus* (aphid)	HiMV	117
Honeysuckle latent *Carlavirus*	HnLV	289
Honeysuckle yellow vein mosaic *Geminivirus* III	HYVMV	—
Hop latent *Carlavirus*	HpLV	261
Hop mosaic *Carlavirus*	HpMV	241
Hop trefoil cryptic virus 1 *Cryptovirus* (A)	HTCV1	—
Hop trefoil cryptic virus 2 *Cryptovirus* (B)	HTCV2	—
Hop trefoil cryptic virus 3 *Cryptovirus* (A)	HTCV3	—
Hydrangea mosaic *Ilavirus*	HdMV	—
Hydrangea ringspot *Potexvirus*	HRSV	114
Hypochoeris mosaic ? *Furovirus*	HyMV	273
Indian peanut clump ? *Furovirus*	IPCV	—
Iris fulva mosaic *Potyvirus* (aphid)	IFMV	310
Iris mild mosaic *Potyvirus* (aphid)	IMMV	324
Iris severe mosaic *Potyvirus* (aphid)	ISMV	338
Kalanchoë latent *Carlavirus*	KLV	—
Kennedya yellow mosaic *Tymovirus*	KYMV	193
Lamium mild mosaic *Fabavirus*	LMMV	—
Leek yellow stripe *Potyvirus* (aphid)	LYSV	240
Lettuce big vein—	LBVV	—
Lettuce infectious yellows ? *Closterovirus*	LIYV	—
Lettuce mosaic *Potyvirus* (aphid)	LMV	9
Lettuce necrotic yellows rhabdovirus (1)	LNYV	26
Lettuce speckles mottle—	LSMV	—
Lilac chlorotic leaf spot *Closterovirus*/? *Capillovirus*	LCLV	202
Lilac ring mottle *Ilarvirus*	LRMV	201
Lily symptomless *Carlavirus*	LSV	96
Lucerne Australian latent *Nepovirus*	LALV	225
Lucerne transient streak *Sobemovirus*	LTSV	224
Lychnis ringspot *Hordeivirus*	LRSV	—
Maclura mosaic ? *Potyvirus* (aphid)	MacMV	239
Maize chlorotic dwarf MCDV group	MCDV	194
Maize chlorotic mottle ? *Sobemovirus*	MCMV	284

(continued)

Table 1 (*Continued*)

Virus name	Acronym	A.A.B. Description No.
Maize dwarf mosaic *Potyvirus* (aphid)	MDMV	—
Maize mosaic rhabdovirus (2)	MMV	94
Maize rayado fino *Marafivirus*	MRFV	220
Maize rough dwarf *Fijivirus*	MRDV	72
Maize sterile stunt rhabdovirus (1)	MSSV	—
Maize streak *Geminivirus* I	MSV	133
Maize stripe *Tenuivirus*	MStpV	300
Maize white line mosaic—	MWLMV	283
Maracuja mosaic *Tobamovirus*	MaMV	—
Melandrium yellow fleck *Bromovirus*	MYFV	236
Melon necrotic spot *Carmovirus*	MNSV	302
Mirabilis mosaic *Caulimovirus*	MiMV	—
Miscanthus streak *Geminivirus* I	MiSV	—
Moroccan pepper *Tombusvirus*	MPV	—
Mulberry ringspot *Nepovirus*	MRSV	142
Mung bean yellow mosaic *Geminivirus* III	MYMV	323
Myrobalan latent ringspot *Nepovirus*	MLRSV	160
Nandina mosaic *Potexvirus*	NaMV	—
Nandina stem-pitting ? *Capillovirus*	NSPV	—
Narcissus latent *Carlavirus*	NLV	170
Narcissus mosaic *Potexvirus*	NMV	45
Narcissus tip necrosis ? *Carmovirus*	NTNV	166
Narcissus yellow stripe *Potyvirus* (aphid)	NYSV	76
Neckar river *Tombusvirus*	NRV	—
Nerine latent *Carlavirus*	NeLV	—
Nerine X *Potexvirus*	NVX	336
Nicotiana velutina mosaic ? *Furovirus*	NVMV	189
Northern ceral mosaic rhabdovirus (1)	NCMV	322
Oat blue dwarf *Marafivirus*	OBDV	123
Oat golden stripe ? *Furovirus*	OGSV	—
Oat mosaic *Potyvirus* (fungus)	OMV	145
Oat necrotic mottle *Potyvirus* (mite)	ONMV	169
Oat sterile dwarf *Fijivirus*	OSDV	217
Odontoglossum ringspot *Tobamovirus*	ORSV	155
Okra mosaic *Tymovirus*	OkMV	128
Olive latent ringspot *Nepovirus*	OLRSV	301
Olive latent virus 1 ? *Sobemovirus*	OLV1	—
Olive latent virus 2—	OLV2	—
Onion yellow dwarf *Potyvirus* (aphid)	OYDV	158
Ononis yellow mosaic *Tymovirus*	OYMV	—
Orchid fleck ? rhabdovirus (?2)	OFV	183
Ourmia melon—	OuMV	—
Pangola stunt *Fijivirus*	PaSV	175
Panicum mosaic ? *Sobemovirus*	PMV	177
Papaya mosaic *Potexvirus*	PapMV	56
Papaya ringspot (=Watermelon mosaic virus 1) *Potyvirus* (aphid)	PRSV	292
Parsnip mosaic *Potyvirus* (aphid)	ParMV	91
Parsnip yellow fleck—	PYFV	129
Passionfruit woodiness *Potyvirus* (aphid)	PWV	122
Pea early-browning *Tobravirus*	PEBV	120
Pea enation mosaic PEMV group	PEMV	257
Pea leaf roll *Luteovirus*	PeLRV	286

Virus name	Acronym	A.A.B. Description No.
Pea seed-borne mosaic *Potyvirus* (aphid)	PSbMV	146
Pea streak *Carlavirus*	PeSV	112
Peach rosette mosaic *Nepovirus*	PRMV	150
Peanut clump *Furovirus*	PCV	235
Peanut mottle *Potyvirus* (aphid)	PeMoV	141
Peanut stunt *Cucumovirus*	PSV	92
Pelargonium flower-break *Carmovirus*	PFBV	130
Pelargonium leaf curl *Tombusvirus*	PLCV	—
Pelargonium line pattern ? *Carmovirus*	PLPV	—
Pelargonium zonate spot—	PZSV	272
Pepino mosaic *Potexvirus*	PepMV	—
Pepper mild mottle *Tobamovirus*	PMMV	330
Pepper mottle *Potyvirus* (aphid)	PepMoV	253
Pepper ringspot *Tobravirus*	PepRSV	—
Pepper veinal mottle *Potyvirus* (aphid)	PVMV	104
Peru tomato *Potyvirus* (aphid)	PTV	255
Petunia asteroid mosaic *Tombusvirus*	PeAMV	—
Physalis mosaic *Tymovirus*	PhyMV	—
Plantago mottle *Tymovirus*	PlMoV	—
Plantain X *Potexvirus*	PlVX	266
Plum pox *Potyvirus* (aphid)	PPV	70
Poa semilatent *Hordeivirus*	PSLV	—
Poinsettia mosaic ? *Tymovirus*	PnMV	311
Pokeweed mosaic *Potyvirus* (aphid)	PkMV	97
Poplar mosaic *Carlavirus*	PopMV	75
Potato A *Potyvirus* (aphid)	PVA	54
Potato aucuba mosaic ? *Potexvirus*	PAMV	98
Potato black ringspot *Nepovirus*	PBRSV	206
Potato leafroll *Luteovirus*	PLRV	291
Potato M *Carlavirus*	PVM	87
Potato mop top *Furovirus*	PMTV	138
Potato S *Carlavirus*	PVS	60
Potato T *Capillovirus*	PVT	187
Potato V *Potyvirus* (aphid)	PVV	316
Potato X *Potexvirus*	PVX	4
Potato Y *Potyvirus* (aphid)	PVY	242
Potato yellow dwarf rhabdovirus (2)	PYDV	35
Prune dwarf *Ilarvirus*	PDV	19
Prunus necrotic ringspot *Ilarvirus*	PNRSV	5
Quail pea mosaic *Comovirus*	QPMV	238
Radish mosaic *Comovirus*	RaMV	121
Radish yellow edge *Cryptovirus* (?A)	RYEV	298
Raspberry bushy dwarf ? Ilarvirus	RBDV	165
Raspberry ringspot *Nepovirus*	RRSV	198
Raspberry vein chlorosis rhabdovirus (1)	RVCV	174
Red clover cryptic 1 *Cryptovirus* (A)	RCCV1	—
Red clover cryptic 2 *Cryptovirus* (B)	RCCV2	—
Red clover mottle *Comovirus*	RCMV	74
Red clover necrotic mosaic *Dianthovirus*	RCNMV	181
Red clover vein mosaic *Carlavirus*	RCVMV	22
Ribgrass mosaic *Tobamovirus*	RMV	152
Rice black-streaked dwarf *Fijivirus* (2)	RBSDV	135

(*continued*)

Table 1   (*Continued*)

Virus name	Acronym	A.A.B. Description No.
Rice dwarf *Phytoreovirus*	RDV	102
Rice gall dwarf *Phytoreovirus*	RGDV	296
Rice grassy stunt *Tenuivirus*	RGSV	320
Rice hoja blanca *Tenuivirus*	RHBV	299
Rice necrosis mosaic *Potyvirus* (fungus)	RNMV	172
Rice ragged stunt plant reovirus (no genus name)	RRSV	248
Rice stripe *Tenuivirus*	RSV	269
Rice transitory yellowing rhabdovirus (2)	RTYV	100
Rice tungro bacilliform—	RTBV	—
Rice tungro spherical ? MCDV group	RTSV	67
Rice yellow mottle *Sobemovirus*	RYMV	149
Robinia mosaic *Cucumovirus*	RbMV	65
Rubus yellow net—	RYNV	188
Ryegrass cryptic *Cryptovirus* (A)	RGCV	—
Ryegrass mosaic *Potyvirus* (mite)	RGMV	86
Saguaro cactus *Carmovirus*	SCV	148
Sammons' opuntia *Tobamovirus*	SOV	—
Satellite of TNV—	STNV	15
Satsuma dwarf ? *Nepovirus*	SDV	208
Scrophularia mottle *Tymovirus*	ScrMV	113
Shallot latent *Carlavirus*	SLV	250
Soilborne wheat mosaic *Furovirus*	SBWMV	77
Solanum nodiflorum mottle *Sobemovirus*	SNMV	318
Sonchus yellow net rhabdovirus (2)	SYNV	205
Southern bean mosaic *Sobemovirus*	SBMV	274
Sow thistle yellow vein rhabdovirus (2)	SYVV	62
Sowbane mosaic *Sobemovirus*	SoMV	64
Soybean chlorotic mottle *Caulimovirus*	SbCMV	331
Soybean dwarf *Luteovirus*	SbDV	179
Soybean mosaic *Potyvirus* (aphid)	SbMV	93
Spinach latent *Ilarvirus*	SpLV	281
Squash leaf curl *Geminivirus* III	SLCV	—
Squash mosaic *Comovirus*	SMV	43
Strawberry crinkle rhabdovirus (1)	SCrV	163
Strawberry latent ring spot ? *Nepovirus*	SLRSV	126
Strawberry veinbanding *Caulimovirus*	SVBV	219
Subterranean clover mottle *Sobemovirus*	SCMoV	329
Subterranean clover stunt—	SCSV	—
Sugarcane mosaic *Potyvirus* (aphid)	SCMV	88
Sunnhemp mosaic *Tobamovirus*	SHMV	152
Sweet clover necrotic mosaic *Dianthovirus*	SCNMV	321
Sweet potato feathery mottle *Potyvirus* (aphid)	SPFMV	—
Sweet potato mild mottle *Potyvirus* (whitefly)	SPMMV	162
Tephrosia symptomless ? *Carmovirus*	TeSV	256
Thistle mottle *Caulimovirus*	ThMoV	—
Tobacco etch *Potyvirus* (aphid)	TEV	258
Tobacco leaf curl *Geminivirus* III	TLCV	232
Tobacco mild green mosaic *Tobamovirus*	TMGMV	—
Tobacco mosaic *Tobamovirus*	TMV	151
Tobacco mottle—	TMoV	—
Tobacco necrosis *Necrovirus*	TNV	14
Tobacco necrotic dwarf *Luteovirus*	TNDV	234
Tobacco rattle *Tobravirus*	TRV	12

Virus name	Acronym	A.A.B. Description No.
Tobacco ringspot *Nepovirus*	TRSV	309
Tobacco streak *Ilarvirus*	TSV	307
Tobacco stunt—	TStV	313
Tobacco vein distorting ? *Luteovirus*	TVDV	—
Tobacco vein mottling *Potyvirus* (aphid)	TVMV	325
Tobacco yellow dwarf *Geminivirus* I	TYDV	278
Tobacco yellow vein—	TYVV	—
Tobacco yellow vein assistor ? *Luteovirus*	TYVAV	—
Tomato aspermy *Cucumovirus*	TAV	79
Tomato black ring *Nepovirus*	TBRV	38
Tomato bushy stunt *Tombusvirus*	TBSV	69
Tomato golden mosaic *Geminivirus* III	TGMV	303
Tomato mosaic *Tobamovirus*	ToMV	156
Tomato ringspot *Nepovirus*	ToRSV	290
Tomato spotted wilt *Tospovirus*	TSWV	39
Tomato yellow leaf curl *Geminivirus* III	TYLCV	—
Tulare apple mosaic *Ilarvirus*	TAMV	42
Tulip breaking *Potyvirus* (aphid)	TBV	71
Tulip X *Potexvirus*	TVX	276
Turnip crinkle *Carmovirus*	TCV	109
Turnip mosaic *Potyvirus* (aphid)	TuMV	8
Turnip rosette *Sobemovirus*	TRoV	125
Turnip yellow mosaic *Tymovirus*	TYMV	230
Ullucus C *Comovirus*	UVC	277
Ullucus mild mottle *Tobamovirus*	UMMV	—
Velvet tobacco mottle *Sobemovirus*	VTMoV	317
Vicia cryptic *Cryptovirus* (A)	VCV	—
Viola mottle *Potexvirus*	VMV	247
Voandzeia necrotic mosaic *Tymovirus*	VNMV	279
Watermelon mosaic virus 1 = Papaya ring-spot virus		
Watermelon mosaic virus 2 *Potyvirus* (aphid)	WMV2	291
Wheat dwarf *Geminivirus* I	WDV	—
Wheat spindle streak mosaic *Potyvirus* (fungus)	WSSMV	167
Wheat streak mosaic *Potyvirus* (mite)	WSMV	48
Wheat yellow leaf *Closterovirus*	WYLV	157
White clover cryptic 1 *Cryptovirus* (A)	WCCV1	—
White clover cryptic 2 *Cryptovirus* (B)	WCCV2	332
White clover cryptic 3 *Cryptovirus* (A)	WCCV3	—
White clover mosaic *Potexvirus*	WC1MV	41
Wild cucumber mosaic *Tymovirus*	WCMV	105
Wineberry latent *Potexvirus*	WLV	304
Wound tumor *Phytoreovirus*	WTV	34
Yam mosaic *Potyvirus* (aphid)	YMV	314
Yucca bacilliform—	YBV	—
Zucchini yellow fleck *Potyvirus* (aphid)	ZYFV	—
Zucchini yellow mosaic *Potyvirus* (aphid)	ZYMV	282

Table 2
Viroids

Viroid name	Acronym	A.A.B. Description No.
Apple scar skin	ASSVd	—
Avocado sun-blotch	ASBVd	254
Burdock stunt	BSVd	—
Carnation stunt-associated	CSAVd	—
Chrysanthemum chlorotic mottle	CCMVd	—
Chrysanthemum stunt	CSVd	—
Citrus cachexia	CCaVd	—
Citrus exocortis	CEVd	226
Coconut cadang-cadang	CCCVd	287
Coconut tinangaja	CTVd	—
Columnea latent	CLVd	—
Cucumber pale fruit	CPFVd	—
Grapevine 1B	GVd1B	—
Grapevine yellow speckle	GYSVd	—
Hop latent	HLVd	—
Hop stunt	HSVd	326
Potato spindle tuber	PSTVd	66
Tomato apical stunt	TASVd	—
Tomato bunchy top	TBTVd	—
Tomato planta macho	TPMVd	—

Bibliography

Aapola, A. I. E., and Rochow, W. F. (1971). Relationships among three isolates of barley yellow dwarf virus. *Virology* **46,** 127–141.

Abad, P., Ponchet, M., Ferrario, S., Poupet, A., and Cardin, L. (1986). Influence of light on the hypersensitive reaction of *Nicotiana tabacum* var Xanthi nc to tobacco mosaic virus; Absence of correlation between the b proteins induction and the viral resistance. *C. R. Seances Acad. Sci. Ser. 3* **302**(16), 603–608.

Abad-Zapatero, C., Abdel-Meguid, S. S., Johnson, J. E., Leslie, A. G. W., Rayment, I., Rossmann, M. G., Such, D., and Tsukihara, T. (1980). The structure of Southern bean mosaic virus at 2.8 Å resolution. *Nature (London)* **286,** 33–39.

Abdel-Meguid, S. S., Yamane, T., Fukuyama, K., and Rossmann, M. G. (1981). The location of calcium ions in southern bean mosaic virus. *Virology* **114,** 81–85.

Abe, H., and Tamada, T. (1986). Association of beet necrotic yellow vein virus with isolates of *Polymyxa betae* Keskin. *Ann. Phytopathol. Soc. Jpn.* **52,** 235–247.

AbouHaidar, M. (1981). Polar assembly of clover yellow mosaic virus. *J. Gen. Virol.* **57,** 199–203.

AbouHaidar, M. G. (1983). The structure of the 5′ and 3′ ends of clover yellow mosaic virus RNA. *Can. J. Microbiol.* **29,** 151–156.

AbouHaidar, M. G. (1988). Nucleotide sequence of the capsid protein gene and 3′ non-coding region of papaya mosaic virus RNA. *J. Gen. Virol.* **69,** 219–226.

AbouHaidar, M. G., and Bancroft, J. B. (1978). The initiation of papaya mosaic virus assembly. *Virology* **90,** 54–59.

AbouHaidar, M. G., and Erickson, J. W. (1985). Structure and *in vitro* assembly of papaya mosaic virus. *In* "Molecular Plant Virology" (J. W. Davies, ed.), Vol. 1, pp. 85–121. CRC Press, Boca Raton, Florida.

AbouHaidar, M. G., and Hirth, L. (1977). 5′-Terminal structure of tobacco rattle virus RNA. Evidence for polarity of reconstitution. *Virology* **76,** 173–185.

AbouHaidar, M. G., and Lai, R. (1989). Nucleotide sequence of the 3′-terminal region of clover yellow mosaic virus RNA. *J. Gen. Virol.* **70,** 1871–1875.

AbouHaidar, M. G., and Paliwal, Y. C. (1988). Comparison of the nucleotide sequences of viroid-like satellite RNA of the Canadian and Australasian strains of lucerne transient streak virus. *J. Gen. Virol.* **69,** 2369–2373.

AbouHaidar, M. G., Pfeiffer, P., Fritsch, C., and Hirth, L. (1973). Sequential reconstitution of tobacco rattle virus. *J. Gen. Virol.* **21,** 83–97.

A'Brook, J. (1964). The effect of planting date and spacing on the incidence of groundnut rosette disease and of the vector *Aphis craccivora,* Koch, at Mokawa, Northern Nigeria. *Ann. Appl. Biol.* **54,** 199–208.

A'Brook, J. (1968). The effect of plant spacing on the numbers of aphids trapped over the groundnut crop. *Ann. Appl. Biol.* **61,** 289–294.

A'Brook, J. (1973). The effect of plant spacing on the number of aphids trapped over cocksfoot and kale crops. *Ann. Appl. Biol.* **74,** 279–285.

Abu Salih, H. S., Ishag, H. M., and Siddig, S. A. (1973). Effect of sowing date on incidence of Sudanese broad bean mosaic virus in, and yield of, *Vicia faba. Ann. Appl. Biol.* **74**, 371–378.

Accotto, G. P., and Boccardo, G. (1986). The coat proteins and nucleic acids of two beet cryptic viruses. *J. Gen. Virol.* **67**, 363–366.

Accotto, G. P., Brisco, M. J., and Hull, R. (1987). *In vitro* translation of the double-stranded RNA genome from beet cryptic virus 1. *J. Gen. Virol.* **68**, 1417–1422.

Accotto, G. P., Donson, J., and Mullineaux, P. M. (1989). Mapping of *Digitaria* streak virus transcripts reveals different RNA species from the same transcription unit. *EMBO J.* **8**, 1033–1039.

Acs, G., Klett, H., Schonberg, M., Christman, J., Levin, D. H., and Silverstein, S. C. (1971). Mechanism of reovirus double-stranded ribonucleic acid synthesis *in vivo* and *in vitro. J. Virol.* **8**, 684–689.

Adam, G., and Hsu, H. T. (1984). Comparison of structural proteins from two potato yellow dwarf viruses. *J. Gen. Virol.* **65**, 991–994.

Adam, G., Sander, E., and Shepherd, R. J. (1979). Structural differences between pea enation mosaic virus strains affecting transmissibility by *Acyrthosiphon pisum* (Harris). *Virology* **92**, 1–14.

Adam, G., Gaedigk, K., and Mundry, K. W. (1983). Alterations of a plant rhabdovirus during successive mechanical transfers. *Z. Pflanzenkr. Pflanzenschutz* **90**, 28–35.

Adams, A. N., and Barbara, D. J. (1982). The use of F(ab′)₂-based ELISA to detect serological relationships among carlaviruses. *Ann. Appl. Biol.* **101**, 495–500.

Adams, D. B., and Kuhn, C. W. (1977). Seed transmission of peanut mottle virus in peanuts. *Phytopathology* **67**, 1126–1129.

Adams, M. J., Jones, P., and Swaby, A. G. (1987). The effect of cultivar used as host for *Polymyxa graminis* on the multiplication and transmission of barley yellow mosaic virus (BaYMV). *Ann. Appl. Biol.* **110**, 321–327.

Adams, M. J., Swaby, A. G., and Jones, P. (1988). Confirmation of the transmission of barley yellow mosaic virus (BaYMV) by the fungus *Polymyxa graminis. Ann. Appl. Biol.* **112**, 133–141.

Adams, R. L. P., Burdon, R. H., Campbell, A. M., Leader, D. P., and Smellie, R. M. S. (1981). "The Biochemistry of the Nucleic Acids," 9th ed. Chapman & Hall, London and New York.

Adams, S. E., Jones, R. A. C., and Coutts, R. H. A. (1986). Effect of temperature on potato virus X infection in potato cultivars carrying different combinations of hypersensitivity genes. *Plant Pathol.* **35**, 517–526.

Adanson, M. (1763). "Familles des Plantes," Vol. 1. Vincent, Paris.

Adejare, G. O., and Coutts, R. H. A. (1981). Eradication of cassava mosaic disease from Nigerian cassava clones by meristem-tip culture. *Plant Cell, Tissue Organ Cult.* **1**, 25–32.

Adolph, K. W., and Butler, P. J. G. (1977). Studies on the assembly of a spherical plant virus. III. Reassembly of infectious virus under mild conditions. *J. Mol. Biol.* **109**, 345–357.

Adomako, D., Lesemann, D. E., Paul, H. L., and Owusu, G. K. (1983). Improved methods for the purification and detection of cocoa swollen shoot virus. *Ann. Appl. Virol.* **103**, 109–116.

Agranovsky, A. A., Dolja, V. V., Kavsan, V. M., and Atabekov, J. G. (1978). Detection of polyadenylate sequences in RNA components of barley stripe mosaic virus. *Virology* **91**, 95–105.

Agranovsky, A. A., Dolja, V. V., Kagramanova, V. K., and Atabekov, J. G. (1979). The presence of a cap structure at the 5′ end of barley stripe mosaic virus RNA. *Virology* **95**, 208–210.

Agranovsky, A. A., Dolja, V. V., Gorbulev, V. G., Kozlov, Y. V., and Atabekov, J. G. (1981). Aminoacylation of barley stripe mosaic virus RNA: Polyadenylate-containing RNA has a 3′ terminal tyrosine-accepting structure. *Virology* **113**, 174–187.

Agranovsky, A. A., Dolja, V. V., and Atabekov, J. G. (1982). Structure of the 3′ extremity of barley stripe mosaic virus RNA: Evidence for internal poly (A) and a 3′-terminal tRNA-like structure. *Virology* **119**, 51–58.

Agranovsky, A. A., Dolja, V. V., and Atabekov, J. G. (1983). Differences in polyadenylate length between individual barley stripe mosaic virus RNA species. *Virology* **129**, 344–349.

Agrios, G. N., Walker, M. E., and Ferro, D. N. (1985). Effect of cucumber mosaic virus inoculation at successive weekly intervals on growth and yield of pepper *(Capsicum annuum)* plants. *Plant Dis.* **69**, 52–55.

Aharoni, N., Marco, S., and Levy, D. (1977). Involvement of gibberellins and abscisic acid in the suppression of hypocotyl elongation in CMV-infected cucumbers. *Physiol. Plant Pathol.* **11**, 189–194.

Ahlquist, P. (1987). Molecular biology and molecular genetics of plant bromoviruses. *In* "Plant Molecular Biology" (D. von Wettstein and N.-H. Chua, eds.), pp. 419–431. Plenum, New York.

Ahlquist, P., and Janda, M. (1984). cDNA cloning and *in vitro* transcription of the complete brome mosaic virus genome. *Mol. Cell. Biol.* **4**, 2876–2882.

Ahlquist, P., and Kaesberg, P. (1979). Determination of the length distribution of poly (A) at the 3′

terminus of the virion RNAs of EMC virus, poliovirus, rhinovirus, RAV-61 and CPMV and of mouse globin mRNA. *Nucleic Acids Res.* **7**, 1195–1204.

Ahlquist, P., Luckow, V., and Kaesberg, P. (1981a). Complete nucleotide sequence of brome mosaic virus RNA3. *J. Mol. Biol.* **153**, 23–38.

Ahlquist, P., Dasgupta, R., and Kaesberg, P. (1981b). Near identity of 3′ RNA secondary structure in bromoviruses and cucumber mosaic virus. *Cell (Cambridge, Mass.)* **23**, 183–189.

Ahlquist, P., Dasgupta, R., and Kaesberg, P. (1984a). Nucleotide sequence of brome mosaic virus genome and its implications for virus replication. *J. Mol. Biol.* **172**, 369–383.

Ahlquist, P., French, R., Janda, M., and Loesch-Fries, L. S. (1984b). Multicomponent RNA plant virus infection derived from cloned viral cDNA. *Proc. Natl. Acad. Sci. U.S.A.* **81**, 7066–7070.

Ahlquist, P., Bujarski, J., Kaesberg, P., and Hall, T. C. (1984c). Localisation of the replicase recognition site within brome mosaic virus RNA by hybrid-arrested RNA synthesis. *Plant Mol. Biol.* **3**, 37–44.

Ahlquist, P., Strauss, E. G., Rice, C. M., Strauss, J. H., Haseloff, J., and Zimmern, D. (1985). Sindbis virus proteins nsP1 and nsP2 contain homology to non-structural proteins from several RNA plant viruses. *J. Virol.* **53**, 536–542.

Ahlquist, P., French, R., and Bujarski, J. J. (1987). Molecular studies of brome mosaic virus using infectious transcripts from cloned cDNA. *Adv. Virus Res.* **32**, 215–242.

Ahlquist, P., Allison, R., Dejong, W., Janda, M., Kroner, P., Pacha, R., and Traynor, P. (1990). Molecular biology of bromovirus replication and host specificity. *In* "Viral Genes and Plant Pathogenesis" (T. P. Pirone and J. G. Shaw, eds.), pp. 144–155. New York, Springer-Verlag.

Ahmed, M. E., Black, L. M., Perkins, E. G., Walker, B. L., and Kummerow, F. A. (1964). Lipid in potato yellow dwarf virus. *Biochem. Biophys. Res. Commun.* **17**, 103–107.

Ajanyi, O., and Dewar, A. M. (1982). The effect of barley yellow dwarf virus on honeydew production by the cereal aphids, *Sitobion avenae* and *Metopolophum dirhodum*. *Ann. Appl. Biol.* **100**, 203–212.

Ajayni, O., and Dewar, A. M. (1983). The effects of barley yellow dwarf virus, aphids and honeydew on *Cladosporium* infection of winter wheat and barley. *Ann. Appl. Biol.* **102**, 57–65.

Ajtkhozhin, M. A., Doschanov, Kh. J., and Akhanov, A. U. (1976). Informosomes as a stored form of mRNA in wheat embryos. *FEBS Lett.* **66**, 124–126.

Al Ani, R., Pfeiffer, P., Lebeurier, G., and Hirth, L. (1979a). The structure of cauliflower mosaic virus. I. pH-induced structural changes. *Virology* **93**, 175–187.

Al Ani, R., Pfeiffer, P., and Lebeurier, G. (1979b). The structure of cauliflower mosaic virus. II. Identity and location of the viral polypeptides. *Virology* **93**, 188–197.

Alblas, F., and Bol, J. F. (1977). Factors influencing the infection of cowpea mesophyll protoplasts by alfalfa mosaic virus. *J. Gen. Virol.* **36**, 175–185.

Albrecht, H., Geldreich, A., Menissier de Murcia, J., Kirchherr, D., Mesnard, J.-M., and Lebeurier, G. (1988). Cauliflower mosaic virus gene 1 product detected in a cell-wall-enriched fraction. *Virology* **163**, 503–508.

Alconero, R. (1973). Mechanical transmission of viruses from sweet potato. *Phytopathology* **63**, 377–380.

Alconero, R., Weeden, N. F., Gonsalves, D., and Fox, D. T. (1985). Loss of genetic diversity in pea germplasm by the elimination of individuals infected by pea seedborne mosaic virus. *Ann. Appl. Biol.* **106**, 357–364.

Aldaoud, R., Dawson, W. O., and Jones, G. E. (1989). Rapid random evolution of the genetic structure of replicating tobacco mosaic virus populations. *Intervirology* **30**, 227–233.

Alhubaishi, A. A., Walkey, D. G. A., Webb, M. J. W., Bolland, C. J., and Cook, A. A. (1987). A survey of horticultural plant virus diseases in the Yemen Arab Republic. *FAO Plant Prot. Bull.* **35**, 135–143.

Allen, R. N., and Dale, J. L. (1981). Application of rapid biochemical methods for detecting avocado sunblotch disease. *Ann. Appl. Biol.* **98**, 451–461.

Allen, W. R. (1981). Dissemination of tobacco mosaic virus from soil to plant leaves under glasshouse conditions. *Can. J. Plant Pathol.* **3**, 163–168.

Allison, R. F., Sorenson, J. C., Kelly, M. E., Armstrong, F. B., and Dougherty, W. G. (1985a). Sequence determination of the capsid protein gene and flanking regions of tobacco etch virus: Evidence for synthesis and processing of a polyprotein in potyvirus genome expression. *Proc. Natl. Acad. Sci. U.S.A.* **82**, 3969–3972.

Allison, R. F., Dougherty, W. G., Parks, T. D., Willis, L., Johnston, R. E., Kelly, M., and Armstrong, F. B. (1985b). Biochemical analysis of the capsid protein gene and capsid protein of tobacco etch virus: N-terminal amino acids are located on the virion's surface. *Virology* **147**, 309–316.

Allison, R. F., Johnston, R. E., and Doughterty, W. G. (1986). The nucleotide sequence of the coding region of tobacco etch virus genomic RNA: Evidence for the synthesis of a single polyprotein. *Virology* **154**, 9–20.

Allison, R. F., Janda, M., and Ahlquist, P. (1988). Infectious *in vitro* transcripts from cowpea chlorotic mottle virus cDNA clones and exchange of individual RNA components with brome mosaic virus. *J. Virol.* **62**, 3581–3588.

Allison, R. F., Janda, M., and Ahlquist, P. (1989). Sequence of cowpea chlorotic mottle virus RNAs 2 and 3 and evidence of a recombination event during bromovirus evolution. *Virology* **172**, 321–330.

Allison, R. F., Thompson, C., and Ahlquist, P. (1990). Regeneration of a functional RNA virus genome by recombination between deletion mutants, and requirement for cowpea chlorotic mottle virus 3a and coat genes for systemic infection. *Proc. Natl. Acad. Sci. U.S.A.* **87**, 1820–1824.

Al Moudallal, Z., Briand, J. P., and Van Regenmortel, M. H. V. (1982). Monoclonal antibodies as probes of the antigenic structure of tobacco mosaic virus. *EMBO J.* **1**, 1005–1010.

Al Moudallal, Z., Altschuh, D., Briand, J. P., and Van Regenmortel, M. H. V. (1984). Comparative sensitivity of different ELISA procedures for detecting monoclonal antibodies. *J. Immunological Methods* **68**, 35–43.

Al Moudallal, Z., Briand, J. P., and Van Regenmortel, M. H. V. (1985). A major part of the polypeptide chain of tobacco mosaic virus protein is antigenic. *EMBO J.* **4**, 1231–1235.

Al Musa, A. M., and Mink, G. I. (1981). Beet necrotic yellow vein virus in North America. *Phytopathology* **71**, 773–776.

Alper, M., Salomon, R., and Loebenstein, G. (1984). Gel electrophoresis of virus-associated polypeptides for detecting viruses in bulbous irises. *Phytopathology* **74**, 960–962.

Altenbach, S. B., and Howell, S. H. (1981). Identification of a satellite RNA associated with turnip crinkle virus. *Virology* **112**, 25–33.

Altenbach, S. B., and Howell, S. H. (1984). Nucleic acid species related to the satellite RNA of turnip crinkle virus in turnip plants and virus particles. *Virology* **134**, 72–77.

Altschuh, D., Hartman, D., Reinbolt, J., and Van Regenmortel, M. H. V. (1983). Immunological studies of tobacco mosaic virus. V. Localization of four epitopes in the protein subunit by inhibition tests with synthetic peptides and cleavage peptides from three strains. *Mol. Immunol.* **20**, 271–278.

Altschuh, D., Al Moudallal, Z., Briand, J. P., and Van Regenmortel, M. H. V. (1985). Immunochemical studies of tobacco mosaic virus. VI. Attempts to localize viral epitopes with monoclonal antibodies. *Mol. Immunol.* **22**, 329–337.

Altschuh, D., Lesk, A. M., Bloomer, A. C., and Klug, A. (1987). Correlation of coordinated amino acid substitutions with function in viruses related to tobacco mosaic virus. *J. Mol. Biol.* **193**, 693–707.

Alvarez, M., and Campbell, R. N. (1976). Influence of squash mosaic virus on yield and quality of cantaloup. *Plant Dis. Rep.* **60**, 636–639.

Amin, P. W. (1985). Apparent resistance of groundnut cultivar Robut 33-1 to bud necrosis disease. *Plant Dis.* **69**, 718–719.

Anderer, F. A. (1959). Reversible Denaturierung des Proteins aus Takakmosaikvirus *Z. Naturforsch. B: Anorg. Chem., Org. Chem., Biochem., Biophys., Biol.* **14**, 642–647.

Anderer, F. A., Uhlig, H., Weber, E., and Schramm, G. (1960). Primary structure of the protein of tobacco mosaic virus. *Nature (London)* **186**, 922–925.

Anderson, M. T., Richardson, K. A., Harbison, S.-A., and Morris, B. A. M. (1988). Nucleotide sequence of the geminivirus chloris striate mosaic virus. *Virology* **164**, 443–449.

Andrews, J. H., and Shalla, T. A. (1974). The origin, development and conformation of amorphous inclusion body components in tobacco etch virus-infected cells. *Phytopathology* **64**, 1234–1243.

Anfinsen, C. B. (1973). Principles that govern the folding of protein chains. *Science* **181**, 223–230.

Angenent, G. C., Linthorst, H. J. M., van Belkum, A. F., Cornelissen, B. J. C., and Bol, J. F. (1986). RNA2 of tobacco rattle virus strain TCM encodes an unexpected gene. *Nucleic Acids Res.* **14**, 4673–4682.

Angenent, G. C., Verbeek, H. B. M., and Bol, J. F. (1989a). Expression of the 16K cistron of tobacco rattle virus in protoplasts. *Virology* **169**, 305–311.

Angenent, G. C., Posthumus, E., Brederode, F. T., and Bol, J. F. (1989b). Genome structure of tobacco rattle virus strain PLB: Further evidence on the occurrence of RNA recombination among tobraviruses. *Virology* **171**, 271–274.

Angenent, G. C., Posthumus, E., and Bol, J. F. (1989c). Biological activity of transcripts synthesised *in vitro* from full-length and mutated DNA copies of tobacco rattle virus RNA 2. *Virology* **173**, 68–76.

Angenent, G. C., van den Ouweland, J. M. W., and Bol, J. F. (1990). Susceptibility to virus infection of transgenic tobacco plants expression structural and non-structural genes of tobacco rattle virus. *Virology* **175**, 191–198.

Anjaneyulu, A., and Bhaktavatsalam, G. (1987). Evaluation of synthetic pyrethroid insecticides for tungro management. *Trop. Pest Manage.* **33**, 323–326.

Anno-Nyako, F. O., Vetten, H. J., Allen, D. J., and Thottappilly, G. (1983). The relation between cowpea golden mosaic and its vector *Bemisia tabaci*. Ann. Appl. Biol. **102**, 319–323.

Antignus, Y., Sela, I., and Harpaz, I. (1971). Species of RNA extracted from tobacco and *Datura* plants and their differential sensitivity to actinomycin D. *Biochem. Biophys. Res. Commun.* **44**, 78–88.

Antoniw, J. F., and White, R. F. (1986). Changes with time in the distribution of virus and PR protein around single local lesions of TMV-infected tobacco. *Plant Mol. Biol.* **6**, 145–149.

Antoniw, J. F., Ritter, C. E., Pierpoint, W. S., and van Loon, L. C. (1980). Comparison of three pathogenesis-related proteins from plants of two cultivars of tobacco infected with TMV. *J. Gen. Virol.* **47**, 79–87.

Antoniw, J. F., Ooms, G., White, R. F., Wullems, G. J., and van Vloten-Doting, L. (1983). Pathogenesis related proteins in plants and tissues of *Nicotiana tabacum* transformed by *Agrobacterium tumefaciens*. *Plant Mol. Biol.* **2**, 317–320.

Anton-Lamprecht, I. (1966). Beitrage zum Problem der Plastidenabänderung. III. Uber das Vorkommen von "Rückmutationen" in einer spontan entstandenen Plastidenschecke von *Epilobium hirsutum*. *Z. Pflanzenphysiol.* **54**, 417–445.

Anzola, J. V., Xu, Z., Asamizu, T., and Nuss, D. L. (1987). Segment-specific inverted repeats found adjacent to conserved terminal sequences in wound tumor virus genome and defective interfering RNAs. *Proc. Natl. Acad. Sci. U.S.A.* **84**, 8301–8305.

Anzola, J. V., Dall, D. J., Xu, Z., and Nuss, D. L. (1989a). Complete nucleotide sequence of wound tumor virus genomic segments encoding nonstructural peptides. *Virology* **171**, 222–228.

Anzola, J. V., Xu, Z., and Nuss, D. L. (1989b). The complete nucleotide sequence of wound tumor virus genomic segment S7. *Nucleic Acids Res.* **17**, 3300.

Aoki, S., and Takebe, I. (1969). Infection of tobacco mesophyll protoplasts by tobacco mosaic virus ribonucleic acid. *Virology* **39**, 439–448.

Appiano, A., and D'Agostino, G. (1983). Distribution of tomato bushy stunt virus in root tips of systemically infected *Gomphrena globosa*. *J. Ultrastruct. Res.* **85**, 239–248.

Appiano, A., and Pennazio, S. (1972). Electron microscopy of potato meristem tips infected with potato virus X. *J. Gen. Virol.* **14**, 273–276.

Appiano, A., Pennazio, S., D'Agostino, G., and Redolfi, P. (1977). Fine structure of necrotic local lesions induced by tomato bushy stunt virus in *Gomphrena globosa* leaves. *Physiol. Plant Pathol.* **11**, 327–332.

Appiano, A., D'Agostino, G., Bassi, M., Barbieri, N., Viale, G., and Dell'Orto, P. (1986). Origin and function of tomato bushy stunt virus-induced inclusion bodies. II. Quantitative electron microscope autoradiography and immunogold cytochemistry. *J. Ultrastruct. Mol. Struct. Res.* **97**, 31–38.

Araya, J. E., and Foster, J. E. (1987). Laboratory study on the effects of barley yellow dwarf virus on the life cycle of *Rhopalosiphum padi* (L). *J. Plant Dis. Prot.* **94**, 578–583.

Argos, P. (1981). Secondary structure prediction of plant virus coat proteins. *Virology* **110**, 55–62.

Argos, P. (1988). A sequence motif in many polymerases. *Nucleic Acids Res.* **16**, 9909–9916.

Argos, P., Kamer, G., Nicklin, M. J. H., and Wimmer, E. (1984). Similarity in gene organization and homology between proteins of animal picornaviruses and a plant comovirus suggest common ancestry of these virus families. *Nucleic Acids Res.* **12**, 7251–7267.

Armour, S. L., Melcher, U., Pirone, T. P., Lyttle, D. J., and Essenberg, R. C. (1983). Helper component for aphid transmission encoded by region II of cauliflower mosaic virus DNA. *Virology* **129**, 25–30.

Asamizu, T., Summers, D., Motika, M. B., Anzola, J. V., and Nuss, D. L. (1985). Molecular cloning and characterization of the genome of wound tumour virus: A tumor-inducing plant reovirus. *Virology* **144**, 398–409.

Ashby, J. W., Teh, P. B., and Close, R. C. (1979). Symptomatology of subterranean clover red leaf virus and its incidence in some legume crops, weed hosts and certain alate aphids in Canterbury, New Zealand. *N.Z. J. Agric. Res.* **22**, 361–365.

Asjes, C. J. (1978). Minerale oliën op lelies om virusverspreiding tegen te gaan. *Bloembollencultuur* **88**, 1046–1047.

Asjes, C. J. (1980). Toepassing van minerale olie om verspreiding van hyacintemozaiekvirus in hyacinten tegen te gaan. *Bloembollencultuur* **90**, 1396–1397.

Asjes, C. J., de Vos, P., and von Slogteren, D. H. M. (1973). Brown ring formation and streak mottle, two distinct syndromes in Lilies associated with complex infections of lily symptomless virus and tulip breaking virus. *Neth. J. Plant Pathol.* **79**, 23–35.

Asselin, A., and Zaitlin, M. (1978). An anomalous form of tobacco mosaic virus RNA observed upon polyacrylamide gel electrophoresis. *Virology* **88**, 191–193.

Atabekov, J. G. (1975). Host specificity of plant viruses. *Annu. Rev. Phytopathol.* **13**, 127–145.

Atabekov, J. G., and Dorokhov, Yu. L. (1984). Plant virus-specific transport function and resistance of plants to viruses. *Adv. Virus Res.* **29**, 313–364.

Atabekov, J. G., and Taliansky, M. E. (1990). Expression of a plant virus-coded transport function by different viral genomes. *Adv. Virus Res.* **38**, 201–248.

Atabekov, J. G., Novikov, V. K., Vishnichenko, V. K., and Javakhia, V. G. (1970a). A study of the mechanisms controlling the host range of plant viruses. II. The host range of hybrid viruses reconstituted *in vitro,* and of free viral RNA. *Virology* **41**, 108–115.

Atabekov, J. G., Schaskolskaya, N. D., Atabekova, T. I., and Sacharovskaya, G. A. (1970b). Reproduction of temperature-sensitive strains of TMV under restrictive conditions in the presence of temperature-resistant helper strain. *Virology* **41**, 397–407.

Atabekov, J. G., Dorokhov, Yu. L., and Taliansky, M. E. (1983). A virus-coded function responsible for the transport of virus genome from infected to healthy cells. *Tag. ungs ber.—Akad. Landwirtschafts wiss. D.D.R.* **216**, 53–58.

Atabekova, T. I., Taliansky, M. E., and Atabekov, J. G. (1975). Specificity of protein–RNA and protein–protein interaction upon assembly of TMV *in vivo* and *in vitro. Virology* **67**, 1–13.

Atchison, B. A. (1973). Division, expansion and DNA synthesis in meristematic cells of French bean (*Phaseolus vulgaris* L.) root-tips invaded by tobacco ringspot virus. *Physiol. Plant Pathol.* **3**, 1–8.

Atchison, B. A., and Francki, R. I. B. (1972). The source of tobacco ringspot virus in root-tip tissue of bean plants. *Physiol. Plant Pathol.* **2**, 105–111.

Athow, K. L., and Bancroft, J. B. (1959). Development and transmission of tobacco ringspot virus in soybean. *Phytopathology* **49**, 697–701.

Athow, K. L., and Laviolette, F. A. (1962). Relation of seed position and pod location to tobacco ringspot virus seed transmission in soybean. *Phytopathology* **52**, 714–715.

Atiri, G. I. (1984). The occurrence and importance of okra mosaic virus in Nigerian weeds. *Ann. Appl. Biol.* **104**, 261–265.

Atiri, G. I., Ekpo, E. J. A., and Thottappilly, G. (1984). The effect of aphid-resistance in cowpea on infestation and development of *Aphis craccivora* and the transmission of cowpea aphid-borne mosaic virus. *Ann. Appl. Biol.* **104**, 339–346.

Atkinson, P. H., and Matthews, R. E. F. (1970). On the origin of dark green tissue in tobacco leaves infected with tobacco mosaic virus. *Virology* **40**, 344–356.

Atkinson, T. G., and Grant, M. N. (1967). An evaluation of streak mosaic losses in winter wheat. *Phytopathology* **57**, 188–192.

Atkinson, T. G., and Slykhuis, J. T. (1963). Relation of spring drought, summer rains, and high fall temperatures to the wheat streak mosaic epiphytotic in southern Alberta, 1963. *Can. Plant Dis. Surv.* **43**, 154–159.

Atreya, C. D., and Siegel, A. (1989). Localization of multiple TMV encapsidation initiation sites in rbcL gene transcripts. *Virology* **168**, 388–392.

Austin, R. B., Ford, M. A., Edrich, J. A., and Hooper, B. E. (1976). Some effects of leaf posture on photosynthesis and yield in wheat. *Ann. Appl. Biol.* **83**, 425–446.

Ausubel, F. M., Brent, R., Kingston, R. E., Moore, D. D., Seidman, J. G., Smith, J. A., and Struhl, K., eds. (1987). "Current Protocols in Molecular Biology," Vol. 1. Wiley, New York.

Ausubel, F. M., Brent, R., Kingston, R. E., Moore, D. D., Seidman, J. G., Smith, J. A., and Struhl, K., eds. (1988). "Current Protocols in Molecular Biology," Vol. 2. Wiley, New York.

Avila-Rincon, M. J., Collmer, C. W., and Kaper, J. M. (1986a). *In vitro* translation of cucumoviral satellites. I. Purification and nucleotide sequence of cucumber mosaic virus-associated RNA5 from cucumber mosaic virus strain S. *Virology* **152**, 446–454.

Avila-Rincon, M. J., Collmer, C. W., and Kaper, J. M. (1986b). *In vitro* translation of cucumovirus satellites. II. CARNA5 from cucumber mosaic virus strain S and SP6 transcripts of cloned (S)CARNA5 cDNA produce electrophoretically comigrating protein products. *Virology* **152**, 455–458.

Baas, P. D., and Jansz, H. S. (1978). Replication of φX174 RF DNA *in vivo. In* "The Single-Stranded DNA Phages" (D. T. Denhardt, D. Dressler, and D. S. Ray, eds.), pp. 215–244. Cold Spring Harbor Lab., Cold Spring Harbor, New York.

Babos, P., and Shearer, G. B. (1969). RNA synthesis in tobacco leaves infected with tobacco mosaic virus. *Virology* **39**, 286–295.

Backus, R. C., and Williams, R. C. (1950). Use of spraying methods and of volatile suspending media in the preparation of specimens for electron microscopy. *J. Appl. Phys.* **21**, 11–15.

Bagnall, R. H. (1988). Epidemics of potato leaf roll in North America and Europe linked to drought and sunspot cycles. *Can. J. Plant. Pathol.* **10**, 193–202.

Baker, P. F. (1960). Aphid behaviour on healthy and on yellow virus-infected sugar beet. *Ann. Appl. Biol.* **48**, 384–391.

Balázs, E., Gáborjányi, R., and Király, Z. (1973). Leaf senescence and increased virus susceptibility in tobacco. The effect of abscisic acid. *Physiol. Plant Pathol.* **3**, 341–346.

Balázs, E., Sziráki, I., and Király, Z. (1977). The role of cytokinins in the systemic acquired resistance of tobacco hypersensitive to tobacco mosaic virus. *Physiol. Plant Pathol.* **11**, 29–37.

Balázs, E., Guilley, H., Jonard, G., and Richards, K. (1982). Nucleotide sequence of DNA from an altered-virulence isolate D/H of the cauliflower mosaic virus. *Gene* **19**, 239–249.

Balázs, E., Bouzoubaa, S., Guilley, H., Jonard, G., Paszkowski, J., and Richards, K. (1985). Chimeric vector construction for higher plant transformation. *Gene* **40**, 343–348.

Bald, J. G., and Tinsley, T. W. (1967). A quasi-genetic model for plant virus host ranges. III. Congruence and relatedness. *Virology* **32**, 328–336.

Bald, J. G., and Tinsley, T. W. (1970). A quasi-genetic model for plant virus host ranges. IV. Cacao swollen-shoot and mottle-leaf viruses. *Virology* **40**, 369–378.

Balint, R., and Cohen, S. S. (1985a). The incorporation of radiolabelled polyamines and methionine into turnip yellow mosaic virus in protoplasts from infected plants. *Virology* **144**, 181–193.

Balint, R., and Cohen, S. S. (1985b). The effects of dicyclohexylamine on polyamine biosynthesis and incorporation into turnip yellow mosaic virus in chinese cabbage protoplasts infected *in vitro*. *Virology* **144**, 194–203.

Ball, E. M. (1973). Solid phase radioimmunoassay for plant viruses. *Virology* **55**, 516–520.

Baltimore, D. (1985). Retroviruses and retrotransposons: The role of reverse transcription in shaping the eukaryotic genome. *Cell (Cambridge, Mass.)* **40**, 481–482.

Bancroft, J. B. (1972). A virus made from parts of the genomes of brome mosaic and cowpea chlorotic mottle viruses. *J. Gen. Virol.* **14**, 223–228.

Bancroft, J. B., and Horne, R. W. (1977). Bromoviruses (brome mosaic virus) group. *In* "The Atlas of Insect and Plant Viruses" (K. Maramorosch, ed.), pp. 287–302. Academic Press, New York.

Bancroft, J. B., and Pound, G. S. (1954). Effect of air temperature on the multiplication of tobacco mosaic virus in tobacco. *Phytopathology* **44**, 481–482.

Bancroft, J. B., Hills, G. J., and Markham, R. (1967). A study of the self-assembly process in a small spherical virus. Formation of organized structures from protein subunits *in vitro*. *Virology* **31**, 354–379.

Bancroft, J. B., McLean, G. D., Rees, M. W., and Short, M. N. (1971). The effect of an arginyl to a cysteinyl replacement on the uncoating behaviour of a spherical plant virus. *Virology* **45**, 707–715.

Banttari, E. E., and Goodwin, P. H. (1985). Detection of potato viruses S, X, and Y by enzyme-linked immunosorbent assay on nitrocellulose membranes (Dot-ELISA). *Plant Dis.* **69**, 202–205.

Banttari, E. E., and Zeyen, R. J. (1969). Chromatographic purification of the oat blue dwarf virus. *Phytopathology* **59**, 183–186.

Banttari, E. E., and Zeyen, R. J. (1972). Ultrastructure of flax with a simultaneous virus and mycoplasma-like infection. *Virology* **49**, 305–308.

Barbara, D. J., and Clark, M. F. (1982). A simple indirect ELISA using F(ab′)₂ fragments of immunoglobulin. *J. Gen. Virol.* **58**, 315–322.

Barbara, D. J., Kawata, E. E., Ueng, P. P., Lister, R. M., and Larkins, B. A. (1987). Production of cDNA clones from the MAV isolate of barley yellow dwarf virus. *J. Gen. Virol.* **68**, 2419–2427.

Barbara, D. J., Morton, A., and Adams, A. N. (1990). Assessment of U.K. hops for the occurrence of hop latent and hop stunt viroids. *Ann. Appl. Biol.* **116**, 265–272.

Bar-Joseph, M., Segev, D., Blickle, V., Yesodi, V., Franck, A., and Rosner, A. (1986). Application of synthetic DNA probes for the detection of viroids and viruses. *In* "Developments and Applications in Virus Testing" (R. A. C. Jones and L. Torrance, eds.), Dev. Appl. Biol. I, pp. 13–23. Association of Applied Biologists, Wellsbourne.

Barker, H. (1987). Invasion of non-phloem tissue in *Nicotiana clevelandii* by potato leafroll luteovirus is enhanced in plants also infected with potato virus Y potyvirus. *J. Gen. Virol.* **68**, 1223–1227.

Barker, H., and Harrison, B. D. (1978). Double infection, interference and superinfection in protoplasts exposed to two strains of raspberry ringspot virus. *J. Gen. Virol.* **40**, 647–658.

Barker, H., and Harrison, B. D. (1986). Restricted distribution of potato leafroll virus antigen in resistant potato genotypes and its effect on transmission of the virus by aphids. *Ann. Appl. Biol.* **109**, 595–604.

Barker, H., and Woodford, J. A. T. (1987). Unusually mild symptoms of potato leafroll virus in the progeny of late-infected mother plants. *Potato Res.* **30**, 345–348.

Barker, R. F., Jarvis, N. P., Thompson, D. V., Loesch-Fries, L. S., and Hall, T. C. (1983). Complete nucleotide sequence of alfalfa mosaic virus RNA3. *Nucleic Acids Res.* **11**, 2881–2891.

Barlow, D. J., Edwards, M. S., and Thornton, J. M. (1986). Continuous and discontinuous protein antigenic determinants. *Nature (London)* **322**, 747–752.

Barnett, O. W., and Fulton, R. W. (1971). Differential response of *Prunus* necrotic ringspot and tulare apple mosaic viruses to stabilizing agents. *Virology* **46**, 613–619.

Bartels, R. (1954). Serologische Untersuchungen über das Verhalten des Kartoffel-A-virus in Tabakpflanzen. *Phytopathol. Z.* **21**, 395–406.

Bassi, M., and Favali, M. A. (1972). Electron microscopy of maize rough dwarf virus assembly sites in maize. Cytochemical and autoradiographic observations. *J. Gen. Virol.* **16**, 153–160.

Bassi, M., Favali, M. A., and Conti, G. G. (1974). Cell wall protrusions induced by cauliflower mosaic virus in chinese cabbage leaves: A cytochemical and autoradiographic study. *Virology* **60**, 353–358.

Bassi, M., Barbieri, N., Appiano, A., Conti, M., D'Agostino, G., and Caciagli, P. (1980). Cytochemical and autoradiographic studies on the genome and site(s) of replication of barley yellow striate mosaic virus in barley plants. *J. Submicrosc. Cytol.* **12**, 201–207.

Bassi, M., Appiano, A., Barbieri, N., and D'Agostino, G. (1985). Chloroplast alterations induced by tomato bushy stunt virus in *Datura* leaves. *Protoplasma* **126**, 233–235.

Bassi, M., Barbieri, N., Appiano, A., and D'Agostino, G. (1986). Origin and function of tomato bushy stunt virus-induced inclusion bodies. I. Differential staining, cytochemistry and serial sectioning. *J. Ultrastruct. Mol. Struct. Res.* **96**, 194–203.

Bastin, M., and Kaesberg, P. (1975). Radioactive labelling of brome mosaic virus. *J. Gen. Virol.* **26**, 321–325.

Bath, J. E., and Tsai, J. H. (1969). The use of aphids to separate two strains of pea enation mosaic virus. *Phytopathology* **59**, 1377–1380.

Baughman, G. A., and Howell, S. H. (1988). Cauliflower mosaic virus 35S RNA leader region inhibits translation of downstream genes. *Virology* **167**, 125–135.

Baughman, G. A., Jacobs, J. D., and Howell, S. H. (1988). Cauliflower mosaic virus gene VI produces a symptomatic phenotype in transgenic tobacco plants. *Proc. Natl. Acad. Sci. U.S.A.* **85**, 733–737.

Baulcombe, D. C., Saunders, G. R., Bevan, M. W., Mayo, M. A., and Harrison, B. D. (1986). Expression of biologically active viral satellite RNA from the nuclear genome of transformed plants. *Nature (London)* **321**, 446–449.

Baur, E. (1904). Zur Aetiologie der infectiösen Panachierung. *Ber. Dtsch. Bot. Ges.* **22**, 453–460.

Bawden, F. C. (1964). "Plant Viruses and Virus Diseases," 4th ed. Ronald Press, New York.

Bawden, F. C., and Kassanis, B. (1945). The suppression of one plant virus by another. *Ann. Appl. Biol.* **32**, 52–57.

Bawden, F. C., and Kassanis, B. (1950). Some effects of host nutrition on the susceptibility of plants to infection by certain viruses. *Ann. Appl. Biol.* **37**, 46–57.

Bawden, F. C., and Pirie, N. W. (1936). Experiments on the chemical behaviour of potato virus X. *Br. J. Exp. Pathol.* **17**, 64–74.

Bawden, F. C., and Pirie, N. W. (1937). The isolation and some properties of liquid crystalline substances from solanaceous plants infected with three strains of tobacco mosaic virus. *Proc. R. Soc. London, Ser. B* **123**, 274–320.

Bawden, F. C., and Pirie, N. W. (1956). Observations on the anomalous proteins occurring in extracts of plants infected with strains of tobacco mosaic virus. *J. Gen. Microbiol.* **14**, 460–477.

Bawden, F. C., and Roberts, F. M. (1947). The influence of light intensity on the susceptibility of plants to certain viruses. *Ann. Appl. Biol.* **34**, 286–296.

Bawden, F. C., and Sheffield, F. M. L. (1939). The intracellular inclusions of some plant virus diseases. *Ann. Appl. Biol.* **26**, 102–115.

Bawden, F. C., Pirie, N. W., Bernal, J. D., and Fankuchen, I. (1936). Liquid crystalline substances from virus-infected plants. *Nature (London)* **138**, 1051–1052.

Bayer, M. E., and Bocharov, A. F. (1973). The capsid structure of bacteriophage lambda. *Virology* **54**, 465–475.

Beachy, R. N. (1988). Virus cross-protection in transgenic plants. *In* "Temporal and Spatial Regulation of Plant Genes" (D. P. S. Verma and R. B. Goldberg, eds.), pp. 313–331. Springer-Verlag, New York.

Beachy, R. N., and Zaitlin, M. (1975). Replication of tobacco mosaic virus. VI. Replicative intermediate and TMV-RNA-related RNAs associated with polyribosomes. *Virology* **63**, 84–97.

Beachy, R. N., and Zaitlin, M. (1977). Characterization and *in vitro* translation of the RNAs from less-than-full-length, virus-related, nucleoprotein rods present in tobacco mosaic virus preparations. *Virology* **81**, 160–169.

Beachy, R. N., Abel, P., Oliver, M. J., De, B., Fraley, R. T., Rogers, S. G., and Horsch, R. B. (1986). Potential for applying genetic information to studies of virus pathogenesis and cross-protection. *In* "Biotechnology in Plant Science: Relevance to Agriculture in the Eighties" (M. Zaitlin, P. Day, and A. Hollaender, eds.), pp. 265–275. Academic Press, Orlando, Florida.

Beale, H. P. (1928). Immunologic reactions with tobacco mosaic virus. *Proc. Soc. Exp. Biol. Med.* **25**, 602.

Bedbrook, J. R. (1972). The effects of TYMV infection on photosynthetic carbon metabolism. M.Sc. Thesis, University of Auckland, Auckland, New Zealand.

Bedbrook, J. R., and Matthews, R. E. F. (1973). Changes in the flow of early products of photosynthetic carbon fixation associated with the replication of TYMV. *Virology* **53**, 84–91.

Bedbrook, J. R., and Matthews, R. E. F. (1976). Location, rate and assemmetry of ds-RNA synthesis during replication of TYMV in Chinese cabbage. *Ann. Microbiol. (Paris)* **127A,** 55–60.

Bedbrook, J. R., Douglas, J., and Matthews, R. E. F. (1974). Evidence for TYMV-induced RNA and DNA synthesis in the nuclear fraction from infected Chinese cabbage leaves. *Virology* **58,** 334–344.

Beemster, A. B. R. (1966). The rate of infection of potato tubers with potato virus Y^N in relation to position of the inoculated leaf. *In* "Viruses of Plants" (A. B. R. Beemster and J. Dijkstra, eds.), pp. 44–47. North-Holland Publ., Amsterdam.

Beemster, A. B. R. (1967). Partial infection with potato virus Y^N of tubers from primarily infected potato plants. *Neth. J. Plant Pathol.* **73,** 161–164.

Beemster, A. B. R. (1979). Acquisition of potato virus Y^N by *Myzus persicae* from primarily infected 'Bintje' potato plants. *Neth. J. Plant Pathol.* **85,** 75–81.

Beer, S. V., and Kosuge, T. (1970). Spermidine and spermine-polyamine components of turnip yellow mosaic virus. *Virology* **40,** 930–938.

Beier, H., and Bruening, G. (1975). The use of an abrasive in the isolation of cowpea leaf protoplasts which support the multiplication of cowpea mosaic virus. *Virology* **64,** 272–276.

Beier, H., and Bruening, G. (1976). Factors influencing the infection of cowpea protoplasts by cowpea mosaic virus RNA. *Virology* **72,** 363–369.

Beier, H., Siler, D. J., Russell, M. L., and Bruening, G. (1977). Survey of susceptibility to cowpea mosaic virus among protoplasts and intact plants from *Vigna sinensis* lines. *Phytopathology* **67,** 917–921.

Beier, H., Mundry, K. W., and Issinger, O.-G. (1980). *In vivo* and *in vitro* translation of the RNAs of four tobamoviruses. *Intervirology* **14,** 292–299.

Beier, H., Barciszewska, M., Krupp, G., Mitnacht, R., and Gross, H. J. (1984a). UAG readthrough during TMV RNA translation: Isolation and sequence of two tRNAs^{tyr} with suppressor activity from tobacco plants. *EMBO J.* **3,** 351–356.

Beier, H., Barciszewska, M., and Sickinger, H.-D. (1984b). The molecular basis for the differential translation of TMV RNA in tobacco protoplasts and wheat germ extracts. *EMBO J.* **3,** 1091–1096.

Beijerinck, M. W. (1898). Over een contagium vivum fluidum als oorzaak van de vlekziekte der tabaksbladen. *Versl. Gewone Vergad. Wis- Natuurkd. Afd., K. Akad. Wet. Amsterdam* **7,** 229–235.

Bell, C. D., Omar, S. A., and Lee, P. E. (1978). Electron microscopic localization of wheat striate mosaic virus in its leafhopper vector, *Endria inimica*. *Virology* **86,** 1–9.

Benda, G. T. A., and Bennett, C. W. (1964). Effect of curly top virus on tobacco seedlings; Infection without obvious symptoms. *Virology* **24,** 97–101.

Bendena, W. G., and Mackie, G. A. (1986). Translational strategies in potexviruses: Products encoded by clover yellow mosaic virus, foxtail mosaic virus, and viola mottle virus RNAs *in vitro*. *Virology* **153,** 220–229.

Bendena, W. G., and Mackie, G. A. (1987). Lack of homology among potexvirus RNAs. *Intervirology* **27,** 112–116.

Bendena, W. G., Bancroft, J. B., and Mackie, G. A. (1987). Molecular cloning of clover yellow mosaic virus RNA: Identification of coat protein coding sequences *in vivo* and *in vitro*. *Virology* **157,** 276–284.

Benfey, P. N., and Chua, N.-H. (1989). Regulated genes in transgenic plants. *Science* **244,** 174–181.

Benfey, P. N., Ren, L., and Chua, N.-H. (1989). The CaMV 35S enhancer contains at least two domains which can confer different developmental and tissue-specific expression patterns. *EMBO J.* **8,** 2195–2202.

Benfey, P. N., Ren, L., and Chua, N.-H. (1990a). Tissue-specific expression from CaMV 35S enhancer subdomains in early stages of plant development. *EMBO J.* **9,** 1677–1684.

Benfey, P. N., Ren, L., and Chua, N.-H. (1990b). Combinatorial and synergistic properties of CaMV 35S enhancer domains. *EMBO J.* **9,** 1685–1696.

Benner, S. A., Ellington, A. D., and Tauer, A. (1989). Modern metabolism as a palimpsest of the RNA world. *Proc. Natl. Acad. Sci. U.S.A.* **86,** 7054–7058.

Bennett, C. W. (1934). Plant tissue relations of sugar beet curly top virus. *J. Agric. Res.* **48,** 665–701.

Bennett, C. W. (1940a). Relation of food translocation to movement of virus of tobacco mosaic. *J. Agric. Res.* **60,** 361–390.

Bennett, C. W. (1940b). Acquisition and transmission of viruses by dodder *(Cuscuta subinclusa)*. *Phytopathology* **30,** 2 (abstr.).

Bennett, C. W. (1944). Latent virus of dodder and its effect on sugar beet plants. *Phytopathology* **34,** 77–91.

Bennett, C. W. (1955). Recovery of water pimpernel from curly top and the reaction of recovered plants to reinoculation with different virus strains. *Phytopathology* **45,** 531–536.

Bennett, C. W. (1960). Sugar beet yellows disease in the United States. *U.S., Dep. Agric., Tech. Bull.* **1218**, 1–63.

Bennett, C. W. (1963). Highly virulent strains of curly top virus in sugar beet in western United States. *J. Am. Soc. Sugar Beet Technol.* **12**, 515–520.

Ben-Sin, C., and Po, T. (1982). Infection of barley protoplasts with barley stripe mosaic virus detected and assayed by immunoperoxidase. *J. Gen. Virol.* **58**, 323–327.

Bentley, G. A., Lewit-Bentley, A., Liljas, L., Skoglund, U., Roth, M., and Unge, T. (1987). Structure of RNA in satellite tobacco necrosis virus. A low resolution neutron diffraction study using $H_2O/^2H_2O$ solvent contrast variation. *J. Mol. Biol.* **194**, 129–141.

Berger, P. H., and Pirone, T. P. (1986). The effect of helper component on the uptake and localization of potyviruses in *Myzus persicae*. *Virology* **153**, 256–261.

Berger, P. H., and Toler, R. W. (1983). Quantitative immunoelectrophoresis of panicum mosaic virus and strains of St. Augustine decline. *Phytopathology* **73**, 185–189.

Berger, P. H., Thornbury, D. W., and Pirone, T. P. (1985). Detection of picogram quantities of potyviruses using a dot blot immunobinding assay. *J. Virol. Methods* **12**, 31–39.

Berger, P. H., Hunt, A. G., Domier, L. L., Hellman, G. M., Stram, Y., Thornbury, D. W., and Pirone, T. P. (1989). Expression in transgenic plants of a viral gene product that mediates insect transmission of potyviruses. *Proc. Natl. Acad. Sci. U.S.A.* **86**, 8402–8496.

Bergeson, G. B., Athow, K. L., Laviolette, F. A., and Thomasine, M. (1964). Transmission, movement, and vector relationships of tobacco ringspot virus in soybean. *Phytopathology* **54**, 723–726.

Bergh, O., Børsheim, K. Y., Bratbak, G., and Heldal, M. (1989). High abundance of viruses found in aquatic environments. *Nature (London)* **340**, 467–468.

Bergh, S. T., and Siegel, A. (1989). Intraviral homology and subgenomic RNAs of pepper ringspot virus. *Virology* **168**, 339–343.

Bergh, S. T., Koziel, M. G., Huang, S.-C., Thomas, R. A., Gilley, D. P., and Siegel, A. (1985). The nucleotide sequence of tobacco rattle virus RNA-2 (CAM strain). *Nucleic Acids Res.* **13**, 8507–8518.

Bergström, G. C., Johnson, M. C., and Kuć, J. (1982). Effects of local infection of cucumber by *Colletotrichum lagenarium*, *Pseudomonas lachrymans* or tobacco necrosis virus on systemic resistance to cucumber mosaic virus. *Phytopathology* **72**, 922–926.

Berlinger, M. J., and Dahan, R. (1987). Breeding for resistance to virus transmission by whiteflies in tomatoes. *Insect Sci. Appl.* **8**, 783–784.

Berna, A., Briand, J.-P., Stussi-Garaud, C., and Godefroy-Colburn, T. (1986). Kinetics of accumulation of the three non-structural proteins of alfalfa mosaic virus in tobacco plants. *J. Gen. Virol.* **67**, 1135–1147.

Bernal, J. D., and Fankuchen, I. (1937). Structure types of protein 'crystals' from virus-infected plants. *Nature (London)* **139**, 923–924.

Bernardi, G. (1971). Chromatography of nucleic acids on hydroxyapatite columns. *In* "Methods in Enzymology" (L. Grossman and K. Moldave, eds.), Vol. 21, Part D, pp. 95–139. Academic Press, New York.

Berzofsky, J. A. (1985). Intrinsic and extrinsic factors in protein antigenic structure. *Science* **229**, 932–940.

Best, R. J. (1936). Precipitation of the tobacco mosaic virus complex at its isoelectric point. *Aust. J. Exp. Biol. Med. Sci.* **14**, 1–13.

Best, R. J. (1937a). The quantitative estimation of relative concentrations of the viruses of ordinary and yellow tobacco mosaics and of tomato spotted wilt by the primary lesion method. *Aust. J. Exp. Biol. Med. Sci.* **15**, 65–79.

Best, R. J. (1937b). Artificially prepared visible paracrystalline fibres of tobacco mosaic virus nucleoprotein. *Nature (London)* **140**, 547–548.

Best, R. J. (1954). Cross protection by strains of tomato spotted wilt virus and a new theory to explain it. *Aust. J. Biol. Sci.* **7**, 415–424.

Best, R. J. (1968). Tomato spotted wilt virus. *Adv. Virus Res.* **13**, 65–146.

Best, R. J., and Katekar, G. F. (1964). Lipid in a purified preparation of tomato spotted wilt virus. *Nature (London)* **203**, 671–672.

Bevan, M. W., Mason, S. E., and Goelet, P. (1985). Expression of tobacco mosaic virus coat protein by a cauliflower mosaic virus promoter in plants transformed by *Agrobacterium*. *EMBO J.* **4**, 1921–1926.

Bharadwaj, R. K., Reddy, D. V. R., and Sinha, R. C. (1966). A reinvestigation of the alimentary canal in the leaf-hopper *Agallia constricta* (Homoptera: Cicadellidae) *Ann. Entomol. Soc. Am.* **59**, 616–617.

Bhargava, K. S. (1951). Some properties of four strains of cucumber mosaic virus. *Ann. Appl. Biol.* **38,** 377–388.

Bigger, M. (1981). The relative abundance of the mealy bug vectors (Hemiptera, Coccidae and Pseudococcidae) of cocoa swollen shoot disease in Ghana. *Bull. Entomol. Res.* **71,** 435–448.

Bilofsky, H. S., Burks, C., Fickett, J. W., Goad, W. B., Lewitter, F. I., Rindone, W. P., Swindel, C. D., and Tung, C. S. (1986). The GenBank[R] genetic sequence data bank. *Nucleic Acids Res.* **14,** 1–4.

Bisaro, D. M., Hamilton, W. D. O., Coutts, R. H. A., and Buck, K. W. (1982). Molecular cloning and characterisation of the two DNA components of tomato golden mosaic virus. *Nucleic Acids Res.* **10,** 4913–4922.

Bisaro, D. M., Sunter, G., Revington, G. N., Brough, C. L., Hormuzdi, S. G., and Hartitz, M. (1990). Molecular genetics of tomato golden mosaic virus replication: Progress towards defining gene functions, transcription units and the origin of DNA replication. *In* "Viral Genes and Plant Pathogens" (T. P. Pirone and J. G. Shaw, eds.), pp. 89–105. Springer-Verlag, New York.

Bisessar, S., and Temple, P. J. (1977). Reduced ozone injury on virus-infected tobacco in the field. *Plant Dis. Rep.* **61,** 961–963.

Bitterlin, M. W., Gonsalves, D., and Scorza, R. (1987). Improved mechanical transmission of tomato ringspot virus to *Prunus* seedlings. *Phytopathology* **77,** 560–563.

Bittner, H., Schenk, G., and Schuster, G. (1987). Chemotherapeutical elimination of potato virus X from potato stem cuttings. *J. Phytopathol.* **120,** 90–92.

Black, D. R., and Knight, C. A. (1970). Ribonucleic acid transcriptase activity in purified wound tumor virus. *J. Virol.* **6,** 194–198.

Black, L. M. (1941). Specific transmission of varieties of potato yellow-dwarf virus by related insects. *Am. Potato J.* **18,** 231–233.

Black, L. M. (1972). Plant tumors of viral origin. *Prog. Exp. Tumor Res.* **15,** 110–137.

Black, L. M. (1979). Vector cell monolayers and plant viruses. *Adv. Virus Res.* **25,** 191–271.

Black, L. M. (1984). The controversy regarding multiplication of some plant viruses in their insect vectors. *Curr. Top. Vector Res.* **2,** 1–30.

Black, L. M., and Brakke, M. K. (1952). Multiplication of wound tumor virus in an insect vector. *Phytopathology* **42,** 269–273.

Blair, P. (1719). "Botanik Essays." W. and J. Innys, London.

Blencowe, J. W., and Tinsley, T. W. (1951). The influence of density of plant population on the incidence of yellows in sugar beet crops. *Ann. Appl. Biol.* **38,** 395–401.

Blok, J., Gibbs, A., and Mackenzie, A. (1987a). The classification of tymoviruses by cDNA–RNA hybridization and other measures of relatedness. *Arch. Virol.* **96,** 225–240.

Blok, J., Mackenzie, A., Guy, P., and Gibbs, A. (1987b). Nucleotide sequence comparisons of turnip yellow mosaic virus isolates from Australia and Europe. *Arch. Virol.* **97,** 283–295.

Bloomer, A. C., Champness, J. N., Bricogne, G., Staden, R., and Klug, A. (1978). Protein disk of tobacco mosaic virus at 2.8 Å resolution showing the interactions within and between subunits. *Nature (London)* **276,** 362–368.

Blum, H., Gross, H. J., and Beier, H. (1989). The expression of the TMV-specific 30-kDa protein in tobacco protoplasts is strongly and selectively enhanced by actinomycin. *Virology* **169,** 51–61.

Boag, B. (1985). The localised spread of virus-vector nematodes adhering to farm machinery. *Nematologica* **31,** 234–235.

Boccara, M., Hamilton, W. D. O., and Baulcombe, D. C. (1986). The organisation and interviral homologies of genes at the 3′ end of tobacco rattle virus RNA1. *EMBO J.* **5,** 223–229.

Boccardo, G., and Accotto, G. P. (1988). RNA dependent RNA polymerase activity in two morphologically different white clover cryptic viruses. *Virology* **163,** 413–419.

Boccardo, G., and Milne, R. G. (1975). The maize rough dwarf virion. I. Protein composition and distribution of RNA in different viral fractions. *Virology* **68,** 79–85.

Boccardo, G., Lisa, V., Lusoni, E., and Milne, R. G. (1987). Cryptic plant viruses. *Adv. Virus Res.* **32,** 171–214.

Bockstahler, L. E. (1967). Biophysical studies on double-stranded RNA from turnip yellow mosaic virus-infected plants. *Mol. Gen. Genet.* **100,** 337–348.

Boedtker, H. (1960). Configurational properties of tobacco mosaic virus ribonucleic acid. *J. Mol. Biol.* **2,** 171–188.

Boege, F., Rohde, W., and Sänger, H. L. (1982). *In vitro* transcription of viroid RNA into full length copies by RNA-dependent RNA polymerase from healthy tomato leaf tissue. *Biosci. Rep.* **2,** 185–194.

Bol, J. F. (1988). Structure and expression of plant genes encoding pathogenesis related proteins. *In*

"Plant Gene Research" (D. P. S. Verma and R. B. Goldberg, eds.), pp. 201–221. Springer-Verlag, New York.

Bol, J. F., and Lak-Kaashoek, M. (1974). Composition of alfalfa mosaic virus nucleoproteins. *Virology* **60,** 476–484.

Bol, J. F., and van Kan, J. A. L. (1988). The synthesis and possible functions of virus-induced proteins in plants. *Microbiol. Sci.* **5,** 47–52.

Bol, J. F., and van Vloten-Doting, L. (1973). Function of top component a RNA in the initiation of infection by alfalfa mosaic virus. *Virology* **51,** 102–108.

Bol, J. F., Brederode, F. T., Janze, G. C., and Rauh, D. K. (1975). Studies on sequence homology between the RNAs of alfalfa mosaic virus. *Virology* **65,** 1–15.

Bol, J. F., Bakhuizen, C. E. G. C., and Rutgers, T. (1976). Composition and biosynthetic activity of polyribosomes associated with alfalfa mosaic virus infections. *Virology* **75,** 1–17.

Bol, J. F., van de Rhee, M. D., van Kan, J. A. L., González Jaén, M. T., and Linthorst, H. J. M. (1989). Characterization of two virus-inducible plant promoters. *In* "Signal Molecules in Plants and Plant-Microbe Interactions" (B. J. J. Lugtenberg, ed.), pp. 169–174. Springer-Verlag, Berlin and Heidelberg.

Bol, J. F., van Rossum, C. M. A., Cornelissen, B. J. C., and Linthorst, H. J. M. (1990). Induction of host genes by the hypersensitive response of tobacco to virus infection. *In* "Viral Genes and Plant Pathogenesis" (T. P. Pirone and J. G. Shaw, eds.), pp. 1–12. Springer-Verlag, New York.

Bonneville, J. M., Sanfaçon, H., Fuetterer, J., and Hohn, T. (1990). Post-transcriptional transactivation in cauliflower mosaic virus. *Cell (Cambridge, Mass.)* **59,** 1135–1143.

Borissenko, S., Schuster, G., and Schmygla, W. (1985). Obtaining a high percentage of explants with negative serological reactions against viruses by combining potato meristem culture with anti-phytoviral chemotherapy. *Phytopathol. Z.* **114,** 185–188.

Borror, D. J., De Long, D. M., and Triplehorn, C. A. (1981). "An Introduction to the Study of Insects," 5th ed. Saunders, Philadelphia, Pennsylvania.

Bos, L. (1970). The identification of three new viruses isolated from *Wisteria* and *Pisum* in The Netherlands, and the problem of variation within the potato virus Y group. *Neth. J. Plant Pathol.* **76,** 8–46.

Bos, L. (1975). The application of TMV particles as an internal magnification standard for determining virus particle sizes with the electron microscope. *Neth. J. Plant Pathol.* **81,** 168–175.

Bos, L. (1978). "Symptoms of Virus Diseases in Plants," 3rd ed. Pudoc, Wageningen.

Bos, L. (1981). Hundred years of Koch's postulates and the history of etiology in plant virus research. *Neth. J. Plant Pathol.* **87,** 91–110.

Bos, L. (1982). Crop losses caused by viruses. *Crop Prot.* **1,** 263–282.

Boswell, K. F. and Gibbs, A. J. (1986). The VIDE data bank for plant viruses. *In* "Developments and Applications in Virus Testing" (R. A. C. Jones and L. Torrance, eds.), Dev. Appl. Biol. I, pp. 283–287. Association of Applied Biologists, Wellsbourne.

Boswell, K. F., Dallwitz, M. J., Gibbs, A. J., and Watson, L. (1986). The VIDE (virus identification data exchange) Project: A data bank for plant viruses. *Rev. Plant Pathol.* **65,** 221–230.

Botstein, D. (1980). A theory of modular evolution for bacteriophages. *Ann. N.Y. Acad. Sci.* **354,** 484–491.

Botstein, D., and Shortle, D. (1985). Strategies and applications in *in vitro* mutagenesis. *Science* **229,** 1193–1201.

Boukema, I. W. (1980). Allelism of genes controlling resistance to TMV in *Capsicum* L. *Euphytica* **29,** 433–439.

Bouley, J.-P., Briand, J.-P., Genevaux, M., Pinck, M., and Witz, J. (1976). The structure of eggplant mosaic virus: Evidence for the presence of low molecular weight RNA in top component. *Virology* **69,** 775–781.

Boulton, M. I., and Markham, P. G. (1986). The use of squash-blotting to detect plant pathogens in insect vectors. *In* "Developments and Applications in Virus Testing" (R. A. C. Jones and L. Torrance, eds.), Dev. Appl. Biol. I, pp. 55–69. Association of Applied Biologists, Wellsbourne.

Boulton, M. I., Buchholz, W. G., Marks, M. S., Markham, P. G., and Davies, J. W. (1989a). Specificity of *Agrobacterium*-mediated delivery of maize streak virus DNA to members of the Graminae. *Plant Mol. Biol.* **12,** 31–40.

Boulton, M. I., Steinkellner, H., Donson, J., Markham, P. G., King, D. I., and Davies, J. W. (1989b). Mutational analysis of the virion-sense genes of maize streak virus. *J. Gen. Virol.* **70,** 2309–2323.

Boulton, R. E., Jellis, G. J., Baulcombe, D. C., and Squire, A. M. (1986). The application of complementary DNA probes to routine virus detection, with particular reference to potato viruses.

In "Developments and Applications in Virus Testing" (R. A. C. Jones and L. Torrance, eds.), Dev. Appl. Biol. I, pp. 41–53. Association of Applied Biologists, Wellsbourne.

Bouquet, A. (1981). Resistance to grapevine fanleaf virus in muscadine grape inoculated with *Xiphinema index. Plant Dis.* **65,** 791–793.

Bouzoubaa, S., Guilley, H., Jonard, G., Richards, K., and Putz, C. (1985). Nucleotide sequence analysis of RNA-3 and RNA-4 of beet necrotic yellow vein virus, isolates F2 and G1. *J. Gen. Virol.* **66,** 1553–1564.

Bouzoubaa, S., Ziegler, V., Beck, D., Guilley, H., Richards, K., and Jonard, G. (1986). Nucleotide sequence of beet necrotic yellow vein virus RNA2. *J. Gen. Virol.* **67,** 1689–1700.

Bouzoubaa, S., Quillet, L., Guilley, H., Jonard, G., and Richards, K. (1987). Nucleotide sequence of beet necrotic yellow vein virus RNA1. *J. Gen. Virol.* **68,** 615–626.

Bové, C., Mocquot, B., and Bové, J. M. (1972). Turnip yellow mosaic virus-RNA synthesis in the plastids: Partial purification of a virus-specific, DNA-independent enzyme–template complex. *Symp. Biol. Hung.* **13,** 43–59.

Bové, J. M., and Bové, C. (1985). Turnip yellow mosaic virus RNA replication on the chloroplast envelope. *Physiol. Vég.* **23,** 741–748.

Bové, J. M., Bové, C., Rondot, M.-J., and Morel, G. (1967). Chloroplasts and virus-RNA synthesis. *In* "The Biochemistry of Chloroplasts" (T. W. Goodwin, ed.), Vol. 2, pp. 329–339. Academic Press, New York.

Boxall, M., and MacNeill, B. H. (1974). Local lesions as sources of biologically pure strains of tobacco mosaic virus. *Can. J. Bot.* **52,** 23–25.

Boyle, J. S., and Bergman, E. L. (1967). Factors affecting incidence and severity of internal browning of tomato induced by tobacco mosaic virus. *Phytopathology* **57,** 354–362.

Bradley, R. H. E. (1956). Effects of depth of stylet penetration on aphid transmission of potato virus Y. *Can. J. Microbiol.* **2,** 539–547.

Bradley, R. H. E. (1959). Loss of virus from the stylets of aphids. *Virology* **8,** 308–318.

Bradley, R. H. E. (1962). Effect of keeping viruliferous aphids on plants or glass on their transmission of sugar beet yellows virus. *Virology* **17,** 571–573.

Bradley, R. H. E., and Rideout, D. W. (1953). Comparative transmission of potato virus Y by four aphid species that infest potato. *Can. J. Zool.* **31,** 333–341.

Bradley, R. H. E., Wade, C. V., and Wood, F. A. (1962). Aphid transmission of potato virus Y inhibited by oils. *Virology* **18,** 327–328.

Brakke, M. K. (1951). Density gradient centrifugation; A new separation technique. *J. Am. Chem. Soc.* **73,** 1847–1848.

Brakke, M. K. (1953). Zonal separations by density gradient centrifugation. *Arch. Biochem. Biophys.* **45,** 275–290.

Brakke, M. K. (1960). Density gradient centrifugation and its application to plant viruses. *Adv. Virus Res.* **7,** 193–224.

Brakke, M. K. (1963). Stabilization of brome grass mosaic virus by magnesium and calcium. *Virology* **19,** 367–374.

Brakke, M. K. (1970). Systemic infections for the assay of plant viruses. *Annu. Rev. Phytopathol.* **8,** 61–84.

Brakke, M. K. (1984). Mutations, the aberrant ratio phenomenon, and virus infection of maize. *Annu. Rev. Phytopathol.* **22,** 77–94.

Brakke, M. K. (1988). Perspectives on progress in plant virology. *Annu. Rev. Phytopathol.* **26,** 331–350.

Brakke, M. K., and Daly, J. M. (1965). Density-gradient centrifugation: Non-ideal sedimentation and the interaction of major and minor components. *Science* **148,** 387–389.

Brakke, M. K., and Rochow, W. F. (1974). Ribonucleic acid of barley yellow dwarf virus. *Virology* **61,** 240–248.

Brakke, M. K., and van Pelt, N. (1970). Properties of infectious ribonucleic acid from wheat streak mosaic virus. *Virology* **42,** 699–706.

Brakke, M. K., Ball, E. M., Hsu, Y. H., and Langenberg, W. G. (1987a). Wheat streak mosaic virus cylindrical inclusion body protein. *J. Gen. Virol.* **68,** 281–287.

Brakke, M. K., White, J. L., Samson, R. G., and Joshi, J. (1987b). Chlorophyll, chloroplast ribosomal RNA, and DNA are reduced by barley stripe mosaic virus systemic infection. *Phytopathology* **78,** 570–574.

Brakke, M. K., Ball, E. M., and Langenberg, W. G. (1988). A non-capsid protein associated with unencapsidated virus RNA in barley infected with barley stripe mosaic virus. *J. Gen. Virol.* **69,** 481–491.

Branch, A. D., and Dickson, E. (1980). Tomato DNA contains no detectable regions complementary to potato spindle tuber viroid as assayed by Southern hybridization. *Virology* **104,** 10–26.

Branch, A. D., and Robertson, H. D. (1984). A replication cycle for viroids and other small infectious RNAs. *Science* **223,** 450–455.

Branch, A. D., Robertson, H. D., and Dickson, E. (1981). Longer-than-unit-length viroid minus strands are present in RNA from infected plants. *Proc. Natl. Acad. Sci. U.S.A.* **78,** 6381–6385.

Branch, A. D., Robertson, H. D., Greer, C., Gegenheimer, P., Peebles, C., and Abelson, J. (1982). Cell-free circularization of viroid progeny RNA by an RNA ligase from wheat germ. *Science* **217,** 1147–1149.

Branch, A. D., Benenfeld, B. J., and Robertson, H. D. (1988a). Evidence for a single rolling circle in the replication of potato spindle tuber viroid. *Proc. Natl. Acad. Sci. U.S.A.* **85,** 9128–9132.

Branch, A. D., Benenfeld, B. J., Franck, E. R., Shaw, J. F., Varban, M. L., Willis, K. K., Rosen, D. L., and Robertson, H. D. (1988b). Interference between coinoculated viroids. *Virology* **163,** 538–546.

Branch, A. D., Levine, B. J., and Robertson, H. D. (1990). The brotherhood of circular RNA pathogens: Viroids, circular satellites, and the delta agent. *Sem. Virol.* **1,** 143–152.

Braudeau, J. (1969). "Le Cacaoyer Techniques agricoles et productions tropicales," XVII. G.-P. Maisonneuve et Larose, Paris.

Brederode, F. T., Koper-Zwarthoff, E. C., and Bol, J. F. (1980). Complete nucleotide sequence of alfalfa mosaic virus RNA4. *Nucleic Acids Res.* **8,** 2213–2223.

Breijo, G., Garro, F. J., and Conejero, V. (1990). C_7 (P32) and C_6 (P34) PR proteins induced in tomato leaves by citrus exocortis viroid infection and chitinases. *Physiol. Molec. Plant Pathol.* **36,** 249–260.

Breyel, E., Casper, R., Ansa, O. A., Kuhn, C. W., Misari, S. M., and Demski, J. W. (1988). A simple procedure to detect a dsRNA associated with groundnut rosette. *J. Phytopathol.* **121,** 118–124.

Briand, J.-P. (1978). Contribution à l'étude de la structure des tymovirus et à l'organisation de leur génome. Thèse D.Sc., l'Université Louis Pasteur de Strasbourg.

Briand, J. P., Al-Moudallal, Z., and Van Regenmortel, M. H. V. (1982). Serological differentiation of tobamoviruses by means of monoclonal antibodies. *J. Virol. Methods* **5,** 293–300.

Briddon, R. W., Watts, J., Markham, P. G., and Stanley, J. (1989). The coat protein of beet curly top virus is essential for infectivity. *Virology* **172,** 628–633.

Briddon, R. W., Pinner, M. S., Stanley, J., and Markham, P. G. (1990). Geminivirus coat protein gene replacement alters insect specificity. *Virology,* **177,** 85–94.

Brisco, M. J., Hull, R., and Wilson, T. M. A. (1985). The effect of extraction protocol on the yield, purity, and translation products of RNA from an isometric plant virus. *J. Virol. Methods* **10,** 195–202.

Brisco, M. J., Hull, R., and Wilson, T. M. A. (1986a). Swelling of isometric and of bacilliform plant virus nucleocapsids is required for virus-specific protein synthesis *in vitro*. *Virology* **148,** 210–217.

Brisco, M. J., Haniff, C., Hull, R., Wilson, T. M. A., and Sattelle, D. B. (1986b). The kinetics of swelling of Southern bean mosaic virus: A study using photon correlation spectroscopy. *Virology* **148,** 218–220.

Brisson, N., Paszkowski, J., Penswick, J. R., Gronenborn, B., Potrykus, I., and Hohn, T. (1984). Expression of a bacterial gene in plants by using a viral vector. *Nature (London)* **310,** 511–514.

Britten, R. J., Graham, D. E., and Neufeld, B. R. (1974). Analysis of repeating DNA sequences by reassociation. *In* "Methods in Enzymology" (L. Grossman and K. Moldare, eds.), Vol. 29, pp. 363–418. Academic Press, New York.

Brlansky, R. H. (1987). Inclusion bodies produced in *Citrus* spp. by citrus tristeza virus. *Phytophylactica* **19,** 211–213.

Brlansky, R. H., Carroll, T. W., and Zaske, S. K. (1986). Some ultrastructural aspects of the pollen transmission of barley stripe mosaic virus in barley. *Can. J. Bot.* **64,** 853–858.

Broadbent, L. H. (1957). "Investigation of Virus Diseases of Brassica Crops," A.R.C. Rep. Ser. No. 14. Cambridge Univ. Press, London and New York.

Broadbent, L. H. (1963). The epidemiology of tomato mosaic. III. Cleaning virus from hands and tools. *Ann. Appl. Biol.* **52,** 225–232.

Broadbent, L. H. (1964). The epidemiology of tomato mosaic. VII. The effect of TMV on tomato fruit yield and quality under glass. *Ann. Appl. Biol.* **54,** 209–224.

Broadbent, L. H. (1965). The epidemiology of tomato mosaic. XI. Seed-transmission of TMV. *Ann. Appl. Biol.* **56,** 177–205.

Broadbent, L. H., and Fletcher, J. T. (1963). The epidemiology of tomato mosaic. IV. Persistence of virus on clothing and glasshouse structures. *Ann. Appl. Biol.* **52,** 233–241.

Broadbent, L. H., Gregory, P. H., and Tinsley, T. W. (1952). The influence of planting date and manuring on the incidence of virus diseases in potato crops. *Ann. Appl. Biol.* **39,** 509–524.

Broadbent, L. H., Heathcote, G. D., McDermott, N., and Taylor, C. E. (1957). The effect of date of planting and of harvesting potatoes on virus infection and on yield. *Ann. Appl. Biol.* **45,** 603–622.

Broadbent, L. H., Green, D. E., and Walker, P. (1963). Narcissus virus diseases. *Daffodil Tulip Yearb.* **28,** 154–160.

Broadbent, L. H., Read, W. H., and Last, F. T. (1965). The epidemiology of tomato mosaic X. Persistence of TMV infected debris in soil and the effects of soil partial sterilisation. *Ann. Appl. Biol.* **55,** 471–483.

Brough, C. L., Hayes, R. J., Morgan, A. J., Coutts, R. H. A., and Buck, K. W. (1988). Effects of mutagenesis *in vitro* on the ability of cloned tomato golden mosaic virus DNA to infect *Nicotiana benthamiana* plants. *J. Gen. Virol.* **69,** 503–514.

Brown, D. J. F. (1986). The transmission of two strains of arabis mosaic virus from England by populations of *Xiphinema diversicaudatum* (Nematoda: Dorylaimoidea) from ten countries. *Rev. Nematol.* **9,** 83–87.

Brown, D. J. F., and Trudgill, D. L. (1984). The spread of carnation ringspot virus in soil with or without nematodes. *Nematologica* **30,** 102–104.

Brown, D. J. F., Murant, A. F., and Trudgill, D. L. (1989). Differences between isolates of the English serotype of tomato black ring virus in their transmissibility by an English population of *Longidorus attennatus* (Nematoda: Dorylaimoidea). *Rev. Nematol.* **12,** 51–56.

Brown, L. M., and Wood, K. R. (1987). Translation of clover yellow mosaic virus RNA in pea mesophyll protoplasts and rabbit reticulocyte lysate. *J. Gen. Virol.* **68,** 1773–1778.

Browning, K. S., Fletcher, L., and Ravel, J. M. (1988). Evidence that requirements for ATP and wheat germ initiation factors 4A and 4F are affected by a region of satellite tobacco necrosis virus RNA that is 3′ to the ribosomal binding site. *J. Biol. Chem.* **263,** 8380–8383.

Bruening, G., Beachy, R. N., Scalla, R., and Zaitlin, M. (1976). *In vitro* and *in vivo* translation of the ribonucleic acids of a cowpea strain of tobacco mosaic virus. *Virology* **71,** 498–517.

Brunt, A. A. (1975). The effects of cocoa swollen-shoot virus on the growth and yield of Amelonado and Amazon cocoa (*Theobroma cacao*) in Ghana. *Ann. Appl. Biol.* **80,** 169–180.

Brunt, A. A., and Kenten, R. H. (1963). The use of protein in the extraction of cocoa swollen-shoot virus from cocoa leaves. *Virology* **19,** 388–392.

Brunt, A. A., and Richards, K. E. (1989). Biology and molecular biology of furoviruses. *Adv. Virus Res.* **36,** 1–32.

Brunt, A. A., and Shikata, E. (1986). Fungus-transmitted and similar labile rod-shaped viruses. *In* "The Plant Viruses" (M. H. V. van Regenmortel and H. Fraenkel-Conrat, eds.), Vol. 2, pp. 305–335. Plenum, New York.

Buchter, H., Hartmann, W., and Stoesser, R. (1987). Anatomical-histological changes in sharka-infected shoots and roots. *Z. Pflanzenkr. Pflanzenschutz* **94,** 46–57.

Buddenhagen, I. W. (1981). Conceptual and practical considerations when breeding for tolerance or resistance. *In* "Plant Disease Control: Resistance and Susceptibility" (R. C. Staples and G. H. Toenniessen, eds.), pp. 221–234. Wiley, New York.

Buddenhagen, I. W. (1983). Crop improvement in relation to virus diseases and their epidemiology. *In* "Plant Virus Epidemiology" (R. T. Plumb and J. M. Thresh, eds.), pp. 25–37. Blackwell, Oxford.

Büttner, C., and Nienhaus, F. (1989). Virus contamination of soils in forest ecosystems of the Federal Republic of Germany. *Eur. J. For. Pathol.* **19,** 47–53.

Büttner, C., Jacobi, V., and Koenig, R. (1987). Isolation of carnation Italian ringspot virus from a creek in a forested area southwest of Bonn. *J. Phytopathol.* **118,** 131–134.

Bujarski, J. J., and Kaesberg, P. (1986). Genetic recombination between RNA components of a multipartite plant virus. *Nature (London)* **321,** 528–531.

Bujarski, J. J., Hardy, S. F., Miller, W. A., and Hall, T. C. (1982). Use of dodecyl-β-D-maltoside in the purification and stabilization of RNA polymerase from brome mosaic virus-infected barley. *Virology* **119,** 465–473.

Bujarski, J. J., Dreher, T. W., and Hall, T. C. (1985). Deletions in the 3′-terminal tRNA-like structure of brome mosaic virus RNA differentially affects aminoacylation and replication *in vitro. Proc. Natl. Acad. Sci. U.S.A.* **82,** 5636–5640.

Burden, R. S., Rowell, P. M., Bailey, J. A., Loeffler, R. S. T., Kemp, M. S., and Brown, C. A. (1985). Debneyol, a fungicidal sesquiterpene from TMV infected *Nicotiana debneyi. Phytochemistry* **24,** 2191–2194.

Burgermeister, W., and Koenig, R. (1984). Electro-blot immunoassay—A means for studying serological relationships among plant viruses? *Phytopathol. Z.* **111,** 15–25.

Burgermeister, W., Koenig, R., Weich, H., Sebald, W., and Lesemann, D.-E. (1986). Diversity of the RNAs in thirteen isolates of beet necrotic yellow vein virus in *Chenopodium quinoa* detected by means of cloned cDNAs. *J. Phytopath.* **115,** 229–242.

Burgess, J., Motoyoshi, F., and Fleming, E. N. (1973). The mechanism of infection of plant protoplasts by viruses. *Planta* **112**, 323–332.

Burgess, J., Motoyoshi, F., and Fleming, E. N. (1974a). Structural changes accompanying infection of tobacco protoplasts with two spherical viruses. *Planta* **117**, 133–144.

Burgess, J., Motoyoshi, F., and Fleming, E. N. (1974b). Structural and autoradiographic observations of the infection of tobacco protoplasts with pea enation mosaic virus. *Planta* **119**, 247–256.

Burgyan, J., Grieco, F., and Russo, M. (1989). A defective interfering RNA molecule in *Cymbidium* ringspot virus infections. *J. Gen. Virol.* **70**, 235–239.

Burk, L. G., Stewart, R. N., and Dermen, H. (1964). Histogenesis and genetics of a plastid-controlled chlorophyll variegation in tobacco. *Am. J. Bot.* **51**, 713–724.

Burnett, P. (1976). Cereal crop pests. *In* "New Zealand Insect Pests" (D. N. Ferro, ed.), pp. 139–152. Lincoln University College of Agriculture, Christchurch.

Burt, P. E., Heathcote, G. D., and Broadbent, L. (1964). The use of insecticides to find when leaf roll and virus Y spread within potato crops. *Ann. Appl. Biol.* **54**, 13–22.

Butler, P. J. G. (1984). The current picture of the structure and assembly of tobacco mosaic virus. *J. Gen. Virol.* **65**, 253–279.

Butler, P. J. G., and Klug, A. (1971). Assembly of the particle of tobacco mosaic virus from RNA and disks of protein. *Nature (London) New Biol.* **229**, 47–50.

Butler, P. J. G., Durham, A. C. H., and Klug, A. (1972). Structures and roles of the polymorphic forms of tobacco mosaic virus protein. IV. Control of mode of aggregation of tobacco mosaic virus protein by proton binding. *J. Mol. Biol.* **72**, 1–18.

Butler, P. J. G., Finch, J. T., and Zimmern, D. (1977). Configuration of tobacco mosaic virus RNA during virus assembly. *Nature (London)* **265**, 217–219.

Buzayan, J. M., Gerlach, W. L., and Bruening, G. (1986a). Satellite tobacco ringspot virus RNA: A subset of the RNA sequence is sufficient for autolytic processing. *Proc. Natl. Acad. Sci. U.S.A.* **83**, 8859–8862.

Buzayan, J. M., Gerlach, W. L., and Bruening, G. (1986b). Non-enzymatic cleavage and ligation of RNAs complementary to a plant virus satellite RNA. *Nature (London)* **323**, 349–353.

Buzayan, J. M., Hampel, A., and Bruening, G. (1986c). Nucleotide sequence and newly formed phosphodiester bond of spontaneously ligated satellite tobacco ringspot virus RNA. *Nucleic Acids Res.* **14**, 9729–9743.

Buzayan, J. M., Gerlach, W. L., Bruening, G., Keese, P., and Gould, A. R. (1986d). Nucleotide sequence of satellite tobacco ringspot RNA and its relationship to multimeric forms. *Virology* **151**, 186–199.

Buzayan, J. M., McNinch, J. S., Schneider, I. R., and Bruening, G. (1987). A nucleotide sequence rearrangement distinguishes two isolates of satellite tobacco ringspot virus RNA. *Virology* **160**, 95–99.

Buzen, F. G., Jr., Niblett, C. L., Hooper, G. R., Hubbard, J., and Newman, M. A. (1984). Further characterisation of panicum mosaic virus and its associated satellite virus. *Phytopathology* **74**, 313–318.

Buzzell, R. I., and Tu, J. C. (1984). Inheritance of soybean resistance to soybean mosaic virus. *J. Hered.* **75**, 82.

Cabautan, P. Q., and Hibino, H. (1984). Incidence of rice tungro bacilliform (RTBV) and rice tungro spherical virus (RTSV) on susceptible rice cultivars. *Int. Rice Newsl.* **9**, 13.

Cadman, C. H., and Harrison, B. D. (1960). Studies on the behavior in soils of tomato black ring, raspberry ringspot, and arabis mosaic viruses. *Virology* **10**, 1–20.

Camacho-Henriquez, A., and Sänger, H. L. (1984). Purification and partial characterisation of the major pathogenesis-related tomato leaf protein P14 from potato spindle tuber viroid (PSTV)-infected tomato leaves. *Arch. Virol.* **81**, 263–284.

Camerini-Otero, R. D., Pusey, P. N., Koppel, D. E., Schaefer, D. W., and Franklin, R. M. (1974). Intensity fluctuation spectroscopy of laser light scattered by solutions of spherical viruses: R 17, Qβ, BSV, PM2 and T7. II. Diffusion coefficients, molecular weights, solvation, and particle dimensions. *Biochemistry* **13**, 960–970.

Campbell, R. N., and Fry, P. R. (1966). The nature of the associations between *Olipidium brassicae* and lettuce big vein and tobacco necrosis viruses. *Virology* **29**, 222–233.

Campbell, R. N., and Grogan, R. G. (1964). Acquisition and transmission of lettuce big-vein virus by *Olpidium brassicae*. *Phytopathology* **54**, 681–690.

Candresse, T., Mouches, C., and Bové, J. M. (1986). Characterisation of the virus encoded subunit of turnip yellow mosaic virus RNA replicase. *Virology* **152**, 322–330.

Candresse, T., Smith, D., and Diener, T. O. (1987a). Nucleotide sequence of a full-length infectious clone of the Indonesian strain of tomato apical stunt viroid (TASV). *Nucleic Acids Res.* **15,** 10, 597.

Candresse, T., Batisti, M., Renaudin, J., Mouches, C., and Bové, J. M. (1987b). Immunodetection of turnip yellow mosaic virus non-structural proteins in infected chinese cabbage leaves and protoplasts. *Ann. Inst. Pasteur./Virol.* **138,** 217–227.

Candresse, T., Macquaire, G., Monsion, M., and Dunez, J. (1988). Detection of chrysanthmum stunt viroid (CSV) using nick translated probes in a dot blot hybridization assay. *J. Virol. Methods* **20,** 185–193.

Candresse, T., Diener, T. O., and Owens, R. A. (1990). The role of the viroid central conserved region in cDNA infectivity. *Virology* **175,** 232–237.

Canova, A. (1966). Richerche sulle malattie da virus delle Graminaceae. III. *Polymyxa graminis* Led. vettore del virus del mosaico del Frumento. *Phytopathol. Mediterr.* **5,** 53–58.

Capoor, S. P. (1949). The movement of tobacco mosaic viruses and potato virus X through tomato plants. *Ann. Appl. Biol.* **36,** 307–319.

Carr, J. P., and Klessig, D. F. (1989). The pathogenesis related proteins. *In* "Genetic Engineering: Principles and Methods" (J. K. Setlow, ed.), Vol. 11, pp. 65–109. Plenum, New York.

Carr, J. P., Beachy, R. N., and Klessig, D. F. (1989). Are the PR1 proteins involved in genetically engineered resistance to TMV? *Virology* **169,** 470–473.

Carr, R. J., and Kim, K. S. (1983). Evidence that bean golden mosaic virus invades non-phloem tissue in double infections with tobacco mosaic virus. *J. Gen. Virol.* **64,** 2489–2492.

Carrington, J. C., and Dougherty, W. G. (1987a). Small nuclear inclusion protein encoded by a plant potyvirus genome is a protease. *J. Virol.* **61,** 2540–2548.

Carrington, J. C., and Dougherty, W. G. (1987b). Processing of the tobacco etch virus 49K protease requires autoproteolysis. *Virology* **160,** 355–362.

Carrington, J. C., and Dougherty, W. G. (1988). A viral cleavage site cassette: Identification of amino acid sequences required for tobacco etch virus polyprotein processing. *Proc. Natl. Acad. Sci. U.S.A.* **85,** 3391–3395.

Carrington, J. C., and Morris, T. J. (1984). Complementary DNA cloning and analysis of carnation mottle virus RNA. *Virology* **139,** 22–31.

Carrington, J. C., and Morris, T. J. (1985). Characterisation of the cell-free translation products of carnation mottle virus genomic and subgenomic RNAs. *Virology* **144,** 1–10.

Carrington, J. C., and Morris, T. J. (1986). High resolution mapping of carnation mottle virus-associated RNAs. *Virology* **150,** 196–206.

Carrington, J. C., and Freed, D. D. (1990). Cap-independent enhancement of translation by a plant potyvirus 5′ non-translated region. *J. Virol.* **63,** 1590–1597.

Carrington, J. C., Morris, T. J., Stockley, P. G., and Harrison, S. C. (1987). Structure and assembly of turnip crinkle virus. IV. Analysis of the coat protein gene and implications of the subunit primary structure. *J. Mol. Biol.* **194,** 265–276.

Carrington, J. C., Cary, S. M., and Dougherty, W. G. (1988). Mutational analysis of tobacco etch virus polyprotein processing: *Cis* and *trans* proteolytic activities of polyproteins containing the 49-kilodalton proteinase. *J. Virol.* **62,** 2313–2320.

Carrington, J. C., Cary, S. M., Parks, T. D., and Dougherty, W. G. (1989a). A second proteinase encoded by a plant potyvirus genome. *EMBO J.* **8,** 365–370.

Carrington, J. C., Heaton, L. A., Zuidema, D., Hillman, B. I., and Morris, T. J. (1989b). The genome structure of turnip crinkle virus. *Virology* **170,** 219–226.

Carrington, J. C., Freed, D. D., and Oh, C-S. (1990). Expression of potyvirus polyproteins in transgenic plants reveals three proteolytic activities required for complete processing. *EMBO J.* **9,** 1347–1353.

Carroll, T. W. (1972). Seed transmissibility of two strains of barley stripe mosaic virus. *Virology* **48,** 323–336.

Carroll, T. W. (1974). Barley stripe mosaic virus in sperm and vegetative cells of barley pollen. *Virology* **60,** 21–28.

Carroll, T. W. (1981). Seedborne viruses: Virus–host interactions. *In* "Plant Diseases and Vectors: Ecology and Epidemiology" (K. Maramorosch and K. F. Harris, eds.), pp. 293–317. Academic Press, New York.

Carroll, T. W. (1983). Certification schemes against barley stripe mosaic. *Seed Sci. Technol.* **11,** 1033–1042.

Carroll, T. W. (1986). Hordeiviruses biology and pathology. *In* "The Plant Viruses" (M. H. V. van Regenmortel and H. Fraenkel-Conrat, eds.), Vol. 2, pp. 373–395. Plenum, New York.

Carroll, T. W., and Chapman, S. R. (1970). Variation in embryo infection and seed transmission of barley stripe mosaic virus within and between two cultivars of barley. *Phytopathology* **60**, 1079–1081.

Carroll, T. W., Gossel, P. L., and Hockett, E. A. (1979). Inheritance of resistance to seed transmission of barley stripe mosaic virus in barley. *Phytopathology* **69**, 431–433.

Carroll, T. W., Hockett, E. A., and Zaske, S. K. (1983). Registration of mobet barley germplasm. *Crop Sci.* **23**, 599–600.

Carsner, E. (1925). Attenuation of the virus of sugar beet curly-top. *Phytopathology* **15**, 745–757.

Carter, M. J., and ter Meulen, V. (1984). The application of monoclonal antibodies in the study of viruses. *Adv. Virus Res.* **29**, 95–130.

Cascone, P. J., Carpenter, C. D., Li, X. H., and Simon, A. E. (1990). Recombination between satellite RNAs of turnip crinkle virus. *EMBO J.* **9**, 1709–1715.

Caspar, D. L. D. (1963). Assembly and stability of the tobacco mosaic virus particle. *Adv. Protein Chem.* **18**, 37–121.

Caspar, D. L. D., and Holmes, K. C. (1969). Structure of dahlemense strain of tobacco mosaic virus: A periodically deformed helix. *J. Mol. Biol.* **46**, 99–133.

Caspar, D. L. D., and Klug, A. (1962). Physical principles in the construction of regular viruses. *Cold Spring Harbor Symp. Quant. Biol.* **27**, 1–24.

Caspar, D. L. D., Dulbecco, R., Klug, A., Lwoff, A., Stoker, M. G. P., Tournier, P., and Wildy, P. (1962). Proposals. *Cold Spring Harbor Symp. Quant. Biol.* **27**, 49.

Cassells, A. C., and Herrick, C. C. (1977a). The identification of mild and severe strains of tobacco mosaic virus in double inoculated tomato plants. *Ann. Appl. Biol.* **86**, 37–46.

Cassells, A. C., and Herrick, C. C. (1977b). Cross protection between mild and severe strains of tobacco mosaic virus in doubly inoculated tomato plants. *Virology* **78**, 253–260.

Castel, A., Kraal, B., De Graaf, J. M., and Bosch, L. (1979). The primary structure of the coat protein of alfalfa mosaic virus strain VRU. A hypothesis on the occurrence of two conformations in the assembly of the protein shell. *Eur. J. Biochem.* **102**, 125–138.

Castillo, M. B., and Orlob, G. B. (1966). Transmission of two strains of cucumber mosaic and alfalfa mosaic viruses by single aphids of *Myzus persicae*. *Phytopathology* **56**, 1028–1030.

Cataldo, D. A. (1974). Vein loading: The role of the symplast in intercellular transport of carbohydrate between the mesophyll and minor veins of tobacco leaves. *Plant Physiol.* **53**, 912–917.

Catherall, P. L. (1966). Effects of barley yellow dwarf virus on the growth and yield of single plants and simulated swards of perennial rye-grass. *Ann. Appl. Biol.* **57**, 155–162.

Catherall, P. L. (1987). Effects of barley yellow dwarf and ryegrass mosaic viruses alone and in combination on the productivity of perennial and Italian ryegrasses. *Plant Pathol.* **36**, 73–78.

Catherall, P. L., and Griffiths, E. (1966a). Influence of cocksfoot streak virus on the growth of single cocksfoot plants. *Ann. Appl. Biol.* **57**, 141–148.

Catherall, P. L., and Griffiths, E. (1966b). Influence of cocksfoot streak virus on the growth of cocksfoot swards. *Ann. Appl. Biol.* **57**, 149–154.

Catherall, P. L., Jones, A. T., and Hayes, J. D. (1970). Inheritance and effectiveness of genes in barley that condition tolerance to barley yellow dwarf virus. *Ann. Appl. Biol.* **65**, 153–161.

Cech, T. R. (1986). A model for the RNA-catalysed replication of RNA. *Proc. Natl. Acad. Sci. U.S.A.* **83**, 4360–4363.

Chalcroft, J. P., and Matthews, R. E. F. (1966). Cytological changes induced by turnip yellow mosaic virus infections. *Virology* **28**, 555–562.

Chalcroft, J. P., and Matthews, R. E. F. (1967a). Virus strains and leaf ontogeny as factors in the production of leaf mosaic patterns by turnip yellow mosaic virus. *Virology* **33**, 167–171.

Chalcroft, J. P., and Matthews, R. E. F. (1967b). Role of virus strains and leaf ontogeny in the production of mosaic patterns by turnip yellow mosaic virus. *Virology* **33**, 659–673.

Chalfant, R. B., and Chapman, R. K. (1962). Transmission of cabbage viruses A and B by the cabbage aphid and the green peach aphid. *J. Econ. Entomol.* **55**, 584–590.

Chamberlain, E. E., Atkinson, J. D., and Hunter, J. A. (1951). Plum mosaic, a virus disease of plums, peaches and apricots in New Zealand. *N. Z. J. Sci. Technol.* **33**(2), 1–16.

Chambers, T. C., and Francki, R. I. B. (1966). Localization and recovery of lettuce necrotic yellows virus from xylem tissues of *Nicotiana glutinosa*. *Virology* **29**, 673–676.

Champness, J. N., Bloomer, A. C., Bricogne, G., Butler, P. J. G., and Klug, A. (1976). The structure of the protein disk of tobacco mosaic virus to 5 Å resolution. *Nature (London)* **259**, 20–24.

Chang, C.-A., Hiebert, E., and Purcifull, D. E. (1988a). Purification, characterisation, and immunological analysis of nuclear inclusions induced by bean yellow mosaic virus and clover yellow vein potyviruses. *Phytopathology* **78**, 1266–1275.

Chang, C.-A., Hiebert, E., and Purcifull, D. E. (1988b). Analysis of *in vitro* translation of bean yellow mosaic virus RNA: Inhibition of proteolytic processing by antiserum to the 49K nuclear inclusion protein. *J. Gen. Virol.* **69,** 1117–1122.

Chauvin, C., Pfeiffer, P., Witz, J., and Jacrot, B. (1978). Structural polymorphism of bromegrass mosaic virus: A neutron small angle scattering investigation. *Virology* **88,** 138–148.

Chen, B., and Francki, R. I. B. (1990). Cucumovirus transmission by the aphid *Myzus persicae* is determined solely by the properties of its coat protein. *J. Gen. Virol.* **71,** 939–944.

Chen, J. M., and Hall, T. C. (1973). Comparison of tyrosyl transfer ribonucleic acid and brome mosaic virus tyrosyl ribonucleic acid as amino acid donors in protein synthesis. *Biochemistry* **12,** 4570–4574.

Chen, L.-C., Durand, D. P., and Hill, J. H. (1982). Detection of pathogenic strains of soybean mosaic virus by enzyme-linked immunosorbent assay with polystyrene plates and beads as the solid phase. *Phytopathology* **72,** 1177–1181.

Chen, Z., Stauffacher, C., Li, Y., Schmidt, T., Bomu, W., Kamer, G., Shanks, M., Lomonossoff, G., and Johnson, J. E. (1989). Protein–RNA interactions in an icosahedral virus at 3.0 Å resolution. *Science* **245,** 154–159.

Chester, K. S. (1935). A serological estimate of the absolute concentration of tobacco mosaic virus. *Science* **82,** 17.

Chester, K. S. (1936). Separation and analysis of virus strains by means of precipitin tests. *Phytopathology* **26,** 778–785.

Chiko, A. W. (1975). Evidence of multiple virion components in leaf-dip preparations of barley stripe mosaic virus. *Virology* **63,** 115–122.

Childress, A. M., and Ramsdell, D. C. (1987). Bee mediated transmission of blueberry leaf mottle virus via infected pollen in highbush blueberry. *Phytopathology* **77,** 167–172.

Childress, S. A., and Harris, K. F. (1989). Localization of virus-like particles in the foreguts of viruliferous *Graminella nigrifrons* leafhoppers carrying the semi-persistent maize chlorotic dwarf virus. *J. Gen. Virol.* **70,** 247–251.

Chiu, R.-J., and Black, L. M. (1969). Assay of wound tumor virus by the fluorescent cell counting technique. *Virology* **37,** 667–677.

Chiu, R.-J., Liu, H.-Y., MacLeod, R., and Black, L. M. (1970). Potato yellow dwarf virus in leafhopper cell culture. *Virology* **40,** 387–396.

Chiykowski, L. N., and Chapman, R. K. (1965). Migration of the six-spotted leaf hopper in central North America. *Res. Bull.—Wis., Agric. Exp. Stn.* **261,** 21–45.

Cho, J. J., Mau, R. F. L., Hamasaki, R. T., and Gonsalves, D. (1988). Detection of tomato spotted wilt virus in individual thrips by enzyme-linked immunosorbent assay. *Phytopathology* **78,** 1348–1352.

Cho, J. J., Mau, R. F. L., German, T. L., Hartmann, R. W., Yudin, L. S., Gonsalves, D., and Provvidenti, R. (1989). A multidisciplinary approach to management of tomato spotted wilt virus in Hawaii. *Plant Dis.* **73,** 375–383.

Choe, I. S., Melcher, U., Richards, K., Lebeurier, G., and Essenberg, R. C. (1985). Recombination between mutant cauliflower mosaic virus DNAs. *Plant Mol. Biol.* **5,** 281–289.

Choudhury, M. M., and Rosenkranz, E. (1983). Vector relationship of *Graminella nigrifrons* to maize chlorotic dwarf virus. *Phytopathology* **73,** 685–690.

Christie, R. G., and Edwardson, J. R. (1986). Light microscopic techniques for detection of plant virus inclusions. *Plant Dis.* **70,** 273–279.

Christie, S. R., Purcifull, D. E., Crawford, W. E., and Ahmed, N. A. (1987). Electron microscopy of negatively stained clarified viral concentrates obtained from small tissue samples with appendices on negative staining techniques. *Bull.—Fla., Agric. Exp. Stn.* **872,** 1–45.

Chu, P. W. G., and Francki, R. I. B. (1979). The chemical subunit of tobacco ringspot virus coat protein. *Virology* **93,** 398–412.

Chu, P. W. G., and Francki, R. I. B. (1982). Detection of lettuce necrotic yellows virus by an enzyme-linked immunosorbent assay in plant hosts and the insect vector. *Ann. Appl. Biol.* **100,** 149–156.

Chu, P. W. G., and Francki, R. I. B. (1983). Chemical and serological comparison of the coat proteins of velvet tobacco mottle and *Solanum nodiflorum* mottle viruses. *Virology* **129,** 350–356.

Chu, P. W. G., Boccardo, G., and Francki, R. I. B. (1981). Requirement of a genome associated protein of tobacco ringspot virus for infectivity but not for *in vitro* translation. *Virology* **109,** 428–430.

Cireli, B. (1976). Observations on the effects of some air pollutants on *Zea mays* leaf tissue. *Phytopathol. Z.* **86,** 233–239.

Cirulli, M., and Alexander, L. J. (1969). Influence of temperature and strain of tobacco mosaic virus on resistance of a tomato breeding line derived from *Lycopersicon peruvianum. Phytopathology* **59,** 1287–1297.

Citovski, V., Knorr, D., Schuster, G., and Zambryski, P. (1990). The P30 movement protein of tobacco mosaic virus is a single-strand nucleic acid binding protein. *Cell* **60**, 637–647.

Clark, M. F., and Adams, A. N. (1977). Characteristics of the microplate method of enzyme-linked immunosorbent assay for the detection of plant viruses. *J. Gen. Virol.* **34**, 475–483.

Clark, M. F., and Barbara, D. J. (1987). A method for the quantitative analysis of ELISA data. *J. Virol. Methods* **15**, 213–222.

Clark, M. F., and Bar-Joseph, M. (1984). Enzyme immunosorbent assays in plant virology. *Methods Virol.* **7**, 51–85.

Clark, M. F., Matthews, R. E. F., and Ralph, R. K. (1964). Ribosomes and polyribosomes in *Brassica pekinensis*. *Biochim. Biophys. Acta* **91**, 289–304.

Clark, R. W. (1976). Calculation of S_{20w} values using ultracentrifuge sedimentation data from linear sucrose gradients, an improved, simplified method. *Biochim. Biophys. Acta* **428**, 269–274.

Clemens, M. (1987). Developments in development. *Nature (London)* **330**, 699–670.

Clement, D. L., Lister, R. M., and Foster, J. E. (1986). ELISA-based studies on the ecology and epidemiology of barley yellow dwarf virus in Indiana. *Phytopathology* **76**, 86–92.

Clinch, P., Loughnane, J. B., and Murphy, P. A. (1938). A study of the infiltration of viruses into seed potato stocks in the field. *Sci. Proc. R. Dublin Soc.* **22**, 18–31.

Close, R. C. (1962). Some interactions between viruses when multiplying together in plants. Ph.D. Thesis, University of London.

Close, R. C. (1964). Some effects of other viruses and of temperature on the multiplication of potato virus X. *Ann. Appl. Biol.* **53**, 151–164.

Close, R. C., and Tomlinson, A. I. (1975). Dispersal of the grain aphid *Macrosiphum miscanthi* from Australia to New Zealand. *N. Z. Entomol.* **6**, 62–65.

Coakley, S. M., Campbell, R. N., and Kimble, K. A. (1973). Internal rib necrosis and rusty brown discoloration of climax lettuce induced by lettuce mosaic virus. *Phytopathology* **63**, 1191–1197.

Cochran, G. W. (1946). Effect of shading techniques on transmission of tobacco mosaic virus through dodder. *Phytopathology* **36**, 396.

Cock, L. J. (1968). Virus diseases of lettuce. *NAAS Q. Rev.* **79**, 126–138.

Cockbain, A. J., Gibbs, A. J., and Heathcoate, G. D. (1963). Some factors affecting the transmission of sugar-beet mosaic and pea mosaic viruses by *Aphis fabae* and *Myzus persicae*. *Ann. Appl. Biol.* **52**, 133–143.

Cockerham, G. (1970). Genetical studies on resistance to potato viruses X and Y. *Heredity* **25**, 309–348.

Cocking, E. C. (1966). An electron microscope study of the initial stages of infection of isolated tomato fruit protoplasts by tobacco mosaic virus. *Planta* **68**, 206–214.

Cohen, J., and Loebenstein, G. (1975). An electron microscope study of starch lesions in cucumber cotyledons infected with tobacco mosaic virus. *Phytopathology* **65**, 32–39.

Cohen, J., Loebenstein, G., and Spiegel, S. (1988). Infection of sweet potato by cucumber mosaic virus depends on the presence of sweet potato feathery mottle virus. *Plant Dis.* **72**, 583–585.

Cohen, S. S. (1981). Reducing the spread of aphid-transmitted viruses in peppers by coarse-net cover. *Phytoparasitica* **9**, 69–76.

Cohen, S. S., and Greenberg, M. L. (1981). Spermidine, an intrinsic component of turnip yellow mosaic virus. *Proc. Natl. Acad. Sci. U.S.A.* **78**, 5470–5474.

Cohen, S. S., and Marco, S. (1973). Reducing the spread of aphid-transmitted viruses in peppers by trapping the aphids on sticky yellow polyethylene sheets. *Phytopathology* **63**, 1207–1209.

Cohen, S. S., Duffus, J. E., Larsen, R. C., Liu, H. Y., and Flock, R. A. (1983). Purification, serology and vector relationships of squash leaf curl virus, a whitefly-transmitted geminivirus. *Phytopathology* **73**, 1669–1673.

Cohen, S. S., Duffus, J. E., and Liu, H. Y. (1989). Acquisition, interference, and retention of cucurbit leaf curl viruses in whiteflies. *Phytopathology* **79**, 109–113.

Coleman, J., Hirashima, A., Inokuchi, Y., Green, P. J., and Inouye, M. (1985). A novel immune system against bacteriophage infection using complementary RNA (mic RNA). *Nature (London)* **315**, 601–603.

Collendavelloo, J., Legrand, M., and Fritig, B. (1982). Plant disease and regulation of enzymes involved in lignification. *De novo* synthesis controls *o*-methyltransferase activity in hypersensitive tobacco leaves infected by tobacco mosaic virus. *Physiol. Plant Pathol.* **21**, 271–281.

Collmer, C. W., and Kaper, J. M. (1985). Double-stranded RNAs of cucumber mosaic virus and its satellite contain an unpaired terminal guanosine: Implications for replication. *Virology* **145**, 249–259.

Collmer, C. W., and Kaper, J. M. (1988). Site-directed mutagenesis of potential protein-coding regions in expressible cloned cDNAs of cucumber mosaic viral satellites. *Virology* **163**, 293–298.

Collmer, C. W., and Zaitlin, M. (1983). The H protein isolated from tobacco mosaic virus reassociates with virions reconstituted *in vitro*. *Virology* **126,** 449–458.

Collmer, C. W., Vogt, V. M., and Zaitlin, M. (1983). H protein, a minor protein of TMV virions, contains sequences of the viral coat protein. *Virology* **126,** 429–448.

Collmer, C. W., Hadidi, A., and Kaper, J. M. (1985). Nucleotide sequence of the satellite of peanut stunt virus reveals structural homologies with viroids and certain nuclear and mitochondrial introns. *Proc. Natl. Acad. Sci. U.S.A.* **82,** 3110–3114.

Comeau, A., and Plourde, A. (1987). Cell, tissue culture and intergeneric hybridization for barley yellow dwarf virus resistance in wheat. *Can. J. Plant Pathol.* **9,** 188–192.

Connolly, M. (1987). Breeding beats maize streak virus. *Span* **30,** 27–28.

Conti, M. (1972). Investigations on the epidemiology of maize rough dwarf virus. I. Overwintering of virus in its planthopper vector. *Actas Congr. Uniao Fitopatol. Mediterr., 3rd, 1972* pp. 11–17.

Conti, M. (1984). Epidemiology and vectors of plant reo-like viruses. *Curr. Top. Vector Res.* **2,** 112–139.

Conti, M., and Appiano, A. (1973). Barley yellow striate mosaic virus and associated viroplasms in barley cells. *J. Gen. Virol.* **21,** 315–322.

Conti, M., and Masenga, V. (1977). Identification and prevalence of pepper viruses in northwest Italy. *Phytopathol. Z.* **90,** 212–222.

Converse, R. H., and Lister, R. M. (1969). The occurrence and some properties of black raspberry latent virus. *Phytopathology* **59,** 325–333.

Converse, R. H., and Martin, R. R. (1982). Use of the Barbara–Clark (F(ab′)$_2$) indirect ELISA test for detection of viruses in small fruits. *Acta Hortic.* **129,** 21.

Converse, R. H., and Tanne, E. (1984). Heat therapy and stolon apex culture to eliminate mild yellow-edge virus from Hood strawberry. *Phytopathology* **74,** 1315–1316.

Cooke, R., and Penon, P. (1990). *In vitro* transcription from cauliflower mosaic virus promoters by a cell free extract from tobacco cells. *Plant Molecular Biology* **14,** 391–405.

Cooper, J. I. (1971). The distribution in Scotland of tobacco rattle virus and its nematode vectors in relation to soil type. *Plant Pathol.* **20,** 51–58.

Cooper, J. I., and Harrison, B. D. (1973). Distribution of potato mop-top-virus in Scotland in relation to soil and climate. *Plant Pathol.* **22,** 73–78.

Cooper, J. I., and Jones, A. T. (1983). Responses of plants to viruses: Proposals for the use of terms. *Phytopathology* **73,** 127–128.

Cooper, J. I., and Mayo, M. A. (1972). Some properties of the particles of three tobravirus isolates. *J. Gen. Virol.* **16,** 285–297.

Cooper, J. I., Jones, R. A. C., and Harrison, B. D. (1976). Field and glasshouse experiments on the control of potato mop-top virus. *Ann. Appl. Biol.* **83,** 215–230.

Cooper, J. I., Massalski, P. R., and Edwards, M.-L. (1984). Cherry leaf roll virus in the female gametophyte and seed of birch and its relevance to vertical transmission. *Ann. Appl. Biol.* **105,** 55–64.

Cope, W. A., Walker, S. K., and Lucas, L. T. (1978). Evaluation of selected white clover clones for resistance to viruses in the field. *Plant Dis. Rep.* **62,** 267–270.

Cornelissen, B. J. C., and Bol, J. F. (1984). Homology between the proteins encoded by tobacco mosaic virus and two tricornaviruses. *Plant Mol. Biol.* **3,** 379–384.

Cornelissen, B. J. C., Brederode, F. T., Moormann, R. J. M., and Bol, J. F. (1983a). Complete nucleotide sequence of alfalfa mosaic virus RNA1. *Nucleic Acids Res.* **11,** 1253–1265.

Cornelissen, B. J. C., Brederode, F. T., Veeneman, G. H., van Boom, J. H., and Bol, J. F. (1983b). Complete nucleotide sequence of alfalfa mosaic virus RNA2. *Nucleic Acids Res.* **11,** 3019–3025.

Cornelissen, B. J. C., Janssen, H., Zuidema, D., and Bol, J. F. (1984). Complete nucleotide sequence of tobacco streak virus RNA3. *Nucleic Acids Res.* **12,** 2427–2437.

Cornelissen, B. J. C., Linthorst, H. J. M., Brederode, F. T., and Bol, J. F. (1986a). Analysis of the genome structure of tobacco rattle virus strain PSG. *Nucleic Acids Res.* **14,** 2157–2169.

Cornelissen, B. J. C., Hooft van Huijsduijnen, R. A. M., and Bol, J. F. (1986b). A tobacco mosaic virus-induced tobacco protein is homologous to the sweet-tasting protein thaumatin. *Nature (London)* **321,** 531–532.

Cornelissen, B. J. C., Horowitz, J., van Kan, J. A. L., Goldberg, R. B., and Bol, J. F. (1987). Structure of tobacco genes encoding pathogenesis-related proteins from the PR-I group. *Nucleic Acids Res.* **15,** 6799–6811.

Correia, J.-J., Shire, S., Yphantis, D. A., and Schuster, T. M. (1985). Sedimentation equilibrium measurements of the intermediate-size tobacco mosaic virus protein polymers. *Biochemistry* **24,** 3292–3297.

Costa, A. S., and Muller, G. W. (1980). Tristeza control by cross-protection: A US–Brazil cooperative success. *Plant Dis.* **64,** 538–541.

Cotten, J. (1979). The effectiveness of soil sampling for virus-vector nematodes in MAFF certification schemes for fruit and hops. *Plant Pathol.* **28,** 40–44.

Couch, H. B. (1955). Studies on the seed transmission of lettuce mosaic virus. *Phytopathology* **45,** 63–70.

Coutts, R. H. A. (1986). New threats to crops from changing farm practices. *Nature (London)* **322,** 594.

Coutts, R. H. A., and Buck, K. W. (1985). DNA and RNA polymerase activities of nuclei and hypotonic extracts of nuclei isolated from tomato golden mosaic virus infected tobacco leaves. *Nucleic Acids Res.* **13,** 7881–7897.

Coutts, R. H. A., Buck, K. W., Roberts, E. J. F., Brough, C. L., Hayes, R. J., Macdonald, H., MacDowell, S. W., Petty, I. T. D., Slomka, M. J., Hamilton, W. D. O., and Bevan, M. W. (1987). Studies on the replication of tomato golden mosaic virus and the construction of gene vectors. *In* "Molecular Strategies for Crop Protection" (C. J. Arntzen and C. Ryan, eds.), pp. 307–318. Alan R. Liss, New York.

Covey, S. N., and Hull, R. (1981). Transcription of cauliflower mosaic virus DNA. Detection of transcripts, properties, and location of the gene encoding the virus inclusion body protein. *Virology* **111,** 463–474.

Covey, S. N., Turner, D., and Mulder, G. (1983). A small DNA molecule containing covalently-linked ribonucleotides originates from the large intergenic region of the cauliflower mosaic virus genome. *Nucleic Acids Res.* **11,** 251–264.

Covey, S. N., Turner, D. S., Lucy, A. P., and Saunders, K. (1990). Host regulation of the cauliflower mosaic virus multiplication cycle. *Proc. Natl. Acad. Sci. U.S.A.* **87,** 1633–1637.

Crane, P. R., and Lidgard, S. (1989). Angiosperm diversification and paleolatitudinal gradients in cretaceous floristic diversity. *Science* **246,** 675–678.

Creamer, R., and Falk, B. W. (1990). Direct detection of transcapsidated barley yellow dwarf luteovirus in doubly-infected plants. *J. Gen. Virol.* **71,** 211–217.

Creaser, E. H., Gibbs, A. J., and Pares, R. D. (1987). The amino acid composition of the coat protein of a tobamovirus from an Australian *Capsicum* crop. *Aust. Plant Pathol.* **16,** 85–87.

Creighton, T. E. (1988a). The protein folding problem. *Science* **240,** 267, 344.

Creighton, T. E. (1988b). Toward a better understanding of protein folding pathways. *Proc. Natl. Acad. Sci. U.S.A.* **85,** 5082–5086.

Cress, D. E., Kiefer, M. C., and Owens, R. A. (1983). Construction of infectious potato spindle tuber viroid cDNA clones. *Nucleic Acids Res.* **11,** 6821–6835.

Crick, F. H. C., and Watson, J. D. (1956). Structure of small viruses. *Nature (London)* **177,** 473–475.

Crill, P., Hagedorn, D. J., and Hanson, E. W. (1971). An artificial system for differentiating strains of alfalfa mosaic virus. *Plant Dis. Rep.* **55,** 127–130.

Cronshaw, J., Hoefert, L., and Esau, K. (1966). Ultrastructural features of *Beta* leaves infected with beet yellows virus. *J. Cell Biol.* **31,** 429–443.

Crosbie, E. S., and Matthews, R. E. F. (1974a). Effects of TYMV infection on leaf pigments in *Brassica pekinensis* Rupr. *Physiol. Plant Pathol.* **4,** 379–387.

Crosbie, E. S., and Matthews, R. E. F. (1974b). Effects of TYMV infection on growth of *Brassica pekinensis* Rupr. *Physiol. Plant Pathol.* **4,** 389–400.

Crowley, N. C. (1959). Studies on the time of embryo infection by seed-transmitted viruses. *Virology* **8,** 116–123.

Crowley, N. C., Davison, E. M., Francki, R. I. B., and Owusu, G. K. (1969). Infection of bean root-meristems by tobacco ringspot virus. *Virology* **39,** 322–330.

Crowson, R. A., Ralfe, W. D. I., Smart, J., Waterson, C. D., Willey, E. C., and Wootton, R. J. (1967). "Arthropoda: Chelicerata, Pyconogonida, Palaeoisopus, Myriapoda and Insecta. The Fossil Record," pp. 499–534. Geological Society of London.

Crowther, R. A., and Klug, A. (1975). Structural analysis of macromolecular assemblies by image reconstruction from electron micrographs. *Annu. Rev. Biochem.* **44,** 161–182.

Crowther, R. A., Geelen, J. L. M. C., and Mellema, J. E. (1974). A three-dimensional image reconstruction of cowpea mosaic virus. *Virology* **57,** 20–27.

Crum, C., Johnson, J. D., Nelson, A., and Roth, D. (1988). Complementary oligodeoxynucleotide mediated inhibition of tobacco mosaic virus RNA translation *in vitro*. *Nucleic Acids Res.* **16,** 4569–4581.

Cuillel, M., Herzog, M., and Hirth, L. (1979). Specificity of *in vitro* reconstitution of bromegrass mosaic virus. *Virology* **95,** 146–153.

Cuillel, M., Jacrot, B., and Zulauf, M. (1981). A T = 1 capsid formed by protein of brome mosaic virus in the presence of trypsin. *Virology* **110,** 63–72.

Cuillel, M., Berthet-Colominas, C., Krop, B., Tardieu, A., Vachette, P., and Jacrot, B. (1983). Self assembly of brome mosaic virus capsids. Kinetic studies using neutron and X-ray solution scattering. *J. Mol. Biol.* **164,** 645–650.

Culver, J. N., and Dawson, W. O. (1989a). Point mutations in the coat protein gene of tobacco mosaic virus induce hypersensitivity in *Nicotiana sylvestris. Mol. Plant-Microbe Interact.* **2,** 209–213.

Culver, J. N., and Dawson, W. O. (1989b). Tobacco mosaic virus coat protein: An elicitor of the hypersensitive reaction, but not required for the development of mosaic symptoms in *Nicotiana sylvestris. Virology* **173,** 755–758.

Cuozzo, M., O'Connell, K. M., Kaniewski, W., Fang, R.-X., Chua, N.-H., and Tumer, N. E. (1988). Viral protection in transgenic tobacco plants expressing the cucumber mosaic virus coat protein or its antisense RNA. *Bio/Technology* **6,** 549–557.

Cusack, S., Miller, A., Krijgsmann, P. C. J., and Mellema, J. E. (1981). An investigation of the structure of alfalfa mosaic virus by small-angle neutron scattering. *J. Mol. Biol.* **145,** 525–543.

Cusack, S., Oostergetel, G. T., Krijgsman, P. C. J., and Mellema, J. E. (1983). Structure of the Top a-t component of alfalfa mosaic virus. A non-icosahedral virion. *J. Mol. Biol.* **171,** 139–155.

Cutt, J. R., Harpster, M. H., Dixon, D. C., Carr, J. P., Dunsmuir, P., and Klessig, D. F. (1989). Disease response to tobacco mosaic virus in transgenic tobacco plants that constitutively express the pathogenesis-related PR1b gene. *Virology* **173,** 89–97.

da Graça, J. V., and Martin, M. M. (1975). Ultrastructural changes in tobacco mosaic virus-induced local lesions in *Nicotiana tabacum* L. cv. "Samsun NN." *Physiol. Plant Pathol.* **7,** 287–291.

da Graça, J. V., and Martin, M. M. (1981). Ultrastructural changes in avocado leaf tissue infected with avocado sunblotch. *Phytopathol. Z.* **102,** 185–194.

Dahl, D., and Knight, C. A. (1963). Some nitrous acid-induced mutants of tomato atypical mosaic virus. *Virology* **21,** 580–586.

Dale, J. L. (1987). Banana bunchy top: An economically important tropical plant virus disease. *Adv. Virus Res.* **33,** 301–325.

Dale, W. T. (1962). Diseases and pests of cocoa. A. Virus diseases. *In* "Agriculture and Land Use in Ghana" (J. B. Wills, ed.), pp. 286–316. Oxford Univ. Press, London and New York.

Dall, D. J., Anzola, J. V., Xu, Z., and Nuss, D. L. (1989a). Complete nucleotide sequence of wound tumor virus genomic segment S11. *Nucleic Acids Res.* **17,** 3599.

Dall, D. J., Randles, J. W., and Francki, R. I. B. (1989b). The effect of alfalfa mosaic virus on productivity of annual barrel medic, *Medicago truncatula. Aust. J. Agric. Res.* **40,** 807–815.

Dall, D. J., Anzola, J. V., Xu, Z., and Nuss, D. L. (1990). Functional role for both terminal domains in the selective binding of transcripts encoded by a segmented RNA virus. *EMBO J.* **9,** in press.

Damsteegt, V. D., and Hewings, A. D. (1986). Comparative transmission of soybean dwarf virus by three geographically diverse populations of *Aulacorthum (=Acyrthosiphon) solani. Ann. Appl. Biol.* **109,** 453–463.

Damsteegt, V. D., and Hewings, A. D. (1987). Relationships between *Aulacorthum solani* and soybean dwarf virus: Effect of temperature on transmission. *Phytopatholology* **77,** 515–518.

Darnell, J. E., and Doolittle, W. F. (1986). Speculations on the early course of evolution. *Proc. Natl. Acad. Sci. U.S.A.* **83,** 1271–1275.

Das, S., and Raski, D. J. (1968). Vector-efficiency of *Xiphinema index* in the transmission of grapevine fanleaf virus. *Nematologica* **14,** 55–62.

Datema, K. P., Spruijt, R. B., Verduin, J. M., and Hemminga, M. A. (1987). Interaction of plant viruses and viral coat proteins with model membranes. *Biochemistry* **26,** 6217–6223.

Daubert, S. D., and Bruening, G. (1984). Detection of genome-linked proteins of plant and animal viruses. *In* "Methods in Virology" (K. Maramorosch and H. Koprowski, eds.), pp. 347–379. Academic Press, Orlando, Florida.

Daubert, S. D., Richins, R., Shepherd, R. J., and Gardner, R. C. (1982). Mapping of the coat protein gene of cauliflower mosaic virus by its expression in a prokaryotic system. *Virology* **122,** 444–449.

Daubert, S. D., Schoelz, J., Debao, L., and Shepherd, R. J. (1984). Expression of disease symptoms in cauliflower mosaic virus genomic hybrids. *J. Mol. Appl. Genet.* **2,** 537–547.

Davies, C., and Symons, R. H. (1988). Further implications for the evolutionary relationships between tripartite plant viruses based on cucumber mosaic virus RNA3. *Virology* **165,** 216–224.

Davies, D. L., and Clark, M. F. (1983). A satellite-like nucleic acid of arabis mosaic virus associated with hop nettlehead disease. *Ann. Appl. Biol.* **103,** 439–448.

Davies, J. W. (1979). Translation of plant virus ribonucleic acids in extracts from eukaryotic cells. *In* "Nucleic Acids in Plants" (T. C. Hall and J. W. Davies, eds.), Vol. 2, pp. 111–149. CRC Press, West Palm Beach, Florida.

Davies, J. W. (1987). Geminivirus genomes. *Microbiol. Sci.* **4,** 18–22.

Davies, J. W., and Stanley, J. (1989). Geminivirus genes and vectors. *Trends Genet.* **5,** 77–81.

Davies, J. W., Stanley, J., Donson, J., Mullineaux, P. M., and Boulton, M. I. (1987a). Structure and replication of geminivirus genomes. *J. Cell Sci., Suppl.* **7**, 95–107.

Davies, J. W., Townsend, R., and Stanley, J. (1987b). The structure, expression, functions and possible exploitation of geminivirus genomes. *Adv. Plant Sci.* **4**, 31–52.

Davis, P. H., and Heywood, V. H. (1963). "Principles of Angiosperm Taxonomy." Oliver & Boyd, Edinburgh.

Davis, R. F., and Shifriss, O. (1983). Natural virus infection in silvery and non-silvery lines of *Cucurbita pepo. Plant Dis.* **67**, 379–380.

Dawson, G. W., Griffiths, D. C., Pickett, J. A., Plumb, R. T., Woodcock, C. M., and Zhong-Ning, Z. (1988). Structure/activity studies on aphid alarm pheromone derivatives and their field use against transmission of barley yellow dwarf virus. *Pestic. Sci.* **22**, 17–30.

Dawson, J. R. O. (1967). The adaptation of tomato mosaic virus to resistant tomato plants. *Ann. Appl. Biol.* **60**, 209–214.

Dawson, W. O. (1972). Enhancement of the infectivity, nucleoprotein concentration, and multiplication rate of cowpea chlorotic mottle virus in manganese-treated cowpea. *Phytopathology* **62**, 1206–1209.

Dawson, W. O. (1978a). Isolation and mapping of replication-deficient, temperature-sensitive mutants of cowpea chlorotic mottle virus. *Virology* **90**, 112–118.

Dawson, W. O. (1978b). Time-course of cowpea chlorotic mottle virus RNA replication. *Intervirology* **9**, 119–128.

Dawson, W. O. (1983). Tobacco mosaic virus protein synthesis is correlated with double-stranded RNA synthesis and not single-stranded RNA synthesis. *Virology* **125**, 314–323.

Dawson, W. O. (1990). Relationship of tobacco mosaic virus gene expression to movement within the plant. *In* "Viral Genes and Plant Pathogenesis" (T. P. Pirone and J. G. Shaw, eds.), pp. 39–52. Springer-Verlag, New York.

Dawson, W. O., and Boyd, C. (1987a). Modifications of nucleic acid precursors that inhibit plant virus multiplication. *Phytopathology* **77**, 477–480.

Dawson, W. O., and Boyd, C. (1987b). TMV protein synthesis is not translationally regulated by heat shock. *Plant Mol. Biol.* **8**, 145–149.

Dawson, W. O., and Jones, G. E. (1976). A procedure for specifically selecting temperature sensitive mutants of tobacco mosaic virus. *Mol. Gen. Genet.* **145**, 307–309.

Dawson, W. O., and Lehto, K. (1989). Mechanisms of TMV symptom formation. *NATO Adv. Res. Workshop. Recognition Response Plant-Virus Interact.*, 1989.

Dawson, W. O., and Korhonen-Lehto, K. M. (1990). Regulation of tobamovirus gene expression. *Adv. Virus Res.* **38**, 307–342.

Dawson, W. O., Schlegel, D. E., and Lung, M. C. Y. (1975). Synthesis of tobacco mosaic virus in intact tobacco leaves systemically inoculated by differential temperature treatment. *Virology* **65**, 565–573.

Dawson, W. O., Beck, D. L., Knorr, D. A., and Grantham, G. L. (1986). cDNA cloning of the complete genome of tobacco mosaic virus and production of infectious transcripts. *Proc. Natl. Acad. Sci. U.S.A.* **83**, 1832–1836.

Dawson, W. O., Bubrick, P., and Grantham, G. L. (1988). Modification of the tobacco mosaic virus coat protein gene affecting replication movement and symptomatology. *Phytopathology* **78**, 783–789.

Dawson, W. O., Lewandowski, D. J., Hilf, M. E., Bubrick, P., Raffo, A. J., Shaw, J. J., Grantham, G. L., and Desjardins, P. R. (1989). A tobacco mosaic virus–hybrid expresses and loses an added gene. *Virology* **172**, 285–292.

Dean, J. L. (1979). Sugarcane mosaic virus: Shape of the inoculum–infection curve near the origin. *Phytopathology* **69**, 179–181.

De Bokx, J. A., ed. (1972). "Viruses of Potatoes and Seed-potato Production." Pudoc, Wageningen.

De Bokx, J. A., and Piron, P. G. M. (1981). Transmission of potato spindle tuber viroid by aphids. *Neth. J. Plant Pathol.* **87**, 31–34.

De Bortoli, M., and Roggero, P. (1985). Electrophoretic desorption of intact virus from immunoadsorbent. *Microbiologica* **8**, 113–121.

De Fazio, G., Kudamatsu, M., and Vicente, M. (1980). Virazole pretreatments for the prevention of tomato spotted wilt virus (TSWV) systemic infection in tobacco plants *Nicotiana tabacum*, L. 'White Burley.' *Fitopatol. Bras.* **5**, 343–394.

de Haan, P., Wagemakers, L., Peters, D., and Goldbach, R. (1989). Molecular cloning and terminal sequence determination of the S and M RNA of tomato spotted wilt virus. *J. Gen. Virol.* **70,** 3469–3473.

de Haan, P., Wagemakers, L., Peters, D., and Goldbach, R. (1990). The SRNA segment of tomato spotted wilt virus has an ambisense character. *J. Gen. Virol.* **71,** 1001–1007.

de Jager, C. P. (1976). Genetic analysis of cowpea mosaic virus mutants by supplementation and reassortment tests. *Virology* **70,** 151–163.

Dekker, E. L., Doré, I., Porta, C., and Van Regenmortel, M. H. V. (1987). Conformational specificity of monoclonal antibodies used in the diagnosis of tomato mosaic virus. *Arch. Virol.* **94,** 191–203.

Dekker, E. L., Pinner, M. S., Markham, P. G., and Van Regenmortel, M. H. V. (1988). Characterisation of maize streak virus isolates from different plant species by polyclonal and monoclonal antibodies. *J. Gen. Virol.* **69,** 983–990.

Dekker, E. L., Porta, C., and Van Regenmortel, M. H. V. (1989). Limitations of different ELISA procedures for localizing epitopes in viral coat protein subunits. *Arch. Virol.* **105,** 269–286.

De Laat, A. M. M., and van Loon, L. C. (1983). The relationship between stimulated ethylene production and symptom expression in virus-infected tobacco leaves. *Physiol. Plant Pathol.* **22,** 261–273.

Delauney, A. J., Tabaeizadeh, Z., and Verma, D. P. S. (1988). A stable bifunctional antisense transcript inhibiting gene expression in transgenic plants. *Proc. Natl. Acad. Sci. U.S.A.* **85,** 4300–4304.

De Leeuw, G. T. N. (1975). An easy and precise method to measure the length of flexuous virus particles from electron micrographs. *Phytopathol. Z.* **82,** 347–351.

Delwiche, C. F., Graham, L. E., and Thomson, N. (1989). Lignin-like compounds and sporopollenin in *Coleochaete,* an algal model for land plant ancestry. *Science* **245,** 399–401.

de Mejia, M. V. G., Hiebert, E., Purcifull, D. E., Thornbury, D. W., and Pirone, T. P. (1985). Identification of potyviral amorphous inclusion protein as a non-structural virus-specific protein related to helper component. *Virology* **142,** 34–43.

Demler, S. A., and De Zoeten, G. A. (1989). Characterization of a satellite RNA associated with pea enation mosaic virus. *J. Gen. Virol.* **70,** 1075–1084.

Dempsey, G. P., Bullivant, S., and Bieleski, R. L. (1975). The distribution of P-protein in mature sieve elements of celery. *Planta* **126,** 45–59.

Demski, J. W., Harris, H. B., and Jellum, M. D. (1971). Effects of time of inoculation with tobacco ringspot virus on the chemical composition and agronomic characteristics of soybean. *Phytopathology* **61,** 308–311.

Demski, J. W., Bays, D. C., and Kahn, M. A. (1986). Simple latex agglutination test for detecting flexuous rod-shaped viruses in forage legumes. *Plant Dis.* **70,** 777–779.

Denloye, A. O., Homer, R. B., and Hull, R. (1978). Circular dichroism studies on turnip rosette virus. *J. Gen. Virol.* **41,** 77–85.

Denna, D. W., and Alexander, M. B. (1975). The isoperoxidases of *Cucurbita pepo* L. *In* "Isozymes" (C. L. Markert, ed.), Vol. 2, pp. 851–864. Academic Press, New York.

Deom, C. M., Oliver, M. J., and Beachy, R. N. (1987). The 30-kilodalton gene product of tobacco mosaic virus potentiates virus movement. *Science* **237,** 389–394.

Deom, C. M., Schubert, K. R., Wolf, S., Holt, C. A., Lucas, W. J., and Beachy, R. N. (1990). Molecular characterisation and biological function of the movement protein of tobacco mosaic virus in transgenic plants. *Proc. Natl. Acad. Sci. U.S.A.* **87,** 3284–3288.

Derrick, K. S. (1973). Quantitative assay for plant viruses using serologically specific electron microscopy. *Virology* **56,** 652–653.

Desjardins, P. R., Drake, R. J., and French, J. V. (1969). Transmission of citrus ringspot virus to citrus and non-citrus hosts by dodder *(Campestris subinclusa)*. *Plant Dis. Rep.* **53,** 947–948.

de Varennes, A., Davies, J. W., Shaw, J. G., and Maule, A. J. (1985). A reappraisal of the effect of actinomycin D and cordycepin on the multiplication of cowpea mosaic virus in cowpea protoplasts. *J. Gen. Virol.* **66,** 817–825.

de Varennes, A., Lomonossoff, G. P., Shanks, M., and Maule, A. J. (1986). The stability of cowpea mosaic virus VPg in reticulocyte lysates. *J. Gen. Virol.* **67,** 2347–2354.

Devic, M., Jaegle, M., and Baulcombe, D. (1989). Symptom production on tobacco and tomato is determined by two distinct domains of the satellite RNA of cucumber mosaic virus (strain Y). *J. Gen. Virol.* **70,** 2765–2774.

De Wijs, J. J. (1975). The distribution of passionfruit ringspot virus in its main host plants in Ivory Coast. *Neth. J. Plant Pathol.* **81,** 144–148.

De Wijs, J. J. (1980). The characteristics of mineral oils in relation to their inhibitory activity on the aphid transmission of potato virus Y. *Neth. J. Plant Pathol.* **86**, 291–300.

De Wijs, J. J., and Suda-Bachmann, F. (1979). The long-term preservation of potato virus Y and watermelon mosaic virus in liquid nitrogen in comparison to other preservation methods. *Neth. J. Plant Pathol.* **85**, 23–29.

De Zoeten, G. A. (1966). California tobacco rattle virus, its intracellular appearance, and the cytology of the infected cell. *Phytopathology* **56**, 744–754.

De Zoeten, G. A. (1968). Application of scanning microscopy in the study of virus transmission of aphids. *J. Virol.* **2**, 745–751.

De Zoeten, G. A., and Fulton, R. W. (1975). Understanding generates possibilities. *Phytopathology* **65**, 221–222.

De Zoeten, G. A., and Gaard, G. (1969a). Distribution and appearance of alfalfa mosaic virus in infected plant cells. *Virology* **39**, 768–774.

De Zoeten, G. A., and Gaard, G. (1969b). Possibilities for inter- and intracellular translocation of some icosahedral plant viruses. *J. Cell Biol.* **40**, 814–823.

De Zoeten, G. A., and Gaard, G. (1984). The presence of viral antigen in the apoplast of systemically virus-infected plants. *Virus Res.* **1**, 713–725.

De Zoeten, G. A., Gaard, G., and Diez, F. B. (1972). Nuclear vesiculation associated with pea enation mosaic virus-infected plant tissue. *Virology* **48**, 638–647.

De Zoeten, G. A., Assink, A. M., and van Kammen, A. (1974). Association of cowpea mosaic virus-induced double-stranded RNA with a cytopathological structure in infected cells. *Virology* **59**, 341–355.

De Zoeten, G. A., Powell, C. A., Gaard, G., and German, T. L. (1976). *In situ* localization of pea enation mosaic virus double-stranded ribonucleic acid. *Virology* **70**, 459–469.

De Zoeten, G. A., Penswick, J. R., Horisberger, M. A., Ahl, P., Schultze, M., and Hohn, T. (1989). The expression localization and effect of a human interferon in plants. *Virology* **172**, 213–222.

Diaco, R., Hill, J. H., Hill, G. K., Tachibana, H., and Durand, D. P. (1985). Monoclonal antibody-based biotin-avidin ELISA for the detection of soybean mosaic virus in soybean seeds. *J. Gen. Virol.* **66**, 2089–2094.

Diaco, R., Hill, J. H., and Durand, D. P. (1986a). Purification of soybean mosaic virus by affinity chromatography using monoclonal antibodies. *J. Gen. Virol.* **67**, 345–351.

Diaco, R., Lister, R. M., Hill, J. H., and Durand, D. P. (1986b). Detection of homologous and heterologous barley yellow dwarf virus isolates with monoclonal antibodies in serologically specific electron microscopy. *Phytopathology* **76**, 225–230.

Dias, H. F. (1970). The relationship between cucumber-necrosis virus and its vector *Olpidium cucurbitacearum*. *Virology* **42**, 204–211.

Diaz-Ruiz, J. R., and Kaper, J. M. (1983). Nucleotide sequence relationships among thirty peanut stunt virus isolates determined by competition hybridization. *Arch. Virol.* **75**, 277–281.

Dickerson, P. E., and Trim, A. R. (1978). Conformational states of cowpea chlorotic mottle virus ribonucleic acid components. *Nucleic Acids Res.* **5**, 987–998.

Diener, T. O. (1971). Potato spindle tuber "virus." IV. A replicating, low molecular weight RNA. *Virology* **45**, 411–428.

Diener, T. O. (1981). Are viroids escaped introns? *Proc. Natl. Acad. Sci. U.S.A.* **78**, 5014–5015.

Diener, T. O. (1987a). Biological properties. *In* "The Viroids" (T. O. Diener, ed.), pp. 9–35. Plenum, New York.

Diener, T. O., ed. (1987b). "The Viroids." Plenum, New York.

Diener, T. O. (1989). Circular RNAs; Relics of precellular evolution? *Proc. Natl. Acad. Sci. U.S.A.* **86**, 9370–9374.

Diener, T. O., and Hadidi, A. (1977). Viroids. *Compr. Virol.* **11**, 285–337.

Diener, T. O., and Jenifer, F. G. (1964). A dependable local lesion assay for turnip yellow mosaic virus. *Phytopathology* **54**, 1258–1260.

Dietzgen, R. G. (1986). Immunological properties and biological function of monoclonal antibodies to tobacco mosaic virus. *Arch. Virol.* **87**, 73–86.

Dietzgen, R. G., and Francki, R. I. B. (1987). Nonspecific binding of immunoglobulins to coat proteins of certain plant viruses in immunoblots and indirect ELISA. *J. Virol. Methods* **15**, 159–164.

Dietzgen, R. G., and Francki, R. I. B. (1988). Analysis of lettuce necrotic yellows virus structural proteins with monoclonal antibodies and concanavalin A. *Virology* **166**, 486–494.

Dietzgen, R. G., and Sander, E.-M. (1982). Monoclonal antibodies against a plant virus. *Arch. Virol.* **74**, 197–204.

Dietzgen, R. G., and Zaitlin, M. (1986). Tobacco mosaic virus coat protein and the large subunit of the host protein ribulose-1,5-bisphosphate carboxylase share a common antigenic determinant. *Virology* **155**, 262–266.

Dietzgen, R. G., Hunter, B. G., Francki, R. I. B., and Jackson, A. O. (1989). Cloning of lettuce necrotic yellows virus RNA and identification of virus-specific polyadenylated RNAs in infected *Nicotiana glutinosa* leaves. *J. Gen. Virol.* **70**, 2299–2307.

Di Franco, A., and Martelli, G. P. (1987). Some observations on the ultrastructure of galinsoga mosaic virus infections. *Phytopathol. Mediterr.* **26**, 54–56.

Di Franco, A., Russo, M., and Martelli, G. P. (1984). Ultrastructure and origin of cytoplasmic multi-vesicular bodies induced by carnation Italian ringspot virus. *J. Gen. Virol.* **65**, 1233–1237.

Dijkstra, J. (1962). On the early stages of infection by tobacco mosaic virus in *Nicotiana glutinosa* L. *Virology* **18**, 142–143.

Dijkstra, J. (1966). Multiplication of TMV in isolated epidermal tissue of tobacco and *Nicotiana glutinosa* leaves. *In* "Viruses of Plants" (A. B. R. Beemster and J. Dijkstra, eds.), pp. 19–21. North-Holland Publ., Amsterdam.

Dijkstra, J., Bruin, G. C. A., Burgers, A. C., van Loon, L. C., Ritter, C., van de Sanden, P. A. C. M., and Wieringa-Brants, D. H. (1977). Systemic infection of some N-gene-carrying *Nicotiana* species and cultivars after inoculation with tobacco mosaic virus. *Neth. J. Plant Pathol.* **83**, 41–59.

Dimock, A. W., Geissinger, C. M., and Herst, R. K. (1971). A new adaptation of tissue implantation for the study of virus and mycoplasma diseases. *Phytopathology* **61**, 429–430.

Ding, S.-W., Keese, P., and Gibbs, A. (1989). Nucleotide sequence of the ononis yellow mosaic tymovirus genome. *Virology* **172**, 555–563.

Ding, S.-W., Gibbs, A., and Keese, P. (1990a). The nucleotide sequence of the genomic RNA of kennedya yellow mosaic tymovirus–Jervis Bay isolate: Relationships with potex- and carlaviruses. *J. Gen. Virol.* **71**, 925–931.

Ding, S.-W., Howe, J., Keese, P., Mackenzie, A., Meek, D., Osorio-Keese, M., Skotnicki, M., Srifah, P., Torronen, M., and Gibbs, A. (1990b). The tymobox, a sequence shared by most tymoviruses: Its use in molecular studies of tymoviruses. *Nucleic Acids Res.* **18**, 1181–1187.

Dingwall, C. (1985). The accumulation of proteins in the nucleus. *Trends Biochem. Sci.* **10**, 64–66.

Dinter-Gottlieb, G. (1986). Viroids and virusoids are related to group I introns. *Proc. Natl. Acad. Sci. U.S.A.* **83**, 6250–6254.

Dixon, L. K., and Hohn, T. (1984). Initiation of translation of the cauliflower mosaic virus genome from a polycistronic mRNA: Evidence from deletion mutagenesis. *EMBO J.* **3**, 2731–2736.

Dixon, L. K., Nyffenegger, T., Delly, G., Martinez-Izquierdo, J., and Hohn, T. (1986). Evidence for replicative recombination in cauliflower mosaic virus. *Virology* **150**, 463–468.

Dodds, J. A. (1986). The potential for using double-stranded RNAs as diagnostic probes for plant viruses. *In* "Developments and Applications in Virus Testing" (R. A. C. Jones and L. Torrance, eds.), Dev. Appl. Biol. I, pp. 71–86. Association of Applied Biologists, Wellsbourne.

Dodds, J. A., and Hamilton, R. I. (1974). Masking of the RNA genome of tobacco mosaic virus by the protein of barley stripe mosaic virus in doubly infected barley. *Virology* **59**, 418–427.

Dodds, J. A., and Hamilton, R. I. (1976). Structural interactions between viruses as a consequence of mixed infections. *Adv. Virus Res.* **20**, 33–86.

Dodds, J. A., Morris, T. J., and Jordan, R. L. (1984). Plant viral double-stranded RNA. *Annu. Rev. Phytopathol.* **22**, 151–168.

Dodds, J. A., Lee, S. Q., and Tiffany, M. (1985). Cross protection between strains of cucumber mosaic virus: Effect of host and type of inoculum on accumulation of virions and double-stranded RNA of the challenge strain. *Virology* **144**, 301–309.

Dodds, J. A., Jarupat, T., Lee, J. G., and Roistacher, C. N. (1987). Effects of strain, host, time of harvest, and virus concentration on double-stranded RNA analysis of citrus tristeza virus. *Phytopathology* **77**, 442–447.

Doi, Y., Teranaka, M., Yora, K., and Asuyama, H. (1967). Mycoplasma or PLT group-like micro-organisms found in the phloem elements of plants infected with mulberry dwarf, potato witches broom, aster yellows, or Paulownia witches broom. *Ann. Phytopathol. Soc. Jpn.* **33**, 259–266.

Doke, N., and Ohashi, Y. (1988). Involvement of an $O_2{}^-$ generating system in the induction of necrotic local lesions on tobacco leaves infected with tobacco mosaic virus. *Physiol. Mol. Plant Pathol.* **32**, 163–175.

Dolja, V. V., Sokolova, N. A., Tjulkina, L. G., and Atabekov, J. G. (1979). A study of barley stripe mosaic virus (BSMV) genome. II. Translation of individual RNA species of two BSMV strains in a homologous cell-free system. *Mol. Gen. Genet.* **175**, 93–97.

Dolja, V. V., Lunina, N. A., Leiser, R.-M., Stanarius, T., Belzhelarskaya, S. N., Kozlov, Yu. V., and Atabekov, J. G. (1983). A comparative study on the *in vitro* translation products of individual

RNAs from two-, three-, and four-component strains of barley stripe mosaic virus. *Virology* **127**, 1–14.

Dollet, M., Accotto, G. P., Lisa, V., Menissier, J., and Boccardo, G. (1986). A geminivirus, serologically related to maize streak virus, from *Digitaria sanguinalis* from Vanuatu. *J. Gen. Virol.* **67**, 933–937.

Domier, L. L., Franklin, K. M., Shahabuddin, M., Hellmann, G. M., Overmeyer, J. H., Hiremath, S. T., Siaw, M. F. E., Lomonossoff, G. P., Shaw, J. G., and Rhoads, R. E. (1986). The nucleotide sequence of tobacco vein mottling virus RNA. *Nucleic Acids Res.* **14**, 5417–5431.

Domier, L. L., Shaw, J. G., and Rhoads, R. E. (1987). Potyviral proteins share amino acid sequence homology with picorna-, como-, and caulimoviral proteins. *Virology* **158**, 20–27.

Domier, L. L., Franklin, K. M., Hunt, A. G., Rhoads, R. E., and Shaw, J. G. (1989). Infectious *in vitro* transcripts from cloned cDNA of a potyvirus, tobacco vein mottling virus. *Proc. Natl. Acad. Sci. U.S.A.* **86**, 3509–3513.

Domingo, E., and Holland, J.-J. (1988). High error rates, population equilibrium, and evolution of RNA replication systems. *In* "RNA Genetics" (E. Domingo, J.-J. Holland, and P. Ahlquist, eds.), Vol. 3, pp. 3–36. CRC Press, Boca Raton, Florida.

Domingo, E., Martínez-Salas, E., Sobrino, F., de la Torre, J. C., Portela, A., Ortín, J., López-Galindez, C., Pérez-Breña, P., Villanueva, N., Nájera, R., VandePol, S., Steinhauer, D., DePolo, N., and Holland, J. (1985). The quasispecies (extremely heterogenous) nature of viral RNA genome populations: Biological relevance—A review. *Gene* **40**, 1–8.

Doncaster, J. P., and Gregory, P. H. (1948). The spread of virus diseases in the potato crop. *G. B., Agric. Res. Counc., Rep. Ser.* **7**, 1–189.

Donofrio, J. C., Kuchta, J., Moore, R., and Kacymarczyk, W. (1986). Properties of a solubilized replicase isolated from corn infected with maize dwarf mosaic virus. *Can. J. Microbiol.* **32**, 637–644.

Donson, J., Morris-Krsinich, B. A. M., Mullineaux, P. M., Boulton, M. I., and Davies, J. W. (1984). A putative primer for second strand DNA synthesis of maize streak virus is virion associated. *EMBO J.* **3**, 3069–3073.

Donson, J., Accotto, G. P., Boulton, M. I., Mullineaux, P. M., and Davies, J.-W. (1987). The nucleotide sequence of a geminivirus from *Digitaria sanguinalis*. *Virology* **161**, 160–169.

Donson, J., Gunn, H. V., Woolston, C. J., Pinner, M. S., Boulton, M. I., Mullineaux, P. M. M., and Davies, J. W. (1988). *Agrobacterium*-mediated infectivity of cloned digitaria streak virus DNA. *Virology* **162**, 248–250.

Doodson, J. K., and Saunders, P. J. W. (1970). Some effects of barley yellow dwarf virus on spring and winter cereals in field trials. *Ann. Appl. Biol.* **66**, 361–374.

Dore, I., Dekker, E. L., Porta, C., and Van Regenmortel, M. H. V. (1987a). Detection by ELISA of two tobamoviruses in orchids using monoclonal antibodies. *J. Phytopathol.* **20**, 317–326.

Dore, I., Altschuh, D., Al-Moudallal, Z., and Van Regenmortel, M. H. V. (1987b). Immunochemical studies of tobacco mosaic virus. VII. Use of comparative surface accessibility of residues in antigenically related viruses for delineating epitopes recognised by monoclonal antibodies. *Mol. Immunol.* **24**, 1351–1358.

Dore, I., Weiss, E., Altschuh, D., and Van Regenmortel, M. H. V. (1988). Visualization by electron microscopy of the location of tobacco mosaic virus epitopes reacting with monoclonal antibodies in enzyme immunoassay. *Virology* **162**, 279–289.

Dore, I., Ruhlmann, C., Oudet, P., Cahoon, M., Caspar, D. L. D., and Van Regenmortel, M. H. V. (1989). Polarity of binding of monoclonal antibodies to tobacco mosaic virus rods and stacked disks. *Virology* **176**, 25–29.

Dore, J.-M., and Pinck, L. (1988). Plasmid DNA containing a copy of RNA3 can substitute for RNA3 in alfalfa mosaic virus RNA inocula. *J. Gen. Virol.* **69**, 1331–1338.

Dorokhov, Y. L., Miroshnichenko, N. A., Alexandrova, N. M., and Atabekov, J. G. (1981). Development of systemic TMV infection in upper non-inoculated tobacco leaves after differential temperature treatment. *Virology* **108**, 507–509.

Dorokhov, Y. L., Alexandrova, N. M., Miroshnichenko, N. A., and Atabekov, J. G. (1983). Isolation and analysis of virus-specific ribonucleoprotein of tobacco mosaic virus-infected tobacco. *Virology* **127**, 237–252.

Dorokhov, Y. L., Alexandrova, N. M., Miroshnichenko, N. A., and Atabekov, J. G. (1984a). Stimulation by aurintricarboxylic acid of tobacco mosaic virus-specific RNA synthesis and production of informosome-like infection-specific ribonucleoprotein. *Virology* **135**, 395–405.

Dorokhov, Y. L., Alexandrova, N. M., Miroschnichenko, N. A., and Atabekov, J. G. (1984b). The informosome-like virus-specific ribonucleoprotein (vRNP) may be involved in the transport of tobacco mosaic virus infection. *Virology* **137**, 127–134.

Dorssers, L., van der Meer, J., van Kammen, A., and Zabel, P. (1983). The cowpea mosaic virus RNA replication complex and the host-encoded RNA-dependent RNA polymerase-template complex are functionally different. *Virology* **125,** 155–174.

Dorssers, L., van der Krol, S., van der Meer, J., van Kammen, A., and Zabel, P. (1984). Purification of cowpea mosaic virus RNA replication complex: Identification of a viral-encoded 110,000-dalton polypeptide responsible for RNA chain elongation. *Proc. Natl. Acad. Sci. U.S.A.* **81,** 1951–1955.

Doudna, J. A., and Szostak, J. W. (1989). RNA catalysed synthesis of complementary-strand RNA. *Nature (London)* **339,** 519–522.

Dougherty, W. G. (1983). Analysis of viral RNA isolated from leaf tissue infected with tobacco etch virus. *Virology* **131,** 473–481.

Dougherty, W. G., and Carrington, J. C. (1988). Expression and function of potyviral gene products. *Annu. Rev. Phytopathol.* **26,** 123–143.

Dougherty, W. G., and Hiebert, E. (1980). Translation of potyvirus RNA in a rabbit reticulocyte lysate: Identification of nuclear inclusion proteins as products of tobacco etch virus RNA translation and cylindrical inclusion protein as a product of the potyvirus genome. *Virology* **104,** 174–182.

Dougherty, W. G., and Hiebert, E. (1985). Genome structure and gene expression of plant RNA viruses. *In* "Plant Molecular Biology" (J. W. Davies, ed.), Vol. 2, pp. 23–81. CRC Press, Boca Raton, Florida.

Dougherty, W. G., Franke, C. A., and Hruby, D. E. (1986). Construction of a recombinant vaccinia virus which expresses immunoreactive plant virus proteins. *Virology* **149,** 107–113.

Dougherty, W. G., Carrington, J. C., Cary, S. M., and Parks, T. D. (1988). Biochemical and mutational analysis of a plant virus polyprotein cleavage site. *EMBO J.* **7,** 1281–1287.

Dougherty, W. G., Cary, S. M., and Parks, T. D. (1989a). Molecular genetic analysis of a plant virus polyprotein cleavage site: A model. *Virology* **171,** 356–364.

Dougherty, W. G., Parks, T. D., Cary, S. M., Bazan, J. F., and Fletterick, R. J. (1989b). Characterization of the catalytic residues of the tobacco etch virus 49-kDa proteinase. *Virology* **172,** 302–310.

Dougherty, W. G., Parks, T. D., Smith, H. A., and Lindbo, J. A. (1990). Expression of the potyvirus genome: The role of proteolytic processing. *In* "Viral Genes and Plant Pathology" (T. P. Pirone and J. G. Shaw, eds.), pp. 124–139. Springer-Verlag, New York.

Douthit, L. B., and McGuire, J. M. (1975). Some effects of temperature on *Xiphinema americanum* and infection of cucumber by tobacco ringspot virus. *Phytopathology* **65,** 134–138.

Dreher, T. W., and Hall, T. C. (1988a). Mutational analysis of the sequence and structural requirements in brome mosaic virus RNA for minus strand promoter activity. *J. Mol. Biol.* **201,** 31–40.

Dreher, T. W., and Hall, T. C. (1988b). Mutational analysis of the tRNA mimicry of brome mosaic virus RNA. Sequence and structural requirements for aminoacylation and 3'-adenylation. *J. Mol. Biol.* **201,** 41–55.

Dreher, T. W., Bujarski, J. J., and Hall, T. C. (1984). Mutant viral RNAs synthesised *in vitro* show altered aminoacylation and replicase template activities. *Nature (London)* **311,** 171–175.

Dreher, T. W., Rao, A. L. N., and Hall, T. C. (1989). Replication *in vivo* of mutant brome mosaic virus RNAs defective in aminoacylation. *J. Mol. Biol.* **206,** 425–438.

Driedonks, R. A., Krijgsman, P. C. J., and Mellema, J. E. (1977). Alfalfa mosaic virus protein polymerization. *J. Mol. Biol.* **113,** 123–140.

Driedonks, R. A., Tjok Joe, M. K. K., and Mellema, J. E. (1980). Application of band centrifugation to the study of the assembly of alfalfa mosaic virus. *Biopolymers* **19,** 575–595.

Dudits, D., Maroy, E., Praznovszky, T., Olah, Z., Gyorgyey, J., and Cella, R. (1987). Transfer of resistance traits from carrot into tobacco by asymmetric somatic hybridization: Regeneration of fertile plants. *Proc. Natl. Acad. Sci. U.S.A.* **84,** 8434–8438.

Duffus, J. E. (1960). Radish yellows, a disease of radish, sugar beet and other crops. *Phytopathology* **50,** 389–394.

Duffus, J. E. (1971). Role of weeds in the incidence of virus diseases. *Annu. Rev. Phytopathol.* **9,** 319–340.

Duffus, J. E. (1973). The yellowing virus diseases of beet. *Adv. Virus Res.* **18,** 347–386.

Duffus, J. E. (1985). Whitefly-borne viruses. *Phytoparasitica* **13,** 274.

Duffus, J. E. (1987). Durability of resistance. *Ciba Found. Symp.* **133,** 196–199.

Duffus, J. E., and Liu, H. Y. (1987). First report of rhizomania of sugar-beet from Texas. *Plant Dis.* **71,** 557.

Duffus, J. E., and Skoyen, I. O. (1977). Relationship of age of plants and resistance to a severe isolate of the beet curly top virus. *Phytopathology* **67,** 151–154.

Duffus, J. E., Larsen, R. C., and Liu, H. Y. (1986). Lettuce infectious yellows virus—A new type of whitefly transmitted virus. *Phytopathology* **76,** 97–100.

Dumas, E., and Gianinazzi, S. (1986). Pathogenesis related (b) proteins do not play a central role in TMV localisation in *Nicotiana rustica*. *Physiol. Plant Pathol.* **28,** 243–250.

Dumas, E., Lherminier, J., Gianinazzi, S., White, R. F., and Antoniw, J. F. (1988). Immunocytochemical location of pathogenesis related b1 protein induced in tobacco mosaic virus-infected or polyacrylic acid-treated tobacco plants. *J. Gen. Virol.* **69,** 2687–2694.

Dumas, P., Moras, D., Florentz, C., Giegé, R., Verlaan, P., van Belkum, A., and Pleij, C. W. A. (1987). 3D graphics modelling of the tRNA-like 3′-end of turnip yellow mosaic virus RNA: Structural and functional implications. *J. Biomol. Struct. Dyn.* **4,** 707–728.

Dunez, J. (1983). "Indexing of Virus and Virus-like Diseases of Fruit Trees." Int. Soc. Hortic. Sci., Castet IMP, Bordeaux.

Dunigan, D. D., Dietzen, R. G., Schoelz, J. E., and Zaitlin, M. (1988). Tobacco mosaic virus particles contain ubiquitinated coat protein subunits. *Virology* **165,** 310–312.

Dunn, D. B., and Hitchborn, J. H. (1965). The use of bentonite in the purification of plant viruses. *Virology* **25,** 171–192.

Dupin, A., Collot, D., Peter, R., and Witz, J. (1985). Comparison between the primary structure of the coat proteins of turnip yellow mosaic virus and eggplant mosaic virus. *J. Gen. Virol.* **66,** 2571–2579.

Du Plessis, D. H., and Smith, P. (1981). Glycosylation of the cauliflower mosaic virus capsid polypeptide. *Virology* **109,** 403–408.

Duran-Vila, N., Roistacher, C. N., Rivera-Bustamante, R., and Semancik, J. S. (1988). A definition of citrus viroid groups and their relationship to the exocortis disease. *J. Gen. Virol.* **69,** 3069–3080.

Durham, A. C. H. (1978). The roles of small ions especially calcium in virus disassembly takeover and transformation. *Biomedicine* **28,** 307–313.

Durham, A. C. H., and Bancroft, J. B. (1979). Cation binding by papaya mosaic virus and its protein. *Virology* **93,** 246–252.

Durham, A. C. H., and Finch, J. T. (1972). Structures and roles of the polymorphic forms of tobacco mosaic virus protein. II. Electron microscope observations of the larger polymers. *J. Mol. Biol.* **67,** 307–314.

Durham, A. C. H., Finch, J. T., and Klug, A. (1971). States of aggregation of tobacco mosaic virus protein. *Nature (London), New Biol.* **229,** 37–42.

Durham, A. C. H., Hendry, D. A., and von Wechmar, M. B. (1977). Does calcium ion binding control plant virus disassembly. *Virology* **77,** 524–533.

Durham, A. C. H., Witz, J., and Bancroft, J. B. (1984). The semipermeability of simple spherical virus capsids. *Virology* **133,** 1–8.

Dynan, W. S. (1989). Modularity in promoters and enhancers. *Cell (Cambridge, Mass.)* **58,** 1–4.

Dzianott, A. M., and Bujarski, J. J. (1989). Derivation of an infectious viral RNA by autolytic cleavage of *in vitro* transcribed viral cDNAs. *Proc. Natl. Acad. Sci. U.S.A.* **86,** 4823–4827.

Eastop, V. F. (1977). Worldwide importance of aphids as virus vectors. *In* "Aphids as Virus Vectors" (K. F. Harris and K. Maramorosch, eds.), pp. 3–62. Academic Press, New York.

Eastop, V. F. (1986). Aphid–plant associations. *In* "Coevolution and Systematics" (A. R. Stone and D. L. Hawksworth, eds.), pp. 35–54. Systematics Association, Clarendon Press, Oxford.

Ebbels, D. L. (1979). A historical review of certification schemes for vegetatively-propagated crops in England and Wales. *ADAS Q. Rev.* **32,** 21–58.

Eecen, H., G., van Dierendonck, J. H., Pleij, C. W. A., Mandel, M., and Bosch, L. (1985). Hydrodynamic properties of RNA: Effect of multivalent cations on the sedimentation behaviour of turnip yellow mosaic virus RNA. *Biochemistry* **24,** 3610–3617.

Ecker, J. R., and Davis, R. W. (1987). Plant defense genes are regulated by ethylene. *Proc. Natl. Acad. Sci. U.S.A.* **84,** 5202–5206.

Edwardson, J. R. (1974). Host ranges of viruses in the PVY group. *Fla., Agric. Exp. Stn., Monogr.* **5,** 1–225.

Edwardson, J. R., and Christie, R. G. (1983). Cytoplasmic cylindrical and nucleolar inclusions induced by potato virus-A. *Phytopathology* **73,** 290–293.

Edwardson, J. R., and Corbett, M. K. (1962). A virus-like syndrome in tomato caused by a mutation. *Am. J. Bot.* **49,** 571–574.

Egan, B. T., and Hall, P. (1983). Monitoring the Fiji disease epidemic in sugar cane at Bundaberg, Australia. *In* "Plant Virus Epidemiology" (R. T. Plumb and J. M. Thresh, eds.), pp. 287–296. Blackwell, Oxford.

Egan, B. T., Ryan, C. C., and Francki, R. I. B. (1989). Fiji disease. *In* "Diseases of Sugar Cane. Major Diseases" (C. Ricaud, B. T. Egan, A. G. Gillespie, Jr., and C. G. Hughes, eds.), pp. 263–287. Elsevier, Amsterdam.

Eggen, R., Kaan, A., Goldbach, R., and van Kammen, A. (1988). Cowpea mosaic virus RNA replication in crude membrane fractions from infected cowpea and *Chenopodium amaranticolor*. *J. Gen. Virol.* **69,** 2711–2720.

Eggen, R., Verver, J., Wellink, J., Pleij, K., van Kammen, A., and Goldbach, R. (1990). Analysis of sequences involved in cowpea mosaic virus RNA replication using site specific mutants. *Virology* **173,** 456–464.

Eglinton, G., and Hamilton, R. J. (1967). Leaf epicuticular waxes. *Science* **156,** 1322–1335.

Ehlers, U., and Paul, H.-L. (1986). Characterisation of the coat proteins of different types of barley yellow mosaic virus by polyacrylamide gel electrophoresis and electro-blot immunoassay. *J. Phytopathol.* **115,** 294–304.

Ehresmann, B., Briand, J.-P., Reinbolt, J., and Witz, J. (1980). Identification of binding sites of turnip yellow mosaic virus protein and RNA by crosslinks induced *in situ*. *Eur. J. Biochem.* **108,** 123–129.

Ehresmann, C., Baudin, F., Mougel, M., Romby, P., Ebel, J.-P., and Ehresmann, B. (1987). Probing the structure of RNA's in solution. *Nucleic Acids Res.* **15,** 9109–9128.

Eichler, W. (1948). Some rules on ectoparasitism. *Ann. Mag. Nat. Hist. Ser. 12* **1,** 588–598.

Ekpo, E. J. A., and Saettler, A. W. (1974). Distribution pattern of bean common mosaic virus in developing bean seed. *Phytopathology* **64,** 269–270.

Eldredge, N. (1985). "Time Frames. The Rethinking of Darwinian Evolution and the Theory of Punctuated Equilibria." Simon & Schuster, New York.

Elliot, M. S., Cromray, H. L., Zettler, F. W., and Carpenter, W. R. (1987). A mosaic disease of wax myrtle associated with a new species of eryophyid mite. *HortScience* **22,** 258–260.

Elmer, J. S., Brand, L., Sunter, G., Gardiner, W. E., Bisaro, D. M., and Rogers, S. G. (1988a). Genetic analysis of tomato golden mosaic virus. II. The product of the AL1 coding sequence is required for replication. *Nucleic Acids Res.* **16,** 7043–7060.

Elmer, J. S., Sunter, G., Gardiner, W. E., Brand, L., Browning, C. K., Bisaro, D. M., and Rogers, S. G. (1988b). *Agrobacterium*-mediated inoculation of plants with tomato golden mosaic virus DNAs. *Plant Mol. Biol.* **10,** 225–234.

Elmer, S., and Rogers, S. G. (1990). Selection for wild type size derivatives of tomato golden mosaic virus during systemic infection. *Nucleic Acids Res.* **18,** 2001–2006.

Elnagar, S., and Murant, A. F. (1976). The role of the helper virus, anthriscus yellows, in the transmission of parsnip yellow fleck virus by the aphid *Cavariella aegopodii*. *Ann. Appl. Biol.* **84,** 169–181.

Emori, Y., Shiba, T., Kanaya, S., Inouye, S., Yuki, S., and Saigo, K. (1985). The nucleotide sequences of *copia* and *copia*-related RNA in *Drosophila* virus-like particles. *Nature (London)* **315,** 773–776.

Erickson, J. W., and Rossmann, M. G. (1982). Assembly and crystallization of a T=1 icosahedral particle from trypsinized Southern bean mosaic virus coat protein. *Virology* **116,** 128–136.

Eşanu, V., and Dumitrescu, M. (1971). A comparative study of isoperoxidases of tobacco as influenced by TMV-infection and genetic constitution. *Acta Phytopathol. Acad. Sci. Hung.* **6,** 31–35.

Esau, K. (1956). An anatomist's view of virus diseases. *Am. J. Bot.* **43,** 739–748.

Esau, K. (1979). Beet yellow stunt virus in cells of *Sonchus oleraceus* L. and its relation to host mitochondria. *Virology* **98,** 1–8.

Esau, K., and Cronshaw, J. (1967). Relation of tobacco mosaic virus with host cells. *J. Cell Biol.* **33,** 665–678.

Esau, K., and Hoefert, L. L. (1972). Ultrastructure of sugarbeet leaves infected with beet western yellows virus. *J. Ultrastruct. Res.* **40,** 556–571.

Esau, K., and Hoefert, L. L. (1973). Particles and associated inclusions in sugarbeet infected with the curly top virus. *Virology* **56,** 454–464.

Esau, K., and Hoefert, L. L. (1978). Hyperplastic phloem in sugarbeet leaves infected with the beet curly top virus. *Am. J. Bot.* **65,** 772–783.

Esau, K., and Hoefert, L. L. (1981). Beet yellow stunt virus in the phloem of *Sonchus oleraceus* L. *J. Ultrastruct. Res.* **75,** 326–338.

Esau, K., Cronshaw, J., and Hoefert, L. L. (1966). Organisation of beet-yellows inclusions in leaf cells of *Beta*. *Proc. Natl. Acad. Sci. U.S.A.* **55,** 486–493.

Esau, K., Cronshaw, J., and Hoefert, L. L. (1967). Relation of beet yellows virus to the phloem and to movement in the sieve tube. *J. Cell Biol.* **32,** 71–87.

Eslick, R. F., and Afanasiev, M. M. (1955). Influence of time of infection with barley stripe mosaic on symptoms, plant yield and seed infection of barley. *Plant Dis. Rep.* **39,** 722–724.

Espinosa, A. M., Markham, P. G., Maule, A. J., and Hull, R. (1988). *In vitro* biological activity associated with the aphid transmission factor of caulimoviruses. *J. Gen. Virol.* **69,** 1819–1830.

Estes, A. P., and Brakke, M. K. (1966). Correlation of *Polymyxa graminis* with transmission of soil-borne wheat mosaic virus. *Virology* **28**, 772–774.

Etessami, P., Callis, R., Ellwood, S., and Stanley, J. (1988). Delimitation of essential genes of cassava latent virus DNA2. *Nucleic Acids Res.* **16**, 4811–4829.

Etessami, P., Watts, J., and Stanley, J. (1989). Size reversion of African cassava mosaic virus coat protein gene deletion mutants during infection of *Nicotiana benthamiana*. *J. Gen. Virol.* **70**, 277–289.

Evans, A. C. (1954). Groundnut rosette disease in Tanganyika. 1. Field studies. *Ann. Appl. Biol.* **41**, 189–206.

Evans, D. (1985). Isolation of a mutant of cowpea mosaic virus which is unable to grow in cowpeas. *J. Gen. Virol.* **66**, 339–343.

Evans, T. A., and Stephens, C. T. (1989). Increased susceptibility to fusarium crown and root rot in virus-infected asparagus. *Phytopathology* **79**, 253–258.

Evered, D., and Harnett, S., eds. (1987). "Plant Resistance to Viruses," Ciba Found. Symp. No. 133. Wiley, Chichester.

Faccioli, G., and Rubies-Autonell, C. (1982). PVX and PVY distribution in potato meristem tips and their eradication by the use of thermotherapy and meristem-tip culture. *Phytopathol. Z.* **103**, 66–75.

Faccioli, G., Rubies-Autonell, C., and Resca, R. (1988). Potato leafroll virus distribution in potato meristem tips and production of virus-free plants. *Potato Res.* **31**, 511–520.

Faed, E. M., and Matthews, R. E. F. (1972). Leaf ontogeny and virus replication in *Brassica pekinensis* infected with turnip yellow mosaic virus. *Virology* **48**, 546–554.

Fairall, L., Finch, J. T., Hui, C.-F., and Cantor, C. R., and Butler, P. J. G. (1986). Studies of tobacco mosaic virus reassembly with an RNA tail blocked by a hydridised and cross-linked probe. *Eur. J. Biochem.* **156**, 459–465.

Falk, B. W., Morris, T. J., and Duffus, J. E. (1979a). Unstable infectivity and sedimentable ds-RNA associated with lettuce speckles mottle virus. *Virology* **96**, 239–248.

Falk, B. W., Duffus, J. E., and Morris, T. J. (1979b). Transmission, host range, and serological properties of the viruses that cause lettuce speckles disease. *Phytopathology* **69**, 612–617.

Falk, B. W., Tsai, J. H., and Lommel, S. A. (1987). Differences in levels of detection for the maize stripe virus capsid and major non-capsid proteins in plant and insect hosts. *J. Gen. Virol.* **68**, 1801–1811.

Falk, B. W., Kim, K. S., and Tsai, J. H. (1988). Electron microscopic and physicochemical analysis of a reo-like virus of the planthopper *Peregrinus maidis Intervirology* **29**, 195–206.

Falk, B. W., Chin, L.-S., and Duffus, J. E. (1989). Complementary DNA cloning and hybridization analysis of beet western yellows luteovirus RNAs. *J. Gen. Virol.* **70**, 1301–1309.

Fang, R.-X., Nagy, F., Sivasubramanian, S., and Chua, N.-H. (1989). Multiple cis regulatory elements for maximal expression of the cauliflower mosaic virus 35S promoter in transgenic plants. *Plant Cell* **1**, 141–150.

Fargette, D., Jenniskens, M. J., and Peters, D. (1982). Acquisition and transmission of pea enation mosaic virus by the individual pea aphid. *Phytopathology* **72**, 1386–1390.

Fargette, D., Fauquet, C., and Thouvenel, J.-C. (1985). Field studies on the spread of African cassava mosaic. *Ann. Appl. Biol.* **106**, 285–294.

Fargette, D., Thouvenel, J.-C., and Fauquet, C. (1987). Virus content of leaves of cassava infected by African cassava mosaic virus. *Ann. Appl. Biol.* **110**, 65–73.

Fargette, D., Fauquet, C. F., and Thouvenel, J.-C. (1988). Yield losses induced by African cassava mosaic virus in relation to the mode and date of infection. *Trop. Pest Manage.* **34**, 89–92.

Faulkner, G., and Kimmins, W. C. (1975). Staining reactions of the tissue bordering lesions induced by wounding, tobacco mosaic virus, and tobacco necrosis virus in bean. *Phytopathology* **65**, 1396–1400.

Fauquet, C., and Beachy, R. N. (1989). International cassava-trans project. Cassava viruses and genetic engineering. *Occas. Publ. ORSTOM*.

Fauquet, C., and Thouvenel, J.-C. (1983). Association d'un nouveau virus en bâtonnet avec la maladie necrose à rayures du Riz en Côte d'Ivoire. *C. R. Seances Acad. Sci., Ser. 3* **296**, 575–580.

Fauquet, C., Dejardin, J., and Thouvenel, J.-C. (1986). Evidence that the amino acid composition of the particle proteins of plant viruses is characteristic of the virus group. I. Multidimensional classification of plant viruses. *Intervirology* **25**, 1–13.

Faustmann, O., Kern, R., Sänger, H. L., and Muehlbach, H.-P. (1986). Potato spindle tuber viroid (PSTV) RNA oligomers of (+) and (−) strand polarity are synthesised in potato protoplasts after liposome-mediated infection with PSTV. *Virus Res.* **4**, 213–227.

Favali, M. A., Bassi, M., and Appiano, A. (1974). Synthesis and migration of maize rough dwarf virus in the host cell: An autoradiographic study. *J. Gen. Virol.* **24**, 563–565.

Fenner, F. (1979). Portraits of viruses: The poxviruses. *Intervirology* **11**, 137–157.

Fenner, F., and Gibbs, A. (1983). Cryptograms—1982. *Intervirology* **19**, 121–128.

Fenoll, C., Black, D. M., and Howell, S. H. (1988). The intergenic region of maize streak virus (MSV) contains promoter elements involved in rightward transcription of the viral genome. *EMBO J.* **7**, 1589–1596.

Fernandez-Gonzalez, O., Renaudin, J., and Bové, J. M. (1980). Infection of chlorophyll-less protoplasts from etiolated chinese cabbage hypocotyls by turnip yellow mosaic virus. *Virology* **104**, 262–265.

Fernow, K. H. (1967). Tomato as a test plant for detecting mild strains of potato spindle tuber virus. *Phytopathology* **57**, 1347–1352.

Filipowicz, W., and Haenni, A.-L. (1979). Binding of ribosomes to 5′ terminal leader sequences of eukaryotic messenger RNAs. *Proc. Natl. Acad. Sci. U.S.A.* **76**, 3111–3115.

Finch, J. T. (1972). The hand of the helix of tobacco mosaic virus. *J. Mol. Biol.* **66**, 291–294.

Finch, J. T., and Holmes, K. C. (1967). Structural studies of viruses. *In* "Methods in Virology" (K. Maramorosch and H. Koprowski, eds.), Vol. 3, pp. 351–474. Academic Press, New York.

Finch, J. T., and Klug, A. (1966). Arrangement of protein subunits and the distribution of nucleic acid in turnip yellow mosaic virus. II. Electron microscopic studies. *J. Mol. Biol.* **15**, 344–364.

Finch, J. T., and Klug, A. (1967). Structure of broad bean mottle virus. 1. Analysis of electron micrographs and comparison with turnip yellow mosaic virus and its top component. *J. Mol. Biol.* **24**, 289–302.

Finch, J. T., Klug, A., and Van Regenmortel, M. H. V. (1967a). The structure of cucumber mosaic virus. *J. Mol. Biol.* **24**, 303–305.

Finch, J. T., Leberman, R., and Berger, J. E. (1967b). Structure of broad bean mottle virus. II. X-ray diffraction studies. *J. Mol. Biol.* **27**, 17–24.

Finlay, K. W. (1953). Inheritance of spotted wilt resistance in the tomato. II. Five genes controlling spotted wilt resistance in four tomato types. *Aust. J. Biol. Sci.* **6**, 153–163.

Fletcher, J. T., and Butler, D. (1975). Strain changes in populations of tobacco mosaic virus from tomato crops. *Ann. Appl. Biol.* **81**, 409–412.

Flore, P. H. (1986). *In vitro* transcription of sonchus yellow net virus RNA by a virus-associated RNA-dependent RNA polymerase. Ph.D. Thesis, Wageningen, The Netherlands.

Florentz, C., Briand, J. P., and Giegé, R. (1984). Possible functional role of viral tRNA-like structures. *FEBS Lett.* **176**, 295–300.

Föglein, F. J., Kalpagam, C., Bates, D. C., Premecz, G., Nyitrai, A., and Farkas, G. L. (1975). Viral RNA synthesis is renewed in protoplasts isolated from TMV-infected Xanthi tobacco leaves in an advanced stage of virus infection. *Virology* **67**, 74–79.

Föglein, F. J., Nyitrai, A., Gulyás, A., Premecz, G., Oláh, T., and Farkas, G. L. (1976). Crystalline inclusion bodies are degraded in protoplasts isolated from TMV-infected tobacco leaves. *Phytopathol. Z.* **86**, 266–269.

Ford, R. E. (1973). Concentration and purification of clover yellow mosaic virus from pea roots and leaves. *Phytopathology* **63**, 926–930.

Forster, A. C., and Symons, R. H. (1987a). Self-cleavage of plus and minus RNAs of a virusoid and a structural model for the active sites. *Cell (Cambridge, Mass.)* **49**, 211–220.

Forster, A. C., and Symons, R. H. (1987b). Self-cleavage of virusoid RNA is performed by the proposed 55-nucleotide active site. *Cell (Cambridge, Mass.)* **50**, 9–16.

Forster, A. C., McInnes, J. L., Skingle, D. C., and Symons, R. H. (1985). Non-radioactive hybridisation probes prepared by the chemical labelling of DNA and RNA with a novel reagent, photobiotin. *Nucleic Acids Res.* **13**, 745–761.

Forster, A. C., Davies, C., Sheldon, C. C., Jeffries, A. C., and Symons, R. H. (1988). Self-cleaving viroid and newt RNAs may only be active as dimers. *Nature (London)* **334**, 265–267.

Forster, R. L. S., and Morris-Krsinich, B. A. M. (1985). Synthesis and processing of the translation products of tobacco ringspot virus in rabbit reticulocyte lysates. *Virology* **144**, 516–519.

Forster, R. L. S., Guilford, P. J., and Faulds, D. V. (1987). Characterisation of the coat protein subgenomic RNA of white clover mosaic virus. *J. Gen. Virol.* **68**, 181–190.

Forster, R. L. S., Bevan, M. W., Harbison, S.-A., and Gardner, R. C. (1988). The complete nucleotide sequences of the potexvirus white clover mosaic virus. *Nucleic Acids Res.* **16**, 291–303.

Foster, J. A., and Ross, A. F. (1975). Properties of the initial tobacco mosaic virus infection sites revealed by heating symptomless inoculated tobacco leaves. *Phytopathology* **65**, 610–616.

Fowlks, E., and Young, R. J. (1970). Detection of heterogeneity in plant viral RNA by polyacrylamide gel electrophoresis. *Virology* **42**, 548–550.

Fraenkel-Conrat, H. (1956). The role of nucleic acid in the reconstitution of active tobacco mosaic virus. *J. Am. Chem. Soc.* **78**, 882–883.

Fraenkel-Conrat, H. (1957). Degradation of tobacco mosaic virus with acetic acid. *Virology* **4**, 1–4.

Fraenkel-Conrat, H., and Singer, B. (1957). Virus reconstitution. II. Combination of protein and nucleic acid from different strains. *Biochim. Biophys. Acta* **24**, 540–548.

Fraenkel-Conrat, H., and Williams, R. C. (1955). Reconstitution of active tobacco mosaic virus from its inactive protein and nucleic acid components. *Proc. Natl. Acad. Sci. U.S.A.* **41**, 690–698.

Fraile, A., and García-Arenal, F. (1990). A classification of the tobamoviruses based on the comparison of their "126K" proteins. *J. Gen. Virol.* **71**, 2223–2228.

Franck, A., Guilley, H., Jonard, G., Richards, K., and Hirth, L. (1980). Nucleotide sequence of cauliflower mosaic virus DNA. *Cell (Cambridge, Mass.)* **21**, 285–294.

Francki, R. I. B. (1973). Plant rhabdoviruses. *Adv. Virus Res.* **18**, 257–345.

Francki, R. I. B. (1980). Limited value of the thermal inactivation point, longevity *in vitro* and dilution endpoint as criteria for the characterization, identification and classification of plant viruses. *Intervirology* **13**, 91–98.

Francki, R. I. B. (1985). Plant virus satellites. *Annu. Rev. Microbiol.* **39**, 151–174.

Francki, R. I. B. (1987). Responses of plant cells to virus infection with special references to the sites of RNA replication. *In* "Positive Strand RNA Viruses" (M. A. Brinton and R. R. Rueckert, eds.). pp. 423–436. Alan R. Liss, New York.

Francki, R. I. B., and Boccardo, G. (1983). The plant reoviridae. *In* "The Reoviridae" (W. K. Joklik, ed.), pp. 505–563. Plenum, New York and London.

Francki, R. I. B., and Habili, N. (1972). Stabilization of capsid structure and enhancement of immunogenicity of cucumber mosaic virus (Q strain) by formaldehyde. *Virology* **48**, 309–315.

Francki, R. I. B., and Hatta, T. (1980). Cucumber mosaic virus—Variation and problems of identification. *Acta Hortic.* **110**, 167–174.

Francki, R. I. B., and Matthews, R. E. F. (1962). Some effects of 2-thiouracil on the multiplication of turnip yellow mosaic virus. *Virology* **17**, 367–380.

Francki, R. I. B., and Miles, R. (1985). Mechanical transmission of sowbane mosaic virus carried on pollen from infected plants. *Plant Pathol.* **34**, 11–19.

Francki, R. I. B., and Randles, J. W. (1972). RNA-dependent RNA polymerase associated with particles of lettuce necrotic yellows virus. *Virology* **47**, 270–275.

Francki, R. I. B., and Randles, J. W. (1973). Some properties of lettuce necrotic yellows virus RNA and its *in vitro* transcription by virion-associated transcriptase. *Virology* **54**, 359–368.

Francki, R. I. B., and Randles, J. W. (1978). Composition of the plant rhabdovirus lettuce necrotic yellows virus in relation to its biological properties. *In* "Negative Strand Viruses and the Host Cell" (B. W. J. Mahy and R. D. Barry, eds.), pp. 223–242.

Francki, R. I. B., and Randles, J. W. (1980). Rhabdoviruses infecting plants. *In* "Rhabdoviruses" (D. H. L. Bishop, ed.), Vol. 3, pp. 135–165. CRC Press, Boca Raton, Florida.

Francki, R. I. B., Gould, A. R., and Hatta, T. (1980a). Variation in the pathogenicity of three viruses of tomato. *Ann. Appl. Biol.* **96**, 219–226.

Francki, R. I. B., Hatta, T., Boccardo, G., and Randles, J. W. (1980b). The composition of chloris striate mosaic virus, a geminivirus. *Virology* **101**, 233–241.

Francki, R. I. B., Milne, R. G., and Hatta, T. (1985a). "Atlas of Plant Viruses," Vol. 1. CRC Press, Boca Raton, Florida.

Francki, R. I. B., Milne, R. G., and Hatta, T., eds. (1985b). "Atlas of Plant Viruses," Vol. 2. CRC Press, Boca Raton, Florida.

Francki, R. I. B., Milne, R. G., and Hatta, T. (1985c). Geminivirus group. *In* "Atlas of Plant Viruses" (R. I. B. Francki, R. G. Milne, and T. Hatta, eds.), Vol. 1, pp. 33–46. CRC Press, Boca Raton, Florida.

Francki, R. I. B., Milne, R. G., and Hatta, T. (1985d). Plant Rhabdoviridae. *In* "Atlas of Plant Viruses" (R. I. B. Francki, R. G. Milne, and T. Hatta, eds.), Vol. 1, pp. 73–100. CRC Press, Boca Raton, Florida.

Francki, R. I. B., Grivell, C. J., and Gibb, K. S. (1986a). Isolation of velvet tobacco mottle virus capable of replication with and without a viroid-like RNA. *Virology* **148**, 381–384.

Francki, R. I. B., Zaitlin, M., and Palukaitis, P. (1986b). *In vivo* encapsidation of potato spindle tuber viroid by velvet tobacco mottle virus particles. *Virology* **155**, 469–473.

Francki, R. I. B., Ryan, C. C., Hatta, T., Rohozinski, J., and Grivell, C. J. (1986c). Serological detection of Fiji disease virus antigens in the planthopper *Perkinsiella saccharricida* and its inefficient ability to transmit the virus. *Plant Pathol.* **35**, 324–328.

Franssen, H., Goldbach, R., Broekhuijsen, M., Moerman, M., and van Kammen, A. (1982). Expression of middle-component RNA of cowpea mosaic virus: *In vitro* generation of a precursor to both capsid proteins by a bottom-component RNA-encoded protease from infected cells. *J. Virol.* **41,** 8–17.

Franssen, H., Moerman, M., Rezelman, G., and Goldbach, R. (1984a). Evidence that the 32,000-dalton protein encoded by bottom-component RNA of cowpea mosaic virus is a proteolytic processing enzyme. *J. Virol.* **50,** 183–190.

Franssen, H., Leunissen, J., Goldbach, R., Lomonossoff, G., and Zimmern, D. (1984b). Homologous sequences in non-structural proteins from cowpea mosaic virus and picornaviruses. *EMBO J.* **3,** 855–861.

Fraser, L. G., and Matthews, R. E. F. (1979a). Efficient mechanical inoculation of turnip yellow mosaic virus using small volumes of inoculum. *J. Gen. Virol.* **44,** 565–568.

Fraser, L. G., and Matthews, R. E. F. (1979b). Strain-specific pathways of cytological change in individual Chinese cabbage protoplasts infected with turnip yellow mosaic virus. *J. Gen. Virol.* **45,** 623–630.

Fraser, L. G., and Matthews, R. E. F. (1981). A rapid transient inhibition of leaf initiation induced by turnip yellow mosaic virus infection. *Physiol. Plant Pathol.* **19,** 325–336.

Fraser, L. G., and Matthews, R. E. F. (1983). A rapid transient inhibition of leaf initiation by abscisic acid. *Plant Sci. Lett.* **29,** 67–72.

Fraser, L. G., Keeling, J., and Matthews, R. E. F. (1984). A reduction in starch accumulation in the apical dome of chinese cabbage seedlings following inoculation with turnip yellow mosaic virus. *Physiol. Plant Pathol.* **24,** 157–162.

Fraser, R. D. B., Furlong, D. B., Trus, B. L., Nibert, M. L., Fields, B. N., and Steven, A. C. (1990). Molecular structure of the cell-attachment protein of *Reovirus*: correlation of computer-processed electron micrographs with sequence-based predictions. *J. Virol.* **64,** in press.

Fraser, R. S. S. (1982). Are "pathogenesis-related" proteins involved in acquired systemic resistance of tobacco plants to tobacco mosaic virus? *J. Gen. Virol.* **58,** 305–313.

Fraser, R. S. S. (1983). Varying effectiveness of the N[1] gene for resistance to tobacco mosaic virus in tobacco infected with virus strains differing in coat protein properties. *Physiol. Plant Pathol.* **22,** 109–119.

Fraser, R. S. S. (1985). Genetics of host resistance to viruses and of virulence. *In* "Mechanisms of Resistance to Plant Disease" (R. S. S. Fraser, ed.), pp. 62–79. Martinus Nijhoff/W. Junk, Dordrecht.

Fraser, R. S. S. (1987a). Resistance to plant viruses. *Oxford Surv. Plant Mol. Cell Biol.* **4,** 1–45.

Fraser, R. S. S. (1987b). "Biochemistry of Virus-Infected Plants." Research Studies Press, Letchworth.

Fraser, R. S. S. (1988). Virus recognition and pathogenicity: Implications for resistance mechanisms and breeding. *Pestic. Sci.* **23,** 267–275.

Fraser, R. S. S., and Gerwitz, A. (1985). A new physical assay method for tobacco mosaic virus using a radioactive virus recovery standard and the first derivative of the ultraviolet absorption spectrum. *J. Virol. Methods* **11,** 289–298.

Fraser, R. S. S., and Gerwitz, A. (1987). The genetics of resistance and virulence in plant virus disease. *In* "Genetics and Plant Pathogenesis" (P. R. Day and G. J. Ellis, eds.), pp. 33–44. Blackwell, Oxford.

Fraser, R. S. S., and Whenham, R. J. (1982). Plant growth regulators and virus infection: A critical review. *Plant Growth Regul.* **1,** 37–59.

Fraser, R. S. S., and Whenham, R. J. (1989). Abscisic acid metabolism in tomato plants infected with tobacco mosaic virus: Relationships with growth, symptoms, and the Tm-1 gene for TMV resistance. *Physiol. Mol. Plant Pathol.* **34,** 215–226.

Fraser, R. S. S., Loughlin, S. A. R., and Connor, J. C. (1980). Resistance to tobacco mosaic virus in tomato: Effects of the Tm-1 gene on symptom formation and multiplication of virus strain 1. *J. Gen. Virol.* **50,** 221–224.

Fraser, R. S. S., Gerwitz, A., and Morris, G. E. L. (1986). Multiple regression analysis of the relationships between tobacco mosaic virus multiplication, the severity of mosaic symptoms, and the growth of tobacco and tomato. *Physiol. Mol. Plant Pathol.* **29,** 239–249.

Frazier, N. W., Voth, V., and Bringhurst, R. S. (1965). Inactivation of two strawberry viruses in plants grown in a natural high temperature environment. *Phytopathology* **53,** 1203–1205.

French, N., and Wilson, W. R. (1976). Influence of crop rotation, weed control and nematicides on spraing in potatoes. *Plant Pathol.* **25,** 167–172.

French, R., and Ahlquist, P. (1987). Intercistronic as well as terminal sequences are required for efficient amplification of brome mosaic virus RNA3. *J. Virol.* **61,** 1457–1465.

French, R., and Ahlquist, P. (1988). Characterisation and engineering of sequences controlling *in vivo* synthesis of brome mosaic virus subgenomic RNA. *J. Virol.* **62,** 2411–2420.

French, R., Janda, M., and Ahlquist, P. (1986). Bacterial gene inserted in an engineered RNA virus: Efficient expression in monocotyledonous plant cells. *Science* **231**, 1294–1297.

Frenkel, M. J., Ward, C. W., and Shukla, D. D. (1989). The use of 3'-noncoding nucleotide sequences in the taxonomy of potyviruses: Application to watermelon mosaic virus 2 and soybean mosaic virus-N. *J. Gen. Virol.* **70**, 2775–2783.

Fribourg, C. E., Jones, R. A. C., and Koenig, R. (1977). Andean potato mottle, a new member of the cowpea mosaic virus group. *Phytopathology* **67**, 969–974.

Fridlund, P. R. (1967a). Effect of time and temperature on the rate of graft transmission of the *Prunus* ringspot virus. *Phytopathology* **57**, 230–231.

Fridlund, P. R. (1967b). The relationship of inoculum–receptor contact period to the rate of graft transmission of twelve *Prunus* viruses. *Phytopathology* **57**, 1296–1299.

Friis, E. M., Chaloner, W. G., and Crane, P. R., eds. (1987). "The Origins of Angiosperms and Their Biological Consequences." Cambridge Univ. Press, London and New York.

Frison, E. A., and Ng, S. Y. (1981). Elimination of sweet potato virus disease agents by meristem tip culture. *Trop. Pest Manage.* **27**, 452–454.

Fritsch, C., Stussi, C., Witz, J., and Hirth, L. (1973). Specificity of TMV RNA encapsidation: *In vitro* coating of heterologous RNA by TMV protein. *Virology* **56**, 33–45.

Fritsch, C., Mayo, M. A., and Hirth, L. (1977). Further studies on the translation products of tobacco rattle virus RNA *in vitro*. *Virology* **77**, 722–732.

Fritsch, C., Mayo, M. A., and Murant, A. F. (1978). Translation of the satellite RNA of tomato black ring virus *in vitro* and in tobacco protoplasts. *J. Gen. Virol.* **40**, 587–593.

Fritsch, C., Mayo, M. A., and Murant, A. F. (1980). Translation products of genome and satellite RNAs of tomato black ring virus. *J. Gen. Virol.* **46**, 381–389.

Frosheiser, F. I. (1969). Variable influence of alfalfa mosaic virus strains on growth and survival of alfalfa and on mechanical and aphid transmission. *Phytopathology* **59**, 857–862.

Frosheiser, F. I. (1974). Alfalfa mosaic virus transmission to seed through alfalfa gametes and longevity in alfalfa seed. *Phytopathology* **64**, 102–105.

Frost, E. H. E. (1977). Radioactive labelling of viruses: An iodination technique preserving biological properties. *J. Gen. Virol.* **35**, 181–185.

Frost, R. R., Harrison, B. D., and Woods, R. D. (1967). Apparent symbiotic interaction between particles of tobacco rattle virus. *J. Gen. Virol.* **1**, 57–70.

Fry, P. R., and Campbell, R. N. (1966). Transmission of a tobacco necrosis virus by *Olpidium brassicae*. *Virology* **30**, 517–527.

Fry, P. R., and Matthews, R. E. F. (1963). Timing of some early events following inoculation with tobacco mosaic virus. *Virology* **19**, 461–469.

Fry, P. R., and Taylor, W. B. (1954). Analysis of virus local lesion experiments. *Ann. Appl. Biol.* **41**, 664–674.

Fuchs, M., Pinck, M., Serghini, M. A., Ravelonandro, M., Walter, B., and Pinck, L. (1989). The nucleotide sequence of satellite RNA in grapevine fanleaf virus strain F13. *J. Gen. Virol.* **70**, 955–962.

Fudl-Allah, A. E. A., Calavan, E. C., and Desjardins, P. R. (1971). Comparative anatomy of healthy and exocortis virus-infected citron plants. *Phytopathology* **61**, 990–993.

Fuetterer, J., and Hohn, T. (1987). Involvement of nucleocapsids in reverse transcription: A general phenomenon? *Trends Biochem. Res.* **12**, 92–95.

Fuetterer, J., Gordon, K., Bonneville, J. M., Sanfaçon, H., Pison, B., Penswick, J., and Hohn, T. (1988). The leading sequence of caulimovirus large RNA can be folded into a large stem-loop structure. *Nucleic Acids Res.* **16**, 8377–8390.

Fujisawa, I., Hayashi, T., and Matsui, C. (1967). Electron microscopy of mixed infections with two plant viruses. I. Intracellular interactions between tobacco mosaic virus and tobacco etch virus. *Virology* **33**, 70–76.

Fukuda, M., and Okada, Y. (1985). Elongation in the major direction of tobacco mosaic virus assembly. *Proc. Natl. Acad. Sci. U.S.A.* **82**, 3631–3634.

Fukuda, M., and Okada, Y. (1987). Bidirectional assembly of tobacco mosaic virus *in vitro*. *Proc. Natl. Acad. Sci. U.S.A.* **84**, 4035–4038.

Fukuda, M., Ohno, T., Okada, Y., Otsuki, Y., and Takebe, I. (1978). Kinetics of biphasic reconstitution of tobacco mosaic virus *in vitro*. *Proc. Natl. Acad. Sci. U.S.A.* **75**, 1727–1730.

Fukuda, M., Okada, Y., Otsuki, Y., and Takebe, I. (1980). The site of initiation of rod assembly on the RNA of a tomato and a cowpea strain of tobacco mosaic virus. *Virology* **101**, 493–502.

Fukumoto, F., and Tochihara, H. (1984). Effect of various additives on the long-term preservation of tobacco ringspot and radish mosaic viruses. *Ann. Phytopathol. Soc. Jpn.* **50**, 158–165.

Fukushi, T. (1933). Transmission of the virus through the eggs of an insect vector. *Proc. Imp. Acad. (Tokyo)* **8,** 457–460.

Fukushi, T. (1940). Further studies on the dwarf disease of rice plant. *J. Fac. Agric., Hokkaido Univ.* **45,** 83–154.

Fukushi, T. (1969). Relationships between propagative rice viruses and their vectors. *In* "Viruses, Vectors, and Vegetation" (K. Maramorosch, ed.), pp. 279–301. Wiley, New York and London.

Fukuyama, K., Abdel-Meguid, S. S., Johnson, J. E., and Rossmann, M. G. (1983). Structure of a T=1 aggregate of alfalfa mosaic virus coat protein seen at 4.5 Å resolution. *J. Mol. Biol.* **167,** 873–894.

Fulton, J. P. (1962). Transmission of tobacco ringspot virus by *Xiphinema americanum. Phytopathology* **52,** 375.

Fulton, J. P. (1969). Transmission of tobacco ringspot virus to the roots of a conifer by a nematode. *Phytopathology* **59,** 236.

Fulton, J. P., Gergerich, R. C., and Scott, H. A. (1987). Beetle transmission of plant viruses. *Annu. Rev. Phytopathol.* **25,** 111–123.

Fulton, R. W. (1952). Mutation in a tobacco necrosis virus strain. *Phytopathology* **42,** 156–158.

Fulton, R. W. (1962). The effect of dilution on necrotic ringspot virus infectivity and the enhancement of infectivity by noninfective virus. *Virology* **18,** 477–485.

Fulton, R. W. (1964). Transmission of plant viruses by grafting, dodder, seed and mechanical inoculation. *In* "Plant Virology" (M. K. Corbett and H. D. Sisler, eds.), pp. 39–67. Univ. of Florida Press, Gainesville.

Fulton, R. W. (1975). The role of top particles in recombination of some characters of tobacco streak virus. *Virology* **67,** 188–196.

Fulton, R. W. (1978). Superinfection by strains of tobacco streak virus. *Virology* **85,** 1–8.

Fulton, R. W. (1980). Biological significance of multicomponent viruses. *Annu. Rev. Phytopathol.* **18,** 131–146.

Fulton, R. W. (1986). Practices and precautions in the use of cross protection for plant virus disease control. *Annu. Rev. Phytopathol.* **24,** 67–81.

Furumoto, W. A., and Mickey, R. (1967). A mathematical model for the infectivity–dilution curve of tobacco mosaic virus; Experimental tests. *Virology* **32,** 224–233.

Furumoto, W. A., and Wildman, S. G. (1963a). Studies on the mode of attachment of tobacco mosaic virus. *Virology* **20,** 45–52.

Furumoto, W. A., and Wildman, S. G. (1963b). The specific infectivity of tobacco mosaic virus. *Virology* **20,** 53–61.

Gáborjányi, R., Balázs, E., and Király, Z. (1971). Ethylene production, tissue senescence and local virus infections. *Acta Phytopathol. Acad. Sci. Hung.* **6,** 51–55.

Gabriel, C. J., and De Zoeten, G. A. (1984). The *in vitro* translation of pea enation mosaic virus RNA. *Virology* **139,** 223–230.

Gaddipati, J. P., and Siegel, A. (1990). Study of TMV assembly with heterologous RNA containing the origin-of-assembly sequence. *Virology* **174,** 337–344.

Gaddipati, J. P., Atreya, C. D., Rochon, D'A., and Siegel, A. (1988). Characterisation of the TMV encapsidation initiation site on 18S rRNA. *Nucleic Acids Res.* **16,** 7303–7313.

Gadh, I. P. S., and Hari, V. (1986). Association of tobacco etch virus related RNA with chloroplasts in extracts of infected plants. *Virology* **150,** 304–307.

Gaedigk, K., Adam, G., and Mundry, K.-W. (1986). The spike protein of potato yellow dwarf virus and its functional role in the infection of insect vector cells. *J. Gen. Virol.* **67,** 2763–2773.

Gairdner, A. E. (1936). The inheritance of factors in *Cheiranthus cheiri. J. Genet.* **32,** 479–486.

Gallagher, W. H., and Lauffer, M. A. (1983). Calcium ion binding by tobacco mosaic virus. *J. Mol. Biol.* **170,** 905–919.

Gallie, D. R., and Kado, C. I. (1989). A translational enhancer derived from tobacco mosaic virus is functionally equivalent to a Shine–Dalgarno sequence. *Proc. Natl. Acad. Sci. U.S.A.* **86,** 129–132.

Gallie, D. R., Sleat, D. E., Watts, J. W., Turner, P. C., and Wilson, T. M. A. (1987a). The 5′ leader sequence of tobacco mosaic virus RNA enhances the expression of foreign gene transcripts *in vitro* and *in vivo. Nucleic Acids Res.* **15,** 3257–3273.

Gallie, D. R., Sleat, D. E., Watts, J. W., Turner, P. C., and Wilson, T. M. A. (1987b). A comparison of eukaryotic viral 5′-leader sequences as enhancers of mRNA expression *in vivo. Nucleic Acids Res.* **15,** 8693–8711.

Gallie, D. R., Sleat, D. E., Watts, J. W., Turner, P. C., and Wilson, T. M. A. (1987c). *In vivo* uncoating and efficient expression of foreign mRNAs packaged in TMV-like particles. *Science* **236,** 1122–1124.

Gallie, D. R., Plaskitt, K. A., and Wilson, T. M. A. (1987d). The effect of multiple dispersed copies of the origin-of-assembly sequence from TMV RNA on the morphology of pseudovirus particles assembled *in vitro*. *Virology* **158,** 473–476.

Gallie, D. R., Sleat, D. E., Watts, J. W., Turner, P. C., and Wilson, T. M. A. (1988a). Mutational analysis of tobacco mosaic virus 5'-leader for altered ability to enhance translation. *Nucleic Acids Res.* **16,** 883–893.

Gallie, D. R., Walbot, V., and Hershey, J. W. B. (1988b). The ribosomal fraction mediates the translational enhancement associated with the 5'-leader of tobacco mosaic virus. *Nucleic Acids Res.* **16,** 8675–8694.

Gallitelli, D., and Hull, R. (1985). Characterisation of satellite RNAs associated with tomato bushy stunt virus and five other definitive tombusviruses. *J. Gen. Virol.* **66,** 1533–1543.

Gallitelli, D., Hull, R., and Koenig, R. (1985). Relationships among viruses in the tombusvirus group: Nucleic acid hybridization studies. *J. Gen. Virol.* **66,** 1523–1531.

Gallitelli, D., Vovlas, C., Martelli, G., Moutasser, M. S., Tousigant, M. E., and Kaper, J. M. (1990). Satellite-mediated protection of tomato against cucumber mosaic virus. II. Field test under natural epidemic conditions in southern Italy. *Plant Dis.* (in press).

Gallo, J., and Čiampor, F. (1977). Transmission of alfalfa mosaic virus through *Nicandra physaloides* seeds and its localization in embryo cotyledons. *Acta Virol.* **21,** 344–346.

Gamez, R., and Léon, P. (1988). Maize raydo fino and related viruses. *In* "The Plant Viruses" (R. Koenig, ed.), Vol. 3, pp. 213–233. Plenum, New York.

García, J. A., Schrijvers, L., Tan, A., Vos, P., Wellink, J., and Goldbach, R. (1987a). Proteolytic activity of the cowpea mosaic virus encoded 24K protein synthesised in *Escherichia coli*. *Virology* **159,** 67–75.

García, J. A., Hille, J., Vos, P., and Goldbach, R. (1987b). Transformation of cowpea *Vigna unguiculata* with a full length DNA copy of cowpea mosaic virus mRNA. *Plant Sci.* **48,** 89–98.

García, J. A., Riechmann, J. L., and Laín, S. (1989). Proteolytic activity of the plum pox Potyvirus NI$_a$-like protein in *Escherichia coli*. *Virology* **170,** 362–369.

García-Arenal, F. (1988). Sequence and structure at the genome 3' end of the U2 strain of tobacco mosaic virus, a histidine accepting tobamovirus. *Virology* **167,** 201–206.

García-Arenal, F., Palukaitis, P., and Zaitlin, M. (1984). Strains and mutants of tobacco mosaic virus are both found in virus derived from single-lesion-passaged inoculum. *Virology* **132,** 131–137.

García-Luque, I., Kaper, J. M., Diaz-Ruiz, J. R., and Rubio-Huertos, M. (1984). Emergence and characterization of satellite RNAs associated with Spanish cucumber mosaic virus isolates. *J. Gen. Virol.* **65,** 539–547.

Gardiner, W. E., Sunter, G., Brand, L., Elmer, J. S., Rogers, S. G., and Bisaro, D. M. (1988). Genetic analysis of tomato golden mosaic virus: The coat protein is not required for systemic spread or symptom development. *EMBO J.* **7,** 899–904.

Gardner, R. C., Howarth, A., Hahn, P., Brown-Leudi, M., Shepherd, R. J., and Messing, J. (1981). The complete nucleotide sequence of an infectious clone of cauliflower mosaic virus by M13mp7 shotgun sequencing. *Nucleic Acids Res.* **9,** 2871–2888.

Gargouri, R., Joshi, R. L., Bol, J. F., Astier-Manifacier, S., and Haenni, A.-L. (1989). Mechanism of synthesis of turnip yellow mosaic virus coat protein subgenomic RNA *in vivo*. *Virology* **171,** 386–393.

Garnier, M., Mamoun, R., and Bové, J. M. (1980). TYMV RNA replication *in vivo*: Replicative intermediate is mainly single stranded. *Virology* **104,** 357–374.

Garnier, M., Candresse, T., and Bové, J. M. (1986). Immunocytochemical localization of TYMV coded structural and non-structural proteins by the protein A-gold technique. *Virology* **151,** 100–109.

Garnsey, S. M., and Randles, J. W. (1987). Biological interactions and agricultural implications of viroids. *In* "Viroids and Viroid-like Pathogens" (J. S. Semancik, ed.), pp. 127–160. CRC Press, Boca Raton, Florida.

Garnsey, S. M., and Whidden, R. (1970). A rapid technique for making leaf-tissue grafts to transmit citrus viruses. *Plant Dis. Rep.* **54,** 907–908.

Garnsey, S. M., and Whidden, R. (1973). Efficiency of mechanical inoculation procedures for citrus exocortis virus. *Plant Dis. Rep.* **57,** 886–889.

Garnsey, S. M., Gonsalves, D., and Purcifull, D. E. (1977). Mechanical transmission of citrus tristeza virus. *Phytopathology* **67,** 965–968.

Garrett, R. G., and O'Loughlin, G. T. (1977). Broccoli necrotic yellows virus in cauliflower and in the aphid *Brevicoryne brassicae* L. *Virology* **76,** 653–663.

Gaspar, J. O., Vega, J., Camargo, I. J. B., and Costa, A. S. (1984). An ultrastructural study of particle distribution during microsporogenesis in tomato plants infected with the Brazilian tobacco rattle virus. *Can. J. Bot.* **62,** 372–378.

Gasser, C. S., and Fraley, R. T. (1989). Genetically engineering plants for crop improvement. *Science* **244,** 1293–1299.

Gay, J. D., Johnson, A. W., and Chalfant, R. B. (1973). Effects of a trapcrop on the introduction and distribution of cowpea virus by soil and insect vectors. *Plant Dis. Rep.* **57,** 684–688.

Gebre Selassie, K., Marchoux, G., Delecolle, B., and Pochard, E. (1985). Varabilitié naturelle des souches du virus Y de la pomme de terre dans le cultures du piment du sud-est de la France. Caractérisation et classification en pathotypes. *Agronomie* **5,** 621–630.

Geelen, J. L. M. C., van Kammen, A., and Verduin, B. J. M. (1972). Structure of the capsid of cowpea mosaic virus. The chemical subunit: Molecular weight and number of subunits per particle. *Virology* **49,** 205–213.

Gehrke, L., Auron, P. E., Quigley, G. J., Rich, A., and Sonenberg, N. (1983). 5′-Conformation of capped alfalfa mosaic virus ribonucleic acid 4 may reflect its independence of the cap structure or of cap-binding protein for efficient translation. *Biochemistry* **22,** 5157–5164.

Geldreich, A., Lebeurier, G., and Hirth, L. (1986). *In vivo* dimerisation of cauliflower mosaic virus DNA can explain recombination. *Gene* **48,** 277–286.

Gera, A., and Loebenstein, G. (1988). An inhibitor of virus replication associated with green island tissue of tobacco infected with cucumber mosaic virus. *Physiol. Mol. Plant Pathol.* **32,** 373–385.

Gera, A., Loebenstein, G., and Raccah, B. (1979). Protein coats of two strains of cucumber mosaic virus affect transmission by *Aphis gossypii*. *Phytopathology* **69,** 396–399.

Gerber, M., and Sarkar, S. (1989). The coat protein of tobacco mosaic virus does not play a significant role for cross-protection. *J. Phytopathol.* **124,** 323–331.

Gergerich, R. C., and Scott, H. A. (1988a). The enzymatic function of ribonuclease determines plant virus transmission by leaf-feeding beetles. *Phytopathology* **78,** 270–272.

Gergerich, R. C., and Scott, H. A. (1988b). Evidence that virus translation and virus infection of non-wounded cells are associated by transmissibility by leaf-feeding beetles. *J. Gen. Virol.* **69,** 2935–2938.

Gergerich, R. C., Scott, H. A., and Fulton, J. P. (1983). Regurgitant as a determinant of specificity in the transmission of plant viruses by beetles. *Phytopathology* **73,** 936–938.

Gergerich, R. C., Scott, H. A., and Fulton, J. P. (1986). Evidence that ribonuclease in beetle regurgitant determines the transmission of plant viruses. *J. Gen. Virol.* **67,** 367–370.

Gerlach, W. L., Miller, W. A., and Waterhouse, P. M. (1987a). *Barley Yellow Dwarf Newsletter* **1,** 17–19.

Gerlach, W. L., Llewellyn, D., and Haseloff, J. (1987b). Construction of a plant resistance gene from the satellite RNA of tobacco ringspot virus. *Nature (London)* **328,** 802–805.

Gerpa, C. P. (1984). Applied and theoretical aspects of virus adsorption to surfaces. *Adv. Appl. Microbiol.* **30,** 133–168.

Geysen, H. M., Meloen, R. H., and Barteling, S. J. (1984). Use of peptide synthesis to probe viral antigens for epitopes to a resolution of a single amino acid. *Proc. Natl. Acad. Sci. U.S.A.* **81,** 3998–4002.

Ghabrial, S. A., and Lister, R. M. (1973). Anomalies in molecular weight determinations of tobacco rattle virus protein by SDS–polyacrylamide gel electrophoresis. *Virology* **51,** 485–488.

Ghabrial, S. A., and Shepherd, R. J. (1980). A sensitive radioimmunosorbent assay for the detection of plant viruses. *J. Gen. Virol.* **48,** 311–317.

Ghosh, A., Rutgers, T., Ke-Qiang, M., and Kaesberg, P. (1981). Characterisation of the coat protein mRNA of Southern bean mosaic virus and its relationship to the genomic RNA. *J. Virol.* **39,** 87–92.

Gianinazzi, S., Martin, C., and Vallée, J.-C. (1970). Hypersensibilité aux virus, température et protéines solubles chez le *Nicotiana* Xanthi n.c. Apparition de nouvelles macromolécules lors de la répression de la synthèse virale. *C. R. Hebd. Seances Acad. Sci., Ser. D.* **270,** 2383–2386.

Gianinazzi, S., Deshayes, A., Martin, C., and Vernoy, R. (1977). Differential reactions to tobacco mosaic virus infection in Samsun 'nn' tobacco plants. I. Necrosis, mosaic symptoms and symptomless leaves following the ontogenic gradient. *Phytopathol. Z.* **88,** 347–354.

Gianinazzi, S., Ahl, P., Cornu, A., Scalla, R., and Cassini, R. (1980). First report of host b-protein appearance in response to a fungal infection in tobacco. *Physiol. Plant Pathol.* **16,** 337–342.

Giband, M., Mensard, J. M., and Lebeurier, G. (1986). The gene III product (P15) of cauliflower mosaic virus is a DNA-binding protein while an immunologically related P11 polypeptide is associated with virions. *EMBO J.* **5,** 2433–2438.

Gibb, K. S., and Randles, J. W. (1988). Studies on the transmission of velvet tobacco mottle virus by the mirid *Cyrtopeltis nicotianae*. *Ann. Appl. Biol.* **112,** 427–437.

Gibbs, A. J. (1962). Lucerne mosaic virus in British lucerne crops. *Plant Pathol.* **11,** 167–171.

Gibbs, A. J. (1969). Plant virus classification. *Adv. Virus Res.* **14,** 263–328.

Gibbs, A. (1986). Tobamovirus classification. *In* "The Plant Viruses" (M. H. V. van Regenmortel and H. Fraenkel-Conrat, eds.), Vol. 2, pp. 167–180. Plenum, New York.

Gibbs, A. (1987). Molecular evolution of viruses; "Trees," "clocks" and "modules." *J. Cell Sci., Suppl.* **7,** 319–337.

Gibbs, A. J., and McIntyre, G. A. (1970). A method for assessing the size of a protein from its composition: Its use in evaluating data on the size of the protein subunits of plant virus particles. *J. Gen. Virol.* **9,** 51–67.

Gibbs, A. J., Harrison, B. D., Watson, D. H., and Wildy, P. (1966). What's in a virus name? *Nature (London)* **209,** 450–454.

Gibson, R. W., and Plumb, R. T. (1977). Breeding plants for resistance to aphid infestation. *In* "Aphids as Virus Vectors" (K. F. Harris and K. Maramorosch, eds.), pp. 473–500. Academic Press, New York.

Gibson, R. W., and Rice, A. D. (1986). The combined use of mineral oils and pyrethroids to control plant viruses transmitted non- and semi-persistently by *Myzus persicae*. *Ann. Appl. Biol.* **109,** 465–472.

Gibson, R. W., Pickett, J. A., Dawson, G. W., Rice, A. D., and Stribley, M. F. (1984). Effects of aphid alarm pherome derivatives and related compounds on non- and semi-persistent plant virus transmission by *Myzus persicae*. *Ann. Appl. Biol.* **104,** 203–209.

Gibson, R. W., Payne, R. W., and Katis, N. (1988). The transmission of potato virus Y by aphids of differing vectoring abilities. *Ann. Appl. Biol.* **113,** 35–43.

Gibson, T. J., and Argos, P. (1990). Protruding domain of tomato bushy stunt virus coat protein is a hitherto unrecognised class of jellyroll conformation. *J. Mol. Biol.* **212,** 7–9.

Giegé, R., Briand, J.-P., Mengual, R., Ebel, J.-P., and Hirth, L. (1978). Valylation of the two RNA components of turnip-yellow mosaic virus and specificity of the tRNA aminoacylation reaction. *Eur. J. Biochem.* **84,** 251–256.

Gierer, A. (1957). Structure and biological function of ribonucleic acid from tobacco mosaic virus. *Nature (London)* **179,** 1297–1299.

Gierer, A. (1958). Die Gross der biologisch aktiven Einheit der Ribosenucleinsäure des Tabakmosaikvirus. *Z. Naturforsch., B: Anorg. Chem., Org. Chem., Biochem., Biophys., Biol.* **13,** 485–488.

Gierer, A., and Mundry, K. W. (1958). Production of mutants of tobacco mosaic virus by chemical alteration of its ribonucleic acid *in vitro*. *Nature (London)* **182,** 1457–1458.

Gierer, A., and Schramm, G. (1956). Infectivity of ribonucleic acid from tobacco mosaic virus. *Nature (London)* **177,** 702–703.

Gilbert, W. (1986). The RNA world. *Nature (London)* **319,** 618.

Gildow, F. E. (1982). Coated-vesicle transport of luteoviruses through salivary glands of *Myzus persicae*. *Phytopathology* **72,** 1289–1296.

Gildow, F. E. (1983). Influence of barley yellow dwarf virus-infected oats and barley on morphology of aphid vectors. *Phytopathology* **73,** 1196–1199.

Gildow, F. E. (1985). Transcellular transport of barley yellow dwarf virus into the hemocoel of the aphid vector, *Rhopalosiphum padi*. *Phytopathology* **75,** 292–297.

Gildow, F. E. (1987). Virus–membrane interactions involved in circulative transmission of luteoviruses by aphids. *Curr. Top. Vector Res.* **4,** 93–120.

Gildow, F. E., and D'Arcy, C. J. (1988). Barley and oats as reservoirs for an aphid virus and the influence on barley yellow dwarf transmission. *Phytopathology* **78,** 811–816.

Gil-Fernandez, C., and Black, L. M. (1965). Some aspects of the internal anatomy of the leaf-hopper *Agallia constricta* (Homoptera: Cicadellidae). *Ann. Ent. Soc. Am.* **58,** 275–284.

Gill, C. C. (1968). Rate of movement of barley yellow dwarf virus out of inoculated cereal leaves. *Phytopathology* **58,** 870–871.

Gill, C. C. (1970). Aphid nymphs transmit an isolate of barley yellow dwarf virus more efficiently than do adults. *Phytopathology* **60,** 1747–1752.

Gill, C. C. (1974). Inclusions and wall deposits in cells of plants infected with oat necrotic mottle virus. *Can. J. Bot.* **52,** 621–626.

Gill, C. C., and Chong, J. (1975). Development of the infection in oat leaves inoculated with barley yellow dwarf virus. *Virology* **66,** 440–453.

Gill, C. C., and Chong, J. (1979). Cytological alterations in cells infected with corn leaf aphid-specific isolates of barley yellow dwarf virus. *Phytopathology* **69,** 363–368.

Gill, C. C., and Chong, J. (1981). Vascular cell alterations and predisposed xylem infection in oats by inoculations with paired barley yellow dwarf viruses. *Virology* **114**, 405–414.

Gillaspie, A. G., and Bancroft, J. B. (1965). The rate of accumulation, specific infectivity and electrophoretic characteristics of bean pod mottle virus in bean and soybean. *Phytopathology* **55**, 906–908.

Gillet, J. M., Marimoto, K. M., Ramsdell, D. C., Baker, K. K., Chaney, W. G., and Esselman, W. J. (1982). A comparison between the relative abilities of ELISA, RIA, and ISEM to detect blueberry shoestring virus in its aphid vector. *Acta Hortic.* **129**, 25–30.

Gilmer, R. M. (1965). Additional evidence of tree-to-tree transmission of sour cherry yellows virus by pollen. *Phytopathology* **55**, 482–483.

Gilmer, R. M., and Brase, K. D. (1963). Nonuniform distribution of prune dwarf virus in sweet and sour cherry trees. *Phytopathology* **53**, 819–821.

Gilmer, R. M., Uyemoto, J. K., and Kelts, L. J. (1970). A new grapevine disease induced by tobacco ringspot virus. *Phytopathology* **60**, 619–627.

Gingery, R. E. (1988). The rice stripe virus group. *In* "The Plant Viruses" (R. G. Milne, ed.), Vol. 4, pp. 297–329. Plenum, New York.

Gingery, R. E., and Louie, R. (1985). A satellite-like virus particle associated with maize white line mosaic virus. *Phytopathology* **75**, 870–874.

Giunchedi, L., and Langenberg, W. G. (1982). Beet necrotic yellow vein virus transmission by *Polymyxa betae* Keskin zoospores. *Phytopathologia Mediterranea* **21**, 5–7.

Givord, L., and Den Boer, L. (1980). Insect transmission of okra mosaic virus in the Ivory Coast. *Ann. Appl. Biol.* **94**, 235–241.

Givord, L., Xiong, C., Giband, M., Koenig, I., Hohn, T., Lebeurier, G., and Hirth, L. (1984). A second cauliflower mosaic virus gene product influences the structure of the viral inclusion body. *EMBO J.* **3**, 1423–1427.

Givord, L., Dixon, L., Rauseo-Koenig, I., and Hohn, T. (1988). Cauliflower mosaic virus ORF VII is not required for aphid transmissibility. *Ann. Inst. Pasteu/Virol.* **139**, 227–231.

Glover, J. F., and Wilson, T. M. A. (1982). Efficient translation of the coat protein cistron of tobacco mosaic virus in a cell free system from *Escherichia coli*. *Eur. J. Biochem.* **122**, 485–492.

Godchaux, W., III, and Schuster, T. M. (1987). Isolation and characterisation of nucleoprotein assembly intermediates of tobacco mosaic virus. *Biochemistry* **26**, 454–461.

Godefroy-Colburn, T., Thivent, C., and Pinck, L. (1985a). Translational discrimination between the four RNA's of alfalfa mosaic virus. A quantitative evaluation. *Eur. J. Biochem.* **147**, 541–548.

Godefroy-Colburn, T., Ravelonandro, M., and Pinck, L. (1985b). Cap accessibility correlates with the initiation efficiency of alfalfa mosaic virus RNAs. *Eur. J. Biochem.* **147**, 549–552.

Godefroy-Colburn, T., Gagey, M.-J., Berna, A., and Stussi-Garaud, C. (1986). A non-structural protein of alfalfa mosaic virus in the walls of infected tobacco cells. *J. Gen. Virol.* **67**, 2233–2239.

Godefroy-Colburn, T., Schoumacher, F., Erny, C., Berna, A., Moser, O., Gagey, M. J., and Stussi-Garaud, C. (1990). The movement protein of some plant viruses. In NATO ASI Series Vol. H41 "Recognition and response in plant-virus interactions. (R. S. S. Fraser, ed.) 207–231. Springer-Verlag, Berlin.

Goelet, P., and Karn, J. (1982). Tobacco mosaic virus induces the synthesis of a family of 3′ coterminal messenger RNAs and their complements. *J. Mol. Biol.* **154**, 541–550.

Goelet, P., Lomonossoff, G. P., Butler, P. J. G., Akam, M. E., Gait, M. J., and Karn, J. (1982). Nucleotide sequence of tobacco mosaic virus RNA. *Proc. Natl. Acad. Sci. U.S.A.* **79**, 5818–5822.

Goffeau, A., and Bové, J. M. (1965). Virus infection and photosynthesis. I. Increased photophosphorylation by chloroplasts from Chinese cabbage infected with turnip yellow mosaic virus. *Virology* **27**, 243–252.

Gokhale, D. V., and Bald, J. G. (1987). Relationship between plant virus concentration and infectivity: A "growth curve" model. *J. Virol. Methods* **18**, 225–232.

Goldbach, R. (1986). Molecular evolution of plant RNA viruses. *Annu. Rev. Phytopathol.* **24**, 289–310.

Goldbach, R. (1987). Genome similarities between plant and animal RNA viruses. *Microbiol. Sci.* **4**, 197–202.

Goldbach, R., and Krijt, J. (1982). Cowpea mosaic virus-encoded protease does not recognise primary translation products of mRNAs from other comoviruses. *J. Virol.* **43**, 1151–1154.

Goldbach, R., and Rezelman, G. (1983). Orientation of the cleavage map of the 200-kilodalton polypeptide encoded by the bottom-component RNA of cowpea mosaic virus. *J. Virol.* **46**, 614–619.

Goldbach, R., and van Kammen, A. (1985). Structure, replication, and expression of the bipartite genome of cowpea mosaic virus. *In* "Molecular Plant Virology" (J. W. Davies, ed.), Vol. 2, pp. 83–120. CRC Press, Boca Raton, Florida.

Goldbach, R., and Wellink, J. (1988). Evolution of plus-strand RNA viruses. *Intervirology* **29**, 260–267.

Goldbach, R., Rezelman, G., and van Kammen, A. (1980). Independent replication and expression of B-component RNA of cowpea mosaic virus. *Nature (London)* **285**, 297–300.

Goldbach, R. W., Schilthuis, J. G., and Rezelman, G. (1981). Comparison of *in vivo* and *in vitro* translation of cowpea mosaic virus RNAs. *Biochem. Biophys. Res. Commun.* **99**, 89–94.

Goldbach, R., Rezelman, G., Zabel, P., and van Kammen, A. (1982). Expression of the bottom-component RNA of cowpea mosaic virus: Evidence that the 60-kilodalton VPg precursor is cleaved into single VPg and a 58-kilodalton polypeptide. *J. Virol.* **42**, 630–635.

Goldberg, K.-B., and Brakke, M. K. (1987). Concentration of maize chlorotic mottle virus increased in mixed infections with maize dwarf mosaic virus strain B. *Phytopathology* **77**, 162–167.

Goldberg, K.-B., Kiernan, J., Shoelz, J. E., and Shepherd, R. J. (1990). Transgenic host responses to gene VI of two cauliviruses. *In* "Viral Genes and Plant Pathogenesis" (T. P. Pirone and J. G. Shaw, eds.), pp. 58–66. Springer-Verlag, New York.

Goldberg, R. B. (1988). Plants. Novel developmental processes. *Science* **240**, 1460–1467.

Golemboski, D. B., Lomonossoff, G. P., and Zaitlin, M. (1990). Plants transformed with a tobacco mosaic virus non-structural gene sequence are resistant to the virus. *Proc. Natl. Acad. Sci. U.S.A.* **87**, 6311–6315.

Gonsalves, D., and Fulton, R. W. (1977). Activation of *Prunus* necrotic ringspot virus and rose mosaic virus by RNA 4 components of some ilarviruses. *Virology* **81**, 398–407.

Golsalves, D., Provvidenti, R., and Edwards, M. C. (1982). Tomato white leaf: The relation of an apparent satellite RNA and cucumber mosaic virus. *Phytopathology* **72**, 1533–1538.

Goodey, T. (1963). "Soil and Freshwater Nematodes," 2nd rev. ed. Methuen, London.

Gooding, G. V., Jr. (1970). Natural serological strains of tobacco ringspot virus. *Phytopathology* **60**, 708–713.

Gooding, G. V., Jr. (1975). Inactivation of tobacco mosaic virus on tomato seed with trisodium orthophosphate and sodium hypochlorite. *Plant Dis. Rep.* **59**, 770–772.

Goodman, M., Czelusniak, J., Koop, B. F., Tagle, D. A., and Slightom, J. L. (1987). Globins: A case study in molecular phylogeny. *Cold Spring Harbor Symp. Quant. Biol.* **52**, 875–890.

Goodman, R. M. (1977a). Single-stranded DNA genome in a whitefly-transmitted plant virus. *Virology* **83**, 171–179.

Goodman, R. M. (1977b). Infectious DNA from a whitefly-transmitted virus of *Phaseolus vulgaris*. *Nature (London)* **266**, 54–55.

Goodman, R. M. (1981). Geminiviruses. *In* "Handbook of Plant Virus Infections and Comparative Diagnosis" (E. Kurstak, ed.), pp. 883–910. Am. Elsevier, New York.

Goodman, R. M., and Bird, J. (1978). Bean golden mosaic virus. *CMI/AAB Descriptions Plant Viruses* No. 192.

Goodman, R. M., and Ross, A. F. (1974a). Enhancement of potato virus X synthesis in doubly infected tobacco occurs in doubly infected cells. *Virology* **58**, 16–24.

Goodman, R. M., and Ross, A. F. (1974b). Independent assembly of virions in tobacco doubly infected by potato virus X and potato virus Y or tobacco mosaic virus. *Virology* **59**, 314–318.

Gorbalenya, A. E., Blinov, V. M., and Koonin, E. V. (1985). Prediction of nucleotide-binding properties of virus specific proteins from their primary structure. *Mol. Genet.* **11**, 30–36.

Gorbalenya, A. E., Donchenko, A. P., and Blinov, V. M. (1986). A possible common origin of poliovirus proteins with different functions. *Mol. Genet. Mikrobiol. Virusol.* **1**, 36–41.

Gorbalenya, A. E., Donchenko, A. P., Blinov, V. M., and Koonin, E. V. (1989a). Cysteine proteases of positive strand RNA viruses and chymotrypsin-like serine proteases. A distinct protein superfamily with a common structural fold. *FEBS Lett.* **243**, 103–114.

Gorbalenya, A. E., Koonin, E. V., Donchenko, A. P., and Blinov, V. M. (1989b). Two related super-families of putative helicases involved in replication, recombination, repair and expression of DNA and RNA genomes. *Nucleic Acids Res.* **17**, 4713–4730.

Gorbalenya, A. E., Blinov, V. M., Donchenko, A. P., and Koonin, E. V. (1989c). An NTP-binding motif is the most conserved sequence in a highly diverged monophyletic group of proteins involved in positive strand RNA viral replication. *J. Mol. Evol.* **28**, 256–268.

Gorbalenya, A. E., Koonin, E. V., and Wolf, Y. I. (1990). A new superfamily of putative NTP-binding domains encoded by genomes of small DNA and RNA viruses. *FEBS Letters* **262**, 145–148.

Gordon, K. H. J., and Symons, R. H. (1983). Satellite RNA of cucumber mosaic virus forms a secondary structure with partial 3′-terminal homology to genomal RNAs. *Nucleic Acids Res.* **11**, 947–960.

Gordon, K. H. J., Pfeiffer, P., Fuetterer, J., and Hohn, T. (1988). *In vitro* expression of cauliflower mosaic virus genes. *EMBO J.* **7**, 309–317.

Gould, A. R., and Symons, R. H. (1983). A molecular biological approach to relationships among viruses. *Annu. Rev. Phytopathol.* **21**, 179–199.

Gould, A. R., Palukaitis, P., Symons, R. H., and Mossop, D. W. (1978). Characterization of a satellite RNA associated with cucumber mosaic virus. *Virology* **84**, 443–455.

Gowda, S., Wu, F. C., Scholthof, H. B., and Shepherd, R. J. (1989). Gene VI of figwort mosaic virus (caulimovirus group) functions in post transcriptional expression of genes on the full-length RNA transcript. *Proc. Natl. Acad. Sci. U.S.A.* **86**, 9203–9207.

Gracia, O., and Shepherd, R. J. (1985). Cauliflower mosaic virus in the nucleus of *Nicotiana*. *Virology* **146**, 141–145.

Graddon, D. J., and Randles, J. W. (1986). Single antibody dot immunoassay—A simple technique for rapid detection of a plant virus. *J. Virol. Methods* **13**, 63–69.

Grandbastien, M.-A., Spielmann, A., and Caboche, M. (1989). Tnt 1, a mobile retroviral-like transposable element of tobacco isolated by plant cell genetics. *Nature (London)* **337**, 376–380.

Granell, A., Belles, J. M., and Conejero, V. (1987). Induction of pathogenesis-related proteins in tomato by citrus exocortis viroid, silver ion, and ethephon. *Physiol. Mol. Plant Pathol.* **31**, 83–90.

Grasmick, M. E., and Slack, S. A. (1986). Effect of potato spindle tuber viroid on sexual reproduction and viroid transmission in true potato seed. *Can. J. Bot.* **64**, 336–340.

Gratia, A. (1933). Pluralité antigénique et identification sérologique des virus de plantes. *C. R. Seances Soc. Biol. Ses Fil.* **114**, 923.

Graybosch, R., Hellmann, G. M., Shaw, J. G., Rhoads, R. E., and Hunt, A. G. (1989). Expression of a potyvirus non-structural protein in transgenic tobacco. *Biochem. Biophys. Res. Commun.* **160**, 425–432.

Gregory, P. H. (1948). The multiple-infection transformation. *Ann. Appl. Biol.* **35**, 412–417.

Grieco, F., Burgyan, J., and Russo, M. (1989). The nucleotide sequence of *Cymbidium* ringspot virus RNA. *Nucleic Acids Res.* **17**, 6383.

Grief, C., Hemmer, O., and Fritsch, C. (1988). Nucleotide sequence of tomato black ring virus RNA-1. *J. Gen. Virol.* **69**, 1517–1529.

Grill, L. K., and Semancik, J. S. (1978). RNA sequences complementary to citrus exocortis viroid in nucleic acid preparations from infected *Gynura aurantiaca*. *Proc. Natl. Acad. Sci. U.S.A.* **75**, 896–900.

Grimsley, N., Hohn, B., Hohn, T., and Walden, R. (1986). "Agroinfection," an alternative route for viral infection of plants by using the Ti plasmid. *Proc. Natl. Acad. Sci. U.S.A.* **83**, 3282–3286.

Grimsley, N., Hohn, T., Davies, J. W., and Hohn, B. (1987). *Agrobacterium*-mediated delivery of infectious maize streak virus into maize plants. *Nature (London)* **325**, 177–179.

Grivell, A. R., Grivell, C. J., Jackson, J. F., and Nicholas, D. J. D. (1971). Preservation of lettuce necrotic yellows and some other plant viruses by dehydration with silica gel. *J. Gen. Virol.* **12**, 55–58.

Grogan, R. G., and Bardin, R. (1950). Some aspects concerning the seed transmission of lettuce mosaic virus. *Phytopathology* **40**, 965.

Grogan, R. G., Welch, J. E., and Bardin, R. (1952). Common lettuce mosaic and its control by the use of disease free seed. *Phytopathology* **42**, 573–578.

Grogan, R. G., Zink, F. W., Hewitt, W. B., and Kimble, K. A. (1958). The association of *Olpidium* with the big-vein disease of lettuce. *Phytopathology* **48**, 292–297.

Gronenborn, B., Gardner, R. C., Schaefer, S., and Shepherd, R. J. (1981). Propagation of foreign DNA in plants using cauliflower mosaic virus as vector. *Nature (London)* **294**, 773–776.

Gröning, B. R., Abouzid, A., and Jeske, H. (1987). Single-stranded DNA from abutilon mosaic virus is present in the plastids of infected *Abutilon sellovianum*. *Proc. Natl. Acad. Sci. U.S.A.* **84**, 8996–9000.

Gross, H. J., Domdey, H., Lossow, C., Jank, P., Raba, M., Alberty, H., and Sänger, H. L. (1978). Nucleotide sequence and secondary structure of potato spindle tuber viroid. *Nature (London)* **273**, 203–208.

Grosset, J., Meyer, I., Chartier, Y., Kauffmann, S., Legrand, M., and Fritig, B. (1990). Tobacco mesophyll protoplasts synthesise 1,3-β-glucanase chitinases and "Osmotins" during *in vitro* culture. *Plant Physiol.* **92**, 520–527.

Gugerli, P. (1976). Different states of aggregation of tobacco rattle virus coat protein. *J. Gen. Virol.* **33**, 297–307.

Gugerli, P. (1977). Untersuchungen über die großräumige und lokale Verbreitung des Tabakrattlevirus (TRV) und seiner Vektoren in der Schweiz. *Phytopathol. Z.* **89**, 1–24.

Gugerli, P. (1984). Isopycnic centrifugation of plant viruses in Nycodenz^R density gradients. *J. Virol. Methods* **9**, 249–258.

Gugerli, P., and Fries, P. (1983). Characterization of monoclonal antibodies to potato virus Y and their use for virus detection. *J. Gen. Virol.* **64,** 2471–2477.

Guilford, P. J. (1989). A molecular analysis of tobacco rattle mosaic virus RNA1. Ph.D. Thesis, University of Cambridge.

Guilford, P. J., and Forster, R. L. S. (1986). Detection of polyadenylated subgenomic RNAs in leaves infected with the potexvirus Daphne virus X. *J. Gen. Virol.* **67,** 83–90.

Guilley, H., Jonard, G., Richards, K. E., and Hirth, L. (1975). Sequence of a specifically encapsidated RNA fragment originating from the tobacco-mosaic-virus coat-protein cistron. *Eur. J. Biochem.* **54,** 135–144.

Guilley, H., Richards, K. E., and Jonard, G. (1983). Observations concerning the discontinuous DNAs of cauliflower mosaic virus. *EMBO J.* **2,** 277–282.

Guilley, H., Carrington, J. C., Balàzs, E., Jonard, G., Richards, K., and Morris, T. J. (1985). Nucleotide sequence and genome organisation of carnation mottle virus RNA. *Nucleic Acids Res.* **13,** 6663–6677.

Guinchedi, L., and Langenberg, W. G. (1982). Beet necrotic yellow vein virus transmission by *Polymyxa betae* Keskin zoospores. *Phytopathol. Mediterr.* **21,** 5–7.

Gumpf, D. J., Cunningham, D. S., Heick, J. A., and Shannon, L. M. (1977). Amino acid sequence in the proteolytic glycopeptide of barley stripe mosaic virus. *Virology* **78,** 328–330.

Gunasinghe, U. B., Irwin, M. E., and Kampmeier, G. E. (1988). Soybean leaf pubescence affects aphid vector transmission and field spread of soybean mosaic virus. *Ann. Appl. Biol.* **112,** 259–272.

Gunnery, S., and Datta, A. (1987). An inhibitor RNA of translation from barley embryo. *Biochem. Biophys. Res. Commun.* **142,** 383–388.

Gunning, B. E. S., and Overall, R. L. (1983). Plasmodesmata and cell-to-cell transport in plants. *BioScience* **33,** 260–265.

Gupta, B. M., Chandra, K., Verma, H. N., and Verma, G. S. (1974). Induction of antiviral resistance in *Nicotiana glutinosa* plants by treatment with *Trichothecium* polysaccharide and its reversal by actinomycin D *J. Gen. Virol.* **24,** 211–213.

Gustafson, F. G., and Armour, S. L. (1986). The complete nucleotide sequence of RNA β from the type strain of barley stripe mosaic virus. *Nucleic Acids Res.* **14,** 3895–3909.

Gustafson, G. D., Larkins, B. A., and Jackson, A. O. (1981). Comparative analysis of polypeptides synthesised *in vivo* and *in vitro* by two strains of barley stripe mosaic virus. *Virology* **111,** 579–587.

Gustafson, G. D., Hunter, B., Hanau, R., Armour, S. L., and Jackson, A. O. (1987). Nucleotide sequence and genetic organisation of barley stripe mosaic virus RNA λ. *Virology* **158,** 394–406.

Gustafson, G. D., Armour, S. L., Gamboa, G. C., Burgett, S. G., and Shepherd, J. W. (1989). Nucleotide sequence of barley stripe mosaic virus RNAα: RNAα encodes a single polypeptide with homology to corresponding proteins from other viruses. *Virology* **170,** 370–377.

Guy, P. L. (1988). Pasture ecology of barley yellow dwarf viruses at Sanford, Tasmania. *Plant Pathol.* **37,** 546–550.

Guy, P. L., and Gibbs, A. J. (1985). Further studies on turnip yellow mosaic tymovirus isolates from an endemic Australian *Cardamine*. *Plant Pathol.* **34,** 532–544.

Gwynn, G. R., Reilly, K. R., Komn, J. J., Burk, L. G., and Reed, S. M. (1986). Genetic resistance to tobacco mosaic virus, cyst nematodes root-knot nematodes, and wildfire from *Nicotiana repanda* incorporated into *N. tabacum*. *Plant Dis.* **70,** 958–962.

Haas, B., Klanner, A., Ramm, K., and Sänger, H. L. (1988). The 7s RNA from tomato leaf tissue resembles a signal recognition particle RNA and exhibits a remarkable sequence complementarity to viroids. *EMBO J.* **7,** 4063–4074.

Haber, S., Polston, J. E., and Bird, J. (1987). Use of DNA to diagnose plant diseases caused by single-stranded DNA plant viruses. *Can. J. Plant Pathol.* **9,** 156–161.

Habili, N., and Symons, R. H. (1989). Evolutionary relationship between luteoviruses and other RNA plant viruses based on the sequence motifs in their putative RNA polymerases and nucleic acid helicases. *Nucleic Acids Res.* **17,** 9543–9555.

Habili, N., McInnes, J. L., and Symons, R. H. (1987). Nonradioactive, photobiotin-labelled DNA probes for the routine diagnosis of barley yellow dwarf virus. *J. Virol. Methods* **16,** 225–237.

Hadidi, A. (1986). Relationship of viroids and certain other plant pathogenic nucleic acids to group I and II introns. *Plant Mol. Biol.* **7,** 129–142.

Hadidi, A. (1988). Synthesis of disease associated proteins in viroid-infected tomato leaves and binding of viroid to host proteins. *Phytopathology* **78,** 575–578.

Hadidi, A., Cress, D. E., and Diener, T. O. (1981). Nuclear DNA from uninfected or potato spindle tuber viroid-infected tomato plants contains no detectable sequences complementary to cloned double-stranded viroid cDNA. *Proc. Natl. Acad. Sci. U.S.A.* **78,** 6932–6935.

Hafner, M. S., and Nadler, S. A. (1988). Phylogenetic trees support the coevolution of parasites and their hosts. *Nature (London)* **332,** 258–259.

Hagborg, W. A. F. (1970). A device for injecting solutions and suspensions into thin leaves of plants. *Can. J. Bot.* **48,** 1135–1136.

Hagen, T. J., Taylor, D. B., and Meagher, R. B. (1982). Rocket immunoelectrophoresis assay for cauliflower mosaic virus. *Phytopathology* **72,** 239–242.

Hagenbüchle, O., Santer, M., Steitz, J. A., and Mans, R. J. (1978). Conservation of the primary structure at the 3′ end of 18S rRNA from eukaryotic cells. *Cell (Cambridge, Mass.)* **13,** 551–563.

Hagiwara, K., Minobe, Y., Nozu, Y., Hibino, H., Kimura, I., and Omura, T. (1986). Component proteins and structure of rice ragged stunt virus. *J. Gen. Virol.* **67,** 1711–1715.

Hahn, P., and Shepherd, R. J. (1980). Phosphorylated proteins in cauliflower mosaic virus. *Virology* **107,** 295–297.

Hahn, P., and Shepherd, R. J. (1982). Evidence for a 58-kilodalton polypeptide as precursor of the coat protein of cauliflower mosaic virus. *Virology* **116,** 480–488.

Haight, E., and Gibbs, A. (1983). Effect of viruses on pollen morphology. *Plant Pathol.* **32,** 369–372.

Hajimorad, M. R., and Francki, R. I. B. (1988). Alfalfa mosaic virus isolates from lucerne in South Australia: Biological variability and antigenic similarity. *Ann. Appl. Biol.* **113,** 45–54.

Halk, E. L., and McGuire, J. M. (1973). Translocation of tobacco ringspot virus in soybean. *Phytopathology* **63,** 1291–1300.

Halk, E. L., Hus, H. T., Aebig, J., and Franke, J. (1984). Production of monoclonal antibodies against three ilarviruses and alfalfa mosaic virus and their use in serotyping. *Phytopathology* **74,** 367–372.

Hall, A. E., and Loomis, R. S. (1972). An explanation for the difference in photosynthetic capabilities of healthy and beet yellows virus-infected sugar beets (*Beta vulgaris* L.). *Plant Physiol.* **50,** 576–580.

Hall, J. D., Barr, R., Al-Abbas, A. H., and Crane, F. L. (1972). The ultrastructure of chloroplasts in mineral-deficient maize leaves. *Plant Physiol.* **50,** 404–409.

Halliwell, R. S., and Gazaway, W. S. (1975). Quantity of microinjected tobacco mosaic virus required for infection of single cultured tobacco cells. *Virology* **65,** 583–587.

Hamilton, R. I., Leung, E., and Nichols, C. (1977). Surface contamination of pollen by plant viruses. *Phytopathology* **67,** 395–399.

Hamilton, R. I., Edwardson, J. R., Francki, R. I. B., Hsu, H. T., Hull, R., Koenig, R. and Milne, R. G. (1981). Guidelines for the identification and characterization of plant viruses. *J. Gen. Virol.* **54,** 223–241.

Hamilton, W. D. O., and Baulcombe, D. L. (1989). Infectious RNA produced by *in vitro* transcription of a full-length tobacco rattle virus RNA-1 cDNA. *J. Gen. Virol.* **70,** 963–968.

Hamilton, W. D. O., Bisaro, D. M., and Buck, K. W. (1982). Identification of novel DNA forms in tomato golden mosaic virus infected tissue. Evidence for a two component viral genome. *Nucleic Acids Res.* **10,** 4901–4912.

Hamilton, W. D. O., Bisaro, D. M., Coutts, R. H. A., and Buck, K. W. (1983). Demonstration of the bipartite nature of the genome of a single-stranded DNA plant virus by infection with the cloned DNA components. *Nucleic Acids Res.* **11,** 7387–7396.

Hamilton, W. D. O., Stein, V. E., Coutts, R. H. A., and Buck, K. W. (1984). Complete nucleotide sequence of the infectious cloned DNA components of tomato golden mosaic virus: Potential coding regions and regulatory sequences. *EMBO J.* **3,** 2197–2205.

Hamilton, W. D. O., Boccara, M., Robinson, D. J., and Baulcombe, D. C. (1987). The complete nucleotide sequence of tobacco rattle virus RNA-1. *J. Gen. Virol.* **68,** 2563–2575.

Hammond, J. (1982). *Plantago* as a host of economically important viruses. *Adv. Virus Res.* **27,** 103–140.

Hammond, J., Lister, R. M., and Foster, J. E. (1983). Purification, identity and some properties of an isolate of barley yellow dwarf virus from Indiana. *J. Gen. Virol.* **64,** 667–676.

Hammond, R. W., and Owens, R. A. (1987). Mutational analysis of potato spindle tuber viroid reveals complex relationships between structure and infectivity. *Proc. Natl. Acad. Sci. U.S.A.* **84,** 3967–3971.

Hammond, R. W., Diener, T. O., and Owens, R. A. (1989). Infectivity of chimeric viroid transcripts reveals the presence of alternative processing sites in potato spindle tuber viroid. *Virology* **170,** 486–495.

Hammond, R. W., Smith, D. R., and Diener, T. O. (1990). Nucleotide sequence and proposed secondary structure of *Columnea* latent viroid: A natural mosaic of viroid sequences. *Nucleic Acids Res.* **17,** 10083–10094.

Hampton, R. O. (1966). Probabilities of failing to detect prune dwarf virus in cherry trees by bud indexing. *Phytopathology* **56,** 650–652.

Hampton, R. O. (1975). The nature of bean yield reduction by bean yellow and bean common mosaic viruses. *Phytopathology* **65,** 1342–1346.

Hampton, R. O. (1983). Plant virus working collections, present and future. *Phytopathology* **73,** 771 (A40).

Hampton, R. O. (1988). Health status (virus) of native North American *Humulus lupulus* in the natural habitat. *J. Phytopathol.* **123,** 353–362.

Hanada, K. (1984). Electrophoretic analysis of virus particles of fourteen cucumovirus isolates. *Ann. Phytopathol. Soc. Jpn.* **50,** 361–367.

Hanada, K., and Francki, R. I. B. (1989). Kinetics of velvet tobacco mottle virus satellite RNA synthesis and encapsidation. *Virology* **170,** 48–54.

Hanada, K., and Harrison, B. D. (1977). Effects of virus genotype and temperature on seed transmission of nepoviruses. *Ann. Appl. Biol.* **85,** 79–92.

Hanada, K., Kusunoki, M., and Iwaki, M. (1986). Properties of virus particles, nucleic acid and coat protein of cycas necrotic stunt virus. *Ann. Phytopathol. Soc. Jpn.* **52,** 422–427.

Hanley-Bowdoin, L., Elmer, J. S., and Rogers, S. G. (1988). Transient expression of heterologous RNAs using tomato golden mosaic virus. *Nucleic Acids Res.* **16,** 10511–10528.

Hanley-Bowdoin, L., Elmer, J. S., and Rogers, S. G. (1990). Expression of functional replication protein from tomato golden mosaic virus in transgenic tobacco plants. *Proc. Natl. Acad. Sci. U.S.A.* **87,** 1446–1450.

Hansen, A. J., and Lane, W. D. (1985). Elimination of apple chlorotic leafspot virus from apple shoot cultures by ribavirin. *Plant Dis.* **69,** 134–135.

Harbison, S.-A., Wilson, T. M. A., and Davies, J. W. (1984). An encapsidated, subgenomic messenger RNA encodes the coat protein of carnation mottle virus. *Biosci. Rep.* **4,** 949–956.

Harbison, S.-A., Davies, J. W., and Wilson, T. M. A. (1985). Expression of high molecular weight polypeptide by carnation mottle virus RNA. *J. Gen. Virol.* **66,** 2597–2604.

Harbison, S.-A., Forster, R. L. S., Guilford, P. J., and Gardner, R. C. (1988). Organisation and interviral homologies of the coat protein gene of white clover mosaic virus. *Virology* **162,** 459–465.

Hare, J. D., and Dodds, J. A. (1987). Survival of the Colorado potato beetle on virus-infected tomato in relation to plant nitrogen and alkaloid content. *Entomol. Exp. Appl.* **44,** 31–35.

Hari, V. (1981). The RNA of tobacco etch virus: Further characterisation and detection of protein linked to the RNA. *Virology* **112,** 391–399.

Hari, V., Siegel, A., Rozek, C., and Timberlake, W. E. (1979). The RNA of tobacco etch virus contains poly(A). *Virology* **92,** 568–571.

Harker, C. L., Woolston, C. J., Markham, P. G., and Maule, A. J. (1987a). Cauliflower mosaic virus aphid transmission factor protein is expressed in cells infected with some aphid non-transmissible isolates. *Virology* **160,** 252–254.

Harker, C. L., Mullineaux, P. M., Bryant, J. A., and Maule, A. J. (1987b). Detection of CaMV gene I and gene VI protein products *in vivo* using antisera raised to COOH-terminal β-galactosidase fusion proteins. *Plant Mol. Biol.* **8,** 275–287.

Harlan, J. R. (1976). Diseases as a factor in plant evolution. *Annu. Rev. Phytopathol.* **14,** 31–51.

Harper, F. R., Nelson, G. A., and Pittman, U. J. (1975). Relationship between leaf roll symptoms and yield in netted gem potato. *Phytopathology* **65,** 1242–1244.

Harrington, R., and Gibson, R. W. (1989). Transmission of potato virus Y by aphids trapped in potato crops in southern England. *Potato Res.* **32,** 167–174.

Harrington, R., Katis, N., and Gibson, R. W. (1987). Monitoring aphids to assess the spread of potato virus Y. *In* "Aphid Migration and Forecasting Euraphid Systems in European Community Countries. Proceedings of the EC Expert Meeting" (R. Cavalloro, ed.), pp. 173–175. Commission of the European Communities Joint Research Centre, Ispra.

Harrington, R., Dewar, A. M., and George, B. (1989). Forecasting the incidence of virus yellows in sugar beet in England. *Ann. Appl. Biol.* **114,** 459–469.

Harris, J. I., and Knight, C. A. (1952). Action of carboxypeptidase on tobacco mosaic virus. *Nature (London)* **170,** 613–614.

Harris, J. I., and Knight, C. A. (1955). Studies on the action of carboxypeptidase on tobacco mosaic virus. *J. Biol. Chem.* **214,** 215–230.

Harris, K. F. (1983). Sternorrhynchous vectors of plant viruses: Virus–vector interactions and transmission mechanisms. *Adv. Virus Res.* **28,** 113–140.

Harris, K. F., and Bath, J. E. (1973). Regurgitation by *Myzus persicae* during membrane feeding: Its likely function in transmission of nonpersistent plant viruses. *Am. Entomol. Soc. Am.* **66,** 793–796.

Harris, K. F., and Bradley, R. H. E. (1973) Tobacco mosaic virus: Can aphids inoculate it into plants with their mouthparts? *Phytopathology* **63,** 1343–1345.

Harris, K. F., and Maramorosch, K. (eds.) (1982). "Pathogens, Vectors, and Plant Diseases: Approaches to Control." Academic Press, New York.

Harris, K. F., Treur, B., Tsai, J., and Toler, R. (1981). Observations on leafhopper ingestion–egestion behaviour: Its likely role in the transmission of non-circulative viruses and other plant pathogens. *J. Econ. Entomol.* **74,** 446–453.

Harrison, B. D. (1958). Ability of single aphids to transmit both avirulent and virulent strains of potato leaf roll virus. *Virology* **6,** 278–286.

Harrison, B. D. (1985). Advances in geminivirus research. *Annu. Rev. Phytopathol.* **23,** 55–82.

Harrison, B. D., and Crockatt, A. A. (1971). Effects of cycloheximide on the accumulation of tobacco rattle virus in leaf discs of *Nicotiana clevelandii*. *J. Gen. Virol.* **12,** 183–185.

Harrison, B. D., and Jones, R. A. C. (1970). Host range and some properties of potato mop-top virus. *Ann. Appl. Biol.* **65,** 393–402.

Harrison, B. D., and Jones, R. A. C. (1971). Effects of light and temperature on symptom development and virus content of tobacco leaves inoculated with potato mop-top virus. *Ann. Appl. Biol.* **67,** 377–387.

Harrison, B. D., and Murant, A. F. (1984). Involvement of plant virus-coded proteins in transmission of plant viruses by vectors. *In* "Vectors in Virus Biology" (M. A. Mayo and K. R. Harrap, eds.), pp. 1–36. Academic Press, London.

Harrison, B. D., and Nixon, H. L. (1959). Separation and properties of particles of tobacco rattle virus with different lengths. *J. Gen. Microbiol.* **21,** 569–581.

Harrison, B. D., and Roberts, I. M. (1968). Association of tobacco rattle virus with mitochondria. *J. Gen. Virol.* **3,** 121–124.

Harrison, B. D., and Winslow, R. D. (1961). Laboratory and field studies on the relation of arabis mosaic virus to its nematode vector *Xiphinema diversicaudatum* Micoletzky. *Ann. Appl. Biol.* **49,** 621–633.

Harrison, B. D., and Woods, R. D. (1966). Serotypes and particle dimensions of tobacco rattle viruses from Europe and America. *Virology* **28,** 610–620.

Harrison, B. D., Peachey, J. E., and Winslow, R. D. (1963). The use of nematicides to control the spread of arabis mosaic virus by *Xiphinema diversicaudatum* (Micol). *Ann. Appl. Biol.* **52,** 243–255.

Harrison, B. D., Finch, J. T., Gibbs, A. J., Hollings, M., Shepherd, R. J., Valenta, V., and Wetter, C. (1971). Sixteen groups of plant viruses. *Virology* **45,** 356–363.

Harrison, B. D., Murant, A. F., and Mayo, M. A. (1972). Two properties of raspberry ringspot virus determined by its smaller RNA. *J. Gen. Virol.* **17,** 137–141.

Harrison, B. D., Murant, A. F., Mayo, M. A., and Roberts, I. M. (1974a) Distribution of determinants for symptom production, host range and nematode transmissibility between the two RNA components of raspberry ringspot virus. *J. Gen. Virol.* **22,** 233–247.

Harrison, B. D., Robertson, W. M., and Taylor, C. E. (1974b). Specificity of retention and transmission of viruses by nematodes. *J. Nematol.* **6,** 155–164.

Harrison, B. D., Kubo, S., Robinson, D. J., and Hutcheson, A. M. (1976). The multiplication cycle of tobacco rattle virus in tobacco mesophyll protoplasts. *J. Gen. Virol.* **33,** 237–248.

Harrison, B. D., Barker, H., Bock, K. R., Guthrie, E. J., Meredith, G., and Atkinson, M. (1977). Plant viruses with circular single-stranded DNA *Nature (London)* **270,** 760–762.

Harrison, B. D., Robinson, D. J., Mowat, W. P., and Duncan, G. H. (1983). Comparison of nucleic acid hybridisation and other tests for detecting tobacco rattle virus in narcissus plants and potato tubers. *Ann. Appl. Biol.* **102,** 331–338.

Harrison, B. D., Mayo, M. A., and Baulcombe, D. C. (1987). Virus resistance in transgenic plants that express cucumber mosaic virus satellite RNA. *Nature (London)* **328,** 799–802.

Harrison, B. D., Barker, H., and Derrick, P. M. (1989). Intercellular movement of potato leafroll luteovirus: Effects of plant resistance and co-infection. *Abstr., NATO Adv. Res. Workshop, Recognition Response Plant-Virus Interact., 1989.*

Harrison, S. C. (1983). Virus structure: High resolution perspectives. *Adv. Virus Res.* **28,** 175–240.

Harrison, S. C., Olson, A. J., Schutt, C. E., Winkler, F. K., and Bricogne, G. (1978). Tomato bushy stunt virus at 2.9 Å resolution. *Nature (London)* **276,** 368–373.

Hartman, K. A., McDonald-Ordzie, P. E., Kaper, J. M., Prescott, B., and Thomas, G. J., Jr. (1978). Studies of virus structure by Laser–Raman spectroscopy. Turnip yellow mosaic virus and capsids. *Biochemistry* **17,** 2118–2123.

Hartman, R. D. (1974). Dasheen mosaic virus and other phytopathogens eliminated from caladium, taro, and cocoyam by culture of shoot tips. *Phytopathology* **64,** 237–240.

Harvey, J. D. (1973). Diffusion coefficients and hydrodynamic radii of three spherical RNA viruses by laser light scattering. *Virology* **56**, 365–368.

Harvey, J. D., Farrell, J. A., and Bellamy, A. R. (1974). Biophysical studies of reovirus type 3. II. Properties of the hydrated particle. *Virology* **62**, 154–160.

Hasegawa, A., Verver, J., Shimada, A., Saito, M., Goldbach, R., van Kammen, A., Miki, K., Kameya-Iwaki, M., and Hibi, T. (1989). The complete nucleotide sequence of soybean chlorotic mottle virus DNA and the identification of a novel promoter. *Nucleic Acids Res.* **17**, 9993–10013.

Haselkorn, R. (1962). Studies on infectious RNA from turnip yellow mosaic virus. *J. Mol. Biol.* **4**, 357–367.

Haseloff, J., and Gerlach, W. (1988). Simple RNA enzymes with new and highly specific endoribonuclease activities. *Nature (London)* **334**, 585–591.

Haseloff, J., and Symons, R. H. (1981). Chrysanthemum stunt viroid: Primary sequence and secondary structure. *Nucleic Acids Res.* **9**, 2741–2752.

Haseloff, J., Mohamed, N. A., and Symons, R. H. (1982). Viroid RNAs of cadang-cadang disease of coconuts. *Nature (London)* **299**, 316–321.

Haseloff, J., Goelet, P., Zimmern, D., Ahlquist, P., Dasgupta, R., and Kaesberg, P. (1984). Striking similarities in amino acid sequence among non-structural proteins encoded by RNA viruses that have dissimilar genomic organisation. *Proc. Natl. Acad. Sci. U.S.A.* **81**, 4358–4362.

Hashimoto, J., and Koganezawa, H. (1987). Nucleotide sequence and secondary structure of apple scar skin viroid. *Nucleic Acids Res.* **15**, 7045–7052.

Hatta, T. (1976). Recognition and measurement of small isometric virus particles in thin sections. *Virology* **69**, 237–245.

Hatta, T., and Francki, R. I. B. (1976). Anatomy of virus-induced galls on leaves of sugarcane infected with Fiji disease virus and the cellular distribution of virus particles. *Physiol. Plant Pathol.* **9**, 321–330.

Hatta, T., and Francki, R. I. B. (1977). Morphology of Fiji disease virus. *Virology* **76**, 797–807.

Hatta, T., and Francki, R. I. B. (1981a). Identification of small polyhedral virus particles in thin sections of plant cells by an enzyme cytochemical technique. *J. Ultrastruct. Res.* **74**, 116–129.

Hatta, T., and Francki, R. I. B. (1981b). Cytopathic structures associated with tonoplasts of plant cells infected with cucumber mosaic and tomato aspermy viruses. *J. Gen. Virol.* **53**, 343–346.

Hatta, T., and Francki, R. I. B. (1981c). Development and cytopathology of virus-induced galls on leaves of sugarcane infected with Fiji disease virus. *Physiol. Plant Pathol.* **19**, 337–346.

Hatta, T., and Francki, R. I. B. (1984). Differences in the morphology of isometric particles of some plant viruses stained with uranyl acetate as an aid to their identification. *J. Virol. Methods* **9**, 237–247.

Hatta, T., and Matthews, R. E. F. (1974). The sequence of early cytological changes in Chinese cabbage leaf cells following systemic infection with turnip yellow mosaic virus. *Virology* **59**, 383–396.

Hatta, T., and Matthews, R. E. F. (1976). Sites of coat protein accumulation in turnip yellow mosaic virus-infected cells. *Virology* **73**, 1–16.

Hatta, T., Nakamoto, T., Takagi, Y., and Ushiyama, R. (1971). Cytological abnormalities of mitochondria induced by infection with cucumber green mottle mosaic virus. *Virology* **45**, 292–297.

Hatta, T., Bullivant, S., and Matthews, R. E. F. (1973). Fine structure of vesicles induced in chloroplasts of Chinese cabbage leaves by infection with turnip yellow mosaic virus. *J. Gen. Virol.* **20**, 37–50.

Hawkes, J. G. (1967). The history of the potato. Part III. *J. R. Hortic. Soc.* **92**, 288–302.

Hayashi, T. (1974). Fate of tobacco mosaic virus after entering the host cell. *Jpn. J. Microbiol.* **18**(4), 279–286.

Hayes, R. J., and Buck, K. W. (1989). Replication of tomato golden mosaic virus DNA B in transgenic plants expressing open reading frames (ORFs) of DNA A: Requirement of ORF AL2 for production of single-stranded DNA. *Nucleic Acids Res.* **17**, 10213–10222.

Hayes, R. J., Buck, K. W., and Brunt, A. A. (1984). Double-stranded and single-stranded subgenomic RNAs from plant tissue infected with tomato bushy stunt virus. *J. Gen. Virol.* **65**, 1239–1243.

Hayes, R. J., Brough, C. L., Prince, V. E., Coutts, R. H. A., and Buck, K. W. (1988a). Infection of *Nicotiana benthamiana* with uncut cloned tandem dimers of tomato golden mosaic virus DNA. *J. Gen. Virol.* **69**, 209–218.

Hayes, R. J., MacDonald, H., Coutts, R. H. A., and Buck, K. W. (1988b). Agroinfection of *Tritium aestivum* with cloned DNA of wheat dwarf virus. *J. Gen. Virol.* **69**, 891–896.

Hayes, R. J., MacDonald, H., Coutts, R. H. A., and Buck, K. W. (1988c). Priming of complementary DNA synthesis *in vitro* by small DNA molecules tightly bound to virion DNA of wheat dwarf virus. *J. Gen. Virol.* **69**, 1345–1350.

Hayes, R. J., Petty, I. T. D., Coutts, R. H. A., and Buck, K. W. (1988d). Gene amplification and expression in plants by a replicating geminivirus vector. *Nature (London)* **334**, 179–182.

Hayes, R. J., Coutts, R. H. A., and Buck, K. W. (1988e). Agroinfection of *Nicotiana* spp. with cloned DNA of tomato golden mosaic virus. *J. Gen. Virol.* **69,** 1487–1496.

Hayes, R. J., Brunt, A. A., and Buck, K. W. (1988f). Gene mapping and expression of tomato bushy stunt virus. *J. Gen. Virol.* **69,** 3047–3057.

Hayes, R. J., Coutts, R. H. A., and Buck, K. W. (1989). Stability and expression of bacterial genes in replicating geminivirus vectors in plants. *Nucleic Acids Res.* **17,** 2391–2403.

Heard, A. J., and Chapman, P. F. (1986). A field study of the pattern of local spread of ryegrass mosaic virus in mown grassland. *Ann. Appl. Biol.* **108,** 341–345.

Hearne, P. Q., Knorr, D. A., Hillman, B. I., and Morris, T. J. (1990). The complete genome structure and synthesis of infectious RNA from clones of tomato bushy stunt virus. *Virology* **177,** in press.

Heathcote, G. D. (1978). Beet mosaic—A declining disease in England. *Plant Pathol.* **22,** 42–45.

Heathcote, G. D. (1978). Review of losses caused by virus yellows in English sugar beet crops and the cost of partial control with insecticides. *Plant Pathol.* **27,** 12–17.

Heathcote, G. D., and Broadbent, L. (1961). Local spread of potato leaf roll and Y viruses. *Eur. Potato J.* **4,** 138–143.

Heaton, L. A., Zuidema, D., and Jackson, A. O. (1987). Structure of the M2 protein gene of Sonchus yellow net virus. *Virology* **161,** 234–241.

Heaton, L. A., Hillman, B. I., Hunter, B. G., Zuidema, D., and Jackson, A. O. (1989a). A physical map of the genome of Sonchus yellow net virus, a plant rhabdovirus with six genes and conserved gene-junction sequences. *Proc. Natl. Acad. Sci. U.S.A.* **86,** 8665–8668.

Heaton, L. A., Carrington, J.-C., and Morris, T. J. (1989b). Turnip crinkle virus infection from RNA synthesized *in vitro*. *Virology* **170,** 214–218.

Hebert, T. T. (1963). Precipitation of plant viruses by polyethylene glycol. *Phytopathology* **53,** 362.

Hecht, E. I., and Bateman, D. F. (1964). Non-specific acquired resistance to pathogens resulting from localised infections by *Thielaviopsis basicola* or viruses in tobacco leaves. *Phytopathology* **54,** 523–530.

Hecker, R., Wang, Z., Steger, G., and Riesner, D. (1988). Analysis of RNA structures by temperature-gradient gel electrophoresis: Viroid replication and processing. *Gene* **72,** 59–74.

Heide, M., and Lange, L. (1988). Detection of potato leaf roll virus and potato viruses M, S, X, and Y by dot immunobinding on plain paper. *Potato Res.* **31,** 367–373.

Heijbroek, W. (1988). Dissemination of rhizomania by soil, beet seeds, and stable manure. *Neth. J. Plant Pathol.* **94,** 9–15.

Heijtink, R. A., and Jaspars, E. M. J. (1974). RNA contents of abnormally long particles of certain strains of alfalfa mosaic virus. *Virology* **59,** 371–382.

Heijtink, R. A., and Jaspars, E. M. J. (1976). Characterization of two morphologically distinct top component a particles from alfalfa mosaic virus. *Virology* **69,** 75–80.

Heijtink, R. A., Houwing, C. J., and Jaspars, E. M. J. (1977). Molecular weights of particles and RNAs of alfalfa mosaic virus. Number of subunits in protein capsids. *Biochemistry* **16,** 4684–4693.

Hein, A. (1971). Zur Wirkung von Öl auf die Virusübertragung durch Blattläuse. *Phytopathol. Z.* **71,** 42–48.

Hein, A. (1987). A contribution to the effect of fungicides and additives for formulation on the rhizomania infection of sugar beets (beet necrotic yellow vein virus). *J. Plant. Dis. Prot.* **94,** 250–259.

Heinrichs, E. A. (1988). Variable resistance of homopterans to rice cultivars. *ISI Atlas of Sci. Anim. Plant Sci.* **1,** 213–220.

Heinrichs, E. A., Rapusas, H. R., Aquino, G. B., and Palis, F. (1986). Integration of host plant resistance and insecticides in the control of *Nephotettix virescens* (Homoptera: Cicadellidae), a vector of rice tungro virus. *J. Econ. Entomol.* **79,** 437–443.

Hellen, C. U. T., and Cooper, J. I. (1987). The genome-linked protein of cherry leaf roll virus. *J. Gen. Virol.* **68,** 2913–2917.

Hellmann, G. M., Shaw, J. G., and Rhoads, R. E. (1988). *In vitro* analysis of tobacco vein mottling virus NI_a cistron: Evidence for a virus-encoded protease. *Virology* **163,** 554–562.

Helms, K., and Wardlaw, I. F. (1976). Movement of viruses in plants: Long distance movement of tobacco mosaic virus in *Nicotiana glutinosa*. In "Transport and Transfer Processes in Plants" (I. F. Wardlaw and J. B. Passioura, eds.), pp. 283–293. Academic Press, New York.

Helms, K., and Zaitlin, M. (1970). Enhancement of infectivity of tobacco mosaic virus particles partially uncoated by alkali. *Virology* **41,** 549–557.

Helms, K., Waterhouse, P. M., and Muller, W. J. (1985). Subterranean clover red leaf virus disease: Effects of temperature on plant symptoms, growth, and virus content. *Phytopathology* **75,** 337–341.

Hemenway, C., Fang, R.-X., Kaniewski, W. K., Chua, N.-H., and Tumer, N. E. (1988). Analysis of the mechanism of protection in transgenic plants expressing the potato virus X coat protein or its antisense RNA. *EMBO J.* **7**, 1273–1280.

Hemenway, C., Weiss, J., O'Connell, K., and Tumer, N. E. (1990). Characterization of infectious transcripts from a potato virus X cDNA clone. *Virology* **175**, 365–371.

Hemida, S. K., Murant, A. F., and Duncan, G. H. (1989). Purification and some particle properties of anthricus yellows virus, a phloem-limited semipersistent aphid-borne virus. *Ann. Appl. Biol.* **114**, 71–86.

Hemmati, K., and McLean, D. L. (1977). Gamete-seed transmission of alfalfa mosaic virus and its effect on seed germination and yield in alfalfa plants. *Phytopathology* **67**, 576–579.

Hemmer, O., Meyer, M., Greif, C., and Fritsch, C. (1987). Comparison of the nucleotide sequences of five tomato black ring virus satellite RNAs. *J. Gen. Virol.* **68**, 1823–1833.

Hennig, W. (1981). "Insect Phylogeny." Wiley, New York.

Henriksson, D., Tanis, R. J., Tashian, R. E., and Nyman, P. O. (1981). Amino acid sequence of the coat protein subunit in satellite tobacco necrosis virus. *J. Mol. Biol.* **152**, 171–179.

Hershey, A. D., and Chase, M. (1952). Independent functions of viral protein and nucleic acid in growth of bacteriophage. *J. Gen. Physiol.* **36**, 39–56.

Hewish, D. R., Shukla, D. D., and Gough, K. H. (1986). The use of biotin-conjugated antisera in immunoassays for plant viruses. *J. Virol. Methods* **13**, 79–85.

Hewitt, W. B., and Chiarappa, L. eds. (1977). "Plant Health and Quarantine in International Transfer of Genetic Resources." CRC Press, Boca Raton, Florida.

Hewitt, W. B., Raski, D. J., and Goheen, A. C. (1958). Nematode vector of soil borne fan-leaf virus of grapevines. *Phytopathology* **48**, 586–595.

Heywood, V. H. (1978). "Flowering Plants of the World." Oxford Univ. Press, London and New York.

Hiatt, A. Cafferkey, R., and Bowdish, K. (1989). Production of antibodies in transgenic plants. *Nature (London)* **342**, 76–78.

Hibi, T., and Saito, Y. (1985). A dot immunobinding assay for the detection of tobacco mosaic virus in infected tissues. *J. Gen. Virol.* **66**, 1191–1194.

Hibi, T., Rezelman, G., and van Kammen, A. (1975). Infection of cowpea mesophyll protoplasts with cowpea mosaic virus. *Virology* **64**, 308–318.

Hibi, T., Omura, T., and Saito, Y. (1984). Double-stranded RNA of rice gall dwarf virus. *J. Gen. Virol.* **65**, 1585–1590.

Hibino, H., and Cabauatan, P. Q. (1987). Infectivity neutralization of rice tungro-associated viruses acquired by vector leafhoppers. *Phytopathology* **77**, 473–476.

Hibino, H., Tiongco, E. R., Cabunagan, R. C., and Flores, Z. M. (1987). Resistance to rice tungro-associated viruses in rice under experimental and natural conditions. *Phytopathology* **77**, 871–875.

Hidaka, S., Hanada, K., Ishikawa, K., and Miura, K.-I. (1988). Complete nucleotide sequence of two new satellite RNAs associated with cucumber mosaic virus. *Virology* **164**, 326–333.

Hiddinga, H. J., Crum, C. J., Hu, J., and Roth, D. A. (1988). Viroid-induced phosphorylation of a host protein related to a dsRNA-dependent protein kinase. *Science* **241**, 451–453.

Hiebert, E., and McDonald, J. G. (1973). Characterisation of some proteins associated with viruses in the potato Y group. *Virology* **56**, 349–361.

Hiebert, E., Bancroft, J. B., and Bracker, C. E. (1968). The assembly *in vitro* of some small spherical viruses, hybrid viruses and other nucleoproteins. *Virology* **34**, 492–508.

Higgins, T. J. V., Goodwin, P. B., and Whitfeld, P. R. (1976). Occurrence of short particles in beans infected with the cowpea strain of TMV. II. Evidence that short particles contain the cistron for coat-protein. *Virology* **71**, 486–497.

Higgins, T. J. V., Whitfeld, P. R., and Matthews, R. E. F. (1978). Size distribution and *in vitro* translation of the RNAs isolated from turnip yellow mosaic virus nucleoproteins. *Virology* **84**, 153–161.

Hilborn, M. T., Hyland, F., and McCrum, R. C. (1965). Pathological anatomy of apple trees affected by the stem-pitting virus. *Phytopathology* **55**, 34–39.

Hill, R. E., and Hastie, N. D. (1987). Accelerated evolution in the reactive centre regions of serine protease inhibitors. *Nature (London)* **326**, 96–99.

Hill, S. A. (1984a). "Methods in Plant Virology." Blackwell, Oxford.

Hills, S. A. (1984b). The ELISA (Enzyme-Linked Immunosorbent Assay) technique for the detection of plant viruses. *Soc. Appl. Bacteriol. Tech. Ser.* **19**, 349–363.

Hill, S. A. (1987). Cereal virus diseases: Contrasting experience. *In* "Populations of Plant Pathogens: Their Dynamics and Genetics" (M. S. Wolfe and C. E. Caten, eds.), 149–159. Blackwell, Oxford.

Hill, S. A., and Torrance, L. (1989). Rhizomania disease of sugar beet in England. *Plant Pathol.* **38**, 114–122.

Hillman, B. I., Morris, T. J., and Schlegel, D. E. (1985). Effects of low-molecular-weight RNA and temperature on tomato bushy stunt virus symptom expression. *Phytopathology* **75**, 361–365.

Hillman, B. I., Carrington, J. C., and Morris, T. J. (1987). A defective interfering RNA that contains a mosaic of a plant virus genome. *Cell (Cambridge, Mass.)* **51**, 427–433.

Hillman, B. I., Hearne, P., Rochon, D'A., and Morris, T. J. (1989). Organisation of tomato bushy stunt virus genome: Characterisation of the coat protein gene and the 3' terminus. *Virology* **169**, 42–50.

Hills, G. J., Plaskitt, K. A., Young, N. D., Dunigan, D. D., Watts, J. W., Wilson, T. M. A., and Zaitlin, M. (1987). Immunogold localisation of the intracellular sites of structural and non-structural tobacco mosaic virus proteins. *Virology* **161**, 488–496.

Hilty, J. W. (1981). Remote sensing of virus diseased corn and sorghum. Virus and virus-like diseases of maize in the United States. *South. Coop. Ser. Bull.* **247**, 124–126.

Hinegardner, R. (1976). Evolution of genome size. *In* "Molecular Evolution" (F. J. Ayala, ed.), pp. 179–199. Sinauer, Sunderland, Massachusetts.

Hiruki, C. (1987). Recovery and identification of tobacco stunt virus from air-dried resting spores of *Olpidium brassicae*. *Plant Pathol.* **36**, 224–228.

Hiruki, C., and Tu, J. C. (1972). Light and electron microscopy of potato virus M lesions and marginal tissue in red kidney bean. *Phytopathology* **62**, 77–85.

Hizi, A., Henderson, L. E., Copeland, T. D., Sowder, R. C., Hixon, C. V., and Oroszlan, S. (1987). Characterisation of mouse mammary tumor virus *gag-pro* gene products and the ribosomal frame-shift site by protein sequencing. *Proc. Natl. Acad. Sci. U.S.A.* **84**, 7041–7045.

Hobbs, H. A., Kuhn, C. W., Papa, K. E., and Brantley, B. B. (1987). Inheritance of non-necrotic resistance to southern bean mosaic virus in cowpea. *Phytopathology* **77**, 1624–1629.

Hodgman, T. C. (1986). An amino acid sequence motif linking viral DNA polymerases and plant viral proteins involved in RNA replication. *Nucleic Acids Res.* **14**, 6769.

Hodgman, T. C. (1988). A new super family of replicative proteins. *Nature (London)* **333**, 22–23.

Hoekema, A., Huisman, M. J., Molendijk, L., van den Elzen, P. J. M., and Cornelissen, B. J. C. (1989). The genetic engineering of two commercial potato cultivars for resistance to potato virus X. *Bio/Technology* **7**, 273–278.

Hogle, J. M., Maeda, A., and Harrison, S. C. (1986). Structure and assembly of turnip crinkle virus 1. X-ray crystallographic structure analysis at 3.2 Å resolution. *J. Mol. Biol.* **191**, 625–638.

Hogue, R., and Asselin, A. (1984). Polyacrylamide–agarose gel electrophoretic analysis of tobacco mosaic virus disassembly intermediates. *Can. J. Bot.* **62**, 2336–2339.

Hohn, T., Hohn, B., and Pfeiffer, P. (1985). Reverse transcription in CaMV. *Trends Biochem. Sci.* **10**, 205–209.

Holland, J., Spindler, K., Horodyski, F., Grabau, E., Nichol, S., and Vande Pol, S. (1982). Rapid evolution of RNA genomes. *Science* **215**, 1577–1585.

Hollings, M. (1965). Disease control through virus-free stock. *Annu. Rev. Phytopathol.* **3**, 367–396.

Hollings, M. (1978). Annual Report of Glasshouse Crops Research Institute, Littlehampton, Great Britain.

Hollings, M., and Stone, O. M. (1970). The long-term survival of some plant viruses preserved by lyophilization. *Ann. Appl. Biol.* **65**, 411–418.

Holmes, F. O. (1929). Local lesions in tobacco mosaic. *Bot. Gaz. (Chicago)* **87**, 39–55.

Holmes, F. O. (1931). Local lesions of mosaic in *Nicotiana tabacum* L. *Contrib. Boyce Thompson Inst.* **3**, 163–172.

Holmes, F. O. (1934). A masked strain of tobacco mosaic virus. *Phytopathology* **24**, 845–873.

Holmes, F. O. (1938). Inheritance of resistance to tobacco mosaic disease in tobacco. *Phytopathology* **28**, 553–561.

Holmes, F. O. (1939). "Handbook of Phytopathogenic Viruses." Burgess, Minneapolis, Minnesota.

Holmes, F. O. (1946). A comparison of the experimental host ranges of tobacco-etch and tobacco mosaic viruses. *Phytopathology* **36**, 643–659.

Holmes, F. O. (1955). Elimination of spotted wilt virus from dahlias by propagation of tip cuttings. *Phytopathology* **45**, 224–226.

Holmes, F. O. (1964). Symptomology of viral diseases in plants. *In* "Plant Virology" (M. K. Corbett and H. D. Sisler, eds.), pp. 17–38. Univ. of Florida Press, Gainesville.

Holness, C. L., Lomonossoff, G. P., Evans, D., and Maule, A. J. (1989). Identification of the initiation codons for translation of cowpea mosaic virus middle component RNA using site-directed mutagenesis of an infectious cDNA clone. *Virology* **172**, 311–320.

Honda, Y., Kajita, S., Matsui, C., Otsuki, Y., and Takebe, I. (1975). An ultrastructural study of the infection of tobacco mesophyll protoplasts by potato virus X. *Phytopathol. Z.* **84**, 66–74.

Hooft van Huijsduijnen, R. A. M., Alblas, S. W., De Rijk, R. H., and Bol, J. F. (1986a). Induction by salicylic acid of pathogenesis related proteins and resistance to alfalfa mosaic virus infection in various plant species. *J. Gen. Virol.* **67,** 2135–2143.

Hooft van Huijsduijnen, R. A. M., and van Loon, L. C., and Bol, J. F. (1986b). cDNA cloning of six mRNAs induced by TMV infection of tobacco and a characterization of their translation products. *EMBO J.* **5,** 2057–2061.

Hooft van Huijsduijnen, R. A. M., Kaufmann, S., Brederode, F. T., Cornelissen, B. J. C., Legrand, M., Fritig, B., and Bol, J. F. (1987). Homology between chitinases that are induced by TMV infection of tobacco. *Plant Mol. Biol.* **9,** 411–420.

Horikoshi, M., Nakayama, M., Yamaoka, N., Furusawa, I., and Shishiyama, J. (1987). Brome mosaic virus coat protein inhibits viral RNA synthesis *in vitro*. *Virology* **158,** 15–19.

Horikoshi, M., Mise, K., Furusawa, I., and Shishiyama, J. (1988). Immunological analysis of brome mosaic virus replicase. *J. Gen. Virol.* **69,** 3081–3087.

Horne, R. W. (1985). The development and application of electron microscopy to the structure of isolated plant viruses. *In* "Molecular Plant Virology" (J. W. Davies, ed.), Vol. 1, pp. 1–41. CRC Press, Boca Raton, Florida.

Horne, R. W., and Wildy, P. (1961). Symmetry in virus architecture. *Virology* **15,** 348–373.

Horst, R. K., Langhans, R. W., and Smith, S. H. (1977). Effects of chrysanthemum stunt, chlorotic mottle, aspermy, and mosaic on flowering and rooting of chrysanthemums. *Phytopathology* **67,** 9–14.

Horváth, J. (1983a). New artificial hosts and non-hosts of plant viruses and their role in the identification and separation of viruses. XVIII. Concluding remarks. *Acta Phytopathol. Acad. Sci. Hung.* **18,** 121–161.

Horváth, J. (1983b). The role of some plants in the ecology of cucumber mosaic virus with special regard to bean. *Acta Phytopathol. Acad. Sci. Hung.* **18,** 217–224.

Houwing, C. J., and Jaspars, E. M. J. (1982). Protein binding sites in nucleation complexes of alfalfa mosaic virus RNA4. *Biochemistry* **21,** 3408–3414.

Houwing, C. J., and Jaspars, E. M. J. (1986). Coat protein blocks the *in vitro* transcription of the virion RNAs of alfalfa mosaic virus. *FEBS Lett.* **209,** 284–288.

Houwing, C. J., and Jaspars, E. M. J. (1987). *In vitro* evidence that the coat protein is the programming factor in alfalfa mosaic virus-induced RNA synthesis. *FEBS Lett.* **221,** 337–342.

Howarth, A. J., and Vandemark, G. J. (1989). Phylogeny of geminiviruses. *J. Gen. Virol.* **70,** 2717–2727.

Howarth, A. J., Gardner, R. C., Messing, J., and Shepherd, R. J. (1981). Nucleotide sequence of naturally occurring deletion mutants of cauliflower mosaic virus. *Virology* **112,** 678–685.

Howarth, A. J., Caton, J., Bossert, M., and Goodman, R. M. (1985). Nucleotide sequence of bean golden mosaic virus and a model for gene regulation in geminiviruses. *Proc. Natl. Acad. Sci. U.S.A.* **82,** 3572–3576.

Howell, S. H. (1984). Physical structure and genetic organisation of the genome of maize streak virus (Kenya isolate). *Nucleic Acids Res.* **12,** 7359–7375.

Howell, S. H., Walker, L. L., and Dudley, R. K. (1980). Cloned cauliflower mosaic virus DNA infects turnips (*Brassica rapa*). *Science* **208,** 1265–1267.

Howell, S. H., Walker, L. L., and Walden, R. M. (1981). Rescue of *in vitro* generated mutants of cloned cauliflower mosaic virus genome in infected plants. *Nature (London)* **293,** 483–486.

Howell, W. E., and Mink, G. I. (1971). The relationship between volunteer sugarbeets and occurrence of beet mosaic and beet western yellow viruses in Washington beet fields. *Plant Dis. Rep.* **55,** 676–678.

Howell, W. E., and Mink, G. I. (1977). The role of weed hosts, volunteer carrots, and overlapping growing seasons in the epidemiology of carrot thin leaf and carrot motley dwarf viruses in central Washington. *Plant. Dis. Rep.* **61,** 217–222.

Howell, W. E., and Mink, G. I. (1988). Natural spread of cherry rugose mosaic disease and two prunus necrotic ringspot virus biotypes in a central Washington sweet cherry orchard. *Plant Dis.* **72,** 636–640.

Hsu, C. H., Sehgal, O. P., and Pickett, E. E. (1976). Stabilizing effect of divalent metal ions on virions of southern bean mosaic virus. *Virology* **69,** 587–595.

Hsu, H. T. (1978). Cell fusion induced by a plant virus. *Virology* **84,** 9–18.

Hsu, H. T., and Black, L. M. (1973). Comparative efficiencies of assays of a plant virus by lesions on leaves and on vector cell monolayers. *Virology* **52,** 284–286.

Hsu, H. T., and Black, L. M. (1974). Multiplication of potato yellow dwarf virus on vector cell monolayers. *Virology* **59,** 331–334.

Hsu, Y.-H. (1984). Immunogold for detection of antigen on nitrocellulose paper. *Anal. Biochem.* **142,** 221–225.

Hsu, Y.-H., and Brakke, M. K. (1985). Cell-free translation of soil-borne wheat mosaic virus RNAs. *Virology* **143,** 272–279.

Huber, R., Rezelman, G., Hibi, T., and van Kammen, A. (1977). Cowpea mosaic virus infection of protoplasts from Samsun tobacco leaves. *J. Gen. Virol.* **34,** 315–323.

Huez, G., Cleuter, Y., Bruck, C., van Vloten-Doting, L., Goldbach, R., and Verduin, B. (1983). Translational stability of plant viral RNAs microinjected into living cells. Influence of a 3-poly (A) segment. *Eur. J. Biochem.* **130,** 205–209.

Hughes, G., Davies, J. W., and Wood, K. R. (1986). *In vitro* translation of the bipartite genomic RNA of pea early browning virus. *J. Gen. Virol.* **67,** 2125–2133.

Hughes, J. d'A., and Thomas, B. J. (1988). The use of protein A-sandwich ELISA as a means for quantifying serological relationships between members of the tobamovirus group. *Ann. Appl. Biol.* **112,** 117–126.

Huguenot, C., van den Dobbelsteen, G., de Haan, P., Wagemakers, C. A. M., Drost, G. A. Osterhaus, A. D. M. E., and Peters, D. (1990). Detection of tomato spotted wilt virus using monoclonal antibodies and riboprobes. *Arch. Virol.* **110,** 47–62.

Huisman, M. J., Sarachu, A. N., Alblas, F., and Bol, J. F. (1985). Alfalfa mosaic virus temperature-sensitive mutants. II. Early functions encoded by RNAs 1 and 2. *Virology* **141,** 23–29.

Huisman, M. J., Sarachu, A. N., Alblas, F., Broxterman, H. J. G., van Vloten-Doting, L., and Bol, J. F. (1986). Alfalfa mosaic temperature sensitive mutants. III. Mutants with a putative defect in cell-to-cell transport. *Virology* **154,** 401–404.

Huisman, M. J., Linthorst, H. J. M., Bol, J. F., and Cornelissen, B. J. C. (1988). The complete nucleotide sequence of potato virus X and its homologies at the amino acid level with various plus-stranded RNA viruses. *J. Gen. Virol.* **69,** 1789–1798.

Hull, R. (1965). control of sugar beet yellows. *Ann. Appl. Biol.* **56,** 345–347.

Hull, R. (1970a). Studies on alfalfa mosaic virus. IV. An unusual strain. *Virology* **42,** 283–292.

Hull, R. (1970b). Large RNA plant-infecting viruses. *In* "The Biology of Large RNA Viruses" (R. D. Barry and B. W. J. Mahy, eds.), pp. 153–164. Academic Press, New York.

Hull, R. (1976). The structure of tubular viruses. *Adv. Virus Res.* **20,** 1–32.

Hull, R. (1977). The banding behaviour of the viruses of southern bean mosaic virus group in gradients of caesium sulphate. *Virology* **79,** 50–57.

Hull, R. (1985). Purification, biophysical and biochemical characterisation of viruses with especial reference to plant viruses. *In* "Virology: A Practical Approach" (B. W. J. Mahy, ed.), pp. 1–14. IRL Press, Oxford.

Hull, R. (1986). The potential for using dot-blot hydridisation in the detection of plant viruses. *In* "Developments and Applications in Virus Testing" (R. A. C. Jones and L. Torrance, eds.), Appl. Biol. I, pp. 3–12. Association of Applied Biologists, Wellsbourne.

Hull, R. (1988). The Sobemovirus group. *In* "The Plant Viruses" (R. Koenig, ed.), Vol. 3, pp. 113–146. Plenum, New York.

Hull, R., and Covey, S. N. (1983). Does cauliflower mosaic virus replicate by reverse transcription? *Trends Biochem. Sci.* **8,** 119–121.

Hull, R., and Covey, S. (1985). Cauliflower mosaic virus: Pathways of infection. *BioEssays* **3,** 160–163.

Hull, R., and Donson, J. (1982). Physical mapping of the DNAs of carnation etched ring and figwort mosaic viruses. *J. Gen. Virol.* **60,** 125–134.

Hull, R., and Heathcote, G. D. (1967). Experiments on the time of application of insecticide to decrease the spread of yellowing viruses of sugar beet. 1954–1966. *Ann. Appl. Biol.* **60,** 469–478.

Hull, R., and Plaskitt, A. (1970). Electron microscopy on the behaviour of two strains of alfalfa mosaic virus in mixed infections. *Virology* **42,** 773–776.

Hull, R., Hills, G. J., and Plaskitt, A. (1970). The *in vivo* behaviour of twenty-four strains of alfalfa mosaic virus. *Virology* **42,** 753–772.

Hull, R., Sadler, J., and Longstaff, M. (1986). The sequence of carnation etched ring virus DNA: Comparison with cauliflower mosaic virus and retroviruses. *EMBO J.* **5,** 3083–3090.

Hull, R., Covey, S. N., and Maule, A. J. (1987). Structure and replication of caulimovirus genomes. *J. Cell Sci., Suppl.* **7,** 213–229.

Hull, R., Brown, F., and Payne, C. (1989). "Virology. Directory and Dictionary of Animal Bacterial and Plant Viruses." Macmillan, London.

Hull, R., Jones, M. C., Dasgupta, I., Cliffe, J. M., Mingins, C., Lee, G., and Davies, J. W. (1990). The molecular biology of rice tungro viruses: evidence for a new retroid virus., in press.

Hull, R., Milne, R. G., and Van Regenmortel, M. H. V. (1991). A list of proposed standard acronyms for plant viruses and viroids. *Intervirology* (in press).

Hull, R. E. (1968). A virus disease of Hart's-tongue fern. *Virology* **35,** 333–335.

Hull, R. E., and Johnson, M. W. (1968). The precipitation of alfalfa mosaic virus by magnesium. *Virology* **34**, 388–390.

Hull, R. E., Rees, M., and Short, M. N. (1969a). Studies on alfalfa mosaic virus. I. The protein and nucleic acid. *Virology* **37**, 404–415.

Hull, R. E., Hills, G. J., and Markham, R. (1969b). Studies on alfalfa mosaic virus. II. The structure of the virus components. *Virology* **37**, 416–428.

Hunt, R. E., Nault, L. R., and Gingery, R. E. (1988). Evidence for infectivity of maize chlorotic dwarf virus and for a helper component in its leafhopper transmission. *Phytopathology* **78**, 499–504.

Hunter, B. G., Smith, J., Fattouh, F., and Jackson, A. O. (1989). Relationships of Lychnis ringspot virus to barley stripe mosaic virus and poa semilatent virus. *Intervirology* **30**, 18–26.

Hunter, T., Jackson, R., and Zimmern, D. (1983). Multiple proteins and subgenomic mRNAs may be derived from a single open reading frame on tobacco mosaic virus RNA. *Nucleic Acids Res.* **11**, 801–821.

Huss, B., Muller, S., Sommermeyer, G., Walter, B., and Van Regenmortel, M. H. V. (1987). Grapevine fanleaf virus monoclonal antibodies: Their use to distinguish different isolates. *J. Phytopathol.* **119**, 358–370.

Hutchins, C. J., Keese, P., Visvader, J. E., Rathjen, P. D., McInnes, J. L., and Symons, R. H. (1985). Comparison of multimeric plus and minus forms of viroids and virusoids. *Plant Mol. Biol.* **4**, 293–304.

Hutchins, C. J., Rathgen, P. D., Forster, A. C., and Symons, R. H. (1986). Self-cleavage of plus and minus transcripts of avocado sunblotch viroid. *Nucleic Acids Res.* **14**, 3627–3640.

Huxley, H. E., and Zubay, G. (1960). The structure of the protein shell of turnip yellow mosaic virus. *J. Mol. Biol.* **2**, 189–196.

Ikegami, M., and Francki, R. I. B. (1973). Presence of antibodies to double-stranded RNA in sera of rabbits immunized with rice dwarf and maize rough dwarf viruses. *Virology* **56**, 404–406.

Ikegami, M., and Francki, R. I. B. (1975). Some properties of RNA from Fiji disease subviral particles. *Virology* **64**, 464–470.

Ikegami, M., and Francki, R. I. B. (1976). RNA-dependent RNA polymerase associated with subviral particles of Fiji disease virus. *Virology* **70**, 292–300.

Inoue, H., and Timmins, P. A. (1985). The structure of rice dwarf virus determined by small-angle neutron scattering measurements. *Virology* **147**, 214–216.

Irvin, J. D. (1975). Purification and partial characterization of the antiviral protein from *Phytolacca americana* which inhibits eukaryotic protein synthesis. *Arch. Biochem. Biophys.* **169**, 522–528.

Irvine, R. F., and Osborne, D. J. (1973). The effect of ethylene on [1-^{14}C] glycerol incorporation into phospholipids of etiolated pea stems. *Biochem. J.* **136**, 1133–1135.

Irwin, M. E., and Ruesink, W. G. (1986). Vector intensity: A product of propensity and activity. *In* "Plant Virus Epidemics" (G. D. McClean, R. G. Garrett, and W. G. Ruesink, eds.), pp. 13–33. Academic Press, Orlando, Florida.

Ishiie, T., Doi, Y., Yora, K., and Asuyama, H. (1967). Suppressive effects of antibiotics of tetracycline group on symptom development of mulberry dwarf disease. *Ann. Phytopathol. Soc. Jpn.* **33**, 267–275.

Ishikawa, M. Meshi, T., Ohno, T., Okada, Y., Sano, T., Ueda, I., and Shikata, E. (1984). A revised replication cycle for viroids: The role of longer than unit length RNA in viroid replication. *Mol. Gen. Genet.* **196**, 421–428.

Ishikawa, M., Meshi, T., Motoyoshi, F., Takamatsu, N., and Okada, Y. (1986). *In vitro* mutagenesis of the putative replicase genes of tobacco mosaic virus. *Nucleic Acids Res.* **14**, 8291–8305.

Ishikawa, M., Meshi, T., Watanabe, Y., and Okada, Y. (1988). Replication of chimeric tobacco mosaic viruses which carry heterologous combinations of replicase genes and 3' non-coding regions. *Virology* **164**, 290–293.

Ismail, I. D., and Milner, J. J. (1988). Isolation of defective interfering particles of Sonchus yellow net virus from chronically infected plants. *J. Gen. Virol.* **69**, 999–1006.

Ismail, I. D., Hamilton, I. D., Robertson, E., and Milner, J. J. (1987). Movement and intracellular location of Sonchus yellow net virus within infected *Nicotiana edwardsonii. J. Gen. Virol.* **68**, 2429–2438.

Ivanovic, M., Macfarlane, I., and Wood, R. D. (1983). Viruses transmitted by fungi: Viruses of sugar beet associated with *Polymyxa betae. Rep. Rothamstead Exp. Stn., 1982*, p. 189.

Iwanowski, D. (1892). Ueber die Mosaikkrankheit der Tabakspflanze. *Bull. Acad. Imp. Sci. St.-Petersbourg [N.W.]* **3**, 65–70.

Jacks, T., Power, M. D., Masiarz, F. R., Luciw, P. A., Barr, P. J., and Varmus, H. E. (1988). Characterisation of ribosomal frameshifting in HIV-1 *gag-pol* expression. *Nature (London)* **331**, 280–283.

Jackson, A. O. (1978). Partial characterization of the structural proteins of sonchus yellow net virus. *Virology* **87**, 172–181.

Jackson, A. O., and Christie, S. R. (1977). Purification and some physicochemical properties of sonchus yellow net virus. *Virology* **77**, 344–355.

Jackson, A. O., Mitchell, D. M., and Siegel, A. (1971). Replication of tobacco mosaic virus. I. Isolation and characterization of double-stranded forms of ribonucleic acid. *Virology* **45**, 182–191.

Jackson, A. O., Zaitlin, M., Siegel, A., and Francki, R. I. B. (1972). Replication of tobacco mosaic virus. III. Viral RNA metabolism in separated leaf cells. *Virology* **48**, 655–665.

Jackson, A. O., Dawson, J. R. O., Covey, S. N., Hull, R., Davies, J. W., McFarland, J. E., and Gustafson, G. D. (1983). Sequence relations and coding properties of a subgenomic RNA isolated from barley stripe mosaic virus. *Virology* **127**, 37–44.

Jackson, A. O., Francki, R. I. B., and Zuidema, D. (1987). Biology structure and replication of plant rhabdoviruses. *In* "The Rhabdoviruses" (R. R. Wagner, ed.), pp. 427–508. Plenum, New York.

Jackson, A. O., Hunter, B. G., and Gustafson, G. D. (1989). Hordeivirus relationships and genome organisation. *Annu. Rev. Phytopathol.* **27**, 95–121.

Jacquemond, M. (1982). Phenomena of interferences between the two types of satellite RNA of cucumber mosaic virus. Protection of tomato plants against lethal necrosis. *C. R. Seances Acad. Sci. Ser.,* 3 **294**, 991–994.

Jacquemond, M., Amselem, J., and Tepfer, M. (1988). A gene coding for a monomeric form of cucumber mosaic virus satellite RNA confers tolerance to CMV. *Mol. Plant-Microbe Interact.* **1**, 311–316.

Jacrot, B. (1975). Studies on the assembly of a spherical plant virus. II. The mechanism of protein aggregation and virus swelling. *J. Mol. Biol.* **95**, 433–446.

Jacrot, B., Pfeiffer, P., and Witz, J. (1976). The structure of a spherical plant virus (bromegrass mosaic virus) established by neutron diffraction. *Philos. Trans. R. Soc. London, Ser. B* **276**, 109–112.

Jacrot, B., Chauvin, C., and Witz, J. (1977). Comparative neutron small-angle scattering study of small spherical RNA viruses. *Nature (London)* **266**, 417–421.

Jadot, R., and Roland, G. (1971). Observations sur les deplacements des aphides à partir des plantes adventices marquées dans un champ de betteraves. *Meded. Fac. Landbouwwet., Rijksuniv. Gent* **36**, 940–944.

Jaegle, M., and Van Regenmortel, M. H. V. (1985). Use of ELISA for measuring the extent of serological cross-reactivity between plant viruses. *J. Virol. Methods* **11**, 189–198.

Jaegle, M., Wellink, J., and Goldbach, R. (1987). The genome-linked protein of cowpea mosaic virus is bound to the 5' terminus of virus RNA by a phosphodiester linkage to serine. *J. Gen. Virol.* **68**, 627–632.

Jaegle, M., Briand, J. P., Burckard, J., and Van Regenmortel, M. H. V. (1988). Accessibility of three continuous epitopes in tomato bushy stunt virus. *Ann. Inst. Pasteur/Virol.* **139**, 39–50.

Jaenicke, R., and Lauffer, M. A. (1969). Determination of hydration and partial specific volume of proteins with the spring balance. *Biochemistry* **8**, 3077–3082.

Jagus, R. (1987a). Translation in cell-free systems. *In* "Methods in Enzymology" (S. L. Berger and A. R. Kimmel, eds.), Vol. **152**, pp. 267–296. Academic Press, San Diego, California.

Jagus, R. (1987b). Characterisation of *in vitro* tranlation products. In "Methods in Enzymology" (S. L., Berger and A. R. Kimmel, eds.) Vol. **152**, pp. 296–304. Academic Press, San Diego, California.

Jakab, G., Kiss, T., and Solymosy, F. (1986). Viroid pathogenicity and pre-rRNA processing: A model amenable to experimental testing. *Biochim. Biophys. Acta* **868**, 190–197.

Jamet, E., and Fritig, B. (1986). Purification and characterisation of 8 of the pathogenesis-related proteins in tobacco leaves reacting hypersensitively to tobacco mosaic virus. *Plant Mol. Biol.* **6**, 69–80.

Janda, M., French, R., and Ahlquist, P. (1987). High efficiency T7 polymerase synthesis of infectious RNA from cloned brome mosaic virus cDNA and effects of 5' extensions of transcript infectivity. *Virology* **158**, 259–262.

Jardetzky, O., Akasaka, K., Vogel, D., Morris, S., and Holmes, K. C. (1978). Unusual segmental flexibility in a region of tobacco mosaic virus coat protein. *Nature (London)* **273**, 564–566.

Jaspars, E. M. J. (1985). Interaction of alfalfa mosaic virus nucleic acid and protein. *In* "Plant Molecular Biology" (J. W. Davies, ed.), Vol. I, pp. 155–221. CRC Press, Boca Raton, Florida.

Jedlinski, H. (1964). Initial infection processes by certain mechanically transmitted plant viruses. *Virology* **22**, 331–341.

Jellison, J. (1987). Exploratory numerical taxonomy based on biochemical and biophysical characters of the tymoviruses. *Intervirology* **27**, 61–68.

Jeng, T.-W., Crowther, R. A., Stubbs, G., and Chiu, W. (1989). Visualization of alpha-helices in tobacco mosaic virus by cryo-protection microscopy. *J. Mol. Biol.* **205**, 251–257.

Jennings, D. L., and Jones, A. T. (1986). Immunity from raspberry vein chlorosis virus in raspberry and its potential for control of the virus through plant breeding. *Ann. Appl. Biol.* **108**, 417–422.

Jenns, A. E., and Kuć, J. (1980). Characteristics of anthracnose resistance induced by localized infection of cucumber with tobacco necrosis virus. *Physiol. Plant Pathol.* **17**, 81–91.

Jensen, S. G. (1973). Systemic movement of barley yellow dwarf virus in small grains. *Phytopathology* **63**, 854–856.

Jepson, P. C., and Green, R. E. (1983). Prospects for improving control strategies for sugar-beet pests in England. *Adv. Appl. Biol.* **7**, 175–250.

Jobling, S. A., and Gehrke, L. (1987). Enhanced translation of chimeric messenger RNAs containing a plant viral untranslated leader sequence. *Nature (London)* **325**, 622–625.

Jobling, S. A., and Wood, K. R. (1985). Translation of tobacco ringspot virus RNA in reticulocyte lysate: Proteolytic processing of the primary translation products. *J. Gen. Virol.* **66**, 2589–2596.

Jockusch, H. (1964). *In vivo-* and *in vitro-*Verhalten temperature sensitiver Mutanten des tabakmosaik-virus. *Z. Vererbungsl.* **95**, 379–382.

Jockusch, H. (1966a). The role of host genes, temperature and polyphenoloxidase in the necrotization of TMV infected tobacco tissue. *Phytopathol. Z.* **55**, 185–192.

Jockusch, H. (1966b). Temperature-sensitive mutation des Tabakmosaikvirus I *in vivo-*verhalten. *Z. Vererbungsl* **98**, 320–343.

John, V. T. (1965). A micro-immuno-osmophoretic technique for assay of tobacco mosaic virus. *Virology* **27**, 121–123.

John, V. T., and Weintraub, M. (1966). Symptoms resembling virus infection induced by high temperature in *Nicotiana glutinosa. Phytopathology* **56**, 502–506.

Johns, M. A., Mottinger, J., and Freeling, M. (1985). A low copy number, *copia-*like transposon in maize. *EMBO J.* **4**, 1093–1102.

Johnson, C. S., and Main, C. E. (1983). Yield/quality tradeoffs of tobacco mosaic virus-resistant tobacco cultivars in relation to disease management. *Plant Dis.* **67**, 886–890.

Johnson, E. M., and Valleau, W. D. (1946). Field strains of tobacco mosaic virus. *Phytopathology* **36**, 112–116.

Johnson, J. (1927). The classification of plant viruses. *Res. Bull.—Wis., Agric. Exp. Stn.*, **76**, 1–16.

Johnson, J. (1947). Virus attenuation and mutation. *Phytopathology* **37**, 12.

Johnson, J., and Hoggan, I. A. (1935). A descriptive key for plant viruses. *Phytopathology* **25**, 328–343.

Johnson, J., and Ogden, W. B. (1929). The overwintering of tobacco mosaic virus. *Wis., Agric. Exp. Stn., Bull.* **95**, 1–25.

Johnson, M. W. (1964). The binding of metal ions by turnip yellow mosaic virus. *Virology* **24**, 26–35.

Johnson, M. W., and Markham, R. (1962). Nature of the polyamine in plant viruses. *Virology* **17**, 276–281.

Johnstone, G. R., Koen, T. B., and Conley, H. L. (1982). Incidence of yellows in sugar beet as affected by variation in plant density and arrangement. *Bull. Entomol. Res.* **72**, 289–294.

Johnstone, G. R., Ashby, J. W., Gibbs, A. J., Duffus, J. E., Thottappilly, G., and Fletcher, J. D. (1984). The host ranges, classification and identification of eight persistent aphid-transmitted viruses causing disease in legumes. *Neth. J. Plant Pathol.* **90**, 225–245.

Jonard, G., Richards, K. E., Guilley, H., and Hirth, L. (1977). Sequence from the assembly nucleation region of TMV RNA. *Cell (Cambridge, Mass.)* **11**, 483–493.

Jones, A. T. (1987). Control of virus infection in crop plants through vector resistance: A review of achievements, prospects and problems. *Ann. Appl. Biol.* **111**, 745–772.

Jones, A. T., and Mayo, M. A. (1984). Satellite nature of the viroid-like RNA-2 of solanum nodiflorum mottle virus and the ability of other plant viruses to support the replication of viroid-like RNA molecules. *J. Gen. Virol.* **65**, 1713–1721.

Jones, A. T., and Mitchell, M. J. (1986). Propagation of black raspberry necrosis virus (BRNV) in mixed culture with solanum nodiflorum mottle virus, and the production and use of BRNV antiserum. *Ann. Appl. Biol.* **109**, 323–336.

Jones, A. T., and Mitchell, M. J. (1987). Oxidising activity in root extracts from plants inoculated with virus or buffer that interferes with ELISA when using the substrate 3,3′, 5,5′ tetramethyl-benzidine. *Ann. Appl. Biol.* **111**, 359–364.

Jones, A. T., Mayo, M. A., and Duncan, G. H. (1983). Satellite-like properties of small circular RNA molecules in particles of lucerne transient streak virus. *J. Gen. Virol.* **64**, 1167–1173.

Jones, G. E., and Dawson, W. O. (1978). Stability of mutations conferring temperature sensitivity on tobacco mosaic virus. *Intervirology* **9**, 149–155.

Jones, I. M., and Reichmann, M. E. (1973). The proteins synthesized in tobacco leaves infected with tobacco necrosis virus and satellite tobacco necrosis virus. *Virology* **52**, 49–56.

Jones, M., Dasgupta, I., Cliffe, J., Davies, J. W., and Hull, R. (1989). An RNA and a DNA virus cause rice tungro disease. *Annu. Rep. John Innes Inst.* 36–38.

Jones, R. A. C. (1975). Systemic movement of potato mop-top virus in tobacco may occur through the xylem. *Phytopathol. Z.* **82**, 352–355.

Jones, R. A. C. (1982). Breakdown of potato virus X resistance gene Nx: Selection of a group four strain from group three. *Plant Pathol.* **31**, 325–331.

Jones, R. A. C. (1985). Further studies on resistance-breaking strains of potato virus X. *Plant Pathol.* **34**, 182–189.

Jones, R. A. C., and Fribourg, C. E. (1977). Beetle, contact and potato true seed transmission of Andean potato latent virus. *Ann. Appl. Biol.* **86**, 123–128.

Jones, R. A. C., and Harrison, B. D. (1969). The behaviour of potato mop-top virus in soil, and evidence for its transmission by *Spongospora subterranea* (Wallr) Lagerh. *Ann. Appl. Biol.* **63**, 1–17.

Jones, R. A. C., and Harrison, B. D. (1972). Ecological studies on potato mop-top virus in Scotland. *Ann. Appl. Biol.* **71**, 47–57.

Jones, R. W., Jackson, A. O., and Morris, T. J. (1990). Defective-interfering RNAs and elevated temperatures inhibit replication of tomato bushy stunt virus in inoculated protoplasts. *Virology* **176**, 539–545.

Jones, T. A., and Liljàs, L. (1984). Structure of satellite tobacco necrosis virus after crystallographic refinement at 2.5 Å resolution. *J. Mol. Biol.* **177**, 735–767.

Jones, T. R., and Hyman, R. W. (1983). Specious hybridization between herpes simplex virus DNA and human cellular DNA. *Virology* **131**, 555–560.

Joshi, S., Chapeville, F., and Haenni, A.-L. (1982). Turnip yellow mosaic virus RNA is aminoacylated *in vivo* in Chinese cabbage leaves. *EMBO J.* **1**, 935–938.

Joshi, S., Pleij, C. W. A., Haenni, A.-L., Chapeville, F., and Bosch, L. (1983). Properties of the tobacco mosaic virus intermediate length RNA-2 and its translocation. *Virology* **127**, 100–111.

Joshi, R. L., Faulhammer, H., Chapeville, F., Sprinzl, M., and Haenni, A.-L., (1984). Aminoacyl RNA domain of turnip yellow mosaic virus Val-RNA interacting with elongation factor Tu. *Nucleic Acids Res.* **12**, 7467–7478.

Joshi, R. L., Chapeville, F., and Haenni, A.-L. (1985). Conformational requirements of tobacco mosaic virus RNA for aminoacylation and adenylation. *Nucleic Acids Res.* **13**, 347–354.

Joubert, J. J., Hahn, J. S., von Wechmar, M. B., and Van Regenmortel, M. H. V. (1974). Purification and properties of tomato spotted wilt virus. *Virology* **57**, 11–19.

Joyce, G. F. (1989). RNA evolution and the origins of life. *Nature (London)* **338**, 217–224.

Juckes, I. R. M. (1971). Fractionation of proteins and viruses with polyethylene glycol. *Biochim. Biophys. Acta* **229**, 535–546.

Juniper, B. E., and Jeffree, C. E. (1983). "Plant Surfaces." Edward Arnold, London.

Jupin, I., Quillet, L., Ziegler-Graff, V., Guilley, H., Richards, K., and Jonard, G. (1988). *In vitro* translation of natural and synthetic beet necrotic yellow vein virus RNA1. *J. Gen. Virol.* **69**, 2359–2367.

Jupin, I., Quillet, L., Niesbach-Klösgen, U., Bouzoubaa, S., Richards, K., Guilley, H., and Jonard, G. (1990). Infectious synthetic transcripts of beet necrotic yellow vein virus RNAs and their use in investigating structure–function relations. *In* "Viral Genes and Plant Pathogenesis" (T. P. Pirone and J. G. Shaw, eds.), pp. 187–204. Springer-Verlag, New York.

Jurík, M., Mucha, V., and Valenta, V. (1980). Intraspecies variability in transmission efficiency of stylet-borne viruses by the pea aphid (*Acyrthosiphon pisum*). *Acta Virol.* **24**, 351–357.

Kaempfer, R. (1984). Regulation of eukaryote translation. *Compr. Virol.* **19**, 99–175.

Kaesberg, P. (1987). Organisation of tripartite plant virus genomes: The genome of brome mosaic virus. *In* "The Molecular Biology of Positive Strand RNA Viruses" (D. J. Rowlands, M. A. Mayo, and B. W. J. Mahy, eds.), pp. 219–235. Academic Press, London.

Kahn, R. P. (1976). Aseptic plantlet culture to improve the phytosanitary aspects of plant introduction for asparagus. *Plant Dis. Rep.* **60**, 459–461.

Kahn, R. P., and Monroe, R. L. (1970). Viruses isolated from arborescent *Datura* species from Bolivia, Ecuador, and Columbia. *Plant Dis. Rep.* **54**, 675–677.

Kaiser, W. J., and Danesh, D. (1971). Etiology of virus-induced wilt of *Cicer arietinum*. *Phytopathology* **61**, 453–457.

Kaiser, W. J., Wyatt, S. D. and Pesho, G. R. (1982). Natural hosts and vectors of tobacco streak virus in eastern Washington. *Phytopathology* **72**, 1508–1512.

Kallender, H., Petty, I. D. T., Stein, V. E., Panico, M., Blench, I. P., Etienne, A. T., Morris, H. R., Coutts, R. H. A., and Buck, K. W. (1988). Identification of the coat protein gene of tomato golden mosaic virus. *J. Gen. Virol.* **69**, 1351–1357.

Kamei, T., Goto, T., and Matsui, C. (1969). Turnip virus multiplication in leaves infected with cauliflower mosaic virus. *Phytopathology* **59**, 1795–1797.

Kamer, G., and Argos, P. (1984). Primary structural comparison of RNA-dependent polymerases from plant, animal and bacterial viruses. *Nucleic Acids Res.* **12**, 7269–7282.

Kan, J. H., Andree, P.-J., Kouijzer, L. C., and Mellema, J. E. (1982). Proton-magnetic-resonance studies on the coat protein of alfalfa mosaic virus. *Eur. J. Biochem.* **126**, 29–33.

Kan, J. H., Cremers, A. F. M., Haasnoot, C. A. G., and Hilbers, C. W. (1987). The dynamical structure of the RNA of alfalfa mosaic virus studied by ^{31}P nuclear magnetic resonance. *Eur. J. Biochem.* **168**, 635–639.

Kaper, J. M. (1972). Experimental analysis of the stabilising interactions of simple RNA viruses. *Proc. FEBS Meet.* **27**, 19–41.

Kaper, J. M. (1975). The chemical basis of virus structure, dissociation and reassembly. *Front. Biol.* **39**, 1–485.

Kaper, J. M., and Collmer, C. W. (1988). Modulation of viral plant diseases by secondary RNA agents. *In* "RNA Genetics" (E. Domingo, J. J. Holland, and P. Ahlquist, eds.), Vol. 3, pp. 171–193. CRC Press, Boca Raton, Florida.

Kaper, J. M., and Diaz-Ruiz, J. R. (1977). Molecular weights of the double-stranded RNAs of cucumber mosaic virus strain S and its associated RNA5. *Virology* **80**, 214–217.

Kaper, J. M., and Tousignant, M. E. (1977). Cucumber mosaic virus-associated RNA5. I. Role of host plant and helper strain in determining amount of associated RNA5 with virions. *Virology* **80**, 186–195.

Kaper, J. M., and Waterworth, H. E. (1977). Cucumber mosaic virus associated RNA5. Causal agent for tomato necrosis. *Science* **196**, 429–431.

Kaper, J. M., Diener, T. O., and Scott, H. A. (1965). Some physical and chemical properties of cucumber mosaic virus (strain Y) and of its isolated ribonucleic acid. *Virology* **27**, 54–72.

Kaper, J. M., Tousignant, M. E., and Lot, H. (1976). A low molecular weight replicating RNA associated with a divided genome plant virus: Defective or satellite RNA? *Biochem. Biophys. Res. Commun.* **72**, 1237–1243.

Kaper, J. M., Tousignant, M. E., and Steen, M. T. (1988a). Cucumber mosaic virus-associated RNA5. XI. Comparison of 14 CARNA5 variants relates ability to induce tomato necrosis to a conserved nucleotide sequence. *Virology* **163**, 284–292.

Kaper, J. M., Tousignant, M. E., and Steger, G. (1988b). Nucleotide sequence predicts circularity and self cleavage of 300-ribonucleotide satellite of arabis mosaic virus. *Biochem. Biophys. Res. Commun.* **154**, 318–325.

Kaper, J. M., Gallitelli, D., and Tousignant, M. E. (1990). Identification of a 334-ribonucleotide viral satellite as principal aetiological agent in a tomato necrosis epidemic. *Res. Virol.* **141**, 81–95.

Kaplan, I. B., Kozlov, Y. V., Pshennikova, E. S., Taliansky, M. E., and Atabekov, J. G. (1982). A study of TMVts mutant Ni 2519. III. Location of the reconstitution sites on Ni 2519 RNA. *Virology* **118**, 317–323.

Karpova, O. V., Tyulkina, L. G., Atabekov, K. J., Rodionova, N. P., and Atabekov, J. G. (1989). Deletion of intercistronic poly(A) tract from brome mosaic virus RNA3 by ribonuclease H and its restoration in progeny of the religated RNA3. *J. Gen. Virol.* **70**, 2287–2297.

Kassanis, B. (1939). Intranuclear inclusions in virus-infected plants. *Ann. Appl. Biol.* **26**, 705–709.

Kassanis, B. (1952). Some effects of high temperature on the susceptibility of plants to infection with viruses. *Ann. Appl. Biol.* **39**, 358–369.

Kassanis, B. (1962). Properties and behaviour of a virus depending for its multiplication on another. *J. Gen. Microbiol.* **27**, 477–488.

Kassanis, B. (1965). Therapy of virus-infected plants. *J. R. Agric. Soc. Engl.* **126**, 105–114.

Kassanis, B., and Bastow, C. (1971). The relative concentration of infective intact virus and RNA of four strains of tobacco mosaic virus as influenced by temperature. *J. Gen. Virol.* **11**, 157–170.

Kassanis, B., and Govier, D. A. (1971). New evidence on the mechanism of aphid transmission of potato C and potato aucuba mosaic viruses. *J. Gen. Virol.* **10**, 99–101.

Kassanis, B., and Macfarlane, I. (1964a). Transmission of tobacco necrosis virus by zoospores of *Olpidium brassicae*. *J. Gen. Microbiol.* **36**, 79–93.

Kassanis, B., and Macfarlane, I. (1964b). Transmission of tobacco necrosis virus to tobacco callus tissues by zoospores of *Olpidium brassicae*. *Nature (London)* **201**, 218–219.

Kassanis, B., and Macfarlane, I. (1968). The transmission of satellite viruses of tobacco necrosis virus by *Olpidium brassicae*. *J. Gen. Virol.* **3**, 227–232.

Kassanis, B., and Phillips, M. P. (1970). Serological relationship of strains of tobacco necrosis virus and their ability to activate strains of satellite virus. *J. Gen. Virol.* **9**, 119–126.

Kassanis, B., and Varma, A. (1967). The production of virus-free clones of some British potato varieties. *Ann. Appl. Biol.* **59,** 447–450.

Kassanis, B., and White, R. F. (1974). A simplified method of obtaining tobacco protoplasts for infection with tobacco mosaic virus. *J. Gen. Virol.* **24,** 447–452.

Kassanis, B., Vince, D. A., and Woods, R. D. (1970). Light and electron microscopy of cells infected with tobacco necrosis and satellite viruses. *J. Gen. Virol.* **7,** 143–151.

Kassanis, B., White, R. F., and Woods, R. D. (1975). Inhibition of multiplication of tobacco mosaic virus in protoplasts by antibiotics and its prevention by divalent metals. *J. Gen. Virol.* **28,** 185–191.

Katis, N., Carpenter, J. M., and Gibson, R. W. (1986). Interference between potyviruses during aphid transmission. *Plant Pathol.* **35,** 152–157.

Katouzian-Safadi, M., and Berthet-Colominas, C. (1983). Evidence for the presence of a hole in the capsid of turnip yellow mosaic virus after RNA release by freezing and thawing. Decapsidation of turnip yellow mosaic virus *in vitro. Eur. J. Biochem.* **137,** 47–55.

Katouzian-Safadi, M., and Haenni, A.-L. (1986). Studies on the phenomenon of turnip yellow mosaic virus RNA release by freezing and thawing. *J. Gen. Virol.* **67,** 557–565.

Katz, D., and Kohn, A. (1984). Immunoabsorbent electron microscopy for detection of viruses. *Adv. Virus Res.* **29,** 169–194.

Kauffmann, S., Legrand, M., Geoffroy, P., and Fritig, B. (1987). Biological function of "pathogenesis-related" proteins: Four PR proteins of tobacco have $1,3,-\beta$-glucanase activity. *EMBO J.* **6,** 3209–3212.

Kauffmann, S., Legrand, M., and Fritig, B. (1989). Isolation and characterisation of six pathogenesis-related (PR) proteins of Samsun NN tobacco. *Plant Mol. Biol.* **14,** 381–390.

Kausche, G. A., Pfankuch, E., and Ruska, A. (1939). Die Sichtbormachung von pflanzlichem Virus im Übermikroskop. *Naturwissenschaften* **27,** 292–299.

Kavanau, J. L. (1949). On the correlation of the phenomena associated with chromosomes, foreign proteins and viruses. III. Virus associated phenomena, characteristics and reproduction. *Am. Nat.* **83,** 113–138.

Kawano, S., Uyeda, I., and Shikata, E. (1984). Particle structure and double-stranded RNA of rice ragged stunt virus. *J. Fac. Agric., Hokkaido Univ.* **61,** 408–418.

Kay, R., Chan, A., Daly, M., and McPherson, J. (1987). Duplication of CaMV 35S promoter sequences creates a strong enhancer for plant genes. *Science* **236,** 1299–1302.

Keeling, J., and Matthews, R. E. F. (1982). Mechanism for release of RNA from turnip yellow mosaic virus at high pH. *Virology* **119,** 214–218.

Keeling, J., Collins, E. R., and Matthews, R. E. F. (1979). Behaviour of turnip yellow mosaic virus nucleoproteins under alkaline conditions. *Virology* **97,** 100–111.

Keese, P., and Symons, R. H. (1985). Domains in viroids: Evidence of intermolecular RNA rearrangements and their contribution to viroid evolution. *Proc. Natl. Acad. Sci. U.S.A.* **82,** 4582–4586.

Keese, P., and Symons, R. H. (1987). The structure of viroids and virusoids. *In* "Viroids and Viroid-like Pathogens" (J. S. Semancik, ed.), pp. 1–47. C.R.C. Press, Boca Raton, Fla.

Keese, P., Bruening, G., and Symons, R. H. (1983). Comparative sequence and structure of circular RNAs from two isolates of lucerne transient streak virus. *FEBS Lett.* **159,** 185–190.

Keese, P., Osario-Keese, M. E., and Symons, R. H. (1988a). Coconut tinangaja viroid: Sequence homology with coconut cadang-cadang viroid and other potato spindle tuber viroid related RNAs. *Virology* **162,** 508–510.

Keese, P., Visvader, J. E., and Symons, R. H. (1988b). Sequence variability in plant viroid RNAs. *In* "RNA Genetics" (E. Domingo, J. Holland, and P. Ahlquist, eds.), Vol. 3, pp. 71–98. CRC Press, Boca Raton, Florida.

Keese, P., Mackenzie, A., and Gibbs, A. (1989). Nucleotide sequence of the genome of an Australian isolate of turnip yellow mosaic tymovirus. *Virology* **172,** 536–546.

Keller, P., Lüttge, U., Wang, X.-C., and Büttner, G. (1989). Influence of rhizomania disease on gas exchange and water relations of a susceptible and a tolerant sugar beet variety. *Physiol. Mol. Plant Pathol.* **34,** 379–392.

Kendall, T. L., Langenberg, W. G., and Lommel, S. A. (1988). Molecular characterization of sorghum chlorotic spot virus, a proposed furovirus. *J. Gen. Virol.* **69,** 2335–2345.

Kennedy, J. S. (1951). Benefits to aphids from feeding on galled and virus-infected leaves. *Nature (London)* **168,** 825–826.

Kenten, R. H., and Lockwood, G. (1977). Studies on the possibility of increasing resistance to cocoa swollen-shoot virus by breeding. *Ann. Appl. Biol.* **85,** 71–78.

Kerlan, C., Robert, Y., Perennec, P., and Guillery, E. (1987). Survey of the level of infection by PVY-O and control methods developed in France for potato seed production. *Potato Res.* **30,** 651–667.

Ketellapper, H. J. (1963). Temperature induced chemical defects in higher plants. *Plants Physiol.* **38,** 175–179.

Khan, I. A., and Jones, G. E. (1989). Selection for a specific tobacco mosaic virus variant during bolting of *Nicotiana sylvestris. Can. J. Bot.* **67,** 88–94.

Khan, M. A., and Maxwell, D. P. (1977). Use of inclusions in the rapid diagnosis of virus diseases of red clover. *Plant Dis. Rep.* **61,** 679–683.

Khan, M. A., and Mukhopadhyay, S. (1985). Studies on the effect of some alternative cultural methods on the incidence of yellow vein mosaic virus (YVMV)disease of okra (*Abelmoschus esculentus* (L.) Moench.). *Indian J. Virol.* **1,** 69–72.

Khan, M. A., Maxwell, D. P., and Maxwell, M. D. (1977). Light microscopic cytochemistry and ultrastructure of red clover vein mosaic virus-induced inclusions. *Virology* **78,** 173–182.

Khan, Z. A., Hiriyanna, K. T., Chavez, F., and Fraenkel-Conrat, H. (1986). RNA-directed RNA polymerases from healthy and from virus infected cucumber. *Proc. Natl. Acad. Sci. U.S.A.* **83,** 2383–2386.

Khetarpal, R. K., Bossennec, J.-M., Burghofer, A., Cousin, A., and Maury, Y. (1988). Effect of pea seed-borne mosaic virus on yield of field pea. *Agronomie* **8,** 811–815.

Kiberstis, P. A., and Zimmern, D. (1984). Translational strategy of *Solanum nodiflorum* mottle virus RNA: Synthesis of a coat protein precursor *in vitro* and *in vivo. Nucleic Acids Res.* **12,** 933–943.

Kiberstis, P. A., Loesch-Fries, L. S., and Hall, T. C. (1981). Viral protein synthesis in barley protoplasts inoculated with native and fractionated brome mosaic virus RNA. *Virology* **112,** 804–808.

Kiberstis, P. A., Pessi, A., Atherton, E., Jackson, R., Hunter, T., and Zimmern, D. (1983). Analysis of *in vitro* and *in vivo* products of the TMV 30 kDa open reading frame using antisera raised against a synthetic peptide. *FEBS Lett.* **164,** 355–360.

Kiefer, M. C., Daubert, S. D., Schneider, I. R., and Bruening, G. (1982). Multimeric forms of satellite of tobacco ringspot virus RNA. *Virology* **121,** 262–273.

Kiefer, M. C., Owens, R. A., and Diener, T. O. (1983). Structural similarities between viroids and transposable genetic elements. *Proc. Natl. Acad. Sci. U.S.A.* **80,** 6234–6238.

Kiho, Y., Machida, H., and Oshima, N. (1972). Mechanism determining the host specificity of tobacco mosaic virus. I. Formation of polysomes containing infecting viral genome in various plants. *Jpn. J. Microbiol.* **16,** 451–459.

Kikkawa, H., Nagata, T., Matsui, C., and Takebe, I. (1982). Infection of protoplasts from tobacco suspension cultures by tobacco mosaic virus. *J. Gen. Virol.* **63,** 451–456.

Kikuchi, Y., Tyc, K., Filipowicz, W., Sanger, H. L., and Gross, H. J. (1982). Circularization of linear viroid RNA via 2′-phosphomonoester, 3′,5′-phosphodiester bonds by a novel type of RNA ligase from wheat germ and *Chlamydomonas. Nucleic Acids Res.* **10,** 7521–7529.

Kim, K. S. (1977). An ultrastructural study of inclusions and disease development in plant cells infected by cowpea chlorotic mottle virus. *J. Gen. Virol.* **35,** 535–543.

Kim, K. S., Fulton, J. P., and Scott, H. A. (1974). Osmiophilic globules and myelinic bodies in cells infected with two comoviruses. *J. Gen. Virol.* **25,** 445–452.

Kimble, K. A., Grogan, R. G., Greathead, A. S., Paulus, A. O., and House, J. K. (1975). Development, application, and comparison of methods for indexing lettuce seed for mosaic virus in California. *Plant Dis. Rep.* **59,** 461–464.

Kimmins, W. C., and Brown, R. G. (1973) Hypersensitive resistance. The role of cell wall glycoproteins in virus localization. *Can. J. Bot.* **51,** 1923–1926.

Kimura, I. (1986). A study of rice dwarf virus in dwarf virus in vector cell monolayers by fluorescent antibody focus counting. *J. Gen. Virol.* **67,** 2119–2124.

Kimura, I., and Black, L. M. (1972). The cell-infecting unit of wound tumor virus. *Virology* **49,** 549–561.

Kimura, I., Minobe, Y., and Omura, T. (1987). Changes in a nucleic acid and a protein component of rice dwarf virus particles associated with an increase in symptom severity. *J. Gen. Virol.* **68,** 3211–3215.

King, A. M. Q. (1988). Genetic recombination in positive strand RNA viruses. *In* "RNA Genetics" (E. Domingo, J. J. Holland, and P. Ahlquist, eds.), Vol. 2, pp. 149–165. CRC Press, Boca Raton, Florida.

King, L., and Leberman, R. (1973). Derivatisation of carboxyl groups of tobacco mosaic virus with cystamine. *Biochim. Biophys. Acta* **322,** 279–293.

Kingsbury, D. W. (1988). Biological concepts in virus classification. *Intervirology* **29,** 242–253.

Kirby, R., Clarke, B. A., and Rybicki, E. P. (1989). Evolutionary relationships of three Southern African streak virus isolates. *Intervirology* **30,** 96–101.

Kirchanski, S. J. (1975). The ultrastructural development of the dimorphic plastids of *Zea mays* L. *Am. J. Bot.* **62,** 695–705.

Kiss, T., Pósfai, J., and Solymosy, F. (1983). Sequence homology between potato spindle tuber viroid and U3B snRNA. *FEBS Lett.* **163**, 217–220.

Kitajima, E. W., and Costa, A. S. (1969). Association of pepper ringspot virus (Brazilian tobacco rattle virus) and host cell mitochondria. *J. Gen. Virol.* **4**, 177–181.

Kitajima, E. W., and Costa, A. S. (1973). Aggregates of chloroplasts in local lesions induced in *Chenopodium quinoa* wild. by turnip mosaic virus. *J. Gen. Virol.* **20**, 413–416.

Kitajima, E. W., and Lauritis, J. A. (1969). Plant virions in plasmodesmata. *Virology* **37**, 681–685.

Kitajima, E. W., and Lovisolo, O. (1972). Mitochondrial aggregates in *Datura* leaf cells infected with henbane mosaic virus. *J. Gen. Virol.* **16**, 265–271.

Klaff, P., Gruner, R., Hecker, R., Sättler, A., Theissen, G., and Riesner, D. (1989). Reconstituted and cellular viroid–protein complexes. *J. Gen. Virol.* **70**, 2257–2270.

Klausner, A. (1987). Immunoassays flourish in new markets. *Bio Technology* **5**, 551–556.

Kleczkowski, A. (1949). The transformation of local lesion counts for statistical analysis. *Ann. Appl. Biol.* **36**, 139–152.

Kleczkowski, A. (1950). Interpreting relationships between concentrations of plant viruses and numbers of local lesions. *J. Gen. Microbiol.* **4**, 53–69.

Kleczkowski, A. (1953). A method for testing results of infectivity tests with plant viruses for compatibility with hypotheses. *J. Gen. Microbiol.* **8**, 295–301.

Klein, M., and Harpaz, I. (1970). Heat suppression of plant-virus propagation in the insect vector's body. *Virology* **41**, 72–76.

Klein, R. E., Wyatt, S. D., and Kaiser, W. J. (1988). Incidence of bean common mosaic virus in USDA *Phaseolus* germ plasm collection. *Plant Dis.* **72**, 301–302.

Kleinhempel, H., and Kegler, G. (1982). Transmission of tomato bushy stunt virus without vectors. *Acta Phytopathol. Acad. Sci. Hung.* **17**, 17–21.

Klinkenberg, F. A., Ellwood, S., and Stanley, J. (1989). Fate of African cassava mosaic virus coat protein deletion mutants after agroinoculation. *J. Gen. Virol.* **70**, 1837–1844.

Klug, A., and Berger, J. E. (1964). An optical method for the analysis of periodicities in electron micrographs, and some observations on the mechanism of negative staining. *J. Mol. Biol.* **10**, 565–569.

Klug, A., and Caspar, D. L. D. (1960). The structure of small viruses. *Adv. Virus Res.* **7**, 225–325.

Klug, A., Longley, W., and Leberman, R. (1966). Arrangement of protein subunits and the distribution of nucleic acid in turnip yellow mosaic virus. I. X-ray diffraction studies. *J. Mol. Biol.* **15**, 315–343.

Kluge, S., Kirsten, U., and Oertel, C. (1983). Infection of *Dianthus* protoplasts with carnation mottle virus. *J. Gen. Virol.* **64**, 2485–2487.

Knorr, D. A., and Dawson, W. O. (1988). A point mutation in the tobacco mosaic virus capsid protein gene induces hypersensitivity in *Nicotiana sylvestris*. *Proc. Natl. Acad. Sci. U.S.A.* **85**, 170–174.

Knowland, J., Hunter, T., Hunt, T., and Zimmern, D. (1975). Translation of tobacco mosaic virus RNA and isolation of the messenger for TMV coat protein. *Colloq.—Inst. Natl. Sante Rech. Med.* **47**, 211–216.

Knudson, D. L., and MacLeod, R. (1972). The proteins of potato yellow dwarf virus. *Virology* **47**, 285–295.

Kobayashi, D. Y., Tamaki, S. J., and Keen, N. T. (1989). Cloned avirulence genes from the tomato pathogen *Pseudomonas syringae* pv tomato confer cultivar specificity on soybean. *Proc. Natl. Acad. Sci. U.S.A.* **86**, 157–161.

Kodama, T., and Suzuki, N. (1973). RNA polymerase activity in purified rice dwarf virus. *Ann. Phytopathol. Soc. Jpn.* **39**, 251–258.

Koenig, R. (1976). A loop-structure in the serological classification system of tymoviruses. *Virology* **72**, 1–5.

Koenig, R. (1981). Indirect ELISA methods for the broad specificity detection of plant viruses. *J. Gen. Virol.* **55**, 53–62.

Koenig, R. (1986). Plant viruses in rivers and lakes. *Adv. Virus Res.* **31**, 321–333.

Koenig, R., and Bercks, R. (1968). Änderungen in heterologen Reaktionsvermögen von Antiseren gegen Vertroter der potato virus X-Gruppe im Laufe des Immunisierungsprozesses. *Phytopathol. Z.* **61**, 382–398.

Koenig, R., and Burgermeister, W. (1986). Applications of immunoblotting in plant virus diagnosis. *In* "Developments and Applications in Virus Testing" (R. A. C. Jones and L. Torrance, eds.), Dev. Appl. Biol. I, pp. 121-137. Association of Applied Biologists, Wellsbourne.

Koenig, R., and Gibbs, A. (1986). Serological relationships among tombusviruses. *J. Gen. Virol.* **67**, 75–82.

Koenig, R., and Givord, L. (1974). Serological interrelationships in the turnip yellow mosaic virus group. *Virology* **58**, 119–125.

Koenig, R., and Paul, H. L. (1982). Variants of ELISA in plant virus diagnosis. *J. Virol. Methods* **5**, 113–125.

Koenig, R., and Torrance, L. (1986). Antigenic analysis of potato virus X by means of monoclonal antibodies. *J. Gen. Virol.* **67**, 2145–2151.

Koenig, R., Tremaine, J. H., and Shepard, J. F. (1978). *In situ* degradation of the protein chain of potato virus X at the N- and C-termini. *J. Gen. Virol.* **38**, 329–337.

Koenig, R., Fribourg, C. E., and Jones, R. A. C. (1979). Symptomatological, serological, and electrophoretic diversity of isolates of Andean potato latent virus from different regions of the Andes. *Phytopathology* **69**, 748–752.

Koenig, R., Burgermeister, W., Weich, H., Sebald, W., and Kothe, C. (1986). Uniform RNA patterns of beet necrotic yellow vein virus in sugar beet roots, but not in leaves from several plant species. *J. Gen. Virol.* **67**, 2043–2046.

Koenig, R., An, D., and Burgermeister, W. (1988a). The use of filter hybridisation techniques for the identification, differentiation and classification of plant viruses. *J. Virol. Methods* **19**, 57–68.

Koenig, R., An, D., Lesemann, D. E., and Burgermeister, W. (1988b). Isolation of carnation ringspot virus from a canal near a sewage plant; cDNA hybridization analysis, serology, and cytopathology. *J. Phytopathol.* **121**, 346–356.

Kohl, R. J., and Hall, T. C. (1974). Aminoacylation of RNA from several viruses. Amino acid specificity and differential activity of plant, yeast and bacterial synthetases. *J. Gen. Virol.* **25**, 257–261.

Kohl, R. J., and Hall, T. C. (1977). Loss of infectivity of brome mosaic virus RNA after chemical modification of the 3′ or 5′ terminus. *Proc. Natl. Acad. Sci. U.S.A.* **74**, 2682–2686.

Köhler, G., and Milstein, C. (1975). Continuous cultures of fused cells secreting antibody of predefined specificity. *Nature (London)* **256**, 495–497.

Koltunow, A. M., and Rezaian, M. A. (1988). Grapevine yellow speckle viroid: Structural features of a new viroid group. *Nucleic Acids Res.* **16**, 849–864.

Koltunow, A. M., and Rezaian, M. A. (1989a). Grapevine viroid 1B, a new member of the apple scar skin viroid group, contains the left terminal region of tomato plants macho viroid. *Virology* **170**, 575–578.

Koltunow, A. M., and Rezaian, M. A. (1989b). A scheme for viroid classification. *Intervirology* **30**, 194–201.

Konarska, M., Filipowicz, W., Domdey, H., and Gross, H. J. (1981). Binding of ribosomes to linear and circular forms of the 5′-terminal leader fragment of tobacco-mosaic-virus RNA. *Eur. J. Biochem.* **114**, 221–227.

Konate, G., and Fritig, B. (1983). Extension of the ELISA method to the measurement of the specific radioactivity of viruses in crude cellular extracts. *J. Virol. Methods* **6**, 347–356.

Konate, G., and Fritig, B. (1984). An efficient microinoculation procedure to study plant virus multiplication at predetermined individual infection sites on the leaves. *Phytopathol. Z.* **109**, 131–138.

Kontzog, H. G., Kleinhempel, H., and Kegler, H. (1988). Detection of plant pathogenic viruses in waters. *Arch. Phytopathol. Pflanzenschutz.* **24**, 171–172.

Koonin, E. V., and Gorbalenya, A. E. (1989). Evolution of RNA genomes: Does the high mutation rate necessitate high rate of evolution of viral proteins? *J. Mol. Evol.* **28**, 524–527.

Koper-Zwarthoff, E. C., and Bol, J. F. (1980). Nucleotide sequence of the putative recognition site for coat protein in the RNA's of alfalfa mosaic virus and tobacco streak virus. *Nucleic Acids Res.* **8**, 3307–3318.

Kopp, M., Geoffroy, P., and Fritig, B. (1981). Studies on tobacco mosaic virus replication by means of radiolabelling the virus under isotonic conditions. *Ann. Phytopathol.* **12**, 314.

Kopp, M., Rouster, J., Fritig, B., Darvill, A., and Albersheim, P. (1989). Host–pathogen interactions. XXXII. A fungal glucan preparation protects *Nicotiana* against infection by viruses. *Plant Physiol.* **90**, 208–216.

Kozak, M. (1981). Possible role of flanking nucleotides in recognition of the AUG initiator codon by eukaryotic ribosomes. *Nucleic Acids Res.* **9**, 5233–5252.

Kozak, M. (1986). Point mutations define a sequence flanking the AUG initiator codon that modulates translation by eucaryotic ribosomes. *Cell (Cambridge, Mass.)* **44**, 283–292.

Krass, C. J., and Schlegel, D. E. (1974). "Motley dwarf" virus disease complex of California carrots. *Phytopathology* **64**, 151–152.

Kroner, P., Richards, D., Traynor, P., and Ahlquist, P. (1989). Defined mutations in a small region of the brome mosaic virus 2a gene cause diverse temperature-sensitive RNA replication phenotypes. *J. Virol.* **63**, 5302–5309.

Kruckelmann, H.-W., and Seyffert, W. (1970). Wechselwirkungen zwischen einem turnip-mosaik-Virus und dem Genom des Wirtes. *Theor. Appl. Genet.* **40**, 121–123.

Krüse, J., Krüse, K. M., Witz, J., Chauvin, C., Jacrot, B., and Tardieu, A. (1982). Divalent ion-dependent reversible swelling of tomato bushy stunt virus and organisation of the expanded virion. *J. Mol. Biol.* **162**, 393–417.

Krüse, J., Timmins, P., and Witz, J. (1987). The spherically averaged structure of a DNA isometric plant virus: Cauliflower mosaic virus. *Virology* **159**, 166–168.

Kryczynski, S., Paduch-Cichal, E., and Skrzeczkowski, L. J. (1988). Transmission of three viroids through seed and pollen of tomato plants. *J. Phytopathol.* **121**, 51–57.

Kubo, S., Harrison, B. D., Robinson, D. J., and Mayo, M. A. (1975a). Tobacco rattle virus in tobacco mesophyll protoplasts: Infection and virus multiplication. *J. Gen. Virol.* **27**, 293–304.

Kubo, S., Harrison, B. D., and Barker, H. (1975b). Defined conditions for growth of tobacco plants as sources of protoplasts for virus infection. *J. Gen. Virol.* **28**, 255–257.

Kummert, J., and Semal, J. (1969). Study of the incorporation of radioactive uridine into virus-infected leaf fragments. *Phytopathol. Z.* **65**, 101–123.

Kunkel, L. O. (1922). Insect transmission of yellow stripe disease. *Hawaii Plant. Rec.* **26**, 58–64.

Kuntz, I. D., Jr., and Kauzmann, W. (1974). Hydration of proteins and polypeptides. *Adv. Protein Chem.* **28**, 239–345.

Kurath, G., and Palukaitis, P. (1989a). Satellite RNAs of cucumber mosaic virus: Recombinants constructed *in vitro* reveal independent functional domains for chlorosis and necrosis in tomato. *Mol. Plant–Microbe Interact.* **2**, 91–96.

Kurath, G., and Palukaitis, P. (1989b). RNA sequence heterogeneity in natural populations of three satellite RNAs of cucumber mosaic virus. *Virology* **173**, 231–240.

Kurath, G., and Palukaitis, P. (1990). Serial passage of infectious transcripts of a cucumber mosaic virus satellite RNA clone results in sequence heterogeneity. *Virology* **176**, 8–15.

Kurkinen, M. (1981). Fidelity of protein synthesis affects the read through translation of tobacco mosaic virus RNA. *FEBS Lett.* **124**, 79–83.

Kurtz-Fritsch, C., and Hirth, L. (1972). Uncoating of two spherical plant viruses. *Virology* **47**, 385–396.

Kyte, J., and Doolittle, R. F. (1982). A simple method for displaying the hydropathic character of a protein. *J. Mol. Biol.* **157**, 105–132.

Lackey, C. F. (1932). Restoration of virulence of attenuated curleytop virus by passage through *Stellaria media. J. Agric. Res.* **44**, 755–765.

Laflèche, D., and Bové, J. M. (1968). Sites d'incorporation de l'uridine tritiée dans les cellules du parenchyme foliare de *Brassica chinensis,* saines ou infectées par le virus de la mosaique jaune du navet. *C.R. Hebd. Seances Acad. Sci.* **266**, 1839–1841.

Leflèche, D., Bové, C., Dupont, G., Mouches, C., Astier, T., Garnier, M., and Bové, J. M. (1972). Site of viral RNA replication in the cells of higher plants; TYMV-RNA synthesis on the chloroplast outer member system. *Proc. FEBS Meet.* **27**, 43–71.

Laidlaw, W. M. R. (1986). Mechanical aids to improve the speed and sensitivity of plant virus diagnosis by the biological test method. *Ann. Appl. Biol.* **108**, 309–318.

Laidlaw, W. M. R. (1987). A new method for mechanical virus transmission and factors affecting its sensitivity. *OEPP/EPPO Bull.* **17**, 81–89.

Lain, S., Riechmann, J. L., and Garcia, J. A. (1989). The complete nucleotide sequence of plum pox potyvirus RNA. *Virus Res.* **13**, 157–172.

Lakshman, D. K., Gonsalves, D., and Fulton, R. W. (1985). Role of *Vigna* species in the appearance of pathogenic variants of cucumber mosaic virus. *Phytopathology* **75**, 751–757.

Lakshman, D. K., Hiruki, C., Wu, X. N., and Leung, W. C. (1986). Use of [^{32}P]RNA probes for the dot-hybridization detection of potato spindle tuber viroid. *J. Virol. Methods* **14**, 309–319.

Lana, A. F. (1981). Prospects of infectivity tests as a tool in plant virus disease diagnosis in the third world. *Trop. Pest Manage.* **27**, 24–28.

Lange, L. (1975). Infection of *Daucus carota* by tobacco necrosis virus. *Phytopathol. Z.* **83**, 136–143.

Lange, L., and Heide, M. (1986). Dot immuno binding (DIB) for detection of virus in seed. *Can. J. Plant Pathol.* **8**, 373–379.

Langenberg, W. G. (1979). Chilling of tissue before glutaraldehyde fixation preserves fragile inclusions of several plant viruses. *J. Ultrastruct. Res.* **66**, 120–131.

Langenberg, W. G. (1982). Fixation of plant inclusions under conditions designed for freeze-fracture. *J. Ultrastruct. Res.* **81**, 184–188.

Langenberg, W. G. (1986a). Virus protein associated with the cylindrical inclusions of two viruses that infect wheat. *J. Gen. Virol.* **67**, 1161–1168.

Langenberg, W. G. (1986b). Deterioration of several rod-shaped wheat viruses following antibody decoration. *Phytopathology* **76**, 339–341.

Langenberg, W. G. (1987). Barley stripe mosaic virus but not brome mosaic virus binds to wheat streak mosaic virus cylindrical inclusions *in vivo*. *Phytopathology* **78**, 589–594.

Langenberg, W. G. (1989). Rapid antigenic modification of wheat streak mosaic virus *in vitro* is prevented in glutaraldehyde fixed tissue. *J. Gen. Virol.* **70**, 969–973.

Langenberg, W. G., and Schroeder, H. F. (1975). The ultrastructural appearance of cowpea mosaic virus in cowpea. *J. Ultrastruct. Res.* **51**, 166–175.

Langereis, K., Neeleman, L., and Bol, J. F. (1986a). Biologically active transcripts of cloned DNA of the coat protein messenger of two plant viruses. *Plant Mol. Biol.* **6**, 281–288.

Langereis, K., Mugnier, M.-A., Cornelissen, B. J. C., Pinck, L., and Bol, J. F. (1986b). Variable repeats and poly (A)-stretches in the leader sequence of alfalfa mosaic virus RNA3. *Virology* **154**, 409–414.

Lapierre, H., Cortillot, M., Kusiak, C., and Hariri, D. (1985). Field resistance of autumn-sown wheat to wheat soil-borne mosaic virus (WSBMV) *Agronomie* **5**, 565–572.

Laquel, P., Ziegler, V., and Hirth, L. (1986). The 80K polypeptide associated with the replication complexes of cauliflower mosaic virus is recognised by antibodies to gene V translation product. *J. Gen. Virol.* **67**, 197–201.

Larson, R. H., Matthews, R. E. F., and Walker, J. C. (1950). Relationships between certain viruses affecting the genus *Brassica*. *Phytopathology* **40**, 955–962.

Lastra, J. R., and Esparza, J. (1976). Multiplication of vesicular stomatitis virus in the leafhopper *Peregrinus maidis* (Ashm), a vector of a plant rhabdovirus. *J. Gen. Virol.* **32**, 139–142.

Lastra, J. R., and Schlegel, D. E. (1975). Viral protein synthesis in plants infected with broadbean mottle virus. *Virology* **65**, 16–26.

Laurence, J. A., Aluiso, A. L., Weinstein, L. H., and McCune, D. C. (1981). Effect of sulphur dioxide on southern bean mosaic and maize dwarf mosaic. *Environ. Poll., Ser. A* **24**, 185–191.

Laviolette, F. A., and Athow, K. L. (1971). Longevity of tobacco ringspot virus in soybean seed. *Phytopathology* **61**, 755.

Lawrence, J. (1714). "The Clergyman's Recreation," 2nd ed. B. Lintott, London.

Lawson, R. H., and Taconis, P. J. (1965). Transfer of dahlia mosaic virus with liquid nitrogen and relation of transfer to symptoms and inclusions. *Phytopathology* **55**, 715–718.

Lawson, R. H., Hearon, S. S., and Smith, F. F. (1971). Development of pinwheel inclusions associated with sweet potato russet crack virus. *Virology* **46**, 453–463.

Lawton, M. A., Tierney, M. A., Nakamura, I., Anderson, E., Komeda, Y., Dubé, P., Hoffman, N., Frayley, R. T., and Beachy, R. N. (1987). Expression of a soybean β-conglycinin gene under the control of the cauliflower mosaic virus 35S and 19S promoters in transformed petunia tissue. *Plant Mol. Biol.* **9**, 315–324.

Lazarowitz, S. G. (1987). The molecular characterisation of geminiviruses. *Plant Mol. Biol. Rep.* **4**, 177–192.

Lazarowitz, S. G. (1988). Infectivity and complete nucleotide sequence of the genome of a South African isolate of maize streak virus. *Nucleic Acids Res.* **16**, 229–249.

Lazarowitz, S. G. (1990). Molecular characterisation of two bipartite geminiviruses causing squash leaf curl disease: Role of transactivation and defective genomic components in determining host range. *Virology* (in press).

Lazarowitz, S. G., Pinder, A. J., Damsteegt, V. D., and Rogers, S. G. (1989). Maize streak virus genes essential for systemic spread and symptom development. *EMBO J.* **8**, 1023–1032.

Leary, J. J., Brigati, D. J., and Ward, D. C. (1983). Rapid and sensitive colorimetric method for visualising biotin-labelled DNA probes hybridised to DNA or RNA immobilised on nitro-cellulose:Bio-blots. *Proc. Natl. Acad. Sci. U.S.A.* **80**, 4045–4049.

Lebeurier, G., and Hirth, L. (1966). Effect of elevated temperatures on the development of two strains of tobacco mosaic virus. *Virology* **29**, 385–395.

Lebeurier, G., Nicolaieff, A., and Richards, K. E. (1977). Inside-out model for self-assembly of tobacco mosaic virus. *Proc. Natl. Acad. Sci. U.S.A.* **74**, 149–153.

Lebeurier, G., Hirth, L., Hohn, B. and Hohn, T. (1982). *In vivo* recombination of cauliflower mosaic virus DNA. *Proc. Natl. Acad. Sci. U.S.A.* **79**, 2932–2936.

Lecoq, H., and Pitrat, M. (1985). Specificity of the helper-component-mediated aphid transmission of three potyviruses infecting muskmelon. *Phytopathology* **75**, 890–893.

Lecoq, H., Pitrat, M., and Labonne, G. (1981). Resistance to virus transmission by aphids in a *Cucumis melo* line presenting non-acceptance to *Aphis gossypii*. *Bull. SROP* **4**, 147–151.

Lee, C. L., and Black, L. M. (1955). Anatomical studies of *Trifolium incarnartum* infected by wound tumor virus. *Am. J. Bot.* **42**, 160–168.

Lee, C. R., and Singh, R. P. (1972). Enhancement of diagnostic symptoms of potato spindle tuber virus by manganese. *Phytopathology* **62**, 516–520.

Lee, P. E., Boerjan, M., and Peters, D. (1972). Electron microscopic evidence for a neuraminic acid in sowthistle yellow vein virus. *Virology* **50,** 309–311.

Lee, R. F., Garnsey, S. M., Brlansky, R. H., and Goheen, A. C. (1987). A purification procedure for enhancement of citrus tristeza virus yields and its application to other phloem limited viruses. *Phytopathology* **77,** 543–549.

Lee, S. Y., Uyeda, I., and Shikata, E. (1987). Characterisation of RNA polymerase associated with rice ragged stunt virus. *Intervirology* **27,** 189–195.

Lee, Y.-S., and Ross, J. P. (1972). Top necrosis and cellular changes in soybean doubly infected by soybean mosaic and bean pod mottle viruses. *Phytopathology* **62,** 839–845.

Legg, J. T., and Lockwood, G. (1977). Evaluation and use of a screening method to aid selection of cocoa (*Theobroma cacao*) with field resistance to cocoa swollen-shoot virus in Ghana. *Ann. Appl. Biol.* **86,** 241–248.

Legrand, M., Kauffmann, S., Geoffroy, P., and Fritig, B. (1987). Biological function of pathogenesis-related proteins: Four tobacco pathogenesis-related proteins are chitinases. *Proc. Natl. Acad. Sci. U.S.A.* **84,** 6750–6754.

Lehto, K., and Dawson, W. O. (1990a). Changing the start codon context of the 30K gene of tobacco mosaic virus from "weak" to "strong" does not increase expression. *Virology* **174,** 169–176.

Lehto, K., and Dawson, W. O. (1990b). Replication, stability, and gene expression of tobacco mosaic virus mutants with a second 30K ORF. *Virology* **175,** 30–40.

Lehto, K., Grantham, G. L., and Dawson, W. O. (1990a). Insertion of sequences containing the coat protein subgenomic RNA promoter and leader in front of the tobacco mosaic virus 30K ORF delays its expression and causes defective cell-to-cell movement. *Virology* **174,** 145–157.

Lehto, K., Bubrick, P., and Dawson, W. O. (1990b). Time course of TMV 30K protein accumulation in intact leaves. *Virology* **174,** 290–293.

Lei, J. D., and Agrios, G. N. (1986). Mechanisms of resistance in corn to maize dwarf mosaic virus. *Phytopathology* **76,** 1034–1040.

Lemaire, O., Merdinoglu, D., Valentin, P., Putz, C., Ziegler-Graff, V., Guilley, H., Jonard, G., and Richards, K. (1988). Effect of beet necrotic yellow vein virus RNA composition on transmission by *Polymyxa betae*. *Virology* **162,** 232–235.

Leonard, D. A., and Zaitlin, M. (1982). A temperature-sensitive strain of tobacco mosaic defective in cell-to-cell movement generates an altered viral-coded protein. *Virology* **117,** 416–424.

Lesemann, D. E., Koenig, R., Torrance, L., Bunton, G., Boonekamp, P. M., Peters, D., and Schots, A. (1990). Electron-microscopical demonstration of different binding sites for monoclonal antibodies on particles of beet necrotic yellow vein virus. *J. Gen. Virol.* **71,** 731–733.

Leung, D. W., Gilbert, C. W., Smith, R. E., Sasavage, N. L., and Clark, J. M., Jr. (1976). Translation of satellite tobacco necrosis virus ribonucleic acid by an *in vitro* system from wheat germ. *Biochemistry* **15,** 4943–4950.

Levy, D., and Marco, S. (1976). Involvement of ethylene in epinasty of CMV-infected cucumber cotyledons which exhibit increased resistance to gaseous diffusion. *Physiol. Plant Pathol.* **9,** 121–126.

Lewis, T., and Dibley, G. C. (1970). Air movement near windbreaks and a hypothesis of the mechanism of the accumulation of airborne insects. *Ann. Appl. Biol.* **66,** 477–484.

Li, W.-H., Wolfe, K. H., Sourdis, J., and Sharp, P. M. (1987). Reconstruction of phylogenetic trees and estimation of divergence times under non-constant rates of evolution. *Cold Spring Harbor Symp. Quant. Biol.* **52,** 847–856.

Li, X. H., Heaton, L. A., Morris, J., and Simon, A. E. (1989). Turnip crinkle defective interfering RNAs intensify viral symptoms and are generated *de novo*. *Proc. Natl. Acad. Sci. U.S.A.* **86,** 9173–9177.

Liljas, L., Unge, T., Alwyn Jones, T., Fridborg, K., Lövgren, S., Skoglund, U., and Strandborg, B. (1982). Structure of satellite tobacco necrosis virus at 3.0Å resolution. *J. Mol. Biol.* **159,** 93–108.

Lim, W. L., and Hagedorn, D. J. (1977). Bimodal transmission of plant viruses. *In* "Aphids as Virus Vectors" (K. F. Harris and K. Maramorosch, eds.), pp. 237–251. Academic Press, New York.

Lin, M. T., Campbell, R. N., Smith, P. R., and Temmink, J. H. M. (1970). Lettuce big-vein virus transmission by single-sporangium isolates of *Olpidium brassicae*. *Phytopathology* **60,** 1630–1634.

Lin, N.-S., and Langenberg, W. G. (1984a). Chronology of appearance of barley stripe mosaic virus protein in infected wheat cells. *J. Ultrastruct. Res.* **89,** 309–323.

Lin, N.-S., and Langenberg, W. G. (1984b). Distribution of barley stripe mosaic virus protein in infected wheat root and shoot tips. *J. Gen. Virol.* **65,** 2217–2224.

Lin, N.-S., and Langenberg, W. G. (1985). Peripheral vesicles in proplastids of barley stripe mosaic virus-infected wheat cells contain double-stranded RNA. *Virology* **142,** 291–298.

Lin, N.-S., Hsu, Y.-H., and Chiu, R.-J. (1987). Identification of viral structural proteins in the nucleoplasm of potato yellow dwarf virus-infected cells. *J. Gen. Virol.* **68,** 2723–2728.

Linstead, P. J., Hills, G. J., Plaskitt, K. A., Wilson, I. G., Harker, C. L., and Maule, A. J. (1988). The subcellular location of the gene 1 product of cauliflower mosaic virus is consistent with a function associated with virus spread. *J. Gen. Virol.* **69,** 1809–1818.

Linthorst, H. J. M., and Kaper, J. M. (1984). Circular satellite-RNA molecules in satellite of tobacco ringspot virus-infected tissue. *Virology* **137,** 206–210.

Linthorst, H. J. M., and Kaper, J. M. (1985). Cucumovirus satellite RNAs cannot replicate autonomously in cowpea protoplasts. *J. Gen. Virol.* **66,** 1839–1842.

Linthorst, H. J. M., Meuwissen, R. L. J., Kayffmann, S., and Bol., J. F. (1989). Constitutive expression of pathogenesis-related PR-1, GRP, and PR-S in tobacco has no effect on virus infection. *Plant Cell* **1,** 285–291.

Lisa, V., Luisoni, E., and Milne, R. G. (1986). Carnation cryptic virus. *AAB Descriptions Plant Viruses* No. 315.

Lister, R. M. (1960). Transmission of soil borne virus through seed. *Virology* **10,** 547–549.

Lister, R. M. (1966). Possible relationship of virus specific products of tobacco rattle virus infections. *Virology* **28,** 350–353.

Lister, R. M. (1968). Functional relationships between virus-specific products of infection by viruses of the tobacco rattle type. *J. Gen. Virol.* **2,** 43–58.

Litvak, S., Carré, D. S., and Chapeville, F. (1970). TYMV RNA as a substrate of the tRNA nucleotidyltransferase. *FEBS Lett.* **11,** 316–319.

Litvak, S., Tarragó, A., Tarragó-Litvak, L., and Allende, J. E. (1973). Elongation factor viral genome interaction dependent on the aminoacylation of TYMV and TMV RNA's *Nature (London), New Biol.* **241,** 88–90.

Lloyd, A. M., Barnason, A. R., Rogers, S. G., Byrne, M. C. Fraley, R. T., and Horsch, R. B. (1986). Transformation of *Arabidopsis thaliana* with *Agrobacterium tumefaciens. Science* **234,** 464–466.

Lockhart, B. E. L. (1990). Evidence for a double-stranded circular DNA genome in a second group of plant viruses. *Phytopathology* **80,** 127–131.

Loebenstein, G., Cohen, J., Shabtai, S., Coutts, R. H. A., and Wood, K. R. (1977). Distribution of cucumber mosaic virus in systemically infected tobacco leaves. *Virology* **81,** 117–125.

Loesch-Fries, L. S., and Hall, T. C. (1980). Synthesis, accumulation and encapsidation of individual brome mosaic virus RNA components in barley protoplasts. *J. Gen. Virol.* **47,** 323–332.

Loesch-Fries, L. S., and Hall, T. C. (1982). *In vivo* aminoacylation of brome mosaic and barley stripe mosaic virus RNAs. *Nature (London)* **298,** 771–773.

Loesch-Fries, L. S., Jarvis, N. P., Krahn, K. J., Nelson, S. E., and Hall, T. C. (1985). Expression of alfalfa mosaic virus RNA4 cDNA transcripts *in vitro* and *in vivo. Virology* **146,** 177–187.

Loesch-Fries, L. S., Merlo, D., Zinnen, T., Burhop, L., Hill, K., Krahn, K., Jarvis, N., Nelson, S., and Halk, E. (1897). Expression of alfalfa mosaic virus RNA4 in transgenic plants confers resistance. *EMBO J.* **6,** 1845–1852.

Logan, A. E., and Zettler, F. W. (1985). Rapid *in vitro* propagation of virus-indexed gladioli. *Acta Hortic.* **164,** 169–180.

Lok, S., and AbouHaidar, M. G. (1986). The nucleotide sequence of the 5 end of papaya mosaic virus RNA: Site of *in vitro* assembly initiation. *Virology* **153,** 289–296.

Lommel, S. A., Weston-Fina, M., Xiong, Z., and Lomonossoff, G. P. (1988). The nucleotide sequence and gene organisation of red clover necrotic mosaic virus RNA2. *Nucleic Acids Res.* **16,** 8587–8602.

Lomonossoff, G. P., and Shanks, M. (1983). The nucleotide sequence of cowpea mosaic virus B RNA. *EMBO J.* **2,** 2253–2258.

Lomonossoff, G. P., and Wilson, T. M. A. (1985). Structure and *in vitro* assembly of tobacco mosaic virus. *In* "Molecular Plant Virology" (J. W. Davies, ed.), Vol. 1, pp. 43–83. CRC Press, Boca Raton, Florida.

Lomonossoff, G. P., Shanks, M., and Evans, D. (1985). The structure of cowpea mosaic virus replicative form RNA. *Virology* **144,** 351–362.

Long, D. G., Borsa, J., and Sargent, M. D. (1976). A potential artifact generated by pelleting viral particles during preparative ultracentrifugation. *Biochim. Biophys. Acta* **451,** 639–642.

Loor, F. (1967). Comparative immunogenicities of tobacco mosaic virus protein subunits, and reaggregated protein subunits. *Virology* **33,** 215–220.

Lot, H., and Kaper, J. M. (1976). Physical and chemical differentiation of three strains of cucumber mosaic virus and peanut stunt virus. *Virology* **74,** 209–222.

Louie, R., and Knoke, J. K. (1975). Strains of maize dwarf mosaic virus. *Plant Dis. Rep.* **59,** 518–522.

Louis, R., and Lorbeer, J. W. (1966). Mechanical transmission of onion yellow dwarf virus. *Phytopathology* **56,** 1020–1023.

Louie, R., Findley, W. R., and Knoke, J. K. (1976). Variation in resistance within corn inbred lines to infection by maize dwarf mosaic virus. *Plant Dis. Rep.* **60,** 838–842.

Louie, R., Knoke, J. K., and Reichard, D. L. (1983). Transmission of maize dwarf mosaic virus with solid-stream inoculum. *Plant Dis.* **67,** 1328–1331.

Lovisolo, O. (1985). International transport of flowers, foliage, nursery stock, and ornamental plants in Europe and the Mediterranean basin. *Acta Hortic.* **164,** 139–151.

Lowe, A. D. (1964). The ecology of the cereal aphid in Canterbury. *Proc. N.Z. Weed Pest Control Conf.* **17,** 175–186.

Lozoya-Saldana, H., and Dawson, W. O. (1982). Effect of alternating temperature regimes on reduction or elimination of viruses in plant tissues. *Phytopathology* **72,** 1059–1064.

Luhn, C. F., and Goheen, A. C. (1970). Viruses in early California grapevines. *Plant Dis. Rep.* **54,** 1055–1056.

Luisoni, E., Milne, R. G., and Boccardo, G. (1975). The maize rough dwarf virion. II. Serological analysis. *Virology* **68,** 86–96.

Lundquist, R. E., Lazar, J. M., Klein, W. H., and Clark, J. M., Jr. (1972). Translation of satellite tobacco necrosis virus ribonucleic acid. II. Initiation of *in vitro* translation in procaryotic and eucaryotic systems. *Biochemistry* **11,** 2014–2019.

Lütke, H. A., Chow, K. C., Mickel, F. S., and Moss, K. A., Kern, H. F., and Scheele, G. A. (1987). Selection of AUG initiation codons differs in plants and animals. *EMBO J.* **6,** 43–48.

MacDonald, R. J. H., Coutts, R. H. A., and Buck, K. W. (1988a). Characterization of a subgenomic DNA isolated from *Triticum aestivium* plants infected with wheat dwarf virus. *J. Gen. Virol.* **69,** 1339–1344.

MacDonald, R. J. H., Coutts, R. H. A., and Buck, K. W. (1988b). Priming of complementary DNA synthesis *in vitro* by small DNA molecules tightly bound to virion DNA of wheat dwarf virus. *J. Gen. Virol.* **69,** 1345–1350.

MacDowell, S. W., MacDonald, R. J. H., Hamilton, W. D. O., Coutts, R. H. A., and Buck, K. W. (1985). The nucleotide sequence of cloned wheat dwarf virus DNA. *EMBO J.* **4,** 2173–2180.

MacDowell, S. W., Coutts, R. H. A., and Buck, K. W. (1986). Molecular characterisation of sub-genomic single-stranded and double-stranded DNA forms isolated from plants infected with tomato golden mosaic virus. *Nucleic Acids Res.* **14,** 7967–7984.

MacFarlane, S. A., Taylor, S. C., King, D. I., Hughes, G., and Davies, J. W. (1989). Pea early browning virus RNA1 encodes four polypeptides including a putative zinc-finger protein. *Nucleic Acids Res.* **17,** 2245–2260.

MacKenzie, D. J., and Tremaine, J. H. (1986). The use of a monoclonal antibody specific for the N-terminal region of southern bean mosaic virus as a probe of virus structure. *J. Gen. Virol.* **67,** 727–735.

MacKenzie, D. J., Tremaine, J. H., and Stace-Smith, R. (1989). Organization and interviral homologies of the 3′-terminal portion of potato virus S. *J. Gen. Virol.* **70,** 1053–1063.

MacKenzie, D. R., Cole, H., Smith, C. B., and Ercegovich, C. (1970). Effects of atrazine and maize dwarf mosaic virus infection on weight and macro and micro element constituents of maize seedlings in the greenhouse. *Phytopathology* **60,** 272–279.

Mackie, G. A., Johnston, R., and Bancroft, J. B. (1988). Single- and double-stranded viral RNAs in plants infected with the potexviruses papaya mosaic virus and foxtail mosaic virus. *Intervirology* **29,** 170–177.

MacLeod, R., Black, L. M., and Moyer, F. H. (1966). The fine structure and intracellular localisation of potato yellow dwarf virus. *Virology* **29,** 540–552.

Macnicol, P. K. (1976). Rapid metabolic changes in the wounding response of leaf discs following excision. *Plant Physiol.* **57,** 80–84.

Madden, L. V., and Campbell, C. L. (1986). Descriptions of virus disease epidemics in time and space. *In* "Plant Virus Epidemics" (G. D. McLean, R. G. Garrett, and W. G. Ruesink, eds.), pp. 273–293. Academic Press, Orlando, Florida.

Madden, L. V., Louie, R., and Knoke, J. K. (1987a). Temporal and spatial analysis of maize dwarf mosaic epidemics. *Phytopathology* **77,** 148–156.

Madden, L. V., Pirone, T. P., and Raccah, B. (1987b). Analysis of spatial patterns of virus-diseased tobacco plants. *Phytopathology* **77,** 1409–1417.

Maeda, T., and Inouye, N. (1985). Insolubilization of cucumber mosaic virus with glutaraldehyde and its use for isolation of specific antibody. *Ann. Phytopath. Soc. Jpn.* **51,** 312–314.

Magyarosy, A. C., and Duffus, J. E. (1977). The occurrence of highly virulent strains of the beet curly top virus in California. *Plant Dis. Rep.* **61,** 248–251.

Magyarosy, A. C., Buchanan, B. B., and Schürmann, P. (1973). Effect of a systemic virus infection on chloroplast function and structure. *Virology* **55,** 426–438.

Maiss, E., Timpe, U., Brisske, A., Jelkmann, W., Casper, R., Himmler, G., Mattanovich, D., and Katinger, H. W. D. (1989). The complete nucleotide sequence of plum pox virus RNA. *J. Gen. Virol.* **70,** 513–524.

Malyshenko, S. I., Lapchic, L. G., Kondakova, O. A., Kuznetzova, L. L., Taliansky, M. E., and Atabekov, J. G. (1988). Red clover mottle comovirus B-RNA spreads between cells in to-bamovirus-infected tissues. *J. Gen. Virol.* **69,** 407–412.

Mandelkow, E., Stubbs, G., and Warren, S. (1981). Structures of the helical aggregates of tobacco mosaic virus protein. *J. Mol. Biol.* **152,** 375–386.

Mandryk, M. (1963). Acquired systemic resistance to tobacco mosaic virus in *Nicotiana tabacum* evoked by stem injection with *Perenospora tabacina.* *Adam. Aust. J. Agric. Res.* **14,** 315–318.

Mang, K., Gosh, A., and Kaesberg, P. (1982). A comparative study of the cowpea and bean strains of southern bean mosaic virus. *Virology* **116,** 264–274.

Manhart, J. R., and Palmer, J. D. (1990). The gain of two chloroplast tRNA introns marks the green algal ancestors of land plants. *Nature, London* **345,** 268–270.

Maramorosch, K., and Harris, K. F., eds. (1981). "Plant Diseases and Vectors: Ecology and Epidemiology." Academic Press, New York.

Maramorosch, K., and Raychaudhuri, S. P., eds. (1988). "Mycoplasma Diseases of Crops. Basic and Applied Aspects." Springer-Verlag, New York.

Marbrook, J., and Matthews, R. E. F. (1966). The differential immunogenicity of plant virus proteins and nucleoproteins. *Virology* **28,** 219–228.

Marcinka, K., and Musil, M. (1977). Disintegration of red clover mottle virus virions under different conditions of storage *in vitro.* *Acta Virol.* **21,** 71–78.

Marco, S., and Levy, D. (1979). Involvement of ethylene in the development of cucumber mosaic virus-induced chlorotic lesions in cucumber cotyledons. *Physiol. Plant Pathol.* **14,** 235–244.

Marco, S., Levy, D., and Aharoni, N. (1976). Involvement of ethylene in the suppression of hypocotyl elongation in CMV-infected cucumbers. *Physiol. Plant Pathol.* **8,** 1–7.

Marinos, N. G. (1967). Multifunctional plastids in the meristematic region of potato tuber buds. *J. Ultrastruct. Res.* **17,** 91–113.

Markham, P. G., and Hull, R. (1985). Cauliflower mosaic virus aphid transmission facilitated by transmission factors from other caulimoviruses. *J. Gen. Virol.* **66,** 921–923.

Markham, P. G., Pinner, M. S., Raccah, B., and Hull, R. (1987). The acquisition of a caulimovirus by different aphid species: Comparison with a potyvirus. *Ann. Appl. Biol.* **111,** 571–587.

Markham, P. G., Pinner, M. S., Mesfin, T., Nebbache, S., Briddon, R., and Medina, V. (1988). Geminiviruses and their interaction with vectors. *AFRC Institute of Plant Science Research and John Innes Institute Annual Report,* 63–67.

Markham, R. (1951). Physicochemical studies on the turnip yellow mosaic virus. *Discuss. Faraday Soc.* **11,** 221–227.

Markham, R. (1962). The analytical centrifuge as a tool for the investigation of plant viruses. *Adv. Virus Res.* **9,** 241–270.

Markham, R., and Smith, J. D. (1951). Chromatographic studies on nucleic acids. 4. The nucleic acid of the turnip yellow mosaic virus, including a note on the nucleic acid of tomato bushy stunt virus. *Biochem. J.* **49,** 401–406.

Markham, R., and Smith, K. M. (1949). Studies on the virus of turnip yellow mosaic. *Parasitology* **39,** 330–342.

Markham, R., Matthews, R. E. F., and Smith, K. M. (1948). Testing potato stocks for virus X. *Farming* February, pp. 40–46.

Markham, R., Frey, S., and Hills, G. J. (1963). Methods for enhancement of image detail and accentuation of structure in electron microscopy. *Virology* **20,** 88–102.

Markham, R., Hitchborn, J. H., Hills, G. J., and Frey, S. (1964). The anatomy of tobacco mosaic virus. *Virology* **22,** 342–359.

Marrou, J., Messiaen, C.-M., and Migliori, A. (1967). Méthode de contrôle de l'état sanitaire des graines de laitice. *Ann. Epiphyt.* **18,** 227–248.

Marrou, J., Quiot, J. B., Duteil, M., Labonne, G., Leclant, F., and Renoust, M. (1979). Ecology and epidemiology of cucumber mosaic virus. III. Interest of the exposure of bait plants in the study of cucumber mosaic virus dissemination. *Ann. Phytopathol.* **11,** 291–306.

Marsh, L. E., and Guifoyle, T. J. (1987). Cauliflower mosaic virus replication intermediates are encapsidated into virion-like particles. *Virology* **161,** 129–137.

Marsh, L. E., and Hall, T. C. (1987). Evidence implicating a tRNA heritage for the promoters of positive strand RNA synthesis in brome mosaic and related viruses. *Cold Spring Harbor Symp. Quant. Biol.* **52**, 331–341.

Marsh, L. E., Dreher, T. W., and Hall, T. C. (1988). Mutational analysis of the core and modulator sequences of the BMV RNA3 subgenomic promoter. *Nucleic Acids Res.* **16**, 981–995.

Marsh, L. E., Pogue, G. P., and Hall, T. C. (1989). Similarities among plant virus (+) and (−) RNA termini imply a common ancestry with promoters of eukaryotic +RNA. *Virology* **172**, 415–427.

Marsh, L. E., Pogue, G. P., Huntley, C. C., and Hall, T. C. (1990b). Insight to replication strategies and evolution of (+) strand RNA viruses provided by brome mosaic virus. *Oxford Surveys of Plant Molecular and Cell Biology,* in press.

Marshall, B., and Matthews, R. E. F. (1981). Okra mosaic virus protein shells in nuclei. *Virology* **110**, 253–256.

Martelli, G. P., and Castellano, M. A. (1971). Light and electron microscopy of the intracellular inclusions of cauliflower mosaic virus. *J. Gen. Virol.* **13**, 133–140.

Martelli, G. P., and Russo, M. (1981). The fine structure of *Cymbidium* ringspot virus in host tissues. *J. Ultrastruct. Res.* **77**, 93–104.

Martelli, G. P., Di Franco, A., and Russo, M. (1984). The origin of multivesicular bodies in tomato bushy stunt virus-infected *Gomphrena globosa* plants. *J. Ultrastruct. Res.* **88**, 275–281.

Martin, R. (1986). Use of double-stranded RNA for detection and identification of virus diseases of *Rubus* species. *Acta Hortic.* **186**, 51–62.

Martin, R. R., and Converse, R. H. (1982). An improved buffer for mechanical transmission of viruses from *Fragaria* and *Rubus*. *Acta Hortic.* **129**, 69–72.

Martin, T. J., Harvey, T. L., Bender, C. G., and Seifers, D. L. (1984). Control of wheat streak mosaic virus with vector resistance in wheat. *Phytopathology* **74**, 963–964.

Martinez-Izquierdo, J. A., and Hohn, T. (1987). Cauliflower mosaic virus coat protein is phosphorylated *in vitro* by a virion-associated protein kinase. *Proc. Natl. Acad. Sci. U.S.A.* **84**, 1824–1828.

Martinez-Izquierdo, J. A., Fütterer, J., and Hohn, T. (1987). Protein encoded by ORF1 of cauliflower mosaic virus is part of the viral inclusion body. *Virology* **160**, 527–530.

Marx, J. L. (1988). Multiplying genes by leaps and bounds. *Science* **240**, 1408–1410.

Marzachi, C., Milne, R. G., and Boccardo, G. (1988). *In vitro* synthesis of double-stranded RNA by carnation cryptic virus-associated RNA-dependent RNA polymerase. *Virology* **165**, 115–121.

Mason, W. S., Taylor, J. M., and Hull, R. (1987). Retroid virus genome replication. *Adv. Virus Res.* **32**, 35–96.

Massalski, P. R., and Harrison, B. D. (1987). Properties of monoclonal antibodies to potato leafroll luteovirus and their use to distinguish virus isolates differing in aphid transmissibility. *J. Gen. Virol.* **68**, 1813–1821.

Massalski, P. R., Cooper, J. I., Hellen, C. U. T., and Edwards, M. L. (1988). The effect of cherry leafroll virus infection on the performance of birch pollen and studies on virus replication in germinating pollen. *Ann. Appl. Biol.* **112**, 415–425.

Masters, P. S., and Banerjee, A. K. (1988). Replication of non-segmented negative strand RNA viruses. *In* "RNA Genetics" (E. Domingo, J. J. Holland, and P. Ahlquist, eds.), Vol. 1, pp. 137–158. CRC Press, Boca Raton, Florida.

Masuta, C., and Takanami, Y. (1989). Determination of sequence and structural requirements for pathogenicity of a cucumber mosaic virus satellite RNA (Y-sat RNA). *Plant Cell* **1**, 1165–1173.

Masuta, C., Zuidema, D., Hunter, B. G., Heaton, L. A., Sopher, D. S., and Jackson, A. O. (1987). Analysis of the genome of satellite panicum mosaic virus. *Virology* **159**, 329–338.

Masuta, C., Kuwata, S., and Takanami, Y. (1988a). Disease modulation on several plants by cucumber mosaic virus satellite RNA (Y strain). *Ann. Phytopathol. Soc. Jpn.* **54**, 332–336.

Masuta, C., Kuwata, S., and Takanami, Y. (1988b). Effects of extra 5′ non-viral bases on the infectivity of transcripts from a cDNA clone of satellite RNA (strain Y) of cucumber mosaic virus. *J. Biochem. (Tokyo)* **104**, 841–846.

Masuta, C., Komari, T., and Takanami, Y. (1989). Expression of cucumber mosaic virus satellite RNA from cDNA copies in transgenic tobacco plants. *Ann. Phytopathol. Soc. Jpn.* **55**, 49–55.

Mathon, M. P., Tavert, G., and Malato, G. (1987). Comparison of three methods for homogenising samples of plant material prior to ELISA testing. *OEPP/EPPO Bull.* **17**, 97–103.

Matisová, J. (1971). Alfalfa mosaic virus in lucerne plants and its transmission by aphids in the course of the vegetation period. *Acta Virol.* **15**, 411–420.

Matsubara, A., Kojima, M., Kawano, S., Narita, M., Hattori, M., Uyeda, I., and Shikata, E. (1985). Purification and serology of a Japanese isolate of barley yellow dwarf virus. *Ann. Phytopathol. Soc. Jpn.* **51**, 152–158.

Matsuoka, M., and Ohashi, Y. (1986). Induction of pathogenesis-related proteins in tobacco leaves. *Plant Physiol.* **80**, 505–510.

Matsuoka, M., Yamamoto, N., Kano-Murakami, Y., Tanaka, Y., Ozeki, Y., Hirano, H., Kagawa, H., Oshima, M., and Ohashi, Y. (1987). Classification and structural comparison of full-length cDNAs for pathogenesis-related proteins. *Plant Physiol.* **85**, 942–946.

Matthews, R. E. F. (1949a). Studies on potato virus X. I. Types of change in potato virus X infections. *Ann. Appl. Biol.* **36**, 448–459.

Matthews, R. E. F. (1949b). Studies on potato virus X. II. Criteria of relationships between strains. *Ann. Appl. Biol.* **36**, 460–474.

Matthews, R. E. F. (1949c). Reactions of *Cyphomandra betacea* to strains of potato virus X. *Parasitology* **39**, 241–244.

Matthews, R. E. F. (1949d). Studies on two plant viruses. Ph.D. Thesis, University of Cambridge.

Matthews, R. E. F. (1953a). Factors affecting the production of local lesions by plant viruses. I. Effect of time of day of inoculation. *Ann. Appl. Biol.* **40**, 377–383.

Matthews, R. E. F. (1953b). Factors affecting the production of local lesions by plant viruses. II. Some effects of light, darkness and temperature. *Ann. Appl. Biol.* **40**, 556–565.

Matthews, R. E. F. (1953c). Incorporation of 8-azaguanine into nucleic acid of tobacco mosaic virus. *Nature (London)* **171**, 1065–1066.

Matthews, R. E. F. (1954). Effects of some purine analogues on tobacco mosaic virus. *J. Gen. Microbiol.* **10**, 521–532.

Matthews, R. E. F. (1957). "Plant Virus Serology." Cambridge Univ. Press, London and New York.

Matthews, R. E. F. (1966). Reconstitution of turnip yellow mosaic virus RNA with TMV protein subunits. *Virology* **30**, 82–96.

Matthews, R. E. F. (1970). "Plant Virology." Academic Press, New York.

Matthews, R. E. F. (1973). Induction of disease by viruses, with special reference to turnip yellow mosaic virus. *Annu. Rev. Phytopathol.* **11**, 147–170.

Matthews, R. E. F. (1974). Some properties of TYMV nucleoproteins isolated in cesium chloride density gradients. *Virology* **60**, 54–64.

Matthews, R. E. F. (1975). A classification of virus groups based on the size of the particle in relation to genome size. *J. Gen. Virol.* **27**, 135–149.

Matthews, R. E. F. (1981). "Plant Virology," 2 ed. Academic Press, New York.

Matthews, R. E. F. (1982). "Classification and Nomenclature of Viruses," 4th Report of the International Committee for Taxonomy of Viruses. Karger, Basel.

Matthews, R. E. F. (1983a). The history of viral taxonomy. *In* "A Critical Appraisal of Viral Taxonomy" (R. E. F. Mathews, ed.), pp. 1–35. CRC Press, Boca Raton, Florida.

Matthews, R. E. F. (1983b). Future prospects for viral taxonomy. *In* "A Critical Appraisal of Viral Taxonomy" (R. E. F. Matthews, ed.), pp. 219–245. CRC Press, Boca Raton, Florida.

Matthews, R. E. F. (1983c). The origin of viruses from cells. *Int. Rev. Cytol., Suppl.* **15**, 245–280.

Matthews, R. E. F. (1985a). Viral taxonomy. *Microbiol. Sci.* **2**, 74–75.

Matthews, R. E. F. (1985b). Viral taxonomy for the non-virologist. *Annu. Rev. Microbiol.* **39**, 451–474.

Matthews, R. E. F. (1987). The changing scene in plant virology. *Annu. Rev. Phytopathol.* **25**, 10–23.

Matthews, R. E. F., and Sarkar, S. (1976). A light-induced structural change in chloroplasts of Chinese cabbage cells infected with turnip yellow mosaic virus. *J. Gen. Virol.* **33**, 435–446.

Matthews, R. E. F., and Witz, J. (1985). Uncoating of turnip yellow mosaic virus *in vivo. Virology* **144**, 318–327.

Matthews, R. E. F., Bolton, E. T., and Thompson, H. R. (1963). Kinetics of labelling of turnip yellow mosaic virus with P^{32} and S^{35}. *Virology* **19**, 179–189.

Mauch, F. C., and Staehelin, L. A. (1988). Subcellular localization of chitinase and β-1,3-glucanase in bean leaves. Functional implications for their involvement in plant–pathogen interactions. *J. Cell. Biochem., Suppl.* **12c**, 269.

Maule, A. J. (1983). Infection of protoplasts from several *Brassica* species with cauliflower mosaic virus following inoculation using polyethylene glycol. *J. Gen. Virol.* **64**, 2655–2660.

Maule, A. J. (1991). Spread of viruses within plants. *Proc. Phytochem. Soc. Eur., Int. Symp. Biochem. Mol. Biol. Plant–Pathog. Interact.* (in press).

Maule, A. J., Boulton, M. I., and Wood, R. K. (1980). Resistance of cucumber protoplasts to cucumber mosaic virus: A comparative study. *J. Gen. Virol.* **51**, 271–279.

Maule, A. J., Espinoza, A. M., Hull, R., Woolston, C. J., Wilson, I. G., and Vlak, J. (1988).

Expression of cauliflower mosaic virus non-structural genes in insect cells using a baculovirus vector. *Annu. Rep., John Innes Inst.*, pp. 61–62.

Maule, A. J., Harker, C. L., and Wilson, I. G. (1989). The pattern of accumulation of cauliflower mosaic virus-specific products in infected turnips. *Virology* **169**, 436–446.

Maurin, J., Ackermann, H. W., Lebeurier, G., and Lwoff, A. (1984). Un système des virus—1983. *Ann. Inst. Pasteur Virol.* **135E,** 105–110.

Maury, Y., Bossennec, J.-M., Boudazin, G., Hampton, R., Pietersen, G., and Macquire, J. (1987). Factors influencing ELISA evaluation of transmission of pea seed-borne mosaic virus in infected pea seed: Seed-group size and seed decortication. *Agronomie* **7,** 225–230.

Mayer, A. (1886). Ueber die Mosaikkrankheit des Tabaks. *Landwirtsch. Vers.-Stn.* **32,** 451–467.

Mayo, M. A., and Reddy, D. V. R. (1985). Translation products of RNA from Indian peanut clump virus. *J. Gen. Virol.* **66,** 1347–1351.

Mayo, M. A., Barker, H., and Harrison, B. D. (1982a). Specificity and properties of the genome-linked proteins of nepoviruses. *J. Gen. Virol.* **59,** 149–162.

Mayo, M. A., Barker, H., Robinson, D. J., Tamada, T., and Harrison, B. D. (1982b). Evidence that potato leafroll virus RNA is positive stranded, is linked to a small protein and does not contain polyadenylate. *J. Gen. Virol.* **59,** 163–167.

Mayo, M. A., Barker, H., and Robinson, D. J. (1982c). Satellite RNA in particles of strawberry latent ringspot virus. *J. Gen. Virol.* **63,** 417–423.

Mayo, M. A., Robinson, D. J., Jolly, C. A., and Hyman, L. (1989). Nucleotide sequence of potato leafroll luteovirus RNA. *J. Gen. Virol.* **70,** 1037–1051.

Mazzolini, L., Bonneville, J. M., Volovitch, M., Magazin, M., and Yot, P. (1985). Strand-specific viral DNA synthesis in purified viroplasms isolated from turnip leaves infected with cauliflower mosaic virus. *Virology* **145,** 293–303.

McClements, W. L., and Kaesberg, P. (1977). Size and secondary structure of potato spindle tuber viroid. *Virology* **76,** 477–484.

McDaniel, L. L., Ammar, E.-D., and Gordon, D. T. (1985) Assembly, morphology and accumulation of a Hawaiian isolate of maize mosaic virus. *Phytopathology* **75,** 1167–1172.

McDonald, J. G., and Hiebert, E. (1974). Ultrastructure of cylindrical inclusions induced by viruses of the potato Y group as visualised by freeze-etching. *Virology* **58,** 200–208.

McDonald, J. G., and Hiebert, E. (1975). Characterization of the capsid and cylindrical inclusion proteins of three strains of turnip mosaic virus. *Virology* **63,** 295–303.

McGovern, M. H., and Kuhn, C. W. (1984). A new strain of southern bean mosaic virus derived at low temperatures. *Phytopathology* **74,** 95–99.

McGuire, J. M., and Douthit, L. B. (1978). Host effect on acquisition and transmission of tobacco ringspot virus by *Xiphinema americanum*. *Phytopathology* **68,** 457–459.

McGuire, J. M., and Schneider, I. R. (1973). Transmission of satellite of tobacco ringspot virus by *Xiphinema americanum*. *Phytopathology* **63,** 1429–1430.

McGuire, J. M., Kim, K. S., and Douthit, L. B. (1970). Tobacco ringspot virus in the nematode *Xiphinema americanum*. *Virology* **42,** 212–216.

McInnes, J. L., and Symons, R. H. (1989). Nucleic acid probes in the diagnosis of plant viruses and viroids. *In* "Nucleic Acids Probes" (R. H. Symons, ed.). CRC Press, Boca Raton, Florida (in press).

McInnes, J. L., Habili, N., and Symons, R. H. (1989). Nonradioactive, photobiotin-labelled DNA probes for routine diagnosis of viroids in plant extracts. *J. Virol. Methods* **23,** 299–312.

McIntyre, J. L., Dodds, J. A., and Hare, J. D. (1981). Effects of localized infections of *Nicotiana tabacum* by tobacco mosaic virus on systemic resistance against diverse pathogens and an insect. *Phytopathology* **71,** 297–301.

McKinney, H. H. (1929). Mosaic diseases in the Canary Islands, West Africa and Gibraltar. *J. Agric. Res.* **39,** 557–578.

McKinney, H. H., Silber, G., and Greeley, L. W. (1965). Longevity of some plant viruses stored in chemically dehydrated tissues. *Phytopathology* **65,** 1043–1044.

McKlusky, D. J., and Stobbs, L. W. (1985). A modified local lesion assay procedure with improved sensitivity and reproducibility. *Can. J. Plant Pathol.* **7,** 347–350.

McLachlan, A. D., Bloomer, A. C., and Butler, P. J. G. (1980). Structural repeats and evolution of tobacco mosaic virus coat protein and RNA. *J. Mol. Biol.* **136,** 203–224.

McLaughlin, M. R., Barnett, O. W., Gibson, P. B., and Burrows, P. M. (1984). Enzyme-linked immunosorbent assay of viruses infecting forage legumes. *Phytopathology* **74,** 965–969.

McLean, D. L., and Kinsey, M. G. (1984). The precibarial valve and its role in the feeding behaviour of the pea aphid, *Acyrthosiphon pisum*. *Bull. Entomol. Soc. Am.* **30,** 26–31.

McLean, G. D., and Francki, R. I. B. (1967). Purification of lettuce necrotic yellows virus by column chromatography on calcium phosphate gel. *Virology* **31,** 585–591.

McLean, G. D., Burt, J. R., Thomas, D. W., and Sproul, A. N. (1982). The use of reflective mulch to reduce the incidence of watermelon mosaic virus in Western Australia. *Crop Prot.* **1,** 491–496.

McLean, G. D., Garrett, R. G., and Ruesink, W. G., eds. (1986). "Plant Virus Epidemics: Monitoring, Modelling and Predicting Outbreaks." Academic Press, Orlando, Florida.

McMullen, C. R., Gardner, W. S., and Myers, G. A. (1977). Ultrastructure of cell-wall thickenings and paramural bodies induced by barley stripe mosaic virus. *Phytopathology* **67,** 462–467.

McNaughton, P., and Matthews, R. E. F. (1971). Sedimentation of small viruses at very low concentrations. *Virology* **45,** 1–9.

McRitchie, J. J., and Alexander, L. J. (1963). Host-specific *Lycopersicon* strains of tobacco mosaic virus. *Phytopathology* **53,** 394–398.

Mead, D. A. Szczesna-Skorupa, E., and Kemper, B. (1986). Single-stranded "blue" T7 promotor plasmids: A versatile tandem promotor system for cloning and protein engineering. *Protein Eng.* **1,** 67–74.

Meints, R. H., Lee, K., and van Etten, J. L. (1986). Assembly site of the virus PBCV-1 in a *Chlorella*-like green alga: Ultrastructural studies. *Virology* **154,** 240–245.

Melander, W. R. (1975). Effect of aggregation on the kinetic properties of aspartate aminotransferase *Biochim. Biophys. Acta* **410,** 74–86.

Melchers, G. (1968). Techniques for the quantitative study of mutation in plant viruses. *Theor. Appl. Genet.* **38,** 275–279.

Melchers, G., Jockusch, H., and Sengbusch, P. V. (1966). A tobacco mutant with a dominant allele for hypersensitivity against some TMV-strains. *Phytopathol. Z.* **55,** 86–88.

Mellema, J. E., and Amos, L. A. (1972). Three-dimensional image reconstruction of turnip yellow mosaic virus. *J. Mol. Biol.* **72,** 819–822.

Mellema, J. E., and van den Berg, H. J. N. (1974). The quaternary structure of alfalfa mosaic virus. *J. Supramol. Struct.* **2,** 17–31.

Mellema, J.-R., Benicourt, C., Haenni, A.-L., Noort, A., Pleij, C. W. A., and Bosch, L. (1979). Translational studies with turnip yellow mosaic virus RNAs isolated from major and minor virus particles. *Virology* **96,** 38–46.

Ménissier, J., Lebeurier, G., and Hirth, L. (1982). Free cauliflower mosaic virus supercoiled DNA in infected plants. *Virology* **117,** 322–328.

Ménissier, J., de Murcia, G., Lebeurier, G., and Hirth, L. (1983). Electron microscopic studies of the different topological forms of the cauliflower mosaic virus DNA: Knotted encapsidated DNA and nuclear minichromosome. *EMBO J.* **2,** 1067–1071.

Ménissier, J., Laquel, P., Lebeurier, G., and Hirth, L. (1984). A DNA polymerase activity is associated with cauliflower mosaic virus. *Nucleic Acids Res.* **12,** 8769–8778.

Ménissier de Murcia, J., Geldreich, A., and Lebeurier, G. (1986). Evidence for a protein kinase activity associated with purified particles of cauliflower mosaic virus. *J. Gen. Virol.* **67,** 1885–1891.

Mernaugh, R. L., Gardner, W. S., and Yocom, K. L. (1980). Three dimensional structure of pinwheel inclusions as determined by analytical geometry. *Virology* **106,** 273–281.

Meshi, T., Ohno, T., Iba, H., and Okada, Y. (1981). Nucleotide sequence of a cloned cDNA copy of TMV (cowpea strain) RNA, including the assembly origin, the coat protein cistron, and the 3' noncoding region. *Mol. Gen. Genet.* **184,** 20–25.

Meshi, T., Kiyama, R., Ohno, T., and Okada, Y. (1983). Nucleotide sequence of the coat protein cistron and the 3' noncoding region of cucumber green mottle mosaic virus (watermelon strain) RNA. *Virology* **127,** 54–64.

Meshi, T., Ishikawa, M., Motoyoshi, F., Semba, K., and Okada, Y. (1986). *In vitro* transcription of infectious RNAs from full-length cDNAs of tobacco mosaic virus. *Proc. Natl. Acad. Sci. U.S.A.* **83,** 5043–5047.

Meshi, T., Watanabe, Y., Saito, T., Sugimoto, A., Maeda, T., and Okada, Y. (1987). Function of the 30 kD protein of tobacco mosaic virus: Involvement in cell-to-cell movement and dispensibility for replication. *EMBO J.* **6,** 2557–2563.

Meshi, T., Motoyoshi, F., Adachi, A., Watanabe, Y., Takamatsu, N., and Okada, Y. (1988). Two concomitant base substitutions in the putative replicase genes of tobacco mosaic virus confer the ability to overcome the effects of a tomato resistance gene, Tm-1. *EMBO J.* **7,** 1575–1581.

Meshi, T., Motoyoshi, F., Maeda, T., Yoshiwoka, S., Watanabe, H., and Okada, Y. (1989). Mutations in the tobacco mosaic virus 30-kD protein gene overcome Tm-2 resistance in tomato. *Plant Cell* **1,** 515–522.

Mesnard, J.-M., Kirchherr, D., Wurch, T., and Lebeurier, G. (1990). The cauliflower mosaic virus gene III product is a non-sequence-specific DNA binding protein. *Virology* **174,** 622–624.

Messieha, M. (1969). Transmission of tobacco ringspot virus by thrips. *Phytopathology* **59,** 943–945.

Meyer, M., Hemmer, O., and Fritsch, C. (1984). Complete nucleotide sequence of a satellite RNA of tomato black ring virus. *J. Gen. Virol.* **65,** 1575–1583.

Meyer, M., Hemmer, O., Mayo, M. A., and Fritsch, C. (1986). The nucleotide sequence of tomato black ring virus RNA-2. *J. Gen. Virol.* **67,** 1257–1271.

Meyer, T. E., Cusanovich, M. A., and Kamen, M. D. (1986). Evidence against use of bacterial amino acid sequence data for construction of all-inclusive phylogenetic trees. *Proc. Natl. Acad. Sci. U.S.A.* **83,** 217–220.

Miller, W. A., and Hall, T. C. (1984). RNA-dependent RNA polymerase isolated from cowpea chlorotic mottle virus-infected cowpeas is specific for bromoviral RNA. *Virology* **132,** 53–60.

Miller, W. A., Dreher, T. W., and Hall, T. C. (1985). Synthesis of brome mosaic virus subgenomic RNA *in vitro* by internal initiation on (−)-sense genomic RNA. *Nature (London)* **313,** 68–70.

Miller, W. A., Bujarski, J. J., Dreher, T. W., and Hall, T. C. (1986). Minus-strand initiation by brome mosaic virus replicase within the 3′ tRNA-like structure of native and modified RNA templates. *J. Mol. Biol.* **187,** 537–546.

Miller, W. A., Waterhouse, P. M., Kortt, A. A., and Gerlach, W. L. (1988a). Sequence and identification of the barley yellow dwarf virus coat protein gene. *Virology* **165,** 306–309.

Miller, W. A., Waterhouse, P. M., and Gerlach, W. L. (1988b). Sequence and organisation of barley yellow dwarf virus genomic RNA. *Nucleic Acids Res.* **16,** 6097–6111.

Mills, P. R. (1987). Comparison of cDNA hybridisation and other tests for the detection of potato mop top virus. *Potato Res.* **30,** 319–327.

Milne, R. G. (1967). Plant viruses inside cells. *Sci. Prog. (Oxford)* **55,** 203–222.

Milne, R. G. (1970). An electron microscope study of tomato spotted wilt virus in sections of infected cells and in negative stain preparations. *J. Gen. Virol.* **6,** 267–276.

Milne, R. G. (1984a). Electron microscopy for the identification of plant viruses in *in vitro* preparations. *Methods Virol.* **7,** 87–120.

Milne, R. G. (1984b). The species problem in plant virology. *Microbiol. Sci.* **1,** 113–118.

Milne, R. G. (1986). New developments in electron microscope serology and their possible applications. *In* "Developments and Applications in Virus Testing" (R. A. C. Jones and L. Torrance, eds.), Dev. Appl. Virol. I, pp. 179–191. Association of Applied Biologists, Wellsbourne.

Milne, R. G. (1988). Species concept should not be universally applied to virus taxonomy—But what to do instead? *Intervirology* **29,** 254–259.

Milne, R. G. (1990). Immunoelectron-microscopy for virus identification. *In* "Electron Microscopy of Plant Pathogens" (K. Mendgen and D. E. Lesemann, eds.). (Springer-Verlag, New York (in press).

Milne, R. G., and Francki, R. I. B. (1984). Should tomato spotted wilt virus be considered as a possible member of the family Bunyaviridae? *Intervirology* **22,** 72–76.

Milne, R. G., and Lesemann, D.-E. (1984). Immunoabsorbent electron microscopy in plant virus studies. *Methods Virol.* **8,** 85–101.

Milne, R. G., and Lovisolo, O. (1977). Maize rough dwarf and related viruses. *Adv. Virus Res.* **21,** 267–341.

Milne, R. G., and Luisoni, E. (1977). Rapid immune electron microscopy of virus preparations. *Methods Virol.* **6,** 265–281.

Milne, R. G., Conti, M., and Lisa, V. (1973). Partial purification, structure and infectivity of complete maize rough dwarf virus particles. *Virology* **53,** 130–141.

Mirkov, T. E., Mathews, D. M., Duplessis, D. H., and Dodds, J. A. (1989). Nucleotide sequence and translation of satellite tobacco mosaic virus RNA. *Virology* **170,** 139–146.

Miroshnichenko, N. A., Karpova, O. V., Morozov, W. Y., Rodionova, N. P., and Atabekov, J. G. (1988). Translation arrest of potato virus X RNA in Krebs-2 cell-free system: RNase H cleavage promoted by complementary oligodeoxynucleotides. *FEBS Lett.* **234,** 65–68.

Miura, K.-I., Kimura, I., and Suzuki, N. (1966). Double-stranded ribonucleic acid from rice dwarf virus. *Virology* **28,** 571–579.

Mizuno, A., Sano, T., Fujii, H., Miura, K., and Yazaki, K. (1986). Supercoiling of the genomic double-stranded RNA of rice dwarf virus. *J. Gen. Virol.* **67,** 2749–2755.

Moghal, S. M., and Francki, R. I. B. (1976). Towards a system for the identification and classification of potyviruses. I. Serology and amino acid composition of six distinct viruses. *Virology* **73,** 350–362.

Mohamed, N. A. (1981). Isolation and characterisation of subviral structures from tomato spotted wilt virus. *J. Gen. Virol.* **53,** 197–206.

Mohier, E., Pinck, L., and Hirth, L. (1974). Replication of alfalfa mosaic virus RNAs. *Virology* **58,** 9–15.

Mohier, E., Hirth, L., LeMeur, M.-A., and Gerlinger, P. (1976). Analysis of alfalfa mosaic virus 17S RNA translational products. *Virology* **71**, 615–618.

Moline, H. E., and Ford, R. E. (1974). Clover yellow mosaic virus infection of seedling roots of *Pisum sativum*. *Physiol. Plant Pathol.* **4**, 219–228.

Momma, T., and Takahashi, T. (1983). Cytopathology of shoot apical meristem of hop plants infected with hop stunt viroid. *Phytopathol. Z.* **106**, 272–280.

Monette, P. L. (1986). Elimination *in vitro* of two grapevine nepoviruses by an alternating temperature regime. *J. Phytopathol.* **116**, 88–91.

Monis, J., Scott, H. A., and Gergerich, R. C. (1986). Effect of beetle regurgitant on plant virus transmission using the gross wounding technique. *Phytopathology* **76**, 808–811.

Montalbini, P., and Lupattelli, M. (1989). Effect of localised and systemic tobacco mosaic virus infection on some photochemical and enzymatic activities of isolated chloroplasts. *Physiol. Mol. Plant Pathol.* **34**, 147–162.

Montasser, M. S., Tousignant, M. E., and Kaper, J. M. (1990). Satellite-mediated protection of tomato against cucumber mosaic virus. I. Greenhouse experiments and simulated epidemic conditions in the field. *Plant Dis.* (in press).

Montelaro, R. C., and Rueckert, R. R. (1975). Radiolabeling of proteins and viruses *in vitro* by acetylation with radioactive acetic anhydride. *J. Biol. Chem.* **250**, 1413–1421.

Montelius, I., Liljas, L., and Unge, T. (1988). Structure of EDTA-treated satellite tobacco necrosis virus at pH 6.5. *J. Mol. Biol.* **201**, 353–363.

Morch, M.-D., Zagórski, W., and Haenni, A.-L. (1982). Proteolytic maturation of the turnip-yellow-mosaic-virus polyprotein coded *in vitro* occurs by internal catalysis. *Eur. J. Biochem.* **127**, 259–265.

Morch, M.-D., and Joshi, R. L., Denial, T. M., and Haenni, A.-L. (1987). A new "sense" RNA approach to block viral RNA replication *in vitro*. *Nucleic Acids Res.* **15**, 4123–4130.

Morch, M.-D., Boyer, J.-C., and Haenni, A.-L. (1988). Overlapping open reading frames revealed by complete nucleotide sequencing of turnip yellow mosaic virus genomic RNA. *Nucleic Acids Res.* **16**, 6157–6173.

Morch, M.-D., Drugeon, G., Szafranski, P., and Haenni, A.-L. (1989). Proteolytic origin of the 15-kilodalton protein encoded by turnip yellow mosaic virus genomic RNA. *J. Virol.* **63**, 5153–5158.

Moreira, A., Jones, R. A. C., and Fribourg, C. E. (1978). A resistance breaking strain of potato virus X that does not cause local lesions in *Gomphrena globosa*. *Proc. Int. Congr. Plant Pathol., 3rd, 1978* Abstr., p. 56.

Mori, K., Hosokawa, D., and Watanabe, M. (1982). Studies on multiplication and distribution of viruses in plants by the use of fluorescent antibody technique. I. Multiplication and distribution of viruses in shoot apices. *Ann. Phytopathol. Soc. Jpn.* **48**, 433–443.

Morinaga, T., Ikegami, M., Arai, T., Yazaki, K., and Miura, K. (1988). Infectivity of cloned tandem dimer DNAs of bean golden mosaic virus. *J. Gen. Virol.* **69**, 897–902.

Morris, B. A. M., Richardson, K. A., Anderson, M. T., and Gardner, R. C. (1988). Cassava latent virus infections mediated by the Ti plasmids of *Agrobacterium tumefaciens* containing either monomeric or dimeric viral DNA. *Plant Mol. Biol.* **11**, 795–803.

Morris, C., Gallois, P., Copley, J., and Kreis, M. (1988). The 5′ flanking region of a barley B hordein gene controls tissue and developmental specific CAT expression in tobacco plants. *Plant Mol. Biol.* **10**, 359–366.

Morris, T. J., and Carrington, J. C. (1988). Carnation mottle virus and viruses with similar properties. *In* "The Plant Viruses" (R. Koenig, ed.), Vol. 3, pp. 73–112. Plenum, New York.

Morris, T. J., and Hillman, B. I. (1989). Defective interfering RNAs of a plant virus. *In* "Molecular Biology of Plant–Pathogen Interactions," pp. 185–197. Alan R. Liss, New York.

Morris-Krsinich, B. A. M., and Forster, R. L. S. (1983). Lucerne transient streak virus RNA and its translation in rabbit reticulocyte lysate and wheat germ extract. *Virology* **128**, 176–185.

Morris-Krsinich, B. A. M., and Hull, R. (1983). Replication of turnip rosette virus RNA in inoculated turnip protoplasts. *J. Gen. Virol.* **64**, 2661–2668.

Morris-Krsinich, B. A. M., Forster, R. L. S., and Mossop, D. W. (1983). The synthesis and processing of the nepovirus grapevine fanleaf virus proteins in rabbit reticulocyte lysate. *Virology* **130**, 523–526.

Morris-Krsinich, B. A. M., Mullineaux, P. M., Donson, J., Boulton, M. I., Markham, P. G., Short, M. N., and Davies, J. W. (1985). Bidirectional translation of maize streak virus DNA, and identification of the coat protein gene. *Nucleic Acids Res.* **13**, 7237–7256.

Mosch, W. H. M., Huttings, H., and Rast, A. T. B. (1973). Some chemical and physical properties of 18 tobacco mosaic virus isolates from tomato. *Neth. J. Plant Pathol.* **79**, 104–111.

Moser, O., Gagey, M.-J., Godefroy-Colburn, T., Stussi-Garaud, C., Ellwart-Tschürtz, M., Nitschko, H., and Mundry, K. W. (1988). The fate of the transport protein of tobacco mosaic virus in systemic and hypersensitive hosts. *J. Gen. Virol.* **69**, 1367–1373.

Mossop, D. W., and Francki, R. I. B. (1977). Association of RNA3 with aphid transmission of cucumber mosaic virus. *Virology* **81**, 177–181.

Mossop, D. W., and Francki, R. I. B. (1978). Survival of a satellite RNA *in vivo* and its dependence on cucumber mosaic virus for replication. *Virology* **86**, 562–566.

Mossop, D. W., and Francki, R. I. B. (1979a). The stability of satellite viral RNAs *in vivo* and *in vitro*. *Virology* **94**, 243–253.

Mossop, D. W., and Francki, R. I. B. (1979b). Comparative studies on two satellite RNAs of cucumber mosaic virus. *Virology* **95**, 395–404.

Motoyoshi, F., and Hull, R. (1974). The infection of tobacco protoplasts with pea enation mosaic virus. *J. Gen. Virol.* **24**, 89–99.

Motoyoshi, F., and Oshima, N. (1975). Infection with tobacco mosaic virus of leaf mesophyll protoplasts from susceptible and resistant lines of tomato. *J. Gen. Virol.* **29**, 81–91.

Motoyoshi, F., and Oshima, N. (1976). The use of tris—HCl buffer for inoculation of tomato protoplasts with tobacco mosaic virus. *J. Gen. Virol.* **32**, 311–314.

Motoyoshi, F., and Oshima, N. (1977). Expression of genetically controlled resistance to tobacco mosaic virus infection in isolated tomato leaf mesophyll protoplasts. *J. Gen. Virol.* **34**, 499–506.

Mouches, C., Candresse, T., and Bové, J. M. (1984). Turnip yellow mosaic virus RNA-replicase contains host and virus-encoded subunits. *Virology* **134**, 78–90.

Mühlbach, H.-P., and Sänger, H. L. (1979). Viroid replication is inhibited by α-amantin. *Nature (London)* **278**, 185–187.

Muller, G. W., and Costa, A. S. (1987). Search for outstanding plants in tristeza infected citrus orchards: The best approach to control the disease by preimmunization. *Phytophylactica* **19**, 197–198.

Müller, H. O. (1942). Die Ausmessung der Tiefe übermikroskopischer Objekte. *Kolloid-Z.* **99**(1), 6–28; *Chem. Abstr.* **37**, 3991 (1943).

Müller, K. O., and Munro, J. (1951). The reaction of virus infected potato plants to *Phytophthora infestans. Ann. Appl. Biol.* **38**, 765–773.

Mullin, R. H., Smith, S. H., Frazier, N. W., Schlegel, D. E., and McCall, S. R. (1974). Meristem culture frees strawberries of mild yellow edge, pallidosis, and mottle diseases. *Phytopathology* **64**, 1425–1429.

Mullineaux, P. M., Donson, J., Morris-Krsinich, B. A. M., Boulton, M. I., and Davies, J. W. (1984). The nucleotide sequence of maize streak virus DNA. *EMBO J.* **3**, 3063–3068.

Mullineaux, P. M., Donson, J., Stanley, J., Boulton, M. I., Morris-Krsinich, B. A. M., Markham, P. G., and Davies, J. W. (1985). Computer analysis identifies sequence homologies between potential gene products of maize streak virus and those of cassava latent virus and tomato golden mosaic virus. *Plant Mol. Biol.* **5**, 125–131.

Mullineaux, P. M., Boulton, M. I., Bowyer, P., van der Vlugt, R., Marks, M., Donson, J., and Davies, J. W. (1988). Detection of a non-structural protein of M_I 11,000 encoded by the virion DNA of maize streak virus. *Plant Mol. Biol.* **11**, 57-66.

Mumford, D. L. (1972). A new method of mechanically transmitting curly top virus. *Phytopathology* **62**, 1217–1218.

Mumford, D. L. (1977). Application of the latex flocculation serological assay to curly top virus. *Phytopathology* **67**, 949–952.

Mumford, D. L., and Thornley, W. R. (1977). Location of curly top virus antigen in bean, sugarbeet, tobacco, and tomato by fluorescent antibody staining. *Phytopathology* **67**, 1313–1316.

Mundry, K. W. (1957). Die abhängigkeit des auftretens neuer virusstämme von der kulturetemperatur der wirtspflanzen. *Z. Indukt. Abstamm.- Vererbungsl.* **88**, 407–426.

Mundry, K. W., Watkins, P. A. C., Ashfield, T., Plaskitt, K. A., Eisele-Walter, S., and Wilson, T. M. A. (1990). Complete uncoating of the 5′-leader sequence of tobacco mosaic virus RNA occurs rapidly and is required to initiate cotranslational virus disassembly *in vitro. J. Gen. Virol.* (in press).

Murakishi, H., Lesney, M., and Carlson, P. (1984). Protoplasts and plant viruses. *Adv. Cell Cult.* **3**, 1–55.

Murakishi, H. H. (1963). Transfer of virus by a direct leaf-to-leaf method. *Nature (London)* **198**, 312–313.

Murakishi, H. H., Hartmann, J. X., Beachy, R. N., and Pelcher, L. E. (1971). Growth curve and yield of tobacco mosaic virus in tobacco callus cells. *Virology* **43**, 62–68.

Murant, A. F. (1984). Helper dependence among persistent and semi-persistent aphid-borne viruses. *Phytoparasitica* **12,** 207.

Murant, A. F., and Kumar, I. K. (1990). Different variants of the satellite RNA of groundnut rosette virus are responsible for the chlorotic and green forms of groundnut rosette disease. *Ann. Appl. Biol.* **117,** in press.

Murant, A. F., and Lister, R. M. (1967). Seed-transmission in the ecology of nematode-borne viruses. *Ann. Appl. Biol.* **59,** 63–76.

Murant, A. F., and Mayo, M. A. (1982). Satellites of plant viruses. *Annu. Rev. Phytopathol.* **20,** 49–70.

Murant, A. F., and Taylor, C. E. (1965). Treatment of soil with chemicals to prevent transmission of tomato black ring and raspberry ringspot viruses by *Longidorus elongatus*. (de Man). *Ann. Appl. Biol.* **55,** 227–237.

Murant, A. F., Mayo, M. A., Harrison, B. D., and Goold, R. A. (1973). Evidence for two functional RNA species and a "satellite" RNA in tomato blackring virus. *J. Gen. Virol.* **19,** 275–278.

Murant, A. F., Chambers, J., and Jones, A. T. (1974). Spread of raspberry bushy dwarf virus by pollination, its association with crumbly fruit, and problems of control. *Ann. Appl. Biol.* **77,** 271–281.

Murant, A. F., Taylor, M., Duncan, G. H., and Raschké, J. H. (1981). Improved estimates of molecular weight of plant virus RNA by agarose gel electrophoresis and electron microscopy after denaturation with glyoxal. *J. Gen. Virol.* **53,** 321–332.

Murant, A. F., Waterhouse, P. M., Raschké, J. H., and Robinson, D. J. (1985). Carrot red leaf and carrot mottle viruses: Observations on the composition of the particles in single and mixed infections. *J. Gen. Virol.* **66,** 1575–1579.

Murant, A. F., Mayo, M. A., and Raschké, J. H. (1986). Some biochemical properties of raspberry bushy dwarf virus. *Acta Hortic.* **186,** 23–30.

Murant, A. F., Rajeshwari, R., Robinson, D. J., and Raschké, J. H. (1988a). A satellite RNA of groundnut rosette virus that is largely responsible for symptoms of groundnut rosette disease. *J. Gen. Virol.* **69,** 1479–1486.

Murant, A. F., Raccah, B., and Pirone, T. P. (1988b). Transmission by vectors. *In* "The Plant Viruses" (R. G. Milne, ed.), Vol. 4, The Filamentous Plant Viruses, pp. 237–273. Plenum, New York and London.

Murdock, D. J., Nelson, P. E., and Smith, S. H. (1976). Histopathological examination of *Pelargonium* infected with tomato ringspot virus. *Phytopathology* **66,** 844–850.

Nagaich, B. B., Upadhya, M. D., Prakash, O., and Singh, S. J. (1968). Cytoplasmically determined expression of symptoms of potato virus X crosses between species of *Capsicum*. *Nature (London)* **220,** 1341–1342.

Nagaraj, A. N. (1965). Immunofluorescence studies on synthesis and distribution of tobacco mosaic virus antigen in tobacco. *Virology* **25,** 133–142.

Nagaraj, A. N., and Black, L. M. (1961). Localisation of wound-tumor virus antigen in plant tumors by the use of fluorescent antibodies. *Virology* **15,** 289–294.

Nagy, F., Odell, J., Morelli, G., and Chua, N. (1986). Properties of expression of the 35S promoter from CaMV in transgenic tobacco plants. *In* "Biotechnology in Plant Science" (M. Zaitlin, P. Day, and A. Hollaender, eds.), pp. 227–236. Academic Press, Orlando, Florida.

Naidu, R. A., Krishnan, M., Nayudu, M. V., and Gnanam, A. (1986). Studies on peanut green mosaic virus infected peanut (*Arachis hypogaea* L.) leaves. III. Changes in the polypeptides of photosystem II particles. *Physiol. Mol. Plant Pathol.* **29,** 53–58.

Namba, K., and Stubbs, G. (1986). Structure of tobacco mosaic virus at 3.6 Å resolution: Implications for assembly. *Science* **231,** 1401–1406.

Namba, K., Caspar, D. L. D., and Stubbs, G. J. (1984). Computer graphics representation of levels of organization in tobacco mosaic virus structure. *Science* **227,** 773–776.

Namba, K., Caspar, D. L. D., and Stubbs, G. (1988). Enhancement and simplification of macromolecular images. *Biophys. J.* **53,** 469–475.

Namba, K., Pattanayek, R., and Stubbs, G. (1989). Visualization of protein–nucleic acid interactions in a virus. Refined structure of intact tobacco mosaic virus at 2.9 Å resolution by X-ray fibre diffraction. *J. Mol. Biol.* **208,** 307–325.

Nameth, S. T., and Dodds, J. A. (1985). Double-stranded RNAs detected in cucurbit varieties not inoculated with viruses. *Phytopathology* **75,** 1293.

Nameth, S. T., Dodds, J. A., Paulus, A. O., and Laemmlen, F. F. (1986). Cucurbit viruses of California: An every-changing problem. *Plant Dis.* **70,** 8–11.

Nassuth, A., and Bol, J. F. (1983). Altered balance of the synthesis of plus- and minus-strand RNAs induced by RNAs 1 and 2 of alfalfa mosaic virus in the absence of RNA3. *Virology* **124,** 75–85.

Nassuth, A., Alblas, F., and Bol, J. F. (1981). Localisation of genetic information involved in the replication of alfalfa mosaic virus. *J. Gen. Virol.* **53**, 207–214.

Nassuth, A., Ten Bruggencate, G., and Bol, J. F. (1983a). Time course of alfalfa mosaic virus RNA and coat protein synthesis in cowpea protoplasts. *Virology* **125**, 75–84.

Nassuth, A., Alblas, F., Van Der Geest, A. J. M., and Bol, J. F. (1983b). Inhibition of alfalfa mosaic virus RNA and protein synthesis by actinomycin D and cycloheximide. *Virology* **126**, 517–524.

Nault, L. R. (1987). Origin and evolution of Auchenorrhyncha-transmitted, plant infecting viruses. *In* "Leafhoppers and Plant Hoppers of Economic Importance" (M. R. Wilson and L. R. Nault, eds.), Proc. 2nd Workshop, pp. 131–149. CIE, London.

Nault, L. R. (1991). Transmission biology, vector specificity and evolution of planthopper transmitted plant viruses. *In* "Planthoppers. Their Ecology, Genetics and Management" (R. F. Denno and T. J. Perfect, eds.). Chapman and Hall, New York.

Nault, L. R., and Ammar, E. D. (1989). Leafhopper and plant hopper transmission of plant viruses. *Annu. Rev. Entomol.* **34**, 503–529.

Nault, L. R., and Gordon, D. T. (1988). Multiplication of maize stripe virus in *Peregrinus maidis.* *Phytopathology* **78**, 991–995.

Nault, L. R., and Madden, L. V. (1988). Phylogenetic relatedness of maize chlorotic dwarf virus leafhopper vectors. *Phytopathology* **78**, 1683–1687.

Nault, L. R., and Styer, W. E. (1972). Effects of sinigrin on host selection by aphids. *Entomol. Exp. Appl.* **15**, 423–437.

Navarro, L., Llacer, G., Cambra, M., Arregui, J. M., and Juarez, J. (1983). Shoot-tip grafting *in vitro* for elimination of viruses in peach plants (*Prunus persica* Batsch). *Acta Hortic.* **130**, 185–192.

Nejidat, A., and Beachy, R. N. (1989). Decreased levels of TMV coat protein in transgenic tobacco plants at elevated temperatures reduce resistance to TMV infection. *Virology* **173**, 531–538.

Nejidat, A., and Beachy, R. N. (1990). Transgenic tobacco plants expressing a coat protein gene of tobacco mosaic virus are resistant to some other tobamoviruses. *Molecular Plant-Microbe Interactions* **3**, 247–251.

Nelson, O., ed. (1988). "Plant Transposable Elements." Plenum, New York and London.

Nelson, R. S., Powell Abel, P., and Beachy, R. N. (1987). Lesions and virus accumulation in inoculated transgenic tobacco plants expressing the coat protein gene of tobacco mosaic virus. *Virology* **158**, 126–132.

Nelson, R. S., McCormick, S. M., Delannay, X., Dubé, P., Layton, J., Anderson, E.-J., Kaniewska, M., Proksch, R. K., Horsch, R. B., Rogers, S. G., Frayley, R. T., and Beachy, R. N. (1988). Virus tolerance, plant growth, and field performance of transgenic tomato plants expressing coat protein from tobacco mosaic virus. *Bio/Technology* **6**, 403–409.

Newcomb, E. H. (1967). Fine structure of protein storing plastids in bean root tips. *J. Cell Biol.* **33**, 143–163.

Niblett, C. L., Dickson, E., Fernow, K. H., Horst, R. K., and Zaitlin, M. (1978). Cross protection among four viroids. *Virology* **91**, 198–203.

Nickerson, K. W., and Lane, L. C. (1977). Polyamine content of several RNA plant viruses. *Virology* **81**, 455–459.

Nilsson-Tillgren, T., Kolehmainen-Sevéus, L., and von Wettstein, D. (1969). Studies on the biosynthesis of TMV. 1. A system approaching a synchronized virus synthesis in a tobacco leaf. *Mol. Gen. Genet.* **104**, 124–141.

Nishiguichi, M., and Motoyoshi, F. (1987). Resistance mechanisms of tobacco mosaic virus strains in tomato and tobacco. *In* "Plant Resistance to Viruses" (D. Evered and S. Harnett, eds.), pp. 38–56. Wiley, Chichester.

Nishiguchi, M., Motoyoshi, F., and Oshima, N. (1978). Behaviour of a temperature sensitive strain of tobacco mosaic virus in tomato leaves and protoplasts. *J. Gen. Virol.* **39**, 53–61.

Nishiguchi, M., Motoyoshi, F., and Oshima, N. (1980). Further investigation of a temperature-sensitive strain of tobacco mosaic virus: Its behaviour in tomato leaf epidermis. *J. Gen. Virol.* **46**, 497–500.

Nishiguchi, M., Kikuchi, S., Kiho, Y., Ohno, T., Meshi, T., and Okada, Y. (1985). Molecular basis of plant viral virulence; The complete nucleotide sequence of an attenuated strain of tobacco mosaic virus. *Nucleic Acids Res.* **13**, 5585–5590.

Nishiguchi, M., Langridge, W. H. R., Szalay, A. A., and Zaitlin, M. (1986). Electroporation-mediated infection of tobacco leaf protoplasts with tobacco mosaic virus RNA and cucumber mosaic virus RNA. *Plant Cell Rep.* **5**, 57–60.

Nitta, N., Takanami, Y., Kuwata, S., and Kubo, S. (1988). Inoculation with RNAs 1 and 2 of cucumber mosaic virus induces viral RNA replicase activity in tobacco mesophyll protoplasts. *J. Gen. Virol.* **69**, 2695–2700.

Nixon, H. L., and Gibbs, A. J. (1960). Electron microscope observations on the structure of turnip yellow mosaic virus. *J. Mol. Biol.* **2**, 197–200.

Noort, A., Van Den Dries, C. L. A. M., Pleij, C. W. A., Jaspars, E. M. J., and Bosch, L. (1982). Properties of turnip yellow mosaic virus in cesium chloride solutions: The formation of high-density components. *Virology* **120**, 412–421.

Nuorteva, P. (1962). Studies on the causes of the phytopathogenicity of *Calligypona pellucida* (F). Hom. Araeopidae). *Ann. Zool. Soc. Zool. Bot. Fenn. Vanamo* **23**(4), 1–58.

Nuss, D. L. (1984). Molecular biology of wound tumor virus. *Adv. Virus Res.* **29**, 57–90.

Nuss, D. L., and Dall, D. J. (1989). Structural and functional properties of plant reovirus genomes. *Adv. Virus Res.* **38**, 249–306.

Nuss, D. L., and Peterson, A. J. (1980). Expression of wound tumour virus gene products *in vivo* and *in vitro. J. Virol.* **34**, 532–541.

Nuss, D. L., and Peterson, A. J. (1981). Resolution and genome assignment of mRNA transcripts synthesized *in vitro* by wound tumor virus. *Virology* **114**, 399–404.

Nuss, D. L., and Summers, D. (1984). Variant dsRNAs associated with transmission-defective isolates of wound tumor virus represent terminally conserved remnants of genome segments. *Virology* **133**, 276–288.

Nutter, R. C., Scheets, K., Panganiban, L. C., and Lommel, S. A. (1989). The complete nucleotide sequence of the maize chlorotic mottle virus genome. *Nucleic Acids Res.* **17**, 3163–3177.

Odell, J. T., and Howell, S. H. (1980). The identification, mapping, and characterization of mRNA for P66, a cauliflower mosaic virus-coded protein. *Virology* **102**, 349–359.

Odell, J. T., Nagy, F., and Chua, N.-H. (1985). Identification of DNA sequences required for activity of the cauliflower mosaic virus 35S promoter. *Nature (London)* **313**, 810–812.

Odell, J. T., Knowlton, S., Lin, W., and Mauvais, C. J. (1988). Properties of an isolated transcription stimulating sequence derived from the cauliflower mosaic virus 35S promoter. *Plant Mol. Biol.* **10**, 263–272.

O'Donnell, I. J., Shukla, D. D., and Gough, K. H. (1982). Electro-blot radioimmunoassay of virus-infected plant sap—A powerful new technique for detecting plant viruses. *J. Virol. Methods* **4**, 19–26.

Odumosu, A. O., Homer, R. B., and Hull, R. (1981). Circular dichroism studies on Southern bean mosaic virus. *J. Gen. Virol.* **53**, 193–196.

Offord, R. E. (1966). Electron microscopic observations on the substructure of tobacco rattle virus. *J. Mol. Biol.* **17**, 370–375.

Ofori, F. A., and Francki, R. I. B. (1983). Evidence that maize wallaby ear disease is caused by an insect toxin. *Ann. Appl. Biol.* **103**, 185–189.

Ofori, F. A., and Francki, R. I. B. (1985). Transmission of leafhopper A virus, vertically through eggs and horizontally through maize in which it does not multiply. *Virology* **144**, 152–157.

Ogawa, M., and Sakai, F. (1984). A messenger RNA for tobacco mosaic virus coat protein in infected tobacco mesophyll protoplasts. *Phytopathol. Z.* **109**, 193–203.

Oh, C-S., and Carrington, J. C. (1989). Identification of essential residues in potyvirus proteinase HC-Pro by site-directed mutagenesis. *Virology* **173**, 692–699.

Ohashi, Y., and Matsuoka, M. (1987). Localization of pathogenesis-related proteins in the epidermis and intercellular spaces of tobacco leaves after their induction by potassium salicylate or tobacco mosaic virus infection. *Plant Cell Physiol.* **28**, 1227–1235.

Ohki, S. T., and Inouye, T. (1987). Use of Gelrite as a gelling agent in immunodiffusion test for identification of plant viral antigens. *Ann. Phytopathol. Soc. Jpn.* **53**, 557–561.

Ohki, S. T., Leps, W. T., and Hiruki, C. (1986). Effects of alfalfa mosaic virus infection on factors associated with symbiotic N_2 fixation in alfalfa. *Can. J. Plant Pathol.* **8**, 277–281.

Ohno, T., Ishikawa, M., Takamatsu, N., Meshi, T., Okada, Y., Sano, T., and Shikata, E. (1983a). *In vitro* synthesis of infectious RNA molecules from cloned hop stunt viroid complementary DNA. *Proc. Jpn. Acad., Ser. B* **59**, 251–254.

Ohno, T., Takamatsu, N., Meshi, T., Okada, Y., Nishiguchi, M., and Kiho, Y. (1983b). Single amino acid substitution in 30K protein of TMV defective in virus transport function. *Virology* **131**, 255–258.

Okada, Y. (1986a). Cucumber green mottle mosaic virus. *In* "The Plant Viruses" (M. H. V. Van Regenmortel and H. Fraenkel-Conrat, eds.), Vol. 2, pp. 267–281. Plenum, New York.

Okada, Y. (1986b). Molecular assembly of tobacco mosaic virus *in vitro. Adv. Biophys.* **22**, 95–149.

Okada, Y., Ohashi, Y., Ohno, T., and Nozu, Y. (1970). Sequential reconstitution of tobacco mosaic virus. *Virology* **42**, 243–245.

Okamoto, S., Machida, Y., and Takebe, I. (1988). Subcellular localization of tobacco mosaic virus minus strand RNA in infected protoplasts. *Virology* **167**, 194–200.

Okuno, T., and Furusawa, I. (1979). RNA polymerase activity and protein synthesis in brome mosaic virus-infected protoplasts. *Virology* **99,** 218–225.

Old, R. W., and Primrose, S. B. (1985). "Principles of Gene Manipulation," 3rd edition. Blackwell, Oxford.

O'Loughlin, G. T., and Chambers, T. C. (1967). The systemic infection of an aphid by a plant virus. *Virology* **33,** 262–271.

Olson, A. J., Bricogne, G., and Harrison, S. C. (1983). Structure of tomato bushy stunt virus. IV. The virus particle at 2.9 Å resolution. *J. Mol. Biol.* **171,** 61–93.

Olson, A. J., Tainer, J. A., and Getsoff, E. D. (1985). Computer graphics in the study of macromolecular interactions. *In* "Crystallography in Molecular Biology" (D. Moras, J. Drenth, B. Strandberg, D. Suck, and K. Wilson, eds.), pp. 131–139. Plenum, New York and London.

Olszewski, N., Hagen, G., and Guilfoyle, T. J. (1982). A transcriptionally active, covalently closed minichromosome of cauliflower mosaic virus DNA isolated from infected turnip leaves. *Cell (Cambridge, Mass.)* **29,** 395–402.

Omar, S. A. M., Bailiss, K. W., Chapman, G. P., and Mansfield, J. W. (1986). Effects of virus infection of faba bean on subsequent infection by *Uromyces viciae-fabae. Plant Pathol.* **35,** 535–543.

Omura, T., Minobe, Y., Matsuoka, M., Nozu, Y., Tsuchizaki, T., and Saito, Y. (1985). Location of structural proteins in particles of rice gall dwarf virus. *J. Gen. Virol.* **66,** 811–815.

Omura, T., Takahashi, Y., Shohara, K., Minobe, Y., Tsuchizaki, T., and Nozu, Y. (1986). Production of monoclonal antibodies against rice stripe virus for the detection of virus antigen in infected plants and viruliferous insects. *Ann. Phytopathol. Soc. Jpn.* **52,** 270–277.

Omura, T., Minobe, Y., and Tsuchizaki, T. (1988a). Nucleotide sequence of segment S10 of the rice dwarf virus genome. *J. Gen. Virol.* **69,** 227–231.

Omura, T., Kimura, I., Tsuchizaki, T., and Saito, Y. (1988b). Infection by rice gall dwarf virus of cultured monolayers of leafhopper cells. *J. Gen. Virol.* **69,** 429–432.

Ooshika, I., Watanabe, Y., Meshi, T., Okada, Y., Igano, K., Inouye, K., and Yoshida, N. (1984). Identification of the 30K protein of TMV by immunoprecipitation with antibodies directed against a synthetic peptide. *Virology* **132,** 71–78.

Oostergetel, G. T., Krijgsman, P. C. J., Mellema, J. E., Cusack, S., and Miller, A. (1981). Evidence for the absence of swelling of alfalfa mosaic virions. *Virology* **109,** 206–210.

Oostergetel, G. T., Mellema, J. E., and Cusack, S. (1983). Solution scattering study on the structure of alfalfa mosaic virus strain VRU. *J. Mol. Biol.* **171,** 157–173.

Orellana, R. G., Fan, F., and Sloger, C. (1978). Tobacco ringspot virus and *Rhizobium* interactions in soybean: Impairment of leghemoglobin accumulation and nitrogen fixation. *Phytopathology* **68,** 577–582.

Orlob, G. B. (1963). Reappraisal of transmission of tobacco mosaic virus by insects. *Phytopathology* **53,** 822–830.

Orlob, G. B. (1966). Feeding and transmission characteristics of *Aceria tulipae* Keifer as a vector of wheat streak mosaic virus. *Phytopathol. Z.* **55,** 218–238.

Orlob, G. B. (1968). Relationships between *Tetranychus urticae* Koch and some plant viruses. *Virology* **35,** 121–133.

Orlob, G. B., and Takahashi, Y. (1971). Location of plant viruses in two-spotted spider mite, *Tetranychus urticae* Koch. *Phytopathol. Z.* **72,** 21–28.

Orr, C. C., and Newton, O. H. (1971). Distribution of nematodes by wind. *Plant Dis. Rep.* **55,** 61–63.

Osaki, T., Yamada, M., and Inouye, T. (1985). Whitefly transmitted viruses from three plant species (abstract in Japanese). *Ann. Phytopathol. Soc. Jpn.* **51,** 82–83.

Osario-Keese, M. E., Keese, P., and Gibbs, A. (1989). Nucleotide sequence of the genome of eggplant mosaic Tymovirus. *Virology* **172,** 547–554.

Osbourn, J. K., Plaskitt, K. A., Watts, J. W., and Wilson, T. M. A. (1989a). Tobacco mosaic virus coat protein and reporter gene transcripts containing the TMV origin-of-assembly sequence do not interact in double-transgenic tobacco plants: Implications for coat protein-mediated protection. *Mol. Plant–Microbe Interact.* **2,** 340–345.

Osbourn, J. K., Watts, J. W., Beachy, R. N., and Wilson, T. M. A. (1989b). Evidence that nucleocapsid disassembly and a later step in virus replication are inhibited in transgenic tobacco protoplasts expressing the TMV coat protein. *Virology* **172,** 370–373.

Osman, T. A. M., and Buck, K. W. (1987). Replication of red clover necrotic mosaic virus RNA in cowpea protoplasts: RNA1 replicates independently of RNA2. *J. Gen. Virol.* **68,** 289–296.

Osman, T. A. M., and Buck, K. W. (1989). Properties of three spontaneous mutants of red clover necrotic mosaic virus. *J. Gen. Virol.* **70,** 491–497.

Osman, T. A. M., Dodd, S. M., and Buck, K. W. (1986). RNA2 of red clover necrotic mosaic virus determines lesion morphology and systemic invasion of cowpea. *J. Gen. Virol.* **67,** 203–207.

Otsuki, Y., and Takebe, I. (1969). Fluorescent antibody staining of tobacco mosaic virus antigen in tobacco mesophyll protoplasts. *Virology* **38**, 497–499.

Otsuki, Y., and Takebe, I. (1976a). Double infection of isolated tobacco mesophyll protoplasts by unrelated plant viruses. *J. Gen. Virol.* **30**, 309–316.

Otsuki, Y., and Takebe, I. (1976b). Interaction of tobacco mosaic virus strains in doubly infected tobacco protoplasts. *Ann. Microbiol. (Paris)* **127**, 21 (abstr.).

Otsuki, Y., and Takebe, I. (1978). Production of mixedly coated particles in tobacco mesophyll protoplasts doubly infected by strains of tobacco mosaic virus. *Virology* **84**, 162–171.

Otsuki, Y., Takebe, I., Honda, Y., and Matsui, C. (1972a). Ultrastructure of infection of tobacco mesophyll protoplasts by tobacco mosaic virus. *Virology* **49**, 188–194.

Otsuki, Y., Shimomura, T., and Takebe, I. (1972b). Tobacco mosaic virus multiplication and expression of the N gene in necrotic responding tobacco varieties. *Virology,* **50**, 45–50.

Otsuki, Y., Takebe, I., Honda, Y., Kajita, S., and Matsui, C. (1974). Infection of tobacco mesophyll protoplasts by potato virus X. *J. Gen. Virol.* **22**, 375–385.

Otsuki, Y., Takebe, I., Ohno, T., Fukuda, M., and Okada, Y. (1977). Reconstitution of tobacco mosaic virus rods occurs bidirectionally from an internal initiation region: Demonstration by electron microscopic serology. *Proc. Nat. Acad. Sci. U.S.A.* **74**, 1913–1917.

Ouchterlony, O. (1962). Diffusion-in-gel methods for immunological analysis. II. *Prog. Allergy* **6**, 30–154.

Ow, D. W., Jacobs, J. D., and Howell, S. H. (1987). Functional regions of the cauliflower mosaic virus 35S RNA promoter determined by use of the firefly luciferase gene as a reporter of promoter activity. *Proc. Natl. Acad. Sci. U.S.A.* **84**, 4870–4874.

Ow, D. W., Wood, K. V., DeLuca, M., De Wet, J. R., Helinski, D. R., and Howell, S. H. (1986). Transient and stable expression of the firefly luciferase gene in plant cells and transgenic plants. *Science* **234**, 856–859.

Owen, J., and Palukaitis, P. (1988). Characterisation of cucumber mosaic virus. I. Molecular heterogeneity mapping of RNA3 in eight CMV strains. *Virology* **166**, 495–502.

Owens, R. A., and Bruening, G. (1975). The pattern of amino acid incorporation into two cowpea mosaic virus proteins in the presence of ribosome-specific protein synthesis inhibitors. *Virology* **64**, 520–530.

Owens, R. A., and Cress, D. E. (1980). Molecular cloning and characterisation of potato spindle tuber viroid cDNA sequences. *Proc. Natl. Acad. Sci. U.S.A.* **77**, 5302–5306.

Owens, R. A., and Diener, T. O. (1981). Sensitive and rapid diagnosis of potato spindle tuber viroid disease by nucleic acid hybridization. *Science* **213**, 670–672.

Owens, R. A., and Diener, T. O. (1982). RNA intermediates in potato spindle tuber viroid replication. *Proc. Natl. Acad. Sci. U.S.A.* **79**, 113–117.

Owens, R. A., and Diener, T. O. (1984). Spot hybridization for detection of viroids and viruses. *Methods Virol.* **7**, 173–187.

Owens, R. A., and Kaper, J. M. (1977). Cucumber mosaic virus-associated RNA5. II. *In vitro* translation in a wheat germ protein-synthesis system. *Virology* **80**, 196–203.

Owens, R. A., and Schneider, I. R. (1977). Satellite of tobacco ringspot virus RNA lacks detectable mRNA activity. *Virology* **80**, 222–224.

Owens, R. A., Hammond, R. W., Gardner, R. C., Kiefer, M. C., Thompson S. M., and Cress, D. E. (1986). Site-specific mutagenesis of potato spindle tuber viroid cDNA: Alterations within premelting region 2 that abolish infectivity. *Plant Mol. Biol.* **6**, 179–192.

Owens, R. A., Candresse, T., and Diener, T. O. (1990). Construction of novel viroid chimeras containing portions of tomato apical stunt and citrus exocortis viroids. *Virology* **175**, 238–246.

Owusu, G. K., Crowley, N. C., and Francki, R. I. B. (1968). Studies of the seed-transmission of tobacco ringspot virus. *Ann. Appl. Biol.* **61**, 195–202.

Oxelfelt, P. (1970). Development of systemic tobacco mosaic virus infection. I. Initiation of infection and time course of virus multiplication. *Phytopathol. Z.* **69**, 202–211.

Oxelfelt, P. (1975). Development of systemic tobacco mosaic virus infection. IV. Synthesis of viral RNA and intact virus and systemic movement of two strains as influenced by temperature. *Phytopathol. Z.* **83**, 66–76.

Oxelfelt, P. (1976) Biological and physicochemical characteristics of three strains of red clover mottle virus. *Virology* **74**, 73–80.

Pacha, R. F., Allison, R. F., and Ahlquist, P. (1990). *Cis*-acting sequences required for *in vivo* amplification of genomic RNA3 are organised differently in related bromoviruses. *Virology* **174**, 436–443.

Paduch-Cichal, E., and Kryczynski, S. (1987). A low temperature therapy and meristem-tip culture for eliminating four viroids from infected plants. *J. Phytopathol.* **118**, 341–346.

Paguio, O. R., Kuhn, C. W., and Boerma, H. R. (1988). Resistance-breaking variants of cowpea chlorotic mottle virus in soybean. *Plant Dis.* **72**, 768–770.

Paje-Manalo, L. L., and Lommel, S. A. (1989). Independent replication of red clover necrotic mosaic virus RNA-1 in electroporated host and nonhost *Nicotiana* species protoplasts. *Phytopathology* **79**, 457–461.

Paliwal, Y. C. (1980). Relationship of wheat streak mosaic and barley stripe mosaic viruses to vector and nonvector eriophyid mites. *Arch. Virol.* **63**, 123–132.

Paliwal, Y. C., and Andrews, C. J. (1979). Effects of barley yellow dwarf and wheat spindle streak mosaic viruses on cold hardiness of cereals. *Can. J. Plant Pathol.* **1**, 71–75.

Palomar, M. K., and Brakke, M. K. (1976). Concentration and infectivity of barley stripe mosaic virus in barley. *Phytopathology* **66**, 1422–1426.

Paludan, N. (1985). Spread of viruses by a recirculating water supply in soilless culture. *Phytoparasitica* **13**, 276.

Palukaitis, P. (1984). Detection and characterisation of subgenomic RNA in plant viruses. *Methods Virol.* **7**, 259–317.

Palukaitis, P. (1987). Potato spindle tuber viroid: Investigation of the long-distance, intra-plant transport route. *Virology* **158**, 239–241.

Palukaitis, P. (1988). Pathogenicity regulation by satellite RNAs of cucumber mosaic virus: Minor nucleotide sequence changes alter host responses. *Mol. Plant–Microbe Interaction.* **1**, 175–181.

Palukaitis, P., and Zaitlin, M. (1984a). Satellite RNAs of cucumber mosaic virus: Characterization of two new satellites. *Virology* **132**, 426–435.

Palukaitis, P., and Zaitlin, M. (1984b). A model to explain the "cross protection" phenomenon shown by plant viruses and viroids. *In* "Plant–Microbe Interactions" (T. Kosuge and E. W. Nester, eds.), Vol. 1, pp. 420–429. Macmillan, New York.

Palukaitis, P., and Zaitlin, M. (1987). The nature and biological significance of linear potato spindle tuber viroid molecules. *Virology* **157**, 199–210.

Palukaitis, P., Rakowski, A. G., Alexander, D. McE., and Symons, R. H. (1981). Rapid indexing of the sunblotch disease of avocados using a complementary DNA probe to avocado sunblotch viroid. *Ann. Appl. Biol.* **98**, 439–449.

Palukaitis, P., García-Arenal, F., Sulzinski, M. A., and Zaitlin, M. (1983). Replication of tobacco mosaic virus. VII. Further characterisation of single- and double-stranded virus-related RNAs from TMV-infected plants. *Virology* **131**, 533–545.

Parejarlarn, A., Lapis, D. B., and Hibino, H. (1984). Reaction of rice varieties to rice ragged stunt virus (RSV)infection by three known plant hopper (BPH) biotypes. *Int. Rice Res. Newsl.* **9**, 7–8.

Pares, R. D. (1988). Serological comparison of an Australian isolate of capsicum mosaic virus with capsicum tobamovirus isolates from Europe and America. *Ann. Appl. Biol.* **112**, 609–612.

Paris, H. S., Cohen, S., Burger, Y., and Yoseph, R. (1988). Single-gene resistance to zucchini yellow mosaic virus in *Cucurbita moschata*. *Euphytica* **37**, 27–29.

Parish, R. W. (1975). The lysosome-concept in plants. I. Peroxidases associated subcellular and wall functions of maize root tips. Implications for vacuole development. *Planta* **123**, 1–13.

Parkinson, J. (1656). "Paradisi in sole paradisus terrestris, or a garden of all sorts of pleasant flowers . . . with a kitchen garden of all manner of herbes etc." London.

Partridge, J. E., Shannon, L., and Gumpf, D. (1976). A barley lectin that binds free amino sugars. 1. Purification and characterization. *Biochim. Biophys. Acta* **451**, 470–483.

Paszkowski, J., Pisan, B., Shillito, R. D., Hohn, T., Hohn, B., and Potrykus, I. (1986). Genetic transformation of *Brassica compestris* var. rapa protoplasts with an engineered cauliflower mosaic virus genome. *Plant Mol. Biol.* **6**, 303–312.

Patterson, S., and Verduin, B. J. M. (1987). Applications of immunogold labelling in animal and plant virology. *Arch. Virol.* **97**, 1–26.

Paul, H. L. (1975). SDS polyacrylamide gel electrophoresis of virion proteins as a tool for detecting the presence of virus in plants. II. Examination of further virus–host combinations. *Phytopathol. Z.* **83**, 303–310.

Pawley, G. S. (1962). Plane groups on polyhedra. *Acta Crystallogr.* **15**, 49–53.

Pead, M. T., and Torrance, L. (1989). Some characteristics of monoclonal antibodies to a British MAV-like isolate of barley yellow dwarf virus. *Ann. Appl. Biol.* **113**, 639–644.

Peden, K. W. C., May, J. T., and Symons, R. H. (1972). A comparison of two plant virus-induced RNA polymerases. *Virology* **47**, 498–501.

Pelcher, L. E., Murakishi, H. H., and Hartmann, J. X. (1972). Kinetics of TMV-RNA synthesis and its correlation with virus accumulation and crystalline viral inclusion formation in tobacco tissue culture. *Virology* **47**, 787–796.

Pelham, H. R. B. (1978). Leaky UAG termination codon in tobacco mosaic virus RNA. *Nature (London)* **272,** 469–471.

Pelham, H. R. B. (1979). Translation of tobacco rattle virus RNAs *in vitro:* Four proteins from three RNAs. *Virology* **97,** 256–265.

Pelletier, J., and Sonenberg, N. (1988). Internal initiation of translation of eukaryotic mRNA directed by a sequence derived from poliovirus RNA. *Nature (London)* **334,** 320–325.

Pennazio, S., and Roggero, P. (1988). Systemic acquired resistance induced in tobacco plants by localized virus infection does not operate against challenging viruses that infect systemically. *J. Phytopathol.* **121,** 255–266.

Pennazio, S., and Roggero, P. (1990). Ethylene stimulation during the hypersensitive reaction of soybean to tobacco necrosis virus under different photo periods. *J. Phytopathol.* **128,** 177–183.

Pennazio, S., and Sapetti, C. (1981). Electrolyte leakage in relation to viral and abiotic stresses inducing necrosis in cowpea leaves. *Biol. Plant.* **24,** 218–255.

Pennazio, S., Colariccio, D., Roggero, P., and Lenzi, R. (1987). Effect of salicylate stress on the hypersensitive reaction of asparagus bean to tobacco necrosis virus. *Physiol. Mol. Plant Pathol.* **30,** 347–357.

Penny, D., Hendy, M. D., and Henderson, I. M. (1987). Reliability of evolutionary trees. *Cold Spring Harbor Symp. Quant. Biol.* **52,** 857–862.

Penny, D., Hendy, M. D., and Steel, M. (1991). Accuracy of tree reconstruction methods. *Trends Ecol. Evol.* (in press).

Pérez de san Román, C., Legorburu, F. J., Pascualena, J., and Gil, A. (1988). Simultaneous detection of potato viruses Y, leaf roll, X and S by DAS-ELISA technique with artificial polyvalent antibodies (APAs). *Potato Res.* **31,** 151–158.

Perham, R. N. (1973). The reactivity of functional groups as a probe for investigating the topography of tobacco mosaic virus. The use of mutants with additional lysine residues in the coat protein. *Biochem. J.* **131,** 119–126.

Perham, R. N., and Wilson, T. M. A. (1976). The polarity of stripping of coat protein subunits from the RNA in tobacco mosaic virus under alkaline conditions. *FEBS Lett.* **62,** 11–15.

Perham, R. N., and Wilson, T. M. A. (1978). The characterization of intermediates formed during the disassembly of tobacco mosaic virus at alkaline pH. *Virology* **84,** 293–302.

Perret, V., Florentz, C., Dreher, T., and Giege, R. (1989). Structural analogies between the 3′ tRNA-like structure of brome mosaic virus RNA and yeast tRNA^{tyr} revealed by protection studies with yeast tyrosyl-tRNA synthetase. *Eur. J. Biochem.* **185,** 331–339.

Perrin, R. M., and Gibson, R. W. (1985). Control of some insect-borne plant viruses with the pyrethroid PP321 (Karate). *Int. Pest Control* Nov./Dec., pp. 142–143.

Pesic, Z., Hiruki, C., and Chen, M. H. (1988). Detection of viral antigen by immunogold cytochemistry in ovules, pollen, and anthers of alfalfa infected with alfalfa mosaic virus. *Phytopathology* **78,** 1027–1032.

Peters, D., and Black, L. M. (1970). Infection of primary cultures of aphid cells with a plant virus. *Virology* **40,** 847–853.

Peterson, A. J., and Nuss, D. L. (1986). Regulation of expression of the wound tumour virus genome in persistently infected vector cells is related to changes in the translational activity of viral transcripts. *J. Virol.* **59,** 195–202.

Peterson, J. F., and Brakke, M. K. (1973). Genomic masking in mixed infections with brome mosaic and barley stripe mosaic viruses. *Virology* **51,** 174–182.

Petty, I. T. D., Coutts, R. H. A., and Buck, K. W. (1988). Transcriptional mapping of the coat protein gene of tomato golden mosaic virus. *J. Gen. Virol.* **69,** 1359–1365.

Petty, I. T. D., Hunter, B. G., Wei, N., and Jackson, A. O. (1989). Infectious barley stripe mosaic virus RNA transcribed *in vitro* from full-length genomic cDNA clones. *Virology* **171,** 342–349.

Pfeiffer, P., and Durham, A. C. H. (1977). The cation binding associated with structural transitions in bromegrass mosaic virus. *Virology* **81,** 419–432.

Pfeiffer, P., and Hohn, T. (1983). Involvement of reverse transcription in the replication of cauliflower mosaic virus: A detailed model and test of some aspects. *Cell (Cambridge, Mass.)* **33,** 781–789.

Pfeiffer, P., and Hohn, T. (1989). Cauliflower mosaic virus as a probe for studying gene expression in plants. *Physiol. Plant.* **77,** 625–632.

Pfeiffer, P., Laquel, P., and Hohn, T. (1984). Cauliflower mosaic virus replication complexes: Characterisation of the associated enzymes and the polarity of DNA synthesised *in vitro*. *Plant Mol. Biol.* **3,** 261–270.

Pfeiffer, P., Gordon, K., Fütterer, J., and Hohn, T. (1987). The life cycle of cauliflower mosaic virus. *In* "Plant Molecular Biology" (D. von Wettstein and N.-H. Chua, eds.), pp. 443–458. Plenum, New York.

Pfitzner, U. M., and Goodman, H. M. (1987). Isolation and characterisation of cDNA clones encoding pathogenesis related proteins from tobacco mosaic virus infected tobacco plants. *Nucleic Acids Res.* **15,** 4449–4465.

Piazzolla, P., Tousignant, M. E., and Kaper, J. M. (1982). Cucumber mosaic virus-associated RNA5. IX. The overtaking of viral RNA synthesis by CARNA5 and dsCARNA5 in tobacco. *Virology* **122,** 147–157.

Piazzolla, P., Rubino, L., Tousignant, M. E., and Kaper, J. M. (1989a). Two different types of satellite RNA associated with chicory yellow mottle virus. *J. Gen. Virol.* **70,** 949–954.

Piazzolla, P., Palmieri, F., and Nuzzaci, M. (1989b). Infectivity studies on cucumber mosaic virus treated with a clay material. *J. Phytopathology* **127,** 291–295.

Pierpoint, W. S. (1966). The enzymic oxidation of chlorogenic acid and some reactions of the quinone produced. *Biochem. J.* **98,** 567–580.

Pierpoint, W. S., Ireland, R. J., and Carpenter, J. M. (1977). Modification of proteins during the oxidation of leaf phenols: Reaction of potato virus X with chlorogenoquinone. *Phytochemistry* **16,** 29–34.

Pietrzak, M., and Hohn, T. (1987). Translation products of cauliflower mosaic virus ORF V, the coding region corresponding to the retrovirus *pol* gene. *Virus Genes* **1,** 83–96.

Pilowski, M., Frankel, R., and Cohen, S. (1981). Studies of the variable reaction at high temperature of F_1 hybrid tomato plants resistant to tobacco mosaic virus. *Phytopathology* **71,** 319–323.

Pinck, L., and Hirth, L. (1972). The replicative RNA and the viral RNA synthesis rate in tobacco infected with alfalfa mosaic virus. *Virology* **49,** 413–425.

Pinck, L., Genevaux, M., Bouley, J. P., and Pinck, M. (1975). Amino acid accepter activity of replicative form from some tymovirus RNAs. *Virology* **63,** 589–590.

Pinck, L., Fuchs, M., Pinck, M., Ravelonandro, M., and Walter, B. (1988). A satellite RNA in grapevine fanleaf virus strain F13. *J. Gen. Virol.* **69,** 233–239.

Pinck, M., Yot, P., Chapeville, F., and Duranton, H. M. (1970). Enzymatic binding of valine to the 3' end of TYMV-RNA. *Nature (London)* **226,** 954–956.

Piron, P. G. M. (1986). New aphid vectors of potato virus Y^N. *Neth. J. Plant Pathol.* **92,** 223–229.

Pirone, T. P. (1967). Acquisition and release of infectious tobacco mosaic virus by aphids. *Virology* **31,** 569–571.

Pirone, T. P. (1977). Accessory factors in nonpersistent virus transmission. *In* "Aphids as Virus Vectors" (K. F. Harris and K. Maramorosch, eds.), pp. 221–235. Academic Press, New York.

Pirone, T. P. (1981). Efficiency and selectivity of the helper-component-mediated aphid transmission of purified potyviruses. *Phytopathology* **71,** 922–924.

Pirone, T. P., and Harris, K. F. (1977). Nonpersistent transmission of plant viruses by aphids. *Annu. Rev. Phytopathol.* **15,** 55–73.

Pirone, T. P., and Thornbury, D. W. (1988). Quantity of virus required for aphid transmission of a potyvirus. *Phytopathology* **78,** 104–107.

Pirone, T. P., Raccah, B., and Madden, L. V. (1988). Suppression of aphid colonization by insecticides: Effect on the incidence of potyviruses in tobacco. *Plant Dis.* **72,** 350–353.

Pitcher, R. S., and McNamara, D. G. (1973). The control of *Xiphinema diversicaudatum*, the vector of arabis mosaic virus in hops. *Ann. Appl. Biol.* **75,** 468–469.

Plaskitt, K. A., Watkins, P. A. C., Sleat, D. E., Gallie, D. R., Shaw, J. G., and Wilson, T. M. A. (1987). Immunogold labelling locates the site of disassembly and transient gene expression of tobacco mosaic virus-like pseudovirus particles *in vivo. Mol. Plant–Microbe Interact.* **1,** 10–16.

Pleij, C. W. A. (1990). Pseudoknots: A new motif in the RNA game. *Trends Biochem. Sci.* **15,** 143–147.

Pleij, C. W. A., Mellema, J. R., Noort, A., and Bosch, L. (1977). The occurrence of the coat protein messenger RNA in the minor components of turnip yellow mosaic virus. *FEBS Lett.* **80,** 19–22.

Pleij, C. W. A., Abrahams, J. P., van Belkum, K. R., Rietveld, K., and Bosch, L. (1987). The spatial folding of the 3' noncoding region of aminoacylatable plant viral RNAs. *In* "Positive Strand RNA Viruses," (M. A. Brinton and R. R. Rueckert, eds.), pp. 299–316. Alan R. Liss, New York.

Pleše, N., Hoxha, E., and Miličić, D. (1975). Pathological anatomy of trees affected with apple stem grooving virus. *Phytopathol. Z.* **82,** 315–325.

Plumb, R. T. (1987). Aphid trapping to forecast virus disease. *Span* **30,** 35–37.

Plumb, R. T., and Macfarlane, I. (1977). Cereal diseases: A "new" virus of oats. *Rep. Rothamsted Exp. Stn.,* p. 256.

Polak, J. (1983). Variability of resistance of white cabbage to turnip mosaic virus. *Tagungsber.—Akad. Landwirtschaftswiss. D.D.R.* **216,** 331–335.

Pollard, D. G. (1977). Aphid penetration of plant tissues. *In* "Aphids as Virus Vectors" (K. F. Harris and K. Maramorosch, eds.), pp. 105–118. Academic Press, New York.

Polson, A., and von Wechmar, M. B. (1980). A novel way to transmit plant viruses. *J. Gen. Virol.* **51**, 179–181.

Ponz, F., and Bruening, G. (1986). Mechanisms of resistance to plant viruses. *Annu. Rev. Phytopathol.* **24**, 355–381.

Ponz, F., Glascock, C. B., and Bruening, G. (1988a). An inhibitor of polyprotein processing with the characteristics of a natural virus resistance factor. *Mol. Plant–Microbe Interact.* **1**, 25–31.

Ponz, F., Russell, M. L., Rowhani, A., and Bruening, G. (1988b). A cowpea line has distinct genes for resistance to tobacco ringspot virus and cowpea mosaic virus. *Phytopathology* **78**, 1124–1128.

Poolpol, P., and Inouye, T. (1986). Ultrastructure of plant cells doubly infected with potyviruses and other unrelated viruses. *Bull. Univ. Osaka Prefect., Ser. B.* **38**, 13–23.

Posnette, A. F. (1947). Virus diseases of cacao in West Africa. 1. Cacao viruses 1A, 1B, 1C, and ID. *Ann. Appl. Biol.* **34**, 388–402.

Posnette, A. F. (1950). Virus diseases of cacao in West Africa. VII. Virus transmission by different vector species. *Ann. Appl. Biol.* **37**, 378–384.

Posnette, A. F. (1969). Tolerance of virus infection in crop plants. *Rev. Appl. Mycol.* **48**, 113–118.

Posnette, A. F., and Todd, J. M. C. A. (1955). Virus diseases of cacao in West Africa. IX. Strain variation and interference in virus 1A. *Ann. Appl. Biol.* **43**, 433–453.

Powell, C. A. (1975). The effect of cations on the alkaline dissociation of tobacco mosaic virus. *Virology* **64**, 75–85.

Powell, C. A., and DeZoeten, G. A. (1977). Replication of pea enation mosaic virus RNA in isolated pea nuclei. *Proc. Natl. Acad. Sci. U.S.A.* **74**, 2919–2922.

Powell, C. A., De Zoeten, G. A., and Gaard, G. (1977). The localization of pea enation mosaic virus-induced RNA-dependent RNA polymerase in infected peas. *Virology* **78**, 135–143.

Powell, P. A., Stark, D. M., Sanders, P. R., and Beachy, R. N. (1989). Protection against tobacco mosaic virus in transgenic plants that express tobacco mosaic virus antisense RNA. *Proc. Natl. Acad. Sci. U.S.A.* **86**, 6949–6952.

Powell, P. A., Saunders, P. R., Tumer, N., Frayley, R. T., and Beachy, R. N. (1990). Protection against tobacco mosaic virus infection in transgenic plants requires accumulation of coat protein rather than coat protein RNA sequences. *Virology* **175**, 124–130.

Powell Abel, P., Nelson, R. S., De, B., Hoffmann, N., Rogers, S. G., Frayley, R. T., and Beachy, R. N. (1986). Delay of disease development in transgenic plants that express the tobacco virus coat protein gene. *Science* **232**, 738–743.

Pratt, M. J., and Matthews, R. E. F. (1971). Non-uniformities in the metabolism of excised leaves and leaf discs. *Planta* **99**, 21–36.

Preece, D. A. (1967). Nested balanced incomplete block designs. *Biometrika* **54**, 479–486.

Preil, W., Koenig, R., Engelhardt, M., and Meier-Dinkel, A. (1982). Elimination of poinsettia mosaic virus (PoiMV) and poinsettia cryptic virus (PoiCV) from *Euphorbia pulcherrima* Willd by cell suspension culture. *Phytopathol. Z.* **105**, 193–197.

Price, W. C. (1966). Flexuous rods in phloem cells of lime plants infected with citrus tristeza virus. *Virology* **29**, 285–294.

Prochiantz, A., and Haenni, A. L. (1973). TYMV RNA as a substrate of the tRNA maturation endonuclease. *Nature (London), New Biol.* **241**, 168–170.

Proctor, V. W. (1968). Long-distance dispersal of seeds by retention in digestive tracts of birds. *Science* **160**, 321–322.

Prody, G. A., Bakos, J. T., Buzayan, J. M., Schneider, I. R., and Bruening, G. (1986). Autolytic processing of dimeric plant virus satellite RNA. *Science* **231**, 1577–1580.

Proeseler, G. (1980). Piesmids. *In* "Vectors of Plant Pathogens" (K. F. Harris and K. Maramorosch, eds.), pp. 97–113. Academic Press, New York.

Proeseler, G., and Kastirr, U. (1988). Research into the effect of fungicides against *Polymyxa graminis* Led. as vector of barley yellow mosaic virus. *Nachrichten bl. Pflanzenschutzdienst DDR* **42**, 116–117.

Provvidenti, R., and Hoch, H. C. (1977). Tomato leaf roll caused by the interaction of the wilty gene and tobacco mosaic virus infection. *Plant Dis. Rep.* **61**, 500–502.

Provvidenti, R., and Schroeder, W. T. (1973). Resistance in *Phaseolus vulgaris* to the severe strain of bean yellow mosaic virus. *Phytopathology* **63**, 196–197.

Puchta, H., Ramm, K., and Sänger, H. L. (1988). The molecular structure of hop latent viroid (HLV) a new viroid occurring worldwide in hops. *Nucleic Acids Res.* **16**, 4197–4216.

Purcifull, D. E., Hiebert, E., and McDonald, J. G. (1973). Immunochemical specificity of cytoplasmic inclusions induced by viruses in the potato Y group. *Virology* **55**, 275–279.

Putz, C., Kuszala, J., Kuszala, M., and Spindler, C. (1974). Variation de pouvoir pathogène des isolats du virus de la mosaïque du concombre associée à la necrose de la tomate. *Ann. Phytopathol.* **6,** 139–154.

Pyne, J. W., and Hall, T. C. (1979). Efficient ribosome binding of brome mosaic virus (BMV) RNA4 contributes to its ability to outcompete the other BMV RNAs for translation. *Intervirology* **11,** 23–29.

Quadt, R., Verbeek, H. J. M., and Jaspars, E. M. J. (1988). Involvement of a non-structural protein in the RNA-synthesis of brome mosaic virus. *Virology* **165,** 256–261.

Quemada, H., Kearney, C., Gonsalves, D., and Slightom, J. L. (1989). Nucleotide sequences of the coat protein genes and flanking regions of cucumber mosaic virus strains C and WL RNA 3. *J. Gen. Virol.* **70,** 1065–1073.

Quesniaux, V., Briand, J.-P., and Van Regenmortel, M. H. V. (1983a). Immunochemical studies of turnip yellow mosaic virus. II. Localisation of a viral epitope in the N-terminal residues of the coat protein. *Mol. Immunol.* **20,** 179–185.

Quesniaux, V., Jaegle, M., and Van Regenmortel, M. H. V. (1983b). Immunological studies of turnip yellow mosaic virus. III. Localisation of two viral epitopes in residues 57–64 and 183–189 of the coat protein. *Biochim. Biophys. Acta* **743,** 226–231.

Quillet, L., Guilley, H., Jonard, G., and Richards, K. (1989). *In vitro* synthesis of biologically active beet necrotic yellow vein virus RNA. *Virology* **172,** 293–301.

Quiot, J. B., Devergne, J. C., Marchoux, G., Cardin, L., and Douine, L. (1979a). Ecology and epidemiology of cucumber mosaic virus (CMV) in South East France. VI. Distribution of two CMV groups in weeds. *Ann. Phytopathol.* **11,** 349–357.

Quiot, J. B., Devergne, J. C., Cardin, L., Verbrugge, M., Marchoux, G., and Labonne, G. (1979b). Ecology and epidemiology of cucumber mosaic virus in the south-east of France. VII. Occurrence of two virus populations in various crops. *Ann. Phytopathol.* **11,** 359–373.

Raccah, B., and Pirone, T. P. (1984). Characteristics of and factors affecting helper-component-mediated aphid transmission of a potyvirus. *Phytopathology* **74,** 305–308.

Raccah, B., Loebenstein, G., and Bar-Joseph, M. (1976). Transmission of citrus tristeza virus by the melon aphid. *Phytopathology* **66,** 1102–1104.

Rackwitz, H.-R., Rohde, W., and Sänger, H. L. (1981). DNA-dependent RNA polymerase II of plant origin transcribes viroid RNA into full-length copies. *Nature (London)* **291,** 297–301.

Rader, W. E., Fitzpatrick, H. F., and Hildebrand, E. M. (1947). A seed-borne virus of musk-melon. *Phytopathology* **37,** 809–816.

Raga, I. N., Ito, K., Matsui, M., and Okada, M. (1988). Effects of temperature on adult longevity, fertility, and rate of transovarial passage of rice stripe virus in the small brown planthopper, *Laodelphax striatellus* Fallen (Homoptera: Delphacidae). *Appl. Entomol. Zool.* **23,** 67–75.

Ragetli, H. W. J., Weintraub, M., and Elder, M. (1973). Effective mechanical inoculation of plant viruses in the absence of water. *Can. J. Bot.* **51,** 1977–1981.

Raghavendra, K., Adams, M. L., and Schuster, T. M. (1985). Tobacco mosaic virus protein aggregates in solution: Structural comparison of 20S aggregates with those near conditions for disk crystallisation. *Biochemistry* **24,** 3298–3304.

Raghavendra, K., Salunke, D. M., Caspar, D. L. D., and Schuster, T. M. (1986). Disk aggregates of tobacco mosaic virus protein in solution: Electron microscopy observations. *Biochemistry* **25,** 6276–6279.

Rajeshwari, R., and Murant, A. F. (1988). Purification and particle properties of groundnut rosette assistor virus and production of a specific antiserum. *Ann. Appl. Biol.* **112,** 403–414.

Rakowski, A. G., and Symons, R. H. (1989). Comparative sequence studies of variants of avocado sunblotch viroid. *Virology* **173,** 352–356.

Ralph, R. K. (1969). Double-stranded viral RNA. *Adv. Virus Res.* **15,** 61–158.

Ralph, R. K., and Wojcik, S. J. (1966). Synthesis of double-stranded viral RNA by cell-free extracts from turnip yellow mosaic virus-infected leaves. *Biochim. Biophys. Acta* **119,** 347–361.

Ralph, R. K., Matthews, R. E. F., Matus, A. I., and Mandel, H. G. (1965). Isolation and properties of double-stranded RNA from virus infected plants. *J. Mol. Biol.* **11,** 202–212.

Ralph, R. K., Bullivant, S., and Wojcik, S. J. (1971). Cytoplasmic membranes, a possible site of tobacco mosaic virus RNA replication. *Virology* **43,** 713–716.

Randles, J. W., and Francki, R. I. B. (1972). Infectious nucleocapsid particles of lettuce necrotic yellows virus with RNA-dependent RNA polymerase activity. *Virology* **50,** 297–300.

Randles, J. W., and Rohde, W. (1990). *Nicotiana velutina* mosaic virus: Evidence for a bipartite genome comprising 3kb and 8kb RNAs. *J. Gen. Virol.* **71,** 1019–1027.

Randles, J. W., Harrison, B. D., Murant, A. F., and Mayo, M. A. (1977). Packaging and biological activity of the two essential RNA species of tomato black ring virus. *J. Gen. Virol.* **36,** 187–193.

Rao, A. L. N., and Francki, R. I. B. (1981). Comparative studies on tomato aspermy and cucumber mosaic viruses. VI. Partial compatibility of genome segments from the two viruses. *Virology* **114**, 573–575.

Rao, A. L. N., and Francki, R. I. B. (1982). Distribution of determinants for symptom production and host range on the three RNA components of cucumber mosaic virus. *J. Gen. Virol.* **61**, 197–205.

Rao, A. L. N., and Hall, T. C. (1990). Requirement for a viral *trans*-acting factor encoded by brome mosaic virus RNA-2 provides strong selection *in vivo* for functional recombinants. *J. Virol.* **64**, 2437–2441.

Rao, A. L. N., and Hiruki, C. (1987). Unilateral compatibility of genome segments from two distinct strains of red clover necrotic mosaic virus. *J. Gen. Virol.* **68**, 191–194.

Rao, A. L. N., Hatta, T., and Francki, R. I. B. (1982). Comparative studies on tomato aspermy and cucumber mosaic viruses. VII. Serological relationships reinvestigated. *Virology* **116**, 318–326.

Rao, A. L. N., Dreher, T. W., Marsh, L. E., and Hall, T. C. (1989). Telomeric function of the t-RNA-like structure of brome mosaic virus RNA. *Proc. Natl. Acad. Sci. U.S.A.* **86**, 5335–5339.

Rao, A. L. N., Sullivan, B. P., and Hall, T. C. (1990). Use of *Chenopodium hybridum* facilitates isolation of brome mosaic virus RNA recombinants. *J. Gen. Virol.* **71**, in press.

Rao, A. S. (1968). Biology of *Polymyxa graminis* in relation to soil-borne wheat mosaic virus. *Phytopathology* **58**, 1516–1521.

Rao, G. P., Shukla, K., and Gupta, S. N. (1987). Effect of cucumber mosaic virus infection on nodulation, nodular physiology and nitrogen fixation of pea plants. *J. Plant Dis. Prot.* **94**, 606–613.

Rao, R. D. V. J. P., and Yaraguntiah, R. C. (1976). Natural occurrence of potato virus-Y on *Datura metel. Curr. Sci.* **45**, 467.

Rappaport, I., and Siegel, A. (1955). Inactivation of tobacco mosaic virus by rabbit antiserum. *J. Immunol.* **74**, 106–116.

Rappaport, I., and Wildman, S. G. (1957). A kinetic study of local lesion growth on *Nicotiana glutinosa* resulting from tobacco mosaic virus infection. *Virology* **4**, 265–274.

Rappaport, I., and Wu, J.-H. (1963). Activation of latent virus infection by heat. *Virology* **20**, 472–476.

Raski, D. J., and Hewitt, W. B. (1963). Plant–parasitic nematodes as vectors of plant viruses. *Phytopathology* **53**, 39–47.

Raski, D. J., Goheen, A. C., Lider, L. A., and Meredith, C. P. (1983). Strategies against grapevine fanleaf virus and its nematode vector. *Plant Dis.* **67**, 335–339.

Rast, A. T. B. (1967a). Yield of glasshouse tomatoes as affected by strains of tobacco mosaic virus. *Neth. J. Plant Pathol.* **73**, 147–156.

Rast, A. T. B. (1967b). Differences in aggressiveness between TMV-isolates from tomato on clones of *Lycopersicum peruvianum. Neth. J. Plant Pathol.* **73**, 186–189.

Rast, A. T. B. (1975). Variability of tobacco mosaic virus in relation to control of tomato mosaic in glasshouse tomato crops by resistance breeding and cross protection. *Agric. Res. Rep. (Wageningen)* **834**, 1–76.

Ready, M. P., Brown, D. T., and Robertus, J. D. (1986). Extracellular localization of pokeweed antiviral protein. *Proc. Natl. Acad. Sci. U.S.A.* **83**, 5053–5056.

Reanney, D. C. (1987). Genetic error and genome design. *Cold Spring Harbor Symp. Quant. Biol.* **52**, 751–757.

Reddy, D. V. R., and Black, L. M. (1972). Increase of wound tumor virus in leafhoppers as assayed on vector cell monolayers. *Virology* **50**, 412–421.

Reddy, D. V. R., and Black, L. M. (1973). Electrophoretic separation of all components of the double-stranded RNA of wound tumor virus. *Virology* **54**, 557–562.

Reddy, D. V. R., and Black, L. M. (1974). Deletion mutations of the genome segments of wound tumor virus. *Virology* **61**, 458–473.

Reddy, D. V. R., and Black, L. M. (1977). Isolation and replication of mutant populations of wound tumor virions lacking certain genome segments. *Virology* **80**, 336–346.

Reddy, D. V. R., and MacLeod, R. (1976). Polypeptide components of wound tumor virus. *Virology* **70**, 274–282.

Reddy, D. V. R., Rhodes, D. P., Lesnaw, J. A., MacLeod, R., Banerjee, A. K., and Black, L. M. (1977). *In vitro* transcription of wound tumor virus RNA by virion-associated RNA transcriptase. *Virology* **80**, 356–361.

Reeck, G. R., de Haën, C., Teller, D. C., Doolittle, R. F., Fitch, W. M., Dickerson, R. E., Chambon, P., McLachlan, A. D., Margoliash, E., Jukes, T. H., and Zuckerkandl, E. (1987). "Homology" in proteins and nucleic acids: A terminology muddle and a way out of it. *Cell (Cambridge, Mass.)* **50**, 667.

Rees, M. W., and Short, M. N. (1965). Variations in the composition of two strains of tobacco mosaic virus in relation to their host. *Virology* **26**, 596–602.

Rees, M. W., and Short, M. N. (1975). The amino acid sequence of the cowpea strain of tobacco mosaic virus protein. *Biochim. Biophys. Acta* **393**, 15–23.

Rees, M. W., and Short, M. N. (1982). The primary structure of cowpea chlorotic mottle virus coat protein. *Virology* **119**, 500–503.

Register, J. C., III, and Beachy, R. N. (1988). Resistance to TMV in transgenic plants results from interference with an early event in infection. *Virology* **166**, 524–532.

Register, J. C., III, and Beachy, R. N. (1989). Effects of protein aggregation state on coat protein-mediated protection against tobacco mosaic virus using a transient protoplast assay. *Virology* **173**, 656–663.

Register, J. C., III, Powell, P. A., Nelson, R. S., and Beachy, R. N. (1988). Genetically engineered cross protection against TMV interferes with initial infection and long distance spread of the virus. *In* "Molecular Biology of Plant–Pathogen Interactions" (B. Staskawicz, P. Ahlquist, and O. Yoder, eds.), pp. 269–282. Alan R. Liss, New York.

Reichenbächer, D., Kalinina, I., Schulze, M., Horn, A., and Kleinhempel, H. (1984). Ultramicro-ELISA with a fluorogenic substrate for detection of potato viruses. *Potato Res.* **27**, 353–364.

Reid, M. S., and Matthews, R. E. F. (1966). On the origin of the mosaic induced by turnip yellow mosaic virus. *Virology* **28**, 563–570.

Reijnders, L., Sloof, P., and Borst, P. (1973). The molecular weights of the mitochondrial-ribosomal RNAs of *Saccharomyces carlsbergensis*. *Eur. J. Biochem.* **35**, 266–269.

Reijnders, L., Aalbers, A. M. J., van Kammen, A., and Thuring, R. W. J. (1974). Molecular weights of plant viral RNAs determined by gel electrophoresis under denaturing conditions. *Virology* **60**, 515–521.

Reinero, A., and Beachy, R. N. (1986). Association of TMV coat protein with chloroplast membranes in virus-infected leaves. *Plant Mol. Biol.* **6**, 291–301.

Reisman, D., and De Zoeten, G. A. (1982). A covalently linked protein at the 5′ end of the genomic RNAs of pea enation mosaic virus. *J. Gen. Virol.* **62**, 187–190.

Renaudin, J., Bové, J. M., Otsuki, Y., and Takebe, I. (1975). Infection of *Brassica* leaf protoplasts by turnip yellow mosaic virus. *Mol. Gen. Genet.* **141**, 59–68.

Resconich, E. C. (1963). Movement of tobacco necrosis virus in systemically infected soybeans. *Phytopathology* **53**, 913–916.

Revington, G. N., Sunter, G., and Bisaro, D. M. (1989). DNA sequences essential for replication of the B genome component of tomato golden mosaic virus. *Plant Cell* **1**, 985–992.

Rezaian, M. (1990). Australian grapevine viroid—evidence for extensive recombination between viroids. *Nucleic Acids Res.* **18**, 1813–1818.

Rezaian, M. A., and Francki, R. I. B. (1973). Replication of tobacco ringspot virus. I. Detection of a low molecular weight double-stranded RNA from infected plants. *Virology* **56**, 238–249.

Rezaian, M. A., and Symons, R. H. (1986). Anti-sense regions in satellite RNA of cucumber mosaic virus form stable complexes with the viral coat protein gene. *Nucleic Acids Res.* **14**, 3229–3239.

Rezaian, M. A., Francki, R. I. B., Chu, P. W. G., and Hatta, T. (1976). Replication of tobacco ringspot virus. III. Site of virus synthesis in cucumber cotyledon cells. *Virology* **74**, 481–488.

Rezaian, M. A., Williams, R. H. V., Gordon, K. H. J., Gould, A. R., and Symons, R. H. (1984). Nucleotide sequence of cucumber-mosaic-virus RNA2 reveals a translation product significantly homologous to corresponding proteins of other viruses. *Eur. J. Biochem.* **143**, 277–284.

Rezaian, M. A., Williams, R. H. V., and Symons, R. H. (1985). Nucleotide sequence of cucumber mosaic virus RNA1—Presence of a sequence complementary to part of the viral satellite RNA and homologies with other viral RNAs. *Eur. J. Biochem.* **150**, 331–339.

Rezaian, M. A., Skene, K. G. M., and Ellis, J. G. (1988). Anti-sense RNAs of cucumber mosaic virus in transgenic plants assessed for control of the virus. *Plant. Mol. Biol.* **11**, 463–471.

Rezelman, G., Goldbach, R., and van Kammen, A. (1980). Expression of bottom component RNA of cowpea mosaic virus in cowpea protoplasts. *J. Virol.* **36**, 366–373.

Rezelman, G., Franssen, H. J., Goldbach, R. W., Ie, T. S., and van Kammen, A. (1982). Limits to the independence of bottom component RNA of cowpea mosaic virus. *J. Gen. Virol.* **60**, 335–342.

Rezelman, G., van Kammen, A., and Wellink, J. (1989). Expression of cowpea mosaic virus M RNA in cowpea protoplasts. *J. Gen. Virol.* **70**, 3043–3050.

Rhodes, D. P., Reddy, D. V. R., MacLeod, R., Black, L. M., and Banerjee, A. K. (1977). *In vitro* synthesis of RNA containing 5′-terminal structure ^{7}mG (5′) ppp (5′) Ap^m . . . by purified wound tumor virus. *Virology* **76**, 554–559.

Rice, R. H. (1974). Minor protein components in cowpea chlorotic mottle virus and satellite of tobacco necrosis virus. *Virology* **61,** 249–255.

Richards, K. E., and Williams, R. C. (1976). Assembly of tobacco mosaic virus *in vitro. Compr. Virol.* **6,** 1–37.

Richards, K. E., Jonard, G., Jacquemond, M., and Lot, H. (1978). Nucleotide sequence of cucumber mosaic virus-associated RNA5. *Virology* **89,** 395–408.

Richards, K. E., Guilley, H., and Jonard, G. (1981). Further characterization of the discontinuities in cauliflower mosaic virus DNA. *FEBS Lett.* **134,** 67–70.

Richardson, J., and Sylvester, E. S. (1968). Further evidence of multiplication of sowthistle yellow vein virus in its aphid vector *Hyperomyzus lactucae. Virology* **35,** 347–355.

Richardson, M., Valdes-Rodriques, S., and Blanco-Labra, A. (1987). A possible function for thaumatin and a TMV-induced protein suggested by homology to a maize inhibitor. *Nature* **327,** 432–434.

Richins, R. D., Scholthof, H. B., and Shepherd, R. J. (1987). Sequence of figwort mosaic virus DNA (caulimovirus group). *Nucleic Acids Res.* **15,** 8451–8466.

Riesner, D., Henco, K., Rokohl, U., Klotz, G., Kleinschmidt, A. K., Domdey, H., Jank, P., Gross, H. J., and Sänger, H. L. (1979). Structure and structure formation of viroids. *J. Mol. Biol.* **133,** 85–115.

Riesner, D., Colpan, M., Goodman, T. C., Nagel, L., Schumacher, J., and Steger, G. (1983). Dynamics and interactions of viroids. *J. Biomol. Struct. Dyn.* **1,** 669–688.

Rivera-Bustamante, R. F., and Semancik, J. S. (1989). Properties of a viroid-replicating complex solubilized from nuclei. *J. Gen. Virol.* **70,** 2707–2716.

Rizvi, S. A. H., and Raman, K. V. (1983). Effect of glandular trichomes on the spread of potato virus Y (PVY) and potato leaf roll virus (PLRV) in the field. *In* "Proceedings of the International Congress on Research for the Potato in the Year 2,000" (W. J. Hooker, ed.), pp. 162–163. International Potato Centre, Lima, Peru.

Rizzo, T. M., and Palukaitis, P. (1988). Nucleotide sequence and evolutionary relationships of cucumber mosaic virus (CMV) strains: CMV RNA2. *J. Gen. Virol.* **69,** 1777–1787.

Rizzo, T. M., and Palukaitis, P. (1989). Nucleotide sequence and evolutionary relationships of cucumber mosaic virus (CMV) strains: CMV RNA1. *J. Gen. Virol.* **70,** 1–11.

Rizzo, T. M., and Palukaitis, P. (1990). Construction of full-length cDNA clones of cucumber mosaic virus RNAs 1, 2, and 3: Generation of infectious transcripts. *Mol. Gen. Genet.* **222,** 249–256.

Robaglia, C., Durand-Tardif, M., Tronchet, M., Boudazin, G., Astier-Manifacier, S., and Casse-Delbart, F. (1989). Nucleotide sequence of potato virus Y (N strain) genomic RNA. *J. Gen. Virol.* **70,** 935–947.

Roberts, D. A. (1982). Systemic acquired resistance induced in hypersensitive plants by non-necrotic localized viral infections. *Virology* **122,** 207–209.

Roberts, D. A., Christie, R. G., and Archer, M. C., Jr. (1970). Infection of apical initials in tobacco shoot meristems by tobacco ringspot virus. *Virology* **42,** 217–220.

Roberts, F. M. (1950). The infection of plants by viruses through roots. *Ann. Appl. Biol.* **37,** 385–396.

Roberts, I. M. (1986). Practical aspects of handling, preparing and staining samples containing plant virus particles for electron microscopy. *In* "Developments and Applications in Virus Testing" (R. A. C. Jones and L. Torrance, eds.), Dev. Appl. Biol. I., pp. 213–243. Association of Applied Biologists, Wellsbourne.

Roberts, I. M. (1988). The structure of particles of tobacco ringspot nepovirus: Evidence from electron microscopy. *J. Gen. Virol.* **69,** 1831–1840.

Roberts, I. M., and Brown, D. J. F. (1980). Detection of six nepoviruses in their nematode vectors by immunosorbent electron microscopy. *Ann. Appl. Biol.* **96,** 187–192.

Roberts, I. M., and Harrison, B. D. (1979). Detection of potato leafroll and potato mop top viruses by immunosorbent electron microscopy. *Ann. Appl. Biol.* **93,** 289–297.

Roberts, I. M., and Mayo, M. A. (1980). Electron microscope studies of the structure of the disk aggregate of tobacco rattle virus protein. *J. Ultrastruct. Res.* **71,** 49–59.

Roberts, I. M., Robinson, D. J., and Harrison, B. D. (1984). Serological relationships and genome homologies among geminiviruses. *J. Gen. Virol.* **65,** 1723–1730.

Roberts, P. L., and Wood, K. R. (1981). Methods for enhancing the synchrony of cucumber mosaic virus replication in tobacco plants. *Phytopathol. Z.* **102,** 114–121.

Robertson, A. D. (1987). The whitefly *Bemisia tabaci* (Gennadius) as a vector of African cassava mosaic virus at the Kenya coast and ways in which the yield losses in cassava, *Manihot esculenta* Crantz, caused by the virus can be reduced. *Insect. Sci. Appl.* **8,** 797–801.

Robertson, H. D., and Branch, A. D. (1987). The viroid replication process. *In* "Viroids and Viroid-like Pathogens" (J. S. Semancik, ed.), pp. 49–69. CRC Press, Boca Raton, Florida.

Robertson, N. L., and Carroll, T. W. (1988). Virus-like particles and a spider mite intimately associated with a new disease of barley. *Science* **240,** 1188–1190.

Robertson, W. M., and Henry, C. E. (1986). An association of carbohydrates with particles of arabis mosaic virus retained within *Xiphinema diversicaudatum. Ann. Appl. Biol.* **109,** 299–305.

Robinson, D. J. (1973). Inactivation and mutagenesis of tobacco rattle virus by nitrous acid. *J. Gen. Virol.* **18,** 215–222.

Robinson, D. J. (1977). A variant of tobacco rattle virus: Evidence for a second gene in RNA-2. *J. Gen. Virol.* **35,** 37–43.

Robinson, D. J., and Harrison, B. D. (1985). Unequal variation in the two genome parts of tobraviruses and evidence for the existence of three separate viruses. *J. Gen. Virol.* **66,** 171–176.

Robinson, D. J., Barker, H., Harrison, B. D., and Mayo, M. A. (1980). Replication of RNA-1 of tomato black ring virus independently of RNA-2. *J. Gen. Virol.* **51,** 317–326.

Robinson, D. J., Mayo, M. A., Fritsch, C., Jones, A. T., and Raschke, J. H. (1983). Origin and messenger activity of two small RNA species found in particles of tobacco rattle virus strain SYM. *J. Gen. Virol.* **64,** 1591–1599.

Robinson, D. J., Hamilton, W. D. O., Harrison, B. D., and Baulcombe, D. C. (1987). Two anomalous tobravirus isolates: Evidence for RNA recombination in nature. *J. Gen. Virol.* **68,** 2551–2561.

Rochon, D'A. M., and Tremaine, J. H. (1989). Complete nucleotide sequence of the cucumber necrosis virus genome. *Virology* **169,** 251–259.

Rochon, D'A., Kelly, R., and Siegel, A. (1986). Encapsidation of 18S RNA by tobacco mosaic virus coat protein. *Virology* **150,** 140–148.

Rochow, W. F. (1970). Barley yellow dwarf virus: Phenotypic mixing and vector specificity. *Science* **167,** 875–878.

Rochow, W. F. (1977). Dependent virus transmission from mixed infections. *In* "Aphids as Virus Vectors" (K. F. Harris and K. Maramorosch, eds.), pp. 253–273. Academic Press, New York.

Rochow, W. F. (1979). Field variants of barley yellow dwarf virus: Detection and fluctuation during twenty years. *Phytopathology* **69,** 655–660.

Rochow, W. F., and Duffus, J. E. (1978). Relationship between barley yellow dwarf and beet western yellows viruses. *Phytopathology* **68,** 51–58.

Rochow, W. F., and Muller, I. (1975). Use of aphids injected with virus-specific antiserum for study of plant viruses that circulate in vectors. *Virology* **63,** 282–286.

Rochow, W. F., Muller, I., and Gildow, F. E. (1983). Interference between two luteoviruses in an aphid: Lack of reciprocal competition. *Phytopathology* **73,** 919–922.

Rodriguez, D., Lopez-Abella, D., and Diaz-Ruiz, J. R. (1987). Viroplasms of an aphid-transmissible isolate of cauliflower mosaic virus contain helper component activity. *J. Gen. Virol.* **68,** 2063–2067.

Rodriguez, J. F., Rodriguez, D., Rodriguez, J.-R., McGowan, E. B., and Esteban, M. (1988). Expression of the firefly luciferase gene in vaccinia virus: A highly sensitive marker to follow virus dissemination in tissues of infected animals. *Proc. Natl. Acad. Sci. U.S.A.* **85,** 1667–1671.

Rodriguez-Cerezo, E., and García-Arenal, F. (1989). Genetic heterogeneity of the RNA genome population of the plant virus U5-TMV. *Virology* **170,** 418–423.

Rodriguez-Cerezo, E., Moya, A., and García-Arenal, F. (1989). Variability and evolution of the plant RNA virus pepper mild mottle virus. *J. Virol.* **63,** 2198–2203.

Rodriguez-Cerezo, E., Fernandez-Elena, S., Moya, A., and García-Arenal, F. (1990). High genetic stability in natural populations of the plant virus tobacco mild green mosaic virus. *J. Mol. Evolution* (in press).

Roenhorst, J. W. (1989). Early stages in cowpea chlorotic mottle virus infection. Ph.D. Thesis, Wageningen Agricultural University, The Netherlands.

Roenhorst, J. W., van Lent, J. W. M., and Verdium, B. J. M. (1988). Binding of cowpea chlorotic mottle virus to cowpea protoplasts and relation of binding to virus entry and expression. *Virology* **164,** 91–98.

Roenhorst, J. W., Verdiun, B. J. M., and Goldbach, R. W. (1989). Virus–ribosome complexes from cell-free translation systems supplemented with cowpea chlorotic mottle virus particles. *Virology* **168,** 138–146.

Rogers, S. G., Bisaro, D. M., Horsch, R. B., Frayley, R. T., Hoffmann, N. L., Brand, L., Elmer, J. S., and Lloyd, A. M. (1986). Tomato golden mosaic virus: A component DNA replicates autonomously in transgenic plants. *Cell (Cambridge, Mass.)* **45,** 593–600.

Rogers, S. G., Elmer, J. S., Sunter, G., Gardiner, W. E., Brand, L., Browning, C. K., and Bisaro, D. M. (1989). Molecular genetics of tomato golden mosaic virus. *In* "Molecular Biology of Plant–Pathogen Interactions," (B. Staskowicz, P. Ahlquist, and O. Yoder, eds.), pp. 199–215. Alan R. Liss, New York.

Roggero, P., and Pennazio, S. (1988). Biochemical changes during the necrotic systemic infection of tobacco plants by potato virus Y, necrotic strain. *Physiol. Mol. Plant Pathol.* **32,** 105–113.

Roggero, P., and Pennazio, S. (1989). The extracellular acidic and basic pathogenesis-related proteins of soybean induced by viral infection. *J. Phytopathol.* **127,** 274–280.

Rohde, W., Randles, J. W., Langridge, P., and Hanold, D. (1990). Nucleotide sequence of a circular single-stranded DNA associated with coconut foliar decay virus. *Virology* **176,** 648–651.

Rohozinski, J., Francki, R. I. B., and Chu, P. W. G. (1986). The *in vitro* synthesis of velvet tobacco mottle-virus-specific double-stranded RNA by a soluble fraction in extracts from infected *Nicotiana clevelandii* leaves. *Virology* **155,** 27–38.

Rohozinski, J., Girton, L. E., and van Etten, J. L. (1989). *Chlorella* viruses contain linear non-permuted double-stranded DNA genomes with covalently closed hairpin ends. *Virology* **168,** 363–369.

Roistacher, C. N., Dodds, J. A., and Bash, J. A. (1987). Means of obtaining and testing protective strains of seedling yellows and stem pitting tristeza virus: A preliminary report. *Phytophylactica* **19,** 199–203.

Romaine, C. P., and Zaitlin, M. (1978). RNA-dependent RNA polymerases in uninfected and tobacco mosaic virus-infected tobacco leaves: Viral-induced stimulation of a host polymerase activity. *Virology* **86,** 241–253.

Romaine, C. P., Newhart, S. R., and Anzola, D. (1981). Enzyme-linked immunosorbent assay for plant viruses in intact leaf tissue disks. *Phytopathology* **71,** 308–312.

Ronald, W. P., Schroeder, B., Tremaine, J. H., and Paliwal, Y. C. (1977). Distorted virus particles in electron microscopy: An artifact of grid films. *Virology* **76,** 416–419.

Ronald, W. P., Tremaine, J. H., and MacKenzie, D. J. (1986). Assessment of southern bean mosaic virus monoclonal antibodies for affinity chromatography. *Phytopathology* **76,** 491–494.

Rønde Kristensen, H. (1983). European fights against fruit tree viruses as organised by EPPO and EEC. *Acta Hortic.* **130,** 19–29.

Roosien, J., and van Vloten-Doting, L. (1982). Complementation and interference of ultraviolet-induced Mts mutants of alfalfa mosaic virus. *J. Gen. Virol.* **63,** 189–198.

Roosien, J., and van Vloten-Doting, L. (1983). A mutant of alfalfa mosaic virus with an unusual structure. *Virology* **126,** 155–167.

Roossinck, M. J., and Palukaitis, P. (1990). Rapid induction and severity of symptoms in zucchini squash (*Cucurbita pepo*) map to RNA 1 of cucumber mosaic virus. *Mol. Plant Microbe Interact.* **3,** 188–192.

Rose, D. J. W. (1974). The epidemiology of maize streak disease in relation to population densities of *Cicadulina* spp. *Ann. Appl. Biol.* **76,** 199–207.

Ross, A. F. (1961a). Localized acquired resistance to plant virus infection in hypersensitive hosts. *Virology* **14,** 329–339.

Ross, A. F. (1961b). Systemic acquired resistance induced by localized virus infections in plants. *Virology* **14,** 340–358.

Ross, A. F. (1966). Systemic effects of local lesion formation. *In* "Viruses of Plants" (A. B. R. Beemster and J. Dijkstra, eds.), pp. 127–150. North-Holland Publ., Amsterdam.

Ross, A. F., and Williamson, C. E. (1951). Physiologically active emanations from virus infected plants. *Phytopathology* **41,** 431–438.

Ross, J. P. (1968). Effect of single and double infections of soybean mosaic and bean pod mottle viruses on soybean yield and seed characters. *Plant Dis. Rep.* **52,** 344–348.

Ross, J. P., and Butler, A. K. (1985). Distribution of bean pod mottle virus in soybeans in North Carolina. *Plant Dis.* **69,** 101–103.

Rossmann, M. G. (1984). Constraints on the assembly of spherical virus particles. *Virology* **134,** 1–11.

Rossmann, M. G. (1985). The structure and *in vitro* assembly of southern bean mosaic virus in relation to that of other small spherical plant viruses. *In* "Molecular Plant Virology" (J. W. Davies, ed.), Vol. 1, pp. 123–153. CRC Press, Boca Raton, Florida.

Rossmann, M. G., and Johnson, J. E. (1989). Icosahedral RNA virus structure. *Annu. Rev. Biochem.* **58,** 533–573.

Rossmann, M. G., and Rueckert, R. R. (1987). What does the molecular structure of viruses tell us about viral functions? *Microbiol. Sci.* **4,** 206–214.

Rossmann, M. G., Abad-Zapatero, C., Murthy, M. R. N., Liljàs, L., Alwyn-Jones, T., and Strandberg, B. (1983). Structural comparisons of some small spherical plant viruses. *J. Mol. Biol.* **165,** 711–736.

Rothstein, S. J., Di Maio, J., Strand, M., and Rice, D. (1987). Stable and heritable inhibition of the expression of nopaline synthase in tobacco expressing antisense RNA. *Proc. Natl. Acad. Sci. U.S.A.* **84,** 8439–8443.

Rott, M. E., Rochon, D. M., and Tremaine, J. H. (1988). A 1.9 kilobase homology in the 3'-terminal regions of RNA-1 and RNA-2 of tomato ringspot virus. *J. Gen. Virol.* **69,** 745–750.

Round, F. E. (1984). The systematics of the Chlorophyta: An historical review leading to some modern concepts (Taxonomy of the Chlorophyta). III. *In* "Systematics of the Green Algae" (D. E. G. Irvine and D. M. John, eds.), pp. 1–27. Academic Press, Orlando, Florida.

Rowhani, A., Mircetich, S. M., Shepherd, R. J., and Cucuzza, J. D. (1985). Serological detection of cherry leafroll virus in English walnut trees. *Phytopathology* **75**, 48–52.

Roy, B. P., AbouHaidar, M. G., Sit, T. L., and Alexander, A. (1988). Construction and use of cloned cDNA biotin and ^{32}P-labelled probes for the detection of papaya mosaic potexvirus RNA in plants. *Phytopathology* **78**, 1425–1429.

Roy, B. P., AbouHaidar, M. G., and Alexander, A. (1989). Biotinylated RNA probes for the detection of potato spindle tuber viroid (PSTV) in plants. *J. Virol. Methods* **23**, 149–156.

Rubin, H. N., and Halim, M. N. (1987). A direct evidence for the involvement of poly(A) in protein synthesis. *Biochem. Biophys. Res. Commun.* **144**, 649–656.

Rubino, L., Tousignant, M. E., and Steger, G., and Kaper, J. M. (1990). Nucleotide sequence and structural analysis of two satellite RNAs associated with chicory yellow mottle virus (CYMV). *J. Gen. Virol.* **71**, 1897–1903.

Rueckert, R. R. (1985). Picornaviruses and their replication. *In* "Virology" (B. N. Fields, ed.), pp. 705–738. Raven Press, New York.

Rufty, R. C., Powell, N. T., and Gooding, G. V., Jr. (1983). Relationship between resistance to *Meloidogyne incognita* and a necrotic response to infection by a strain of potato virus Y in tobacco. *Phytopathology* **73**, 1418–1423.

Rupasov, V. V., Morozov, S. Yu., Kanyuka, K. V., and Zavriev, S. K. (1989). Partial nucleotide sequence of potato virus M RNA shows similarities to potexviruses in gene arrangement and the encoded amino acid sequences. *J. Gen. Virol.* **70**, 1861–1869.

Rushing, R. E., Sunter, G., Gardiner, W. E., Dute, R. R., and Bisaro, D. M. (1987). Ultrastructural aspects of tomato golden mosaic virus infection in tobacco. *Phytopathology* **77**, 1231–1236.

Russell, G. E. (1958). Sugar beet yellows: A preliminary study of the distribution and interrelationships of viruses and virus strains found in East Anglia 1955–57. *Ann. Appl. Biol.* **46**, 393–398.

Russell, G. E. (1960). Breeding for resistance to sugar beet yellows. *Br. Sugar Beet Rev.* **28**, 163–170.

Russell, G. E. (1966). The control of *Alternaria* species on leaves of sugar beet infected with yellowing viruses. II. Experiments with two yellowing viruses and virus-tolerant sugar beet. *Ann. Appl. Biol.* **57**, 425–434.

Russo, M., and Martelli, G. P. (1972). Ultrastructural observations on tomato bushy stunt virus in plant cells. *Virology* **49**, 122–129.

Russo, M., Kishtah, A. A., and Martelli, G. P. (1979). Unusual intracellular aggregates of broad bean wilt virus particles. *J. Gen. Virol.* **43**, 453–456.

Russo, M., Di Franco, A., and Martelli, G. P. (1987). Cytopathology in the identification and classification of tombusviruses. *Intervirology* **28**, 134–143.

Russo, M., Burgyan, J., Carrington, J. C., Hillman, B. I., and Morris, T. J. (1988). Complementary DNA cloning and characterisation of cymbidium ringspot virus RNA. *J. Gen. Virol.* **69**, 401–406.

Rybicki, E. P., and von Wechmar, M. B. (1982). Enzyme-assisted immune detection of plant virus proteins electroblotted onto nitrocellulose paper. *J. Virol. Methods* **5**, 267–278.

Rydén, K., and Gerhardson, B. (1978). Rödklövermosaikvirus sprids med slättermaskiner. *Vaextskyddsnotiser* **42**, 112–115.

Ryder, E. J. (1964). Transmission of common lettuce mosaic virus through the gametes of the lettuce plant. *Plant Dis. Rep.* **48**, 522–523.

Saayer-Rieps, J. D., and De Jager, C. P. (1988). Inhibition of different cowpea viruses by non-virulent cowpea mosaic virus is dependent on the type of immunity of the plant to the inhibitory virus. *Neth. J. Plant Pathol.* **94**, 253–256.

Sacher, R., and Ahlquist, P. (1989). Effect of deletions in the N-terminal basic arm of brome mosaic virus coat protein on RNA packaging and systemic infection. *J. Virol.* **63**, 4545–4552.

Sacher, R., French, R., and Ahlquist, P. (1988). Hybrid brome mosaic virus RNAs express and are packaged in tobacco mosaic virus coat protein *in vivo*. *Virology* **167**, 15–24.

Sachs, A. B., and Davis, R. W. (1989). The poly(A) binding protein is required for poly(A) shortening and 60S ribosomal subunit-dependent translation initiation. *Cell (Cambridge, Mass.)* **58**, 857–867.

Saedler, H., Gierl, A., Sommer, H., and Schwarz-Sommer, Z. (1987). Plant transposable elements and their role in the evolution of regulatory units and proteins. *J. Cell Sci., Suppl.* **7**, 139–144.

Saiki, R. K., Gelfand, D. H., Stoffel, S., Scharf, S. J., Higuchi, R., Horn, G. T., Mullis, K. B., and Erlich, H. A. (1988). Primer directed enzymatic amplification of DNA with a thermostable DNA polymerase. *Science* **239**, 487–491.

Saito, T., Meshi, T., Takamatsu, N., and Okada, Y. (1987a). Coat protein gene sequence of tobacco mosaic virus encodes a host response determinant. *Proc. Natl. Acad. Sci. U.S.A.* **84**, 6074–6077.

Saito, T., Hosokawa, D., Meshi, T., and Okada, Y. (1987b). Immunocytochemical localisation of the

130K and 180K proteins (putative replicase components) of tobacco mosaic virus. *Virology* **160,** 477–481.

Saito, T., Imai, Y., Meshi, T., and Okada, Y. (1988). Interviral homologies of the 30K proteins of tobamoviruses. *Virology* **167,** 653–656.

Saito, T., Yamanaka, K., Watanabe, Y., Takamatsu, N., Meshi, T., and Okada, Y. (1989). Mutational analysis of the coat protein gene of tobacco mosaic virus in relation to hypersensitive response in tobacco plants with the N' gene. *Virology* **173,** 11–20.

Saito, T., Yamanaka, K., and Okada, Y. (1990). Long-distance movement and viral assembly of tobacco mosaic virus mutants. *Virology* **176,** 329–336.

Saitou, N., and Nei, M. (1986). Polymorphism and evolution of influenza virus genes. *Mol. Biol. Evol.* **3,** 57–74.

Sakai, F., and Takebe, I. (1970). RNA and protein synthesis in protoplasts isolated from tobacco leaves. *Biochim. Biophys. Acta* **224,** 531–540.

Sakai, F., and Takebe, I. (1974). Protein synthesis in tobacco mesophyll protoplasts induced by tobacco mosaic virus infection. *Virology* **62,** 426–433.

Sakai, F., Dawson, J. R. O., and Watts, J. W. (1979). Synthesis of proteins in tobacco protoplasts infected with brome mosaic virus. *J. Gen. Virol.* **42,** 323–328.

Sakai, F., Dawson, J. R. O., and Watts, J. W. (1983). Interference in infections of tobacco protoplasts with two bromoviruses. *J. Gen. Virol.* **64,** 1347–1354.

Sakimura, K. (1962). The present status of thrips-borne viruses. *In* "Biological Transmission of Disease Agents" (K. Maramorosch, ed.), pp. 33–40. Academic Press, New York.

Sako, N., and Ogata, K. (1980). Different helper factors associated with aphid transmission of some potyviruses. *Virology* **112,** 762–765.

Salaman, R. N. (1933). Protective inoculation against a plant virus. *Nature (London)* **131,** 468.

Salaman, R. N. (1938). The potato virus "X," its strains and reactions. *Philos. Trans. R. Soc. London, Ser. B* **229,** 137-217.

Salazar, L. F., and Jones, R. A. C. (1975). Some studies on the distribution and incidence of potato mop-top virus in Peru. *Am. Potato J.* **52,** 143–150.

Salazar, L. F., Balbo, I., and Owens, R. A. (1988a). Comparison of four radioactive probes for the diagnosis of potato spindle tuber viroid by nucleic acid spot hybridization. *Potato Res.* **31,** 431–442.

Salazar, L. F., Hammond, R. W., Diener, T. O., and Owens, R. A. (1988b). Analysis of viroid replication following *Agrobacterium*-mediated inoculation of non-host species with potato spindle tuber viroid cDNA. *J. Gen. Virol.* **69,** 879–889.

Salinas Calvete, J., and Wieringa-Brants, D. H. (1984). Infection and necrosis of cowpea mesophyll cells by tobacco necrosis virus and two strains of tobacco mosaic virus. *Neth. J. Plant Pathol.* **90,** 71–78.

Salomon, R., and Bar-Joseph, M. (1982). Translational competition between related virus RNA species in cell-free systems. *J. Gen. Virol.* **62,** 343–347.

Salvato, M. S., and Fraenkel-Conrat, H. (1977). Translation of tobacco necrosis virus and its satellite in a cell-free wheat germ system. *Proc. Natl. Acad. Sci. U.S.A.* **74,** 2288–2292.

Samac, D. A., Nelson, S. E., and Loesch-Fries, L. S. (1983). Virus protein synthesis in alfalfa mosaic virus infected alfalfa protoplasts. *Virology* **131,** 455–462.

Sambrook, J., Fritsch, E. F., and Maniatis, T., (1989). Molecular cloning: a laboratory manual. 2d ed., Vols. I, II, and III, Cold Spring Harbor Press.

Samuel, G. (1934). The movement of tobacco mosaic virus within the plant. *Ann. Appl. Biol.* **21,** 90–111.

Samuel, G., and Bald, J. G. (1933). On the use of primary lesions in quantitative work with two plant viruses. *Ann. Appl. Biol.* **20,** 70–99.

Sander, E.-M., and Dietzgen, R. G. (1984). Monoclonal antibodies against plant viruses. *Adv. Virus Res.* **29,** 131–168.

Sander, E.-M., and Mertes, G. (1984). Use of protoplasts and separate cells in plant virus research. *Adv. Virus Res.* **29,** 215–262.

Sander, E.-M., Dietzgen, R. G., Cranage, M. P., and Coombs, R. R. A. (1989). Rapid and simple detection of plant virus by reverse passive haemagglutination. I. Comparison of ELISA (enzyme-linked immunosorbent assay) and RPH (reverse passive haemagglutination) for plant virus diagnosis. *J. Plant Dis. Prot.* **96,** 113–123.

Sanders, P. R., Winter, J. A., Barnason, A. R., Rogers, S. G., and Fraley, R. T. (1987). Comparison of cauliflower mosaic virus 35S and nopaline synthase promoters in transgenic plants. *Nucleic Acids Res.* **15,** 1543–1558.

Sänger, H. L. (1969). Functions of the two particles of tobacco rattle virus. *J. Virol.* **3,** 304–312.

Sänger, H. L. (1987). Viroid function. Viroid replication. *In* "The Viroids" (T. O. Diener, ed.), pp. 117–166. Plenum, New York and London.

Sänger, H. L., Klotz, G., Riesner, D., Gross, H. J., and Kleinschmidt, A. K. (1976). Viroids are single stranded covalently closed circular RNA molecules existing as highly base-paired rod-like structures. *Proc. Natl. Acad. Sci. U.S.A.* **73,** 3852–3856.

Sanger, M., Daubert, S., and Goodman, R. M. (1990). Characteristics of a strong promoter from figwort mosaic virus: Comparison with the analogous 35S promoter from cauliflower mosaic virus and the regulated mannopine synthase promoter. *Plant. Mol. Biol.* **14,** 433–443.

Sano, T., Hataya, T., and Shikata, E. (1988a). Complete nucleotide sequence of a viroid isolated from etrog citron, a new member of hop stunt viroid group. *Nucleic Acids Res.* **16,** 347.

Sano, T., Kudo, H., Sugimoto, T., and Shikata, E. (1988b). Synthetic oligonucleotide hybridization probes to diagnose hop stunt viroid strains and citrus exocortis viroid. *J. Virol. Methods* **19,** 109–120.

Sano, T., Hataya, T., Terai, Y., and Shikata, E. (1989). Hop stunt viroid strains from dapple fruit disease of plum and peach in Japan. *J. Gen. Virol.* **70,** 1311–1319.

Sarachu, A. N., Nassuth, A., Roosien, J., van Vloten-Doting, L., and Bol, J. F. (1983). Replication of temperature-sensitive mutants of alfalfa mosaic virus in protoplasts. *Virology* **125,** 64–74.

Sarachu, A. N., Huisman, M. J., van Vloten-Doting, L., and Bol, J. F. (1985). Alfalfa mosaic virus temperature-sensitive mutants. I. Mutants defective in viral RNA and protein synthesis. *Virology* **141,** 14–22.

Saric, A., and Wrischer, M. (1975). Fine structure changes in different host plants induced by grapevine fanleaf virus. *Phytopathol. Z.* **84,** 97–104.

Sarkar, S. (1969). Evidence of phenotypic mixing between two strains of tobacco mosaic virus. *Mol. Gen. Genet.* **105,** 87–90.

Sarkar, S. (1986). Tobacco mosaic virus: Mutants and strains. *In* "The Plant Viruses" (M. H. V. van Regenmortel and H. Fraenkel-Conrat, eds.), Vol. 2, pp. 59–77. Plenum, New York.

Saunders, K., Lucy, A. P., and Covey, S. N. (1989). Characterization of cDNA clones of host RNAs isolated from cauliflower mosaic virus-infected turnip leaves. *Physiol. Mol. Plant Pathol.* **35,** 339–346.

Savithri, H. S., and Erickson, J. W. (1983). The self assembly of the cowpea strain of southern bean mosaic virus. Formation of $T=1$ and $T=3$ nucleoprotein particles. *Virology* **126,** 328–335.

Savithri, H. S., Munshi, S. K., Suryanarayana, S., Divakar, S., and Murthy, M. R. N. (1987). Stability of belladonna mottle virus particles: The role of polyamines and calcium. *J. Gen. Virol.* **68,** 1533–1542.

Sawyer, L., Tollin, P., and Wilson, W. R. (1987). A comparison between the predicted secondary structures of potato virus X and papaya mosaic virus coat proteins. *J. Gen. Virol.* **68,** 1229-1232.

Saxena, R. C., and Khan, Z. R. (1985). Electronically recorded disturbances in feeding behaviour of *Nephotettix virescens* (Homoptera: Cicadellidae) on neem oil-treated rice plants. *J. Econ. Entomol.* **78,** 222–226.

Scalla, R., Boudon, E., and Rigaud, J. (1976) Sodium dodecyl sulphate–polyacrylamide-gel electrophoretic detection of two high molecular weight proteins associated with tobacco mosaic virus infection in tobacco. *Virology* **69,** 339–345.

Scalla, R., Romaine, P., Asselin, A., Rigaud, J., and Zaitlin, M. (1978). An *in vivo* study of a nonstructural polypeptide synthesized upon TMV infection and its identification with a polypeptide synthesized *in vitro* from TMV RNA. *Virology* **91,** 182–193.

Schachman, H. K. (1959). "Ultracentrifugation in Biochemistry." Academic Press, New York.

Schalk, H.-J., Matzeit, V., Schiller, B., Schell, J., and Gronenborn, B. (1989). Wheat dwarf virus, a geminivirus of graminaceous plants needs splicing for replication. *EMBO J.* **8,** 359–364.

Schaskolskaya, N. D., Atabekov, J. G., Sacharovskaya, G. N., and Javachia, V. G. (1968). Replication of temperature-sensitive strain of tobacco mosaic virus under nonpermissive conditions in the presence of helper strain. *Biol. Sci. USSR* **8** 101–105.

Schell, J. St. (1987). Transgenic plants as tools to study the molecular organisation of plant genes. *Science* **237,** 1176–1183.

Scheller, H. V., and Shukle, R. H. (1986). Feeding behaviour and transmission of barley yellow dwarf virus by *Sitobion avenae* on oats. *Entomol. Exp. Appl.* **40,** 189–195.

Schlegel, D. E., and Delisle, D. E. (1971). Viral protein in early stages of clover yellow mosaic virus infection of *Vicia faba. Virology* **45,** 747–754.

Schmelzer, K., Schmidt, H. E., and Schmidt, H. B. (1966). Viruskrankheiten und Virusverdächtige Erscheinungen an Forstgehölzen. *Arch. Forstwes.* **15,** 107–120.

Schmidt, H. E., and Zobywalski, S. (1984). Determination of pathotypes of bean yellow mosaic virus using *Phaseolus vulgaris* L. as a differential host. *Arch. Phytopathol. Pflanzenschutz* **20,** 95–96.

Schmidt, T., and Johnson, J. E. (1983). The spherically averaged structures of cowpea mosaic virus components by X-ray solution scattering. *Virology* **127,** 65–73.

Schneider, I. R. (1964). Difference in the translocatability of tobacco ringspot and southern bean mosaic viruses in bean. *Phytopathology* **54,** 701–705.

Schneider, I. R., and Thompson, S. M. (1977). Double-stranded nucleic acids found in tissue infected with the satellite of tobacco ringspot virus. *Virology* **78,** 453–462.

Schneider, I. R., and Worley, J. F. (1959a). Upward and downward transport of infectious particles of southern bean mosaic virus through steamed portions of bean stems. *Virology* **8,** 230–242.

Schneider, I. R., and Worley, J. F. (1959b). Rapid entry of infectious particles of southern bean mosaic virus into living cells following transport of the particles in the water stream. *Virology* **8,** 243–249.

Schneider, I. R., Hull, R., and Markham, R. (1972a). Multidense satellite of tobacco ringspot virus: A regular series of components of different densities. *Virology* **47,** 320–330.

Schneider, I. R., White, R. M., and Gooding, G. V., Jr. (1972b). Two new isolates of the satellite of tobacco ringspot virus. *Virology* **50,** 902–905.

Schneider, R. J., and Shenk, T. (1987). Impact of virus infection on host cell protein synthesis. *Annu. Rev. Biochem.* **56,** 317–332.

Schnölzer, M., Haas, B., Ramm, K., Hofmann, H., and Sänger, H. L. (1985). Correlation between structure and pathogenicity of potato spindle tuber viroid (PSTV). *EMBO J.* **4,** 2181–2190.

Schoelz, J. E., and Shepherd, R. J. (1988). Host range control of cauliflower mosaic virus. *Virology* **162,** 30–37.

Schoelz, J. E., and Zaitlin, M. (1989). Tobacco mosaic virus RNA enters chloroplasts *in vivo. Proc. Natl. Acad. Sci. U.S.A.* **86,** 4496–4500.

Schoelz, J. E., Shepherd, R. J., and Daubert, S. D. (1986). Region VI of cauliflower mosaic virus encodes a host range determinant. *Mol. Cell. Biol.* **6,** 2632–3627.

Schroeder, W. T., and Provvidenti, R. (1968). Resistance of bean (*Phaseolus vulgaris*) to the PV2 strain of bean yellow mosaic virus conditioned by the single dominant gene By. *Phytopathology* **58,** 1710.

Schroeder, W. T., and Provvidenti, R. (1971). A common gene for resistance to bean yellow mosaic virus and watermelon mosaic virus 2 in *Pisum sativum. Phytopathology* **61,** 846–848.

Schuchalter-Eicke, G., and Jeske, H. (1983). Seasonal changes in the chloroplast ultrastructure in *Abutilon* mosaic virus (AbMV) infected *Abutilon* spec. (Malvaceae). *Phytopathol. Z.* **108,** 172–184.

Schultze, M., Jiricny, J., and Hohn, T. (1990). Open reading frame VIII is not required for viability of cauliflower mosaic virus. *Virology* **176,** 662–664.

Schumacher, J., Meyer, N., Riesner, D., and Weidemann, H. L. (1986). Diagnostic procedure for detection of viroids and viruses with circular RNAs by "return"-gel electrophoresis. *J. Phytopathol.* **115,** 332–343.

Schuster, A. M., Burbank, D. E., Meister, B., Skrdla, M. P., Meints, R. H., Hattman, S., Swinton, D., and van Etten, J. L. (1986). Characterization of viruses infecting a eukaryotic *Chlorella*-like green alga. *Virology* **150,** 170–177.

Schwinghamer, M. W., and Broadbent, P. (1987). Association of viroids with a graft-transmissible dwarfing symptom in Australian orange trees. *Phytopathology* **77,** 205–209.

Scotti, P. D. (1985). The estimation of virus density in isopycnic cesium chloride gradients. *J. Virol. Methods* **12,** 149–160.

Sdoodee, R., and Teakle, D. S. (1987). Transmission of tobacco streak virus by *Thrips tabaci:* A new method of plant virus transmission. *Plant Pathol.* **36,** 377–380.

Sehnke, P. C., Mason, A. M., Hood, S. J., Lister, R. M., and Johnson, J. E. (1989). A "zinc-finger"-type binding domain in tobacco streak virus coat protein. *Virology* **168,** 48–56.

Sela, I., Reichman, M., and Weissbach, A. (1984). Comparison of dot molecular hybridization and enzyme-linked immunosorbent assay for detecting tobacco mosaic virus in plant tissues and protoplasts. *Phytopathology* **74,** 385–389.

Sela, I., Grafi, G., Sher, N., Edelbaum, O., Yagev, H., and Gerassi, E. (1987). Resistance systems related to the N gene and their comparison with interferon. *In* "Plant Resistance to Viruses" (D. Evered and S. Harnell, eds.), pp. 109–119. Wiley, Chichester.

Selling, B. H., Allison, R. F., and Kaesberg, P. (1990). Genomic RNA of an insect virus directs synthesis of infectious virions in plants. *Proc. Natl. Acad. Sci. U.S.A.* **87,** 434–438.

Selman, B. J. (1973). Beetles—Phytophagous Coleoptera. *In* "Viruses and Invertebrates" (A. J. Gibbs, ed.), pp. 157–177. North-Holland Publ., Amsterdam.

Selsky, M. I., and Black, L. M. (1962). Effect of high and low temperatures on the survival of wound-tumor virus in sweet clover. *Virology* **16,** 190–198.

Selstam, E., and Jackson, A. O. (1983). Lipid composition of sonchus yellow net virus. *J. Gen. Virol.* **64,** 1607–1613.

Semal, J., and Kummert, J. (1969). Effects of actinomycin D on the incorporation of uridine into virus-infected leaf fragments. *Phytopathol. Z.* **65,** 364–372.

Semancik, J. S., ed. (1987). "Viroids and Viroid-like Pathogens." CRC Press, Boca Raton, Florida.

Semancik, J. S., and Vanderwoude, W. J. (1976). Exocortis viroid: Cytopathic effects at the plasma membrane in association with pathogenic RNA *Virology* **69,** 719–726.

Semancik, J. S., Roistacher, C. N., Rivera-Bustamente, R., and Duran-Vila, N. (1988). Citrus cachexia viroid, a new viroid of citrus: Relationship to viroid of the exocortis disease complex. *J. Gen. Virol.* **69,** 3059–3068.

Serwer, P. (1977) Flattening and shrinkage of bacteriophage T7 after preparation for electron microscopy by negative staining. *J. Ultrastruct. Res.* **58,** 235–243.

Shahabuddin, M., Shaw, J. G., and Rhoads, R. E. (1988). Mapping of the tobacco vein mottling virus VPg cistron. *Virology* **163,** 635–637.

Shalla, T. A. (1959). Relations of tobacco mosaic virus and barley stripe mosaic virus to their host cells as revealed by ultrathin tissue-sectioning for the electron microscope. *Virology* **7,** 193–219.

Shalla, T. A., and Petersen, L. J. (1973). Infection of isolated plant protoplasts with potato virus X. *Phytopathology* **63,** 1125–1130.

Shalla, T. A., and Shepard, J. F. (1970). A virus induced soluble antigen associated with potato virus-X infection. *Virology* **42,** 1130–1132.

Shalla, T. A., and Shepard, J. F. (1972). The structure and antigenic analysis of amorphous inclusion bodies induced by potato virus X. *Virology* **49,** 654–667.

Shalla, T. A., Shepherd, R. J., and Petersen, L. J. (1980). Comparative cytology of nine isolates of cauliflower mosaic virus. *Virology* **102,** 381–388.

Shalla, T. A., Petersen, L.-J., and Zaitlin, M. (1982). Restricted movement of a temperature-sensitive virus in tobacco leaves is associated with a reduction in numbers of plasmodesmata. *J. Gen. Virol.* **60,** 355–358.

Shanks, M., and Lomonossoff, G. P. (1990). The primary structure of the 24K protease from red clover mottle virus: Implications for the mode of action of comovirus proteases. *J. Gen. Virol.* **71,** 735–738.

Shanks, M., Lomonossoff, G. P., and Evans, D. (1985). Double-stranded, replicative form RNA molecules of cowpea mosaic virus are not infectious. *J. Gen. Virol.* **66,** 925–930.

Shanks, M., Stanley, J., and Lomonossoff, G. P. (1986). The primary structure of red clover mottle virus middle component RNA. *Virology* **155,** 697–706.

Shanks, M., Maule, A. J., Wilson, I. G., Lomonossoff, G. P., Huskison, N., and Tomenius, K. (1988). RCMV gene expression. *Annu. Rep. John Innes Inst.* **73.**

Shanks, M., Tomenius, K., Chapham, D., Huskisson, N. S., Barker, P. J., Wilson, I. G., Maule, A. J., and Lomonossoff, G. P. (1989). Identification and subcellular localization of a putative cell-to-cell transport protein from red clover mottle virus. *Virology* **173,** 400–407.

Shaw, J. G. (1969). *In vivo* removal of protein from tobacco mosaic virus after inoculation of tobacco leaves. II. Some characteristics of the reaction. *Virology* **37,** 109–116.

Shaw, J. G. (1973). *In vivo* removal of protein from tobacco mosaic virus after inoculation of tobacco leaves. III. Studies on the location on virus particles for the initial removal of protein. *Virology* **53,** 337–342.

Shaw, J. G., Plaskitt, K. A., and Wilson, T. M. A. (1986). Evidence that tobacco mosaic virus particles disassemble cotranslationally *in vivo*. *Virology* **148,** 326–336.

Shaw, J. G., Hunt, A. G. Pirone, T. P., and Rhodes, R. E. (1990). Organisation and expression of potyviral genes. *In* "Viral Genes and Plant Pathogenesis" (T. P. Pirone and J. G. Shaw, eds.), pp. 107–123. Springer-Verlag, New York.

Sheffield, F. M. L. (1939). Micrurgical studies on virus-infected plants. *Proc. R. Soc. London, Ser. B* **126,** 529–538.

Sheldon, C. C., and Symons, R. H. (1989). Mutagenesis analysis of a self-cleaving RNA. *Nucleic Acids Res.* **17,** 5679–5685.

Sheldon, C. C., Jeffries, A. C., Davies, C., and Symons, R. H. (1990). RNA self-cleavage by the hammerhead structure. *Nucleic Acids Mol. Biol.* **4,** 227–242.

Shepard, J. F. (1975). Regeneration of plants from protoplasts of potato virus X-infected tobacco leaves. *Virology* **66,** 492–501.

Shepard, J. F. (1981). Protoplasts as sources of disease resistance in plants. *Annu. Rev. Phytopathol.* **19,** 145–166.

Shepard, J. F., and Uyemoto, J. K. (1976). Influence of elevated temperatures on the isolation and proliferation of mesophyll protoplasts from PVX- and PVY-infected tobacco tissue. *Virology* **70**, 558–560.

Shepardson, S., Esau, K., and McCrum, R. (1980). Ultrastructure of potato leaf phloem infected with potato leafroll virus. *Virology* **105**, 379–392.

Shepherd, R. J. (1972). Transmission of viruses through seed and pollen. *In* "Principles and Techniques in Plant Virology" (C. I. Kado and H. O. Agrawal, eds.), pp. 267–292. Van Nostrand-Reinhold, Princeton, New Jersey.

Shepherd, R. J. (1976). DNA viruses of higher plants. *Adv. Virus Res.* **20**, 305–339.

Shepherd, R. J., and Hills, F. J. (1970). Dispersal of beet yellows and beet mosaic viruses in the inland valleys of California. *Phytopathology* **60**, 798–804.

Shepherd, R. J., Wakeman, R. J., and Romanko, R. R. (1968). DNA in cauliflower mosaic virus. *Virology* **36**, 150–152.

Shepherd, R. J., Bruening, G. E., and Wakeman, R. J. (1970). Double-stranded DNA from cauliflower mosaic virus. *Virology* **41**, 339–347.

Sherwood, J. L. (1987). Demonstration of the specific involvement of coat protein in tobacco mosaic virus (TMV) cross protection using a TMV coat protein mutant. *J. Phytopathol.* **118**, 358–362.

Sherwood, J. L., and Fulton, R. W. (1982). The specific involvement of coat protein in tobacco mosaic virus cross protection. *Virology* **119**, 150–158.

Sherwood, J. L., Sanborn, M. R., Keyser, G. C., and Meyers, L. D. (1989). Use of monoclonal antibodies in detection of tomato spotted wilt virus. *Phytopathology* **79**, 61–64.

Shewmaker, C. K., Caton, J. R., Houck, C. M., and Gardner, R. C. (1985). Transcription of cauliflower mosaic virus integrated into plant genomes. *Virology* **140**, 281–288.

Shields, S. A., and Wilson, T. M. A. (1987). Cell-free translation of turnip mosaic virus RNA. *J. Gen. Virol.* **68**, 169–180.

Shields, S. A., Brisco, M. J., Wilson, T. M. A., and Hull, R. (1989). Southern bean mosaic virus RNA remains associated with swollen virions during translation in wheat germ cell-free extracts. *Virology* **171**, 602–606.

Shih, D.-S., and Kaesberg, P. (1973). Translation of brome mosaic viral ribonucleic acid in a cell-free system derived from wheat embryo. *Proc. Natl. Acad. Sci. U.S.A.* **70**, 1799–1803.

Shih, D.-S., and Kaesberg, P. (1976). Translation of the RNAs of brome mosaic virus: The monocistronic nature of RNA1 and RNA2. *J. Mol. Biol.* **103**, 77–88.

Shih, D.-S., Bu, M., Price, M. A., and Shih, C.-Y. T. (1987). Inhibition of cleavage of a plant viral polyprotein by an inhibitor activity present in wheat germ and cowpea embryos. *J. Virol.* **61**, 912–915.

Shikata, E. (1987). Hop stunt. *In* "The Viroids" (T. O. Diener, ed.), pp. 279–290. Plenum, New York and London.

Shikata, E., and Maramorosch, K. (1966). Electron microscopy of pea enation mosaic virus in plant cell nuclei. *Virology* **30**, 439–454.

Shikata, E., and Maramorosch, K. (1967). Electron microscopy of wound tumor virus assembly sites in insect vectors and plants. *Virology* **32**, 363–377.

Shinshi, H., Mohnen, D., and Meins, F., Jr. (1987). Regulation of a plant pathogenesis-related enzyme: Inhibition of chitinase and chitinase mRNA accumulation in cultured tobacco tissues by auxin and cytokinin. *Proc. Natl. Acad. Sci. U.S.A.* **84**, 89–93.

Shintaku, M., and Palukaitis, P. (1990). Genetic mapping of cucumber mosaic virus. *In* "Viral Genes and Plant Pathogenesis" (T. P. Pirone and J. G. Shaw, eds.), pp. 156–164. Springer-Verlag, New York.

Shirako, Y., and Brakke, M. K. (1984). Spontaneous deletion mutation of soil-borne wheat mosaic virus RNA II. *J. Gen. Virol.* **65**, 855–858.

Short, J. M., Fernandez, J. M., Sorge, J. A., and Huse, W. D. (1988). λ ZAP: A bacteriophage λ expression vector with *in vivo* excision properties. *Nucleic Acids Res.* **16**, 7583–7600.

Short, M. N., and Davies, J. W. (1983). Narcissus mosaic virus: A potexvirus with an encapsidated subgenomic messenger RNA for coat protein. *Biosci. Rep.* **3**, 837–846.

Short, M. N., Turner, D. S., March, J. F., Pappin, D. J. C., Parente, A., and Davies, J. W. (1986). The primary structure of papaya mosaic virus coat protein. *Virology* **152**, 280–283.

Shukla, D. D., and Schmelzer, K. (1973). Studies on viruses and virus disease of cruciferous plants. XIV. Cucumber mosaic virus in ornamental and wild species. *Acta Phytopathol. Acad. Sci. Hung.* **8**, 149–155.

Shukla, D. D., and Schmelzer, K. (1975). Studies on viruses and virus diseases of cruciferous plants. XIX. Analysis of the results obtained with ornamental and wild species. *Acta Phytopathol. Acad. Sci. Hung.* **10**, 217–229.

Shukla, D. D., and Ward, C. W. (1988). Amino acid sequence homology of coat proteins as a basis for identification and classification of the potyvirus group. *J. Gen. Virol.* **69,** 2703–2710.

Shukla, D. D., and Ward, C. W. (1989a). Structure of potyvirus and coat proteins and its application in the taxonomy of the potyvirus group. *Adv. Virus Res.* **36,** 273–314.

Shukla, D. D., and Ward, C. W. (1989b). Identification and classification of potyviruses on the basis of coat protein sequence data and serology. *Arch. Virol.* **106,** 171–200.

Shukla, D. D., O'Donnell, I. J., and Gough, K. H. (1983). Characteristics of the electro-blot radioimmunoassay (EBRIA) in relation to the identification of plant viruses. *Acta Phytopathol. Acad. Sci. Hung.* **18,** 79–84.

Shukla, D. D., Gough, K. H., and Ward, C. W. (1987). Coat protein of potyviruses 3. Comparison of amino acid sequences of four Australian strains of sugarcane mosaic virus. *Arch. Virol.* **96,** 59–74.

Shukla, D. D., McKern, N. M., Gough, K. H., Tracey, S. L., and Letho, S. G. (1988a). Differentiation of potyviruses and their strains by high performance liquid chromatographic peptide profiling of coat proteins. *J. Gen. Virol.* **69,** 493–502.

Shukla, D. D., Strike, P. M., Tracy, S. L., Gough, K. H., and Ward, C. W. (1988b). The N and C termini of the coat proteins of potyviruses are surface located and the N terminus contains the major virus-specific epitopes. *J. Gen. Virol.* **69,** 1497–1508.

Shukla, D. D., Jilka, J., Tosic, M., and Ford, R. E. (1989a). A novel approach to the serology of potyviruses involving affinity-purified polyclonal antibodies directed towards virus specific N-termini of coat proteins. *J. Gen. Virol.* **70,** 13–23.

Shukla, D. D., Tosic, M., Jilka, J., Ford, R. E., Toler, R. W., and Langham, M. A. C. (1989b). Taxonomy of potyviruses infecting maize, sorghum and sugarcane in Australia and the United States as determined by reactivities of polyclonal antibodies directed towards virus-specific N-termini of coat proteins. *Phytopathology* **79,** 223–229.

Shukla, D. D., Tribbick, G., Mason, T.-J., Hewish, D. R., Geysen, H. M., and Ward, C. W. (1989c). Localization of virus-specific and group-specific epitopes of plant potyviruses by systematic immunochemical analysis of overlapping peptide fragments. *Proc. Natl. Acad. Sci. U.S.A.* **86,** 8192–8196.

Shukla, D. D., Ford, R. E., Tosic, M., Jilka, J., and Ward, C. W. (1989d). Possible members of the potyvirus group transmitted by mites or whiteflies share epitopes with aphid-transmitted definitive members of the group. *Arch. Virol.* **105,** 143–151.

Shukla, V. D., and Anjaneyulu, A. (1981). Plant spacing to reduce rice tungro incidence. *Plant Dis.* **65,** 584–586.

Shukla, V. D., and Anjaneyulu, A. (1982). Evaluation of systemic insecticides to reduce tungro disease incidence in rice nursery. *Indian Phytopathol.* **35,** 502–504.

Shukle, R. H., Lampe, D. J., Lister, R. M., and Foster, J. E. (1987). Aphid feeding behaviour: Relationship to barley yellow dwarf virus resistance in *Agropyron* species. *Phytopathology* **77,** 725–729.

Siaw, M. F. E., Shahabuddin, M., Ballard, S., Shaw, J., and Rhoads, R. E. (1985). Identification of a protein covalently linked to the 5′ terminus of tobacco vein mottling virus RNA. *Virology* **142,** 134–143.

Sidwell, R. W., Huffman, J. H., Khare, G. P., Allen, L. B., Witkowski, J. T., and Robins, R. K. (1972). Broad-spectrum antiviral activity of virazole: 1-β-D-ribofuranosyl-1,2,4-triazole-3-carboxamide. *Science* **177,** 705–706.

Siegel, A. (1960). Studies on the induction of tobacco mosaic virus mutants with nitrous acid. *Virology* **11,** 156–167.

Siegel, A. (1971). Pseudovirions of tobacco mosaic virus. *Virology* **46,** 50–59.

Siegel, A. (1985). Plant-virus-based vectors for gene transfer may be of considerable use despite a presumed high error frequency during RNA synthesis. *Plant Mol. Biol.* **4,** 327–329.

Siegel, A., Hari, V., and Kolacz, K. (1978). The effect of tobacco mosaic virus infection on host and virus-specific protein synthesis in protoplasts. *Virology* **85,** 494–503.

Siitari, H., and Kurppa, A. (1987). Time resolved fluoroimmunoassay in the detection of plant viruses. *J. Gen. Virol.* **68,** 1423–1428.

Silber, G., and Burk, L. G. (1965). Infectivity of tobacco mosaic virus stored for fifty years in extracted "unpreserved" plant juice. *Nature (London)* **206,** 740–741.

Silberklang, M., Prochiantz, A., Haenni, A.-L., and Rajbhandary, U. L. (1977). Studies on the sequence of the 3′ terminal region of turnip yellow mosaic virus RNA. *Eur. J. Biochem.* **72,** 465–478.

Silva, A. M., and Rossmann, M. G. (1987). Refined structure of southern bean mosaic virus at 2.9 Å resolution. *J. Mol. Biol.* **197,** 69–87.

Silva, J. L., and Weber, G. (1988). Pressure-induced dissociation of brome mosaic virus. *J. Mol. Biol.* **199,** 149–159.

Simon, A. E., and Howell, S. H. (1986). The virulent satellite RNA of turnip crinkle virus has a major domain homologous to the 3' end of the helper virus genome. *EMBO J.* **5,** 3423–3428.

Simon, A. E., and Howell, S. (1987). Synthesis *in vitro* of infectious RNA copies of the virulent satellite of turnip crinkle virus. *Virology* **156,** 146–152.

Simon, A. E., Engel, H., Johnson, R. P., and Howell, S. H. (1988). Identification of regions affecting virulence, RNA processing and infectivity in the virulent satellite of turnip crinkle virus. *EMBO J.* **7,** 2645–2651.

Simpkins, I., Walkey, D. G. A., and Neely, H. A. (1981). Chemical suppression of virus in cultured plant tissues. *Ann. Appl. Biol.* **99,** 161–169.

Singh, M. N., Khurana, S. M. P., Nagaich, B. B., and Agrawal, H. O. (1988). Environmental factors influencing aphid transmission of potato virus Y and potato leafroll virus. *Potato Res.* **31,** 501–509.

Singh, R. P. (1984). *Solanum* X *berthaultii,* a sensitive host for indexing potato spindle tuber viroid from dormant tubers. *Potato Res.* **27,** 163–172.

Singh, R. P., and Slack, S. A. (1984). Reactions of tuber-bearing *Solanum* species to infection with potato spindle tuber viroid. *Plant Dis.* **68,** 784–787.

Singh, R. P., and Somerville, T. H. (1987). New disease symptoms observed on field-grown potato plants with potato spindle tuber viroid and potato virus Y infections. *Potato Res.* **30,** 127–132.

Singh, R. P., Lee, C. R., and Clark, M. C. (1974). Manganese effect on the local lesion symptom of potato spindle tuber "virus" in *Scopolia sinensis. Phytopathology* **64,** 1015–1018.

Singh, R. P., Boucher, A., and Seabrook, J. E. A. (1988). Detection of the mild strains of potato spindle tuber viroid from single true potato seed by return electrophoresis. *Phytopathology* **78,** 663–667.

Singh, S. K., Anjaneyulu, A., and Lapierre, H. (1984). Use of pectino-cellulolytic enzymes for improving extraction of phloem-limited plant viruses as exemplified by the rice tungro virus complex. *Agronomie* **4,** 479–484.

Sinha, R. C. (1967). Response of wound tumor virus infection in insects to vector age and temperature. *Virology* **31,** 746–748.

Sinijärv, R., Järvekülg, L., Andreeva, E., and Saarma, M. (1988). Detection of potato virus X by one incubation europium time-resolved fluoroimmunoassay and ELISA. *J. Gen. Virol.* **69,** 991–998.

Sisler, W. W., and Timian, R. G. (1956). Inheritance of the barley stripe mosaic resistance of Modjo (CT3212) and CT33121-1. *Plant Dis. Rep.* **40,** 1106–1108.

Sit, T. L., AbouHaidar, M. G., and Holy, S. (1989). Nucleotide sequence of papaya mosaic virus RNA. *J. Gen. Virol.* **70,** 2325–2331.

Skaria, M., Lister, R. M., and Foster, J. E. (1984). Lack of barley yellow dwarf virus dosage effects on virus content in cereals. *Plant. Dis.* **68,** 759–761.

Skotnicki, A., Gibbs, A., and Wrigley, N. G. (1976). Further studies on *Chara corallina* virus. *Virology* **75,** 457–468.

Skryabin, K. G., Kraev, A. S., Morozov, S. Yu, Rozanov, M. N., Chernov, B. K., Lukasheva, L. I., and Atabekov, J. G. (1988a). The nucleotide sequence of potato virus X RNA. *Nucleic Acids Res.* **16,** 10929–10930.

Skryabin, K. G., Morozov, S. Yu, Kraev, A. S., Rozanov, M. N., Chernov, B. K., Lukasheva, L. I., and Atabekov, J. G. (1988b). Conserved and variable elements in RNA genomes of potexviruses. *FEBS Lett.* **240,** 33–40.

Slack, C. R. (1969). Localization of certain photosynthetic enzymes in mesophyll and parenchyma sheath chloroplasts of maize and *Amaranthus palmeri. Photochemistry* **8,** 1387–1391.

Sleat, D. E., and Palukaitis, P. (1990a). Site-directed mutagenesis of a plant viral satellite RNA changes its phenotype from ameliorative to necrogenic. *Proc. Natl. Acad. Sci. U.S.A.* **87,** 2946–2950.

Sleat, D. E., and Palukaitis, P. (1990b). Induction of tobacco chlorosis by cucumber mosaic virus satellite RNAs is specific to subgroup II helper strains. *Virology* **176,** 292–295.

Sleat, D. E., Turner, P. C., Finch, J. T., Butler, P. J. G., and Wilson, T. M. A. (1986). Packaging of recombinant RNA molecules into pseudovirus particles directed by the origin-of-assembly sequences from tobacco mosaic virus RNA. *Virology* **155,** 299–308.

Sleat, D. E., Gallie, D. R., Jefferson, R. A., Bevan, M. W., Turner, P. C., and Wilson, T. M. A. (1987). Characterisation of the 5' leader sequence of tobacco mosaic virus RNA as a general enhancer of translation *in vitro. Gene* **60,** 217–225.

Sleat, D. E., Hull, R., Turner, P. C., and Wilson, T. M. A. (1988a). Studies on the mechanism of translational enhancement by the 5'-leader sequence of tobacco mosaic virus RNA. *Eur. J. Biochem.* **175,** 75–86.

Sleat, D. E., Plaskitt, K. A., and Wilson, T. M. A. (1988b). Selective encapsidation of CAT gene transcripts in TMV-infected transgenic tobacco inhibits CAT synthesis. *Virology* **165**, 609–612.

Sleat, D. E., Gallie, D. R., Watts, J. W., Deom, C. M., Turner, P. C., Beachy, R. N., and Wilson, T. M. A. (1988c). Selective recovery of foreign gene transcripts as virus-like particles in TMV-infected transgenic tobaccos. *Nucleic Acids Res.* **16**, 3127–3140.

Skomka, M. J., Buck, K. W., and Coutts, R. H. A. (1988). Characterisation of multimeric DNA forms associated with tomato golden mosaic virus infection. *Arch. Virol.* **100**, 99–108.

Slykhuis, J. T. (1955). *Aceria tulipae* Keifer (Acarina: Eriophyidae) in relation to the spread of wheat streak mosaic. *Phytopathology* **45**, 116–128.

Slykhuis, J. T. (1962). Mite transmission of plant viruses. *In* "Biological Transmission of Disease Agents" (K. Maramorosch, ed.), pp. 41–61. Academic Press, New York.

Slykhuis, J. T. (1963). Vector and host relations of North American wheat striate mosaic virus. *Can. J. Bot.* **41**, 1171–1185.

Slykhuis, J. T. (1965). Mite transmission of plant viruses. *Adv. Virus Res.* **11**, 97–137.

Slykhuis, J. T. (1974). Differentiation of transmission and incubation temperatures for wheat spindle streak mosaic virus. *Phytopathology* **64**, 554–557.

Slykhuis, J. T. (1976). Stimulating effects of transplanting on the development of wheat spindle streak mosaic virus in wheat plants infected from soil. *Phytopathology* **66**, 130-131.

Slykhuis, J. T., and Barr, D. J. S. (1978). Confirmation of *Polymyxa graminis* as a vector of wheat spindle streak mosaic virus. *Phytopathology* **68**, 639–643.

Slykhuis, J. T., Andrews, J. E., and Pittmann, U. J. (1957). Relation of date of seeding winter wheat in southern Alberta to losses from wheat streak mosaic, root rot and rust. *Can. J. Plant Sci.* **37**, 113–127.

Smart, T. E., Dunigan, D. D., and Zaitlin, M. (1987). *In vitro* translation products of mRNAs derived from TMV-infected tobacco exhibiting a hypersensitive response. *Virology* **158**, 461–464.

Smirnyagina, E. V., Karpova, O. V., Miroshnichenko, N. P., Rodinova, N. P., and Atabekov, J. G. (1989). Translational *in vitro* activity of the 3a gene and the coat protein gene derived from brome mosaic virus RNA3 by site-specific cleavage with RNase H (FEB 07513). *FEBS Lett.* **254**, 66–68.

Smit, C. H., Roosien, J., van Vloten-Doting, L., and Jaspars, E. M. J. (1981). Evidence that alfalfa mosaic virus infection starts with three RNA–protein complexes. *Virology* **112**, 169–173.

Smith, A. E., and Kenyon, D. H. (1972). The origin of viruses from cellular genetic material. *Enzymologia* **43**, 13–18.

Smith, D. B., and Inglis, S. C. (1987). The mutation rate and variability of eukaryotic viruses: An analytical review. *J. Gen. Virol.* **68**, 2729–2740.

Smith, F. D., and Banttari, E. E. (1987). Dot-ELISA on nitrocellulose membranes for detection of potato leafroll virus. *Plant. Dis.* **71**, 795–799.

Smith, H. C. (1967). The effect of aphid numbers and stage of plant growth in determining tolerance to barley yellow dwarf virus in cereals. *N.Z. J. Agric. Res.* **10**, 445–466.

Smith, K. M. (1931). On the composite nature of certain potato virus diseases of the mosaic group as revealed by the use of plant indicators and selective methods of transmission. *Proc. R. Soc. London, Ser. B* **109**, 251–266.

Smith, K. M. (1937). "A Text Book of Plant Virus Diseases." Churchill, London.

Smith, K. M. (1972). "A Textbook of Plant Virus Diseases," 3rd ed. Longmans, Green, New York.

Smith, P. R. (1983). The Australian Fruit Variety Foundation and its role in supplying virus-tested planting material to the fruit industry. *Acta Hortic.* **130**, 263–266.

Smith, S. H., and Schlegel, D. E. (1964). The distribution of clover yellow mosaic virus in *Vicia faba* root tips. *Phytopathology* **54**, 1273–1274.

Smith, T. A. (1985). Polyamines. *Annu. Rev. Plant Physiol.* **36**, 117–143.

Smith, T. A. (1987). The isolation of the two electrophoretic forms of cowpea mosaic virus using fast protein liquid chromatography. *J. Virol. Methods* **16**, 263–269.

Sneath, P. H. A. (1962). The construction of taxonomic groups. *Symp. Soc. Gen. Microbiol.* **12**, 289–332.

Söber, J., Järvekülg, L., Toots, I., Radavsky, J., Villems, R., and Saarma, M. (1988). Antigenic characterisation of potato virus X with monoclonal antibodies. *J. Gen. Virol.* **69**, 1799–1807.

Sol, H. H. (1963). Some data on the occurrence of rattle virus at various depths in the soil and on its transmission. *Tijdschr. Plantenziekten* **69**, 208–214.

Solis, I., and García-Arenal, F. (1990). The complete nucleotide sequence of the genomic RNA of the tobamovirus tobacco mild green mosaic virus. *Virology* **177**, 553–558.

Sorger, P. K., Stockley, P. G., and Harrison, S. C. (1986). Structure and assembly of turnip crinkle virus II. Mechanism of reassembly *in vitro*. *J. Mol. Biol.* **191**, 639–658.

Soto, P. E., Buddenhagen, I. W., and Asnani, V. L. (1982). Development of streak virus-resistant maize populations through improved challenge and selection methods. *Ann. Appl. Biol.* **100**, 539–546.

Spirin, A. S. (1961). The "temperature effect" and macromolecular species of high polymer ribonucleic acids of various origins. *Biokhimiya* **26,** 454–463.

Sprague, G. F., and McKinney, H. H. (1966). Aberrant ratio: An anomaly in maize associated with virus infection. *Genetics* **54,** 1287–1296.

Sprague, G. F., and McKinney, H. H. (1971). Further evidence on the genetic behavior of AR in maize. *Genetics* **67,** 533–542.

Sprague, G. F., McKinney, H. H., and Greeley, L. W. (1963). Virus as a mutagenic agent in maize. *Science* **141,** 1052–1053.

Stace-Smith, R., and Hamilton, R. I. (1988). Inoculum thresholds of seed borne pathogens: Viruses. *Phytopathology* **78,** 875–880.

Stack, J. P., and Tattar, T. A. (1978). Measurement of transmembrane electropotentials of *Vigna sinensis* leaf cells infected with tobacco ringspot virus. *Physiol. Plant Pathol.* **12,** 173–178.

Stahl, E. (1894). Einige Versuche über Transpiration und Assimilation. *Bot. Z.* **6/7,** 117–145.

Stanley, J. (1983). Infectivity of the cloned geminivirus genome requires sequences from both DNAs. *Nature (London)* **305,** 643–645.

Stanley, J. (1985). The molecular biology of geminiviruses. *Adv. Virus Res.* **30,** 139-177.

Stanley, J., and Davies, J. W. (1985). Structure and function of the DNA genome of geminiviruses. *In* "Molecular Plant Virology" (J. W. Davies, ed.), Vol. 2, pp. 191–218. CRC Press, Boca Raton, Florida.

Stanley, J., and Gay, M. R. (1983). Nucleotide sequence of cassava latent virus DNA. *Nature (London)* **301,** 260-262.

Stanley, J., and Townsend, R. (1985). Characterisation of DNA forms associated with cassava latent virus infection. *Nucleic Acids Res.* **13,** 2189–2206.

Stanley, J., Hanau, R., and Jackson, A. O. (1984). Sequence comparison of the 3' ends of a subgenomic RNA and the genomic RNAs of barley stripe mosaic virus. *Virology* **139,** 375–383.

Stanley, J., Townsend, R., and Curson, S. J. (1985). Pseudorecombinants between cloned DNAs of two isolates of cassava latent virus. *J. Gen. Virol.* **66,** 1055–1061.

Stanley, J., Markham, P. G., Callis, R. J., and Pinner, M. S. (1986). The nucleotide sequence of an infectious clone of the geminivirus beet curly top virus. *EMBO J.* **5,** 1761–1767.

Stanley, W. M. (1935). Isolation of a crystalline protein possessing the properties of tobacco-mosaic virus. *Science* **81,** 644–645.

Stanley, W. M. (1936). Chemical studies on the virus of tobacco mosaic virus. VI. The isolation from diseased turkish tobacco plants of a crystalline protein possessing the properties of tobacco mosaic virus. *Phytopathology* **26,** 305–320.

Stauffacher, C. V., Usha, R., Harrington, M., Schmidt, T., Hosur, M. V., and Johnson, J. E. (1985). The structure of cowpea mosaic virus at 3.5 Å resolution. *NATO Adv. Sci. Inst. Ser., Ser. A* **126,** 293–308.

Steckert, J. J., and Schuster, T. M. (1982). Sequence specificity of trinucleoside diphosphate binding to polymerised tobacco mosaic virus protein. *Nature (London)* **299,** 32–36.

Steen, M. T., Kaper, J. M., Pleij, C. W. A., and Hansen, J. N. (1990). *In vitro* translation of cucumoviral satellites, III. Translational efficiencies of cucumber mosaic virus-associated RNA sequence variants can be related to the predicted secondary structures of their first 55 nucleotides. *Virus Genes* **4,** 41–52.

Steere, R. L. (1969). Freeze-etching simplified. *Cryobiology* **5,** 306–323.

Steger, G., Hofmann, H. Förtsch, J., Gross, H. J., Randles, J. W., Sänger, H. L., and Riesner, D. (1984). Conformational transitions in viroids and virusoids: Comparison of results from energy minimization algorithms and from experimental data. *J. Biomol. Struct. Dyn.* **2,** 543–571.

Steger, G., Tabler, M., Brüggermann, W., Colpan, M., Klotz, G., Sänger, H.-L., and Riesner, D. (1986). Structure of viroid replicative intermediates; Physico-chemical studies on SP6 transcripts of cloned oligomeric potato spindle tuber viroid. *Nucleic Acids Res.* **14,** 9613–9630.

Stenger, D. C., Richardson, J., Sylvester, E. S., Jackson, A. O., and Morris, T. J. (1988). Analysis of sowthistle yellow vein virus-specific RNAs in infected hosts. *Phytopathology* **78,** 1473–1477.

Steven, A. C., Trus, B. L., Putz, C., and Wurtz, M. (1981). The molecular organisation of beet necrotic yellow vein virus. *Virology* **113,** 428–438.

Stevens, C., and Gudauskas, R. T. (1983). Effects of maize dwarf mosaic virus infection of corn on inoculum potential of *Helminthosporium maydis* race 0. *Phytopathology* **73,** 439–441.

Stewart, C.-B., and Wilson, A. C. (1987). Sequence convergence and functional adaptation of stomach lysozymes from foregut fermenters. *Cold Spring Harbor Symp. Quant. Biol.* **52,** 891–899.

Stewart, R. N., and Dermen, H. (1975). Flexibility in ontogeny as shown by the contribution of the shoot apical layers to leaves of periclinal chimeras. *Am. J. Bot.* **62,** 935–947.

Stobbs, L. W., and McNeill, B. H. (1980). Increase of tobacco mosaic virus in graft-inoculated TMV resistant tomatoes. *Can. J. Plant Pathol.* **2,** 217–221.

Stobbs, L. W., Manocha, M. S., and Dias, H. F. (1977). Histological changes associated with virus localization in TMV-infected pinto bean leaves. *Physiol. Plant Pathol.* **11,** 87–94.

Stoeckle, M. Y., Shaw, M. W., and Choppin, P. W. (1987). Segment-specific and common nucleotide sequences in the noncoding regions of influenza B virus genome RNAs. *Proc. Natl. Acad. Sci. U.S.A.* **84,** 2703–2707.

Stollar, B. D. (1975). The specificity and applications of antibodies to helical nucleic acids. *CRC Crit. Rev. Biochem.* **3,** 45–69.

Stouffer, R. F. (1969). Transmission of apple viruses by leaf tissue implantation. *Phytopathology* **59,** 274–278.

Stouffer, R. F., and Ross, A. F. (1961). Effect of infection by potato virus Y on the concentration of potato virus X in tobacco plants. *Phytopathology* **51,** 740–744.

Stratford, R., and Covey, S. N. (1988). Changes in turnip leaf messenger RNA populations during systemic infections by severe and mild strains of cauliflower mosaic virus. *Mol. Plant-Microbe Interact.* **1,** 243–249.

Stratford, R., and Covey, S. N. (1989). Segregation of cauliflower mosaic virus symptom genetic determinants. *Virology* **172,** 451–459.

Stratford, R., Plastkitt, K. A., Turner, D. S., Markham, P. G., and Covey, S. N. (1988). Molecular properties of Bari 1, a mild strain of cauliflower mosaic virus. *J. Gen. Virol.* **69,** 2375–2386.

Stubbs, G., Warren, S., and Holmes, K. (1977). Structure of RNA and RNA binding site in tobacco mosaic virus from a 4 Å map calculated from X-ray fibre diagrams. *Nature (London)* **267,** 216–221.

Stubbs, L. L. (1956). Motley dwarf virus disease of carrot in California. *Plant Dis. Rep.* **40,** 763–764.

Stussi-Garaud, C., Garaud, J.-C., Berna, A., and Godefroy-Colburn, T. (1987). *In situ* location of an alfalfa mosaic virus non-structural protein in plant cell walls: Correlation with virus transport. *J. Gen. Virol.* **68,** 1779–1784.

Sugimura, Y., and Matthews, R. E. F. (1981). Timing of the synthesis of empty shells and minor nucleoproteins in relation to turnip yellow mosaic virus synthesis in *Brassica* protoplasts. *Virology* **112,** 70–80.

Sulzinski, M. A., and Zaitlin, M. (1982). Tobacco mosaic virus replication in resistant and susceptible plants: In some resistant species virus is confined to a small number of initially infected cells. *Virology* **121,** 12–19.

Sulzinski, M. A., Gabard, K. A., Palukaitis, P., and Zaitlin, M. (1985). Replication of tobacco mosaic. VIII. Characterisation of a third subgenomic TMV RNA. *Virology* **145,** 132–140.

Sumner, J. B. (1926). The isolation and crystallisation of the enzyme urease. *J. Biol. Chem.* **69,** 435–441.

Sun, W., and Gong, Z.-X. (1988). *In vitro* studies on the nucleocapsid-associated RNA polymerase of wheat rosette stunt virus. *Intervirology* **29,** 154–161.

Sunter, G., and Bisaro, D. M. (1989). Transcription map of the B genome component of tomato golden mosaic virus and comparison with A component transcripts. *Virology* **173,** 647–655.

Sunter, G., Coutts, R. H. A., and Buck, K. W. (1984). Negatively supercoiled DNA from plants infected with a single-stranded DNA virus. *Biochem. Biophys. Res. Commun.* **118,** 747–752.

Sunter, G., Buck, K. W., and Coutts, R. H. A. (1985). S1-sensitive sites in the supercoiled double-stranded form of tomato golden mosaic virus DNA component B; Identification of regions of potential alternative secondary structure and regulatory function. *Nucleic Acids Res.* **13,** 4645–4659.

Sunter, G., Gardiner, W. E., Rushing, A. E., Rogers, S. G., and Bisaro, D. M. (1987). Independent encapsidations of tomato golden mosaic virus A component DNA in transgenic plants. *Plant Mol. Biol.* **8,** 477–484.

Sunter, G., Gardiner, W. E., and Bisaro, D. M. (1989). Identification of tomato golden mosaic virus-specific RNAs in infected plants. *Virology* **170,** 243–250.

Suzuki, N., Kudo, T., Shirako, Y., Ehara, Y., and Tachibana, T. (1989). Distribution of cylindrical inclusion, amorphous inclusion, and capsid proteins of watermelon mosaic virus 2 in systemically infected pumpkin leaves. *J. Gen. Virol.* **70,** 1085–1091.

Swaans, H., and van Kammen, A. (1973). Reconsideration of the distinction between the severe and yellow strains of cowpea mosaic virus. *Neth. J. Plant Pathol.* **79,** 257–265.

Sweet, J. B. (1975). Soil-borne viruses occurring in nursery soils and infecting some ornamental species of Rosaceae. *Ann. Appl. Biol.* **79,** 49–54.

Sweet, J. B. (1980). Hedgerow hawthorn (*Crataegus* spp.) and blackthorn (*Prunus spinosa*) as hosts of fruit tree viruses in Britain. *Ann. Appl. Biol.* **94,** 83–90.

Swinkels, P. P. H., and Bol, J. F. (1980). Limited sequence variation in the leader sequence of RNA4 from several strains of alfalfa mosaic virus. *Virology* **106**, 145–147.

Sylvester, E. S. (1969). Evidence for transovarial passage of sowthistle yellow vein virus in the aphid *Hyperomyzus lactucae*. *Virology* **38**, 440–446.

Sylvester, E. S. (1973). Reduction of excretion, reproduction, and survival in *Hyperomyzus lactucae* fed on plants infected with isolates of sowthistle yellow vein virus. *Virology* **56**, 632–635.

Sylvester, E. S., and Richardson, J. (1970). Infection of *Hyperomyzus lactucae* by sowthistle yellow vein virus. *Virology* **42**, 1023–1042.

Sylvester, E. S., and Richardson, J. (1971). Decreased survival of *Hyperomyzus lactucae* inoculated with serially passed sowthistle yellow vein virus. *Virology* **46**, 310–317.

Sylvester, E. S., and Richardson, J. (1981). Inoculation of the aphids *Hyperomyzus lactucae* and *Chaelosiphon jacobi* with isolates of sowthistle yellow vein virus and strawberry crinkle virus. *Phytopathology* **71**, 598–602.

Sylvester, E. S., Richardson, J., and Frazier, N. W. (1974). Serial passage of strawberry crinkle virus in the aphid *Chaetosiphon jacobi*. *Virology* **59**, 301–306.

Symons, R. H. (1978). The two-step purification of ribosomal RNA and plant viral RNA by polyacrylamide slab gel electrophoresis. *Aust. J. Biol. Sci.* **31**, 25–37.

Symons, R. H. (1979). Extensive sequence homology at the 3′-termini of the four RNAs of cucumber mosaic virus. *Nucleic Acids Res.* **7**, 825–837.

Symons, R. H. (1981). Avocado sunblotch viroid; Primary sequence and proposed secondary structure. *Nucleic Acids Res.* **9**, 6527–6537.

Symons, R. H. (1989). Self-cleavage of RNA in the replication of small pathogens of plants and animals. *Trends Biochem. Sci.* **14**, 445–450.

Symons, R. H. (1990). Encapsidated circular satellite RNAs (virusoids) of plants. *Curr. Top. Microbiol. Immunol.* (in press).

Symons, R. H., Haseloff, J., Visvader, J. E., Keese, P., Murphy, P. J., Gill, D. S., Gordon, K. H. J., and Bruening, G. (1985). On the mechanism of replication of viroids, virusoids, and satellite RNAs. *In* "Subviral Pathogens of Plants and Animals: Viroids and Prions" (K. Maramorosch and J. J. McKelvey, Jr., eds.), pp. 235–263. Academic Press, Orlando, Florida.

Symons, R. H., Hutchins, C. J., Forster, A. C., Rathjen, P. D., Keese, P., and Visvader, J. E. (1987). Self cleavage of RNA in the replication of viroids and virusoids. *J. Cell Sci., Suppl.* **7**, 303–318.

Szatmari-Goodmann, G., and Nault, L. R. (1983). Test of oil-sprays for suppression of aphid-borne maize dwarf mosaic virus in Ohio sweet corn. *J. Econ. Entomol.* **76**, 144–149.

Takahashi, T. (1971). Studies on viral pathogenesis in plant hosts. I. Relation between host leaf age and the formation of systemic symptoms induced by tobacco mosaic virus. *Phytopathol. Z.* **71**, 275–284.

Takahashi, T. (1972a). Studies on viral pathogenesis in plant hosts. II. Changes in developmental morphology of tobacco plants infected systemically with tobacco mosaic virus. *Phytopathol. Z.* **74**, 37–47.

Takahashi, T. (1972b). Studies on viral pathogenesis in plant hosts. III. Leaf age-dependent susceptibility to tobacco mosaic virus infection in 'Samsun NN' and 'Samsun' tobacco plants. *Phytopathol. Z.* **75**, 140-155.

Takahashi, T., and Diener, T. O. (1975). Potato spindle tuber viroid. XIV. Replication in nuclei isolated from infected leaves. *Virology* **64**, 106–114.

Takahashi, W. N. (1956). Increasing the sensitivity of the local-lesion method of virus assay. *Phytopathology* **46**, 654–656.

Takahashi, W. N., and Rawlins, T. E. (1932). Method for determining shape of colloidal particles. Application in study of tobacco mosaic virus. *Proc. Soc. Exp. Biol. Med.* **30**, 155–157.

Takamatsu, N., Ohno, T., Meshi, T., and Okada, Y. (1983). Molecular cloning and nucleotide sequence of the 30K and the coat protein cistron of TMV (tomato strain) genome. *Nucleic Acids Res.* **11**, 3767–3778.

Takamatsu, N., Ishikawa, M., Meshi, T., and Okada, Y. (1987). Expression of bacterial chloramphenicol acetyltransferase gene in tobacco plants mediated by TMV-RNA. *EMBO J.* **6**, 307–311.

Takamatsu, N., Watanabe, Y., Meshi, T., and Okada, Y., (1990a). Mutational analysis of the pseudoknot region in the 3′ non-coding region of TMVRNA *J. Virology* **64**, 3686–3693.

Takamatsu, N., Watanabe, Y., Yanagi, H., Meshi, T., Shiba, T., and Okada, Y. (1990b). Production of enkephalin in tobacco protoplasts using tobacco mosaic virus RNA vector. *FEBS Lett.,* in press.

Takanami, Y. (1981). A striking change in symptoms on cucumber mosaic virus-infected tobacco plants induced by a satellite RNA. *Virology* **109**, 120–126.

Takanami, Y., Kubo, S., and Imaizumi, S. (1977). Synthesis of single- and double-stranded cucumber mosaic virus RNAs in tobacco mesophyll protoplasts. *Virology* **80**, 376–389.

Takanami, Y., Nitta, N., and Kubo, S. (1989). A marked improvement of efficiency in infection of tobacco mesophyll protoplasts with plant viruses and virus RNAs by using polyethyleneimine as a polycation. *Ann. Phytopathol. Soc. Jpn.* **55,** 324–329.

Takatsuji, H., Hirochika, H., Fukushi, T., and Ikeda, J.-E. (1986). Expression of cauliflower mosaic virus reverse transcriptase in yeast. *Nature (London)* **319,** 240–243.

Takebe, I. (1977). Protoplasts in the study of plant virus replication. *Compr. Virol.* **11,** 237–283.

Takebe, I., and Otsuki, Y. (1969). Infection of tobacco mesophyll protoplasts by tobacco mosaic virus. *Proc. Natl. Acad. Sci. U.S.A.* **64,** 843–848.

Takebe, I., Otsuki, Y., and Aoki, S. (1968). Isolation of tobacco mesophyll cells in intact and active state. *Plant Cell Physiol.* **9,** 115–124.

Takebe, I., Otsuki, Y., Honda, Y., and Matsui, C. (1975). Penetration of plant viruses into isolated tobacco leaf protoplasts. *In* "Proceedings of the First Intersectional Congress of the International Association of Microbiological Societies" (T. Hasegawa, ed.), Vol. 3, pp. 55–64. Science Council of Japan, Tokyo.

Taliansky, M. E., Atabekova, T. I., and Atabekov, J. G. (1977). The formation of phenotypically mixed particles upon mixed assembly of some tobacco mosaic virus (TMV) strains. *Virology* **76,** 701–708.

Taliansky, M. E., Atabekova, T. I., Kaplan, I. B., Morozov, S. Yu., Malyshenko, S. I., and Atabekov, J. G. (1982a). A study of TMV ts mutant NI2519. I. Complementation experiments. *Virology* **118,** 301–308.

Taliansky, M. E., Malyshenko, S. I., Pshennikova, E. S., Kaplan, I. B., Ulanova, E. F., and Atabekov, J. G. (1982b). Plant virus-specific transport function. I. Virus genetic control required for systemic spread. *Virology* **122,** 318–326.

Taliansky, M. E., Malyshenko, S. I., Pshennikova, E. S., and Atabekov, J. G. (1982c). Plant virus-specific transport function. II. A factor controlling virus host range. *Virology* **122,** 327–331.

Talley, J., Warren, F. H. J. B., Torrance, L., and Jones, R. A. C. (1980). A simple kit for detection of plant viruses by the latex serological test. *Plant Pathol.* **29,** 77–79.

Tamada, T., Harrison, B. D., and Roberts, I. M. (1984). Variation among British isolates of potato leafroll virus. *Ann. Appl. Biol.* **104,** 107–116.

Tamburro, A. M., Guantieri, V., Piazzolla, P., and Gallitelli, D. (1978). Conformational studies on particles of turnip yellow mosaic virus. *J. Gen. Virol.* **40,** 337–344.

Tanzi, M., Betti, L., De Jager, C. P., and Canova, A. (1986). Isolation of an attenuated virus mutant obtained from a TMV pepper strain after treatment with nitrous acid. *Phytopathol. Mediterr.* **25,** 119–124.

Tas, P. W. L., Boerjan, M. L., and Peters, D. (1977). The structural proteins of tomato spotted wilt virus. *J. Gen. Virol.* **36,** 267–279.

Tatchell, G. M., Plumb, R. T., and Carter, N. (1988). Migration of alate morphs of the bird cherry aphis (*Rhopalosiphum padi*) and implications for the epidemiology of barley yellow dwarf virus. *Ann. Appl. Biol.* **112,** 1–11.

Taylor, C. E., and Murant, A. F. (1968). Chemical control of raspberry ringspot and tomato black ring viruses in strawberry. *Plant Pathol.* **17,** 171–178.

Taylor, C. E., and Thomas, P. R. (1968). The association of *Xiphinema diversicaudatum* (Micoletsky) with strawberry latent ringspot and arabis mosaic viruses in a raspberry plantation. *Ann. Appl. Biol.* **62,** 147–157.

Taylor, G. A., Lewis, G. D., and Rubatzky, V. E. (1969). The influence of time of tobacco mosaic virus inoculation and stage of fruit maturity upon the incidence of tomato internal browning. *Phytopathology* **59,** 732–736.

Taylor, L. R. (1986). The distribution of virus disease and the migrant vector aphid. *In* "Plant Virus Epidemics" (G. D. McLean, R. G. Garrett, and W. G. Ruesink, eds.), pp. 35–57. Academic Press, Orlando, Florida.

Teakle, D. S. (1962). Transmission of tobacco necrosis virus by a fungus, *Olpidum brassicae*. *Virology* **18,** 224–231.

Teakle, D. S. (1986). Abiotic transmission of southern bean mosaic virus in soil. *Aust. J. Biol. Sci.* **39,** 353–359.

Tecott, L. H., Barchas, J. D., and Eberwine, J. H. (1988). *In situ* transcription: Specific synthesis of complementary DNA in fixed tissue sections. *Science* **240,** 1661–1664.

Temin, H. M. (1976). The DNA provirus hypothesis. *Science* **192,** 1075–1080.

Temmink, J. H. M., and Campbell, R. N. (1969a). The ultrastructure of *olpidium brassicae*. II. Zoo-spores. *Can. J. Bot.* **47,** 227–231.

Temmink, J. H. M., and Campbell, R. N. (1969b). The ultrastructure of *Olpidium brassicae*. III. Infection of host roots. *Can. J. Bot.* **47,** 421–424.

Temmink, J. H. M., and Campbell, R. N., and Smith, P. R. (1970). Specificity and site of *in vitro* acquisition of tobacco necrosis virus by zoospores of *Olpidium brassicae. J. Gen. Virol.* **9,** 201–213.

Thomas, C. M., Hull, R., Bryant, J. A., and Maule, A. J. (1985). Isolation of a fraction from cauliflower mosaic virus-infected protoplasts which is active in the synthesis of (+) and (−) strand viral DNA and reverse transcription of primed RNA templates. *Nucleic Acids Res.* **13,** 4557–4576.

Thomas, J. E., Massalski, P. R., and Harrison, B. D. (1986). Production of monoclonal antibodies to African cassava mosaic virus and differences in their reactivities with other whitefly-transmitted geminiviruses. *J. Gen. Virol.* **67,** 2739–2748.

Thomas, P. E. (1970). Isolation and differentiation of five strains of curly top virus. *Phytopathology* **60,** 844–848.

Thomas, P. E., Hassan, S., and Mink, G. I. (1988). Influence of light quality on translocation of tomato yellow top virus and potato leafroll virus in *Lycopersicon peruvianum* and some of its tomato hybrids. *Phytopathology* **78,** 1160–1164.

Thompson, D. L., and Hebert, T. T. (1970). Development of maize dwarf mosaic symptoms in eight phytotron environments. *Phytopathology* **60,** 1761–1764.

Thompson, S., Fraser, R. S. S., and Barnden, K. L. (1988). A beneficial effect of trypsin on the purification of turnip mosaic virus (TuMV) and other potyviruses. *J. Virol. Methods* **20,** 57–64.

Thompson, W. R., Meinwald, J., Aneshansley, D., and Eisner, T. (1972). Flavonols: Pigments responsible for ultraviolet absorption in nectar guide of flower. *Science* **177,** 528–530.

Thompson, W. W., Weier, T. E., and Drever, H. (1964). Electron-microscopic studies on chloroplasts from phosphorus deficient plants. *Am. J. Bot.* **51,** 933–938.

Thompson. A. D. (1961). Effect of tobacco mosaic virus and potato virus Y on infection by potato virus X. *Virology* **13,** 262–264.

Thomson, A. D. (1976). Virus diseases of *Solanum laciniatum* Ait. in New Zealand. *N.Z.J. Agric. Res.* **19,** 521–527.

Thomson, A. D., and Ferguson, J. D. (1976). Effect of varying the nutrient supply on response of pea plants to pea leaf roll. *N.Z.J. Agric. Res.* **19,** 529–533.

Thongmeearkom, P., Ford, R. E., and Jedlinski, H. (1976). Aphid transmission of maize dwarf mosaic virus strains. *Phytopathology* **66,** 332–335.

Thornbury, D. W., and Pirone, T. P. (1983). Helper components of two potyviruses are serologically distinct. *Virology* **125,** 487–490.

Thornbury, D. W., Hellmann, G. M., Rhoads, R. E., and Pirone, T. P. (1985). Purification and characterisation of potyvirus helper component. *Virology* **144,** 260–267.

Thouvenel, J.-C., and Fauquet, C. (1981). Further properties of peanut clump virus and studies on its natural transmission. *Ann. Appl. Biol.* **97,** 99–107.

Thresh, J. M. (1958). The spread of virus diseases in Cacao. *West Afr. Cacao Res. Inst., Tech. Bull.* **5,** 1–36.

Thresh, J. M. (1966). Field experiments on the spread of blackcurrant reversion virus and its gall mite vector (*Phytoptus ribis.* Nal). *Ann. Appl. Biol.* **58,** 219–230.

Thresh, J. M. (1980). The origins and epidemiology of some important plant viruses diseases. *In* "Applied Virology" (T. H. Coaker, ed.), Vol. VII, pp. 1–65. Academic Press, London.

Thresh, J. M. (1982). Cropping practices and virus spread. *Annu. Rev. Phytopathol.* **20,** 193–218.

Thresh, J. M. (1983a). Progress curves of plant virus disease. *Adv. Appl. Biol.* **8,** 1–85.

Thresh, J. M. (1983b). The long range dispersal of plant viruses by arthropod vectors. *Philos. Trans. R. Soc. London, Ser. B* **302,** 497–528.

Thresh, J. M. (1988a). Rice viruses and "the green revolution." *Aspects Appl. Biol.* **17,** 187–194.

Thresh, J. M. (1988b). Eradication as a virus disease control measure. *In* "Control of Plant Diseases: Costs and Benefits" (B. C. Clifford and E. Lester, eds.), pp. 155–194. Blackwell, Oxford.

Thresh, J. M. (1989a). Insect-borne viruses of rice and the green revolution. *Trop. Pest Manage.* **35,** 264–272.

Thresh, J. M. (1989b). Plant virus epidemiology: The battle of the genes. *NATO Adv. Res. Workshop: Recognition Response Plant Virus Interact., 1989.*

Tien, P., Zhang, X., Qiu, B., Qin, B., and Wu, G. (1987). Satellite RNA for the control of plant diseases caused by cucumber mosaic virus. *Ann. Appl. Biol.* **111,** 143–152.

Tilney-Bassett, R. A. E. (1986). "Plant Chimeras." Edward Arnold, London.

Timian, R. G. (1974). The range of symbiosis of barley and barley stripe mosaic virus. *Phytopathology* **64,** 342–345.

Timian, R. G. (1975). Barley stripe mosaic virus and the world collection of barleys. *Plant Dis. Rep.* **59,** 984–988.

Timmis, J. N., and Steele Scott, N. (1984). Promiscuous DNA: Sequence homologies between DNA of separate organelles. *Trends Biochem. Sci.* **9,** 271–273.

Tinsley, T. W. (1953). The effects of varying the water supply of plants on their susceptibility to infection with viruses. *Ann. Appl. Biol.* **40,** 750–760.

Todd, J. M. (1958). Spread of potato virus X over a distance. *Proc. Conf. Potato Virus Dis., 3rd, 1957* pp. 132–143.

Todd, J. M. (1961). The incidence and control of aphid-borne potato virus diseases in Scotland. *Eur. Potato J.* **4,** 316–329.

Toh, H., Hayashida, H., and Miyata, T. (1983). Sequence homology between retroviral reverse transcriptase and putative polymerases of hepatitis B virus and cauliflower mosaic virus. *Nature (London)* **305,** 827–829.

Tollin, P., and Wilson, H. R. (1971). Some observations on the structure of the Campinas strain of tobacco rattle virus. *J. Gen. Virol.* **13,** 433–440.

Tomenius, K., Clapham, D., and Meshi, T. (1987). Localization by immunogold cytochemistry of the virus-coded 30K protein in plasmodesmata of leaves infected with tobacco mosaic virus. *Virology* **160,** 363–371.

Tomita, K., and Rich, A. (1964). X-ray diffraction investigations of complementary RNA. *Nature (London)* **201,** 1160–1164.

Tomlinson, J. A. (1962). Control of lettuce mosaic by the use of healthy seed. *Plant Pathol.* **11,** 61–64.

Tomlinson, J. A. (1981). Chemotherapy of plant viruses and virus diseases. *In* "Pathogens, Vectors, and Plant Diseases: Approaches to Control" (K. F. Harris and K. Maramorosch, eds.), pp. 23–44. Academic Press, New York.

Tomlinson, J. A., and Faithfull, E. M. (1979). Effects of fungicides and surfactants on the zoospores of *Olpidium brassicae. Ann. Appl. Biol.* **93,** 13–19.

Tomlinson, J. A., and Faithfull, E. M. (1984). Studies on the occurrence of tomato bushy stunt virus in English rivers. *Ann. Appl. Biol.* **104,** 485–495.

Tomlinson, J. A., and Hunt, J. (1987). Studies on watercress chlorotic leaf spot virus and on control of the fungus vector (*Spongospora subterranea*) with zinc. *Ann. Appl. Biol.* **110,** 75–88.

Tomlinson, J. A., and Walker, V. M. (1973). Further studies on seed transmission in the ecology of some aphid-transmitted viruses. *Ann. Appl. Biol.* **73,** 293–298.

Tomlinson, J. A., and Ward, C. M. (1978). The reactions of swede (*Brassica napus*) to infection by turnip mosaic virus. *Ann. Appl. Biol.* **89,** 61–69.

Tomlinson, J. A., and Ward, C. M. (1982). Selection for immunity in swede (*Brassica napus*) to infection by turnip mosaic virus. *Ann. Appl. Biol.* **101,** 43–50.

Tomlinson, J. A., Carter, A. L., Dale, W. T., and Simpson, C. J. (1970). Weed plants as sources of cucumber mosaic virus. *Ann. Appl. Biol.* **66,** 11–16.

Tomlinson, J. A., Faithfull, E. M., and Ward, C. M. (1976). Chemical suppression of the symptoms of two virus diseases. *Ann. Appl. Biol.* **84,** 31–41.

Tomlinson, J. A., Faithfull, E., Flemett, T. H., and Beards, G. (1982). Isolation of infective tomato barley stunt virus after passage through the human alimentary tract. *Nature (London)* **300,** 637–638.

Torbet, J. (1983). Internal structural anisotropy of spherical viruses studied with magnetic birefringence. *EMBO J.* **2,** 63–66.

Torbet, J., Timmins, P. A., and Lvov, Y. (1986). Packaging of DNA in cauliflower mosaic virus and bacteriophage Sd studied with magnetic birefringence. *Virology* **155,** 721–725.

Toriyama, S. (1986a). An RNA-dependent RNA polymerase associated with the filamentous nucleoproteins of rice stripe virus. *J. Gen. Virol.* **67,** 1247–1255.

Toriyama, S. (1986b). Rice stripe virus: Prototype of a new group of viruses that replicate in plants and insects. *Microbiol. Sci.* **3,** 347–351.

Toriyama, S. (1987). Ribonucleic acid polymerase activity in filamentous nucleoproteins of rice grassy stunt virus. *J. Gen. Virol.* **68,** 925–929.

Toriyama, S., and Peters, D. (1980). *In vitro* synthesis of RNA by dissociated lettuce necrotic yellows virus particles. *J. Gen. Virol.* **50,** 125–134.

Toriyama, S., and Peters, D. (1981). Differentiation between broccoli necrotic yellows virus and lettuce necrotic yellows virus by their transcriptase activities. *J. Gen. Virol.* **56,** 59–66.

Torrance, L. (1980). Use of protein A. to improve sensitisation of latex particles with antibodies to plant viruses. *Ann. Appl. Biol.* **96,** 45–50.

Torrance, L. (1987). Use of enzyme amplification in an ELISA to increase sensitivity of detection of barley yellow dwarf virus in oats and in individual vector aphids. *J. Virol. Methods* **15,** 131–138.

Torrance, L., and Dolby, C. A. (1984). Sampling conditions for reliable routine detection by enzyme-linked immunosorbent assay of three ilarviruses in fruit trees. *Ann. Appl. Biol.* **104,** 267–276.

Torrance, L., and Jones, R. A. C. (1982). Increased sensitivity of detection of plant viruses obtained by using a fluorogenic substrate in enzyme-linked immunosorbent assay. *Ann. Appl. Biol.* **101**, 501–509.

Torrance, L., Larkins, A. P., and Butcher, G. W. (1986a). Characterisation of monoclonal antibodies against potato virus X and comparison of serotypes with resistance groups. *J. Gen. Virol.* **67**, 57–67.

Torrance, L., Pead, M. T., Larkins, A. P., and Butcher, G. W. (1986b). Characterisation of monoclonal antibodies to a UK isolate of barley yellow dwarf virus. *J. Gen. Virol.* **67**, 549–556.

Toruella, M., Gordon, K., and Hohn, T. (1989). Cauliflower mosaic virus produces an aspartic proteinase to cleave its polyproteins. *EMBO J.* **8**, 2819–2825.

Toussaint, A., Dekegel, D., and Vanheule, G. (1984). Distribution of *Odontoglossum* ringspot virus in apical meristems of infected *Cymbidium* cultivars. *Physiol. Plant Pathol.* **25**, 297–305.

Townsend, R., Stanley, J., Curson, S. J., and Short, M. N. (1985). Major polyadenylated transcripts of cassava latent virus and location of the gene encoding the coat protein. *EMBO J.* **4**, 33–37.

Townsend, R., Watts, J., and Stanley, J. (1986). Synthesis of viral DNA forms in *Nicotiana plumbaginifolia* protoplasts inoculated with cassava latent virus (CLV); Evidence for the independent replication of one component of the CLV genome. *Nucleic Acids Res.* **14**, 1253–1265.

Toyoda, H., Oishi, Y., Matsuda, Y., Chatani, K., and Hirai, T. (1985). Resistance mechanism of cultured plant cells to tobacco mosaic virus. IV. Changes in tobacco mosaic virus concentrations in somaclonal tobacco callus tissues and production of virus-free plantlets. *Phytopathol. Z.* **114**, 126–133.

Traynor, P., and Ahlquist, P. (1990). Use of bromovirus RNA2 hybrids to map *cis* and *trans*-acting functions in a conserved RNA replication gene. *J. Virol.* **64**, 69–77.

Tremaine, J. H., Ronald, W. P., and Valcic, A. (1976). Aggregation properties of carnation ringspot virus. *Phytopathology* **66**, 34–39.

Trevathan, L. E., Tolin, S. A., Moore, L. D., and Orcutt, D. M. (1982). Total lipid, free sterol, free fatty acid, and triacylglycerol fatty acid content of tobacco mosaic virus-infected tobacco. *Can. J. Plant Sci.* **62**, 771–776.

Tsagris, M., Tabler, M., and Sänger, H. L. (1987a). Oligomeric potato spindle tuber viroid (PSTV) RNA does not process autocatalytically under conditions where other RNAs do. *Virology* **157**, 227–231.

Tsagris, M., Tabler, M., Mühlbach, H.-P., and Sänger, H. L. (1987b). Linear oligomeric potato spindle tuber viroid (PSTV) RNAs are accurately processed *in vitro* to the monomeric circular viroid proper when incubated with a nuclear extract from healthy potato cells. *EMBO J.* **6**, 2173–2183.

Tsugita, A., and Fraenkel-Conrat, H. (1960). The amino acid composition and C-terminal sequence of a chemically evoked mutant of tobacco mosaic virus. *Proc. Natl. Acad. Sci. U.S.A.* **46**, 636–642.

Tsugita, A., Gish, D. T., Young, J., Fraenkel-Conrat, H., Knight, C. A., and Stanley, W. M. (1960). The complete amino acid sequence of the protein of tobacco mosaic virus. *Proc. Natl. Acad. Sci. U.S.A.* **46**, 1463–1469.

Tu, J. C., (1973). Electron microscopy of soybean root nodules infected with soybean mosaic virus. *Phytopathology* **63**, 1011–1017.

Tu, J. C. (1977). Effects of soybean mosaic virus infection on ultrastucture of bacteroidal cells in soybean root nodules. *Phytopathology* **67**, 199–205.

Tu, J. C., and Buzzell, R. I. (1987). Stem-tip necrosis: A hypersensitive, temperature dependent dominant gene reaction of soybean to infection by soybean mosaic virus. *Can. J. Plant Sci.* **67**, 661–665.

Tu, J. C., Ford, R. E., and Quiniones, S. S. (1970). Effects of soybean mosaic virus and/or bean pod mottle virus infection on soybean nodulation. *Phytopathology* **60**, 518–523.

Tucker, E. B. (1982). Translocation in staminal hairs of *Setcreasea purpurea*. I. A study of cell ultrastructure and cell-to-cell passage of molecular probes. *Protoplasma* **113**, 193–201.

Tumer, N. E., O'Connell, K. M., Nelson, R. S., Sanders, P. R., Beachy, R. N., Fraley, R. T., and Shah, D. M. (1987). Expression of alfalfa mosaic virus coat protein gene confers cross-protection in transgenic tobacco and tomato plants. *EMBO J.* **6**, 1181–1188.

Turner, D. R., and Butler, P. J. G. (1986). Essential features of the assembly origin of tobacco mosaic virus RNA as studied by directed mutagenesis. *Nucleic Acids Res.* **14**, 9229–9242.

Turner, D. R., Mondragon, A., Fairall, L., Bloomer, A. C., Finch, J. T., van Boom, J. H., and Butler, P. J. G. (1986). Oligonucleotide binding to the coat protein disk of tobacco mosaic virus. Possible steps in the assembly mechanism. *Eur. J. Biochem.* **157**, 269–274.

Turner, D. R., Joyce, L. E., and Butler, P. J. G. (1988). The tobacco mosaic virus assembly origin RNA; Functional characteristics defined by directed mutagenesis. *J. Mol. Biol.* **203**, 531–547.

Turner, D. R., McGuigan, C. J. and Butler, P. J. G. (1989). Assembly of hybrid RNAs with tobacco mosaic virus coat protein. Evidence for incorporation of disks in 5′ elongation along the major RNA tail. *J. Mol. Biol.* **209**, 407–422.

Turner, D. S., and Covey, S. N. (1984). A putative primer for the replication of cauliflower mosaic virus by reverse transcription is virion-associated. *FEBS Lett.* **165,** 285–289.

Turner, D. S., and Covey, S. N. (1988). Discontinuous hairpin DNAs synthesized *in vivo* following specific and non-specific priming of cauliflower mosaic virus DNA (+) strands. *Virus. Res.* **9,** 49–62.

Turner, Ph. C., Watkins, P. A. C., Zaitlin, M., and Wilson, T. M. A. (1987). Tobacco mosaic virus particles uncoat and express their RNA in *Xenopus laevis* oocytes: Implications for early interactions between plant cells and viruses. *Virology* **160,** 515–517.

Tyc, K., Konarska, M., Gross, H. J., and Filipowicz, W. (1984). Multiple ribosome binding to the 5'-terminal leader sequence of tobacco mosaic virus RNA. *Eur. J. Biochem.* **140,** 503–511.

Uegaki, R., Kubo, S., and Fujimori, T. (1988). Stress compounds in the leaves of *Nicotiana undulata* induced by TMV inoculation *Phytochemistry* **27,** 365–368.

Uhlenbeck, O. C. (1987). A small catalytic oligoribonucleotide. *Nature (London)* **328,** 596–600.

Ukai, Y., and Yamashita, A. (1984). Induced mutation for resistance to barley yellow mosaic virus. *JARQ* **17,** 255–259.

Ullman, D. E., Qualset, C. O., and McLean, D. L. (1988). Feeding responses of *Rhopalosiphum padi* (Homoptera: Aphidae) to barley yellow dwarf virus resistant and susceptible barley varieties. *Environ. Entomol.* **17,** 988–991.

Unge, T., Montelius, I., Liljàs, L., and Ofverstedt, L.-G. (1986). The EDTA-treated expanded sattelite tobacco necrosis virus: Biochemical properties and crystallisation. *Virology* **152,** 207–218.

Uppal, B. N. (1934). The movement of tobacco mosaic virus in leaves of *Nicotiana sylvestris*. *Indian J. Agric. Sci.* **4,** 865–873.

Ushiyama, R., and Matthews, R. E. F. (1970). The significance of chloroplast abnormalities associated with infection by turnip yellow mosaic virus. *Virology* **42,** 293–303.

Uyeda, I., and Shikata, E. (1984). Characterisation of RNAs synthesized by the virion-associated transcriptase of rice dwarf virus *in vitro*. *Virus Res.* **1,** 527–532.

Uyeda, I., Matsumura, T., Sano, T., Ohshima, K., and Skikata, E. (1987a). Nucleotide sequence of rice dwarf virus genome segment 10. *Proc. Jpn. Acad., Ser. B* **63,** 227–230.

Uyeda, I., Lee, S. Y., Yoshimoto, H., and Shikata, E. (1987b). RNA polymerase activity of rice ragged stunt and rice black-streaked dwarf viruses. *Ann. Phytopathol. Soc. Jpn.* **53,** 60–62.

Uyeda, I., Kudo, H., Takahashi, T., Sano, T., Oshima, K., Matsumura, T., and Shikata, E. (1989). Nucleotide sequence of rice dwarf virus genome segment 9. *J. Gen. Virol.* **70,** 1297–1300.

Uyemoto, J. K., and Gilmer, R. M. (1972). Properties of tobacco necrosis virus strains isolated from apple. *Phytopathology* **62,** 478–481.

Uyemoto, J. K., Grogan, R. G., and Wakeman, J. R. (1968). Selective activation of satellite virus strains by strains of tobacco necrosis virus. *Virology* **34,** 410–418.

Vaeck, M., Reynaerts, A., Höfte, H., Jansens, S., De Beuckeleer, M., Dean, C., Zabeau, M., Van Montagu, M., and Leemans, J. (1987). Transgenic plants protected from insect attack. *Nature (London)* **328,** 33–37.

Valdez, R. B. (1972). A micro-container technique for studying virus transmission by nematodes. *Plant Pathol.* **21,** 114–117.

Valleau, W. D. (1946). Breeding tobacco varieties resistant to mosaic. *Phytopathology* **36,** 412.

Valleau, W. D. (1952). Breeding tobacco for disease resistance. *Econ. Bot.* **6,** 69–102.

Vallega, V., and Rubies Autonell, C. (1985). Reactions of Italian *Triticum durum* cultivars to soilborne wheat mosaic. *Plant Dis.* **69,** 64–66.

Valverde, R. A., and Dodds, J. A. (1986). Evidence for a satellite RNA associated naturally with the U5 strain and experimentally with the U1 strain of tobacco mosaic virus. *J. Gen. Virol.* **67,** 1875–1884.

Valverde, R. A., and Dodds, J. A. (1987). Some properties of isometric virus particles which contain the satellite RNA of tobacco mosaic virus. *J. Gen. Virol.* **68,** 965–972.

Valverde, R. A., Dodds, J. A., and Heick, J. A. (1986). Double-stranded ribonucleic acid from plants infected with viruses having elongated particles and undivided genomes. *Phytopathology* **76,** 459–465.

van Beek, N. A. M., Lohuis, D., Dijkstra, J., and Peters, D. (1985). Morphogenesis of Festuca leaf streak virus in cowpea protoplasts. *J. Gen. Virol.* **66,** 2485–2489.

van Beek, N. A. M., Derksen, A. C. G., and Dijkstra, J. (1986). Synthesis of sonchus yellow net virus proteins in infected cowpea protoplasts. *J. Gen. Virol.* **67,** 1701–1709.

van Belkum, A., Cornelissen, B., Linthorst, H., Bol, J., Pleij, C., and Bosch, L. (1987). tRNA-like properties of tobacco rattle virus RNA. *Nucleic Acids Res.* **15,** 2837–2850.

van Belkum, A., Verlaan, P., Bing Kun, J., Pleij, C., and Bosch, L. (1988). Temperature dependent chemical and enzymatic probing of the tRNA-like structure of TYMV RNA. *Nucleic Acids Res.* **16,** 1931–1950.

van Beynum, G. M. A., de Graaf, J. M., Castel, A., Kraal, B., and Bosch, L. (1977). Structural studies on the coat protein of alfalfa mosaic virus. *Eur. J. Biochem.* **72,** 63–78.

Vance, V. B., and Beachy, R. N. (1984). Detection of genomic-length soybean mosaic virus RNA on polyribosomes of infected soybean leaves. *Virology* **138,** 26–36.

van der Broek, L. J., and Gill, G. C. (1980). The median latent periods for three isolates of barley yellow dwarf virus in aphid vectors. *Phytopathology* **70,** 644–646.

van der Krol, A. R., Lenting, P. E., Veenstra, J., van der Meer, I. M., Koes, R. E., Gerats, A. G. M., Mol, J. N. M., and Stuitje, A. R. (1988). An anti-sense chalcone synthase gene in transgenic plants inhibits flower pigmentation. *Nature (London)* **333,** 866–869.

van der Kuyl, A. C., Langereis, K., Houwing, C. J., Jaspars, E. M. J., and Bol, J. F. (1990). Cis-acting elements involved in replication of alfalfa mosaic virus RNAs *in vitro. Virology* **176,** 346–354.

Van der Lubbe, J. L. M., Hatta, T., and Francki, R. I. B. (1979). Structure of the antigen from Fiji disease virus particles eliciting antibodies specific to double-stranded polyribonucleotides. *Virology* **95,** 405–414.

van der Meer, J., Dorssers, L., van Kammen, A., and Zabel, P. (1984). The RNA-dependent RNA polymerase of cowpea is not involved in cowpea mosaic virus RNA replication: Immunological evidence. *Virology* **132,** 413–425.

Van der Plank, J. E. (1946). A method for estimating the number of random groups of adjacent diseased plants in a homogeneous field. *Trans. R. Soc. S. Afr.* **31,** 269–278.

Van der Plank, J. E. (1948). The relation between the size of fields and the spread of plant disease into them. Part I. Crowd diseases. *Emp. J. Exp. Agric.* **16,** 134–142.

Vanderveken, J. J. (1977). Oils and other inhibitors of nonpersistent virus transmission. *In* "Aphids as Virus Vectors" (K. F. Harris and K. Maramorosch, eds.), pp. 435–454. Academic Press, New York.

van der Vlugt, R., Allefs, S., de Haan, P., and Goldbach, R. (1989). Nucleotide sequence of the 3'-terminal region of potato virus Y^N RNA. *J. Gen. Virol.* **70,** 229–233.

Van der Want, J. P. H., Boerjan, M. L., and Peters, D. (1975). Variability of some plant species from different origins and their suitability for virus work. *Neth. J. Plant Pathol.* **81,** 205–216.

van der Wilk, F., Huismann, M. J., Cornelissen, B. J. C., Huttinga, H., and Goldbach, R. (1989). Nucleotide sequence and organisation of potato leaf roll virus genomic RNA. *FEBS Lett.* **245,** 51–56.

van Dijk, P., van der Meer, F. A., and Piron, P. G. M. (1987). Accessions of Australian *Nicotiana* species suitable as indicator hosts in the diagnosis of plant virus diseases. *Neth. J. Plant Pathol.* **93,** 73–85.

van Dun, C. M. P., and Bol, J. F. (1988). Transgenic tobacco plants accumulating tobacco rattle virus coat protein resist infection with tobacco rattle virus and pea early browning virus. *Virology* **167,** 649–652.

van Dun, C. M. P., Bol, J. F., and van Vloten-Doting, L. (1987). Expression of alfalfa mosaic virus and tobacco rattle virus coat protein genes in transgenic tobacco plants. *Virology* **159,** 299–305.

van Dun, C. M. P., van Vloten-Doting, L., and Bol, J. F. (1988a). Expression of alfalfa mosaic virus cDNA 1 and 2 in transgenic tobacco plants. *Virology* **163,** 572–578.

van Dun, C. M. P., Overduin, B., van Vloten-Doting, L., and Ból, J. F. (1988b). Transgenic tobacco expressing tobacco streak virus or mutated alfalfa mosaic virus coat protein does not cross-protect against alfalfa mosaic virus infection. *Virology* **164,** 383–389.

van Emmelo, J., Ameloot, P., Plaetinck, G., and Fiers, W. (1984). Controlled synthesis of the coat protein of satellite tobacco necrosis virus in *Escherichia coli. Virology* **136,** 32–40.

van Emmelo, J., Ameloot, P., and Fiers, W. (1987). Expression in plants of the cloned satellite tobacco necrosis virus genome and of derived insertion mutants. *Virology* **157,** 480–487.

van Etten, J. L., Burbank, D. E., Kuczmarski, D., and Meints, R. H. (1983). Virus infection of culturable *Chlorella*-like algae and development of a plaque assay. *Science* **219,** 994–996.

van Etten, J. L., Xia, Y., and Meints, R. H. (1987). Viruses of a *Chlorella*-like green alga. *In* "Plant–Microbe Interactions" (T. Kosuge and E. W. Nester, eds.), Vol. 2, pp. 307-325. Macmillan, New York.

van Etten, J. L., Schuster, A. M., and Meints, R. H. (1988). Viruses of eukaryotic *Chlorella*-like alga. *In* "Viruses of Fungi and Simple Eukaryotes" (Y. Koltin and M. J. Leibowitz, eds.), pp. 411–428. Dekker, New York.

van Griensven, L. J. L. D., van Kammen, A., and Rezelman, G. (1973). Characterization of the double-stranded RNA isolated from cowpea mosaic virus-infected *Vigna* leaves. *J. Gen. Virol.* **18,** 359–367.

van Hoof, H. A. (1970). Some observations on retention of tobacco rattle virus in nematodes. *Neth. J. Plant Pathol.* **76,** 329–330.

van Hoof, H. A. (1976). The bait leaf method for determining soil infestation with tobacco rattle virus-transmitting trichodorids. *Neth. J. Plant Pathol.* **82**, 181–185.

van Hoof, H. A. (1977). Determination of the infection pressure of potato virus Y^N. *Neth. J. Plant Pathol.* **83**, 123–127.

van Hoof, H. A., and Silver, C. N. (1976). Natural elimination of tobacco rattle virus in tulip "Apeldoorn." *Neth. J. Plant Pathol.* **82**, 255–256.

van Kammen, A. (1968). The relationship between the components of cowpea mosaic virus. 1. Two ribonucleoprotein particles necessary for the infectivity of CPMV. *Virology* **34**, 312–318.

van Kammen, A. (1985). The replication of plant virus RNA. *Microbiol. Sci.* **2**, 170–174.

van Kammen, A., and Eggen, H. I. L. (1986). The replication of cowpea mosaic virus. *BioEssays* **5**, 261–266.

van Kan, J. A. L., Cornelissen, B. J. C., and Bol, J. F. (1988). A virus-inducible tobacco gene encoding a glycine-rich protein shares putative regulatory elements with the ribulose bisphosphate carboxylase small subunit gene. *Mol. Plant-Microbe Interact.* **1**, 107–112.

van Kan, J. A. L., van de Rhee, M. D., Zuidema, D., Cornelissen, B. J. C., and Bol. J. F. (1989). Structure of the tobacco genes encoding thaumatin-like proteins. *Plant Mol. Biol.* **12**, 153–155.

van Kooten, O., Meurs, C., and van Loon, L. C. (1990). Photosynthetic electron transport in tobacco leaves infected with tobacco mosaic virus. *Physiol. Plant.* **80**, 446–452.

van Lent, J. W. M., and Verduin, B. J. M. (1985). Specific gold-labelling of antibodies bound to plant viruses in mixed suspensions. *Neth. J. Plant Pathol.* **91**, 205–213.

van Lent, J. W. M., and Verduin, B. J. M. (1986). Detection of viral protein and particles in thin sections of infected plant tissue using immunogold labelling. *In* "Developments and Applications in Virus Testing" (R. A. C. Jones and L. Torrance, eds.), Dev. Appl. Biol. I, pp. 193–211. Association of Applied Biologists, Wellsbourne.

van Lent, J. W. M., and Verduin, B. J. M. (1987). Detection of viral antigen in semi-thin sections of plant tissue by immunogold-silver staining and light microscopy. *Neth. J. Plant Pathol.* **93**, 261–272.

van Lent, J. W. M., Wellink, J., and Goldbach, R. (1990). Evidence for the involvement of the 58K and 48K proteins in the intercellular movement of cowpea mosaic virus. *J. Gen. Virol.* **71**, 219–223.

van Loon, L. C. (1977). Induction by 2-chloroethylphosphonic acid of viral-like lesions, associated proteins, and systemic resistance in tobacco. *Virology* **88**, 417–420.

van Loon, L. C. (1987). Disease induction by plant viruses. *Adv. Virus Res.* **33**, 205–255.

van Loon, L. C. (1989). Stress proteins in infected plants. *In* "Plant–Microbe Interactions" (T. Kosuge and E. W. Nester, eds.), pp. 198–237. McGraw-Hill, New York.

van Loon, L. C., and Antoniw, J. F. (1982). Comparison of the effects of salicylic acid and ethephon with virus-induced hypersensitivity and acquired resistance in tobacco. *Neth. J. Plant Pathol.* **88**, 237–256.

van Loon, L. C., and Dijkstra, J. (1976). Virus-specific expression of systemic acquired resistance in tobacco mosaic virus- and tobacco necrosis virus-infected "Samsun NN" and "Samsun" tobacco. *Neth. J. Plant Pathol.* **82**, 231–237.

van Loon, L. C., and van Kammen, A. (1970). Polyacrylamide disc electrophoresis of the soluble leaf proteins from *Nicotiana tabacum* var. "Samsun" and "Samsun NN." II. Changes in protein constitution after infection with tobacco mosaic virus. *Virology* **40**, 199–211.

van loon. L. C., Gerritsen, Y. A. M., and Ritter, C. E. (1987). Identification, purification, and characterisation of pathogenesis-related proteins from virus-infected Samsun NN tobacco leaves. *Plant Mol. Biol.* **9**, 593–609.

van Oosten, H. J., Meijneke, C. A. R., and Peerbooms, H. (1983). Growth, yield and fruit quality of virus-infected and virus-free Golden Delicious apple trees 1968–1982. *Acta Hortic.* **130**, 213–217.

van Pelt-Heerschap, H., Verbeek, H., Slot, J. W., and van Vloten-Doting, L. (1987a). The location of coat protein and viral RNAs of alfalfa mosaic virus in infected tobacco leaves and protoplasts. *Virology* **160**, 297–300.

van Pelt-Heerschap, H., Verbeek, H., Huisman, M. J., Loesch-Fries, L. S., and van Vloten-Doting, L. (1987b). Non-structural proteins and RNAs of alfalfa mosaic virus synthesized in tobacco and cowpea protoplasts. *Virology* **161**, 190–197.

Van Regenmortel, M. H. V. (1975). Antigenic relationships between strains of tobacco mosaic virus. *Virology* **64**, 415–420.

Van Regenmortel, M. H. V. (1982). "Serology and Immunochemistry of Plant Viruses." Academic Press, New York.

Van Regenmortel, M. H. V. (1984a). Recent advances in immunodiagnosis of viral diseases of crops. *In* "Applied Virology" (E. Kurstak, ed.), pp. 463–477. Academic Press, New York.

Van Regenmortel, M. H. V. (1984b). Molecular dissection of antigens by monoclonal antibodies. *In*

"Hybridoma Technology in Agricultural and Veterinary Research" (N. J. Stern and H. R. Gamble, eds.), pp. 43–82. Rowman and Allanheld, Totowa, New Jersey.

Van Regenmortel, M. H. V. (1984c). Monoclonal antibodies in plant virology. *Microbiol. Sci.* **1**, 73–78.

Van Regenmortel, M. H. V. (1986). The potential for using monoclonal antibodies in the detection of plant viruses. *In* "Developments and Applications in Virus Testing" (R. A. C. Jones and L. Torrance, eds.), Dev. Appl. Biol. I, pp. 89–101. Association of Applied Biologists, Wellsbourne.

Van Regenmortel, M. H. V. (1989a). Applying the species concept to plant viruses. *Arch. Virol.* **104**, 1–17.

Van Regenmortel, M. H. V. (1989b). Structural and functional approaches to the study of protein antigenicity. *Immunol. Today* **10**, 266–272.

Van Regenmortel, M. H. V., and von Wechmar, M. B. (1970). A reexamination of the serological relationship between tobacco mosaic virus and cucumber virus 4. *Virology* **41**, 330–338.

van Telgen, H. J., Goldbach, R. W., and van Loon, L. C. (1985). The 126,000 molecular weight protein of tobacco mosaic virus is associated with host chromatin in mosaic-diseased tobacco plants. *Virology* **143**, 612–616.

Vawter, L., and Brown, W. M. (1986). Nuclear and mitochondrial DNA comparisons reveal extreme rate variation in the molecular clock. *Science* **234**, 194–196.

van Tol, R. G. L., and van Vloten-Doting, L. (1979). Translation of alfalfa-mosaic-virus RNA1 in the mRNA-dependent translation system from rabbit reticulocyte lysates. *Eur. J. Biochem.* **93**, 461–468.

van Tol, R. G. L., and van Vloten-Doting, L. (1981). Lack of serological relationship between the 35K non-structural protein of alfalfa mosaic virus and the corresponding proteins of three other plant viruses with tripartite genomes. *Virology* **109**, 444–447.

van Tol, R. G. L., van Gemeren, R., and van Vloten-Doting, L. (1980). Two leaky termination codons in AMV RNA 1. *FEBS Lett.* **118**, 67–71.

van Vloten-Doting, L. (1968). Verdeling van de genetische informatie over de natuurlijke componenten van een plantvirus. Ph.D. Thesis, University of Leiden.

van Vloten-Doting, L. (1975). Coat protein is required for infectivity of tobacco streak virus: Biological equivalence of the coat proteins of tobacco streak and alfalfa mosaic viruses. *Virology* **65**, 215–225.

van Vloten-Doting, L. (1976). Similarities and differences between viruses with a tripartite genome. *Ann. Microbiol. (Paris)* **127**, 119–129.

van Vloten-Doting, L. (1978). Early events in the infection of tobacco with alfalfa mosaic virus. *J. Gen. Virol.* **41**, 649–652.

van Vloten-Doting. L., and Bol, J. F. (1988). Variability, mutant selection and mutant stability in plant RNA viruses. *In* "RNA Genetics" (E. Domingo, J. J. Holland, and P. Ahlquist, eds.), vol. 3, pp. 37–51. CRC Press, Boca Raton, Florida.

van Vloten-Doting, L., and Jaspars, E. M. J. (1972). The uncoating of alfalfa mosaic virus by its own RNA. *Virology* **48**, 699–708.

van Vloten-Doting, L., and Jaspars, E. M. J. (1977). Plant covirus systems: Three-component systems. *Compr. Virol.* **11**, 1–53.

van Vloten-Doting, L., and Neeleman, L. (1980). Translation of plant viral RNAs. *NATO Adv. Study Inst. Ser., Ser. A* **29**, 511–527.

van Vloten-Doting, L., and Neeleman, L. (1982). Translocation of plant viral RNA's. *Encycl. Plant Physiol., New Ser.* **14B**, 337–367.

van Vloten-Doting, L., Kruseman, J., and Jaspars, E. M. J. (1968). The biological function and mutual dependence of bottom component and top component of alfalfa mosaic virus. *Virology* **34**, 728–737.

van Vloten-Doting, L., Hasrat, J. A., Oosterwijk, E., van'T Sant, P., Schoen, M. A., and Roosien, J. (1980). Description and complementation analysis of 13 temperature sensitive mutants of alfalfa mosaic virus. *J. Gen. Virol.* **46**, 415–426.

van Vloten-Doting, L., Francki, R. I. B., Fulton, R. W., Kaper, J. M., and Lane, L. C. (1981). Tricornaviridae—A proposed family of plant viruses with tripartite, single-stranded RNA genomes. *Intervirology* **15**, 198–203.

van Vloten-Doting, L., Bol, J. F., and Cornelissen, B. (1985). Plant virus-based vectors for gene transfer will be of limited use because of the high error frequency during viral RNA synthesis. *Plant Mol. Biol.* **4**, 323–326.

van Wezenbeek, P., Verver, J., Harmsen, J., Vos, P., and van Kammen, A. (1983). Primary structure and gene organisation of the middle component RNA of cowpea mosaic virus. *EMBO J.* **2**, 941–946.

Varveri, C., Candresse, T., Cugusi, M., Ravelonandro, M., and Dunez, J. (1988). Use of ^{32}P-labelled transcribed RNA probe for dot hybridization detection of plum pox virus. *Phytopathology* **78**, 1280–1283.

Veidt, I., Lot, H., Leiser, M., Scheidecker, D., Guilley, H., Richards, K. and Jonard, G. (1988). Nucleotide sequence of beet western yellows virus RNA. *Nucleic Acids Res.* **16**, 9917–9932.

Vela, C., Cambra, M., Cortes, E., Moreno, P., Miguet, J.-G., Pérez de san Román, C., and Sanz, A. (1986). Production and characterisation of monoclonal antibodies specific for citrus tristeza virus and their use for diagnosis. *J. Gen. Virol.* **67**, 91–96.

Veldee, S., and Fraenkel-Conrat, H. (1962). The characterisation of tobacco mosaic virus strains by their productivity. *Virology* **18**, 56–63.

Venekamp, J. H., and Beemster, A. B. R. (1980). Mature plant resistance of potato against some virus diseases. I. Concurrence of development of mature plant resistance against potato virus X, and decrease of ribosome and RNA content. *Neth. J. Plant Pathol.* **86**, 1–10.

Vera, P., and Conejero, V. (1989). The induction and accumulation of the pathogenesis-related P69 proteinase in tomato during citrus exocortis viroid infection and in response to chemical treatments. *Physiol. Mol. Plant Pathol.* **34**, 323–334.

Verduin, B. J. M., Prescott, B. and Thomas, G. J., Jr. (1984). RNA–protein interactions and secondary structures of cowpea chlorotic mottle virus for *in vitro* assembly. *Biochemistry* **23**, 4301–4308.

Verhagen, W., Van Boxsel, J. A. M., Bol, J. F., van Vloten-Doting, L., and Jaspars, E. M. J. (1976). RNA–protein interactions in alfalfa mosaic virus. *Ann. Microbiol. (Paris)* **127A**, 165–172.

Verkleij, F. N., and Peters, D. (1983). Characterization of a defective form of tomato spotted wilt virus. *J. Gen. Virol.* **64**, 677–686.

Vértesy, J., and Nyéki, J. (1974). Effect of different ringspot viruses on the flowering period and fruit set of Montmorency and Pándy sour cherries. I. *Acta Phytopathol. Acad. Sci. Hung.* **9**, 17–22.

Verver, J., Goldbach, R., Garcia, J. A., and Vos, P. (1987). *In vitro* expression of a full-length DNA copy of cowpea mosaic virus B-RNA: Identification of the B-RNA encoded 24-kd protein as a viral protease. *EMBO J.* **6**, 549–554.

Vetten, H. J., and Allen, D. J. (1983). Effects of environment and host on vector biology and incidence of two whitefly-spread diseases of legumes in Nigeria. *Ann. Appl. Biol.* **102**, 219–227.

Vetten, H. J., Lesemann, D.-E., and Dalchow, J. (1987). Electron microscopical and serological detection of virus-like particles associated with lettuce big vein disease. *J. Phytopathol.* **120**, 53–59.

Virudachalum, R., Harrington, M., Johnson, J. E., and Markley, J. L. (1985). ^{1}H, ^{13}C and ^{31}P nuclear magnetic resonance studies of cowpea mosaic virus: Detection and exchange of polyamines and dynamics of the RNA. *Virology* **141**, 43–50.

Visvader, J. E., and Symons, R. H. (1985). Eleven new sequence variants of citrus exocortis viroid and the correlation of sequence with pathogenicity *Nucleic Acids Res.* **13**, 2907–2920.

Visvader, J. E., and Symons, R. H. (1986). Replication of *in vitro* constructed viroid mutants: Location of the pathogenicity-modulating domain of citrus exocortis viroid. *EMBO J.* **5**, 2051–2055.

Vogel, R. H., and Provencher, S. W. (1988). Three-dimensional reconstruction from electron micrographs of disordered specimens. II. Implementation and results. *Ultramicroscopy* **25**, 223–240.

Vos, P., Jaegle, M., Wellink, J., Verver, J., Eggen, R., van Kammen, A., and Goldbach, R. (1988a). Infectious RNA transcripts derived from full-length DNA copies of the genomic RNAs of cowpea mosaic virus. *Virology* **164**, 33–41.

Vos, P., Verver, J., Jaegle, M., Wellink, J., van Kammen, A., and Goldbach, R. (1988b). Two viral proteins involved in the proteolytic processing of the cowpea mosaic virus polyproteins. *Nucleic Acids Res.* **16**, 1967–1985.

Vriend, G., Verduin, B. J. M., and Hemminga, M. A. (1986). Role of the N-terminal part of the coat protein in the assembly of cowpea chlorotic mottle virus. A 500 MHZ proton nuclear magnetic resonance study and structural calculations. *J. Mol. Biol.* **191**, 453–460.

Wagih, E. E., and Coutts, R. H. A. (1982). Peroxidase, polyphenoloxidase and ribonuclease in tobacco necrosis virus infected or mannitol osmotically-stressed cowpea and cucumber tissue. I. Quantitative Alterations. *Phytopathol. Z.* **104**, 1–12.

Wagner, G. W., and Bancroft, J. B. (1968). The self-assembly of spherical viruses with mixed coat proteins. *Virology* **34**, 748–756.

Wagner, R. R., Prevec, L., Brown, F., Summers, D. F., Sokol, F., and MacLeod, R. (1972). Classification of rhabdovirus proteins: A proposal. *J. Virol.* **10**, 1228–1230.

Wagner, W. H. (1984). A comparison of taxonomic methods in biosystematics. *In* "Plant Biosystematics" (W. F. Grant, ed.), pp. 643–654. Academic Press, Orlando, Florida.

Wakarchuk, D. A., and Hamilton, R. I. (1985). Cellular double-stranded RNA in *Phaseolus vulgaris*. *Plant Mol. Biol.* **5**, 55–63.

Walden, R., and Howell, S. H. (1982). Intergenomic recombination events among pairs of defective cauliflower mosaic virus genomes in plants. *J. Mol. Appl. Genet.* **1**, 447–456.

Waldrop, M. M. (1989). Did life really start out in an RNA world? *Science* **246**, 1248–1249.

Walker, H. L., and Pirone, T. P. (1972a). Particle numbers associated with mechanical and aphid transmission of some plant viruses. *Phytopathology* **62**, 1283–1288.

Walker, H. L., and Pirone, T. P. (1972b). Number of TMV particles required to infect locally or systemically susceptible tobacco cultivars. *J. Gen. Virol.* **17**, 241–243.

Walkey, D. G. A. (1968). The production of virus-free rhubarb by apical tip-culture. *J. Hortic. Sci.* **43**, 283–287.

Walkey, D. G. A. (1985). "Applied Plant Virology." Heinemann, London.

Walkey, D. G. A., and Antill, D. N. (1989). Agronomic evaluation of virus-free and virus-infected garlic (*Allium sativum* L). *J. Hortic. Sci.* **64**, 53–60.

Walkey, D. G. A., and Cooper, J. (1976). Heat inactivation of cucumber mosaic virus in cultured tissues of *Stellaria media. Ann. Appl. Biol.* **84**, 425–428.

Walkey, D. G. A., and Dance, M. C. (1979). The effect of oil sprays on aphid transmission of turnip mosaic, beet yellows, bean common mosaic and bean yellow mosaic viruses. *Plant Dis. Rep.* **63**, 877–881.

Walkey, D. G. A., and Freeman, G. H. (1977). Inactivation of cucumber mosaic virus in cultured tissues of *Nicotiana rustica* by diurnal alternating periods of high and low temperature. *Ann. Appl. Biol.* **87**, 375–382.

Walkey, D. G. A., and Webb, M. J. W. (1968). Virus in plant apical meristems. *J. Gen. Virol.* **3**, 311–313.

Walkey, D. G. A., and Webb, M. J. W. (1970). Tubular inclusion bodies in plants infected with viruses of the NEPO type. *J. Gen. Virol.* **7**, 159–166.

Walkey, D. G. A., and Whittingham-Jones, S. G. (1970). Seed transmission of strawberry latent ringspot virus in celery (*Apium graveolens* var. Dulce). *Plant Dis. Rep.* **54**, 802–803.

Walkey, D. G. A., Fitzpatrick, J., and Woolfitt, J. M. G. (1969). The inactivation of virus in cultured shoot tips of *Nicotiana rustica* L. *J. Gen. Virol.* **5**, 237–241.

Walkey, D. G. A., Brocklehurst, P. A., and Parker, J. E. (1985). Some physiological effects of two seed-transmitted viruses on flowering, seed production, and seed vigour in *Nicotiana* and *Chenopodium* plants. *New Phytol.* **99**, 117–128.

Walkey, D. G. A., Webb, M. J. W., Bolland, C. J., and Miller, A. (1987). Production of virus-free garlic (*Allium sativum* L.) and shallot (*Allium ascalonicum* L.) by meristem tip culture. *J. Hortic. Sci.* **62**, 211–220.

Wallin, J. R., and Loonan, D. V. (1971). Low-level jet winds, aphid vectors, local weather, and barley yellow dwarf virus outbreaks. *Phytopathology* **61**, 1068–1070.

Walsh, J. A. (1986). Virus diseases of oilseed rape and their control. *Br. Crop Prot. Conf.—Pests Dis.* **7A-3**, 737–743.

Walsh, J. A., and Tomlinson, J. A. (1985). Viruses infecting winter oilseed rape (*Brassica napus* ssp. *oleifera*). *Ann. Appl. Biol.* **107**, 485–495.

Wang, M. C., Lin, J. J., Duran-Vila, N., and Semancik, J. S. (1986). Alteration in cell wall composition and structure in viroid-infected cells. *Physiol. Plant Pathol.* **28**, 107–124.

Ward, A., Etessami, P., and Stanley, J. (1988). Expression of a bacterial gene in plants mediated by infectious geminivirus DNA. *EMBO J.* **7**, 1583–1587.

Ward, C. M., Walkey, D. G. A., and Phelps, K. (1987). Storage of samples infected with lettuce or cucumber mosaic viruses prior to testing with ELISA. *Ann. Appl. Biol.* **110**, 89–95.

Warmke, H. E., and Edwardson, J. R. (1966). Electron microscopy of crystalline inclusions of tobacco mosaic virus in leaf tissue. *Virology* **30**, 45–57.

Watanabe, Y., and Okada, Y. (1986). *In vitro* viral RNA synthesis by a subcellular fraction of TMV-inoculated tobacco protoplasts. *Virology* **149**, 64–73.

Watanabe, Y., Ohno, T., and Okada, Y. (1982). Virus multiplication in tobacco protoplasts inoculated with tobacco mosaic virus RNA encapsulated in large unilamellar vesicle liposomes. *Virology* **120**, 478–480.

Watanabe, Y., Meshi, T., and Okada, Y. (1984a). The initiation site for transcription of the TMV 30-kDa protein messenger RNA. *FEBS Lett.* **173**, 247–250.

Watanabe, Y., Emori, Y., Ooshika, I., Meshi, T., Ohno, T., and Okada, Y. (1984b). Synthesis of TMV-specific RNAs and proteins at the early stage of infection in tobacco protoplasts: Transient expression of the 30k protein and its mRNA. *Virology* **133**, 18–24.

Watanabe, Y., Meshi, T., and Okada, Y. (1987a). Infection of tobacco protoplasts with *in vitro* transcribed tobacco mosaic virus RNA using an improved electroporation method. *FEBS Lett.* **219**, 65–69.

Watanabe, Y., Morita, N., Nishiguchi, M., and Okada, Y. (1987b). Attenuated strains of tobacco mosaic virus. Reduced synthesis of a viral protein with a cell-to-cell movement function. *J. Mol. Biol.* **194,** 699–704.

Waterhouse, P. M., and Helms, K. (1985). *Metopalophium dirhodum* (Walker): A newly arrived vector of barley yellow dwarf virus in Australia. *Australas. Plant Pathol.* **14,** 64–66.

Waterworth, H. E., Tousignant, M. E., and Kaper, J. M. (1978). A lethal disease of tomato experimentally induced by RNA-5 associated with cucumber mosaic virus isolated from *Commelina* from El Salvador. *Phytopathology* **68,** 561–566.

Waterworth, H. E., Kaper, J. M., and Tousignant, M. E. (1979). CARNA 5, the small cucumber mosaic virus-dependent replicating RNA, regulates disease expression. *Science* **204,** 845–847.

Watson, L., and Gibbs, A. J. (1974). Taxonomic patterns in the host ranges of viruses among grasses, and suggestions on generic sampling for host-range studies. *Ann. Appl. Biol.* **77,** 23–32.

Watson, M. A., and Heathcote, G. D. (1965). The use of sticky traps and the relation of their catches of aphids to the spread of viruses in crops. *Rep., Rothamsted Exp. Stn.* pp. 292–300.

Watson, M. A., and Plumb, R. T. (1972). Transmission of plant-pathogenic viruses by aphids. *Annu. Rev. Entomol.* **17,** 425–452.

Watson, M. A., Heathcote, G. D., Lauckner, F. B., and Sowray, P. A. (1975). The use of weather data and counts of aphids in the field to predict the incidence of yellowing viruses of sugar-beet crops in England in relation to the use of insecticides. *Ann. Appl. Biol.* **81,** 181–198.

Watts, J. W., King, J. M., and Stacey, N. J. (1987). Inoculation of protoplasts with viruses by electroporation. *Virology* **157,** 40–46.

Watts, L. E. (1975). The response of various breeding lines of lettuce to beet western yellows virus. *Ann. Appl. Biol.* **81,** 393–397.

Way, R. D., and Gilmer, R. M. (1963). Reductions in fruit sets on cherry trees pollinated with pollen from trees with sour cherry yellows. *Phytopathology* **53,** 399–401.

Weber, P. V. V. (1951). Inheritance of a necrotic-lesion reaction to a mild strain of tobacco mosaic virus. *Phytopathology* **41,** 593–609.

Webley, D. P., and Stone, L. E. W. (1972). Field experiments on potato aphids and virus spread in South Wales 1966/9. *Ann. Appl. Biol.* **72,** 197–203.

Weidemann, H. L., Lesemann, D., Paul, H. L., and Koenig, R. (1975). Das Broad Bean Wilt-Virus als Ursache für eine neue Vergilbungskrankheit des Spinats in Deutschland. *Phytopathol. Z.* **84,** 215–221.

Weiland, J. J., and Dreher, T. W. (1989). Infectious TYMV RNA from cloned cDNA: Effects *in vitro* and *in vivo* of point substitutions in the initiation codons of two extensively overlapping ORFs. *Nucleic Acids Res.* **17,** 4675–4687.

Weiner, A. M. (1987). Summary. *Cold Spring Harbor Symp. Quant. Biol.* **52,** 933–941.

Weiner, A. M., and Maizels, N. (1987). tRNA-like structures tag the 3′ ends of genomic RNA molecules for replication: Implications for the origin of protein synthesis. *Proc. Natl. Acad. Sci. U.S.A.* **84,** 7383–7387.

Weiner, H. L., Drayna, D., Averill, D. R., Jr., and Fields, B. N. (1977). Molecular basis for reovirus virulence: Role of the S1 gene. *Proc. Natl. Acad. Sci. U.S.A.* **74,** 5744–5748.

Weintraub, M., and Ragetli, H. W. J. (1970). Electron microscopy of the bean and cowpea strains of southern bean mosaic virus within leaf cells. *J. Ultrastruct. Res.* **32,** 167–189.

Weising, K., Schell, J., and Kahl, G. (1988). Foreign genes in plants: Transfer, structure, expression and applications. *Annu. Rev. Genet.* **22,** 421–477.

Welkie, G. W., and Pound, G. S. (1958). Temperature influence on the rate of passage of cucumber mosaic virus through the epidermis of cowpea leaves. *Virology* **5,** 362–371.

Welkie, G. W., Young, S. F., and Miller, G. W. (1967). Metabolite changes induced by cucumber mosaic virus in resistant and susceptible strains of cowpea. *Phytopathology* **57,** 472–475.

Wellink, J., and van Kammen, A. (1988). Proteases involved in the processing of viral polyproteins. *Arch. Virol.* **98,** 1–26.

Wellink, J., and van Kammen, A. (1989). Cell-to-cell transport of cowpea mosaic virus requires both the 58K/48K proteins and the capsid proteins. *J. Gen. Virol.* **70,** 2279–2286.

Wellink, J., Rezelman, G., Goldbach, R., and Beyreuther, K. (1986). Determination of the proteolytic processing sites in the polyprotein encoded by the bottom component RNA of cowpea mosaic virus. *J. Virol.* **59,** 50–58.

Wellink, J., Jaegle, M., and Goldbach, R. (1987a). Detection of a novel protein encoded by the bottom-component RNA of cowpea mosaic virus, using antibodies raised against a synthetic peptide. *J. Virol.* **61,** 236–238.

Wellink, J., Jaegle, M., Prinz, H., van Kammen, A., and Goldbach, R. (1987b). Expression of the middle component RNA of cowpea mosaic virus *in vivo*. *J. Gen. Virol.* **68,** 2577–2585.

Wellink, J., van Lent, J., and Goldbach, R. (1988). Detection of viral proteins in cytopathic structures in cowpea protoplasts infected with cowpea mosaic virus. *J. Gen. Virol.* **69**, 751–755.

Westheimer, F. H. (1986). Polyribonucleic acids as enzymes. *Nature (London)* **319**, 534–536.

Wetter, C. (1975). Tabakmosaikvirus und Para-Tabakmosaikvirus in Zigaretten. *Naturwissenshaften* **62**, 533.

Wetter, C., and Bernard, M. (1977). Identifizierung, Reinigung und serologischer Nachweis von Tabakmosaikvirus und Par-Tabakmosaikvirus aus Zigaretten. *Phytopathol. Z.* **90**, 257–267.

Whalen, M. C., Stall, R. E., and Staskawicz, B. J. (1988). Characterisation of a gene from a tomato pathogen determining hypersensitive resistance in non-host species and genetic analysis of this resistance in bean. *Proc. Natl. Acad. Sci. U.S.A.* **85**, 6743–6747.

Whenham, R. J. (1989). Effect of systemic tobacco mosaic virus infection on endogenous cytokinin concentration in tobacco (*Nicotiana tabacum* L.) leaves: Consequences for the control of resistance and symptom development. *Physiol. Mol. Plant Pathol.* **35**, 85–95.

Whenham, R. J., and Fraser, R. S. S. (1981). Effect of systemic and local-lesion-forming strains of tobacco mosaic virus on abscisic acid concentration in tobacco leaves: Consequences for the control of leaf growth. *Physiol. Plant Pathol.* **18**, 267–278.

Whenham, R. J., and Fraser, R. S. S. (1982). Does tobacco mosaic virus RNA contain cytokinins? *Virology* **118**, 263–266.

Whenham, R. J., Fraser, R. S. S., and Snow, A. (1985). Tobacco mosaic virus-induced increase in abscisic acid concentration in tobacco leaves: Intracellular location and relationship to symptom severity and to extent of virus multiplication. *Physiol. Plant Pathol.* **26**, 379–387.

Whitcomb, R. F., and Black, L. M. (1961). Synthesis and assay of wound-tumor soluble antigen in an insect vector. *Virology* **15**, 136–145.

White, J. L. (1982). Regeneration of virus-free plants from yellow-green areas and TMV-induced enations of *Nicotiana tomentosa*. *Phytopathology* **72**, 866–867.

White, J. L., and Brakke, M. K. (1983). Protein changes in wheat infected with wheat streak mosaic virus and in barley infected with barley stripe mosaic virus. *Physiol. Plant Pathol.* **22**, 87–100.

White, J. L., and Kaper, J. M. (1987). Absence of lethal stem necrosis in select *Lycopersicon* spp. infected by cucumber mosaic virus strain D and its necrogenic satellite CARNA5. *Phytopathology* **77**, 808–811.

White, J. L., and Kaper, J. M. (1989). A simple method for detection of viral satellite RNAs in small plant tissue samples. *J. Virol. Methods* **23**, 83–94.

White, J. L., Wu, F.-S., and Murakishi, H. H. (1977). The effect of low-temperature pre-incubation treatment of tobacco and soybean callus cultures on rates of tobacco- and southern bean mosaic virus synthesis. *Phytopathology* **67**, 60–63.

White, R. F., and Antoniw, J. F. (1983). Direct control of virus diseases. *Crop Prot.* **2**, 259–271.

White, R. F., Dumas, E., Shaw, P., and Antoniw, J. F. (1986). The chemical induction of PR (b) proteins and resistance to TMV infection in tobacco. *Antiviral Res.* **6**, 177–185.

White, R. F., Rybicki, E. P., von Wechmar, M. B., Dekker, J. L., and Antoniw, J. F. (1987). Detection of PR 1-type proteins in Amaranthaceae, Chenopodiaceae, Graminae and Solanaceae by immunoblotting. *J. Gen. Virol.* **68**, 2043–2048.

Whitmoyer, R. E., Nault, L. R., and Bradfute, O. E. (1972). Fine structure of *Aceria tulipae* (Acarina: Eriophyidae). *Ann. Entomol. Soc. Am.* **65**, 201–215.

Wijdeveld, M. M. G., Goldbach, R. W., Verduin, B. J. M., and van Loon, L. C. (1989). Association of viral 126kDa protein-containing X-bodies with nuclei in mosaic-diseased tobacco leaves. *Arch. Virol.* **104**, 225–239.

Wilcoxson, R. D., Johnson, L. E. B., and Frosheiser, F. I. (1975). Variation in the aggregation forms of alfalfa mosaic virus strains in different alfalfa organs. *Phytopathology* **65**, 1249–1254.

Williams, R. C., and Wycoff, R. G. W. (1944). The thickness of electron microscopic objects. *J. Appl. Phys.* **15**, 712–716.

Williamson, K. I., and Taylor, W. B. (1958). The analysis of particle counts by the spray-drop method. *Br. J. Appl. Phys.* **9**, 264–267.

Willison, J. H. M. (1976). The hexagonal lattice spacing of intracellular crystalline tobacco mosaic virus. *J. Ultrastruct. Res.* **54**, 176–182.

Wilson, T. M. A. (1984a). Cotranslational disassembly of tobacco mosaic virus *in vitro*. *Virology* **137**, 255–265.

Wilson, T. M. A. (1984b). Cotranslational disassembly increases the efficiency of expression of TMV RNA in wheat germ cell-free extracts. *Virology* **138**, 353–356.

Wilson, T. M. A. (1986). Expression of the large 5′ proximal cistron of tobacco mosaic virus by 70S ribosomes during cotranslational disassembly in a prokaryotic cell-free system. *Virology* **152**, 277–279.

Wilson, T. M. A. (1989). Plant viruses: A tool-box for genetic engineering and crop protection. *Bio-Essays* **10,** 179–186.

Wilson, T. M. A., and Glover, J. F. (1983). The origin of multiple polypeptides of molecular weight below 110,000 encoded by tobacco mosaic virus RNA in the messenger-dependent rabbit reticulocyte lysate. *Biochim. Biophys. Acta* **739,** 35–41.

Wilson, T. M. A., and Perham, R. N. (1985). Modification of the coat protein charge and its effect on the stability of the U1 strain of tobacco mosaic virus at alkaline pH. *Virology* **140,** 21–27.

Wilson, T. M. A., and Watkins, P. A. C. (1985). Cotranslational disassembly of a cowpea strain (Cc) of TMV: Evidence that viral RNA–protein interactions at the assembly origin block ribosome translocation *in vitro. Virology* **145,** 346–349.

Wilson, T. M. A., and Watkins, A. C. (1986). Influence of exogenous viral coat protein on the cotranslational disassembly of tobacco mosaic virus (TMV) particles *in vitro. Virology* **149,** 132–135.

Wilson, T. M. A., Perham, R. N., Finch, J. T., and Butler, P. J. G. (1976). Polarity of the RNA in the tobacco mosaic virus particle and the direction of protein stripping in sodium dodecyl sulphate. *FEBS Lett.* **64,** 285–289.

Wilson, T. M. A., Plaskitt, K. A., Watts, J. W., Osbourn, J. K., and Watkins, P. A. C. (1990). Signals and structures involved in early interactions between plants and viruses or pseudoviruses. *NATO Adv. Res. Workshop: Recognition Response Plant-Virus Interact., 1989.*

Windsor, I. M., and Black, L. M. (1973). Evidence that clover club leaf is caused by a rickettsia-like organism. *Phytopathology* **63,** 1139–1148.

Winter, E., Yamamoto, F., Almoguera, C., and Perucho, M. (1985). A method to detect and characterise point mutations in transcribed genes: Amplification and overexpression of the mutant C-K1-*ras* allele in human tumor cells. *Proc. Natl. Acad. Sci. U.S.A.* **82,** 7575–7579.

Wittmann, H. G., and Wittmann-Liebold, B. (1966). Protein chemical studies of two RNA viruses and their mutants. *Cold Spring Harbor Symp. Quant. Biol.* **31,** 163–172.

Wolanski, B. S., and Chambers, T. C. (1971). The multiplication of lettuce necrotic yellows virus. *Virology* **44,** 582–591.

Wolf, S., Deom, C. M., Beachy, R. N., and Lucas, W. J. (1989). Movement protein of tobacco mosaic virus modifies plasmodesmatal size exclusion limit. *Science* **246,** 377–379.

Wolfe, K. H., Li, W.-H., and Sharp, P. M. (1987). Rates of nucleotide substitution vary greatly among plant mitochondrial, chloroplast and nuclear DNAs. *Proc. Natl. Acad. Sci. U.S.A.* **84,** 9054–9058.

Wolfe, K. H., Gouy, M., Yang, Y.-W., Sharp, P. M., and Li, W.-H. (1989). Date of the monocot–dicot divergence estimated from chloroplast DNA sequence data. *Proc. Natl. Acad. Sci. U.S.A.* **86,** 6201–6205.

Wolstenholme, G. E. W., and O'Connor, M., eds. (1971). "Strategy of the Viral Genome." Churchill-Livingstone, Edinburgh and London.

Wood, G. A. (1983). Problems associated with the introduction of virus- and mycoplasma-free apple trees to New Zealand orchards. *Acta Hortic.* **130,** 257–262.

Wood, H. A. (1973). Viruses with double-stranded RNA genomes. *J. Gen. Virol.* **20,** Suppl., 61–85.

Woodford, J. A. T., and Gordon, S. C. (1978). Virus-like symptoms in red raspberry leaves caused by fenitrothion. *Plant Pathol.* **27,** 77–81.

Woodford, J. A. T., Gordon, S. C., and Foster, G. N. (1988). Sideband application of systemic granular pesticides for the control of aphids and potato leaf roll virus. *Crop Prot.* **7,** 96–105.

Woolston, C. J., Covey, S. N., Penswick, J. R., and Davies, J. W. (1983). Aphid transmission and a polypeptide are specified by a defined region of the cauliflower mosaic virus genome. *Gene* **23,** 15–23.

Woolston, C. J., Czaplewski, L. G., Markham, P. G., Goad, A. S., Hull, R., and Davies, J. W. (1987). Location and sequence of a region of cauliflower mosaic virus gene 2 responsible for aphid transmissibility. *Virology* **160,** 246–251.

Woolston, C. J., Reynolds, H. V., Stacey, N. J., and Mullineaux, P. M. (1989). Replication of wheat dwarf virus DNA in protoplasts and analysis of coat protein mutants in protoplasts and plants. *Nucleic Acids Res.* **17,** 6029–6041.

Worley, J. F. (1965). Translocation of southern bean mosaic virus in phloem fibres. *Phytopathology* **55,** 1299–1302.

Wrischer, M. (1973). The effect of ethionine on the fine structure of bean chloroplasts. *Cytobiologie* **7,** 211–214.

Wu, F. S., and Murakishi, H. H. (1979). Synthesis of virus and virus-induced RNA in southern bean mosaic virus-infected soybean cell cultures. *J. Gen. Virol.* **45,** 149–160.

Wu, G., Kang, L., and Tien, P. (1989). The effect of satellite RNA on cross-protection among cucumber mosaic virus strains. *Ann. Appl. Biol.* **114,** 489–496.

Wu, G.-J., and Bruening, G. (1971). Two proteins from cowpea mosaic virus. *Virology* **46**, 596–612.

Wu, J.-H., (1964). Release of inhibited tobacco mosaic virus infection by ultraviolet irradiation as a function of time and temperature after inoculation. *Virology* **24**, 441–445.

Wu, J.-H. (1973). Wound-healing as a factor in limiting the size of lesions of *Nicotiana glutinosa* leaves infected by the very mild strain of tobacco mosaic virus (TMV-VM). *Virology* **51**, 474–484.

Wu, S., Rinehart, C. A., and Kaesberg, P. (1987). Sequence and organisation of Southern bean mosaic virus genomic RNA. *Virology* **161**, 73–80.

Wutscher, H. K., and Shull, A. V. (1975). Machine-hedging of citrus trees and transmission of exocortis and xyloporosis viruses. *Plant Dis. Rep.* **59**, 368–369.

Wyatt, S. D., and Kuhn, C. W. (1977). Highly infectious RNA isolated from cowpea chlorotic mottle virus with low specific infectivity. *J. Gen. Virol.* **35**, 175–180.

Wyatt, S. D., and Shaw, J. G. (1975). Retention and dissociation of tobacco mosaic virus by tobacco protoplasts. *Virology* **63**, 459–465.

Wyatt, S. D., Seybert, L. J., and Mink, G. (1988). Status of the barley yellow dwarf problem of winter wheat in eastern Washington. *Plant Dis.* **72**, 110–113.

Xia, Y., Burbank, D. E., Uher, L., Rabussay, D., and van Etten, J. L. (1987). IL-3A virus infection of a *Chlorella*-like green alga induces a DNA restriction endonuclease with novel sequence specificity. *Nucleic Acids Res.* **15**, 6075–6090.

Xiong, C., Muller, S., Lebeurier, G., and Hirth, L. (1982). Identification by immunoprecipitation of cauliflower mosaic virus *in vitro* major translation product with a specific antiserum against viroplasm protein. *EMBO J.* **1**, 971–976.

Xiong, C., Lebeurier, G., and Hirth, L. (1984). Detection *in vivo* of a new gene product (gene III) of cauliflower mosaic virus. *Proc. Natl. Acad. Sci. U.S.A.* **81**, 6608–6612.

Xiong, Z., and Lommel, S. A. (1989). The complete nucleotide sequence and genome organization of red clover necrotic mosaic virus RNA1. *Virology* **171**, 543–554.

Xiong,, Z., Hiebert, E., and Purcifull, D. E. (1988). Characterisation of the peanut mottle virus genome by *in vitro* translation. *Phytopathology* **78**, 1128–1134.

Xu, Z., Anzola, J. V., and Nuss, D. L. (1989a). Assignment of wound tumor virus non-structural polypeptides to cognate dsRNA genome segments by *in vitro* expression of tailored full-length cDNA clones. *Virology* **168**, 73–78.

Xu, Z., Anzola, J. V., Nalin, C. M., and Nuss, D. L. (1989b). The 3'-terminal sequence of a wound tumour virus transcript can influence conformational and functional properties associated with the 5' terminus. *Virology* **170**, 511–522.

Yamaguchi, A., and Hirai, T. (1967). Symptom expression and virus multiplication in tulip petals. *Phytopathology* **57**, 91–92.

Yamamoto, K., and Yoshikura, H. (1986). Relation between genomic and capsid structures in RNA viruses. *Nucleic Acids Res.* **14**, 389–396.

Yamaoka, N., Furusawa, I., and Yamamoto, M. (1982a). Infection of turnip protoplasts with cauliflower mosaic virus DNA. *Virology* **122**, 503–505.

Yamaoka, N., Morita, T., Furusawa, I., and Yamamoto, M. (1982b). Effect of temperature on the multiplication of cauliflower mosaic virus. *J. Gen. Virol.* **61**, 283–287.

Yamaya, J., Yoshioka, M., Meshi, T., Okada, Y., and Ohno, T. (1988a). Expression of tobacco mosaic virus RNA in transgenic plants. *Mol. Gen. Genet.* **211**, 520–525.

Yamaya, J., Yoshioka, M., Meshi, T., Okada, Y., and Ohno, T. (1988b). Cross protection in transgenic tobacco plants expressing a mild strain of tobacco mosaic virus. *Mol. Gen. Genet.* **215**, 173–175.

Yamazaki, H., and Kaesberg, P. (1963a). Isolation and characterisation of a protein subunit of broadbean mottle virus. *J. Mol. Biol.* **6**, 465–473.

Yamazaki, H., and Kaesberg, P. (1963b). Degradation of bromegrass mosaic virus with calcium chloride and the isolation of its protein and nucleic acid. *J. Mol. Biol.* **7**, 760–762.

Yang, A. F., and Hamilton, R. I. (1974). The mechanism of seed transmission of tobacco ringspot virus in soybean. *Virology* **62**, 26–37.

Yarwood, C. E. (1951). Associations of rust and virus infections. *Science* **114**, 127–128.

Yarwood, C. E. (1952). The phosphate effect in plant virus inoculations. *Phytopathology* **42**, 137–143.

Yarwood, C. E. (1955a). Deleterious effects of water in plant virus inoculations. *Virology* **1**, 268–285.

Yarwood, C. E. (1968). Sequence of supplements in virus inoculations. *Phytopathology* **58**, 132–136.

Yarwood, C. E. (1969). Sulfite in plant virus inoculations. *Virology* **39**, 74–78.

Yarwood, C. E. (1973). Quick drying versus washing in virus inoculations. *Phytopathology* **63**, 72–76.

Yassin, A. M. (1983). A review of factors influencing control strategies against tomato leafcurl virus disease in the Sudan. *Trop. Pest Manage.* **29**, 253–256.

Ye, X. S., Pan, S. Q., and Kuć, J. (1989). Pathogenesis-related proteins and systemic resistance to blue mould and tobacco mosaic virus induced by tobacco mosaic virus, *Peronospora tabacina* and aspirin. *Physiol. Mol. Plant Pathol.* **35**, 161–175.

Yeh, S.-D., and Gonsalves, D. (1984). Purification and immunological analyses of cylindrical-inclusion protein induced by papaya ringspot virus and watermelon mosaic virus 1. *Phytopathology* **74**, 1273–1278.

Yeh, S.-D., Gonsalves, D., Wang, H.-L., Namba, R., and Chiu, R.-J. (1988). Control of papaya ringspot virus by cross protection. *Plant Dis.* **72**, 375–380.

Yi-Li, R., Wen-Li, C., and Rui-Fen, L. (1981). Studies on the rice virus vector small brown planthopper *Laodelphax striatella* Fallen. *Acta Entomol. Sin.* **24**, 290.

Yokoyama, M., Nozu, Y., Hashimoto, J., and Omura, T. (1984). *In vitro* transcription by RNA polymerase associated with rice gall dwarf virus. *J. Gen. Virol.* **65**, 533–538.

Yoshida, K., Goto, T., and Iizuka, N. (1985). Attenuated isolates of cucumber mosaic virus produced by satellite RNA and cross protection between attenuated isolates and virulent ones. *Ann. Phytopathol. Soc. Jpn.* **51**, 238–242.

Yoshikawa, N., Poolpol, P., and Inouye, I. (1986). Use of a dot immunobinding assay for rapid detection of strawberry pseudo mild yellow edge virus. *Ann. Phytopathol. Soc. Jpn.* **52**, 728–731.

Yot, P., Pinck, M., Haenni, A.-L., Duranton, H. M., and Chapeville, F. (1970). Valine-specific tRNA-like structure in turnip yellow mosaic virus RNA. *Proc. Natl. Acad. Sci. U.S.A.* **67**, 1345–1352.

Young, M. J., Daubert, S. D. and Shepherd, R. J. (1987). Gene I products of cauliflower mosaic virus detected in extracts of infected tissue. *Virology* **158**, 444–446.

Young, N., Forney, J. and Zaitlin, M. (1987). Tobacco mosaic virus replicase and replicative structures. *J. Cell Sci. Suppl.* **7**, 277–285.

Young, N. D., and Zaitlin, M. (1986). Analysis of tobacco mosaic virus replicative structures synthesized *in vitro*. *Plant Mol. Biol.* **6**, 455–465.

Ysebaert, M., van Emmelo, J., and Fiers, W. (1980). Total nucleotide sequence of a nearly full-size DNA copy of satellite tobacco necrosis virus RNA. *J. Mol. Biol.* **143**, 273–287.

Yu, M. H., Frenkel, M. J., McKern, N. M., Shukla, D. D., Strike, P. M., and Ward, C. W. (1989). Coat proteins of potyviruses. 6. Amino acid sequences suggest that watermelon mosaic virus 2 and soybean mosaic virus -N are strains of the same potyvirus. *Arch. Virol.* **105**, 55–64.

Yung, K.-H., and Northcote, D. H. (1975). Some enzymes present in the walls of mesophyll cells of tobacco leaves. *Biochem. J.* **151**, 141–144.

Zabel, P., Moerman, M., van Straaten, F., Goldbach, R., and van Kammen, A. (1982). Antibodies against the genome-linked protein VPg of cowpea mosaic virus recognize a 60,000 dalton precursor polypeptide. *J. Virol.* **41**, 1083–1088.

Zabel, P., Moerman, M., Lomonossoff, G., Shanks, M., and Beyreuther, K. (1984). Cowpea mosaic virus VPg-sequencing of radiochemically modified protein allows mapping of the gene on B-RNA. *EMBO J.* **3**, 1629-1634.

Zagorski, W., Morch, M.-D., and Haenni, A.-L. (1983). Comparison of three different cell-free systems for turnip yellow mosaic virus RNA translation. *Biochimie* **65**, 127–133.

Zaitlin, M. (1962). Graft transmissibility of a systemic virus infection to a hypersensitive host—An interpretation. *Phytopathology* **52**, 1222–1223.

Zaitlin, M. (1976). Viral cross protection: More understanding is needed. *Phytopathology* **66**, 382–383.

Zaitlin, M., and Ferris, W. R. (1964). Unusual aggregation of a nonfunctional tobacco mosaic virus protein. *Science* **143**, 1451–1452.

Zaitlin, M., and Hull, R. (1987). Plant–virus–host interactions. *Annu. Rev. Plant Physiol.* **38**, 291–315.

Zaitlin, M., and Israel, H. W. (1975). Tobacco mosaic virus (type strain). *CMI/AAB Descriptions Plant Viruses* No. **151**, pp. 1–5.

Zaitlin, M., Duda, C. T., and Petti, M. A. (1973). Replication of tobacco mosaic virus V. Properties of the bound and solubilized replicase. *Virology* **53**, 300-311.

Zaitlin, M., Beachy, R. N., and Bruening, G. (1977). Lack of molecular hybridization between RNAs of two strains of TMV. A reconsideration of the criteria for strain relationships. *Virology* **82**, 237–241.

Zaitlin, M., Niblett, C. L., Dickson, E., and Goldberg, R. B. (1980). Tomato DNA contains no detectable regions complementary to potato spindle tuber viroid as assayed by solution and filter hybridization. *Virology* **104**, 1–9.

Zaumeyer, W. J., and Meiners, J. P. (1975). Disease resistance in beans. *Annu. Rev. Phytopathol.* **13**, 313–334.

Zech, H. (1952). Untersuchungen über den Infektionsvorgang und die Wanderung des Tabakmosaikvirus im Pflanzenkörper. *Planta* **40**, 461–514.

Zelcer, A., Weaber, K. F., Balázs, E., and Zaitlin, M. (1981). The detection and characterization of viral-related double-stranded RNAs in tobacco mosaic virus-infected plants. *Virology* **113**, 417–427.

Zelcer, A., Zaitlin, M., Robertson, H. D., and Dickson, E. (1982). Potato spindle tuber viroid-infected tissues contain RNA complementary to the entire viroid. *J. Gen. Virol.* **59**, 139–148.

Zettler, F. W. (1967). A comparison of species of Aphididae with species of three other aphid families regarding virus transmission and probe behaviour. *Phytopathology* **57**, 398–400.

Zettler, F. W., and Hartman, R. D. (1987). Dasheen mosaic virus as a pathogen of cultivated aroids and control of the virus by tissue culture. *Plant Dis.* **71**, 958–963.

Zeyen, R. J., and Banttari, E. E. (1972). Histology and ultrastructure of oat blue dwarf virus infected oats. *Can. J. Bot.* **50**, 2511–2519.

Zeyen, R. J., Stromberg, E. L., and Kuehnast, E. L. (1987). Long range aphid transport hypothesis for maize dwarf mosaic virus: History and distribution in Minnesota, USA. *Ann. Appl. Biol.* **111**, 325–336.

Zhou, G. H., and Rochow, W. F. (1984). Differences among five stages of *Schizaphis graminum* in transmission of a barley yellow dwarf luteovirus. *Phytopathology* **74**, 1450–1453.

Zhuravlev, Yu. N., Reifman, V. G., Shumilova, L. A., Yudakova, Z. S., and Pisetskaya, N. F. (1975). Absorption of P^{32}-labelled tobacco mosaic virus by isolated tobacco protoplasts and deproteinization of the virus in them. *Sov. Plant Physiol. (Engl. Transl.)* **22**, 941–943.

Ziegler, V., Richards, K., Guilley, H., Jonard, G., and Putz, C. (1985). Cell-free translation of beet necrotic yellow vein virus: Readthrough of the coat protein cistron. *J. Gen. Virol.* **66**, 2079–2087.

Ziegler-Graff, V., Bouzoubaa, S., Jupin, I., Guilley, H., Jonard, G., and Richards, K. (1988). Biologically active transcripts of beet necrotic yellow vein virus RNA-3 and RNA-4. *J. Gen. Virol.* **69**, 2347–2357.

Ziemiecki, A., and Peters, D. (1976). The proteins of sowthistle yellow vein virus: Characterization and location. *J. Gen. Virol.* **32**, 369–381.

Ziemiecki, A., and Wood, K. R. (1976). Proteins synthesized by cucumber cotyledons infected with two strains of cucumber mosaic virus. *J. Gen. Virol.* **31**, 373–381.

Zimmermann, D., and Van Regenmortel, M. H. V. (1989). Spurious cross-reactions between plant viruses and monoclonal antibodies can be overcome by saturating ELISA plates with milk proteins. *Arch. Virol.* **106**, 15–22.

Zimmern, D. (1977). The nucleotide sequence at the origin for assembly on tobacco mosaic virus RNA. *Cell (Cambridge, Mass.)* **11**, 463–482.

Zimmern, D. (1982). Do viroids and RNA viruses derive from a system that exchanges genetic information between eukaryotic cells? *Trends Biochem. Sci.* **7**, 205–207.

Zimmern, D. (1983a). Homologous proteins encoded by yeast mitochondrial introns and by a group of RNA viruses from plants. *J. Mol. Biol.* **171**, 345–352.

Zimmern, D. (1983b). An extended secondary structure model for the TMV assembly origin, and its correlation with protection studies and an assembly defective mutant. *EMBO J.* **2**, 1901–1907.

Zimmern, D. (1988). Evolution of RNA viruses. *In* "RNA Genetics" (E. Domingo, J. J. Holland, and P. Ahlquist, eds.), Vol. 2, pp. 211–240. CRC Press, Boca Raton, Florida.

Zimmern, D., and Wilson, T. M. A. (1976). Location of the origin for viral reassembly on tobacco mosaic virus RNA and its relation to stable fragment. *FEBS Lett.* **71**, 294–298.

Zink, F. W., and Duffus, J. E. (1972). Association of beet western yellows and lettuce mosaic viruses with internal rib necrosis of lettuce. *Phytopathology* **62**, 1141–1144.

Zink, F. W., Grogan, R. G., and Welch, J. E. (1956). The effect of percentage of seed transmission upon subsequent spread of lettuce mosaic virus. *Phytopathology* **46**, 622–624.

Zinnen, T., and Fulton, R. W. (1986). Cross-protection between sunn-hemp mosaic and tobacco mosaic viruses. *J. Gen. Virol.* **67**, 1679–1687.

Zrein, M., Burckard, J., and Van Regenmortel, M. H. V. (1986). Use of the biotin–avidin system for detecting a broad range of serologically related plant viruses by ELISA. *J. Virol. Methods* **13**, 121–128.

Zuidema, D., and Jaspars, E. M. J. (1985). Specificity of RNA and coat protein interaction in alfalfa mosaic virus and related viruses. *Virology* **140**, 342–350.

Zuidema, D., Bierhuizen, M. F. A., and Jaspars, E. M. J. (1983). Removal of the N-terminal part of alfalfa mosaic virus coat protein interferes with the specific binding to RNA1 and genome activation. *Virology* **129**, 255–260.

Zuidema, D., Heaton, R. H., Hanau, R., and Jackson, A. O. (1986). Detection and sequence of plus-strand leader RNA of sonchus yellow net virus, a plant rhabdovirus. *Proc. Natl. Acad. Sci. U.S.A.* **83**, 5019–5023.

Zuidema, D., Heaton, L. A., and Jackson, A. O. (1987). Structure of the nucleocapsid protein gene of sonchus yellow net virus. *Virology* **159,** 373–380.

Zuidema, D., Linthorst, H. J. M., Huisman, M. J., Asjes, C. J., and Bol, J. F. (1989). Nucleotide sequence of narcissus mosaic virus RNA. *J. Gen. Virol.* **70,** 267–276.

Zulauf, M. (1977). Swelling of brome mosaic virus as studied by intensity fluctuation spectroscopy. *J. Mol. Biol.* **114,** 259–266.

Index